DC/AC Principles: Analysis and Troubleshooting

DC/AC Principles: Analysis and Troubleshooting

RON WALLS
Okaloosa-Walton Community College

WES JOHNSTONE
Okaloosa-Walton Community College

WEST PUBLISHING COMPANY
ST. PAUL NEW YORK
LOS ANGELES SAN FRANCISCO

Composition: G&S Typesetters, Inc.
Copyeditor: Sylvia Dovner, Technical Texts
Design: Diane Beasley
Dummy Artist: Diane Beasley
Artist: Ed Rose, Visual Graphics Systems
Indexer: Marcia Carlson
Technical Problem Checker: Diane Smith
Production, Prepress, Printing and Binding: West Publishing Company

Printed in the United States of America
99 98 97 96 95 94 93 92 8 7 6 5 4 3 2 1 0

Library of Congress Cataloging-in-Publication Data

Walls, Ron.
DC/AC principles : analysis and troubleshooting / Ron Walls, Wes Johnstone.
p. cm.—(West series in electronics technology)
Includes index.
ISBN 0-314-88350-9
1. Electronic circuits. 2. Electric circuit analysis.
I. Johnstone, Wes. II. Title. III. Series.
TK7867.W275 1992
621.381′5—dc20 91-34945
CIP

Photo Credits

1 © Jay Freis, The Image Bank; **2** Courtesy Tektronix; **3** IBM; **21** IBM; **23** Tony Stone WorldWide; **62** IBM; **65** NSP; **88** Motorola; **91** © Steve Schneider, Frozen Images; **141** © Steve Schneider, Frozen Images; **182** Honeywell; **185** Courtesy of Northwest Airlines; **232** Courtesy Hummingbird; **235** IMB; **272** IMB; **275** © Grant V. Faint, Image Bank; **303** © Grant V. Faint, Image Bank; **330** Sencore; **339** Sencore; **373** Courtesy Air Force Armament and Development Test Center Eglin Air Force Base; **355** Mallory Capacitor Company; **403** Tektronix; **423** Tektronix; **427** © 1990 Steve Schneider, Frozen Images; **457** Tektronix; **464** Ford Motor Company; **467** NSP; **493** NASA; **497** © Alfred Pasieka, Image Bank; **533** © Steven Pratt/Stockphotos, Inc., Frozen Images; **562** Hank Delespinasse, Image Bank; **565** Chris Gilbert; **599** IBM; **635** © Alvis Upitis, Image Bank; **657** Courtesy of St. Paul Ramsey Hospital; **684** © R. Stephen Marks, Stockphotos, Inc., Image Bank

Custom Photography by Thomas Perry

3 Figure 1-2; **4** Figure 1-4, 1-5 and 1-6; **5** Figure 1-7; **15** Figure 1-8; **16** Figure 1-9a&b and 1-10a&b; **20** Figure 1-11, 1-12 and 1-13; **46** Figure 2-23; **51** Figure 2-29; **52** Figure 2-31; **53** Figure 2-33a&b; **54** Figure 2-35a-c; **55** Figure 2-37a-c; **56** Figure 2-39a-c; **80** Figure 3-10; **82** Figure 3-13a&b; **88** Figure F3-1; **89** Figure F3-2; **98** Figure 4-17; **128, 129** Figure 4-75b_1-d_1; **130** Figure 4-76b_1; **131** Figure 4-76c_1&d_1; **132** Figure 4-76e_1&f_1; **133** Figure 4-77b_1; **134** Figure 4-77c_1&d_1; **135** Figure 4-77e_1; **136** Figure 4-78b_1&c_1; **137** Figure 4-78d_1&e_1; **138** Figure FO4; **139** Figure F4-3; **212** Figure 6-27; **232** Figure FO6; **297** Figure 8-20; **305** Figure CO9; **316** Figure 9-9, 9-10 and 9-12; **327** Figure 9-12; **329** Figure 9-23 a&b; **335** Figure 9-45 and 9-46; **337** Figure FO9; **340** Figure 10-1a; **353** Figure 10-11, 10-12 and 10-13; **354** Figure 10-15; **355** Figure 10-16 and 10-17; **372** Figure 10-53; **375** Figure CO11; **385** Figure 11-12a&b; **391** Figure 11-19; **394** Figure 11-25; **395** Figure 11-27 and 11-28; **402** Figure 11-30; **403** Figure 11-32; **405** Figure 11-36; **406** Figure 11-38; **407** Figure 11-39; **408** Figure 11-40; **409** 11-42; **420** Figure 11-64; **421** Figure 11-65 and 11-67; **482** Figure 13-24; **484** Figure 13-30b; **531** Figure FO14; **633** Figure FO15; **Inside back cover**

Contents

Preface to the Instructor

This textbook, *DC/AC Principles: Analysis and Troubleshooting,* has been written with the needs of both the students and the instructor as our primary consideration. We have presented the fundamentals of dc and ac circuits in a style and depth suitable for community colleges, technical colleges, and technical institutes. This text, which is an electron flow version, is intended for electronics technician training programs. For those instructors who prefer conventional flow and want more mathematical analysis, we have written another text, *Introduction to Circuit Analysis,* that contains the many features found in this text.

Why do we need another dc/ac circuits text, you might ask? We have tried to write a textbook that will make a difference in getting students to read and understand the important concepts of electricity and electronics. The underlying theme of this book is one of *explaining* concepts, rather than simply presenting material for the student to memorize. Mathematics is never used to *explain* a concept, it is used only to *support* concepts.

The most basic concept of the book is the fact that all electronic systems function by controlling the flow of current. The flow of current is controlled through the circuit properties of resistance, capacitance, and inductance in series, parallel, and series-parallel circuits. The fundamental laws of these circuits are Ohm's law, Kirchhoff's voltage law, and Kirchhoff's current law. These fundamental properties and laws are not just presented, but are continuously stressed and reinforced throughout the text.

Text Organization

The text contains nineteen chapters and may be used in a two-semester dc/ac sequence. Chapters 1, 2, and 3 are introductory in nature, covering the basics of electricity through Ohm's law and power. Included in Chapter 1 are a review of powers-of-ten principles and an introduction of the scientific calculator.

Analysis of dc circuits is presented in Chapters 4, 5, 6, and 7. It is done in the traditional manner that includes series, parallel, and series-parallel circuits, as well as the network theorems applied to dc circuits. Magnetism, its effects, and its applications are covered in Chapter 8.

Chapters 9 and 10, inductance and capacitance in dc circuits, may be presented as part of the first semester or reserved for the second semester at the discretion of the instructor. These subjects are reviewed briefly in Chapters 12 and 14, respectively, for students who study them during the first semester.

Chapters 11 and through 18 cover alternating current and ac circuits in a fairly traditional order. Complex notation is not used in these chapters. Chapter 19, which is the last chapter, covers complex numbers. Thus, it may be ignored entirely or introduced earlier at the instructor's discretion.

Chapter Organization

Each chapter begins with a set of measureable performance objectives. Next follows a brief introduction that defines the scope of the chapter and how it relates to prior learning and that which is yet to come. The body of the text is made up of numbered sections of convenient length, each followed by a brief set of review questions.

At the end of the chapter is a summary of key points, a list of key equations and the variables used in them, and a list of key terms. Each chapter concludes with four sets of questions and problems designed to allow students to test their knowledge at several levels. First is a set of test-your-knowledge questions that require short answers. Second is a set of basic review problems, each keyed to the section of the text in which it was introduced. Third is a set of more challenging problems. Finally, there is a set of problems that require the use of troubleshooting principles for solution.

Pedagogy

We believe this text has several unique features that, when combined, make it a better partner in the learning process. First, like your students, this text is very *practical*. Each major topic is addressed with these four questions in mind:

1. *What is it?* This simple question is often ignored in other texts. For each topic, we take pains to let the student know what it is they must understand.
2. *What does it do?* In order to know when to use a device, equation, or tool, the student must understand its intended function. It is not enough to be able to recite Ohm's law; the student must learn to picture the laws of electronics and apply them where needed.
3. *How does it work?* In order to truly master a subject or concept, an understanding of the mechanics of that topic is necessary. An electronics student may completely understand what a device is and what it does, but unless he or she knows the basic concept of how it does it, the student will have no flexibility in applying it to a variety of situations.
4. *What happens if it fails?* This question is especially relevant to the electronics technician. It is our belief that all learning should be geared to the goal of preparing students for the work force. All the theoretical knowledge a student may possess is of little value if he or she is unable to apply it in the troubleshooting and repair of electronic equipment.

While it may not be practical for the answers to these questions to be spelled out for each topic, we take this practical approach wherever possible on a concept-by-concept basis.

The second feature of this text is an *active learning format*. It has been our experience that too few students make the effort to actually read their textbooks, and when they do, that effort is hampered by a dry, formal style of writing. With this in mind, we have developed an active learning format that will draw the student into the book and actively engage them in the learning process.

It all begins with our writing style, which is intended to be conversational and encouraging. Explanations are made in the same tone as used by classroom instructors, thereby extending the lecture to the students' home study.

A question format is used throughout each chapter to alert students to look for important ideas as they read the material. These questions serve to develop an interactive learning experience and help to develop critical thinking skills.

Several other elements are important pedagogical features of the book.

Study Skill boxes are placed throughout the early chapters. Each study skill briefly discusses some aspect of studying electronics, doing homework, or preparing for and taking exams. The intent of these tips is to provide a brief handbook of college survival skills.

Focus on Technology articles are found at the end of each chapter. These articles show

how electronics is used in everyday life and provide the students with a glimpse of the potential careers available in electronics.

Troubleshooting, which is a main component in many electronics careers and also a problem-solving method in electronic circuits, is heavily emphasized throughout the text, both conceptually and in practice. It is intoduced early in Chapter 2 and carried through each chapter. For practical application, a section of the end-of-chapter problems, specifically devoted to troubleshooting, are provided in each chapter beginning with Chapter 2.

Four-color artwork is used throughout the text, not for cosmetic or aesthetic purposes, but to make the illustrations more useful and easier to understand. This color is not in a special section, but on the page in which it is needed. A good example is in Chapter 6, series-parallel circuits, in which color is used to show clearly the various currents flowing in the circuits. The color art is mostly computer generated so the colors are true and will not be confusing to the students.

Finally, numerous charts, tables, graphs, pictures, and illustrations are used to demonstrate and explain concepts. Numerous examples are included to demonstrate mathematical analysis, many of which include the calculator sequence for arriving at the solution. Kirchhoff's laws are recalled and reviewed each time they are applicable to the discussion.

■ End-of-Text Learning Resources

1. Appendices with supporting texts, tables, and graphs.
2. A glossary containing an alphabetical listing of key terms and their definitions.
3. Answers to section review questions.
4. Answers to odd-numbered end-of-chapter problems including illustrations where necessary.

■ Supplementary Materials

For the instructor, the following are available:

Instructor's Resource Manual, containing worked out solutions to all problems, teaching tips for each chapter, extra problem worksheets (which the instructor may copy) for use by the students, lab exercise solutions, and a test file of over 350 questions.

Transparency masters, containing over 50 figures and circuits taken from the text.

Computerized testing service

The following materials are available for students:

Laboratory Manual, written by the authors and containin 50 experiments that are directly related to the text and that are designed to work with most existing lab equipment.

Interactive software by Thompson and Mauser that contains troubleshooting exercises in an exciting circuit simulation format.

The Interactive Computer Simulations Troubleshooting software is available free to adopters of the text. This user-friendly, menu-driven program was developed by Jerry Thompson and Liz Mauser of Catawba Valley Community College.

The program is a 13-disk package (3½″ or 5¼″) using Icon Author. It runs on IBM PC, XT, AT, PS/2 or 100% compatible machines. A hard disk with 640K Ram and MS-DOS version 3.1 or higher is required. A color monitor is preferred (CGA or EGA recommended), but a monochrome monitor with a graphics card will also work. The program includes a run-time version of "Windows," or Microsoft "Windows" version 2.1 can be used. While a mouse is not necessary, the program will run with a mouse that is Microsoft compatible.

Students can use this program to solve simulated problems on the computer. The main menu of the program corresponds to the table of contents in the text. Each problem contains a reference screen option that refers students back to particular sections of the textbook if they become "stuck" on a problem.

Each chapter has a demonstration problem that "walks" the student through a particular problem. The student may go through the demonstration problem or proceed directly to the chapter problems. For many of the circuit analysis problems, students have the option to solve multiple problems using the same schematic. This feature gives the student a drill-and-practice option if they need to work on a particular area.

The software is accompanied by a documentation package that contains directions for loading the software onto the hard disk, and some basic directions for how to move around the screen. In addition, students can go through the introductory segment of the program, which provides instructions on how to select options from the menus, take measurements, select answers and so on.

Acknowledgments

It took the efforts of a great number of people to bring this text from an idea to a reality. The entire professional staff at West Educational Publishing has been most helpful, courteous, and supportive. Our special thanks go to acquisition editor Chris Conty, production editor Stacy Lenzen, and copy editor Sylvia Dovner. A special thanks goes to Janet Walls for the sandwiches, countless cups of coffee, and many hours of work on the art program. Thanks also to Evelyn Floyd who supplied the inspiration when things got tough.

This text would not have been possible without the extensive and conscientious suggestions from fellow instructors, including many not listed here who graciously took time to answer questions over the phone. We do wish to express our great appreciation and thanks to the following people who reviewed all or part of the manuscript:

Don Abernathy, DeVry Institute of Technology–Irving; Venkata S. Anandu, Southwest Texas State University; Ernest Arney, ITT Technical Institute–Indianapolis; A. Duane Bailey, Southern Alberta Institute of Technology; George Borchers, ITT Technical Institute–Salt Lake City; Gary Bryan, DeVry Institute of Technology–Phoenix; Gary Cardinale, DeVry Technical Institute–Woodbridge; Neil Catone, Pima Community College; Steven R. Coe, DeVry Institute of Technology–Phoenix; Doug Douty, Fresno City College; Donald Embree, Seminole Community College; Burton A. Fierstine, C. S. Mott Community College; F. Douglas Fuller, Humber College; Jim Gee, ITT Technical Institute–Salt Lake City; Dietmar Goetz, Meridian Community College; John Hamilton, National Education Center; Jill Harlamert, DeVry Institute of Technology–Columbus; Gene D. Hart, DeVry Institute of Technology–Chicago; Ron Hessman, ITT Technical Institute, Indianapolis; Shirley Hickman, Lamar University; Gary House, DeVry Institute of Technology–Decatur; Chester Howard, American River College; Michael Howard, IVY Tech–Evansville; Edward Kaufenberg, Blackhawk Technical College (Wisconsin); Bruce Koller, Diablo Valley College; Stanley Lawrence, Salt Lake Community College; Robert Lehman, Clackamas Community College; Arlo "Chip" Lusby, Mt. San Antonio College; Dave Marshall, Texas State Technical Institute; Ron McBride, Schoolcraft College; Mary McNamara, Lorain County Community College; David L. Meine, Texas State Technical Institute; Charles Merten, Milwaukee Area Technical College; William Mowbray, Community College of Rhode Island; Frank Naujokas, DeVry Institute of Technology; Darwin L. Pace; Gateway Electronics Institute; James Predko, Lansing Community College; Greg Rasmussen, St. Paul Technical College; Ronald Reis, Los Angeles Trade and Technical College; Bill Robertson, ITT Technical Institute; Bruce Sargent, Middlesex Community College (Massachusetts); Kate Sawyer, Community College of Southern Nevada; Terry Schulz, Mt. Hood Community College; Michael F. Siemion, Madison Area Technical College; Diane

Smith, Vincennes University; Paul Svatik, Owens Technical College; Robert Van Elsen, National Education Center; Mary Waller, Milwaukee Area Technical College; Steve Yelton, Cincinnati Technical College.

We also want to acknowledge the following people for their help with the project:

Ernest Arney, Jim Gee, John Hamilton, Ronald Reis, and Robert Lehman for their detailed suggestions, which helped shape the manuscript.
Greg Rasmussen and St. Paul Technical College for helping with the color photographs.
Diane Smith for reviewing the entire manuscript and checking all answers for accuracy.

Finally we wish to extend our sincere appreciation to our friend Tom Tucker for all the work and dedication he put into this text.

DC/AC Principles: Analysis and Troubleshooting

CHAPTER

1 Introduction

UPON COMPLETION OF THIS CHAPTER, YOU WILL BE ABLE TO

1. Define the term *electronics.*
2. List five applications of electronics.
3. List six careers in electronics.
4. Know good safety practices for electricity, and demonstrate their use.
5. Write numbers in scientific notation.
6. Perform mathematical manipulations with numbers expressed in scientific notation.
7. Replace powers of ten with the appropriate metric prefixes.
8. Demonstrate competence in the use of the scientific calculator.

STUDY SKILLS

How Much Study Time Is Enough?

Most college courses expect you to spend 2–4 hours of outside study time for every classroom hour. It is important that this time be spent wisely and efficiently if it is to be of maximum benefit.

To do this, you should distribute your studying over a period of time. Don't try to do all of your studying in one session. You will find your ability to understand is greater in short sessions than in long ones.

In addition, if your study sessions do exceed an hour in length, it is a good idea to take a ten-minute break each hour. This break will help to clear your mind and allow you to think more clearly.

1–1 WHAT IS ELECTRONICS?

Welcome to the fascinating world of electronics. A formal introduction is, of course, unnecessary. Throughout your day, you have many encounters with electronics. In the home are radios, televisions, microwave ovens, "smart" coffee pots, and a host of other appliances. In the world of business and commerce, there are computers, communication satellites, robots, and "smart machines." It is indeed difficult to find minutes of your day in which electronics does not play a part. Up to this point, you have associated electronics with its many applications. Now you will begin a new relationship, based on the laws and theories that apply to electronics.

A television, a microwave oven, or other piece of electronic equipment is actually an **electronic system**—that is, an organized collection of electronic parts that operate through the control of electric current. A typical electronic system, a digital computer, is shown in Figure 1–1. In any electronic system, the electric current is the movement of subatomic particles known as electrons (hence, the name "electronics"). Anything in

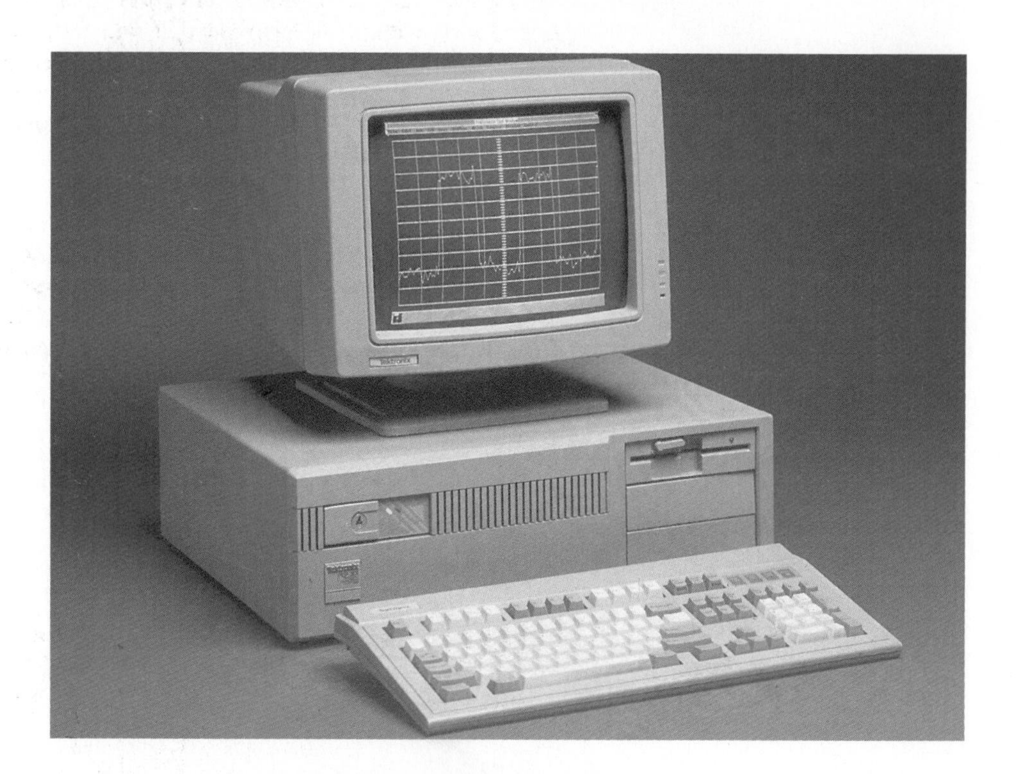

FIGURE 1–1 A typical electronic system: a digital computer

FIGURE 1–2 Typical electronic devices

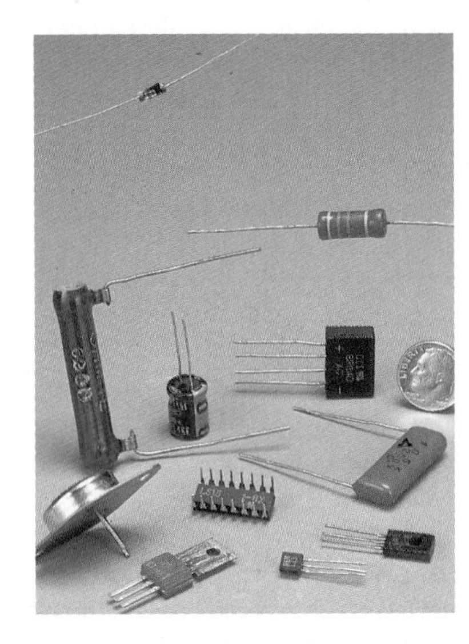

FIGURE 1–3 A typical printed circuit board showing installed devices and components

motion, be it electrons or a tractor, can be made to do work. For this work to be useful, however, the movement must be controlled.

Electronics is the science dealing with the control of electron flow. Control of current is accomplished through the use of such devices as diodes, transistors, and resistors—a collection of which is shown in Figure 1–2. These devices are incorporated into circuits, such as amplifer and oscillator circuits, that are the components of an electronic system. Figure 1–3 shows a typical circuit board composed of these devices and components.

Not long ago, "electronics" was synonymous with some form of radio communication. Today, with the development of small, fast, inexpensive, and reliable components, electronics has found its way into a host of other fields. It has become the facilitating force behind research and development. More and more, problems are first defined and then solved through electronic means. It is truly an exciting time to be entering the electronics field.

Today's trend seems to be toward even more applications of electronics. Large amounts of time and money are being invested in developing smaller, more powerful, and less expensive electronic components. These will be used as the building blocks of even faster, more powerful computers. These computers, in turn, will assist in the discovery of even more sophisticated applications for electronics. What is most remarkable is that all this sophistication is based upon the principles presented in this text.

SECTION 1–1 REVIEW

1. What is electronics?
2. List five applications of electronics.

1–2 CAREERS IN ELECTRONICS

Preparation

Preparation for employment in an electronics career must begin with a thorough study of the theory of electricity as well as electronics. These studies include the theory and operation of electronic devices, electronic circuits, and the test equipment used in circuit testing and troubleshooting. Equipment typically used in electronics training is shown in

FIGURE 1–4 A typical electronics training station

Figure 1–4. Anyone acquiring the knowledge involved in this equipment is said to be at "entry level" for employment in the electronics industry. In electronics, the completion of school training is not the end of learning, but the beginning of a lifetime of learning.

Although the applications of electronics seem limitless, the basic, underlying principles are the same. Students must not merely absorb the facts of electronics theory, they must also develop their skills in reasoning and analysis. Electronics technicians must be capable of, and indeed eager, to grow with the industry. Otherwise, as the technology changes, they may find their skills to be inadequate.

As broad as the electronics field is, the student often asks: "What am I being trained to do?" In a program providing a thorough grounding in the basics, the answer is "Anything!" The technician enters the job market with entry level skills, and the employer provides the training on specific items of equipment, usually during work hours and at full pay.

There are many job classifications in electronics, each employing technicians with specific skills. Some are common to the industry. Others are designated by a specific company in order to meet its own needs. The careers described in the following sections are generic in nature, and the list is not meant to be all inclusive.

Service Store Technician

A person in this classification is involved in the servicing and repair of consumer products such as televisions, video cassette recorders (VCRs), and sound systems. An entry level technician usually trains on the job under the tutelage of an experienced technician. Instruction on the more sophisticated equipment is usually accomplished in schools and at seminars sponsored by the manufacturer.

Although capable of servicing many brands, the store may be franchised for the sale and service of a specific brand. Items to be repaired are usually brought to the store. Some stores do, however, make housecalls in which minor repairs are made on the spot. In these situations, the technician must deal directly with the customer. Thus, not only technical skills, but also those in good interpersonal relationships must be developed. Some technicians begin their careers working in a serivce store and then become self-employed, owning a service store themselves. A service store technician is shown at work in Figure 1–5.

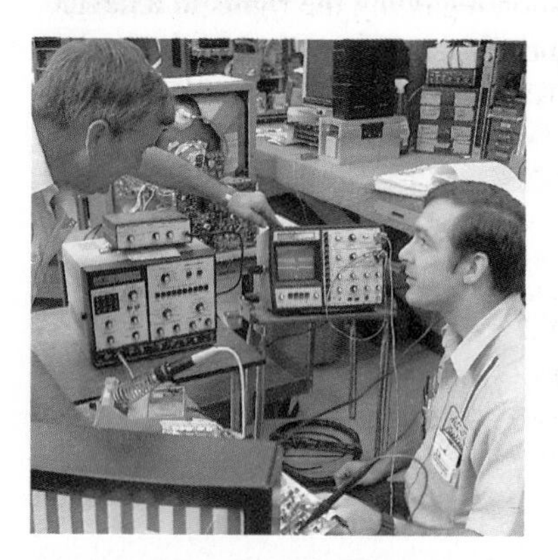
FIGURE 1–5 A service store technician at work

Field Service Technician

A field service technician works for an equipment manufacturer and provides service and repair at the equipment user's location. The equipment may include automatic bank teller machines, environmental control systems, large computers, copying machines, and radar systems. Training on the equipment is usually provided at the manufacturer's school. The technician assigned to service a given system may, in fact, assist in its installation and testing. Field service technicians may work out of the corporate headquarters or any one of a number of remote service centers. Technicians from each center service equipment within a given territory, usually making short trips of less than one day. The company often provides tools and, in many instances, transportation for technicians. A field service technician is shown at work in Figure 1–6.

FIGURE 1–6 A field service technician at work

Technologist

In this classification, the technician works closely with scientists and engineers involved in research and development. As new circuit devices and concepts are developed, they must be tested, modified as needed, and refined before they can be incorporated into electronic systems. Usually, basic devices of circuits, such as transistors, are being developed, and the technician may assist in fabricating them as well as testing them in electronic circuits. The test results must be reported to the scientists, usually in written form. Many colleges and universities have research programs in which technicians can find work. Once again, more than merely technical skills are required. The scientists and technicians must communicate well in order to define goals and develop test procedures and schedules. The reporting requires skills in written communication. Thus, anyone entering this area of endeavor should take courses in oral and written communication.

Engineering Assistant

FIGURE 1–7 An electronics technician at work

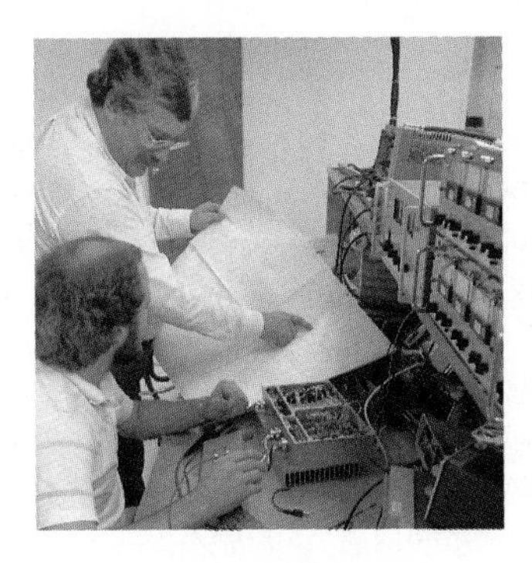

Engineers are assigned tasks in the areas of system development and modification. Each engineer selects a technician to assist in the task. Working from drawings supplied by the engineer, the technician procures the parts, builds and tests the circuits, and reports the results. The technician is usually involved in the project from its inception and continues through the early stages of the manufacturing process. An engineering assistant must maintain a daily log documenting all of his/her activities in a bound notebook, containing numbered pages upon which every test and its results must be logged. No erasures are allowed; incorrect entries are lined through, but must be readable. None of the numbered pages may be removed from the log book. These stringent requirements are necessary to protect the employer's interests in the event of litigation regarding the rights to a device. Where two or more parties claim patent rights, this notebook is often the deciding factor. Thus, once again skill in written communication is very important. Figure 1–7 shows an engineering assistant at work.

Technical Writer

A technical writer must assemble the technical data on a system and then write manuals covering its theory, operation, and servicing. In addition to technical skills, the technical writer must possess the ability to develop and clearly communicate all material needed by users and service technicians. This material forms the system's documentation and is a critical part of the development process. A poorly documented system is extremely frustrating for the user. A well-documented system is much easier to sell than one in which the instructions are vague or nonexistent!

Sales Representative

Technical sales can be a very rewarding career. A technician aspiring to such a career must possess not only the required technical knowledge, but also the ability to communicate with prospective customers. The successful salesperson is one who makes a study of human nature and possesses the attributes of ambition, friendliness, a cheerful personality, and mental toughness. Courses not only in communication, but also in salesmanship are helpful here. Some positions in technical sales may involve a great deal of travel.

SECTION 1–2 REVIEW

1. What preparation is necessary for a career in electronics?
2. List five careers in electronics.
3. In addition to technical skills, list at least three other skills that a technician should develop in order to be successful.

STUDY SKILLS

Previewing Material

Before attending class, preview the material to be covered in that session. First look at major headings and try to get a feeling for what the section is about. Then go back and read the material carefully.

By reading the material before class, you will be able to follow the instructor more easily during the lecture. Also, if there are points you did not understand while reading, you will be prepared to ask the instructor during class time.

1–3 SAFETY

Your course of study in electronics technology will consist of two types of learning: (1) the theory and underlying principles and (2) the practices followed in the applications of theory. For example, in the lab, you will perform experiments in which circuits are built and tested in order to demonstrate the theory presented in the classroom. Consequently, one of the first learning experiences must be the safety factors involved in working with electricity.

The key to electronics safety is knowledge. With knowledge, any fear of electricity is lost and is replaced with a healthy respect. Electronics safety is a matter of good news and bad news! The bad news is that contact with electricity can be devastating to the human body. The good news is that it is a simple, common-sense matter to avoid contact with electricity and still function well as an electronics technician. Hundreds of volts may be present and scores of amperes of current may be flowing, but so long as electrical contact is not made, you will be perfectly safe.

There are two basic factors in electronics safety: (1) avoiding contact with conductors of electric current and (2) minimizing the effects if contact is inadvertently made. Before these two factors are explored in detail in the following subsections, one additional point must be raised: Knowledge of safety procedures is essential, but it is of no value unless put into practice. You are the only one who can ensure your safety!

■ Avoiding Contact

The very best way to avoid contact with electricity is to de-energize the circuit while working on it. This practice is especially important in the laboratory where changes must often be made in the circuit under test. *Never make changes in a circuit with the power on.* Doing so is unsafe for both you and the electronic devices involved.

If the circuit must be energized, then avoiding contact becomes a matter of touching insulated points only. Insulated points are exposed conductors that have been covered with such materials as plastic or rubber that are not conductors of electricity. The insulating material acts as a barrier between you and the electricity in the circuit. You must be especially careful, for example, when using the test leads of a voltmeter. The ends of the probes are metal in order to make contact with points in the circuit. Care must be taken to be sure that you do not touch the metal ends and that you hold the leads only by the insulators provided. Test leads must be inspected periodically to ensure that the insulation has not become frayed or damaged in any way.

■ Miniziming Effects of Electric Shock

By staying fully alert and concentrating on what is being done, the chance of getting an electric shock is almost nonexistant. However, Murphy's Law states that anything that can happen eventually will, so it is best to be aware of the possibility. Most technicians have, at one time or another, experienced electric shock, and according to the overwhelming majority who have survived, it seems that the effects can be minimized. Nevertheless, remember that avoiding shock is the intelligent thing to do.

Human skin is a relatively poor conductor of electricity. The oils present on the skin also supply some insulation. When the skin is wet with water or perspiration, however, the skin resistance drops remarkably. This lower skin resistance allows a much greater current to flow in the event of electric shock, which greatly increases the danger. So one way to minimize the effect of electric shock is to keep your hands dry.

When you work on live circuits, it is a good safety practice to work with one hand and to place the other hand in your pocket or behind your back when making measurements. If both hands are used, a current path may be set up from one hand to the other. This path would be, of course, through vital organs such as the heart. Working with one hand prevents such a current path from forming and thus greatly reduces the damage done in the event of electric shock.

Another safety practice that should be observed whenever working around live circuits or mechanical systems is the removal of all jewelry. As innocent as watches, rings, or necklaces may seem, they pose a threat to both the technician and the equipment. As with all safety procedures, an ounce of prevention is worth a pound of cure.

A safety practice when working with high voltages is to work with another person, not alone. Another person nearby could disconnect the equipment or trip a circuit breaker and prevent a serious or even fatal accident.

How much current is required to cause serious damage to the human body is very difficult to determine. Many factors are involved in determining how much current an individual is able to withstand. A general rule of thumb is that any flow of current through the body greater than .01 amperes can cause considerable pain. A current of .1 amperes or greater can cause serious nerve damage and even death.

Electricity, like any other tool, should not be feared but should be used with respect and proper caution.

1–4 SCIENTIFIC NOTATION

■ Rationale

In the analysis of electronic circuits, both very large and very small numbers are encountered. For example, they may be as large as 3,000,000,000 and as small as 0.000,000,000,001. Such numbers must be manipulated using all the operations of mathematics. They are awkward to write and to work with, and the number of digits may exceed the capacity of a calculator's display.

Each of the numbers given as examples in the preceding paragraph has only one significant digit (3 and 1, respectively). The zeros are used merely to place the decimal points. Without them, the numbers would be much easier to write and manipulate. The difficulty of working with all of the zeros can be overcome by expressing the numbers in scientific notation.

■ Converting Numbers Using Positive Powers of Ten

Scientific notation is a method for expressing large numbers as small numbers between one and ten, multiplied by a power of ten. Expressing a number in scientific notation is a two-step process that involves moving the decimal point to the left, until only one nonzero digit is to its left. The next step involves selecting an appropriate power of ten, which, when multiplied by the reduced number, will return the reduced number to its original value.

Table 1–1 contains several numbers and their representation as positive powers of ten. The ten is known as the *base* and the number to which it is raised is known as the *exponent* or *power*. There are several things to note in this table. First of all, notice that as the

■ TABLE 1–1 Positive powers of ten

Number	Power of Ten
1	10^0
10	10^1
100	10^2
1,000	10^3
10,000	10^4
100,000	10^5
1,000,000	10^6

numbers increase from one to one million, the number of digits *increases,* but in scientific notation the number of digits remains the *same.* Second, notice that the exponent is the number of places the decimal point will move to the *right* of 1 when multiplying in order to write the number in its original form. (Recall that each time the decimal place is moved to the right, the number is multiplied by ten.)

The two steps for expressing a number in scientific notation are as follows:

1. Convert the number to a number between one and ten.
2. Select the appropriate power of ten as the multiplier.

For example, the number 3,000,000,000 can be expressed in scientific notation as follows:

1. To convert to a number between one and ten, move the decimal point to the left until only one significant digit remains to its left:

$$\begin{matrix} 3 & 000 & 000 & 000. \\ 3. & 000 & 000 & 000 \end{matrix}$$

2. Next, to choose the appropriate power of ten for the multiplier, count the number of places the decimal point was moved to the left, which is equal to the number of places the power of ten must move it to the right to restore the original value. The exponent indicates the number of places that it will move to the right, which in this case must be nine. Thus,

$$3{,}000{,}000{,}000 = 3 \times 10^9$$

EXAMPLE 1–1 Express 80,400 in scientific notation.

■ Solution

1. Convert to a number between one and ten:

$$\begin{matrix} 8 & 0 & 4 & 0 & 0. \\ 8. & 0 & 4 & 0 & 0 \end{matrix}$$

2. Choose the appropriate power of ten for the multiplier. The decimal point was moved four places to the left so the appropriate power of ten is 10^4. Thus,

$$80{,}400 = 8.04 \times 10^4$$

CAUTION

The first zero to the right of the eight is a significant digit just as the four is. Do not drop any such digits and do not perform rounding on them.

EXAMPLE 1–2 Express 3,650 in scientific notation?

■ Solution

1. Convert to a number between one and ten:

3	6	5	0.
3.	6	5	0

2. Choose the appropriate power of ten for the multiplier. Since the decimal point was moved three places to the left, the proper power of ten is 10^3. Thus,

$$3{,}650 = 3.65 \times 10^3$$

Converting Numbers Using Negative Powers of Ten

Table 1–2 contains several *decimal fractions,* or numbers having a value of less than one, and their representation as powers of ten. A negative power of ten indicates the number of places the decimal point will move to the *left* when multiplying. In the following example for using negative powers of ten, the number 0.000,000,000,001 will be expressed in scientific notation.

TABLE 1–2 Negative powers of ten

Number	Power of Ten
.1	10^{-1}
.01	10^{-2}
.001	10^{-3}
.0001	10^{-4}
.00001	10^{-5}
.000001	10^{-6}
.0000001	10^{-7}

1. To convert the number to a number between one and ten, move the decimal point to the right one point past the first nonzero digit.

0.000	000	000	001
000	000	000	001.

2. Next, to choose the appropriate power of ten for the multiplier, count the number of places the decimal point was moved the right. The answer is 12 places, so the power of ten must move it 12 places to the left, Thus,

$$0.000{,}000{,}000{,}001 = 1 \times 10^{-12}$$

EXAMPLE 1–3 Express 0.000 000 38 in scientific notation.

Solution

1. Convert to a number between one and ten:

0.0 0 0	0 0 0	3 8
0 0 0	0 0 0	3.8

2. Choose the appropriate power of ten for the multiplier. The decimal point was moved seven places to the right, so the proper power of ten is 10 – 7. Thus,

$$0.000\ 000\ 38 = 3.8 \times 10^{-7}$$

EXAMPLE 1–4 Express 0.009 55 in scientific notation.

Solution

1. Convert to a number between one and ten:

0.0	0	9 5	5
0	0	9.5	5

2. Choose the appropriate power of ten for the multiplier. The decimal point was moved three places to the right, so the proper power of ten is 10^{-3}. Thus,

$$0.009\ 55 = 9.55 \times 10^{-3}$$

■ Arithmetic with Powers of Ten

As mentioned previously, the purpose of scientific notation is to make very large or very small numbers easier to manipulate. The arithmetic operations of addition, subtraction, multiplication, and division can be performed more easily when the numbers are in scientific notation. In the following subsections, the rules for these operations are reviewed.

Addition and Subtraction

In order to add or subtract numbers in scientific notation, the powers of ten must be the same. If not, one must be converted to match the other. The rule for addition or subtraction is as follows: **Add or subtract the numbers and assign the common power of ten to the sum or difference.**

EXAMPLE 1–5 Add the following numbers: 2×10^3 and 5×10^2

■ Solution

The exponent in 2×10^3 indicates that the decimal point in the original number was moved three places to the left. The exponent in 5×10^2 indicates it was moved two places to the left. To make the exponents match, move the decimal point in 5×10^2 one additional place to the left. Thus,

$$5 \times 10^2 = 0.5 \times 10^3$$

Now add the numbers:

$$(2 \times 10^3) + (0.5 \times 10^3) = 2.5 \times 10^3$$

EXAMPLE 1–6 Add the following numbers: 5.3×10^6 and 4×10^4.

■ Solution

The exponent in 5.3×10^6 indicates that the decimal point in the original number was moved six places to the left. The exponent in 4×10^4 indicates it was moved four places to the left. To make the exponents match, move the decimal point in 4×10^4 two additional places to the left. Thus,

$$4 \times 10^4 = 0.04 \times 10^6$$

Add the numbers:

$$(5.3 \times 10^6) + (0.04 \times 10^6) = 5.34 \times 10^6$$

EXAMPLE 1–7 Subtract 2.66×10^4 from 5.89×10^5.

■ Solution

The exponent in 5.89×10^5 indicates the decimal point in the original number was moved five places to the left. To make the exponent in 5.89×10^5 match the exponent in 2.66×10^4, move the decimal point one place to the right for a new total of four places to the left in the original number. Thus,

$$5.89 \times 10^5 = 58.9 \times 10^4$$

and

$$(58.9 \times 10^4) - (2.66 \times 10^4) = 5.624 \times 10^5$$

EXAMPLE 1–8 Subtract 5.89×10^{-4} from 2.66×10^{-3}.

■ Solution

The exponent in 5.89×10^{-4} indicates the decimal point in the original number was moved four places to the right. Thus,

$$5.89 \times 10^{-4} = 0.589 \times 10^{-3}.$$

and

$$(2.66 \times 10^{-3}) - (0.589 \times 10^{-3}) = 2.071 \times 10^{-3}$$

Multiplication

The rule for multiplication of numbers in scientific notation is as follows: **Multiply the numbers and algebraically add the powers of ten.**

EXAMPLE 1–9 Multiply 1.88×10^2 and 3×10^3.

■ Solution

Multiply the numbers:

$$1.88 \times 3 = 5.64$$

Algebraically add the exponents:

$$10^2 \times 10^3 = 10^{(2+3)} = 10^5$$

Thus,

$$(1.88 \times 10^2)(3 \times 10^3) = 5.64 \times 10^5$$

EXAMPLE 1–10 Multiply 1.8×10^{-1} and 320×10^{-3}.

■ Solution

Multiply the numbers:

$$1.8 \times 320 = 576$$

Algebraically add the exponents:

$$10^{-1} \times 10^{-3} = 10^{[-1+(-3)]} = 10^{-4}$$

Thus,

$$(1.8 \times 10^{-1})(320 \times 10^{-3}) = 5.76 \times 10^{-2}$$

Division

The rule for division of numbers in scientific notation is as follows: **Divide the numbers and algebraically subtract the exponent of the divisor from the exponent of the dividend.**

EXAMPLE 1–11 Divide 8.5×10^3 by 2.5×10^{-2}.

■ Solution

Divide the numbers:

$$8.53 \div 2.5 = 3.4$$

Algebraically subtract the exponents:

$$10^3 \div 10^{-2} = 10^{[3-(-2)]} = 10^5$$

Thus,

$$(8.5 \times 10^3) \div (2.5 \times 10^{-2}) = 3.4 \times 10^5$$

EXAMPLE 1–12 Divide 120×10^4 by 3.9×10^2.

■ Solution

Divide the numbers:

$$120 \div 3.9 = 30.77$$

Algebraically subtract the exponents:

$$10^4 \div 10^2 = 10^{(4-2)} = 10^2$$

Thus,

$$(120 \times 10^4) \div (3.9 \times 10^2) = 3.077 \times 10^3$$

■ SECTION 1–4 REVIEW

1. Express each of the following numbers in scientific notation: a. 1,100 b. 0.000,85 c. 2,250,000.
2. Write each of the following numbers without powers of ten: a. 3×10^6 b. 2.08×10^{-9} c. 2.25×10^{12}.
3. Add 2.35×10^3 to 9.33×10^6.
4. Subtract 5.6×10^{-9} from 4.8×10^{-6}.
5. Multiply 6×10^3 and 7.55×10^4.
6. Divide 49×10^6 by 7×10^3.

TABLE 1–3 Electrical quantities

Quantity	Symbol	Unit	Abbreviation
Capacitance	C	Farad	F
Charge	Q	Coulomb	C
Conductance	G	Siemen	S
Current	I	Ampere	A
Energy	W	Joule	J
Frequency	f	Hertz	Hz
Impedance	Z	Ohm	Ω
Inductance	L	Henry	H
Power	P	Watt	W
Reactance	X	Ohm	Ω
Resistance	R	Ohm	Ω
Time	t	Second	s
Voltage	V	Volt	V

1–5 ELECTRICAL QUANTITIES

Table 1–3 is a list of the most common electrical quantities, with their symbols, abbreviations, and units of measure. They are presented now for your information and will be covered in detail in succeeding chapters. In troubleshooting and analysis of electronic circuits, these quantities give the "clues" as to the condition of the circuit.

The symbols and units in Table 1–3 are from the **International System** (in French, *Système Internationale*), which is known as the SI system. For example, notice that lower case t is the symbol for time and its unit is the second symbolized by lower case s. These quantities and units will become second nature to you as you proceed through the electronics curriculum.

1–6 METRIC PREFIXES

In electronics, certain powers of ten are used more often than others. These have been given names in order to avoid confusion in written and oral communication. Since the base is ten, the names have been chosen from the metric system. These are shown in Table 1–4. They are standard for expressing electronic quantities and should be committed to memory.

The usefulness of metric prefixes can be demonstrated by the following examples. Suppose the quantity to be expressed is 1,500 volts (V). (The volt is the unit of electrical pressure and will be introduced in the next chapter.) Using scientific notation, this quantity can be expressed as

$$1{,}500 \text{ V} = 1.5 \times 10^3 \text{V}$$

Table 1–4 shows that 10^3 is equivalent to the metric prefix *kilo* (abbreviated k). Thus, the quantity can now be written as

$$1.5 \times 10^3 \text{ V} = 1.5 \text{ kV}$$

TABLE 1–4 Metric prefixes used in electronics

Prefix	Power of Ten	Mathematical Ratio	Symbol
Multiples			
giga	10^{9}	Billions	G
mega	10^{6}	Millions	M
kilo	10^{3}	Thousands	k (K)
Fractions			
milli	10^{-3}	Thousandths	m
micro	10^{-6}	Millionths	μ
nano	10^{-9}	Billionths	n
pico	10^{-12}	Trillionths	p

As a second example, suppose the quantity to be expressed is 0.000635 amperes (A). (The ampere is the unit of electric current flow and will also be introduced in the next chapter.) It can be expressed as

$$0.000635 \text{ A} = 635 \times 10^{-6} \text{ A}$$

Table 1–4 shows that 10^{-6} is equivalent to the metric prefix *micro* (abbreviated μ). The number can be expressed as

$$635 \times 10^{-6} = 635\mu\text{A}$$

EXAMPLE 1–13 Express 1,350 volts using metric prefixes.

■ Solution

Using powers of ten, the quantity is expressed as

$$1{,}350 \text{ V} = 1.35 \times 10^{3} \text{ V}$$

From Table 1–4, 10^{3} is equivalent to kilo so the quantity is expressed as

$$1.35 \times 10^{3} \text{ V} = 1.35 \text{ kV}$$

EXAMPLE 1–14 Express 0.000 000 000 015 siemens (S) using metric prefixes.

■ Solution

$$0.000\,000\,000\,015 \text{ S} = 15 \times 10^{-12} \text{ S}$$

From table 1–4, 10^{-12} is equivalent to the metric prefix *pico* (p), so the quantity is expressed as

$$15 \times 10^{-12} = 15 \text{ pS}$$

SECTION 1–6 REVIEW

1. Express 150,000 V using the appropriate metric prefix.
2. Express 10,000,000 Hz using the appropriate metric prefix.
3. Express 0.000,050 A using the appropriate metric prefix.
4. Express 0.015,5 S using the appropriate metric prefix.

STUDY SKILLS

Doing Exercises

After reviewing the appropriate material in class, you are ready to begin to do the exercises. Your goal at this point should be accuracy. When working problems with new concepts or equations, it is always a good idea to take your time. By concentrating on the process or the equation that you are using, your comprehension will be enhanced, and the concept will be reinforced.

Don't worry about speed! It takes time and practice to become proficient. It is better at this point to go at a slower pace and be accurate. Once you thoroughly understand what you are doing, speed will follow.

1–7 USE OF THE SCIENTIFIC CALCULATOR

In all phases of the electronics technology program, it is necessary to make calculations. Since the calculation itself is not as important as the significance of the results, it is a good idea to acquire a calculator. It will relieve you of a task that is merely time consuming, leaving more time for important learning. One thing must be clearly understood: The calculator is merely a tool, and it will do only what you tell it to do. So if your input is faulty, the answer will be faulty. To have confidence in the results obtained from the calculator, you must become completely familiar with it.

You will need what is known as a **scientific calculator,** which has, in addition to the standard arithmetic functions, scientific notation, trigonometric functions, and logarithms. A calculator without these functions is of little value in an electronics curriculum. You may wish to get other features such as coordinate and number base conversions. While these are nice to have, they are not really necessary. Some calculators are programmable and have battery back-up in order to allow them to retain what is stored in memory. Be aware that if a programmable calculator is used on tests, you may have to clear the memory prior to use. In this section, only the calculator functions needed now are considered; the others will be introduced as needed.

FIGURE 1–8 A typical scientific calculator

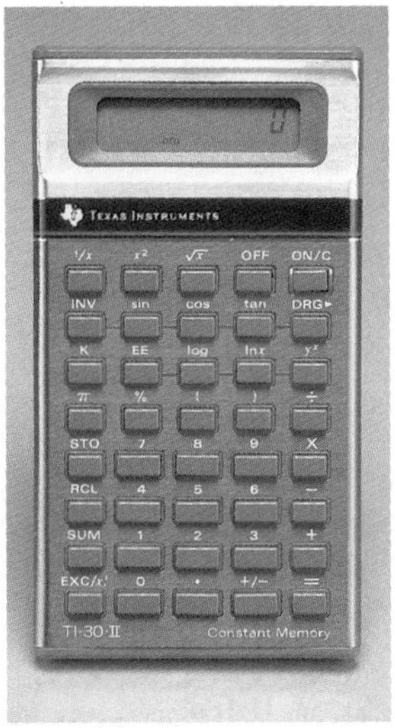

Figure 1–8 shows a typical calculator keyboard. The keys for the arithmetic functions of addition, subtraction, multiplication, and division are self-explanatory. The 1/X key provides the reciprocal of the number entered in the display. As the key shows, the reciprocal of a number is one divided by that number. The X^2 key, when depressed, provides the square of the number entered in the display while the $\sqrt{X}$ key does the opposite by providing the square root of the number. The STO key stores the number from the display in memory while the RCL key will return the number from memory to the display. (Some calculators may have more than one location for storing data.) Consult the manual provided with your particular calculator in order to learn of any special functions or features.

The EE, or EXP, key allows the entry of numbers in scientific notation. Suppose you wished to enter the number 6.25×10^6. The keystrokes and display will be as follows:

Enter: 6.25	*Display reads:* 6.25
Depress: EE	*Display reads:* 6.25 00
Enter: 6	*Display reads:* 6.25 06

The display contains the number 6,250,000 entered in scientific notation. The displays for steps 1 and 3 of this calculator sequence are shown in Figure 1–9a and b.

Note: All examples assume the use of a scientific calculator in the scientific (SCI) display mode

FIGURE 1–9 Keystrokes for entering 6.25×10^6

a. Enter 6.25. Depress EE.

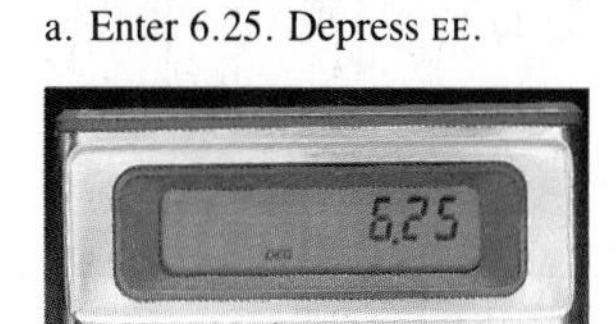

b. Enter 6.

6.25 06

EXAMPLE 1–15 Enter the number 0.00453 in scientific notation.

■ Solution

$$0.00453 = 4.53 \times 10^{-3}$$

Enter: 4.53 *Display reads:* 4.53
Depress: EE *Display reads:* 4.53 00
Depress: +/– *Display reads:* 4.53 –00
Enter: 3 *Display reads:* 4.53 –03

The display should indicate 4.53 –03, which is 4.53×10^{-3}. The displays for steps 1 and 4 of this entry are shown in Figure 1–10a and b.

FIGURE 1–10 Keystrokes for entering 4.53×10^{-3}

a. Enter 4.53. Depress EE. Depress +/–.

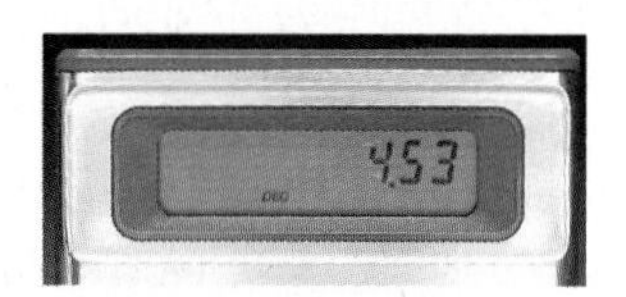

b. Enter 3.

Notice in the example above that the +/– key was used to change the sign of the exponent from positive to negative. A positive exponent will be indicated by no sign between the number and the exponent.

EXAMPLE 1–16 Enter the number 6,250,000,000,000,000,000 into the calculator in scientific notation.

■ Solution

$$6{,}250{,}000{,}000{,}000{,}000{,}000 = 6.25 \times 10^{18}$$

Enter: 6.25 *Display reads:* 6.25
Depress: EE *Display reads:* 6.25 00
Enter: 18 *Display reads:* 6.25 18

The display should indicate 6.25 18, which is 6.25×10^{18}.

Example 1–16 demonstrates clearly the need for scientific notation in using the calculator. Numbers of this size would never fit the display of a calculator. Although a number of this magnitude is not usually encountered in electronics, some numbers will exceed the capacity of the display and will have to be manipulated.

EXAMPLE 1–17 Add the numbers 975,000,000 and 250,000,000.

■ **Solution**

$$975{,}000{,}000 = 9.75^{8}$$

$$250{,}000{,}000 = 2.75^{8}$$

Enter: 9.75 *Display reads:* 9.75
Depress: EE *Display reads:* 9.75 00
Enter: 8 *Display reads:* 9.75 08
Depress: + *Display reads:* 9.75 08
Enter: 2.5 *Display reads:* 2.5
Depress: EE *Display reads:* 2.5 00
Enter: 8 *Display reads:* 2.5 08
Depress: = *Display reads:* 1.225 09

The display should now read 1.225 09, which is 1.225×10^{9}.

EXAMPLE 1–18 Subtract the number 0.000,625 from 0.009,50.

■ **Solution**

$$0.000625 = 6.25 \times 10^{-4}$$

$$0.00950 = 9.5 \times 10^{-3}$$

Enter: 9.5 *Display reads:* 9.5
Depress: EE *Display reads:* 9.5 00
Depress: +/− *Display reads:* 9.5 −00
Enter: 3 *Display reads:* 9.5 −03
Depress: − *Display reads:* 9.5 −03
Enter: 6.25 *Display reads:* 6.25
Depress: EE *Display reads:* 6.25 00
Depress: +/− *Display reads:* 6.25 −00
Enter: 4 *Display reads:* 6.25 −04
Depress: = *Display reads:* 8.875 −03

The display should read 8.875 −03, which is 8.875×10^{-3}.

EXAMPLE 1–19 Multiply 65,000 times 0.003.

■ **Solution**

$$65{,}000 = 6.5 \times 10^{4}$$

$$0.003 = 3.0 \times 10^{-3}$$

Enter: 6.5	*Display reads:* 6.5
Depress: EE	*Display reads:* 6.5 00
Enter: 4	*Display reads:* 6.5 04
Depress: ×	*Display reads:* 6.5 04
Enter: 3	*Display reads:* 3
Depress: EE	*Display reads:* 3 00
Enter: +/−	*Display reads:* 3 −00
Enter: 3	*Display reads:* 3 −03
Depress: =	*Display reads:* 195

The display should read 1.95 02.

EXAMPLE 1–20 Divide 250,000 by 6,400.

■ Solution

$$250{,}000 = 2.50 \times 10^5$$

$$6{,}400 = 6.4 \times 10^3$$

Enter: 2.5	*Display reads:* 2.5
Depress: EE	*Display reads:* 2.5 00
Enter: 5	*Display reads:* 2.5 05
Depress: ÷	*Display reads:* 2.5 05
Enter: 6.4	*Display reads:* 6.4
Depress: EE	*Display reads:* 6.4 00
Enter: 3	*Display reads:* 6.4 03
Depress: =	*Display reads:* 3.9 01

STUDY SKILLS

Following Directions

In electronics, as in all disciplines of science, a systematic approach to problem solving must be followed. When using equations, it is necessary that quantities be entered in their base units (volts, amperes, etc.) in the proper sequence. Failure to adhere to this rule will result in incorrect answers.

In the case of laboratory exercises, failure to follow the proper sequence of steps may result in inaccurate data or possibly unsafe conditions.

It is better to proceed at a slower pace and concentrate on accuracy than to rush ahead with poor results. In all cases, whenever a question arises that might possibly result in equipment damage or danger to the student, an instructor should be consulted.

■ CHAPTER SUMMARY

1. Electronics is the science dealing with the control of electric current.
2. An electronics technology program prepares graduates for employment in any of the various areas of the electronics field.
3. Careers in electronics include service technician, technologist, engineering assistant, field service technician, technical writer, and sales representative.
4. In scientific notation, a number is expressed as a number between one and ten times a power of ten.

5. A positive power of ten indicates how many places the decimal point will move to the *right* when multiplying.
6. A negative power of ten indicates how many places the decimal point must be moved to the *left* when multiplying.
7. Adding or subtracting numbers in scientific notation requires that they have the same power of ten.
8. To add numbers represented as powers of ten, add the numbers and assign the common power of ten to the result.
9. To subtract numbers represented as powers of ten, subtract the numbers and assign the common power of ten to the result.
10. To multiply two numbers represented as powers of ten, multiply the numbers and algebraically add the powers of ten.
11. To divide two numbers represented as powers of ten, divide the numbers and algebraically subtract the powers of ten.
12. Certain powers of ten are used more often than others in electronics and can be replaced by metric prefixes.

KEY TERMS

electronic system
electronics
scientific notation
International System
scientific calculator

TEST YOUR KNOWLEDGE

1. What is the function of an engineering assistant?
2. Why is it important to be able to write and manipulate numbers in scientific notation?
3. What use is made of metric prefixes in writing electrical quantities?
4. What are the metric prefixes for the most often used powers of ten.

PROBLEM SET: BASIC

Section 1–4

1. Express the following numbers in scientific notation:
   a. 65,000 **b.** 320 **c.** 8750 **d.** 37
2. Express the following numbers in scientific notation:
   a. 0.002 68 **b.** 0.000 007 9 **c.** 0.99 **d.** 0.0039
3. Express the following numbers in decimal form:
   a. 2.5×10^{-6} **b.** 300×10^{3}
   c. 1×10^{-9} **d.** 25×10^{6}
4. Express the following numbers in decimal form:
   a. 2.5×10^{-3} **b.** 300×10^{6}
   c. 1×10^{5} **d.** 25×10^{-7}
5. Add the following numbers:
   a. $(5 \times 10^{3}) + (7.5 \times 10^{5})$
   b. $(4.9 \times 10^{-3}) + (16 \times 10^{-6})$
   c. $(465 \times 10^{2}) + (132 \times 10^{3})$
6. Find the difference of the following numbers:
   a. $(35 \times 10^{-6}) - (24 \times 10^{-3})$
   b. $(925 \times 10^{3}) - (337 \times 10^{2})$
   c. $(2.25 \times 10^{12}) - (81 \times 10^{9})$
7. Multiply the following:
   a. $(6 \times 10^{-6})(3 \times 10^{-6})$
   b. $(8.35 \times 10^{3})(4.41 \times 10^{-6})$
   c. $(0.55 \times 10^{6})(0.18 \times 10^{-3})$
8. Divide the following:
   a. $(45 \times 10^{3}) \div (9 \times 10^{-3})$
   b. $(0.95 \times 10^{6}) \div (40 \times 10^{-2})$
   c. $(5.25 \times 10^{-3}) \div (10 \times 10^{-3})$

Section 1–6

9. Express the following quantities using metric prefixes:
   a. 3,100,000,000 Hz **b.** 1,800,000 Ω **c.** 1,500 V
10. Express the following quantities using metric prefixes:
    a. 0.0035 A **b.** 0.000,050 F
    c. 0.000 000 385 A **d.** 0.000 000 000 015 F
11. Write the following quantities without the metric prefix using powers of ten:
    a. 9000 MHz **b.** 1.5 kV **c.** 4.8 GHz
12. Write the following quantities without the metric prefix using powers of ten:
    a. 12 mA **b.** 450 μA **c.** 6000 nS **d.** 50 pF

Section 1–7

13. The display of a scientific calculator reads 18 03. (Figure 1–11) Write the decimal equilavent of this reading.
14. The display of a scientific calculator reads 6.25 18. (Figure 1–12) Write the decimal equivalent of this reading.

FIGURE 1–11 Calculator display for Problem 13

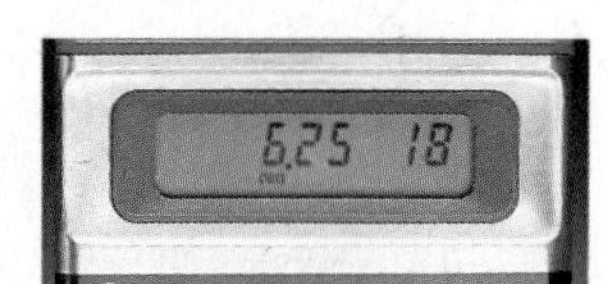

FIGURE 1–12 Calculator display for Problem 14

FIGURE 1–13 Calculator display for Problem 15

15. The display of a scientific calculator reads 35 − 03. (Figure 1–13) Write the decimal equivalent of this reading.

16. How would the number 0.000 042 5 appear on the calculator display in scientific notation?

17. Which key on the calculator allows a power of ten to be entered?

■ PROBLEM SET: CHALLENGING

18. Convert the quantity 1200 microvolts to the following metric prefixes:
a. ? mV **b.** ? pV **c.** ? nV **d.** ? V

19. Convert the quantity 1.4 MHz to the following metric prefixes:
a. ? kHz **b.** ? Hz **c.** ? GHz

20. Convert the quantity 100 pF to the following metric prefixes:
a. ? nF **b.** ? μF **c.** ? mF **d.** ? F

A Day In the Life of an Electronics Technician

Ed is employed as a technician in a successful company that has recently moved into its new and modern facility in the city's industrial park. The firm, which has been in business for about ten years, designs, builds, and markets custom-made power supplies. Although much of their business is with U.S. and foreign military agencies, they also do a great deal of business in the civilian market. The company is comprised of four divisions: (1) administration, (2) marketing, (3) engineering, and (4) production. Ed works in the production division, where the tasks are producing, testing, and certifying the power supplies contracted for by marketing and designed by engineering. This division also determines problems with power supplies returned by customers for warranty service. Ed's particular job is to test the completed power supplies that have come off the production line, repair any not meeting specifications, retest them, and certify them for shipment.

Ed's day begins, as usual, with an informal meeting in the break room (with coffee and pastry furnished by the company), during which supervisors discuss work assignments and production schedules with the employees. Ed learns that he and another technician are to begin inspecting a new run of power supplies that will be arriving at their benches within the hour. Ed is already involved in another task, but his supervisor informs him that the new job is of higher priority.

Ed goes to his work area and begins to clear the decks for action. His

work area, which is in the shape of a "U," has a bench along one side for test equipment, a large bulletin board in the rear, and a work bench along the other side. Ed removes all the old drawings, notes, and schematics from the bulletin board in order to make room for the schematic of the new units. He next makes a check of the test equipment at his bench to make certain that it is operational.

Ed notes that in the next area, Evelyn, the other technician assigned to the task, is making similar preparations. At this point, the QA (quality assurance) manager arrives with the new schematics, the test procedures, and the certification sheets. The three confer about how to approach the task. They study the schematic of the power supply and discuss possible trouble points. They also review the electrical specifications that the equipment must meet. A projected time table for completion of the task is discussed.

With the arrival of the first two dozen of the power supplies, Ed and Evelyn begin the inspecting and testing procedures. First, Ed visually inspects a unit. Is it complete and intact? Are any screws or washers missing? Are there any dents or abrasions in the cover, which could indicate that the unit had been dropped or mishandled in assembly? He next observes the printed circuit board. Are there any obvious breaks? Are there any broken or damaged components? Are all the holes in the board filled?

After the physical inspection, Ed begins the series of electrical checks, that must be performed for each unit before power is applied to the unit. Ed completes these checks for the first unit without detecting problems that would have to be corrected before he inserted the power cord. Ed fills in the results of each check on the inspection sheet for the unit.

Next, Ed applies power to the unit and watches it intently. Is there any smoke? Is there any odor of burning insulation or resistors? Does any part of the surface feel hot to the touch? There are no apparent problems, so Ed connects the unit to an ATE (Automatic Test Equipment) system for a dynamic test, which will determine whether the unit's components are operating properly. The automated test system facilitates and greatly speeds up this phase of testing process. Ed finds that all results are acceptable, and he signs the inspection form to certify that the unit is ready for "cook-in" testing, which will be

continued

continued

done by another technician. In the cook-in test, maximum power will be applied to the unit for 12 hours, a period judged to be long enough to determine whether the unit will or will not fail when used by the customer.

Ed continues testing the units without detecting a problem until the fourteenth unit. In troubleshooting this unit, Ed finds two resistors have been interchanged on the printed circuit board. He repairs the problem and continues testing units. Thus it goes throughout the day.

At the end of the day, Ed and Evelyn review their progress to see if the task is on schedule. They find that they have completed testing and repairing 36 units. They also compare notes to see whether they have found any recurring problems to report to the QA supervisors, who can then inform the production line supervisors.

In all, it has been an eventful day for Ed, filled with moments of both exultation and frustration as the units were tested. As he leaves for home, he reflects on his career. Ed feels that his job is both useful and satisfying. The company trusts him to place its stamp of approval on the products through which they stay in business. (The 36 units completed today represents a total sales price of over $5000.) Thus, he has a very responsible job. And besides, the work is clean, performed in climate-controlled environment, interesting, and challenging. And, it pays well too!

CHAPTER

2 Basic Electricity

■ UPON COMPLETION OF THIS CHAPTER, YOU WILL BE ABLE TO

1. Describe the basic structure of matter to include the atom, element, compound, and molecule.
2. Define and give the symbols and units of measure for charge, voltage, current, resistance, and conductance.
3. Describe the characteristics of conductors, semiconductors, and insulators.
4. Describe the construction of various fixed and variable resistors.
5. Determine the values of fixed resistors using the color code.
6. Draw a schematic of a typical electric circuit and explain the purpose of each component.
7. Diagram the position of the meter in measuring voltage, current, and resistance.
8. Describe a general approach to troubleshooting for a simple electric circuit.

STUDY SKILLS

Comparing and Contrasting Examples

Most of what you are learning is totally new and difficult to relate to any previous experience. This unfamiliarity makes it easy to get confused and to treat things that are different as though they are the same because they "look" similar. Circuit analysis can be especially confusing because of the many variables involved.

It is very important that you learn to make decisions based on comparing and contrasting concepts that look almost identical but are not. It is important that you identify the ways in which they are similar and the ways in which they are not. For example, current through a resistor may be found by dividing the voltage across it by its value of resistance. In some circuits, however, the voltage across any one resistor is not equal to the voltage applied to the entire circuit. Unless the circuit is properly identified, an incorrect answer will result.

When working exercises, ask yourself these questions: "What examples or concepts in the chapter are similar to those I am now doing? How are they similar? How are they different?" By asking such questions, you will help yourself avoid making careless errors and eliminate needless confusion.

This chapter presents a logical approach to the understanding of what electricity is, how it is produced, and how it is put to work. It begins deep within the structure of matter in a building block known as the atom. Inside the atom are found the two basic particles of electric charge: the proton and the electron. Separation of these two particles produces electricity. Several methods of producing electricity are presented in this chapter.

In electronics, charge, voltage, current, resistance, and conductance are basic quantities. Their definitions, symbols, and units of measure are presented. Each quantity has a specific function in an electric circuit. If one is to master the studies that follow, these quantities must be *understood,* not just memorized.

In this chapter, the electric circuit and its operation is introduced. Although the circuits presented in chapters that follow will become more complex, the basic facts presented in this chapter will still apply. One component of an electric circuit, an electronic device known as the resistor, is also introduced in this chapter. It is probably the most widely used electronic device and the first of many that you will meet in your studies.

2–1 STRUCTURE OF MATTER

Question: What is matter?

Matter is anything that has weight or occupies space. As such, it is everything of which our world is composed. Matter may be found in any of three physical states: liquid, solid, or gas. The properties of any specific matter are determined by the basic substances of which it is formed. These substances, of which there are over 100, are known as elements.

An *element* is a unique substance that cannot be chemically decomposed. An element can stand alone as matter, or it can combine with another element to form a substance known as a *compound*.

An example of a compound is water, which is a combination of two elements, hydrogen and oxygen. The process of chemical bonding between these elements results in two parts of hydrogen uniting with one part of oxygen to form a single molecule of water, shown in Figure 2–1. A *molecule* is the smallest part of a compound that retains all of the compound's chemical properties. Other examples of compounds are table salt (sodium and chlorine) and rust (iron and oxygen).

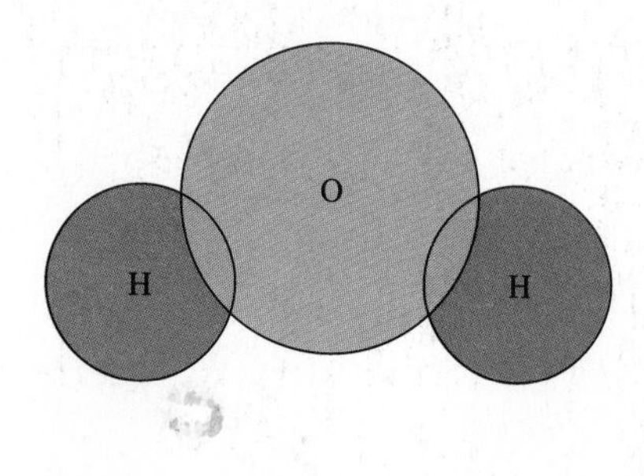

FIGURE 2–1 Chemical bonding of two elements. Two parts of hydrogen combine with one part of oxygen to form a single molecule of the compound water

2–2 ATOMS AND ATOMIC STRUCTURE

Question: What is an atom?

An **atom** is the smallest particle of an element that can retain that element's properties and characteristics. Since each element has a unique atomic structure, there are as many different atoms as there are elements. The atom is made up of many particles according to the modern concept, but this study is concerned with only three: the proton, the neutron and the electron.

■ Atomic Structure

An atom has a structure similar to the planetary arrangement of the solar system. The atom contains three types of particles, and the interaction of these particles is referred to as an electric charge. The center of the atom, known as the *nucleus,* is composed of two particle types: The **proton** is a particle of positive charge, while the **neutron** is a neutral particle and has no charge. In orbits around the nucleus are negatively charged particles known as **electrons.** Figure 2–2 illustrates this arrangement.

To provide a perspective of the atom, Table 2–1 shows the mass in grams and charge given in a unit of measure called coulombs (C) for the three particles. The mass of the electron is much less than that of the proton, but they have equal electrical charges. Another fact of atomic structure is that, on an atomic scale, the distance from nucleus to electron is vast. The radius of the electron's orbit is about 0.000,000,05 centimeters. This is 5,555,000,000,000,000,000 times greater than its mass! The neutron, which does little more than add weight to the nucleus, has about the same mass as the proton.

■ Atoms of the Elements

As previously noted, each element has individual characteristics that are the result of its unique atomic structure. The differences between elements arise from basically two things: (1) the number of protons in the nucleus and (2) the way in which the orbital electrons are arranged around the nucleus.

Question: What are the differences that exist among the atoms of the elements?

In the field of chemistry, the elements are arranged in what is known as a *periodic table.* The periodic table groups into families of elements those elements with similar properties and characteristics. Table 2–2 shows the symbol, atomic number, atomic weight for each. The **atomic number** of an element is the number of protons in the nucleus. In the periodic table, the elements are arranged in ascending order of their atomic numbers. The actual weight of an atom is so small as to be inconvenient for practical use; thus the **atomic weight** is not the true weight, but a number proportional to it. For example, helium is

FIGURE 2–2 The atom. Protons and neutrons make up the nucleus, and electrons orbit the nucleus.

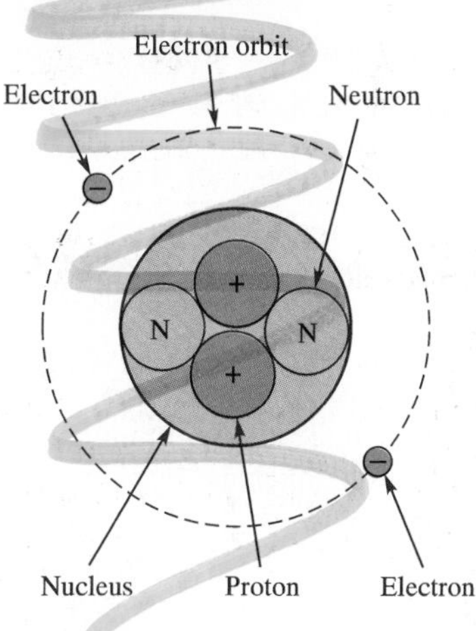

■ **TABLE 2–1 Charge and mass of atomic particles**

Particle	Charge	Mass
Electron	1.6×10^{-19} C (−)	9.097×10^{-28} g
Proton	1.6×10^{-19} C (+)	1.6729×10^{-24} g
Neutron	None	1.6758×10^{-24} g

■ **TABLE 2–2 Periodic table of the elements**

Group	Element	Symbol	Atomic Number	Atomic Wgt.	1	2	3	4	5	6	7
IA	Hydrogen	H	1	1.008	1						
	Lithium	Li	3	6.941	2	1					
	Sodium	Na	11	22.99	2	8	1				
	Potassium	K	19	39.10	2	8	8	1			
	Rubidium	Rb	37	85.47	2	8	18	8	1		
	Cesium	Cs	55	132.9	2	8	18	18	8	1	
	Francium	Fr	87	223	2	8	18	32	18	8	1
IIA	Berylium	Be	4	9.012	2	2					
	Magnesium	Mg	12	24.3	2	8	2				
	Calcium	Ca	20	40.08	2	8	8	2			
	Strontium	Sr	38	87.62	2	8	18	8	2		
	Barium	Ba	56	137.3	2	8	18	18	8	2	
	Radium	Ra	88	226.02	2	8	18	32	18	8	2
IIIA	Boron	B	5	10.81	2	3					
	Aluminum	Al	13	26.98	2	8	3				
	Gallium	Ga	31	69.72	2	8	18	3			
	Indium	In	49	114.8	2	8	18	18	3		
	Thallium	Tl	81	204.3	2	8	18	32	18	3	
IVA	Carbon	C	6	12.01	2	4					
	Silicon	Si	14	28.08	2	8	4				
	Germanium	Ge	32	72.5	2	8	18	4			
	Tin	Sn	50	118.6	2	8	18	18	4		
	Lead	Pb	82	207.2	2	8	18	32	18	4	
VA	Nitrogen	N	7	14	2	5					
	Phosphorus	P	15	30.97	2	8	5				

four times heavier than hydrogen, sulphur is twice as heavy as oxygen, calcium is about the same weight as potassium, and so on.

Figure 2–3(a) shows the nucleus of a hydrogen atom. This simplest of all atoms has 1 proton in the nucleus and 1 orbital electron. The proton's charge of +1 is exactly balanced by the electron's charge of −1, making the atom neutral. This *balancing of charges* is true for all atoms. The next element in the table, helium, has 2 protons and 2 orbital electrons and is illustrated in Figure 2–3(b). Following helium in the table is lithium with an atomic number of 3. In the lithium atom, 3 electrons orbit the nucleus, but notice in Figure 2–3(c) their positioning. Two share an orbit near the nucleus, while the other is in an orbit of its own at a greater distance from the nucleus. A final example, the sodium atom, is shown in Figure 2–3(d). Its atomic number is 11.

Group	Element	Symbol	Atomic Number	Atomic Wgt.	1	2	3	4	5	6	7
	Arsenic	As	33	74.92	2	8	18	5			
	Antimony	Sb	51	121.7	2	8	18	18	5		
	Bismuth	Bi	83	208.98	2	8	18	32	18	5	
VIA	Oxygen	O	8	15.99	2	6					
	Sulfur	S	16	32.06	2	8	6				
	Selenium	Se	34	78.9	2	8	18	6			
	Tellurium	Te	52	127.6	2	8	18	18	6		
	Polonium	Po	84	210	2	8	18	32	18	6	
VIIA	Fluorine	F	9	18.99	2	7					
	Chlorine	Cl	17	35.45	2	8	7				
	Bromine	Br	35	79.9	2	8	18	7			
	Iodine	I	53	126.9	2	8	18	18	7		
	Astatine	At	85	210	2	8	18	32	18	7	
VIIIA	Helium	He	2	4	2						
	Neon	Ne	10	20.17	2	8					
	Argon	Ar	18	39.94	2	8	8				
	Krypton	Kr	36	83.8	2	8	18	8			
	Xenon	Xe	54	131.3	2	8	18	18	8		
	Radon	Rn	86	222	2	8	18	32	18	8	
IB	Copper	Cu	29	63.54	2	8	18	1			
	Silver	Ag	47	107.87	2	8	18	18	1		
	Gold	Au	79	196.97	2	8	18	32	18	1	
IIB	Zinc	Zn	30	65.38	2	8	18	2			
	Cadmium	Cd	48	112.4	2	8	18	18	2		
	Mercury	Hg	80	200.5	2	8	18	32	18	2	

Arrangement of Electrons

Question: What is an orbital shell?

Notice the arrangement of the 11 orbital electrons in Figure 2–3(d). They are arranged in what are known as *orbital shells*. The shell nearest the nucleus contains 2 electrons. The second shell contains 8, and the third shell contains 1 electron. There is a maximum number of electrons that can occupy any given shell. This value can be computed using the following equation:

$$\text{maximum electrons in shell} = 2N^2$$

The N in this equation is the shell number. It has a value of 1 for the shell nearest the nucleus. The second shell has a value of 2, the third shell has a value of 3, and so on, outward.

FIGURE 2–3 Atomic numbers. The atomic number of an element is determined by the number of protons in the nucleus of that element's atom. The number of positively charged protons is counterbalanced by an equal number of negatively charged electrons.

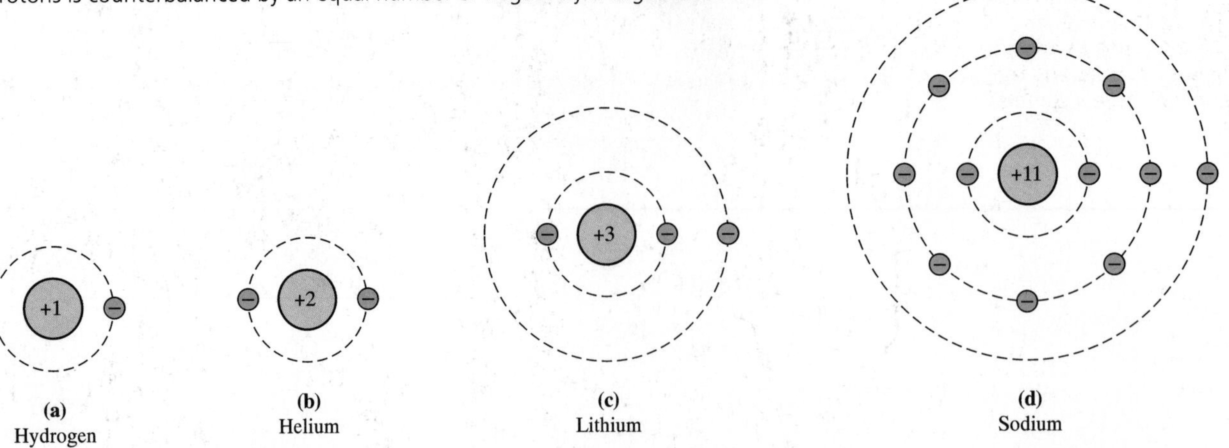

EXAMPLE 2–1 Compute the maximum number of electrons that may occupy the fourth shell in the nucleus of an atom having numerous shells.

■ Solution:

$$\text{maximum electrons in shell} = 2N^2$$
$$= 2 \times 4^2 = 32 \text{ electrons}$$

EXAMPLE 2–2 Using Table 2–2, determine the number of electrons in each shell of the copper atom.

■ Solution:

The symbol for copper is Cu. Finding copper in Table 2–2 shows that the first three shells contain their maximum number of electrons, thus leaving 1 electron in the outer shell. The shells then contain

Shell 1: 2
Shell 2: 8
Shell 3: 18
Shell 4: 1

The Valence Shell

Question: What is special about the outer shell?

The number of electrons in the outer shell of the atom, which is known as the **valence shell,** determines the element's chemical and electrical properties. Eight electrons, which is the maximum that will be found in any valence shell, makes the element relatively inert. The periodic table (Table 2–2) shows the number of valence electrons in all the elements. Notice the elements in Group VIIIA, which, with the exception of helium, all have 8 electrons in their valence shells. This group of elements, known as the inert or noble gases, is the most stable of all the families of elements. The complement of 8 electrons and helium's complete outer shell account for their stability.

FIGURE 2–4 Bonding. The bonding process takes place as the single valence electron of the sodium atom transfers to the chlorine atom in order that the chlorine atom will have 8 valence electrons and the sodium atom will have a complete outer shell.

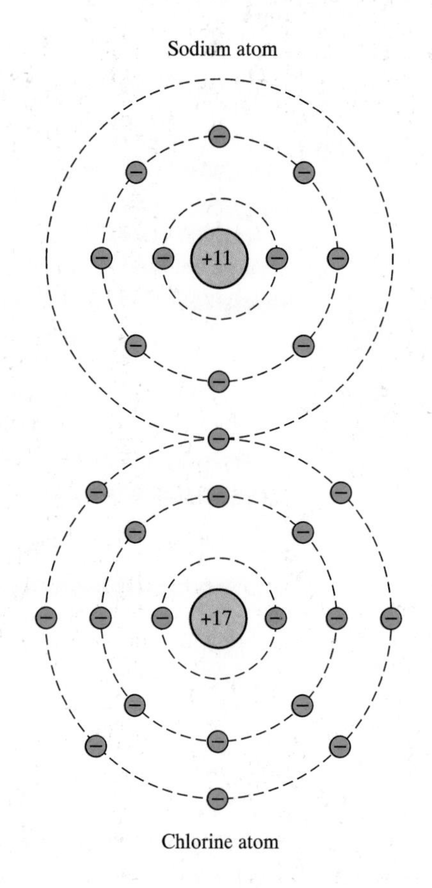

All other elements have less than 8 valence electrons. In seeking stability, they form compounds, sharing their valence electrons so as to "see" 8 electrons in the valence shell. As illustrated in Figure 2–4, the sodium atom has 1 valence electron while the chlorine atom has 7. The single electron from the sodium atom actually transfers to the chlorine atom, so that the sodium atom sees a complete outer shell, and the chlorine atom sees 8 electrons in its valence shell. The result is a stable compound known as sodium chloride or, more commonly, table salt.

A different type of bond forms the molecules of water. Two atoms of hydrogen, each having 1 valence electron, combine with 1 atom of oxygen, which has 6 valence electrons. No electrons are transferred between atoms, but they are shared. In this sharing, each atom once again sees 8 electrons and a stable compound is formed. This type of bond is known as a covalent bond. Covalent bonds occur in the semiconductor elements, which will be introduced later in the electronics curriculum.

■ SECTION 2–2 REVIEW

1. What is an atom?
2. What is the charge of an electron? A proton?
3. What does the atomic number of an element indicate?
4. What is the maximum number of electrons in the valence shell of any atom?
5. What is a compound?

2–3 STATIC ELECTRICITY

Question: Is it possible for an atom to gain or lose an electron?

Although an atom is neutral, it may lose or gain an electron. If an atom gains an electron, then the total number of electrons no longer equals the number of protons. As a result, there is no longer a complete cancellation of charges. Since there is one extra electron, the atom will assume a net charge of -1. In other words, the atom will have a negative charge of 1 electron. An atom of this type is known as a *negative ion.* In a similar manner, an atom that loses an electron assumes a charge of $+1$ and is known as a *positive ion.*

Charged Bodies

From the concept of the atom presented here, it would appear that all matter should be uncharged or neutral. Under certain conditions, however, some forms of matter will assume a charge. That is, matter will become a *charged body,* either by losing or by gaining electrons and thus contain a deficient or an excess number of electrons. Materials such as glass, styrofoam, and most plastics, when handled in a manner that produces friction, will become charged. A *charge* results from one type of material giving up electrons to another due to the energy of friction. The body that gains electrons becomes negatively charged because it has an excess of electrons.

Nearly everyone has seen examples of the behavior of charged bodies. The small pieces of styrofoam used to pack delicate items for shipment are good examples (These are known as "peanuts" in the shipping industry). The energy generated by the continual movement of the pieces of packing material causes gains and losses of electrons to take place. The result is charged bodies that store positive or negative charges according to whether an excess or deficient number of electrons are present. A charged body is said to possess *static electricity* when the electrons are stored, not in motion (thus the term *static* is used). When an item is unpacked, the charged pieces of packing material will cling to hands and clothing. (Try taking a handful of the "peanuts" and throwing them!) This clinging is due to the fundamental law of electric charges.

Known as the law of attraction and repulsion, the law of electric charges may be stated as follows:

Unlike charges attract one another, while like charges repel one another.

As illustrated in Figure 2–5, this law can be demonstrated by the standard pith-ball experiment, which you may have witnessed in a physical science class. When like charges are induced in the pith-balls, whether negative or positive, they repel one another. When unlike charges are induced, the pith-balls attract one another.

Another familiar example of the law of charges occurs when a person walks across a rug on a dry day. The friction between shoes and rug cause a transfer of electrons and the storage of charge. When the person touches some object, such as a doorknob, the electrons move in a manner so as to balance the charges, and the person experiences a slight shock. This shock is the result of electrical discharge, similar on an infinitely smaller scale, to lightning.

SECTION 2–3 REVIEW

1. What is meant by static electricity?
2. What is the law of repulsion and attraction for electric charges?
3. What causes a body to be negatively charged?
4. What causes the pieces of styrofoam used as packing to become charged?

FIGURE 2–5 Pith-ball example of the fundamental law of electric charges

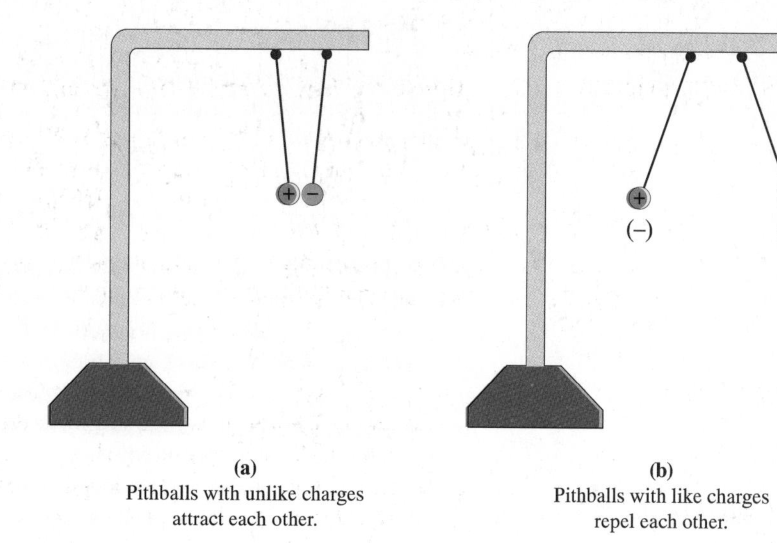

STUDY SKILLS

Reviewing Prior Material

One of the most difficult areas in the study of electronics is that each concept or skill is dependent on those previously learned. If a certain skill has not been learned well enough, it will most likely affect your ability to learn the next concept or skill.

Simply passing an exam or making it through a topic or chapter is no cause for celebration! Sooner or later the skills and concepts presented in that chapter will reappear. If those skills have not been mastered, it may not be possible to comprehend subsequent topics.

Constant review of all previous skills and equations will allow new concepts to be understood with less effort and help place things in a more meaningful perspective.

2–4 ELECTRIC CHARGE

Question: What is the result of transferring electrons from one body to another?

An **electric charge** is an excess or deficiency of electrons in a material. After electrons are moved from one body to another, a static charge is stored because there is no further movement of electrons. In order to move the electrons to store charge, energy is required, as for example that produced by friction.

The symbol for charge is Q, meaning quantity. The unit of measure for charge is the **coulomb** (abbreviated C), which is defined as an excess or deficiency of 6.25×10^{18} electrons. For example, the body in Figure 2–6(a) has an excess of 6.25×10^{18} electrons and is said to have a negative stored charge of 1 C or

$$Q = -1\text{C}$$

The other body (Figure 2–6(b)) has a deficiency of the same number of electrons and is said to have a positive stored charge of 1 C or

$$Q = +1\text{C}$$

A force field, known as an *electrostatic field*, exists between the charged bodies as shown in Figure 2–6(c). The arrows indicate the direction of movement of an electron

FIGURE 2–6 Charged bodies. (a) has an excess of electrons, and (b) has a deficiency of electrons. (c) is the electrostatic field between the two charged bodies.

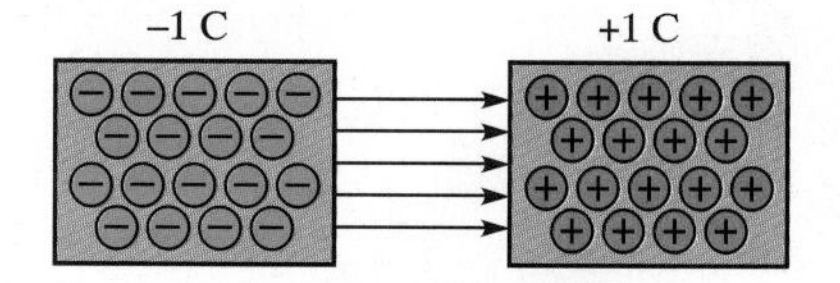

FIGURE 2–7 Electrostatic force fields between charged bodies

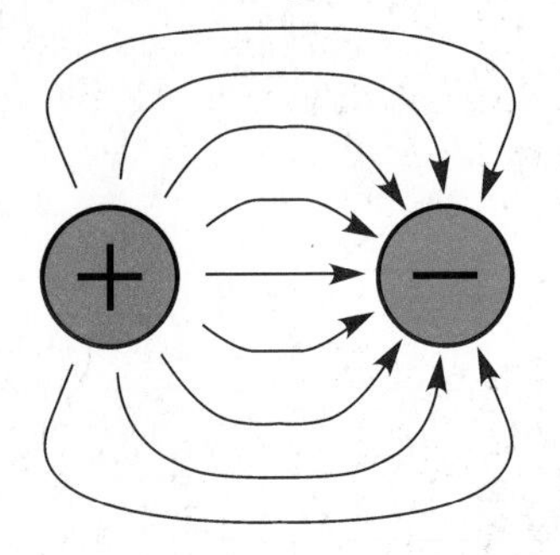

(a)
Unlike charges

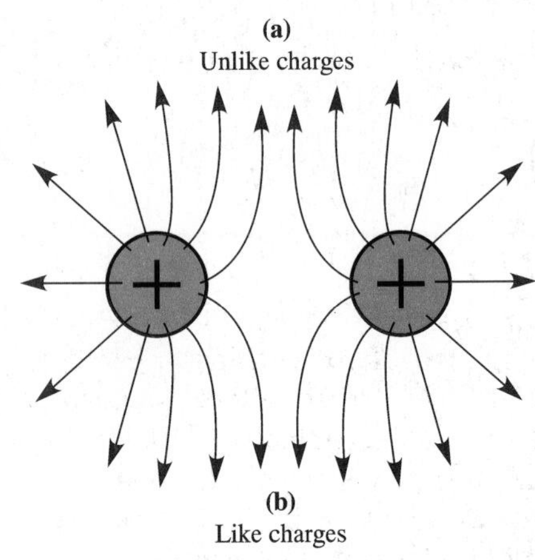

(b)
Like charges

within this field: The positively charged protons are attracted by positively charged electrons in the other body. Figure 2–7 illustrates the electrostatic fields between bodies with unlike and like charges. Note the force of attraction for unlike charges and repulsion for like charges.

Question: What are the effects of stored charge?

The entire concept of electricity and electronics is based upon the fact that negatively and positively charged particles can be separated. This separation is accomplished by *moving electrons, never protons.* (To move protons would require breaking up the nucleus of the atom, which is what is done in producing nuclear energy, not electricity!) Separating particles produces an imbalance: more electrons than normal in one body and fewer than normal in another. This unbalanced condition results in an electrical pressure.

Recall the previous example of how charge was stored on the body after walking across a rug. Then, when the doorknob was touched, a small electric shock occurred. This shock was caused by the movement of electrons from the body with an excess to the body with a deficiency. Once the charges were equal, the flow of electrons stopped. Newton's laws of motion tell us that no inanimate object moves unless an outside force is applied to it. Thus, there must be a pressure produced by stored charge.

The following example from the practical, everyday world illustrates the concept of electrical pressure. One purpose of municipal government is to provide water service for its citizens. The water will not, of course, leap from the wells and gush from faucets upon demand. It must somehow be placed under pressure. For example, by pumping the water into huge containers placed high above the ground, the force of gravity will supply the necessary pressure. As illustrated in Figure 2–8, the water stored under pressure in the tank will flow downward seeking its lowest level when the valve is opened. Of course, the city water main would be connected to this valve, and pipes distributing the water would go to each house and business. The pressure developed due to the water's height above the ground would cause the water to flow from faucets and hydrants. The higher the water is raised, the greater the pressure.

FIGURE 2–8 Water analogy for stored charge. Water stored in a tank high above the ground has pressure due to gravity. When the pressure is released by opening the valve, the water flows.

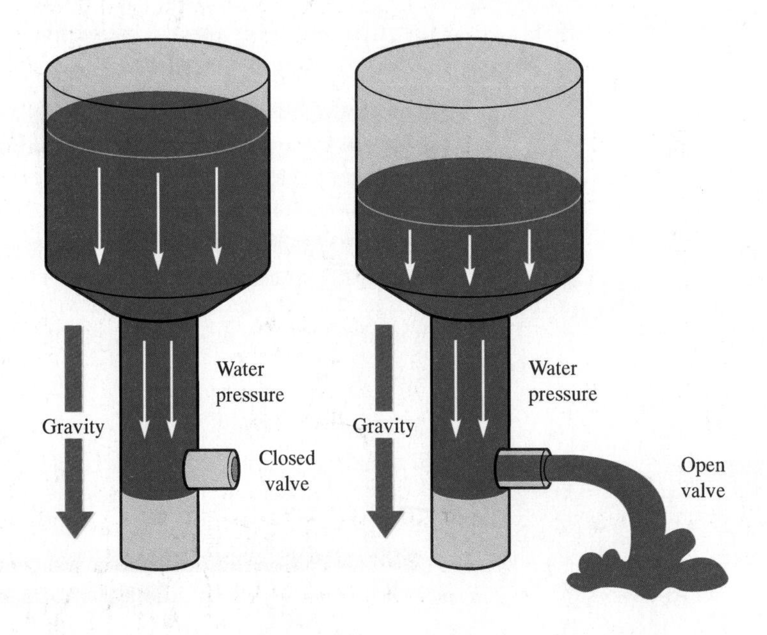

Putting water under pressure is analogous to storing electric charge. Stored charge results in an electrical pressure between the charged bodies. This electrical pressure is known as electromotive force, abbreviated EMF. It is, however, more commonly known as voltage. In a manner similar to the water analogy, the stored electrons, if given a suitable conducting path, will move in a way to correct the imbalance, as shall be seen in the next section.

■ SECTION 2–4 REVIEW

1. Define static charge.
2. What is the symbol and unit of measure for charge?
3. What is one result of stored charge?
4. Four coulombs of charge are stored. How many electrons have been added or removed from a material?

2–5 VOLTAGE

Question: What is the difference between energy and work?

Before discussing voltage, it is necessary to gain some knowledge of energy and work. **Energy** is defined as the capacity for doing work. The law of conservation of energy states: Energy can be neither created nor destroyed but only changed in form. In doing work, the form of the energy is changed. Take for example the automobile. The fuel, whether gasoline or diesel, has energy due to its composition. When ignited, it supplies energy through the power train of the car to the wheels in order to do the work of setting the car in motion. Once set in motion, the car has a portion of the energy that was released from the fuel. This energy of motion is known as *kinetic energy.*

A body of matter may possess energy due to its position. An illustration is given in our previous example in which water was stored in high towers in order to produce pressure. This type of energy is *potential energy* because it has the potential for doing work.

Work can be defined as a force acting through a distance. If one pushes against a brick wall, no work is done because nothing moves. If a brick is moved from the ground and placed atop the wall, work is done, energy is dissipated in doing this work, and potential energy is stored in the brick. The work done is the product of the weight of the item and the distance through which it is moved. In equation form, work is defined as follows:

(2–1) $$W = w \times D \quad [\text{work} = \text{weight} \times \text{distance}]$$

WHERE: W = work done (in ft · lb)
w = weight (in lb)
D = distance moved (in ft)

The raised brick will have the same amount of potential energy as the work that was done in raising it to that point. (If you doubt that the brick has stored energy, stand near the wall and have someone kick it off onto you!) The unit of energy is the same as that for work, the foot · pound.

■ Definition of Voltage

Question: What is the result of separating charges?

Work must be done in order to separate charges. That is, electrons must be moved. The result is potential electrical energy equal to the amount of work done in separating the charges. Once moved, the excess of electrons in the negatively charged body feel a pressure to return to the positively charged body and restore the balance. This electrical pressure from stored charge is known as **voltage,** denoted by V. (As we noted previously, this pressure is also known as the electromotive force, denoted E. Thus, voltage may be denoted by either V or E, but in this text, the symbol used for voltage is V.)

The unit of voltage is known as the volt (abbreviated V), which is defined in terms of the work done and number of coulombs of charge moved. The practical unit of electronic work is the **joule** (J), which is also the unit of electrical energy and which, as a frame of reference, is equivalent to 0.736 foot · pounds of mechanical work. **A volt** is defined as the amount of electrical pressure, or potential difference, that exists between two points when 1J of work is done in moving 1C of charge from one point to another. Figure 2–9 illustrates the concept of voltage according to this definition of the volt. In equation form, voltage is

(2–2)

$$V = \frac{W}{Q} \quad \left[\text{voltage} = \frac{\text{work}}{\text{charge}}\right]$$

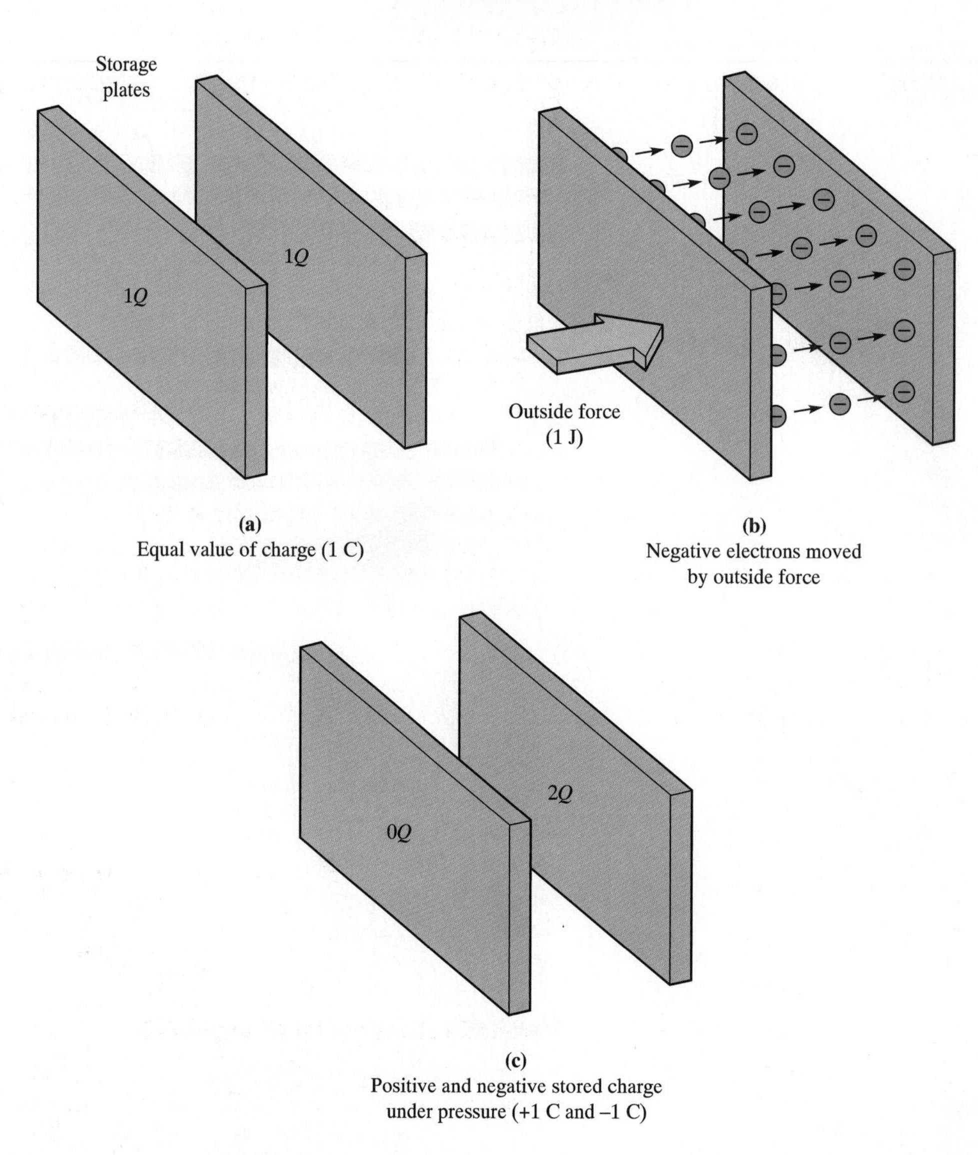

FIGURE 2–9 Voltage. Voltage is the energy stored as result of work done in moving a charge.

WHERE: V = voltage (V)
W = work (J)
Q = charge (C)

EXAMPLE 2–3 Compute the voltage developed if 0.03 J of work are done in moving 300 microcoulombs (μC) of charge from one point to another.

■ **Solution:**

$$V = \frac{W}{Q} \quad \left[\text{voltage} = \frac{\text{work}}{\text{charge}}\right]$$
$$= \frac{0.03}{300 \times 10^{-6}} = 100 \text{ V}$$

Calculator sequence for Example 2–3

[.] [0] [3] [÷] [3] [0] [0] [EXP] [±] [6] [=]

EXAMPLE 2–4 How many joules of work would be required to produce 300 V in moving 2000 μC of charge?

■ **Solution:**

Solving Eq. 2–2 for W yields

$$W = V \times Q \quad [\text{work} = \text{voltage} \times \text{charge}]$$
$$= (300)(2000 \times 10^{-6}) = 0.6 \text{ J}$$

Calculator sequence for Example 2–4

[3] [0] [0] [×] [2] [0] [0] [0] [EXP] [±] [6] [=]

EXAMPLE 2–5 If 0.1 J of work produce 10 V, how much charge was moved?

■ **Solution:**

Solving for Q gives

$$Q = \frac{W}{V} \quad \left[\text{charge} = \frac{\text{work}}{\text{voltage}}\right]$$
$$= \frac{0.1}{10} = 10 \text{ mC}$$

Calculator sequence for Example 2–5

[.] [1] [÷] [1] [0] [=]

NOTE

Note carefully, in this example, how the basic equation was transposed to obtain equations for all the variables. Thus, it is only necessary to know the basic equation for a quantity in order to find the other two. You will encounter numerous situations in the electronics curriculum where this skill is a necessity. It will greatly reduce the amount of memory work required.

SECTION 2–5 REVIEW

1. Define energy and work.
2. Define voltage and give its symbol and unit of measure.
3. What is the unit of electrical energy?

2–6 SOURCES OF VOLTAGE

Question: How is voltage produced?

As was stated previously, energy cannot be created or destroyed, but its form can be changed. Electrical energy is produced according to this law of energy. For example, heat and light are two forms of energy that can be converted to electrical energy. The energy of chemical reactions can also be converted to electrical energy. As examples of each method, this section describes the generator, the solar cell, and the battery.

Electric Generator

In a generator, electricity is produced by rotating loops of copper wire in a strong magnetic field. This process is known as electromagnetic induction. (Electromagnetic induction will be covered in detail in Chapter 8.) An outside source must provide the energy needed to rotate the generator. The process described here is one that is used in commercial power production.

The process begins with the burning of a fossil fuel such as coal or oil in order to produce heat. The heat turns water into steam, which is used to power a turbine engine. The engine in turn rotates the generator producing electricity. Figure 2–10 illustrates this process.

FIGURE 2–10 Converting energy in fossil fuel to electricity

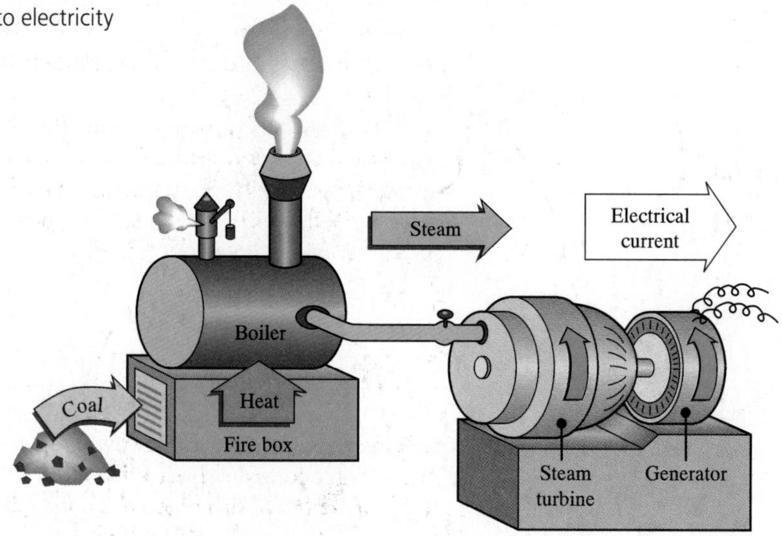

FIGURE 2–11 Producing electricity by chemical action

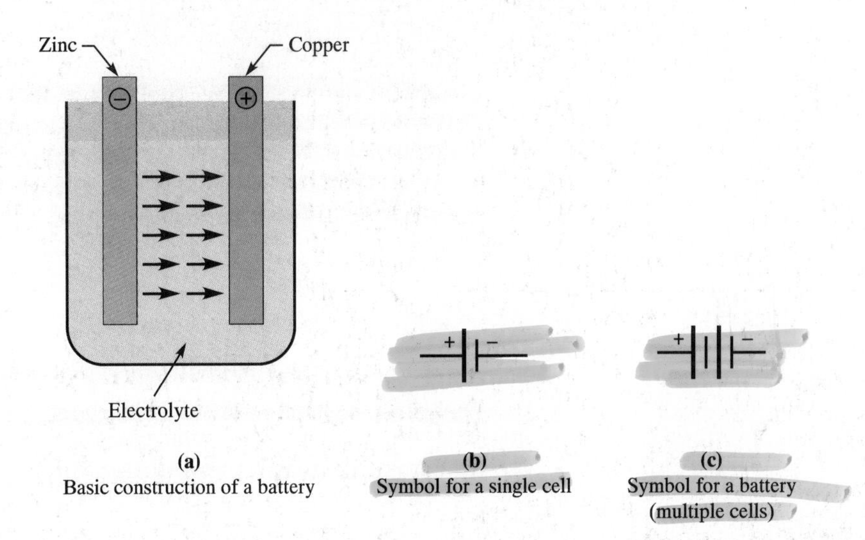

(a) Basic construction of a battery

(b) Symbol for a single cell

(c) Symbol for a battery (multiple cells)

■ Solar Cell

Solar cells get their name from the fact that sunlight is used as their energy source. They operate on what is known as the photovoltaic principle. *Photo* is a combining form meaning "produced by light"; hence, *photovoltaic* means voltage produced by light. These cells find applications where small amounts of power are required such as in hand-held calculators. They have also been used in space to provide electrical power for satellites.

The solar cell is composed of a junction of two types of processed silicon. When the junction is exposed to light, electrons in one of the materials gain enough energy to cross the junction into the other. The result is that one of the materials has an excess of electrons while the other has a deficiency. Charge is thus stored, and voltage produced as long as the junction is illuminated.

■ Battery

A **battery,** known as a voltaic cell, is a device that produces electricity through chemical action. Figure 2–11(a) illustrates the basic construction of a battery. Two dissimilar metals are immersed in an *electrolyte,* or liquid that will conduct electricity. Chemical action between the electrolyte and the metals causes one to gain electrons while the other loses electrons. In a practical battery, the metals are attached to metal posts, or *electrodes.* The electrode that has lost electrons will be marked with a plus sign (+), while the one gaining electrons will be marked with a minus sign (−). Once again, charge has been stored and voltage produced.

A battery is composed of units known as *cells.* The symbol for a cell is shown in Figure 2–11(b). The symbol for a battery is shown in Figure 2–11(c). Cells for the various types of batteries are composed of different chemicals and metals. The amount of voltage produced by a cell and its other characteristics are determined by the type of metals and electrolyte used. The types of batteries, their characteristics, and their ratings are covered in Appendix E.

■ SECTION 2–6 REVIEW

1. Describe the process for converting the energy of a fossil fuel into electrical energy.
2. List three forms of energy that can be converted to electrical energy.
3. What are the basic components of a voltaic cell?

2–7 CONDUCTORS, SEMICONDUCTORS, AND INSULATORS

Storage of charge produces an electrical pressure. The next step in causing electricity to do useful work is to make it flow through some device such as a toaster, lamp, or blow dryer. If given a suitable conducting path, the pressure will cause electrons to flow from the negative electrode, through the device, and back to the positive electrode. This flow of electrons is known as electric current, which will be discussed in more detail in the next section.

Question: Is the flow of electric current the same in all materials?

How well electric current flows in a material is determined by its properties. Current flows easily through some materials, not so easily through others, and in some, only under extreme circumstances. With regard to how easily electric current will flow, there are three classes of materials: conductors, semiconductors, and insulators. Each is covered in this section.

■ Conductors

Recall that the number of electrons in the valence shell of an element determines its electrical characteristics. The stable elements—the so-called noble gases—each have either a complete outer shell or one containing 8 electrons. These electrons are very tightly bound to the nucleus and would require a tremendous amount of energy to free them from their orbits.

Copper, as Figure 2–12 shows, has three complete inner shells and a single electron in the valence shell. In order to gain stability, the copper atom has two options: either gain 7 electrons for its valence shell or lose the single electron from it. The single electron in its valence shell is not tightly bound. Thus it is far easier for the atom to give up this loosely bound electron than to gain 7 electrons.

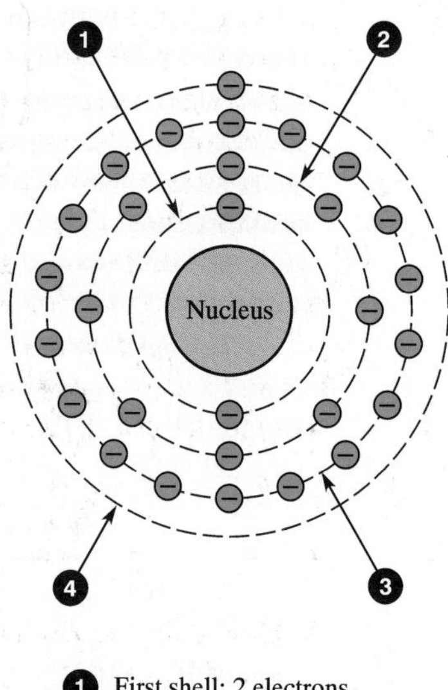

FIGURE 2–12 Atomic structure of copper, with a single valence electron

Ordinary conditions allow these electrons that are not tightly bound to move freely from atom-to-atom within a material, and they are thus known as **free electrons.** Any material with many free electrons (and having less than 4 valence electrons) is a **conductor** of electricity. In addition to copper, good conductors of electricity are silver and gold. Notice in Table 2–2 that they are in the same family of elements as copper. Like copper, they have 1 valence electron and thus many free electrons. Gold and silver are not, of course, used in commercial wiring applications due to their high cost and lack of strength. Gold is what is known as a noble metal. It doesn't rust or corrode in any manner. It is used in applications of electronics where corrosion must be avoided to ensure good electrical conductivity.

Insulators

An **insulator** is a material that has few free electrons (and has more than 5 valence electrons); it thus is a poor conductor of electricity. In these materials, then, all orbital electrons are tightly bound within their atoms, and few free electrons are available as charge carriers. Glass, paper, rubber, and plastic are good examples of insulators. Within a circuit, insulators prevent electrical contact between conductors. They also protect technicians and operators of electronic equipment from electric shock.

In certain uses, an insulator is known as a dielectric. A **dielectric,** though not capable of conducting electricity, is capable of storing electrical energy in the form of an electrostatic field. The electrostatic lines of force between charged bodies are developed in a dielectric. This term will be seen again in the study of devices known as capacitors.

Semiconductors

Between the conductivity of conductors and the nonconductivity of insulators lies a class of elements known as semiconductors. **Semiconductors** have 4 valence electrons, and as the name implies, they are neither good conductors nor insulators. The semiconductor elements are carbon, silicon, germanium, tin, and lead. In order to gain stability, these elements can either lose or gain 4 valence electrons. These elements, however, do not gain or lose valence electrons; they share them. This sharing of electrons, as noted previously, is known as a covalent bond. The semiconductors are the only family of elements that gain stability in this manner.

In a *covalent bond,* each atom shares one of its 4 valence electrons with the 4 atoms on both sides and immediately above and below. Each of these 4 share 1 of their electrons with this atom. In this manner, each atom "sees" 8 valence electrons. The result is fewer free electrons than a conductor but more than an insulator. Figure 2–13 illustrates the covalent bond.

FIGURE 2–13 Covalent bonds in pure semiconductor material. In this illustration, which shows valence shell electrons only, each atom "sees" 8 valence electrons through sharing with adjacent atoms.

The conductivity of all elements of the semiconductor family is not the same. Tin and lead, of which common solder is composed, have a great deal more conductivity than silicon and germanium. The same can be said for carbon of which some electronic devices are made. It is only the elements silicon and germanium that have the degree of conductivity, or lack of it, that make them useful in modern electronic devices such as transistors.

SECTION 2–7 REVIEW

1. How does the number of electrons in the valence shell affect the conductivity of an element?
2. What is meant by a "free" electron?
3. Why will there be few free electrons in an insulator?
4. What is the significance of the name semiconductor?
5. Describe a covalent bond.

2–8 ELECTRIC CURRENT

Question: Is all electron movement within a conductor electric current?

Free electrons within a conductor are in continuous movement from one atom to another. This motion is random, however, and does not constitute electric current. It is not the type of electron movement that can perform useful work. Figure 2–14(a) illustrates this type of movement.

Definition of Electric Current

Electric current, denoted by I for intensity, is the flow of electrons in a conductor under the influence of a voltage. If a source of voltage, such as a battery, is placed across a conductor, one end will be at a positive potential and the other at a negative potential. The negative pole will repel free electrons in the conductor while the positive pole will attract them. The result is a net flow of electrons toward the positive potential. This flow of electrons is what constitutes electric current and is illustrated in Figure 2–14(b).

The use of electricity involves the flow of electrons. There are two schools of thought on how this flow is represented. The electric current in Figure 2–14(b) represents the view known as *electron flow.* The other view is *conventional flow,* which shows positive charges moving from the positive to the negative side of the source. Texts in electrical engineering usually use conventional flow in explanations of circuit behavior.

FIGURE 2–14 Flow of electrons in a conductor. Electric current results when the electrons flow toward the positive potential.

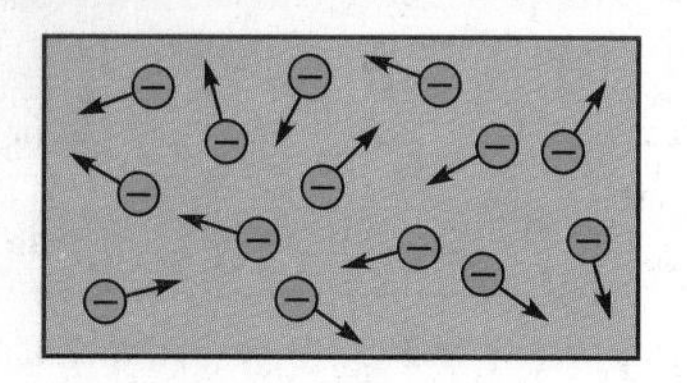

(a)
Random movement of electrons within a conductor

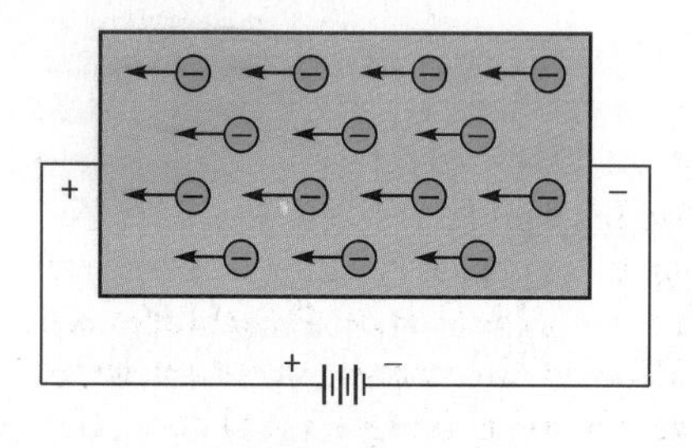

(b)
Flow of electrons from negative to positive poles

This text uses electron flow exclusively because it makes explanations simpler from a technology standpoint. Indeed, it is really not the direction of current flow that is important, but simply the fact that it does move. Any circuit can be successfully analyzed, with the same results, using either view.

In Figure 2–14(b), it is assumed that the polarity of the voltage source is fixed value and that current flows in only one direction. This type of current is known as direct current, or dc. There is another type of current, known as alternating current, or ac, in which the current changes directions periodically. This ability to change direction means, of course, that the polarity of the source must also change periodically. Alternating current and its circuits will be covered in the second part of this text. This section is concerned with the flow of direct current.

The unit of measure for the intensity of the flow of current is the **ampere** (abbreviated A), which is defined as being equal to 1 C of flow per second. Thus, if 1 C of charge flows past a given point in 1 s, the intensity of the current flow is 1 A. In equation form, current is defined as

(2–3)
$$I = \frac{Q}{t} \qquad \left[\text{current} = \frac{\text{charge}}{\text{time}} \right]$$

WHERE: I = the current flow (A)
Q = the charge (C)
t = time of flow (s)

The ampere is a rather large unit of current flow. In electronics, current will normally be measured in milliamperes (mA) or microamperes (μA). However, in high-power circuits, currents in the ampere range may be encountered.

Control of Current

Electronic systems, no matter what their purpose, have one thing in common: They operate through the systematic control of electric current. Electronic control devices such as resistors, transistors, diodes, and so forth, cause the current to increase, decrease, or drop to zero in response to some input signal. This basic principle lies behind radio, television, sound systems, and a host of other applications. In becoming an electronics technician, one must learn the operation of a great number of electronic control devices. If it is kept in mind that the purpose of all these devices is to control current, understanding will be greatly enhanced.

SYNOPSIS

At some point in certain chapters of this text, you will arrive at a point where the presentation of theory is complete and application begins. At these points a synopsis, or review, will be presented. Its purpose is to reinforce your knowledge of the important points covered and provide a checklist of essential information. You are at such a point in this chapter.

Up to now, charge, voltage, and current have been discussed. Stored charge produces a potential difference known as a voltage. Voltage is pressure that causes an excess of electrons to move toward a deficiency of electrons if a conducting path is provided. This flow of electrons is known as electric current.

Although it is sometimes heard, it is never correct to say that voltage "flows." The voltage *pushes* and the current *flows*. This usage is part of the language of electronics and should be made a part of your vocabulary.

SECTION 2–8 REVIEW

1. Does the random movement of electrons within a conductor constitute electric current?
2. What is the definition of electric current?
3. What is the symbol and unit of measure for current?
4. Define the ampere.
5. What is the purpose of most electronic control devices?

2–9 RESISTANCE

Question: What is meant by resistance?

When electrons move through a conductor, they collide with other electrons and atoms. Their movement is slowed by these collisions, causing them to give up energy in the form of heat. This slowing in the electron movement is known as resistance. **Resistance** (abbreviated R), then, is the opposition to current flow. The unit of measure for resistance is the **ohm,** which is symbolized by Ω, the upper case omega of the Greek alphabet. The ohm is named for George Simon Ohm who discovered the relationship between current, voltage, and resistance.

Many factors are involved in how much resistance a conductor has, including its length, cross-sectional area, and the element(s) of which it is made. A discussion of resistivity of conductors and a wire gauge table are contained in Appendix A.

Any conductive material contains some resistance, which is a property of the material known as *distributed resistance*. Resistance that is introduced into a circuit in order to limit current to some maximum value is *lumped resistance*. Lumped resistance is introduced into a circuit using a device known, appropriately enough, as a resistor. There are two types of resistors: fixed and variable. The symbols for these resistors are shown in Figure 2–15.

FIGURE 2–15 Resistor symbols

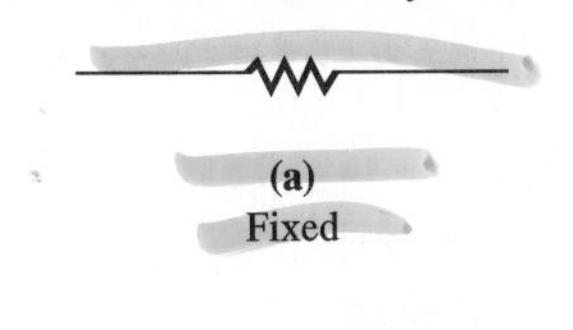

(a) Fixed

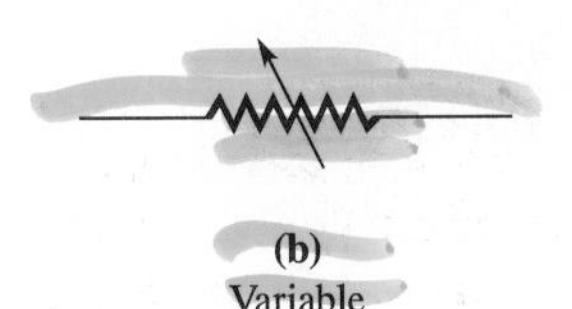

(b) Variable

■ Fixed Resistors

Fixed resistors are manufactured in a wide range of ohmic values and tolerances. It is not convenient, or in most cases possible, to alter this manufactured value. There are several varieties of fixed resistors including carbon composition, metal or carbon film, and wire-wound. Each has specific characteristics and applications. Figure 2–16 shows examples of each type.

Carbon composition resistors are made of a mixture of finely ground (powdered) carbon and insulating material held together by a resin binder. The value of the resistance is set by the ratio of carbon to insulating material. The mixture is formed into a cylindrical bar and is covered with a plastic insulating material, which provides for safety and gives mechanical strength. The ends of the bar have metal caps for connecting axial leads used in making circuit connections. A list of standard, or preferred, values for carbon composition resistors is contained in Appendix B.

Carbon composition resistors are usually manufactured with a precision of no better than ±5%. This measure of precision means that a 100 Ω, ±5% resistor could be any value between 95 and 105 Ω. Carbon composition resistors are also made with precision values of ±10% and ±20%.

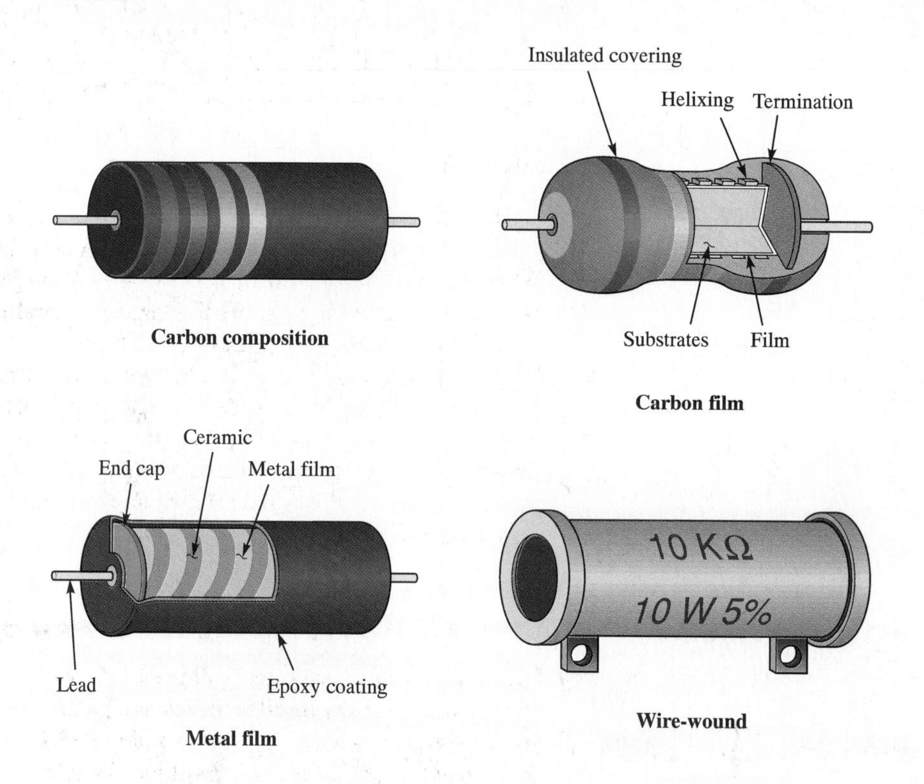

FIGURE 2–16 Varieties of fixed resistors

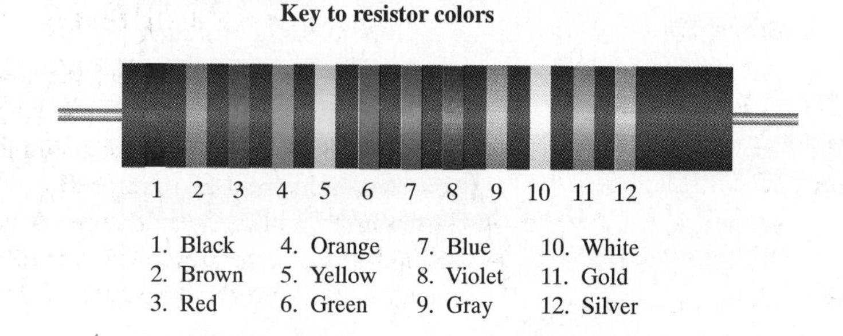

Caution: The numbers 1–12 in this figure are used to identify each color only. They do not represent the numerical value of that color as applied to the resistor color code.

TABLE 2–3 Resistor color codes

Color	Value	Multiplier	Tolerance
Black	0	10^0	
Brown	1	10^1	
Red	2	10^2	
Orange	3	10^3	
Yellow	4	10^4	
Green	5	10^5	
Blue	6	10^6	
Violet	7	10^7	
Gray	8	10^8	
White	9	10^9	
Gold		10^{-1}	5%
Silver		10^{-2}	10%

Film resistors are made by depositing a resistive element on a high-grade ceramic rod. There are a number of films used. Carbon or nickel chromium may be used. The film may also be a mixture of metal and glass or metal and metal oxide. The precise resistance is achieved by removing part of the film in a spiral pattern around the rod. Tolerances of ±1% can be achieved in metal film resistors.

Wire-wound resistors are made by wrapping what is known as resistance wire around an insulating core. Wire made of tungsten and manganin is commonly used for this purpose. The resistor is made by wrapping bare wire on an insulating core and encasing the entire unit in an insulator such as ceramic. Resistance wire will have a resistance value for a given length. Thus the value of a wire-wound resistor is determined by the length of the resistance wire.

Resistor Color Codes

Question: Do all resistors have their resistance value printed on them?

Resistance values are often indicated by coded colors. Table 2–3 contains a listing of these colors and the discrete values assigned to each. Two colors—gold and silver—are used to indicate tolerance or as the multiplier for resistor values of less than 10 Ω. A four-color band system is used for resistors with tolerances of ±5% and ±10%. A three-band system is used for those with tolerances of ±20%. This system is illustrated in Figure 2–17.

FIGURE 2–17 Resistor color code bands

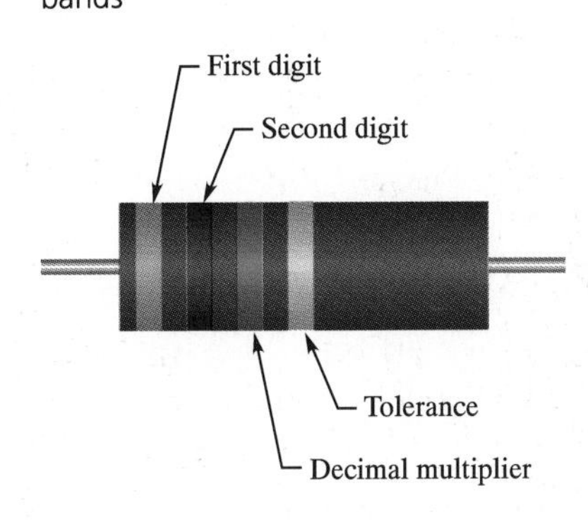

The resistance is decoded according to the band system in the following manner:

1. The color band nearest the end of the resistor indicates the first digit.
2. The second band indicates the second digit.
3. The third band is the decimal multiplier. It indicates the power of ten by which the first two digits must be multiplied.
4. The fourth band is the tolerance. A gold band means the resistor has a tolerance of ±5% while a silver band means ±10%. Lack of a fourth band (the three-band system mentioned) indicates a tolerance of ±20%.

FIGURE 2–18 Color-coded resistor

As an example, the resistor with color bands of red-yellow-orange-gold shown in Figure 2–18 is decoded as follows:

1. The first color band is red, so the first digit is 2.
2. The second color band is yellow, so the second digit is 4.
3. The third color band is orange, so the multiplier is 10^3, or 1000.
4. The fourth color band is gold, so the tolerance is $\pm 5\%$.

The resistance is 24 times 1000 or 24,000 Ω (24 kΩ). The tolerance is ±5%, and the range of resistances is computed in the following manner:

$$24{,}000 + (24{,}000 \times .05) = 25{,}200$$

$$24{,}000 - (24{,}000 \times .05) = 22{,}800$$

Thus, if the resistance value is between 22,800 Ω and 25,200 Ω the resistor is acceptable. The color coded value of resistance is its *nominal value.*

EXAMPLE 2–6 Determine the value of the resistor color coded brown-black-red-silver shown in Figure 2–19.

FIGURE 2–19 Resistor for Example 2–6

■ Solution:

1. First digit (brown) = 1.
2. Second digit (black) = 0.
3. Third digit (red) = 10^2 (or 100).
4. Tolerance (silver) = ±10%.

The resistance is 10 × 100 Ω or 1 kΩ. The tolerance is ±10%, so the range of acceptable values is

$$1000 + (1000 \times 0.1) = 1100\ \Omega$$

$$1000 - (1000 \times 0.1) = 900\ \Omega$$

The resistor is acceptable if its value is between 900 and 1100 Ω.

EXAMPLE 2–7 What is the color code of a 4.7 kΩ resistor with a tolerance of ±20%?

■ Solution:

1. First color band (4) = yellow.
2. Second color band (7) = violet.
3. Multiplier band (10^2) = red.
4. Tolerance, 20% = no band.

The color code is yellow-violet-red with no tolerance band.

EXAMPLE 2–8 Determine the value of the resistor color coded green-blue-black-gold shown in Figure 2–20.

■ Solution:

1. First digit (green) = 5.

FIGURE 2–20 Resistor for Example 2–8

2. Second digit (blue) = 6.
3. Multiplier (black) = 10^0 (or 1).
4. Tolerance (gold) = ±5%.

The value of the resistor is 56 × 1 or 56 Ω. The tolerance is ±5%, so the range of acceptable values is

$$56 + (56 \times .05) = 58.8\ \Omega$$

$$56 - (56 \times .05) = 53.2\ \Omega$$

One variation of this coding is for resistors with a value less than 10 Ω. In these cases, the multiplier band is either gold or silver. If gold, the multiplier is 10^{-1} and if silver 10^{-2}. The first two digits will be multiplied by 0.1 in the case of gold and 0.01 in the case of silver. Except for the values of the multipliers, the decoding is done in the same manner as in Example 2–8.

EXAMPLE 2–9 Determine the value of the resistor color coded blue-gray-gold-silver shown in Figure 2–21.

FIGURE 2–21 Resistor for Example 2–9

■ Solution:

1. First digit (blue) = 6.
2. Second digit (gray) = 8.
3. Multiplier (gold) = 10^{-1} (or 0.1).
4. Tolerance (silver) = ±10%.

The value of this resistor is 68 × 0.1 or 6.8 Ω. The tolerance is ±10%, so the range of acceptable values is

$$6.8 + (6.8 \times 0.1) = 7.48\ \Omega$$

$$6.8 - (6.8 \times 0.1) = 6.12\ \Omega$$

The resistor is acceptable if its value is between 6.12 and 7.48 Ω.

NOTE

It is possible that at some time you may encounter a resistor that is marked by a single black band. Recall that the color black represents a value of zero. As odd as it may seem, this is indeed a "zero ohm" resistor! Why a resistor with no resistance? Most electronic circuits manufactured today use a printed circuit board (PCB) to hold the various components and provide the required electrical connections. Often these circuits are modified to be used in several different pieces of equipment, or they are updated by the manufacturer to include improvements in design.

As a result, a resistor may have to be removed from the circuit, which presents a problem in that the path for current is broken when the resistor is removed. In order to restore a complete path, a component the same size as the resistor is inserted in its place. This replacement acts as a conductor with "zero ohms" of resistance and completes the circuit, providing the needed bridge to allow current to flow.

Precision Resistors

Fixed resistors with greater precision, those with tolerances of ±1% to ±0.1% tolerance, may be coded with five color bands. A table of preferred values and color code is provided in Appendix C.

■ Variable Resistors

Question: Are there resistors whose value may be adjusted?

FIGURE 2–22 Three types of variable resistors

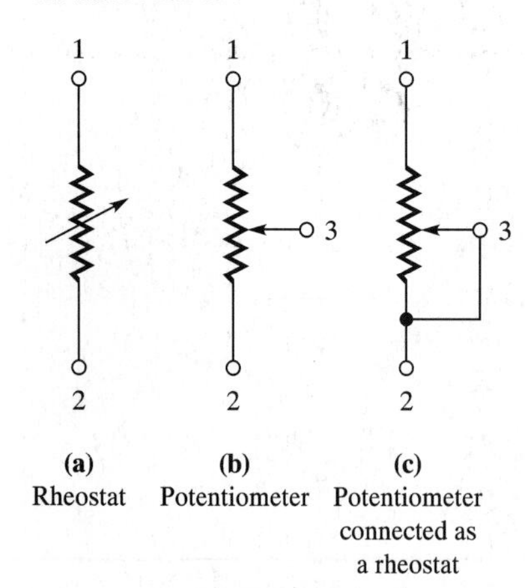

FIGURE 2–23 A variety of variable resistors

A *variable resistor,* as the name implies, is one whose value is easily changed. It is basically a resistor with a mechanical slider, known as a "wiper," which can be moved from one end of the resistor to the other. The resistor may be wirewound or made of carbon composition. The resistive element is usually formed in a circular fashion, and the wiper is moved by an attached shaft. Enough insulation is removed from the resistive element to allow electrical contact with the wiper. There are two basic types of variable resistors: the rheostat and the potentiometer. Their symbols are shown in Figure 2–22 and several varieties are shown in the color photograph Figure 2–23.

A *potentiometer* is a three-terminal device in which the ratio of two resistances to the total resistance can be varied. For example, the total resistance from two terminals—terminals 1 and 2 in Figure 2–22(b)—is a constant value regardless of the position of the wiper. The position of the wiper represents the third terminal, and it divides the resistor into two resistances: one from terminal 1 to terminal 3 and one from terminal 2 to terminal 3. As shown in Figure 2–24(a), if the wiper is set at midpoint, the ratio of the two resistances is the same. That is, if the total resistance were 10 kΩ, the two resistances would be 5 kΩ each. If, as in Figure 2–24(b), the wiper were moved toward terminal 1, the resistance between terminals 1 and 3 would decrease and resistance between 2 and 3 increase. As shown in Figure 2–24(c), movement of the wiper toward terminal 2 will have the opposite effect. Table 2–4 shows the resistances that result from positioning the wiper at various points along the resistor. The values are for a potentiometer with a resistance from terminals 1 to 2 of 10 kΩ. A device with this value is known as a 10k potentiometer or a "10k pot."

The variable resistors spoken of here are said to have a linear taper. (*Taper* refers to the physical shape of the resistor, or the distribution of the resistance along its length.) Thus, the resistance varies in a linear manner according to the angle of rotation of the shaft that turns the wiper (such as shown in Table 2–4). Others may have a nonlinear taper. An example would be one in which the resistance varied in a logarithmic manner in relation to the angle of shaft rotation. In this case, setting the wiper at its midpoint would not result in an equal division of the resistance.

A *rheostat* (Figure 2–22(a)) is a two-terminal device whose resistance changes as the wiper is moved. As shown in Figure 2–22(c), a potentiometer can be converted to a rheostat by connecting the wiper contact to one end of the resistor, thereby shorting a portion of it. (*Shorting* refers to a path of extremely low resistance—a "shortcut" for the current—around a resistor.) As the wiper is moved toward terminal 1, more of the resistor is shorted and the resistance decreases. Conversely, as the wiper is moved toward terminal 2, less of the resistor is shorted and the resistance increases. Thus, the resistance can be varied from zero to the maximum value of the resistor.

■ Conductance

Conductance is a measure of a materials ability to pass electric current. It is equal to the reciprocal of resistance, and as such, it is often referred to as the opposite of resistance. Its symbol is G and its unit of measure is the **siemen,** which is abbreviated S. (Note that siemen is abbreviated with capital "S"; lower case "s" is the abbreviation for seconds.) In equation form, conductance is defined as

(2–4)
$$G = \frac{1}{R} \quad \left[\text{conductance} = \frac{1}{\text{resistance}}\right]$$

Conductance, though useful and well worth studying, is not used nearly as often as resistance. Its major use is in the analysis of parallel circuits introduced in Chapter 5. You

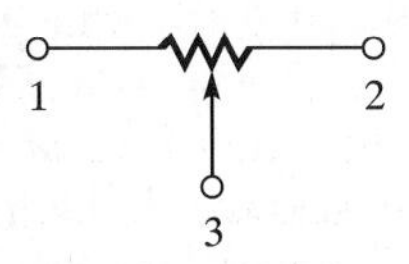

FIGURE 2–24 Operation of a potentiometer. The ratio of two resistances from terminals 1 to 3 and from terminals 2 to 3 is altered by positioning the wiper.

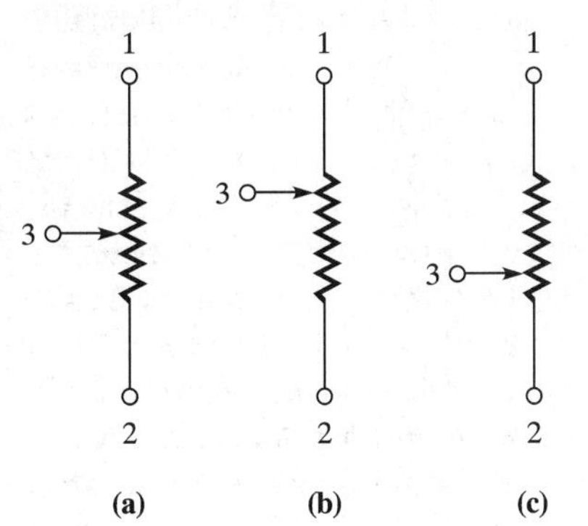

■ **TABLE 2–4 Potentiometer resistances**

Wiper Position from Terminals 1 to 2	Resistance Terminals 1 to 2	Resistance Terminals 1 to 3	Resistance Terminals 2 to 3
At terminal 1	10 kΩ	0	10 kΩ
One-tenth	10 kΩ	1 kΩ	9 kΩ
Two-tenths	10 kΩ	2 kΩ	8 kΩ
Three-tenths	10 kΩ	3 kΩ	7 kΩ
Four-tenths	10 kΩ	4 kΩ	6 kΩ
One-half	10 kΩ	5 kΩ	5 kΩ
Six-tenths	10 kΩ	6 kΩ	4 kΩ
Seven-tenths	10 kΩ	7 kΩ	3 kΩ
Eight-tenths	10 kΩ	8 kΩ	2 kΩ
Nine-tenths	10 kΩ	9 kΩ	1 kΩ
At terminal 2	10 kΩ	10 kΩ	0

will also encounter conductance later in the electronics curriculum in the analysis of field-effect transistors.

EXAMPLE 2–10 Compute the conductance of the following resistances: (a) 2 kΩ, (b) 100 kΩ, (c) 1.5 megaohms (MΩ).

■ Solution:

a. $G = \frac{1}{R} = \frac{1}{2000} = 5 \times 10^{-4}\text{S} = 500\ \mu\text{S}.$

b. $G = \frac{1}{R} = \frac{1}{100{,}000} = 1 \times 10^{-5}\text{S} = 10\ \mu\text{S}.$

c. $G = \frac{1}{R} = \frac{1}{1{,}500{,}000} = 6.667 \times 10^{-7}\text{S} = 0.6667\ \mu\text{S}.$

Calculator Sequence for Example 2–10

a. [2] [0] [0] [0] [1/x] [=]

b. [1] [0] [0] [0] [0] [0] [1/x] [=]

c. [1] [5] [0] [0] [0] [0] [0] [1/x] [=]

EXAMPLE 2–11 A resistance has a conductance of 2 μS. What is its value in ohms?

■ Solution:

Solving Eq. 2–4 for R gives

$$R = \frac{1}{G} \quad \left[\text{resistance} = \frac{1}{\text{conductance}}\right]$$

$$= \frac{1}{2 \times 10^{-6}} = 500\ \text{k}\Omega$$

Calculator sequence for Example 2–11

[2] [EXP] [±] [6] [1/x] [=]

■ Temperature Coefficient

Question: Do changes in temperature affect conductors and resistors?

An increase in the temperature of a conductor causes random motion of free electrons to increase. In turn, this increase causes more collisions between electrons and more interference with the flow of electrons; hence, the resistance is greater. The opposite is also true. A decrease in temperature results is less random motion and a lower resistance. Thus, the resistance of a conductor is directly proportional to its temperature. If a rise in temperature results in an increase in resistance, a device has a positive temperature coefficient.

The **temperature coefficient** of a material is the factor by which its resistance varies per degree Celsius temperature change. This number is designated α, lowercase alpha from the Greek alphabet. Table 2–5 contains the temperature coefficients for several materials.

Semiconductors have a negative temperature coefficient. Thus, an increase in temperature will cause a decrease in resistance. Recall that semiconductors get their stability through the formation of covalent (shared electron) bonds. The application of heat breaks some of these bonds, releasing more free electrons for current flow. The result is a decrease in resistance.

Some materials such as constantin (an alloy of copper and nickel) and manganin (an alloy of copper, manganese, and nickel) have a zero, or at least small, temperature coefficient of resistance. As such, they can be used to make precision, wire-wound resistors whose value will not change significantly with a change in temperature. (See Table 2–5)

■ **TABLE 2–5 Temperature coefficients**

Material	Temperature Coefficient (α) (20°C)
Silver	3.8×10^{-3}
Carbon	-5×10^{-4}
Constantin	8×10^{-7}
Copper	3.93×10^{-3}
Nickel	6×10^{-3}
Tungsten	4.5×10^{-3}

■ SECTION 2–9 REVIEW

1. Define resistance. What is its symbol and unit of measure?
2. What is the nominal value of a resistor color coded red-red-green-gold?
3. What is the maximum allowed value of a resistor color coded yellow-violet-yellow-gold?
4. What is the value of each resistor in Figure 2–25.
5. What is the difference between a rheostat and a potentiometer?
6. Define conductance. What is its symbol and unit of measure?

2–10 THE ELECTRIC CIRCUIT

Question: What are the requirements of an electric circuit?

An **electric circuit** has three essential parts: (1) a voltage source, (2) a conducting path for current, and (3) an electrically operated device, or **load.** The voltage source may be any of those mentioned in Section 2–6. Two wires, one connected to the positive and one

FIGURE 2–25 Resistors for Section Review Problem 4

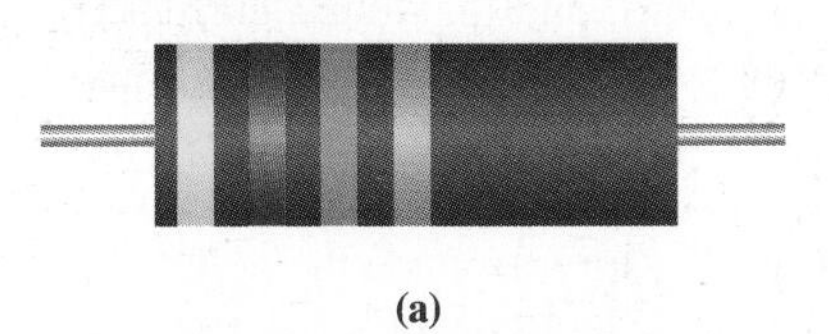

(a)

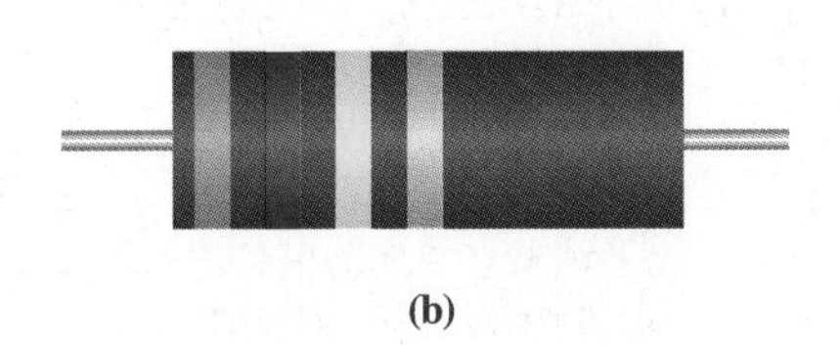

(b)

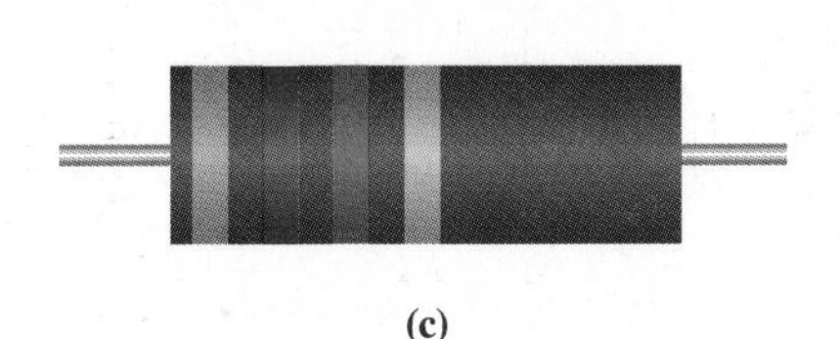

(c)

FIGURE 2–26 Circuit drawings

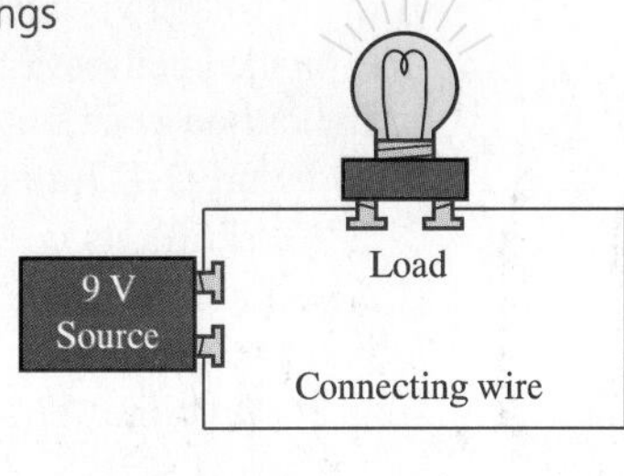

(a)

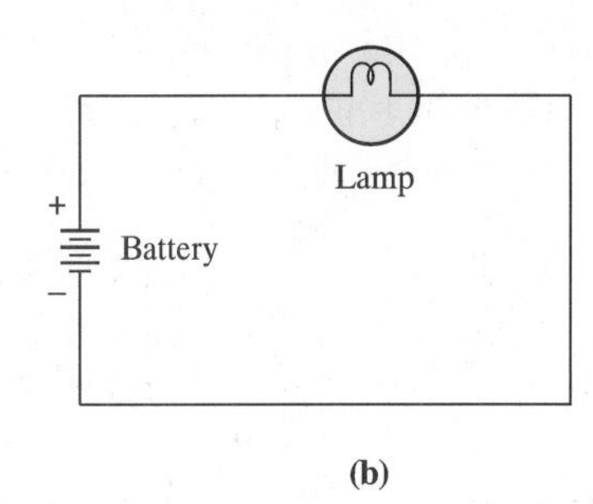

(b)

connected to the negative terminal, provide a conducting path for current. For purposes of this explanation, the load will be an electric lamp.

The circuit is connected as shown in Figure 2–26(a). This type of drawing is known as a *pictorial diagram*. In it, the components are shown as they actually appear. The source provides the pressure (voltage) to move the stored charge (electrons) from the negative terminal, through the conducting wires and load, and back to the positive terminal. This flow of electrons constitutes electric current. Within the lamp is a filament made up of resistance wire. The current flowing through this resistance produces heat, causing the filament to glow and emit light.

Figure 2–26(b) shows the same circuit as a *schematic diagram*. In a schematic diagram, a symbol is used for each component rather than an actual drawing of it. This is the type of diagram technicians deal with most often. Notice that as the name implies, in a "circuit" there is a complete path for current from one side of the source to the other. Note further that every basic electrical quantity presented earlier in this chapter comes into play in the circuit.

STUDY SKILLS

Reflecting

After you have finished working a problem or reading, it is always good to take a few minutes and consider what you have just covered. Consider how the examples relate to what you have learned previously. How are the examples or problems similar or different from ones you have already learned?

Practical Electric Circuits

In addition to the three basic elements of any electric circuit, a *practical circuit* has at least two more components: a switch with which to open and close the circuit and a device to protect the source and conductors in the event of an overload. An *overload* is a condition that occurs when a load attempts to draw more current than the source is designed to supply. The protective device may be either a fuse or a circuit breaker.

■ Switches

A switch has two positions—open or closed—which are usually referred to as ON and OFF. In the open position, the switch's resistance will be infinite, and no current will flow. In the closed position, its resistance will be practically zero, and maximum current will flow. Figure 2–27(a) shows the condition of the circuit with the switch closed and Figure 2–27(b) shows its condition with the switch open. The switch used in these illustrations is a *single pole-single throw* (SPST) switch. An SPST switch has the input voltage attached to a single pole, and moving the switch to the on position applies the voltage to a single output line. There are several other types of switches that must be considered.

FIGURE 2–27 Operation of a switch. In (a), the switch is ON, the current path is complete, and the lamp glows. In (b), the switch is OFF, the current path is broken, no current flows, and the lamp is extinguished.

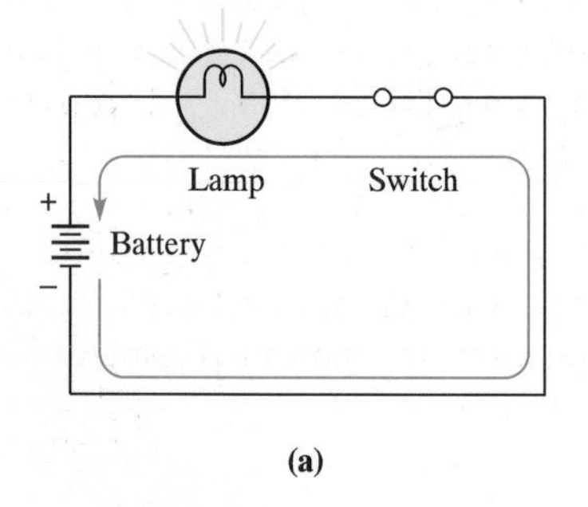

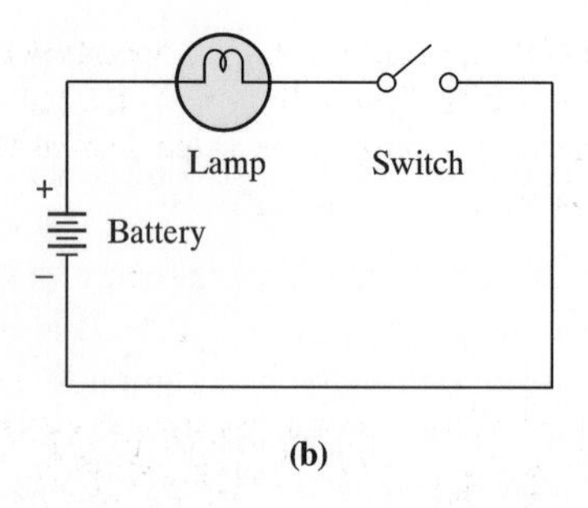

Single Pole-Double Throw (SPDT)

The SPDT switch, shown in Figure 2–28(a), has a single input line to the pole but can be switched to one of two output lines. It is used to switch the electric current from the source to either of two loads. This type switch may also have a center OFF position that de-energizes both loads.

Double Pole-Single Throw (DPST)

The DPST switch, shown in Figure 2–28(b), is really two SPST switches in the same package. The dashed line means that the contacts are mechanically linked and they will open and close as one. This switch allows the opening and closing of two conducting paths simultaneously.

Double Pole-Double Throw (DPDT)

The DPDT switch, shown in Figure 2–28(c), is the same as the DPST but now there are two ON positions rather than one. Input lines are connected to the poles, which can be switched to either of two sets of contacts. This switch may also have a center OFF position.

Normally Open Push-Button (NOPB)

The NOPB, shown in Figure 2–28(d), is one in which a connection is made between contacts when the switch is depressed and broken when the switch is released. This type of switch is sometimes constructed to mechanically lock in the ON position and is released by depressing it a second time.

Normally Closed Push-Button (NCPB)

The NCPB, shown in Figure 2–28(e), is one in which the contact is broken when the switch is depressed and made when the switch is released. It may also have the mechanical locking feature of the NOPB.

■ Protective Devices

Question: Why are protective devices needed in electric circuits?

A source and its conductors are constructed so that they can safely supply a certain amount of current to a load. If a load attempts to draw more than this rated current, serious damage may be done to the voltage source. Devices are installed in the current path that open, becoming an infinite resistance, if the rated current is exceeded. Two devices are used for this purpose: the fuse and the circuit breaker.

Fuses

A *fuse* is made of a metal link encapsulated in glass or some other nonconductor. Heat is produced by current flowing through the link; the higher the current the greater the heat produced. When a certain current, specified by the design of the fuse, is exceeded, the link will melt or "fuse," thereby opening the circuit and stopping current flow. For example, if the maximum rated current of the voltage source is 1 A, the fuse placed in the

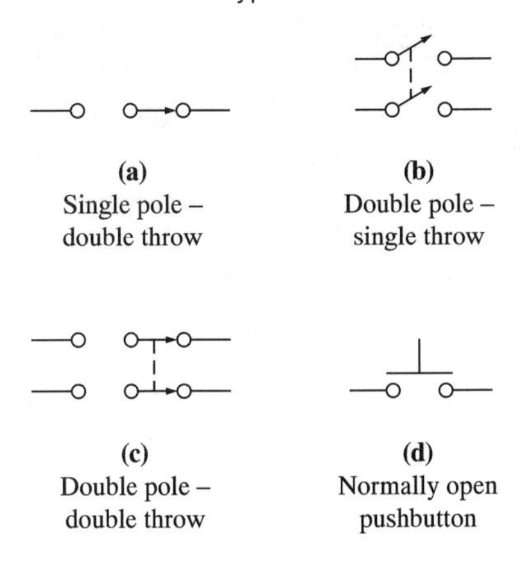

FIGURE 2–28 Types of switches

current path would be designed to open when 1 A of current was exceeded. That is, the fuse has a current rating of 1 A. Most fuses that fail in electronic circuits are replaced and discarded. In some high current applications, the fuse may be repairable by replacing the link.

In addition to a current rating, fuses have a voltage rating. For example, a blown fuse creates an open circuit, and the source voltage will be found across the fuse. Even though the current path is open, the electrical pressure is still present. If the voltage is great enough, and the gap across the fuse is short enough, current may arc across it creating an unsafe condition. Therefore, the fuse must have a voltage rating below which arcing will not occur. Examples of a variety of fuses are shown in the color photograph of Figure 2–29. Figure 2–30 shows the symbol for a fuse.

Circuit Breakers

A *circuit breaker* is a mechanical device that opens the circuit when an overload occurs. There are two types of circuit breakers: those actuated by heat and those actuated by magnetic action. In the thermal type, an excess current produces heat that causes a metal strip to expand and bow. This bowing action trips a spring-loaded switch, opening the circuit. In the magnetic type, an electromagnet, whose strength is directly proportional to the amount of current flowing, is used to actuate a switch. When the safe current is exceeded, the magnetic field becomes strong enough to open the circuit. In either case, the circuit breaker can be reset and operation restored when the reason for the excess current is corrected. Examples of a variety of circuit breakers are shown in Figure 2–31. The symbol for a circuit breaker is shown in Figure 2–32.

CAUTION

A circuit breaker is not designed to be used as an ON-OFF switch! Continued operation of these devices can seriously alter the current value at which they will trip. If this value increases an appreciable amount, a serious safety hazard can exist. If the value decreases, the circuit may not operate even though there is no fault.

FIGURE 2–29 Various types of fuses

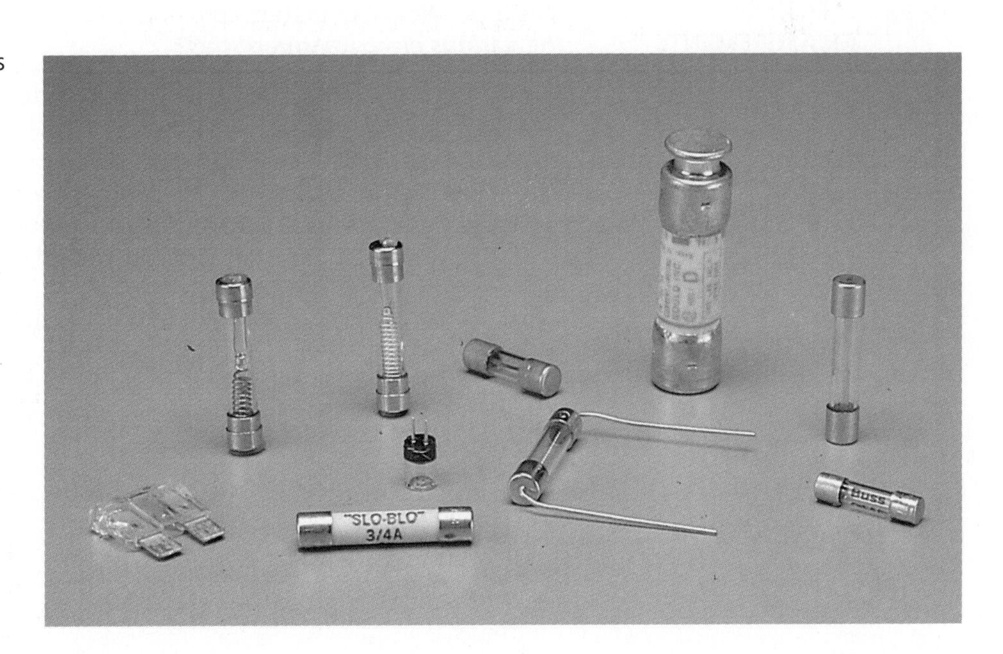

FIGURE 2–30 Symbol for a fuse

FIGURE 2–31 A variety of circuit breakers

FIGURE 2–32 Symbol for a circuit breaker

■ SECTION 2–10 REVIEW

1. What are the essential parts of an electric circuit?
2. What is the difference between a pictorial and a schematic diagram?
3. What is the purpose of a fuse or circuit breaker?
4. Describe the action of a double pole-double throw switch.
5. Why does a fuse have a voltage rating as well as a current rating?

2–11 ELECTRICAL MEASUREMENTS

Question: How are meters connected to measure the various electrical quantities?

■ Measuring Instruments

The most common measurements in electronics are those of voltage, current, and resistance. In equipment tests, alignments, and troubleshooting, these quantities must be routinely and accurately made. A single instrument, a *multimeter* or VOM (volt-ohm-milliampere meter), is used to measure the three quantities, as its name implies. The multimeter is equipped with a function switch for selecting the quantity to be measured.

There are two types of multimeters: analog and digital. An analog meter is shown in Figure 2–33(a). Its rotating pointer indicates the value on the scale beneath it. Figure 2–33(b) shows a digital multimeter (DMM). The DMM indicates the value on a display as discrete digits. A DMM or VOM may be conditioned through switching to measure any one of the quantities of voltage, current, or resistance. It may be referred to as a *voltmeter* when conditioned to measure voltage, an *ammeter* when conditioned to measure current, or an *ohmmeter* when conditioned to measure resistance.

■ Measuring Voltage

Voltage, because it is a difference in potential, must always be measured between two points in a circuit. Voltage is measured by placing the leads of the voltmeter on either side of the load, as illustrated in Figure 2–34. Figure 2–35(a) shows an analog voltmeter

FIGURE 2–33 Two types of multimeters

a. Analog multimeter

b. Digital multimeter

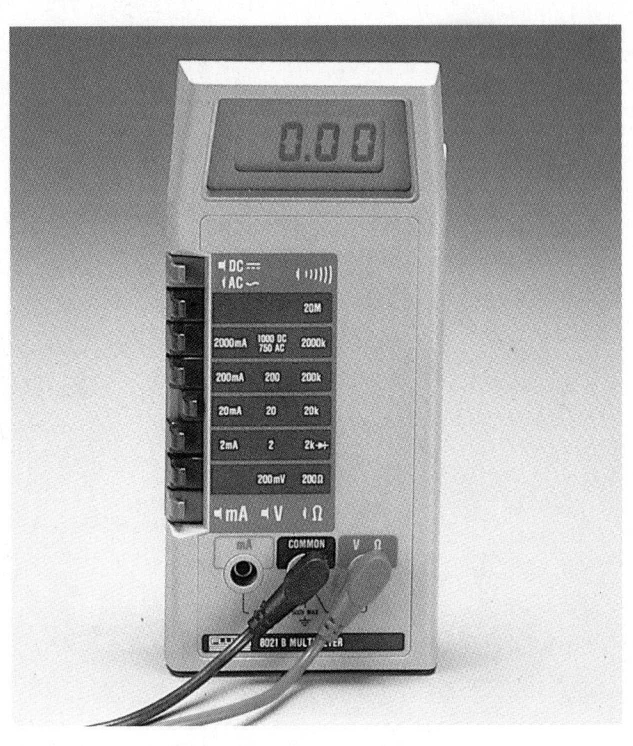

FIGURE 2–34 Schematic diagram of a voltmeter properly connected to measure voltage.

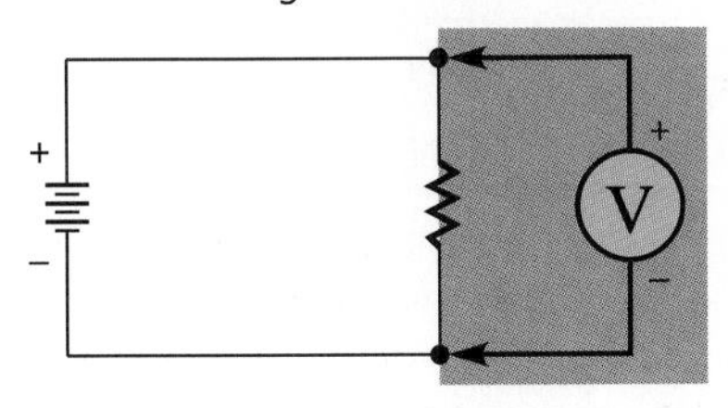

and 2–35(b) a digital voltmeter connected across a circuit. Devices connected in this manner are said to be in parallel, a circuit configuration covered in detail in Chapter 5. A practical voltmeter will have several voltage scales. One popular voltmeter in use today has 5 scales: zero to 2.5, 10, 50, 250, and 1000 volts.

When measuring unknown voltages, it is wise to begin with the voltmeter set to its highest scale and then move lower to the most convenient scale. Otherwise, too great a voltage could damage the meter. In addition, proper polarity must be observed when connecting the meter. Otherwise the pointer would try to move down-scale and possibly cause damage to the meter. Usually the lead to the red jack of the meter must be connected to the positive side of the source and the black one to the negative side.

Measuring Current

Current is the flow of electrons in a conductor. Thus, the ammeter must be placed within the line of current flow. This method of connecting electronic components is known as a series connection. Series circuits are covered in detail in Chapter 4. As shown in Figure 2–36, the current leaves the circuit, is metered while flowing through the meter, and then returns to the circuit. Note the symbol for the ammeter. Figure 2–37 shows analog and digital ammeters in operation within a circuit. Once again, polarity must be observed in connecting the meter. It is also good practice in measuring current to begin with the highest scale. Having to open the circuit to insert the meter can be difficult in some instances. For this reason, current measurements are made only when absolutely necessary. Some power sources have a permanently mounted ammeter for continuous monitoring of current.

FIGURE 2–35 Measuring voltage with analog and digital meters.

a. Analog multimeter measuring voltage

b. Digital multimeter measuring voltage

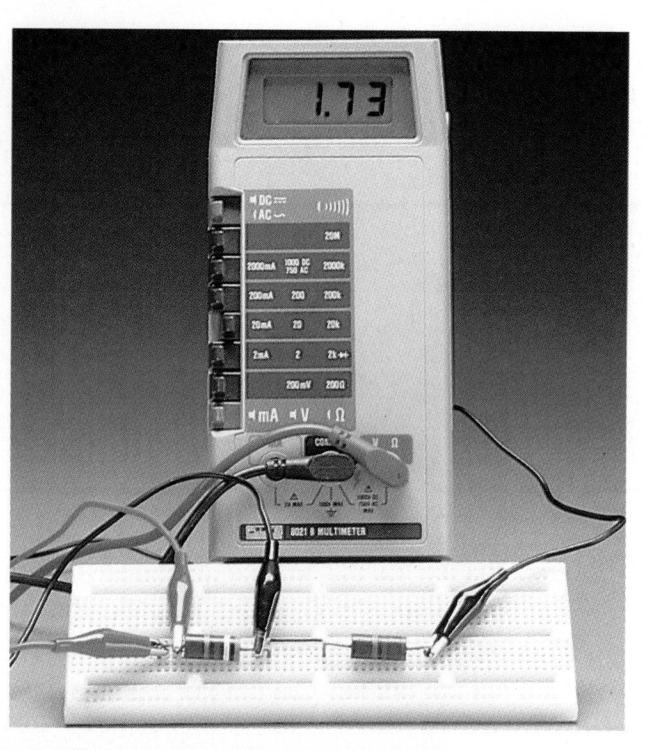

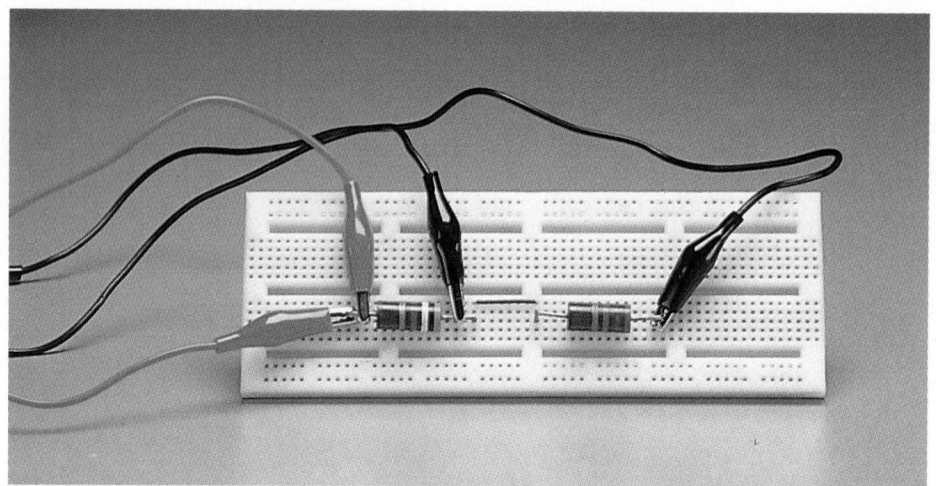

FIGURE 2–36 Schematic diagram of an ammeter properly connected to measure current.

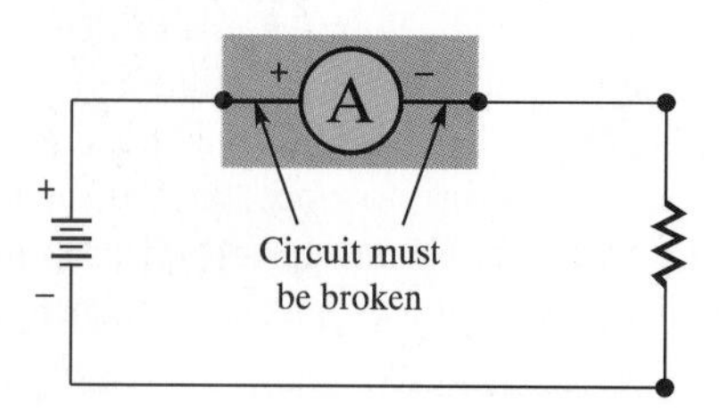

FIGURE 2–37 Measuring current with analog and digital meters.

a. Analog multimeter measuring current

b. Digital multimeter measuring current

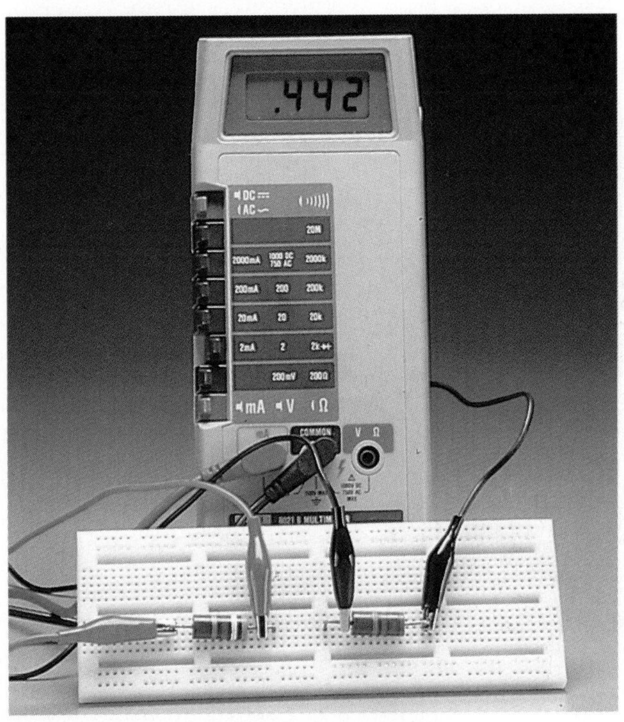

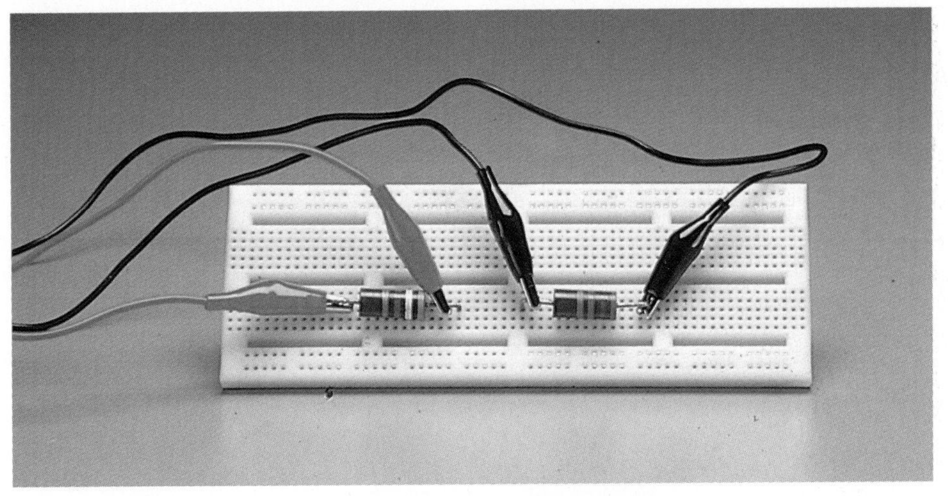

FIGURE 2–38 Schematic diagram of an ohmmeter properly connected to measure resistance. Note that the power must be removed from the circuit or the device removed.

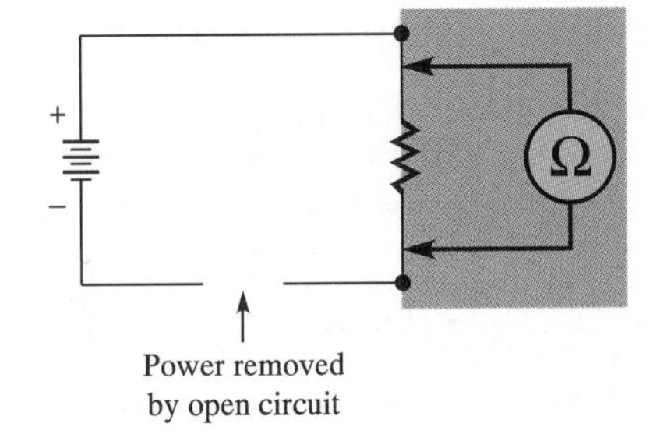

FIGURE 2–39 Measuring resistance with analog and digital meters.

a. Analog multimeter measuring resistance

b. Digital multimeter measuring resistance

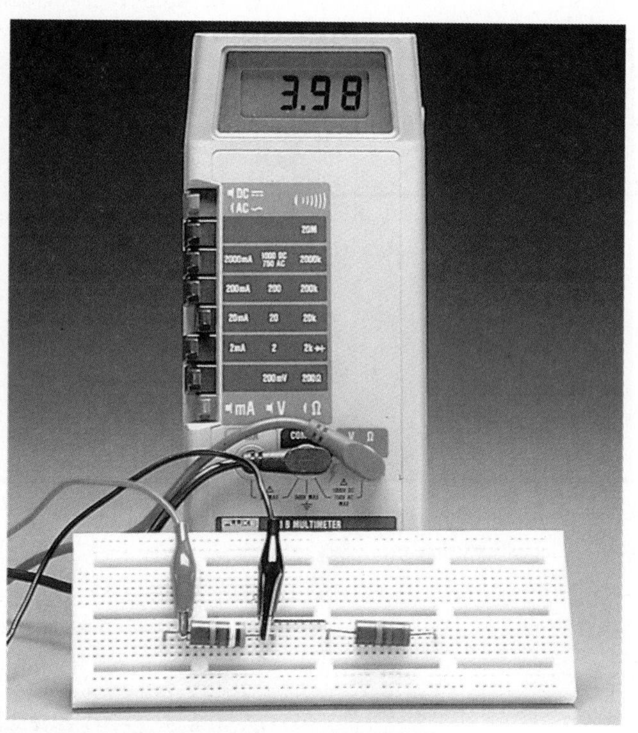

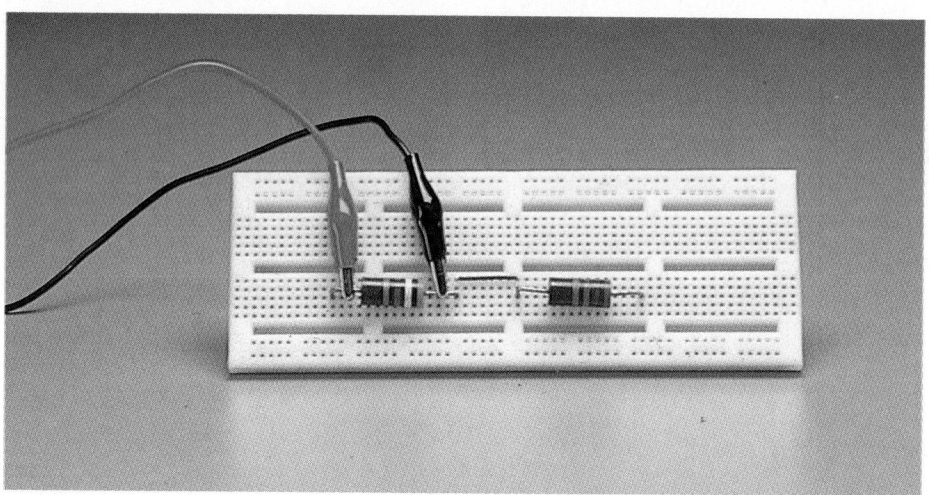

■ Measuring Resistance

Resistance is measured as shown in Figure 2–38. Figure 2–39 shows analog and digital ohmmeters in use. Note that *severe damage* to the ohmmeter can result if resistance is measured when voltage is applied to the circuit. Consequently, the power switches in the circuit must be placed in the OFF position prior to making resistance measurements. Other precautions that must be taken in the interpretation of ohmmeter indications when making in circuit tests of resistors will be pointed out in the Troubleshooting sections of appropriate chapters. There is no polarity to be observed in resistance measurements.

SECTION 2–11 REVIEW

1. What are the two types of meters used in electrical measurements?
2. Why are current measurements made only when absolutely necessary?
3. How is an ammeter connected in a circuit?
4. What precaution must be taken prior to measuring resistance?

2–12 TROUBLESHOOTING

Modern electronic systems are highly reliable, but are still subject to failure. An electronics technician must be capable of quickly and accurately analyzing a problem and making necessary repairs, and must often do so under conditions of great stress. Thus, it is necessary that a technician develop a logical, step-by-step approach to fault analysis.

To begin with a simple example, consider an electric lantern. This well-known device contains, with the exception of a fuse, all the elements of a basic circuit as given earlier. Its schematic is shown in Figure 2–40(a). The source of electrical energy is a 6 V battery. (Note the symbol for a battery.) When the switch is closed and the circuit completed, the voltage pushes current through the load, which is the electric lantern, shown by the lamp symbol. Inside the lamp is a filament made of resistance wire. Current flowing through this filament heats it to incandescence. (*Incandescence* means heating so intense that the material glows white, emitting light.) The useful work done by the current in this circuit is the production of light for the user's convenience.

FIGURE 2–40 Circuits for troubleshooting the lantern

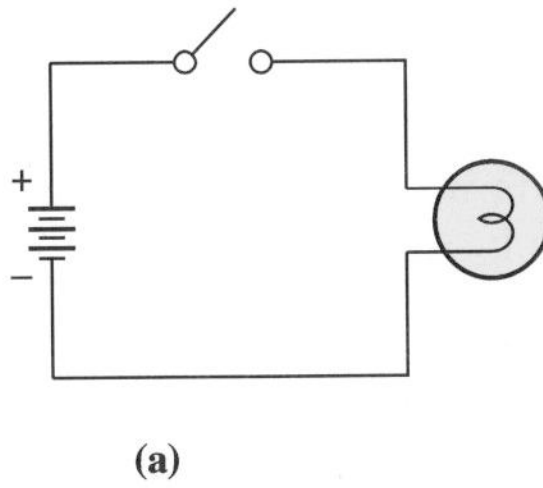

(a)
Circuit for lamp

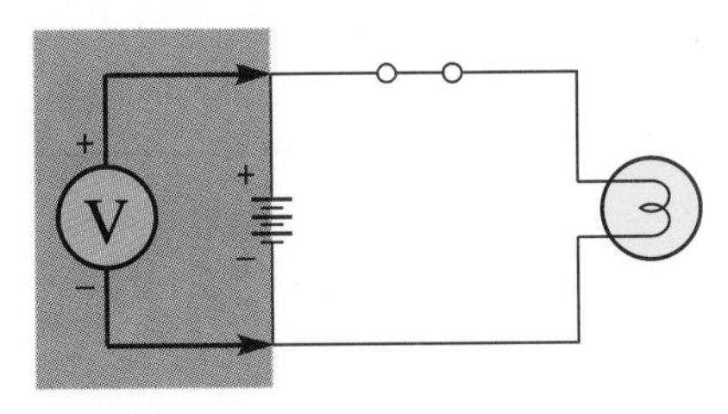

(b)
Measuring source voltage

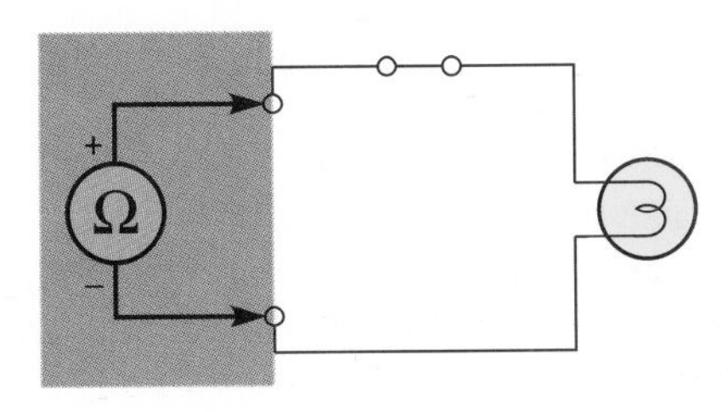

(c)
Measuring circuit resistance

The first step in troubleshooting is to determine that a fault does in fact exist. Just as a medical doctor looks for symptoms of illness in a patient, a technician must know the purpose of the electronic system and look for symptoms of trouble. In the case of the electric lantern, a symptom could be a reduced light level or no light at all. Once the symptom is determined, all possible causes must be considered and tested until the actual fault is found.

One point that must be emphasized here can be stated in terms of a maxim of the electronics industry: "If it is not broken, do not try to fix it." Thus, if no symptoms exist, testing is needless and should not be done, because it invites mistakes and errors of judgement that can put a perfectly operating system out of commission.

Suppose, in the case of the electric lantern, the symptom of no light does exist. Recalling what it takes to produce light, one would conclude that there is no current flowing through the filament. There are two possible causes of this problem: (1) the voltage source has failed or (2) the conduction path for current has been broken.

The lantern's voltage source can be checked as shown in Figure 2–40(b). (The switch should be ON during this test.) A reading of near 6 V indicates normal voltage, while a reading of near 0 V indicates no current and thus no light. A voltage reading of less than 6 V indicates reduced current and reduced light. In either the case of low voltage or no voltage, the battery must be replaced.

The most probable cause of a break in the current path would be the burn out of the lantern's filament. This problem can be checked by removing the lantern and testing it with an ohmmeter. A very low resistance reading indicates a good lantern, while a reading of infinite ohms indicates a burned out lantern, which must be replaced.

A defective switch may also be the cause of an open current path. The switch can be checked by measuring its resistance (Figure 2–40(c)) in both the ON and OFF positions. (After first removing the battery, of course.) In the ON position, it should read near zero ohms, and in the OFF position, it should read infinite ohms. The switch may become resistive due to corroded contacts or mechanical wear. In this event, it adds resistance to the circuit, which thus lowers the flow of current and results in a lower light level. A significantly greater resistance than zero ohms would be read by the ohmmeter.

An open conducting path may also be caused by an open interconnecting wire, which can be checked by measuring the resistance between the leads that connect to the battery's electrodes. With the switch ON the resistance should be near zero ohms. (Assuming of course that there is a good lamp installed.) Figure 2–40(c) illustrates this check.

SECTION 2–12 REVIEW

1. What is the first check to be made in fault analysis?
2. What two things could cause the current in the electric lantern to be zero?
3. What two things could cause the current in the electric lantern to be too low?
4. Explain the procedure for testing the switch of the electric lantern?
5. What is the meaning of the term *incandescence*?

CHAPTER SUMMARY

1. An atom is the smallest particle of an element that will retain that element's properties and characteristics.
2. A proton is a particle of positive charge and is found in the nucleus of the atom.
3. An electron is a particle of negative charge and is found in orbit around the nucleus of the atom.
4. A neutron is a particle with no charge and is found in the nucleus of the atom.
5. The number of positive charges in an atom equals the number of negative charges making it electrically neutral.
6. An atom that loses or gains an electron is no longer neutral and is known as an ion.
7. The number of electrons in the outer, or valence shell, of an element's atom determines its electrical characteristics.
8. Charge must be stored in order to produce a potential difference or voltage.
9. A potential difference is electrical pressure, known as voltage or electromotive force.
10. A voltage applied to a conducting path causes current to flow in a circuit.
11. Current flow in a circuit is opposed by resistance.
12. The basic electric circuit consists of a voltage source, a conducting path, and a load.
13. A closed circuit is one in which there is a complete path for current from the negative side to the positive side of the source.
14. An open circuit is one in which there is a break in the conducting path which results in no current flow.
15. A voltmeter is connected across the load or voltage source in order to measure a difference in potential.
16. An ammeter is placed in line with the current path in order to measure the intensity of current flow.
17. To avoid damage to an ohmmeter, voltage must be removed from the circuit or component under test.
18. A switch is a mechanical device that provides a convenient way of opening the circuit.
19. Fuses and circuit breakers are devices that open when excessive current flows in a circuit.
20. In fault analysis, the symptom of the problem must be observed and all possible causes considered and tested until repairs are made.

KEY TERMS

atom
proton
neutron
electron
valence shell
coulomb
energy
work
joule
volt
battery
free electrons
conductor
insulator
dielectric
semiconductors
electric current
ampere

resistance
ohm
conductance
sieman
temperature coefficient
electric circuit
load

EQUATIONS

(2–1) $W = w \times D$

(2–2) $V = \frac{W}{Q}$

(2–3) $I = \frac{Q}{t}$

(2–4) $G = \frac{1}{R}$

Variable Quantities

W = work
w = weight
D = distance
V = voltage
Q = charge
t = time
G = conductance
R = resistance

TEST YOUR KNOWLEDGE

1. What are free electrons? What type of elements have many of them?
2. What is static electricity? How is it produced?
3. What are the laws of repulsion and attraction for charged bodies?
4. What name is given the field that exists between charged bodies?
5. What is a positive ion?
6. What is the common name for potential difference?
7. What must be done in order to produce a potential difference?
8. What is the definition of an ampere?
9. What characteristic makes an element a conductor?
10. What is produced when current flows through a resistance?
11. How does a potentiometer differ from a rheostat?
12. What is the polarity of charge of a single proton? A single electron?
13. What are the parts of a basic electric circuit?
14. What are two purposes of insulation?
15. In making voltage measurements, how is the meter connected?
16. In making current measurements, how is the meter connected?
17. Is it possible to connect a voltage source between two points and have no current flow? Explain your answer.
18. Why are voltage measurements simpler than current measurements?

PROBLEM SET: BASIC

Section 2–2

1. From the periodic table, determine what element has an atomic number of 14.
2. What is the maximum number of electrons allowed in the third orbital shell of an atom?

Section 2–4

3. Compute the charge in coulombs if 18.75×10^{18} electrons are stored.
4. How many electrons must be stored in order to produce 800 μC of charge?
5. What is the charge of a material that has lost 6.25×10^{18} electrons?

Section 2–5

6. If 10^2 J of work is done in moving 5^{-3} C of charge, how much voltage is produced?
7. What is the work done in joules if 50 V is produced in moving 300^{-3} C of charge?
8. If 100 V are produced when 500^{-6} C of charge is moved, how much work was done?

Section 2–8

9. A current with an intensity of 500 mA flows for 10 s. How many coulombs of charge pass a point during this period of time?

10. During a period of 5 s, 25 C of charge pass a given point. What is the intensity of the current in amperes?
11. A current has an intensity of 750 mA. How long will it take 500 μC of charge to pass a given point?

Section 2–9

12. What is the nominal resistance value of each resistor in Figure 2–41?
13. For each resistor of Figure 2–41, state the limits of resistance its value must be between in order to be acceptable.
14. What is the conductance of each resistor in Figure 2–42?
15. Determine the color code for each of the following resistance values:
    a. 47 Ω ± 5% **b.** 160 Ω ± 10%
    c. 470 Ω ± 20% **d.** 22 Ω ± 5%
    e. 130 Ω ± 5% **f.** 390 Ω ± 5%
16. Determine the color code for each of the following resistance values:
    a. 2.2 Ω ± 5% **b.** 0.56 Ω ± 10%
    c. 510 Ω ± 5% **d.** 12 MΩ ± 10%
    e. 12 k Ω ± 10% **f.** 240 kΩ ± 5%
17. For each of the following resistor color codes, determine the limits of resistance its value must be between in order to be acceptable:
    a. Red-black-brown-gold **b.** Brown-black-yellow-silver
    c. Yellow-violet-black-silver **d.** Blue-gray-orange-gold
    e. Red-red-silver-silver **f.** Brown-black-green-gold

FIGURE 2–41 Resistors for Problems 12 and 13

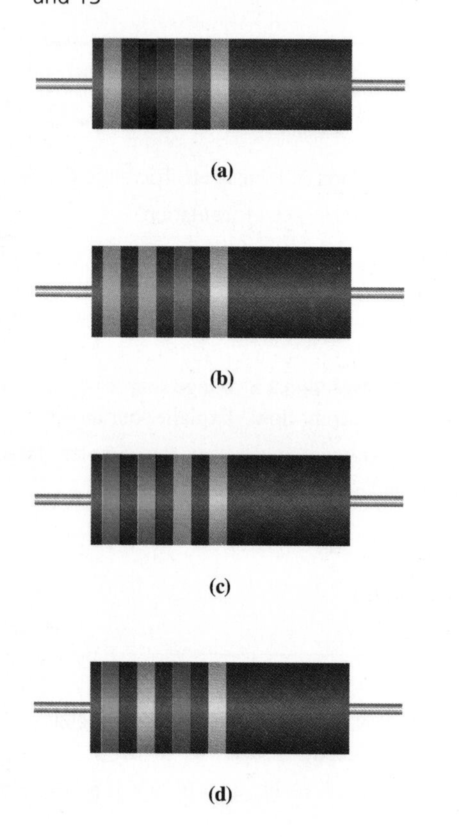

FIGURE 2–42 Resistors for Problem 14

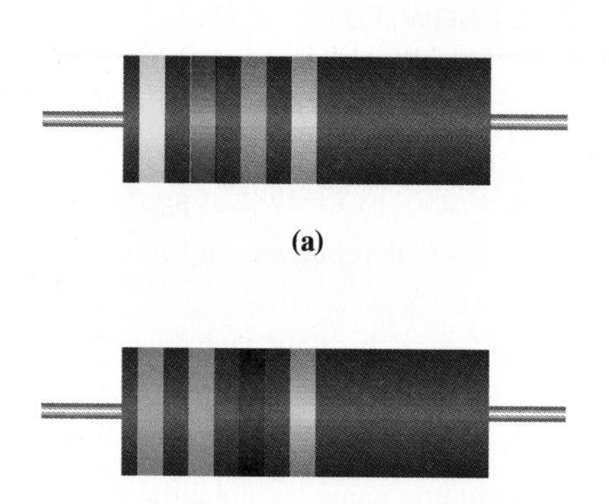

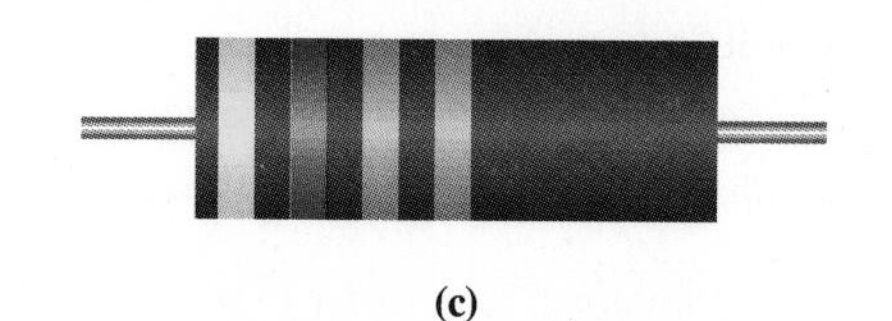

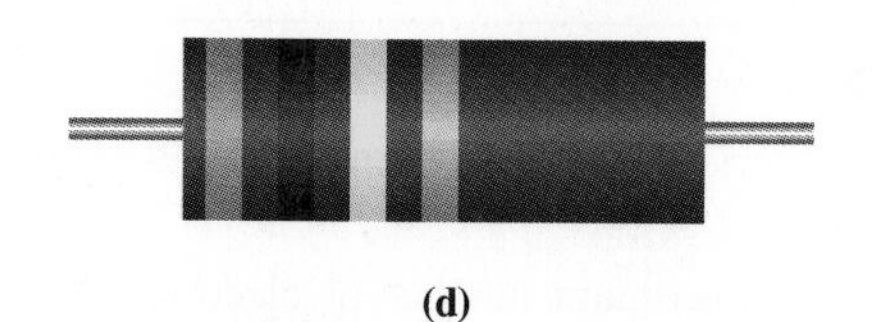

PROBLEM SET: CHALLENGING

18. For the following conductances, find the color code of the closest standard value resistor for each, assuming all have a ±5% tolerance:

a. 13.33 μS **b.** 555 μS **c.** 9.09 mS
d. 66.66 μS **e.** 27.77 μS

19. A calculation reveals that 32,385 Ω of resistance are required in a circuit. If a tolerance of ±10% is acceptable, what standard value of resistance could be chosen?

20. Only the orange multiplier band and silver tolerance band of a resistor are visible. If the resistor is within tolerance, what is the maximum value it could be?

PROBLEM SET: TROUBLESHOOTING

21. A security light controlled by a switch from within the house refuses to light. Its circuit consists of the power source, switch, wiring, and lamp. List four possible causes of this fault.

22. An electric heater refuses to operate. Its circuit, as shown in Figure 2–43, consists of the power source, the wiring, a fuse, a resistive heating element, an ON-OFF switch, and a safety switch. The safety switch opens the circuit if the heater is not in an upright position. List six possible causes of this fault.

FIGURE 2–43 Circuit for Problem 22

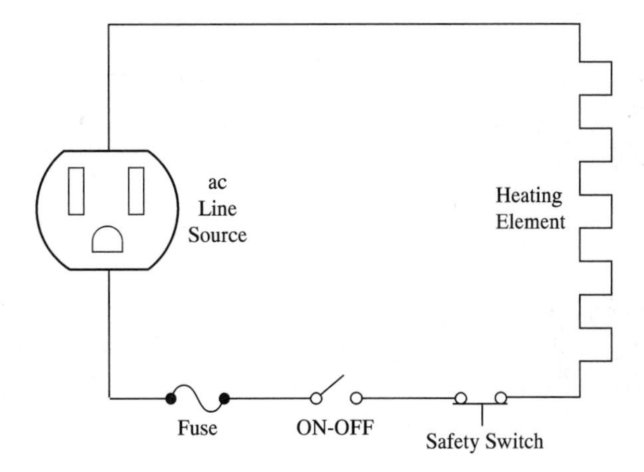

Superconductors

Browse through a few science or electronics publications and you'll more than likely come across an article about superconductors. Much effort is currently being put forth in superconductor research. Various articles explain that the almost universal goal of research scientists is to create a room temperature superconductor. The question for many, however, may be "What is a superconductor?"

Simply put, a superconductor is a conductor that presents no resistance to the flow of electrical current. This phenomenon was first observed in 1911 by the Dutch scientist Kamerlingh Onnes. Onnes discovered that if a mercury crystal was cooled to a temperature just above absolute zero (4 kelvins (K) or −452°F), it lost all resistance to the flow of electricity. For many years, scientists tried to explain what was happening to cause superconductivity. Only recently has this process begun to be understood.

From the original discovery in 1911 until 1986, superconduction was observed only in certain metals when they were cooled to temperatures near absolute zero. Then in 1986 in Zurich, K. Alex Muller observed superconduction in some ceramic compounds at −408°F. Shortly thereafter, C. W. Paul Chu of Houston, Texas, announced that he had achieved superconduction at −284°F. Suddenly the idea of room temperature superconductors seemed a real possibility.

The basic idea of the superconductor is simple. A normal conductor presents resistance to electrical current as it moves through a conductor. This resistance converts energy to heat. If current is to continue to flow, a constant source of energy must be present to replace energy lost as heat. A superconductor has no resistance and therefore does not generate heat. If an electrical current is set flowing in a superconductive loop, it will flow literally forever.

It can easily be seen how superconductive power lines could save enormous amounts of energy in the transmission of commercial power to homes and businesses. Typically 15% to 20% of the energy produced by electric utility companies is lost in transmission to the user. Superconducting transmission lines would not only lower the cost of producing electricity but would conserve natural resources as well. Other less obvious applications that have been proposed for superconductors include medical instruments such as magnetic imaging devices, high-speed computer applications, high-speed trains levitated by magnetic fields created around superconductive rails, and even superlong-life batteries.

Just what causes a material to become a superconductor is still a mystery to scientists, although many facts are known. In a normal conductor, electrons are continually colliding with atoms which make up the conductor. Each collision causes energy to be lost as well as heat to be generated. This process is illustrated in Figure 1. In a superconductor each passing electron causes a small vibration in the atomic

FIGURE 1 Action of electrons in a normal conductor

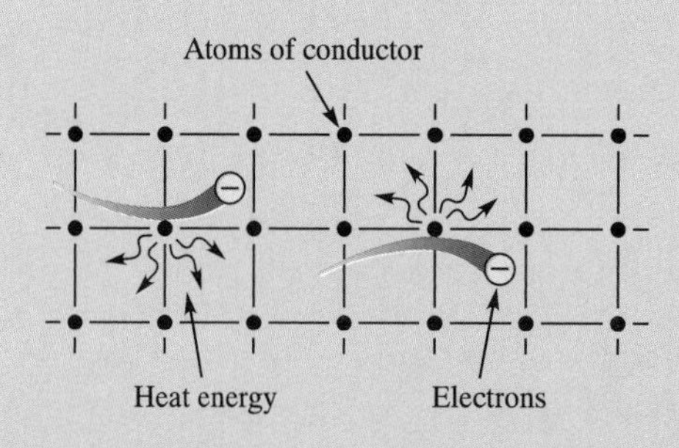

FIGURE 2 Action of electrons in a superconductor

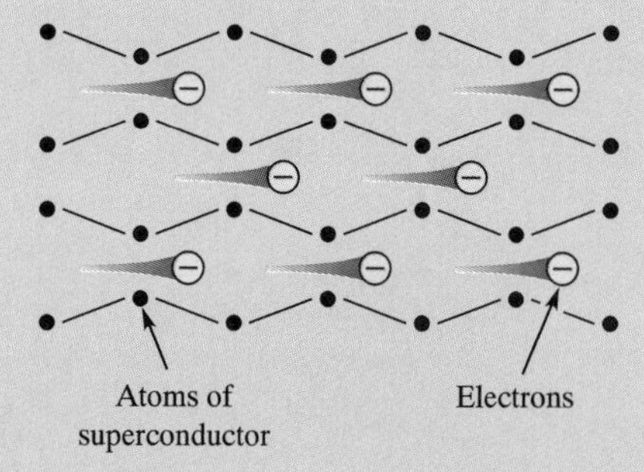

continued

latticework of the conductor. This action clears the way for a second electron, which is pulled along behind (Figure 2). Between each bound pair of electrons may be millions of other electrons, all being pulled along by an invisible tow line. As temperature rises, the motion of the conductor's atoms increases and more and more atomic obstacles present themselves. Eventually the bond between electrons is broken and superconduction stops. For this reason superconduction requires an extremely low operating temperature.

Scientists are currently puzzled over the new superconductor. It is not certain how these newly discovered compounds are able to superconduct at such high temperatures. As it stands now experimentation has produced superconductors that science cannot explain.

CHAPTER

3 Ohm's Law and Power

■ UPON COMPLETION OF THIS CHAPTER, YOU WILL BE ABLE TO

1. Explain how voltage, current, and resistance are related in an electrical circuit.
2. State the three equations expressing Ohm's law.
3. Explain the relationship between energy, work, and power.
4. Explain how power is developed in an electrical circuit.
5. State the three forms of the power equation.
6. Explain how power is measured.

STUDY SKILLS

Preparing for Exams: Is Homework Enough?

Doing homework will help you to develop your skills and better understand the material, but it is not a guarantee that you will do well on an examination for several reasons. First, unlike homework, exams must be completed in a limited amount of time. Second, the fact that your performance is being graded may make you anxious and, therefore, more prone to careless errors. Third, homework usually covers a limited amount of material while exams normally cover much more, which increases the chances of your forgetting or confusing the skills, rules, and concepts involved. Finally, your notes and text book are available while doing homework, but not during an exam.

If you believe that you understand the material and/or you do well on your homework, but your exam grades are not what you feel they should be, then the Study Skill sections presented throughout the text should be helpful.

This chapter examines the ways in which the three basic quantities of voltage, current, and resistance are related. Although the relationships may be expressed in the form of mathematical equations, true knowledge of the way in which the quantities interact is more important than simply the memorization of the equations.

How well a technician understands the voltage-current-resistance relationships will ultimately determine how effective he/she will be in the work force. The basic relationships, first developed by George Simon Ohm (1789–1854), must be mastered early if continued progress in the chapters that follow is expected.

FIGURE 3–1 Basic circuit demonstrating interaction of voltage and current when resistance is held constant

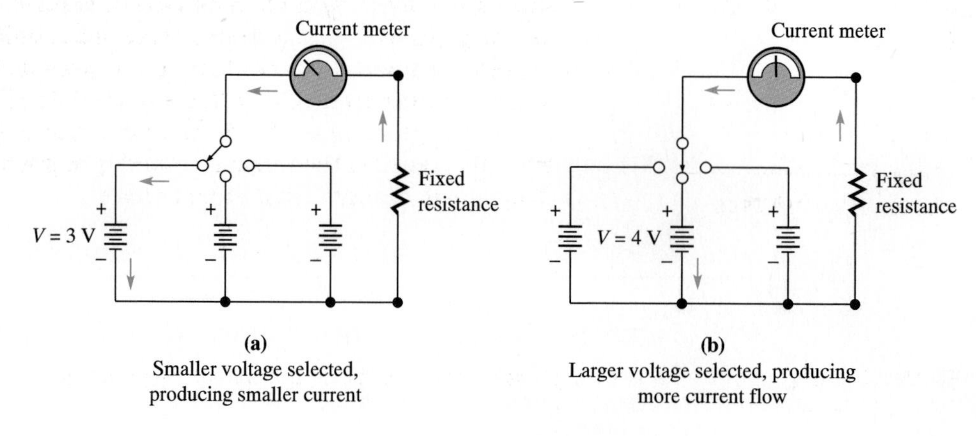

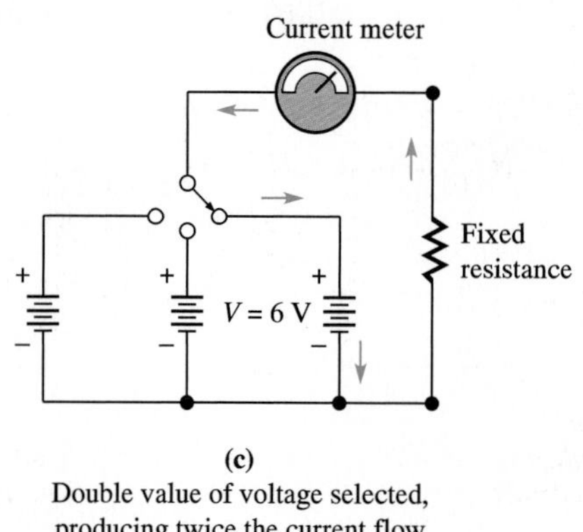

3–1 THE RELATIONSHIP BETWEEN VOLTAGE, CURRENT, AND RESISTANCE

Question: How do voltage, current, and resistance interact in an electrical circuit?

Although several different methods may be used to demonstrate how voltage, current, and resistance interact, the method shown by Figure 3–1 uses a constant value of resistance and a selectable value of voltage. As the selector switch is moved from one position to another to change the voltage, the ammeter indicates the resulting values of current flow in the circuit for each. Note that changing the value of the voltage does *not* change the value of resistance.

It can be observed that as a smaller value of voltage is selected, a smaller value of current is measured. As a greater value of voltage is selected, a greater amount of current is measured. If the voltage is increased to exactly double its original value, the resulting current is exactly double. If the value of voltage is made one half, the value of current is one half. The voltage-to-current ratio is constant and thus represents a **proportional relationship.** It also represents a **direct relationship** in that an increase in one quantity results in a proportional increase in the other.

FIGURE 3–2 Graph showing linear relationship of voltage and current when resistance is held constant

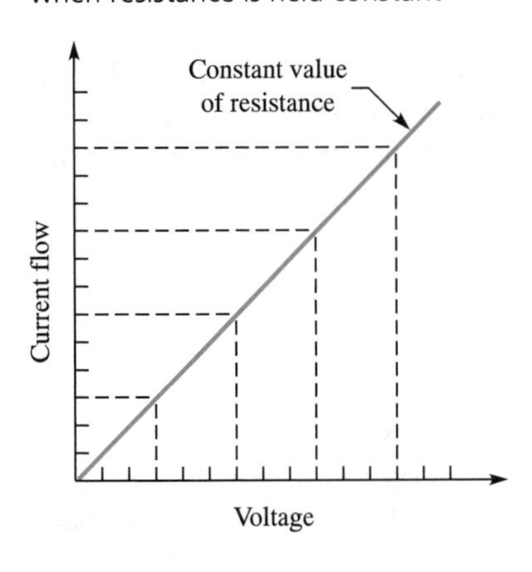

In graph form, this directly proportional relationship is a **linear relationship,** forming a straight line, as Figure 3–2 shows. The graph in this figure plots voltage on the horizontal axis and current on the vertical, and the intersection of all of the voltage values, plotted against the resulting current value, produces a straight line. The slope of this line is determined by dividing the *horizontal value* (voltage) by the *vertical value* (current), which results in a constant value.

Another method of demonstrating the interaction between voltage, current, and resistance is illustrated in Figure 3–3. In this example, the value of the voltage is held constant, and different values of resistance are selected. As the selector switch is placed in various positions to change the resistance, the resulting values of current change. In this example, however, it can be observed that when the value of resistance is made larger, the resulting current is smaller. If the value of the resistor is made smaller, the resulting current is larger in value. The resistance-to-current ratio forms a proportional relationship, but in this case it is an **inverse relationship** in that a change in one quantity results in a change in the opposite direction for the other.

FIGURE 3–3 Basic circuit demonstrating interaction of resistance and current when voltage is held constant

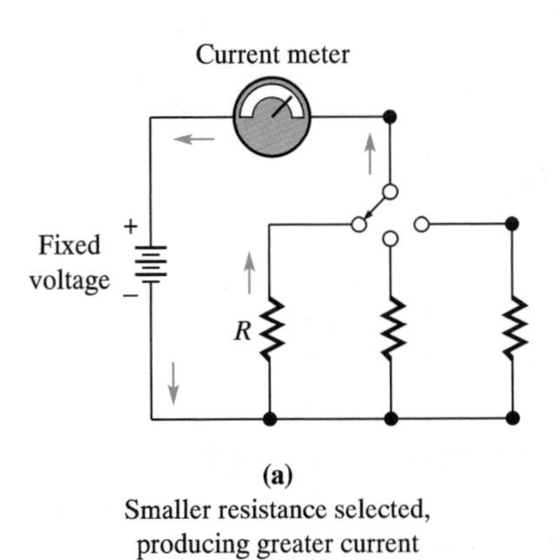

(a)
Smaller resistance selected, producing greater current

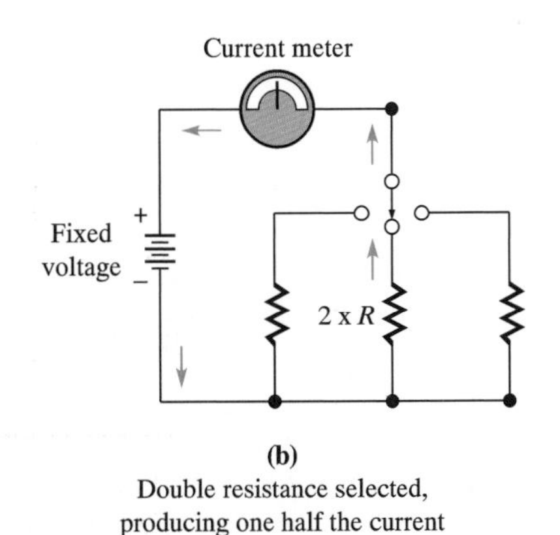

(b)
Double resistance selected, producing one half the current

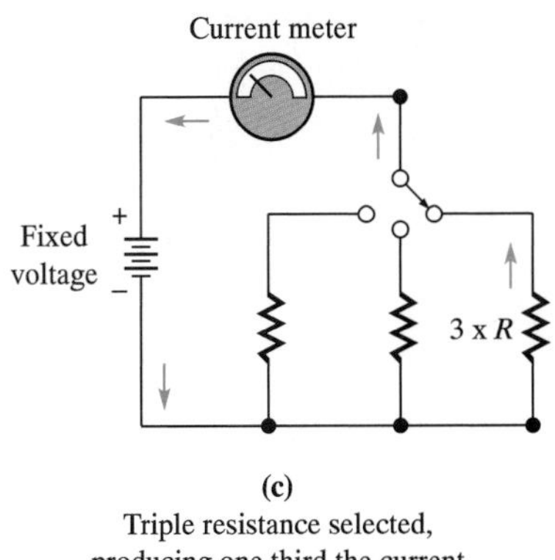

(c)
Triple resistance selected, producing one third the current

Ohm's Law

Question: How did George Simon Ohm express the voltage-current-resistance relationships in the form of his law?

The voltage-current-resistance relationships known as **Ohm's law** can be summarized as follows:

The current in a circuit is directly proportional to the voltage and inversely proportional to the resistance.

According to Ohm's law, the manner in which voltage, current, and resistance are related is both orderly and predictable and can be expressed in the mathematical equations given in the following section. As will be seen, when one of the quantities is unknown, it can quickly and accurately be determined if the other two quantities are known. Applying the Ohm's law equations, however, requires that the technician have a clear understanding of the direct and inverse relationships of the quantities.

Question: Why are the direct and inverse relationships in Ohm's law so important?

No matter how deeply you may become involved in electronics and no matter how sophisticated the electronic systems, you will never outgrow Ohm's law! The relationships stated in Ohm's law always apply—no matter what electronic devices are involved.

As has been stated previously, all electronic systems operate through the systematic control of electric current. Only three things can be done to current: The current flow can be increased, decreased, or made zero. Thus, if a fault occurs in a circuit, it can have only one of three causes: The current is too great, the current is too small, or the current is zero when it shouldn't be. Thus, being able to picture the Ohm's law relationships greatly facilitates understanding and fault analysis of a circuit.

In Chapters 4 and 5, you will learn two new laws, known as Kirchhoff's voltage and current laws. Along with Ohm's law, these two laws will be the basis for your understanding of electric circuits, both basic and advanced.

SECTION 3–1 REVIEW

1. An increase in voltage causes an (increase, decrease) in current.
2. An increase in resistance causes an (increase, decrease) in current.
3. State Ohm's law in words.

3–2 THE OHM'S LAW EQUATIONS

Question: Is there a simple way to state the many ways in which voltage, resistance, and current are related?

A better understanding of how voltage, current, and resistance are related in an electrical circuit can be gained by expressing the relationships as mathematical equations.

Because three variable quantities are involved, any single unknown quantity can be found from the two known values.

Determining Voltage

Ohm's law states that the voltage (V) can be determined at any point within a circuit by multiplying the value of current (I) flowing through that point times the value of resistance (R). Expressed in equation form, then, Ohm's law for voltage is

(3–1) $V = IR$ [voltage = current × resistance]

This mathematical expression shows that voltage is the product of resistance and current. In other words, it takes a predictable amount of pressure (V) to cause a specific amount of current to flow through a particular resistance. An increase in either the resistance to the flow of current, or the value of current flowing through a resistance, causes a directly proportional increase in the resultant voltage at that point in the circuit.

NOTE

Whenever using the Ohm's law equations, the basic units of volts (V), ohms (Ω), and amperes (A) must be used. Further, use of a metric prefix requires converting back to the base unit or use of the appropriate scientific notation.

EXAMPLE 3–1 Verify that an increase in either resistance or current will cause a directly proportional increase in voltage.

Given: $R = 100\ \Omega \qquad I = 2\ \text{A}$

■ **Solution**

1. Use Eq. 3–1 to find the unknown quantity:

$$V = IR$$
$$= (2\ \text{A})(100\ \Omega) = 200\ \text{V}$$

2. Double the resistance, and use Eq. 3–1 again to find the new voltage value:

$$V = (2\ \text{A})\ 200\ \Omega \qquad \text{[double resistance: } 100\ \Omega \times 2]$$
$$= 400\ \text{V} \qquad \text{[double voltage]}$$

3. Double the current, and again find the new voltage value

$$V = (4\ \text{A})(100\ \Omega)\ \text{[double current: } 2\ \text{A} \times 2]$$
$$= 400\ \text{V [double voltage]}$$

Conclusion: Changing the value of either resistance or current causes a directly proportional change in the resultant voltage.

Calculator sequence for Example 3–1

1. [2] [×] [1] [0] [0] [=]
2. [2] [×] [2] [0] [0] [=]
3. [2] [×] [2] [×] [1] [0] [0] [=]

■ Determining Resistance

If the voltage and current in a circuit are known, resistance may be found by the following form of Ohm's law:

(3–2) $R = \dfrac{V}{I} \qquad \left[\text{resistance} = \dfrac{\text{voltage}}{\text{current}}\right]$

EXAMPLE 3–2 Assume the circuit shown in Figure 3–4 has a voltage of 15 V and 3 A of current are measured. What is the value of resistance?

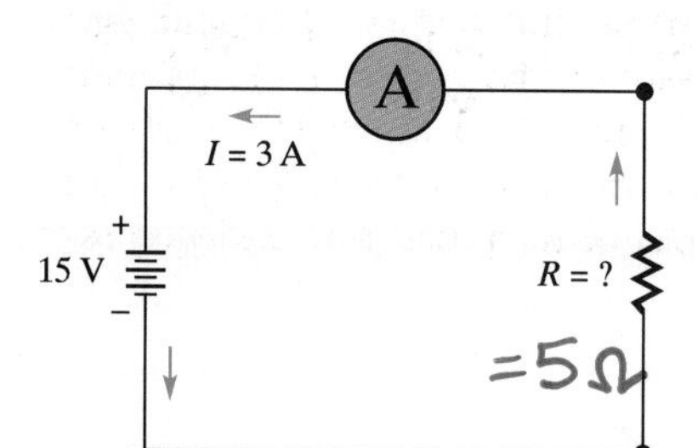

FIGURE 3–4 Circuit for Example 3–2

■ Solution:

Use Eq. 3–2 to find the unknown quantity of resistance:

$$R = \frac{V}{I}$$

$$= \frac{15\text{ V}}{3\text{ A}} = 5\ \Omega$$

Calculator sequence for Example 3–2

[1] [5] [÷] [3] [=]

■ Determining Current

Similarly, Ohm's law may be arranged to give current when voltage and resistance are known:

(3–3)
$$I = \frac{V}{R} \qquad \left[\text{current} = \frac{\text{voltage}}{\text{resistance}}\right]$$

EXAMPLE 3–3 How much current will flow in the circuit of Figure 3–5 with 10 Ω of resistance and an applied voltage of 50 V?

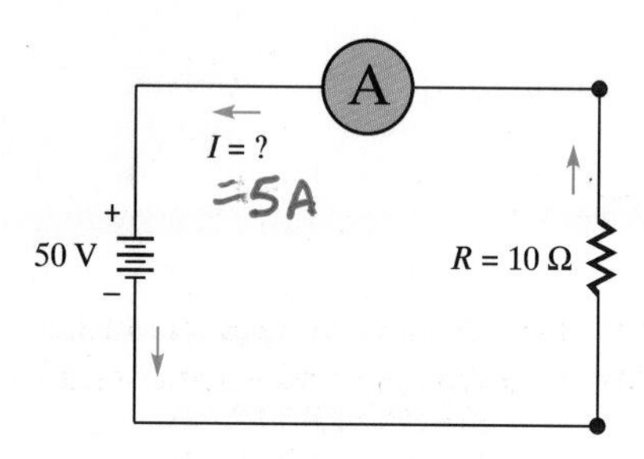

FIGURE 3–5 Circuit for Example 3–3

■ Solution

Use Eq. 3–3 to find the unknown quantity of current:

$$I = \frac{V}{R}$$

$$= \frac{50\text{ V}}{10\ \Omega} = 5\text{ A}$$

Current is Amps
Resistance is Ohms

Calculator sequence for Example 3–3

[5] [0] [÷] [1] [0] [=]

■ Using Ohm's Law Equations

Question: What is the importance of Ohm's law to a technician?

Remember that true mastery of Ohm's law comes when the relationships it represents are thoroughly understood, not when the three forms of the mathematical equation are merely memorized! The analysis of even the most complex circuit can be accomplished by the skillful use and manipulation of Ohm's law. As you will learn throughout your studies of electronics, due to many factors not yet explained, the majority of measurements made in actual equipment will be voltage measurements. As a result, the technician must use voltage as the key to understanding what is happening in the circuit. To do that quickly and accurately requires understanding the Ohm's law relationships, not just memorizing "Ohm's law."

As noted previously, when a fault exists in a circuit, it results in one of three things: (1) the current is too low, (2) the current is too high, or (3) the current is zero when it shouldn't be. The Ohm's law relationship expressed in Eq. 3–3 can be used to determine the fault in the circuit. For example, consider the situation in which the result of a fault is that the current is too low. From the Eq. 3–3, it can be seen that there are two possibilities. Either the voltage is too low

$$I\downarrow = \frac{V\downarrow}{R_{\text{constant}}}$$

or the resistance is too large

$$I\downarrow = \frac{V_{\text{constant}}}{R\uparrow}$$

If the current is too high there are also two possibilities. Either the voltage is too high

$$I\uparrow = \frac{V\uparrow}{R_{\text{constant}}}$$

or the resistance is too small

$$I\uparrow = \frac{V_{\text{constant}}}{R\downarrow}$$

If the current is zero when it is not supposed to be, there are again two possibilities. Either the voltage is zero

$$I_{\text{zero}} = \frac{V_{\text{zero}}}{R\infty}$$

or the resistance is infinitely large

$$I_{\text{zero}} = \frac{V}{R\infty}$$

Thus, troubleshooting begins with the ability to picture the effects of voltage and resistance upon circuit current. Always keep the Ohm's law equations in mind when considering possible circuit faults!

SECTION 3–2 REVIEW

1. If $V = 25$ V and $I = 2$ A, find R. 12.5
2. Assume $V = 35$ V and $R = 250\ \Omega$, what is I? .14 Amps
3. A power supply can provide 50 V at a maximum current of 3 A. What is the smallest value of resistor that may be used? 16.6
4. How much current will flow through a 1 kΩ resistor if 20 V is applied to it? .02 Amps
5. A 330 Ω resistor has 250 mA of current flowing through it. What is the value of voltage across it? .25A 82.5V

STUDY SKILLS

Preparing for Exams: When to Study

When to start studying and how to distribute your time are critical factors in good study habits. The technique of "all night cram sessions" seldom produce the desired results. As with any other skill, electronics skills must be developed by consistent effort over a period of time. It is usually best to begin studying for an exam early. Beginning one to two weeks before the exam allows time to perfect your skills and work out any problems that may prevent your understanding the material.

It is also wise to distribute your study sessions over an extended period of time. Don't study eight hours in one day. Instead, study two hours each day for four days. This practice will allow you to be more alert and help to keep your mind clear. Taking a ten minute break each hour will also help prevent fatigue and keep your mind sharp.

3–3 ENERGY, POWER, AND WORK

Question: Is there a difference between energy and power?

One of the fundamental laws of physics states that **energy,** or the capacity to do work, can be neither created nor destroyed. It can, however, be converted from one form to another. For example, recall the process that takes place each time your car starts, as described in Chapter 2. In order to cause the vehicle to move, it must have an engine, which, in most cases, is an internal combustion engine that uses gasoline as its source of energy. Inside the engine, gasoline is burned under carefully controlled conditions to produce heat. This heat in turn causes expansion of the air within the engine's cylinders, forcing the piston downward and rotating the crankshaft. This force ultimately is transmitted through the auto's drive train to the wheels causing motion. In this process, some of the energy released from the fuel was transferred to the car in the form of kinetic energy, the energy of motion.

Stopping the car is somewhat less complex. Pressure is applied to the brake pedal, which in turn causes pressure to be applied to the brake pads. As the pads rub against the metallic surface of the brake drum or rotor, friction causes another transfer of energy to the rotor, which converts the energy into heat. Once the energy of motion has been converted to heat energy, the car stops because it no longer has the energy of motion.

In both cases, energy was converted from one form to another. The process of converting energy from one form to another is called **work.** The rate at which work is accomplished, or at which energy is converted, is called **power.** Power may be expressed in an energy unit called the **Watt** (W), which is an SI unit, or in the power unit called *horsepower* (HP), which is a non-SI unit. For electronic circuits, the watt is used exclusively. However, many electric motors are rated by horsepower and must be converted from horsepower to watts, which is accomplished by the following factor:

$$1 \text{ HP} = 746 \text{ W}$$

Question: How do energy, power, and work relate to each other?

Put in the form of a mathematical equation, the energy-power-work relationship, given in units of watts for power and in joules for work, is as follows:

(3–4)
$$P = \frac{W}{t} \quad \left[\text{power} = \frac{\text{energy}}{\text{time}}\right]$$

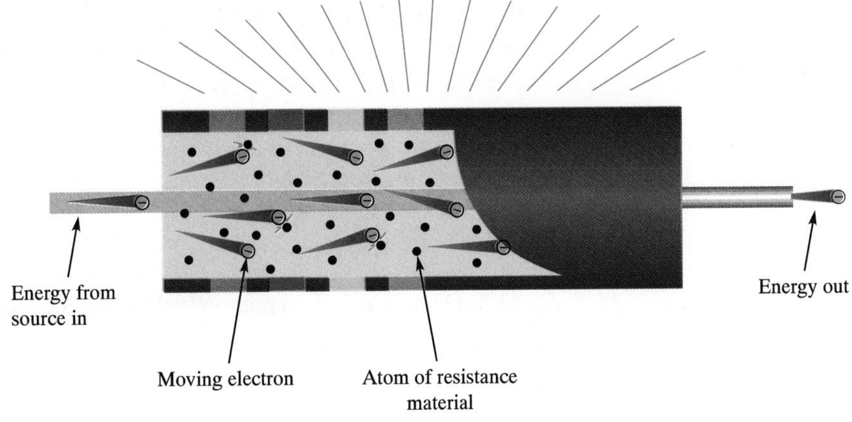

FIGURE 3–6 Power development in a circuit

WHERE: P = power (W)
W = work (energy converted) (J)
t = time (s)

As shown by this equation, 1 W of power is equal to 1 J of energy converted in 1 s of time. If the same amount of work is accomplished in less time, then more power is produced. If time is increased for a given amount of work, then the result is less power produced.

Question: How does power relate to electrical circuits?

In electrical circuits, power is developed when current flows through a resistance. Like the energy in an automobile, the source of energy in a circuit provides the energy to do the work of setting the electrons in motion and thus making the current flow. As shown in Figure 3–6, when the electrons pass through a resistance, the moving electrons collide with the atoms in the resistor creating friction. In the process, energy is transferred from the electrons to the resistor, and the resistor must give off, or **dissipate,** a corresponding amount of energy. Some energy is dissipated in the form of heat, and the rate at which dissipation occurs determines the power developed for the circuit.

To develop a useful equation for this process, the basic definition of voltage must be recalled. This definition states that voltage is equal to work energy in joules, divided by the charge in coulombs. This equation is expressed as

$$V = \frac{W}{Q} \quad \left[\text{voltage} = \frac{\text{work (energy)}}{\text{charge}}\right]$$

Recall also that current is the charge that flows, divided by time, which is expressed as

$$I = \frac{Q}{t} \quad \left[\text{current} = \frac{\text{charge}}{\text{time}}\right]$$

Transposing each of these equations gives

$$W = VQ \quad [\text{work} = \text{voltage} \times \text{charge}]$$

and

$$t = \frac{Q}{I} \quad \left[\text{time} = \frac{\text{charge}}{\text{current}}\right]$$

Power, as given by Eq. 3–4, is

$$P = \frac{W}{t}$$

Substituting from the transposed equations yields

$$P = (VQ)\left(\frac{I}{Q}\right) \qquad \left[\text{power} = (\text{work} \times \text{charge}) \times \left(\frac{\text{current}}{\text{charge}}\right)\right]$$

Canceling terms gives

(3–5) $$P = VI \qquad [\text{power} = \text{voltage} \times \text{current}]$$

Power can thus be determined if the voltage and the amount of current flowing in the circuit are known.

EXAMPLE 3–4 How much power is developed by a 12 V voltage source delivering 200 mA of current in the circuit shown in Figure 3–7?

FIGURE 3–7 Circuit for Example 3–4

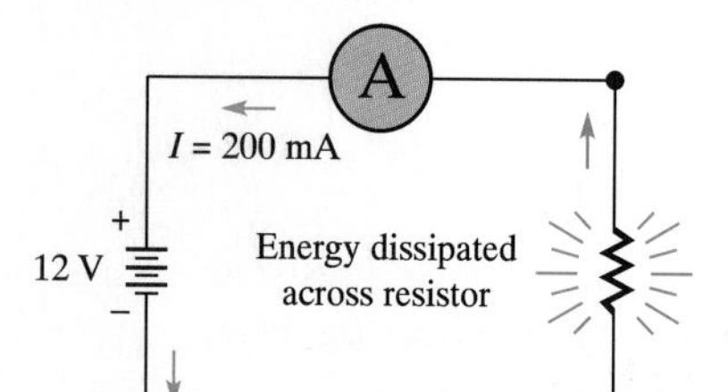

■ Solution:

Use Eq. 3–5 to find the unknown quantity of power:

$$P = VI$$
$$= (12\text{ V})(0.2\text{ A}) = 2.4\text{ W}$$

Calculator sequence for Example 3–4

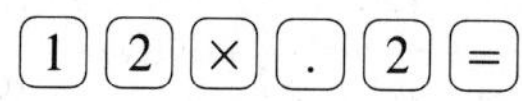

Question: Must voltage and current always be known in order to determine power?

In some cases, it is desirable to determine the amount of power developed by a resistance when only its value and the amount of current flow are known. In this case, the quantity V must be eliminated from the equation and an equivalent statement substituted. Thus, to determine power when the resistance and current are known, use Eq. 3–5:

$$P = VI$$

with

$$V = IR$$

and substitute:

$$P = (I \times R)I$$

Simplifying yields

$$P = (I \times I)R$$

Since the value $I \times I$ is more commonly stated as *I squared,* this equation is written as

$$P = I^2R \quad \text{[power = (current)}^2\text{(resistance)]} \tag{3-6}$$

EXAMPLE 3–5

How much power is developed by the 220 Ω resistor in the circuit of Figure 3–8, when 375 mA of current is flowing?

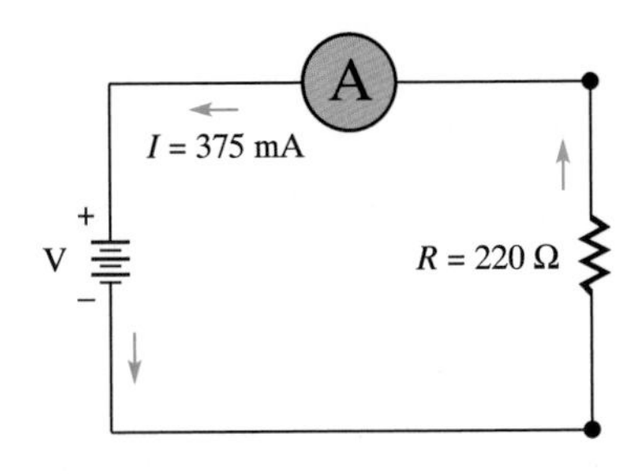

FIGURE 3–8 Circuit for Example 3–5

■ **Solution**

Use Eq. 3–6 to find the unknown quantity of power:

$$P = I^2R$$
$$= (0.375\text{ A})(0.375\text{ A})220\ \Omega = 30.94\text{ W}$$

As mentioned earlier, most measurements made in analyzing electronic equipment are voltage measurements. As a result, the two values frequently available for calculating power are voltage and resistance. Thus, to determine power when voltage and resistance are known, use Eq. 3–5:

$$P = VI$$

with

$$I = \frac{V}{R}$$

and substitute:

$$P = V\frac{V}{R}$$

Simplifying yields

$$P = \frac{V^2}{R} \quad \left[\text{power} = \frac{(\text{voltage})^2}{\text{resistance}}\right] \tag{3-7}$$

Calculator sequence for Example 3–5

[.] [3] [7] [5] [X²] [×] [2] [2] [0] [=]

EXAMPLE 3–6

In Figure 3–9, the 330 Ω resistor has 18 V across it. How much power is developed by the resistor?

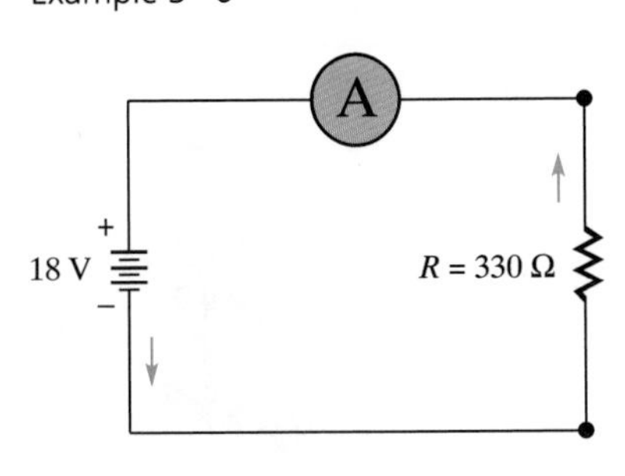

FIGURE 3–9 Circuit for Example 3–6

■ **Solution**

Use Eq. 3–7 to find the unknown quantity of power:

$$P = \frac{V^2}{R}$$
$$= \frac{(18\text{ V})(18\text{ V})}{330\ \Omega} = 0.982\text{ W}$$

Calculator sequence for Example 3–6

[1] [8] [X²] [÷] [3] [3] [0] [=]

STUDY SKILLS

Preparing for Exams: Study Activities

If you are to truly master the subject of electronics, you must concern yourself not only with manipulating mathematical equations, but with understanding what you are doing and why.

All too many times, students memorize procedures for solving a particular type of problem without ever truly understanding the reasons for the procedures. While this type of learning may work for a while and even lead to a good grade-point average, it will eventually cause serious problems!

Concentrating on understanding what a method is and why it works is important. Neither the teacher nor the textbook can cover every possible way a concept may present itself in a problem. If you understand the concept, however, you should be able to recognize it in any type of situation.

In order to achieve the goal of developing both skill and understanding, your study habits should include four things: (1) practicing problems, (2) reviewing the textbook and your notes, (3) drilling with study cards (to be discussed in the next Study Skill), and (4) reflecting on material being reviewed and on exercises being done. It is best to do a little of the first three of these activities *during* each study session and to save some time for reflection at the end of the session.

■ Measuring Power

In Example 3–6, the total power developed by the resistor was less than 1 W. Many electronic circuits are made with small integrated circuit chips (IC chips) and operate at low voltage and current values. As a result, power dissipation in these devices is small and normally rated in *milliwatts* (mW). In radio and other forms of communications receivers, the input signal power is very small and is rated in *microwatts* (μW).

On the other side of the spectrum, utility companies deal with very large values of power. The power used in the average home is typically in the thousands of watts and, for the sake of convenience, requires a larger unit of measure. Consequently, when dealing with commercial power, the prefix *kilo* ($\times$ 1000) is used, and power is measured in a unit called a **kilowatt** (kW), or a thousand watts.

Since power usage is an ongoing factor, it is convenient to have a method of metering energy over a given period of time. Thus, the **watthour** (Wh) is the unit of measure for 1 W of power developed for 1 h. Again, since a more convenient unit is needed when dealing with large values of energy, the **kilowatthour** (kWh) measures thousands of watts per hour and is obtained by multiplying the unit of kilowatts by the time period (hours). Thus, 1 kW for 1 h equals 1 kWh. Similarly, 2 kW for 1 h equals 2 kWh, and 1 kW for 2 h is also equal to 2 kWh.

EXAMPLE 3–7 How many kilowatthours are developed by a 60 W light bulb burning 24 h per day for 30 days?

■ Solution

1. Begin by finding the kilowatt equivalent for the 60W bulb:

$$60 \text{ W} \div 1000 = 0.06 \text{ kW}$$

2. Find the total hours in the time period:

$$(24 \text{ h})(30 \text{ days}) = 720 \text{ h}$$

3. Multiply to find the total kilowatthours:

$$(720\ \text{h})(0.06\ \text{kW}) = 43.2\ \text{kWh}$$

Calculator sequence for Example 3–7

1. [6] [0] [÷] [1] [0] [0] [0] [=]
2. [2] [4] [×] [3] [0] [=]
3. [7] [2] [0] [×] [.] [0] [6] [=]

EXAMPLE 3–8 A 270 W yard light burns 10 h per day for 30 days. What is the cost of this light if electricity is 8 cents per kilowatthour?

■ Solution

1. 270 W ÷ 1000 = 0.270 kW
2. (10 h)(30 days) = 300 h
3. (300 h)(0.270 kW) = 81 kWh
4. (81 kWh)($0.08) = $6.48

Calculator sequence for Example 3–8

1. [2] [7] [0] [÷] [1] [0] [0] [0] [=]
2. [1] [0] [×] [3] [0] [=]
3. [3] [0] [0] [×] [.] [2] [7] [0] [=]
4. [8] [1] [×] [.] [0] [8] [=]

■ SECTION 3–3 REVIEW

1. How much power is developed by a 100 Ω resistor if 500 mA of current are flowing through it?
2. What is the process of converting energy from one form to another called?
3. How much power will be developed by a 470 Ω resistor placed across a 12 V battery?
4. How much current will flow in a 60 W, 120 V light bulb?
5. What is the difference between energy and power?

3–4 EFFICIENCY

Question: What is efficiency and how is it applied to electronic equipment?

The operation of any electronic system is based on the development of electrical power. The development of power in an electronic system manifests itself in the form of heat. For example, in an electric toaster, current flowing through the resistance wire produces heat that is imparted to the bread. The heat imparted to the bread is the *usable* power developed by the toaster. Some of the power developed by the toaster is not imparted to the bread, however, but to the surrounding air, which also becomes quite hot. This latter power is not usable power, but rather represents lost, or *wasted,* power. The combination of usable and wasted power is the *total* power developed by the toaster.

In operating an electronic system, it is this total power for which the customer must pay. How economical it is to operate a system depends upon the amount of input power that it takes to obtain the useful output power desired. The ratio of the usable output power to the total input power of the system is known as its **efficiency.** This ratio, when multiplied by 100, becomes the *percent of efficiency* of the system. In the form of an equation, we have

(3–8)

$$\%\ \text{efficiency} = \frac{P_{out}}{P_{in}}\ 100$$

The greater the efficiency value, the cheaper it will be to operate the system. While the cost of operation is a major concern, it is not the only aspect of efficiency that is important to the user. Over a period of time, the heat of wasted power will cause deterioration of the materials making up the system and can thus increase the number of system failures and shorten its usable life. Sometimes it is even necessary to provide electric fans to carry the excess heat away, which adds to the cost of manufacturing the equipment as well as to the initial cost for the user.

EXAMPLE 3–9 How efficient is a stereo amplifier that operates from 120 V at 500 mA and delivers 20 W per channel output?

■ Solution

1. Use Eq. 3–5, $P = VI$, to find the input power:

$$P_{in} = (120\ \text{V})(0.5\ \text{A}) = 60\ \text{W}$$

2. Find the power output:

$$P_{out} = (20\ \text{W})(2\ \text{channels}) = 40\ \text{W}$$

3. Use Eq. 3–8 to find the efficiency of the system:

$$\%\ \text{efficiency} = \frac{40\text{W}}{60\text{W}}\ 100 = 66.666\%$$

Calculator sequence for Example 3–9

1. [1] [2] [0] [×] [.] [5] [=]
2. [2] [0] [×] [2] [=]
3. [4] [0] [÷] [6] [0] [=] [×] [1] [0] [0] [=]

EXAMPLE 3–10 What would be the required input power to a system whose efficiency was 60% if the required output power is 10 W?

■ Solution

The best way to solve a problem of this type is to "think" your way through it. The system is 60% efficient and the output power is 10 W. According to Eq. 3–6, this output power is 60% of the total input power. Thus, to find the total input power when the output power and the efficiency are known, begin by finding the value of 1% of the total input power:

$$10 \text{ W} \div 60\% = 0.167 \text{ W}$$

Multiply to find 100% or the total input power:

$$(0.167 \text{ W})(100\%) = 16.7 \text{ W}$$

Calculator sequence for Example 3–10

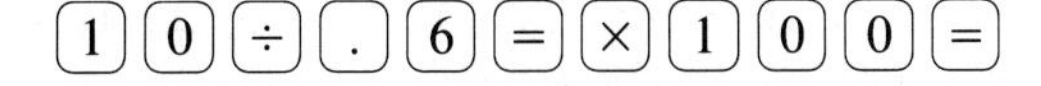

EXAMPLE 3–11 A circuit has an input power of 450 mW. If its efficiency is 85%, what is the output power?

■ Solution

To find the output power when the input power and the efficiency are known, use Eq. 3–8, and transpose to find the unknown quantity:

$$\% \text{ efficiency} = \frac{P_{out}}{P_{in}} 100$$

$$P_{out} = (P_{in})(\text{efficiency})$$

$$= (450 \text{ mW})(0.85) = 382.5 \text{ mW}$$

Calculator sequence for Example 3–11

[4] [5] [0] [EXP] [3] [±] [×] [.] [8] [5] [=]

■ SECTION 3–4 REVIEW

1. A circuit has an input power of 60 W. If the output power is 57 W, compute the efficiency.
2. A circuit with an efficiency of 80% has an output power of 580 mW. Compute the input power.
3. Define what is meant by the efficiency of an electric circuit.

3–5 POWER RATING OF RESISTORS

Question: Why must resistors be rated for a certain value of power dissipation?

Recall that the result of current flow through a resistance is the creation of heat energy, which must be given off, or dissipated, by the resistor to avoid overheating. Carbon, the

FIGURE 3–10 Carbon resistors arranged according to size and power rating

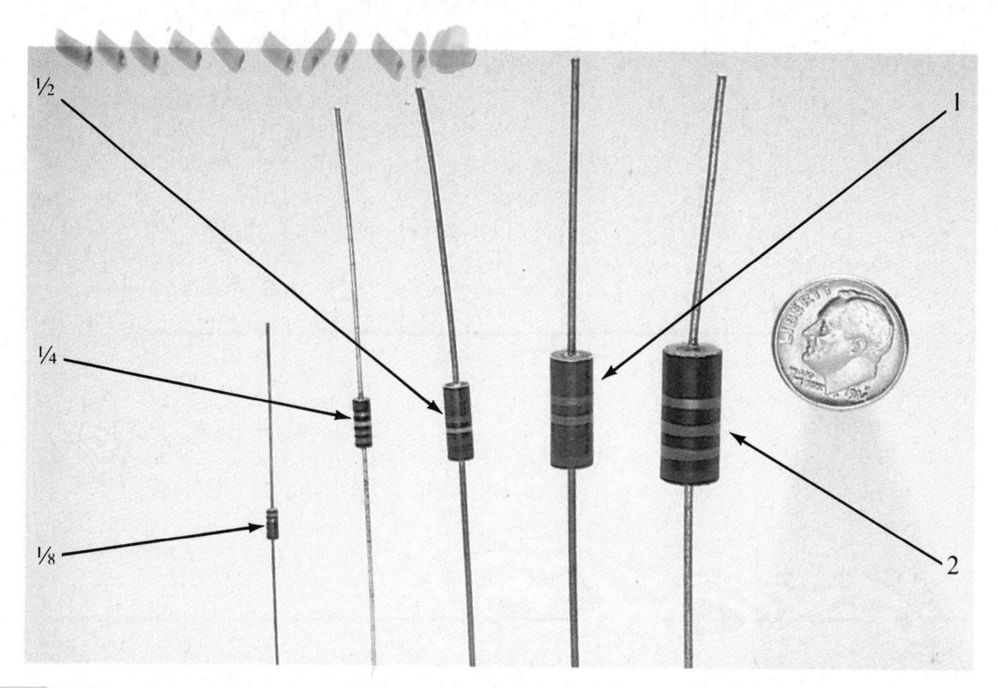

material from which most resistors are made, has a negative temperature coefficient. The ability of a particular resistor to dissipate heat energy is directly related to its mass. Therefore, the larger a resistor is physically, the greater its ability to dissipate heat.

As a resistor nears its maximum ability to dissipate heat energy, its internal temperature begins to rise. Since carbon, the most common material used for resistors, has a negative temperature coefficient, the resistance begins to drop as the temperature rises. As the resistor's resistance drops, the value of current flowing through it increases proportionally, resulting in more power being dissipated, which, in turn, causes the internal temperature to rise again, and the entire process is repeated. This condition is referred to as *thermal runaway* and is, of course, undesirable.

To ensure that thermal runaway does not occur, it is standard practice to double the value of power calculated when selecting resistors. The **power rating** of carbon and carbon film resistors, given in watts, is directly related to their physical size, with larger sizes having greater ability to dissipate heat and thus develop more power. Resistors are available in the standard power ratings of 1/8, 1/4, 1/2, 1, and 2 W. Figure 3–10 shows resistors of each rating and depicts their relative sizes.

When resistors of a wattage rating greater than 2 W are required, the wirewound variety must be used. These resistors, as described and illustrated in Chapter 2, are available in values up to several hundred watts.

EXAMPLE 3–12 For the circuit of Figure 3–11, what is the power rating of the carbon resistor selected?

■ Solution

1. Compute the current by Ohm's law given in Eq. 3–3:

$$I = \frac{V}{R}$$

$$= \frac{10\ \text{V}}{100\ \Omega} = 0.1\ \text{A}$$

FIGURE 3–11 Circuit for Example 3–11

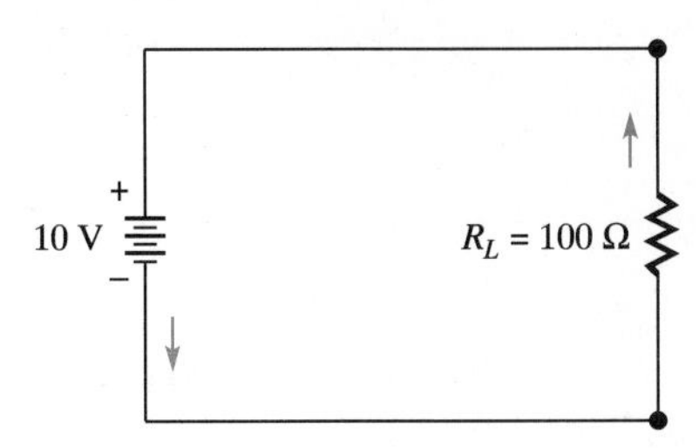

2. Compute the power by Eq. 3–6:

$$P = I^2R$$
$$= (0.1\text{ A})^2(100\ \Omega) = 1\text{ W}$$

The power that the resistor must develop is 1 W. In order to avoid thermal runaway, a 100 Ω, 2 W resistor should be selected.

Calculator sequence for Example 3–12

1. [1] [0] [÷] [1] [0] [0] [=]
2. [.] [1] [X²] [×] [1] [0] [0] [=]

SECTION 3–5 REVIEW

1. How much power is produced in a 400 Ω resistor with 25 mA of current?
2. What would the correct power rating be for a 400 Ω resistor that has 10 V across it?

3–6 TROUBLESHOOTING

Troubleshooting, as introduced in Chapter 2, requires certain definite steps in the process. By way of review, these steps consist of (1) determining from the operation of the equipment that a fault does in fact exist; (2) checking the source voltage for its proper value; (3) determining the cause of zero, high, or reduced current; (4) making repairs to the system; and (5) checking the system for proper operation.

With the knowledge gained in this chapter, you can now take a more analytical approach to troubleshooting. Through Ohm's law, you now know the relationships between voltage, current, and resistance in a circuit. The power equations that were presented in this chapter make it possible to determine the amount of power developed in a circuit or in a component. Mathematical calculations of these quantities make it possible to determine the circuit operation. Measurements in the circuit make it possible to determine if these operational values are being met.

FIGURE 3–12 Simple circuit for troubleshooting circuit operation

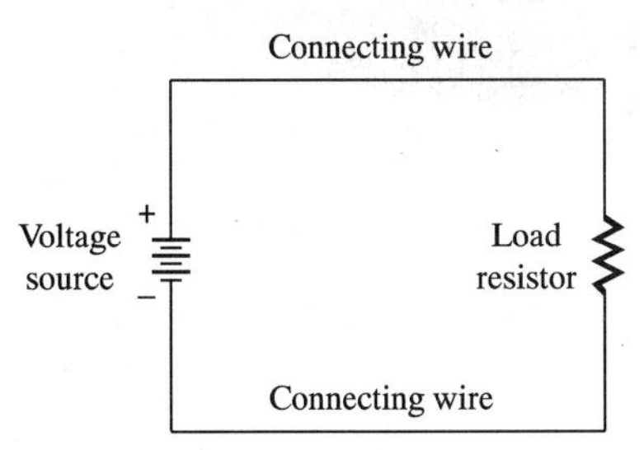

Figure 3–12 shows a simple electronic circuit. A resistor is the load connected to a voltage source. Current flows from the source through the connecting wires and the load. How much current will flow is a function of both the voltage and the resistance of the load ($I = V/R$). Thus, from the values of the voltage and resistance, you can predict what current will flow. To do so, you must determine the values of the voltage source and the resistance. The voltage is found by measuring the source with a voltmeter. You could determine the resistance of the resistor using an ohmmeter. This measurement must, however, be made with the power off, and it is often necessary to desolder one end of the resistor from the circuit. In troubleshooting a practical circuit, it is not usually necessary to know the exact value of the resistance such as is obtained with a measurement. Its nominal value, which can be obtained through the resistor color code, is usually sufficient to predict the circuit operation.

Now, once the values of voltage and resistance are known, the circuit current can be computed using Ohm's law. This current is the current for which the circuit was designed. The voltage source must be capable of supplying this amount of current, the connecting wires must be capable of passing this current without overheating, and the

FIGURE 3–13 Effects of overheating

(a) Photo of discolored printed circuit board

(b) Photo of a burned resistor

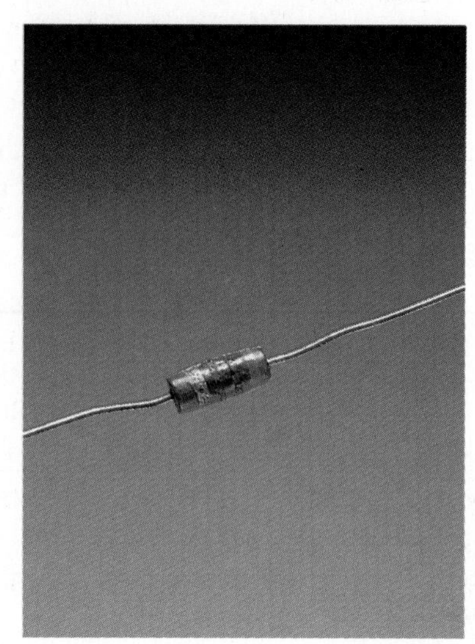

resistor must be capable of developing the required power without overheating. Troubles occur when any of these circuit parameters are exceeded.

In troubleshooting problems involving power, two of the bodies senses, sight and smell, are very helpful. Overheated wiring or printed circuit boards often take on a darkened appearance. This discoloration may or may not be an indication of trouble, but any such appearance should be carefully scrutinized as a possible by-product of some circuit problem. Figure 3–13(a) shows an example of a discolored printed circuit board, and Figure 3–13(b) shows an example of a burned resistor. Resistors may overheat to the point where the surface looks charred, which often makes it impossible to determine the color codes. A burned resistor may even disintegrate completely. Once again, a burned resistor may not be the actual trouble in a circuit, but merely the symptom of the trouble.

Normal heating in an electronic system does not produce the charring or discoloration. Thus, as a technician, it is necessary for you to learn the difference between normal heating and overheating. As you will learn in later portions of the electronics curriculum, some circuits, in normal operation, operate at rather high temperatures. Examples are output circuits controlling such devices as electric motors or speakers.

When operation or troubleshooting electronic equipment, it is often the sense of smell that gives the first indication of trouble. Overheated insulation or resistors have a pungent, telltale odor with which all technicians very quickly become familiar. When this odor is perceived, it is a good idea to also look for wisps of light smoke and then shut the equipment off as quickly as possible! If this is not done, serious damage to the equipment may occur. Once again, components burning or smoking are usually the symptom of a trouble, rather than the ultimate problem.

Since power is equal to the product of the current squared and the resistance, overheating occurs in a circuit when a current greater than the design current flows. Too great a current is produced when either the source voltage is too great or the resistance is too small, as can be confirmed by Ohm's law. The source voltage can be measured to determine its value. If it is correct, then something has happened to reduce the resistance. A possibility is a short path being developed around the resistor. Once the trouble is determined, it must be corrected, and the circuit must be checked.

SECTION 3–6 REVIEW

1. What are the steps in troubleshooting?
2. What are two possible indications of overheating in a circuit?
3. What are two possible causes of overheating of components in a circuit?

CHAPTER SUMMARY

1. Ohm's law is a relationship that exists between voltage, current, and resistance.
2. Ohm's law expresses voltage-current-resistance relationships in a circuit.
3. The relationship between voltage and current, holding resistance constant, is directly proportional.
4. The relationship between resistance and current, holding voltage constant, is inversely proportional.
5. Work is the process of converting energy from one form to another.
6. The mathematical expressions of Ohm's law are critical in circuit analysis.
7. Energy is the capacity to do work, and power is the rate at which the work is done.
8. The unit of electrical power is the watt.
9. The unit of energy is the watthour.
10. The ratio of power output to power input is known as efficiency.
11. The power rating of a carbon resistor relates directly to its physical size.

KEY TERMS

direct relationship
dissipate
efficiency
energy
inverse relationship
kilowatt
kilowatthour
linear relationship
load
Ohm's law
power
power rating
proportional relationship
watt
watthour
work

EQUATIONS

(3–1) $V = IR$

(3–2) $R = \frac{V}{I}$

(3–3) $I = \frac{V}{R}$

(3–4) $P = \frac{W}{t}$

(3–5) $P = VI$

(3–6) $P = I^2R$

(3–7) $P = \frac{V^2}{R}$

(3–8) $\%\ efficiency = \frac{P_{out}}{P_{in}} 100$

Variable Quantities

V = voltage
I = current
R = resistance
P = power
W = work
t = time
P_{in} = input power
P_{out} = output power

TEST YOUR KNOWLEDGE

1. Given a constant voltage source, how does a decrease in resistance affect current flow?
2. How is the voltage across a fixed value of resistance affected by an increase in current flow?

3. If the voltage applied to a circuit is tripled and the circuit resistance remains constant, what happens to current flow in the circuit?
4. When current flow through a resistor is doubled, what happens to power dissipated by the resistor?
5. Doubling the voltage across a resistor will do what to power dissipated by the resistor?
6. How does a decrease in efficiency of a circuit affect its operating temperature?
7. What unit is normally used by utility companies for computing customers' energy usage?
8. What is meant by a directly proportional relationship?
9. What type of measurements are most commonly made in the analysis of electronic equipment?
10. What unit takes into account how much power is used in a specific amount of time?
11. In your own words, state Ohm's law.
12. Write the equation for finding power, given resistance and current.
13. Write the equation for finding current, given resistance and voltage.
14. Write the equation for finding power, given voltage and current.
15. Write the equation for determining % of efficiency, given input and output power.

PROBLEM SET: BASIC

Section 3–1

1. If the voltage is increased in the circuit of Figure 3–14, what is the effect upon current?

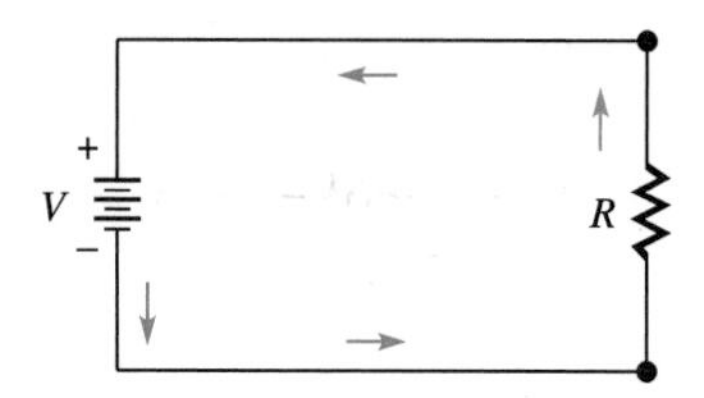

FIGURE 3–14 Circuit for Problems 1 and 2

2. If the resistance is decreased in the same circuit as Problem 1, what is the effect upon current?

Section 3–2

3. For each circuit of Figure 3–15, compute the value of the current.

FIGURE 3–15 Circuits for Problem 3

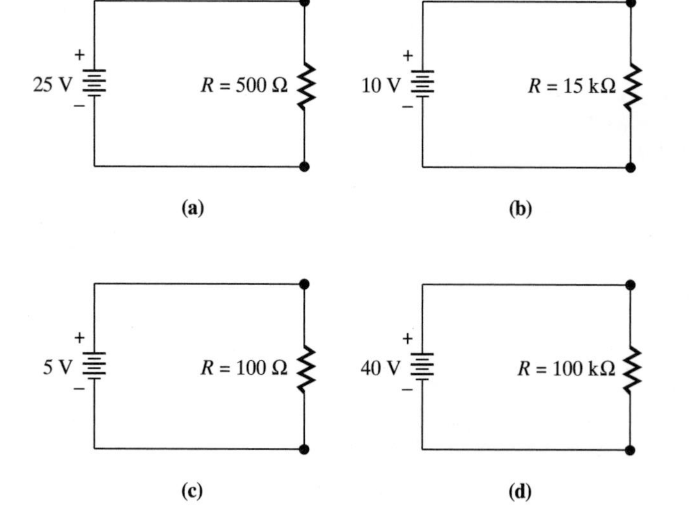

4. For each circuit of Figure 3–16, compute the value of the current.

FIGURE 3–16 Circuits for Problem 4

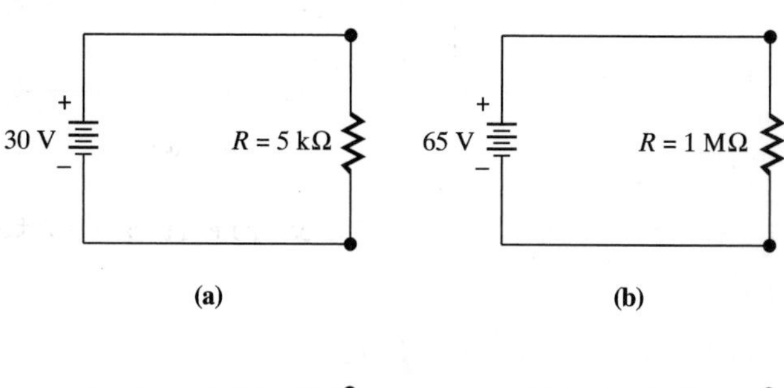

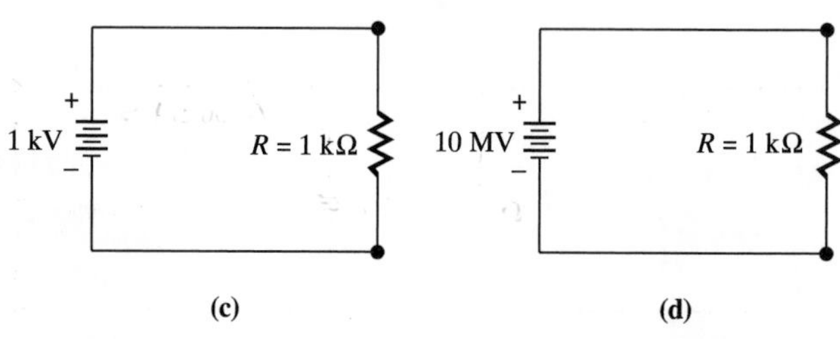

5. For the circuit of Figure 3–17, compute the current that will flow through the lamp.

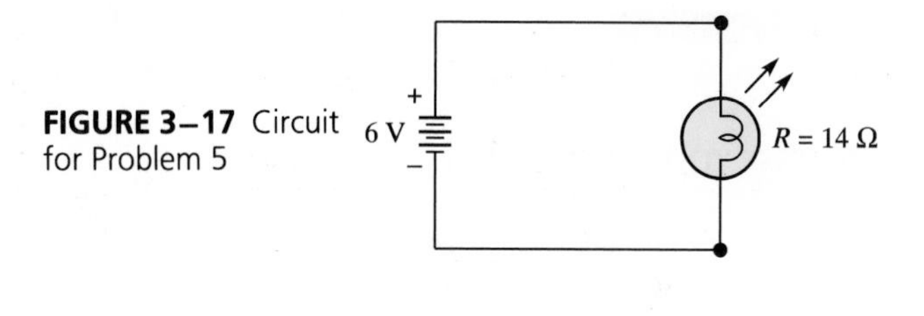

FIGURE 3–17 Circuit for Problem 5

6. A resistor color coded green-blue-red is connected across a 14 V source. How much current flows?
7. If the resistor in Problem 6 has a tolerance of 10%, what is the maximum acceptable current for the circuit?
8. For each circuit of Figure 3–18, compute the value of the applied voltage.

FIGURE 3–18 Circuits for Problem 8

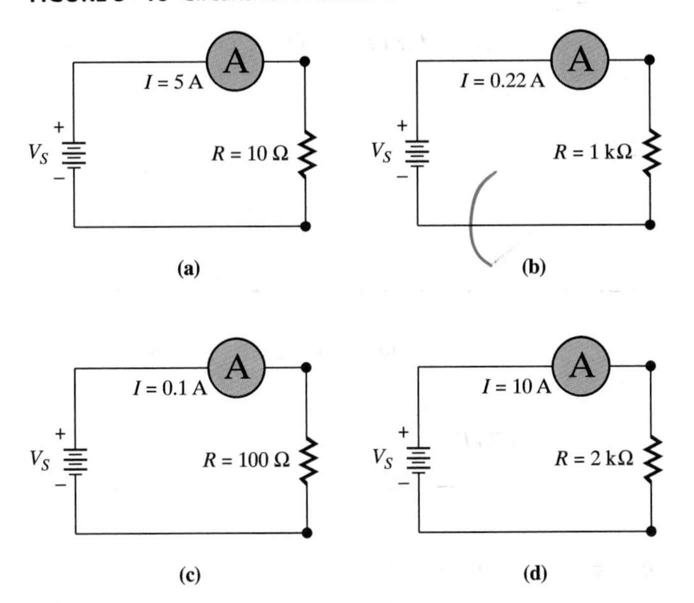

9. For each circuit of Figure 3–19, compute the value of the applied voltage.

FIGURE 3–19 Circuits for Problem 9

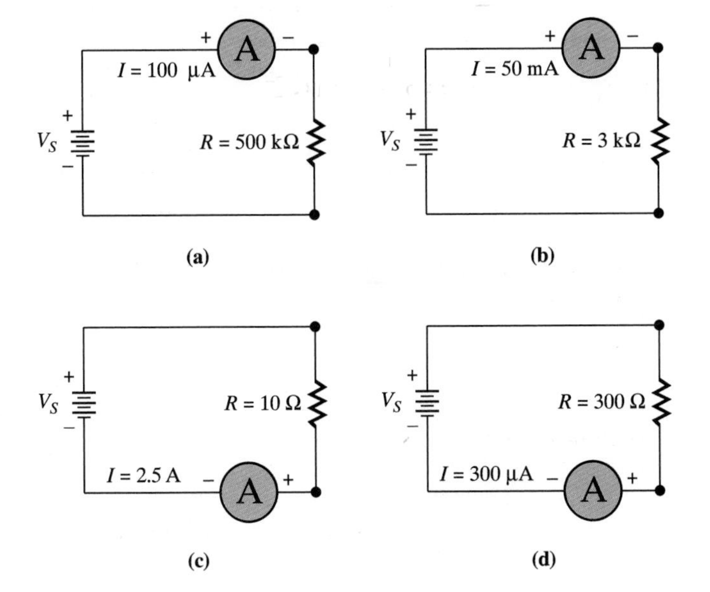

10. An electric toaster draws 12 A of current and has 10 Ω of resistance. How much voltage is required?
11. It is necessary to push 150 mA of current through a 500 Ω load. How much voltage is required?
12. For each circuit of Figure 3–20, compute the value of the resistance.

FIGURE 3–20 Circuits for Problem 12

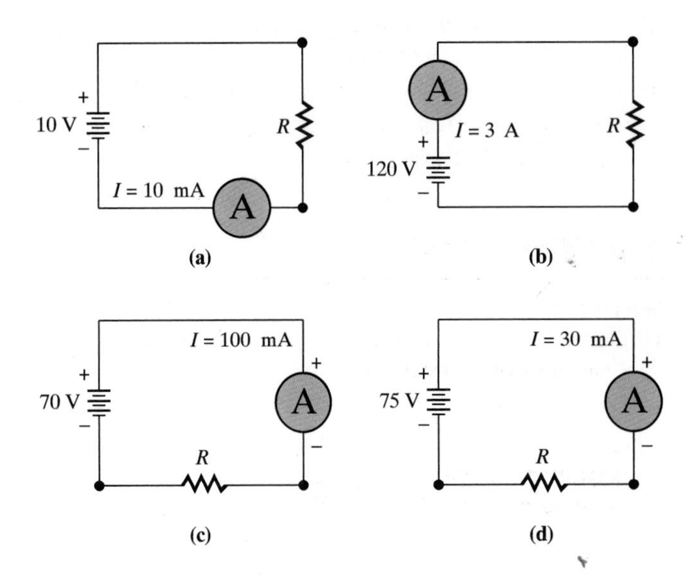

13. For each circuit of Figure 3–21, compute the value of the resistance:

14. A certain resistance has 5 V measured across it. If the current is 70 mA what is the resistance?

15. A current of 5 A flows through a resistance with 120 V applied. What is the value of the resistor?

Section 3–3

16. Compute the power developed for each of the following:
   a. $V = 10$ V; $R = 1000\ \Omega$ **b.** $V = 10$ V; $I = 1.5$ A
   c. $I = 2$ A; $R = 10\ \Omega$ **d.** $V = 5$ V, $I = 10$ A

FIGURE 3–21 Circuits for Problem 13

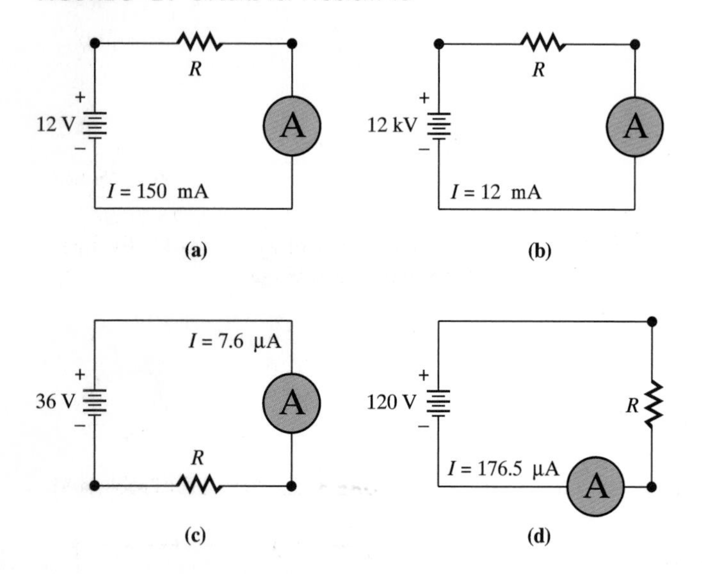

17. Compute the power for each of the following:
   a. $V = 120$ V; $R = 5\ \text{k}\Omega$ **b.** $V = 100$ mV; $I = 250\ \mu$A
   c. $I = 80$ mA; $R = 50\ \Omega$ **d.** $V = 15$ V; $I = 150$ mA

18. For the circuit of Figure 3–22, how much power is developed in the resistor?

Section 3–4

19. A circuit has an input power of 400 W and an output power of 100 W. Compute its efficiency.

20. A circuit is 80% efficient. If the output power is 500 mW, what is the input power?

21. A circuit has an input power of 350 W. If its efficiency is 95%, what is the output power?

Section 3–5

22. A resistor must be capable of developing 350 mW of power. What standard power rating should be chosen?

FIGURE 3–22 Circuit for Problem 18

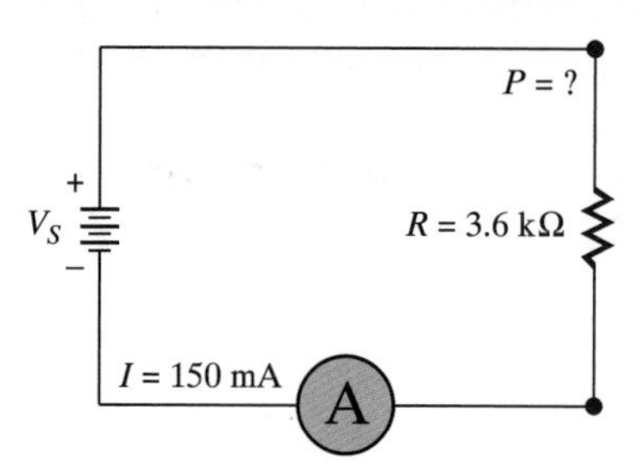

PROBLEM SET: CHALLENGING

23. Make the following conversions of metric prefixes:
   a. 1500 Ω = ______ kΩ
   b. 1,500,000 Ω = ______ MΩ
   c. 0.555 mA = ______ μA
   d. 10.4 GHz = ______ Hz
   e. 10,000 W = ______ kW
   f. 1000 pF = ______ μF
   g. 30,000 V = ______ kV
   h. 0.0006 mA = ______ μA
   i. 6000 μA = ______ mA

24. How much current is required to produce 3.6 W of power in a 1000 Ω resistor?

25. In Problem 24, how much voltage is required?

26. A 500 W yard lamp burns for an average of 8 h per day. If electricity costs 7¢ per kilowatthour, what is the cost to operate this lamp for 30 days?

27. A 60 W bulb burns for 10 h. Compute the energy consumed in joules.

28. A radio transmitter and receiver requires 500 W of power from a 12 V battery. How many amperes of current will flow when the system is in operation?

29. For the circuit of Figure 3–23, what is the source voltage?

FIGURE 3–23 Circuit for Problem 29

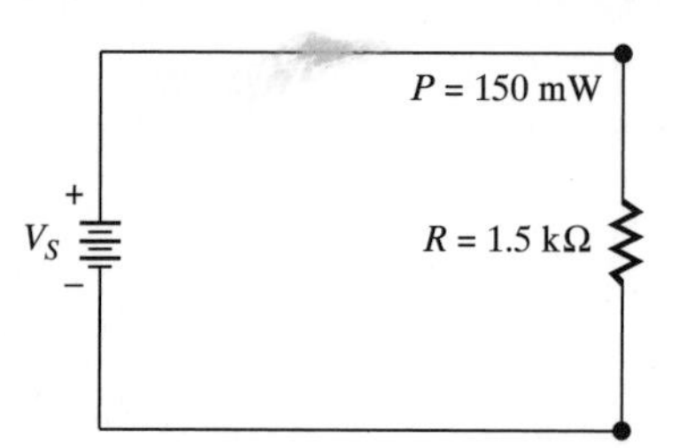

30. For the circuit of Figure 3–24, what is the circuit current?

FIGURE 3–24 Circuit for Problem 30

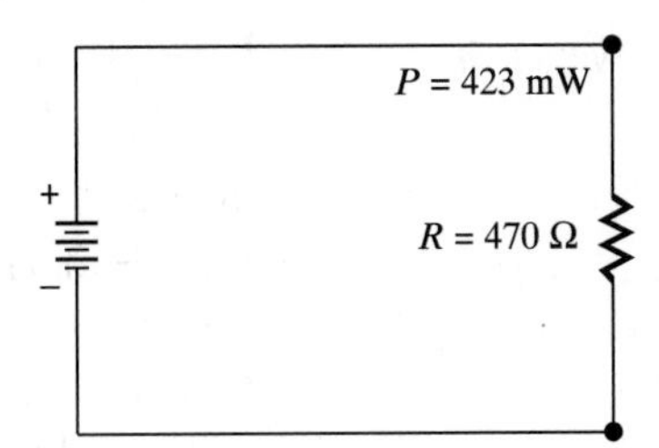

PROBLEM SET: TROUBLESHOOTING

31. Determine whether the current in the circuit of Figure 3–25 is the correct value for the source voltage and load resistance shown?

32. In Problem 31, the source voltage is measured and found to be correct. What is the cause of the fault in this circuit? (*Hint:* Picture Ohm's law!)

33. Determine whether the current in the circuit of Figure 3–26 is the correct value for the source voltage and load resistance shown?

34. In Problem 33, the resistor value is correct. What is the probable cause of the fault in this circuit? (*Hint:* Picture Ohm's law!)

35. Given the current in the circuit of Figure 3–27, list three possible causes for a fault in this circuit.

FIGURE 3–25 Circuit for Problem 31

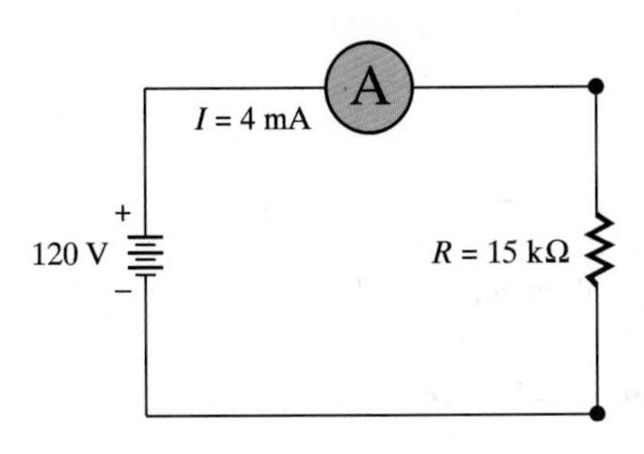

FIGURE 3–26 Circuit for Problem 33

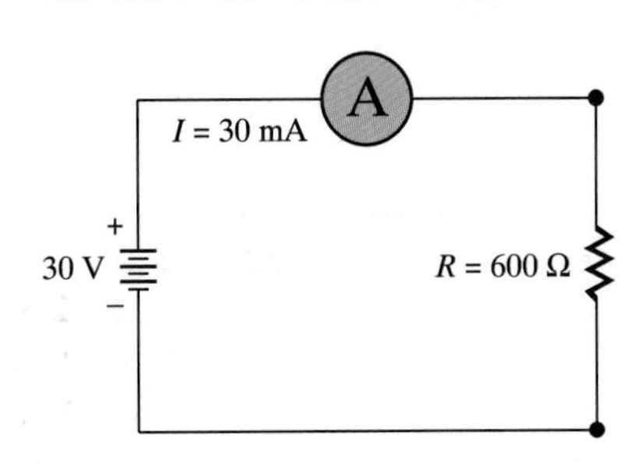

FIGURE 3–27 Circuit for Problem 35

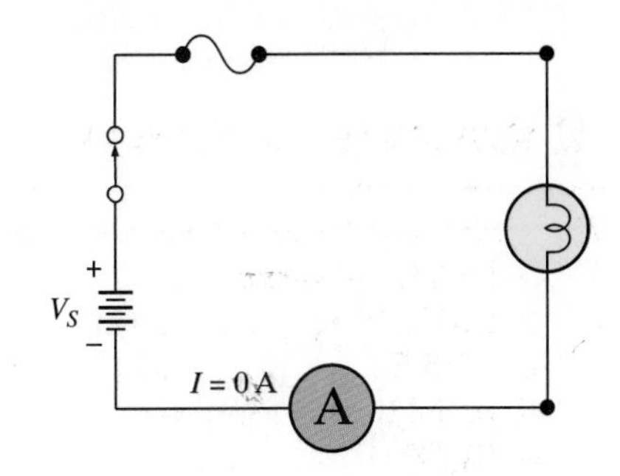

The Digital Computer and IC Chips

The powerful, compact, and relatively cheap digital computer is truly one of the "miracles" of modern technology. Its basic principles were proven in the 1930s and the 1940s, but the early computers were huge, ungainly devices that were difficult and expensive to operate. For example, the ENIAC computer contained over 18,000 vacuum tubes. Just getting rid of the heat from these tubes was a major undertaking and expense in itself!

In the early computers, all of the circuits were *discrete,* which means they were made up of individually mounted and wired components. Software was not readily available, and the computer had to be programmed by arranging groups of cables. Such systems were definitely not to be used for playing video games or keeping family records!

In the 1950s, technological advances greatly reduced the size and power requirements of electronic systems. However, the circuits were still discrete, and the degree of miniaturization possible was limited. Then, the late 1950s and early 1960s saw the birth of the *integrated circuit* (IC), in which all components are inseparably assembled in a single structure. In this assembly, all circuits and devices are integrated on a single piece, or "chip," of silicon material, whose size is on the order of 0.15 to 0.25 inches!

Several typical IC chips are shown in Figure 1. Wires from circuit points on the chip are terminated in pins for connection to the "outside world" components.

The electronics industry has devoted a tremendous amount of energy and money to the development of smaller and more versatile ICs. As a result, ICs have grown steadily smaller, while at the same time more and more components have been integrated. The packing density, which relates to how many components can be integrated into a given space, has increased rapidly over the years. Today, an IC can contain hundreds of thousands of components. An example is the microprocessor, which has been touted as a "computer on a chip." The competition among designers and manufacturers is quite intense, and each year, more varieties of ICs are added to those already available.

FIGURE 1 Examples of integrated circuit (IC) chips

If the digital computer is a miracle of modern technology, the integrated circuit is the cause of the miracle. Without the IC, it would be impossible to build relatively cheap computers that not only are reliable but are compact, have lower power consumption, and are capable of running a multitude of software packages. Figure 2 shows a typical circuit board used in a modern digital computer.

Software packages are groups of standard programs that take advantage of the large amount of memory available in today's computers as the result of high-density ICs. These programs make the computer more effective and easier to use and have thus contributed to its popularity in businesses, schools, and homes. In contrast, owners of early computers were forced to either hire a programmer or learn to do the programming themselves.

The IC has three important features compared to discrete circuits. First, ICs are more reliable, because the connections are made only at the pins of the chip. Discrete circuits require many more connections to wire the individual components. Second, the cost of an IC is many times lower, be-

continued

continuing

cause ICs are made in a single process, while discrete circuits must be assembled individually. In addition, in discrete circuits, individual components must be bought, cataloged, and stored while awaiting assembly, all of which contribute to the higher cost of a discrete circuit. Third, the IC is much faster in its operation. All the components on an IC are much smaller and much closer together than their discrete counterparts. Speed is thus enhanced because the electrons, which cause the circuit action, have a smaller distance to travel.

While the future for ICs is bright, questions arise concerning the effect IC development will have on the future for electronics technicians. Like the popular television commercial that shows the loneliness of a repairman who has nothing to do because the product he services is so reliable, the use of reliable ICs will be so widespread, according to some people, that the electronics technician will become obsolete. If, for example, an entire TV or stereo were to be integrated on a single chip, it won't be repaired but rather discarded if it does fail. With integrated circuits, the needs for system alignment will also be minimal. These are, of course, issues of proper concern.

FIGURE 2 Typical circuit board for a modern digital computer

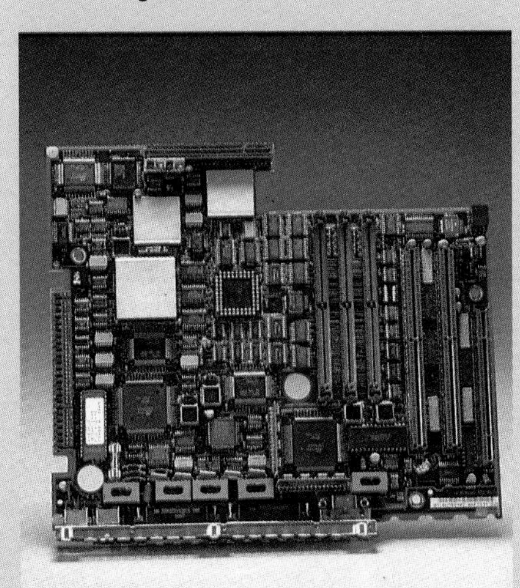

The idea of the obsolescence of electronics technicians has been advanced each time major advances have been made in electronics technology. The events following each advance, however, have proven just the opposite. Each has provided more opportunities for employment, not less. More versatile IC chips invariably allow applications of electronics in areas not before considered. Thus, they have a way of expanding, not limiting, the industry. The nature of the jobs available may change, but the numbers will increase. For their part, electronics technicians must grow with the industry and keep their technical knowledge current.

CHAPTER

4 Series Circuits

■ UPON COMPLETION OF THIS CHAPTER, YOU WILL BE ABLE TO

1. Identify series circuits.
2. Compute the total resistance of a series circuit.
3. Compute the current in a series circuit.
4. Explain why the current is the same at all points in a series circuit.
5. State and apply Kirchhoff's voltage law in analysis of series circuits.
6. Explain why the series circuit is known as a voltage divider.
7. Compute the voltage drops in a series circuit using Ohm's Law.
8. Compute the power developed by each resistor and the total power of a series circuit.
9. Explain the difference between series-aiding and series-opposing voltage sources.
10. Perform troubleshooting on simple series circuits.

STUDY SKILLS

Preparing for Exams: Reviewing Your Notes and Text

Another activity that is strongly suggested in preparing for a test is reviewing your notes and the textbook. Normally, the things you record in your notes indicate that a concept or equation was important—that's why you wrote it down. Your notes should thus serve as a study guide and may help bring to light some important concept or idea that may otherwise be overlooked or was perhaps missed the first time.

Reviewing explanations of problems in the text and in your notes gives you a better perspective and helps tie the material together. Concepts will begin to make more sense when you review and consider how they are interrelated. Reviewing old homework problems, quizzes, and exams will also help you organize your thoughts and maintain the skills that you've already learned. Pay special attention to problems that were done incorrectly on homework or quizzes. These offer an excellent opportunity to improve your skills and increase speed and accuracy.

As you review the material, ask yourself the questions in the Study Skill section "Comparing and Contrasting Examples" given in Chapter 2. Identify as clearly as possible the distinctions that exist in areas that tend to cause you confusion.

The basic electrical circuit presented in Chapter 2 consisted of a source and a single load. A practical electronic circuit may have more than one load. These multiple loads can be connected in two ways: in series or in parallel, producing respectively, series or parallel circuits. In this chapter, the theory and analysis of series circuits is presented. (The other circuit configurations, parallel and series-parallel, will be presented in subsequent chapters.)

In addition, another law of electricity, Kirchhoff's voltage law, is introduced in this chapter. Often abbreviated KVL, it is the basic law of the series circuit. Along with the Ohm's Law and power equations, KVL is used in the analysis of series circuits. The analysis of series circuits, as well as the methods for performing it, must be thoroughly understood. It is the basis for the analysis of many of the electronic circuits and systems that follow in later chapters.

Series circuits have a distinct purpose in electronic systems that is explored in this chapter. Troubleshooting of the series circuit is also presented. Keep in mind that although the circuits considered in this chapter are quite simple, the troubleshooting techniques presented will still apply when the circuits become more complex.

4–1 THE SERIES CONNECTION

Question: How are resistors connected in series?

A *series* is anything in which one thing or event follows another. Some television shows, for example, have stories that continue from one week to another and are known as a series. In October of each year, a series of baseball games are played in order to determine the champion of major league baseball. In an automobile engine, the cylinders fire one after another in an order that provides a smooth flow of power to the wheels. And thus it is in electric series circuits, with one load resistance following another.

FIGURE 4–1 Electronic circuit with three loads

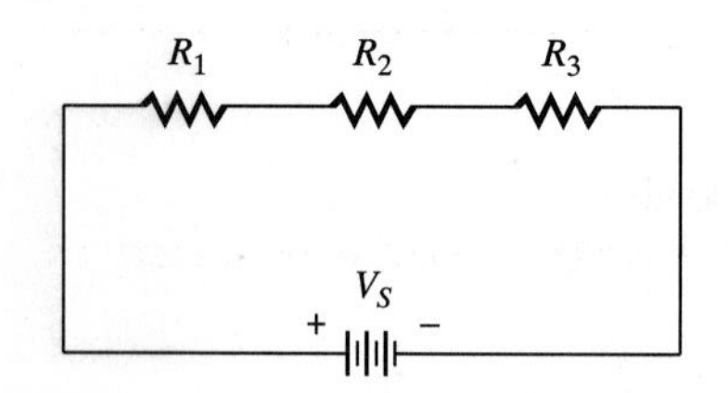

As noted previously, most electronic circuits are composed of multiple loads rather than a single load. A circuit with three loads is illustrated in Figure 4–1, where subscripts (read as R-one, R-two, R-three) are used in order to differentiate the three loads. The resistors are connected end-to-end and are said to be connected in series. The physical orientation of the resistor has no bearing on whether or not the connection is series. For

FIGURE 4–2 Examples of resistors connected in series

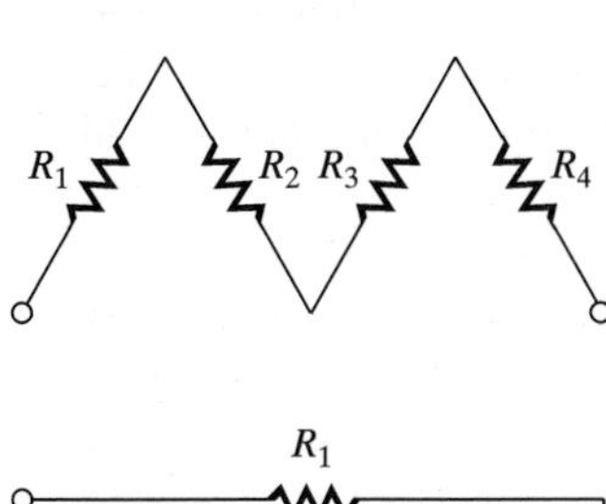

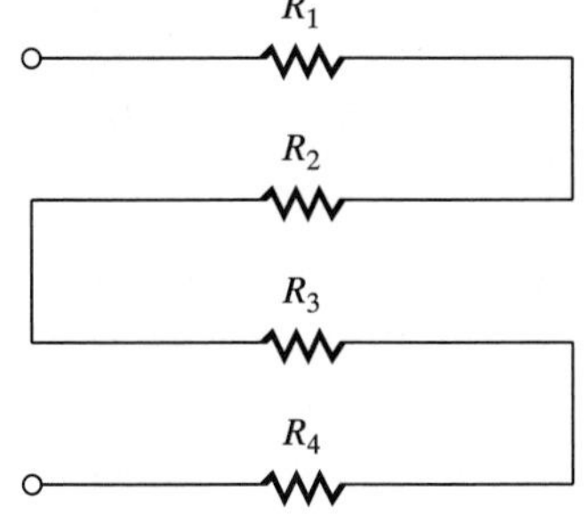

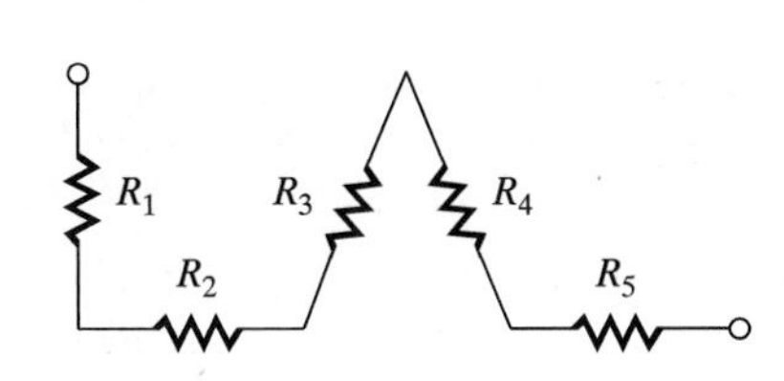

FIGURE 4–3 Resistors for Example 4–1

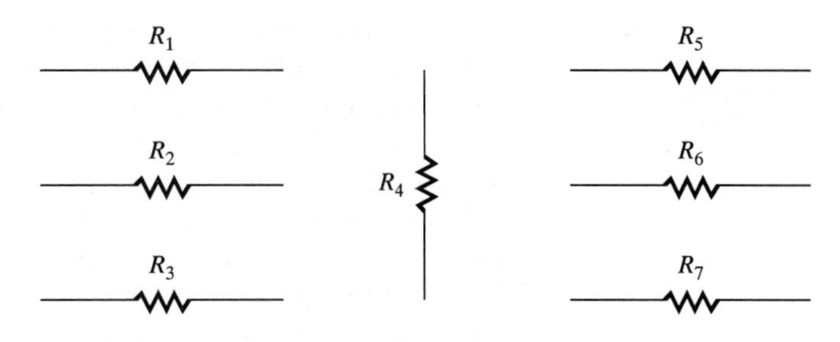

FIGURE 4–4 Series connections for resistors in Example 4–1

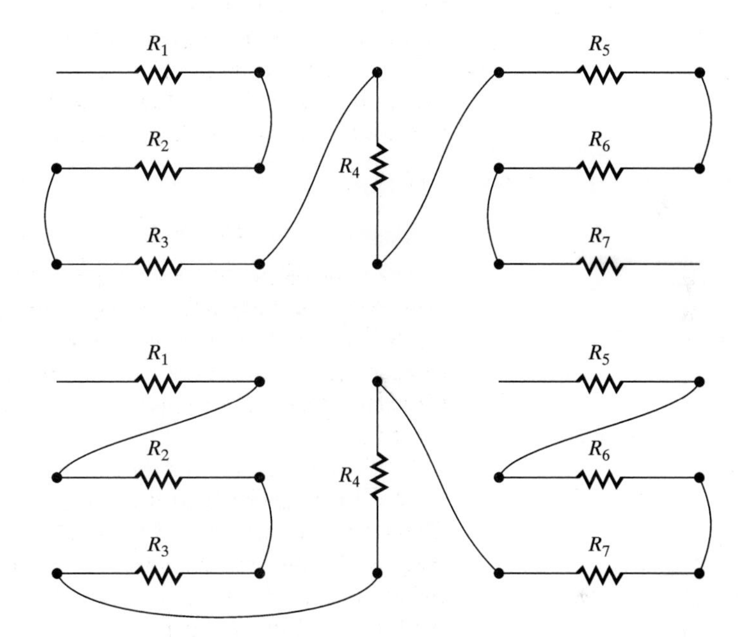

example, as each circuit of Figure 4–2 shows, the resistors are connected end-to-end and are thus in series.

EXAMPLE 4–1 Without changing their physical orientation, connect the resistors of Figure 4–3 in series. (Begin with R_1 and continue in numerical order.)

■ Solution:

The resistors can be connected in series in several ways. Two are illustrated in Figure 4–4.

■ **SECTION 4–1 REVIEW**

1. How are resistors connected in series?
2. Are most circuits composed of single or multiple loads?
3. Does the physical orientation of resistors have any bearing on whether or not they are in series?

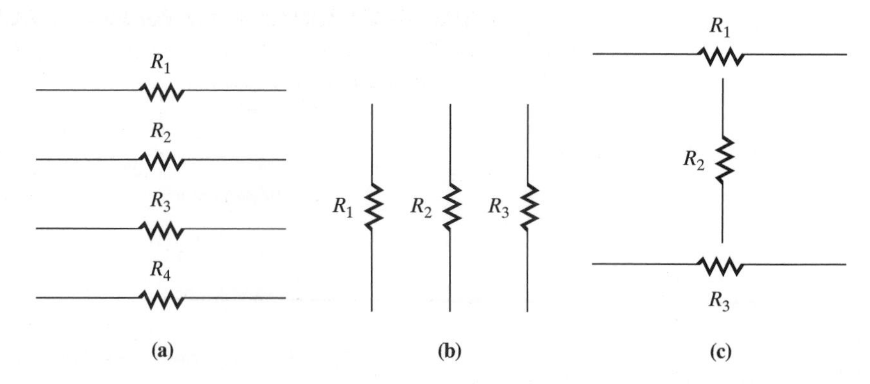

FIGURE 4–5 Resistors for Section Review Problem 4

4. Without changing their physical orientation, connect the resistors of Figure 4–5 in series. (Begin with R_1 and continue in numerical order.)

4–2 CURRENT IN A SERIES CIRCUIT

Question: How does current travel in a series circuit?

In Figure 4–6, a voltage source has been placed across the **string,** or group of resistors connected in series. As the arrows show, current flows from the negative side of the source through each resistor, and back to the positive side. Note carefully that current *cannot* get from the negative to the positive terminal without flowing through *each* resistor. This fact defines the basic **series circuit** as one in which there is only one path for current, from one point to another, resulting in the same current flow through each component. Thus any two components in a circuit will be in series if they have the same current flowing through them.

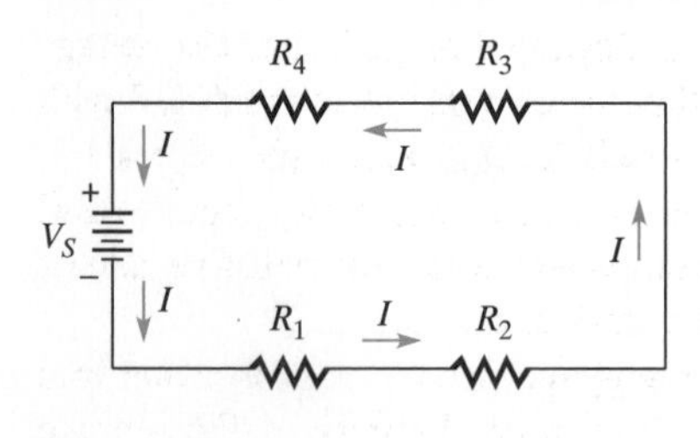

FIGURE 4–6 Current flow in a series circuit

As illustrated in Figure 4–7, four ammeters are placed in the circuit in order to measure the current into and out of each resistor. The voltage source forces current into one end of the resistor, through it, and out the other end. The current flowing out of each resistor must equal the current flowing into it. As Figure 4–7 shows, the current out of one resistor is the current into the next. Thus, it can be said that the same current flows in all parts of a series circuit.

FIGURE 4–7 Measuring current flow in a series circuit

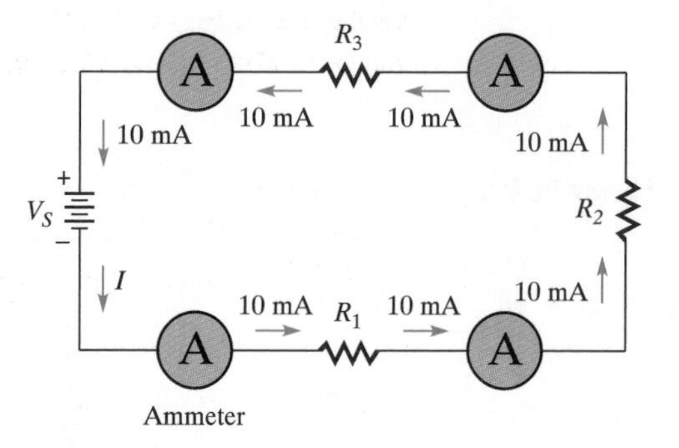

■ **SECTION 4–2 REVIEW**

1. What determines whether or not two electronic devices are in series?
2. In Figure 4–8, if 100 mA of current flows through R_1, how much current is flowing through R_2?
3. In Figure 4–9, what will each ammeter read?

FIGURE 4–8 Circuit for Section Review Problem 2

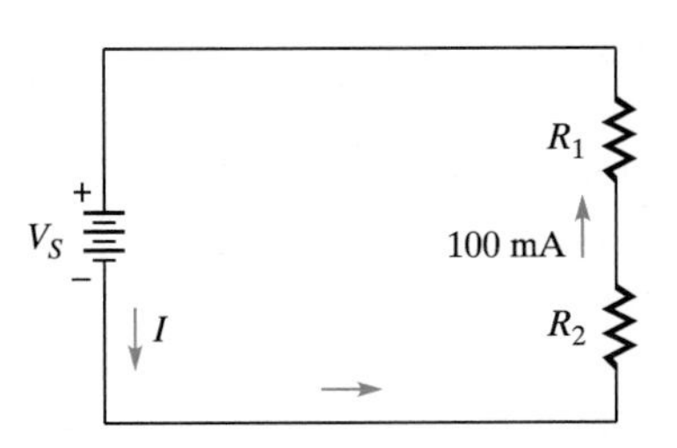

FIGURE 4–9 Circuit for Section Review Problem 3

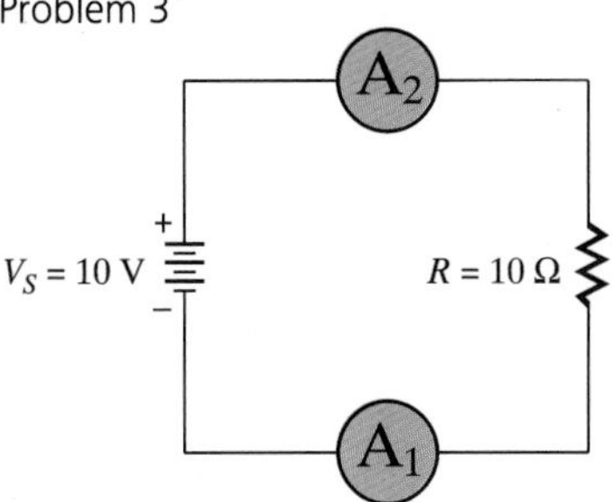

4–3 VOLTAGE IN A SERIES CIRCUIT

Question: What is a voltage drop?

When current flows through a resistor, the electrons emerging have less energy than when they entered. This loss of energy, or difference in potential, creates a voltage across the resistor, known as a **voltage drop,** often referred to as an *IR* drop. (The latter name comes from the Ohm's Law equation for voltage, $V = IR$.) There will be a voltage drop across each resistor in a series circuit.

In a manner similar to the naming of resistors, the voltage drops are labeled V_1, V_2, V_3, and so on, as shown in Figure 4–10. Note also that the source, or supply, voltage is labeled V_S. As each voltage drop occurs, it is subtracted from the source voltage. The distribution or *division* of voltage is often the purpose of a series circuit, and for this reason, a series circuit is referred to as a **voltage divider.**

Figure 4–10 illustrates the series circuit, the current flowing through it, and the voltage drops produced. Notice that the end of the resistor that the current enters is labeled with a minus sign (−) while the end from which it emerges is labeled with a plus sign (+). Even though resistors have no polarity, it is said that the current enters the negative end and emerges from the positive. The end of a series component nearer the negative side of the source will *always* be negative with respect to the other end.

Notice, however, that the current flows out of the negative side of the source and into the positive, giving it a polarity opposite that of a voltage drop. In order to differentiate between the two voltages, the source voltage is referred to as a **voltage rise.** You should notice one other point concerning the voltage source and the voltage drop. The voltage source is always present, even if the circuit is broken. Voltage drops, on the other hand,

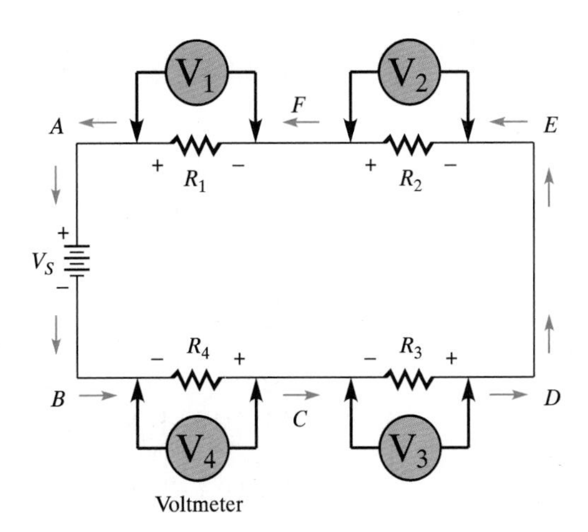

FIGURE 4–10 Current flow, resistors, and voltage drops in a series circuit

FIGURE 4–11 Voltage drops in a series circuit. Source voltage is always present, but current must flow through the circuit for voltage drops to occur.

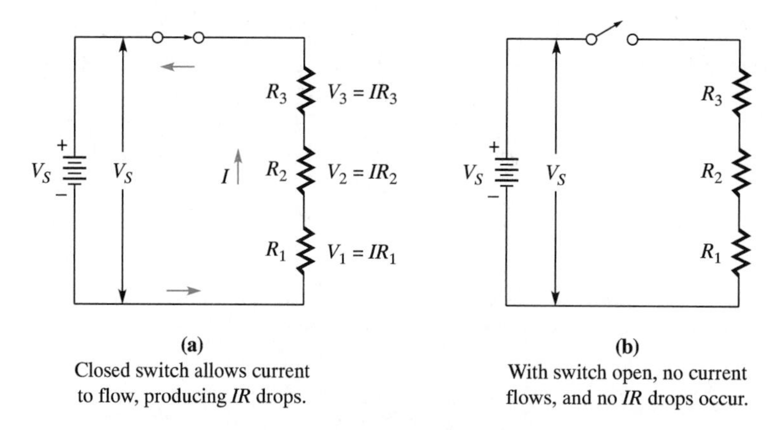

occur *only* when current flows through a resistor, as illustrated in Figure 4–11. As will be demonstrated later in the chapter, this point is important in troubleshooting series circuits.

■ Kirchhoff's Voltage Law

A voltage drop means just that—a drop in potential energy. Thus, as stated previously, as each voltage drop occurs it is subtracted from the source voltage (Figure 4–10). The voltage measured from point A to point B is the source voltage. As subsequent measurements are made from point A to points C, D, and E, in that order, the voltage will decrease. Thus the following mathematical relationship can be stated:

FIGURE 4–12 Relationship of voltage drops to voltage source

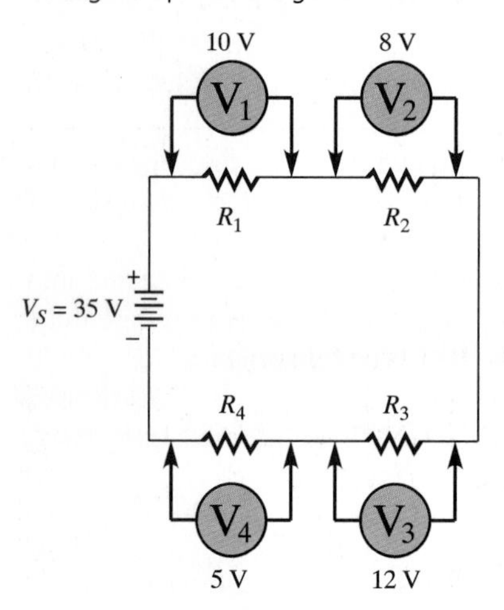

(4–1) $$V_S - V_1 - V_2 - V_3 - V_4 = 0$$

This equation represents the basic law of series circuits, formulated by the German physicist Gustav Kirchhoff. Known as **Kirchhoff's voltage law** (KVL), this basic law can be stated as follows:

The algebraic sum of the voltage drops and the source voltage in a series circuit is equal to zero.

Notice in Figure 4–12 that the source voltage and the voltage drops have been assigned values. The sum of these voltage drops is *always* equal to the source voltage, which can be stated in equation form as

(4–2) $$V_S = V_1 + V_2 + V_3 + V_4$$

This equation shows the source voltage divided into a number of smaller voltages by the series resistors. Thus, **KVL** can be stated in an alternate manner:

The sum of the voltage drops around a series circuit is equal to the source voltage.

CAUTION

What you have been exposed to in this section is something indispensable to your understanding of all electronic concepts and circuits. Kirchhoff's voltage law may seem simple and self-evident, but failure to understand, remember, and apply it at the appropriate times can make subsequent learning very difficult. Kirchhoff's voltage law is *the* basic law for all series circuits. Series circuits are found in all electronic systems. It doesn't matter of what components they may be composed, any series circuit can be analyzed through application of KVL. In this text, Kirchhoff's voltage law will be recalled, when needed, to aid in the understanding of some circuit action. Take time now not to just learn it, but to thoroughly understand it!

EXAMPLE 4–2

For the circuit of Figure 4–13, compute the value of the source voltage.

FIGURE 4–13 Circuit for Example 4–2

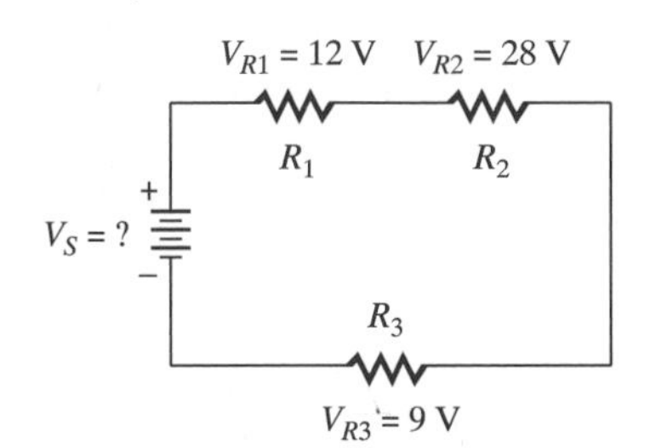

■ **Solution:**

Using Eq. 4–2 gives

$$V_S = V_{R1} + V_{R2} + V_{R3}$$
$$= 12\text{ V} + 28\text{ V} + 9\text{ V} = 49\text{ V}$$

[1] [2] [+] [2] [8] [+] [9] [=]

Example 4–3

For the circuit of Figure 4–14, compute the voltage across R_3.

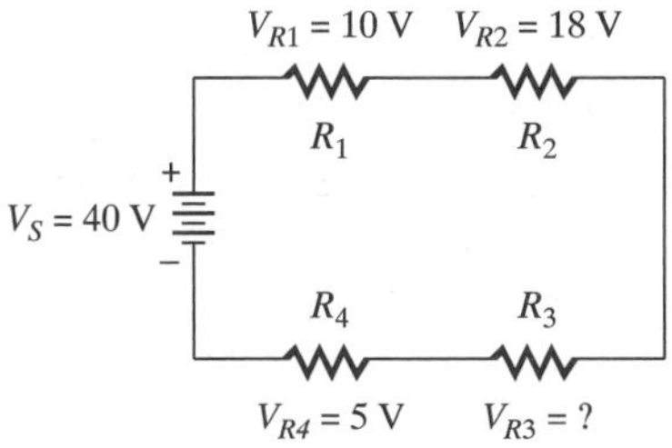

FIGURE 4–14 Circuit for Example 4–3

■ **Solution:**

Use Eq. 4–2 and transpose to solve for V_{R3}:

$$V_S = V_{R1} + V_{R2} + V_{R3} + V_{R4}$$
$$V_{R3} = V_S - (V_{R1} + V_{R2} + V_{R4})$$
$$= 40\text{ V} - (10\text{ V} + 18\text{ V} + 5\text{ V})$$
$$= 40\text{ V} - 33\text{ V} = 7\text{ V}$$

■ SECTION 4–3 REVIEW

1. Write Kirchhoff's Voltage Law.
2. What is the value of V_{R3} in Figure 4–15?
3. What is the value of V_S in Figure 4–16?

FIGURE 4–15 Circuit for Section Review Problem 2

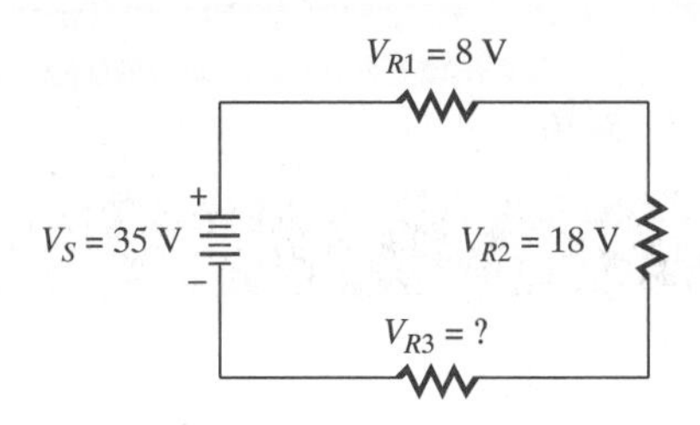

FIGURE 4–16 Circuit for Section Review Problem 3

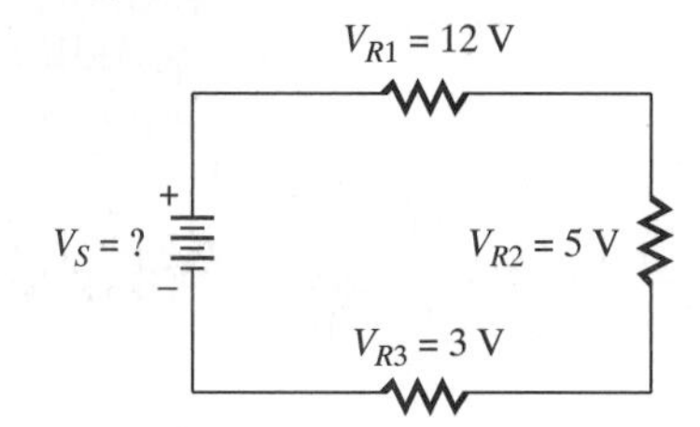

4–4 VOLTAGE SOURCES IN SERIES

Voltage sources, such as batteries, can be placed in series in order to increase the total voltage. A good example of this is the battery pack used in systems such as hand-held radio transceivers. As shown in Figure 4–17, the batteries are placed so the negative terminal of one contacts the positive terminal of the other. The total voltage will be the sum of the voltage of the individual batteries.

STUDY SKILLS

Taking an Exam: Just Before the Exam

During the exam you'll need to think clearly and be alert. In order to be at your best, you should get plenty of rest the night before. A good nourishing breakfast is also essential to an alert mind.

It is not generally a good idea to study up until the last minute. You may find something that you have missed and become anxious because there is not enough time to learn it. This anxiety may result in poor performance on the entire exam, rather than just missing one or two problems. It is better to spend some time before the exam doing something that will have a relaxing and calming effect. Just before the exam, you may want to review key equations and concepts.

Be sure to come prepared. Bring two pencils and scratch paper, and have fresh batteries in your calculator. Also allow plenty of time to get to the exam so you won't start off feeling rushed and anxious.

FIGURE 4–17 Battery pack as an example of voltage sources in series

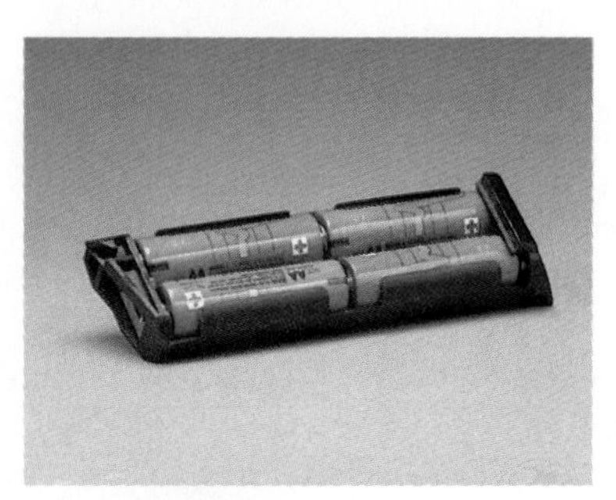

FIGURE 4–18 Configurations of voltage sources in series

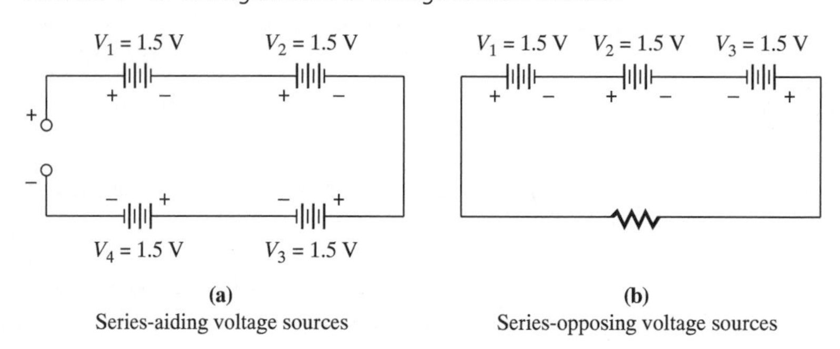

When batteries are connected negative to positive (− to +), the configuration is known as **series-aiding voltage sources,** schematically shown in Figure 4–18(a). When the batteries are connected negative to negative (− to −) and positive to positive (+ to +) so that the voltages subtract one from the other, the configuration is known as **series-opposing** voltage sources, as shown in Figure 4–18(b). Although not encountered in power sources, the series-opposing concept will be apparent in many circuits making up electronic systems.

Example 4–4

What is the total voltage supplied by the sources as connected in Figure 4–19?

FIGURE 4–19 Circuit for Example 4–4

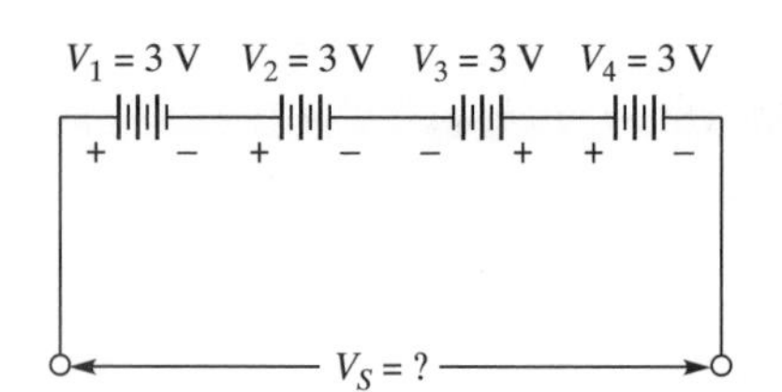

■ Solution:

Sources V_1, V_2, and V_4 are connected series aiding, thus producing a voltage of 9 V. Source V_3 is connected series opposing. The total voltage is found by

$$(V_1 + V_2 + V_4) - V_3$$

Thus,

$$3\text{ V} + 3\text{ V} + 3\text{ V} - 3\text{ V} = 6\text{ V}$$

(3 + 3 + 3) − 3 =

SECTION 4–4 REVIEW

1. Define what is meant by series aiding voltage sources.
2. Define what is meant by series opposing voltage sources.
3. Two voltage sources, one of 15 V and one of 25 V are connected series aiding. Draw this arrangement and compute the total voltage.
4. If the same two voltage sources of Problem 3 were connected series opposing, draw the arrangement and compute the total voltage.

4–5 TOTAL RESISTANCE OF A SERIES CIRCUIT

Question: How is the total resistance of a series circuit determined?

When a voltage source is connected across a series circuit, it doesn't feel the effects of each individual resistor, but rather a value representing their total. Using the equation for Kirchhoff's voltage law, an equation representing the total resistance can be derived.

From Kirchhoff's voltage law,

$$V_S = V_{R1} + V_{R2} + V_{R3} \cdots V_n$$

where n is any number.

From Ohm's Law and substituting,

$$IR_T = IR_1 + IR_2 + IR_3 \cdots IR_n$$

where T is the total.

Now dividing both sides of the equation by current I gives the total resistance:

(4–3)
$$R_T = R_1 + R_2 + R_3 \cdots R_n$$

Equation 4–3 shows that the **total resistance** of a series circuit is *always* equal to the sum of the individual resistance values making up the circuit.

FIGURE 4–20 Circuit for Example 4–5

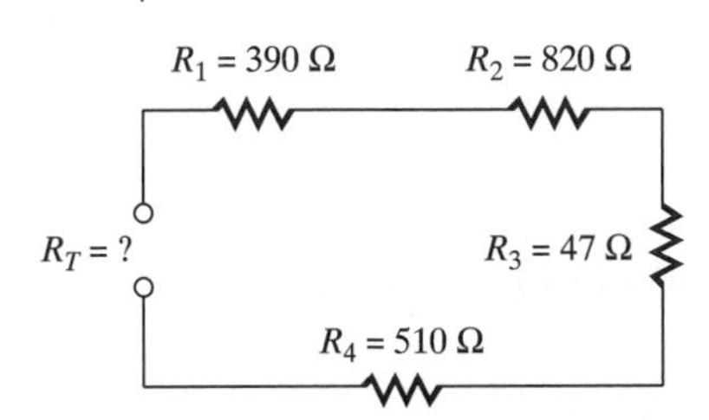

EXAMPLE 4–5 What is the total resistance of the circuit in Figure 4–20?

■ Solution:

Using Eq. 4–3,

$$R_T = R_1 + R_2 + R_3 + R_4$$
$$= 390\ \Omega + 820\ \Omega + 47\ \Omega + 510\ \Omega = 1.767\ \text{k}\Omega$$

EXAMPLE 4–6 If the total resistance in Figure 4–21 is 9.8 kΩ, what is the value of R_3?

FIGURE 4–21 Circuit for Example 4–6

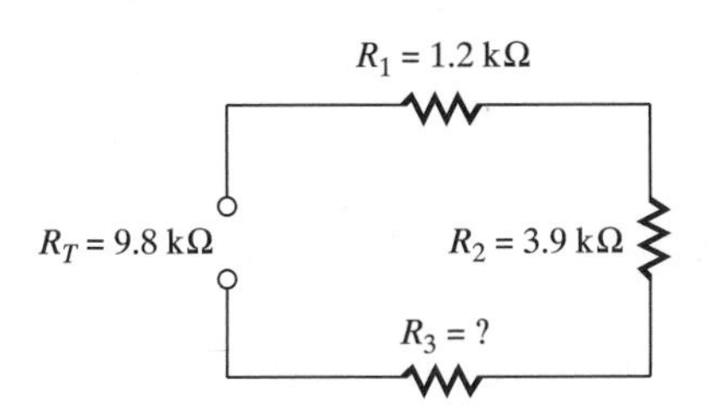

■ Solution:

Solving Eq. 4–3 for R_3 yields

$$R_3 = R_T - (R_1 + R_2)$$
$$= 9.8\ \text{k}\Omega - (1.2\ \text{k}\Omega + 3.9\ \text{k}\Omega) = 4.7\ \text{k}\Omega$$

9 · 8 EXP 3 − (1 · 2 EXP 3 + 3 · 9 EXP 3) =

SECTION 4–5 REVIEW

1. For the circuit of Figure 4–22, compute the total resistance.
2. Three series resistors have a total resistance of 167 Ω. A resistor is added in series that makes the total resistance value 317 Ω. What is the value of the resistor added?
3. In Problem 2, what happened to the circuit current when the resistor was added? Why?

FIGURE 4–22 Circuit for Section Review Problem 1

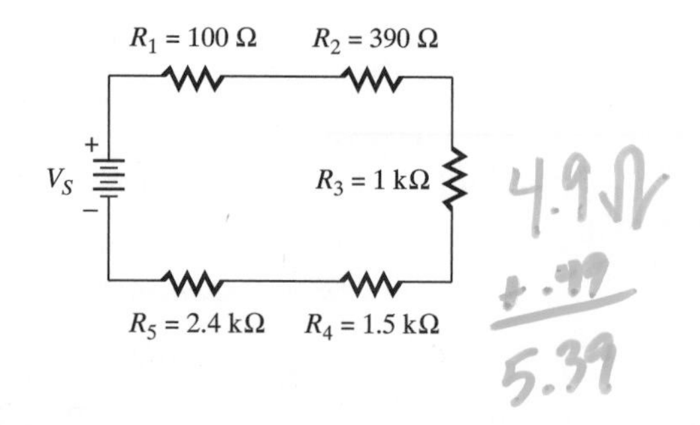

4–6 POWER IN A SERIES CIRCUIT

As stated previously, power is developed in an electric circuit when current flows through a resistance. The **total power** in a series circuit is the sum of the powers developed by each resistor. In equation form, total power is written as

(4–4) $$P_T = P_1 + P_2 + P_3 + \cdots P_n$$

The total power may also be computed using the total values of current, resistance, and voltage:

$$P_T = V_S I$$

$$P_T = \frac{V_S^2}{R_T} \quad \left(\text{remember: } P = \frac{V^2}{R}\right)$$

$$P_T = I^2 R_T \quad (\text{remember: } P = I^2 R)$$

EXAMPLE 4–7

Compute the total power developed in the circuit of Figure 4–23.

FIGURE 4–23 Circuit for Example 4–7

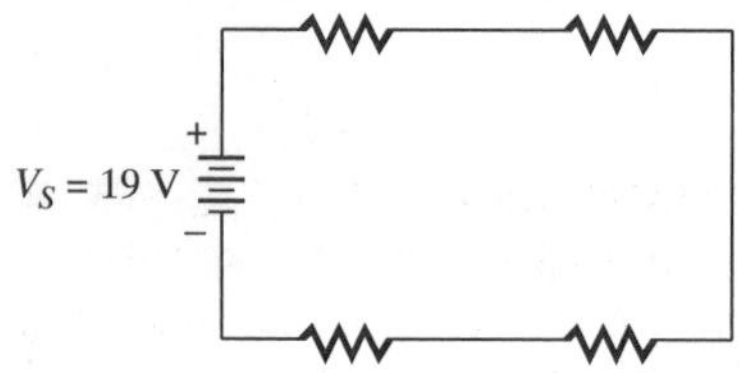

$R_1 = 100\ \Omega$, $R_2 = 330\ \Omega$, $V_S = 19$ V, $R_4 = 1$ kΩ, $R_3 = 470\ \Omega$

■ Solution:

Compute the total resistance:

$$R_T = R_1 + R_2 + R_3 + R_4$$

$$= 100\ \Omega + 330\ \Omega + 470\ \Omega + 1000\ \Omega = 1.9\ \text{k}\Omega$$

Compute the circuit current:

$$I = \frac{V_S}{R_T}$$

$$= \frac{19\ \text{V}}{1900\ \Omega} = 10\ \text{mA}$$

Compute the power developed by each resistor:

$$P_{R1} = I^2 R_1 = (10^{-3} \times 10^{-3})(100\ \Omega) = 10\ \text{mW}$$

$$P_{R2} = I^2 R_2 = (10^{-3} \times 10^{-3})(330\ \Omega) = 33\ \text{mW}$$

$$P_{R3} = I^2R_3 = (10^{-3} \times 10^{-3})(470\ \Omega) = 47\text{ mW}$$

$$P_{R4} = I^2R_4 = (10^{-3} \times 10^{-3})(1000\ \Omega) = 100\text{ mW}$$

Compute total power by adding the individual power values:

$$P_T = P_{R1} + P_{R2} + P_{R3} + P_{R4}$$

$$= 10\text{ mW} + 33\text{ mW} + 47\text{ mW} + 100\text{ mW} = 190\text{ mW}$$

[1] [0] [EXP] [±] [3] [+] [3] [3] [EXP] [±] [3] [+] [4] [7] [EXP] [±] [3] [+]

[1] [0] [0] [EXP] [±] [3] [=]

EXAMPLE 4–8 Compute the total power of Example 4–7 using the alternate methods.

$$P_T = I^2R_T = (10^{-3} \times 10^{-3})(1900\ \Omega) = 190\text{ mW}$$

$$P_T = \frac{V_S^2}{R_T} = \frac{(19\text{ V} \times 19\text{ V})}{1900\ \Omega} = 190\text{ mW}$$

$$P_T = V_S I_T = (19)(10^{-3} \times 10^{-3}) = 190\text{ mW}$$

SYNOPSIS

Before beginning a study of the analysis of series circuits, it would be wise to review their characteristics. The basic facts of a series circuit may seem to be self-evident, and in truth, they are. However, failure to understand them now and apply them later will make the understanding and analysis of electronic systems much more difficult. Time taken now to learn and *thoroughly* understand these facts will be time well spent!

A series circuit is one in which there is only one path for current. As a result, the same current flows in all parts of a series circuit. Each resistor in the circuit will have an associated voltage drop, also known as an IR drop. IR drops occur only when current flows through a resistance. The sum of the IR drops is equal to the source voltage. Finally, the total resistance of a series circuit is the sum of the individual resistances.

4–7 ANALYSIS OF SERIES CIRCUITS

Question: How are series circuits analyzed?

The analysis of any circuit involves determining the current through and the voltage across each component. It is especially important to be able to determine the voltage drops. Recall that voltages can be measured without disturbing the circuit, while measuring current requires opening the circuit. Using voltage measurements, then, will greatly speed and facilitate troubleshooting in electronic systems. Ohm's Law, Kirchoff's voltage law, and the facts of series circuits are the tools of analysis. Always keep in mind that the series connection is one of only two ways of connecting electronic components. In the analysis of series circuits, which are a major part of complex electronic systems, you will use the procedures that are presented in this section. To analyze a complex system, then, all that will be necessary is to learn the operation of the devices, other than resistors, of which it is composed.

STUDY SKILLS

Taking an Exam: Beginning the Exam

It is vital that you listen carefully to all instructions given by the instructor. As soon as you are allowed to begin, write down the equations that you think you might need. Also write down any warnings of common errors you know you may be likely to make. By writing these down first you will be relieved of the burden of worrying about whether or not you will remember them when you need them, thus allowing you to concentrate on the actual problems.

You should refer back to these equations and warnings as you go through the exam to be sure that you have not overlooked an error or misused an equation.

There is a step-by-step procedure to be followed in circuit analysis:

1. Calculate the total resistance.
2. Calculate the value of the current.
3. Calculate the voltage drop across each resistor.
4. Calculate the power developed in each resistor.
5. Calculate the total power developed in the circuit.

The current is the most important and useful quantity in the analysis of series circuits because it is the same through each component.

NOTE

There is, of course, a preliminary step to be done before attempting to solve circuit analysis problems: Make it a habit of carefully reading the problem and writing out all information given. It is very easy to overlook some fact that is a key to the solution, which will also be true in analysis of electronic systems. Remember that solving the problems in this text is not just an academic drill; it is also practice for analysis of practical electronic circuits.

EXAMPLE 4–9 For the circuit of Figure 4–24, Compute (a) the total resistance, (b) the current, (c) the voltage drop across each resistor, (d) the power developed in each resistor, and (e) the total power developed in the circuit.

FIGURE 4–24 Circuit for Example 4–9

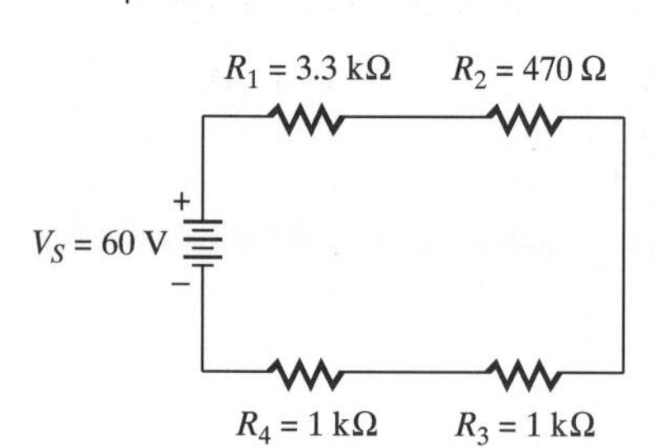

■ **Solution:**

a. Compute the total resistance:

$$R_T = R_1 + R_2 + R_3 + R_4$$
$$= 3.3\text{ k}\Omega + 470\ \Omega + 1\text{ k}\Omega + 1\text{ k}\Omega = 5.77\text{ k}\Omega$$

[3] [·] [3] [EXP] [3] [+] [4] [7] [0] [+] [1] [EXP] [3] [+] [1] [EXP] [3] [=]

b. Calculate the current:

$$I = \frac{V_S}{R_T} = \frac{60\text{ V}}{5.77\text{ k}\Omega} = 10.4\text{ mA}$$

[6] [0] [÷] [5] [·] [7] [7] [EXP] [3] [=]

c. Calculate the voltage across each resistor:

$$V_{R1} = I_{R1}R_1 = (10.4 \text{ mA})(3.3\text{k } \Omega) = 34.32 \text{ V}$$

[1] [0] [·] [4] [EXP] [±] [3] [×] [3] [·] [3] [EXP] [3] [=]

$$V_{R2} = I_{R2}R_2 = (10.4 \text{ mA})(470 \ \Omega) = 4.89 \text{ V}$$

[1] [0] [·] [4] [EXP] [±] [3] [×] [4] [7] [0] [=]

$$V_{R3} = I_{R3}R_3 = (10.4 \text{ mA})(1 \text{ k}\Omega) = 10.4 \text{ V}$$

[1] [0] [·] [4] [EXP] [±] [3] [×] [1] [EXP] [3] [=]

V_{R4} will equal V_{R3} since R_4 and R_3 are the same value and have the same current flowing through them.

d. Calculate the power developed in each resistor. The power developed in each resistor is calculated using one of the power equations. Remember to use the voltage or resistance for a particular resistor when calculating its power.

$$P_{R1} = V_{R1}I = (34.32 \text{ V})(10.4 \text{ mA}) = 357 \text{ mW}$$

[3] [4] [·] [3] [2] [×] [1] [0] [·] [4] [EXP] [±] [3] [=]

$$P_{R2} = V_{R2}I = (4.89 \text{ V})(10.4 \text{ mA}) = 50.8 \text{ mW}$$

[4] [·] [8] [9] [×] [1] [0] [·] [4] [EXP] [±] [3] [=]

$$P_{R3} = V_{R3}I = (10.4 \text{ V})(10.4 \text{ mA}) = 108 \text{ mW}$$

[1] [0] [·] [4] [×] [1] [0] [·] [4] [EXP] [±] [3] [=]

P_{R4} will equal P_{R3} because the voltage drops are the same and they have the same current flow.

e. Calculate total power developed in the circuit:

$$P_T = P_{R1} + P_{R2} + P_{R3} + P_{R4}$$
$$= 357 \text{ mW} + 50.8 \text{ mW} + 108 \text{ mW} + 108 \text{ mW} = 623.8 \text{ mW}$$

[3] [5] [7] [EXP] [±] [3] [+] [5] [0] [·] [8] [EXP] [±] [3] [+] [1] [0] [8]

[EXP] [±] [3] [+] [1] [0] [8] [EXP] [±] [3] [=]

Total power may also be computed using total values of resistance and voltage:

$$P_T = I^2R_T = (10.4^{-3} \text{ mA} \times 10.4^{-3} \text{ mA})(5.77 \times 10^3 \ \Omega) = 624 \text{ mW}$$

[1] [0] [·] [4] [EXP] [±] [3] [X²] [×] [5] [·] [7] [7] [EXP] [3] [=]

$$P_T = V_SI = (60 \text{ V})(10.4^{-3} \text{ mA}) = 624 \text{ mW}$$

[6] [0] [×] [1] [0] [·] [4] [EXP] [±] [3] [=]

$$P_T = \frac{V_S^2}{R_T} = \frac{(60 \text{ V})(60 \text{ V})}{5.77 \times 10^3 \ \Omega} = 624 \text{ mW}$$

[6] [0] [X²] [÷] [5] [·] [7] [7] [EXP] [3] [=]

NOTE

The values of total power obtained by four different means obtained in Example 4–8e agree very closely. The question is "How close is close enough?" For classroom work, your instructor will undoubtedly give you guidance as to what degree of accuracy is required. Calculations in this text are carried to three significant figures.

EXAMPLE 4–10

For the circuit of Figure 4–25, compute the voltage drop across each resistor and the value of the source voltage.

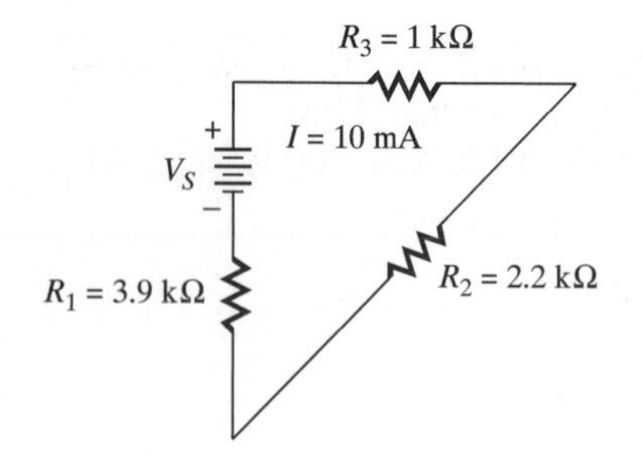

FIGURE 4–25 Circuit for Example 4–10

■ **Solution:**

Circuit current is given as 10 mA. Since this is a series circuit, each resistor will have 10 mA of current flowing through it. So by Ohm's Law,

$$V_{R1} = IR_1 = (10\ \text{mA})(3.9\ \text{k}\Omega) = 39\ \text{V}$$

$$V_{R2} = IR_2 = (10\ \text{mA})(2.2\ \text{k}\Omega) = 22\ \text{V}$$

$$V_{R3} = IR_3 = (10\ \text{mA})(1\ \text{k}\Omega) = 10\ \text{V}$$

Then, by Kirchhoff's voltage law,

$$V_S = V_{R1} + V_{R2} + V_{R3}$$

$$= 39\ \text{V} + 22\ \text{V} + 10\ \text{V} = 71\ \text{V}$$

EXAMPLE 4–11

For the circuit of Figure 4–26, compute the values of R_2, R_3, and R_4.

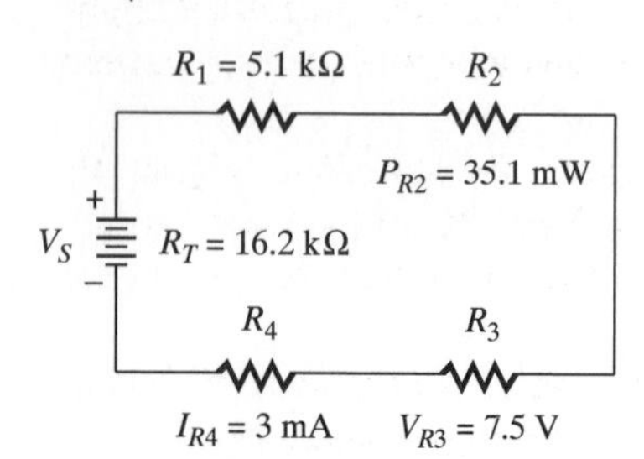

FIGURE 4–26 Circuit for Example 4–11

■ **Solution:**

1. The current through all resistors will be the same as that through R_4 (3 mA).
2. The power developed in R_2 is given. The power equation can be transposed in order to compute R_2:

$$P_{R2} = I^2R_2$$

Then,

$$R_2 = \frac{P_{R2}}{I^2} = \frac{35.1 \times 10^{-3}}{(3 \times 10^{-3}\ \text{A})^2} = 3.9\ \text{k}\Omega$$

3. The voltage drop across R_3 is given. The Ohm's Law equation for voltage can be transposed in order to find R_3:

$$V_{R3} = I_{R3}$$

and

$$R_3 = \frac{V_{R3}}{I} = \frac{7.5\text{V}}{3 \times 10^{-3}\ \text{A}} = 2.5\ \text{k}\Omega$$

4. R_4 can be computed using the equation for total resistance.

$$R_4 = R_T - (R_1 + R_2 + R_3)$$

$$= 16.2\ \text{k}\Omega - (5.1\ \text{k}\Omega + 3.9\ \text{k}\Omega + 2.5\ \text{k}\Omega) = 4.7\ \text{k}\Omega$$

These examples demonstrate the methods of series circuit analysis. Remember that you are not memorizing how to analyze a particular circuit, but you are learning how to approach any such problem. Later in this chapter, these principles will be used in troubleshooting of series circuits.

SECTION 4–7 REVIEW

1. A series circuit is composed of four resistors across a 15 V source. The total resistance is 2.447 kΩ. If three of the resistors are 47 Ω, 390 Ω, and 1200 Ω, what is the value of the fourth?
2. Compute the voltage drop across each resistor in Problem 1.
3. A resistor has a voltage drop of 25 V and a value of 100 Ω. Compute the power developed by it and its minimum power rating.
4. A four-resistor series circuit has a source voltage of 10 V. Three of the voltage drops are 1.5 V, 3.5 V, and 4 V. Compute the value of the fourth voltage drop.
5. In Problem 4, the resistor with a 1.5 V drop has a value of 150 Ω. Compute the value of the other three resistors.

4–8 THE SERIES CIRCUIT AS A VOLTAGE DIVIDER

Question: What is the difference between a loaded and an unloaded voltage divider?

As you have already seen, a series circuit is a *voltage divider* in that it divides a source voltage among several loads. Each of the voltage drops can now be thought of as a voltage source. Thus, a series circuit can be used to supply voltages, from a common voltage source, to loads with different voltage requirements. Few electronic systems require only one voltage in their operation. In the early days of radio, the receivers were powered from at least three different batteries, each with a different voltage rating. Later, series circuits were used as voltage dividers to supply these voltages. This extremely important application of series circuits is still used in modern amplifiers and a host of other electronic circuits.

In this section, the unloaded voltage divider is presented. By **unloaded voltage divider,** it is meant that the voltage has been divided, but no loads have been placed across those voltages produced. (In Chapter 6, loads will be placed across the voltages and the loaded voltage divider developed.)

In Figure 4–27, the source voltage provides current to the three resistors producing three voltage drops V_{R1}, V_{R2}, and V_{R3}. The voltage drops may be computed as follows:

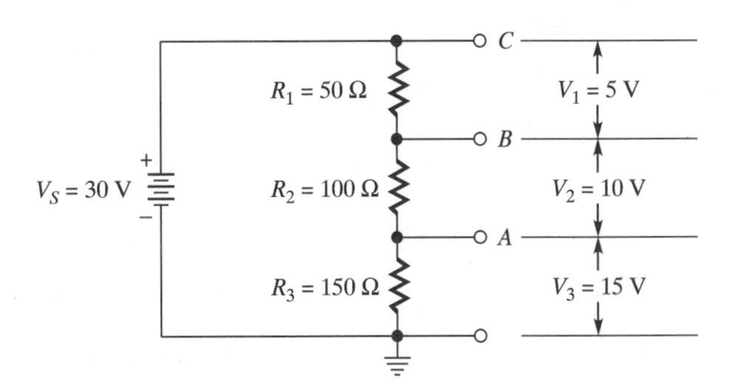

FIGURE 4–27 Unloaded voltage divider circuit

1. $R_T = R_1 + R_2 + R_3 = 50\ \Omega + 100\ \Omega + 150\ \Omega = 300\ \Omega$

2. $I = \dfrac{V_S}{R_T} = \dfrac{30\ \text{V}}{300\ \Omega} = 0.1\ \text{A}$

3. $V_1 = I_{R1}R_1 = (0.1\ \text{A})(50\ \Omega) = 5\ \text{V}$
   $V_2 = I_{R2}R_2 = (0.1\ \text{A})(100\ \Omega) = 10\ \text{V}$
   $V_3 = I_{R3}R_3 = (0.1\ \text{A})(150\ \Omega) = 15\ \text{V}$

The source voltage has been divided into three voltages of 5 V, 10 V, and 15 V. Notice that since the current through all resistors is the same, each individual voltage drop is *directly proportional* to the value of the resistor producing it. These are not the only values of resistors that will divide this source into the same three voltages. *Any* three values, with the same ratios one to another, will give the same result.

This directly proportional relationship may be summarized in the form of **resistance–voltage ratios** as follows: The ratio of each resistance to the total resistance is equal to the ratio of its voltage drop to the total voltage. The following example demonstrates this relationship.

First the resistance ratios are computed:

$$\frac{R_1}{R_T} = \frac{50\ \Omega}{300\ \Omega} = 16.67\%$$

$$\frac{R_2}{R_T} = \frac{100\ \Omega}{300\ \Omega} = 33.33\%$$

$$\frac{R_3}{R_T} = \frac{150\ \Omega}{300\ \Omega} = 50\%$$

Next the voltage ratios are computed:

$$\frac{V_{R1}}{V_S} = \frac{5\ \text{V}}{30\ \text{V}} = 16.67\%$$

$$\frac{V_{R2}}{V_S} = \frac{10\ \text{V}}{30\ \text{V}} = 33.33\%$$

$$\frac{V_{R3}}{V_S} = \frac{15\ \text{V}}{30\ \text{V}} = 50\%$$

Notice that the voltage ratios are the same as their resistance ratios, from which the following useful equation can be derived:

$$\frac{V_x}{V_S} = \frac{R_x}{R_T}$$

where x is an unknown. Rearranging gives

$$V_x \times R_T = V_S \times R_x$$

and solving yields

(4–5)
$$V_x = \frac{R_x}{R_T} V_S$$

Equation 4–5 is known as the *voltage divider equation,* and is quite useful in certain instances. Notice that it can be rearranged as follows:

$$V_x = \frac{V_S}{R_T} R_x$$

Since $V_S/R_T = I$,

$$V_x = IR_x$$

You will recognize this equation as current times the individual resistance, which is the Ohm's law equation for voltage!

EXAMPLE 4–12 Design an unloaded voltage divider to provide voltages of 6 V, 12 V, and 18 V. Do not include the voltage source as one of the voltages provided by the divider.

■ Solution:

The purpose of this example is to provide another example of series circuit analysis. It can be solved using principals presented in previous paragraphs.

1. Since there are three voltages required, the divider must contain three resistors.
2. A current must be chosen for the divider, which is possible since the same current flows through each resistor. Note that 10 mA will be chosen strictly for convenience. Since the same current flows through each resistor, it is the ohmic value of the individual resistor that determines its voltage drop. Thus, any value of current could have been chosen.
3. The resistor values can now be computed:

$$R_1 = \frac{V_{R1}}{I} = \frac{6}{10 \times 10^{-3}\text{ A}} = 600\ \Omega$$

$$R_2 = \frac{V_{R2}}{I} = \frac{12}{10 \times 10^{-3}\text{ A}} = 1200\ \Omega$$

$$R_3 = \frac{V_{R3}}{I} = \frac{18}{10 \times 10^{-3}\text{ A}} = 1800\ \Omega$$

4. The voltage source, by Kirchhoff's voltage law must be

$$V_S = V_{R1} + V_{R2} + V_{R3}$$

$$= 6\text{ V} + 12\text{ V} + 18\text{ V} = 36\text{ V}$$

Notice Example 4–12 that if a different current had been chosen, a different set of resistances would have been computed. You will find that when the loaded voltage divider is presented in a later chapter, there is a definite way of choosing the divider current for best results.

■ SECTION 4–8 REVIEW

1. If a voltage source must produce 5 voltages, how many resistors must the divider contain?
2. Why is it possible to assume a current in designing a voltage divider?
3. Design an unloaded voltage divider to produce 4.5 V, 8 V, and 20 V.

STUDY SKILLS

Taking an Exam: What to Do First

Exams are not always arranged either in ascending order of difficulty or in chronological order. Since time is usually a factor, it is not wise to spend too much time on problems that you find difficult to understand or cannot recall the procedure to solve. Therefore, you should first look over the entire exam and then follow this procedure: (1) Start with problems that you know how to solve quickly; (2) Go back and solve problems that you know how to solve but that take longer; (3) Work on problems that you find difficult but have a general knowledge of how to solve; and (4) Finally, divide the remaining time between the problems you find most difficult and checking your solutions. (And don't forget to check the warnings you wrote down at the beginning of the exam.)

4–9 THE POTENTIOMETER AS A VARIABLE VOLTAGE DIVIDER

Question: Can a potentiometer be used as a variable voltage divider?

Recall from Chapter 2 that a potentiometer is a fixed resistor with a variable wiper arm that can be positioned at any point along it. The ratio of the resistance from the wiper to each end can be varied. You now know that current through a resistor produces a voltage drop whose value is directly proportional to the resistance. Since the ratio of the two resistances of the potentiometer can be varied, then the ratio of the two voltages can be varied, which thus makes the potentiometer a **variable voltage divider.**

In Figure 4–28(a), a 10 kΩ linear potentiometer with a 50 V source is set to its midpoint. The circuit formed is effectively one that has two 5 kΩ resistors in series, as shown in Figure 4-28(b). The voltage divides equally yielding 25 V across each. In Figure 4–28(c), the wiper is moved three-quarters of the way toward point A, and the circuit is effectively the same as the one in Figure 4–28(d). The voltage from A to C is now 12.5 V and the voltage from point B to C is 37.5 V. Any movement of the wiper arm results in a

FIGURE 4–28 Potentiometer as a variable voltage divider

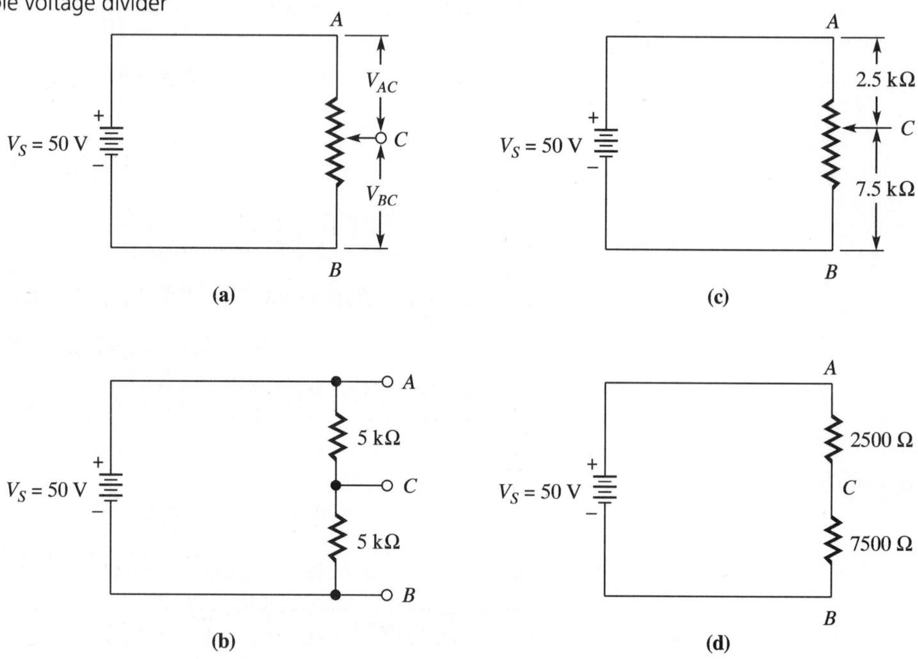

■ **TABLE 4–1 Resistance and voltage ratios for potentiometer in Figure 4–29**

Wiper Arm Position	R_{AC}	R_{CB}	V_{AC}	V_{CB}
At *A*	0 Ω	10 kΩ	0 V	50 V
One-quarter	2.5 kΩ	7.5 kΩ	12.5 V	37.5 V
One-third	3.3 kΩ	6.7 kΩ	16.5 V	33.5V
One-half	5 KΩ	5 kΩ	25 V	25 V
Two-thirds	6.7 kΩ	3.3 kΩ	33.5 V	16.5 V
Three-quarters	7.5 kΩ	2.5 kΩ	37.5 V	12.5 V
At *B*	10 kΩ	0 Ω	50 V	0 V

different resistance ratio and hence a different voltage ratio. Table 4–1 shows the ratios for the potentiometer for various settings of the wiper arm.

CAUTION

The preceding discussion and circuit are merely meant to show the operation of a potentiometer. Do not attempt to build this circuit in order to prove its operation. To do so could result in damage to the potentiometer.

EXAMPLE 4–13 The wiper arm of a 100 kΩ potentiometer is positioned as shown in Figure 4–29. What would be the voltage measured from the wiper to both ends of the potentiometer?

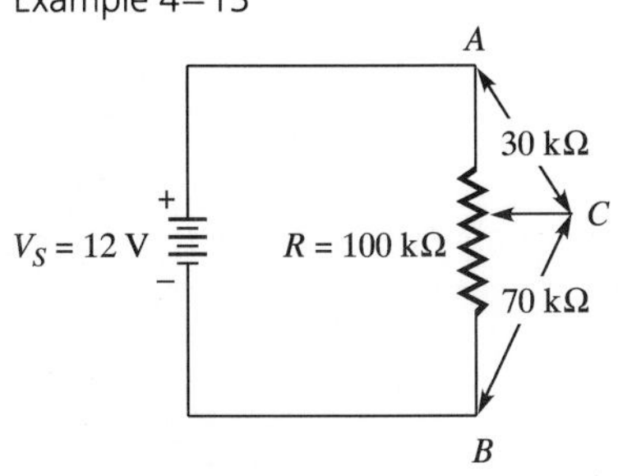

FIGURE 4–29 Circuit for Example 4–13

■ Solution:

The current flowing through the potentiometer is found by dividing the source voltage by the total resistance. Thus,

$$I = \frac{V_S}{R_T}$$

$$= \frac{12\text{ V}}{100\text{ k}\Omega} = 120\ \mu\text{A}$$

[1] [2] [÷] [1] [0] [0] [EXP] [3] [=]

The resistance from points *A* to *C* is 30 kΩ; so the voltage between these two points is

$$V_{AC} = IR_A = (120\ \mu\text{A})(30\text{ k}\Omega) = 3.6\text{ V}$$

[1] [2] [0] [EXP] [±] [6] [×] [3] [0] [EXP] [3] [=]

And

$$V_{CB} = IR_B$$

$$= (120\ \mu\text{A})(70\text{ k}\Omega) = 8.4\text{ V}$$

[1] [2] [0] [EXP] [±] [6] [×] [7] [0] [EXP] [3] [=]

EXAMPLE 4–14 At what point must the wiper arm of a 5 kΩ potentiometer be set in order to provide 1.8 V and 8.2 V from a 10 V source? The circuit is shown in Figure 4–30.

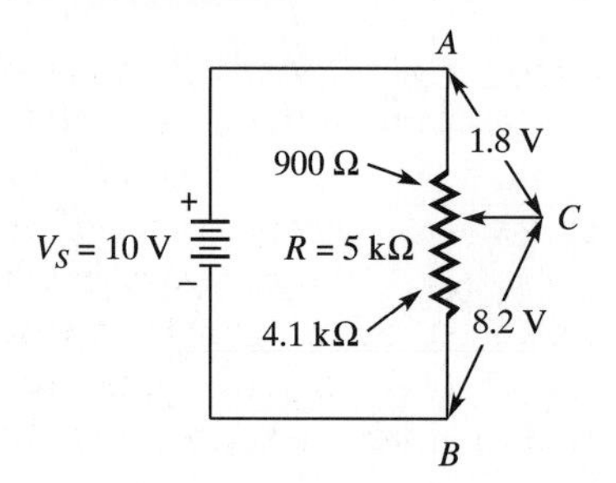

FIGURE 4–30 Circuit for Example 4–14

■ **Solution:**

A 10 V source produces 2 mA of current through the entire potentiometer. Thus, the resistances required to produce 1.8 V and 8.2 V are

$$R_A = \frac{V_{AC}}{I}$$

$$= \frac{1.8 \text{ V}}{2 \text{ mA}} = 900 \ \Omega$$

$$R_B = \frac{V_{CB}}{I}$$

$$= \frac{8.2 \text{ V}}{2 \text{ mA}} = 4.1 \text{ k}\Omega$$

■ SECTION 4–9 REVIEW

1. Why can a potentiometer be considered to be a variable voltage divider?
2. The wiper of a 500 kΩ potentiometer is positioned at three-fourths of its full movement. What are the two resistance values formed?
3. If a 100 V source is placed across the potentiometer of Problem 2, what two voltages will be produced?

4–10 GROUND AS A REFERENCE

In Chapter 2, voltage was defined as a difference in potential electric energy between two points. Thus, in an electronic system, if a voltage is specified, the two points between which it will be measured must also be specified. Specifying these points is often done on schematic diagrams of electronic systems as an aid in troubleshooting. For example, as illustrated in Figure 4–31, the voltage measured between points *A* and *B* is specified and written as V_{AB}. The first letter indicates the point of measurement and the second indicates the reference point. As another example, in the same circuit, the voltage at point *C* with reference to point *A* is labeled V_{CA}.

FIGURE 4–31 Reference points for measuring voltage

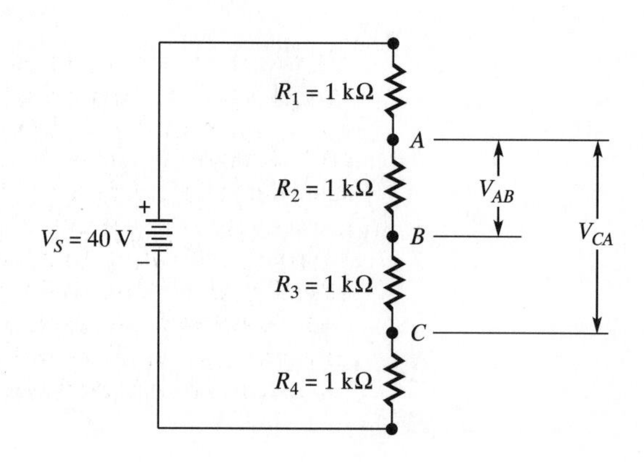

It is more convenient, in most cases within an electronic system, to specify a *common reference*. This common reference is usually given the name ground. Ground, however, can have several meanings.

Question: What does ground mean?

Ground is an often misunderstood subject in electronics. **Ground** is a reference point in a circuit that provides a common return point to the voltage source. A true ground reference, however, is one where the common return is the earth, which is a conductive medium. If two iron copper rods are driven into the earth, *continuity*, or zero ohms will be measured between them. This type of reference is often supplied by commercial power companies. At a generating plant, copper rods are driven deeply into the ground. A wire attached to these rods accompanies the power lines to each customer, and a metal rod driven into the ground at the customer's location establishes a common reference with the power company. Some types of soil are more conductive than others; a salt marsh in Louisiana, for example, is more conductive than dry sand in Arizona. In the latter case, the rod must be driven much more deeply into the ground.

When the earth is used to measure continuity in a circuit, the reference point is known as an **earth ground.** Its symbol is illustrated in Figure 4–32(a).

Question: Is earth ground the only reference encountered in electronic systems?

Earth ground is not the only reference used in electronic systems. A **chassis ground** is one in which the common return point is now the chassis, or metal box, on which the circuit components are mounted. It may also be a large piece of copper foil on a printed circuit board, which is known as a *trace*. Any point of common return to the voltage source is then connected to this ground point. The symbol for a chassis ground is shown in Figure 4–32(c).

FIGURE 4–32 Circuit ground symbols

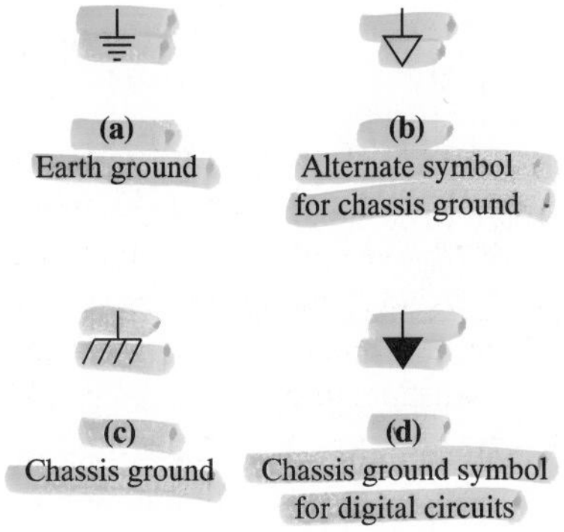

(a) Earth ground

(b) Alternate symbol for chassis ground

(c) Chassis ground

(d) Chassis ground symbol for digital circuits

Chassis ground may or may not be attached to earth ground, so the term is somewhat of a misnomer in these cases. The only way to know for sure is to measure the resistance from the chassis or printed circuit board trace to the power line ground. A reading of zero ohms would indicate that they are common, while a reading of infinity would indicate that they are not. A better name for chassis ground, thus, is **circuit common,** which does not imply that the reference is connected to earth ground. The alternate symbol for chassis ground is shown in Figure 4–32(b). Figure 4–32(d) is the symbol often used for the reference points in digital circuits. (Digital circuits are those used extensively in computers.)

CAUTION

When two or more electronic systems must be used together, they must, for safety reasons, be tied to a common reference or ground. If not, a voltage may exist between the two. A good example is the patient monitors used in recovery rooms of hospitals. Many different systems may be in use at any given time. If any two of these systems were not tied to a common reference, or ground, a voltage might exist between them that could be devastating to the patient or the technician operating them. This voltage and resulting damage could also occur if the ground wire to a system opened for some reason. Thus, tests for ground faults, utilizing a device known as a ground-fault monitor, are conducted daily in hospitals. A technician can determine if two systems share a common reference using a voltmeter. If two systems share a common reference or ground, a voltmeter between the two will indicate zero volts.

FIGURE 4–33 Alternate configurations of resistive series circuit

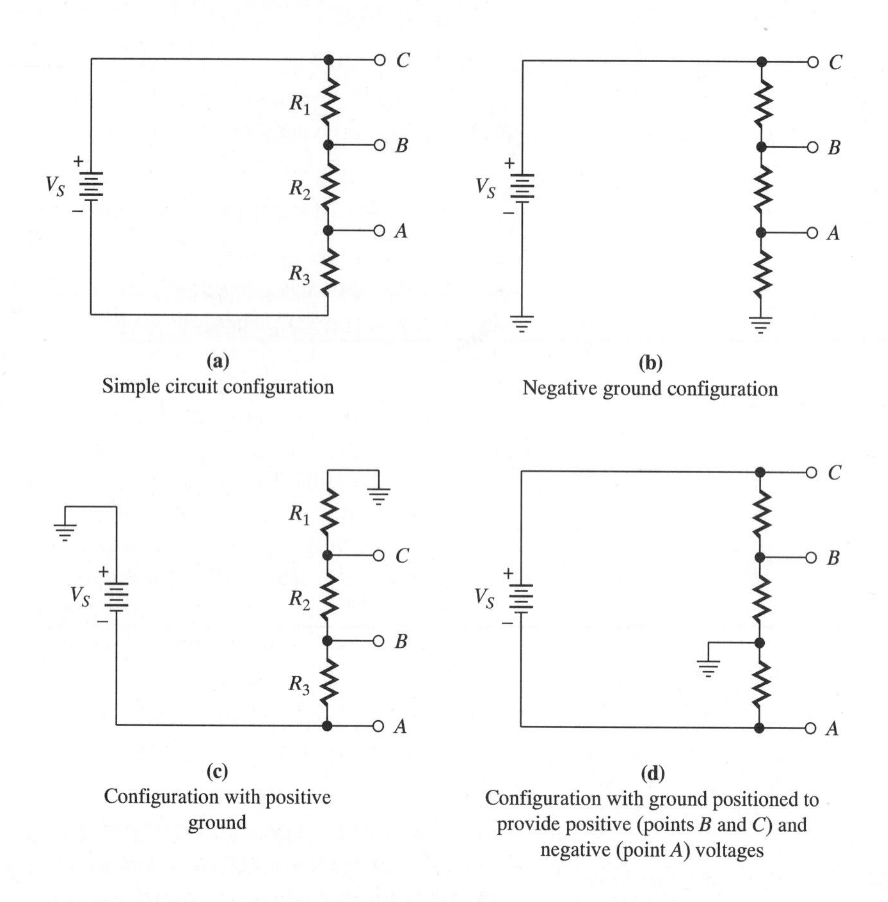

In Figure 4–33(a), you see one way of drawing a simple series resistive circuit. In Figure 4–33(b), you see an alternate configuration. Since the reference points—grounds—are common, the circuit is actually the same as in 4–33(a). Any one point in a circuit can be designated as the reference point. As shown in Figure 4–33(c), the grounds or reference points are now the positive side of the source. The latter system was once used in some American and European cars; it is still used in some foreign-made consumer electronics products. The reference may also be placed at some intermediate point as shown in Figure 4–33(d).

In your studies, you will encounter positive and negative voltages. Voltages are usually considered to be positive or negative with respect to circuit common. In Figure 4–32(b), the reference is the negative side of the source, so the voltages, measured from points *A*, *B* and *C* with respect to it, are all positive. In Figure 4–32(c), the reference is the positive side of the source, so the voltages, measured from points *A*, *B*, and *C* with respect to circuit common, are all negative. In Figure 4–33(d), the placement of circuit common causes point *A* to be negative while points *B* and *C* are positive with respect to it.

EXAMPLE 4–15 For the circuit of Figure 4–34, find the value and polarity of each voltage with respect to the reference. The resistors are of equal value.

■ Solution:

Since the resistors are of equal value and in series, the voltage drops will also be equal. Thus, each voltage drop will be 3 V. Point *C* is the reference. The voltage from point *A*

FIGURE 4–34 Circuit for Example 4–15

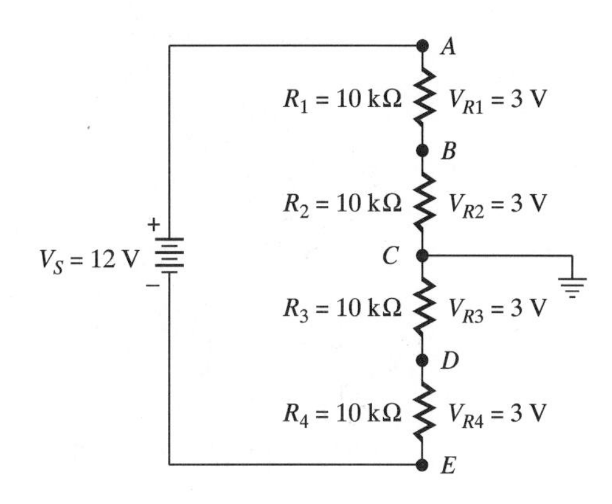

to point C is the sum of the IR drops of R_1 and R_2. Point A lies to the positive side of point C. Thus the voltage from point A to point C is $+6$ V. Point B also lies to the positive side of point C. So the voltage from point B to point C is $+3$ V. Points D and E both lie to the negative side of point C. Thus, the voltage from point E to point C is -6 V, and from point D to point C is -3 V.

■ SECTION 4–10 REVIEW

1. What is the symbol for ground?
2. What are the two symbols used for circuit common?
3. Why is more than one symbol necessary for the circuit reference point?
4. Draw a simple series circuit that will produce two negative and two positive voltages with respect to the circuit common.

4–11 APPLICATIONS OF SERIES CIRCUITS

Applications of series circuits are actually applications of voltage dividers. As stated previously, the use of voltage dividers greatly simplifies power supplies in electronic systems. A portable radio may operate from a single, usually 9 V, battery. Through the use of voltage dividers, the many lower voltages required in its operation are developed. There are, of course, hundreds of applications of voltage dividers, a representative few of which will be presented in this section.

■ Transistor Amplifier

A voltage divider can be used to *bias,* or set the operating point, of a transistor amplifier. Bear in mind that the following discussion is not intended to make you an expert in transistors, but merely to show an application of a voltage divider. Figure 4–35(a) shows a transistor with a voltage divider providing its operating voltage. The voltage divider is made up of R_1 and R_2. The top of R_2, which is connected to the base of the transistor, provides the proper polarity and value of dc voltage to set the operating point of the transistor. (You will see this circuit many times in courses that follow in electronics.) By varying the values of R_1 and R_2, the voltage from the base to ground can be changed. In this manner, the operating point of the transistor can be varied.

FIGURE 4–35 Series circuit applications

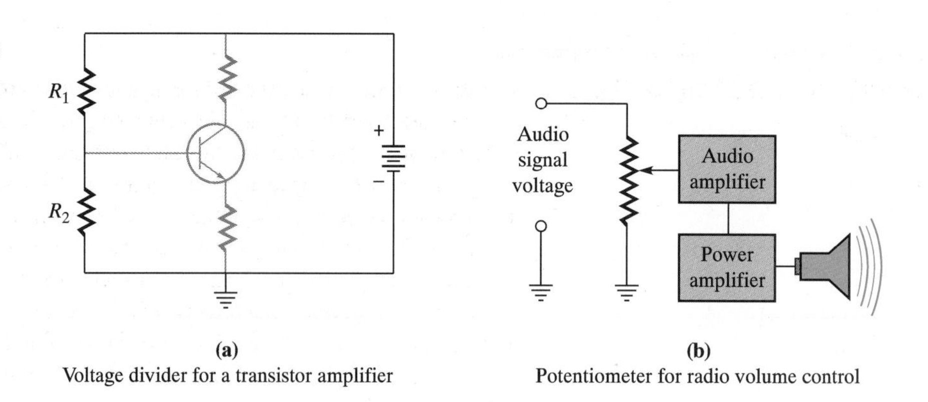

(a)
Voltage divider for a transistor amplifier

(b)
Potentiometer for radio volume control

■ Radio Volume Control

A volume control, such as in a radio, is a potentiometer. As shown in Figure 4–35(b), the input to the potentiometer is a signal voltage, which is analogous to an audio signal generated at the radio transmitter. The volume control knob on the radio moves the wiper arm of the potentiometer. The signal voltage is divided between the upper and lower part of the potentiometer. It is the voltage from the lower part of the potentiometer that enters the audio amplifier and eventually drives the speaker. The greater this voltage the louder the sound from the speaker. The sound level will increase if the wiper is moved upward and decreased if it is moved downward toward circuit common.

■ Liquid-Level Sending Unit

In some industrial processes, it is necessary to meter the liquid level in a container. The unit that is used to detect the level of the liquid is known as a *sending unit* and is illustrated in Figure 4–36. Once again, a potentiometer can be used in this application. As shown on Figure 4–36(b), the wiper of the potentiometer is connected to the arm of a float. As the liquid level rises, the wiper moves upward and the output voltage increases proportionally. The schematic representation is shown in Figure 4–36(c). As the liquid level drops, the wiper arm moves downward toward ground and the voltage decreases. This potentiometer is linear so the number of gallons in the tank is proportional to the voltage.

FIGURE 4–36 Series circuit application: Potentiometer for detecting liquid level in a tank

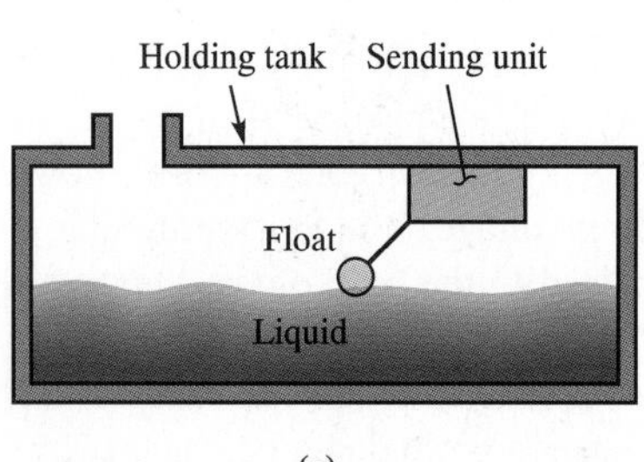

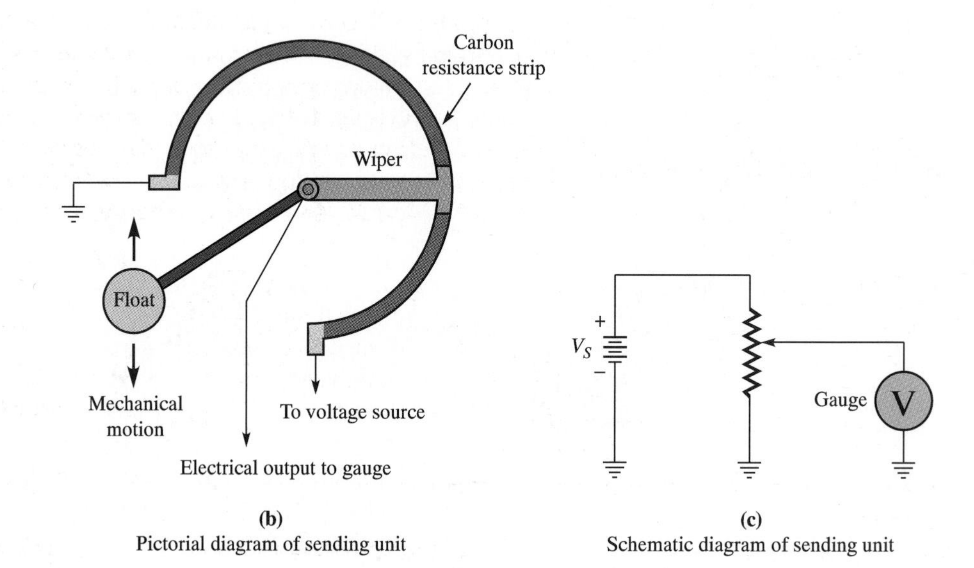

(a)
Setup for detecting liquid level in a tank

(b)
Pictorial diagram of sending unit

(c)
Schematic diagram of sending unit

4–12 THE DC VOLTMETER

The dc voltmeter is a very common and important application of series circuits. Voltage dividers are used to extend the measuring range of voltmeters. In studying this application, you will become more familiar with the analog meters that you have probably used many times in the electronics curriculum. As you know, an analog meter has a pointer that rotates to indicate a value of voltage, current, or resistance. Before covering dc voltmeter circuits, however, this section describes the operation and characteristics of the portion of the meter known as the **meter movement,** which is the portion of an analog meter with the moving parts to which the pointer is attached. While the meter movement will be covered in detail in Chapter 8, at this point, only its characteristics essential in the understanding of the theory and operation of a voltmeter are presented.

■ Meter Movement

Question: What characteristics of the meter movement must be considered in understanding dc voltmeter circuits?

The two characteristics of a meter movement that must be considered in voltmeter circuits are its full-scale current I_{FS} and its internal resistance R_M. The current flowing through the movement causes the pointer to rotate. The amount that the pointer moves is *directly proportional* to the amount of current flowing through the meter movement. The full-scale current is the current that causes the pointer to move from zero to maximum. The internal resistance is the resistance within the meter through which this current must flow. The current flowing through the internal resistance will produce a voltage drop according to Ohm's law:

(4–6) $$V_M = I_M R_M$$

FIGURE 4–37 Meter movement in voltmeter circuit

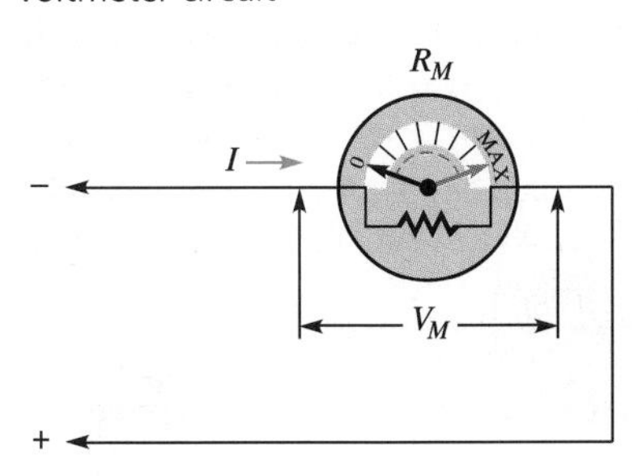

This relationship is illustrated in Figure 4–37, where, for purposes of this discussion, the meter movement is considered to be merely a resistance with a pointer attached. A certain amount of current through this resistance causes the pointer to move full scale. Both the resistance and the amount of current required for full-scale deflection are set quantities. Thus, at full-scale deflection, the meter has a certain voltage drop across it.

For example, a particular meter movement has a full-scale current I_{FS} of 1 mA and an internal resistance R_M of 300 Ω. The meter voltage drop, by Ohm's law, is (1 mA)(300 Ω) = 300 mV. As illustrated in Figure 4–38(a), if 300 mV is placed across the meter leads, the pointer will move to just full scale (since 1 mA of current is produced). If, as shown in Figure 4–38(b), a voltage source greater than 300 mV is placed across the leads, more than 1 mA of current will be produced. The pointer will now overtravel the scale and "peg," as it is called. (Has this ever happened to you in the electronics lab?) Pegging the meter should be avoided since meter damage might result. If, as shown in Figure 4–38(c), the polarity of the voltage between the leads is reversed, the pointer will move downscale instead of upscale. This downscale pegging is corrected by reversing the leads.

FIGURE 4–38 Meter movement operating principles

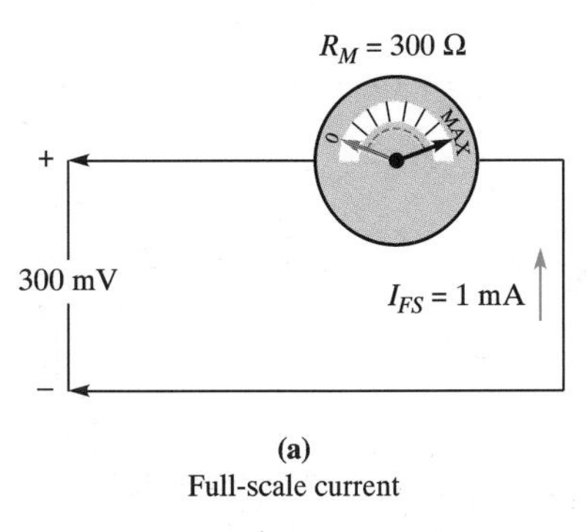

(a)
Full-scale current

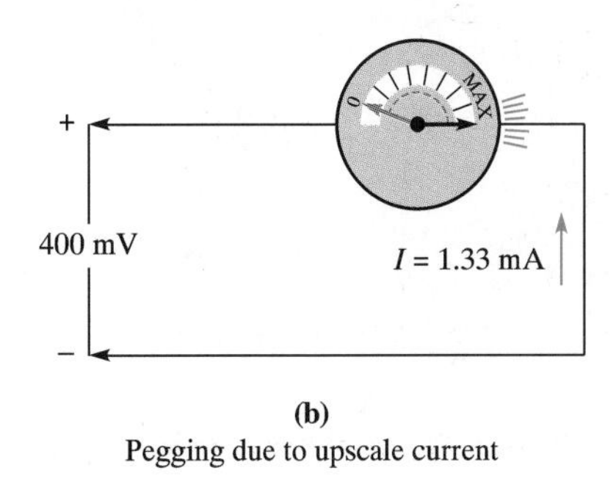

(b)
Pegging due to upscale current

(c)
Downscale pegging with reversed polarity

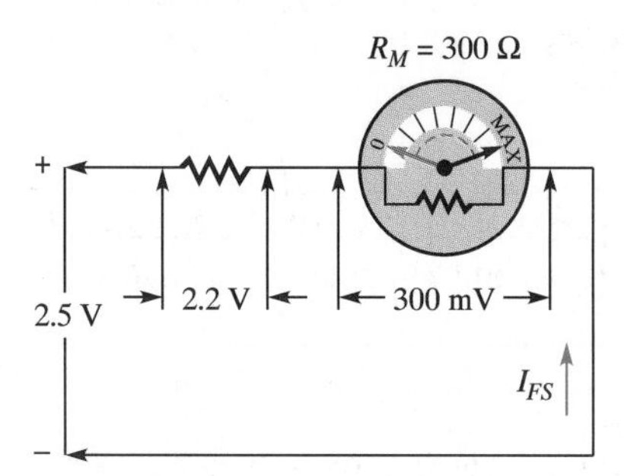

FIGURE 4–39 Multiplier resistor in series with meter movement

The meter movement, by itself, is not a very useful meter. The one described here could be used to measure only voltages as large as 300 mV. In order to extend the scale, it is necessary to place resistors, known as *multipliers,* in series with the movement. What is formed is a series circuit composed of the internal resistance of the movement and the multiplier resistor. At full-scale deflection, 300 mV will be dropped across the meter movement and the rest of the scale voltage across the multiplier resistor. For example, if the voltage scale is 2.5 V, 300 mV will be dropped across the meter, and by Kirchhoff's voltage law, the remaining 2.2 V will be dropped across the multiplier. The combination of the meter resistance and multiplier resistor limits the total current to the full-scale current of the meter, as illustrated in Figure 4–39.

EXAMPLE 4–16 A meter movement has a full-scale current of 1 mA and an internal resistance of 500 Ω. Compute the value of multiplier resistor that will provide a scale of 0–10 V.

■ **Solution:**

At full-scale deflection of the meter, 1 mA of current flows through the movement. The voltage drop across the movement is (1 mA)(500 Ω) = 0.5 V. Thus, if 10 V are applied across the leads, by Kirchhoff's voltage law the remaining 9.5 V must be dropped across the multiplier. The value of the multiplier resistor is found using Ohm's law:

$$R_{\text{mult}} = \frac{V_{\text{mult}}}{I}$$

$$= \frac{9.5\ \text{V}}{1\ \text{mA}} = 9.5\ \text{k}\Omega$$

Notice that the total meter resistance is the sum of the meter and multiplier resistances, or 10 kΩ. With the scale voltage of 10 V applied, the current is 10/10,000, or 1 mA, and the pointer shows full-scale deflection.

Meter Movement Sensitivity

Question: What is meant by the sensitivity of a meter movement?

The sensitivity of a meter movement is determined by its full-scale current. The lower the full-scale current, the higher the sensitivity. In fact, the sensitivity of a meter movement is often given as its full-scale current. It is more often expressed, however, as the number of ohms of resistance required between the leads of a voltmeter in order to measure 1 V. Thus, it can be computed as the reciprocal of the full-scale current:

(4–7)
$$\text{sensitivity} = \frac{1}{I_{FS}}$$

The 1 in this equation means "one volt" and the units are expressed as *ohms per volt.* Thus, the sensitivity of the meter movement of the Example 4–16 is

$$\text{sensitivity} = \frac{1}{I_{FS}}$$

$$= \frac{1}{1\ \text{mA}}$$

$$= 1000\ \Omega/\text{V}$$

In order to measure 10 V with this meter movement, a total of 10 × 1000 Ω of resistance is necessary between the leads, which in fact is what the 9.5 kΩ multiplier and 500 Ω of internal resistance provide. As the following example shows, using the sensitivity figure can simplify the computation of multiplier resistor values.

EXAMPLE 4–17 Design a circuit to provide the following voltage ranges: 2.5 V, 10 V, 100 V, and 250 V. The meter movement has a sensitivity of 20,000 Ω/V, and an internal resistance of 1000 Ω.

FIGURE 4–40 Circuit for Example 4–17

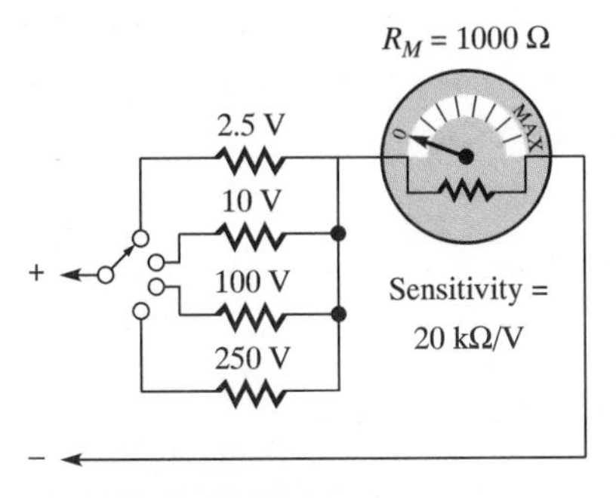

2.5 V = 49 kΩ	100 V = 1.99 MΩ
10 V = 199 kΩ	250 V = 4.999 MΩ

■ Solution:

The circuit is shown in Figure 4–40. Notice that a rotary switch is used to select different values of multiplier resistors. The meter must have 20 kΩ between its leads for each volt to be measured. So for 2.5 V, 50 kΩ is required. The meter already has 1000 Ω resistance, so

$$R_{\text{mult}} = (sensitivity)(scale) - R_M$$

$$= (20{,}000\ \Omega/\text{V})(2.5\text{V}) - 1000\ \Omega = 49\ \text{k}\Omega$$

For 10 V,

$$R_{\text{mult}} = (20{,}000\ \Omega/\text{V})(10\text{V}) - 1000\ \Omega = 199\ \text{k}\Omega$$

For 100 V,

$$R_{\text{mult}} = (20{,}000\ \Omega/\text{V})(100\text{V}) - 1000\ \Omega = 1.999\ \text{M}\Omega$$

For 250 V,

$$R_{\text{mult}} = (20{,}000\ \Omega/\text{V})(250\text{V}) - 1000\ \Omega = 4.999\ \text{M}\Omega$$

■ Accuracy of a Voltmeter

Question: How much faith can a technician place in the value indicated by a given voltmeter?

A voltmeter has an accuracy expressed as a percentage of its full-scale value. For example, this accuracy may be ±5%. Thus, the value of voltage indicated by the voltmeter *may* be in error by as much as plus or minus 5 percent of the full scale value upon which it is read. For example, at the 100 V range, the actual value may vary from the indicated value by plus or minus 5 V, or (100 V)(0.05) = ±5 V. Accordingly, a reading of 45 V can mean anywhere between 40 V and 50 V. This difference represents a possible error of 11.1%. If the indication on the voltmeter were 35 V, the actual value could be anywhere between 30 V and 40 V, which represents a possible error of 14.3%!

Notice that the percentage of error increases as a lower value is read on this scale. If the readings given in the preceding paragraph were made on a lower scale, such as the 50 V range, the percentage of error would decrease. In this range, the actual value of the voltage could be in error by as much as ±2.5 V. Thus, a reading of 45 V can mean between 42.5 V and 47.5 V, which represents a possible error of 5.6%. For a reading of 30 V, the percentage of error is 7.14%. Thus, for greatest accuracy, voltage readings should be made on the lowest possible scale.

EXAMPLE 4–18 A voltage reading of 120 V is made on the 250 V range of a voltmeter. If the voltmeter has an accuracy of ±2%, the actual voltage is between what values?

■ Solution:

The accuracy is measured at full scale, so the error is ±2% of 250 V:

$$\text{error} = 0.02 \times 250 = \pm 5 \text{ V}$$

Thus, the actual voltage read by the meter is between 115 V and 125 V (120 V ± 5 V). This result represents a possible error of ±4.17% (5 V ÷ 120 V × 100).

EXAMPLE 4–19 If this same voltage measurement was made at the 150 V range of this meter, the actual voltage is between what values?

■ Solution:

$$\text{error} = 0.02 \times 150 = \pm 3 \text{ V}$$

Thus, the actual voltage may be between 117 V and 123 V (120 V ± 3 V). The possible error is now ±2.5%.

■ SECTION 4–12 REVIEW

1. What two characteristics of a meter movement are used in computing voltmeter multipliers?
2. What is the unit of measure of sensitivity?
3. Compute the value of a multiplier resistor to produce a 250 V scale. The meter movement has a full-scale current of 1 mA and a resistance of 300 Ω. Use Ohm's law and Kirchhoff's voltage law.
4. A meter movement has a sensitivity of 5000 Ω/V. Its internal resistance is 600 Ω. Compute the value of a multiplier to provide a scale of 100 V.

4–13 TROUBLESHOOTING SERIES CIRCUITS

All electronic circuits operate through the systematic control of electric current. As previously noted, faults occur that cause the current to be too low, too high, or zero. Once the fault is determined, all possible causes must be explored. Ohm's law, Kirchhoff's voltage law, and the facts of series circuits are used in such analyses. In this section, the principles used in troubleshooting any series circuit are demonstrated. Be sure that you understand the purpose and result of each step.

The first step in troubleshooting any circuit or system is checking the voltage source. All circuit current flows through the source so it is a likely trouble point. Also, it stands to reason that no circuit will operate without the proper source voltage. In checking this voltage, especially if the source is a battery, be sure that the power switch is on. Batteries must be tested under full load conditions. In the following analyses, the voltage source is assumed to be correct.

Zero current in a series circuit indicates that the circuit is open. As shown in Figure 4–41, an open circuit can be found using a voltmeter. Recall that a voltage, or *IR*, drop is developed when current flows through a resistor. So with no current flow, all *IR* drops are zero. Since all the voltage drops are zero, then, by Kirchhoff's voltage law, the voltage across the open resistor must equal the source voltage. In the example of Figure 4–41, resistor R_2 is open, and the source voltage is measured across it. Notice that even though the current is under discussion, it is *voltage* that is measured. Since current must flow in

FIGURE 4–41 Zero current resulting from open circuit. V_{R1} and V_{R3} measure zero volts; the source voltage is measured across R_2, the open component.

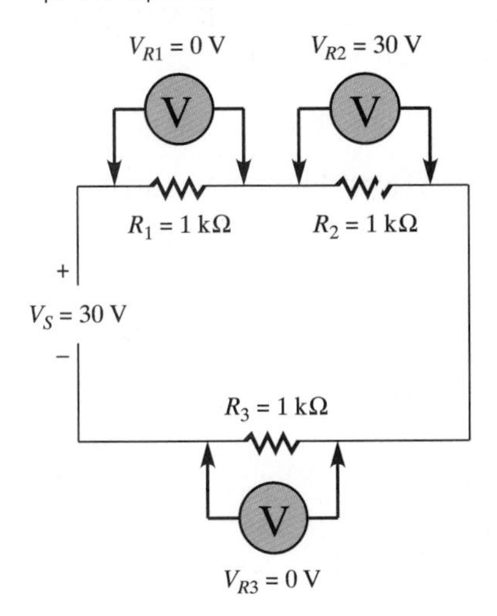

FIGURE 4–42 Short circuit with decreased resistance and increased current

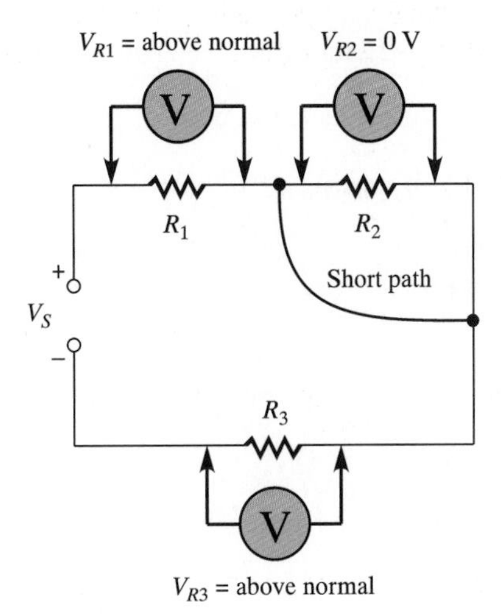

FIGURE 4–43 Circuit with high resistance and low current due to faulty resistor

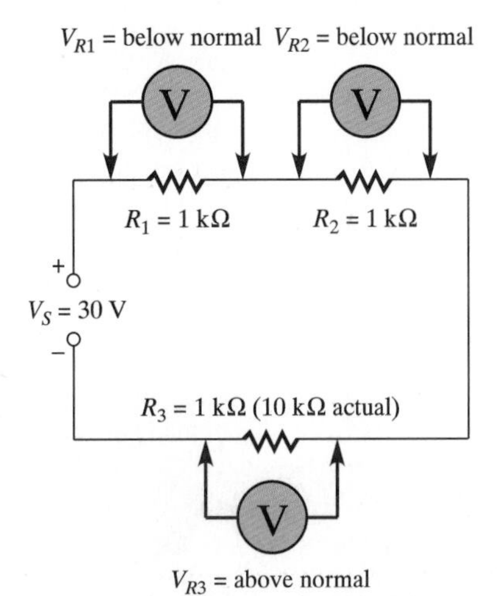

order to get a voltage drop, it is far more convenient to detect current flow with a voltage measurement. An open circuit may be the result of a broken wire or printed circuit board trace. It may also be caused by a component that has burned open (note that a component can burn open and look perfectly normal). This trouble is corrected by replacing the open component or repairing the break in the printed circuit board trace.

If the current current is *too high,* the resistance, by Ohm's law, must be too low. Low resistance can be caused by either a shorted component or the fact that the current has found an alternate path around one or more components. A low-resistance alternate path may be caused by corrosion or some foreign substance, such as solder bits or wire clippings. A short circuit is shown in Figure 4–42. The total resistance is decreased, and the current is greater than normal. The shorted resistor has no voltage drop, while the other components have voltage drops larger than normal. This problem can be corrected by replacing the shorted resistor or removing the alternate path. Note that components such as transistors are more likely to short than resistors. Resistors are more likely to burn open.

If the voltmeter indicates the current is *too low,* the resistance must be too high. The most probable cause is a resistor that has increased in value. This increase can happen through overheating or when internal connections in the resistor develop a resistance. The circuit shown in Figure 4–43 has a resistor whose value has changed. It has a larger voltage drop than normal, and the others have lower voltage drops. This trouble is cleared by replacing the component that has changed value.

Note that if a resistor must be replaced in a circuit, care must be taken that the circuit design parameters are not upset. The replacement item should be exactly like the original, if possible. If an exact replacement is not available, care must be taken in picking a substitute. The original circuit was designed to operate within certain limits, and the resistors were chosen with tolerances that would ensure this. Therefore, never replace a resistor with one that has greater tolerance. The replacement should also have the same wattage rating as the original. If not available in the correct wattage, one with a higher rating may be substituted.

TROUBLESHOOTING PRACTICE

In Figure 4–44, a single fault exists. One of the resistors, R_1 through R_4, is either shorted or open. Your task is to identify which of these faults exists and which component is involved. The first step is to determine the expected values of the voltages. On the job, you would make the necessary measurements *one at a time*. For this practice, after deciding on a measurement, you can determine its value from the table given in Figure 4–44. Then use these measurements, Ohm's law, Kirchhoff's voltage law, and the characteristics of series circuits in order to determine the trouble. Try to determine the trouble in as few steps as possible.

Rate yourself as follows:

Supertech: 2 or less steps
Technician: 3 steps
Apprentice: 4 steps
Need review: 5 or more steps

FIGURE 4–44 Circuit for Section Review Problem 6

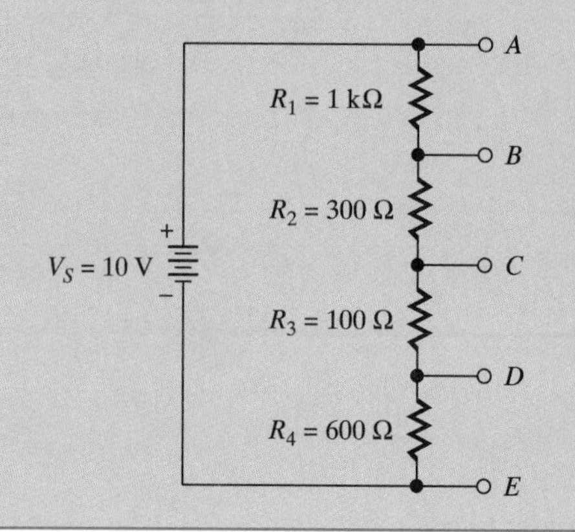

Measurement	Value
V_{AB}	0 V
V_{AC}	0 V
V_{AD}	10 V
V_{AE}	10 V
V_{BC}	0 V
V_{BD}	10 V
V_{CD}	10 V
V_{CE}	10 V
V_{DE}	0 V
V_{BE}	10 V

SECTION 4–13 REVIEW

1. What check is always the first to be made in fault analysis?
2. What three problems may be associated with current in an electric circuit?
3. With power applied, how can a voltmeter be used to determine if a component, such as a resistor is shorted?
4. How can a voltmeter be used to check if the circuit current is too low?
5. What precautions must be observed when replacing a resistor?

CHAPTER SUMMARY

1. Components connected end to end are in series.
2. A series circuit has only one path for current.
3. The current is the same at all points in a series circuit.
4. A series circuit is a voltage divider.
5. The total resistance in a series circuit is the sum of all the series resistors.
6. The sum of the voltage drops around a series loop is equal to the source voltage.
7. A voltage drop is the result of current flowing through a resistor.
8. A voltage drop is negative on the side where the current enters and positive on the side where it exits.
9. The voltage source is known as a rise and is opposite in polarity to a voltage drop.
10. Total power in a series circuit is the sum of the powers developed in each resistor.
11. The reference point for voltage measurements is known as circuit common.
12. In an open circuit, all voltage drops will be zero, and the current is zero.
13. In a series circuit, the largest resistor has the largest voltage drop.

STUDY SKILLS

Taking an Exam: Dealing with Panic

If you have followed the advice given thus far on how to study and how to prepare for an exam, and if you have put your study time to good use, you should feel fairly confident and less anxious about taking an exam. However, you may still find that during the course of the exam, you get stuck or "draw a blank." Such a situation can easily lead to panic and irrational thoughts like . . . "I can't do this" . . . "I don't know any of this stuff" . . . "I'm going to fail this exam." Your pulse rate increases and you begin to breathe faster. Your thinking has become irrational, and you are entering a panic cycle!

Getting stuck on a few problems does not mean that you cannot do well on the rest of the exam. This type of thinking can only serve to interfere with your performance on the entire exam—even the portions that you would have otherwise had no problem with! How do you stop the panic cycle? First, put the exam aside and silently say to your self, "*Stop*!" Now try to relax, clear your mind, and say to yourself . . . "this is only one problem" . . . "this isn't the whole test" . . . "I've done this before, I can do it now."

Now take a couple of deep breaths, and start looking for some problem that you know how to solve. Build your confidence and concentration up steadily with more problems that you know you can do. If time permits, you can go back and try to work those problems that you were stuck on. But don't forget to allow time to check your work.

14. An open circuit has the source voltage across it.
15. A short circuit reduces the total resistance and increases the current.

KEY TERMS

string
series circuit
voltage drop
voltage divider
voltage rise
Kirchhoff's voltage law (KVL)
series-aiding voltage sources
series-opposing voltage sources
total resistance
total power
unloaded voltage divider
resistance–voltage ratios
variable voltage divider
ground
earth ground
chassis ground
circuit common
meter movement

EQUATIONS

(4–1) $V_S - V_1 - V_2 - V_3 - V_4 = 0$ (for any number of voltage drops)

(4–2) $V_S = V_1 + V_2 + V_3 + V_4$ (for any number of voltage drops)

(4–3) $R_T = R_1 + R_2 + R_3 + \cdots R_n$

(4–4) $P_T = P_1 + P_2 + P_3 + \cdots P_n$

(4–5) $V_x = \frac{R_x}{R_T} V_S$

(4–6) $V_M = I_M R_m$

(4–7) $\text{sensitivity} = \frac{1}{I_{FS}}$

Variable Quantities

n = any number
x = unknown value
V_S = source voltage
R_T = total circuit resistance
P_T = total circuit power
V_M = voltage of meter movement
I_M = current of meter movement
R_M = resistance of meter movement
I_{FS} = current required to cause full-scale deflection of the meter movement

TEST YOUR KNOWLEDGE

1. In your own words, explain why current is the same at all points of a series circuit.
2. In your own words, explain why current can be measured at a single point in a circuit, but voltage must be measured between two points.
3. Why are all voltage drops zero when an "open" occurs in a series circuit?
4. Explain why a voltage drop is known as an *IR* drop.
5. What voltage will be measured across an open point in a series circuit? Explain your answer.
6. List all the effects of a shorted resistor in a series circuit.
7. List all the effects of an open resistor in a series circuit.
8. List some conditions that may cause a short circuit.
9. What is the first check to be made in troubleshooting?
10. What is meant by an unloaded voltage divider? How does it differ from a loaded voltage divider?
11. Draw a four-resistor series circuit and place a ground symbol at a point that will make all voltages, measured with respect to it, negative.
12. What effect will a blown fuse have upon the circuit?
13. List three applications of series circuits.
14. Place four 1.5 V cells in series in a manner that will produce a total of 6 V.
15. Place the same four 1.5 V cells in series in a manner that will produce a total of 3 V.

PROBLEM SET: BASIC LEVEL

Section 4–1

1. Draw the schematic diagram of a series circuit consisting of resistors of 10 Ω, 16 Ω, and 39 Ω across a 25 V source.
2. Without changing their orientation, connect the resistors of Figure 4–45 in series.

FIGURE 4–45 Resistors for Problem 2

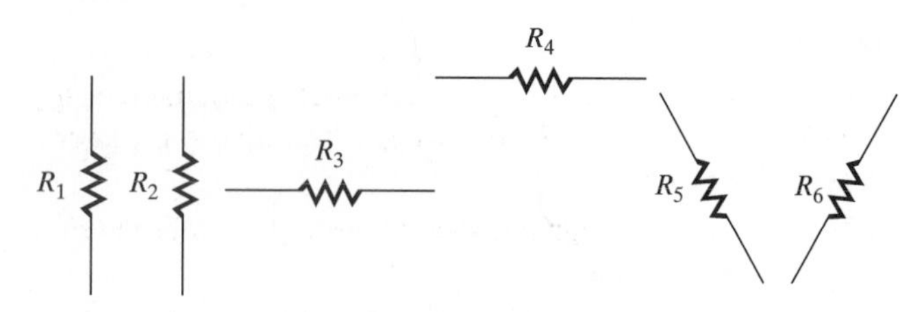

3. Do the same as in Problem 2 for the resistors of Figure 4–46.

FIGURE 4–46 Resistors for Problem 3

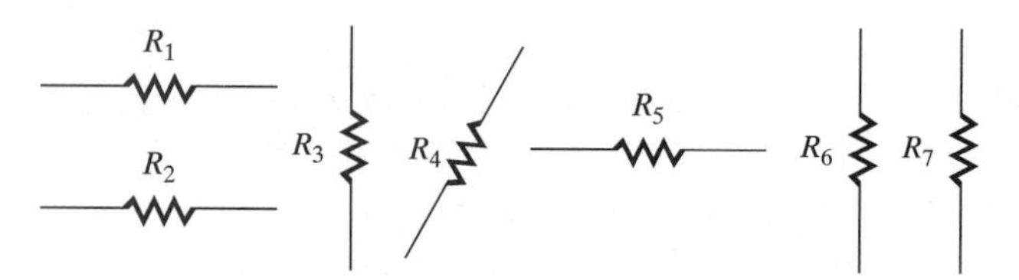

Section 4–2

4. For the circuit of Figure 4–47, compute the current that will be indicated by each ammeter.
5. Two resistors are in series. If one has 150 mA of current flowing through it, how much current flows through the other?

FIGURE 4–47 Circuit for Problem 4

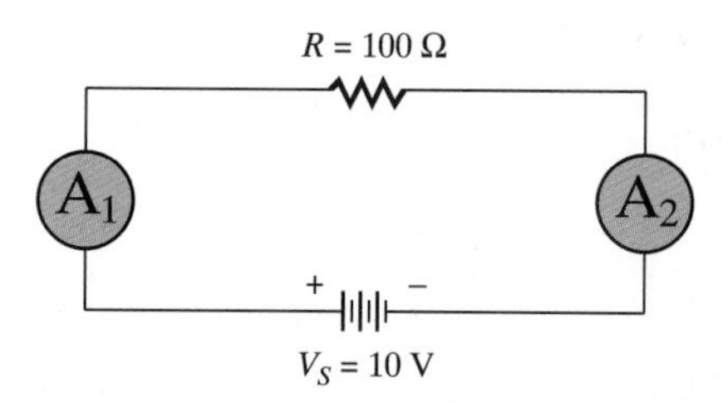

6. Two resistors, 4.7 kΩ (±10%) and 3.3 kΩ (±5%) are connected in series across a 40 V source. Taking into account their tolerances, what would be the minimum acceptable current in this circuit?

Section 4–3

7. For the circuit of Figure 4–48, compute the source voltage V_S.

FIGURE 4–48 Circuit for Problem 7

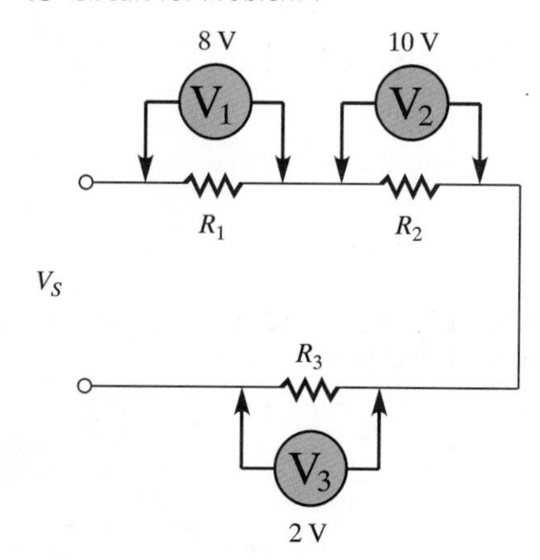

8. For the circuit of Figure 4–49, what is the voltage drop across R_3?

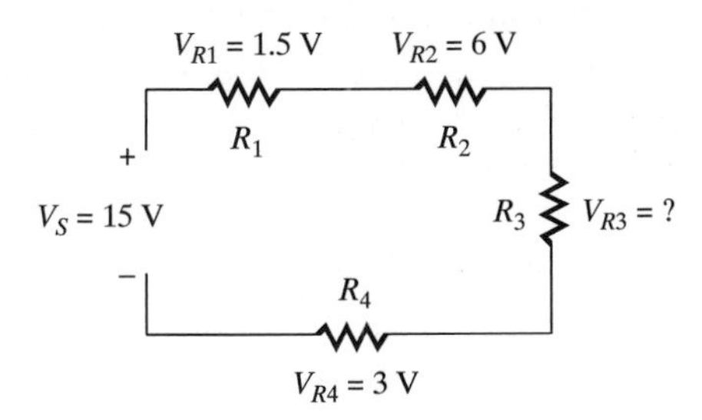

FIGURE 4–49 Circuit for Problem 8

9. A circuit consists of two resistors, 3.9 kΩ and 8.1 kΩ, in series. If the 3.9 kΩ resistor has a voltage drop of 19.5 V, compute the voltage drop across the 8.1 kΩ resistor and the value of the source voltage V_S.

Section 4–4

10. In Figure 4–50, compute the total voltage that will be developed by each series combination of cells. (V_1 and V_2.) Assume each cell is 1.5 V.

FIGURE 4–50 Circuit for Problem 10

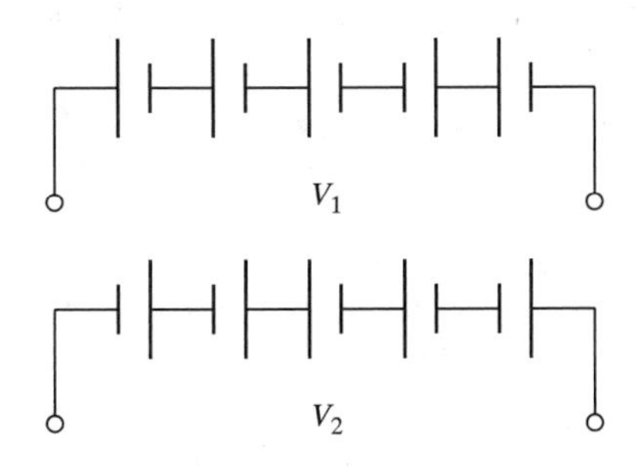

11. In Figure 4–51, indicate the polarity of the output voltage between points A and B. All cells are 1.5 V.

FIGURE 4–51 Circuit for Problem 11

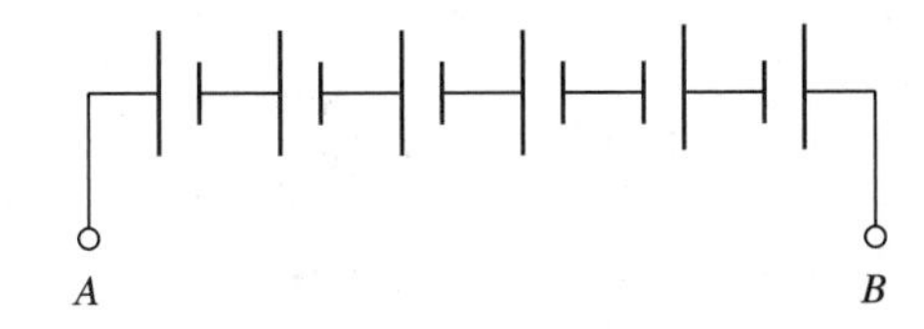

Section 4–5

12. For the circuit of Figure 4–52, compute the total resistance seen by the source voltage.

FIGURE 4–52 Circuit for Problem 12

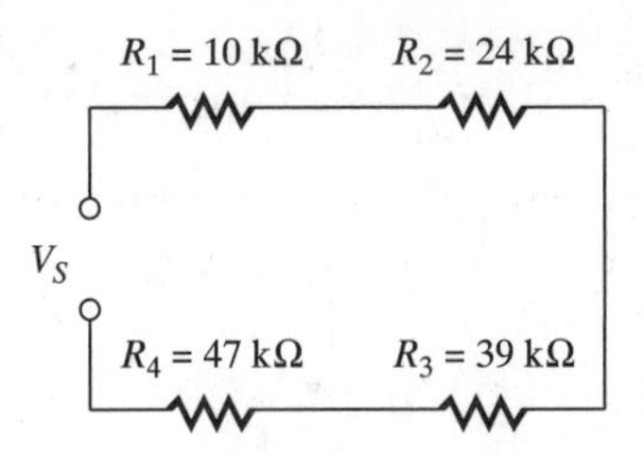

13. For the circuit of Figure 4–53, compute the total resistance between points A and B.

FIGURE 4–53 Circuit for Problem 13

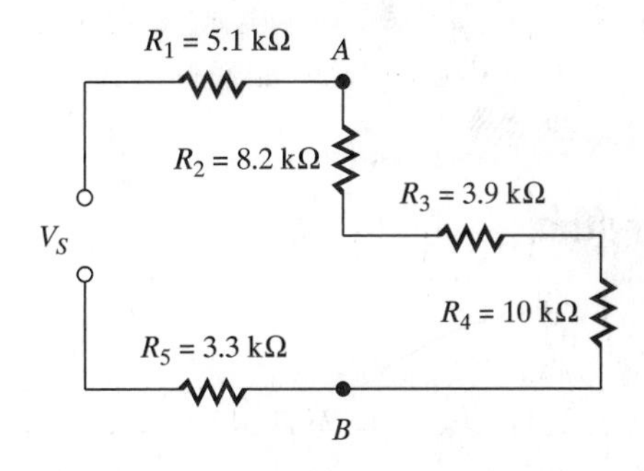

14. A four-resistor series circuit has a total resistance of 2.3 kΩ. Three of the values are 100 kΩ, 390 kΩ, and 1000 Ω. What is the value of the fourth resistor?

Section 4–6

15. Three resistors of 8 Ω, 18 Ω, and 39 Ω are connected in series across a 15 V source. What is the minimum power rating for each resistor?

16. For the circuit of Figure 4–54, compute the power developed in each resistor.

17. Using four different methods, compute the total power developed in the circuit of Figure 4–54.

FIGURE 4–54 Circuit for Problems 16 and 17

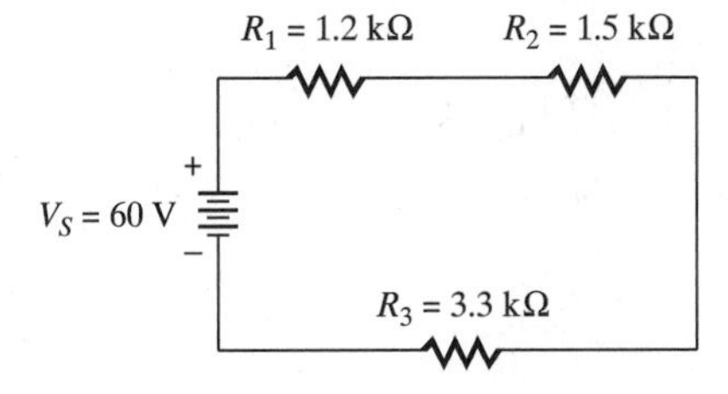

Section 4–8

18. For the circuit of Figure 4–55, compute the current through and the voltage drop across each resistor.

FIGURE 4–55 Circuit for Problem 18

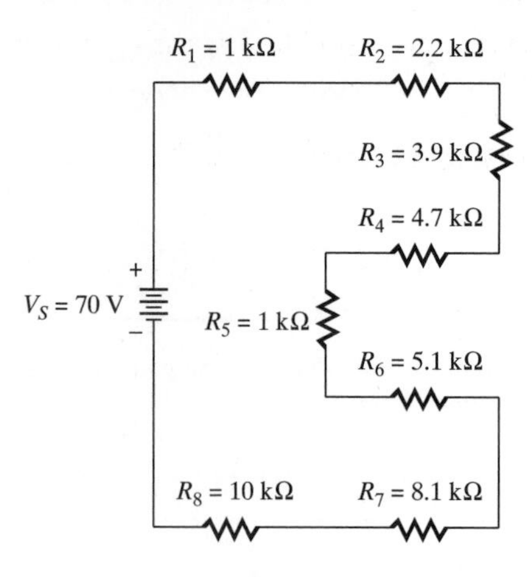

19. In the circuit of Figure 4–56, compute the voltage between each point and circuit common.

FIGURE 4–56 Circuit for Problem 19

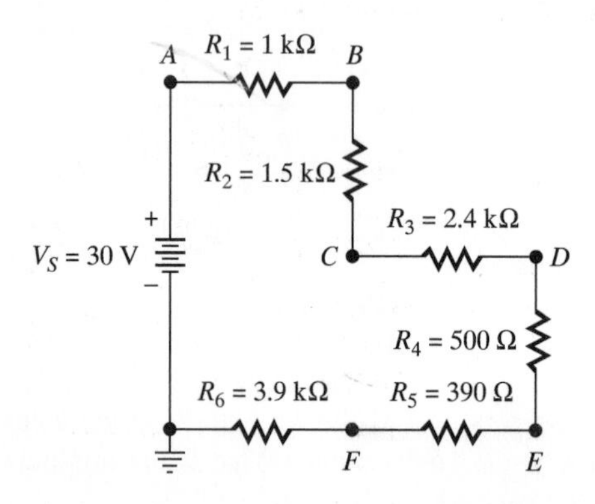

20. In the circuit of Figure 4–57, determine the voltage between the following points: A to C, B to D, C to E, and A to E.

FIGURE 4–57 Circuit for Problem 20

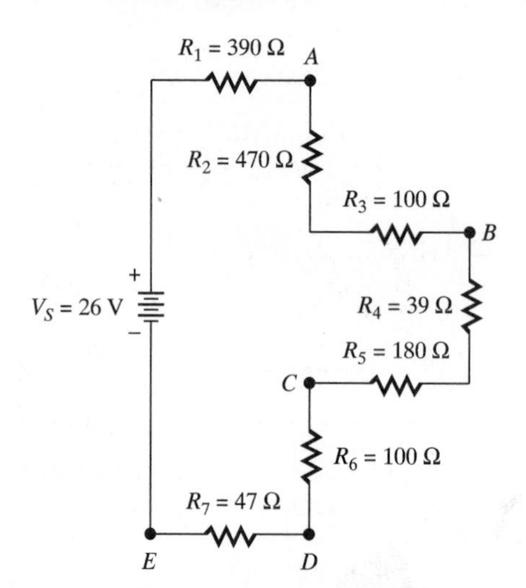

Section 4–9

21. Design an unloaded voltage divider that will provide the following approximate voltages with respect to the negative side of a 120 V source: 25.4 V, 56.1 V, and 86.74 V. Limit the divider current to no more than 1 mA. Use standard values of resistors listed in Appendix B. Draw the schematic and indicate the value of each resistor.

22. For the circuit of Figure 4–58, what percent of the total resistance is each resistor in the circuit?

FIGURE 4–58 Circuit for Problems 22 and 23

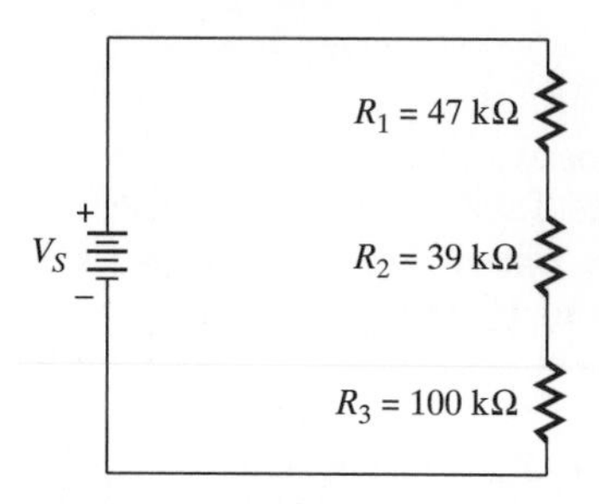

23. In Problem 22, what percentage of the source voltage will be dropped across each resistor?

24. Using the voltage divider equation, compute the voltage across each resistor in Figure 4–59.

FIGURE 4–59 Circuit for Problem 24

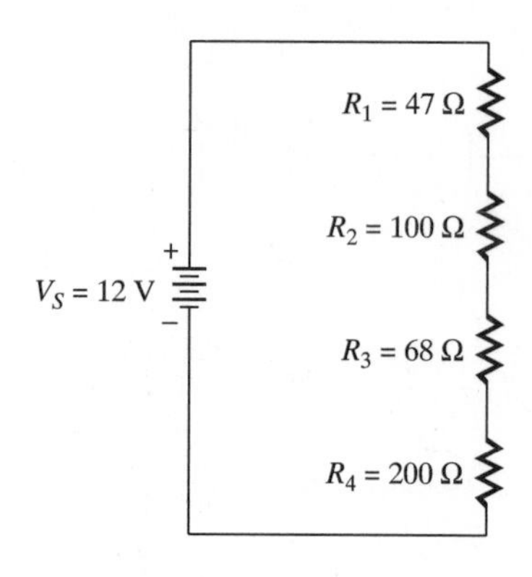

Section 4–10

25. In Figure 4–60, a 10 kΩ potentiometer is across a 100 V source. If the voltage between points A and B is 18.35 V, compute the voltage from point B to point C.

26. For the circuit in Problem 25, compute the resistances from points A to B and from points B to C.

27. Figure 4–61 shows a 5 kΩ potentiometer with the output taken between point C and circuit common. If 8 V is measured between point C and circuit common, what are the resistances between points A and C and B and C?

FIGURE 4–60 Circuit for Problems 25 and 26

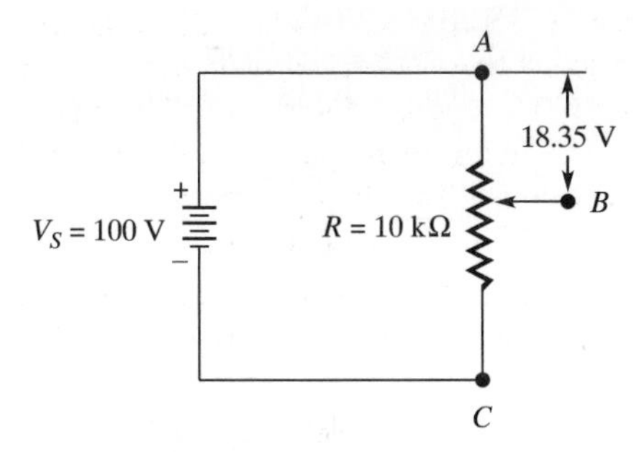

FIGURE 4–61 Circuit for Problem 27

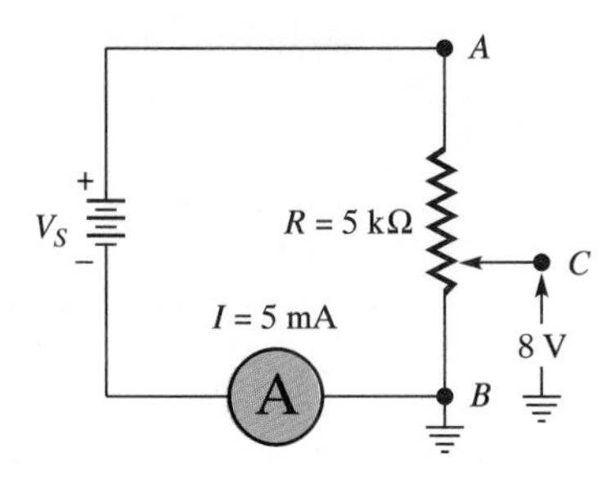

Section 4–11

28. For the circuit of Figure 4–62, compute the voltages at points A, B, C, and D with respect to common.

FIGURE 4–62 Circuit for Problems 28 and 29

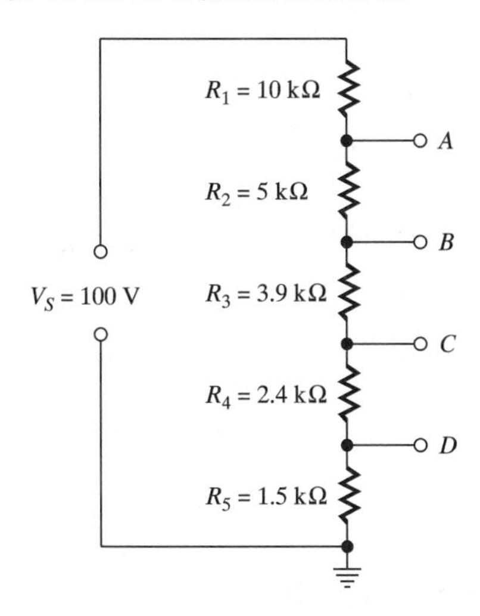

29. For the circuit in Problem 28, compute the power developed in the circuit between the same points and common.

Section 4–12

30. What is the sensitivity of the meter movement in the circuit of Figure 4–63?

FIGURE 4–63 Circuit for Problem 30

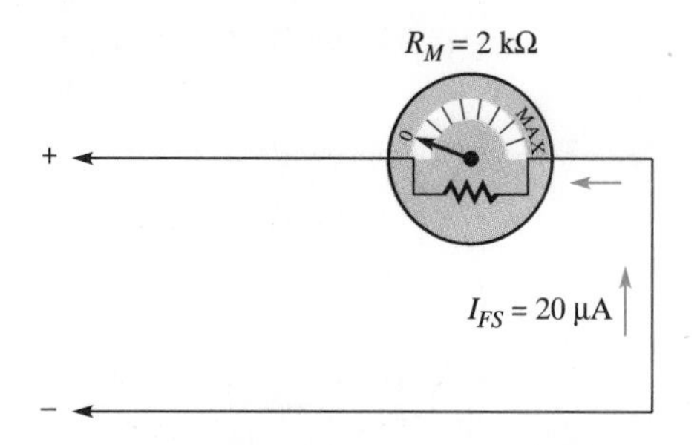

31. In the circuit of Figure 4–64, compute the value of multiplier resistors for the following voltage ranges:
a. 2.5 V **b.** 10 **c.** 50 V **d.** 500 V

FIGURE 4–64 Circuit for Problem 31

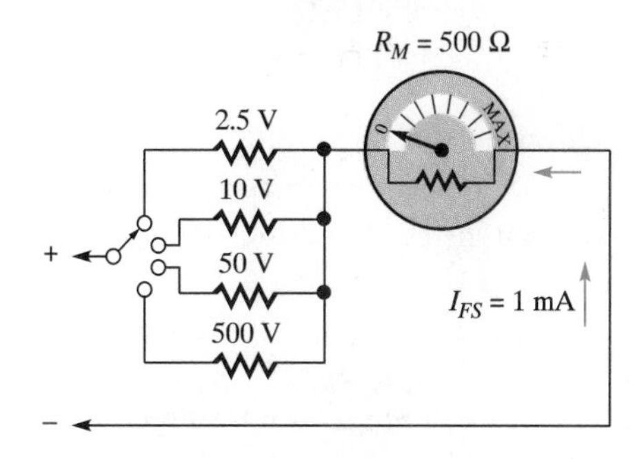

32. In the circuit of Figure 4–65, compute the value of multiplier resistors for the following voltage ranges:
a. 15 V **b.** 50 V **c.** 150 V **d.** 500 V

FIGURE 4–65 Circuit for Problem 32

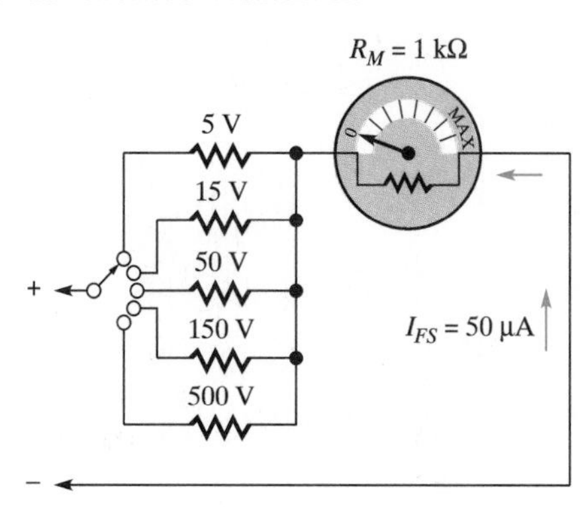

PROBLEM SET: Challenging

33. For the circuit of Figure 4–66, compute:
a. V_S **b.** R_T **c.** R_1, R_2, R_3, R_4, and R_5
d. P_1, P_2, P_3, P_4, P_5 **e.** P_T

FIGURE 4–66 Circuit for Problem 33

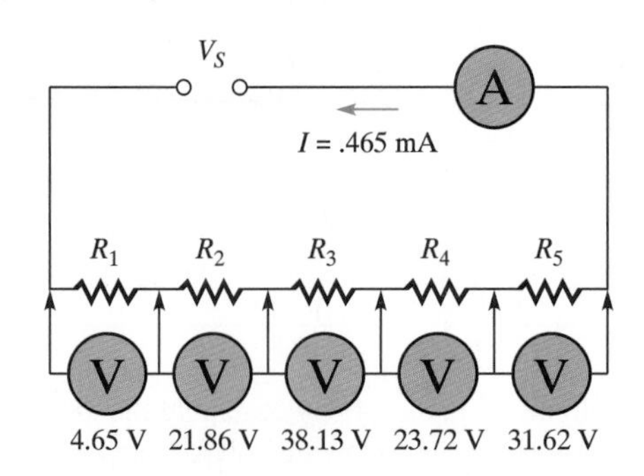

34. In Figure 4–67, R_D is known as a dropping resistor (the source voltage must be "dropped" to 8 V for the load). Compute the value of R_D that will cause 8 V to be developed across the load.

FIGURE 4–67 Circuit for Problem 34

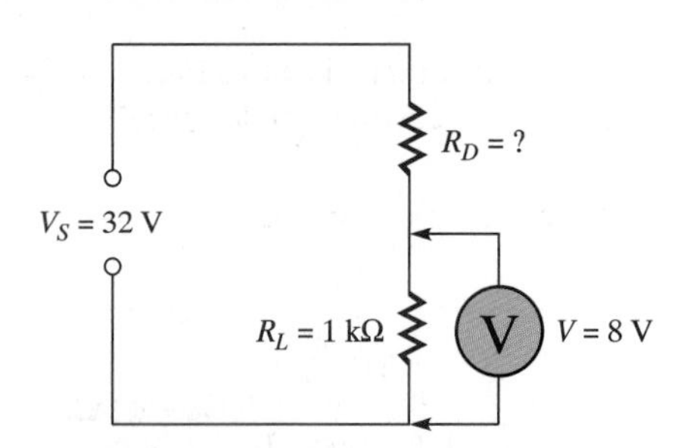

35. In Figure 4–68, compute the value of R_1 that will cause 1.96 mA of current to flow in the circuit.

FIGURE 4–68 Circuit for Problem 35

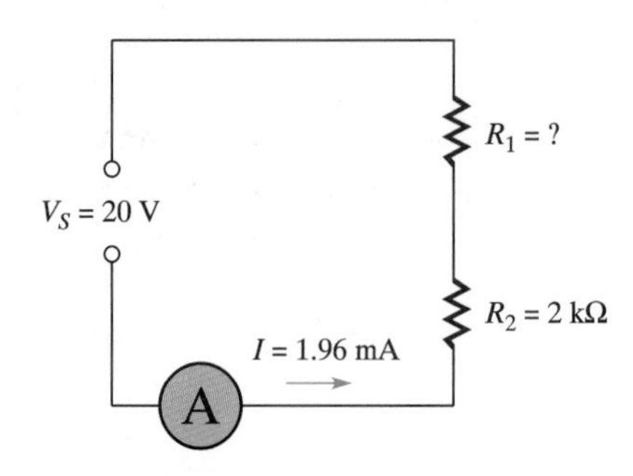

36. A circuit has 150 mA flowing through a total resistance of 1 kΩ. How much resistance must be removed from the circuit in order to increase the current by 25%?

37. In the circuit of Figure 4–69, what would be all of the effects upon the circuit if the 18 Ω resistor were to short?

FIGURE 4–69 Circuit for Problem 37

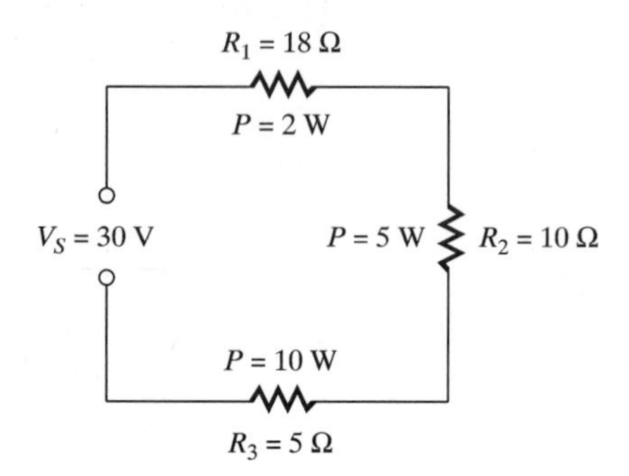

38. In Figure 4–70, Q_1 is a transistor. It is in series with R_C and R_E across the voltage source. Explain in detail how you could determine the current through the transistor without breaking the circuit and installing an ammeter.

FIGURE 4–70 Circuit for Problem 38

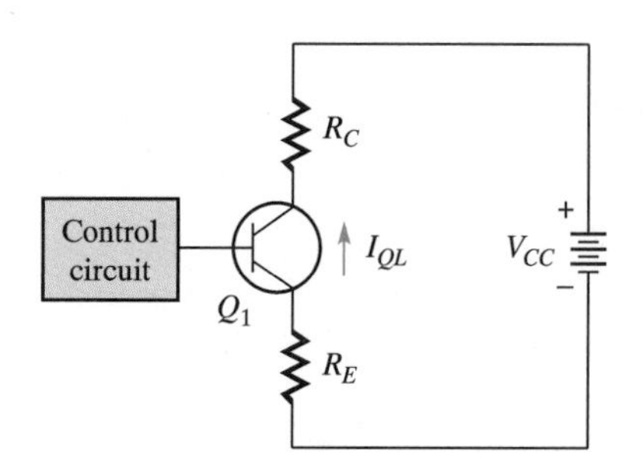

39. For the circuit of Figure 4–71, compute the following:
a. R_2 **b.** R_3 **c.** R_4 **d.** V_S

FIGURE 4–71 Circuit for Problem 39

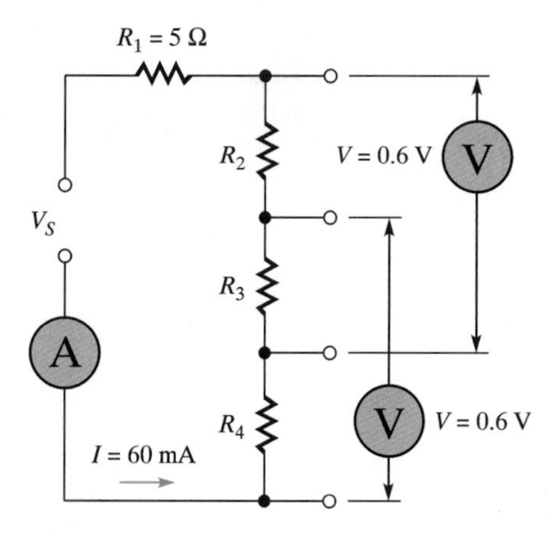

40. For the circuit of Figure 4–72, compute the values of
a. R_1 **b.** R_2 **c.** R_3 **d.** R_4

FIGURE 4–72 Circuit for Problem 40

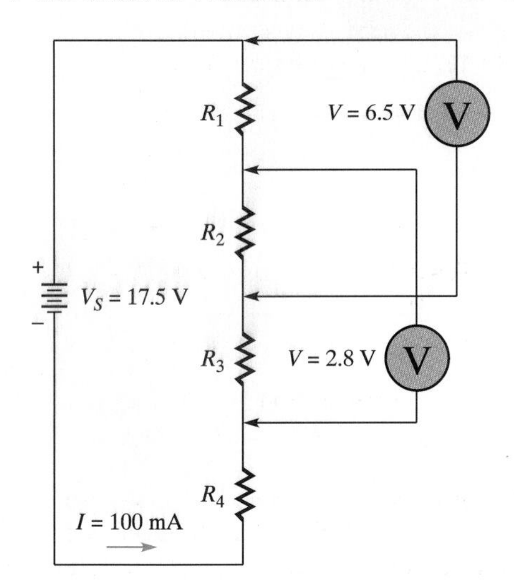

PROBLEM SET: TROUBLESHOOTING

41. One trouble exists in the circuit of Figure 4–73. From the meter readings given, determine the problem.

FIGURE 4–73 Circuit for Problem 41

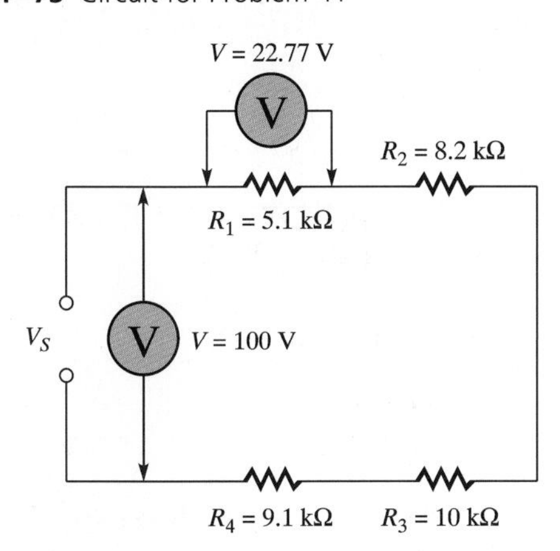

42. One trouble exists in the circuit of Figure 4–74. From the meter readings given, determine the problem.

FIGURE 4–74 Circuit for Problem 42

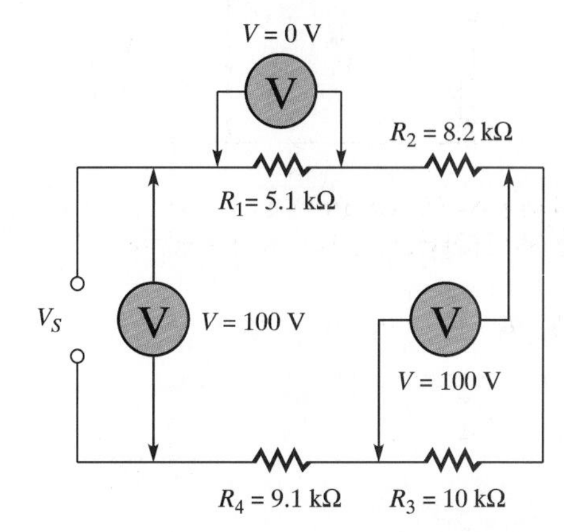

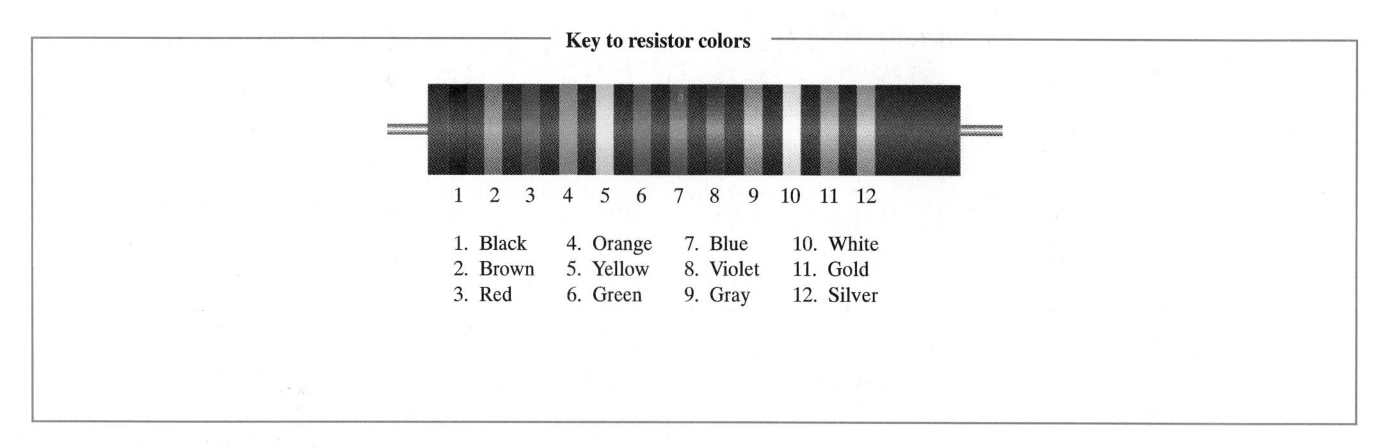

Caution: The numbers 1–12 in this figure are used to identify each color only. They do not represent the numerical value of that color as applied to the resistor color code.

FIGURE 4–75 Illustrations for Figure 4–75 a–d are circuit and meter readings for Problem 43

43. From the voltage readings shown, does a trouble exist in the circuit in Figure 4–75? If a problem exists, what is its probable cause?

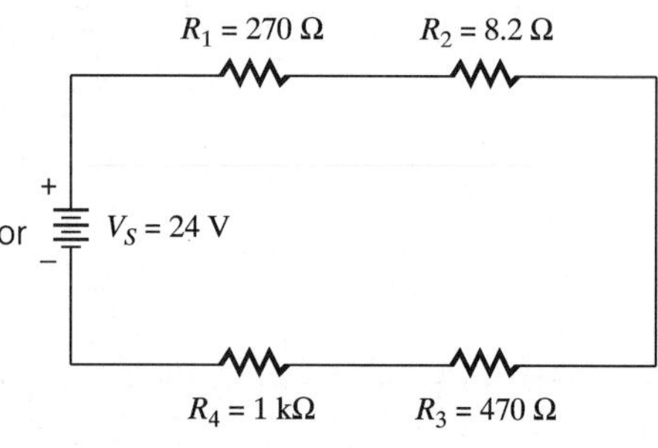

(a) Schematic for Problem 43

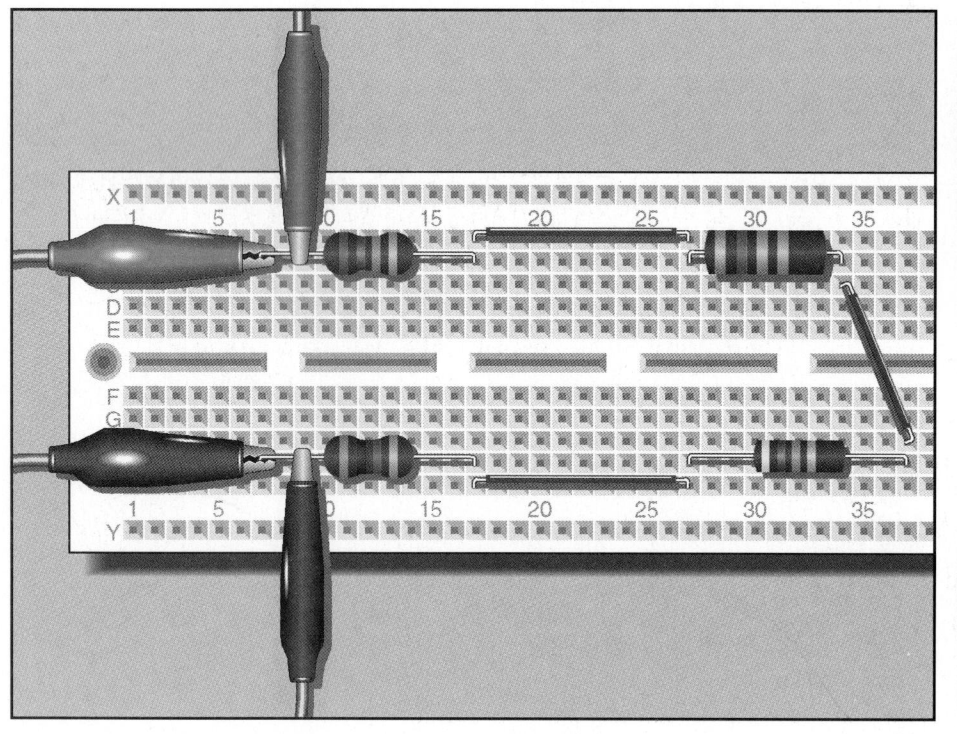

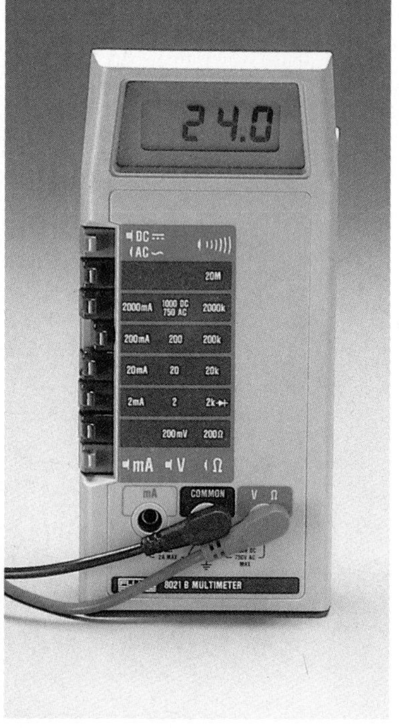

(b) Voltmeter connected to measure V_S

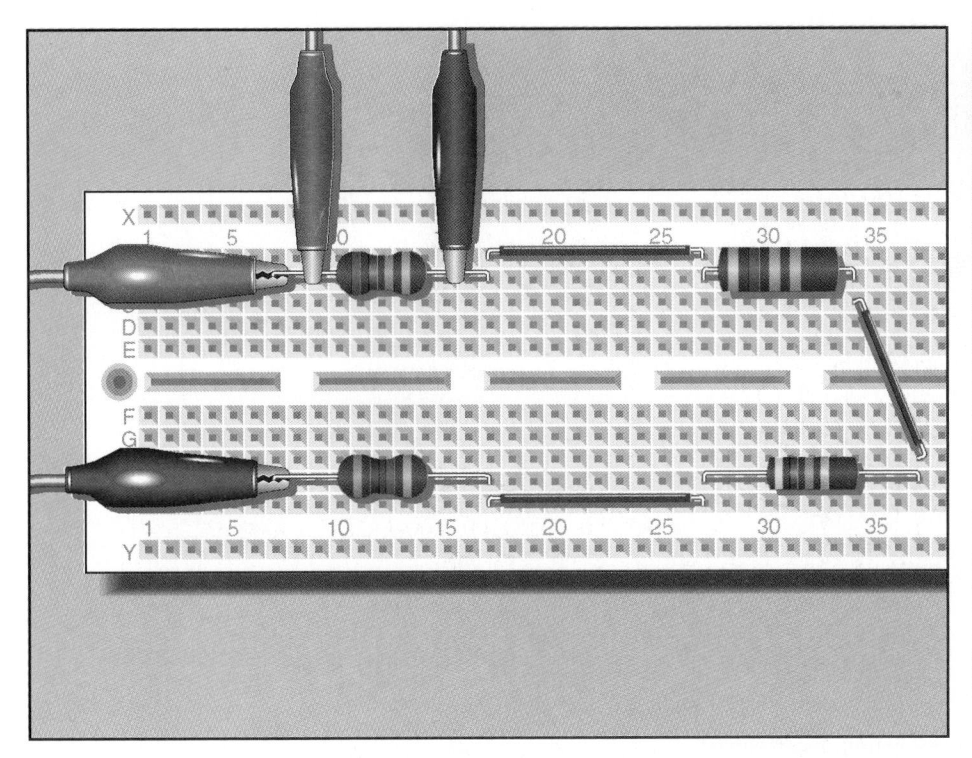

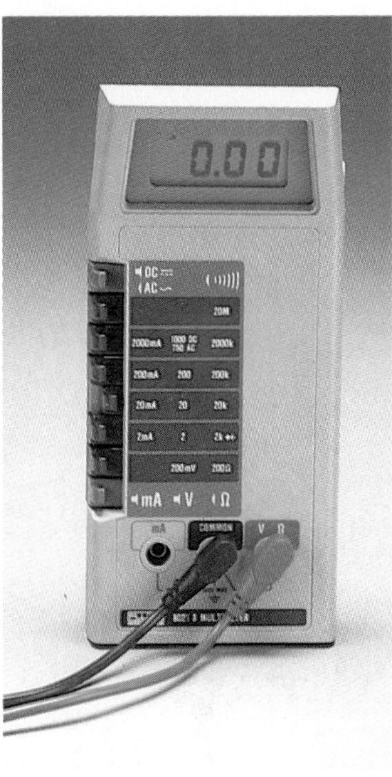

(c) Voltmeter connected to measure V_{R1}

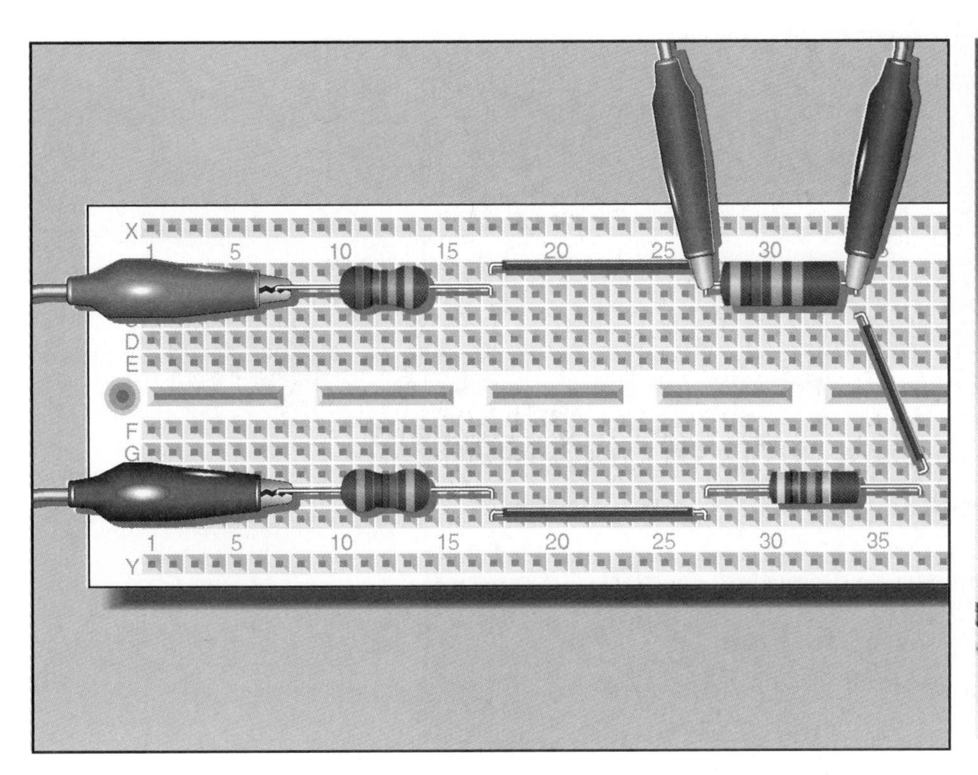

(d) Voltmeter connected to measure V_{R2}

FIGURE 4–76 Illustrations for Figure 4–76 a–f are circuit and meter readings for Problem 44

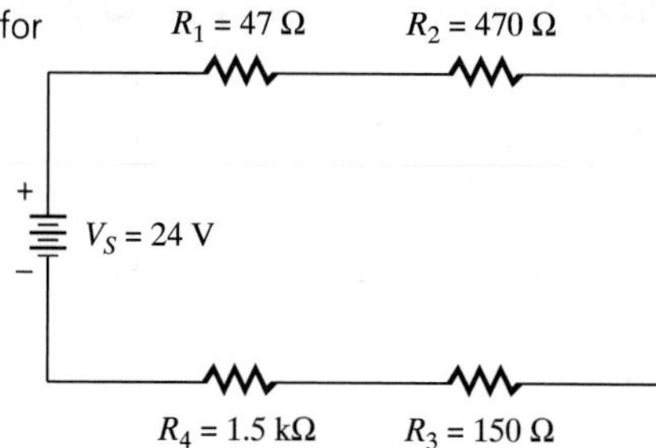

(a) Schematic for problem 44

44. From the current and voltage readings shown, does trouble exist in the circuit in Figure 4–76? If a problem exists, what is its probable cause?

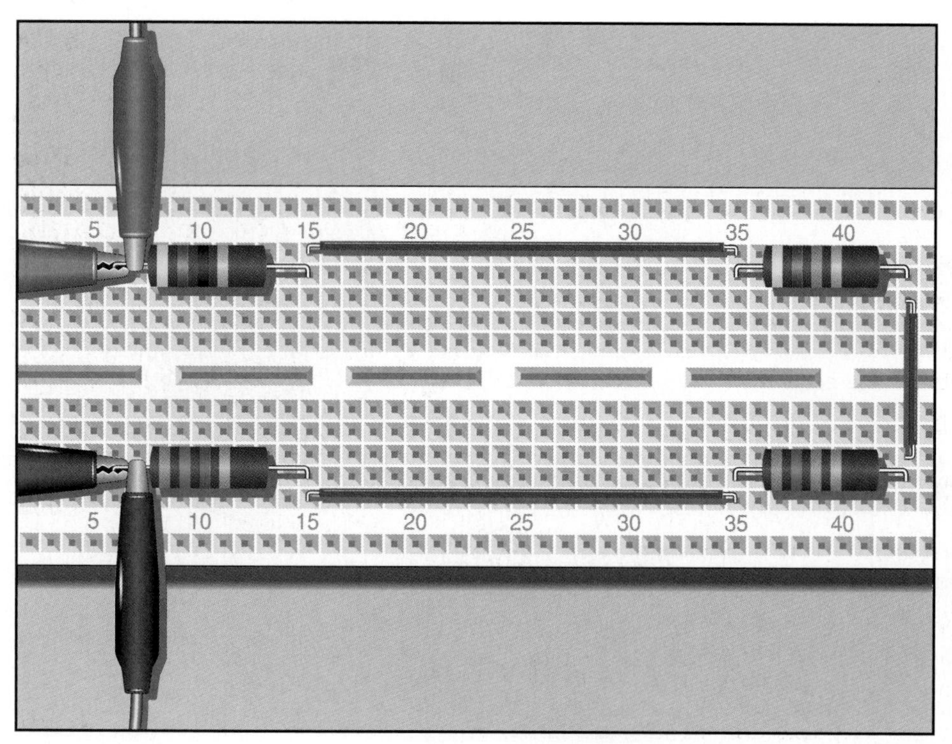

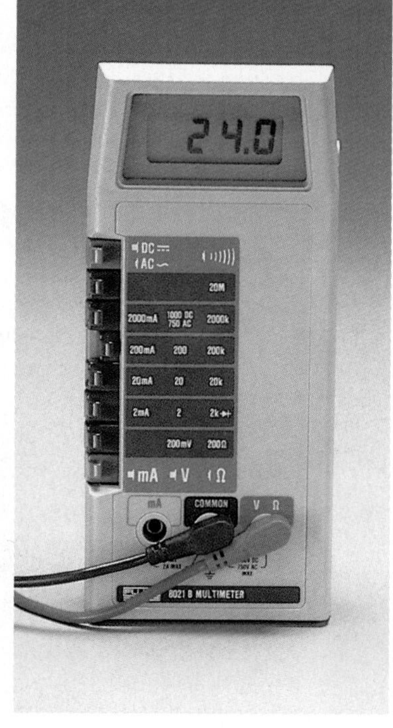

(b) Voltmeter connected to measure V_S

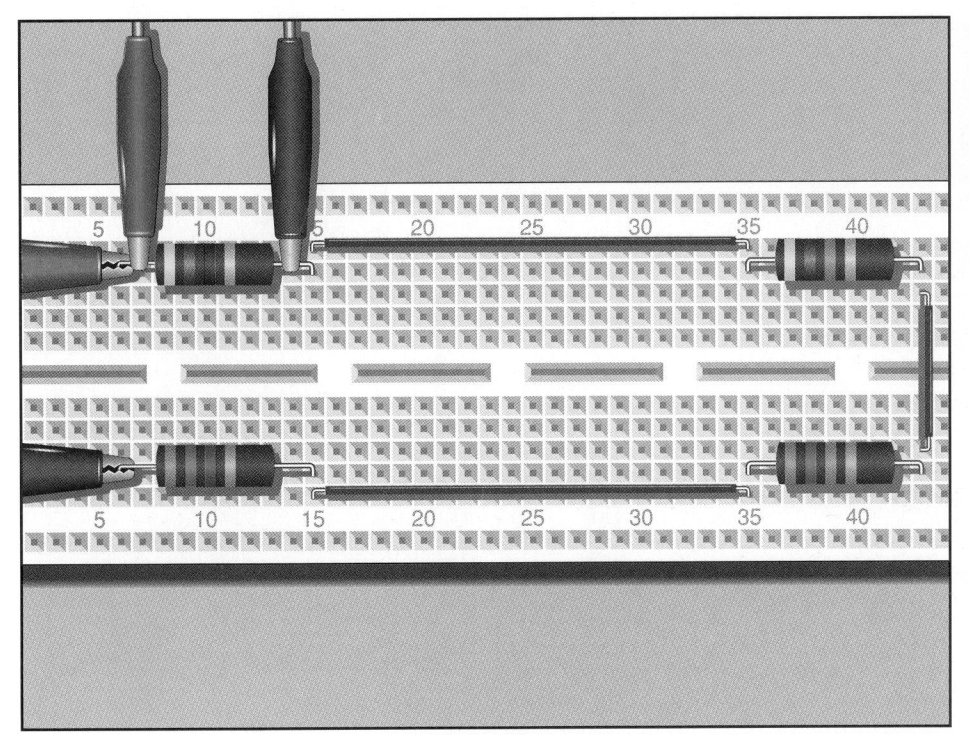

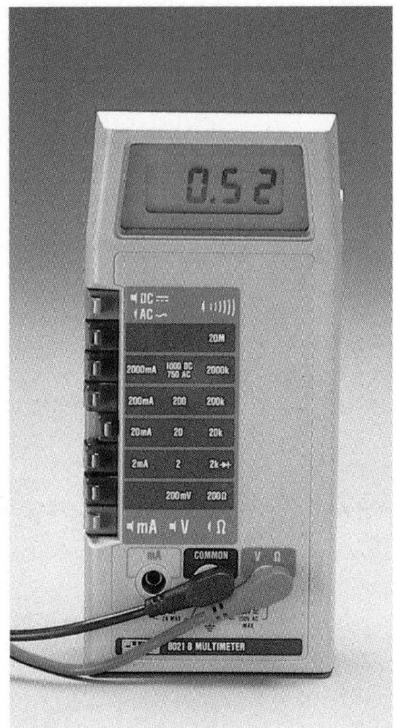

(c) Voltmeter connected to measure V_{R1}

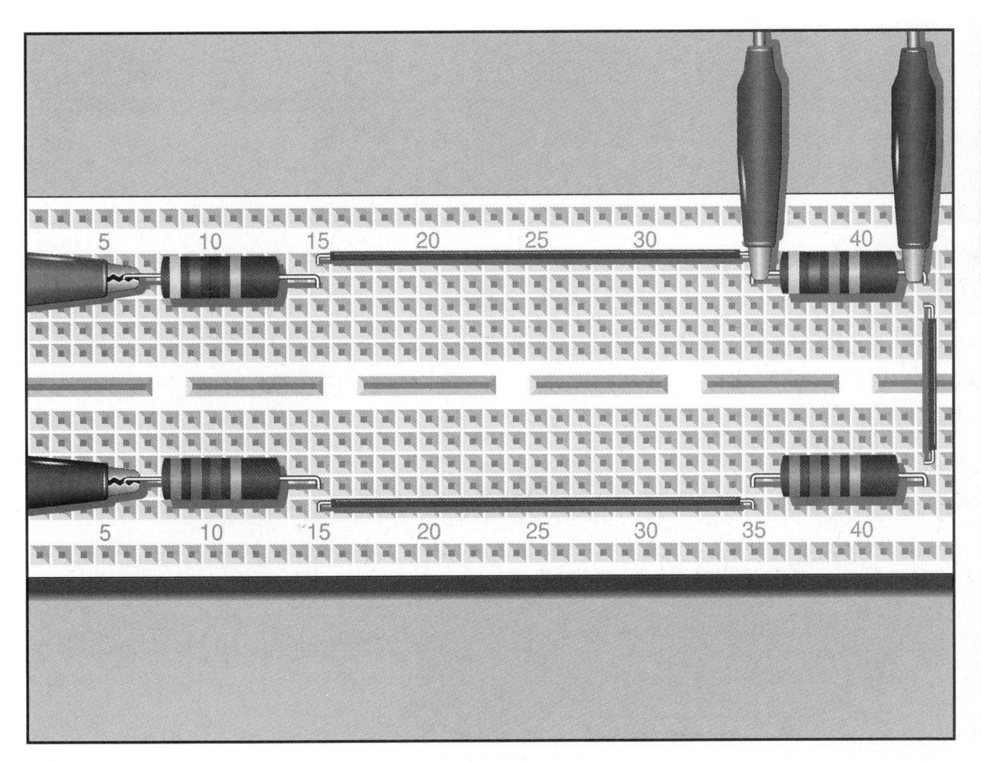

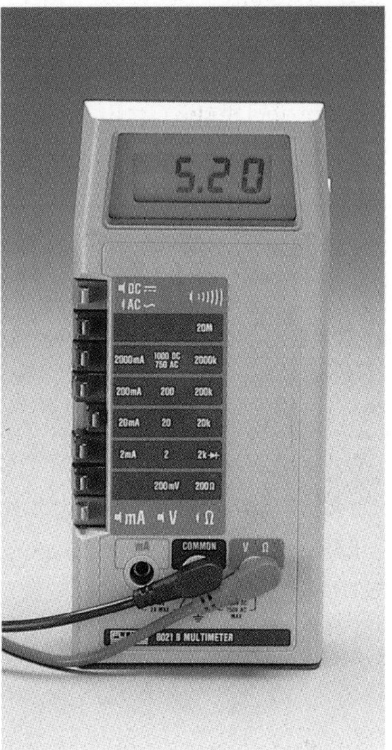

(d) Voltmeter connected to measure V_{R2}

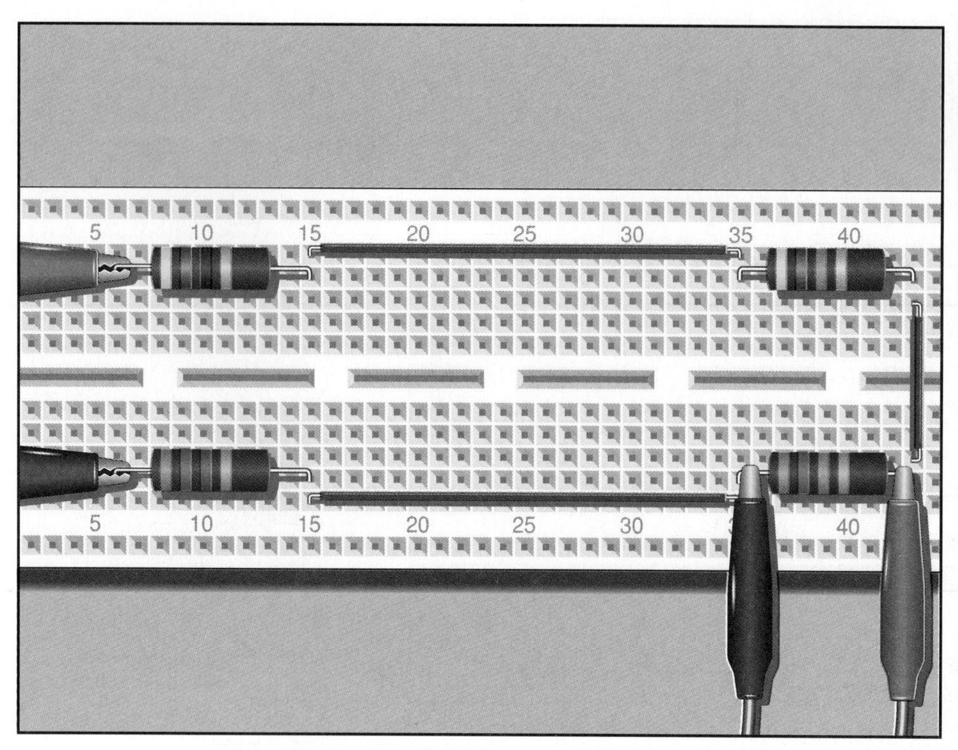

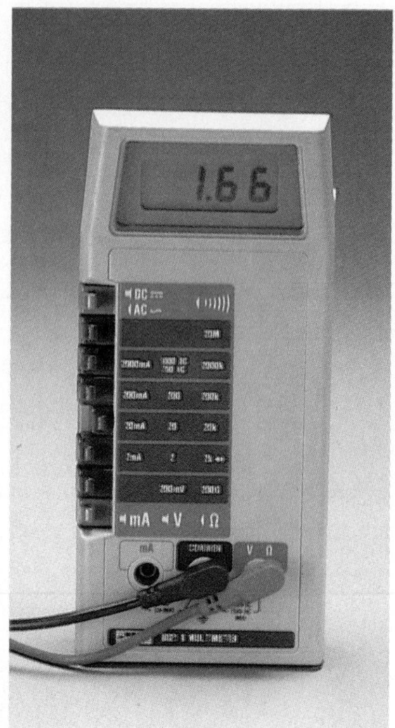

(e) Voltmeter connected to measure V_{R3}

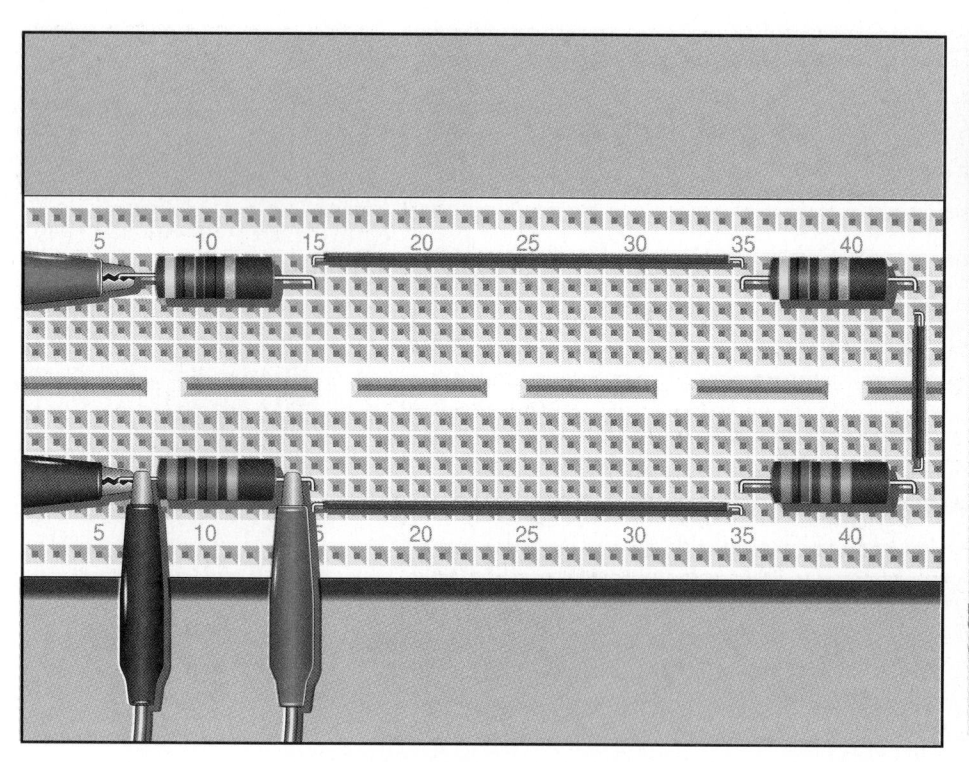

(f) Voltmeter connected to measure V_{R4}

FIGURE 4–77 Illustrations for Figure 4–77 a–e are circuit and meter readings for Problem 45

$R_1 = 9.1\ \text{k}\Omega$ $R_2 = 1\ \text{k}\Omega$

$V_S = 25\ \text{V}$

$R_4 = 5.1\ \text{k}\Omega$ $R_3 = 1.8\ \text{k}\Omega$

(a) Schematic for problem 45

45. From the current and voltage readings shown, does trouble exist in the circuit in Figure 4–77? If a problem exists, what is its probable cause?

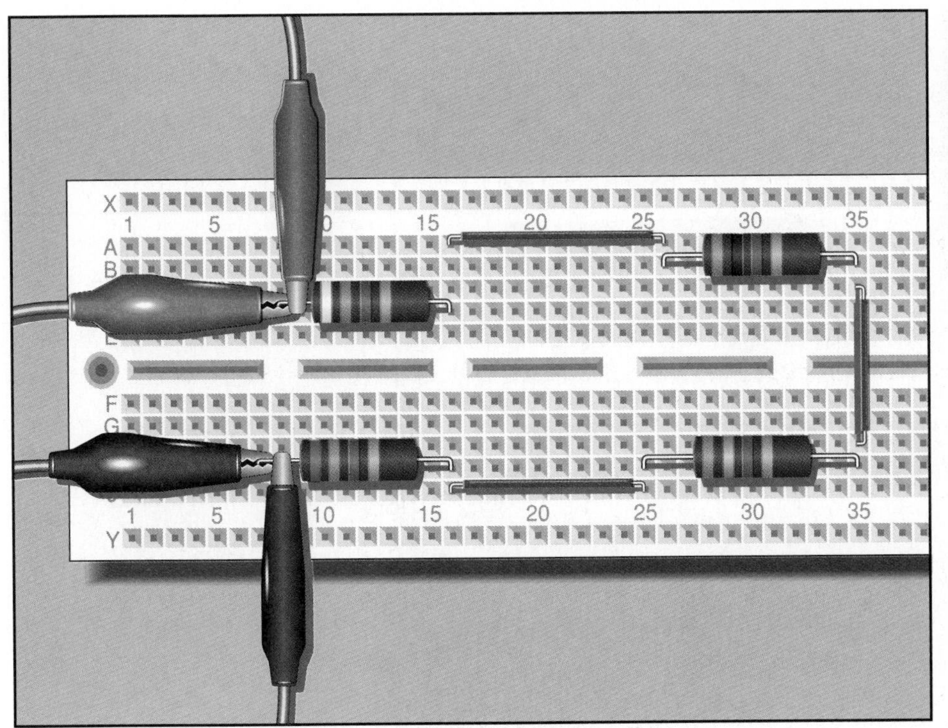

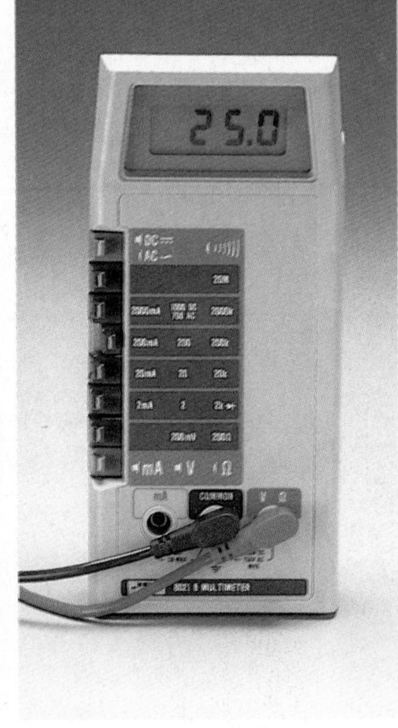

(b) Voltmeter connected to measure V_S

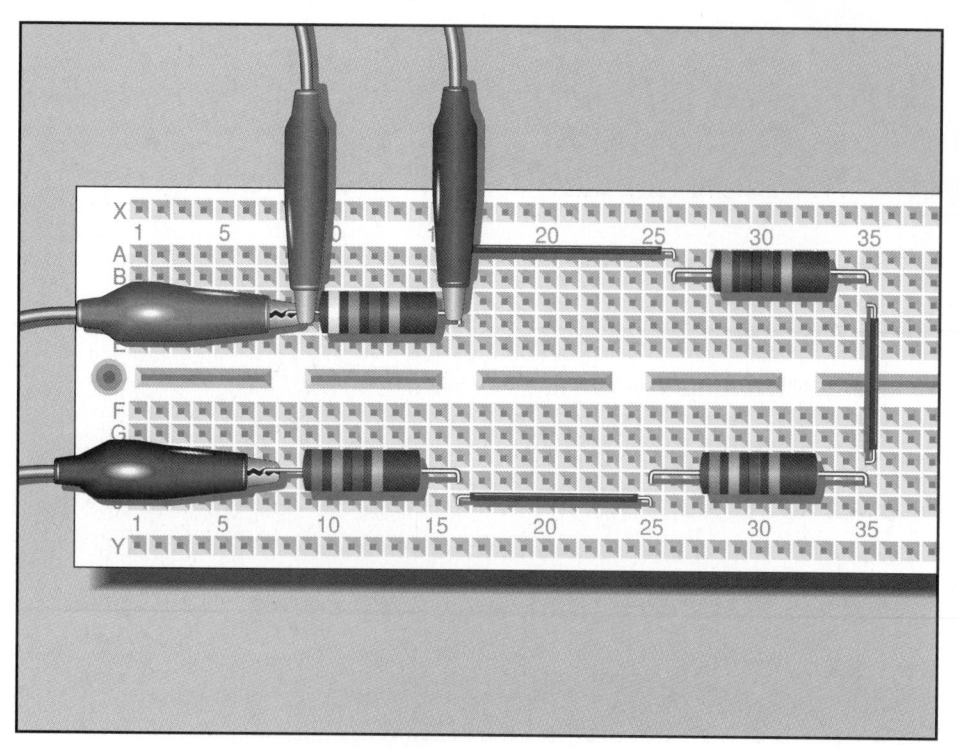

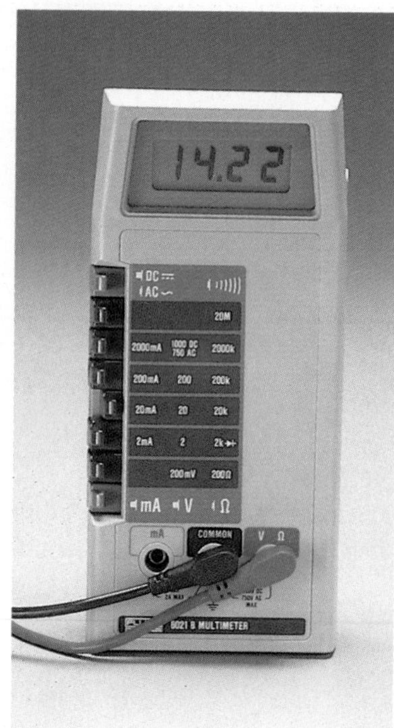

(c) Voltmeter connected to measure V_{R1}

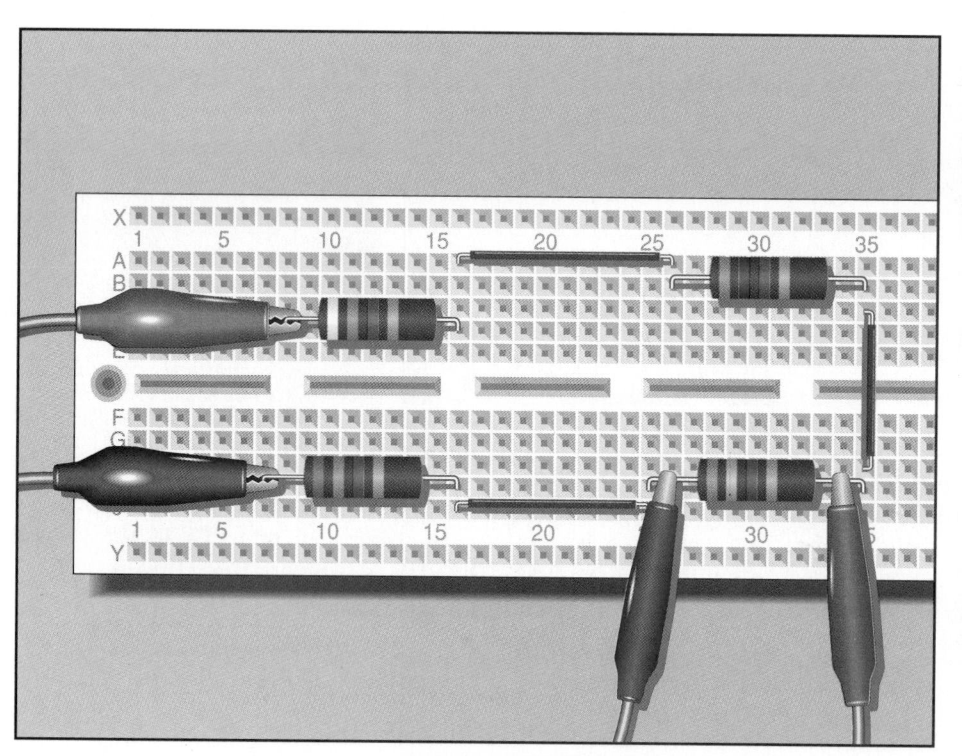

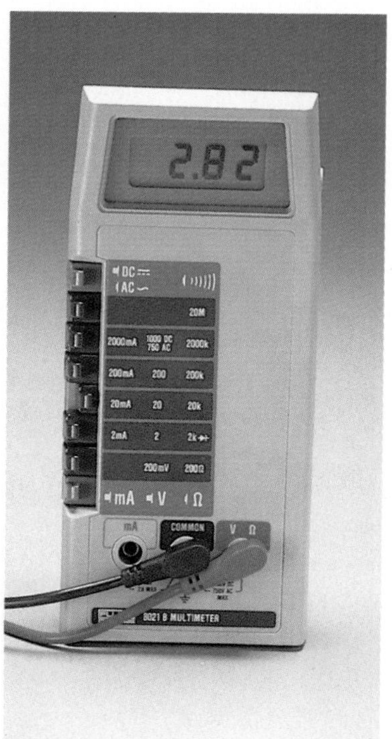

(d) Voltmeter connected to measure V_{R3}

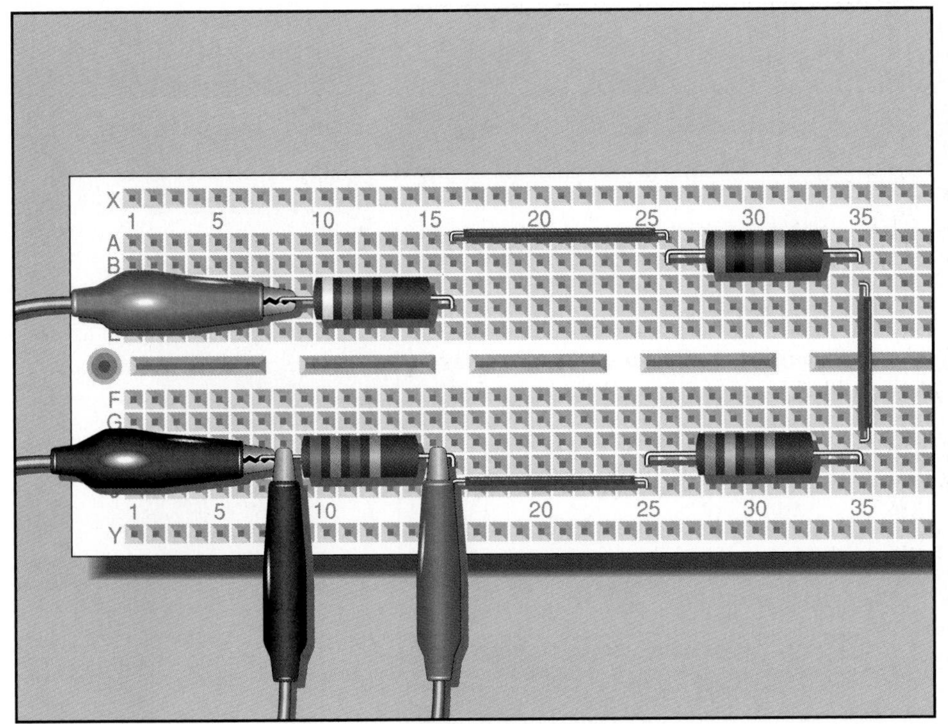

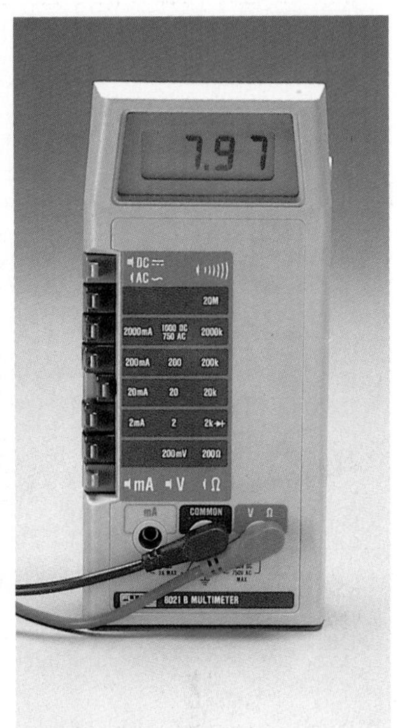

(e) Voltmeter connected to measure V_{R4}

FIGURE 4–78 Illustrations for Figure 4–78 a–e are circuit and meter readings for Problem 46

46. In the circuit in Figure 4–78, the voltage drops are as shown. What is the probable fault or faults in this circuit?

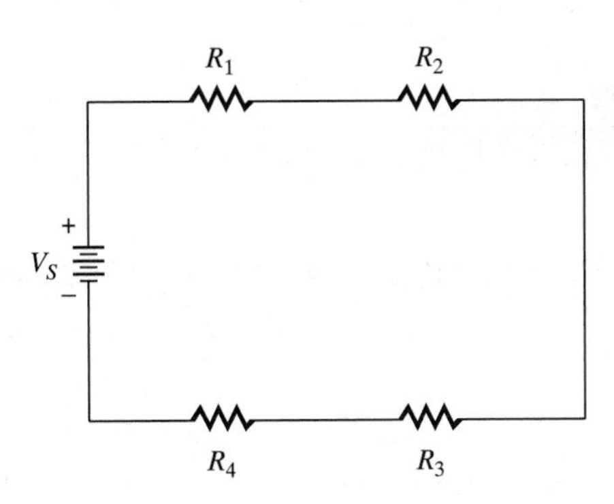

(a) Schematic for problem 46

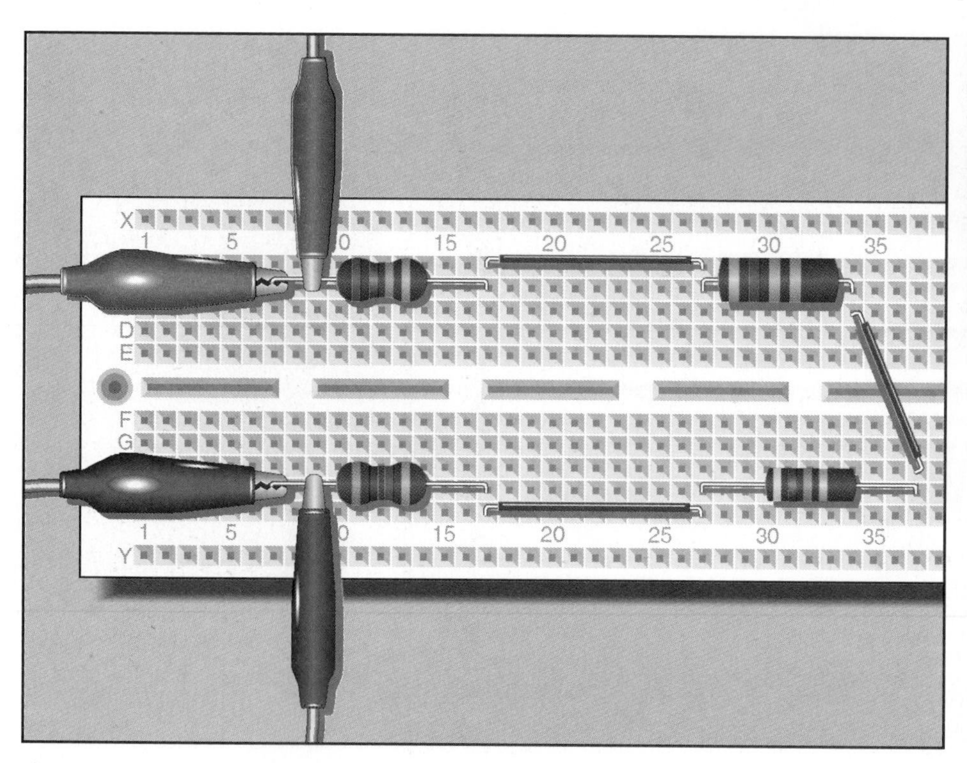

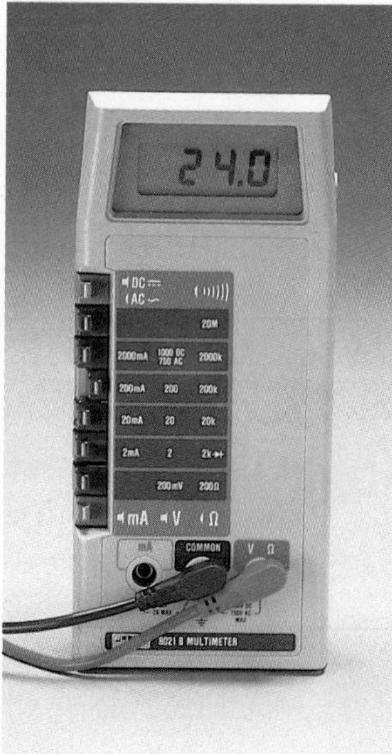

(b) Voltmeter connected to measure V_S

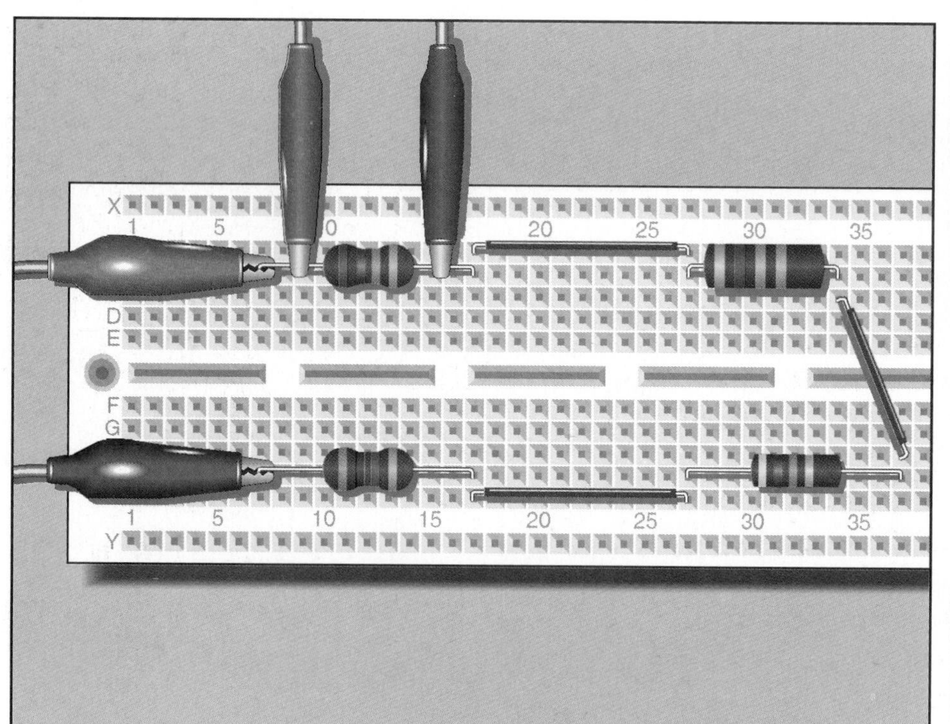

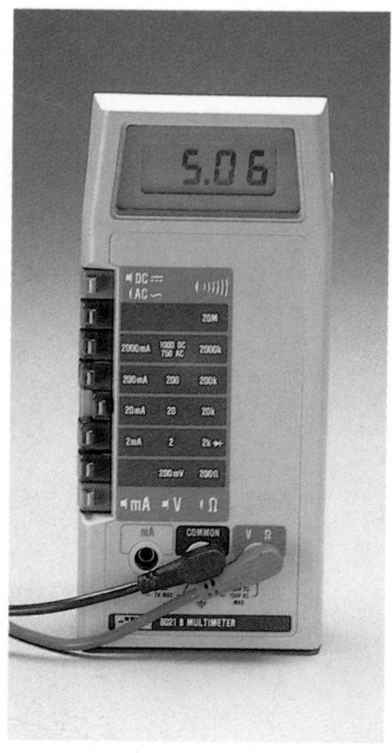

(c) Voltmeter connected to measure V_{R1}

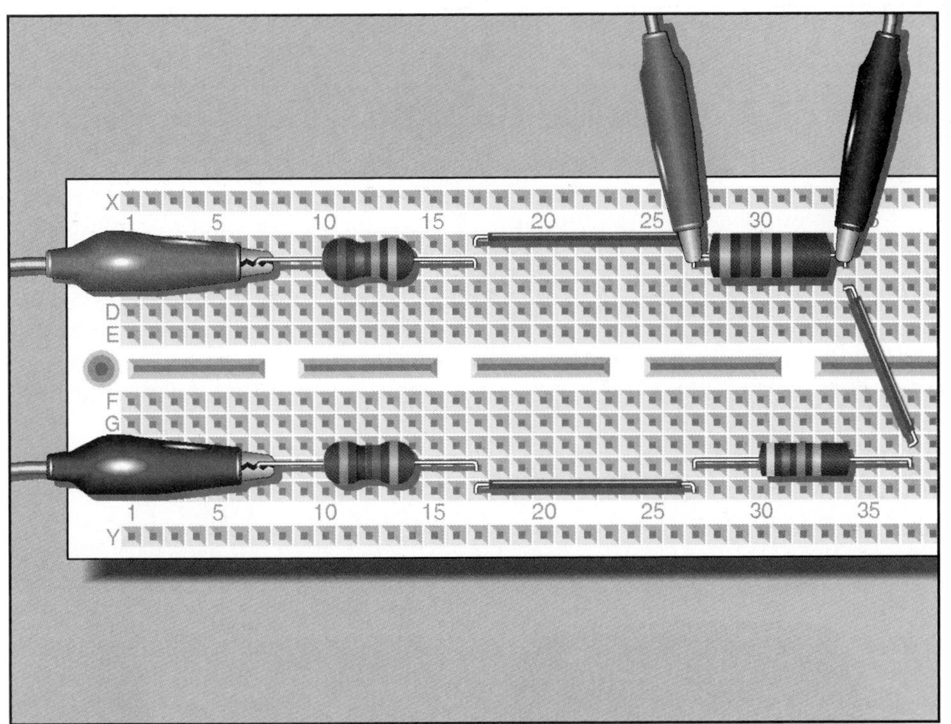

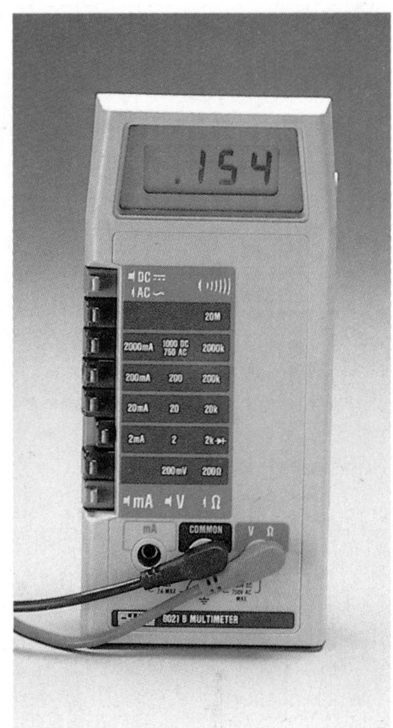

(d) Voltmeter connected to measure V_{R2}

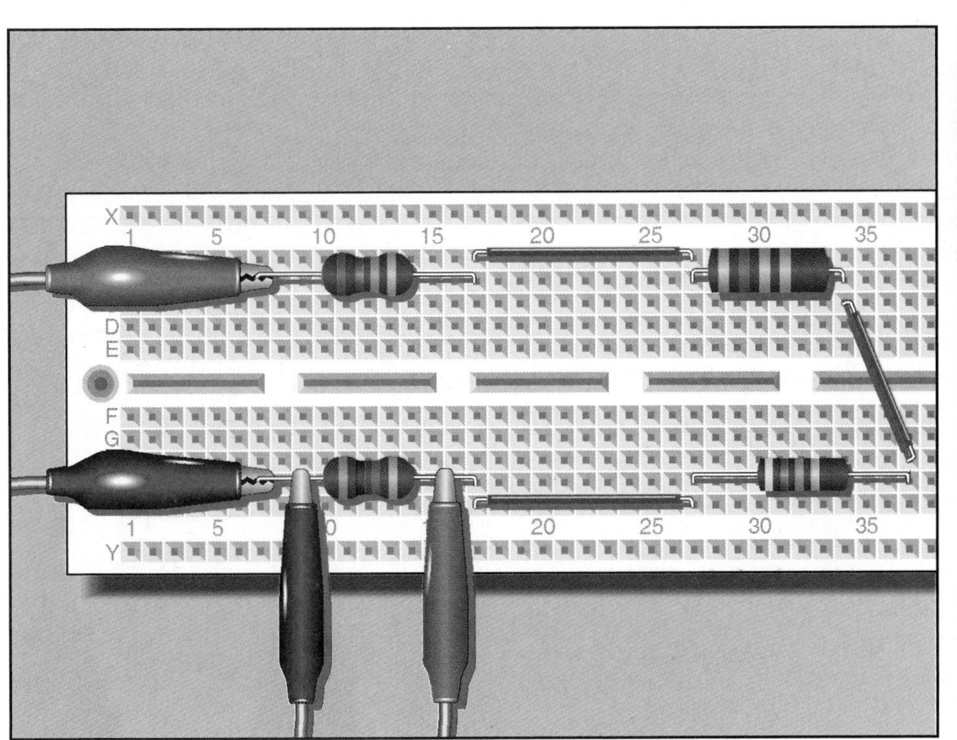

(b) Voltmeter connected to measure V_{R4}

Converting Silicon to IC Chips

In an integrated circuit, all components are integrated on a single piece or "chip" of silicon. There are several different types of IC chips, as they are called, including monolithic, thin film, thick film, and hybrid. (Hybrid is a combination of monolithic and thin-film units.) This discussion focuses on the monolithic type of IC, in which the components are formed on a single piece of silicon.

The two semiconductor elements most often used in the manufacture of electronic components and devices are silicon and germanium. Integrated circuits, however, are made almost exclusively of silicon. Silicon is the second most abundant element in the earth's crust. It is found in sand in the form of silicon dioxide (SiO). Recall that the semiconductor elements are characterized by having 4 valence electrons. In attaining stability, as was explained in Chapter 2, they share their valence electrons, and as a result effectively "see" 8 electrons in their valence shell. This sharing is known as covalent bonding. The stability makes the semiconductor materials, in their pure state, rather poor conductors of electricity. In order to make them conductive, controlled amounts of impurities known as *doping elements* are introduced. Doping elements are those with either 5 or 3 valence electrons. The doping process can produce two types of semiconductors: *p*-type and *n*-type.

An *n*-type semiconductor is one doped with an element, such as arsenic or phosphorus, that has 5 valence electrons. When introduced into pure silicon, the atoms of these elements take their place within the covalent bonds. But where the doping elements fill the covalent bond, there is an extra electron. These electrons are charge carriers that thus make the material conductive. Since the charge carriers are negative, the material is known as *n*-type. If doped with elements such as boron or gallium that have only 3 valence electrons, a *p*-type material is formed. The doping atoms enter the covalent bond, but now each such bond is short one electron. These *holes,* as they are called, are now the charge carriers, and since they are positive, the material is known as *p*-type. The results of the doping process is illustrated in Figure 1.

The manufacturing process for IC chips begins with pure silicon. It is doped with *p*-type impurities and then melted in a crucible (a bowl made of material capable of withstanding the high heat required to melt metals and ores). A large crystal pulled from this melt is then sliced into wafers approximately 15 mils thick (0.038 cm or 0.015 inches). This wafer, which is known as the substrate, is about 1.5″ to 4″ in diameter. The process of manufacturing the wafer is illustrated in Figure 2. The substrate layer (Figure 2(a)) is then highly polished to prepare it for the rest of the manufacturing process. A great number of integrated circuits are made from a single wafer of silicon. The wafer is scribed into squares, about 0.05 inches square, each of which will become an IC. Thus the ICs are made in

FIGURE 1 Results of the doping process

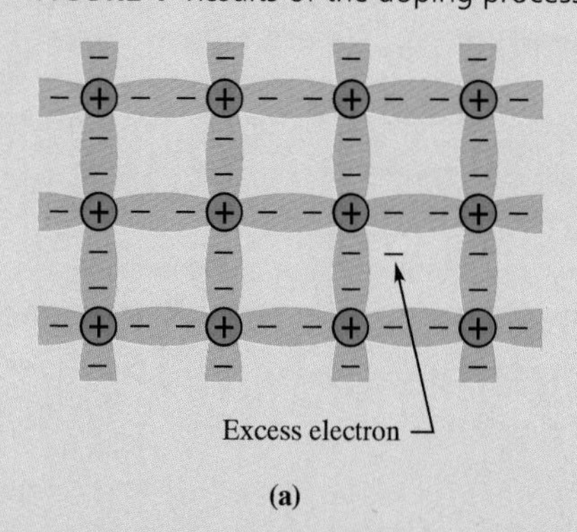

(a)

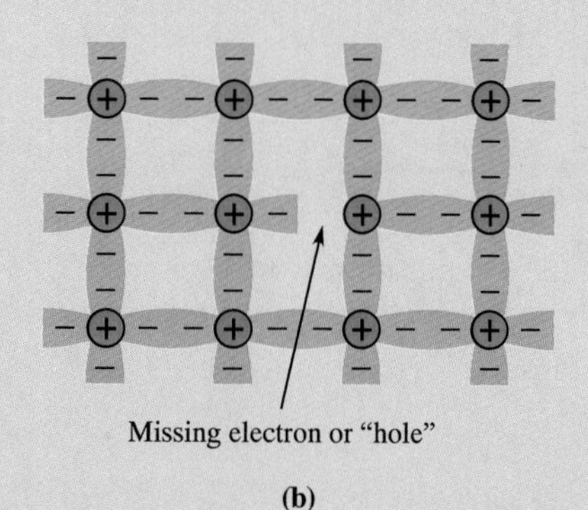

(b)

continued

continuing

batches, which reduces the manufacturing costs.

The next step in the process is to form an oxide coating on the wafer. This step is accomplished by heating the wafer in the presence of water. Next the oxide surface is coated with a material, known as *photoresist,* which hardens when exposed to ultraviolet light. These layers are illustrated in Figure 2(b). The next steps involve what is known as the photolithographic process of IC fabrication. A photomask containing the outline of the circuit to be produced is placed over the wafer (Figure 2(c)). It is then exposed to ultraviolet light to harden the exposed areas. It is washed in chemicals to remove the unhardened areas of photoresist and expose the oxide coating (Figure 2(d)). The oxide coating is removed next (Figure 2(e)), and the substrate is exposed, allowing it to be doped, to produce the desired *p*-type or *n*-type material (Figure 2(f)). Several such masking and developing operations are required to complete the entire integrated circuit.

When the masking, exposing, and etching processes are all complete, each IC is tested. Testing is done by machines, with a computer keeping track of which ICs meet specifications and which do not. The wafer is then scribed using a diamond point tool or a gas laser to separate the individual ICs. The wafer is then stressed, causing it to break along the scribed lines. The ICs not meeting specifications are discarded, and the others are mounted in their cases or packages. There are basically three types of packages: the dual in-line package (DIP), the TO-5, and the flat pack. An example of each is shown in Figure 3. Either aluminum or gold wires are used to make the connections between the IC and the external pins. The cases for the DIP and flat pack are usually made of plastic or ceramic. The TO-5 package is metal.

Some of the advantages of ICs have already been mentioned. They reduce the amount of space needed for an electronic system and serve to make the system more reliable. Since IC components are all made simultaneously from the same material, they are better matched in their characteristics than are those of the discrete variety. There are, however, some disadvantages. The resistors, made as a part of the IC, are limited in their precision. The great number of components on a single piece of silicon limit their power handling capability. Also, it is not possible to integrate inductors on the IC.

FIGURE 3 IC chip packages

FIGURE 2 Manufacture of *p*-type silicon IC chips

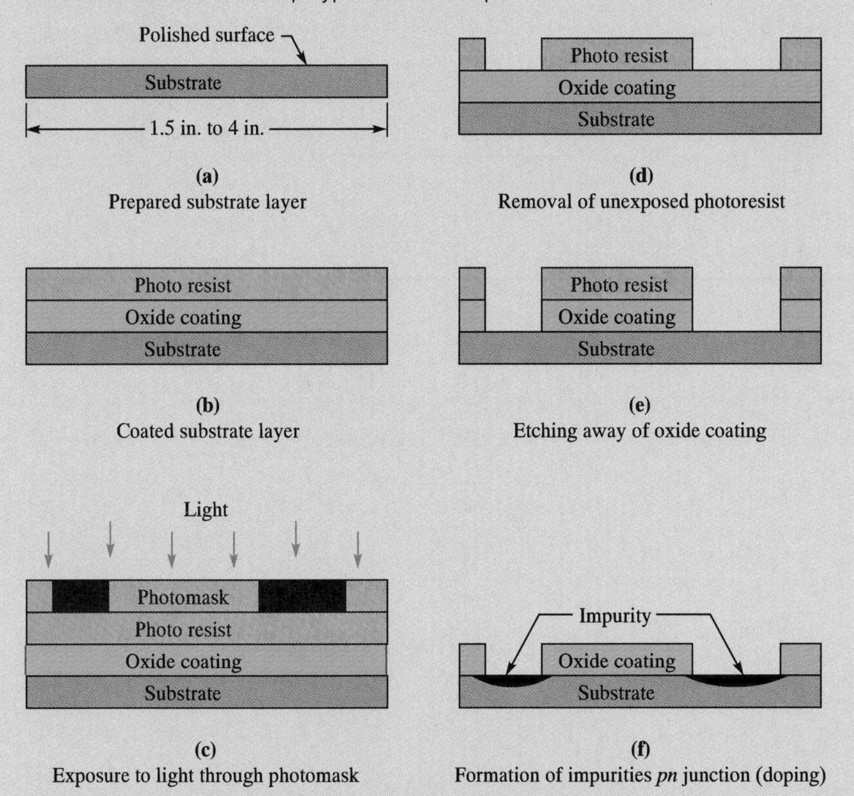

CHAPTER

5 Parallel Circuits

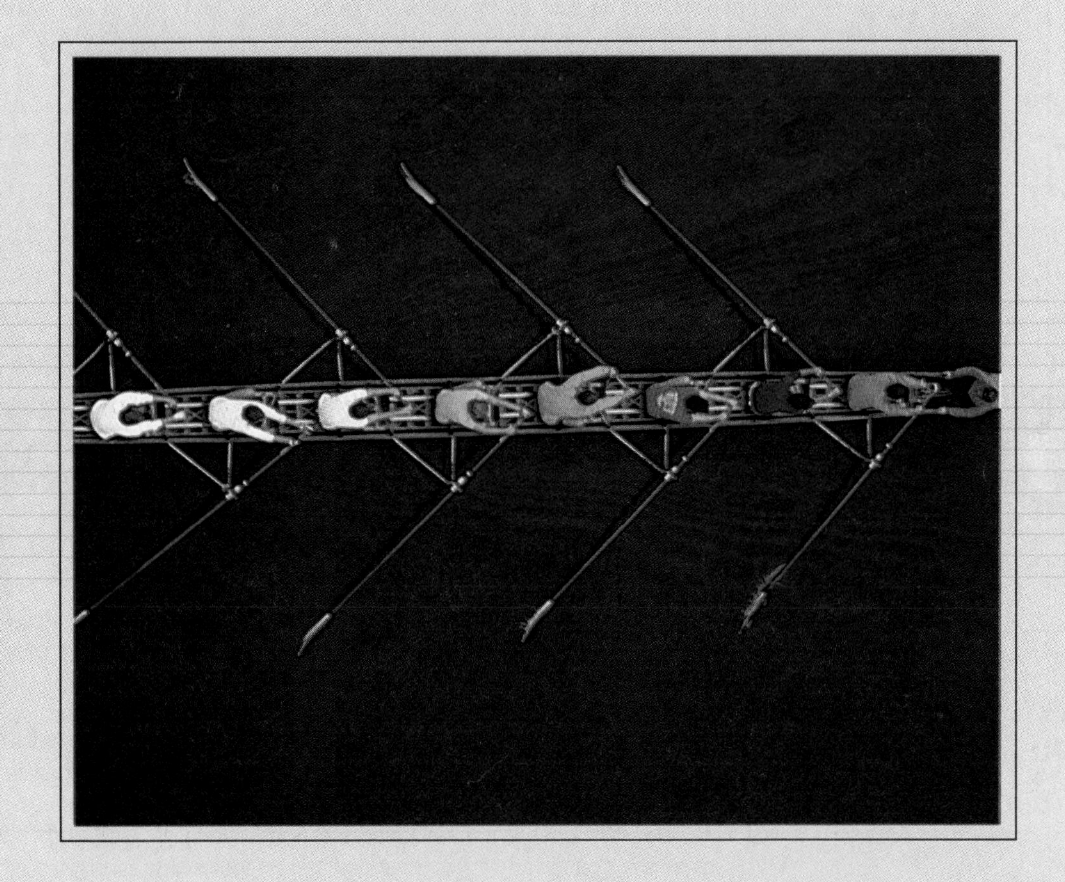

UPON COMPLETION OF THIS CHAPTER, YOU WILL BE ABLE TO

1. Identify parallel circuits.
2. Use conductance to compute the total resistance of a parallel resistive circuit.
3. Define resistor bank.
4. Explain why voltage is the same across all branches of a parallel circuit.
5. State and apply Kirchhoff's current law in the analysis of parallel circuits.
6. Explain why a parallel circuit is a current divider.
7. Compute branch currents in a parallel resistive circuit using Ohm's law.
8. Compute the branch currents in a parallel circuit using the current divider equation.
9. Compute the power developed in each branch of a parallel circuit and the total power.
10. Explain why homes and factories are wired in parallel rather than series.
11. Perform troubleshooting on parallel resistive circuits.

In this chapter, a second way of connecting multiple loads is presented. These loads are connected "side-by-side," or in parallel, in what is known as a parallel circuit. Parallel and series connections are the only ways of connecting electronic components. You will find, however, that most practical electronic circuits are combinations of the two. (Combination circuits, known as series-parallel circuits, will be presented in Chapter 6.)

One thing that should become apparent as you study this chapter is that series and parallel circuits are opposites. What can be said of one type of circuit in almost all cases, the opposite can be said of the other. Watch for this fact as you progress through the chapter. It will also increase your understanding if you determine why, in each instance, the circuits are opposites.

This chapter covers theory and analysis of parallel circuits. The basic law of parallel circuits, Kirchhoff's current law, Ohm's law, and the power equations are used in the analysis of parallel circuits. As was the case with series circuits, your knowledge of parallel circuit theory and analysis must be complete and thorough. This knowledge is essential to your understanding of the electronic systems that follow.

5–1 THE PARALLEL CONNECTION

FIGURE 5–1 Resistors in parallel

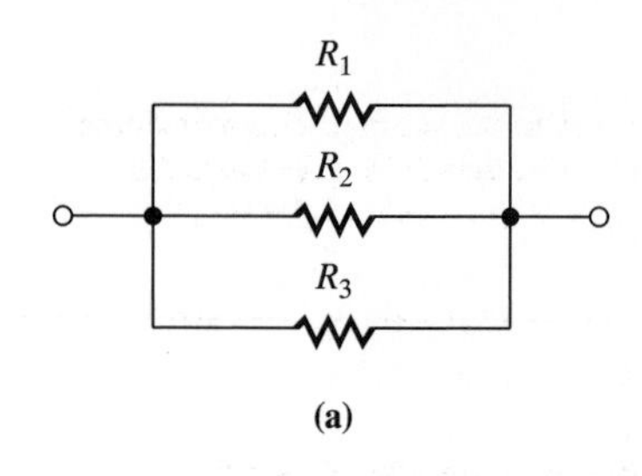

(a)

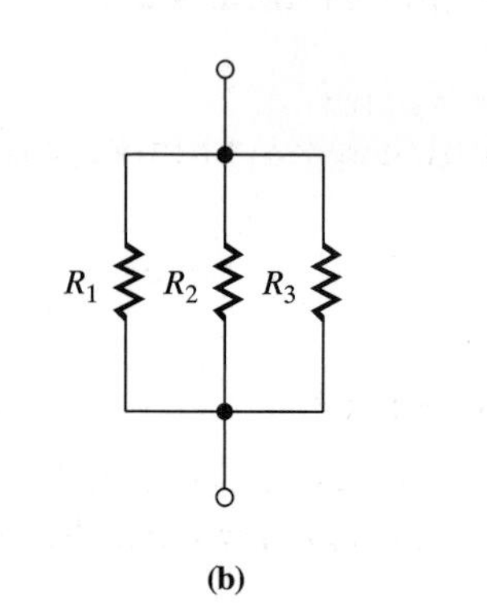

(b)

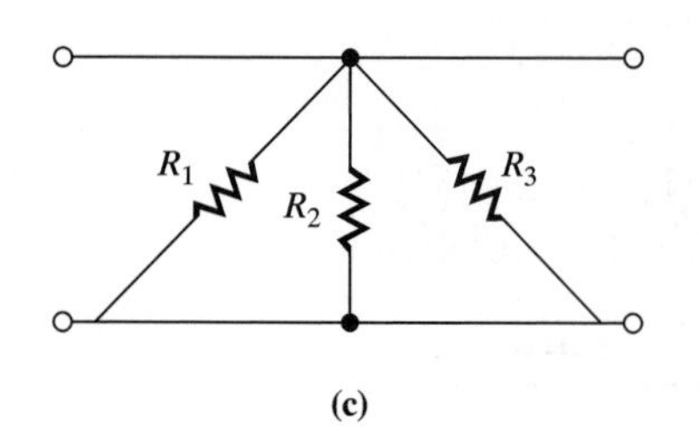

(c)

Question: How are electronic components connected in parallel?

If two or more electronic components share the same voltage source, they are said to be in *parallel*, or "side by side." Configurations of components thus connected are shown in Figure 5–1. The physical orientation of the resistors has no bearing in the matter. The ends are connected to the same electrical point, so they are in parallel. As was the case with series circuits, the resistors connected in parallel are given the names R_1, R_2, R_3, and so forth.

Question: What defines, from an electrical standpoint, electronic devices being connected in parallel?

A group of resistors connected in parallel is often referred to as a **resistor bank** across a common source, as shown in Figure 5–2(a). The top of each resistor load in the figure is connected to a common electrical point, point A. In a like manner, the lower part of each resistor is connected to a common electrical point, point B. The positive side of the voltage source is connected to point A, and the negative side is connected to point B. Thus, the positive side of the source is common to the top of each resistor, and the negative side is common to the bottom of each resistor. Figures 5–2(b), (c), and (d) show the source across each resistor individually, while Figure 5–2(a) shows it across them collectively. Figure 5–2 also shows that the same voltage source is across each branch of the circuit. It can be said, then, for devices to be connected in parallel, the same voltage will be across each load. This last statement defines devices in parallel and must be thoroughly understood.

Another way of illustrating the common voltage source across parallel resistors is shown in Figure 5–3(a). The tops of the resistors are connected to points A, B, and C, respectively, with all these connected to point D, which is the positive side of the source. The bottoms of the resistors connect to points E, F, and G, respectively, with all of these connected to point H, which is the negative side of the source. Figure 5–3(b) shows the resistors connected in an alternate manner. Since points A, B, and C are common, they can be connected to point D. The common points E, F, and G are connected to common point H. Thus, the same voltage is across each resistor.

FIGURE 5–2 Common reference points for resistors connected in parallel

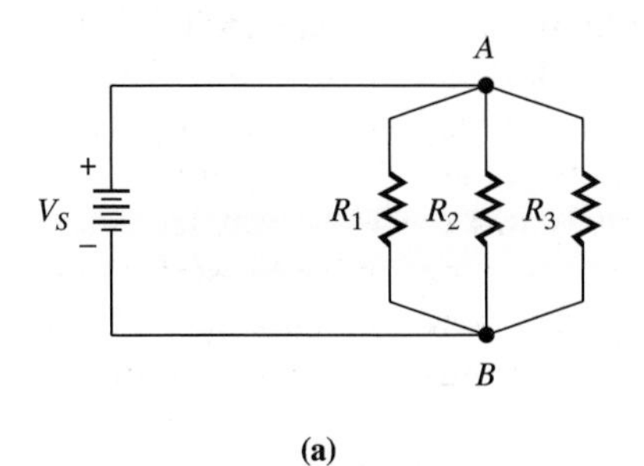

(a)
Resistor bank connected to points A and B

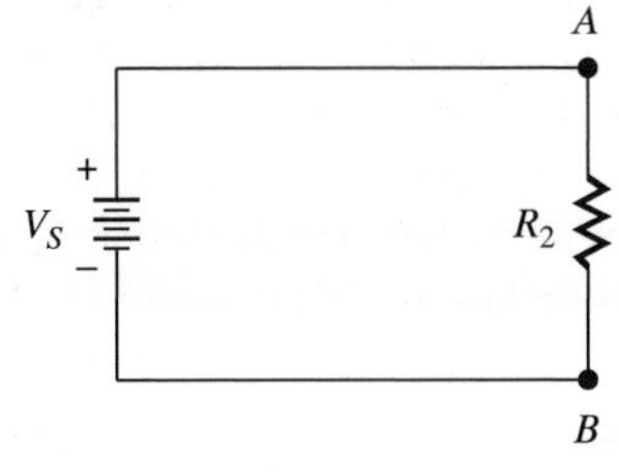
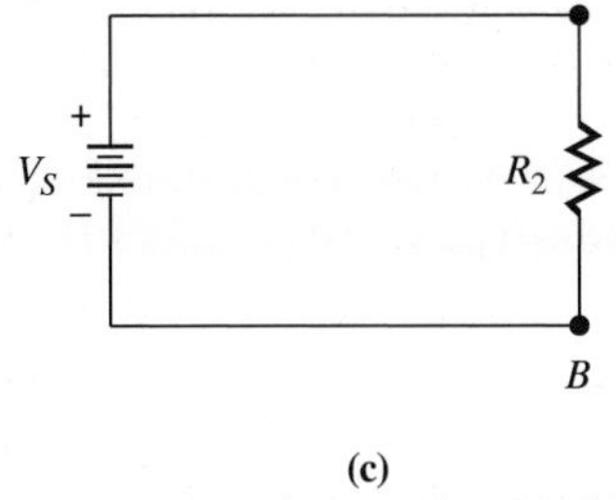

(c)
R_2 connections at points A and B

(b)
R_1 connections at points A and B

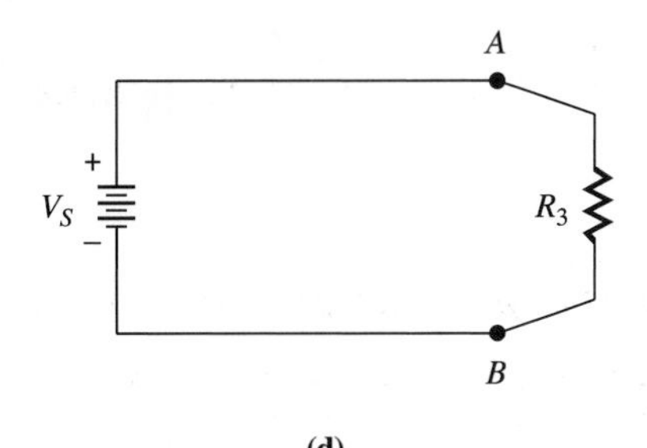

(d)
R_3 connections at points A and B

FIGURE 5–3 Alternate configurations for common reference points

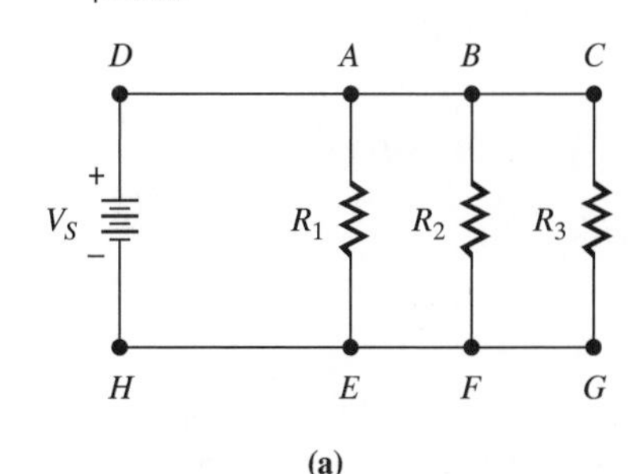

(a)
Points A, B and C referenced to D; points E, F, and G referenced to H

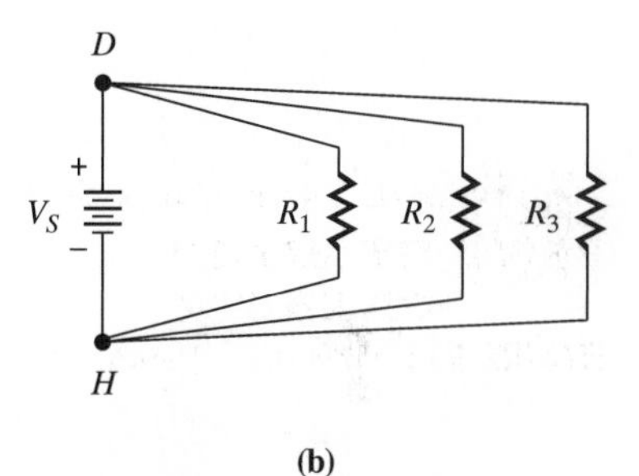

(b)
Alternate configuration of resistors referenced to points D and H

SECTION 5–1 REVIEW

1. What is meant by a parallel connection?
2. What is meant by components being connected to the same electrical point?
3. What is a resistor bank?
4. What defines two or more electronic devices being in parallel?
5. Explain, using a drawing if necessary, why parallel loads have the same voltage across them.

5–2 CURRENT IN A PARALLEL CIRCUIT

Question: What is the basic purpose of a parallel circuit?

The basic purpose of a parallel circuit is to provide the same voltage to loads that may draw different currents. As shown in Figure 5–4, the total current I_T flows from the source

FIGURE 5–4 Division of total current into each branch of a parallel circuit

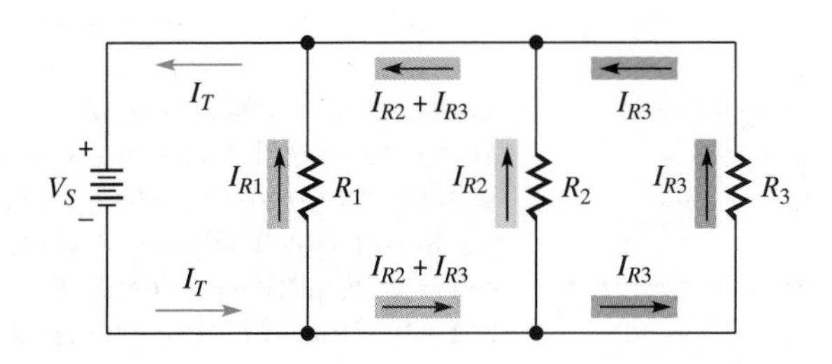

and then divides among the parallel resistors. Thus, a **parallel circuit** is one in which there is more than one path for current and the same voltage is across each load. For these reasons, a parallel circuit is often referred to as a current divider.

Branch Currents

Each current path in the parallel circuit is known as a **branch,** and the current flowing through each branch is the **branch current.** Using Ohm's law, branch current is computed as follows:

$$I_{R1} = \frac{V_S}{R_1}$$

$$I_{R2} = \frac{V_S}{R_2}$$

$$I_{R3} = \frac{V_S}{R_3}$$

As shown, the same voltage is found across each branch. Notice that by Ohm's law, the current in each branch is *inversely proportional* to the resistance of the branch.

EXAMPLE 5–1 Compute the current flowing in each branch of the circuit of Figure 5–5.

Solution:

$$I_{R1} = \frac{V_S}{R_1} = \frac{50}{10\ \text{k}\Omega} = 5\ \text{mA}$$

$$I_{R2} = \frac{V_S}{R_2} = \frac{50}{5\ \text{k}\Omega} = 10\ \text{mA}$$

$$I_{R3} = \frac{V_S}{R_3} = \frac{50}{25\ \text{k}\Omega} = 2\ \text{mA}$$

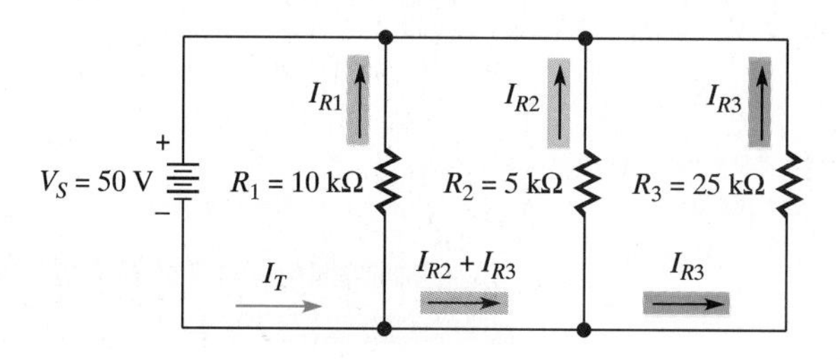

FIGURE 5–5 Circuit for Example 5–1

Kirchhoff's Current Law

Recall from Chapter 4 that the current flowing out of a resistor is the same as the current flowing into it. It is also true that the current flowing out of any point in a circuit equals the current flowing into that point. **Total current** I_T in a parallel circuit leaves the negative side of the source, divides when entering each of the loads, and recombines after flowing through them to return to the positive side of the source.

For example, Point A of Figure 5–6 is known as a *branch point,* and the current flowing into this point is the total current from the source. The current leaving point A

FIGURE 5–6 Schematic diagram showing Kirchhoff's current law

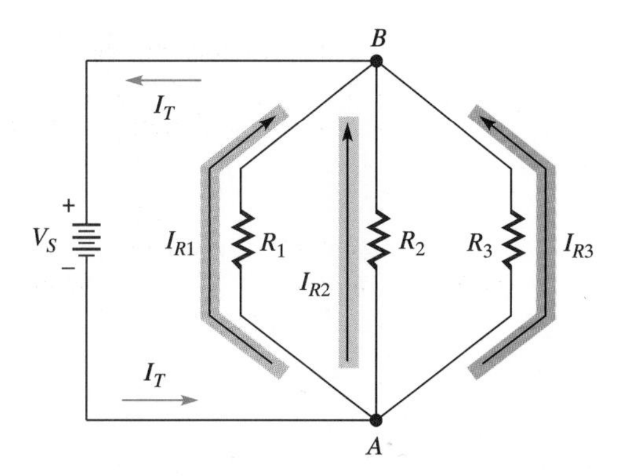

goes three ways—through R_1, R_2, and R_3. After flowing through these resistors, the current combines once more at point B and returns to the positive side of the source. The same current flows at points A and B, so the sum of the currents through resistors R_1, R_2, and R_3, known as I_1, I_2, and I_3, equals the total current. This relationship is, in general, **Kirchhoff's current law** (KCL):

The total current in a parallel circuit is equal to the sum of the branch currents.

In equation form, KCL is written as

(5–1) $$I_T = I_1 + I_2 + I_3 + \cdots I_n$$

If the current flowing into a point is considered positive and that flowing out of the point negative, the current law can be stated in an alternate manner: The algebraic sum of the currents flowing into and out of a point is equal to zero. In equation form, this version of KCL is

$$I_T + (-I_1) + (-I_2) + (-I_3) + \cdots (-I_n) = 0$$

As shown in Figure 5–6, the current into point A, the total current, is given a positive sign while the three currents leaving are given negative signs. It is important to understand that the negative sign does not imply a negative value, only that the current is flowing away from the point of reference. When added with regard to their signed values, the sum is zero.

NOTE

This point within your study of electronics is another at which you have been exposed to something of crucial importance. Kirchhoff's current law is *the basic law* of parallel circuits. Thus, it will be used to explain the action of parallel circuits of every type in subsequent chapters of this text, as will be called to your attention when these points occur. KCL will also be used in explaining the action of some of the electronic circuits making up radios, computers, and so forth. Be sure that when you have completed this section, you not only can solve the problems, but can also "picture" the action of the circuit from the standpoint of Kirchhoff's current law!

EXAMPLE 5–2 For the circuit of Figure 5–7, compute the total current.

■ Solution:

Use Eq. 5–1 and substitute:

$$I_T = I_1 + I_2 + I_3 + I_4$$
$$= 5 \text{ mA} + 18 \text{ mA} + 7.6 \text{ mA} + 3.2 \text{ mA} = 33.8 \text{ mA}$$

[5] [EXP] [±] [3] [+] [1] [8] [EXP] [±] [3] [+] [7] [·] [6] [EXP] [±] [3] [+]

[3] [·] [2] [EXP] [±] [3] [=]

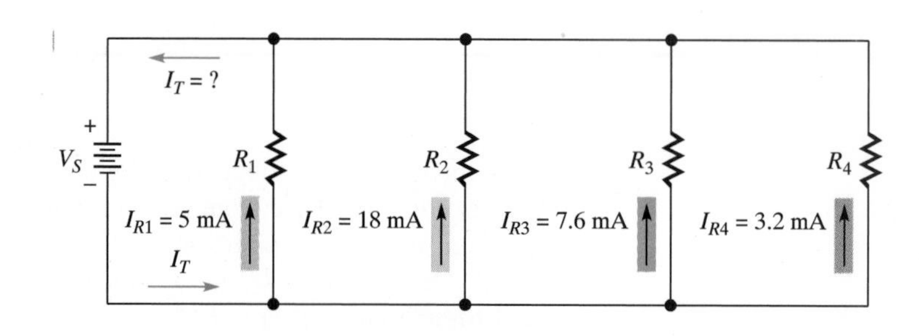

FIGURE 5–7 Circuit for Example 5–2

EXAMPLE 5–3 For the circuit of Figure 5–8, compute the current through branch 3.

■ Solution:

Rearranging and substituting into Eq. 5–1 yields

$$I_3 = I_T - (I_1 + I_2)$$
$$= 36 \text{ ma} - (12 \text{ mA} + 7 \text{ mA}) = 17 \text{ mA}$$

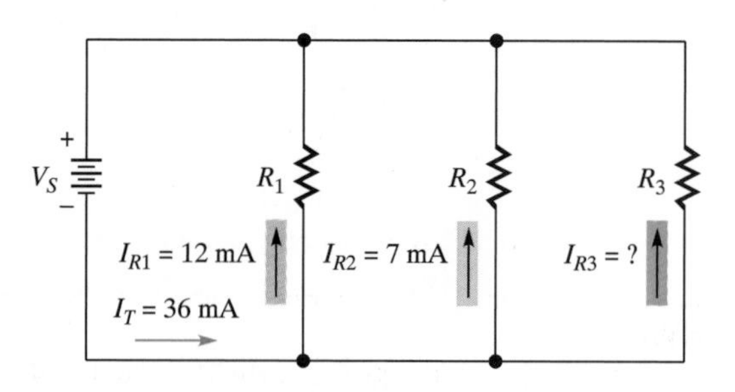

FIGURE 5–8 Circuit for Example 5–3

EXAMPLE 5–4 For the circuit of Figure 5–9, compute the current through each branch and the total current.

■ Solution:

Use Ohm's law to find the current in each branch:

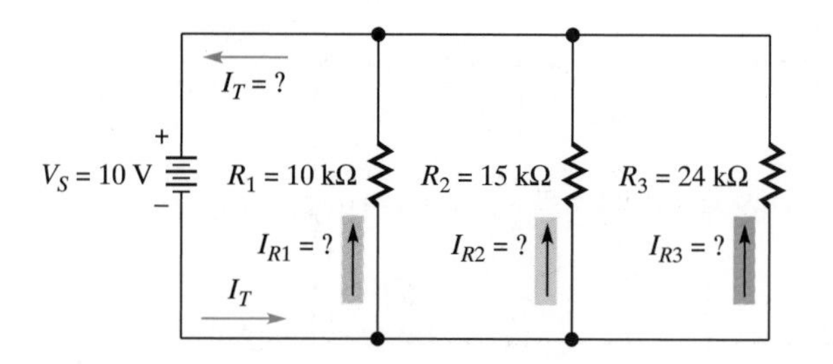

FIGURE 5–9 Circuit for Example 5–4

$$G_4 = G_T - (G_1 + G_2 + G_3)$$

$$= 1.44 \text{ mS} - (.667 \text{ mS} + .417 \text{ mS} + .256 \text{ mS})$$

$$= 1.44 \text{ mS} - 1.34 \text{ mS} = 0.1 \text{ mS}$$

[1] [.] [4] [4] [EXP] [±] [3] [−] [(] [·] [6] [6] [7] [EXP] [±] [3] [+]

[.] [4] [1] [7] [EXP] [±] [3] [+] [·] [2] [5] [6] [EXP] [±] [3] [)] [=]

Then,

$$R_4 = \frac{1}{G_4}$$

$$= \frac{1}{0.1 \text{ mS}} = 10 \text{ k}\Omega$$

Note that more than three significant figures are used in this example in order that the calculation be exact.

■ Total Resistance of Equal Resistors in Parallel

Question: Are there methods of quickly and accurately determining or estimating total resistance?

If all the resistors in a parallel resistor bank are of equal value, there is a simplified equation that may be used in computing total resistance. It may be derived as follows:

1. Use the equation for total conductance:

$$G_T = G_1 + G_2 + G_3$$

2. Substitute reciprocal resistance values:

$$\frac{1}{R_T} = \frac{1}{R_1} + \frac{1}{R_2} + \frac{1}{R_3}$$

3. Since all resistance values are equal, rewrite the equation:

$$\frac{1}{R_T} = \frac{1}{R_x} + \frac{1}{R_x} + \frac{1}{R_x} = \frac{3}{R_x}$$

4. Multiply by 3:

$$3R_T = R_x$$

5. Rearrange to find the simplified equation for total resistance:

$$R_T = \frac{R_x}{3}$$

Called the *equal resistance equation,* this equation shows that if all resistance values are equal in a parallel bank, the total resistance is the ohmic value of one resistor divided by the number of resistors. This principle can be quite helpful as shown in the following examples.

EXAMPLE 5–8 Five 10 kΩ resistors are connected in parallel. Compute the total resistance.

■ **Solution:**

Since all the resistance values are equal,

$$R_T = \frac{R_x}{N} \quad \left[\text{total resistance} = \frac{\text{value of one resistor}}{\text{number of resistors}}\right]$$

$$= \frac{10 \times 10^3}{5}$$

$$= 2\ \text{k}\Omega$$

EXAMPLE 5–9 For the circuit of Figure 5–14, compute the total resistance.

■ **Solution:**

This problem shows the advantage of the equal resistance equation in computing total resistance. Notice that there are three 6 kΩ and two 4 kΩ resistors in parallel. The total of the three 6 kΩ resistors can first be determined: 6000/3 = 2000 Ω. Next the total of the two 4 kΩ resistors can be determined: 4000/2 = 2000 Ω. This result is the equivalent of two 2 kΩ resistances in parallel. Their total can now be determined: 2000/2 = 1000 Ω. Thus, the total of all five resistors is 1 kΩ. As you can see, this type problem can be performed without resorting to pencil, paper, or calculator!

FIGURE 5–14 Circuit for Example 5–9

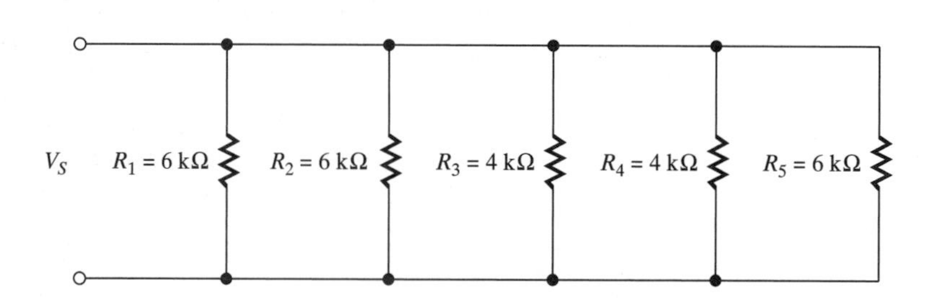

EXAMPLE 5–10 Two parallel resistors have a total resistance of 480 Ω. If the value of one is 1200 Ω, compute the value of the second.

■ **Solution:**

In this example, the total resistance equation for parallel resistors is transposed in order to obtain an equation for the second resistor.

Equation 5–4: $$R_T = \frac{1}{(1/R_1) + (1/R_2)}$$

Cross multiplying: $$R_T[(1/R_1) + (1/R_2)] = 1$$

Converting: $$\frac{R_T}{R_2} + \frac{R_T}{R_1} = 1$$

$$\frac{R_T}{R_2} = \frac{1 - R_T}{R_1}$$

$$\frac{R_T}{R_2} = \frac{R_1 - R_T}{R_1}$$

$$R_2(R_1 - R_T) = R_T R_1$$

Solving:

$$R_2 = \frac{R_T R_1}{R_1 - R_T}$$

$$= \frac{(480)(1200)}{1200 - 480} = 800\ \Omega$$

The solution of Example 5–10 involved a rather heavy amount of arithmetic. A simpler method that demonstrates the utility of using conductance rather than resistance, is shown in Example 5–11.

EXAMPLE 5–11 Two parallel resistors have a total resistance of 800 Ω. If the value of one is 1000 Ω, compute the value of the second.

■ Solution:

Equation 5–2: $G_T = G_1 + G_2$

Transposing: $G_2 = G_T - G_1$

Substituting:

$$G_2 = \frac{1}{R_T} - \frac{1}{R_1}$$

$$= \frac{1}{800\ \Omega} - \frac{1}{1000\ \Omega}$$

$$= 0.00125\ \text{S} - 0.001\ \text{S} = 0.00025\ \text{S}$$

Solving:

$$R_2 = \frac{1}{G_2}$$

$$= \frac{1}{0.00025\ \text{S}} = 4\ \text{k}\Omega$$

■ Approximating Total Resistance

In addition to the methods presented here for calculating total resistance, there are ways in which the total can be approximated. In every calculation made in the previous paragraphs, one thing stands out: The total resistance of resistors in parallel is always smaller than the value of the smallest resistor. For example, even when a 1 MΩ resistor parallels a 100 Ω resistor, the total will be slightly less than 100 Ω. This fact is useful in checking your answers in total resistance computations.

In instances where a resistor is paralleled with a much smaller resistor, a rule-of-thumb may be applied in determining the total. In cases such as these, if the value of one parallel resistor is at least ten times larger than the other, the total may be considered the value of the smaller. Take, for example, a 1000 Ω and a 100 Ω resistor in parallel. The total resistance of this combination is 91 Ω ohms. Taking resistor tolerances into consideration, the approximation of 100 Ω as the total resistance would be, in most cases, satisfactory. The greater the difference, 100 times, 1000 times, and so on, the closer the approximation will come to being the actual total resistance.

NOTE

Approximating values of not just resistance, but also of voltage and current, is a skill you should work at perfecting. In your studies you have, and will continue to, make many calculations of these quantities. Some idea of what the answer will be is essential if human errors are to be avoided. (You may already be aware of how easy it is to make the wrong entry or depress the wrong key in using the calculator.)

Approximating will also be necessary on the job. You will be making measurements of the electrical quantities when performing troubleshooting on electronic systems. For each measurement, a decision will be necessary as to whether or not it is correct. You may not have the time nor inclination to compute the exact value of what the quantity should be. It will then be necessary to make an approximation and decide: "That's about right," or "That couldn't be right!"

SECTION 5–4 REVIEW

1. Explain why the total resistance is reduced each time another load is added to a parallel resistor bank.
2. Compute the total conductance for the circuit of Figure 5–15.
3. Compute the total resistance for the circuit of Figure 5–15.
4. Using the most efficient method, compute the total resistance of six 4.7 kΩ resistors in parallel.
5. Without using pencil, paper, or calculator, what is the approximate total resistance of a 10 Ω and a 200 Ω resistor connected in parallel?
6. What standard value resistor must be added in parallel with 4.7 kΩ in order to produce a total of approximately 3.2 kΩ?

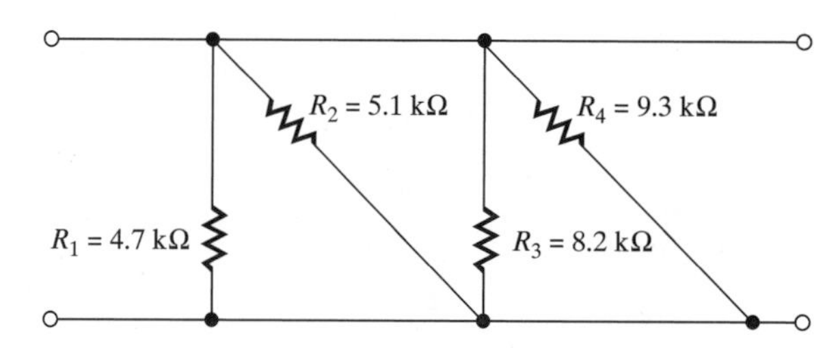

FIGURE 5–15 Circuit for Section Review Problems 2 and 3

5–5 POWER IN A PARALLEL CIRCUIT

Question: How is the total power in a parallel circuit computed?

As previously noted, electrical power is developed when current flows through a resistance. It manifests itself in a circuit in the form of heat. Regardless of the circuit configuration—whether in series or parallel—power is considered in the same manner. Thus, **total power** in a parallel circuit is equal to the sum of the powers developed by the individual loads. In equation form, total power in a parallel circuit is

(5–5)
$$P_T = P_{R1} + P_{R2} + P_{R3} + \cdots P_{Rn}$$

Total power may also be computed using total voltage and total resistance:

$$P_T = V_S I_T$$

$$P_T = \frac{V_S^2}{R_T}$$

$$P_T = I_T^2 R_T$$

EXAMPLE 5–12 Compute the total power developed in the circuit of Figure 5–16.

■ Solution:

1. Using Ohm's law, compute the current in each branch:

$$I_1 = 12 \text{ mA} \qquad I_2 = 6 \text{ mA} \qquad I_3 = 3 \text{ mA}$$

2. Compute the power developed in each branch:

$$P_{R1} = I_1^2R_1 = 0.864 \text{ W}$$

[1] [2] [EXP] [±] [3] [X²] [×] [6] [EXP] [3] [=]

$$P_{R2} = I_2^2R_2 = 0.432 \text{ W}$$

[6] [EXP] [±] [3] [X²] [×] [1] [2] [EXP] [3] [=]

$$P_{R3} = I_3^2R_3 = 0.216 \text{ W}$$

[3] [EXP] [±] [3] [X²] [×] [2] [4] [EXP] [3] [=]

3. Compute the total power using Eq. 5–5:

$$P_T = P_{R1} + P_{R2} + P_{R3}$$

$$= 0.864 \text{ W} + 0.432 \text{ W} + 0.216 \text{ W} = 1.512 \text{ W}$$

[·] [8] [6] [4] [+] [·] [4] [3] [2] [+] [.] [2] [1] [6] [=]

Computing total power by an alternate method gives

$$P_T = I_T^2R_T$$

$$= (21 \times 10^{-3})^2(3429) = 1.512 \text{ W}$$

[2] [1] [EXP] [±] [3] [X²] [×] [3] [4] [2] [9] [=]

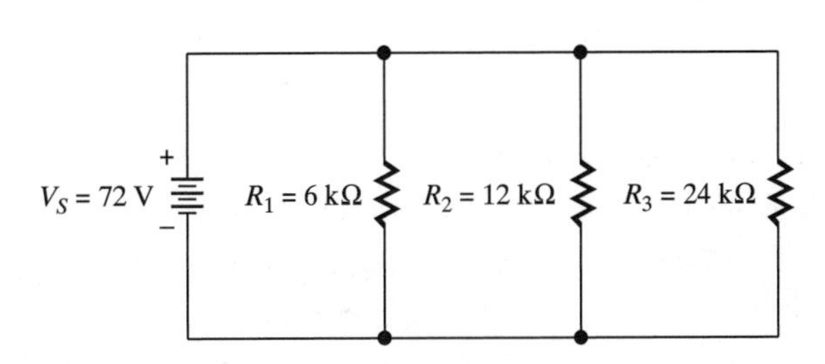

FIGURE 5–16 Circuit for Example 5–12

CAUTION

TECH TIP: CONNECTING SPEAKERS

Often the owner of a home stereo will want to add a second set of speakers in another room or on a patio. While some systems have this capability built in, others do not allow for connecting more than one pair of speakers.

Typically, speakers have an impedance value of 8 Ω. (*Impedance* is the opposition to the flow of alternating current and is used to determine the value of current flow in a circuit just as resistance is used in dc circuits.) The desired extra speakers are added in parallel, thus adding more paths for current and reducing the total resistance. For example, if the speakers have a resistance of 8 ohms, and a second set is added, the total resistance is 4 ohms (equal resistances in parallel). Recall also that when current increases, resulting power increases.

It can be seen that adding the extra set of speakers requires that the amplifier be capable of delivering twice as much power at any particular listening level. To avoid damage, great care must be taken to insure that the amplifier is able to safely supply the extra power before adding any additional speakers.

SECTION 5–5 REVIEW

1. Why can total power in a parallel circuit be computed in the same manner as that in a series circuit?
2. What method other than the sum of each resistor's power can be used to compute total power?

SYNOPSIS

Before beginning a study of parallel circuit analysis, it would be wise to stop and review the basic facts of parallel circuits. In a parallel circuit, all loads connect directly across the voltage source. As a result, all loads have the same voltage across them. The total current leaves the negative side of the source and divides among the branches. For this reason, a parallel circuit is known as a current divider. Each path for current is known as a branch, and the currents that flow through them are known as branch currents. The branch currents can be computed using Ohm's law, and the total current can be computed using Kirchhoff's current law. The total resistance of a parallel circuit is found by first computing the total conductance and then taking its reciprocal. It may also be found by dividing the source voltage by the total current.

As was the case with series circuits, these facts are the basis for circuit analysis and must be understood, not just memorized. After having studied both series and parallel circuits, comparing the opposite characteristics of the two is a good memory aid in that it brings both circuits to mind at the same time. In a parallel circuit, the voltage is the same across all branches, and the current divides. In a series circuit, the current is the same at all points, and the voltage divides. Adding loads to a series circuit increases the total resistance, while adding loads in parallel decreases total resistance.

5–6 PARALLEL CIRCUIT ANALYSIS

Question: Is there a step-by-step procedure for analysis of a parallel circuit?

The analysis of any circuit involves determining the voltage across the current through each component. The preferred measurement in any power-on situation is voltage, since the circuit does not need to be broken in order to make the measurement. The tools of analysis of parallel circuits are Ohm's law, Kirchhoff's current law, and the facts of parallel circuits. Always keep in mind that electronic systems are combinations of series and parallel circuits. The parallel circuits contained within them will be analyzed using the procedures in this section.

The step-by-step procedure for the analysis of parallel circuits is as follows:

1. Calculate the total resistance of the circuit.
2. Calculate the value of the total current.
3. Calculate the value of the current through each branch.
4. Calculate the power developed in each resistor.
5. Calculate the total power developed by the circuit.

FIGURE 5–17 Circuit for parallel circuit analysis

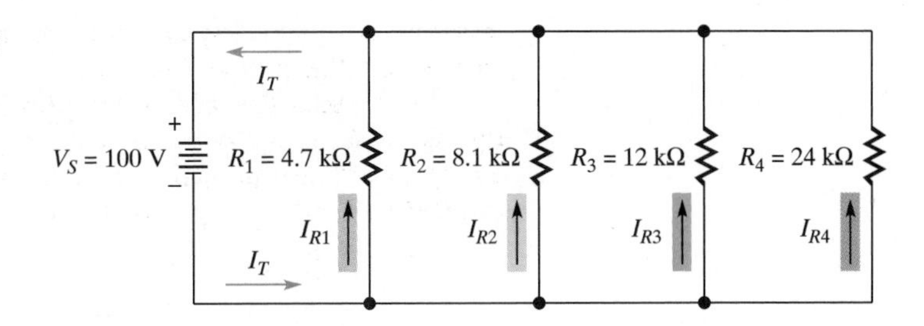

In this analysis procedure, the voltage is the most useful quantity since it is the same across each branch.

The circuit of Figure 5–17 will be used as an example of parallel circuit analysis according to the procedure. Since the source voltage is given, the voltage across each resistor is known: $V_S = V_{R1} = V_{R2} = V_{R3} = V_{R4} = 100$ V.

1. Compute total resistance:

$$G_T = G_1 + G_2 + G_3 + G_4$$

$$G_T = \frac{1}{4.7\text{ k}} + \frac{1}{8.1\text{ k}} + \frac{1}{12\text{ k}} + \frac{1}{24\text{ k}}$$

$$= 212.8\ \mu\text{S} + 123.5\ \mu\text{S} + 83.3\ \mu\text{S} + 41.7\ \mu\text{S} = 461.3\ \mu\text{S}$$

$$R_T = \frac{1}{G}$$

$$= \frac{1}{461.3\ \mu\text{S}} = 2168\ \Omega$$

2. Calculate total current:

$$I_T = \frac{V_S}{R_T}$$

$$= \frac{100\text{ V}}{2168\ \Omega} = 46.1\text{ mA}$$

3. Calculate current through each branch:

$$I_{R1} = \frac{V_S}{R_1} = \frac{100\text{ V}}{4.7\text{ k}\Omega} = 21.28\text{ mA}$$

$$I_{R2} = \frac{V_S}{R_2} = \frac{100\text{ V}}{8.1\text{ k}\Omega} = 12.35\text{ mA}$$

$$I_{R3} = \frac{V_S}{R_3} = \frac{100\text{ V}}{12\text{ k}\Omega} = 8.33\text{ mA}$$

$$I_{R4} = \frac{V_S}{R_4} = \frac{100\text{ V}}{24\text{ k}\Omega} = 4.17\text{ mA}$$

Note that when the branch currents are added, the sum is 46.13 mA, which is within 0.03 mA of the total computed in step 2. The slight difference is due to rounding done in the calculations.

4. Compute power developed in each resistor:

$$P_{R1} = I_{R1}^2R_1 = (21.28 \text{ mA})^2(4.7 \text{ k}\Omega) = 2.13 \text{ W}$$

$$P_{R2} = I_{R2}^2R_2 = (12.35 \text{ mA})^2(8.1 \text{ k}\Omega) = 1.24 \text{ W}$$

$$P_{R3} = I_{R3}^2R_3 = (8.33 \text{ mA})^2(12 \text{ k}\Omega) = 0.833 \text{ W}$$

$$P_{R4} = I_{R4}^2R_4 = (4.17 \text{ mA})^2(24 \text{ k}\Omega) = 0.417 \text{ W}$$

5. Compute total power developed in the circuit:

$$P_T = I_T^2R_T = (46.1 \text{ mA})(2.168 \text{ k}) = 4.61 \text{ W}$$

Total power may also be calculated as the sum of the power values developed in each resistor:

$$P_T = P_{R1} + P_{R2} + P_{R3} + P_{R4}$$

$$P_T = 2.13 \text{ W} + 1.24 \text{ W} + 0.833 \text{ W} + 0.417 \text{ W}$$

$$P_T = 4.62 \text{ W}$$

Note that the two values of total power agree within 0.01 W. Once again, the slight difference is due to rounding done in the calculations.

The preceding example illustrates the general method of circuit analysis when the source voltage and the resistance values are given. It is the simple, straightforward way of performing parallel circuit analysis. The examples that follow show alternate approaches that may be used when other circuit parameters are given.

EXAMPLE 5–13 For the circuit of Figure 5–18, compute R_T, I_T, and V_S.

■ **Solution:**

Equation 5–1:

$$I_T = I_{R1} + I_{R2} + I_{R2}$$
$$= 2 \text{ mA} + .667 \text{ mA} + 0.5 \text{ mA} = 3.167 \text{ mA}$$

[2] [EXP] [±] [3] [+] [·] [6] [6] [7] [EXP] [±] [3] [+] [·] [5] [EXP] [±] [3] [=]

Equation 5–4:

$$R_T = \frac{1}{G_1 + G_2 + G_3}$$
$$= \frac{1}{0.0001 \text{ S} + 0.0000333 \text{ S} + 0.000025 \text{ S}}$$
$$= \frac{1}{0.0001583 \text{ S}} = 2.183 \text{ k}\Omega$$

[·] [1] [EXP] [±] [3] [+] [3] [3] [3] [EXP] [±] [7] [+] [2] [5] [EXP] [±] [6] [=] [1/x]

Ohm's law:

$$V_S = I_TR_T$$
$$= (3.167 \text{ mA})(6.317 \text{ k}\Omega) = 20 \text{ V}$$

[3] [·] [1] [6] [7] [EXP] [±] [3] [×] [6] [·] [3] [1] [7] [EXP] [3] [=]

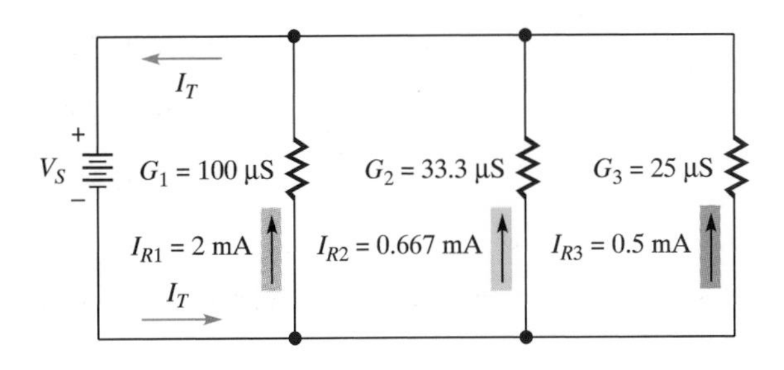

FIGURE 5–18 Circuit for Example 5–13

EXAMPLE 5–14 For the circuit of Figure 5–19, compute the current through each branch.

■ **Solution:**

1. Compute the total resistance of the three branches:

$$R_T = \frac{1}{G_1 + G_2 + G_3}$$

$$= \frac{1}{455\ \mu\text{S} + 123\ \mu\text{S} + 55.6\ \mu\text{S}}$$

$$= \frac{1}{633.6\ \mu\text{S}} = 1.578\ \text{k}\Omega$$

2. Compute the source voltage for the total current given and the total resistance computed:

$$V_S = I_T R_T$$

$$= (76\ \text{mA})(1.578\ \text{k}\Omega) = 120\ \text{V}$$

3. Compute each branch current using Ohm's law:

$$I_1 = \frac{V_S}{R_1} = \frac{120\ \text{V}}{2.2\ \text{k}\Omega} = 54.55\ \text{mA}$$

$$I_2 = \frac{V_S}{R_2} = \frac{120\ \text{V}}{8.1\ \text{k}\Omega} = 14.81\ \text{mA}$$

$$I_3 = \frac{V_S}{R_3} = \frac{120\ \text{V}}{18\ \text{k}\Omega} = 6.67\ \text{mA}$$

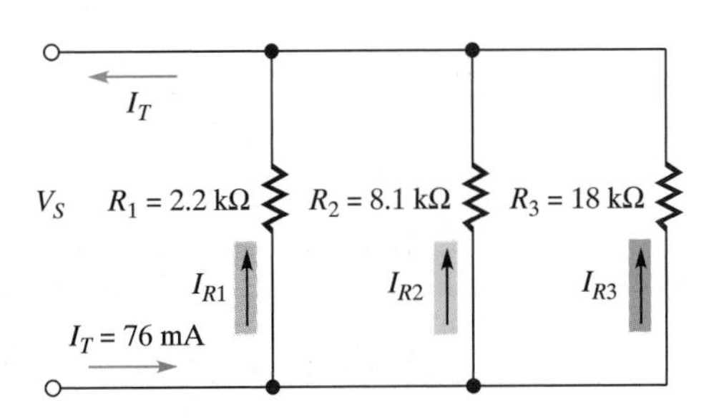

FIGURE 5–19 Circuit for Example 5–14

EXAMPLE 5–15 For the circuit of Figure 5–20, compute the source voltage and the values of R_1 and R_3.

■ Solution:

The value of R_4 and its current are both known. Thus, the voltage across it is the source voltage, or

$$V_{R4} = I_{R4}R_4$$
$$= (1.5 \text{ mA})(10 \text{ k}\Omega) = 15\text{V}$$

Since this is a parallel circuit, the same voltage is across all branches. Thus, the source voltage is 15 V.

Notice that the current at point B is, by KCL, the sum of the currents through R_3 and R_4. Thus, the current through R_3 is

$$I_{R3} = I_{R3+R4} - I_{R4}$$
$$= 1.88 \text{ mA} - 1.5 \text{ mA} = 0.38 \text{ mA}$$

The value of R_3 can now be computed using Ohm's law:

$$R_3 = \frac{V_S}{I_{R3}}$$
$$= \frac{15 \text{ V}}{0.38 \text{ mA}} = 39.5 \text{ k}\Omega$$

The current at point A is the total current. Thus, the current through R_1 is the total current minus that flowing through the other resistances. The current through R_2 is first determined:

$$I_{R2} = \frac{V_S}{R_2}$$
$$= \frac{15}{12 \text{ k}\Omega}$$
$$= 1.25 \text{ mA}$$

Then,

$$I_{R2} + I_{R3} + I_{R4} = 3.13 \text{ mA}$$
$$I_{R1} = I_T - (I_{R2} + I_{R3} + I_{R4})$$
$$= 4.99 \text{ mA} - (1.25 \text{ mA} + 0.38 \text{ mA} + 1.5 \text{ mA}) = 1.86 \text{ mA}$$
$$R_1 = \frac{V_S}{I_{R1}}$$
$$= \frac{15 \text{ V}}{1.86 \text{ mA}} = 8.065 \text{ k}\Omega$$

■ The Parallel Current Divider

In Example 5–14, the problem encountered was computing branch currents when only the branch resistances and total current are known. The calculation required several steps before arriving at the answers. In this section, an equation, known as the *current divider equation,* will be developed and will greatly simplify the solution of such problems. It is

FIGURE 5–20 Circuit for Example 5–15

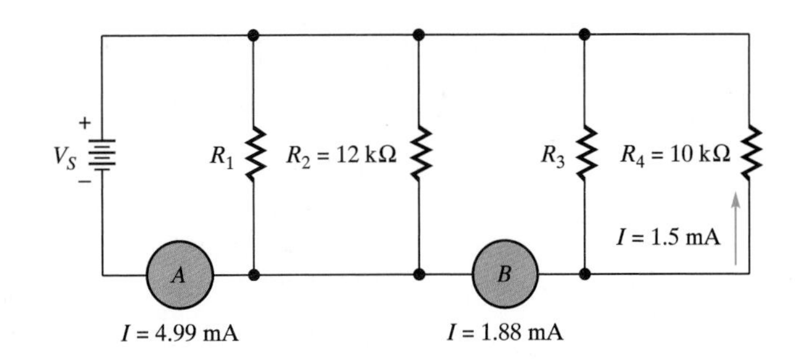

FIGURE 5–21 Circuit for demonstration of current divider equation

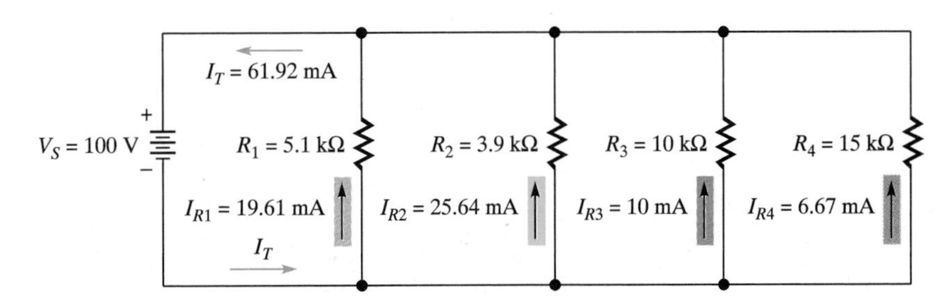

similar to the voltage divider equation presented in Chapter 4.

In the parallel circuit of Figure 5–21, the branch currents are computed using Ohms law:

$$I_{R1} = \frac{V_S}{R_1} = \frac{100\ \text{V}}{5.1\ \text{k}\Omega} = 19.61\ \text{mA}$$

$$I_{R2} = \frac{V_S}{R_2} = \frac{100\ \text{V}}{3.9\ \text{k}\Omega} = 25.64\ \text{mA}$$

$$I_{R3} = \frac{V_S}{R_3} = \frac{100\ \text{V}}{10\ \text{k}\Omega} = 10\ \text{mA}$$

$$I_{R4} = \frac{V_S}{R_4} = \frac{100\ \text{V}}{15\ \text{k}\Omega} = 6.67\ \text{mA}$$

The total current is then computed using Kirchhoff's current law:

$$I_T = I_{R1} + I_{R2} + I_{R3} + I_{R4}$$
$$= 19.61\ \text{mA} + 25.64\ \text{mA} + 10\ \text{mA} + 6.67\ \text{mA} = 61.92\ \text{mA}$$

Because there is a direct relationship between conductance and current in parallel circuits, the conductance of each branch is computed next:

$$G_1 = \frac{1}{R_1} = \frac{1}{5.1\ \text{k}\Omega} = 196\ \mu\text{S}$$

$$G_2 = \frac{1}{R_2} = \frac{1}{3.9\ \text{k}\Omega} = 256\ \mu\text{S}$$

$$G_3 = \frac{1}{R_3} = \frac{1}{10\ \text{k}\Omega} = 100\ \mu\text{S}$$

$$G_4 = \frac{1}{R_4} = \frac{1}{15\ \text{k}\Omega} = 66.6\ \mu\text{S}$$

The total conductance is then computed:

$$G_T = G_1 + G_2 + G_3 + G_4$$
$$= 196\ \mu S + 256\ \mu S + 100\ \mu S + 66.6\ \mu S = 618.6\ \mu S$$

From these computations, the current divider equation can now be derived. (This derivation is similar to that used for the voltage divider equation.) It is based upon the fact that the ratio of a branch current to the total current is *equal* to the ratio of that branch conductance to the total conductance. This relationship will be proven mathematically in the following example.

First the ratio of each branch current to the total current is computed:

$$\frac{I_{R1}}{I_T} = \frac{19.61\ \text{mA}}{61.92\ \text{mA}} = 0.317$$

$$\frac{I_{R2}}{I_T} = \frac{25.64\ \text{mA}}{61.92\ \text{mA}} = 0.414$$

$$\frac{I_{R3}}{I_T} = \frac{10\ \text{mA}}{61.92\ \text{mA}} = 0.161$$

$$\frac{I_{R4}}{I_T} = \frac{6.67\ \text{mA}}{61.92\ \text{mA}} = 0.108$$

Next the ratio of each branch conductance to the total conductance is computed:

$$\frac{G_1}{G} = \frac{196\ \mu S}{618.6\ \mu S} = 0.317$$

$$\frac{G_2}{G} = \frac{256\ \mu S}{618.6\ \mu S} = 0.414$$

$$\frac{G_3}{G} = \frac{100\ \mu S}{618.6\ \mu S} = 0.161$$

$$\frac{G_4}{G} = \frac{66.6\ \mu S}{618.6\ \mu S} = 0.108$$

Since the ratios are in fact equal, the following mathematical relationship can be established:

$$\frac{I_{R1}}{I_T} = \frac{G_1}{G_T}$$

Then,

$$I_{R1} G_T = I_T G_1$$

and

$$I_1 = \frac{G_1}{G_T} \times I_T$$

This relationship can be established regardless of which branch current and conductance is used.

In general, then, it can be expressed as the current divider equation as follows:

(5–6) $$I_x = \frac{G_x}{G_T} \times I_T$$

In electronics, most people do not think in terms of conductance but rather in terms of resistance. The current divider equation thus established can be manipulated in order to convert it to resistance:

$$I_x = \frac{G_x}{G_T} \times I_T$$

$$= \frac{(1/R_x)}{(1/R_T)} \times I_T$$

$$= [(1/R_x)(R_T/1)]I_T$$

Thus,

$$I_x = \frac{R_T}{R_x} \times I_T \qquad \text{(5–7)}$$

Equation 5–7 is often helpful and may be used in parallel circuit analysis where appropriate. For example, in Example 5–14, if this equation is used in its solution rather than the voltage divider equation, notice that the positions of R_T and R_x are reversed. This reversal can be explained by the fact that in a series circuit voltage drops are directly proportional to resistance, while in a parallel circuit branch currents are inversely proportional to resistance.

EXAMPLE 5–16 For the circuit of Figure 5–22, compute the value of the current through each branch using the resistance current divider equation.

■ Solution:

1. Compute total conductance using Eq. 5–2:

$$G_T = \frac{1}{R_1} + \frac{1}{R_2} + \frac{1}{R_3}$$

$$= \frac{1}{1\text{ k}\Omega} + \frac{1}{5\text{ k}\Omega} + \frac{1}{10\text{ k}\Omega} = 1.3\text{ mS}$$

[1] [EXP] [3] [1/x] [+] [5] [EXP] [3] [1/x] [+] [1] [0] [EXP] [3] [1/x]

2. Solve for total resistance using Eq. 5–3:

$$R_T = \frac{1}{G_T}$$

$$= \frac{1}{1.3\text{ mS}} = 769\ \Omega$$

FIGURE 5–22 Circuit for Example 5–16

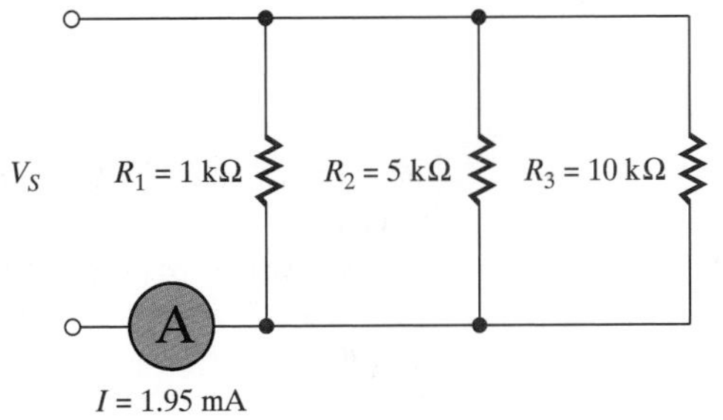

= 1/x

3. Find the current for each branch using the current divider equation:

$$I_1 = \frac{R_T}{R_1} \times I_T$$

$$= \frac{769\ \Omega}{1000\ \Omega} \times 1.95\ \text{mA} = 1.5\ \text{mA}$$

7 6 9 ÷ 1 EXP 3 × 1 . 9 5 EXP ± 3 =

$$I_2 = \frac{R_T}{R_2 I_T}$$

$$= \frac{769\ \Omega}{5000\ \Omega} \times 1.95\ \text{mA} = 0.3\ \text{mA}$$

7 6 9 ÷ 5 EXP 3 × 1 . 9 5 EXP ± 3 =

$$I_3 = \frac{R_T}{R_3} \times I_T$$

$$= \frac{769\ \Omega}{10000\ \Omega} \times 1.95\ \text{mA} = 0.15\ \text{mA}$$

7 6 9 ÷ 1 0 EXP 3 × 1 . 9 5 EXP ± 3 =

EXAMPLE 5–17 For the circuit of Figure 5–23, compute the current through each branch using the conductance current divider equation.

■ Solution:

1. Compute individual and total conductances:

$$G_1 = \frac{1}{R_1} = 100\ \text{mS}$$

$$G_2 = \frac{1}{R_2} = 66.7\ \text{mS}$$

$$G_3 = \frac{1}{R_3} = 10\ \text{mS}$$

$$G_T = G_1 + G_2 + G_3 = 176.7\ \text{mS}$$

2. Compute the branch currents:

$$I_1 = \frac{G_1}{G_T} \times I_T$$

$$= \frac{100\ \text{mS}}{176.7\ \text{mS}} \times 800\ \text{mA} = 452.7\ \text{mA}$$

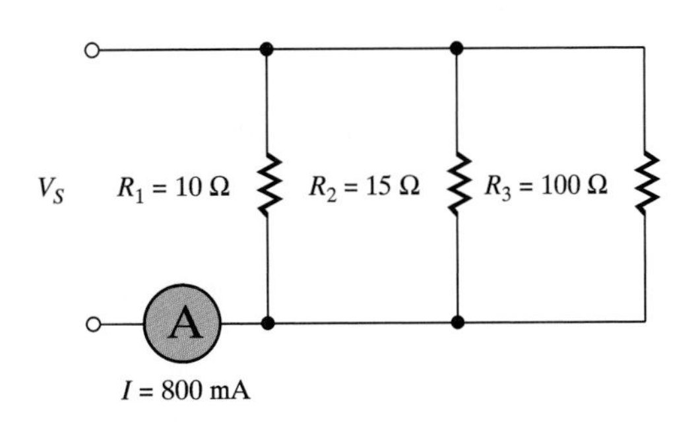

FIGURE 5–23 Circuit for Example 5–17

$$I_2 = \frac{G_2}{G_T} \times I_T$$

$$= \frac{66.7 \text{ mS}}{176.7 \text{ mS}} \times 800 \text{ mA} = 302 \text{ mA}$$

$$I_3 = \frac{G_3}{G_T} \times I_T$$

$$= \frac{10 \text{ mS}}{176.7 \text{ mS}} \times 800 \text{ mA} = 45.3 \text{ mA}$$

■ Voltage Sources in Parallel

In addition to a voltage rating, a source also has a current rating. This value is the maximum current that it can safely supply to a load. If this current is exceeded, the heat produced may be great enough to damage the source. If the voltages are the same, voltage sources can be placed in parallel and thus increase the available current.

Figure 5–24 shows three voltage sources connected in parallel. Each has a terminal voltage of 18 V and a current rating of 2 A. Because the sources are connected in parallel, the voltage is the same across each, and the available voltage is the same as that of one source. With each source capable of supplying 2 A, the three sources can supply a total of 6 A to a load.

If, as shown in Figure 5–25, the sources connected in parallel do not have the same voltages, the lower value source draws excessive current from the larger. The result is a lower voltage supplied to the load. This situation must be avoided, because lower voltage sources discharge higher voltage sources in parallel circuits.

FIGURE 5–24 Voltage sources connected in parallel

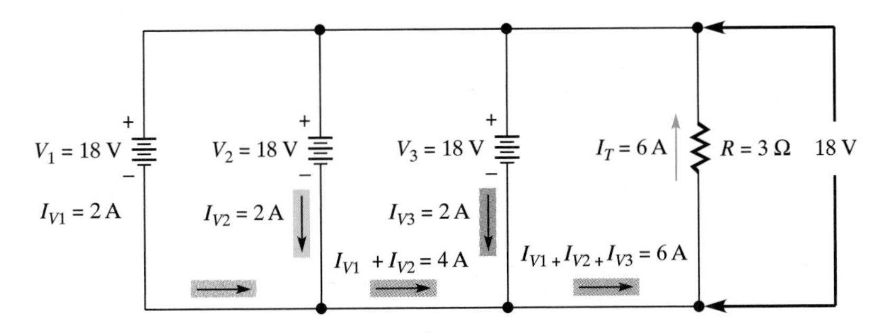

FIGURE 5–25 Results of paralleling unequal cells

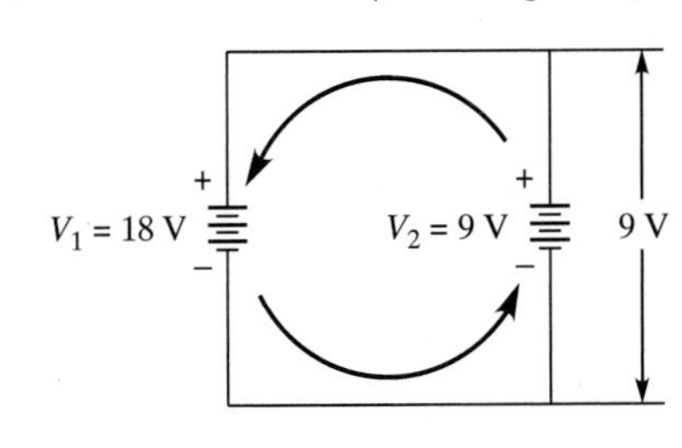

SECTION 5–6 REVIEW

1. A parallel circuit with 100 mA of current flowing has branches of 150 Ω, 390 Ω, and 1000 Ω. Using conductance current divider equation, compute the current through each resistor.
2. A parallel circuit has 150 mA of current flowing and branches of 1.5 kΩ, 3.9 kΩ, and 10 kΩ. Compute the current through each branch using the resistance current divider equation.
3. Four voltage sources, each having a current rating of 5A and a voltage of 48 V are connected in parallel. What value of total current can they safely supply to a load?
4. In Figure 5–26, if the two sources have the same voltage value, what current must each be capable of delivering to the load?

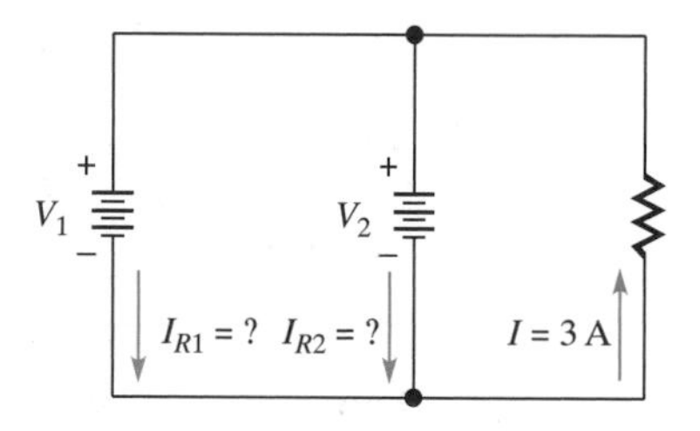

FIGURE 5–26 Circuit for Section Review Problem 4

5–7 APPLICATIONS OF PARALLEL CIRCUITS

Home Wiring

The most common application of parallel circuits is in home wiring. Most of the familiar items such as toasters, electric lamps, radios, televisions, and so forth require 120 V for their operation. So the requirement exists that they have the same voltage. A second requirement is that they must be capable of independent operation, even when all the others are de-energized. The requirements of independent load currents and identical source voltages are satisfied by parallel circuit operation.

In Figure 5–27, the main voltage source, supplied by the local electric utility company, is paralleled into many circuits within the home. One circuit, as shown, may be used for lighting. Note that each lamp is placed across the line and can operate independently. Another circuit may be used to supply power to wall outlets. The wall outlets are the "sockets" into which the power cords of radios, TVs, and electric clocks are inserted. Once again, all outlets have the same voltage across them and are capable of independent operation.

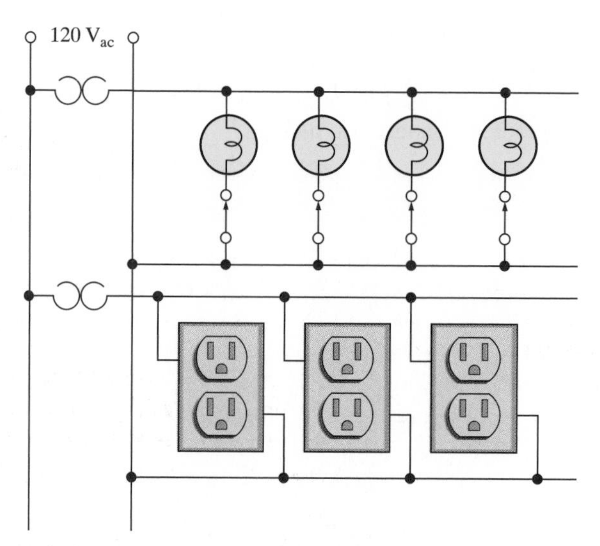

FIGURE 5–27 Home wiring example

FIGURE 5–28 Automotive lighting example

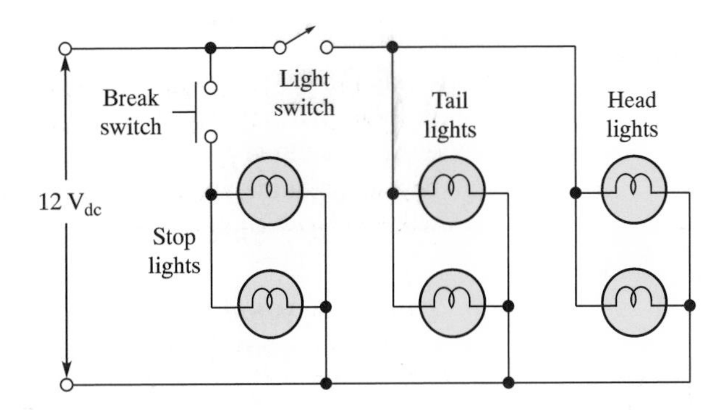

■ Automotive Lighting

Figure 5–28 shows the simplified wiring diagram of the lighting system in an automobile. Notice that the head lights and tail lights are wired in parallel. This arrangement is desirable for night driving. If the lights were wired in series and one light failed, all lights would fail. With the parallel connection, all lights operate independently of one another, so that if one fails, the others will remain operational.

■ Field Expedient

A *field expedient* is something done to keep equipment operating, for example, to avoid a stoppage in production. On the job, you will often be called upon to find ways of maintaining operation, even when some needed replacement part is not available. Suppose, for instance, that you must replace a 5 kΩ, 2 W resistor that has burned open. In checking, you find that the only 5 kΩ resistors available have a 1 W rating. It is not good practice to replace a resistor with one of lower wattage, as it may not be capable of developing the required power without burning open. In checking, you find several 10 kΩ, 1 W resistors available. A possible field expedient is to replace the single 5 kΩ, 2 W resistor with two 10 kΩ, 1 W resistors in parallel, as shown in Figure 5–29. The resistance value will be correct and the current will divide evenly, developing no more than 1 W of power. This field expedient makes use of the current dividing action of a parallel circuit.

FIGURE 5–29 Field expedient example

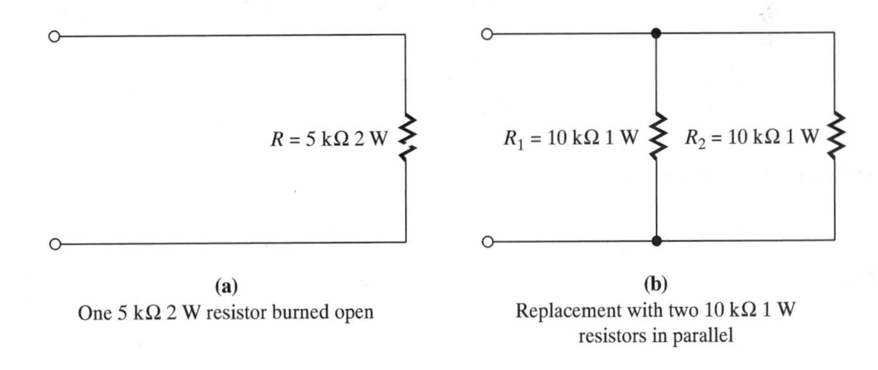

(a)
One 5 kΩ 2 W resistor burned open

(b)
Replacement with two 10 kΩ 1 W resistors in parallel

■ SECTION 5–7 REVIEW

1. Why are the lights of a car wired in parallel?
2. Why are electrical circuits within the home wired in parallel?
3. Why is it dangerous to plug multiple outlet receptacles into wall sockets within the home?

5–8 THE dc AMMETER

Similar to the presentation in Chapter 4, where multiplier resistors of a dc voltmeter were introduced as an application of series circuits, this chapter introduces the dc ammeter as an application of parallel circuits. The same meter movement introduced in Chapter 4 will be used with the ammeter. (Recall that the theory and operation of this meter movement will be discussed in Chapter 8.)

Question: What are the characteristics of the meter movement that are important in study of the dc ammeter?

Recall that a meter movement requires a certain amount of current in order to cause a full-scale indication, called, appropriately enough, the *full-scale current* I_{FS}. A greater current flowing through the movement will cause it to overrun the scale, or to "peg." Pegging must be avoided because it could cause damage to the meter. The current flowing through an internal meter resistance R_M produces a voltage drop V_M, as illustrated in Figure 5–30. For a quality meter movement, both the full-scale current and internal resistance must be kept as low as possible. (The reasons for this will become apparent later in this section.) As was the case previously, the meter movement is shown as this internal resistance with a pointer attached.

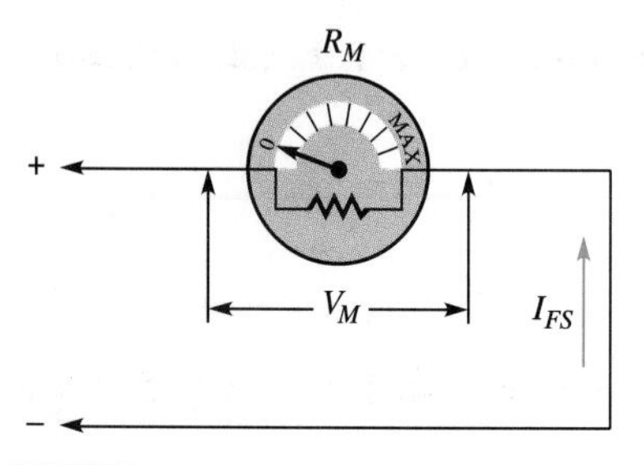

FIGURE 5–30 Current flow through internal meter resistance

By itself, the meter movement is rather useless as a current measuring device. As stated previously, the full-scale current must be kept small and cannot be exceeded, so measurement is limited to one rather small scale. It is necessary to extend the range to measure greater values of current.

Question: How is the range of an ammeter extended?

The range of an ammeter is extended by placing a resistor in parallel with the movement, as is illustrated in Figure 5–31. In this arrangement, the resistor is known as a **shunt,** abbreviated R_{shunt}, because some of the current is bypassed, or shunted, around the movement. In order to find the value required for the shunt, it is necessary to select a scale for the ammeter. Notice in Figure 5–31 that the full-scale current is 1 mA and the resistance is 100 Ω. By Ohm's law, the voltage drop across the meter at full-scale deflection is

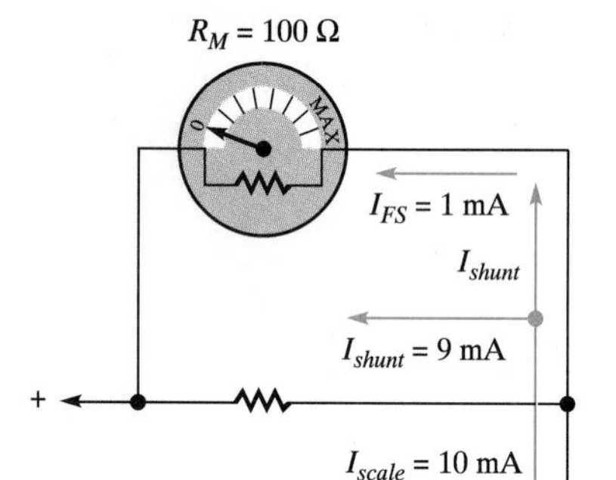

FIGURE 5–31 Circuit for demonstration of shunt resistance

$$V_M = I_{FS}R_M$$
$$= (1\text{ mA})(100\ \Omega) = 0.1\text{ V}$$

Suppose a full-scale of 10 mA is desired. Since only 1 mA can flow through the meter movement, by Kirchhoff's current law, 9 mA must be shunted around it. A general equation for the *shunt current,* is

(5–8)
$$I_{\text{shunt}} = I_{\text{scale}} - I_{FS}$$

where I_{scale} is the current is the ammeter's range. Since the shunt resistor is in parallel with the movement, it will have the same voltage (0.1 V) across it. Thus, the value of the *shunt resistor* can be determined using Ohm's law:

(5–9)
$$R_{\text{shunt}} = \frac{V_{\text{shunt}}}{I_{\text{shunt}}}$$
$$= \frac{0.1\text{ V}}{9\text{ mA}} = 11.1\ \Omega$$

Note that since R_{shunt} is in parallel with R_M, $V_M = V_{\text{shunt}}$.

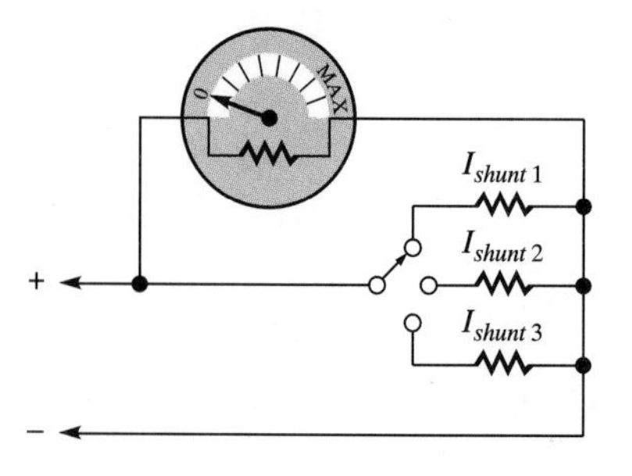

FIGURE 5–32 Ampere ranges selectable by rotary switch

As shown in Figure 5–32, a multiple range ammeter can be constructed using a rotary switch. Each position of the switch selects a different value of shunt resistor and thus a different scale. A rotary switch of the "make-before-break" variety would be used in this application. In this type of switch, contact is made with the next shunt resistor before contact with the first is broken. Otherwise, there might be an instant in switching in which there is no shunt in place, and the pointer would move wildly. The following example shows the process of designing such a meter.

EXAMPLE 5–18 A meter movement has a full-scale current of 50 μA and a resistance of 1000 Ω. Compute the value of shunt resistors for the following scales: (a) 1 mA, (b) 10 mA, (c) 100 mA, and (d) 500 mA. Draw the schematic of the completed circuit.

■ **Solution:**

First it is necessary to compute the voltage drop across the meter at full-scale deflection:

$$V_M = I_{FS}R_M$$

$$= (50\ \mu\text{A})(1000\ \Omega) = 50\ \text{mV}$$

[5] [0] [EXP] [±] [6] [×] [1] [EXP] [3] [=]

Now the value of the shunt resistor for each scale can be found using Eq. 5–8 and Eq. 5–9:

a. 1 mA scale

$$I_{\text{shunt}} = I_{\text{scale}} - I_{FS}$$

$$= 1\ \text{mA} - 50\ \mu\text{A} = 0.95\ \text{mA}$$

[1] [EXP] [±] [3] [−] [5] [0] [EXP] [±] [6] [=]

$$R_{\text{shunt}} = \frac{V_M}{I_{\text{shunt}}}$$

$$= \frac{50\ \text{mV}}{0.95\ \text{mA}} = 52.6\ \Omega$$

[5] [0] [EXP] [±] [3] [÷] [·] [9] [5] [EXP] [±] [3] [=]

b. 10 mA scale:

$$I_{\text{shunt}} = I_{\text{scale}} - I_{FS}$$

$$= 10\ \text{mA} - 50\ \mu\text{A} = 9.95\ \text{mA}$$

[1] [0] [EXP] [±] [3] [−] [5] [0] [EXP] [±] [6] [=]

$$R_{\text{shunt}} = \frac{V_M}{I_{\text{shunt}}}$$

$$= \frac{50\ \text{mV}}{9.95\ \text{mA}} = 5.03\ \Omega$$

[5] [0] [EXP] [±] [3] [÷] [9] [·] [9] [5] [EXP] [±] [3] [=]

c. 100 mA scale:

$$I_{\text{shunt}} = I_{\text{scale}} - I_{FS}$$
$$= 100 \text{ mA} - 50\ \mu\text{A} = 99.95 \text{ mA}$$

[1] [0] [0] [EXP] [±] [3] [−] [5] [0] [EXP] [±] [6] [=]

$$R_{\text{shunt}} = \frac{V_M}{I_{\text{shunt}}}$$
$$= \frac{50 \text{ mV}}{99.95 \text{ mA}} = 0.5\ \Omega$$

[5] [0] [EXP] [±] [3] [÷] [9] [9] [·] [9] [5] [EXP] [±] [3] [=]

d. 500 mA scale:

$$I_{\text{shunt}} = I_{\text{scale}} - I_{FS}$$
$$= 500 \text{ mA} - 50\ \mu\text{A} = 499.95 \text{ mA}$$

[5] [0] [0] [EXP] [±] [3] [−] [5] [0] [EXP] [±] [6] [=]

$$R_{\text{shunt}} = \frac{V_M}{I_{\text{shunt}}}$$
$$= \frac{50 \text{ mV}}{499.95 \text{ mA}} = 0.1\ \Omega$$

[5] [0] [EXP] [±] [3] [÷] [4] [9] [9] [·] [9] [5] [EXP] [±] [3] [=]

The schematic is shown in Figure 5–33.

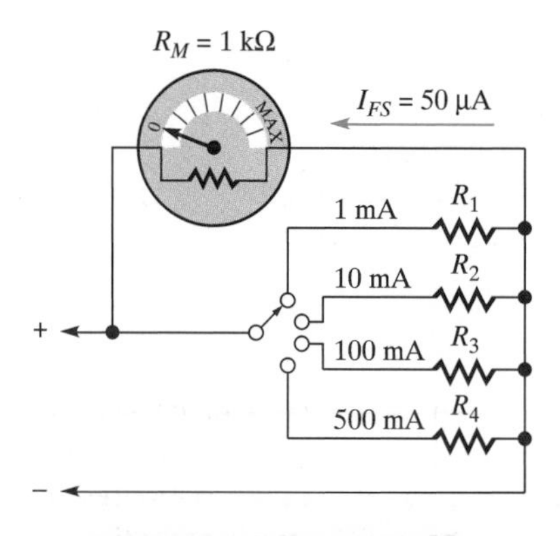

FIGURE 5–33 Circuit for Example 5–18

$R_1 = 52.6\ \Omega$	$R_3 = 0.5\ \Omega$
$R_2 = 5.03\ \Omega$	$R_4 = 0.1\ \Omega$

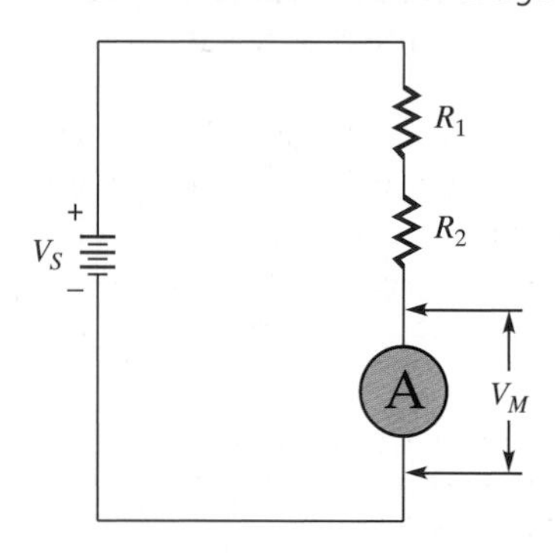

FIGURE 5–34 Meter voltage subtracted available source voltage

When the ammeter is placed in series with the circuit under test, its voltage drop must be *subtracted* from the available source voltage, as shown in Figure 5–34. Whether or not this drop is significant depends upon the values of the circuit components and the applied voltage. In a portion of the circuit where the voltage is relatively high, a 50 mV drop would be insignificant. In a portion of the circuit where the voltage is low, however, the voltage drop could result in false readings. Thus, a quality meter movement has low values of both resistance and full-scale current that result in a small voltage drop across the meter.

■ Ammeter Accuracy

The accuracy of an ammeter is considered in a like manner to that of the voltmeter. If the accuracy of the ammeter is specified as ±5%, then the reading taken may be in error by this percentage of the full-scale value. For example, if the reading is being taken on the 10 mA range, the error could be 0.5 mA (5% of 10 mA = 0.5 mA). Thus, if the meter indicates 4 mA, the actual circuit current is somewhere between 3.5 mA and 4.5 mA. This range represents a much greater possibility of error than 5%. Thus, it is a good policy to take readings as close to the upper end of the scale as possible.

EXAMPLE 5–19 An ammeter has an accuracy of ±2%. If a reading of 25 mA is taken in the 100 mA scale, what is the possible error?

The ±2% accuracy is of the full-scale value, so the error could be as much as 0.02 × 100 mA or 2 mA. Thus, the actual circuit current could be as much as 27 mA or as little as 23 mA.

SECTION 5–8 REVIEW

1. What is the name given the resistors used to extend the range of an ammeter?
2. These resistors are in (series, parallel) with the meter movement.
3. A meter movement has a full-scale current of 1 mA and a resistance of 200 Ω. Compute the value of the resistor required for a current range of 50 mA.

5–9 TROUBLESHOOTING PARALLEL CIRCUITS

The principles of troubleshooting that were applied to series circuits can be applied equally well to parallel circuits. The trouble must be determined, the problem corrected, and the circuit tested for correct operation. Although the parallel circuits considered in this section are quite simple, the principles involved will be used in troubleshooting of the more complex circuits to come.

A parallel circuit failure results in a current that is too high, too low, or zero. As with series circuits the first step in troubleshooting is to check the voltage source. Sometimes the most obvious things are overlooked: a blown fuse, a tripped circuit breaker, or a disconnected power cord. Check all possibilities. No circuit will operate properly if the value of the voltage source is not within design limits. If the voltage source is correct, another cause for the trouble must be found.

Zero current is the result of an "open" in the circuit. The zero current in the circuit of Figure 5–35 indicates the open is between the source and the first resistor. If the open were after the first resistor in the bank, then some current would flow through R_1, and zero volts would be measured across R_2 and R_3. The only other cause of zero current flow would be *all* loads being open.

High current is the result of resistance that is too low. The resistance of a parallel circuit decreases as paths for current are added. If any component in a parallel circuit "shorts," then, a short is across *all* components, as is illustrated in Figure 5–36. Under this circumstance, the current becomes extremely high, and the circuit protection device,

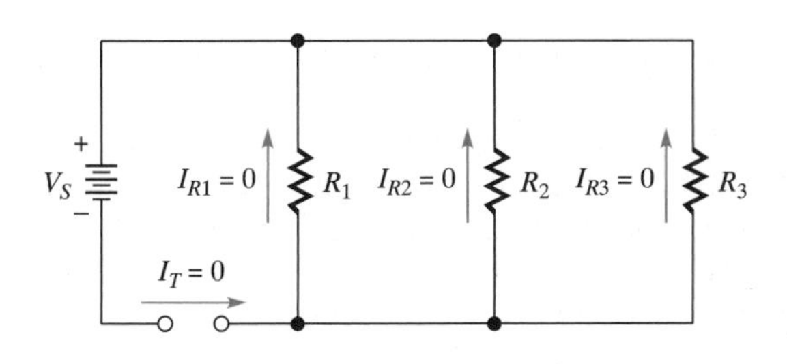

FIGURE 5–35 Zero current. When an "open" occurs prior to the first branch, the result is zero current flow.

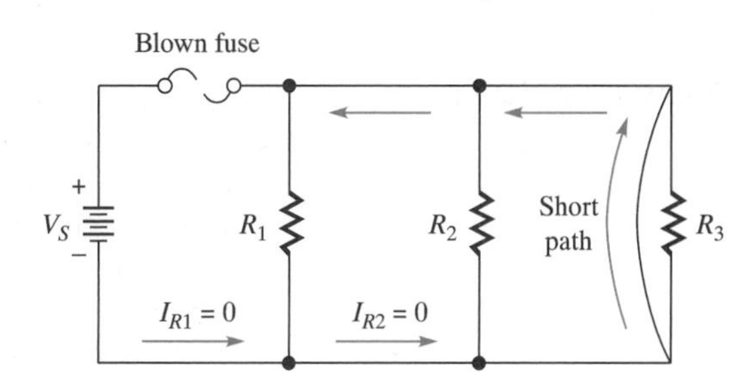

FIGURE 5–36 High current due to a shorted branch. Lowered branch resistance causes extremely high circuit current that can cause the fuse to open.

circuit breaker or fuse, opens to protect the source from damage. As mentioned in Chapter 4, shorted resistors are uncommon. (Solid state control devices such as transistors or diodes are more likely to short.) Thus, a cause of higher current than normal could be a resistor that has significantly decreased in value. In this case, one branch conducts much greater current than normal, thereby increasing the total current, since the lower resistance results in a higher conductance and a higher current. As illustrated in Figure 5–37, it is necessary to measure the branch currents to determine which resistor has changed value. The resistor can then be replaced in order to correct the trouble.

Low current is likely to be caused by a resistor that has opened for some reason. Recall that the source voltage is measured across an open resistor. This measurement is useless, however, in a parallel circuit where the source voltage is across *all* resistors. It is necessary, therefore, to make current measurements to locate the fault. In Figure 5–38, where one of the resistors has opened, the branch is drawing no current, and by Kirchhoff's current law, the total current is lower than normal. The computed total current is 65 mA. The measured total current is 15 mA. The difference between the two is 50 mA, which is the current that should flow through the 1 kΩ resistor. Thus, it can be concluded that the most probable cause of the fault is that the 1 kΩ resistor is open.

When measuring the resistance of a branch, it is necessary to disconnect one end of the resistor, as shown in Figure 5–39(a). Figure 5–39(b) shows the problem that will develop if the resistor is not disconnected. The parallel combination of the other branches gives a false reading of about 3.3 kΩ on the ohmmeter. This point is a good one to remember later on when you are working with actual systems and checking resistance. If the reading is not that expected, look for parallel paths that may give a false reading!

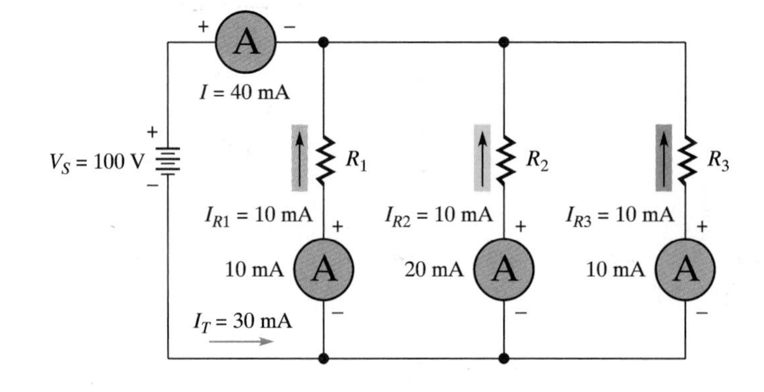

FIGURE 5–37 High current due to decreased resistance in one branch. Arrows indicate computed values and meters indicate measured values. Total current measures higher than normal, and current measurement in R_2 is high, indicating that it has decreased in value.

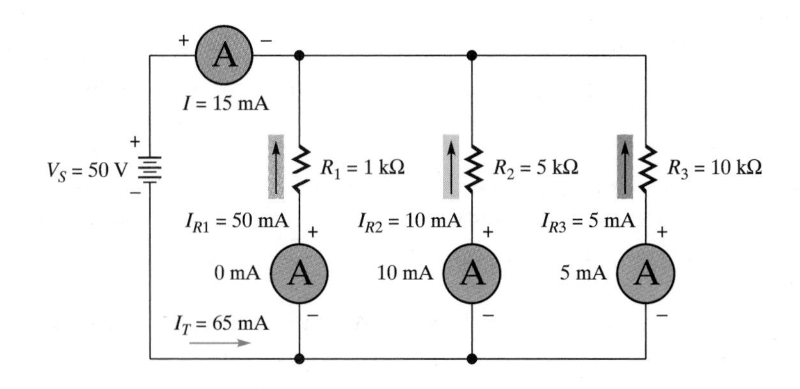

FIGURE 5–38 Low current due to opened resistor. Arrows indicate computed values and meters indicate measured values. The zero reading in branch two indicates that it is open.

FIGURE 5–39 Measuring branch resistance

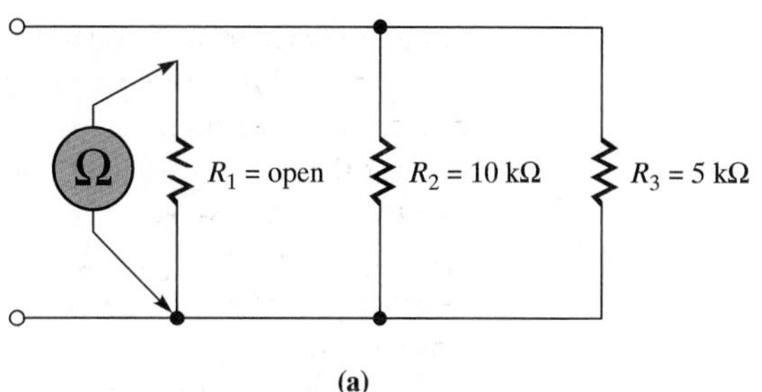

(a)
Correct measurement obtained by raising one end of R_1

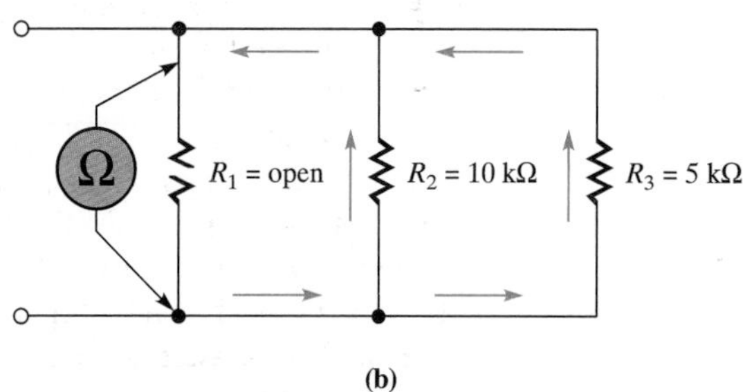

(b)
False reading obtained if R_1 remains in parallel with R_2 and R_3

SECTION 5–9 REVIEW

1. In a parallel circuit, what are two possible causes of the current being too high?
2. In a parallel circuit, what are two possible causes of the current being too low?
3. What precaution must be taken when measuring the resistance of a resistor that is paralleled by other resistors?
4. In a parallel circuit, why is it impossible to detect an open resistor with a voltage measurement?

CHAPTER SUMMARY

1. Components connected side-by-side across a common voltage source are connected in parallel.
2. The same voltage is impressed across all branches of a parallel circuit.
3. A parallel circuit is a current divider.
4. The total current divides among the branches in inverse proportion to their resistance value and in direct proportion to their conductance value.
5. The total conductance of a parallel circuit is equal to the sum of the individual branch conductances.
6. The total resistance of a parallel circuit is equal to the reciprocal of the total conductance.
7. The total current in a parallel circuit is equal to the sum of the branch currents.
8. Total power in a parallel circuit is equal to the sum of the individual powers developed by each resistor.
9. Adding another branch to a parallel circuit decreases the total resistance and increases the total current.
10. Removing a branch from a parallel circuit increases the total resistance and decreases the total current.
11. Sources of the same voltage may be paralleled in order to increase the current supplying capability of the source.
12. Parallel circuits find applications with loads that require the same voltage, draw different currents, and require individual control.
13. A shorted branch is one in which the resistance is zero.
14. An open branch is one whose resistance is infinite.

TROUBLESHOOTING PRACTICE

In Figure 5–40, a single fault exists. There is either a short between two points or an open at some point. Your task is to identify which of these faults exists. The first step is to determine the values of current and voltage that the meters should indicate. On the job, you would make the necessary measurements one at a time. For this practice, after deciding on a measurement, you can determine its value from the table given in Figure 5–40. Then use the measurements, Ohm's law, Kirchhoff's current law, and the facts of parallel circuits to determine what fault exists. Try to determine the trouble in as few steps as possible. Rate yourself as follows:

Supertech: 2 or less steps
Technician: 3 steps
Apprentice: 4 steps
Need review: 5 or more steps

FIGURE 5–40

Measurement	Value	Measurement	Value
I_T	13.18 mA	V_1	10 V
I_1	10 mA	V_2	10 V
I_2	0	V_3	10 V
I_3	1.96 mA	V_4	10 V
I_4	1.219 mA		
I_{2+3+4}	3.179 mA		
I_{3+4}	3.179 mA		

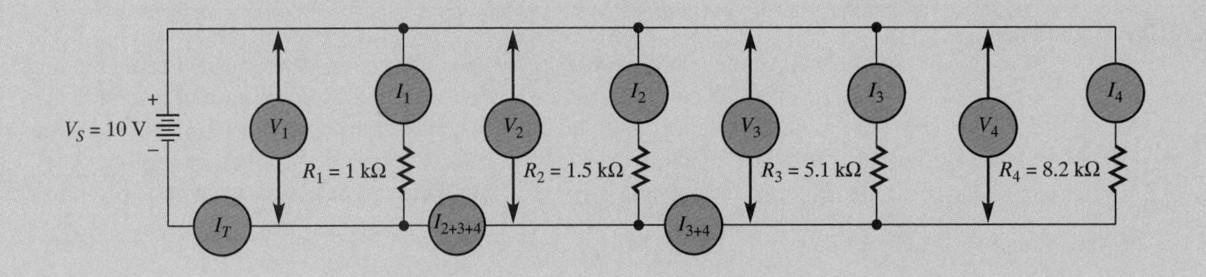

■ KEY TERMS

resistor bank
parallel circuit
branch
branch current
total current
Kirchhoff's current law
total conductance
total power
shunt

■ EQUATIONS

(5–1) $$I_T = I_1 + I_2 + I_3 + \cdots I_n$$

(5–2) $$G_T = G_1 + G_2 + G_3 + \cdots G_n$$

(5–3) $$R_T = \frac{1}{G_T}$$

(5–4) $$R_T = \frac{1}{(1/R_1) + (1/R_2) + (1/R_3) + \cdots (1/R_n)}$$

(5–5) $$P_T = P_{R1} + P_R + P_{R3} + \cdots P_{Rn}$$

(5–6) $$I_x = \frac{G_x}{G_T} \times I_T.$$

(5–7) $$I_x = \frac{R_T}{R_x} \times I_T$$

(5–8) $$I_{\text{shunt}} = I_{\text{scale}} - I_{FS}$$

(5–9) $$R_{\text{shunt}} = \frac{V_{\text{shunt}}}{I_{\text{shunt}}}$$

Variable Quantities

n = any number

x = unknown value

I_T = total circuit current

R_T = total circuit resistance

G_T = total circuit conductance

P_T = total circuit power

I_M = current of meter movement

I_{shunt} = current through the parallel (shunt) resistor

I_{scale} = current through an ammeter's scale range

R_{shunt} = resistance of the parallel (shunt) resistor

TEST YOUR KNOWLEDGE

1. In your own words, explain why the same voltage is across all branches of a parallel circuit.
2. In your own words, explain why a parallel circuit is known as a current divider.
3. Explain why branch currents are inversely proportional to the resistance of the branch.
4. Explain why conductance is used in the computation of total resistance.
5. What effect will a shorted branch have on the operation of a parallel circuit?
6. What effect will an open branch have on the operation of a parallel circuit?
7. Explain in detail the rationale behind the current divider equation.
8. List three applications of parallel circuits.
9. What is the first check that must be made in troubleshooting?
10. If a branch is added to a parallel circuit, what happens to the total resistance and total current?
11. If a branch is deleted from a parallel circuit, what happens to the total resistance and total current?
12. If one branch of a parallel circuit opens, what happens to the other branch currents?
13. Discuss all the differences that exist between series and parallel circuits.
14. Why can an open in a parallel branch component not be detected with a voltage measurement?

PROBLEM SET: BASIC LEVEL

Section 5–1

1. Without changing their physical position, connect the resistors of Figure 5–41 in parallel.

FIGURE 5–41 Resistors for Problem 1

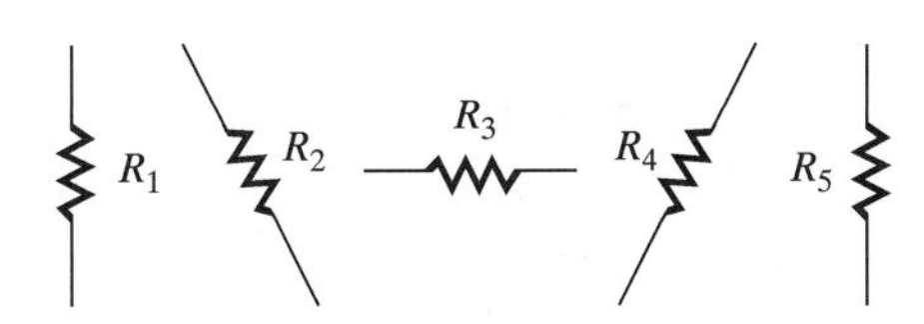

2. Move the power source of Figure 5–42 to three different positions within the branches without changing the circuit's electrical characteristics.

FIGURE 5–42 Circuit for Problem 2

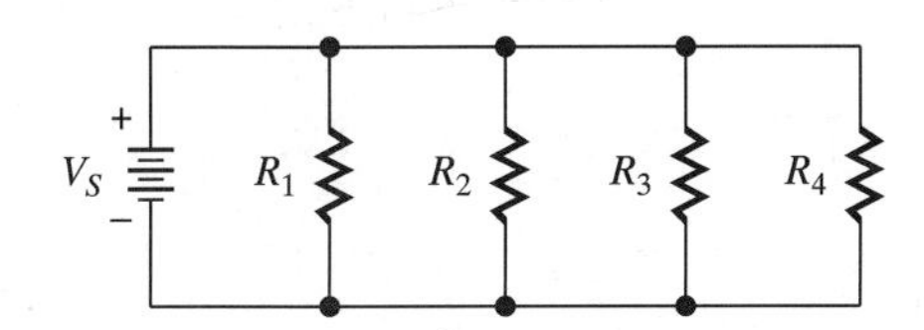

Section 5–2

3. In the circuit of Figure 5–43, what is the voltage across R_1? Across R_3?

FIGURE 5–43 Circuit for Problem 3

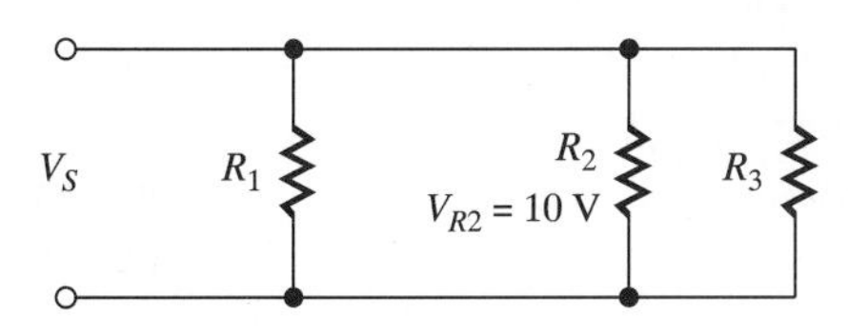

4. For the circuit of Figure 5–44, what is the voltage across R_2?

FIGURE 5–44 Circuit for Problem 4

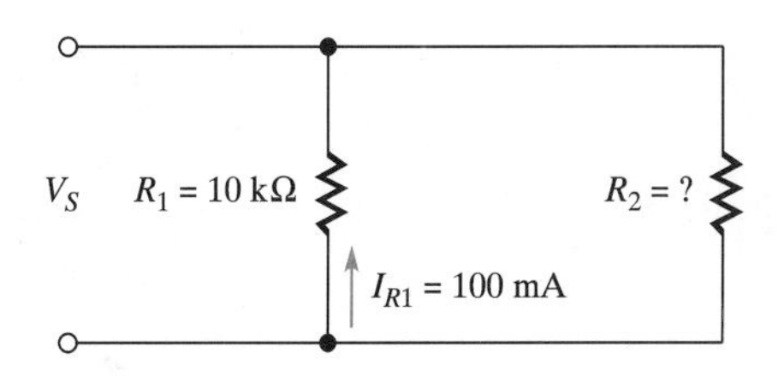

5. In the circuit of Figure 5–45, the voltage across R_2 is measured and found to be 12 V. What is the current through R_1?

FIGURE 5–45 Circuit for Problem 5

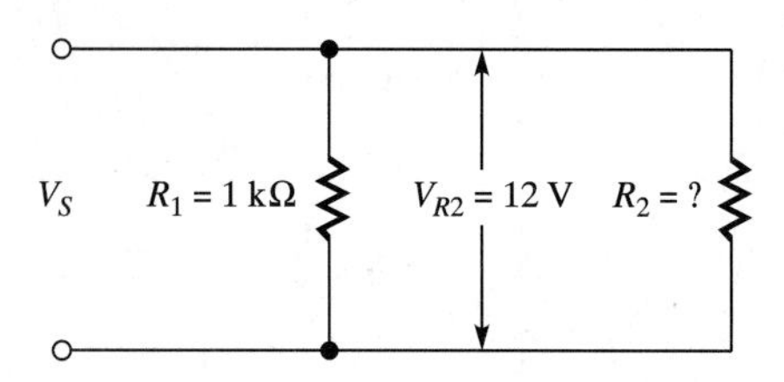

Section 5–3

6. Five resistors of equal value are connected in parallel. If the total current is 10 mA, what is the current through each branch?

7. Compute the total current for the circuit of Figure 5–46.

FIGURE 5–46 Circuit for Problem 7

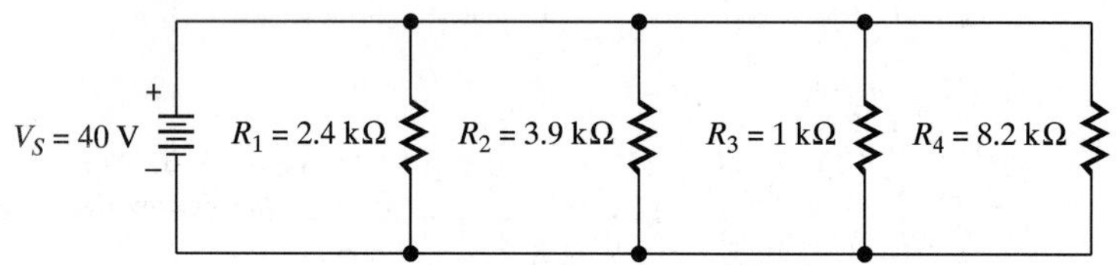

8. For the circuit of Figure 5–47, what is the current through R_3?

FIGURE 5–47 Circuit for Problem 8

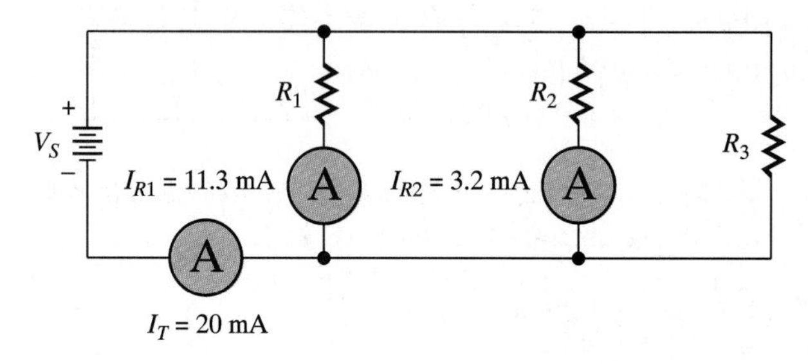

9. For the circuit of Figure 5–48, what is the current through R_1?

FIGURE 5–48 Circuit for Problem 9

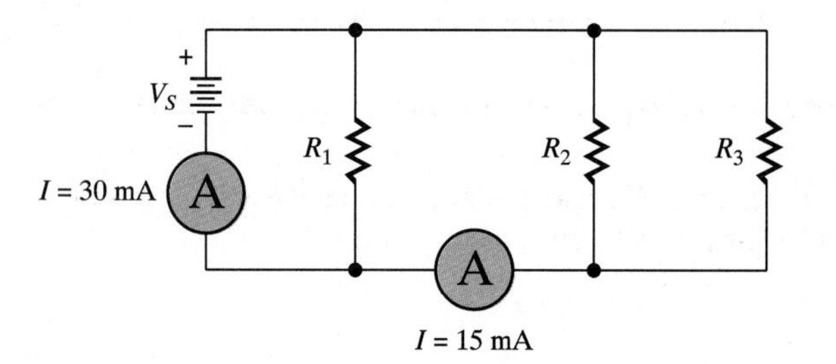

Section 5–4

10. For each of the circuits of Figure 5–49, compute the total conductance.

FIGURE 5–49 Circuits for Problems 10 and 11

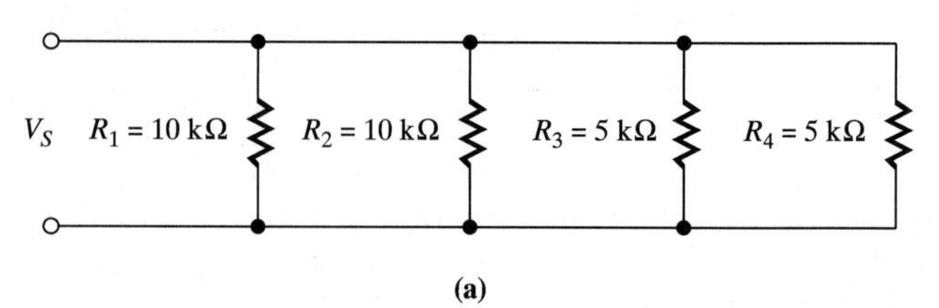

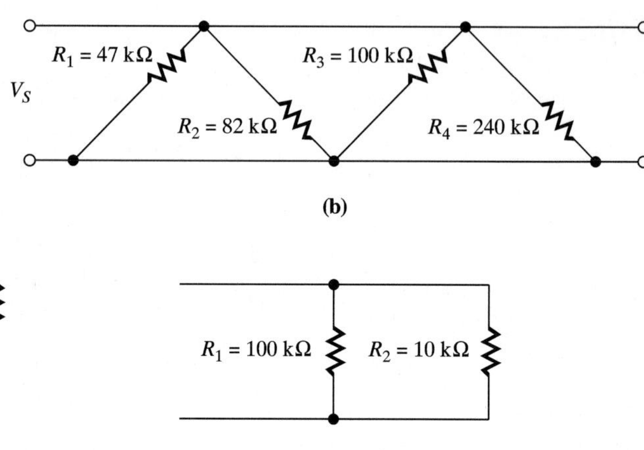

11. In Problem 10, if all resistance values are doubled, what will be the total conductance?

12. In the circuit of Figure 5–50, compute the conductance of branch 4.

FIGURE 5–50 Circuit for Problem 12

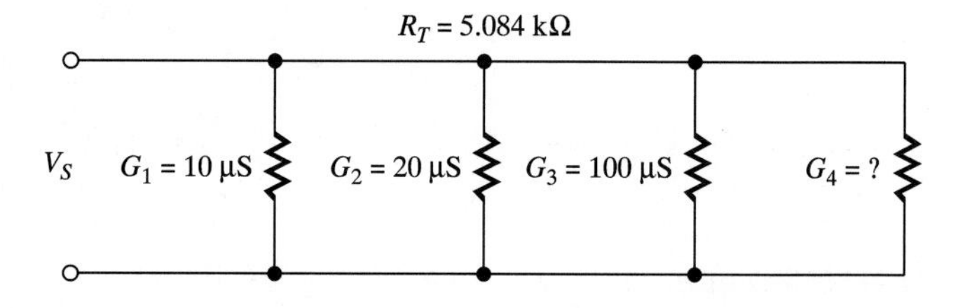

Section 5–5

13. Compute the total resistance of each circuit in Figure 5–51.

14. Five 20 kΩ resistors are connected in parallel. Compute the total resistance.

15. A two-resistor parallel combination circuit consists of a 10 kΩ and a 1 kΩ resistor. Without resorting to the use of pencil, paper, or calculator, what is the approximate total resistance of this combination?

FIGURE 5–51 Circuits for Problem 13

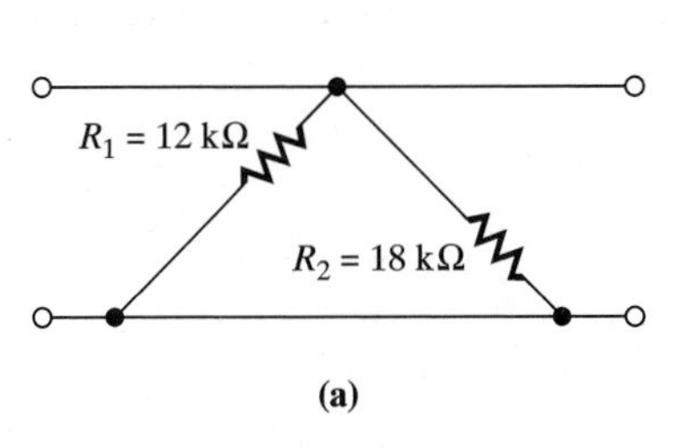

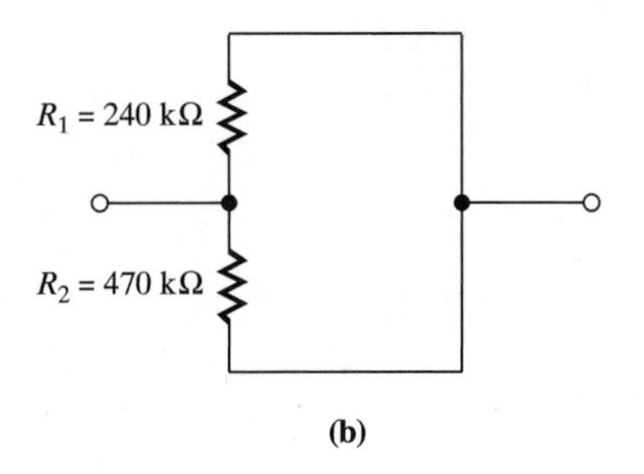

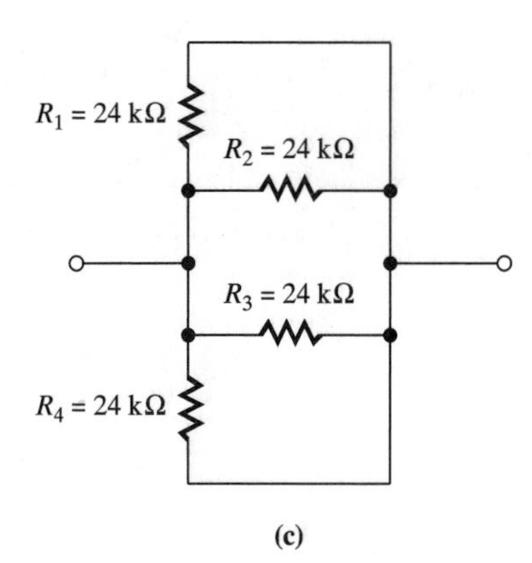

16. A parallel circuit consists of resistances as shown in Figure 5–52. Without resorting to pencil, paper, or calculator, compute the value of the total resistance.

FIGURE 5–52 Circuit for Problem 16

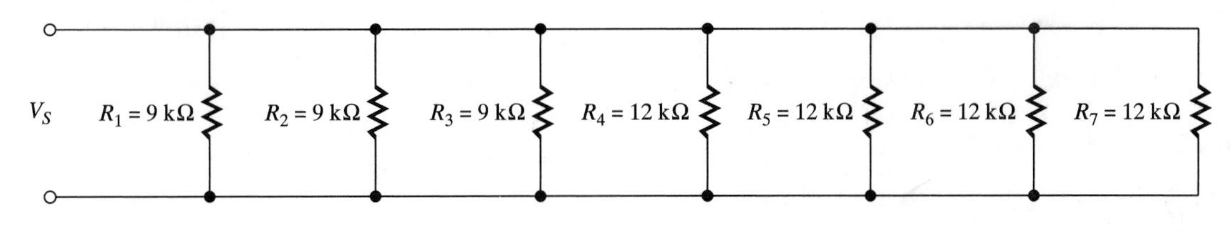

17. A two-resistor parallel combination has a total resistance of 3 kΩ. If the value of one resistor is 12 kΩ, what is the value of the other?

Section 5–6

18. Compute the power developed in each resistor and the total power developed in the circuit of Figure 5–53.

FIGURE 5–53 Circuit for Problem 18

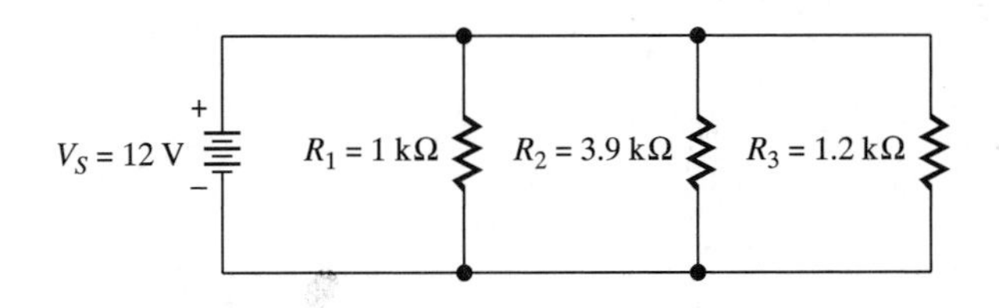

19. What is the total power developed in the circuit of Figure 5–54?

FIGURE 5–54 Circuit for Problem 19

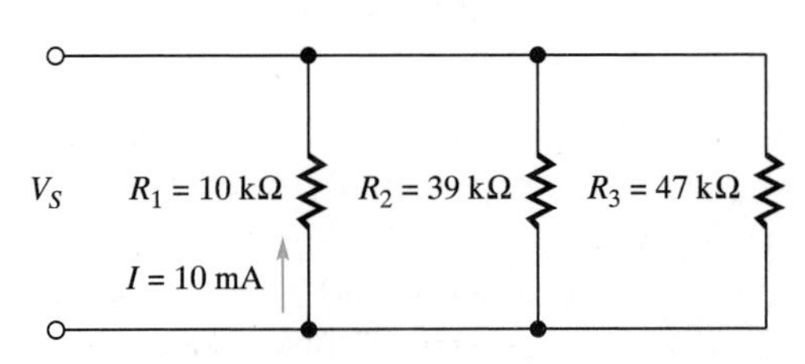

Section 5–7

20. In the circuit of Figure 5–55, compute the value of the current through each resistor, the total current, the power developed in each resistor, and the total power.

FIGURE 5–55 Circuit for Problem 20

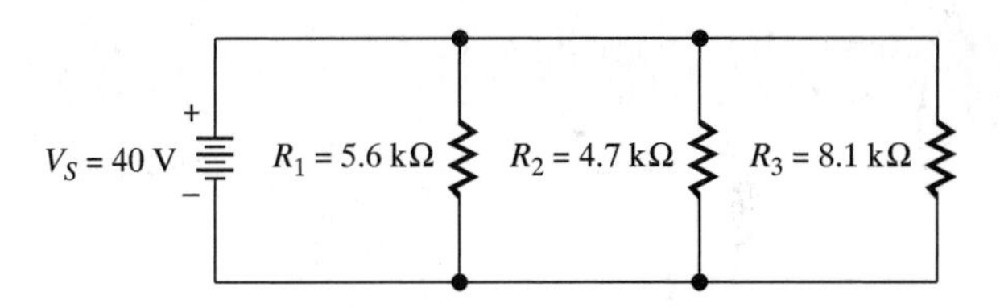

21. The total current in the circuit of Figure 5–56 is 4.8 mA. Compute the value of the source voltage.

FIGURE 5–56 Circuit for Problem 21

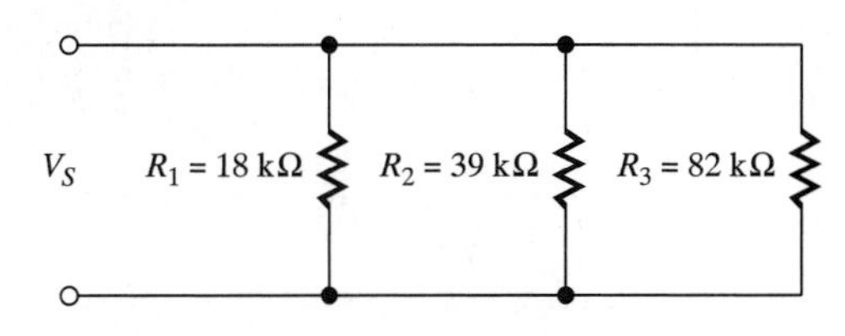

22. What is the voltage across each branch of the circuit in Figure 5–57?

23. In Figure 5–57, compute the currents through R_1 and R_2. Compute the total current.

FIGURE 5–57 Circuit for Problems 22 and 23

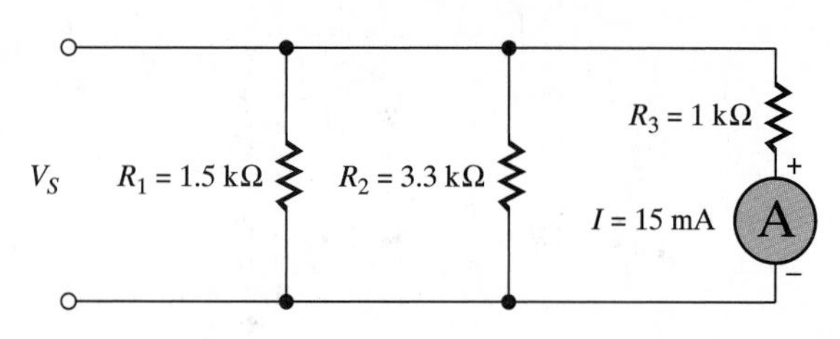

24. For the circuit of Figure 5–58, compute the current through each branch.

FIGURE 5–58 Circuit for Problem 24

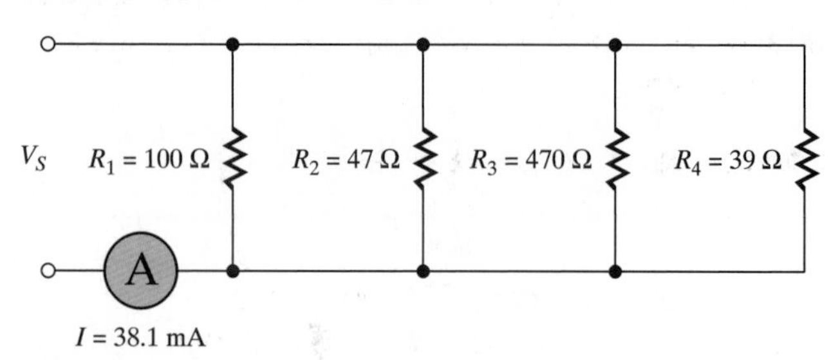

25. Four identical resistors are connected in parallel across a 12 V source. Compute the value of each if the total current is 20 mA.

26. If R_3 were to open in the circuit of Figure 5–59, compute the current through the other two resistors and the total current.

FIGURE 5–59 Circuit for Problem 26

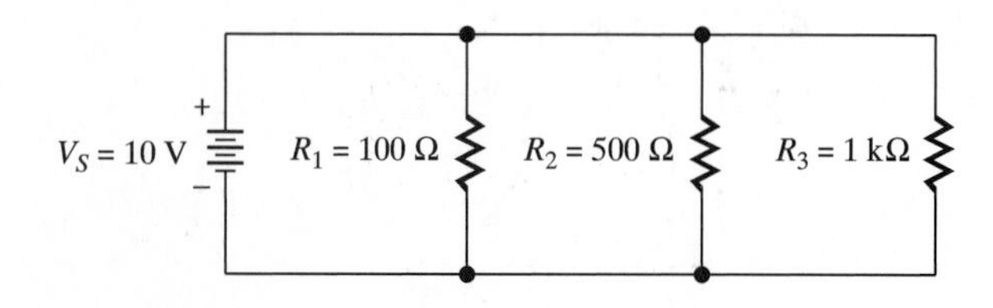

27. The resistors of Figure 5–60 have a $\pm 5\%$ tolerance. Taking into account the resistor tolerances, what would be the maximum current expected in this circuit?

FIGURE 5–60 Circuit for Problem 27

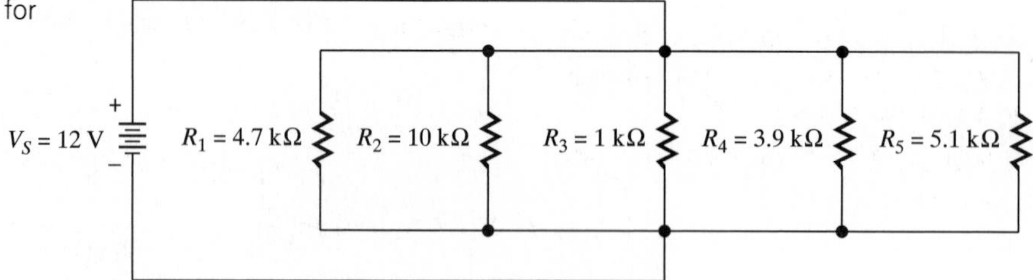

Section 5–8

28. In the circuit of Figure 5–61, what percentage of the total current will flow through each branch?

29. In Figure 5–61, the total current is 35 mA. Use the current divider equation to compute the value of the current in each branch.

30. A fourth branch of 1 kΩ is added to the circuit in Figure 5–61. Use the values from Problem 28 to compute the current through each branch and the total current.

FIGURE 5–61 Circuit for Problems 28, 29, and 30

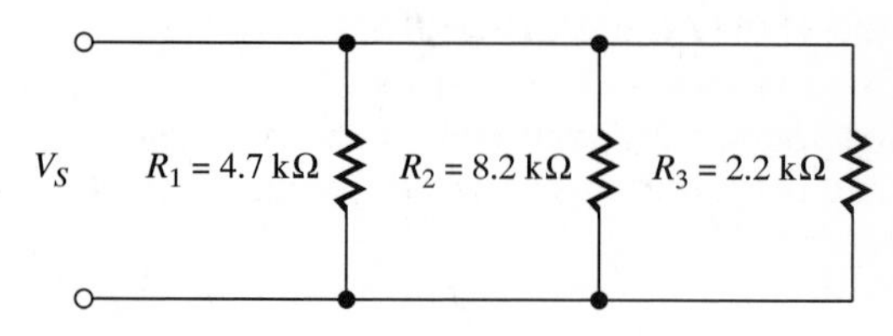

Section 5–9

31. For the ammeter circuit of Figure 5–62, compute the value of shunt resistor needed for each current range.

FIGURE 5–62 Circuit for Problem 31

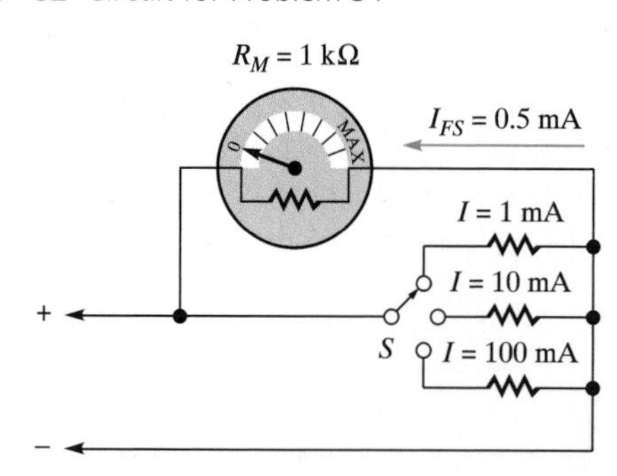

32. If the ammeter of Problem 31 is placed in the circuit of Figure 5–63 as shown, what will be its loading effect?

33. Compute the value of current you would expect the ammeter of Problem 31 to indicate. If the accuracy of the ammeter is ±5%, what are the maximum and minimum current values that may be indicated?

FIGURE 5–63 Circuit for Problems 32 and 33

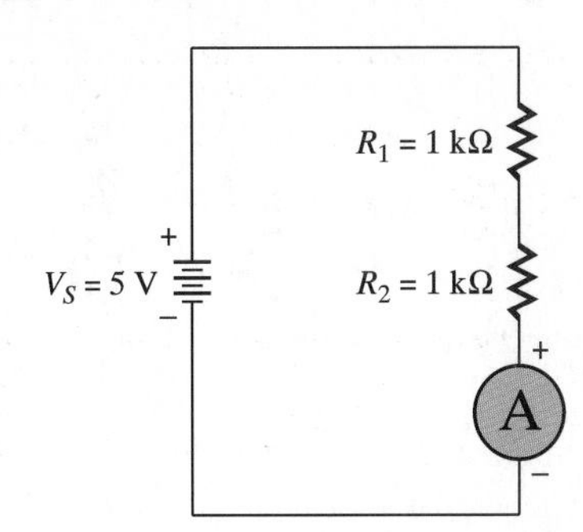

PROBLEM SET: CHALLENGING

34. In the circuit of Figure 5–64, what value resistor must be added in parallel with R_1 in order to make the total current equal 8 mA?

FIGURE 5–64 Circuit for Problem 34

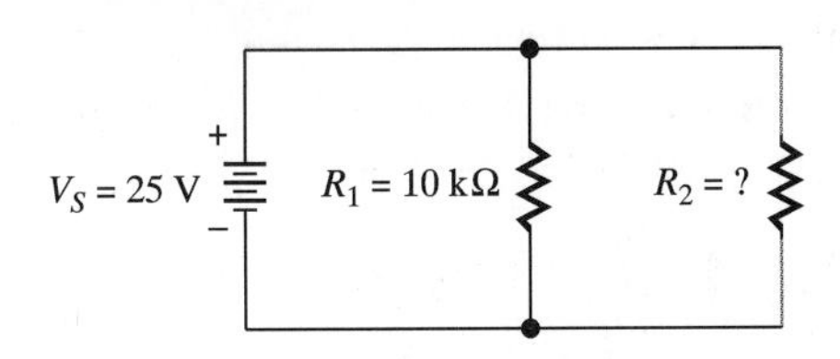

35. For the circuit of Figure 5–65, compute the following: (a) the resistance of each branch, (b) the current in each branch, (c) the power dissipated in each branch, and (d) the total power developed in the circuit.

FIGURE 5–65 Circuit for Problem 35

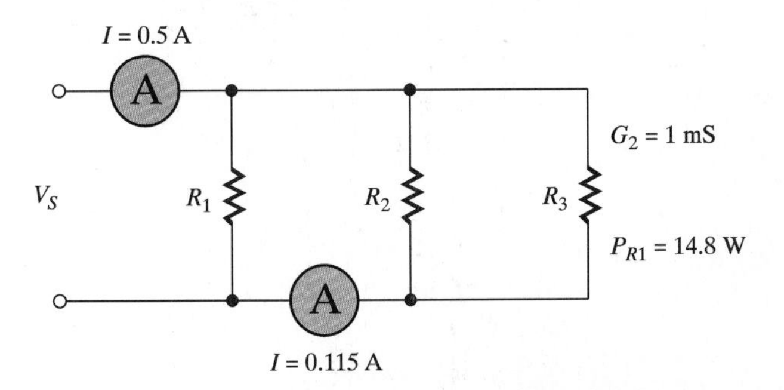

36. The total current in a parallel circuit is 200 mA and the total resistance is 500 Ω. What value resistance must be added in parallel in order to increase the current by 50%?

37. In Problem 36, by what percentage would the total power increase?

38. For the circuit of Figure 5–66, compute the values of R_1, R_2, and V_S.

FIGURE 5–66 Circuit for Problem 38

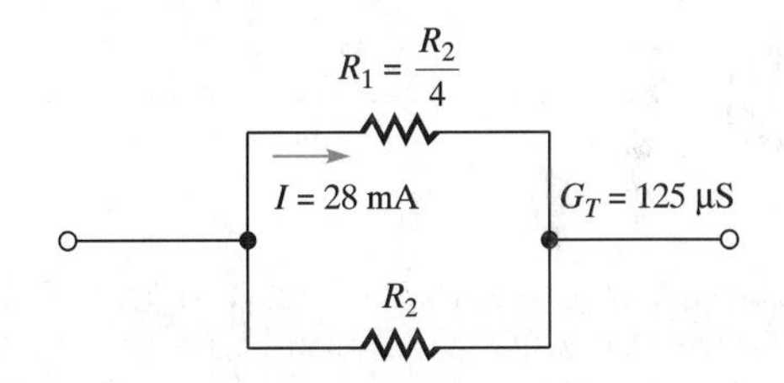

39. In Figure 5–67, compute the values of R_1 and R_2.

FIGURE 5–67 Circuit for Problem 39

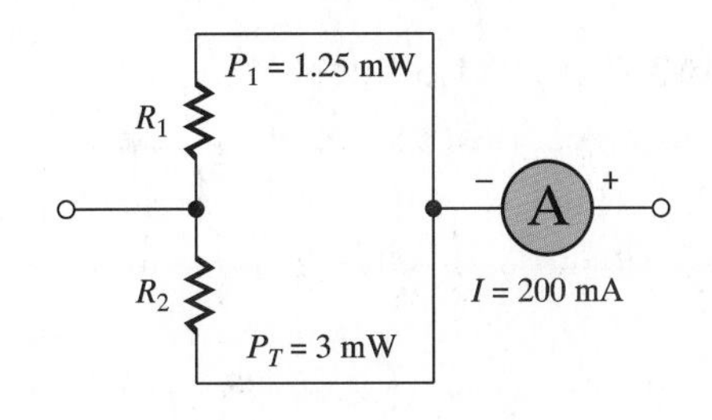

40. In Figure 5–68, compute the current at points A through E.

FIGURE 5–68 Circuit for Problem 40

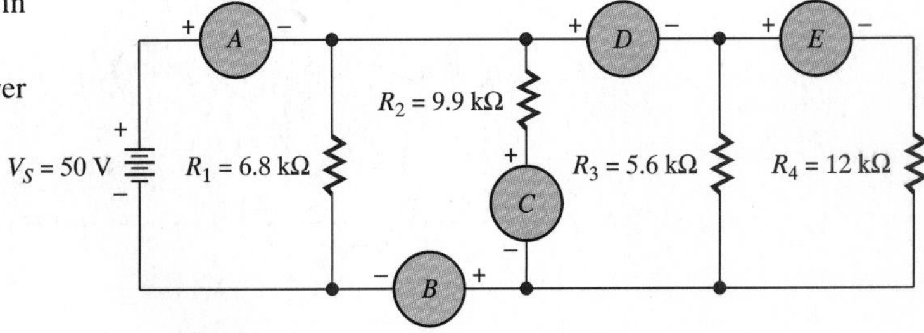

41. Five light bulbs are connected in parallel across a 120 V source. If each is rated at 60 W compute: (a) the resistance of each bulb, (b) the current flowing through each bulb, (c) the total current, and (d) the total power developed.

42. In the circuit of Figure 5–69, compute: (a) I_1, (b) I_2, (c) I_3, (d) R_4, and (e) R_5.

FIGURE 5–69 Circuit for Problem 42

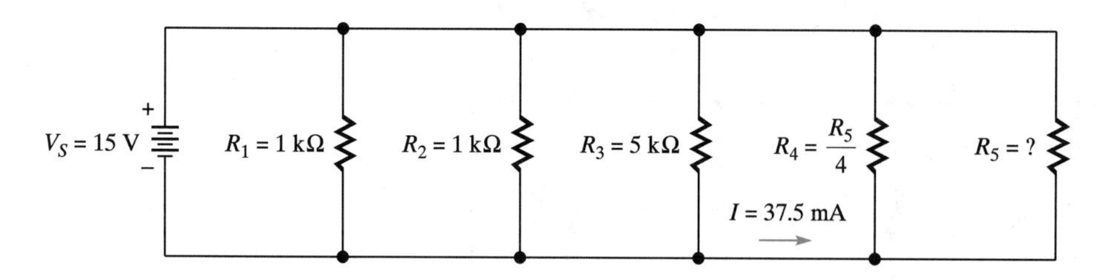

43. Analyze the circuit of Figure 5–70 for all voltages, currents, and component values not specified.

FIGURE 5–70 Circuit for Problem 43

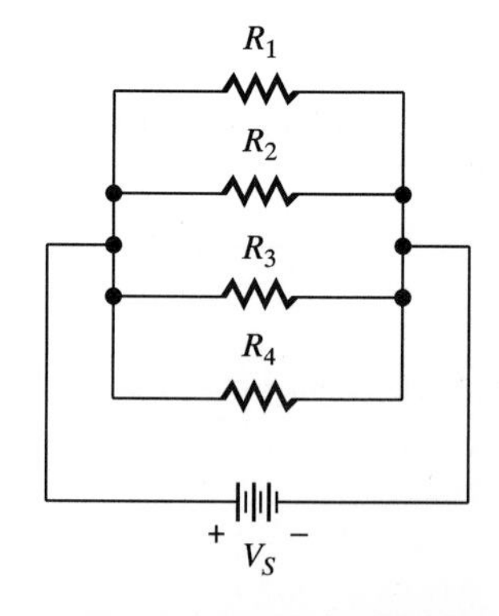

$R_1 = 10$ kΩ	$I_{R1} = 5$ mA
$P_{R2} = 50$ mW	$R_T = 2.586$ kΩ
$I_{R3} = 10$ mA	

PROBLEM SET: TROUBLESHOOTING

Refer to Figure 5–71 for Problems 44 through 46.

44. Determine the fault in the circuit if the meter readings are as follows: A_1 reads 6 mA, V_2 reads 60 V, and V_1 reads 0 V.

45. Determine the fault in the circuit if the meter readings are as follows: V_1 and V_2 both read 60 V and A_1 reads 7.54 mA.

46. Determine the fault in the circuit if the meter readings are as follows: A_1 reads zero, V_1 reads zero, and V_2 reads zero.

47. If Figure 5–72, the normal supply voltage is 10 V. If one fault exists, from the meter readings shown, determine the probable cause.

FIGURE 5–71 Circuit for Problems 44, 45, and 46

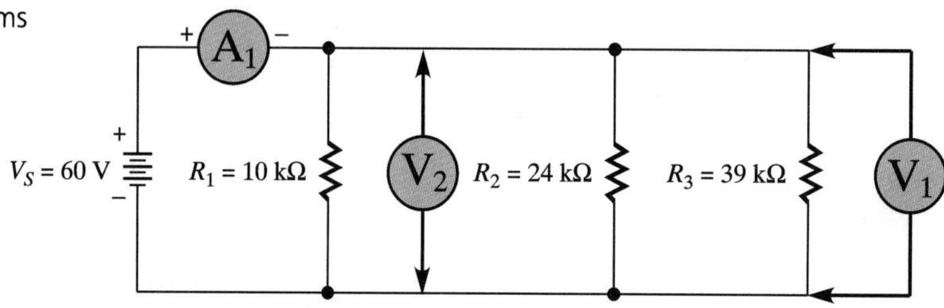

FIGURE 5–72 Circuit for Problem 47

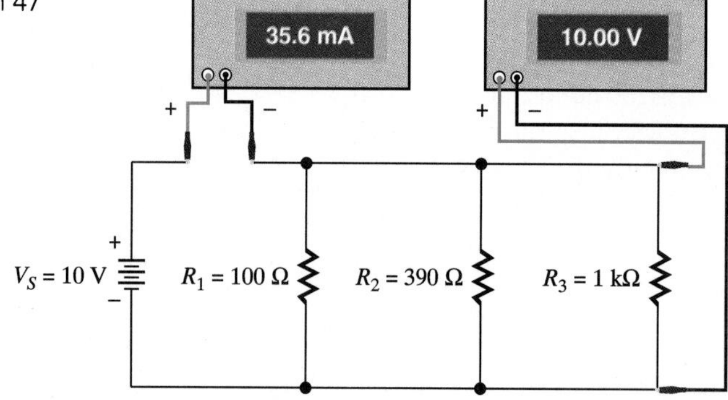

48. In Figure 5–73, once again the correct supply voltage is 10 V. If one fault exists, from the meter readings shown, determine the probable cause.

49. The symptom of a fault in Figure 5–73 is that the power supply shows an overload and shuts down immediately when the power switch is closed. What is the probable cause of this fault? You may assume that the power supply is operating normally.

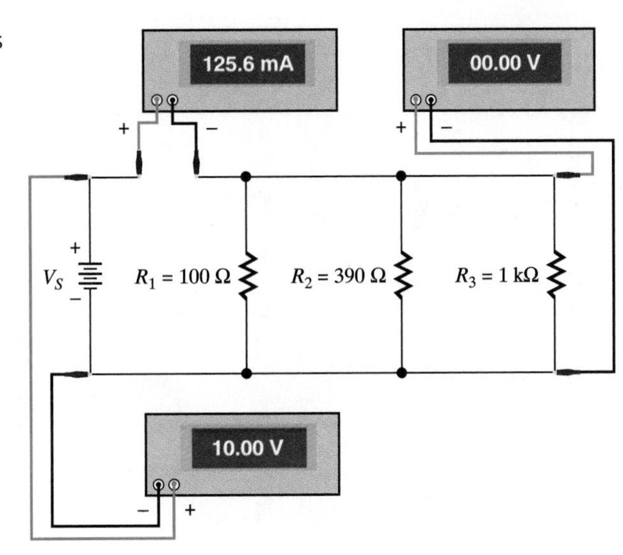

FIGURE 5–73 Circuit for Problems 48 and 49

Electronic Security Systems

Over the past few years, one of the fastest growing applications of electronics has been in the area of security systems. Businesses in particular have been taking advantage of this new technology. In fact, a security system is often a condition in their obtaining burglary insurance. Private citizens in growing numbers have also found it advisable to protect their homes from illegal entry and burglary. Once again, it is often cheaper to insure a home and its possessions if certain security precautions are taken. Many of these precautions taken by both businesses and homeowners are mechanical devices such as window latches and dead bolts, but others are electronic systems designed to alert the police of a burglary in progress.

One type of electronic security system uses magnetically operated switches on each window and door. As shown in Figure 1(a), a magnet mounted on the window holds a switch closed to complete a circuit. If the window is forced open, as shown in Figure 1(b), the magnet will be moved away from the switch and the circuit opened. Opening the circuit triggers the operation of a central system, which dials the emergency police number and a tape announces the illegal entry. The same sequence would occur if any of the windows or doors were opened. The system must be armed by closing a master switch each time the building is secured. Once the system is armed, a certain number of seconds, possibly 30, are allowed for closing the final door. The same is true upon entering the building; the same amount of time is allowed for disarming the system.

In this type of system, the call is automatically made through the telephone system to the police emergency number (911). The call is patched into a special system used by the police in order to locate the building as well as patrol cars in the area. The police computer brings the location of the building up on a CRT, and the dispatcher alerts the police cars in the area to investigate the call.

Other systems use different means of detecting entry. One such system detects motion in a room or vault. A transmitter emits sound waves above the range of human hearing and detects entry by a change in the waves returned from an object. The waves returned by a stationary object are different than those returned by a moving object. Thus, an intrusion can be detected. Since the motion of any object is detected, this type of system requires careful installation. The system is monitored by personnel at a central facility, often operated by a private security service.

Another type of intrusion detection system uses heat sensors to detect the body heat of an intruder. Once again, the system is tied into a central facility monitored by security personnel.

There are also automatic security systems that allow entry to selected individuals only. One such system relies upon fingerprint recognition. A would-be entrant's hand is scanned by sensors, and the result is compared to fingerprint information on file. If the fingerprints match that of an authorized person, the person is allowed entry. In a similar manner, the entry can be allowed through voice recognition. Here, the voice patterns are analyzed and compared to those on file.

Security systems are also used to detect and alert the proper agencies in cases of fire in a business or home. Sensors monitor the temperatures in various areas of the building. If abnormal temperatures are detected, the central system automatically alerts the fire department through the phone system. This system may also activate sprinkler systems and close fire doors. These systems are now standard equipment in large office buildings and apartments. They are often a part of a much larger, electronically controlled, environmental regulating system.

What employment opportunities are there in the security systems industry? There are many opportunities for employment in the security systems industry. Smaller companies may cater to the needs of individual homeowners and small businesses. These systems are usually intrusion alarms that protect the premises when no one is present. The company charges for the installation and then guarantees its operation for a small monthly fee. Thus, employment opportunities exist in both system installation and maintenance.

When large apartment or business buildings are constructed, the builder contracts with someone to install and maintain a comprehensive system of security, fire protection, and environmental control. This contractor may subcontract some of the work to smaller companies. Once again, employment opportunities for electronics technicians are available in installation and maintenance of the systems. These smaller contracts are often the vehicle through which an entrepreneur gains entry into the field.

FIGURE 1 Magnetic switch arrangement used electronic security system

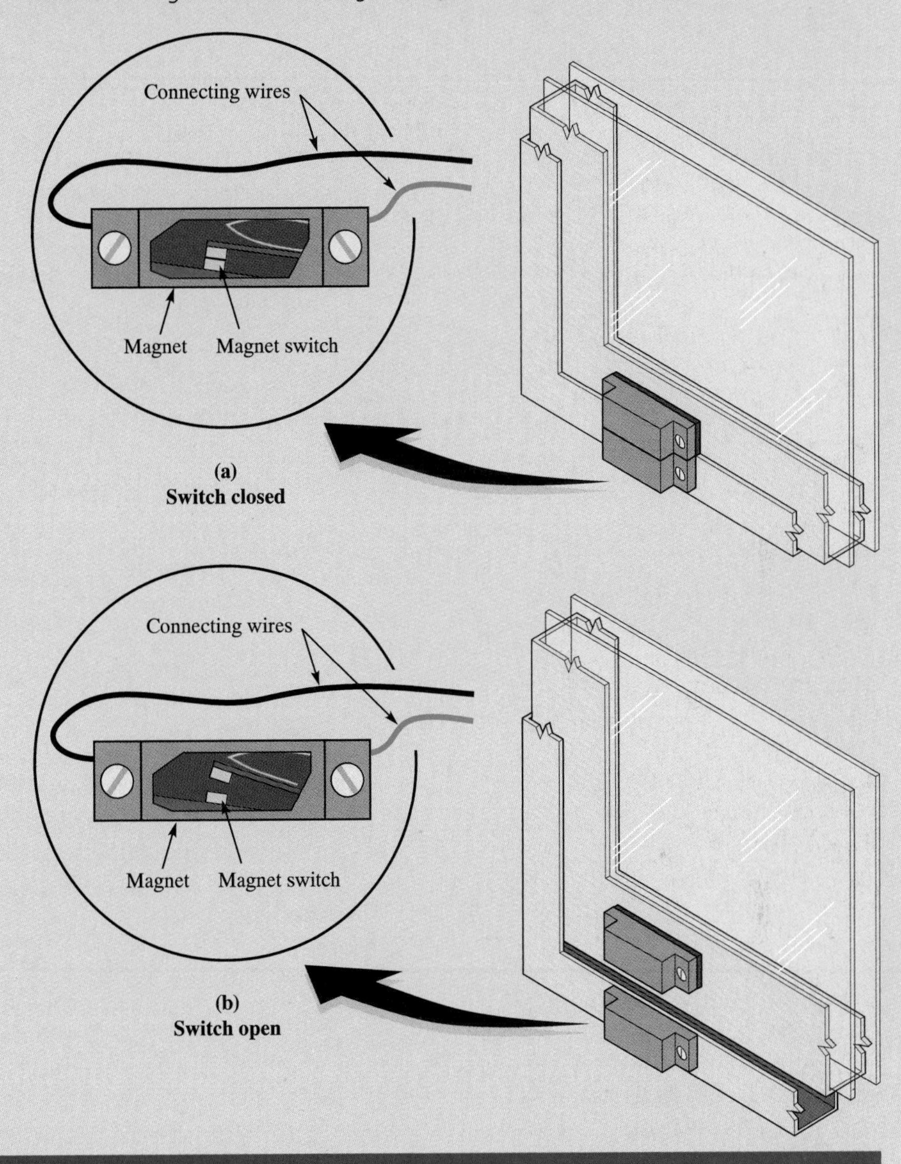

CHAPTER

6 Series-Parallel Circuits

UPON COMPLETION OF THIS CHAPTER, YOU WILL BE ABLE TO

1. Recognize series-parallel resistive circuits.
2. Recognize series components in series-parallel circuits.
3. Recognize parallel components in series-parallel circuits.
4. Compute the total resistance of a series-parallel circuit.
5. Analyze series-parallel circuits for the current through and the voltage across each component.
6. Analyze series-parallel circuits when the circuit ground is placed at various points.
7. Design and analyze loaded voltage dividers.
8. Explain the operation and applications of a Wheatstone bridge.
9. Perform troubleshooting on series-parallel resistive circuits.

Chapters 4 and 5 covered the operation, analysis, and applications of series and parallel circuits. In this chapter, the two will be combined to form series-parallel circuits. This type of circuit is, of course, the most common configuration used in electronic systems. Within a series-parallel circuit, devices such as resistors, transistors, and diodes are placed in series when a common current, but different voltages, are required; other devices that require a common voltage, but draw different currents, are placed in parallel. These characteristics represent the two methods of connecting electronic components.

It will be necessary to become adept at recognizing series and parallel combinations. This task should not be difficult if you recall and apply the facts about both types of circuits. Then, it is a matter of applying the principles of circuit analysis presented in Chapters 4 and 5. Practice is needed in learning to apply these principles individually to series and parallel portions of the circuit.

Some students have difficulty with series-parallel circuits even after doing very well in their studies of the individual circuits. The problem is usually a result of not seeing the overall circuit picture or their failure to apply some fact of the individual circuits. This text tries to avoid these problems by pointing out such pitfalls as they come along. Watch for them!

6–1 THE SERIES-PARALLEL CIRCUIT

Question: What is a series-parallel circuit?

FIGURE 6–1 Series-parallel circuit

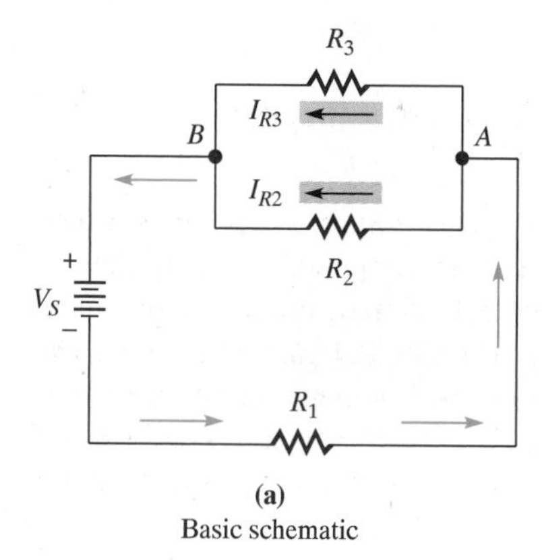

(a)
Basic schematic

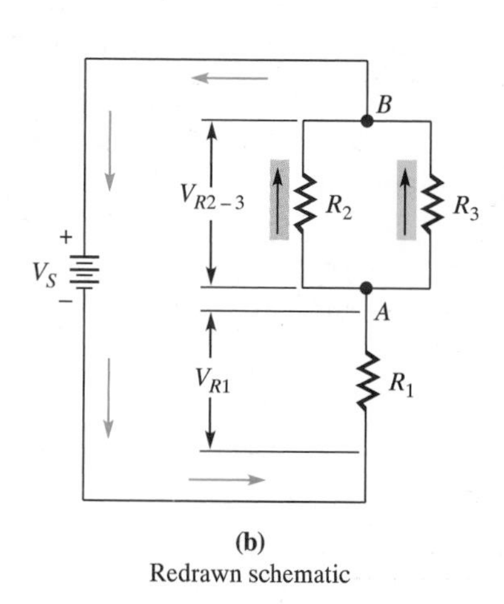

(b)
Redrawn schematic

In Chapters 4 and 5, it was shown that a series circuit is a voltage divider and a parallel circuit is a current divider. A **series-parallel circuit** contains both series and parallel combinations of components, and there is a division of both current and voltage. Consider the circuit in Figure 6–1(a). The current leaves the negative side of the source and flows through R_1. At point *A*, the current branches, with some going through R_2 and the rest through R_3. The current recombines at point *B* and returns to the positive side of the source. Thus, current division takes place in the parallel combination of R_2 and R_3. In the overall circuit, the parallel combination of R_2 and R_3 is in series with R_1.

The same circuit is redrawn in Figure 6–1(b) to illustrate the division of voltage. There are two voltage drops: V_{R_1} across R_1 and $V_{R_{2\text{-}3}}$ across the parallel combination R_2 and R_3. The sum of these two voltages equals the source voltage. Thus, voltage division takes place between R_1 and the common voltage drop of R_2 and R_3. Since both current and voltage division takes place, this circuit is a series-parallel circuit.

Another arrangement of a series-parallel circuit is shown in Figure 6–2(a). Notice in this circuit that R_2 and R_3 are in series, and the voltage across R_1 is divided between them. Thus, voltage divider action takes place. The current branches at point *A*, with some going through R_1 and the rest going through the series combination of R_2 and R_3 and recombining at point B. Once again, both current and voltage division have taken place, making this a series-parallel circuit. In the overall circuit, the series combination of R_2 and R_3 is in parallel with R_1.

One more example is shown in Figure 6–2(b). In this circuit, the total current flows through the series combination of R_1 and R_2. At point *A* it branches, with portions of it flowing through R_3, R_4, and R_5. It recombines at point *B* and immediately branches again, with some going through R_6 and the rest through R_7. It recombines again at point *C* and returns to the positive side of the source. Notice that the circuit has divided the source voltage into V_{R_1}, V_{R_2}, $V_{R_{3\text{-}4\text{-}5}}$, and $V_{R_{6\text{-}7}}$. The total current has been divided twice: in the parallel combination of R_3, R_4, and R_5 and in the parallel combination of R_6 and R_7. In this circuit, the parallel combination of R_6 and R_7 and the parallel combination of R_3, R_4, and R_5, are in series with R_1 and R_2.

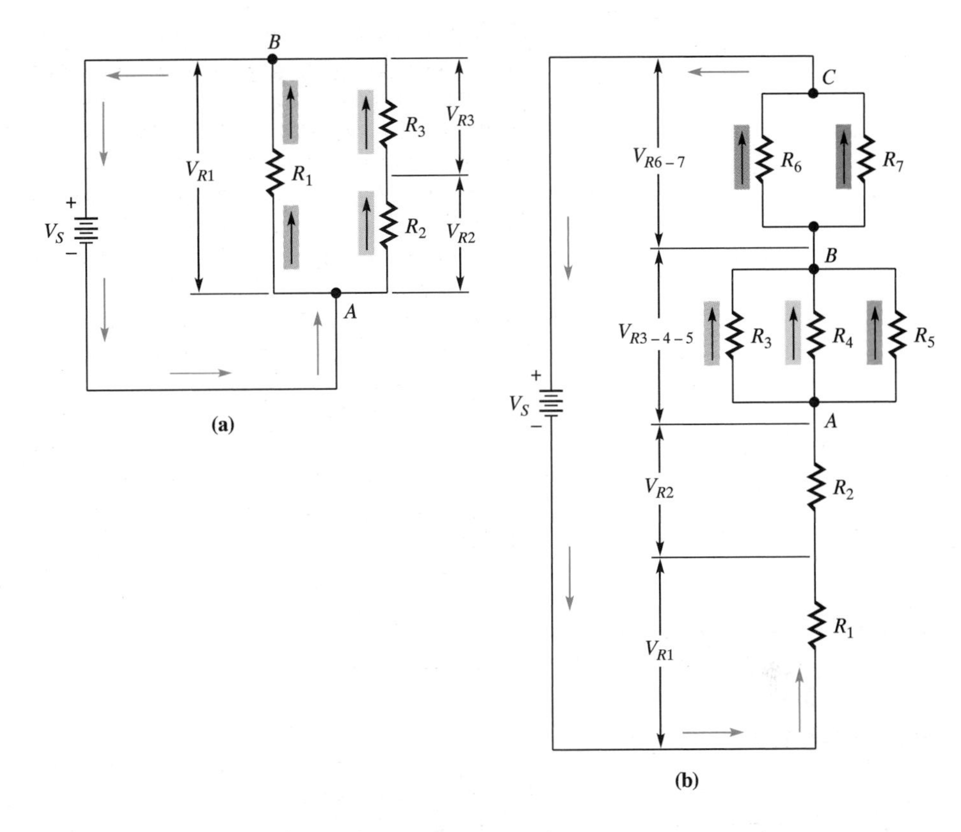

FIGURE 6–2 Additional configurations of series-parallel circuits

Question: Why is it necessary to accurately and quickly recognize series and parallel components in a series-parallel circuit?

In previous chapters, circuit applications that were presented utilized either series or parallel circuits. Real-world electronic systems, however, are composed of many devices connected in complex series-parallel combinations. Recognition of the individual series and parallel combinations is made by using the facts of series and parallel circuits. Devices with a common current are in series, while those with a common voltage are in parallel. It is important that you take time now to learn to recognize series and parallel combinations quickly and accurately. You will need this skill in the analysis of complex electronic systems.

EXAMPLE 6–1 State the series and parallel combinations that exist in Figure 6–3.

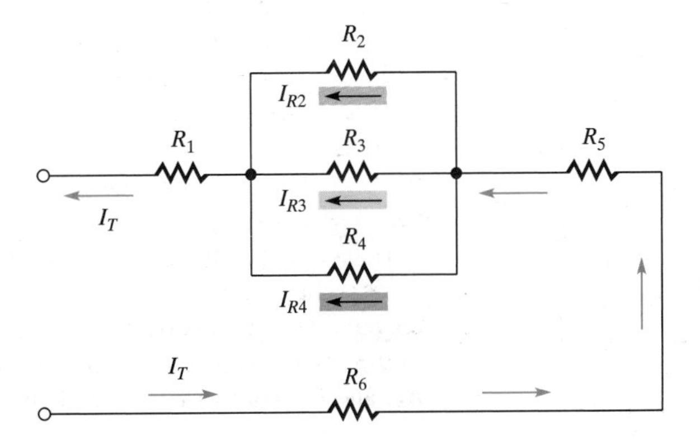

FIGURE 6–3 Circuit for Example 6–1

■ Solution:

R_2, R_3, and R_4 are in parallel, and this parallel combination is in series with R_1, R_5, and R_6.

EXAMPLE 6–2 State the series and parallel combinations of Figure 6–4 between each set of points: (a) A to B, (b) B to C, (c) C to D, and (d) D to A.

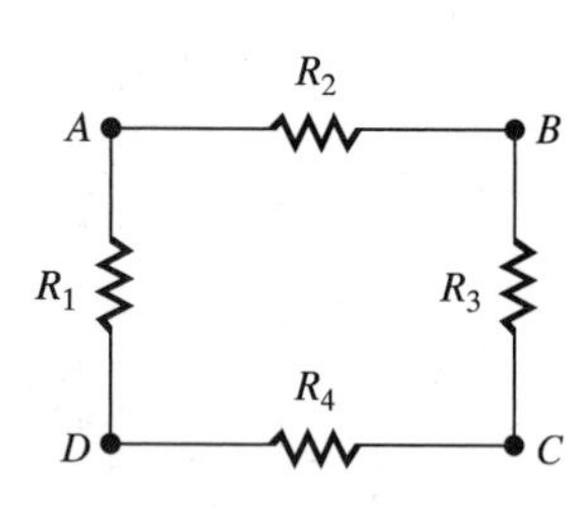

FIGURE 6–4 Circuit for Example 6–2

■ Solution:

a. From points A to B, resistor R_2 is in parallel with the series combination of R_1, R_3, and R_4.
b. From points B to C, resistor R_3 is in parallel with the series combination of R_1, R_2, and R_4.
c. From points C to D, resistor R_4 is in parallel with the series combination of R_1, R_2, and R_3.
d. From points D to A, resistor R_1 is in parallel with the series combination of R_2, R_3, and R_4.

■ SECTION 6–1 REVIEW

1. What is a series-parallel circuit?
2. How do you determine series and parallel combinations in a series-parallel circuit?
3. List the series and parallel combinations of the circuits in Figure 6–5.

FIGURE 6–5 Circuit for Section Review Problem 3

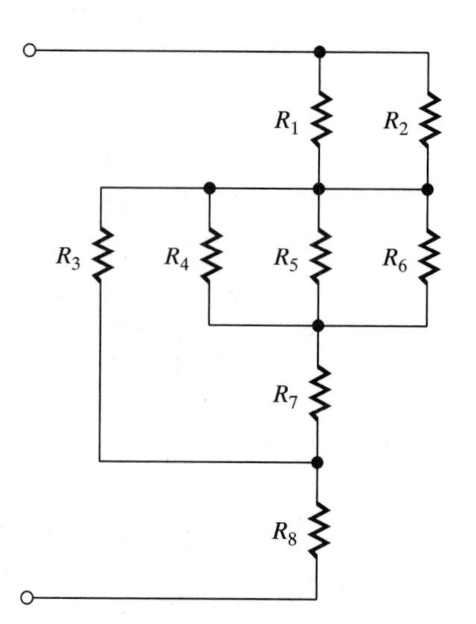

6–2 REDRAWING OF SERIES-PARALLEL CIRCUITS

Question: Why is it usually a good idea to redraw a series parallel circuit prior to analysis?

Redrawing, or altering the orientation of components in a schematic, often makes it easier to determine what is in parallel and what is in series. These determinations must be made by viewing the position of each component in relation to other components and to the

FIGURE 6–6 Series-parallel circuits showing original placement and repositioning of all components

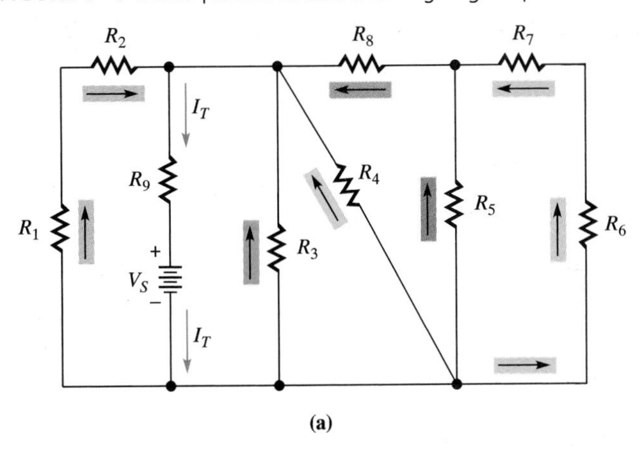

(a)

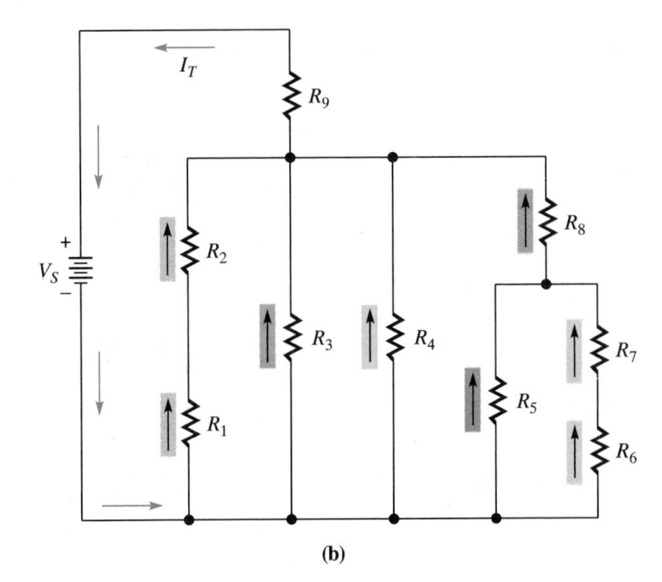

(b)

FIGURE 6–7 Illustration of procedure for redrawing series-parallel circuits

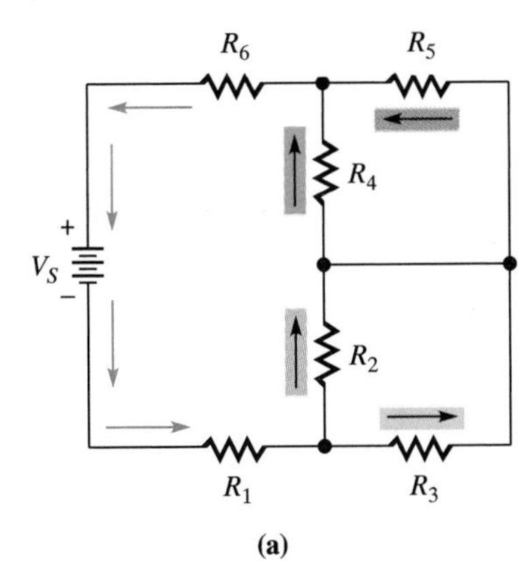

(a)

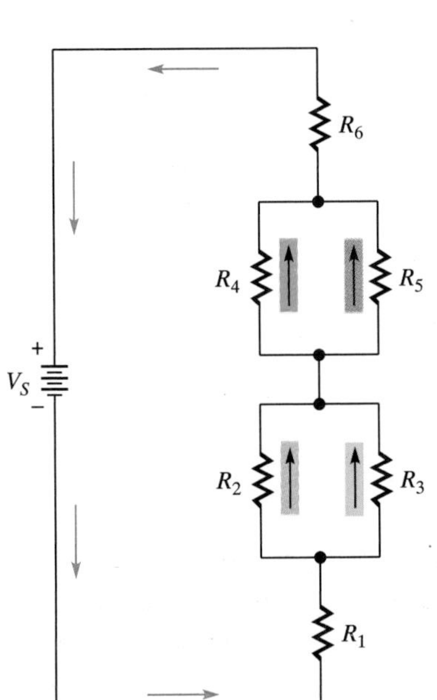

(b)

source. For example, notice the positioning of the components in Figure 6–6(a). The source is shown in the midst of the components, and the components are drawn with several different orientations. Drawn as they are, the series and parallel combinations may not be readily apparent. Now consider Figure 6–6(b), which shows the same circuit redrawn with the source repositioned and the resistors drawn vertically. In this schematic, the series and parallel combinations are more readily apparent, and the divisions of the source voltage stand out more clearly.

The steps for redrawing a series-parallel circuit can be summarized as follows:

1. Draw the source by itself, with the positive side to the top on what will be the left side of the redrawn schematic.
2. Moving from the positive (top) side of the source, draw in each resistor as it is encountered. The resistors should be drawn vertically with the top being the end first encountered.
3. Continue this process until all resistors are accounted for in the schematic, and the negative side of the source is encountered.

Redrawing circuits according to the steps in this list is good practice and makes recognition of the series components and the parallel components much easier. The circuit in Figure 6–7(a) is redrawn here to illustrate the procedure and to determine the series and parallel combinations. The redrawn diagram is shown in Figure 6–7(b). Proceeding from the positive side of the source, the top of R_6 is encountered. The bottom of R_6 contacts the tops of R_4 and R_5. The lower sides of R_4 and R_5 are connected, which places them in parallel. The bottom of this parallel combination contacts the tops of R_2 and R_3. The lower sides of R_2 and R_3 are connected, placing them in parallel. The bottom of this parallel combination contacts the top of R_1. The bottom of R_1 then returns to the negative side of the source. Thus, the parallel combinations of R_4–R_5 and R_2–R_3 are in series with R_1 and R_6.

EXAMPLE 6–4 Redraw the circuit of Figure 6–8(a), and determine the series and parallel combinations.

■ Solution:

The redrawn circuit is shown in Figure 6–8(b). R_1 and R_2 are in series. R_7 and R_8 are in series. These two series combinations are in parallel with R_3, R_5, and R_6. This entire combination is in series with R_4.

FIGURE 6–8 Circuits for Example 6–4

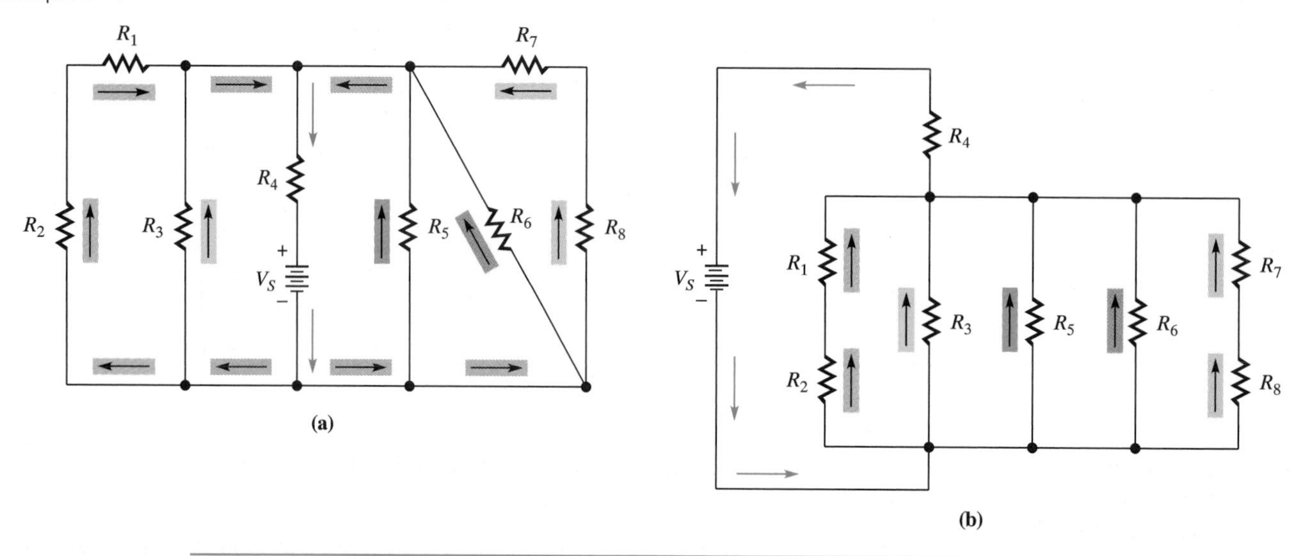

EXAMPLE 6–5 Redraw the circuit of Figure 6–9(a), and determine the series and parallel combinations.

■ Solution:

The redrawn circuit is shown in Figure 6–9(b). In this circuit, R_7 and R_8 are in series. This series combination is in parallel with R_5 and R_6. This parallel combination is in series with R_4. The entire combination is denoted as A on the drawing. R_1 and R_2 are in series, and the combination is denoted as B on the drawing. Combinations A and B are in parallel and are in series with R_3.

FIGURE 6–9 Circuit for Example 6–5

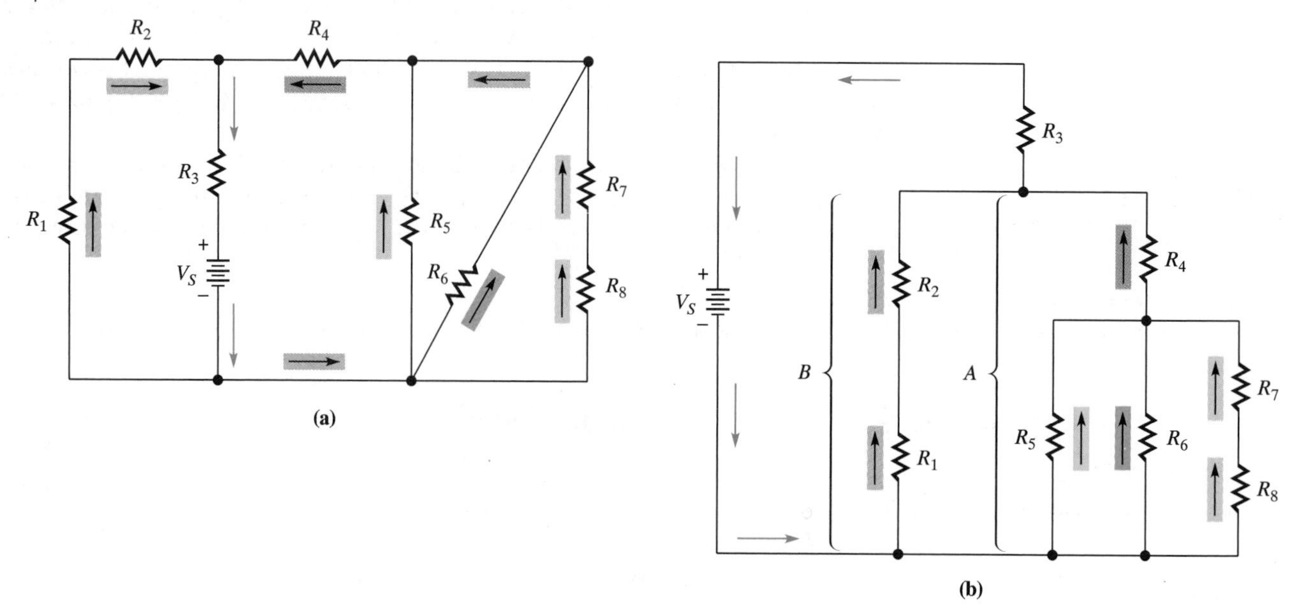

SECTION 6–2 REVIEW

1. Why is it often a good idea to redraw a series-parallel circuit?
2. Where should the source be placed in a redrawn circuit?
3. What should the orientation of the components be in a redrawn circuit?
4. Redraw the circuit of Figure 6–10 using the principles presented in this section.

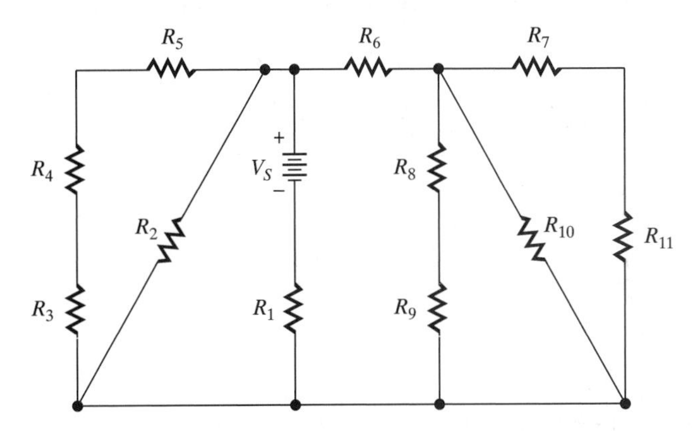

FIGURE 6–10 Circuit for Section Review Problem 4

6–3 TOTAL RESISTANCE OF A SERIES-PARALLEL CIRCUIT

Question: How is the total resistance of a series-parallel circuit "seen" by the voltage source?

The voltage source of an electric circuit does not "see" each individual resistor that makes up the circuit. It sees, rather, a resistance equivalent to the total resistance. The source supplies a total current equal to the source voltage divided by the total resistance. The first step in the analysis of a series-parallel circuit is to determine this total resistance. This value is obtained by starting as far from the source as possible and working back toward the source, combining each series and parallel combination into an equivalent resistance. An **equivalent resistance** is one that can replace a parallel bank or a series string without changing the total current.

An **equivalent circuit** is one in which some of the resistors have been replaced with equivalent resistances. It is good practice, each time an equivalent resistance is formed, to redraw the equivalent circuit. Doing so will make it much easier for you to keep track of what you have already done. Also, as you will find later in series-parallel circuit analysis, this procedure will make the determination of currents and voltages at points in the circuit much easier.

The circuit in Figure 6–11(a) will be used as an example to illustrate the use of equivalent resistance to determine total resistance. Any point at which you look into a circuit can be considered the position of the voltage source, which in this case is between points

FIGURE 6–11 Circuits for determining total resistance in series-parallel circuits

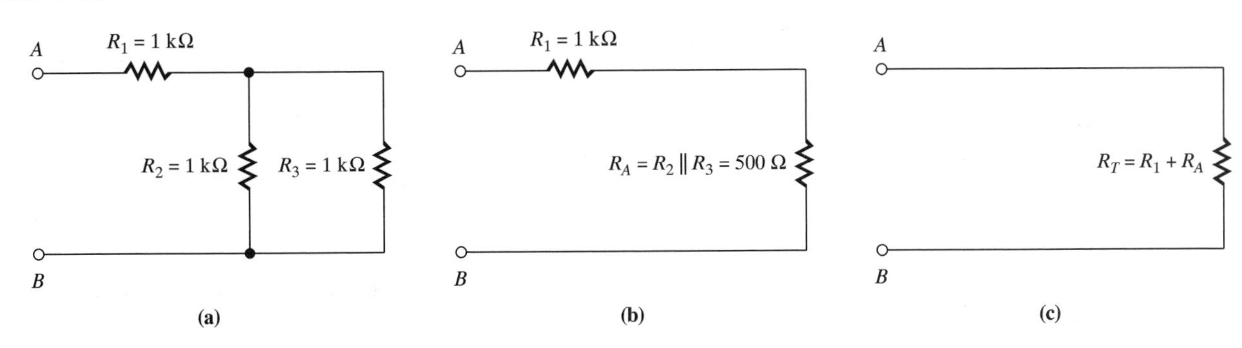

A and B. Beginning as far from the source as possible, notice R_2 and R_3 in parallel. The first step is to form an equivalent resistance for this combination, which will be called R_A:

$$R_A = \frac{R_2}{2} \quad \text{[equal resistors in parallel]}$$

$$= \frac{1000}{2} = 500\ \Omega$$

The result of the first transformation is illustrated in Figure 6–11(b). The parallel combination of R_2 and R_3, which may be written as $R_1 \| R_2$ as shown on the figure, has been replaced with equivalent resistance R_A. Notice now that R_A is in series with R_1. An equivalent resistance can now be formed from this combination:

$$R_T = R_1 + R_A$$

$$= 1000\ \Omega + 500\ \Omega = 1.5\ \text{k}\Omega$$

Notice that with this last transformation (Figure 6–11 c), there is a single resistance across the source, which is known as the total resistance R_T.

EXAMPLE 6–6 Compute the total resistance seen by the source in Figure 6–12(a). Note that even though R_1 and R_5 are separated within the circuit, they are in series.

FIGURE 6–12 Circuits for Example 6–6

■ Solution:

1. Combine R_1 and R_5 as series components:

$$R_A = R_1 + R_5$$

$$= 1.5\ \text{k}\Omega + 1.5\ \text{k}\Omega = 3\ \text{k}\Omega$$

[1] [.] [5] [EXP] [3] [+] [1] [.] [5] [EXP] [3] [=]

2. Notice within the parallel combination, that R_2 and R_3 are in series and can be combined to become R_B:

$$R_B = R_2 + R_3$$

$$= 5\ \text{k}\Omega + 5\ \text{k}\Omega = 10\ \text{k}\Omega$$

[5] [EXP] [3] [+] [5] [EXP] [3] [=]

(a)

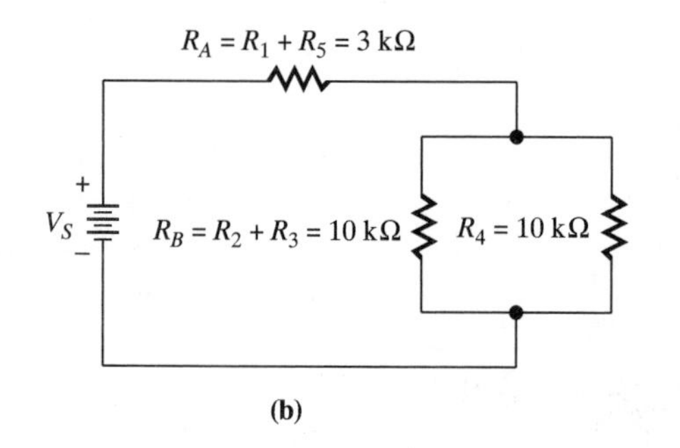

(b)

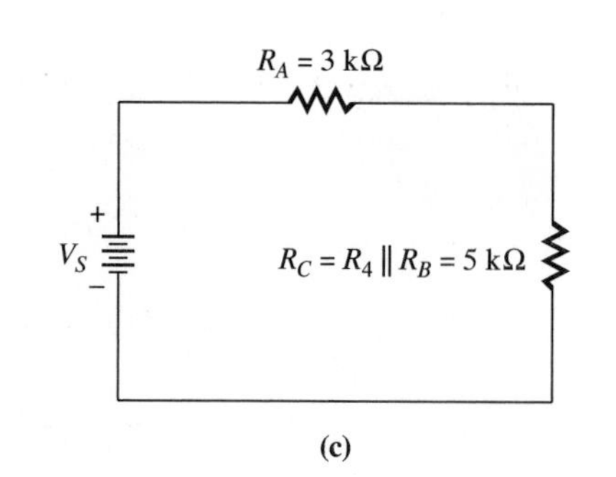

(c)

(d)

At this point, the equivalent circuit appears as in Figure 6–12(b). The equivalent resistor R_A could have been placed on either side of the parallel circuit.

3. Combine the parallel resistance R_B and R_4 to form R_C.

$$R_C = \frac{R_B}{2} \quad \text{[equal resistors in parallel]}$$

$$= \frac{10\ \text{k}\Omega}{2} = 5\ \text{k}\Omega$$

4. Finally, since R_A and R_C are seen in series as shown in Figure 6–12(c), combine them to yield R_T, as shown in Figure 6–12(d):

$$R_T = R_A + R_C$$

$$= 5\ \text{k}\Omega + 3\ \text{k}\Omega = 8\ \text{k}\Omega$$

As stated before, the key to determining total resistance of a series-parallel circuit is the recognition of its series and parallel combinations. In the following example, the circuit will be redrawn in order to make these combinations more apparent.

EXAMPLE 6–7 Compute the total resistance seen by the source in Figure 6–13(a).

■ Solution:

1. First redraw the circuit and mark the current paths as shown in Figure 6–13(b). Begin as far from the source as possible, and form an equivalent resistance R_A for the R_2 and R_3 series combination:

$$R_A = R_2 + R_3$$

$$= 100\ \Omega + 100\ \Omega = 200\ \Omega$$

2. Notice in the redrawn of Figure 6–13(c) that there are now two parallel circuits: R_4 in parallel with R_A and R_5 in parallel with R_6. Combine these to yield R_B (Figure 6–13(d)):

$$R_B = \frac{R_A}{2}$$

$$= \frac{200\ \Omega}{2} = 100\ \Omega$$

Next, use the conductance method to combine R_5 and R_6 and find the value of R_C:

$$G_C = G_5 + G_6$$

$$= 3.8\ \text{mS}$$

$$R_C = \frac{1}{G_C}$$

$$= \frac{1}{3.8 \times 10^{-3}} = 263\ \Omega$$

FIGURE 6–13 Circuits for Example 6–7

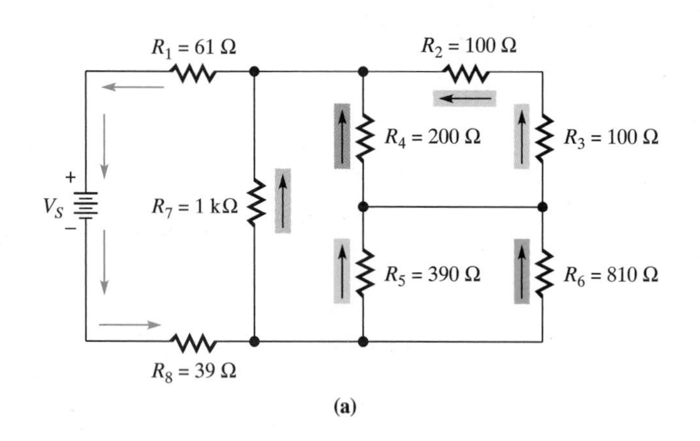

(a)

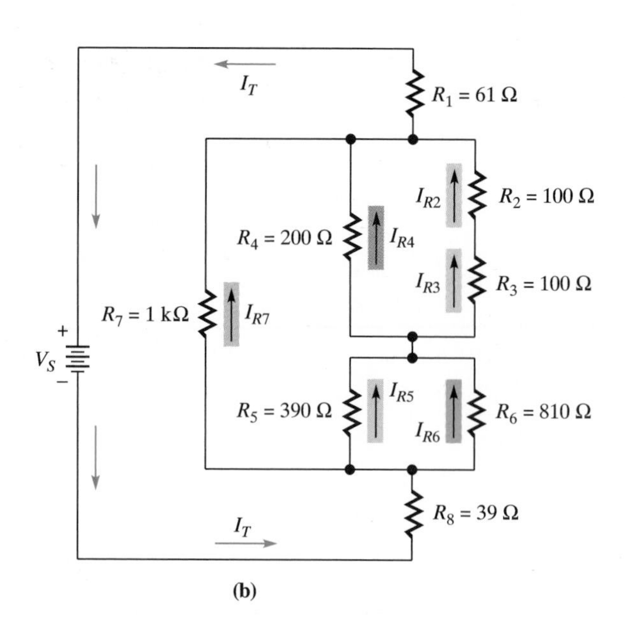

(b)

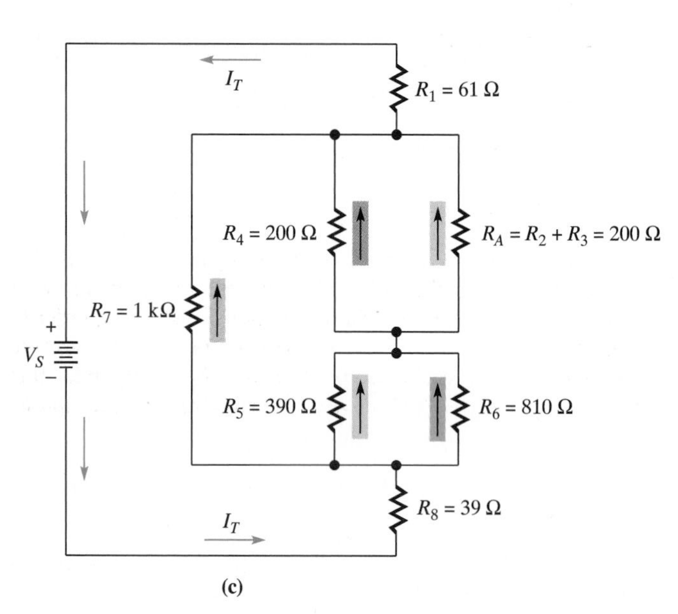

(c)

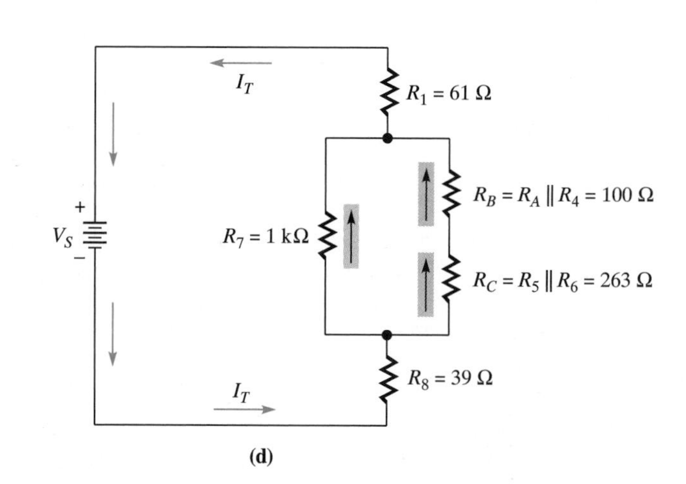

(d)

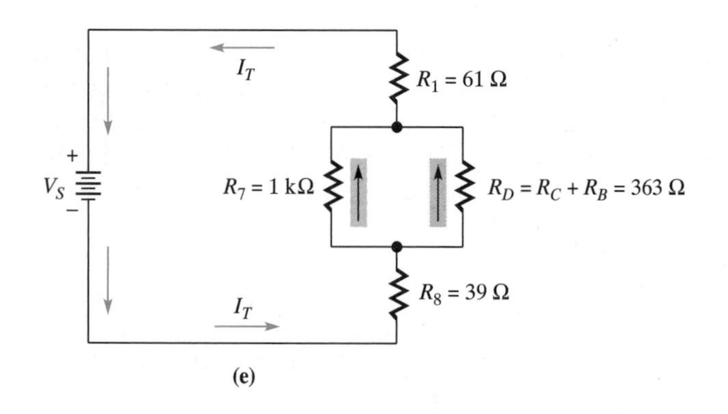

(e)

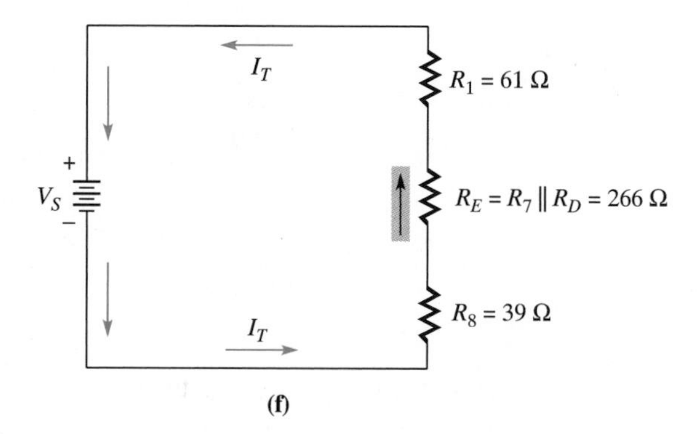

(f)

3. Combine the R_B and R_C series combination to yield R_D (Figure 6–13(e)):

$$R_D = R_B + R_C$$
$$= 100\ \Omega + 263\ \Omega$$
$$= 363\ \Omega$$

4. Next, combine the resistances R_7 and R_D to yield R_E:

$$G_E = G_7 + G_D$$
$$= \frac{1}{1 \times 10^3\Omega} + \frac{1}{363\Omega} = 3.776 \text{ mS}$$
$$R_E = \frac{1}{G_E}$$
$$= \frac{1}{3.776\ \Omega \times 10^{-3}\text{S}} = 266\ \Omega$$

Finally, the total resistance seen by the source is the series combination of R_1, R_E, and R_8 as shown in Figure 6–13(f).

$$R_T = R_1 + R_E + R_8$$
$$= 61\ \Omega + 266\ \Omega + 39\ \Omega = 366\ \Omega$$

FIGURE 6–14 Circuit with R_1 and R_2 series in parallel with R_3

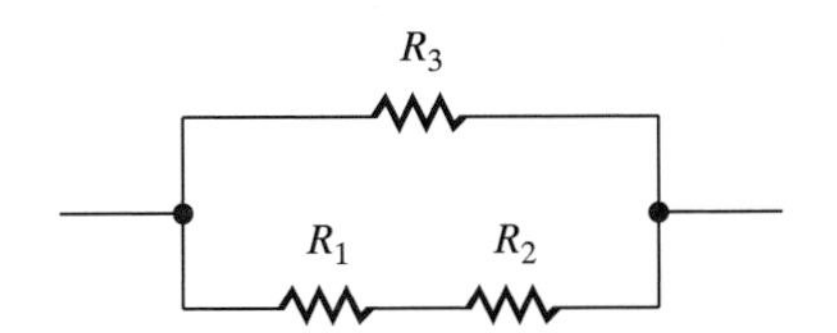

NOTE

In Figure 6–14, some students have been known to see R_1 or R_2 in parallel with R_3, which results, of course, from not looking at the entire picture of the circuit. The resistors appear to be side by side, so the student may see them in parallel. However, in order to be in parallel, devices must have the same voltage across them. In this case, the voltage drop across R_3 is the source for the series combination of R_1 and R_2. R_1 and R_2 divide this voltage and neither of them are in parallel with R_3. The series combination of R_1 and R_2 is in parallel with R_3, however.

EXAMPLE 6–8 In Figure 6–15(a), compute the total resistance seen between A and E.

■ **Solution:**

The easiest way to approach this type of problem is to pretend a voltage source exists between each set of points in question, which is, in this case, between points A and E. The circuit with the pretend source is shown redrawn in Figure 6–15(b). All resistors are a part of the circuit. R_2 and R_3 are in parallel with a total resistance of 500 Ω. This equivalent circuit is shown in Figure 6–15(c). The total is then the sum of the resistances: $R_1 + R_2 \| R_3 + R_1 + R_4 + R_5 = 1.037\text{ k}\Omega$.

EXAMPLE 6–9 Using the circuit in Figure 6–15(a), compute the total resistance seen between points B and D.

FIGURE 6–15 Circuits for Examples 6–8 and 6–9

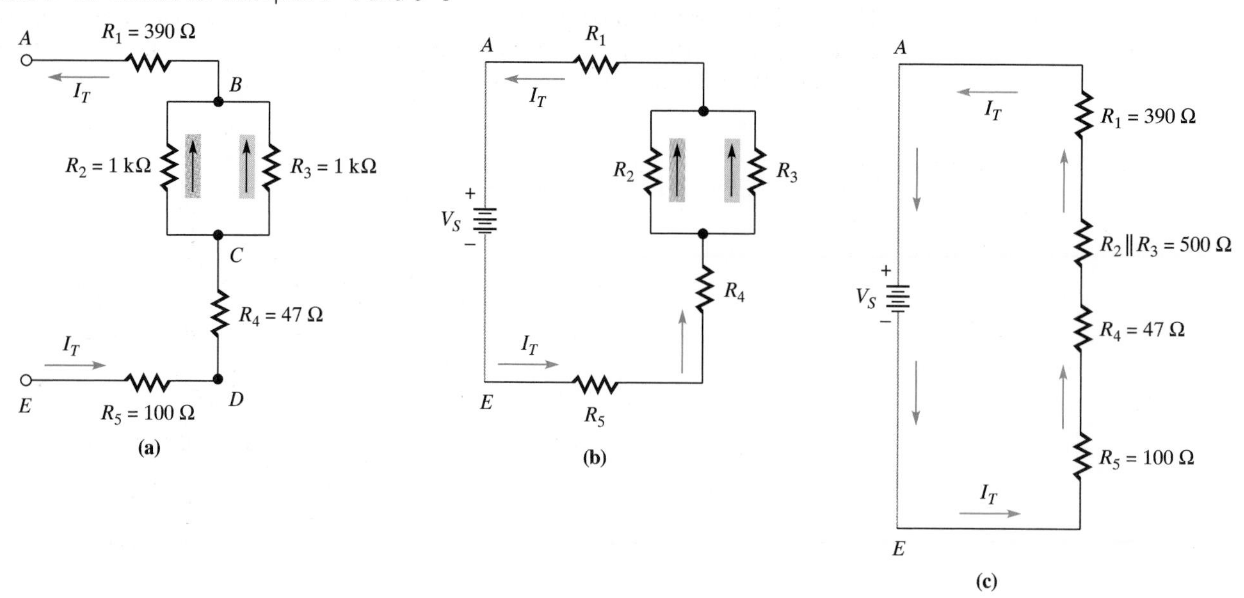

FIGURE 6–16 Circuits for Example 6–9

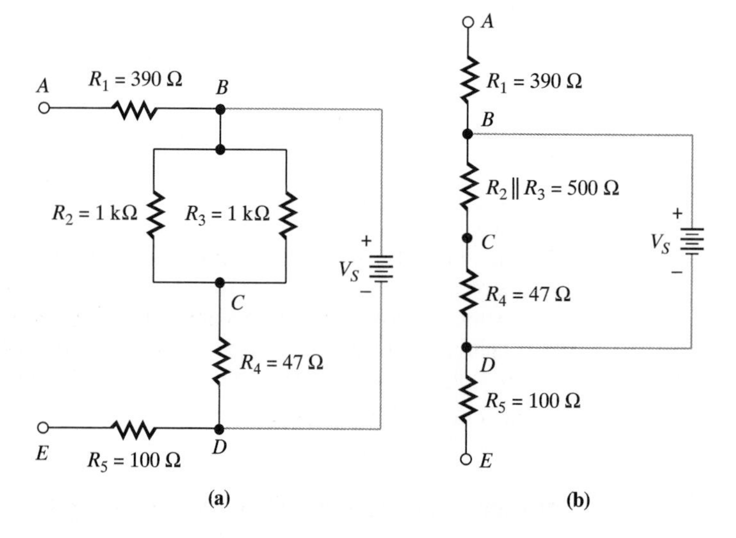

■ Solution:

Once again, pretend a source exists between the points B and D. This redrawn circuit is shown in Figure 6–16(a). Notice that now resistors R_1 and R_5 are *no longer* in the circuit. This circuit is shown redrawn in Figure 6–16(b). The total resistance is now the sum of the parallel combination of R_2 and R_3 plus R_4, or $R_2 \| R_3 + R_4 = 547\ \Omega$.

EXAMPLE 6–10 Again use the circuit in Figure 6–15(a), and compute the total resistance seen between points B and C.

■ Solution:

The circuit is shown redrawn in Figure 6–17. When a source is pretended between points B and C, notice that only resistors R_2 and R_3 are part of the circuit. Thus, the total resistance is their parallel combination, $R_2 \| R_3$, or 500 Ω.

FIGURE 6–17 Circuits for Example 6–10

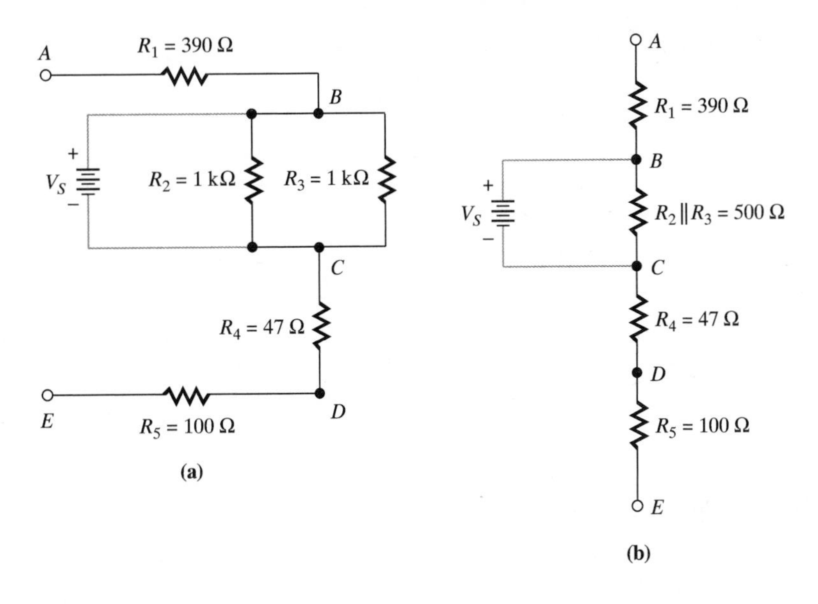

■ SECTION 6–3 REVIEW

1. In your own words, state a procedure for determining the total resistance of a series parallel circuit.
2. Explain what is meant by the term equivalent resistance.
3. Compute the total resistance seen by the source in Figure 6–18.
4. In Figure 6–19, compute the total resistance seen between each of the following sets of points: A to B, B to C, A to D.

FIGURE 6–18 Circuit for Section Review Problem 3

FIGURE 6–19 Circuit for Section Review Problem 4

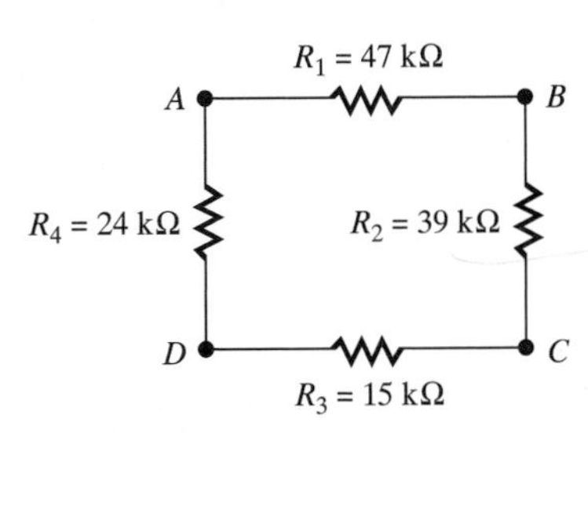

6–4 SERIES-PARALLEL CIRCUIT ANALYSIS

Question: Is there a logical, step-by-step procedure for the analysis of series-parallel circuits?

In the analysis of series-parallel circuits, you are once again seeking to determine the current through and the voltage across each component. As with the individual circuits, a step-by-step procedure is used for series-parallel circuit analysis:

1. Determine the total resistance as demonstrated in the preceding section. Be sure to redraw the circuit each time an equivalent resistance is formed.
2. Beginning with the last conversion, which is R_T across the source, work backward computing the current through and the voltage across each resistor and equivalent resistance.
3. Compute the power developed in each resistor.
4. Compute the total power developed by the circuit.

Once again, the tools of circuit analysis are Ohms's law, Kirchhoff's voltage and current laws, and the facts of series and parallel circuits. Study carefully the following examples and apply the methods presented in the practice that follows later in this chapter. Remember that these are the same methods used in the analysis of complex electronic circuit systems.

As an example, the steps of circuit analysis will be applied to the circuit of Figure 6–20(a). The first step is to determine the total resistance. Combining the series resistors R_3 and R_4 to yield R_A (Figure 6–20(b)):

$$R_A = R_3 + R_4$$

$$= 1.2\ \text{k}\Omega + 3.9\ \text{k}\Omega = 5.1\ \text{k}\Omega$$

Next combining the two parallel combinations R_1 and R_2 and R_5 and R_6 to yield R_B and R_C (Figure 6–20(c)) gives

$$R_B = \frac{R_1}{2}$$

$$= \frac{1.5\ \text{k}\Omega}{2} = 750\ \Omega$$

and

$$G_C = G_5 + G_6$$

$$= \frac{1}{4.7 \times 10^3\ \Omega} + \frac{1}{47 \times 10^3\ \Omega}$$

$$R_C = \frac{1}{G_C}$$

$$= \frac{1}{234\ \mu\text{S}} = 4.273\ \text{k}\Omega$$

The equivalent series circuit composed of R_A, R_B, and R_C, can now be combined to obtain R_T (Figure 6–20(d)):

$$R_T = R_A + R_B + R_C$$

$$= 5.1\ \text{k}\Omega + 0.75\ \text{k}\Omega + 4.723\ \text{k}\Omega$$

$$= 10.123\ \text{k}\Omega$$

FIGURE 6–20 Illustration of circuit analysis procedure

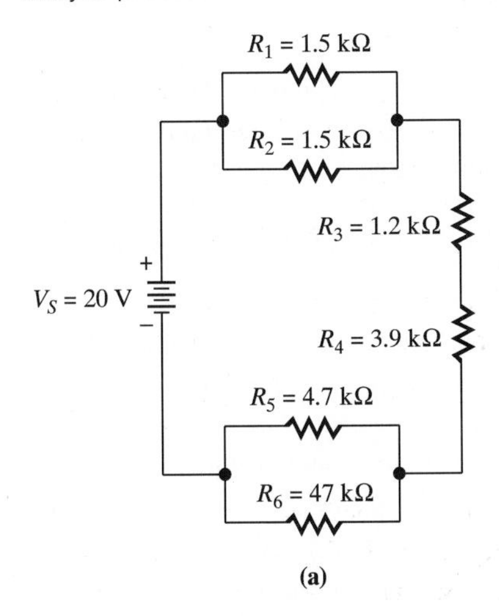

(a)

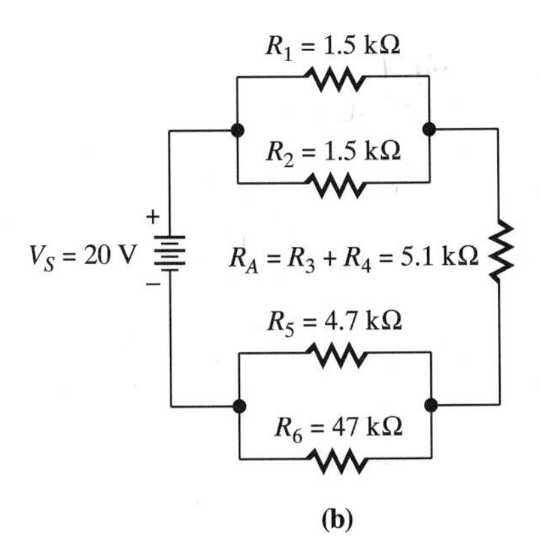

(b)

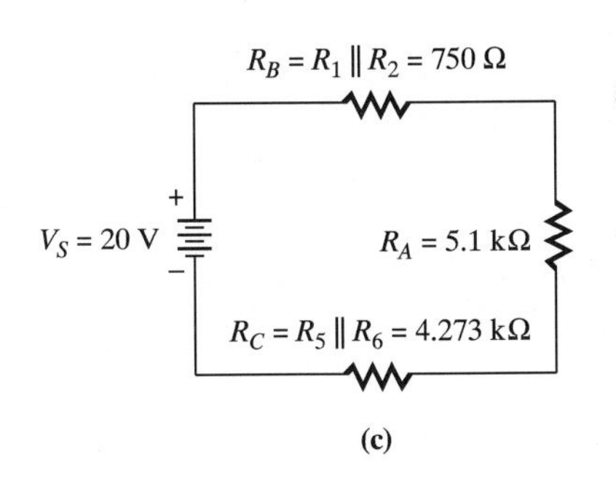

(c)

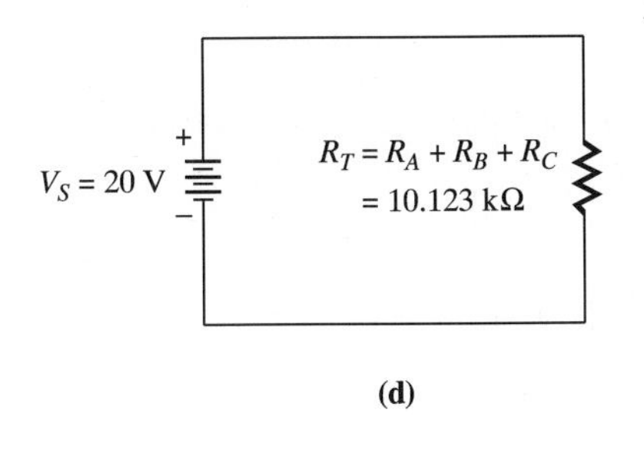

(d)

The next step is to compute the total current using Ohm's law:

$$I_T = \frac{V_S}{R_T}$$

$$= \frac{20\ \text{V}}{10.123\ \text{k}\Omega} = 1.98\ \text{mA}$$

This current value is the total current that flows through R_A, R_B, and R_C in Figure 6–20(c). It has been rounded to three significant figures, which, unless otherwise noted, is the precedent followed throughout the text where currents are computed.

Next, compute the voltage drops across these equivalent resistances:

$$V_{RA} = I_T R_A$$

$$= (1.98\ \text{mA})(5.1\ \text{k}\Omega) = 10.098\ \text{V}$$

$$V_{RB} = I_T R_B$$

$$= (1.98\ \text{mA})(750\ \Omega) = 1.485\ \text{V}$$

$$V_{RC} = I_T R_C$$

$$= (1.98\ \text{mA})(4.273\ \text{k}) = 8.4605\ \text{V}$$

These voltages have great significance to the rest of the analysis. V_{RA} is the total voltage across the series combination of R_3 and R_4. V_{RB} is the total voltage across the parallel combination of R_1 and R_2. V_{RC} is the total voltage across the parallel combination of R_5 and R_6. These voltages will be used in the computation of the individual currents and voltage drops for the components.

The computations for the series combination of R_3 and R_4 are made up as follows: Equivalent resistance ${}_{BA}$ is made up of the series combination of R_3 and R_4. Since the total current flows through R_A, then the total current also flows through the resistors from which it is made. Thus, the total current flows through R_3 and R_4, confirming what you can see through the inspection of Figure 6–20(a).

The definition of devices in series can now be written as

$$I_{R3} = I_{R4} = I_T = 1.98\ \text{mA}$$

and

$$V_{R3} = I_T R_3$$

$$= (1.98\ \text{mA})(1.2\ \text{k}\Omega) = 2.376\ \text{V}$$

$$V_{R4} = I_T R_4$$

$$= (1.98\ \text{mA})(3.9\ \text{k}\Omega) = 7.722\ \text{V}$$

Note that these voltages could have been computed using the voltage divider equation introduced in Chapter 4:

$$V_{R3} = \frac{R_3}{R_T} V_{RA}$$

$$= \frac{1200\ \Omega}{5100\ \Omega} 10.098\ \text{V} = 2.376\ \text{V}$$

$$V_{R4} = \frac{R_4}{R_T} V_{RA}$$

$$= \frac{3900\ \Omega}{5100\ \Omega} 10.098\ \text{V} = 7.722\ \text{V}$$

CAUTION

The voltage divider equation calls for the use of the total resistance and total voltage in computing a voltage drop. Be sure to use the total resistance and total voltage of the *combination* in question, which is not necessarily the total resistance and total voltage of the *circuit*. Each series combination will be analyzed separately in this manner.

The total current flows through resistance R_B, which is the equivalent resistance of the parallel combination R_1 and R_2. In Figure 6–20(c), the voltage across R_B was computed to be 1.485 V. This result leads to the following very helpful fact. The voltage across a parallel combination is the same as the voltage across its equivalent resistance. Thus, the voltage across both R_1 and R_2 are the same as that across R_B:

$$V_{R_1} = V_{R_2} = V_{R_B} = 1.485\ \text{V}$$

Since the resistance of R_1 and R_2 are equal, there is an equal division of the total current between them. So,

$$I_{R_1} = I_{R_2} = \frac{I_T}{2} = \frac{1.98\ \text{mA}}{2} = 0.99\ \text{mA}$$

The currents through R_5 and R_6 can be found in several ways. Once again, the voltage across these resistors is the same as that across their equivalent resistance. The voltage across R_5 and R_6 is the same as the voltage across R_C. Using Ohm's law, their currents are found by

$$I_{R_5} = \frac{V_{R_C}}{R_5}$$

$$= \frac{8.46\ \text{V}}{4.7\ \text{k}\Omega} = 1.8\ \text{mA}$$

$$I_{R_6} = \frac{V_{R_C}}{R_6}$$

$$= \frac{8.46\ \text{V}}{47\ \text{k}\Omega} = 0.18\ \text{mA}$$

Both these currents could be found using the current divider equation introduced in Chapter 5:

$$I_{R_5} = \frac{R_T}{R_5} I_T$$

$$= \frac{4.273\ \text{k}\Omega}{4.7\ \text{k}\Omega} 1.98\ \text{mA} = 1.8\ \text{mA}$$

$$I_{R_6} = \frac{R_T}{R_6} I_T$$

$$= \frac{4.273\ \text{k}\Omega}{47\ \text{k}\Omega} 1.98\ \text{mA} = 0.18\ \text{mA}$$

NOTE

You will find points within any problem of series-parallel analysis at which you can check the accuracy of the results. An example is the computations just made of I_{R5} and I_{R6}. Since they divided the total current, their sums should equal I_T. In a similar fashion, the sum of voltage drops, such as V_{R3} and V_{R4}, should equal the total voltage across them. These uses of Kirchhoff's current and voltage laws can be quite helpful in checking for errors.

The next step in the circuit analysis of the circuit in Figure 6–20(a) is to compute the power developed in each resistor:

$$P_{R1} = I_{R1}{}^2 R_1$$

$$= (0.99 \times 10^{-3} \text{ mA})(1.5 \text{ k}\Omega) = 1.47 \text{ mW}$$

Since R_2 has the same value as R_1,

$$P_{R2} = P_{R1} = 1.47 \text{ mW}$$

$$P_{R3} = I_T{}^2 R_3$$

$$= (1.98 \times 10^{-3} \text{ mA})^2(1.2 \text{ k}\Omega) = 4.7 \text{ mW}$$

$$P_{R4} = I_T{}^2 R_4$$

$$= (1.98 \times 10^{-3} \text{ mA})^2(3.9 \text{ k}\Omega) = 15.3 \text{ mW}$$

$$P_{R5} = I_{R5}{}^2 R_5$$

$$= (1.8 \times 10^{-3} \text{ mA})^2(4.7 \text{ k}\Omega) = 15.2 \text{ mW}$$

$$P_{R6} = I_{R6}{}^2 R_6$$

$$= (0.18 \times 10^{-3} \text{ mA})^2(47 \text{ k}\Omega) = 1.52 \text{ mW}$$

The final step in the circuit analysis procedure is computing total power. The total power developed by the circuit is the sum of the power developed by each resistor. Thus,

$$P_T = P_{R1} + P_{R2} + P_{R3} + P_{R4} + P_{R5} + P_{R6}$$

$$= 1.47 \text{ mW} + 1.47 \text{ mW} + 4.7 \text{ mW} + 15.3 \text{ mW} + 15.2 \text{ mW} + 1.52 \text{ mW}$$

$$= 39.66 \text{ mW}$$

The total power may also be computed using the values for the total resistances and total current:

$$P_T = I_T{}^2 R_T$$

$$= (1.98 \times 10^{-3} \text{ mA})^2(10.123 \text{ k}\Omega)$$

$$= 39.66 \text{ mW}$$

EXAMPLE 6–11 Analyze the series parallel circuit of Figure 6–21(a).

■ **Solution:**

The circuit is more easily grasped if redrawn as in Figure 6–21(b).

1. Determine total resistance, beginning from as far from the source as possible and working backward to combine all series and parallel combinations until only R_T is left.

FIGURE 6–21 Circuits for Example 6–11

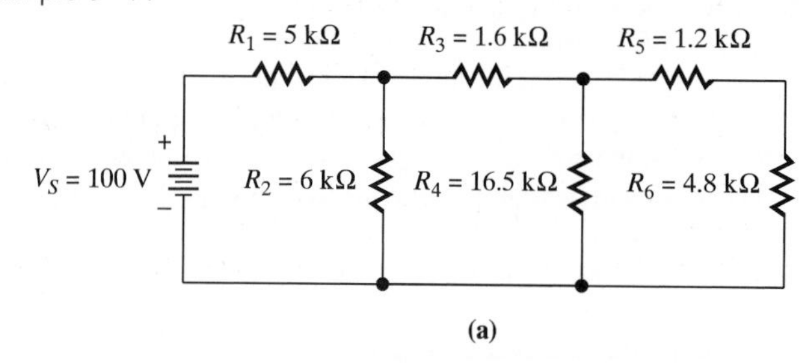

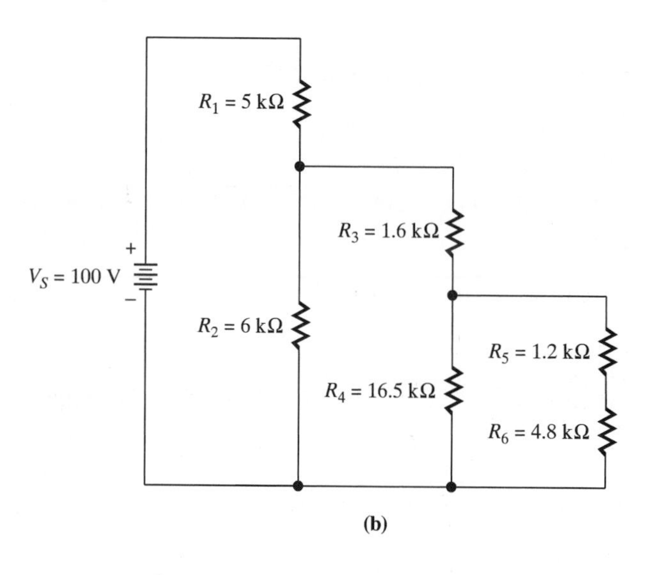

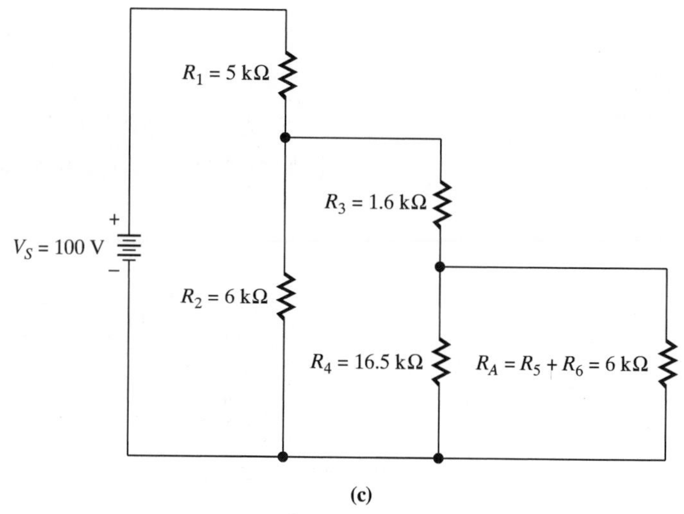

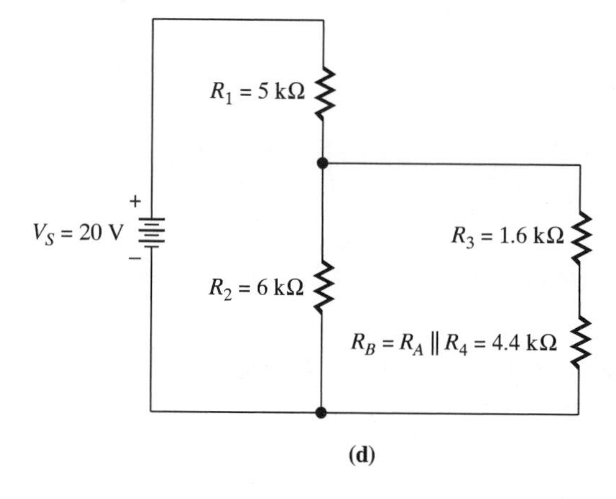

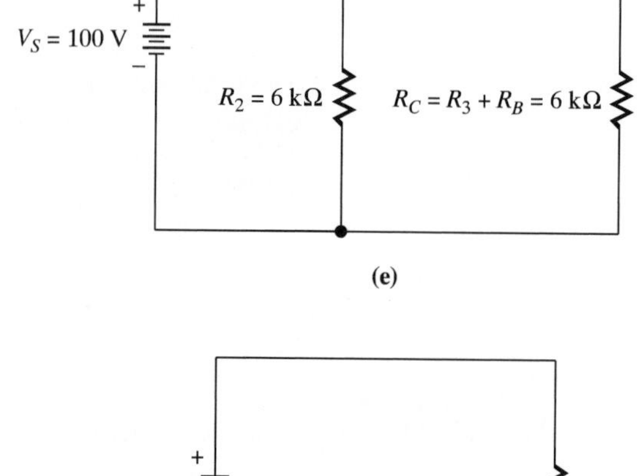

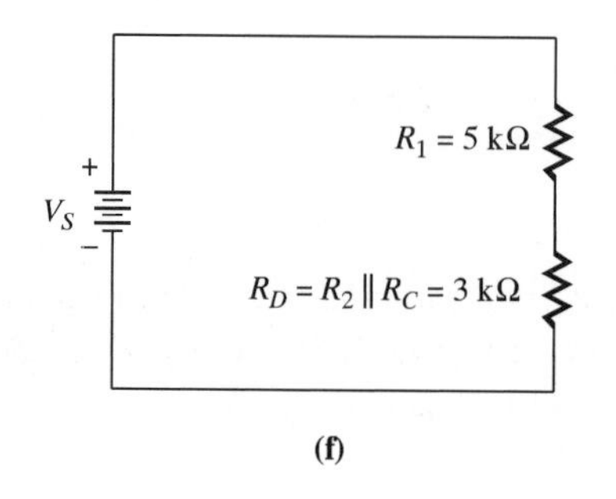

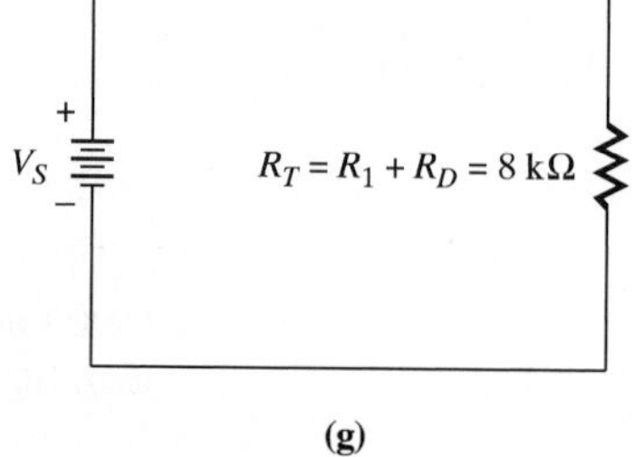

a. Combine R_5 and R_6 in series (Figure 6–21(c)):

$$R_A = R_5 + R_6$$
$$= 1.2\ \text{k}\Omega + 4.8\ \text{k}\Omega = 6\ \text{k}\Omega$$

[1] [·] [2] [EXP] [3] [+] [4] [·] [8] [EXP] [3] [=]

b. Combine the parallel resistance of R_A and R_4 to yield R_B (Figure 6–21(d)):

$$R_B = \frac{1}{(1/R_A) + (1/R_4)}$$
$$= \frac{1}{(1/6\ \text{k}\Omega) + (1/16.5\ \text{k}\Omega)} = 4.4\ \text{k}\Omega$$

[6] [EXP] [3] [1/x] [+] [1] [6] [·] [5] [EXP] [3] [1/x] [=] [1/x]

c. Next combine the series combination of R_3 and R_B to find R_C (Figure 6–21(e)):

$$R_C = R_3 + R_B$$
$$= 1.6\ \text{k}\Omega + 4.4\ \text{k}\Omega = 6\ \text{k}\Omega$$

[1] [·] [6] [EXP] [3] [+] [4] [·] [4] [EXP] [3] [=]

d. Combine the R_C combination, which is in parallel with R_2, to yield R_D (Figure 6–21(f)):

$$R_D = \frac{R_2}{2} \quad \text{[equal resistances]}$$
$$= \frac{6\ \text{k}\Omega}{2} = 3\ \text{k}\Omega$$

[6] [EXP] [3] [÷] [2] [=]

e. Finally, combine R_1 and R_D in series to form R_T (Figure 6–21(g)):

$$R_T = R_1 + R_D$$
$$= 5\ \text{k}\Omega + 3\ \text{k}\Omega = 8\ \text{k}\Omega$$

[5] [EXP] [3] [+] [3] [EXP] [3] [=]

2. Compute the current through and voltage across each resistance.

a. Begin by computing the total current value using R_T and V_S:

$$I_T = \frac{V_S}{R_T}$$
$$= \frac{100\ \text{V}}{8\ \text{k}\Omega} = 12.5\ \text{mA}$$

[1] [0] [0] [÷] [8] [EXP] [3] [=]

b. Now work outward from the source, computing the current through and the voltage across each resistor and resistance. Since I_T flows through R_1 (Figure 6–21(e)), compute the voltage drop across R_1 by

$$V_{R1} = I_T R_1$$

$$= (12.5 \text{ mA})(5 \text{ k}\Omega) = 62.5 \text{ V}$$

1 2 . 5 EXP ± 3 × 5 EXP 3 =

c. Next compute the voltage drop across resistance R_D:

$$V_{RD} = I_T R_D$$

$$= (12.5 \text{ mA})(3 \text{ k}\Omega) = 37.5 \text{ V}$$

1 2 . 5 EXP ± 3 × 3 EXP 3 =

Note that this 37.5 V is the voltage across the parallel combination of R_2 and R_C. (Do you see why?)

d. Now compute the currents, which can be done most simply by recognizing that the resistances are equal and the current divides evenly:

$$I_{RC} = I_{R2} = \frac{I_T}{2} = \frac{12.5 \text{ mA}}{2} = 6.25 \text{ mA}$$

e. Since the current through R_C is the same current flowing through the series combination of R_3 and R_B from which it is made, compute the voltage drops as follows:

$$V_{R3} = I_{RC} R_3$$

$$= (6.25 \text{ mA})(1.6 \text{ k}\Omega) = 10 \text{ V}$$

6 . 2 5 EXP ± 3 × 1 . 6 EXP 3 =

f. Next compute the voltage across R_B, which, by Kirchhoff's voltage law, is the difference between the drop across R_3 and the total across the two:

$$V_{RB} = V_{RD} - V_{R3}$$

$$= 37.5 \text{ V} - 10 \text{ V} = 27.5 \text{ V}$$

3 7 . 5 − 1 0 =

g. Since the voltage across R_B is the same as the voltage across the parallel combination of R_4 and R_A from which it is made, compute the currents through R_4 and R_A as follows:

$$I_{R4} = \frac{V_{RB}}{R_4}$$

$$= \frac{27.5 \text{ V}}{16.5 \text{ k}\Omega} = 1.67 \text{ mA}$$

2 7 . 5 ÷ 1 6 . 5 EXP 3 =

h. Use Kirchhoff's current law to compute the current through I_{RA}

$$I_{RA} = I_{RB} - I_{R4}$$

$$= 6.25 \text{ mA} - 1.67 \text{ mA} = 4.58 \text{ mA}$$

6 · 2 5 EXP ± 3 − 1 · 6 7 EXP ± 3 =

i. Since the current through R_A is the same as the current through the series combination of R_5 and R_6 from which it is made, compute the voltage drop across R_5 and R_6 by

$$V_{R5} = I_{RA}R_5$$
$$= (4.58 \text{ mA})(1.2 \text{ k}\Omega) = 5.5 \text{ V}$$

4 · 5 8 EXP ± 3 × 1 · 2 EXP 3 =

j. Use Kirchhoff's voltage law to compute the voltage across R_6:

$$V_{R6} = V_{RB} - V_{R5}$$
$$= 27.5 \text{ V} - 5.5 \text{ V} = 22 \text{ V}$$

2 7 · 5 − 5 · 5 =

3. Compute the power developed in each resistance. This step in the analysis procedure and Step 4, computing the total power, will be left to the student.

SECTION 6–4 REVIEW

1. What is the object of circuit analysis?
2. What are the steps in circuit analysis of a series-parallel circuit?
3. Perform analysis of the circuit shown in Figure 6–22.

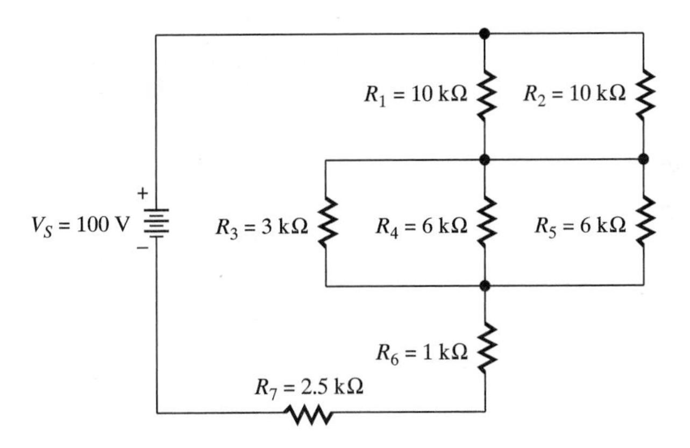

FIGURE 6–22 Circuit for Section Review Problem 3

MAKING IN-CIRCUIT MEASUREMENTS

When working in an actual circuit and making resistance measurements, many parallel and series-parallel paths exist. In order to accurately measure the value of a particular resistor, one end must be removed from the circuit.

To simplify troubleshooting, many schematics diagrams include "resistance to ground" at various test points. These values are typical of what would be measured in a normally functioning circuit. In this way, resistance checks may be made without removing any components, thus saving time and eliminating wear and tear on the circuit.

SYNOPSIS

A series-parallel circuit is one in which there are both strings of series resistors and banks of parallel resistors. There are branch points at which current divides. The voltage divides across each resistor bank and each series resistor. Thus a series-parallel circuit is one in which both current and voltage division take place.

The first step in analysis of series-parallel circuits is computation of the total resistance seen by the source. This computation is done by combining all series resistors and parallel banks into equivalent resistances. This procedure results, eventually, in a series string of equivalent resistances that can be combined to form the total resistance.

Next, the total current is computed using the source voltage and the total resistance. Then, working outward from the source, the current through and the voltage across each resistor or resistance is determined using Ohm's law and Kirchhoff's current and voltage laws.

As stated previously, the circuits encountered in electronic systems will be of the series-parallel variety. Regardless of their sophistication, the circuits comprising these systems can be analyzed in a manner similar to that outlined in this chapter. The only difference will be in the devices making up the circuits. These devices may include, in addition to resistors, inductors, capacitors, diodes, transistors, and thyristors, which will be covered later in the electronics curriculum. For now, keep in mind that the purpose of each of these devices is the same as that of the resistor: to control electric current. All that will be necessary in understanding circuits containing such devices is knowledge of how each device operates. The circuit operation, be it series or parallel, will be the same.

6–5 APPLICATIONS OF SERIES-PARALLEL CIRCUITS

Question: What are the major applications of series-parallel circuits?

As stated previously, nearly all electronic systems are composed of circuits of the series-parallel variety. In many instances, although the applications seem different, a series-parallel circuit acts as what is known as a loaded voltage divider. The design and operation of a loaded voltage divider is explored in this section. Another application of the series-parallel circuit, the Wheatstone bridge circuit, is also presented.

Loaded Voltage Divider

Question: How does a loaded voltage divider differ from an unloaded voltage divider?

As introduced in Chapter 4, a voltage divider is basically a series circuit that divides a source voltage into several smaller voltages. It thus allows a single source to be used in supplying the various voltages required in electronic circuit operation. In the voltage divider of Chapter 4, the voltages were divided, but *no loads* were placed across them.

Figure 6–23(a) illustrates an unloaded voltage divider. From a single voltage source, there are now three voltages with respect to circuit common: 20 V from point C to common, 12 V from point B to common, and 4 V from point A to common. These voltages can be used as sources for various parts of an electronic circuit.

In Figure 6–23(b), a load requiring 4 V is to be placed from point A to common. If the load draws 1.6 mA of current, its resistance can be computed as follows:

$$R_L = \frac{V_{R_L}}{I}$$

$$= \frac{4\text{ V}}{0.0016\text{ A}} = 2.5\text{ k}\Omega$$

FIGURE 6–23 Series-parallel circuits as voltage dividers

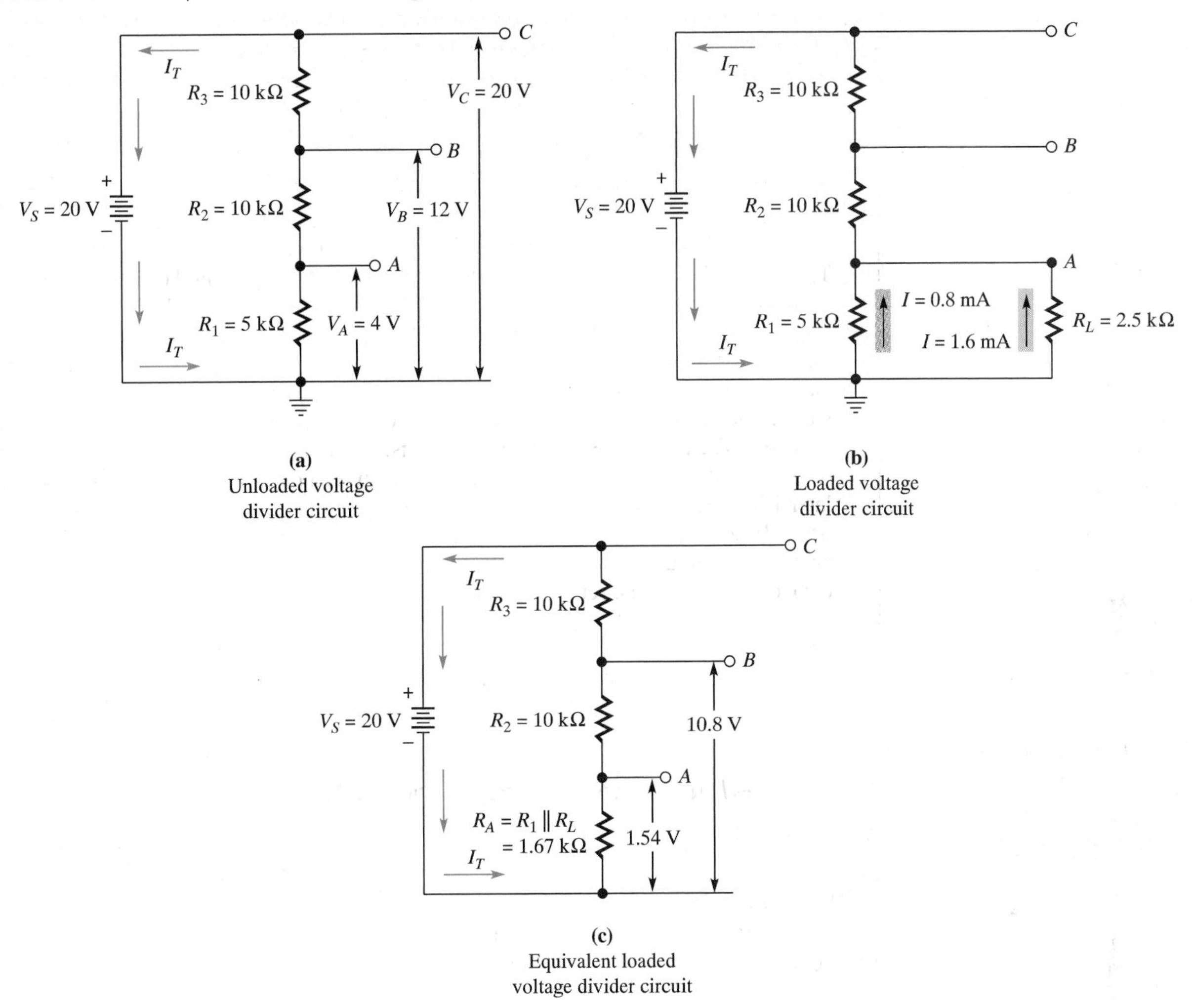

With the load in place, the circuit is the parallel combination of R_1 and R_L in series with R_2 and R_3. The equivalent circuit is shown in Figure 6–23(c). Notice that point A to common is only 1.67 kΩ with a voltage drop of 1.54 V. Obviously, this voltage will not suffice for the 4 V load requirement. It is said that the low resistance placed across the source has loaded it. A voltage divider circuit with loads in place is what is known as a **loaded voltage divider.**

Referring once again to Figure 6–23(c), loading also has an effect on the voltage from point B to common. With the load in place, the resistance from point B to common is

$$R_B = R_2 + R_A$$
$$= 10\ \text{k}\Omega + 1.67\ \text{k}\Omega = 11.67\ \text{k}\Omega$$

The voltage from point B to common, now known as V_B, is

$$V_B = \frac{R_B}{R_T} V_S$$
$$= \frac{11.67\ \text{k}\Omega}{21.67\ \text{k}\Omega} 20\ \text{V} = 10.8\ \text{V}$$

Notice that this voltage is well below its original design value. The only voltage that will not change is V_C, which equals V_S, the source itself.

The load to be applied to a circuit must be taken into consideration when a voltage divider is designed. That is, the voltage required for each load and the amount of current that it will draw must be known. Then resistors can be chosen for the divider to supply the desired voltages under loaded conditions. The loaded voltage divider designed in the example that follows shows these principles and also provides a review of series-parallel circuit analysis.

A voltage divider is to be designed having a source of 180 V that will divide across three loads. Load 3 requires 120 V and will draw 24 mA of current. Load 2 requires 48 V and will draw 19.2 mA of current. Load 1 requires 28 V and will draw 5.6 mA of current. The circuit is shown in Figure 6–24, where the divider is made up of resistors R_1, R_2, R_3, and R_4.

Recall that in the design of an unloaded divider, the same current flowed through each resistor. In the case of the loaded voltage divider, in Figure 6–24, however, the divider current flows through only R_4. R_3 has the divider current, plus that flowing through R_{L_1}; R_2 has the current that flows through R_3, plus that which flows through R_{L_2}; and R_1 has the current that flows through R_2, plus that flowing through R_{L_3}. Thus, each resistor has a different current that must be determined in order to compute its value.

The first step in the design is to decide upon a divider current. Any current could be chosen, but a good rule-of-thumb is to use *approximately one-tenth* of the total load current. The required total load current in this case is 48.8 mA, so a divider current of 5 mA is chosen. This amount of current flows through R_4. The voltage from point A to common must be 28 V, which is the voltage across R_4. The value of R_4 can thus be computed:

$$R_4 = \frac{V_A}{I}$$

$$= \frac{28\text{ V}}{5 \times 10^{-3}\text{A}} = 5.6\text{ k}\Omega$$

The voltage from point B to common must be 48 V. The voltage across R_3 is the difference between the voltages from points B and A to common.

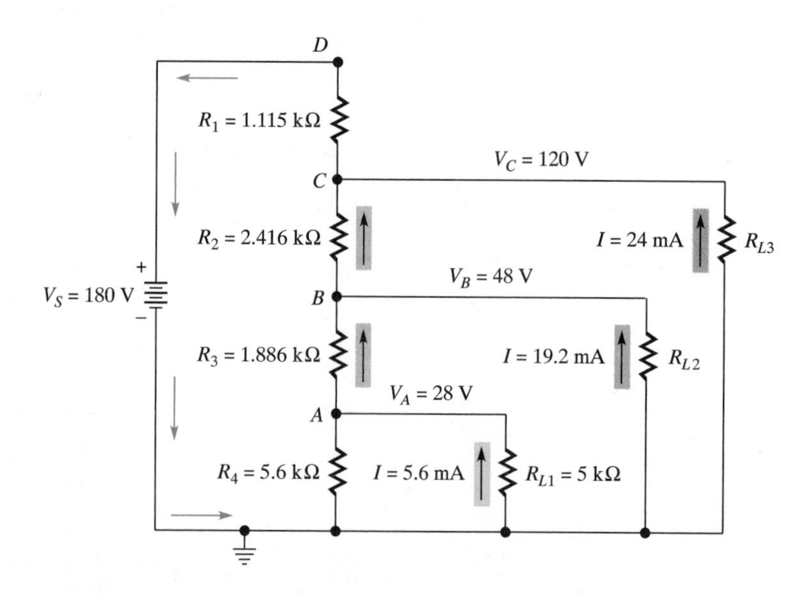

FIGURE 6–24 Circuit to illustrate design of voltage divider with three loads

$$V_{R3} = V_B - V_A$$
$$= 48\text{ V} - 28\text{ V} = 20\text{ V}$$

Notice that the divider current flowing through R_4 and the current through R_{L1} combine at point A and flow through R_3. Thus,

$$I_{R3} = I_{R4} + I_{RL1}$$
$$= 5\text{ mA} + 5.6\text{ mA} = 10.6\text{ mA}$$

The value of R_3 can now be computed:

$$R_3 = \frac{V_{R3}}{I_{R3}}$$
$$= \frac{20\text{ V}}{10.6\text{ mA}} = 1.886\text{ k}\Omega$$

The voltage from the point C to common must be 120 V. The voltage across R_2 is the difference in the voltages from points C to B to common:

$$V_{R2} = V_C - V_B$$
$$= 120\text{ V} - 48\text{ V} = 72\text{ V}$$

The current through R_2 is the current through R_3, plus the current through R_{RL2}. Thus,

$$I_{R2} = I_{R3} + I_{RL2}$$
$$= 10.6\text{ mA} + 19.2\text{ mA} = 29.8\text{ mA}$$

The values of R_2 can now be computed:

$$R_2 = \frac{V_{R2}}{I_{R2}}$$
$$= \frac{72\text{ V}}{29.8\text{ mA}} = 2.416\text{ k}\Omega$$

The voltage from point D to common is the source voltage, 180 V. The voltage across R_1 is the difference in the voltages from points D and C to common:

$$V_{R1} = V_D - V_C$$
$$= 180\text{ V} - 120\text{ V} = 60\text{ V}$$

The current through R_1 is the current through R_2, plus the current through R_{L3}. Thus,

$$I_{R1} = I_{R2} + I_{RL3}$$
$$= 29.8\text{ mA} + 24\text{ mA} = 53.8\text{ mA}$$

The value of R_1 can now be computed:

$$R_1 = \frac{V_{R1}}{I_{R1}}$$
$$= \frac{60\text{ V}}{53.8\text{ mA}} = 1.115\text{ k}\Omega$$

The resistors making up the divider have voltage values of

$$R_1 = 1.115\ \text{k}\Omega$$

$$R_2 = 2.416\ \text{k}\Omega$$

$$R_3 = 1.886\ \text{k}\Omega$$

$$R_4 = 5.6\ \text{k}\Omega$$

The divider voltages thus have tolerances dependent upon the needs of the loads. Standard values of resistors would be chosen that develop voltages within this tolerance.

This completes the design of the loaded voltage divider. Carefully review each step and understand why and how it was done. This example is an excellent demonstration of the principles of series-parallel circuit analysis!

The Wheatstone Bridge

The **Wheatstone bridge,** often referred to as a resistance bridge, is used extensively in instruments employed in industrial processes for measuring temperature and weight. A Wheatstone bridge is shown schematically in Figure 6–25(a) and is redrawn in Figure 6–25(b) in order to help simplify the explanation of its operation.

The bridge has two sets of terminals: one for the input and one for the output. Points A and B are the input terminals and have either an ac or dc voltage across them. Points C and D are the output terminals. The bridge has two conditions: balanced and unbalanced. In balanced condition, the voltage between the output terminals is zero. In the unbalanced

FIGURE 6–25 Wheatstone bridge circuit operation

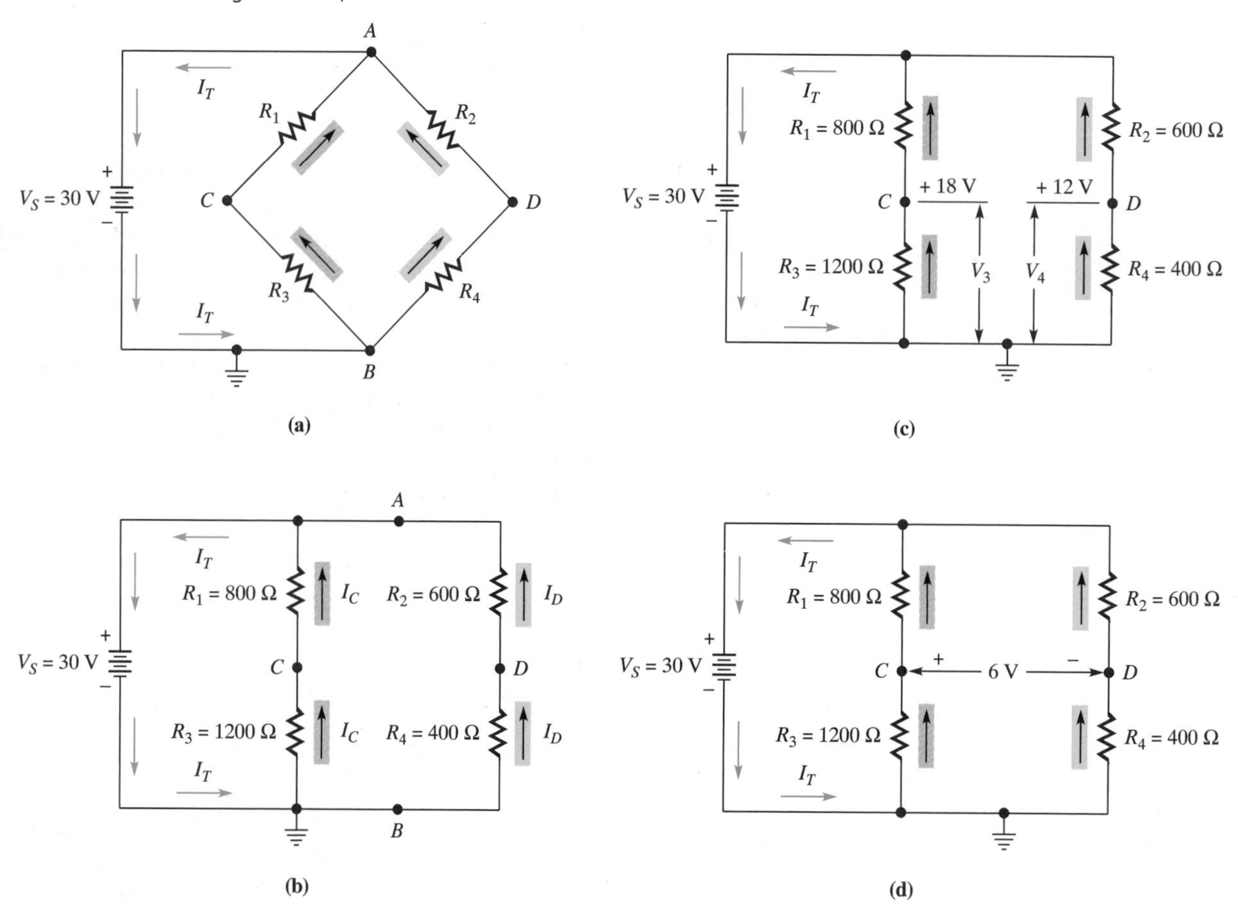

condition, a voltage exists across the output terminals, and its value and polarity are dependent on the degree of imbalance. The conditions establishing a balanced or unbalanced condition are discussed in the following paragraphs.

Refer to Figure 6–25(b). The bridge is a series-parallel circuit. The series circuit, composed of R_1 and R_3, is in parallel with the series circuit made up of R_2 and R_4. Current flows through both branches, producing voltage drops across all resistors. Point C, with respect to common, is positive, and its value is the drop across R_3. Point D, with respect to common, is positive also, and its value is the drop across R_4.

Recall that a voltage is a difference in potential between two points. Thus, if these two voltages are *equal* and the *same polarity,* there is no difference in potential between points C and D, and a voltmeter placed between them will indicate zero voltage. Under these conditions, the output of the bridge is zero, and it is said to be balanced. This balanced condition can be explained using the ratio of the voltages within the bridge.

If V_{R3} is equal to V_{R4} in the balanced condition, then it is also true that V_{R1} is equal to V_{R2}. The ratios of the two voltages in each branch are equal:

$$\frac{V_{R1}}{V_{R3}} = \frac{V_{R2}}{V_{R4}}$$

Substituting yields

$$\frac{I_{R1}R_1}{I_{R3}R_3} = \frac{I_{R2}R_2}{I_{R4}R_4}$$

Canceling gives the conditions for a balanced bridge:

(6–1)
$$\frac{R_1}{R_3} = \frac{R_2}{R_4}$$

Thus, if the ratios of the resistances in each branch are equal, then the output of the bridge is zero, and it is balanced.

If the resistances ratios in a Wheatstone bridge are not equal, then the bridge is in the unbalanced condition and has some output. Referring once again to Figure 6–25(b), assume the voltage drops across R_3 and R_4 are not equal. Thus, the voltage from points C and D to common are also unequal, as illustrated in Figure 6–25(c). Points C and D measure, respectively, 18 V and 12 V with respect to common. Both voltages are positive but point C is more positive than point D. Thus, point C is positive with respect to D, and conversely, point D is negative with respect to C. The net result is shown in Figure 6–25(d) where 6 V is between the two points with polarities. If the voltage from point D to common were larger than point C, then the polarities would be reversed.

One use made of the Wheatstone bridge circuit is in determining the value of an unknown resistance. Recall that the condition of a balanced bridge is expressed in the following equation:

$$\frac{R_1}{R_3} = \frac{R_2}{R_4}$$

This equation can be transposed in order to derive an equation for any of the resistors. For example, R_3 can be found by cross multiplying

$$\frac{R_1}{R_3} = \frac{R_2}{R_4}$$

$$R_2R_3 = R_1R_4$$

FIGURE 6–26 Wheatstone bridge circuit for determining unknown resistance value

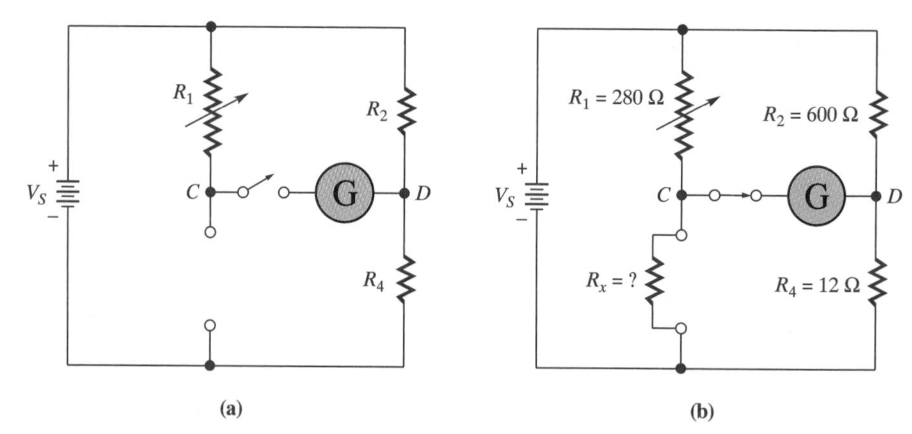

and dividing by R_2

$$R_3 = \frac{(R_1R_4)}{R_2}$$

This procedure can be applied to the circuit of Figure 6–26(a) in order to determine the value of the unknown resistance. Notice three differences in this circuit: (1) R_3 has been omitted, (2) resistor R_1 is now a variable resistor, and (3) a galvanometer has been placed between points C and D. A **galvanometer,** one of which is shown in Figure 6–27, is a very sensitive, center reading, microammeter. Its zero point is in the center of the scale, and it deflects one way or the other, depending on the polarity of the applied voltage. It is used to indicate a difference in potential between points C and D of the bridge. A center reading would mean no voltage between points C and D, and the bridge would be balanced.

FIGURE 6–27 Galvanometer

An unknown resistor, placed across the open part of the bridge as shown in Figure 6–26(b), will cause the bridge to be unbalanced by creating a voltage from points C to D and thus causing the galvanometer to deflect. Variable resistor R_1 is adjusted to make the galvanometer read zero. A zero reading on the galvanometer indicates zero volts between points C and D, which occurs when the resistance ratios of the two branches are equal and the bridge is balanced. The resistance of the variable resistor is measured and inserted, along with the fixed values of R_2 and R_4 into the equation in order to compute the value of the unknown resistance. Precision instruments for making such measurements are available. They are made of precision resistors and contain a calibrated dial for reading the value of the variable resistor. The following example demonstrates this process.

EXAMPLE 6–12 In Figure 6–26(b), the fixed resistors are R_2 (600 Ω) and R_4 (12 Ω). If the variable resistor must be adjusted to 280 Ω in order to balance the bridge, what is the value of the unknown resistance?

■ Solution:

$$R_x = \frac{(R_1R_4)}{R_2}$$

$$= \frac{(280\ \Omega)(12\ \Omega)}{600\ \Omega} = \frac{3360\ \Omega}{600\ \Omega} = 5.6\ \Omega$$

[2] [8] [0] [×] [1] [2] [÷] [6] [0] [0] [=]

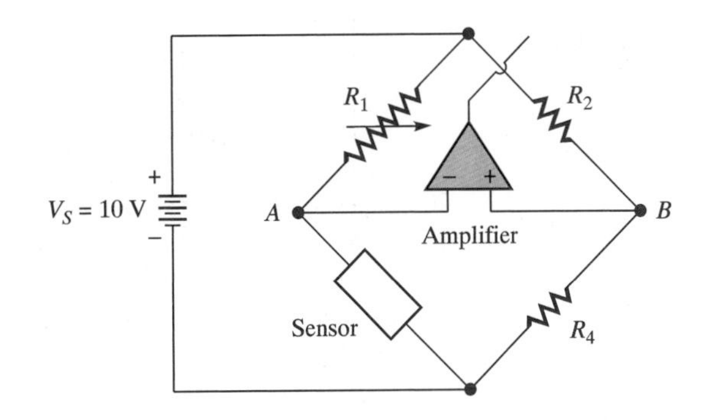

FIGURE 6–28 Wheatstone bridge circuit used for sensor measurements

Another use of the Wheatstone bridge is in measurements of quantities such as temperature, weight, and light. As shown in Figure 6–28, in these applications a sensor replaces one of the resistors in a leg of the bridge. The *sensor* is a device whose resistance varies in response to heat, light, or pressure. The bridge is balanced at some reference value of the sensor resistance. As the measured quantity changes, the resistance of the sensor changes, and the bridge goes out of balance. A voltage is produced across the output of the bridge that is proportional to the change in the sensor's resistance. The voltage is amplified and applied to circuitry that displays or records the quantity.

EXAMPLE 6–13 In Figure 6–28, the sensor has a resistance of 100 Ω. The variable resistor R_1 is set to 135 Ω. R_2 and R_4 have values of 200 and 100 Ω, respectively. What is the voltage developed between points A and B?

■ **Solution:**

The voltages from points A and B to the reference are found using the voltage divider equation:

$$V_A = \frac{R_{\text{sensor}}}{R_{\text{sensor}} + R_1} V_S$$

$$= \frac{100\ \Omega}{100\ \Omega + 135\ \Omega} \times 10$$

$$= \frac{100\ \Omega}{235\ \Omega} 10\ \text{V} = 4.26\ \text{V}$$

$$V_B = \frac{R_4}{R_2 + R_4} V_S$$

$$= \frac{100\ \Omega}{100\ \Omega + 200\ \Omega} \times 10\ \text{V}$$

$$= \frac{100\ \Omega}{300\ \Omega} 10\ \text{V} = 3.33\ \text{V}$$

Then,

$$V_{AB} = V_A - V_B$$

$$= 4.26\ \text{V} - 3.33\ \text{V} = 0.93\ \text{V}$$

SECTION 6–5 REVIEW

1. What is the difference between a loaded and an unloaded voltage divider?
2. Why must the load currents be known in designing a loaded voltage divider?
3. What are two uses of the Wheatstone bridge?
4. What condition causes the Wheatstone bridge to be in balance?
5. In Figure 6–29, compute the voltage between points C and D.
6. For the circuit of Figure 6–30, what value of resistor will cause the bridge to be balanced?

FIGURE 6–29 Circuit for Section Review Problem 5

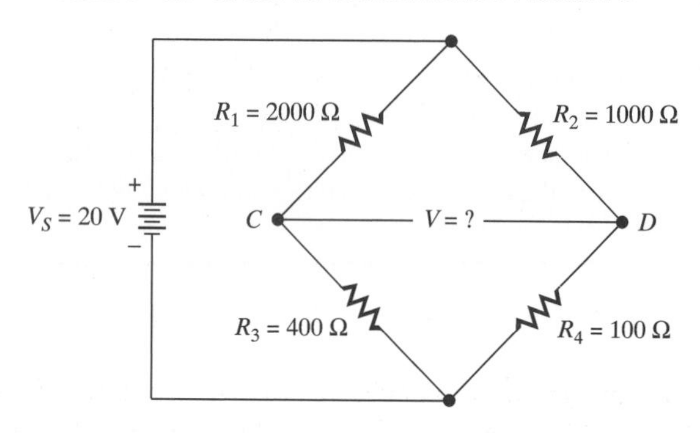

FIGURE 6–30 Circuit for Section Review Problem 6

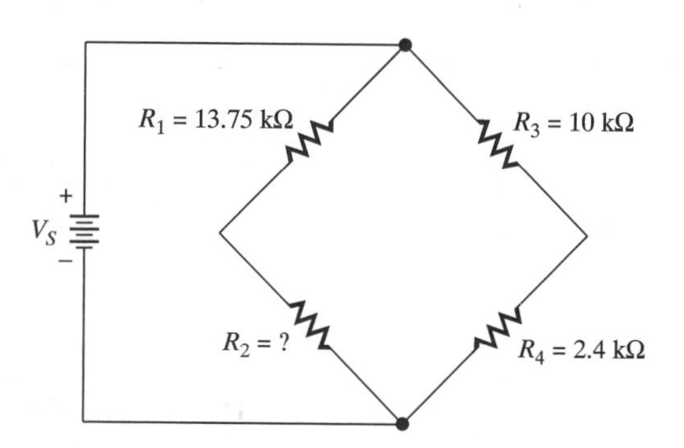

6–6 VOLTAGE MEASUREMENT ACROSS A RESISTOR NOT CONNECTED TO CIRCUIT COMMON

Question: Why are special procedures often necessary when measuring the voltage across a resistor not connected to circuit common?

It is often necessary to measure the voltage across resistors positioned such as R_2 of Figure 6–31. Notice that neither end of this resistor is connected to circuit common. If neither lead of the voltmeter is connected to circuit common, its input is known as a *floating input,* and the measurement can be made as in Figure 6–31(a). If, on the other hand, the neutral lead of the meter and circuit common are referenced together, the circuit would appear as in 6–31(b). Notice now that circuit common has been moved to the bottom of R_2, shorting out R_3. In this situation, a correct reading cannot be obtained.

Question: What special procedures are necessary in measuring the voltage across a nonreferenced resistor?

Figure 6–32 illustrates a method of measuring a resistor's voltage drop without placing the meter across it. Two readings are taken, one from each side of the resistor to circuit common. The voltage drop is the difference between these two voltages. Thus, the voltage across R_2 may be computed as follows:

$$V_{R2} = V_A - V_B$$

$$= 6.32\text{ V} - 1.78\text{ V} = 4.54\text{ V}$$

Notice that the neutral side of the meter connects to circuit common only. Thus, the circuit configuration cannot be changed by the meter.

FIGURE 6–31 Floating input meter

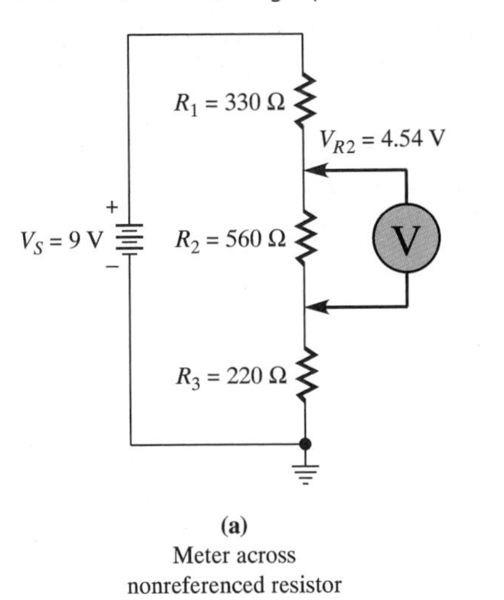

(a)
Meter across
nonreferenced resistor

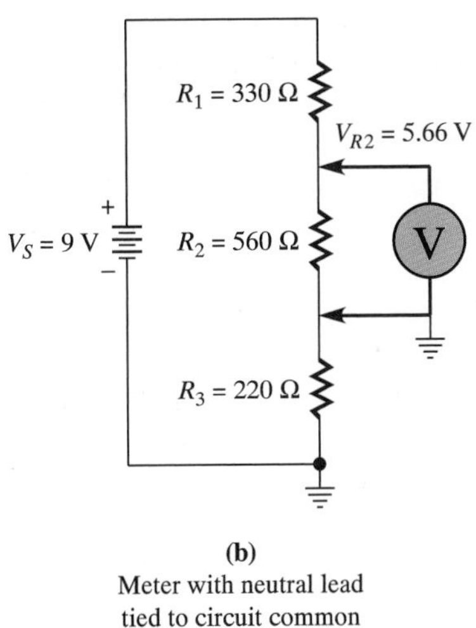

(b)
Meter with neutral lead
tied to circuit common

FIGURE 6–32 Method for measuring voltage across a non-referenced resistor

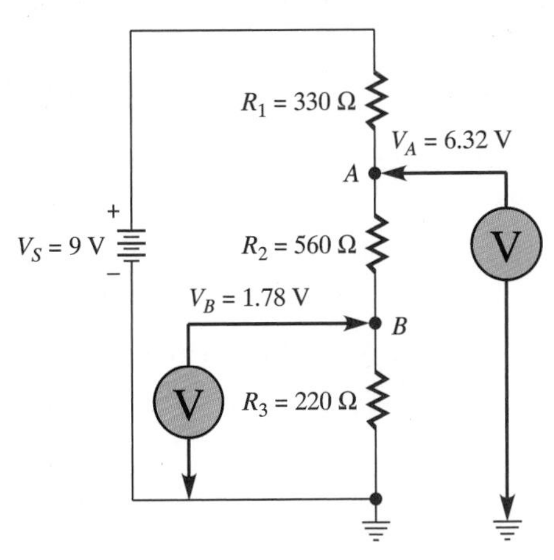

6–7 THE OHMMETER

The discussion here of the ohmmeter and its circuits completes your study of the basic multifunction meter. In Chapter 4, the meter movement and the circuits making up the voltmeter were discussed. In Chapter 5, the circuits of a basic ammeter were discussed. Taken together, these three—voltmeter, ammeter, and ohmmeter—make up the basic functions of what is known as a *multimeter.*

■ The Basic Ohmmeter Circuit

The basic ohmmeter circuit is shown in Figure 6–33. It consists of a meter movement, a current-limiting resistor (R_S), a zero-adjusting rheostat (R_z), and a battery (V_M). The battery provides the current to drive the meter movement. The resistor R_S has a chosen value that allows full-scale current to flow when the leads are shorted together. This full-scale current, which drives the pointer to full scale, also indicates zero ohms of resistance. It is difficult enough to get a value for R_S that allows exactly full-scale current when the leads are shorted; it is impossible to maintain this relationship as the battery ages. Thus, a zeroing rheostat, about one-tenth the value of the current-limiting resistor, is included to finely adjust the zero of the meter. When the leads are open, no current flows. The pointer does not move, which is an indication of infinite ohms.

FIGURE 6–33 Basic ohmmeter

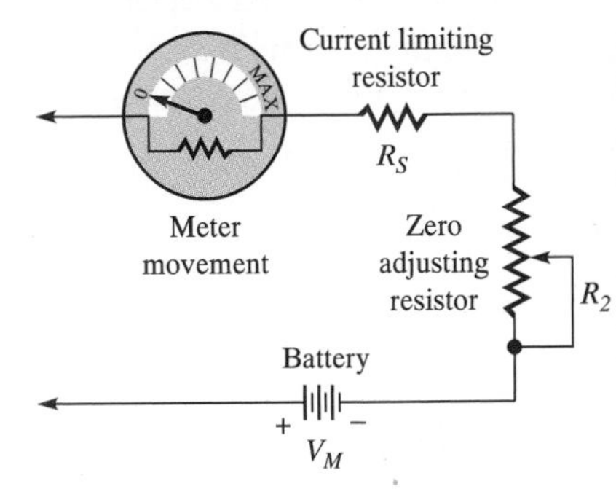

As you may recall, the voltmeter and ammeter have scales that indicate zero on the left and maximum on the right. The ohmmeter scale is opposite: Infinite is on the left, and zero is on the right. Thus, it's a common error, especially among beginning technicians, to see "zero ohms" when the pointer of the ohmmeter doesn't move. This reading is not zero but infinity. Although this error is common, its one you want to avoid!

■ The Ohmmeter Scale

As you know from laboratory exercises, the scale of an ohmmeter is *nonlinear.* That is, the scale is wide at lower values of resistance and compresses at larger values. To understand this type of scale, consider the circuit of Figure 6–34(a). R_S is a lumped resistance of 30 kΩ representing the sum of the internal resistance of the movement, the current-limiting resistor, and the rheostat. The meter movement has a sensitivity of 50 μA for full-scale deflection. Thus, when the leads are shorted, the 1.5 V battery supplies full-scale current, and the pointer deflects from infinity to zero. If the pointer does not reach full-scale deflection, the rheostat of Figure 6–33 is adjusted to allow full-scale current to flow.

Question: How is the nonlinear ohmmeter scale developed?

If a resistor is placed between the ohmmeter leads, the total resistance will increase, and the amount of deflection of the pointer will decrease. The amount of decrease in deflection is proportional to the value of resistance. Thus, a resistance value can be assigned to the amount of deflection of the meter. For example, in Figure 6–34(b) if a resistance of 30 kΩ is placed between the leads, the total resistance doubles, the current is halved, and the pointer moves to the middle of the scale, which is marked "30 kΩ." If a resistance of 60 kΩ is placed between the leads, the resistance is three times the original value, and the current is one-third. The pointer thus moves to the one-third position on the scale, which is marked "60 kΩ." This process could be continued by placing a 120 kΩ resistor between the leads, making the total resistance five times the original value, and the current one-fifth of it. Thus, the pointer only moves to the one-fifth position on the scale, which is marked "120 kΩ."

The development of the complete nonlinear ohms scale can be demonstrated by carrying out the process for a number of resistors. The results for additional resistors are shown in Figure 6–35. The amount of deflection is computed in the table and indicated on a representative scale. Notice that the value of resistance decreases from left to right. Also notice that the scale is wide on the side near zero and compresses as infinity is approached.

FIGURE 6–34 Nonlinear ohmmeter scale

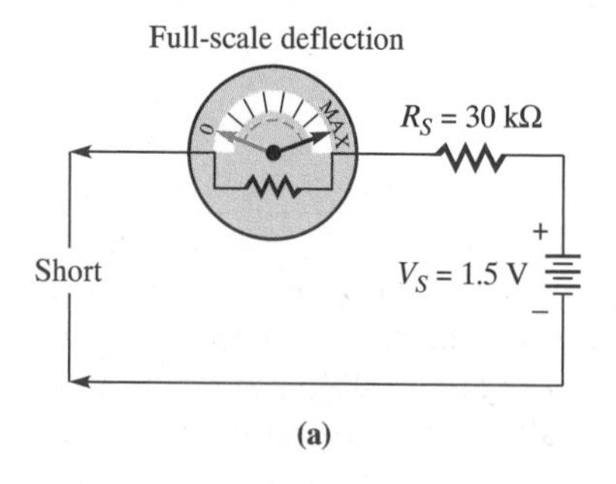

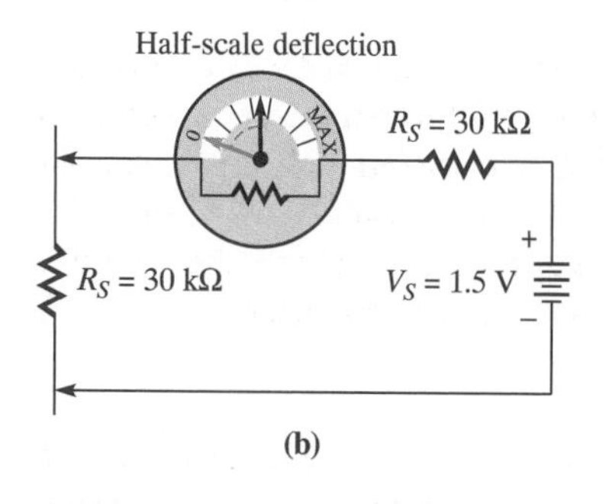

Question: Are there limits on how large or small a resistance can be measured with certain meters?

There are limits to how large or small a resistance can be accurately measured with the ohmmeter in Figure 6–36. Infinite ohms can mean that the circuit or device is open. This

FIGURE 6–35 Calculations for nonlinear ohmmeter scale

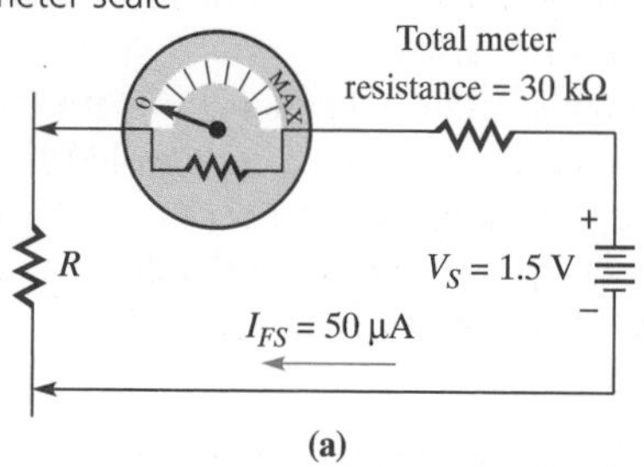

(a)

R	240 kΩ	120 kΩ	60 kΩ	30 kΩ	20 kΩ	10 kΩ	5 kΩ	1 kΩ
I_M	5.6 μA	10 μA	16.7 μA	25 μA	30 μA	37.5 μA	42.9 μA	48.4 μA
% I_{MF}	11 %	20 %	33 %	50 %	60 %	75 %	86 %	97 %
% scale	11 %	20 %	33 %	50 %	60 %	75 %	86 %	97 %

(b)

∞ 240 120 60 30 20 10 5 1

(c)

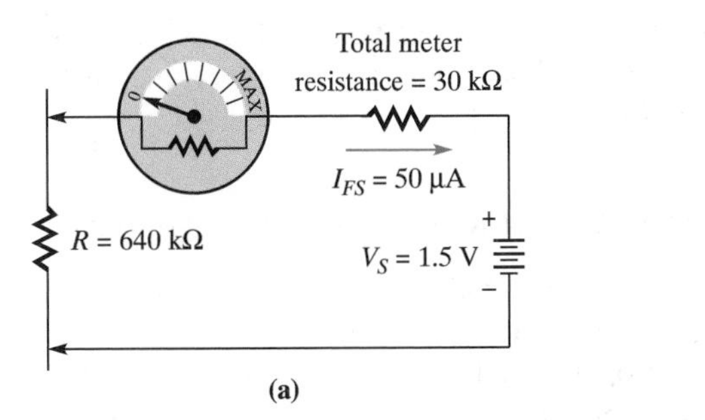

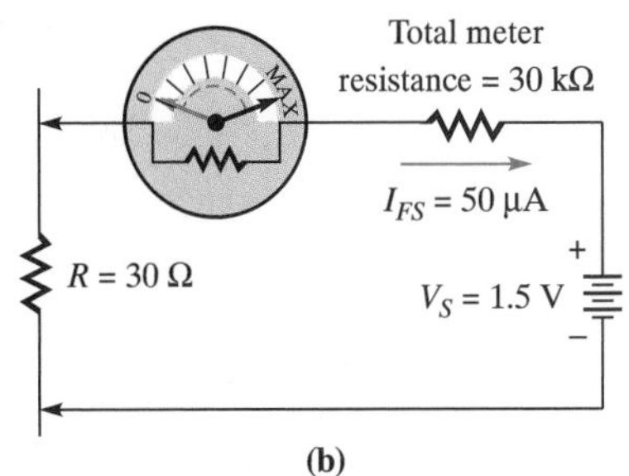

FIGURE 6–36 Ohmmeter scale limitations. The 640 kΩ resistor will only cause about 4% movement of pointer, which could be mistaken for an open circuit.

finding could be the significance of such a reading. The reading can also mean, however, that not enough current is flowing through the meter in order to make it deflect enough to be noticeable. This finding is not the result of an open circuit, but merely a very large resistance in relation to the value of the meter resistance. For example, if a resistance of 640 kΩ is placed between the leads of the ohmmeter of Figure 6–36(a), the deflection would only be four one-hundredths of full scale. For all practical purposes the meter would indicate infinity. A higher voltage or a more sensitive meter movement is required to measure larger resistors.

A resistance many times smaller than the internal resistance of the meter can also cause an ohmmeter reading to be erroneously interpreted. Suppose the resistance to be measured is 30 Ω. The current produced would be 99% of that required for full-scale deflection. As illustrated in Figure 6–36(b) the pointer would move, for all appearances, to full scale. Thus, it is probable that the 30 Ω resistance would be interpreted as being a short circuit. These two paragraphs point out the need for multiple scales in an ohmmeter.

■ Multiple Resistance Ranges

Practical ohmmeters have several ranges that make it possible to measure resistances from less than ohm to several megohms. The ranges of an ohmmeter, however, differ greatly from those of the ammeter and voltmeter. The latter two have a full-scale value while each ohmmeter range extends from zero to infinity. The differences in the ohmmeter ranges comes from their individual *multiplying factor.*

Figures 6–37(a) and (b) show an ohmmeter scale and a range switch. Notice that the ranges are shown as $R \times 1$, $R \times 10$, and $R \times 100$. These ranges indicate the number by which the scale reading must be multiplied in order to obtain the resistance value. On the $R \times 1$ range, the reading is taken directly from the scale. On the $R \times 10$ range, the scale reading is multiplied by 10 for the resistance value, and so on.

Question: How are the various scale values produced in an ohmmeter?

The ranges are produced as illustrated in Figure 6–37(c). As shown in the figure, on the $R \times 1$ scale, the total internal resistance of the meter is shunted by a 30 Ω resistor. Since the internal resistance is 1000 times larger than the shunt, the total resistance is nearly 30 Ω. Thus, if a resistance of 30 Ω is placed between the leads, the current is halved and the pointer moves to the half-scale position, which is marked "30 Ω." The rest of the scale is marked to indicate the resistance directly. On the $R \times 10$ range, the meter is shunted by a 300 Ω resistor. For practical purposes, the total resistance of the meter's internal resistance and the shunt is 300 Ω. It would now take a 300 Ω resistor in series with the meter in order to obtain half-scale deflection. Thus, on this scale it is necessary to multiply the value read by 10. In a similar manner, the $R \times 100$ scale can be developed by using a 3000 Ω shunt.

FIGURE 6–37 Ohmmeter with multiple resistance ranges

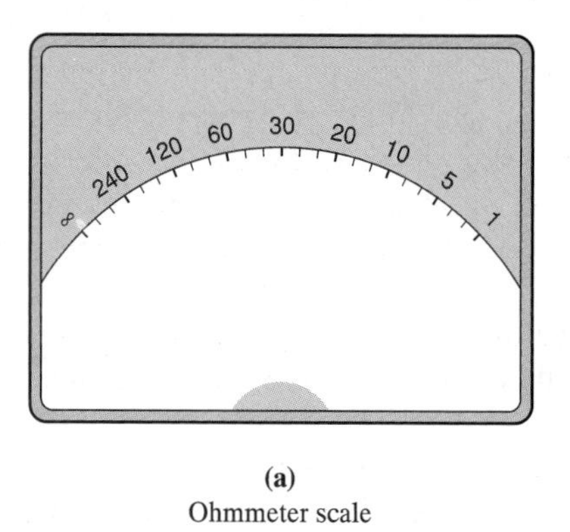

(a)
Ohmmeter scale

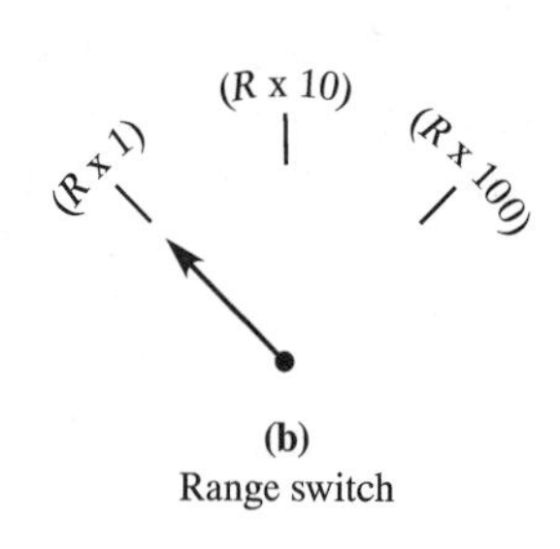

(b)
Range switch

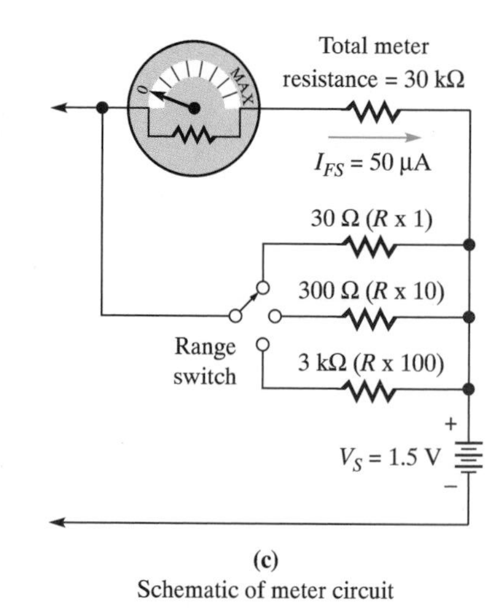

(c)
Schematic of meter circuit

Question: Is there any limitation on how large a scale can be developed?

For this particular meter movement, the $R \times 100$ range is largest for which reasonably accurate readings can be expected. Recall from Chapter 5 that when two resistors are in parallel and one is one-tenth the value of the other, the total resistance is the value of the smaller. In this ohmmeter, on the $R \times 100$ range, the value of the shunt is exactly one-tenth the value of the total resistance of the meter movement. If an $R \times 1000$ range is desired, the shunt has to be 30 kΩ. Notice now that the total resistance is only 15 kΩ. The center scale reading no longer is only 30 or some power of ten greater. If higher resistance ranges are required, it is necessary to use a meter movement with a higher sensitivity or a larger battery voltage. Either of these requires a larger value of current-limiting resistor, increasing the internal resistance of the meter. Thus a larger value of shunt resistor could be used without loading the circuit.

■ SECTION 6–7 REVIEW

1. What is different about the ohmmeter scale as compared to those of the voltmeter and ammeter?
2. What is the purpose of the current-limiting resistor in the basic ohmmeter circuit?
3. Why is a zero-adjusting rheostat needed?
4. What is the purpose of the battery in the ohmmeter?
5. If the meter movement has an internal resistance of 60 kΩ, what is the maximum value of shunt that would not cause appreciable loading?

6–8 TROUBLESHOOTING SERIES-PARALLEL CIRCUITS

Question: In troubleshooting series-parallel circuits, is it necessary merely to apply the principles learned for series and parallel circuits?

You studied troubleshooting for series circuits in Chapter 4 and for parallel circuits in Chapter 5. For series-parallel circuits, it is merely a matter of putting the principles learned together and applying them within the same circuit. The possibilities are the same: Current may be too high, too low, or zero. Each can be caused by the same problems: shorts and opens. Recall that a short, or "short path" for current, reduces resistance and causes the current to increase. An open circuit, on the other hand, causes the resistance to increase and the current to decrease.

A short circuit around a device has several possible causes. It may be caused by small pieces of conductive material, such as bits of solder or metal filings lodging between the traces of a printed circuit board (PCB). A short may also be caused by a breakdown in the insulation of a conductor. The resistance of a short is practically zero, and the current, following the path of least resistance, bypasses the load. A device itself may become a short circuit when subjected to too large a current, especially such devices as transistors and diodes. When checked with an ohmmeter when the power is off, a shorted device should measure near zero ohms. With the power on, the voltage drop across a short should be near zero.

An open circuit may actually be the result of a short circuit. A short circuit causes a large increase in current. If the current exceeds the rating of the fuse, it opens to protect the circuit from further damage. Thus, a blown fuse causes an open circuit. The fuse is usually in the main current line, and when it fails, the total current drops to zero. All voltage drops also will be zero in this case. If a fuse does not blow, a component such as a transistor, diode, or resistor may overheat and burn open. Any device in series with the open one has no current and thus no voltage drop. The source voltage is read across an open device. The total current is lower than normal, and some voltage drops may be lower.

Care must be taken in using an ohmmeter to detect an open circuit. Infinite resistance should be read across an open. If, however, the device is part of a parallel circuitry, the ohmmeter may give a lower, even normal, reading, due to the fact that it is reading the resistance of the parallel components. One end of the suspected device should be disconnected in order to eliminate parallel paths.

Troubleshooting will be demonstrated in the following examples. A circuit and the values of certain quantities will be given. It will be necessary to determine if the values are correct and, if not, what is the probable cause.

EXAMPLE 6–14

In Figure 6–38, one fault exists. From the readings given, determine the probable cause.

FIGURE 6–38 Circuit for Example 6–14

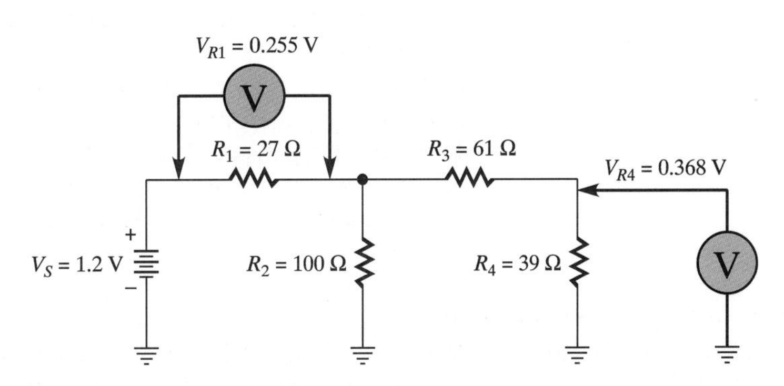

■ Solution:

Determine what the voltage readings should be. First compute the total resistance by finding the combined resistance of R_3 and R_4:

$$\begin{aligned} R_A &= R_3 + R_4 \\ &= 61\ \Omega + 39\ \Omega = 100\ \Omega \end{aligned}$$

This combination is in parallel with R_2. Thus,

$$R_B = \frac{R_2}{2}$$

$$= \frac{100\ \Omega}{2} = 50\ \Omega$$

Since the total resistance is R_B in series with R_1,

$$R_T = R_1 + R_B$$

$$= 27\ \Omega + 50\ \Omega = 77\ \Omega$$

The total current should be

$$I_T = \frac{V_S}{R_T}$$

$$= \frac{1.2\ \text{V}}{77\ \Omega} = 15.6\ \text{mA}$$

The total current flows through R_1, and since R_1's voltage drop is given, the actual total current can be computed:

$$I_T = \frac{V_{R_1}}{R_1}$$

$$= \frac{0.255\ \text{V}}{27\ \Omega} = 9.44\ \text{mA}$$

This last calculation shows that the total current is too low. It follows then that since V_S is given as 1.2 V, the total resistance is too high. One leg of the parallel combination is probably open.

Resistor R_4 has a voltage drop of 0.368 V and a value of 39 Ω. The current flowing through it is

$$I_{R4} = \frac{V_{R4}}{R_4}$$

$$= \frac{0.368\ \text{V}}{39\ \Omega} = 9.44\ \text{mA}$$

Notice that the current through R_1 and R_4 are the same. Thus, they must be in series. The only way they could be in series is if resistor R_2 is open. Replace resistor R_2 to correct the fault.

EXAMPLE 6–15 From the voltmeter readings in Figure 6–39, determine if a fault exists in the circuit, and if so, the probable cause.

■ Solution:

First of all, determine if voltage V_C is correct. To do so, you must determine exactly what the voltmeter is across. One lead is at the top of R_7 and the other is at circuit common. Notice, too, that the bottom of R_7 is at common. So the voltmeter is across R_7. The series

FIGURE 6–39 Circuit for Example 6–15

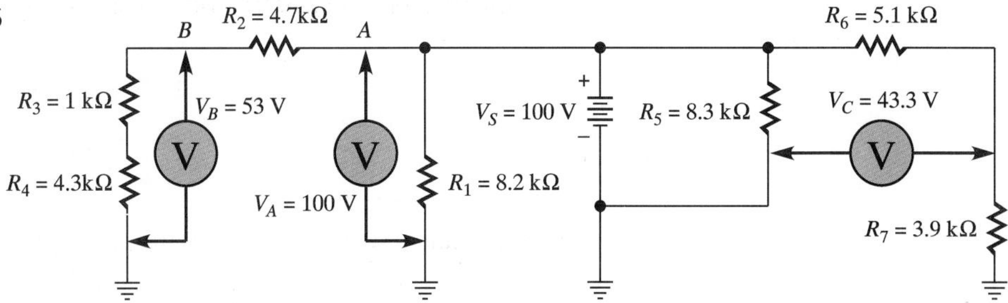

combination of R_6 and R_7 is across the voltage source, so the voltage divider equation can be used to find the voltage across R_7:

$$V_{R7} = \frac{R_7}{R_6 + R_7} V_S$$

$$= \frac{3.9\text{ k}\Omega}{9\text{ k}\Omega} 100\text{ V} = 43.3\text{ V}$$

This result indicates that the voltmeter reading V_C is correct.

Next, notice that the voltage from two points, A and B, to common is given. The difference between these two points is the drop across R_2:

$$V_{R2} = V_A - V_B$$

$$= 100\text{ V} - 53\text{ V} = 47\text{ V}$$

The voltage at point A is correct since R_1 is directly across 100 volt source. The voltage from point B to common is across series combination R_3 and R_4. It too can be computed using the voltage divider equation:

$$R_B = R_3 + R_4$$

$$= 1\text{ k}\Omega + 4.3\text{ k}\Omega = 5.3\text{ k}\Omega$$

$$V_B = \frac{R_B}{R_T} V_S$$

$$= \frac{5.3\text{ k}\Omega}{10\text{ k}\Omega} 100\text{ V} = 53\text{ V}$$

These two voltage readings are also correct, indicating that no fault exists in this circuit.

EXAMPLE 6–16 From the meter readings in Figure 6–40(a), determine if a fault exists in the circuit, and if so, its possible cause.

■ Solution:

This circuit is more easily analyzed if redrawn as in Figure 6–40(b). Notice that it is a series-parallel circuit with three parallel legs. Voltmeter V_1 is actually positioned across R_3. Voltmeter V_2 is across the parallel combination of R_5 and R_6. Notice that this parallel combination is in series with R_4. The correct voltages must be computed and compared to the actual readings.

FIGURE 6–40 Circuits for Example 6–16

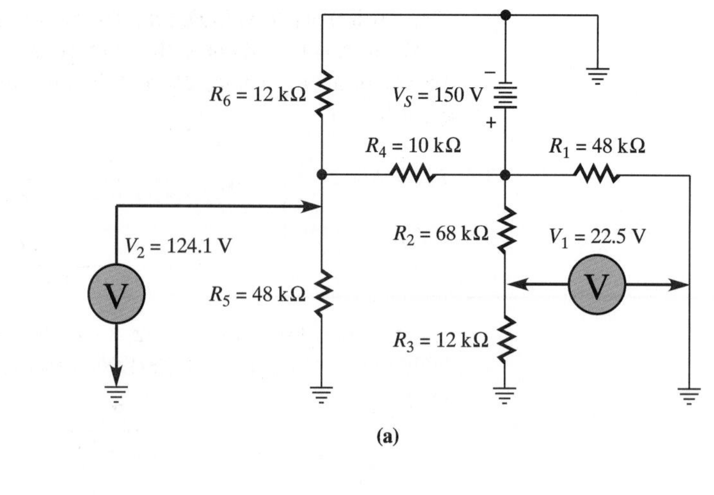

(a)

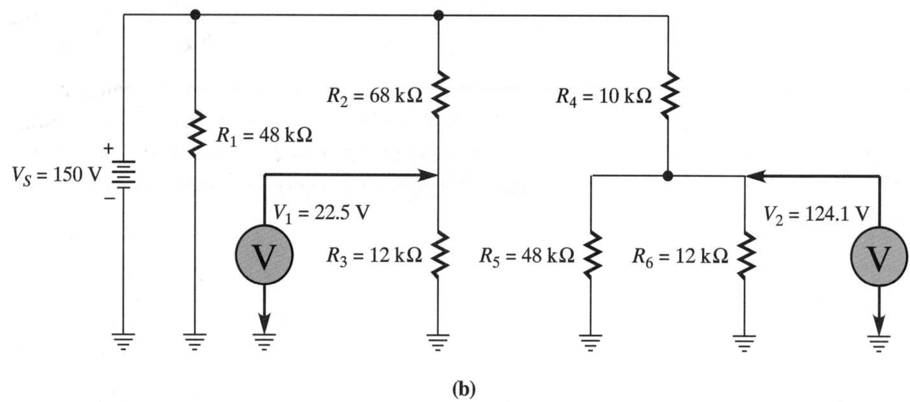

(b)

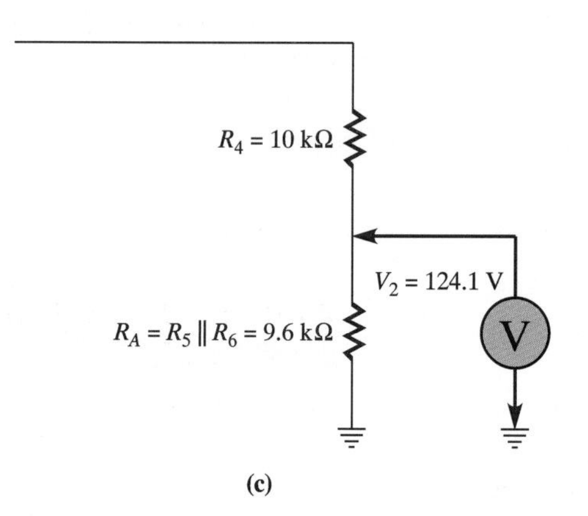

(c)

Notice that R_2 and R_3 are in series across the voltage source. The correct voltage across R_3, which should be indicated by voltmeter V_1, can be computed quite simply using the voltage divider equation:

$$V_{R_3} = \frac{R_3}{R_2 + R_3} V_S$$

$$= \frac{12\ k\Omega}{68\ k\Omega + 12\ k\Omega} 150\ V = 22.5\ V$$

Thus, voltmeter V_1 is indicating the correct voltage.

Voltmeter V_2 indicates the voltage across the parallel combination of R_5 and R_6. To compute this value, R_5 and R_6 are combined into equivalent resistances R_A (Figure 6–40(c)):

$$R_A = \frac{1}{(1/R_5) + (1/R_6)}$$

$$= \frac{1}{(1/48000\ \Omega) + (1/12000\ \Omega)} = 9.6\ \text{k}\Omega$$

This equivalent resistance forms a series circuit with R_4 across the source. Its voltage drop is computed using the voltage divider equation:

$$V_A = \frac{R_A}{R_A + R_4} V_S$$

$$= \frac{9.6\ \text{k}\Omega}{9.6\ \text{k}\Omega + 10\ \text{k}\Omega} 150\ \text{V} = 73.47\ \text{V}$$

The actual voltage reading is much greater than the correct value! The voltmeter shows 124.1 V dropped across the parallel combination. The remainder of the source voltage will be dropped across R_4 (25.9 V). The current flowing in the series combination can now be computed, using the value of R_4 and the voltage drop across it:

$$I_{R_4} = \frac{V_{R_4}}{R_4}$$

$$= \frac{25.9\ \text{V}}{10\ \text{k}\Omega} = 2.59\ \text{mA}$$

The actual resistance of the parallel combination can now be computed using this current and indicated voltage drop. (Recall that current is the same in all parts of the series circuit.) Thus,

$$R_{\text{actual}} = \frac{V_S}{I}$$

$$= \frac{124.1\ \text{V}}{2.59\ \text{mA}} = 48\ \text{k}\Omega$$

The last calculation indicates that 12 kΩ resistor R_6 is open. The problem can be corrected by replacing R_6.

SECTION 6–8 REVIEW

1. List two of the most common faults occurring in an electric circuit.
2. List the symptoms of each of these types of faults.
3. List two possible causes of each type of fault listed.
4. Refer to Figure 6–41. Considering opens and shorts only list all faults that will cause point A (a) to measure zero volts to common; (b) to measure a larger than correct voltage to common; and (c) to measure a near-normal voltage to common.

FIGURE 6–41 Circuit for Section Review Problem 4

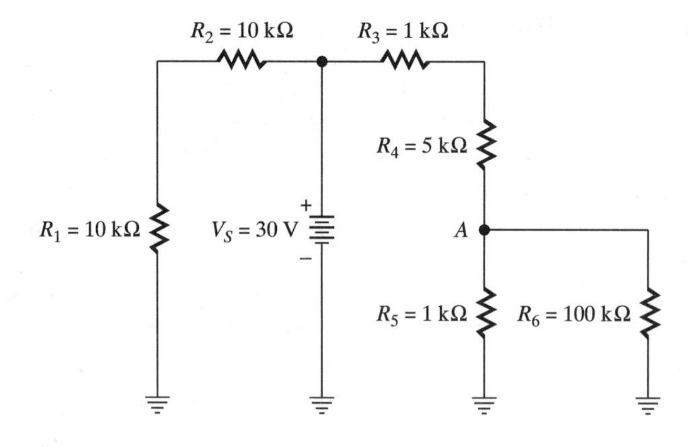

CHAPTER REVIEW

1. A series-parallel circuit is one in which both voltage and current division takes place.
2. A series-parallel circuit contains all characteristics of series and parallel circuits.
3. The source of a series-parallel circuit "sees" a total resistance made up of the combined series and parallel resistances.
4. Analysis of a series-parallel circuit is accomplished by applying analysis procedures used in individual series and parallel analysis.
5. Applications of series-parallel circuits are usually some type of loaded voltage divider.
6. A Wheatstone bridge is a type of series-parallel circuit used in making resistance measurements.
7. The Wheatstone bridge may also be used in instruments used in industrial applications to measure, heat, light, and weight.
8. When the Wheatstone bridge is in the balanced condition, its output voltage is zero.
9. The Wheatstone bridge is balanced when ratios of the resistance in the two legs are equal.
10. The scale of the ohmmeter is nonlinear.
11. The ohmmeter scale indicates infinite ohms on the left edge and zero ohms on the right edge.
12. The zero-adjust on an ohmmeter is used to compensate for aging of the internal battery.

KEY TERMS

series-parallel circuit
equivalent resistance
equivalent circuit
loaded voltage divider
Wheatstone bridge
galvanometer
unloaded voltage divider

EQUATIONS

Equations used in this chapter may be found in Chapters 3, 4, and 5.

(6–1) $$\frac{R_1}{R_2} = \frac{R_3}{R_4}$$

(condition for balanced bridge)

TEST YOUR KNOWLEDGE:

1. Draw the series parallel circuit composed of the following: R_1 and R_2 in parallel and this combination in series with R_3.
2. Draw the series parallel circuit composed of the following: a parallel bank composed of R_1, R_2, and R_3 that is in parallel with the series combination of R_4 and R_5.
3. In your own words, explain the procedure for finding the total resistance in problem 2.
4. A 56 Ω and a 100 Ω resistor are in parallel, and the combination is in series with a 2.2 kΩ resistor. If placed across a voltage source, which will conduct the most current? Why?
5. What determines the amount of output in an unbalanced bridge?
6. If two voltages are both positive with respect to common, explain how one can be negative with respect to the other.
7. Under what condition is a Wheatstone bridge balanced?
8. Which will have more loading effect on a voltage divider, a 1 kΩ or a 1 MΩ resistor? Why?
9. Explain how to measure the voltage drop across a resistor without placing the meter across it?
10. Under what circumstances is the procedure in Problem 9 necessary?
11. What is the effect upon circuit current of a short? Why?
12. What is the effect upon circuit current of an open? Why?
13. What precaution must be taken when measuring the resistance of a resistor while it is installed in a circuit? (Besides ensuring that the power is off.)
14. What is a galvanometer?
15. In your own words, explain how to determine the value of an unknown resistor using the Wheatstone bridge.

PROBLEM SET: BASIC

Section 6–1

1. For the circuits in Figure 6-42, identify the series and parallel combinations.

FIGURE 6–42 Circuits for Problem 1

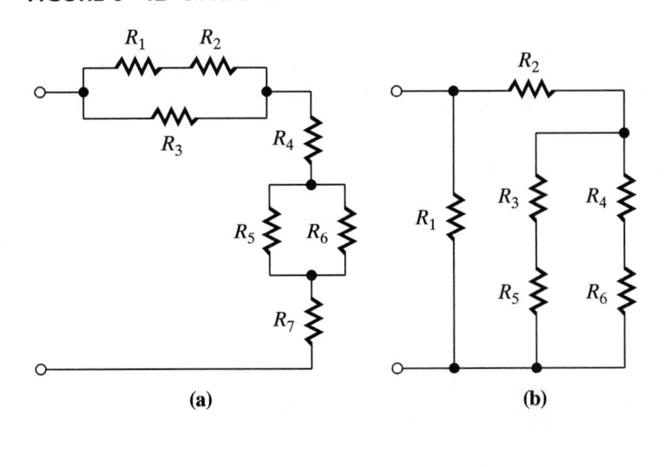

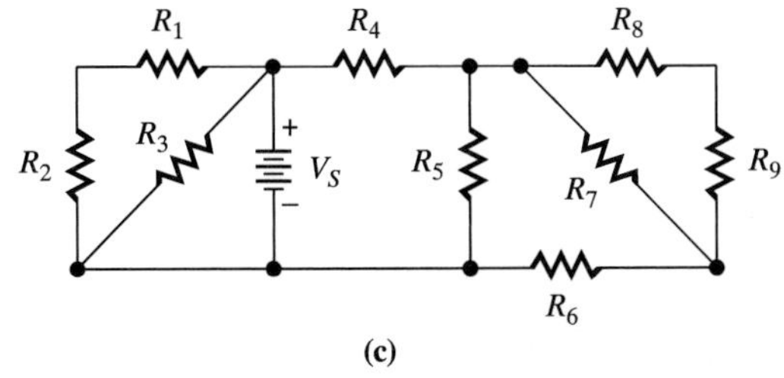

2. For the circuits of Figure 6–43, identify the series and parallel combinations.

FIGURE 6–43 Circuits for Problem 2

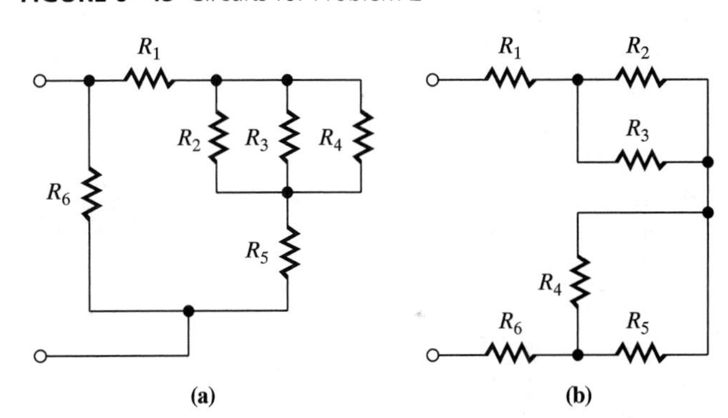

FIGURE 6–44 Circuit for Problem 3

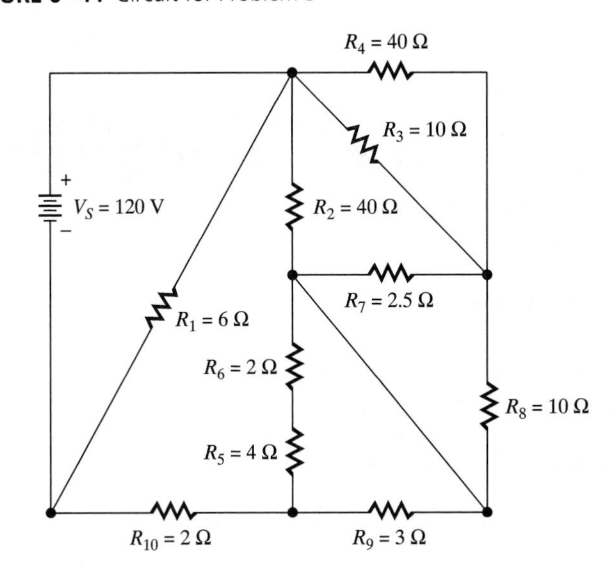

Section 6–2

3. Redraw the series-parallel circuit of Figure 6–44. The source must be at the left of the drawing with all resistors oriented vertically.

4. Using the same procedure as in Problem 3, redraw the circuit of Figure 6–45.

FIGURE 6–45 Circuit for Problem 4

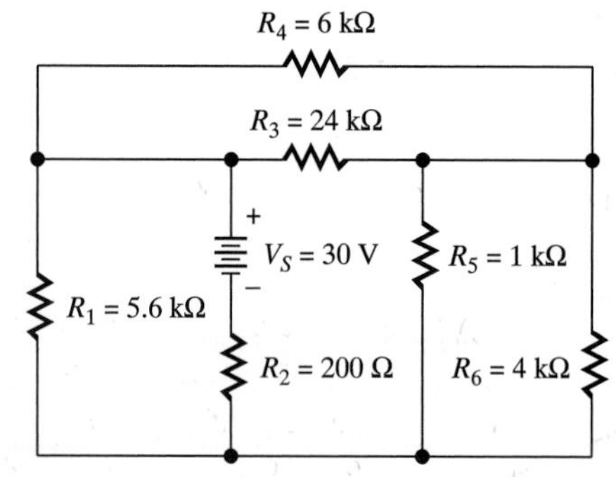

Section 6–3

5. For the circuits of Figure 6–46, compute the total resistance seen by the source.

FIGURE 6–46 Circuits for Problem 5

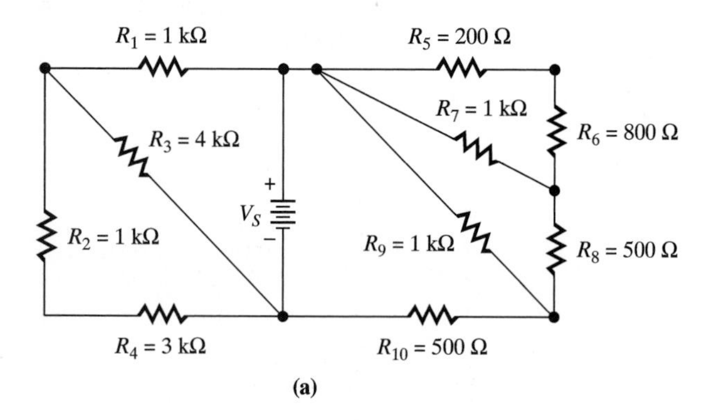

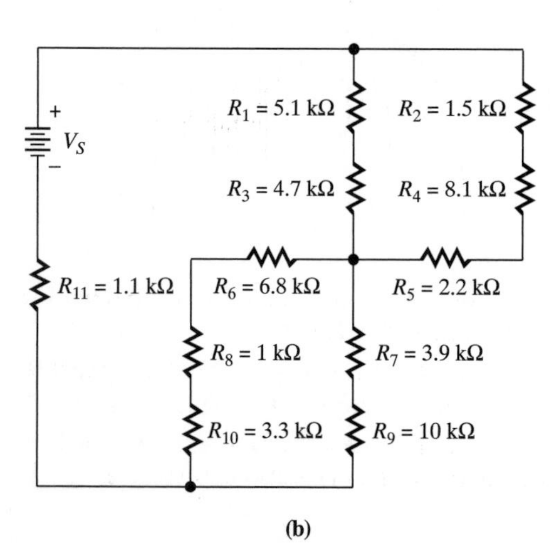

6. For the circuit of Figure 6–47, compute the total resistance seen by the source.

FIGURE 6–47 Circuit for Problem 6

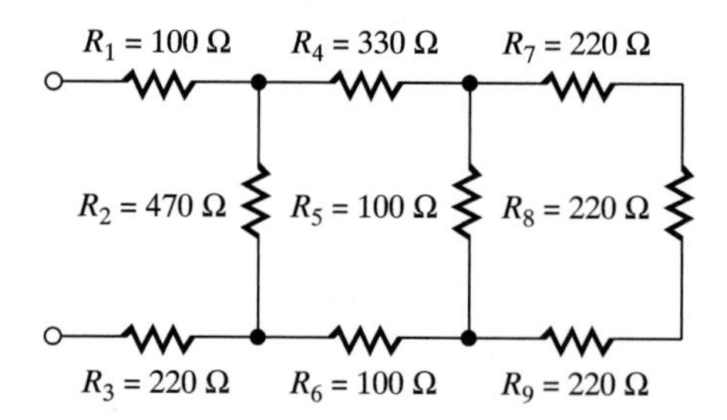

7. For the circuit of Figure 6–48, compute the resistance between points A and B and between points A and C.

FIGURE 6–48 Circuit for Problem 7

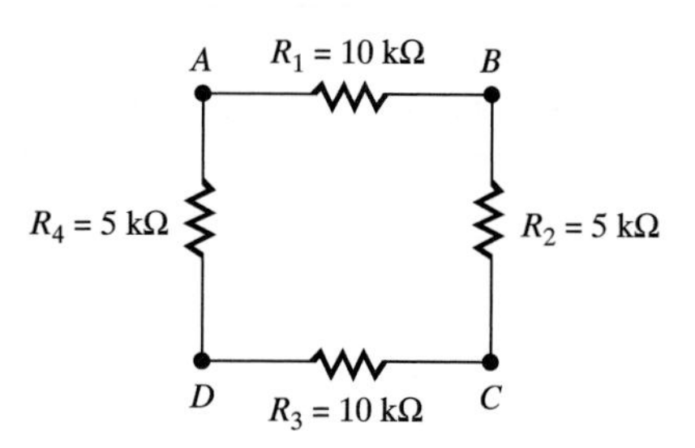

8. In Figure 6–49, what value resistor must be placed in parallel with the resistor bank made up of R_2, R_3, and R_4 in order to reduce the total resistance to 775 Ω?

FIGURE 6–49 Circuit for Problem 8

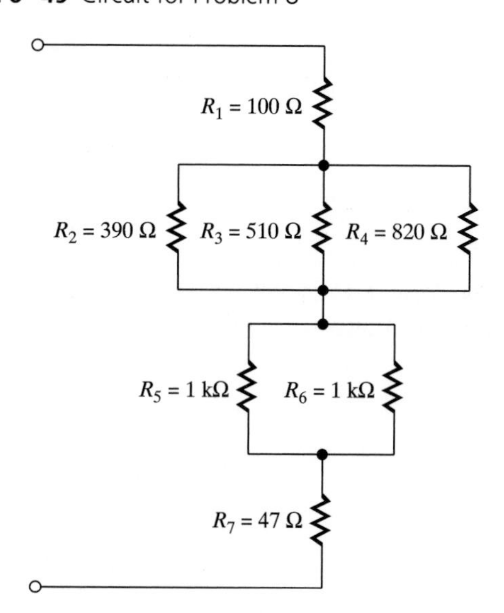

Section 6–4

9. For the circuit of Figure 6–50, compute the current through and the voltage across each resistor.

FIGURE 6–50 Circuit for Problem 9

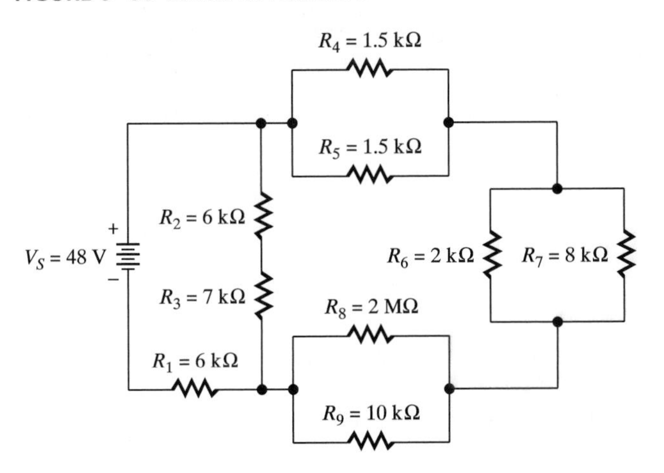

10. For the circuit of Figure 6–51, compute the voltage from each indicated point to common.

FIGURE 6–51 Circuit for Problem 10

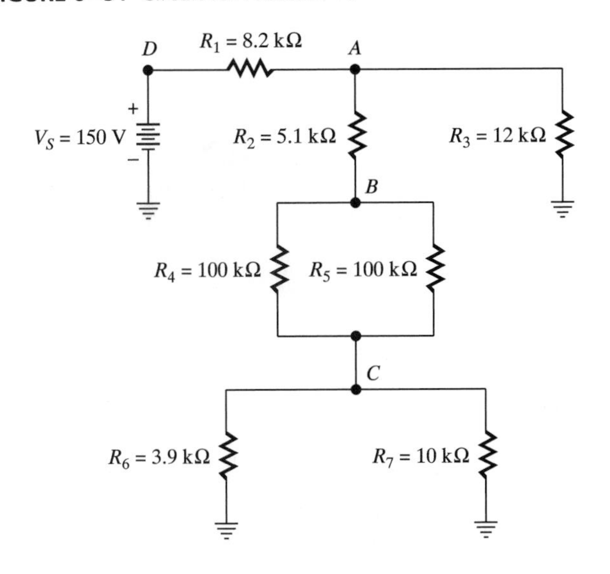

11. For the circuit of Figure 6–52, compute the voltage from each indicated point to common.

FIGURE 6–52 Circuit for Problem 11

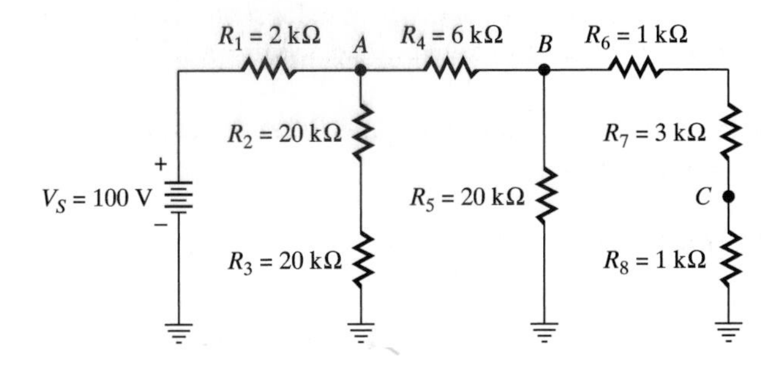

Section 6–6

12. What three voltages are produced by an unloaded voltage divider made up of three 4.7 kΩ resistors across a 15 V source?

13. If a load of 10 kΩ is placed across the smallest voltage in Problem 12, what will the three voltages of the divider be?

14. In Problem 13, would the loading effect be less or greater if the load placed across the divider was 100 kΩ instead of 10 kΩ?

15. An unloaded voltage divider containing two resistors draws 10 mA from a 20 V source. If the voltage across one resistor is 8 V, compute the value of each resistor.

16. Using the voltages given at each point in Figure 6–53, compute the voltage drops for each resistor.

FIGURE 6–53 Circuit for Problem 16

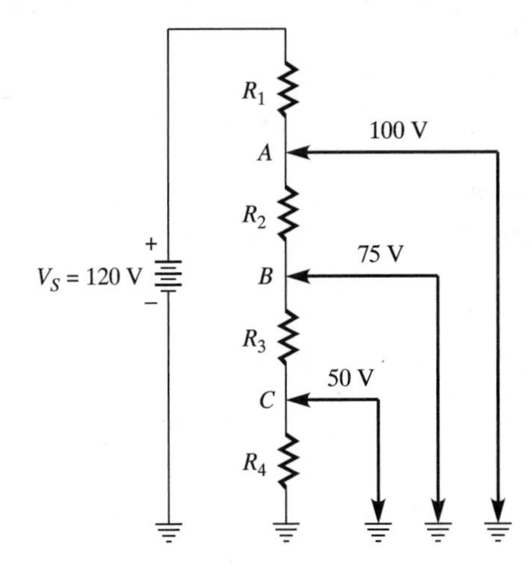

17. For the Wheatstone bridge of Figure 6–54, compute the voltage between points C and D.

18. In Problem 17, what polarity is point D with respect to ground? What polarity is point D with respect to point C?

19. Compute the value of resistance with which you would replace R_4 in order to balance the bridge in Problem 17.

FIGURE 6–54 Circuit for Problem 17, 18 and 19.

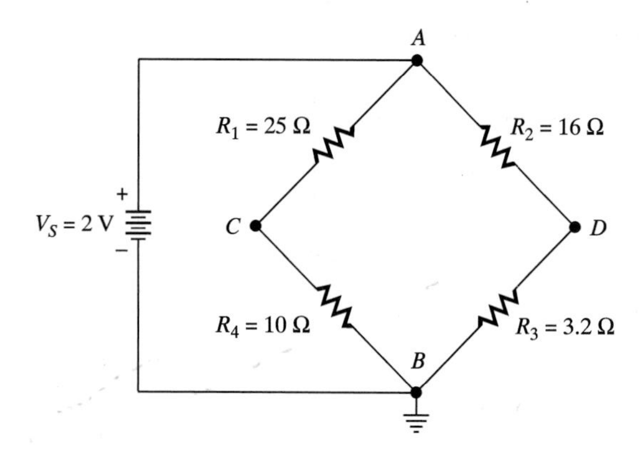

FIGURE 6–55 Circuit for Problem 20

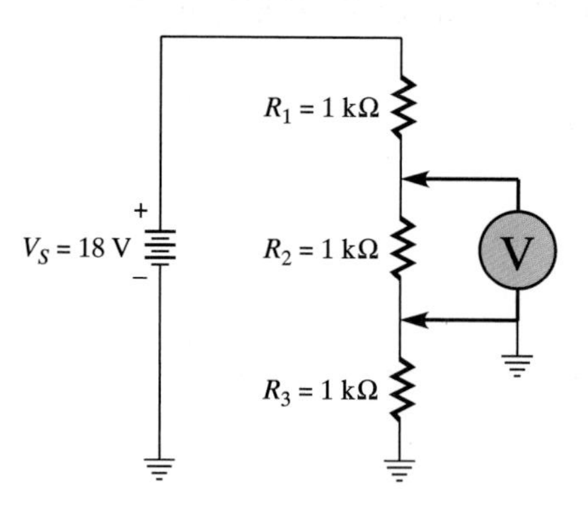

Section 6–7

20. What voltage would the voltmeter of Figure 6–55 indicate?

21. From the voltage readings shown, what is the voltage drop across R_2 of Figure 6–56? What is the value of V_S?

FIGURE 6–56 Circuit for Problem 21

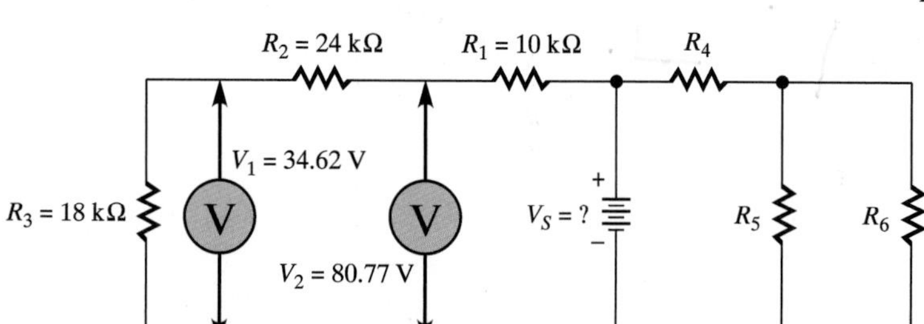

Section 6–8

22. An ohmmeter has a total internal resistance of 10 kΩ. What value of shunt resistor would be used for $R \times 1$ scale?

23. In Problem 22, what value of shunt resistor would be needed for the $R \times 10$ scale?

24. For the ohmmeter of Problem 22, what is the largest value shunt resistor that could be used for accurate resistance readings?

PROBLEM SET: CHALLENGING

25. If the total resistance of the circuit in Figure 6–57 is 3.1 k ohms, what is the value of R_1?

FIGURE 6–57 Circuit for Problem 25

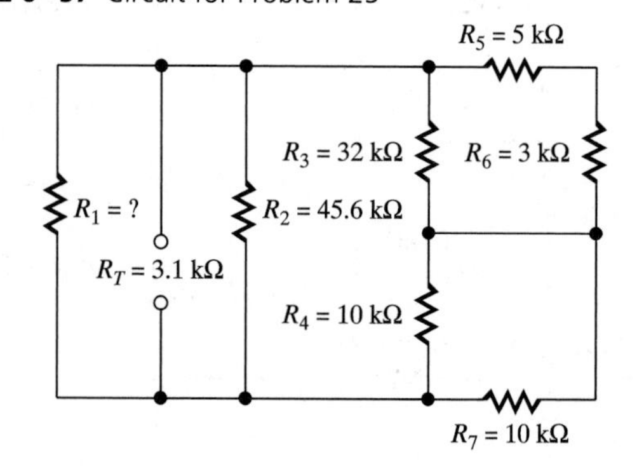

26. Redraw the circuit of Figure 6–58 by placing the source on the left and drawing all resistors vertically.

27. Compute the current through and voltage across each resistor in the circuit of Figure 6–58.

FIGURE 6–58 Circuit for Problems 26 and 27

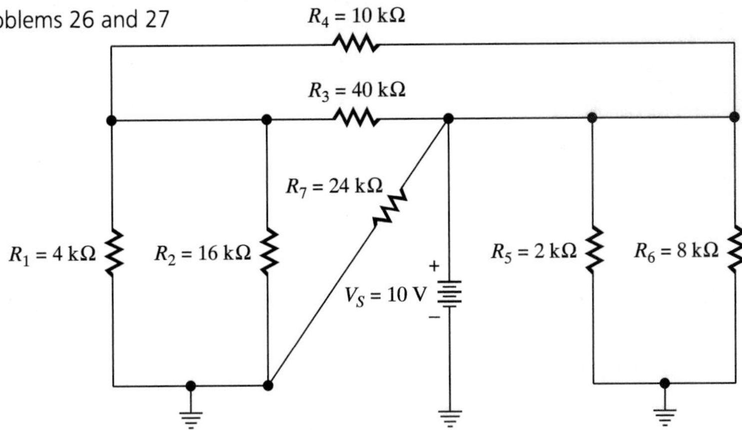

28. For the circuit of Figure 6–59, compute the value of voltages, currents, and resistances not specified.

FIGURE 6–59 Circuit for Problem 28

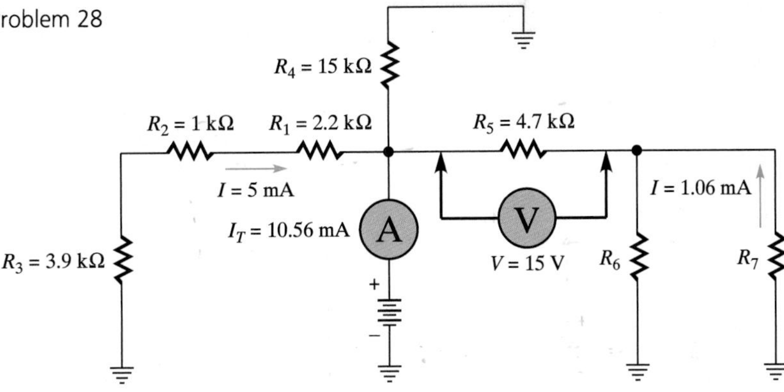

29. Compute the voltages from points A, B, C, and D to circuit common for the circuit of Figure 6–60.

FIGURE 6–60 Circuit for Problem 29

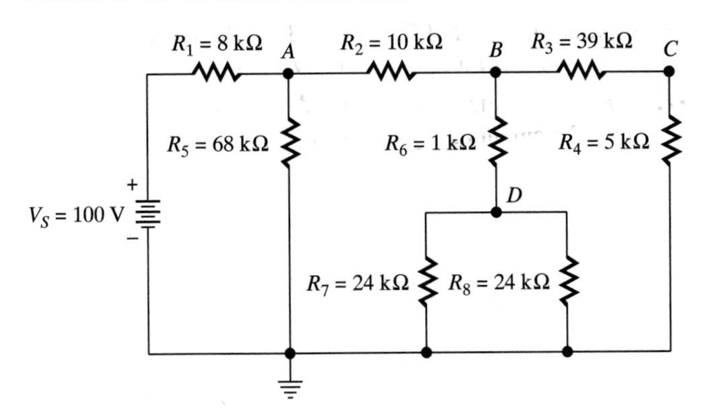

30. In Problem 29, if R_4 were to be open, what would the voltages then be from points A, B, C, and D to circuit common?

31. In Problem 29, if R_8 were to short, what would the voltages then be from points A, B, C, and D to circuit common?

32. Redraw the circuit of Figure 6–61, placing the source on the left and all resistors vertically.

FIGURE 6–61 Circuit for Problems 32 and 33

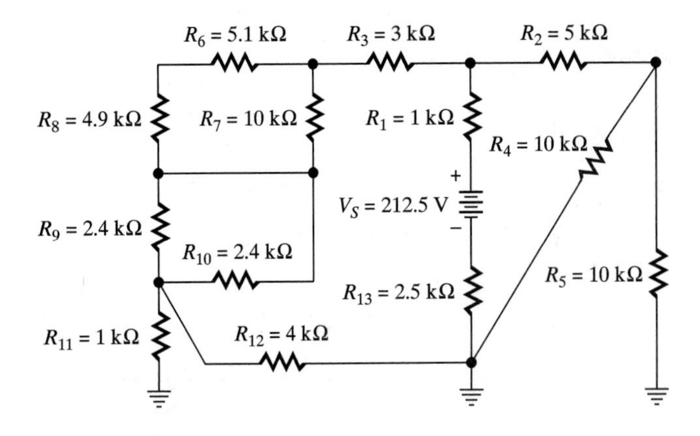

33. Compute the current through and the voltage across each resistor in the circuit of Figure 6–61.

34. Compute the value of the resistors required for the loaded voltage divider of Figure 6–62.

FIGURE 6–62 Circuit for Problem 34

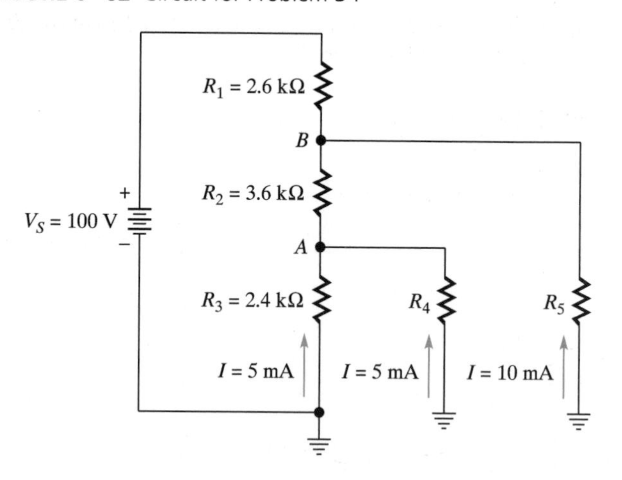

PROBLEM SET: TROUBLESHOOTING

35. Are the voltmeter readings correct in the circuit of Figure 6–63? If not, what is the probable fault?

FIGURE 6–63 Circuit for Problem 35

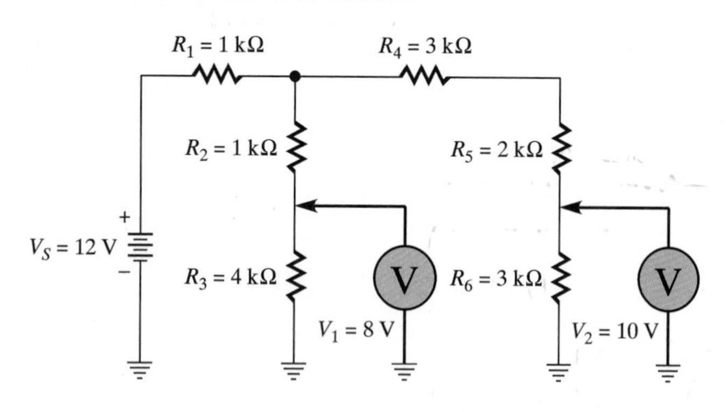

FIGURE 6–64 Circuit for Problem 36

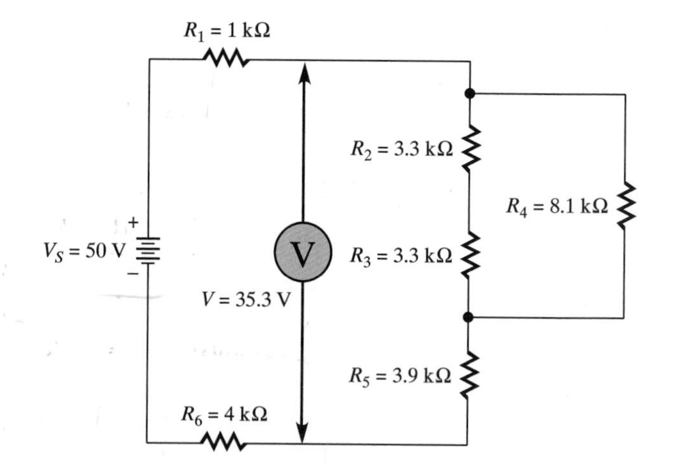

36. Is the voltmeter reading correct in the circuit of Figure 6–64? If not, what is the probable fault?

37. Are the ammeter readings correct in the circuit of Figure 6–65? If not, what is the probable fault?

FIGURE 6–65 Circuit for Problem 37

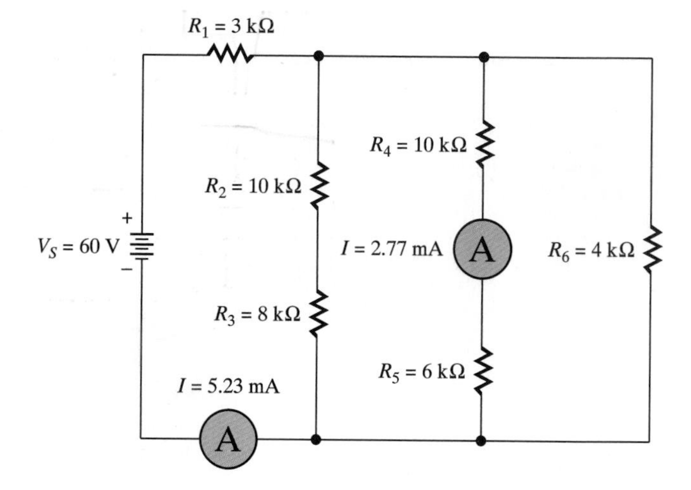

38. In the circuit of Figure 6–66, which resistor, by shorting, would have greatest effect upon voltages from points A, B, C, and D to circuit common?

FIGURE 6–66 Circuit for Problem 38

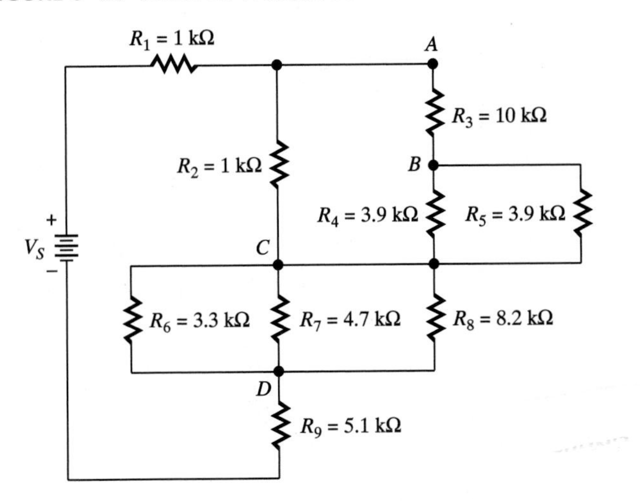

39. For the circuit of Figure 6–67, one fault exists. From the meter readings shown, what is the probable cause?

FIGURE 6–67 Circuit for Problem 39

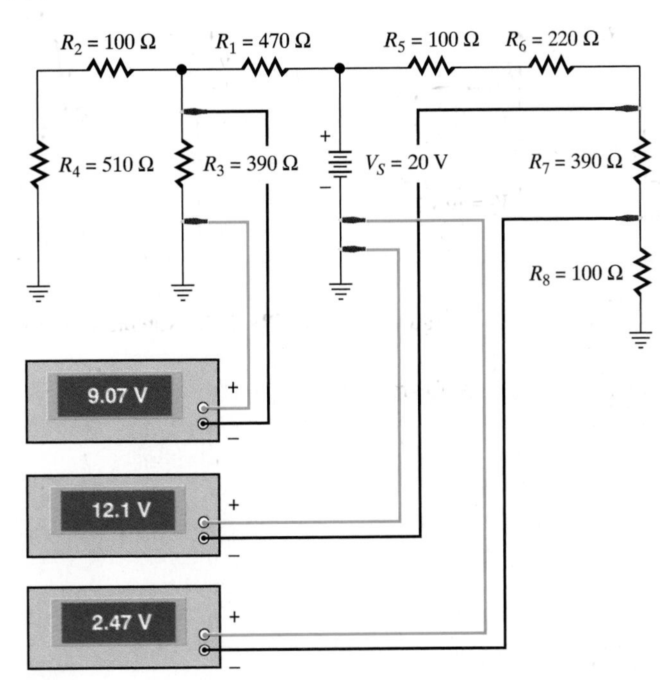

40. For the circuit of Figure 6–68, one fault exists. From the meter readings shown, what is probable cause?

FIGURE 6–68 Circuit for Problem 40

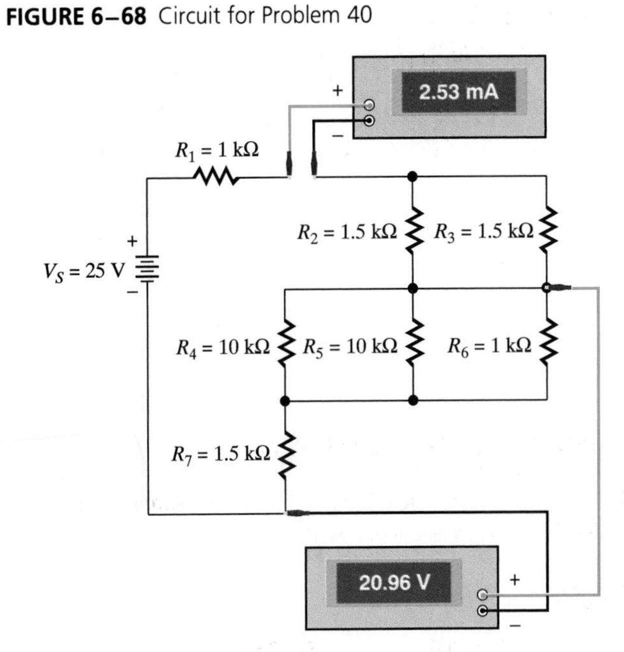

Marine Electronics

From the days of the ancient Egyptians to the present day, people have been fascinated by boats. They have been used to explore the world, provide the means of helping to feed the population, and provided a means of transportation. In modern times, however, they have become more and more a means of leisure pursuit. There are areas in which the increase in the number of boats in use has outstripped the increase in population! Some of these boats are used strictly for commercial fishing. Others are charter boats that take parties for daylong fishing expeditions. The vast majority are small boats, however, that are used strictly for the pleasure of the owners. This includes boating trips, parties, and noncommercial fishing.

Like any life experience, boating can be quite hazardous. Dangers come from rough weather and seas, uncharted shoals, sailing in darkness or fog, and becoming lost at sea. Thus the sailor needs a means of avoiding hazards presented by these conditions. It probably will not surprise you to learn that electronics is playing an ever-expanding role in making boating safer. The development of integrated circuits has made electronic systems small and cheap enough for use on even the smaller personal craft.

One such development is radar. A radar system makes navigating in rain, fog, or darkness much safer. In spite of poor visibility, potential hazards, such as shorelines, bridges, other boats, and so on, show up on the radar screen and can be avoided. Radar is especially helpful when operating in narrow channels where other traffic is also present. Thus, radar is becoming standard equipment on many small pleasure boats. The systems used on boats are small, lightweight, relatively inexpensive, and require very little maintenance. They are also quite easy to operate.

Sailors are not always familiar with all the waterways in which they will sail. Thus the exact depth of the water may not be known. Today many boats are equipped with devices known as depth finders which indicate the depth of the water beneath the boat. They operate on the principle of the sonar used on submarines. Sound waves are transmitted downward to be reflected from the bottom. Electronic circuits determine the time between the transmission and return of the sound waves and convert this time to distance. Thus, the depth of the water is known, and the rocks and shoals can be avoided. On the lighter side, some boats are equipped with what are known as fish finders. These devices, similar in construction and operation to depth finders, can be used to locate schools of fish. This is quite important, of course, to commercial fishermen and charter boat captains.

Another electronic system found on many boats is a navigation aid known as LORAN (Long Range Aid to Navigation). The LORAN system of radio transmitters is constructed and operated by the government; the receiver and presentation system must be provided by the user. Located at various locations throughout the United States, LORAN transmitters send signals to LORAN receivers aboard boats. The signals received from the various LORAN transmitters are processed, and the position of the boat can be quite accurately determined. This system benefits boats making rather long trips along the coast or to islands off the coast. It is also used extensively onboard both commercial and sport fishing boats to locate known reefs and fishing areas.

Marine electronics is thus an area in which an entrepreneur can establish a profitable business. There are many small companies dealing in the marine electronic systems located in populous areas along the coast and on large in-

land waters. They deal in the marketing, installation, and servicing of marine electronic systems on small boats. These companies also provide training for the user in operation of the systems.

It is usually not difficult for an entry level electronics technician to gain employment with a marine electronics company. The technician is expected to learn quickly in an on-the-job setting. It would probably begin with deliveries and installation of the electronic systems. As the technician's learning progresses, he/she can work up to bench repair work, user training, and even sales.

CHAPTER

7 Network Theorems

■ UPON COMPLETION OF THIS CHAPTER, YOU WILL BE ABLE TO

1. Draw and explain the action of ideal and real voltage sources and current sources.
2. Convert a voltage source to a current source, and convert a current source to a voltage source.
3. Write and explain the superposition theorem.
4. Perform analysis of circuits with multiple voltage sources using the superposition theorem.
5. Write and explain Thevenin's theorem.
6. Reduce series-parallel resistive circuits to their Thevenin's equivalent.
7. Analyze complex series-parallel circuits with multiple load values utilizing Thevenin's theorem.
8. Write and explain Norton's theorem.
9. Reduce series-parallel resistive circuits to their Norton's equivalent.
10. Perform conversions between Thevenin and Norton equivalent circuits.
11. Explain under what conditions a source transfers maximum power to a load.
12. Explain under what conditions practically all of the source voltage is developed across the load.
13. Write the procedure for determining the Thevenin equivalent of an actual electronic circuit.

In the previous four chapters, a great deal of time was devoted to the analysis of resistive dc circuits using Ohm's law, Kirchhoff's current and voltage laws, and the facts of series and parallel circuits. As basic and as useful as these tools are, they have their limitations. For instance, circuits with multiple voltage sources cannot be analyzed using only Ohm's law, and they are very difficult to analyze using Kirchhoff's laws.

In this chapter, a new set of tools for circuit analysis is presented. These tools are known as the network theorems. The superposition theorem makes it possible to analyze multiple source circuits using Ohm's law. Thevenin's and Norton's theorems make the analysis of circuits with multiple loads, no matter how complex they may be, a very simple process. The nice thing is that the theorems apply to any electronic circuit, not just the resistor networks that follow. Thus, a knowledge of them will be very helpful in later studies of complex circuits.

In electronics, there are times when it is necessary to transfer as much power as possible to the load. One example is the output of the final amplifier driving the speaker of a radio. Large amounts of power are required in order to operate the speaker. The conditions for maximum transfer of power are contained in the maximum transfer of power theorem, which is also presented in this chapter. In other instances, however, it is desirable to transfer as much voltage as possible to the load. Here, an example is the transfer of voltage developed in the antenna of a radio to the first amplifier. In this case, only a small amount of voltage is generated and as much as possible must be transferred. The conditions under which a relatively large voltage is transferred to the load is covered in the Chapter.

In addition, the concepts of voltage and current sources are introduced in this Chapter. An ideal voltage source is one in which the load voltage does not change when the load resistance varies. The ideal current source is one in which the output current does not change appreciably as the load varies. You will learn, as is the case with most things in our physical world, that real voltage or current sources fall somewhat short of the ideal.

All the concepts and theorems presented in this chapter are related in some way, so the understanding of one will aid you in understanding the others. You will apply them in your study and analysis of the more complex electronic circuits that will follow. A thorough knowledge of them will greatly enhance your learning and understanding.

7–1 THE VOLTAGE SOURCE

Question: What is the action of an ideal voltage source?

An **ideal voltage source** is one in which the load voltage remains constant regardless of changes in load current. Figure 7–1(a) is the symbol for the ideal voltage source. (You will recognize this symbol as that used for a battery in previous chapters to indicate a voltage source.) In Figure 7–1(b), a load resistance is placed across its output terminals. Since there is no other resistance, the entire source voltage is applied to the load. Thus, the action of an ideal voltage source may be stated as follows: With the exception of a short circuit (zero ohms), any value of load resistance has the entire source voltage across it.

Question: Are ideal voltage sources the rule in practical electronics?

The ideal voltage source does not exist in practice. A practical voltage source composed of metals, chemicals, or both has some resistance. Since this resistance is present in several different points in the voltage source, it is known as *distributed resistance*. In an actual circuit, the total of this distributed resistance is lumped, as shown in Figure 7–1(c), and is known as the **source resistance** R_S. It is often referred to as *internal resistance*.

FIGURE 7–1 Voltage sources

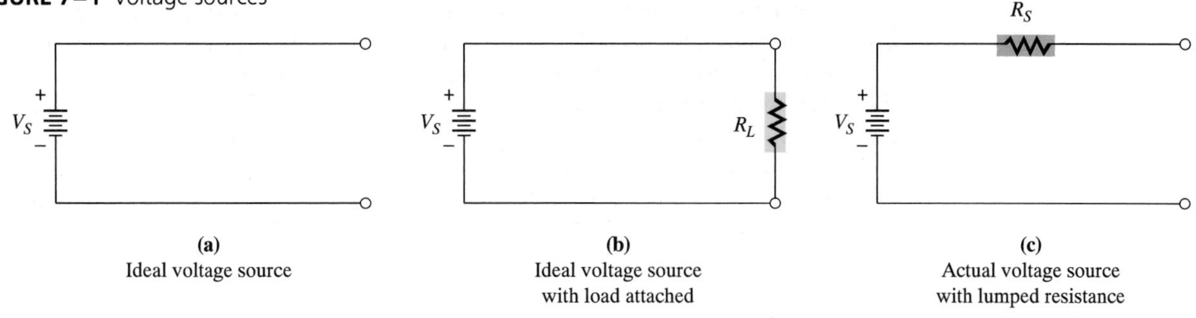

FIGURE 7–2 Practical voltage source with internal resistance in series with load resistance.

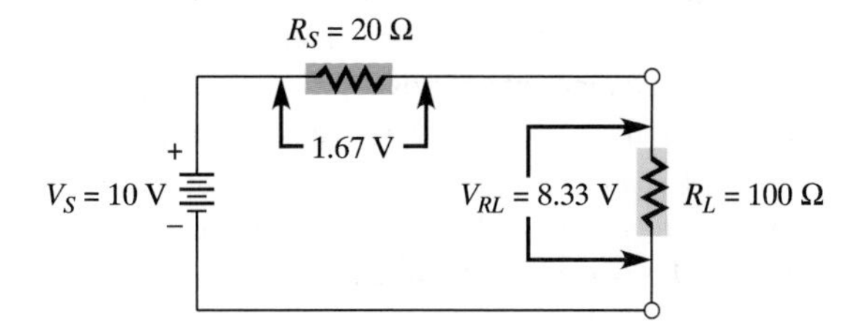

The load resistor is in series with R_S, and voltage divider action takes place, as illustrated in the following example.

In the circuit of Figure 7–2, the circuit has a 10 V source with 20 Ω resistance driving a 100 Ω load. The load voltage can be computed using the voltage divider equation:

$$V_{RL} = \frac{R_L}{R_T} V_S$$

$$= \frac{100\ \Omega}{120\ \Omega} 10\ \text{V} = 8.33\ \text{V}$$

Notice that with the addition of the load, current flows through R_S, a voltage drop of 1.67 V is produced and reduces the voltage available for the load resistor. A lower value of load resistance would mean a lower total resistance, a higher current, a larger voltage drop across R_S and even less load voltage.

EXAMPLE 7–1 A 10 V source with 20 Ω resistance drives a load of 50 Ω. Compute the load voltage.

■ Solution:

$$V_{R1} = \frac{R_1}{R_T} V_S$$

$$= \frac{50\ \Omega}{70\ \Omega} 10\ \text{V} = 7.14\ \text{V}$$

If the load resistance in Example 7–1 were dropped to 20 Ω, then only one-half the source voltage would be felt across it. Thus, any change in load resistance causes a variation in load voltage. This change in load voltage is known as *loading* the voltage source.

Question: Under what condition does the real voltage source approach the ideal?

A voltage source approaches the ideal if the source resistance R_S is very small in relation to the load resistance R_L. In such a case, the small R_S has very little voltage drop, and nearly all the supply voltage appears across the load resistor. How much smaller is considered "very small"? A rule of thumb states that a voltage source approaches ideal if the value of the source resistance is no more than 10% of the load resistance. The smaller this percentage, the smaller the change in load voltage with changes in resistance.

SECTION 7–1 REVIEW

1. Explain what is meant by an ideal voltage source.
2. What is meant by source resistance? By what other name is it known?
3. Explain why a practical voltage source is not ideal.
4. Under what conditions does a voltage source approach the ideal?

7–2 THE CURRENT SOURCE

Question: What is the action of an ideal current source?

An **ideal current source** is one in which the load current remains constant regardless of variations in the load resistance. The symbol for a constant current source is shown in Figure 7–3(a). The arrow indicates the direction of the source current I_S. It must be understood that the value of I_S is a fixed amount not subject to change. Figure 7–3(b) shows a resistance placed across the output terminals. The internal resistance of the source is now looked upon as being in parallel with the load. If this resistance is infinite, then, by current divider action, all the source current passes through the load regardless of its value. (The voltage across the load does not change.) Thus, the action of an ideal current source may be stated as follows: Any value of load resistance, with the exception of an open circuit (infinite ohms), will have the entire source current through it.

Question: Does the ideal current source exist in practical electronics?

Just as the ideal voltage source can only be approached in practice, so it is with the ideal current source. The internal resistance will never be infinite. A current source approaches the ideal if the internal resistance, in parallel with the load is at least 10 times greater than the load resistance, as demonstrated in the following example. For the practical current

FIGURE 7–3 Current sources

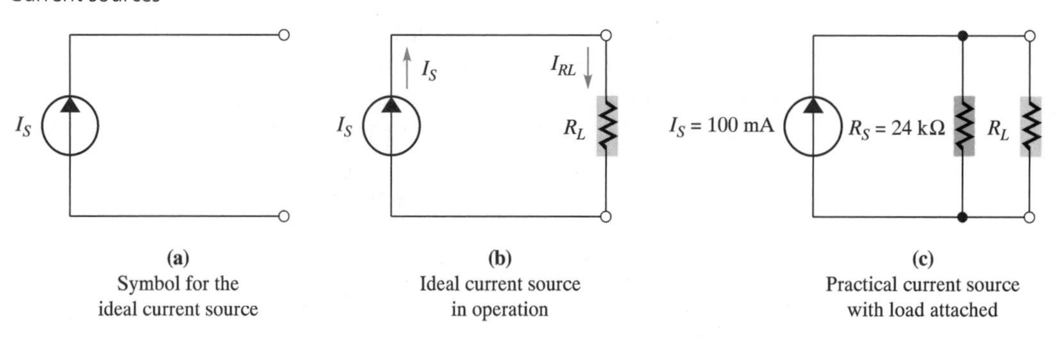

source shown in Figure 7–3(c), where $I_S = 100$ mA, the current divider equation is used to compute the load current for each of the following values of resistance: 200 Ω, 500 Ω, and 1000 Ω.

For 200 Ω,

$$I_{RL} = \frac{R_T}{R_L} I_T = \frac{198\ \Omega}{200\ \Omega} 100\ \text{mA} = 99\ \text{mA}$$

For 500 Ω,

$$I_{RL} = \frac{R_T}{R_L} I_T = \frac{489\ \Omega}{500\ \Omega} 100\ \text{mA} = 97.8\ \text{mA}$$

For 1000 Ω,

$$I_{RL} = \frac{R_T}{R_L} I_T = \frac{960\ \Omega}{1000\ \Omega} 100\ \text{mA} = 96\ \text{mA}$$

Notice that for the values given, the load current is within 5% of the source current. The smaller the value of the load resistance relative to the source resistance, the closer the ideal is approached.

SECTION 7–2 REVIEW

1. Explain what is meant by an ideal current source.
2. Explain why a practical current source is not ideal.
3. Under what conditions does a current source approach the ideal?

7–3 CONVERSION BETWEEN VOLTAGE AND CURRENT SOURCES

Question: Is it possible, and sometimes desirable, to convert between one source and another?

As previously stated, current and voltage sources will be used in the analysis of electronic circuits. The analysis is often easier if the source in question is converted to its opposite (current to voltage source or voltage to current source). The two sources must be equivalent. Thus, if the same value of load resistor is connected to each, the same voltage and current is produced.

The equivalent current source has the same source resistance as the voltage source. The source current is found by dividing the source voltage by the source resistance:

$$I_S = \frac{V_S}{R_S}$$

This equation is used to convert the voltage source to an equivalent current source, as illustrated in Figure 7–4(a). Notice that the source resistance of the equivalent current source is the same as that of the voltage source, which is 15 Ω. The source current I_S is computed by dividing the source voltage V_S by the source resistance R_S.

$$I_S = \frac{V_S}{R_S}$$

$$= \frac{30\ \text{V}}{15\ \Omega} = 2\ \text{A}$$

FIGURE 7–4 Conversion of voltage source to equivalent current source

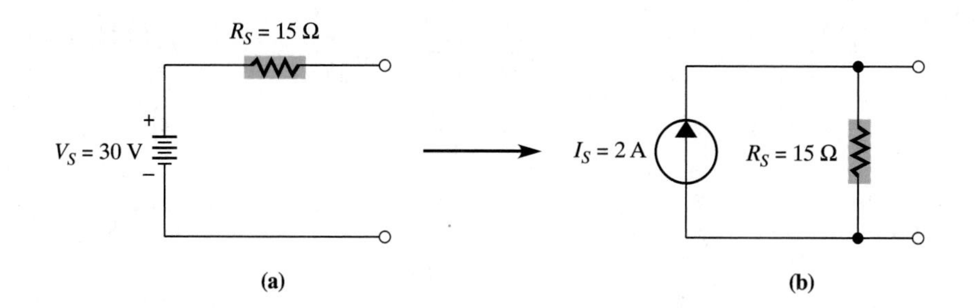

The equivalent current source is shown in Figure 7–4(b).

Equivalence can be proven by placing the same load across each source and computing the resulting load current and load voltage, as shown in the following example.

EXAMPLE 7–2 Prove the equivalence of the sources in Figure 7–4.

■ Solution:

To prove equivalence, install the same value load resistor across each source and solve for the resulting load current and voltage. Use a value of 150 Ω (although any value could be chosen).

1. For the voltage source, use the voltage divider equation:

$$V_{RL} = \frac{R_L}{R_T} V_S$$

$$= \frac{150\ \Omega}{165\ \Omega} 30\ \text{V} = 27.3\ \text{V}$$

[1] [5] [0] [÷] [1] [6] [5] [×] [3] [0] [=]

$$I_{RL} = \frac{V_S}{R_T}$$

$$= \frac{30\ \text{V}}{165\ \Omega} = 182\ \text{mA}$$

[3] [0] [÷] [1] [6] [5] [=]

2. For the current source, again use the current divider equation:

$$I_{RL} = \frac{R_T}{R_L} I_S$$

$$= \frac{13.65\ \Omega}{150\ \Omega} 2\ \text{A} = 182\ \text{mA}$$

[1] [3] [·] [6] [5] [÷] [1] [5] [0] [×] [2] [=]

$$V_{RL} = I_{RL} R_1$$

$$= (182\text{mA})(150\ \Omega) = 27.3\ \text{V}$$

1 8 2 EXP ± 3 × 1 5 0 =

Thus, the two sources are proven to be equivalent.

The procedure for converting a current to a voltage source is similar. Once again, the source resistances are the same. The source voltage V_S is found by multiplying the source current I_S by the source resistance R_S.

EXAMPLE 7–3 Convert the current source of Figure 7–5(a) to an equivalent voltage source.

■ Solution:

Using a source resistance of 2.5 k Ω, compute the source voltage by multiplying I_S and R_S:

$$V_S = I_S R_S$$
$$= (5 \text{ mA})(2.5 \text{ k } \Omega) = 12.5 \text{ V}$$

The equivalent voltage source is shown in Figure 7–5(b). Its equivalence may be proven as in Example 7–2.

FIGURE 7–5 Circuits for Example 7–3

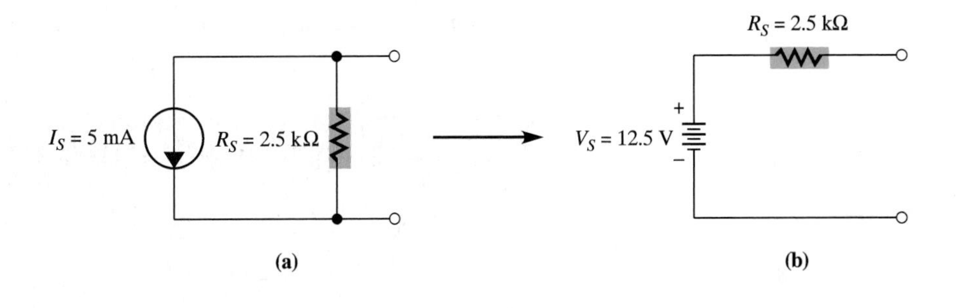

■ SECTION 7–3 REVIEW

1. What is the procedure for converting a voltage source to a current source?
2. What is the procedure for converting a current source to a voltage source?
3. What is meant by equivalence?
4. Convert the voltage source of Figure 7–6 to an equivalent current source.
5. Using a 100 Ω load, prove the two sources of Problem 4 are equivalent.

FIGURE 7–6 Circuit for Section Review Problem 4

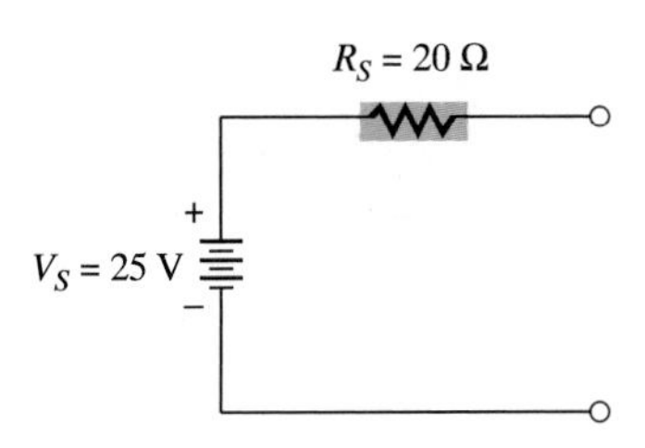

7–4 THE SUPERPOSITION THEOREM

Question: What is the importance of the superposition theorem?

In Chapters 4, 5, and 6, the analysis of series, parallel, and series-parallel circuits was presented. The prime tools for this analysis were Ohm's law, Kirchhoff's voltage law, and Kirchhoff's current law. These laws work well for circuits with one voltage source, but are inadequate for multiple source circuits. When they are used in conjunction with the superposition theorem, however, circuits with any number of voltage sources can be analyzed. The superposition theorem, then, is another tool of circuit analysis.

Question: How is the superposition theorem used in multiple voltage source circuit analysis?

The **superposition theorem** basically states that the current in any branch of a multiple source circuit is the algebraic sum of the currents that would be produced by each source acting separately. The following step-by-step procedure may be used in computing the value of these currents:

1. Remove all sources but one. Replace those removed with their internal resistance or with a "short" if the source is considered to be ideal.
2. Compute the current that the remaining source produces in each branch. Mark the direction and value of each current on a schematic near the branch in which it is produced.
3. Repeat steps 1 and 2 for the remaining sources, always keeping one source connected in the circuit. There will be a current in all branches when each source is considered.
4. Mark the branch currents on a common schematic, observing the direction obtained for each source.
5. Mark currents flowing in one direction with a plus (+) sign and those flowing in the other with a minus (−) sign. Add them algebraically.
6. For each branch, determine the current by finding the algebraic sum of the currents in the branch, and assign its direction according to that of the largest.

The superposition theorem makes it possible to analyze any circuit, no matter how many sources are involved. It merely needs to be solved the same number of times as there are voltage or current sources. The following example uses the superposition theorem to compute the current through each branch of the circuit in Figure 7–7(a). To simplify the analysis, the sources are considered to be ideal—that is, having zero internal resistance.

First, source V_{S2} is replaced with a short, as shown in Figure 7–7(b). Notice the arrows drawn to indicate the direction of current flowing through each resistor produced by the remaining source.

Next, the total resistance is computed:

$$R_T = \frac{1}{(1/R_2) + (1/R_3)} + R_1$$

$$= \frac{1}{(1/2.2\ \text{k}\Omega) + 1/1\ \text{k}\Omega)} + 390\ \Omega$$

$$= 688\ \Omega + 390\ \Omega = 1.078\ \text{k}\Omega$$

Computing the value of the total current gives

$$I_T = \frac{V_{S1}}{R_T}$$

FIGURE 7–7 Circuit analysis with the superposition theorem

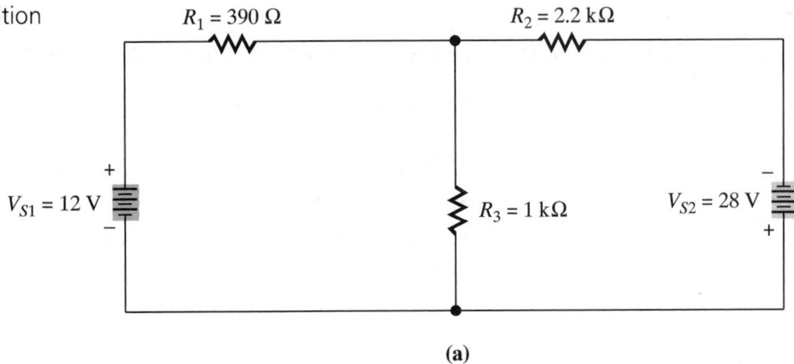

(a)

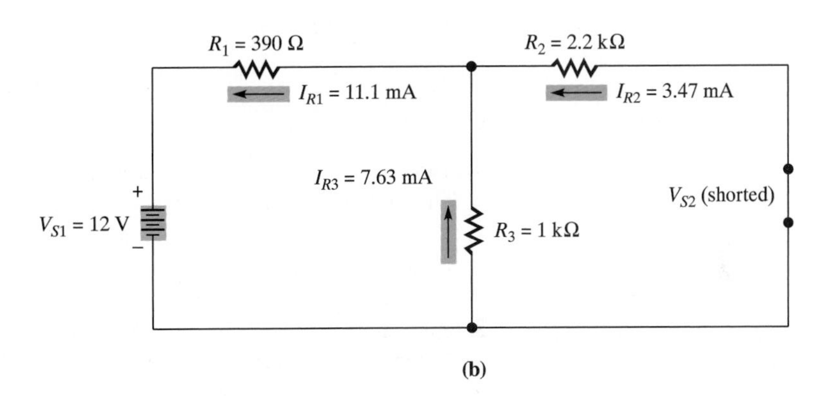

(b)

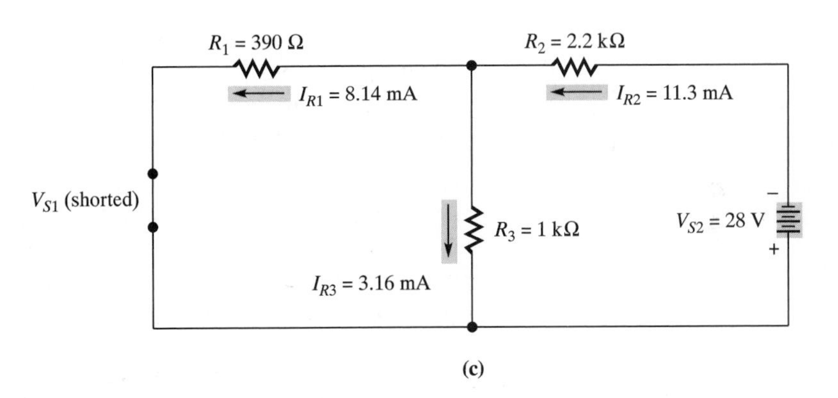

(c)

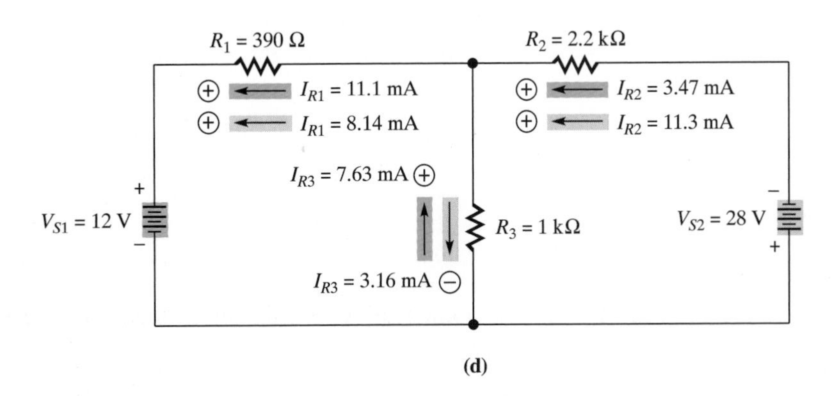

(d)

$$= \frac{12\text{ V}}{1078\ \Omega} = 11.1\text{ mA}$$

This result is the current through R_1 and is recorded near this resistor.

Using the current divider equation, the current through R_2 and R_3 is now computed:

$$I_{R2} = \frac{R_T}{R_2} I_T$$

$$= \frac{688\ \Omega}{2200\ \Omega} 11.1\text{ mA} = 3.47\text{ mA}$$

$$I_{R3} = I_T - I_{R2}$$

$$= 11.1\text{ mA} - 3.47\text{ mA} = 7.63\text{ mA}$$

Each of these values is also recorded near the appropriate resistor in Figure 7–7(b).

The process is repeated with V_{S2} reinstalled and with V_{S1} replaced by a short, as shown in Figure 7–7(c). The arrows again show the direction of the current flow through each resistor.

The total resistance with only V_{S2} connected is computed:

$$R_T = \frac{1}{(1/R_1) + (1/R_3)} + R_2$$

$$= \frac{1}{(1/390\ \Omega) + (1/1\text{ k}\Omega)} + 2.2\text{ k}\Omega$$

$$= 281\ \Omega + 2200\ \Omega = 2.481\text{ k}\Omega$$

Next, the total current with only V_{S2} installed is computed:

$$I_T = \frac{V_{S2}}{R_T}$$

$$= \frac{28\text{V}}{2.481\text{ k}\Omega} = 11.3\text{ mA}$$

This value is the current through R_2 and is recorded near this resistor.

Computing the current through R_1 and R_3 gives

$$I_{R1} = \frac{R_T}{R_1} I_T$$

$$= \frac{281\ \Omega}{390\ \Omega} 11.3\text{ mA} = 8.14\text{ mA}$$

$$I_{R3} = I_T - I_{R1}$$

$$= 11.3\text{ mA} - 8.14\text{ mA} = 3.16\text{ mA}$$

On the original schematic, the value and direction of each current is marked near the appropriate resistor, as shown in Figure 7–7(d). This process is similar to "superimposing" the set of currents from V_{S1} over those of V_{S2}. (Could this be the origin of the term "superposition"?) The currents flowing in one direction are indicated by a plus (+) sign, and those in the other, by a minus (−) sign. For each branch, the currents are added algebraically to determine the current through each resistor with both sources active:

$$I_{R1} = 11.1 \text{ mA} + 8.14 \text{ mA} = 19.2 \text{ mA}$$

$$I_{R2} = 11.3 \text{ mA} + 3.47 \text{ mA} = 14.77 \text{ mA}$$

$$I_{R3} = 7.63 \text{ mA} - 3.16 \text{ mA} = 4.47 \text{ mA}$$

EXAMPLE 7–4 Using the superposition theorem, compute the current through resistor R_1 in Figure 7–8(a). Notice that the sources are current sources.

■ Solution:

An ideal current source, when removed, is replaced by an open rather than a short circuit. For simplicity, consider the internal resistances to be infinite.

1. Replace source I_{S1} with an open circuit, as shown in Figure 7–8(b).
2. Indicate the value and direction of the current supplied by I_{S2}, now flowing through R_1, R_2, and R_3 as shown.
3. Reinstall source I_{S1}, and replace I_{S2} by an open circuit Figure 7–8(c).
4. Indicate the value and direction of the current I_{S1} flowing through R_1 as shown.
5. Superimpose the currents on the original schematic and mark with plus (+) and minus (−) signs as in Figure 7–8(d). Since only current I_{S2} flows through R_2 and R_3, the current through each of these resistors is the value of I_{S2}.
6. Find the current through R_1, which has both I_{S1} and I_{S2} flowing through it, with the smaller current flowing downward and the larger flowing upward. Compute the current through R_1 by subtracting the smaller current from the larger:

$$I_{R1} = I_{S1} - I_{S2}$$

$$= 180 \text{ mA} - 38 \text{ mA} = 142 \text{ mA}$$

Thus, the current through R_1 due to both sources is 142 mA.

FIGURE 7–8 Circuits for Example 7–4

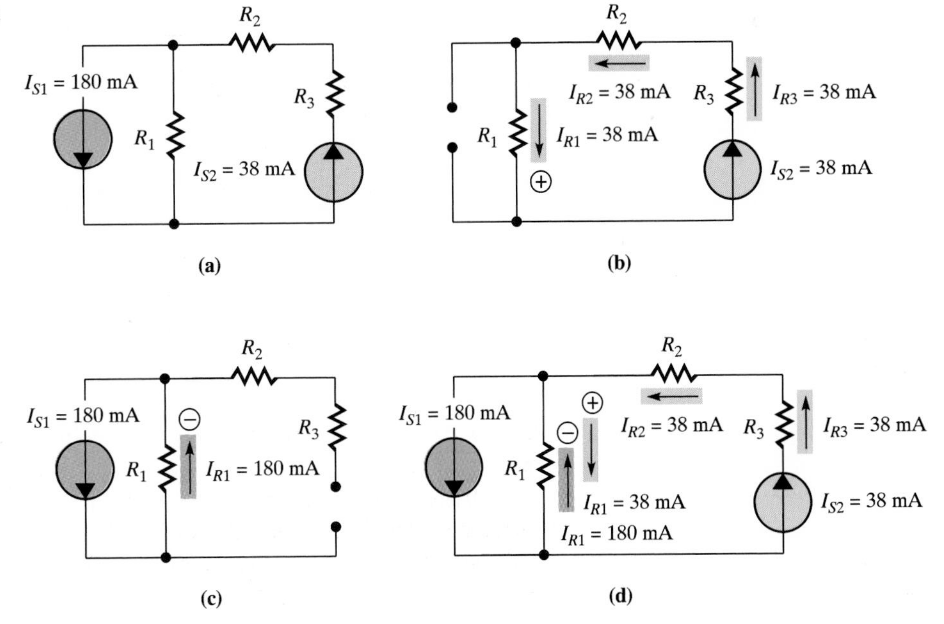

EXAMPLE 7–5 Using the superposition theorem, compute the current through each resistor of the circuit of Figure 7–9(a).

■ Solution:

1. Replace source V_{S1} with a short as in Figure 7–9(b).
2. For the polarity of the source as shown, draw arrows showing the direction of the current through each resistor.
3. Compute the total resistance for this circuit:

$$R_T = \frac{1}{(1/R_1) + (1/R_2)} + R_3$$

$$= \frac{1}{(1/60\ \Omega) + (1/24\ \Omega)} + 10\ \Omega = 27.10\ \Omega$$

[6] [0] [1/x] [+] [2] [4] [1/x] [=] [1/x] [+] [10] [=]

4. Compute the total current for this combination:

$$I_T = \frac{V_{S2}}{R_T}$$

$$= \frac{21\ \text{V}}{27.10\ \Omega} = 0.774\ \text{A}$$

[2] [1] [÷] [2] [7] [·] [1] [=]

Note that this result is the current through R_3. Mark this value near it.

5. Compute the current through R_1 using the current divider equation:

$$I_{R1} = \frac{R_T}{R_1} I_T$$

$$= \frac{17.10\ \Omega}{60\ \Omega} 0.774\ \text{A} = 220\ \text{mA}$$

6. Compute the current through R_2:

$$I_{R2} = I_T - I_{R1}$$

$$= 774\ \text{mA} - 220\ \text{mA} = 554\ \text{mA}$$

[7] [7] [4] [EXP] [±] [3] [−] [2] [2] [0] [EXP] [±] [3] [=]

Mark both of these current values near the appropriate resistor as shown in Figure 7–9(b).

7. Next, replace source V_{S2} with a short circuit, as shown in Figure 7–9(c). (Note that the circuit is shown redrawn in Figure 7–9(d) simplification.)
8. Draw arrows near each resistor showing the direction of current flow.
9. Compute the total resistance for this combination:

$$R_T = \frac{1}{(1/R_1) + (1/R_3)} + R_2$$

$$= \frac{1}{(1/60\ \Omega) + (1/10\ \Omega)} + 24\ \Omega = 32.6\ \Omega$$

6 0 1/x + 1 0 1/x = 1/x + 2 4 =

10. Compute the total current:

$$I_T = \frac{V_{S1}}{R_T}$$

$$= \frac{8\ \text{V}}{32.6\ \Omega} = 245\ \text{mA}$$

8 ÷ 3 2 · 6 =

This result is the current through R_2. Mark the value near this resistor in Figure 7–9(c).

FIGURE 7–9 Circuits for Example 7–5

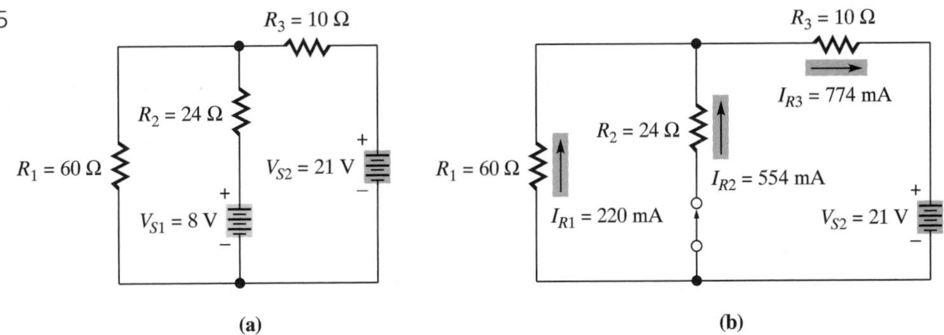

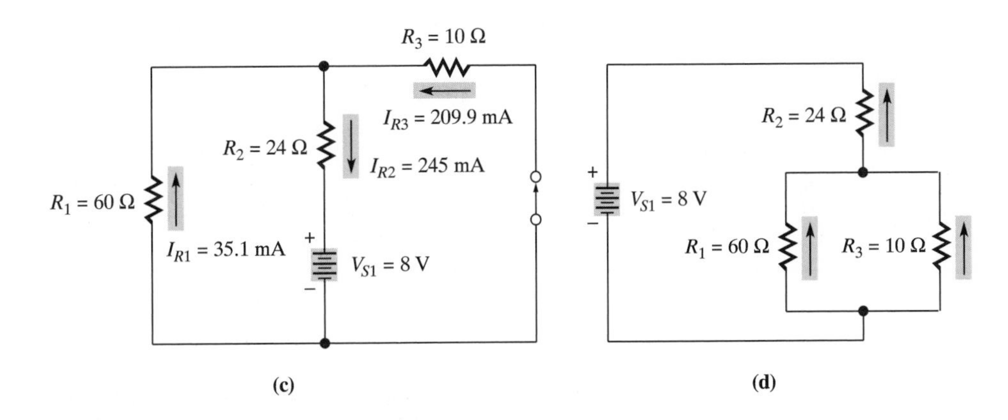

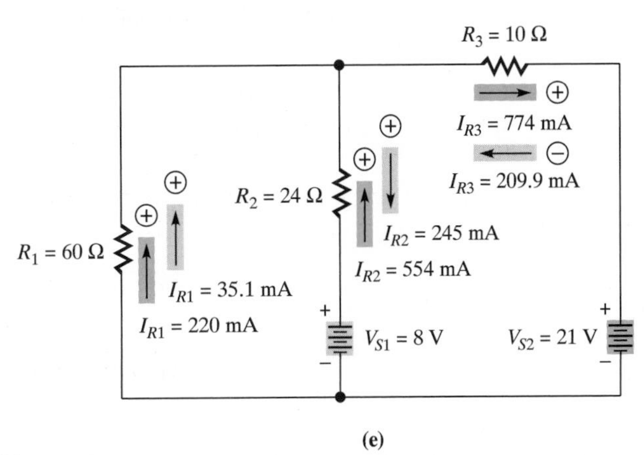

12. Compute the current through R_1 using the current divider equation:

$$I_{R1} = \frac{R_T}{R_1} I_T$$

$$= \frac{8.6\ \Omega}{60\ \Omega} 245\ \text{mA} = 35.1\ \text{mA}$$

[8] [·] [6] [÷] [6] [0] [×] [2] [4] [5] [EXP] [±] [3] [=]

13. Find the current through R_3:

$$I_{R3} = I_T - I_{R1}$$

$$= 245\ \text{mA} - 35.1\ \text{mA} = 209.9\ \text{mA}$$

[2] [4] [5] [EXP] [±] [3] [−] [3] [5] [·] [1] [EXP] [±] [3] [=]

Mark both these current values near the appropriate resistors as shown in Figure 7–9(c).

14. Mark the value and direction of the current flow for each resistor on the original schematic shown in Figure 7–9(e).

15. Add the currents when they are flowing in the same direction and subtract them when the directions are opposite to determine the current through each resistor with both sources active:

$$I_{R1} = 220\ \text{mA} + 35.1\ \text{mA} = 255.1\ \text{mA}$$

[2] [2] [0] [EXP] [±] [3] [+] [3] [5] [·] [1] [EXP] [±] [3] [=]

$$I_{R2} = 554\ \text{mA} - 245\ \text{mA} = 309\ \text{mA}$$

[5] [5] [4] [EXP] [±] [3] [−] [2] [4] [5] [EXP] [±] [3] [=]

$$I_{R3} = 774\ \text{mA} - 209.9\ \text{mA} = 564.1\ \text{mA}$$

[7] [7] [4] [EXP] [±] [3] [−] [2] [0] [9] [·] [9] [EXP] [±] [3] [=]

SECTION 7–4 REVIEW

1. Write the superposition theorem.
2. Write the step-by-step procedure for utilizing the superposition theorem in multiple source circuit analysis.
3. Using the superposition theorem, compute the current through each resistor of Figure 7–10.

FIGURE 7–10 Circuit for Section Review Problem 3

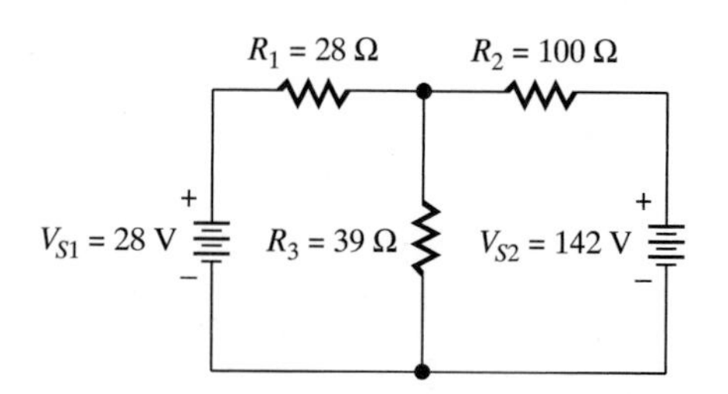

7–5 THEVENIN'S THEOREM

Question: What is Thevenin's theorem?

Thevenin's theorem is a method of reducing a circuit, no matter how complex, to a single voltage source V_{Th} and a single series resistance R_{Th}. Figure 7–11(a) shows a rather complex series-parallel circuit containing two sources. Consider how long it would take to compute the voltage across and the current through R_L utilizing the tools of analysis presented to this point. Figure 7–11(b) is the same circuit, reduced to an equivalent, through the application of Thevenin's theorem. Now consider how long the analysis would take using the reduced circuit. Needless to say, the time needed for the first would be "very long" while for the second it would be "very short"! The only qualification placed upon the use of Thevenin's theorem is that the circuit must be linear. A **linear circuit** is one in which any change in voltage produces a proportional change in current. When plotted, these variables produce a straight line.

Question: How is a circuit reduced by the application of Thevenin's theorem?

As shown in Figure 7–12(a), the first step in developing the Thevenin equivalent of a circuit is to remove the load resistor. Doing so actually makes the load resistance infinite, and the maximum possible voltage is developed across the points where it was removed.

FIGURE 7–11 Thevenin's theorem: Reducing a complex circuit to single voltage and resistance sources

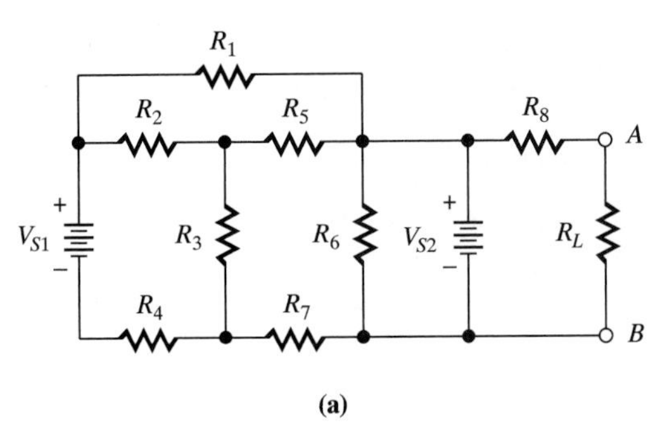

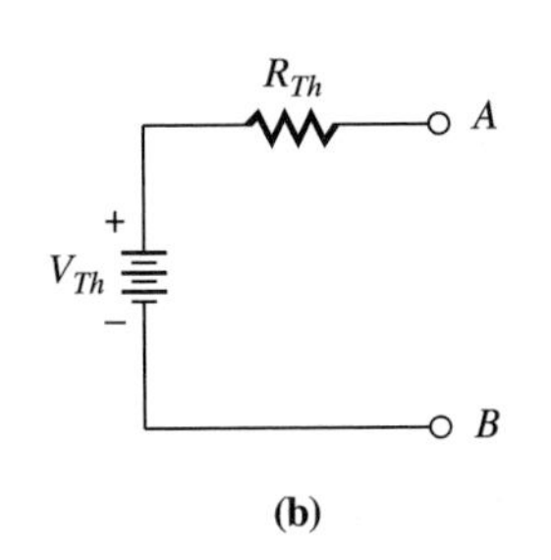

FIGURE 7–12 Demonstration of Thevenin's theorem

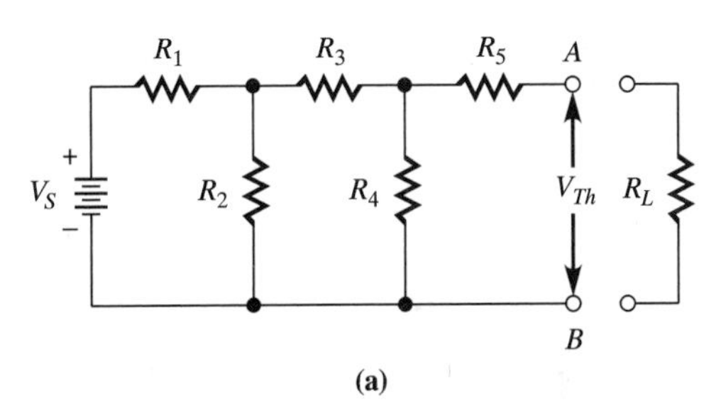

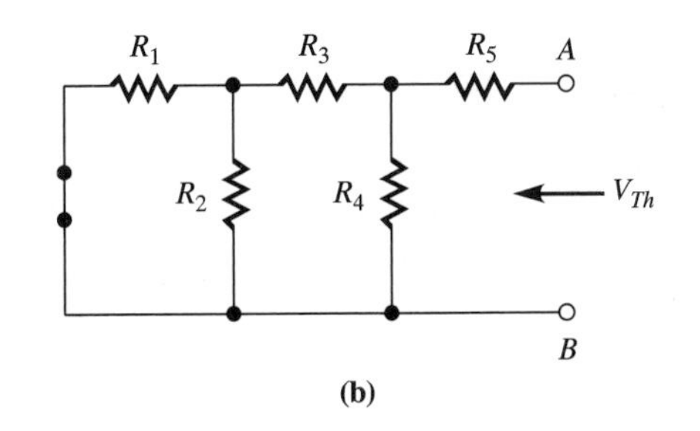

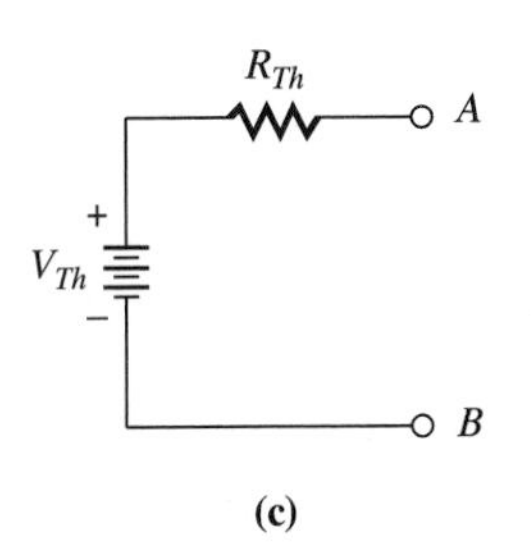

This voltage, known as the Thevenin voltage V_{Th} is the source voltage of the reduced circuit. Next, the total resistance across the open circuit is computed. As shown in Figure 7–12(b), all sources are replaced with their internal resistances, and the total resistance is computed looking into the circuit from where the load was removed. The Thevenin equivalent circuit will be this resistance R_{Th} in series with the Thevenin voltage V_{Th} (Figure 7–12(c)).

Once the Thevenin equivalent of the circuit has been developed, the load resistance, which had been previously removed, is reinstalled in series with R_{Th}. This procedure results in the simplest of series circuits, and the voltage divider equation can be used to solve for the load current and voltage. Any resistor, of any value, when installed in the Thevenin equivalent circuit, will produce the same current and voltage as it would if installed in the original circuit.

The procedure for determining the Thevenin equivalent circuit is summarized as follows:

1. Remove the load resistor.
2. Solve for the open circuit voltage across the points where the load was removed.
3. Replace the voltage or current source with its internal resistance, and solve for the total resistance looking into the circuit from the load.
4. Place the resistance found in step 3 in series with the voltage found in step 2, leaving the circuit open.
5. Install the original load in the equivalent Thevenin circuit and solve for its current and voltage drop.

The following example demonstrates this procedure for using Thevenin's theorem in circuit analysis. For the simple circuit of Figure 7–13(a), conventional means of circuit analysis are used to solve for the load current and voltage. Then the Thevenin equivalent circuit is developed to verify that using it yields the same values for load current and voltage.

In solving by conventional means, the total resistance is computed as follows: R_3 and R_L are in series with a combined total of 10 kΩ. This total is in parallel with R_2, and their total is, by the equal resistors in parallel equation, 5 kΩ. This combined total is in series with R_1 for a total circuit resistance of 15 kΩ. (A calculator is not always necessary, is it?)

Then,

$$I_T = \frac{V_S}{R_T}$$

$$= \frac{36\ \text{V}}{15{,}000\ \Omega} = 2.4\ \text{mA}$$

This total current divides evenly between R_2 and the series combination of R_3 and R_L. Thus, the current through R_L is 1.2 mA, $I_T/2$, and its voltage is

$$V_{RL} = I_{RL}R_L$$

$$= (1.2\ \text{mA})(5000\ \Omega) = 6\ \text{V}$$

The Thevenin equivalent of the circuit is determined by removing the load, as shown in Figure 7–13(b), and solving for the open circuit voltage. (Ask yourself, "If I placed a voltmeter in parallel with R_2, what would I measure?") Since the circuit is open from point A to point B, there is no current flowing through R_3. Thus, the open circuit voltage is that across R_2. R_1 and R_2 are in series and equal, so the source voltage divides between them. The Thevenin voltage is thus $V_{Th} = 36/2 = 18$ V.

FIGURE 7–13 Circuit analysis with Thevenin's theorem

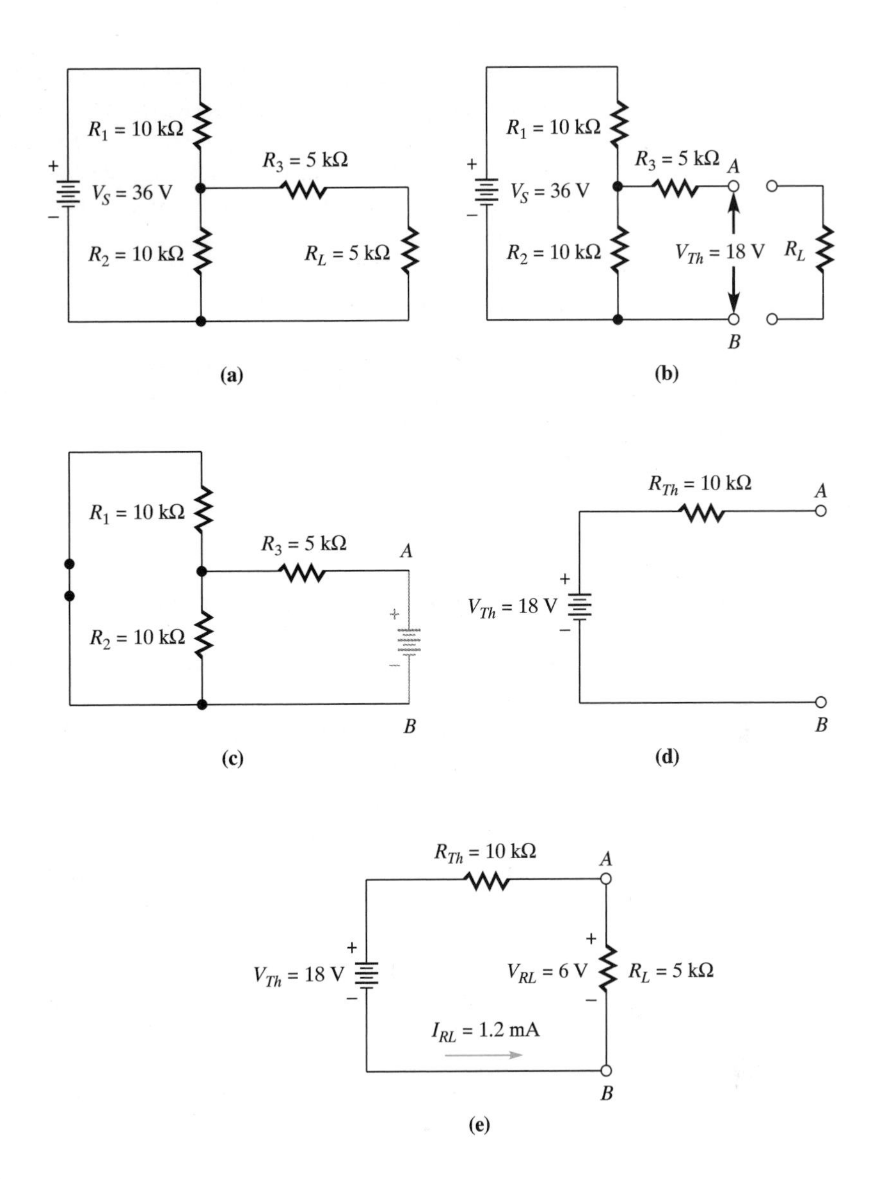

The Thevenin resistance can be found by looking into the circuit from the open. (To simplify the calculation, the internal resistance of the source is assumed to be zero.) Once again, in order to determine series and parallel combinations of resistors, it is a good idea to picture a voltage source where the load was removed, as shown in Figure 7–13(c). Resistors R_1 and R_2 form a parallel combination in series with R_3. Thus, by the equal resistors in parallel equation, the combined resistance of R_1 and R_2 is 5 kΩ. Since this combination is in series with R_3, the total resistance is 10 kΩ, which is the Thevenin resistance R_{Th}.

The Thevenin equivalent circuit is shown in Figure 7–13(d). Installing the load resistor (Figure 7–13(e)) and solving for V_{RL} and I_{RL} yields

$$V_{RL} = \frac{R_L}{R_T} V_{Th}$$

$$= \frac{5000\ \Omega}{15000\ \Omega}\ 18\ \text{V} = 6\ \text{V}$$

$$I_{RL} = \frac{V_{RL}}{R_L}$$

$$= \frac{6 \text{ V}}{5000 \ \Omega} = 1.2 \text{ mA}$$

The results, using either conventional means or Thevenin's theorem, are in fact equal! The two circuits, the original and the Thevenin equivalent, are not physically the same. But if the same value of load resistor is placed across each, the same voltage and current is developed in the load resistor. Herein lies the advantage of the use of Thevenin's theorem. In analyzing a complicated circuit, the difficult, time-consuming work is done *only once*. The action of the rest of the loads involved are computed from a simple, two-resistor voltage divider. When computed by conventional means, the difficult work must be done each time.

EXAMPLE 7–6 Find the Thevenin equivalent of the circuit of Figure 7–14(a), and compute the current through and voltage across R_3.

■ Solution:

1. Remove the load to produce the circuit shown in Figure 7–14(b). Notice that there are two voltage sources in this example. V_{AB} can be computed using the superposition theorem or conventional means. The former will be used here.
2. Replace the source V_{S2} with a short, as in Figure 7–14(c). Notice that the voltage from point A to point B is the voltage drop across R_3. Also notice that point A is positive with respect to point B.
3. Use the voltage divider equation:

$$V_{AB1} = \frac{R_2}{R_T} V_{S1}$$

$$= \frac{40 \ \Omega}{100 \ \Omega} 20 \text{ V} = 8\text{V}$$

4. Replace V_{S1} by a short, as shown in Figure 7–14(d). Notice that point A is now negative with respect to point B, because point B was chosen as the reference.
5. Compute the voltage from points A to B with only V_{S2} connected:

$$V_{AB2} = \frac{R_1}{R_T} V_{S2}$$

$$= \frac{60 \ \Omega}{100 \ \Omega} (-20 \text{ V}) = -12 \text{ V}$$

6. Compute the voltage between points A and B with both sources active by finding the algebraic sum of the two voltages:

$$V_{AB} = V_{AB2} + V_{AB1}$$

$$= -12 \text{ V} + 8 \text{ V} = -4 \text{ V}$$

7. Find the Thevenin equivalent circuit by replacing both sources with shorts and looking into the circuit from where the load was removed. As Figure 7–14(e) shows, R_1 and R_2 are in parallel, and a source is pretended at the point the load was removed in order to better see series and parallel combinations. Thus, compute the Thevenin resistance:

FIGURE 7–14 Circuits for Example 7–6

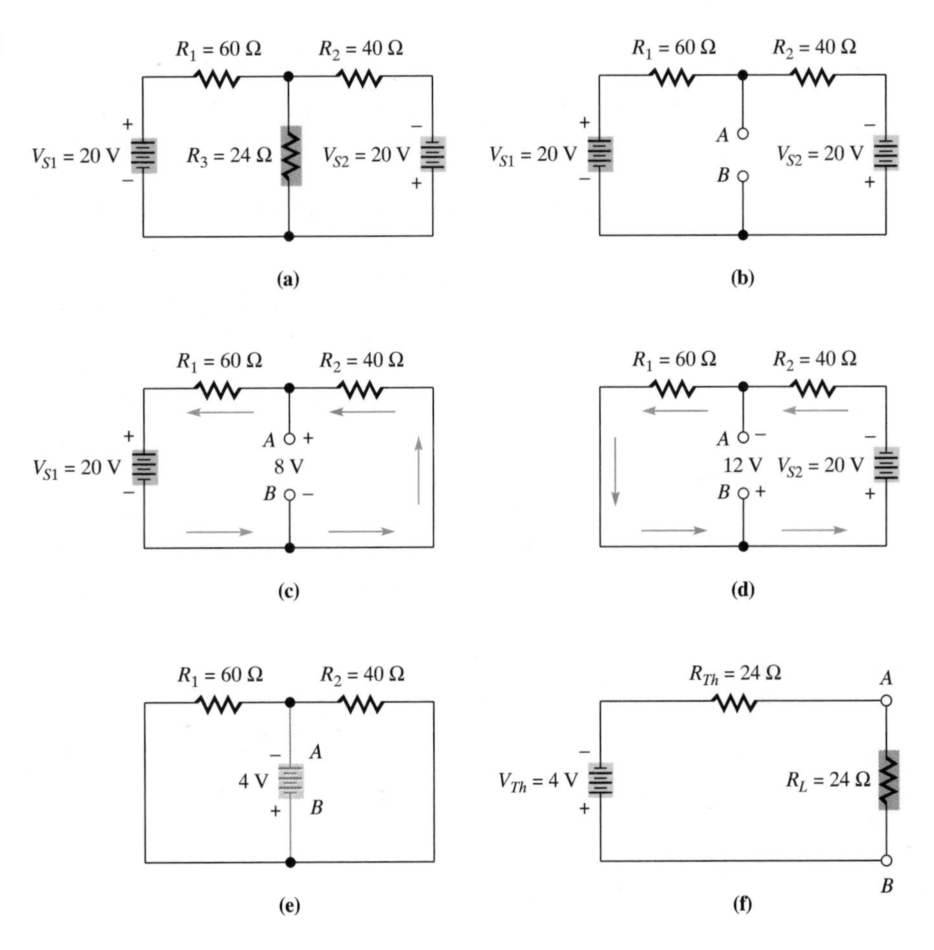

$$R_{Th} = \frac{1}{(1/R_1) + (1/R_2)}$$

$$= \frac{1}{(1/60\ \Omega) + (1/40\ \Omega)} = 24\ \Omega$$

The Thevenin equivalent circuit is shown in Figure 7–14(f), where R_3 is now installed across the terminals of the equivalent circuit. Since R_{Th} and R_L are equal and in series, V_{Th} is divided equally between them.

8. Solve for current through and voltage across R_3:

$$V_{R3} = -2\text{ V}$$

$$I_{R3} = \frac{V_{RL}}{R_L}$$

$$= \frac{-2\text{ V}}{24\ \Omega} = -83.3\text{ mA}$$

The minus sign on the current means merely that the current is flowing downward toward the reference.

EXAMPLE 7–7 The classic example of a problem made for the application of Thevenin's Theorem is shown in Figure 7–15(a). This circuit is the Wheatstone bridge circuit introduced in Chapter 6 with a load resistor across the output. This circuit would be very difficult to analyze by conventional means, because it is not a straight-forward series-parallel circuit. (If you doubt this, try redrawing the circuit to identify series and parallel combinations!) Analysis is simple, however, using the Thevenin's Theorem.

■ Solution:

1. Remove the load resistor to produce the circuit shown in Figure 7–15(b). Notice that V_{AB} is the voltage between points A and B and, as in the bridge circuit, will be the difference in the voltage drops across R_2 and R_4. Also notice that R_1 is in series with R_2, and R_3 is in series with R_4.
2. Compute the voltage between points A and B:

$$V_{R2} = \frac{R_2}{R_1 + R_2} V_S$$

$$= \frac{500\ \Omega}{1000\ \Omega + 500\ \Omega} 18\ \text{V} = 6\ \text{V}$$

$$V_{R4} = \frac{R_4}{R_3 + R_4} V_S$$

$$= \frac{360\ \Omega}{3600\ \Omega + 360\ \Omega} 18\ \text{V} = 1.64\ \text{V}$$

$$V_{AB} = V_{R2} - V_{R4}$$

$$= 6 - 1.64 = 4.36\ \text{V}$$

3. Compute R_{Th} by looking into the circuit from where the load was removed after redrawing the circuit as in Figure 7–15(c):

$$R_{Th} = \frac{1}{(1/R_1) + (1/R_2)} + \frac{1}{(1/R_3) + (1/R_4)}$$

$$= \frac{1}{(1/1\ \text{k}\Omega) + 1/500\ (\Omega)} + \frac{1}{(1/3.6\ \text{k}\Omega) + (1/360\ \Omega)}$$

$$= 333\ \Omega + 327\ \Omega = 660\ \Omega$$

4. Install the 1 kΩ load in the Thevenin equivalent circuit shown in Figure 7–15(d), and solve for V_{RL} and I_{RL}:

$$V_{RL} = \frac{R_L}{R_L + R_{Th}} V_{Th}$$

$$= \frac{1000\ \Omega}{1000\ \Omega + 660\ \Omega} 4.36\ \text{V} = 2.63\ \text{V}$$

$$I_{RL} = \frac{V_{RL}}{R_L}$$

$$= \frac{2.63\ \text{V}}{1000\ \Omega} = 2.63\ \text{mA}$$

FIGURE 7–15 Circuit for Example 7–7

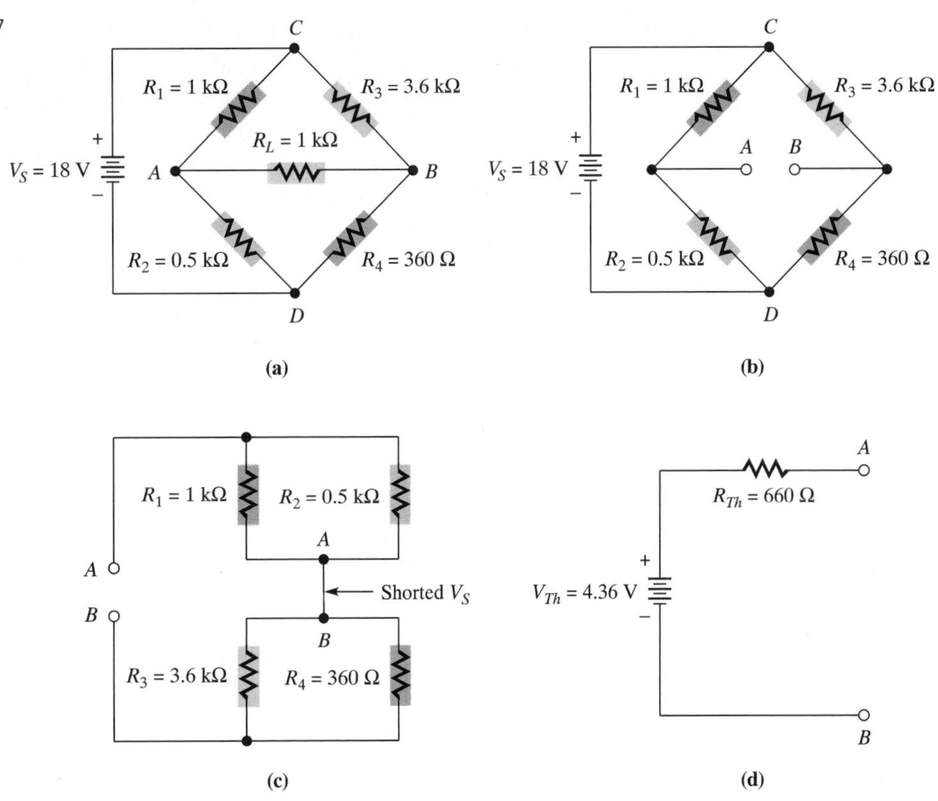

SECTION 7–5 REVIEW

1. What does Thevenin's theorem state?
2. What is the advantage of Thevenin's theorem in circuit analysis?
3. List the steps for finding the Thevenin equivalent of a circuit.
4. Why is the load removed in order to find V_{Th}?
5. Using conventional means, compute the current through and voltage across R_L in Figure 7–16. Then solve for the same values using Thevenin's theorem.

FIGURE 7–16 Circuit for Section Review Problem 5

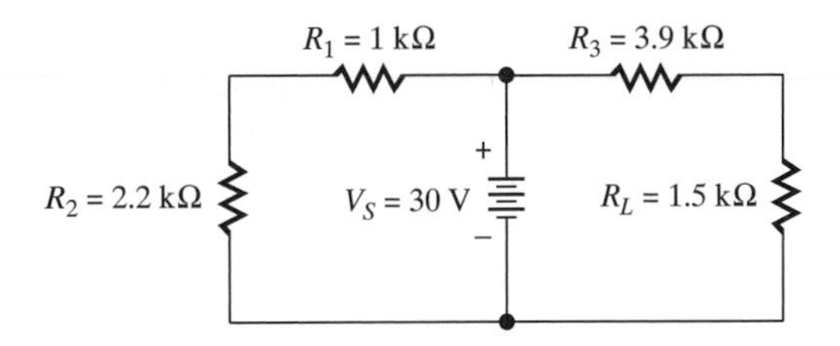

7–6 NORTON's THEOREM

Question: What is Norton's theorem?

Norton's theorem states that any electrical circuit, no matter how complex, can be reduced to a single current source and a single parallel resistance. This reduced circuit, as illustrated in Figure 7–17(a), is the current source introduced earlier in this chapter. A

FIGURE 7–17 Norton's theorem

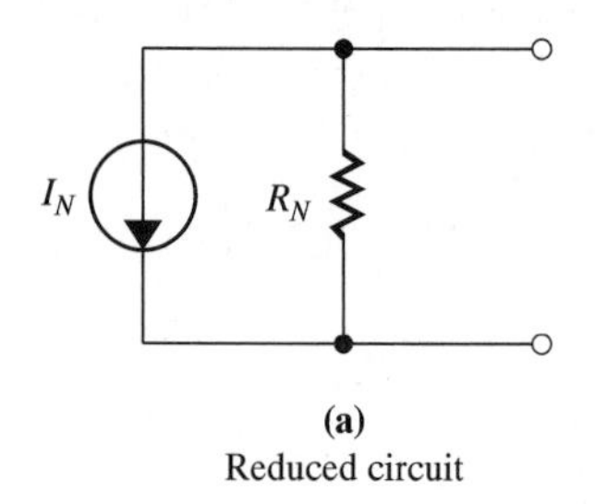

(a)
Reduced circuit

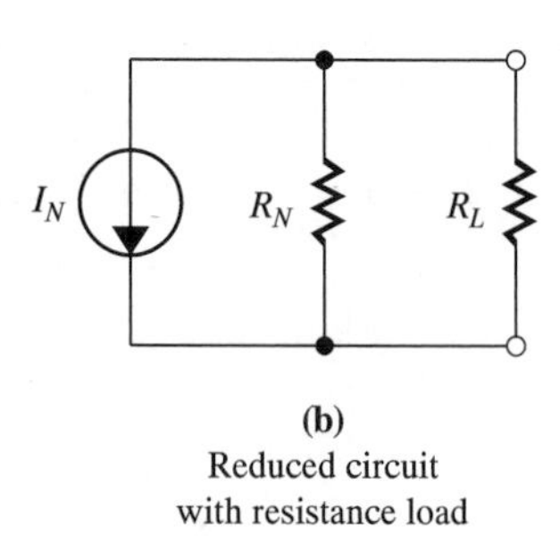

(b)
Reduced circuit
with resistance load

load placed in parallel with R_N will produce the same current and voltage as it would in the original circuit from which the Norton equivalent circuit was derived. Once again, the Norton equivalent circuit will be much simpler to analyze than the original.

Question: How is a circuit reduced through the application of Norton's theorem?

The application of Norton's theorem begins with placing a short around the load resistor. The short reduces the load resistance to zero, producing maximum current in the circuit. The current through this short, often referred to as the short circuit current, is known as the *Norton current* I_N. It is the source current of a Norton equivalent circuit. The Norton resistance R_N is found in the same manner as R_{Th}. The circuit is opened at the load, and the total resistance is computed by looking into the circuit from this open. The Norton equivalent circuit is the Norton resistance in parallel with the Norton current.

Once the circuit is reduced to its Norton equivalent, the load resistor, which was previously removed, is reinstalled in parallel with the Norton resistance. As illustrated in Figure 7–17(b), the results is a simple current divider that can be solved for the load current and voltage. Any resistor, no matter what its value, when installed in the Norton equivalent circuit, produces the same current and voltage as it would in the original circuit.

The steps in finding the Norton equivalent circuit are summarized as follows:

1. Place a short around the load resistance.
2. Compute the current, which is the Norton current, through this short.
3. Replace all sources with their internal resistances.
4. Remove the load and compute the total resistance, which is the Norton resistance, by looking into the circuit from where the load was removed.
5. Place the current source computed in step 2 in parallel with the resistance computed in step 4.
6. Place the load resistance in parallel with R_N and compute I_{RL} and V_{RL}.

The following example demonstrates the application of Norton's Theorem in circuit analysis.

EXAMPLE 7–8 Reduce the circuit of Figure 7–18(a) to a Norton equivalent, and compute the current through and voltage across the load.

■ Solution:

Removing R_L and replacing it with a short produces the circuit of Figure 7–18(b). The Norton current flows through the short. It can be determined by computing the current through R_3, which is in series with the short.

1. Compute R_T of this circuit:

$$R_T = \frac{1}{(1/R_2 + 1/R_3)} + R_1$$

$$= \frac{1}{(1/10\ \text{k}\Omega) + (1/5\ \text{k}\Omega)} + 5.1\ \text{k}\Omega = 8.433\ \text{k}\Omega$$

2. Compute the total current:

FIGURE 7–18 Circuits for Example 7–8

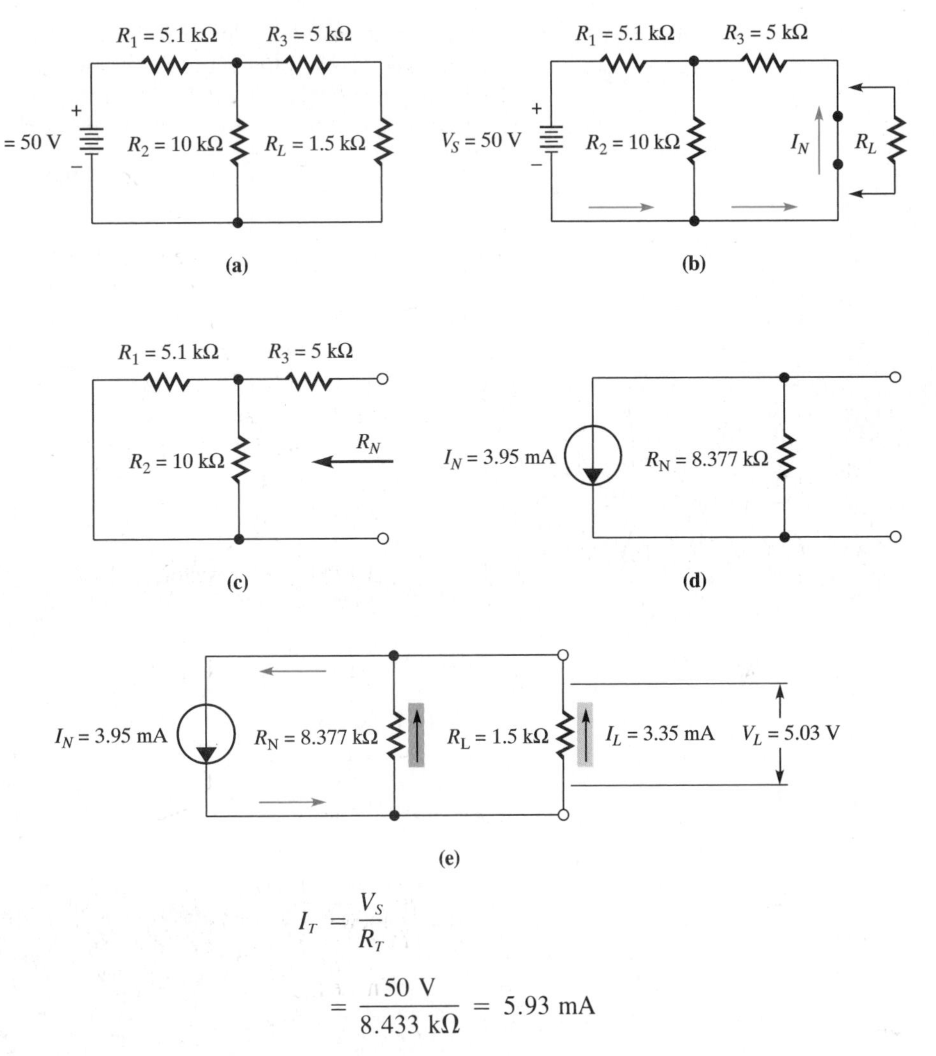

$$I_T = \frac{V_S}{R_T}$$

$$= \frac{50 \text{ V}}{8.433 \text{ k}\Omega} = 5.93 \text{ mA}$$

3. Apply the current divider equation to the parallel combination of R_2 and R_3. Determine the current through the 5 kΩ resistor and thus through the short:

$$I_N = \frac{R_T}{R_3} I_T$$

$$= \frac{3333 \ \Omega}{5000 \ \Omega} 5.93 \text{ mA} = 3.95 \text{ mA}$$

4. Now compute the Norton resistance in the same manner as the Thevenin resistance in the preceding section: As shown in Figure 7–18(c), remove the source and replace it with its internal resistance. (The internal resistance is assumed to be zero in order to simplify the calculations.) With the load and short removed, look into the circuit from the open to compute the Norton resistance:

$$R_N = \frac{1}{(1/R_1) + (1/R_2)} + R_3$$

$$= \frac{1}{(1/5.1 \text{ k}\Omega + (1/10 \text{ k}\Omega)} + 5 \text{ k}\Omega = 8.377 \text{ k}\Omega$$

The Norton equivalent circuit is as shown in Figure 7–18(d).

5. Place the previously removed load resistor in parallel with R_N to compute the load current and voltage, as shown in Figure 7–18(e).

6. Apply the current divider equation to the parallel combination of R_N and R_L to compute the load current:

$$I_{RL} = \frac{R_T}{R_L} I_N$$

$$= \frac{1272\ \Omega}{1500\ \Omega} 3.95\ \text{mA} = 3.35\ \text{mA}$$

7. Compute the load voltage using Ohm's law:

$$V_{RL} = I_L R_L$$

$$= (3.35\ \text{mA})(1.5\ \text{k}\Omega) = 5\ \text{V}$$

The Norton equivalent circuit can now be used to compute the load voltage and current that would be developed by any resistor installed in the original circuit.

The advantages in the use of Norton's theorem in circuit analysis are the same as those for Thevenin's theorem. The hard work is done only once in reducing the circuit to a simple current source and single parallel resistor. The analysis involves merely inserting the load into this circuit and applying the current divider equation.

SECTION 7–6 REVIEW

1. What does Norton's theorem state?
2. What is the advantage of Norton's theorem in circuit analysis?
3. List the steps for finding the Norton equivalent of a circuit.
4. Why is the load replaced with a short in computing I_N?
5. Compute I_{RL} and V_{RL} in the circuit of Figure 7–19 using conventional means. Confirm that the same values are obtained using Norton's theorem.

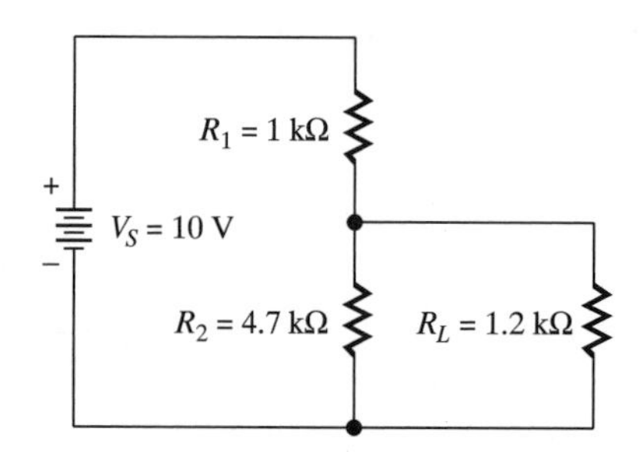

FIGURE 7–19 Circuit for Section Review Problem 5

7–7 MAXIMUM TRANSFER OF POWER THEOREM

Question: What type load develops relatively large quantities of power in its operation?

Recall that power is equal to the square of the current times the resistance through which it flows. Thus, any load that draws a large amount of current develops a large amount of power. This situation exists at the output of an electronic system. In the early stages of an electronic system, which are composed of control circuits, only small amounts of current flow. At the output, however, power to do work must be developed because something must be moved or illuminated. This power requires much larger currents. Most of the output devices, such as speakers, electric motors, and displays, have a low resistance and thus draw relatively large amounts of current. They must be supplied with enough power to operate properly, but it must be supplied as efficiently as possible.

Question: What is the problem in transferring power to a load?

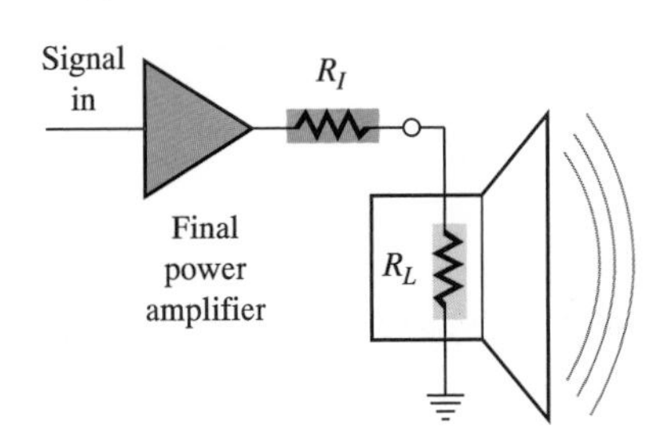

FIGURE 7–20 Transferring power to a load

The problem of transferring power to a load is illustrated in Figure 7–20. The final amplifier of the system must transfer maximum power to the load, which in this case is a speaker. The speaker is known as a *transducer,* a device that converts one form of energy to another. In this case, electrical energy is converted to sound waves. As with any electrical source, the amplifier has an internal resistance R_I. This resistance is in series with the resistance of the load R_L. The output current flows through both resistances, and power is developed in each. The power developed in the internal resistance is, of course, wasted and must be kept to a minimum, while at the same time, maximum power must be developed in the load. This problem must be addressed when designing the output circuit.

Question: Under what circumstances will the maximum possible power be developed in a load?

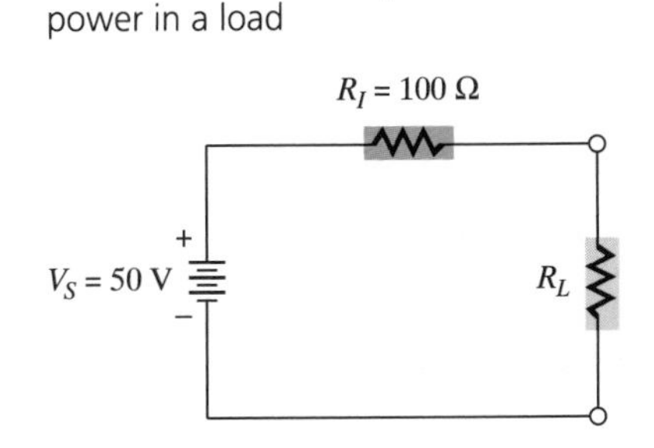

FIGURE 7–21 Development of power in a load

The development of maximum power in a load can best be explained by an example in which the load power is calculated for various values of load resistance. For the circuit of Figure 7–21, the power developed in the load is computed for the following values of load resistance R_L: 10 Ω, 30 Ω, 50 Ω, 80 Ω, 100 Ω, 200 Ω, and 500 Ω.

For 10 Ω,

$$R_T = R_I + R_L = 100\ \Omega + 10\ \Omega = 110\ \Omega$$

$$I_L = \frac{V_{RL}}{R_L} = \frac{50\text{ V}}{100\ \Omega} = 455\text{ mA}$$

$$P_{RL} = I_{RL}{}^2\text{R}_L = (0.455)^2(10\ \Omega) = 2.07\text{ W}$$

For 30 Ω,

$$R_T = R_I + R_L = 100\ \Omega + 30\ \Omega = 130\ \Omega$$

$$I_{RL} = \frac{V_{RL}}{R_L} = \frac{50\text{ V}}{130\ \Omega} = 385\text{ mA}$$

$$P_{RL} = I_{RL}{}^2\text{R}_L = (0.385)^2(30\ \Omega) = 4.45\text{ W}$$

For 50 Ω,

$$R_T = R_I + R_L = 100\ \Omega + 50\ \Omega = 150\ \Omega$$

$$I_{RL} = \frac{V_{RL}}{R_L} = \frac{50\text{ V}}{150\ \Omega} = 333\text{ mA}$$

$$P_{RL} = I_{RL}{}^2\text{R}_{RL} = (0.333)^2 50\ \Omega = 5.54\text{ W}$$

For 80 Ω,

$$R_T = R_I + R_L = 100\ \Omega + 80\ \Omega = 180\ \Omega$$

$$I_{RL} = \frac{V_{RL}}{R_{RL}} = \frac{50\text{ V}}{180\ \Omega} = 278\text{ mA}$$

$$P_{RL} = I_{RL}{}^2\text{R}_L = (0.278)^2 80\ \Omega = 6.18\text{ W}$$

For 100 Ω,

$$R_T = R_I + R_L = 100\ \Omega + 100\ \Omega = 200\ \Omega$$

$$I_{RL} = \frac{V_{RL}}{R_L} = \frac{50\text{ V}}{200\ \Omega} = 250\text{ mA}$$

$$P_{RL} = I_{RL}{}^2\text{R}_L = (0.25)^2(100\ \Omega) = 6.25\text{ W}$$

For 200 Ω,

$$R_T = R_I + R_L = 100\ \Omega + 200\ \Omega = 300\ \Omega$$

$$I_{RL} = \frac{V_{RL}}{R_L} = \frac{50\text{ V}}{300\ \Omega} = 167\text{ mA}$$

$$P_{RL} = I_{RL}{}^2\text{R}_L = (0.167)^2(200\ \Omega) = 5.58\text{ W}$$

For 300 Ω,

$$R_T = R_I + R_L = 100\ \Omega + 300\ \Omega = 400\ \Omega$$

$$I_{RL} = \frac{V_{RL}}{R_L} = \frac{50\text{ V}}{400\ \Omega} = 125\text{ mA}$$

$$P_{RL} = I_{RL}{}^2\text{R}_L = (0.125)^2(300\ \Omega) = 4.69\text{ W}$$

These calculations of power show that as the load resistance increases, the power developed in it increases. In this example, the power developed in the load peaked when the load resistance was equal to the internal resistance. As the load resistance was increased further, the load power began to decrease. This data is shown on a graph in Figure 7–22. From these results, the **maximum transfer of power theorem** can be stated as follows:

FIGURE 7–22 Transfer of power curve

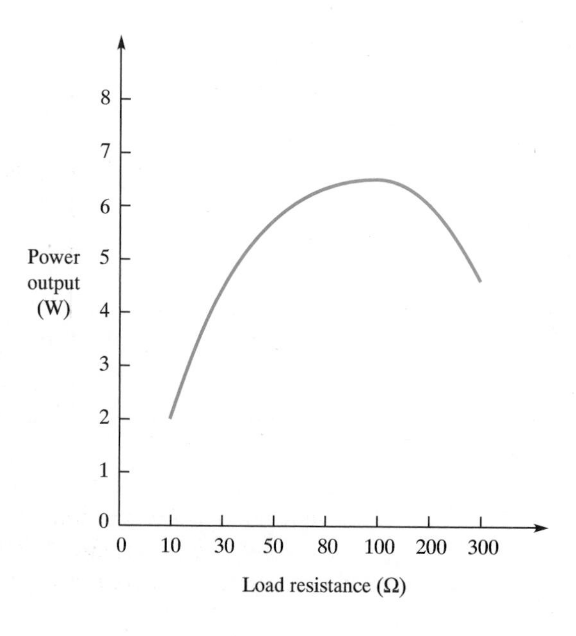

Maximum power is transferred from a source to a load when the load resistance is equal to the resistance of the source. In equation form, this relationship is written as

(7–1) $$R_I = R_L$$

This equation is a very important consideration in the design of the output circuits of electronic systems.

Question: Does the maximum transfer of power mean that it is impossible to drive a load from a source when their resistances are not the equal?

It is not impossible to drive a load efficiently with a source when their resistances are not equal, but it does require special circuitry in order to achieve maximum transfer of power to the load. If the load and source resistances are not equal, they are said to be *mismatched.* Circuits can be developed that, when placed between the load and source, can make them appear equal to one another. These circuits are known as *impedance matching circuits.* (*Impedance* is a term used in ac circuits and has a meaning similar to resistance.) An example of a device that can be used for this purpose is the transformer, which will be covered in Chapter 13.

Question: What is the circuit efficiency when maximum power is transferred from source to load?

Unfortunately, when maximum power is being developed in the load, the efficiency of the circuit is not maximum, as can be demonstrated from the load resistance data of Figure 7–22. The circuit **input power** is the total developed in the source and load resistances. The **usable output power** is that developed in the load resistance only. Recall from Chapter 3 that the circuit efficiency is computed as the ratio of the output power to the input power. In equation form, the percentage of efficiency is

$$\% \text{ efficiency} = \frac{P_{out}}{P_{in}} 100$$

Applying this equation to the data gives the following list of efficiencies:

Load Resistance	Efficiency
10 Ω	9%
30 Ω	23%
50 Ω	33%
80 Ω	44%
100 Ω	50%
200 Ω	66%
300 Ω	74%

Notice that when maximum power is transferred, the circuit efficiency is 50%. As the load resistance is increased, notice that even though the amount of power transferred decreases, the efficiency increases.

SECTION 7–7 REVIEW

1. State the maximum transfer of power theorem.
2. When is it necessary to transfer maximum power from source to load?
3. Under what circumstances will maximum power be transferred from source to load?
4. For the circuit of Figure 7–23, compute the maximum power that can be developed in the load and the circuit efficiency.

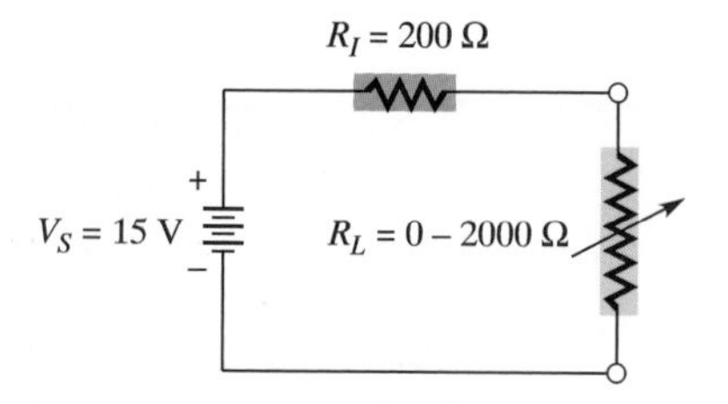

FIGURE 7–23 Circuit for Section Review Problem 4

7–8 MAXIMUM TRANSFER OF VOLTAGE

Question: Why is it impossible to transfer all of the voltage developed by a source to a load?

As stated previously, a practical voltage source, whether a battery, power supply, or transducer, has an internal resistance that is in series with any attached load. A voltage divider is thus formed, and the source voltage, which is the practical voltage source, is divided between the source resistance and load. If the internal resistance is high, a low resistance load has only a small portion of the source voltage developed across it. This drop in load voltage is known as loading the source. Thus, no matter what the value of the load resistance, it will load the source to some degree.

Question: When is it necessary to transfer as much of the source voltage as possible to the load?

Recall that maximum power must be transferred to a load at the output of an electronic system. Maximum voltage, on the other hand, must be developed *early* in an electronic system in what are known as the control circuits. An example is an audio system in which sound waves are converted to electrical signals in a transducer known as a microphone. As Figure 7–24 illustrates, sound waves striking the microphone produce voltages that vary in direct proportion to their amplitude. The voltages, being quite small in magnitude, must be amplified. The input resistance R_{in} of the amplifier is in series with the internal resistance of the source, and the two divide the signal generated by the microphone in direct proportion to their magnitude. It is necessary to develop as much of the signal voltage as possible across the input resistance of the amplifier. Thus, the input resistance must be as large as possible in relation to the source resistance.

Question: What relationship must exist in order to develop a relatively large proportion of the signal voltage across the input of the amplifier?

As previously stated, a voltage source approaches the ideal when the load resistance is much greater than the internal resistance. In equation form, this relationship is written

(7–2)
$$R_L \gg R_I$$

FIGURE 7–24 Transferring voltage to a load

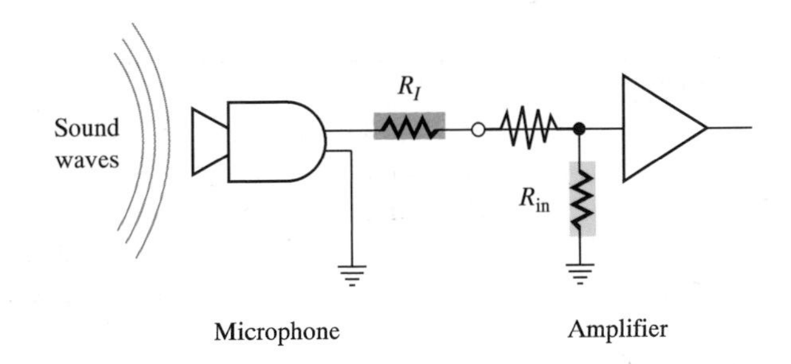

As a rule of thumb, the load resistance must be at least 10 times the internal resistance. If the input resistance of the amplifier in Figure 7–24 is at least 10 times the source resistance, almost the entire source voltage will be developed across it. This problem of amplifier design will be studied in detail in later portions of the electronics curriculum. It is sufficient at the moment to understand that maximum voltage is applied to the load when its impedance is as high as possible.

TECH TIP: STARTING YOUR CAR IN COLD WEATHER

Sometimes on a very cold morning it seems like the battery in your car just doesn't have what it takes to turn the engine over and get it fired up. Here's a tip to help you get it started a little easier.

First it is assumed that the car is properly tuned and that the battery is not defective. There is no substitute for proper maintenance!

The reason that the battery appears not to be providing enough power to the starter is because it's not! That doesn't mean its defective, it's simply unable to. As the outside temperature drops, the internal resistance of a storage cell rises. As the internal resistance rises, so does the associated voltage drop. Kirchhoff's law demands that the sum of the voltage drops must equal the applied voltage. In other words V_{internal} plus V_{starter} must equal V_{source}. Therefore, as the battery's resistance rises, its internal voltage drop rises, and the voltage applied to the starter falls. Of course, a lower starter voltage means less power.

One way to lower the internal resistance of the battery is to heat it. That's great if you have a heated garage, but what if your car sits outside?

Remember that I^2R means *heat!* Simply turn on the headlights for about 30 seconds before attempting to start the car. The heat generated by the current flow will warm the battery and lower its internal resistance, thus allowing a higher starting voltage.

SECTION 7–8 REVIEW

1. Why is it impossible to develop across the load all voltage produced by a source?
2. In what portion of an electronic system is necessary to transfer as much of the source voltage as possible to the load?
3. Under what conditions will practically all of the source voltage be developed across the load?

7–9 APPLICATIONS OF NETWORK THEOREMS

The network theorems are applied almost exclusively in the analysis of complex electronic circuits. The superposition theorem makes it possible to analyze circuits with more than one voltage source using only Ohm's and Kirchhoff's laws. Thevenin's and Norton's theorems allow the most complicated linear circuit to be reduced to a single source and resis-

FIGURE 7–25 Using Thevenin's theorem to measure internal resistance

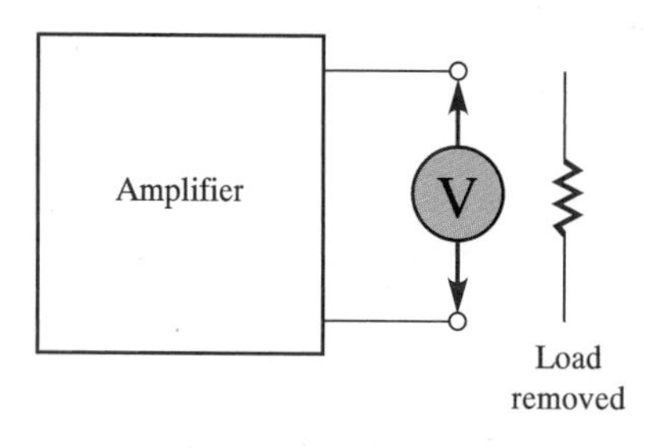

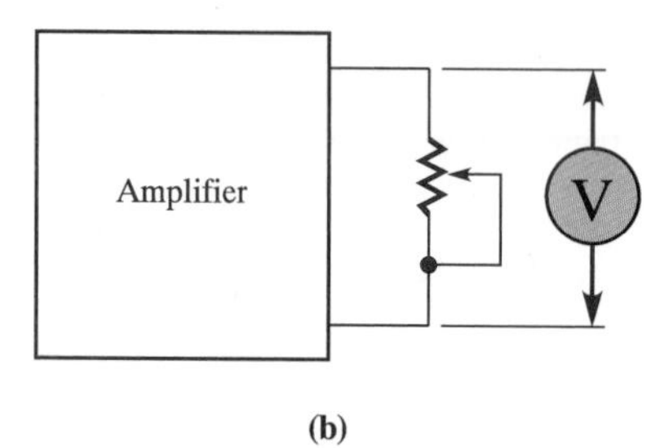

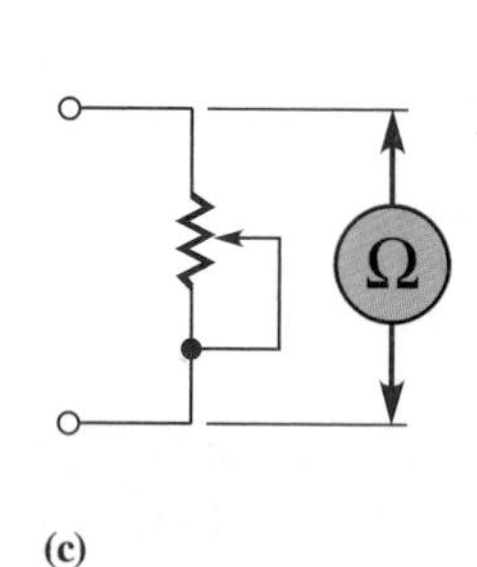

tance for any value of load. The maximum transfer of power and voltage theorems make it possible to predict the action of circuits when they are interfaced one to another. The only qualifier placed upon their use was that the circuit to be analyzed must be linear. Circuits containing a transistor or diode cannot be analyzed because they are not linear devices.

The input and output circuits of transistor amplifiers can be represented by Thevenin or Norton equivalent circuits in order to demonstrate their operation under varying loads.

In certain circuits, physical measurements can be made through which the Thevenin's equivalent of an actual circuit can be obtained. Once the Thevenin's equivalent is obtained, the action of the circuit under any load condition can be determined using Ohm's law. The procedure is illustrated in Figure 7–25. As was the first step in finding the Thevenin equivalent, the entire load must be removed from the source. Then the open circuit voltage is measured, as shown in Figure 7–25(a), to determine V_{Th}. Next a variable resistor (rheostat) is placed across the open circuit, as shown in Figure 7–25(b), and adjusted until one-half the source voltage is measured across it. Next, the variable resistor is removed from the circuit taking care not to disturb its setting. The resistance is measured, as shown in Figure 7–25(c), to determine the value of R_L.

What has been discussed in the previous paragraph is known as the "black box" approach to circuit analysis. A *black box* is the name often given an electronic circuit in which you have access to only the input and output terminals. You know the purpose of the circuit, but not the components of which it is made or the way in which they are connected. You cannot see into the circuit, hence, the name "black box." If its internal resistance can be determined, however, its response to a given value of load can be predicted. This determination of internal resistance was what was demonstrated in the previous paragraph. It is a task you will encounter often in a career as an electronics technician.

Two notes of caution are in order in this procedure for determining the Thevenin's equivalent circuit. First, ensure that the source is capable of delivering the current required by the variable resistor. The source may overload or be damaged if too much current is drawn. Also begin with the rheostat at its maximum resistance value. This practice avoids beginning with maximum current being drawn from the source and once again risking overload. Second, the voltage reading should be made with a high impedance meter in order to avoid loading of the source.

SECTION 7–9 REVIEW

1. What is the major application of the network theorems?
2. What are the steps in determining the Thevenin's equivalent of an actual circuit?
3. Explain why the resistance measured in this section is the Thevenin resistance.

CHAPTER REVIEW

1. Some circuits are very difficult and some are impossible to analyze using only Ohm's and Kirchhoff's laws.
2. The superposition theorem makes it possible to analyze a circuit with two or more voltage sources using Ohm's and Kirchhoff's laws.
3. The application of Thevenin's theorem reduces a complex circuit to a single voltage source and a single series resistor.
4. The application of Norton's theorem reduces a complex circuit to a single current source and a single parallel resistor.
5. In an ideal voltage source, the load voltage does not change appreciably as the load resistance varies.

6. In an ideal current source, the source current does not change appreciably as the load resistance varies.
7. Maximum power is developed in a load when the load resistance is equal to the source resistance.
8. The voltage developed across the load is approximately equal to the source voltage when the load resistance is at least 10 times the source resistance.
9. Circuit efficiency is 50% when maximum power is developed in the load.
10. The major application of the network theorems is in the analysis of complex linear circuits.

KEY TERMS

ideal voltage source
source resistance
ideal current source
Superposition theorem
Thevenin's theorem
Norton's theorem
maximum transfer of power
input power
usable output power
maximum transfer of voltage theorem

EQUATIONS

(7–1) $R_I = R_L$

(7–2) $R_L \gg R_I$

Variable Quantities

R_I = internal resistance of source device

R_L = resistance of the load

TEST YOUR KNOWLEDGE

1. Write the steps in analyzing a circuit with more than one voltage source using the superposition theorem.
2. In your own words, write Thevenin's theorem.
3. List the steps for finding the Thevenin equivalent of a circuit.
4. In your own words, write Norton's theorem.
5. List the steps for finding the Norton equivalent of a circuit.
6. What is a linear circuit?
7. Why is it impossible to develop all the circuit power in the load?
8. In what portion of an electronic system is it necessary to develop maximum power in the load?
9. Under what conditions is maximum power developed in the load?
10. What is the circuit efficiency when maximum power is developed in the load?
11. Why is it impossible to develop all the source voltage across the load?
12. In what portion of an electronic circuit is it necessary to develop as much of the source voltage as possible across the load?
13. Under what conditions will practically all the source voltage be developed across the load?
14. What application is made of the network theorems?
15. What are the steps in determining the Thevenin's equivalent of a practical electronic circuit?

PROBLEM SET: BASIC

Section 7–1

1. Is the circuit of Figure 7–26 an approximation of an ideal voltage source? Why?

FIGURE 7–26 Circuit for Problem 1

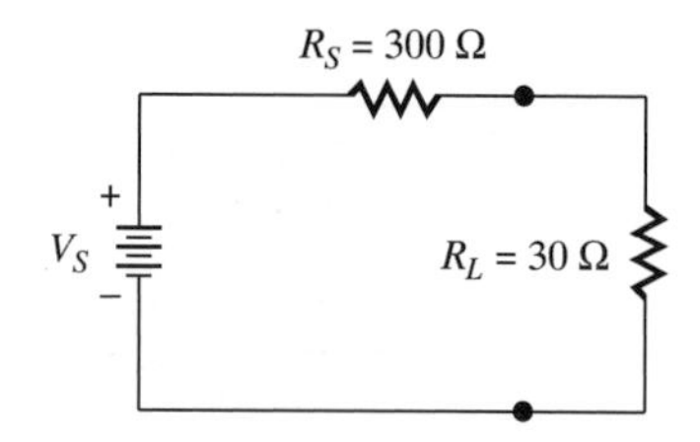

2. In the circuit of Figure 7–27, what is the minimum value of load resistor that could be attached to make the source an ideal voltage source?

FIGURE 7–27 Circuit for Problem 2

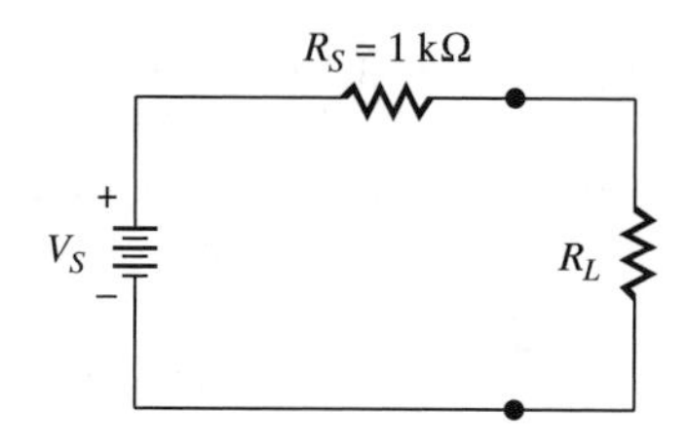

Section 7–2

3. Is the circuit of Figure 7–28 an approximation of an ideal current source? Why?

FIGURE 7–28 Circuit for Problem 3

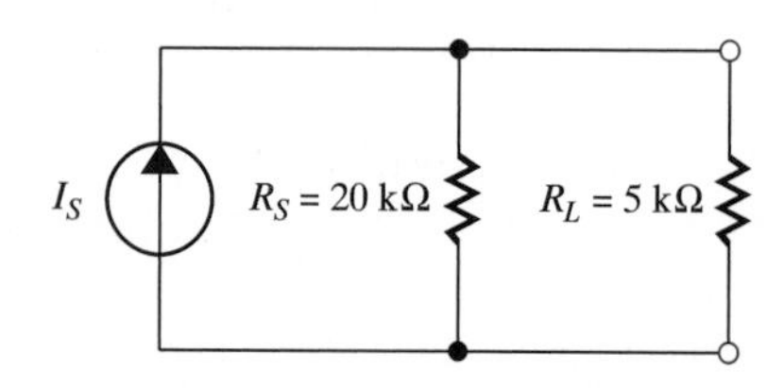

4. In the circuit of Figure 7–29, what is the maximum value of load resistor that could be attached to make the source an ideal current source?

FIGURE 7–29 Circuit for Problem 4

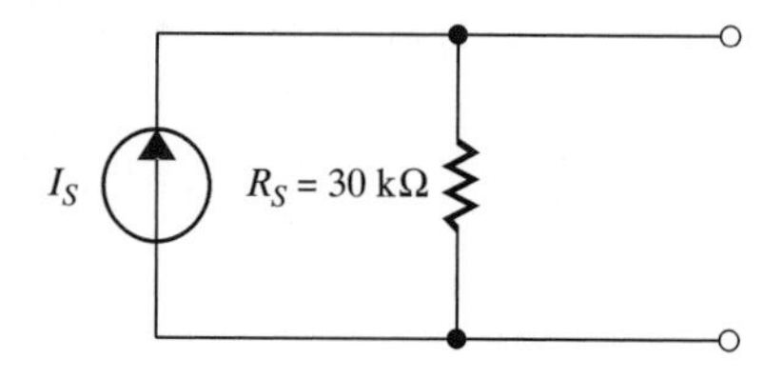

Section 7–3

5. Convert the voltage sources of Figure 7–30(a) and 7–30(b) to equivalent current sources.

FIGURE 7–30 Circuits for Problem 5

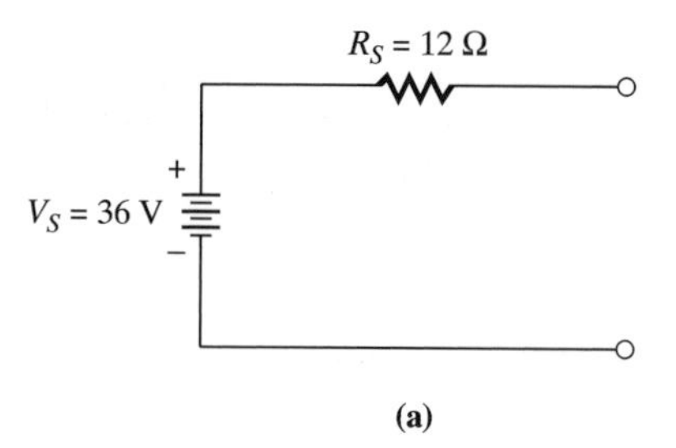

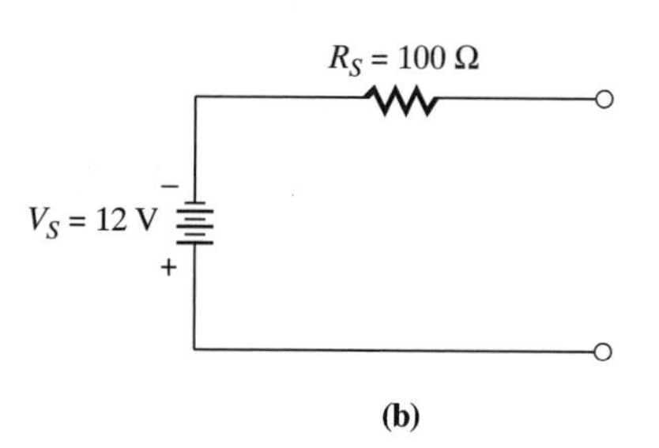

6. Convert the current sources of Figure 7–31(a) and 7–31(b) to equivalent voltage sources.

FIGURE 7–31 Circuits for Problem 6

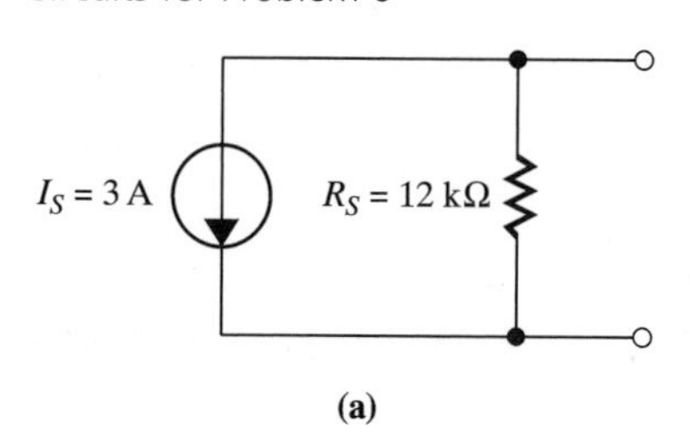

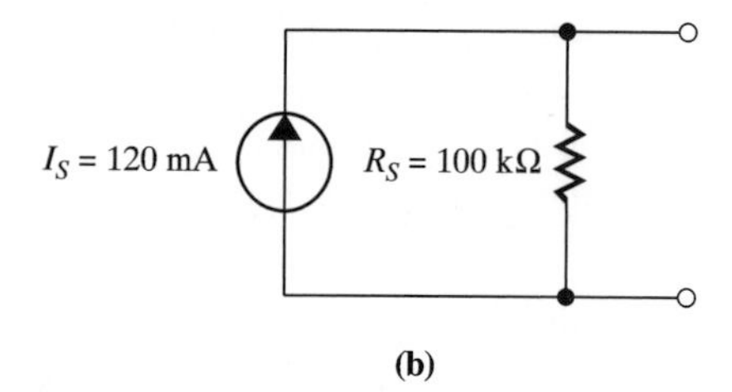

Section 7–4

7. Using the superposition theorem, compute the current through and the voltage across R_L in Figure 7–32.

FIGURE 7–32 Circuit for Problem 7

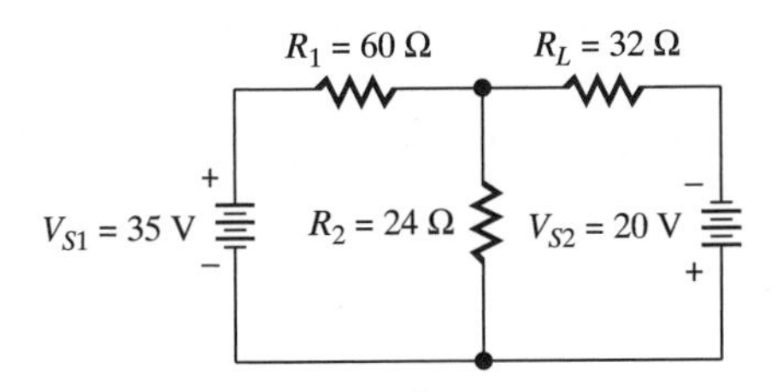

8. Using the superposition theorem compute the voltage between points A and B in Figure 7–33.

FIGURE 7–33 Circuit for Problem 8

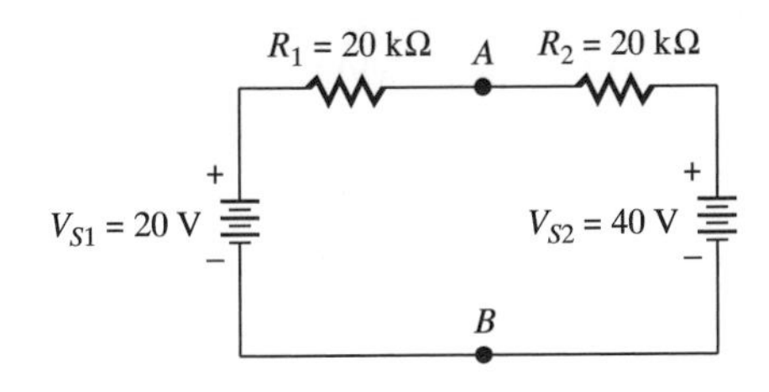

9. Using the superposition theorem, compute the current through R_6 of Figure 7–34.

FIGURE 7–34 Circuit for Problem 9

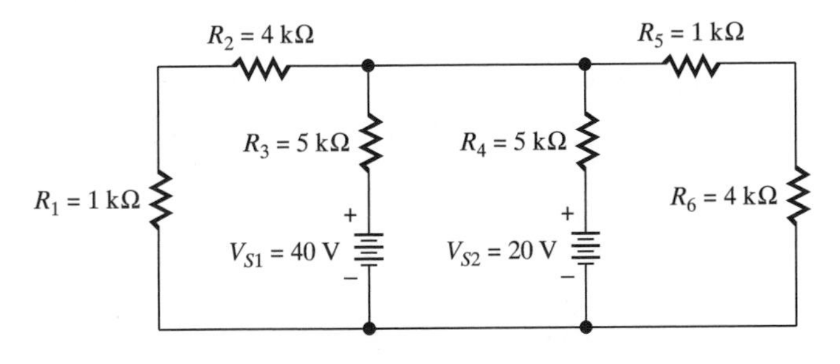

Section 7–5

10. Compute the current through and the voltage across R_L of Figure 7–35 using conventional means.

FIGURE 7–35 Circuit for Problems 10 and 11

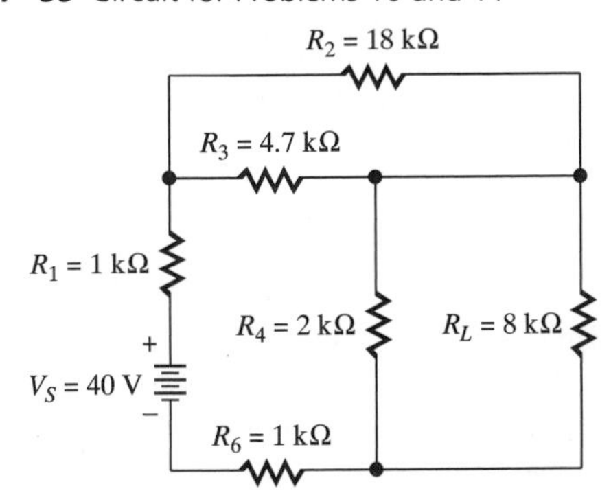

11. Convert the circuit of Figure 7–35 to its Thevenin equivalent. Compute the current through and the voltage across R_L using the Thevenin equivalent circuit. Are the results the same?

12. Convert the circuit of Figure 7–36 to its Thevenin equivalent circuit. Determine the current through and the voltage across R_L using the Thevenin equivalent.

FIGURE 7–36 Circuit for Problem 12

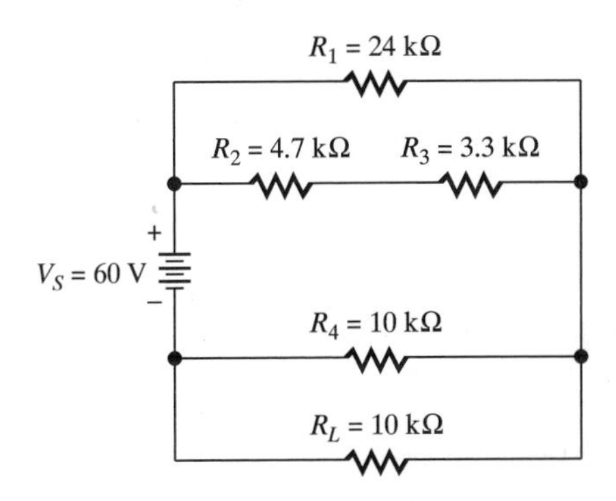

13. The values of a Thevenin equivalent circuit are: $R_{Th} = 234\ \Omega$ and $V_{Th} = 12.5$ V. Draw the circuit.

14. Looking into the circuit of Figure 7–37 from points A and B, determine the Thevenin equivalent.

FIGURE 7–37 Circuit for Problems 14 and 15

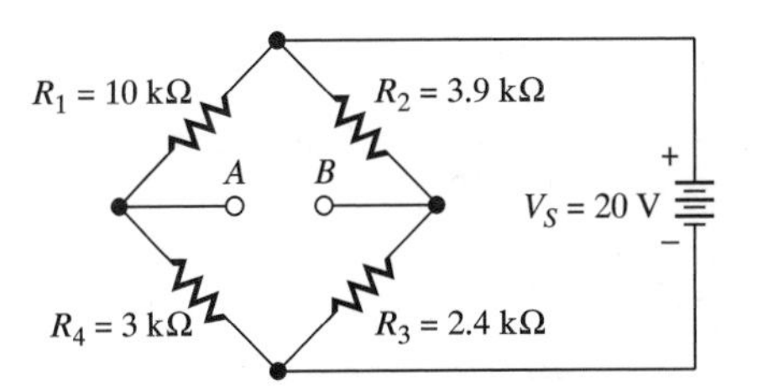

15. Convert the Thevenin equivalent of Problem 14 to its Norton equivalent.

Section 7–6

16. Using conventional means, compute the value of the current through and the voltage across R_L in the circuit of Figure 7–38.

FIGURE 7–38 Circuit for Problems 16 and 17

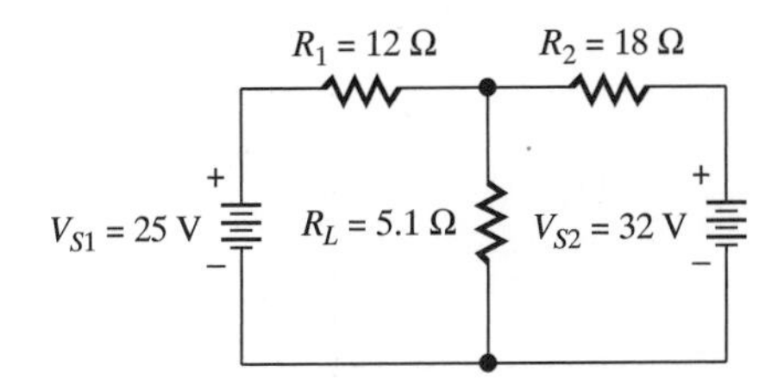

17. Determine the Norton equivalent of the circuit of Figure 7–38. Compute the current through and the voltage across R_L using the Norton equivalent. Are the results the same?

18. For the circuit of Figure 7–39, determine the current through R_L using a Norton equivalent.

FIGURE 7–39 Circuit for Problems 18 and 19

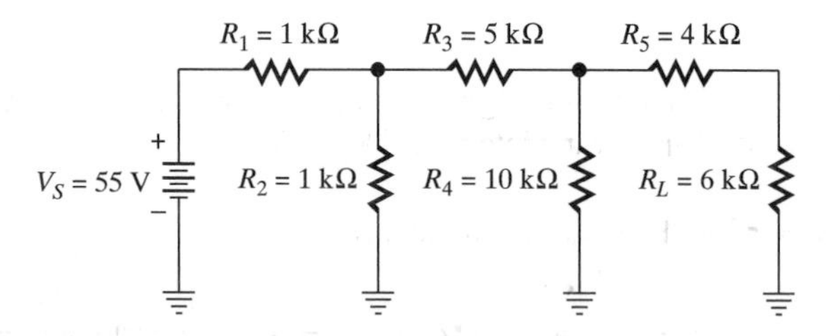

19. Convert the Norton equivalent of Problem 18 to its Thevenin's equivalent.

Section 7–7

20. In the circuit of Figure 7–40, what setting of the variable resistor R_L will cause maximum transfer of power?

FIGURE 7–40 Circuit for Problems 20 and 21

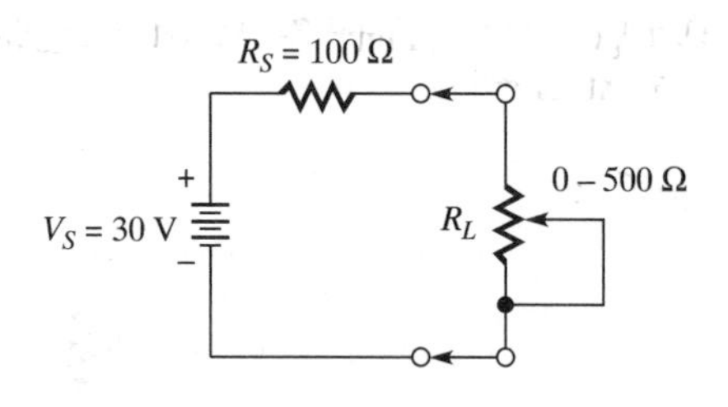

21. In Problem 20, compute the maximum power that can be developed in the load and also the circuit efficiency.

Section 7–8

22. The source of Figure 7–41 has an internal resistance of 60 Ω and a maximum open circuit voltage of 10 V. What is the minimum value of load resistance that can be placed across the source and still transfer maximum voltage to the load?

23. If the load resistor in Problem 22 is 6000 Ω, what is the load voltage?

FIGURE 7–41 Circuit for Problems 22 and 23

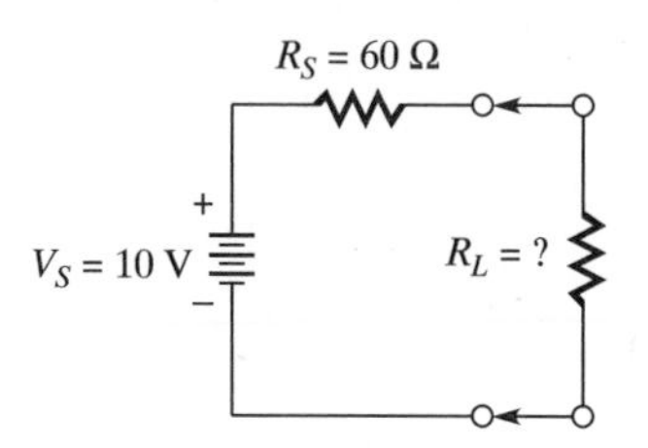

PROBLEM SET: CHALLENGING

24. A power source has an open circuit voltage of 7 V. In Figure 7–42, a 600 Ω load placed across output causes the voltage to drop to 5 V. Compute the internal resistance of the source.

FIGURE 7–42 Circuit for Problem 24

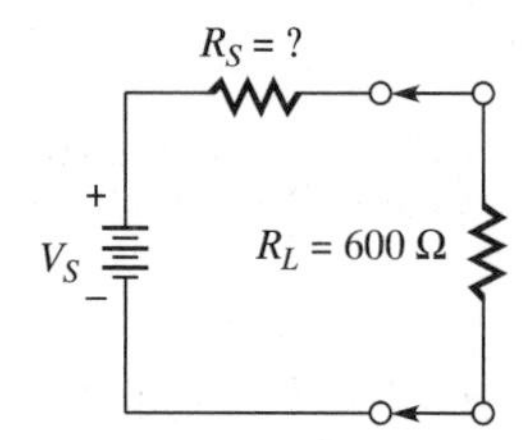

25. A power source has an open circuit voltage of 10 V and an internal resistance of 100 Ω, as in Figure 7–43. What value of load resistor will cause the output voltage to drop to 8 V?

FIGURE 7–43 Circuit for Problem 25

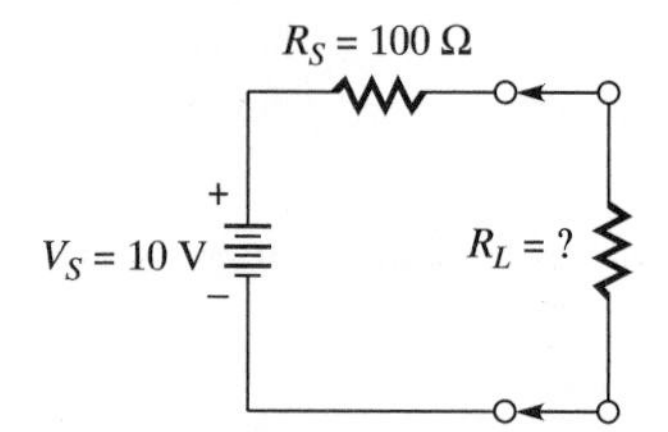

26. An amplifier has an open circuit output voltage of 16 Volts and an internal resistance of 1000 Ω, as in Figure 7–44. What value of load resistor will cause the output voltage to decrease by 10%?

FIGURE 7–44 Circuit for Problem 26

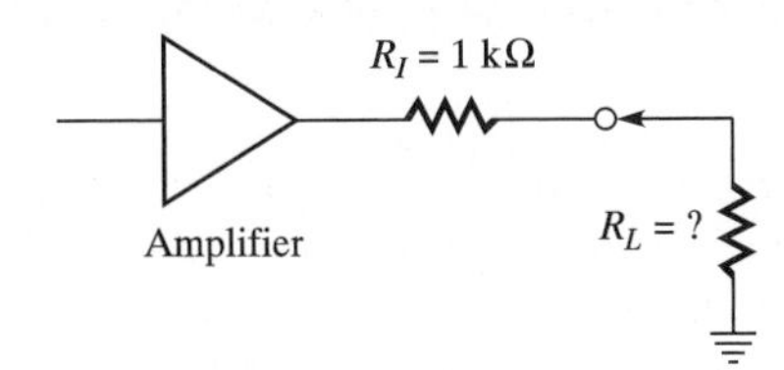

27. The load in Figure 7–45 requires 2.4 V for its operation and draws 300 mA of current. If maximum power is to be transferred, what must be the value of the internal resistance?

FIGURE 7–45 Circuit for Problem 27

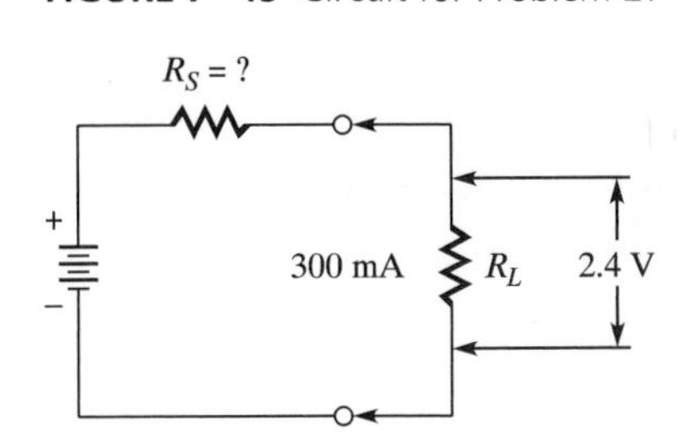

28. Compute the values for and draw the Thevenin equivalent of the circuit in Figure 7–46. Compute V_{RL} and I_{RL} using the Thevenin equivalent circuit.

FIGURE 7–46 Circuit for Problem 28

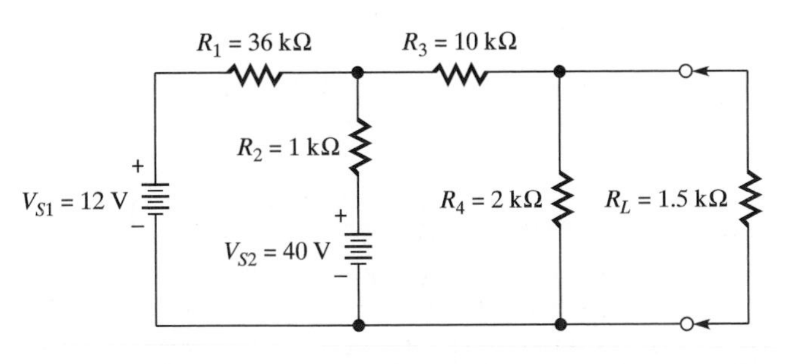

29. What value of load resistance will cause the circuit of Figure 7–47 to act as a voltage source?

FIGURE 7–47 Circuit for Problem 29

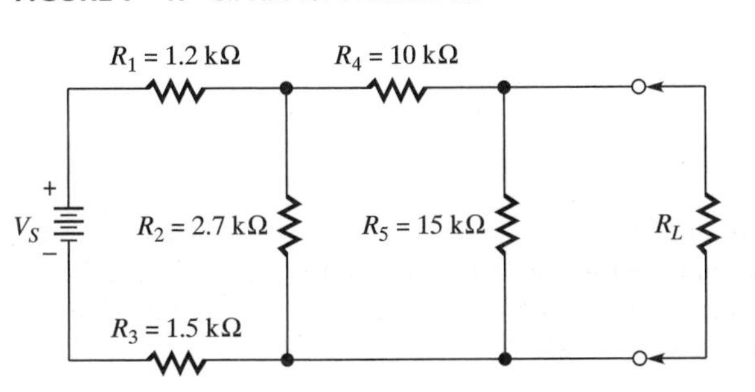

30. For the circuit of Figure 7–48, determine the Thevenin equivalent looking into points *A* and *B*.

FIGURE 7–48 Circuit for Problem 30

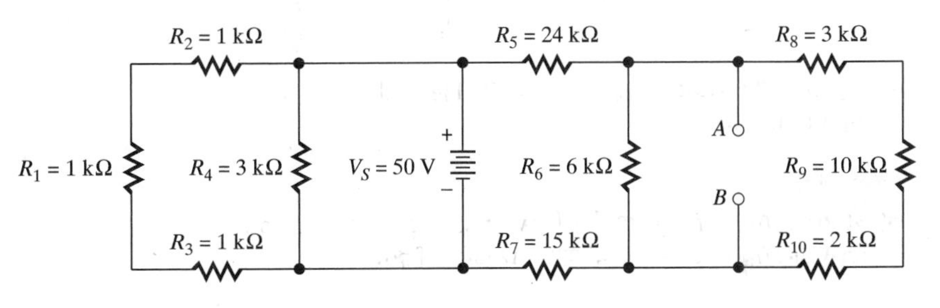

31. For the bridge circuit of Figure 7–49, determine the voltage between points *A* and *B*.

FIGURE 7–49 Circuit for Problem 31

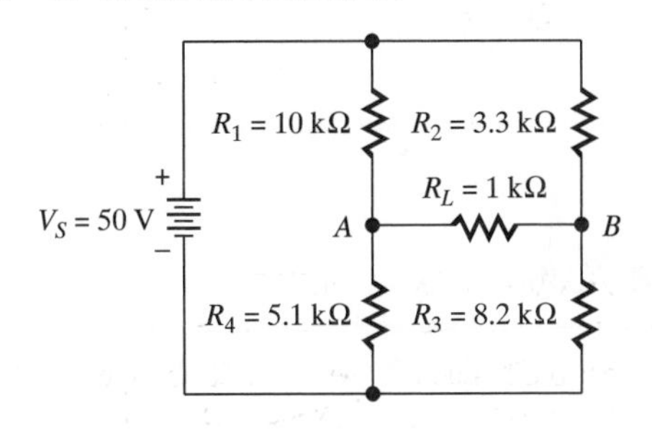

32. Using the superposition theorem, determine the current through R_3 in Figure 7–50.

FIGURE 7–50 Circuit for Problem 32

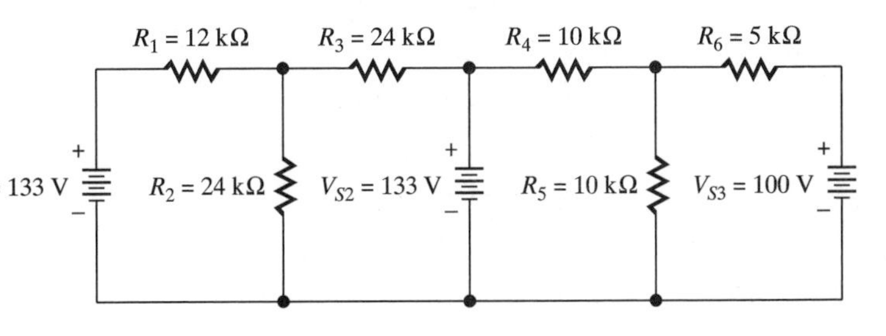

33. Using the superposition theorem, determine the current through R_5 in Figure 7–51.

FIGURE 7–51 Circuit for Problem 33

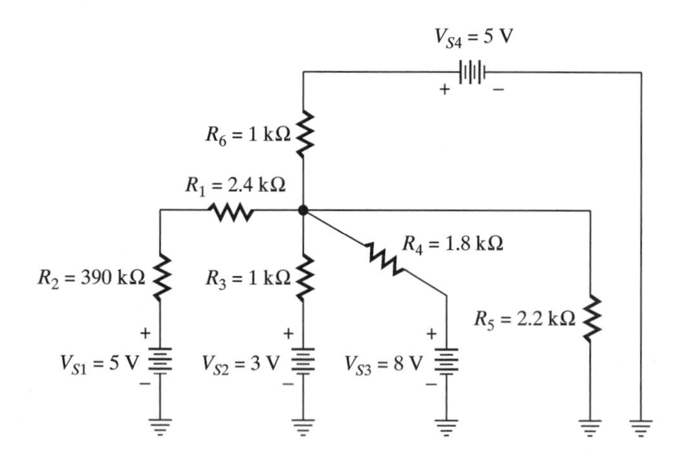

34. Determine the Norton equivalent of the circuit in Figure 7–52. Determine the current through R_L using the Norton equivalent.

35. Convert the circuit of Problem 34 to its Thevenin equivalent. Find the current through R_L using the Thevenin equivalent.

FIGURE 7–52 Circuit for Problems 34 and 35

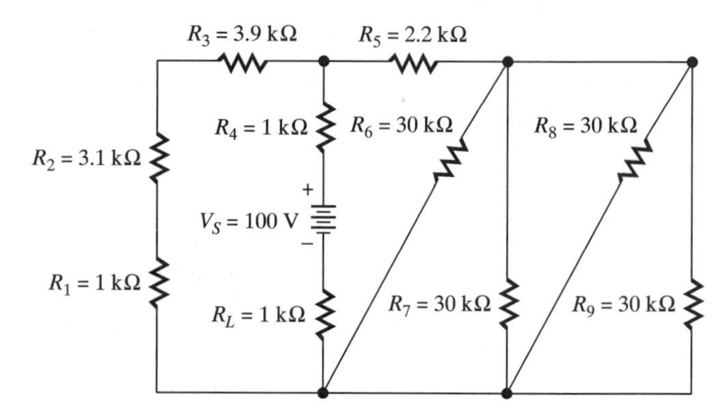

PROBLEM SET: TROUBLESHOOTING

36. Determine the Thevenin equivalent of the circuit of Figure 7–53(a). Using the equivalent circuit, compute the current through R_L.

37. Using the procedures introduced in Section 7–9, the actual Thevenin equivalent is determined and found to be as shown in Figure 7–53(b). If one fault exists in the circuit, what is its probable cause?

FIGURE 7–53 Circuits for Problems 36 and 37

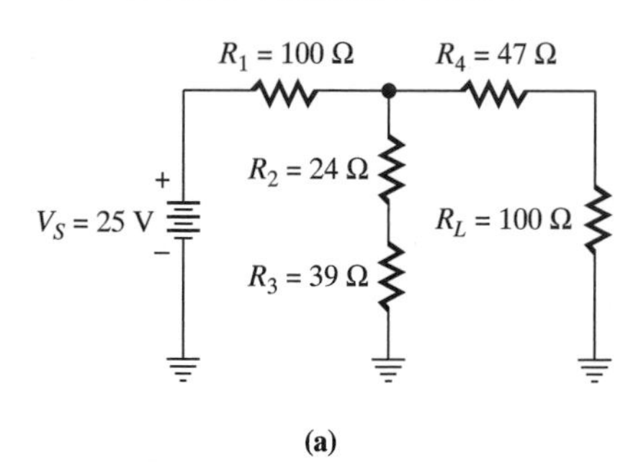

(a)

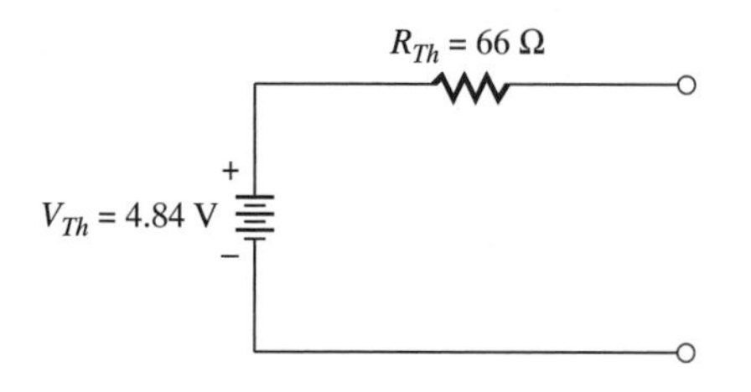

(b)

38. Determine the Norton equivalent of the circuit of Figure 7–54. Using the equivalent circuit, compute the current through R_L.

FIGURE 7–54 Circuit for Problems 38 and 39

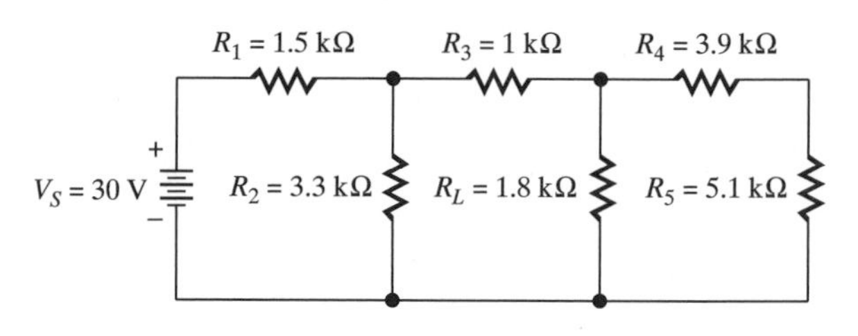

39. The current through R_L of Figure 7–54 is measured and found to be 6.25 mA. If the load resistance is correct, determine what fault exists.

40. From the readings shown in Figure 7–55, determine the internal resistance of the circuit contained within the block labeled "black box."

FIGURE 7–55 Circuit for Problem 40

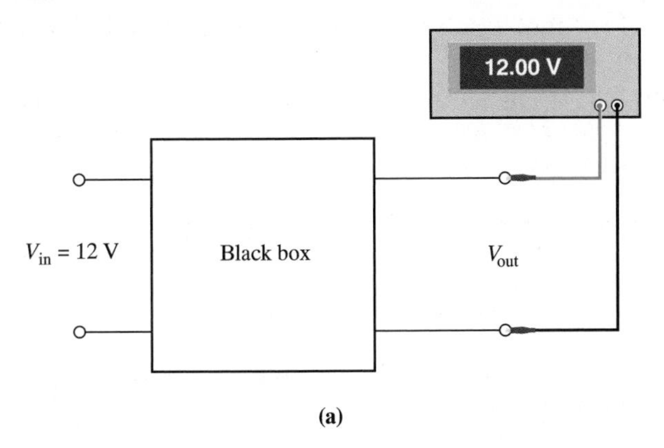

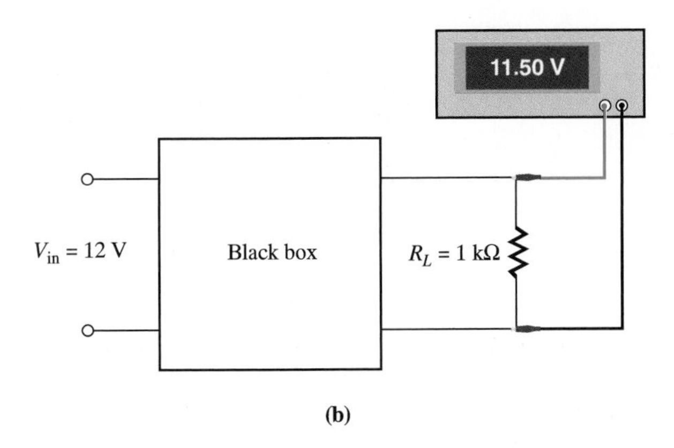

Quality Assurance

Practically everyone has, at one time or another, purchased some item that has turned out to be a "lemon." Examples of "lemons," or defective items, are many: the car that leaks around the rear window, the toaster that burns your bread to a cinder, the TV that only works sometimes, the calculator that puts 8's across the whole display when you press the 8 key, the computer that sometimes won't talk to the terminal, and even the space capsule door that won't open! Usually the manufacturer or distributor makes good on the warranty for such defective products, and the consumer is satisfied.

In the case of a true "lemon," however, the item keeps finding new and novel ways to fail. During the warranty period, the manufacturer assumes responsibility for the repairs; after the warranty period, the dissatisfied customer bears the expense.

Regardless of what consumers may think, manufacturers do not want to produce lemons. A manufacturer does not wish to do warranty work. It is a waste of manpower and facilities that must be avoided if at all possible. And every manufacturer wants the word "quality" associated with its name.

The vast majority of workers want to do quality work, too. The problem is that they are not robots, and they make human mistakes. A master assembler can make a poor solder connection or place a component in the wrong position on a printed circuit board. An assembler can fail to notice incorrectly wired plugs, frayed insulation, nicked wires, and numerous other problems.

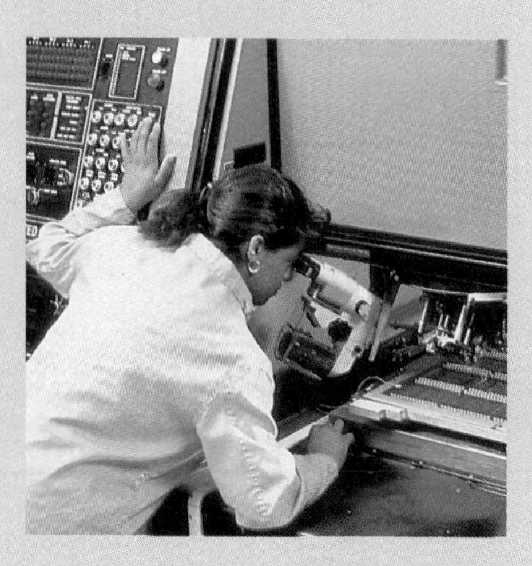

Such considerations have led manufacturers to institute quality assurance QA programs. QA programs have several purposes: (1) to detect problems before items are shipped to customers and (2) to ensure that workmanship conforms to standards of good practice. Although inspectors are employed specifically to inspect and certify work, QA programs involve all workers in the process, raising their awareness of the need for quality.

Quality assurance inspectors are basically critics. They must inspect work and either pass or reject it. The work is then redone to the proper standards, inspected once more, and accepted. Occasionally, however, personalities creep into the process and hard feelings develop between QA inspectors and other employees. In these cases, leadership must be provided by top management to ensure that the QA program has positive rather than negative results.

There are times when an honest difference of opinion exists on what is and what is not acceptable. There are also times when a production line supervisor may feel that he/she must "stand up for the employees on the line" against a "picky" QA inspector. Situations such as these are not uncommon, and they usually end up being arbitrated by higher management, as illustrated by the actual case from industry that follows.

Dixie has been a QA inspector for 8 years with a company that specializes in the design and manufacture of instrumented tracking radars. She has risen steadily in the program through her knowledge of standards and her ability to deal with people at all levels in the production process. In working with production employees, she has earned their respect through letting neither personalities nor friendships enter into her decisions. In short, she has learned the art of disagreeing without being disagreeable. She is enhancing her knowledge of electronics theory through night classes in the Electronics Engineering Technology program at the local community college.

This is not to say that Dixie's job is always a smooth one. As she enters the plant on this particular day, she thinks of a problem called to her attention the day before. Certain resistors in some printed circuit boards already in production are being mounted in a nonstandard manner. As she understands it, the spacing on the printed circuit boards, which have already been manufactured, will not allow standard mounting of precision resistors. The required precision

continued

continued

resistors are larger than the nonprecision variety and these will not fit the allotted space. These resistors are supposed to be mounted as shown in Figure 1. The resistor leads are bent to a certain length and inserted through the board and then soldered. However, to avoid the delay and expense of manufacturing new boards, the production staff is having the resistors installed as shown in Figure 2. Dixie knows that it will be her decision to either pass or reject this procedure.

Arriving at her office, Dixie finds that Dennis, the production engineer in charge, is waiting to see her concerning the problem. After apologizing for not informing her of the problem when it was first discovered, he points out that the procedure should cause no problems in circuit operation, and so long as the equipment functions properly, no one will even notice the nonstandard installation. The solution of condemning all the boards and manufacturing new ones would be a waste of time and money, he feels. Dixie neither agrees or disagrees with this analysis, but does promise to keep it in mind in making her decision.

After Dennis leaves, Dixie ponders the problem and all its ramifications. On the one hand, she knows that the standards have not been met and that Dennis is probably right in thinking circuit operation will not be affected. Further, she knows that the project is running rather close to the deadline and that the contract is a lucrative one that management most certainly wants to see finished on time. Finally, she knows that Dennis is quite determined to see that the problem does not delay or add expense to the project.

On the other hand, she must consider other issues such as the customer's opinion and the company's position in the industry. The customer is a valued one, and she also knows it is made up of shrewd business-people who will not tolerate shoddy work. Her company has earned its enviable reputation through pleasing just such customers. She further knows that such reputations hang by a thread in such a competitive business as electronics! Weighing carefully all the factors on both sides, Dixie decides to reject the printed circuit boards.

After lunch she informs Dennis of her decision. They discuss the matter at length, but cannot agree that the decision is satisfactory to both QA and production. Dennis feels that they should both express their views to the general manager. This plan is satisfactory to Dixie, and a meeting is scheduled for the following morning. The manager listens to the views of both parties, never seeming to agree or disagree. After asking a few questions, the manager decides that the work has to be redone to conform to appropriate standards. He suggests that engineering re-evaluate the circuits to ascertain if nonprecision resistors can be used.

Walking down the hall after the meeting, Dixie feels satisfaction at being vindicated, but she is also apprehensive. Dennis is an old and dear friend, and she hopes their friendship will survive this difference of opinion. As they walk down the hall, she is relieved to feel an arm across her shoulder and hear Dennis say, rather ruefully, "Well, Dixie, I guess we win some and we lose some." With a great feeling of relief, Dixie returns to her office to confront the next problem.

FIGURE 1 Precision resistor soldered to PC board with proper spacing

FIGURE 2 Precision resistor soldered to PC board with too little spacing

CHAPTER

8 Magnetism and Electromagnetism

■ UPON COMPLETION OF THIS CHAPTER, YOU WILL BE ABLE TO

1. Sketch a bar magnet and its magnetic field.
2. Write the laws of repulsion and attraction for magnetic poles.
3. Define and explain the differences between magnetic flux and flux density.
4. Explain how materials become magnetized.
5. Define residual magnetism.
6. Define retentivity, permeability, and reluctance.
7. Explain the operation of an electromagnet.
8. Define magnetomotive force and field intensity.
9. Draw and explain the graph of the *BH* magnetization curve.
10. Define magnetic hysteresis.
11. Explain several applications of magnetism and electromagnetism, including motor action, generator action, analog meter movement, and the Hall effect.
12. List precautions to be taken in the handling and storage of magnets.

Magnetism is a subject that has fascinated mankind from its discovery to the present day. Over 2500 years ago, in an area of Asia Minor known as Magnesia, mineral rocks were dug from the ground that had the property of attracting iron. These rocks were the subject of both Greek and Roman writers who referred to them as magnetite, named for the area in which they were found. You can well imagine the mystic and magic properties assigned to these rocks by the ancients!

Centuries later, around 1300 A.D., it was discovered that if the magnetite was shaped into a bar and suspended so that it was free to rotate, it always aligned itself in the same direction. This type of device was the forerunner of the magnetic compass and was used by early mariners and land travelers in their journeys of exploration and trade. The early devices "led the way," so to speak, and were known as lodestones or "leading stones."

Over the years, science has unlocked many of the secrets of magnetism. Magnetite has been determined to be an oxide of iron that has gained "spontaneous" magnetism. The action of the compass is explained by the fact that the earth itself has a magnetic field and is, in fact, a rather large natural magnet. Permanent magnets in many shapes and sizes are manufactured today and find a host of uses. These applications range from the mundane such as magnetic latches on cabinet doors to the sophisticated such as position sensors used in automated factories.

In the nineteenth century, the work of many scientists indicated a link between magnetism and electricity. This discovery, which many rank in importance with the discovery of fire and invention of the wheel, has had profound effects upon modern society. The telegraph, electric generator, and electric motor are all applications of early research in magnetism and electricity. In the course of this research, the basic principles that led to radio and television broadcasting were discovered. Today, electricity is being used to produce very powerful magnetic fields, one application of which is to propel trains at tremendous speeds. These trains promise a great deal of comfort because they don't ride on the rails, but levitate on a magnetic cushion above them! All of these accomplishments have occurred even though there is still a tremendous amount to be learned about magnetism. The future is an exciting one in this area!

In the first part of this chapter, the basics of magnetism and permanent magnets are covered in order to provide an understanding of electromagnetism and electromagnetic induction, which is presented in the second part of the chapter. The chapter provides the basic knowledge needed for the understanding of inductance, a magnetic effect, presented in the next chapter. Thus, the material covered in this chapter is an important part of your technical knowledge.

8–1 BASICS OF MAGNETISM

There are three basic sources of magnetism: (1) **natural magnets,** which are dug from the ground in the form of magnetite, (2) **permanent magnets,** which are manufactured through a process known as magnetic induction, and (3) **electromagnets,** which are the result of a flow of electric current through a material. The latter two sources of magnetism find many applications in electronic equipment and systems.

Question: What is the nature of the magnetic field?

A *bar magnet,* which gets its name from the metal bar of which it is made, is illustrated in Figure 8–1, along with its associated magnetic field. The magnet has a north magnetic pole (N) at one end and a south magnetic pole (S) at the other end. Invisible lines of magnetic force emanate from the north pole, move through the space surrounding the magnet, and re-enter at the south pole. They form complete loops, never crossing one

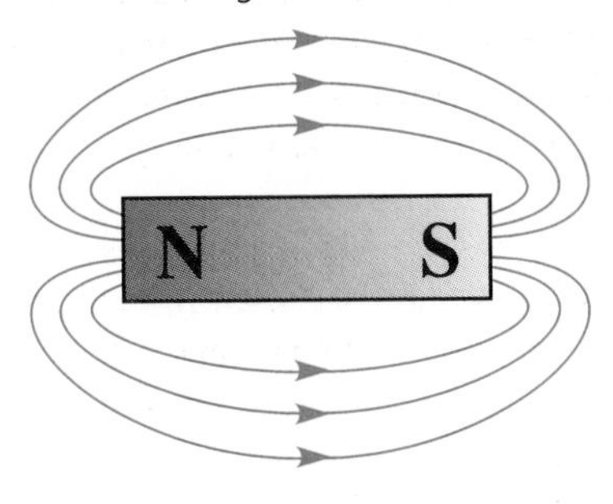

FIGURE 8–1 Bar magnet and its associated magnetic field

another, and move through the shortest possible route. The entire group of magnetic lines of force surrounding a magnet are known as a **magnetic field.** Recall that a voltage source has a polarity determined by the position of its positive (+) and negative (−) poles. In a similar manner, the magnet has a magnetic polarity determined by the position of its north and south pole.

NOTE

Notice the similarity between an electric and a magnetic circuit. Several more similarities will be pointed out as they arise. Watch for them. They will aid you in your understanding of magnetic circuits.

Question: How are the north and south poles of a magnet designated?

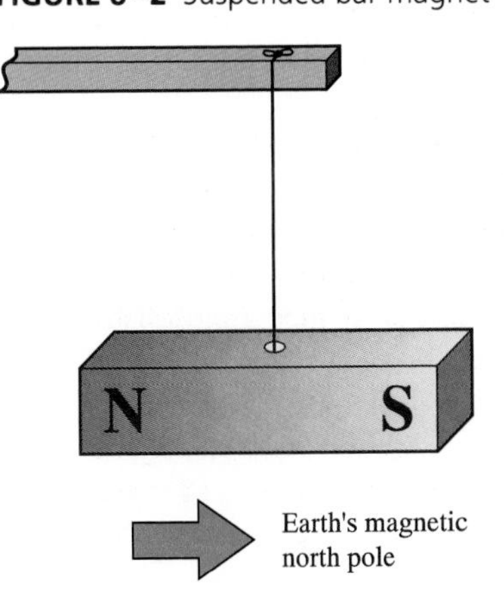

FIGURE 8–2 Suspended bar magnet

The north and south poles are not designated arbitrarily. The earth is itself a large natural magnet. It has north and south magnetic poles. These poles differ from geographic north and south by a number of degrees, known as *declination.* As illustrated in Figure 8–2, if a bar magnet is suspended and allowed to rotate, it will align itself with the earth's magnetic field. The pole pointing north is the magnet's south pole, which is called the *north-seeking pole.* The other pole, the magnet's north pole, is its south-seeking pole.

Recall that two bodies with opposite electrical charges attract one another, while bodies with like electrical charges repel. In a similar manner, unlike magnetic poles attract one another, while like magnetic poles repel. This principle is illustrated in Figure 8–3. The repulsion or attraction of magnetic poles is used in many applications of magnetism.

Question: Can north and south poles be separated in order to produce monopoles?

A magnet is said to be a *dipole,* meaning two poles. It would seem that if a bar magnet were cut at its exact center into two equal pieces, the result would be two pieces, with

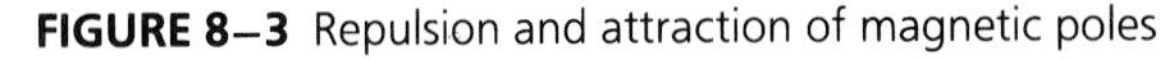

FIGURE 8–3 Repulsion and attraction of magnetic poles

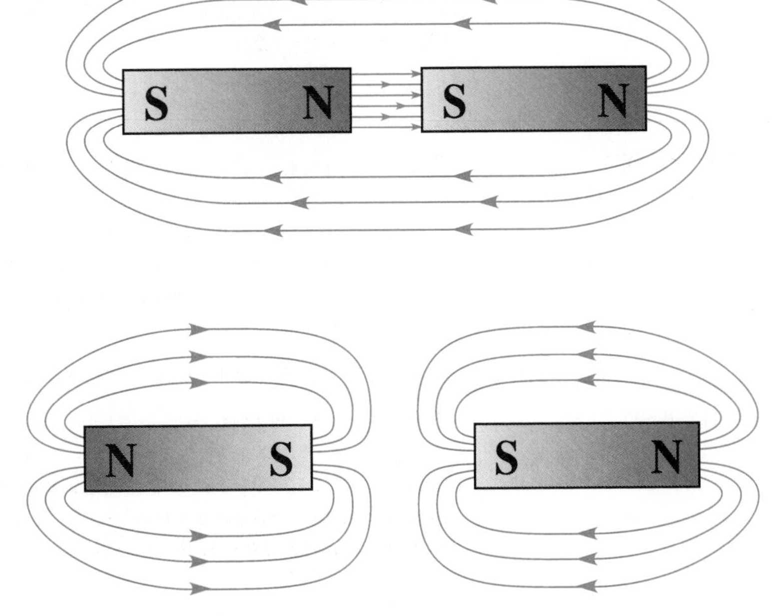

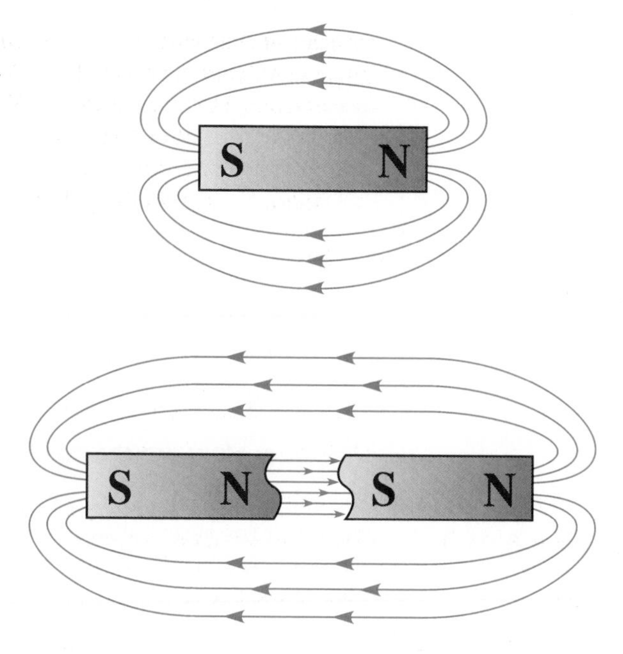

FIGURE 8–4 The bar magnet as a dipole. Splitting a bar magnet results in two sets of north and south poles.

one having only a north pole and the other having only a south pole. However, as Figure 8–4 shows, the result is not two pieces with single poles, but rather two magnets, each having both a north and a south pole. Two magnets result *each* time a piece is divided. The search for the monopole has been long, extensive, and fruitless. The thinking within scientific circles today is that the monopole does not exist.

NOTE

The reason that the search for the elusive monopole has been so diligent is that there are those who feel it may be an energy source. Recall that electricity is produced by the separation of negative and positive charges. The electrons moving in a manner to neutralize the charge are made to do useful work. If monopoles were known to exist, then it may be possible to separate north and south poles. This imbalance could cause a pressure that could cause one or the other to move. Could the moving monopoles be made to do useful work? Perhaps. An energy source of this type does excite the imagination, but to date, there is no evidence that the poles can be separated to create monopoles.

SECTION 8–1 REVIEW

1. How is the magnetic polarity of a magnet determined?
2. State two characteristics of the lines of force that make up the magnetic field.
3. What are the laws of repulsion and attraction for magnetic poles?

8–2 FLUX AND FLUX DENSITY

The magnetic lines of force making up a magnetic field are known as **flux lines.** The entire group of these flux lines is known as **magnetic flux** and is symbolized by the Greek letter phi (ϕ). The number of flux lines produced by a magnetic source is one of the factors that determine the strength of the magnetic field.

The **maxwell** (Mx), named for Scottish physicist James Clerk Maxwell, is the unit of measure of magnetic flux and is equal to 1 line of magnetic flux. A strong one-pound magnet produces approximately 5000 Mx of flux. Another and much larger unit of measure of magnetic flux, the **weber** (Wb), named for the German physicist Wilhelm Weber, is equal to 100 million Mx (1 Wb = 1×10^8 Mx).

The 5000 Mx of flux produced by the one-pound magnet can be converted to webers by dividing the maxwells by 100 million. Thus,

$$\text{Wb} = \frac{\text{Mx}}{100 \times 10^6}$$

$$= \frac{5000}{100 \times 10^6} = 50\ \mu\text{Wb}$$

Since the weber is a very large unit, it is usually expressed in milli (m) or micro (μ) units.

EXAMPLE 8–1 Convert 8000 Mx to webers.

■ Solution:

$$\text{Wb} = \frac{\text{Mx}}{100 \times 10^6}$$

$$= \frac{8000}{100 \times 10^6} = 80\ \mu\text{Wb}$$

In addition to the number of lines of flux generated, the strength of the magnetic field is also related to the concentration, or density, of these lines of flux. As shown in Figure 8–5, the flux lines are most concentrated near the surface of the magnet and diverge as they move outward. There are more flux lines in the unit area near the magnet than in the same unit area further from it. Thus, the magnetic field strength and flux density are

FIGURE 8–5 Flux density and magnetic strength

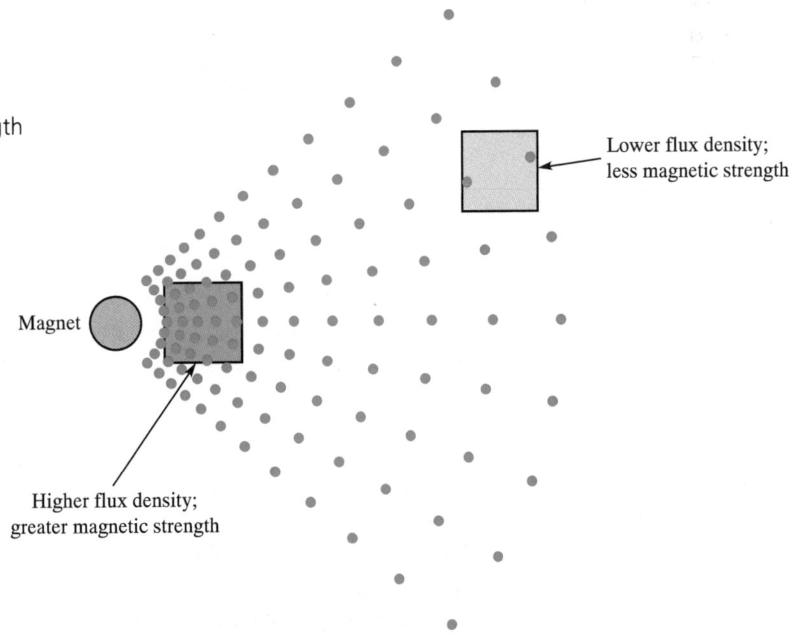

Magnet

Higher flux density; greater magnetic strength

Lower flux density; less magnetic strength

greatest near the surface of the magnet and decrease as the distance from the magnet increases. It is especially strong in the area of each magnetic pole where the lines are most concentrated and the field strength is greatest.

Flux density, symbolized by B, is defined as the number of flux lines in a given area A within the magnetic field. Just as was the case with flux, there are two units of measure, one small and one large, for flux density. The **gauss** (G) named for Karl Friedrich Gauss, the German mathematician, is the number of maxwells per square centimeter. In equation form, flux density in gauss is written as

(8–1)

$$B = \frac{\text{Mx}}{A\ (\text{cm}^2)}$$

The larger unit of flux density, the **tesla** (T), named for the Austrian-born American inventor Nikola Tesla, is the number of webers of flux per square meter. In equation form, flux density in teslas is written as

(8–2)

$$B = \frac{\text{Wb}}{A\ (\text{m}^2)}$$

One tesla is equal to 10,000 G ($1\ \text{T} = 1 \times 10^4\ \text{G}$). Thus, the flux density in gauss can be converted to teslas by dividing by 1×10^4.

EXAMPLE 8–2

3000 maxwells of flux are found to occupy 5 cm^2. What is the flux density in gauss?

■ Solution:

Use Eq. 8–1:

$$B = \frac{\text{Mx}}{A\ (\text{cm}^2)}$$

$$= \frac{3000}{5} = 600\ \text{G}$$

EXAMPLE 8–3

Convert the value of flux density computed in Example 8–2 to teslas.

■ Solution:

Use Eq. 8–2:

$$1\ \text{T} = \frac{\text{G}}{1 \times 10^4}$$

Thus,

$$B = \frac{600}{1 \times 10^4} = 60\ \text{mT}$$

■ SECTION 8–2 REVIEW

1. What are the two units of measure for magnetic flux?
2. State the difference between flux and flux density.

3. Where is the strength of the magnetic field the greatest? Why?
4. Define magnetic flux.

8–3 PERMANENT MAGNETS

Question: What gives a material magnetic properties?

Recall that it is the *position* of the electrons in the atoms of elements that determined their electrical properties. Copper, with only one electron in its valence shell, is a good conductor of electricity. Copper, however, is a nonmagnetic material because of the way the electrons in its atoms *move*. Electrons not only orbit the nucleus of an atom, they also rotate on their axis in much the same manner as the earth does. Each produces a weak magnetic field. In copper and other nonmagnetic materials, electrons with opposite spins are paired and the fields cancel. The result is thus a lack of magnetic properties.

In magnetic materials such as steel, there is not an even pairing, and more electrons rotate one direction than the other, which creates a net magnetic field with a definite north–south dipole. Groups of these atoms form what are known as *magnetic domains*. These domains can be looked upon as minute bar magnets, each being a magnetic dipole. In a material that is not magnetized, they are in a state of disarray as shown in Figure 8–6. The presence of these domains give the material magnetic properties in varying degrees. These materials are attracted by magnets and can themselves be magnetized.

FIGURE 8–6 Random alignment of magnetic domains within a nonmagnetized material

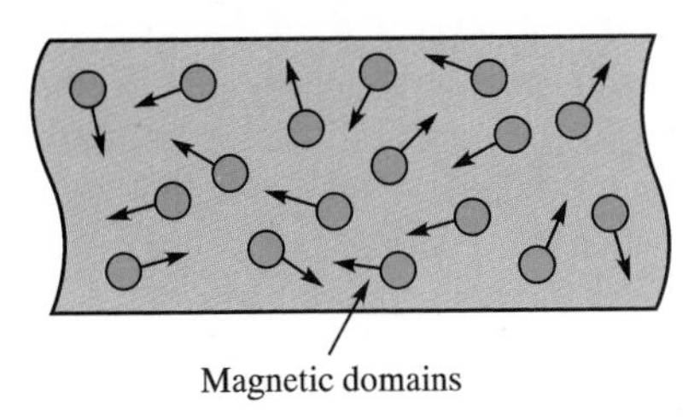

Question: What are the properties of magnetic materials?

The most important magnetic properties of a material are retentivity, permeability, and reluctance. Each will be discussed in the following subsections.

■ Permeability

The **permeability** of a material is a measure of the ease with which magnetic flux can be established within it. It is also often defined as the ability of a material to concentrate lines of magnetic flux. It is symbolized by the Greek letter μ (mu).

The action of a highly permeable material in distorting a magnetic field is illustrated in Figure 8–7. A piece of soft iron, which is highly permeable, is placed near a magnetic field. Notice that now the lines of flux detour through the iron on their way back to the other pole of the magnet in spite of the fact that the path is longer. The iron more easily allows the establishment of magnetic flux than air. Thus, the flux lines follow the path of highest permeability. Notice the similarity to electric current that follows the path of least resistance.

FIGURE 8–7 Distortion of magnetic field as flux lines detour to follow path of high permeability

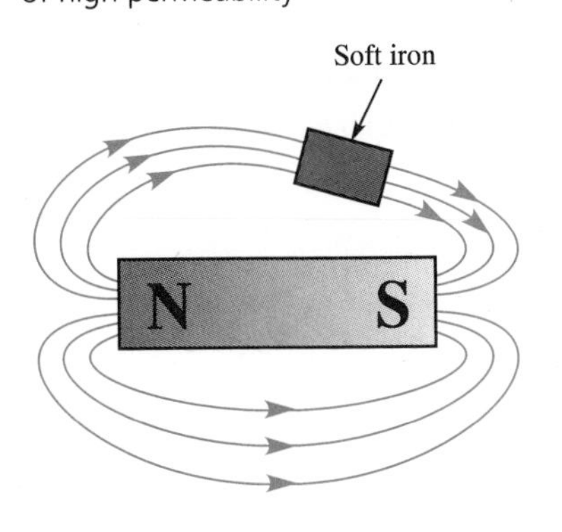

■ Reluctance

Reluctance is basically the opposite of permeability in that it is the opposition a material presents to the establishment of magnetic flux within itself. It is symbolized by $\mathcal{R}$. Reluctance is similar in its action to that of resistance in an electric circuit. Thus, it is very difficult to magnetize a material that has a high reluctance. Looking once again at Figure 8–7, the magnetic flux detours through the iron because the air path has a much higher reluctance than the iron.

Retentivity

If a magnetic material is acted upon by an external magnetic field, it in itself becomes a magnet. The iron in Figure 8–7 actually becomes a magnet because of the action of the external magnetic field. When the external field is removed, some magnetism remains in the material. The amount of magnetism retained by the material when a magnetizing force is removed is a measure of the material's **retentivity.** Metal screwdrivers often become slightly magnetized after long periods of use around electronic equipment. For the manufacture of permanent magnets, materials with high retentivity are desirable.

Question: How are materials magnetized?

FIGURE 8–8 Magnetic induction. Magnetic domains align with the external field when a magnetizing force is appled to a magnetic material, causing it to become a magnet.

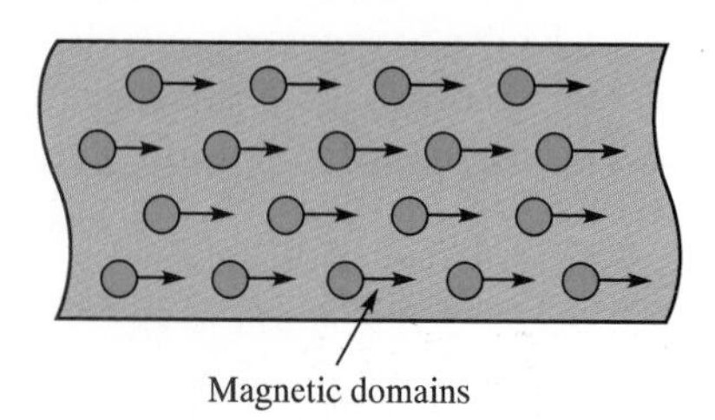

Permanent magnets are manufactured through a process known as **magnetic induction,** in which the material is placed in a strong magnetic field. As you have seen, a magnetic material with high permeability is easily magnetized. Recall that in a nonmagnetized material, as shown in Figure 8–6, the magnetic domains are in a state of disarray. As illustrated in Figure 8–8, when a magnetizing force is applied, the domains align themselves with the external magnetic field. Thus, one end of the material develops a north pole and the other a south pole, and the material becomes a magnet. If the material has a high retentivity, it will retain most of this magnetism when the external field is removed. If its retentivity is low, however, external heat and vibration cause the domains to once again go into a state of disarray. When this occurs, the material loses its magnetism.

Most permanent magnets are made of some alloy of iron. A material commonly used in permanent magnets is **ALNICO,** which is composed of aluminum, nickel, cobalt, and iron. It produces strong magnets with high flux density and with high retentivity. A major use of these magnets is in permanent magnet loudspeakers. Another material, *permalloy,* which is made of nickel and iron, also yields magnets with high values of flux density and retentivity.

Question: How does the air gap affect the strength of a magnet?

Figure 8–9(a) shows once again the magnetic field surrounding a bar magnet. Notice that the *air gap,* which the flux must pass through, is the distance from the north to the south pole. Figure 8–9(b) shows the same magnet shaped into what is known as a horseshoe magnet. Notice that now the air gap is much shorter. The flux density between the poles with the shorter air gap is much greater than that with the longer air gap. This narrowing of the air gap can be done only when the required shape of the magnet allows it.

SECTION 8–3 REVIEW

1. Describe a magnetic domain.
2. What are three characteristics of magnetic materials?
3. Which is more difficult to magnetize, a material with high permeability or high reluctance?
4. How is reluctance similar in its action to an electric circuit?
5. What is meant by magnetic induction?

8–4 ELECTROMAGNETISM

A current-carrying conductor has an associated magnetic field along its entire length. The strength of this magnetic field is directly proportional to the amount of current flow in the conductor. Its magnetic polarity depends upon the direction of current flow. Looking

FIGURE 8–9 Effect of the air gap on magnetic strength

FIGURE 8–10 Magnetic polarity and the left-hand rule

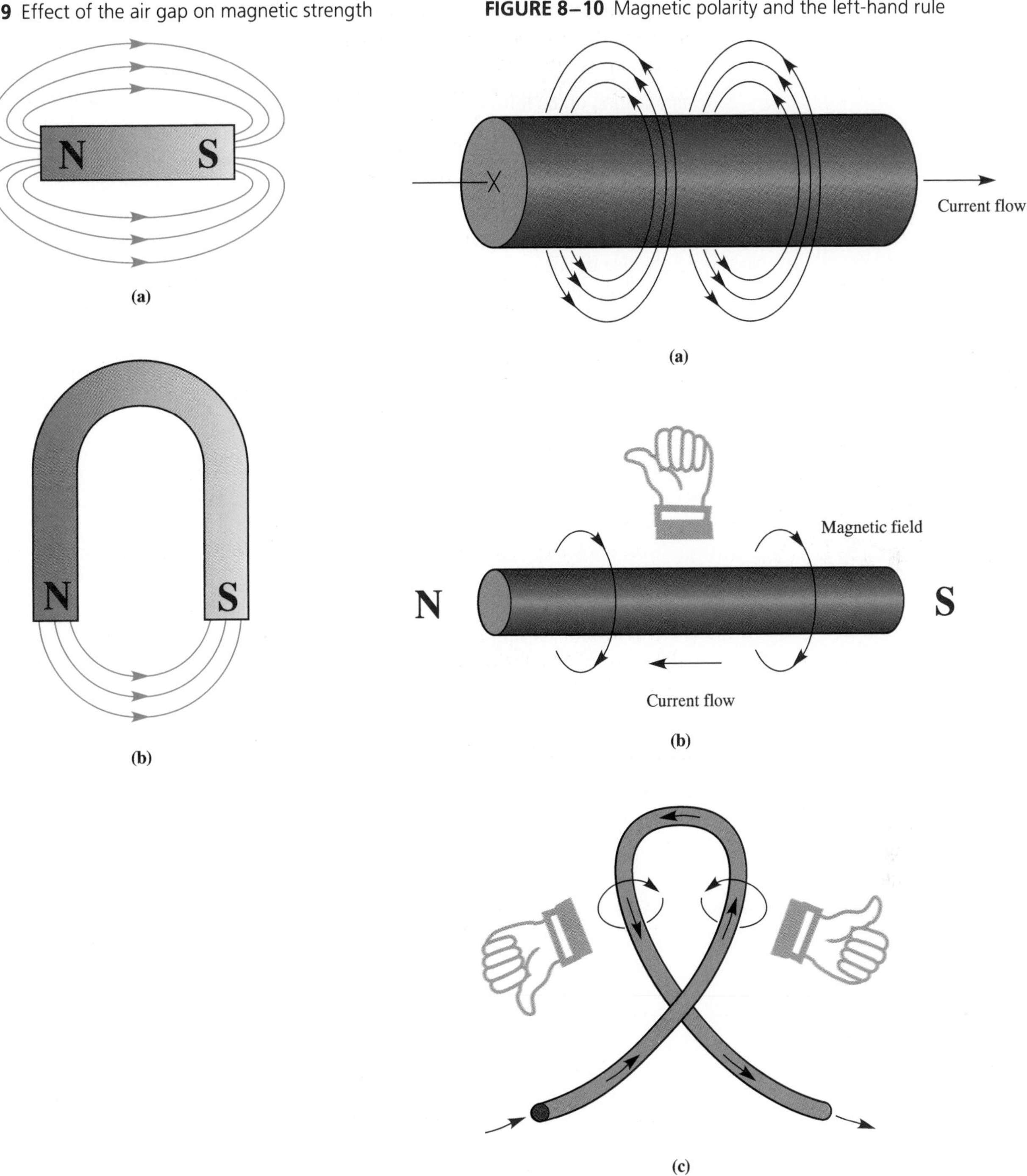

down the conductor in the same direction as the current flow, the magnetic field is seen as being *counterclockwise,* as shown in Figure 8–10(a). The direction of magnetic flux can also be determined using what is known as the **left-hand rule:** If the thumb of the left hand is pointed in the direction of current flow and the fingers are wrapped around the wire, the fingers point in the direction of the magnetic flux, as shown in Figure 8–10(b).

Question: What is the effect upon the magnetic flux of winding the conductor into a coil?

A current-carrying conductor is formed into a loop as shown in Figure 8–10(c). Through application of the left-hand rule, it can be seen that the flux lines have directions as shown. Notice that in the center of the loop, all the flux lines are moving in the same direction. They reinforce one another, thus creating a stronger magnetic flux in the center. If as shown in Figure 8–11(a), the conductor is wound into a coil, the lines of flux in the center of the coil merge to form one large magnetic field. It has a definite north and south pole and becomes what is known as an electromagnet.

Notice in this drawing that the windings of the coil go up the front and down the back of the coil. With the polarity of the source as shown, the current also travels up the front and down the back. The magnetic polarity of the electromagnet can be determined using an *alternate left-hand rule* that states: If the fingers of the left hand are placed around the coil in the direction of current flow, the thumb points toward the north pole. If either the polarity of the source or the direction of the coil windings is reversed, the magnetic polarity is also reversed, as illustrated in Figure 8–11(b).

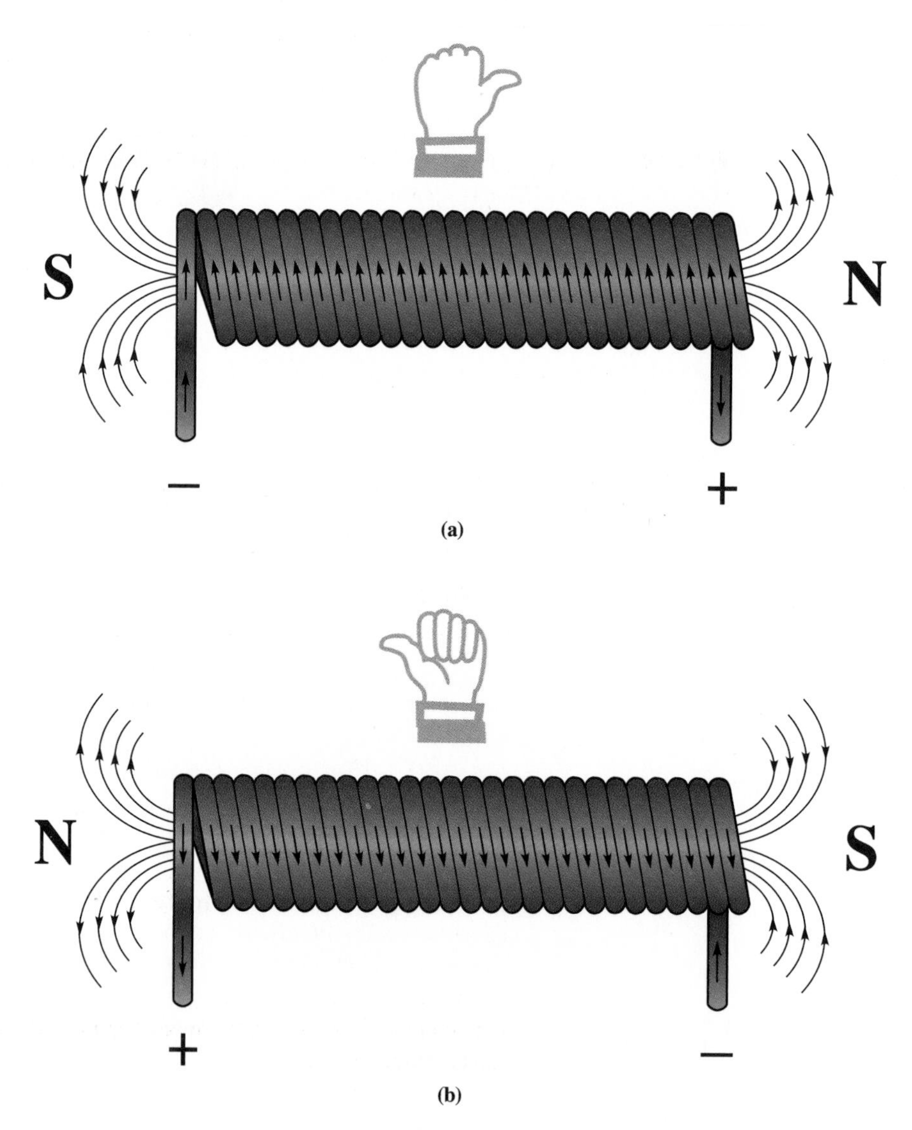

FIGURE 8–11 Magnetic polarity in a coiled conductor and the alternate left-hand rule

Question: What determines the strength of the magnetic field produced by an electromagnet?

As previously stated, a coil with current flowing through it becomes an electromagnet. The strength of the magnetic field is determined by four things: (1) the amount of current flowing through the coil, (2) the number of turns on the coil, (3) the length of the coil, and (4) the permeability of the core material.

■ Magnetomotive Force

The magnetizing force producing the magnetic field in an electromagnet is known as **magnetomotive force,** abbreviated F_{mm}. Its action is similar to that of the voltage in an electric circuit. The strength of the magnetic field is directly proportional to the magnetomotive force. It is equal to the product of the current and the number of turns on the coil. Its unit of measure is the **ampere-turn** (A · t). In equation form, magnetomotive force is written as

(8–3)
$$F_{mm} = NI$$

where: N = the number of turns on the coil

I = current in amperes

EXAMPLE 8–4 A coil has 300 turns of wire and 4 A of current flowing through it. Compute the magnetomotive force.

■ Solution:

Use Eq. 8–3:

$$F_{mm} = NI$$
$$= 300(4) = 1200 \text{ A} \cdot \text{t}$$

EXAMPLE 8–5 A current of 2 A is required to produce a magnetizing force of 6000 A · t. How many turns of wire are there on the coil?

■ Solution:

Use Eq. 8–3, and rearrange to solve for N:

$$F_{mm} = NI$$
$$N = \frac{F_{mm}}{I}$$
$$= \frac{6000}{2} = 3000$$

Thus, 3000 turns on the coil produce 6000 A · t of F_{mm}.

EXAMPLE 8–6 What value of current must flow through a 500-turn coil in order to produce 5000 A · t of magnetomotive force?

■ Solution:

Rearrange Eq. 8–3 to solve for I:

$$F_{mm} = NI$$

$$I = \frac{F_{mm}}{N}$$

$$= \frac{5000}{500} = 10\text{A}$$

■ Field Intensity

Field intensity, abbreviated H, is the flux developed at a given point on a magnet, and its unit of measure is the ampere-turn per meter (At/m). The magnetomotive force produces the total flux, but the intensity of the field depends upon the length of the coil. If the coil is longer, then the turns are more widely spaced. The lines of flux of the more widely spaced loops do not reinforce each other as greatly as they do under tight spacing. Thus, the field intensity is lower. The field intensity is equal to the magnetomotive force in ampere-turns divided by the length l of the coil. In equation form, magnetic field intensity is written as

(8–4)
$$H = \frac{F_{mm}}{l}$$

EXAMPLE 8–7 The length of a coil between its poles is 0.08 m. If the coil has 25 turns and 0.6 A flowing through it, compute the field intensity.

■ Solution:

1. Use Eq. 8–3 to compute the magnetomotive force:

$$F_{mm} = NI$$

$$= 25(0.6) = 15 \text{ A} \cdot \text{t}$$

2. Use Eq. 8–4 to compute the field intensity:

$$H = \frac{F_{mm}}{l}$$

$$= \frac{15}{0.08} = 187.5 \text{ A} \cdot \text{t/m}$$

■ Permeability

It is in the core upon which the coil is wound that permeability comes into play. Recall that permeability is a material's ability to concentrate lines of magnetic flux. This ability directly affects the flux density and thus the strength of the magnetic field. A coil with only air as its core produces a certain field strength dependent upon the permeability of air. In SI units, the permeability of free space (air or vacuum) μ_0 is equal to $4\pi \times 10^{-7}$. Rounded for calculations, the value of μ_0 is 1.26×10^{-6}. If the coil is wound on a core of some highly permeable material, more lines of flux are concentrated and the magnetic field is stronger.

The **relative permeability** μ_r of a material is the ratio of the flux density produced by a magnetic force acting on a material relative to the flux density that would be produced by the same force in air. For example, if a magnetic force produces 3000 G in a material and only 6 G in air, the μ_r is 3000/6 = 500. Notice that the units in this ratio cancel. Thus, relative permeability has no units.

It is not the material's relative permeability but its absolute permeability that comes into play here. **Absolute permeability** μ is equal to the ratio of the flux density B in teslas to the field intensity H in ampere-turns per square meter. The greater this ratio, the greater the absolute permeability of the core. In equation form, absolute permeability is written as

(8–5)
$$\mu = \frac{B}{H}$$

The absolute permeability of a material has the units of teslas per ampere-turn per meter (T/A · t/m).

Recall that relative permeability is the ratio of the flux density produced in a material by a magnetizing force to the flux density it would produce in air. Thus, the relative permeability μ_r is also the ratio of its absolute permeability μ to the permeability of air μ_0. Written in equation form, this relationship is

$$\mu_r = \frac{\mu}{\mu_0}$$

Transposing gives

(8–6)
$$u = \mu_r \mu_0$$
$$= \mu_r (1.26 \times 10^{-6})$$

Thus, it can be seen that the absolute permeability of a material is the product of its relative permeability and the absolute permeability of air.

EXAMPLE 8–8 What is the absolute permeability of a material with a relative permeability of 12,000?

■ Solution:

Use Eq. 8–6:

$$\mu = \mu_r \mu_0$$
$$= 12{,}000(1.26 \times 10^{-6})$$
$$= 1.512 \times 10^{-2} \text{ At/m}$$

[1] [2] [EXP] [3] [×] [1] [·] [2] [6] [EXP] [±] [6] [=]

EXAMPLE 8–9 A core material has an absolute permeability of 3.78×10^{-4}. Compute its relative permeability.

■ Solution:

Use Eq. 8.6 and transpose to solve for μ_r:

$$\mu = \mu_r \mu_0$$

$$\mu_r = \frac{\mu}{\mu_0}$$

$$= \frac{3.78 \times 10^{-4}}{1.26 \times 10^{-6}} = 300 \quad \text{[units cancel]}$$

SECTION 8–4 REVIEW

1. What factors affect the strength of the field of an electromagnet?
2. State the left-hand rule for determining the magnetic polarity of a coil.
3. What is the difference between relative and absolute permeability.
4. A coil has 500 turns of wire and 1.5 A of current flowing through it. Compute the F_{mm}.
5. If the coil in problem 4 is 0.05 m long, compute its field intensity H.

8–5 MAGNETIC HYSTERESIS

The term *hysteresis* applies to some effect lagging behind its cause. In magnetism, when a magnetizing force H is applied, the magnetic domains within the material must align themselves in order to produce magnetic flux. Thus, the magnetic field that is formed in the material lags the magnetizing force that produces it. This lagging of the flux density B behind the magnetizing force H is known as **magnetic hysteresis,** and it occurs as flux density and field strength is increased in an electromagnet.

Question: Will the flux density in an electromagnet continue to increase without end as the field intensity is increased?

Like almost anything in the physical world, the amount of flux lines that can be induced in the core of an electromagnet has its limits. As has already been established, the flux density of a magnetic field is dependent upon the number of lines of force and their concentration. In an electromagnet, once it is built, the only practical way of increasing the flux density and thus the magnetic field strength is by increasing the current through the coil. As the current is increased, more of the magnetic domains within the core material align, and the flux density increases. However, a point, known as **magnetic saturation,** is reached at which there are no more domains to align, and any increase in current will cause little or no increase in the flux density. Figure 8–12 shows a graph of this relationship.

FIGURE 8–12 Magnetic saturation

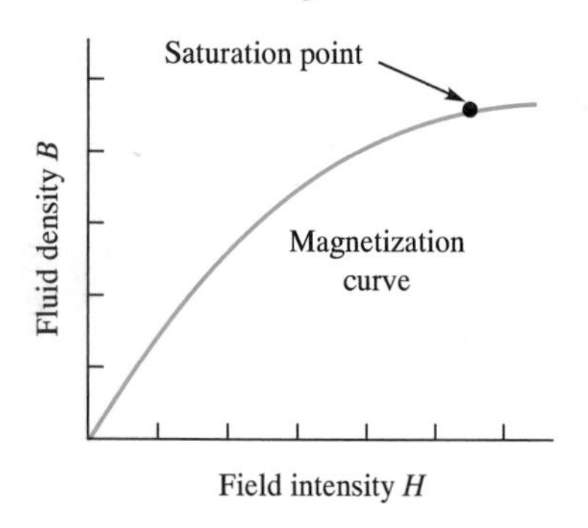

Recall that flux density is the product of the absolute permeability of the core and the field intensity. The field intensity, which is the ratio of the magnetomotive force and the length of the coil, continues to increase as long as the current through the coil increases. So, if the field intensity increases with no change in flux density, the absolute permeability decreases. As the domain alignment progresses, there are fewer and fewer to add to the magnetization of the core. Thus, the core becomes more difficult to magnetize and its permeability decreases.

■ The Hysteresis Loop

Figure 8–13 is a graph showing a *hysteresis loop,* in which the flux lags the magnetic force. The horizontal axis of the graph represents the field intensity. Notice that it begins at the reference, point O, and extends in both directions, $+H$ to the right and $-H$ to the left. Recall that the magnetic polarity of a coil is dependent upon the direction of current

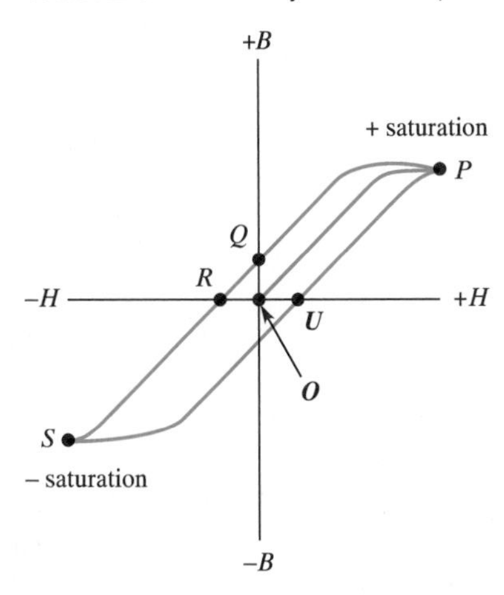

FIGURE 8–13 The hysteresis loop

flow in it. Thus $+H$ means the current flows one way, and $-H$ means the current flows the other. When the current flows to form $+H$, a flux density known as $+B$ is produced. When the current flows to form $-H$, a flux density known as $-B$ is produced.

To begin the discussion, assume that all magnetic fields are zero, which is point O on the graph. You are already familiar with what occurs when the coil is magnetized by $+H$: The flux density builds up along the curve from point O to point P, which is the $+B$ magnetic saturation point. If the current producing $+H$ is returned to zero, the flux density does not return to zero. Due to the retentivity of the core, it drops only to point Q. The flux that remains, represented by point O to point Q on the graph, is the residual magnetism of the core. What must be done in order to return the flux to zero? The current in the coil must be reversed to produce enough $-H$, known as coercive force, to remove the residual magnetism. The flux is now zero, but notice that this is at point R on the graph.

If the current in the coil is increased to produce more $-H$, the flux density now builds to point S, which is the $-B$ magnetic saturation. Reducing $-H$ to zero does not reduce the flux to zero. Because of its retentivity, the flux drops only to point T on the graph. The remaining flux, represented by point O to point T on the graph, is once again the residual magnetism in the core. To remove the residual magnetism and reduce the magnetic field to zero, the current must once again be reversed and enough $+H$, coercive force, developed to do so, which is point U on the graph. A further increase in $+H$ will move the magnetization back to point P and the cycle could be repeated.

What this discussion and graph of the hysteresis loop shows is that the flux is *always* lagging the magnetizing force. This lag results from the fact that when a magnetizing force is removed, the magnetic domains do not return to their original positions. They must be physically moved by the magnetizing force. This movement requires electrical energy, which is converted to heat in the core. Thus, magnetic hysteresis results in a loss of power to any system in which it is a factor.

The power loss due to magnetic hysteresis occurs each time the current changes directions. It is not a problem, then, where direct current is involved. However, in the case of alternating current, which is discussed in the second portion of this text, this power loss is a concern.

■ SECTION 8–5 REVIEW

1. What is the meaning of hysteresis?
2. Why does magnetic saturation take place?
3. Which parameter causes magnetic saturation: magnetomotive force, field intensity, or core permeability?
4. What is the force called that is required to return the flux to zero after the core has been magnetized?
5. What causes the power loss due to the magnetic hysteresis of the core?

8–6 APPLICATIONS

There are literally thousands of uses of magnetism, electromagnetism, and electromagnetic induction. They range from magnetic switches used in entry alarm systems to the electromagnetic waves which make radio and television broadcasting possible. In this section only a few of the more important applications are presented.

Generator Action

Generator action is one of the major uses of magnetism and electromagnetism. A **generator** produces electricity through the interaction between a conductor and a magnetic field. The process of producing electricity with a generator is known as **electromagnetic induction.** There are three requirements for electromagnetic induction: (1) a conductor, (2) a magnetic field, and (3) relative motion between the conductor and magnetic field. When this relative motion between conductor and flux lines takes place, voltage is induced on the conductor. The magnetic field may be supplied by either a permanent magnet or an electromagnet.

NOTE

In electromagnetic induction, there is no physical contact between the components. The stem word for *induction* is *induce*. To induce means to bring about something. In electronics, it is a voltage that is brought about without any physical contact between components of the system. Thus, an induced voltage is one that has been brought about through induction.

As shown in Figure 8–14(a), a conductor is moved downward through the field of a horseshoe magnet. This movement causes electrons to move from one end of the conductor and to collect at the other. Thus, charges are separated, and a voltage, with polarity as shown, is developed across the conductor. (Recall that a separation of charges produces a voltage.) If as shown in Figure 8–14(b) the conductor is moved upward through the field, a voltage with the opposite polarity is induced. If a load were connected to the conductor, the voltage would produce current through it. The current flow reverses direction each time the voltage changes polarity. Thus, an up-down motion of the conductor produces an *alternating current*.

CAUTION

Study the preceding paragraphs and illustrations very carefully. The requirements, methods, and results of electromagnetic induction must be committed not just to memory, but to understanding. It is not only basic to the study of inductance that will be presented in the next chapter, but will be encountered time and again in your study of electronics!

Question: What determines the amount of voltage that is induced in electromagnetic induction?

The amount of voltage induced in electromagnetic induction is determined by basically four factors: (1) the strength of the magnetic field, (2) the number of turns of wire in the conductor, (3) the rate at which the lines of flux are cut, and (4) the angle at which the conductor cuts the flux lines.

Strength of Magnetic Field

The amount of induced voltage is directly proportional to the number of flux lines making up the field. Therefore, a strong permanent or electromagnet produces more voltage than a weaker one.

Number of Turns on the Conductor

Although the conductor was shown straight in Figure 8–14, it can be wound into a coil to increase its length, as is done in the practical generator, where the loop is rotated rather than moved in a linear manner. Voltage is induced in each loop. The loops are in series and the individual voltages add. Thus, the voltage induced is directly proportional to the number of turns in the conductor.

■ Rate at Which the Flux Lines are Cut

The voltage induced is directly proportional to the rate at which the lines of flux are cut, which is done by increasing the speed of the moving conductor. At faster speeds, more flux lines are cut over a specific time frame, and the induced voltage increases.

■ Angle at Which the Flux Lines Are Cut

One of the requirements for electromagnetic induction is relative motion between the conductor and magnetic field. Relative motion means that the conductor must cut across the lines of force at some angle other than zero. If the conductor cuts the lines of flux at

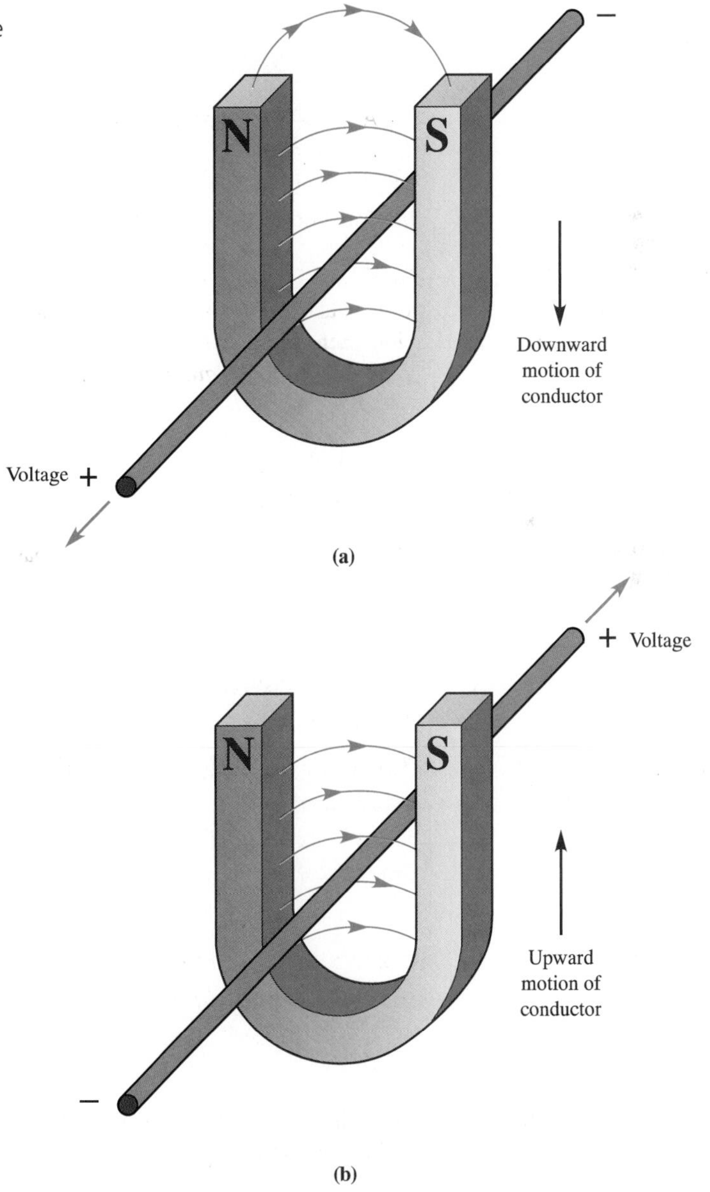

FIGURE 8–14 Electromagnetic induction. The three requirements are a magnetic field, a conductor, and relative motion between the two.

an angle of 90°, maximum voltage is induced. As the angle decreases, the amount of the voltage induced decreases. At an angle of zero degrees the conductor is moving parallel to the lines of flux, and no voltage is induced, as is illustrated in Figure 8–15.

■ Faraday's Law of Induced Voltage

Named for Michael Faraday, the British physicist credited with the discovery of electromagnetic induction, **Faraday's law** states: when the magnetic flux linking a coil changes, a voltage proportional to the number of turns and the rate of change in flux is induced in the coil.

FIGURE 8–15 Percentage of maximum voltage relative to cutting angle

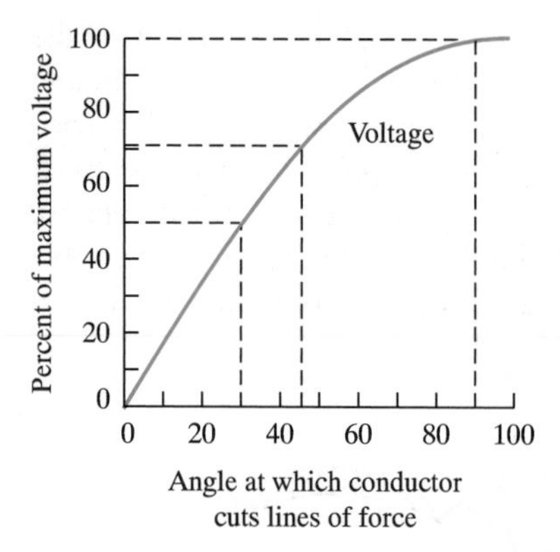

In equation form, Faraday's law is written

(8–7)

$$V_{ind} = N\frac{\Delta\phi}{\Delta t}$$

WHERE V_{ind} = the induced voltage

N = number of turns

$\frac{\Delta\phi}{\Delta t}$ = rate at which the flux cuts across the conductor (Wb/s)

Thus, if the number of flux lines is expressed in maxwells, it will be necessary to convert this value to webers. (Recall that 1 Mx is equal to 1×10^{-8}Wb.)

EXAMPLE 8–10 What is the amount of voltage produced if the magnetic flux cuts 500 turns of wire at the rate of 0.5 Wb/s?

■ Solution:

Use Eq. 8–8:

$$V_{ind} = N\frac{\Delta\phi}{\Delta t}$$

$$= 500(0.5 \text{ Wb/s}) = 250 \text{ V}$$

EXAMPLE 8–11 If 1000 Mx of flux cuts 200 turns in 2 μs, how much voltage is induced in the coil?

■ Solution:

Since 1 Mx is equal to 1×10^{-8} Wb, the flux is equal to $1000(1 \times 10^{-8})$ Wb. Find the time rate of cutting:

$$\frac{1000 \times 10^{-8}}{2 \ \mu\text{s}} = 5 \text{ Wb/s}$$

Then,

$$V_{ind} = N\frac{\Delta\phi}{\Delta t}$$

$$= 200(5) = 1 \text{ kV}$$

■ Motor Action

Another well-known application of magnetism and electromagnetism is the electric motor. **Motor action** means that movement is produced through the interaction of two magnetic fields. In generator action, you will recall, a *current* is produced when a loop of wire is rotated within a magnetic field. In motor action, *motion* is produced when current flows through a conductor in a magnetic field.

Motor action is illustrated in Figure 8–16. In Figure 8–16(a), a loop of wire is within a strong magnetic field. A voltage with polarity as shown is applied to the loop producing a current that flows into the loop on the left and out of the loop on the right. Thus, the left portion of the loop has a counterclockwise magnetic field, while the right portion has a clockwise magnetic field. The interaction of these magnetic fields produces *torque*, or a twisting motion, on the loop.

This interaction of magnetic fields is more easily explained and understood using the two dimensional view of Figure 8–16(b). Notice the relationship of the flux lines around the left portion of the loop to the flux lines of the fixed field. They are of the *same* magnetic polarity *below* the loop but of the *opposite* magnetic polarity *above* the loop. The flux lines aid one another below the loop producing a strong field, but they oppose one another above the loop producing a weak field. The loop is forced upward on this side toward the weak field. In the right portion of the loop, the opposite occurs. Its

FIGURE 8–16 Motor action

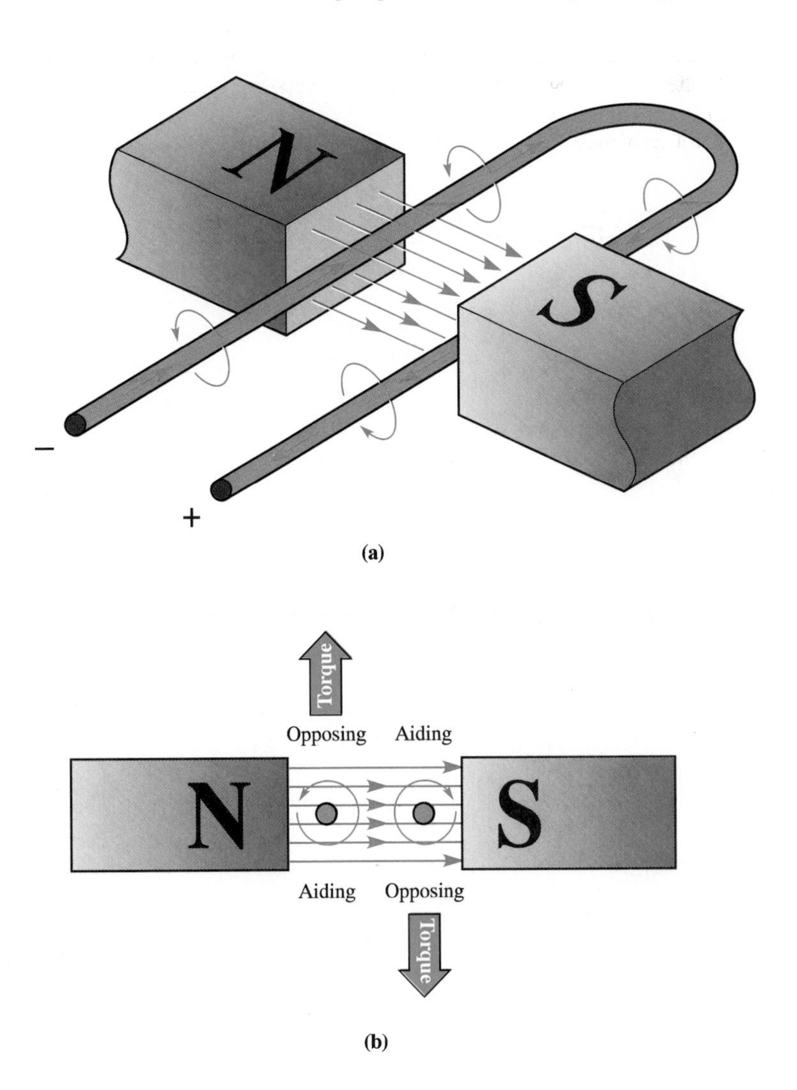

clockwise magnetic field produces a strong field above the loop and a weak field below the loop. The loop is forced downward on this side toward the weak field. Thus, torque is produced and imparts a rotating motion to the loop.

■ dc Meter Movement

In Chapter 4, the dc voltmeter was introduced. At that time, only the bare essentials of the meter movement were presented. Its theory of operation can now be presented since it is an example of motor action. Recall that the meter movement contained the moving

FIGURE 8–17 Motor action in dc meter movement

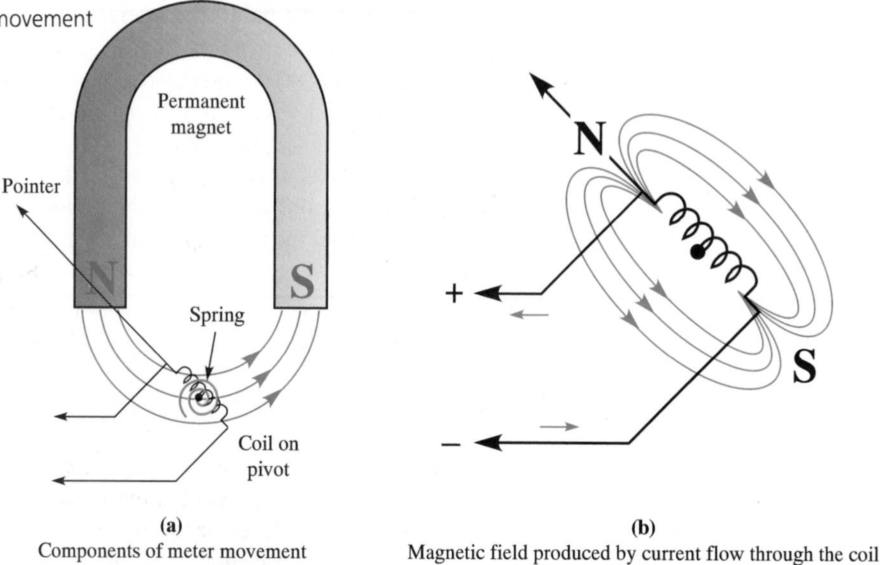

(a)
Components of meter movement

(b)
Magnetic field produced by current flow through the coil

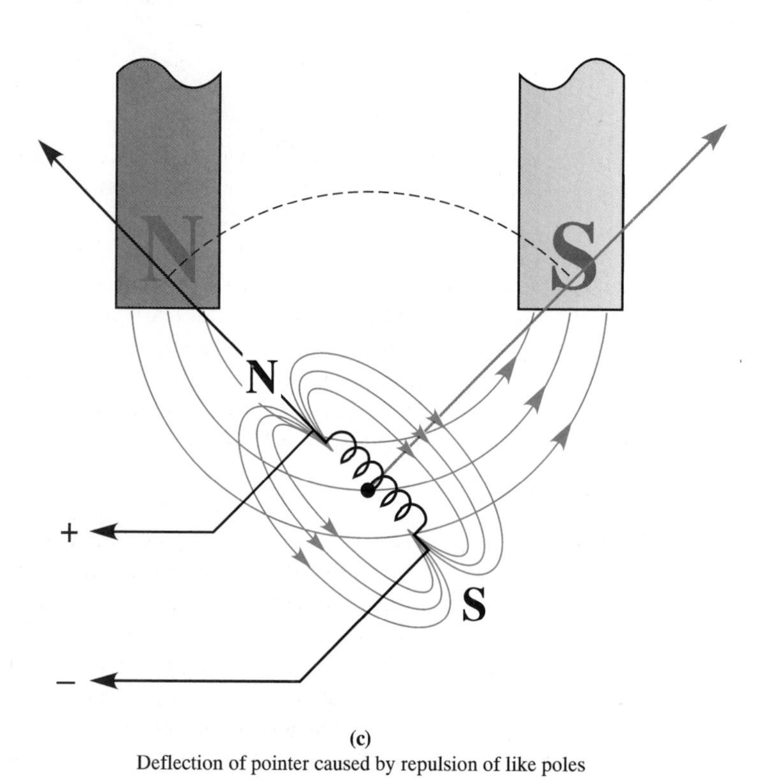

(c)
Deflection of pointer caused by repulsion of like poles

parts of an analog meter. Figure 8–17(a) shows the basic components of the meter movement. These are a permanent magnet, a coil with a pointer attached, a spiral spring, and a meter scale.

As shown in Figure 8–17(b), the coil is connected to a set of contacts in which the meter leads are inserted. These leads are the points of contact into the circuit that is to be measured. Current flows as shown through the coil producing a magnetic field around it. The strength of this magnetic field is directly proportional to the current through the coil, and its magnetic polarity is determined by the direction of the current flow. Thus, the intensity of the current flowing through the coil is now represented by the strength of the magnetic field.

As illustrated in Figure 8–17(c), the current in the coil produces a magnetic field with a polarity the same as that of the permanent magnet. The like magnetic poles repel and cause the coil and pointer to pivot against the pressure of the spiral spring. The amount of deflection of the pointer is then directly proportional to the amount of current flowing in the coil. If the magnetic field is strong enough, the pointer moves to the full-scale position. This current is known as the full-scale current I_{FS}. When the current is removed from the coil, the coil, and pointer return to the zero position by the spring.

When the full-scale current is multiplied by the meter resistance R_M, the result is a voltage V_M that is developed across the movement. Recall that these characteristics of the meter movement were used in the computation of resistance values in analog meter circuits (Chapters 4, 5, and 6).

■ The Hall Effect

In addition to the generator, magnetism can be used in still another way to produce a voltage. When a magnetic field is placed at right angles to a current-carrying conductor, a small voltage is produced across the conductor and is known as the **Hall effect.** Named for its discoverer American physicist E. M. Hall, it finds many applications in switching and measurement operations. Voltages produced by this process are quite small, in the microvolt region, when conductors are used. When a semiconductor material such as indium arsenide is used, however, the voltage produced can be as much as 100 mV.

The Hall effect is illustrated in Figure 8–18. A block of indium arsenide has a small current flowing through it. If a magnetic field is placed perpendicular to the direction of current flow, a voltage is produced across the width of the semiconductor. The amount of voltage produced is directly proportional to the flux density of the magnetic field.

A natural use for the Hall effect is as a sensor in instruments designed to measure the strength of magnetic fields. One such device is known as a gauss meter. Within its attached probe is a Hall effect device. When placed within a magnetic field, it produces a voltage directly proportional to the flux density of the field. This voltage is scaled and displayed on the meter, which is calibrated in gauss. There are many other uses for Hall effect devices, including position sensors for industrial machines, switches for computer keyboards, and trigger generators in electronic ignition systems in automobiles.

■ Magnetic Tape Recording and Reproduction

Another major and very popular application of magnetism is the recording and replaying of audio tapes. Sound waves produced as voice or music are first converted into varying electronic signals using a microphone. The signals vary in amplitude with the sound waves striking the microphone, as illustrated in Figure 8–19(a). The varying currents then produce a varying magnetic field around an electromagnet known as a *record head.* Tape coated with a magnetic material is drawn across the record head by a constant speed motor. The field produced by the record head magnetizes the coating, and the signals are recorded in the form of varying magnetic fields, as shown in Figure 8–19(b).

FIGURE 8–18 Generation of Hall effect voltage

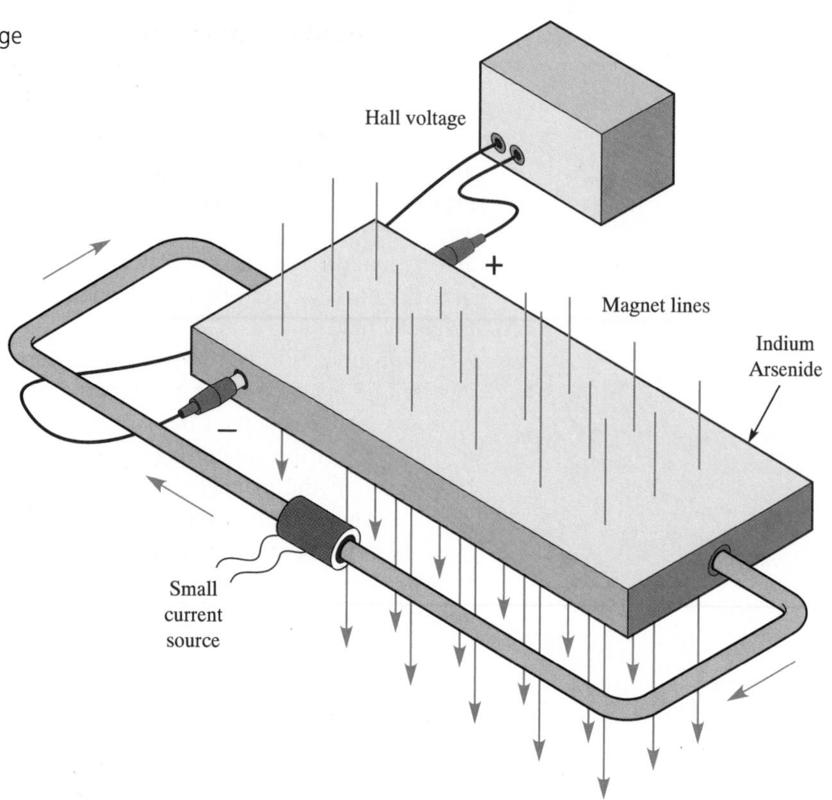

The signals are reproduced by drawing the tape across a coil known as a *reproduce head.* The magnetic field of the tape cuts across the windings of the reproduce head, inducing currents in it (recall the three things required for electromagnetic induction). The currents reproduced are nearly identical to those that magnetized the tape when it passed the record head. These signals can now be amplified and reproduced as sound by the speaker, as illustrated in Figure 8–19(c). Care must be taken to protect magnetic tapes from external magnetic fields that would effectively erase the recorded audio.

■ Magnetic Shielding

Moving magnetic fields are prevalent in and around electronic systems. As you have learned, they induce a voltage in any conductor that they may cut across, which is sometimes undesirable in that the induced voltages may impair system operation. Thus, some means of magnetic shielding must be applied to the affected device.

As illustrated in Figure 8–20, a coil is to be shielded from moving magnetic fields. Magnetic shielding can be accomplished by surrounding the device with a box made of some very low reluctance material such as soft iron. As shown, the lines of flux follow the path of lowest reluctance, which in this case is the walls of the soft iron box. Thus, they will not cut across the coil and no unwanted voltages will be induced.

8–7 TROUBLESHOOTING

■ Permanent Magnets

High-quality permanent magnets are made of elements with relatively high reluctance that require a very large magnetizing force in order to align the domains. Once magnetized, however, magnets made from these elements maintain their field strength almost indefinitely. There are certain conditions that must be avoided if the magnetic field is to remain strong.

FIGURE 8–19 Principle of magnetic tape recording and reproduction

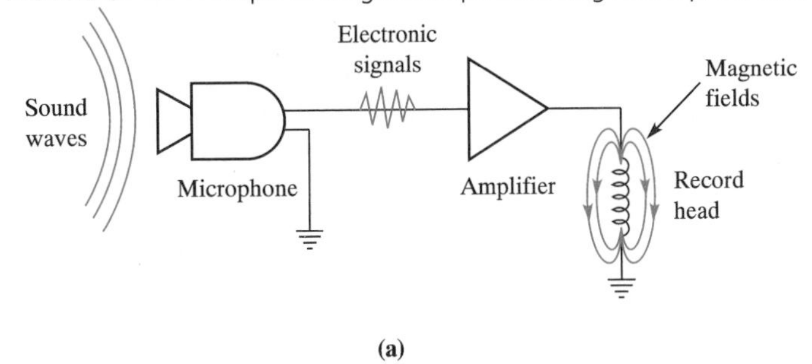

(a)

Record head

Magnetized areas on tape

Magnetic tape

(b)

Speaker

Amplifier

Record head

Magnetized areas on tape

Magnetic tape

(c)

FIGURE 8–20 Shielded coil

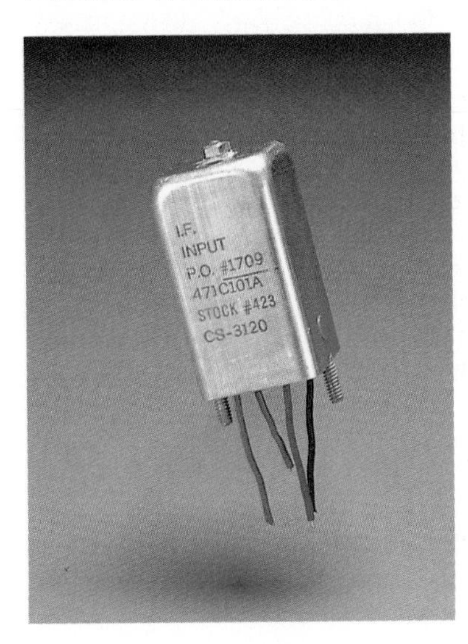

Problems can occur if permanent magnets are exposed to high temperatures and mechanical vibration and shock. For example, if a magnet, even one of high quality, is heated with a blowtorch, it will lose its magnetism. Thermal energy supplied by the flame causes the domains to return to the disarray that they were in prior to magnetization. Even when cooled, the magnet does not regain its magnetism and has to be remagnetized. Heating to this degree is both extreme and unrealistic, but in industrial settings, very high surrounding temperatures can be encountered. Thus, in any setting in which a magnet is to be used, the surrounding temperature must be taken into consideration.

Extreme vibration and shock affect a magnet in basically the same way as high temperatures. The vibrations cause some of the domains to move out of alignment, weakening the magnetic field. A sharp blow to the magnet has this effect. Care should be taken when handling magnets to avoid dropping them, especially those made with relatively low reluctance elements.

To avoid problems, certain precautions can be taken when storing permanent magnets. Permanent magnets should be stored as shown in Figure 8–21. In Figure 8–21(a), bar magnets are placed with opposite poles touching. As you know, they attract one another and will thus stay in this position. A horseshoe magnet is stored with a magnetic keeper across the air gap as shown in Figure 8–21(b). In both cases, a low reluctance path is supplied for the lines of magnetic flux. If not stored in this manner, their magnetic polarity could be reversed by a strong, constant magnetic field, while a rapidly varying external magnetic field could cause them to become demagnetized.

Electromagnets

In troubleshooting an electromagnet that has lost its magnetic field strength, recall the factors affecting the field intensity of an electromagnet: the amount of current through the coil, the number of turns on the coil, the length of the core, and the permeability of the core. Changes in any of these factors can affect the magnetic field strength. The first two are more likely possibilities than the latter two.

Even though changes in core length are not likely, the effects are considered here. Magnetic field strength is inversely proportional to the length of the core. The core would have to be lengthened in order to reduce the field strength, which could only be accomplished by its removal and replacement with one of greater length. Thus, the length of the core would not be the first check you would make in troubleshooting!

FIGURE 8–21 Methods for storing magnets to contain flux line in low reluctance paths

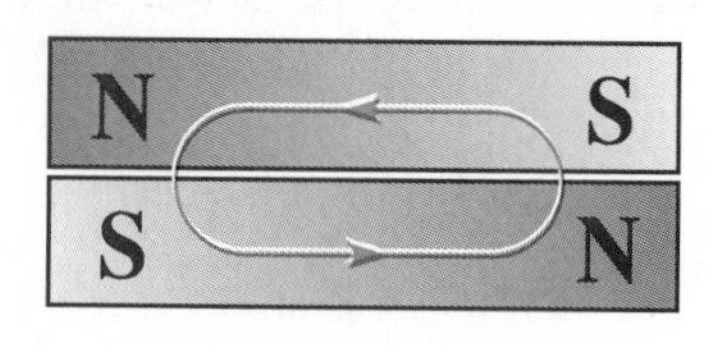

(a)
Bar magnets

(b)
Horseshoe magnet with magnetic keeper

NOTE

While considering the length of the core may seem to be a ridiculous waste of time, it does make an important point in electronics: Never reject any possibility when performing troubleshooting! All possibilities must be considered, even though the more likely ones should be checked first. Don't be a victim of "Murphy's Law" in your electronics career. It can be very embarrassing! For example, many high frequency amplifiers use coils, referred to as free-standing coils, that have no coil form. Through one process or another, these coils sometimes become deformed and stretched. This change in length results in a change in their value of inductance and may cause the circuit to be "untunable." A little "tweaking" and a lot of luck may make the circuit operational again. More likely, the coil will have to be replaced.

The permeability of the core is also not likely to change. The only change in permeability occurs when the core begins to saturate making magnetization more difficult. In a practical sense, it would take replacement of the core in order to effectively change the permeability.

The flux density is directly proportional to the number of turns on the coil. Thus, anything that causes a reduction in the number of turns lowers the magnetic field strength. Such a reduction could be caused if a portion of the turns were shorted, as illustrated in Figure 8–22. A coil with 2000 turns, but shorted as shown, would have effectively only 1000 turns. The flux density would also be halved. This problem could be detected by measuring the resistance of the coil windings and noting any decrease in the resistance of the wire making it up.

FIGURE 8–22 Loss in flux density due to shorted windings

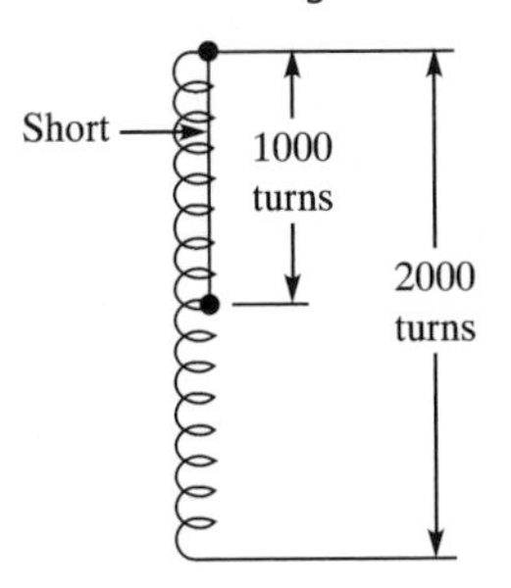

The flux density is also directly proportional to the current flowing through the coil. Thus, anything causing a decrease in current would reduce the magnetic field strength. This could be the result of a decrease in the source voltage or an increase in source resistance. The source voltage is always the first check in troubleshooting. An increase in resistance would most likely be the result of a loose or deteriorated solder or screw terminal connection.

Demagnetizing

Sometimes components in electronic systems become magnetized causing a deterioration in the system performance. Two quite common examples are the picture tube (CRT) in a television set and the record or play heads on a tape recorder. Both of these operate by being magnetized by a signal current, so if they become permanently magnetized, their performance will be impaired. It is often necessary to demagnetize these devices. This process is also often referred to as degaussing. The process and equipment for accomplishing it are shown in Figure 8–23.

FIGURE 8–23 Demagnetizing process

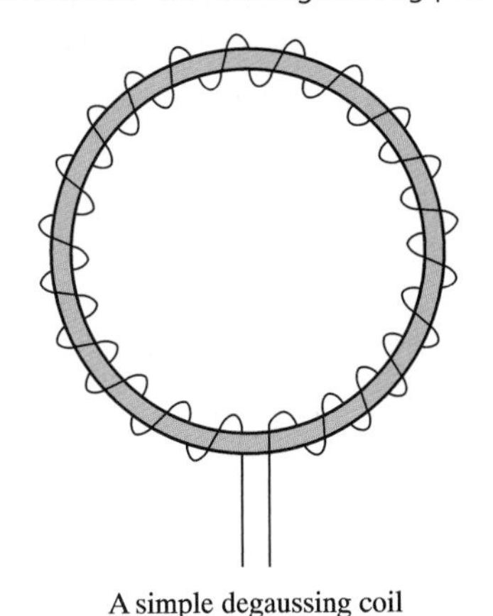

A simple degaussing coil

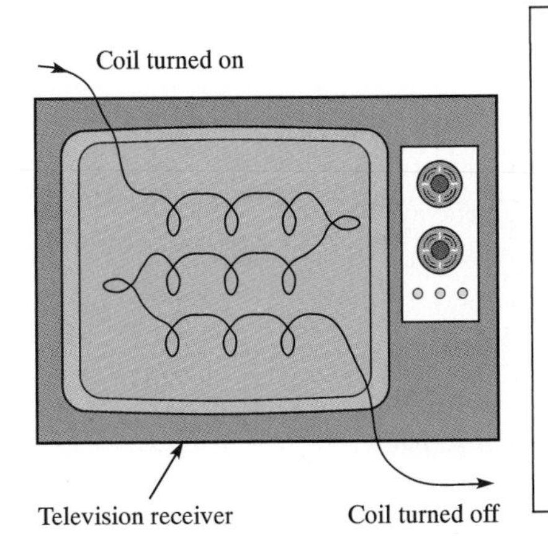

The CRT of a television receiver is demagnetized using a large coil to which a rapidly reversing alternating current is applied. This current produces a rapidly reversing magnetic field. The coil is placed near the side of the television and slowly drawn away while being moved in a rotary fashion. The rapidly reversing magnetic field causes the domains in the magnetized coating inside the CRT to revert to a condition of disarray, and the magnetism is removed. The same is done for the heads on a tape deck. The device used has a long, protruding core that is held close to the head and slowly withdrawn while being moved in a rotary fashion.

TECH TIP: AUDIO-VISUAL SPEAKER SYSTEMS

You're probably aware of the current trend toward Audio-Visual (AV) systems. Many modern manufacturers advertise their speakers as being "AV speakers." You might ask: "What are AV speakers? Aren't they just regular speakers used with a TV?" Surprisingly the answer is no.

A true AV speaker has one feature that is significantly different than a normal speaker designed for a stereo system. An AV speaker is designed to sit next to, or even on top of, a color television receiver. Remember that speakers use large permanent magnets that produce strong magnetic fields. These fields can severely distort the color purity of the television if not properly shielded. Also setting magnetic recording tape, such as cassettes, on these speakers can cause the audio or video recorded upon them to be erased.

Speakers designated as AV speakers employ special shielding that allows them to be placed close to a TV receiver with out affecting it. If you're not sure if a speaker is AV safe or not, use caution when placing it near a television or near magnetic recording tape.

SECTION 8–7 REVIEW

1. Why does excessive heat or vibration cause a decrease in the strength of an electromagnet?

2. How does a decrease in the number of turns of an electromagnet cause a decrease in magnetic field strength?
3. What could happen to cause a decrease in the number of turns on an electromagnet?
4. How does a decrease in the coil current cause a decrease in magnetic field strength?
5. What could happen to cause a decrease in the coil current of an electromagnet?

CHAPTER SUMMARY

1. A magnet is a device that attracts iron or other magnetic materials.
2. The three sources of magnetism are natural magnets, manufactured permanent magnets, and electromagnets.
3. A magnet is surrounded by an invisible field made up of magnetic flux lines.
4. Like magnetic poles repel one another while unlike poles attract.
5. The strength of a magnetic field depends on the number of flux lines and their concentration.
6. A material is magnetized through the alignment of magnetic domains.
7. The most important properties of a magnetic material are retentivity, permeability, and reluctance.
8. An electromagnet is made up of a coil wound on a magnetic core.
9. The strength of an electromagnetic field depends upon the current through the coil, the length of the core, and permeability of the core.
10. The core of an electromagnet can only be magnetized to a certain strength at which time magnetic saturation occurs.
11. Power must be developed in the core in order to reverse the domains when the current changes directions.
12. Generator action produces voltage when a conductor is rotated in a magnetic field.
13. Motor action produces motion when a current flows through a conductor in a magnetic field.
14. In the Hall effect, a voltage is produced across the width of a conductor when a magnetic field is applied perpendicular to it.
15. The Hall effect is useful as a sensor device for measuring the strength of a magnetic field.
16. Extremes of heat and vibration must be avoided in order to maintain the strength of a magnet.
17. Demagnetizing is accomplished by subjecting the magnetized component to a decreasing, rapidly varying magnetic field.
18. Magnetic shielding is accomplished by surrounding the device with a box made of a low reluctance material such as soft iron.

KEY TERMS

natural magnets
permanent magnets
electromagnets
magnetic field
flux lines
magnetic flux
Maxwell
Weber
flux density
gauss
tesla
permeability
reluctance
retentivity
magnetic induction
left-hand rule
magnetomotive force
ampere-turn
field intensity
magnetic hysteresis
magnetic saturation
generator
electromagnetic induction
Faraday's law
motor action
Hall effect

EQUATIONS

(8–1) $B = \dfrac{\text{Mx}}{A\ (\text{cm}^2)}$

(8–2) $B = \dfrac{\text{Wb}}{A\ (\text{m}^2)}$

(8–3) $F_{mm} = NI$

(8–4) $H = \dfrac{F_{mm}}{l}$

(8–5) $\mu = \dfrac{B}{H}$

(8–6) $\mu = \mu_r \times \mu_0$

(8–7) $V_{\text{ind}} = N\dfrac{d\phi}{dt}$

Variable Quantities

B = flux density

F_{mm} = magnetomotive force

μ = absolute permeability

μ_0 = permeability of air

μ_r = relative permeability

H = magnetizing force

$A\ (\text{cm}^2)$ = area in square centimeters

$A\ (\text{m}^2)$ = area in square meters

N = number of turns

I = current flow

$\dfrac{\Delta\phi}{\Delta t}$ = rate of change of flux with respect to time

TEST YOUR KNOWLEDGE

1. State the laws of repulsion and attraction for like and unlike magnetic poles.
2. What factors determine the strength of a magnetic field?
3. Describe what happens in magnetic induction.
4. What are the most important characteristics of a magnetic material?
5. What are two alloys from which magnets are made?
6. What determines the strength of the field of an electromagnet?
7. Why does magnetic saturation occur in an electromagnet?
8. What causes magnetic saturation in an electromagnet?
9. Describe generator action.
10. Describe motor action.
11. What factors must be avoided in ensuring that a magnet remains strong?
12. Explain how demagnetizing is accomplished.
13. Explain the need for and the method of magnetic shielding.
14. Diagram the magnetic hysteresis loop.
15. Diagram and explain the Hall effect.

PROBLEM SET: BASIC

Section 8–1

1. For the bar magnet of Figure 8–24, sketch the lines of force making up the magnetic field around it.

FIGURE 8–24 Bar magnet for Problem 1

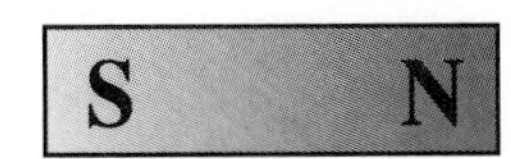

2. For the bar magnets shown in Figure 8–25, sketch the lines of magnetic force to show either repulsion or attraction.

FIGURE 8–25 Bar magnets for Problem 2

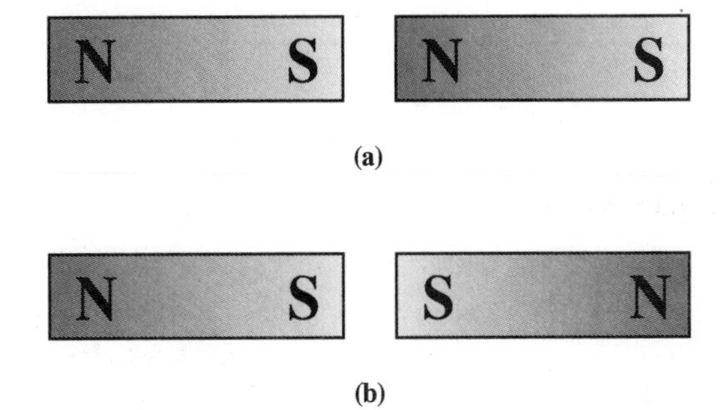

Section 8–2

3. A magnetic field has 25,000 flux lines. How many maxwells is this?
4. A magnetic field consists of 8000 Mx. How many webers is this?
5. A magnetic field has 2800 lines of flux in 8 cm². Compute the flux density.
6. How many gauss are there in 0.5 T?
7. How many teslas are there in 5000 G?

Section 8–4

8. A current of 0.2 A flows through a 250-turn coil. Compute the magnetomotive force.

9. A current of 0.15 A flows through a 300-turn coil. If the distance between the poles is 0.05 m, compute the field intensity.

10. A current of 0.5 A flows through a 300-turn coil. The distance between the poles is 0.15 m. If the absolute permeability of the core is 6×10^{-2} T/A · t/m, compute the flux density B.

Section 8–5

11. If 500×10^{-5} Wb cut across a 300-turn coil in 5 μs, how much voltage is produced?

12. If 1200 Mx of flux cut across a 400-turn coil in 10 μs, how much voltage is produced?

13. In Figure 8–26, the magnet is moved in the direction shown, causing the magnetic field to cut the coil at the rate of 10 Wb/s. What is the voltage produced?

FIGURE 8–26 Set up for Problem 13

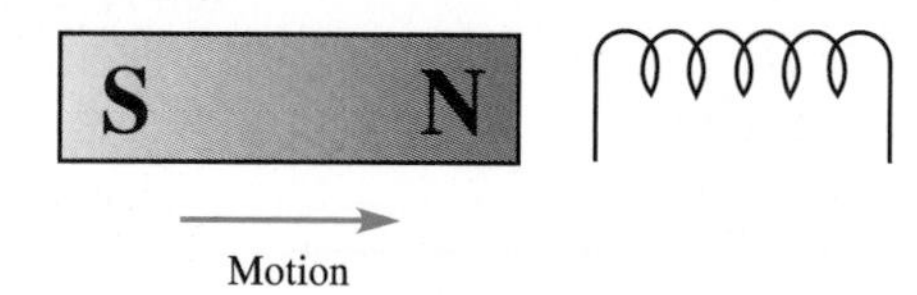

PROBLEM SET: CHALLENGING

14. An alloy has a relative permeability μ_r of 350. Compute its absolute permeability μ.

15. Compute the amount of magnetomotive force F_{mm} required to produce 420 A · t/m of field intensity H on a 0.25m core.

16. Compute the relative permeability of a material that has an absolute permeability of 1.26 T/A · t/m.

17. A flux density of 200 G/cm² would equal how many teslas?

18. For the circuit of Figure 8–27, compute: (a) I, (b) F_{mm}, (c) H, and (d) B.

FIGURE 8–27 Circuit for Problem 18

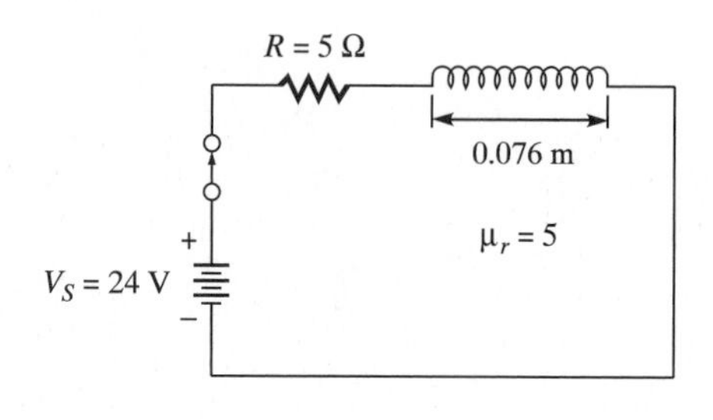

19. In Problem 18, compute the same values if the iron core were removed.

20. In Figure 8–28, determine which is the north and which is the south pole.

FIGURE 8–28 Circuit for Problem 20

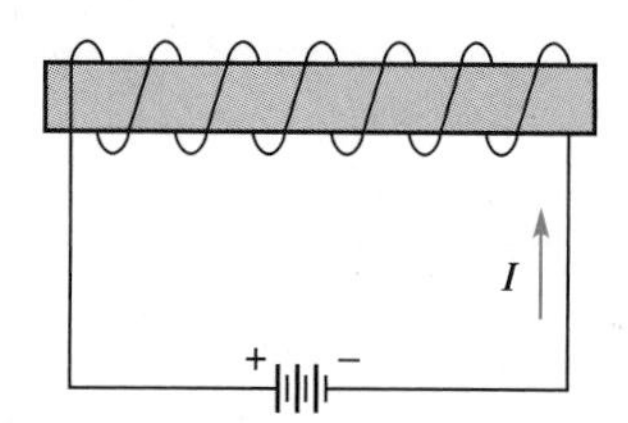

21. The strength of a magnetic field is measured and found to be 2000 G. What is the value of the flux density in teslas?

PROBLEM SET: TROUBLESHOOTING

22. The recording process involves transferring sound waves to the magnetic coating on a tape. If an audio cassette tape is played and no audio is found, list one possible cause.

23. A permanent magnet is found to be very weak when put into use. List two possible causes for it losing its magnetic strength.

24. An electromagnet, when put into use, is found to produce a very weak magnetic field. List two possible causes of this fault.

Magnetic Levitation

Levitation, the seeming defiance of the law of gravity, is a subject that has fascinated humankind for generations. One of the most amazing acts in the street carnivals of bygone days was the fast talker who could make somebody or something float in the air (with no strings attached, of course)! People who saw these acts went away in a state of wonder about how it was done. While some of the feats that scientists have performed in the last few years have been just as amazing, no one has, as yet, truly levitated. It is true that the airplane has gotten us into the air, but this is not levitation. It is merely one law of science compensating for another. We are still subject to the law of gravity.

Much has been made lately of something known as magnetic levitation, or "maglev." Using this technology, the Japanese and several European nations are developing trains that do not ride upon rails, but "levitate" above the rails on a magnetic cushion. Experimental models of this type of train have reached speeds of 300 miles per hour with an incredibly smooth ride. The forward motion of the train is produced by the attraction of fields with opposite magnetic polarities and the repulsion of fields with like magnetic polarities. For this type of an application, the magnetic fields must be incredibly strong!

Magnetic levitation, on a much smaller scale can be demonstrated in the laboratory with a rather simple device. The components of the device are inexpensive and available almost anywhere. The parts list is as follows:

1. An 8′ × 6″ × ¾″ piece of plywood for the base.
2. A strip of 8′ × 1″ × ⅜″ wide aluminum for the center spacer.
3. Two 8′ lengths of 1″ × 1″ × ⅜″ angle aluminum for side rails.
4. 200 1″ flat magnets, polarized top and bottom.

The track is assembled as shown in Figure 1. Be sure to use recessed screws in mounting the aluminum to the wooden platform. The magnets are special ceramic magnets that have their north and south poles on the surface, not the ends. They are mounted on the base with the poles facing the

FIGURE 1 Assembly diagram for maglev track

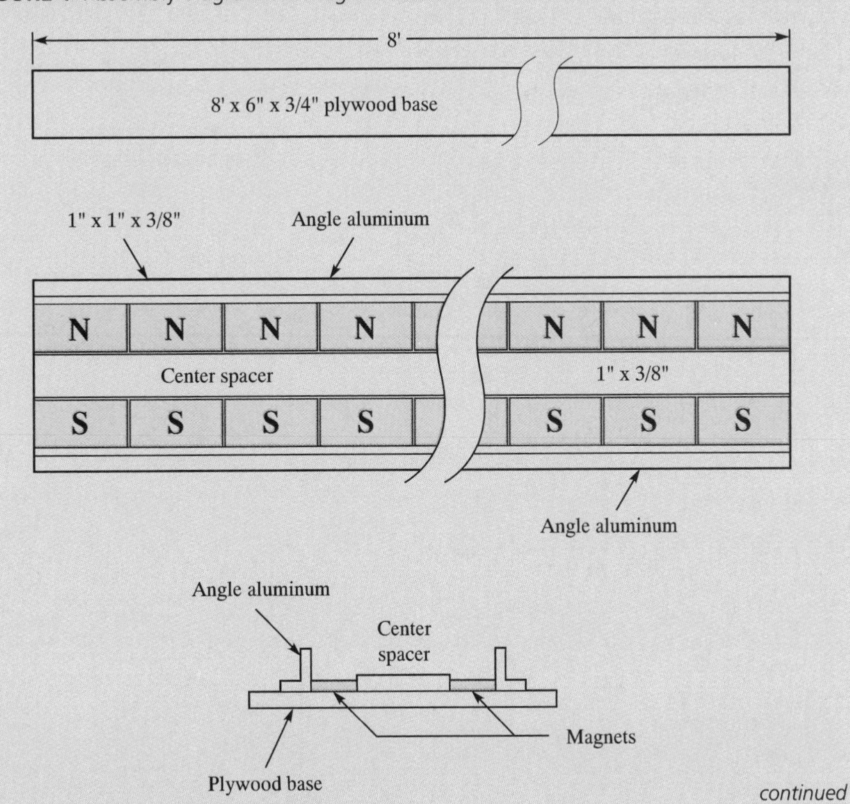

continued

continued

same direction. The polarity of the two strips of magnets must be reversed, however. The magnets are attached to the base using glue. A hot glue gun gives best results here. The magnets must be mounted flush to the center spaces and to each other, as shown in Figure 1. Care must be taken here because they will try to repell one another. They must be held in place for a few seconds until the glue sets. Be sure to check each magnet for proper polarity prior to gluing. Also make sure the two rows have opposite polarization.

After all the magnets have been installed, mount the aluminum side rails. The riser side of the rails should be mounted flush with the magnets. The distance between the side rails must be kept constant throughout their length. The screws mounting the rails should be loosely installed and tightened as the center spacer is moved down the base. The maglev track is now ready for operation!

You can construct a vehicle to test the track quite simply. Use a piece of wood slightly smaller than the spacer board. Glue magnets on both edges observing the same polarities as was observed for the track magnets. If the vehicle is placed properly on the track, it should levitate. If not, all you must do is reverse it. It can be propelled down the track on its magnetic cushion.

Vehicles with more sophisticated means of propulsion can then be conceived. They can be propelled by a number of means, but sling-shots and CO_2 capsules should be avoided. If an electric motor and a propellor is used, the side rails can be electrified to supply the voltage. Finding new and unique means of propulsion can become a class challenge, and a contest can be held to find whose vehicle completes the course in the shortest time. The degree of difficulty can be increased by raising one end of the maglev track slightly. Eight feet was chosen in these plans, but the track can be made longer if you so desire. All that is required is a longer base and aluminum rails and more magnets.

CHAPTER

9 Inductance in dc Circuits

■ UPON COMPLETION OF THIS CHAPTER, YOU WILL BE ABLE TO

1. Define inductance in two ways, and explain what is meant by self-inductance.
2. State Lenz' law.
3. Describe the characteristics of the voltage produced by self-inductance.
4. Write the symbols for inductance and inductors.
5. List the factors that determine the inductance of an inductor, and state whether each is an inverse or a direct relationship.
6. Compute the total inductance of inductors in series and inductors in parallel.
7. Define the henry in terms of the induced voltage and the rate of change of current with respect to time.
8. Draw the equivalent circuit of an inductor to include a pure inductance and the winding resistance.
9. Explain the power considerations for an inductor in a dc circuit.
10. Compute the time constant for an *RL* dc circuit, and graph the energizing and de-energizing current.
11. List examples of applications of inductors.
12. Describe methods of troubleshooting inductors.

In the preceding chapters, the circuit property of resistance was considered. As you know, resistance is introduced into the circuit in the form of devices known as resistors. Resistors are devices used in the control of electric current. In this chapter, a second circuit property and a second device is considered. The property is known as inductance, and it is introduced into circuits in the form of devices known as inductors. The purpose of an inductor is the same as that of a resistor: to control current. It does so in a quite different manner, however. A resistor affects a current uniformly at all times, but an inductor has an effect only upon currents that are changing in value. This effect, you will find, is greatest when the current is changing most rapidly. In fact, an inductor may act like an open circuit to a very rapidly changing current, but act like a short circuit for a steady value of current.

Inductors are used quite extensively in electronic circuits and systems. In an automobile, an inductor known as the ignition coil, takes the 12 V of the battery and produces over 20,000 V in order to make the spark plugs fire. A special arrangement of inductors, known as the transformer, can either increase or decrease the value of ac signals. Inductors are the heart of both electric motors and generators. Inductors are used in conjunction with other components in selecting one of the many TV or radio signals impinging upon an antenna. The list could go on, but these examples illustrate the importance and many uses of inductors.

Along with resistance and inductance, the only other electrical property to be mastered is capacitance, which will be presented in the next chapter. These three—resistance, inductance, and capacitance—are the basis for all electrical circuits. Needless to say, then, inductance is well worth any effort expended in mastering its characteristics!

9–1 INDUCTANCE

Review of Electromagnetic Induction

Recall that a current-carrying conductor is surrounded by a magnetic field and that the strength of this field is dependent upon the amount of current flow. Any increase in current results in an increase in the number of flux lines around the conductor, while a decrease in current has the opposite effect. As their numbers increase, the flux lines move outward from the center of the conductor. A decrease in the number of flux lines sees them moving back toward the center of the conductor, as illustrated in Figure 9–1. Thus, a *varying* current produces a *moving* magnetic field in the space around a conductor.

Three things are required in order to induce a voltage through electromagnetic induc-

FIGURE 9–1 Principle of electromagnetic current

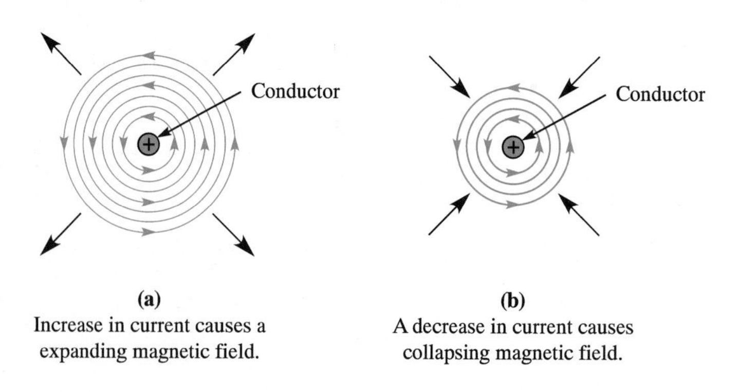

tion: a conductor, a magnetic field, and relative motion between the two. (Relative motion means that the lines of magnetic force must cut across the conductor, not move parallel to it.) An electronic circuit is made up of devices connected together by conductors. The values of the currents flowing in these conductors are not static, but are usually varying in response to some input. Thus, everything needed for induction—a conductor, a magnetic field, and relative motion between the two—are present in an electric circuit.

Question: What is inductance?

Inductance is the property of a circuit that induces a voltage within itself when the current changes. Its symbol is L. The unit of measure for inductance is the **henry** (H) named for the American physicist Joseph Henry. Varying currents within a circuit produce varying magnetic fields. These varying fields cut across circuit conductors inducing voltages across them. This type of inductance is known as *distributed inductance*, or stray inductance, because it can occur anywhere within the circuit. Inductance is introduced into the circuit through a device that is basically a coil, known as an **inductor.** The symbol for an inductor is shown in Figure 9–2(a). An inductor is known as a lumped inductance because the inductance is concentrated rather than being distributed throughout the circuit.

When the current through an inductor varies, the magnetic flux surrounding it also varies. It increases when the current increases, and it decreases when the current decreases. In either case, the moving magnetic field induces a voltage across the inductor. This induction, demonstrated in Figures 9–2(b) and (c) is known as *self-inductance*, or just *inductance*. In Figure 9–2(b), the switch is closed, and current flows through the inductor. The magnetic field builds outward from the center of the inductor, inducing a voltage with polarity as shown. In Figure 9–2(c), when the switch is opened, the current flow ceases. The magnetic field collapses back into the center of the inductor. Once again voltage is induced, but now it is of the opposite polarity.

FIGURE 9–2 Operation of an inductor

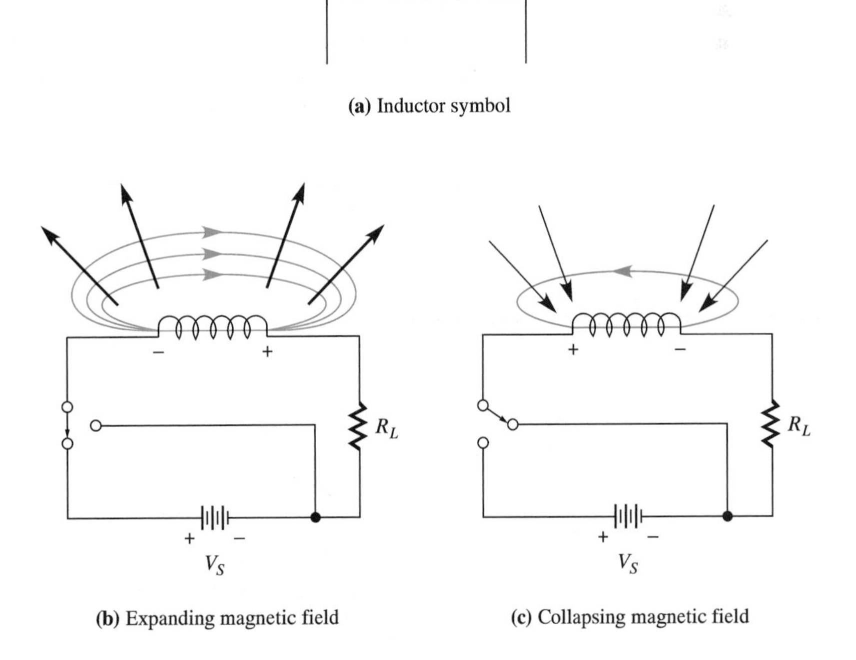

SECTION 9–1 REVIEW

1. Define inductance.
2. What is meant by a distributed inductance?
3. Explain what action causes a voltage to be induced across an inductor.
4. What is the basic inductor?
5. What is the symbol and unit of measure for inductance?

9–2 NATURE OF THE INDUCED VOLTAGE

When the circuit current varies, the inductor induces a voltage within its own windings that is known as a **self-induced voltage.** This voltage, and the effects produced by it, are what inductance contributes to circuit action. The following subsections explore how this voltage is produced, the polarity of the induced voltage, and its effect upon the current that produces it.

Polarity of the Induced Voltage

Question: What is Lenz' law?

Heinrich Lenz, a German-born scientist, made a very important contribution to the understanding of electromagnetic induction. **Lenz' law** states:

The magnetic field of an induced current has a polarity that opposes the external magnetic field producing it.

This relationship is illustrated in Figure 9–3. Notice in Figure 9–3(a) that as the magnet is moved into the coil, a current is induced, producing a south pole at the side of entry. The result is *opposing* magnetic fields, in which the north pole of the magnet opposes the north pole produced around the coil by the induced current. In Figure 9–3(b), as the magnet is moved out of the coil, the poles produced by the induced current are reversed. Since there are opposing magnetic fields, it takes mechanical energy to overcome the opposition of the unlike magnetic poles. This mechanism is the means for converting mechanical to electrical energy.

FIGURE 9–3 Polarity in an induced current

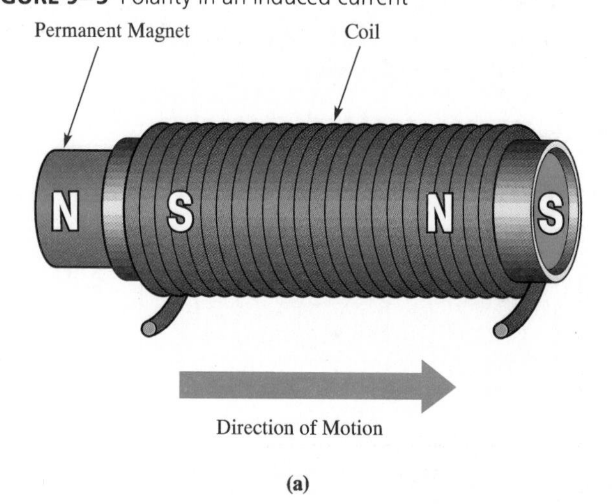

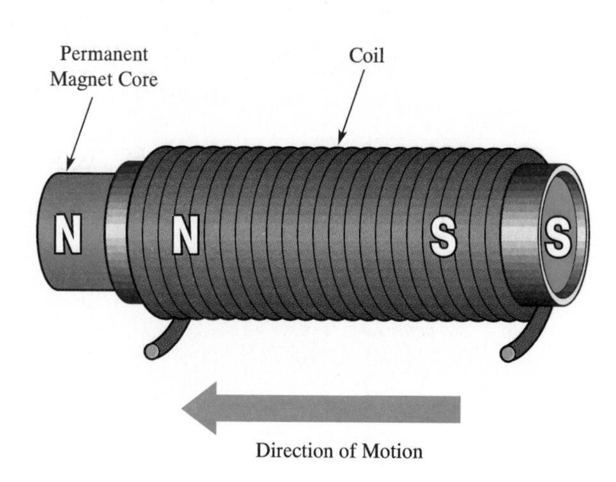

FIGURE 9–4 Polarity in an induced voltage

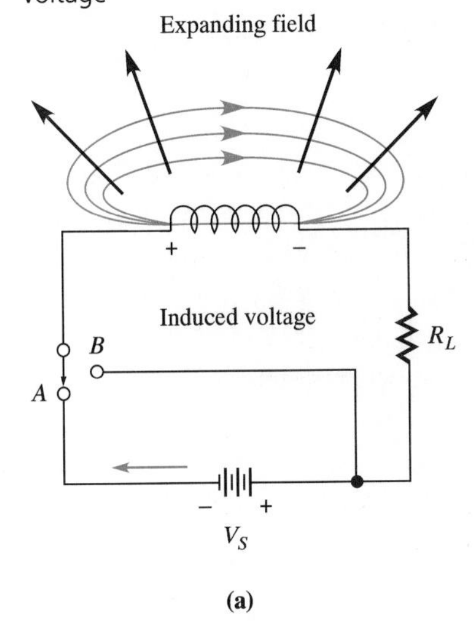

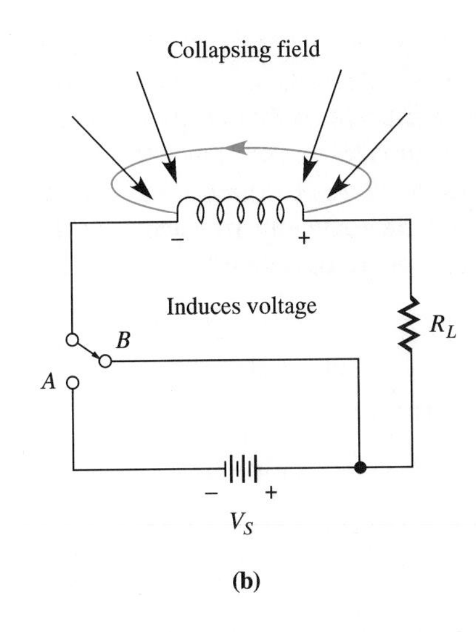

Question: What determines the polarity of a self-induced voltage?

An induced voltage is required to produce an induced current, such as the one in Figure 9–4. The change in direction of the current was produced by a change in polarity of the induced voltage. The movement of a permanent magnet and its magnetic field were used to produce induction. When moved in one direction, a certain polarity of voltage and certain direction of current flow were produced. When moved in the opposite direction, the opposite polarity of voltage and the opposite direction of current were produced. Thus, the *direction of movement* of the magnetic field relative to the conductor determines the *polarity* of the induced voltage, as illustrated in Figure 9–4.

In Figure 9–4(a), notice that a source of dc voltage is applied to a coil through a two-way switch. When the switch is in position *A*, current flows through the coil in the direction shown, producing a magnetic field around it. This magnetic field does not appear instantly, but must build up from the center of the coil and move outward as shown. In moving across the coil, the magnetic field induces a voltage across it. This induced voltage, known as a *counter electromotive force,* or CEMF, opposes the establishment of an energizing current in the inductor. (Recall that electromotive force is another name for voltage.) Thus, the induced voltage must be of a polarity opposite that of the source voltage.

As shown in Figure 9–4(b), when the switch is moved to position *B*, the current stops and the magnetic field is no longer supported. Thus, it collapses and once again induces a voltage in the inductor. Since the direction of movement of the magnetic field has reversed, the polarity of the induced voltage is also reversed. Its polarity is, in fact, the same as the source voltage. The polarity of the CEMF now opposes the decrease in the energizing current in the inductor.

Question: What are the effects of the self-induced voltage?

The effect of the self-induced voltage is to *oppose* any change in current in the circuit. At the instant the source voltage is applied, the magnetic field is expanding at its greatest rate. Thus, by Faraday's law, maximum counter voltage is induced at this time, and the current is instantaneously zero. So at the instant the switch is closed, the inductor appears as an *open* circuit. As the magnetic flux lines move outward, the rate at which they cross the loops of the inductor decreases, resulting in a decrease in the CEMF. This reduced counter voltage allows the current to build up more rapidly. This process continues until the magnetic field is at its maximum strength, the CEMF is zero, and the current is maximum. The inductor is now acting practically as a *short* circuit. The graph of current build-up within the inductor is illustrated in Figure 9–5. Notice that the inductor did not stop the change in current, but it definitely slowed it down.

FIGURE 9–5 Current build-up within an inductor

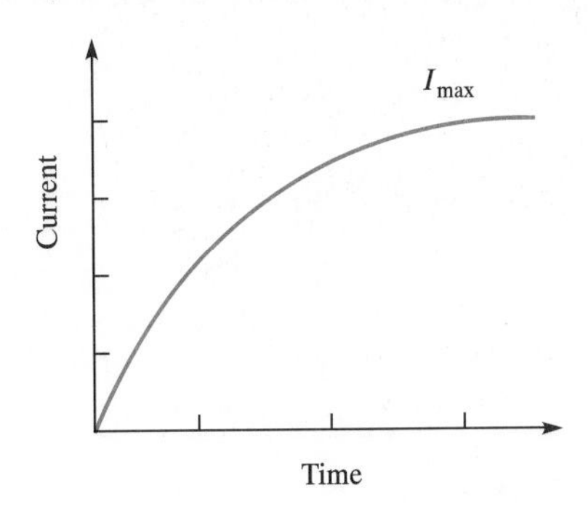

One thing to be noted carefully is that inductance has *no* effect on steady-state direct current. Without a change in current, there is no change in the magnetic field and thus no counter voltage produced. Thus, the inductive effect upon direct current occurs only when the circuit is first energized, de-energized, or when the current is made to vary. During steady-state conditions, the current is limited only by the resistance of the wire making up the inductor.

SECTION 9–2 REVIEW

1. State Lenz' law.
2. What produces the self-induced voltage in an inductor?
3. What determines the polarity of the self-induced voltage in an inductor?
4. Why is the self-induced voltage called a counter electromotive force?
5. What effect does this counter electromotive force have when the current changes?
6. When is direct current affected by inductance?

9–3 MAGNITUDE OF THE INDUCED VOLTAGE

Question: What factors determine the magnitude of the self-induced voltage?

The value of the self-induced voltage is determined by the value of the inductance and the rate at which the flux lines cut across the windings of the inductor. The rate of change in flux is determined by the rate at which the current in the inductor increases or decreases. Thus, the *magnitude* of the self-induced voltage is determined by the inductance value and the rate of change of current with respect to time. Recall that the unit of inductance is the henry. An inductor has an inductance of 1 H if a change of 1 A in 1 s induces 1 V across the coil. In equation form, the magnitude of a self-induced voltage is written

$$L = \frac{V_I}{\Delta i/\Delta t}$$

WHERE L = inductance of 1 H
V_I = induced voltage of 1 V
Δi = change in current of 1 A
Δt = change in time of 1 s

Multiplying both sides of this equation by $\Delta i/\Delta t$ yields an equation for the value of the induced voltage:

(9–1)
$$V_I = \mathrm{L}\frac{\Delta i}{\Delta t}$$

Notice in Eq. 9–1 that a lowercase i was used to indicate current. Lower case letters are used to indicate currents or voltages which have *dynamic* rather than *static* values. This convention is used throughout electronics and in the remainder of this text.

To evaluate the significance of Eq. 9–1, you must recall Faraday's law presented in Chapter 8. It stated that a voltage induced through electromagnetic induction is directly proportional to the number of turns on the coil and the rate at which the flux lines cut across it. In Eq. 9–1, notice that the self-induced voltage is directly proportional to the inductance value. Because inductance is a magnetic effect, anything increasing the magnetic effect also increases the induced voltage. The induced voltage also varies in direct proportion to the second factor in the equation, $\Delta i/\Delta t$. Thus, the induced voltage varies in direct proportion to the rate of change of current with respect to time.

EXAMPLE 9–1 Compute the voltage induced across a 300 mH inductor when the current changes at a rate 2000 A/s.

■ Solution:

Use Eq. 9–1:

$$V_l = L\frac{\Delta i}{\Delta t}$$

$$= (300 \times 10^{-3}\ \text{H})(2 \times 10^{3}\ \text{A/s}) = 600\ \text{V}$$

EXAMPLE 9–2 A change of 300 mA in 1 μs produces 3 kV of induced voltage, what is the value of the inductor?

■ Solution:

Use Eq. 9–1 and rearrange to solve for L:

$$V_l = L\frac{\Delta i}{\Delta t}$$

$$L = \frac{V_l}{\Delta i/\Delta t}$$

$$= \frac{3 \times 10^{3}\ \text{V}}{(300 \times 10^{-3}\ \text{A})/(1 \times 10^{-6}\ \text{s})} = 10\ \text{mH}$$

A definition of a henry is often given in terms of the factors making up Eq. 9–1. An inductance of 1 H exists if a change of 1 A in 1 s produces 1 V of induced voltage. Although you may not often need this relationship in practical situations, it does help to fix in your mind the factors involved in a value of inductance and their relationship one to another.

■ SECTION 9–3 REVIEW

1. What factors determine the value of the induced voltage?
2. What will be the effect on the voltage induced across an inductor by a decrease in inductance L?
3. What will be the effect on the induced voltage of an inductor by an increase in the time t of cutting?
4. What will be the effect on the induced voltage of an inductor by an increase in current I?
5. Write the equation for self-induced voltage relating inductance and the rate of change of current with respect to time.

9–4 INDUCTANCE AND POWER

Question: Is electrical power developed in an inductor?

The concept of a pure inductor must be introduced to answer this question. A pure inductor would be a device containing the property of inductance only. An inductor, however, is made of coiled wire that has some resistance. Thus, the pure inductor, shown in Figure 9–6, consists of a pure inductance in series with a lumped resistance R_W that has the value of the distributed resistance of the coil. It should be made clear that R_W is actually a part of the inductor and not a separate resistor. This R_W, or winding resistance, can be measured with an ohmmeter, however.

FIGURE 9–6 Resistance within an inductor

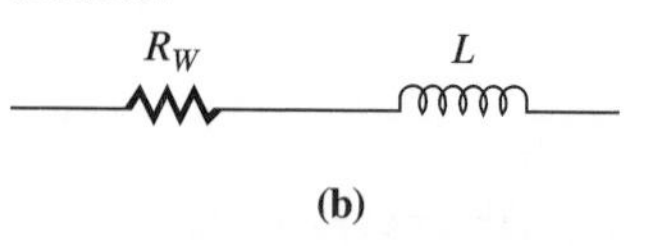

Heat
L
I
I^2R_W
(b)

Recall that power is developed in an electric circuit when current flows through a resistance. Thus, as shown in Figure 9–6(b), power is developed in the winding resistance in the form of heat. The inductance, on the other hand, produces a magnetic field when current flows through it. The electrical energy needed to produce this field is not dissipated, but is actually stored within it. When the current ceases to flow, the magnetic field collapses and returns the energy to the source. Thus, in a practical inductor, energy is dissipated in the winding resistance, but no energy is dissipated in the inductance.

The amount of energy stored in the magnetic field of an inductor is in direct proportion to its inductance value and the amount of current flowing through it. It can be computed using the following equation:

(9–2)

$$E = \frac{L \cdot I^2}{2}$$

WHERE: E = energy (J)
L = inductance (H)
I = current (A)

EXAMPLE 9–3 Compute the energy stored by an 5 H inductor when 1.5 A of current flows.

▪ Solution:

Use Eq. 9–2:

$$E = \frac{L \cdot I^2}{2}$$

$$= \frac{(5 \text{ H})(1.5 \text{ A})^2}{2} = 5.625 \text{ J}$$

EXAMPLE 9–4 A 100 mH inductor stores 3 mJ of electrical energy. What is the amount of current flow in the inductor?

▪ Solution:

1. Use Eq. 9–2:

$$E = \frac{LI^2}{2}$$

2. Cross multiply:

$$2E = LI^2$$

3. Solve for I:

$$I = \sqrt{\frac{2E}{L}}$$

$$= \sqrt{\frac{2 \times 3 \times 10^{-3}\ \text{mJ}}{100 \times 10^{-3}\ \text{H}}} = 244\ \text{mA}$$

SECTION 9–4 REVIEW

1. What is meant by a pure inductor?
2. Where is the energy stored in inductor action?
3. Is there any true power developed in an inductor?
4. If 250 mA of current flows through a 30 mH inductor, how much energy is stored?

9–5 CHARACTERISTICS OF INDUCTORS

Inductance is a physical property with four physical factors that determine its value for any given inductor.

Question: What are the physical factors affecting the inductance value of an inductor?

The physical factors that determine the inductance of an inductor are (1) the number of turns of wire on the coil, (2) the absolute permeability of the core, (3) the area of the core, and (4) the length of the coil. Each of these is indicated in Figure 9–7. The equation relating these physical factors to inductance is

(9–3)
$$L = \frac{N^2 A \mu}{l}$$

WHERE: N = number of turns
A = area of the core
μ = permeability of the core
l = length of the core

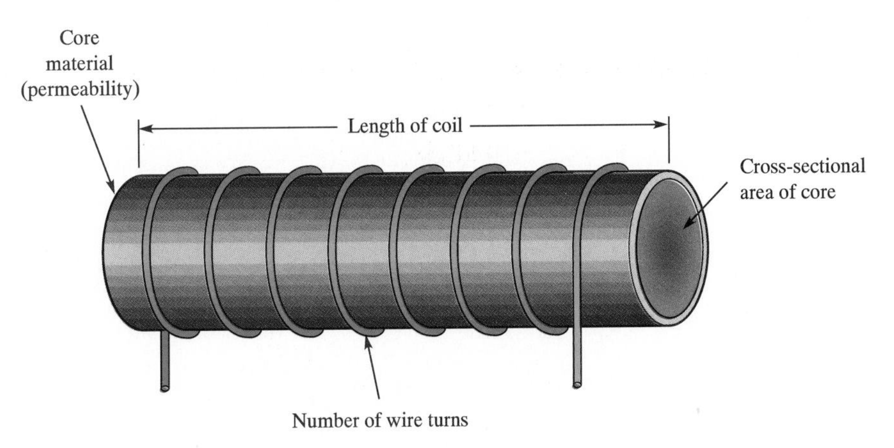

FIGURE 9–7 Inductance factors

This equation shows that inductance is directly proportional to the absolute permeability, the square of the number of turns, and the area of the core. These direct relationships exist because an increase in any one of them increases the magnetic effect, and any increase in the magnetic effect increases the inductance. Inductance is inversely proportional to the length of the coil.

■ Number of Turns

Since this quantity is squared, the number of turns have a great effect upon the inductance. Recall that magnetomotive force is equal to the current times the number of turns on the coil. Thus, a greater number of turns means a stronger mangetic effect and greater inductance.

■ Area of the Core

The area of the core is one determinant in how many lines of force can be developed when the coil is magnetized. Recall that as the magnetomotive force is increased, a point is reached at which the core is saturated, and no more flux lines are developed. A larger core allows the development of a greater number of flux lines and, hence, a stronger magnetic field. Thus, inductance is directly proportional to the area of the core.

■ Absolute Permeability

Recall that the relative permeability of the core is a measure of the ease with which it can be magnetized. A higher permeability means that for a given magnetizing force, a greater number of flux lines are developed, resulting in a stronger magnetic field and greater inductance. Note that in Eq. 9–3, the *absolute permeability* is used. The relative permeability must be multiplied by the absolute permeability of air, 1.26×10^{-6}, in order to convert it to absolute permeability.

■ Length of the Coil

If the turns of a coil are closely spaced, the lines of flux are more concentrated, and the field is more intense. Conversely, if the same number of turns are spaced over a greater length, the field is less concentrated and less intense. Thus, all other things being equal, the inductance is inversely proportional to the length of the coil.

EXAMPLE 9–5 For the values given, compute the inductance of the inductor in Figure 9–7.

■ Solution:

Use Eq. 9–3:

$$L = \frac{N^2 A \mu}{l}$$

$$= \frac{8^2(3.14 \times 10^{-4})(6.3 \times 10^{-4})}{0.15} = 84.4\ \mu\text{H}$$

[8] [X²] [×] [3] [·] [1] [4] [EXP] [±] [4] [×] [6] [·] [3] [EXP] [±] [4] [=]

[÷] [·] [1] [5] [=]

EXAMPLE 9–6 Compute the inductance if the number of turns on the inductor of Figure 9–7 is doubled.

■ **Solution:**

$$L = \frac{N^2 A \mu}{l}$$

$$= \frac{16^2(3.14 \times 10^{-4})(6.3 \times 10^{-4})}{0.15} = 337.6\ \mu\text{H}$$

EXAMPLE 9–7 If the permeability of the core of the inductor of Example 9–1 is lowered to 1, what is the new value of inductance?

■ **Solution:**

$$L = \frac{N^2 A(1.26 \times 10^{-6})}{l}$$

$$= \frac{8^2(3.14 \times 10^{-4})(1.26 \times 10^{-6})}{0.15} = 0.169\ \mu\text{H}$$

Notice that lowering the permeability of the core has resulted in a very large decrease in inductance.

NOTE

Notice that when the number of turns was doubled, the inductance was increased by a factor of four. If the number of turns were halved, the inductance would be one-quarter its original value. (Try the computation again using four turns to verify this statement.) This increase by a factor of four is true anytime the square of a factor is involved. It is mentioned here to help you avoid making snap and incorrect judgments (especially on certification tests)! Watch for squared quantities in equations, and find ways to prove your decisions by substitution of values.

■ Types of Inductor Cores

As previously mentioned, the inductor is basically a coil of wire. The inductor core may be made of air, iron, or ferrite. The different varieties of inductors are made to satisfy the special requirements of electronic circuits. The symbols for the various types of inductors are shown in Figure 9–8. Notice that the symbols relate to the type of core and whether or not the value of inductance can be varied. The following subsections focus on these characteristics and the construction of each type of core.

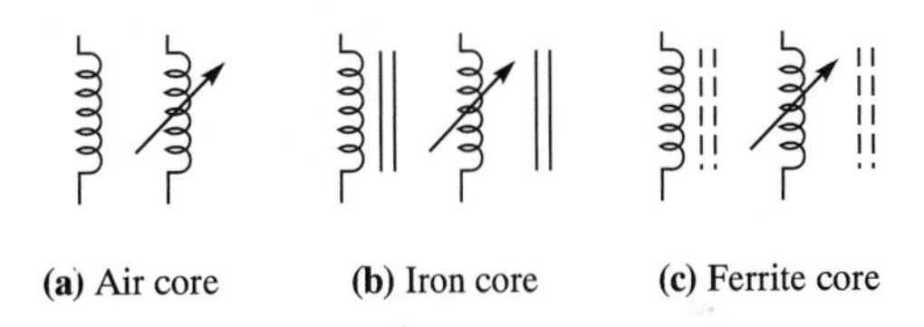

FIGURE 9–8 Symbols for various types of fixed and variable inductors

FIGURE 9–9 Typical air core inductors

FIGURE 9–10 Typical iron core inductors

FIGURE 9–12 Typical of ferrite core inductor

Question: How does the type of core affect the inductance?

Air Core

An inductor with an air core has no magnetic material in its flux path. The loops of wire are wound on a tubular form made of an insulating material. If the wire used is heavy or the coil is very small, it may be self-supporting. A typical air core inductor is shown in Figure 9–9. Without a core, there is very little energy required in order to reverse the magnetic field when the current changes direction. Thus, magnetic hysteresis losses are low and this type inductor can be used in what are known as high frequency applications such as radio and television. (*Frequency* is the number of reversals per second of alternating current, which will be discussed in detail in Chapter 11.) All other factors being equal, air core inductors have a lower inductance than others because of the low absolute permeability of air. The value of small air core inductors is in the microhenry range.

Iron Core

Iron is often used as a core in an inductor. Its high permeability makes it possible to manufacture inductors with much higher inductance values than those with an air core. Their values range from a few microhenries to many henrys. A typical iron core inductor is shown in the photograph of Figure 9–10. They are not without their disadvantages, however. One disadvantage is the bulk and weight added by the iron. It also takes a great deal more energy to reverse the magnetic field in iron than in air. As the number of current reversals per second increases, this problem looms larger. Thus, iron core inductors are not used in high frequency applications.

Iron cores are constructed of thin sheets of iron held together with an insulating adhesive. This type of construction is used to prevent losses due to the flow of what are known as eddy currents. **Eddy currents** are those induced in the iron core by the flux lines cutting across it. They flow at right angles to the movement of the flux lines, as shown in Figure 9–11. The insulated laminations provide smaller paths for eddy currents and greatly reduce their power losses. (Eddy currents are covered in detail in Chapter 13.)

Ferrite Core

A **ferrite core** is a mixture of iron oxide and ceramic, and it is characterized by high permeability and low losses due to hysteresis and eddy currents. A typical ferrite core inductor is shown in Figure 9–12. Cores made of this material have the high permeability of an iron core, but not nearly as much power loss due to magnetic hysteresis and eddy

FIGURE 9–11 Eddy currents

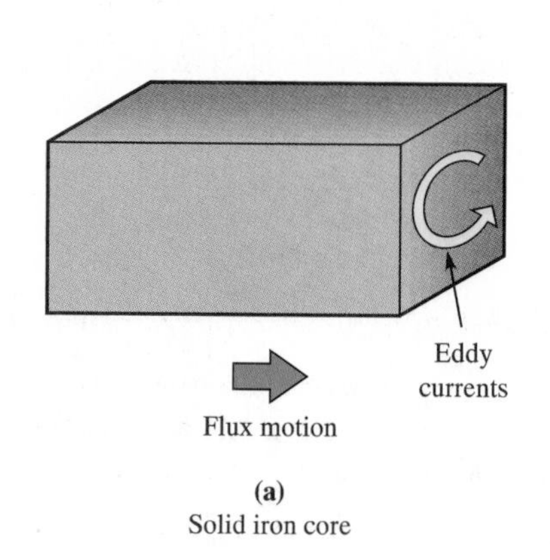

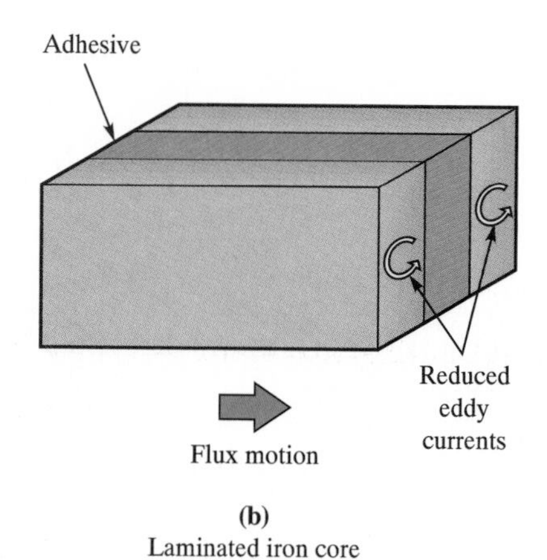

currents. The ceramic material acts as an insulator and breaks up the paths for eddy currents. Thus, ferrite core inductors find applications in alternating current circuits where the frequency, or number of alternations per second, is very high. (Recall that this was the same application as for the air core inductor.) The ferrite core, with its higher permeability, has a higher value of inductance than a comparable air core.

SECTION 9–5 REVIEW

1. What are the physical factors affecting the inductance of an inductor?
2. How do absolute and relative permeability differ?
3. Why are hysteresis losses low in air core inductors?
4. What is ferrite material made of?
5. Why are laminated iron cores not good for high frequency applications?
6. Compute the inductance of the inductor shown in Figure 9–13.

FIGURE 9–13 Inductor for Section Review Problem 6

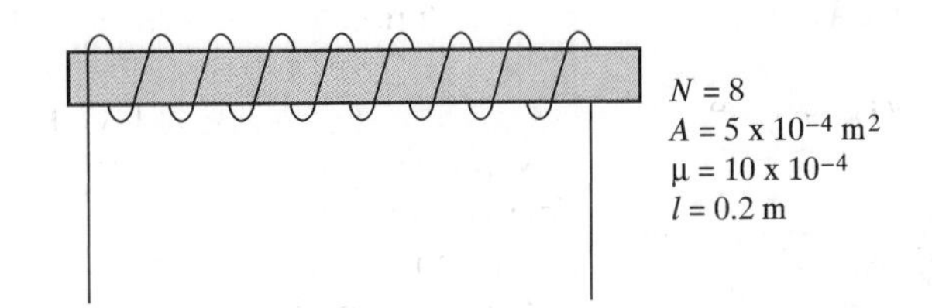

SYNOPSIS

At this point, it is a good idea to summarize the characteristics and effects of inductance. Inductance is a magnetic property that has an effect only when the current within a circuit changes. This effect is due to the change in mangetic flux that accompanies any change in current. The flux cuts across the windings of the inductor, self-inducing a voltage across it. This self-induced voltage has a polarity that opposes the voltage producing the current change. Thus, this counter electromotive force opposes the change in current. This opposition will be present whether the current is increasing or decreasing.

Inductance, then, is the property of a circuit that induces a voltage when the current changes. The polarity of this voltage is such that it always opposes the change in current. The symbol for inductance is L and the unit of measure is the henry. Inductance is directly proportional to the square of the number of turns, the relative permeability of the core, and the area of the core. It is inversely proportional to the length of the core. The amount of voltage induced is directly proportional to the inductance and to the rate of change of current with respect to time.

9–6 INDUCTORS IN SERIES

Question: Why are inductors sometimes connected in series in a dc circuit?

Inductors are connected in series in order to increase the total effective inductance. A series string of inductors is shown in Figure 9–14. When connected in series, the equation for computing the total inductance L_T is

(9–4)
$$L_T = L_1 + L_2 + L_3 \cdots + L_n$$

FIGURE 9–14 Series string of inductors

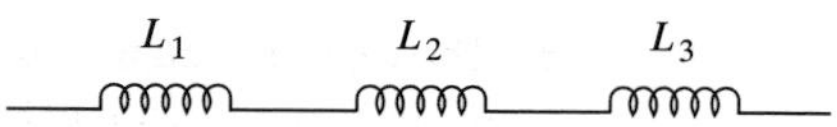

L_1 L_2 L_3

Notice that this equation for total inductance is similar to the equation used in computing the total resistance of resistors in series.

NOTE

Be aware at this point that Eq. 9–4 is only valid if there is no magnetic coupling between the inductors. If the inductors are close to one another, their magnetic fields cross and what is known as *mutual inductance* takes place. (Mutual inductance is covered in detail in the Chapter 13.)

EXAMPLE 9–8 Three inductors with values of 10 mH, 2 mH, and 250 μH are connected in series. Compute the total inductance.

■ Solution:

Use Eq. 9–4:

$$L_T = L_1 + L_2 + L_3$$
$$= (10 \times 10^{-3}) + (2 \times 10^{-3}) + (250 \times 10^{-6}) = 12.25 \text{ mH}$$

[1] [0] [EXP] [±] [3] [+] [2] [EXP] [±] [3] [+] [2] [5] [0] [EXP] [±] [6] [=]

EXAMPLE 9–9 For the inductor string of Figure 9–15, if L_T equals 37 mH, compute the value of L_4.

■ Solution:

Since the values of L_1, L_2, and L_3 are known, transpose Eq. 9–4 to solve for L_4:

$$L_T = L_1 + L_2 + L_3 + L_4$$
$$= (L_1 + L_2 + L_3) + L_4$$
$$L_4 = L_T - (L_1 + L_2 + L_3)$$
$$= 37 \text{ mH} - (10 \text{ mH} + 2 \text{ mH} + 5 \text{ mH}) = 20 \text{ mH}$$

FIGURE 9–15 Inductors in series for Example 9–9

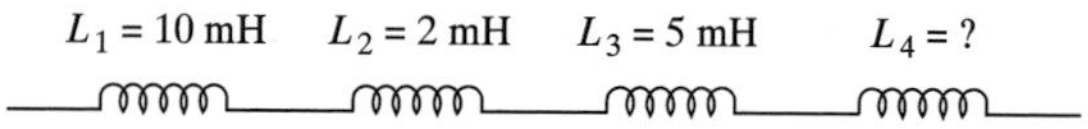

$L_1 = 10$ mH $L_2 = 2$ mH $L_3 = 5$ mH $L_4 = ?$

SECTION 9–6 REVIEW

1. Why are inductors often connected in series?
2. Compute the total inductance of the following five inductors connected in a series string: 109 μH, 59 mH, 0.109 mH, 250 μH, and 36 mH.

3. The total inductance of a three-inductor series string is 200 mH. If two have the values of 30 mH and 150 mH, compute the value of the third.

9–7 INDUCTORS IN PARALLEL

Question: Why are inductors sometimes connected in parallel in a dc circuit?

Inductors may be connected in parallel in order to reduce the total inductance of the circuit. A parallel bank of inductors is shown in Figure 9–16. When connected in parallel, the equation for total inductance is

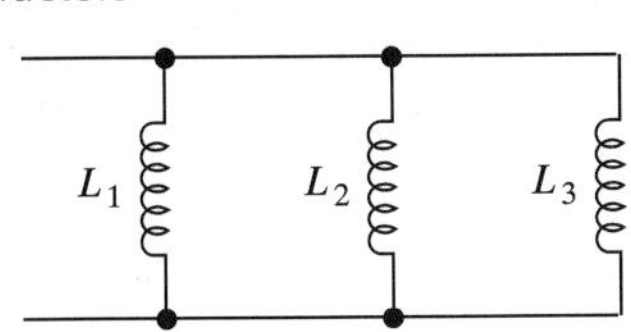

FIGURE 9–16 Parallel bank of inductors

(9–5)
$$L_T = \frac{1}{(1/L_1) + (1/L_2) + (1/L_3) \cdots + (1/L_n)}$$

Notice that this equation for total inductance is similar to the equation for computing the total resistance of resistors in parallel. Once again, this equation is valid only if no magnetic coupling takes place between the inductors.

EXAMPLE 9–10 Compute the total inductance of 4 inductors with values of 2 H, 5 H, 5 H, and 15 H when connected in parallel.

■ Solution:

Use Eq. 9–5:

$$L_T = \frac{1}{(1/L_1) + (1/L_2) + (1/L_3) + (1/L_4)}$$

$$= \frac{1}{(1/2\text{ H}) + (1/5\text{ H}) + (1/5\text{ H}) + (1/15\text{ H})} = \frac{1}{0.9667\text{ H}} = 1.03\text{ H}$$

[2] [1/x] [+] [5] [1/x] [+] [5] [1/x] [+] [1] [5] [1/x] [=] [1/x]

EXAMPLE 9–11 What value of inductance must be placed in parallel with a 30 mH inductor in order to produce 5 mH?

■ Solution:

1. Use the reciprocal form of Eq. 9–5:

$$L_T = \frac{1}{(1/L_1) + (1/L_2)}$$

$$\frac{1}{L_T} = \frac{1}{L_1} + \frac{1}{L_2}$$

2. Solve for L_2:

$$\frac{1}{L_2} = \frac{1}{L_T} - \frac{1}{L_1}$$

$$= \frac{1}{5 \times 10^{-3}} - \frac{1}{30 \times 10^{-3}} = \frac{1}{166.67} = 6\text{ mH}$$

If a number of equal value inductors are placed in parallel, the total can be found by dividing the value of one by the number of inductors. Once again notice that this method is similar to that used in finding the total resistance of equal-value resistors in parallel. Thus,

(9–6)
$$L_T = \frac{L}{N}$$

EXAMPLE 9–12 What is the total inductance of six 30 mH inductors when they are placed in parallel?

■ Solution:

$$L_T = \frac{L}{N}$$
$$= \frac{30 \times 10^{-3}}{6} = 5 \text{ mH}$$

■ SECTION 9–7 REVIEW

1. Why are inductors sometimes placed in parallel in a dc circuit?
2. What is the total inductance of the following inductors when placed in parallel: 250 μH, 109 mH, and 4 μH?
3. What is the total inductance of five 250 mH inductors when placed in parallel?

9–8 THE *RL* TIME CONSTANT

Recall that the current through an inductor and the magnetic field surrounding it do not rise to maximum instantaneously when the voltage is applied. As the current begins to flow, the expanding magnetic field induces a voltage, the polarity of which opposes the increase in current. This opposition decreases until the current has reached its steady-state value. By the same token, the current and magnetic field cannot drop to zero instantaneously when the voltage source is removed. The collapsing magnetic field induces a voltage whose polarity opposes the decrease in current. The *time* involved in either case depends upon the amount of inductance and resistance in the circuit.

Question: How does the value of the inductance affect the time required for the current to build to its steady-state value?

From a purely mathematical standpoint, Eq. 9–1 demonstrates that the self-induced voltage of an inductor is directly proportional to its inductance value. The larger the inductance the greater the counter voltage produced by the inductor. A larger counter voltage presents a greater opposition to the build up in current. Thus, the larger the inductance, the greater the time period required for the current to reach its steady-state value. (The *steady-state value* is the value at which direct current stabilizes after the magnetic field is fully developed.) Another way of considering the relationship between the inductance value and time of current build-up is to recall that inductance is a magnetic effect. Any factor increasing the inductance value of an inductor also increases its magnetic effect. Anything increasing its magnetic effect causes it to induce a larger counter voltage for a given current variation.

Question: How does the circuit resistance affect the time required for the current to build to its steady-state value?

The concept of a "perfect inductor" is purely theoretical. All inductors contain some resistance from the wire used in the coil windings. As was stated previously, this winding resistance, R_W, is effectively in series with the inductor. Actual resistors may be placed in series with the inductor, and all will have an effect on the time taken for the current to reach its steady-state value.

Recall that when the voltage is first applied to an inductor, it appears as an open circuit. This situation is shown in Figure 9–17(a). No current flows, so all the voltage is across the inductor, with none across the resistor. After the magnetic field is fully developed, there will be no more counter voltage, and the inductor will appear as a short circuit to dc. As shown in Figure 9–17(b), all the voltage is across the resistor, and the current has stabilized at its steady-state value. This current value is determined according to Ohm's law by the ratio of the applied voltage to the total resistance.

The steady-state current produced is inversely proportional to the value of the resistance. Thus, a relatively large resistance produces a relatively small steady state current. Since the change in current from zero to maximum is small, the inductive effect of the inductor is also small, and the time taken to reach this current value is short. A relatively small resistance, on the other hand, produces a relatively large steady-state current. In this case, the change from zero to maximum is large, and the inductive effect is greater. Thus, the time required for the current to reach its steady state value is inversely proportional to the resistance.

Question: How is the energizing time of an inductor calculated?

The energizing of an inductor refers to the storage of energy in its magnetic field. This storage of energy in an inductor requires a current known as **energizing current.** When the energizing current is removed and the magnetic fields collapses, the energy stored in the field is returned to the circuit. A current, known as a **de-energizing current,** may flow at this time.

FIGURE 9–17 Inductance and resistance in a dc circuit

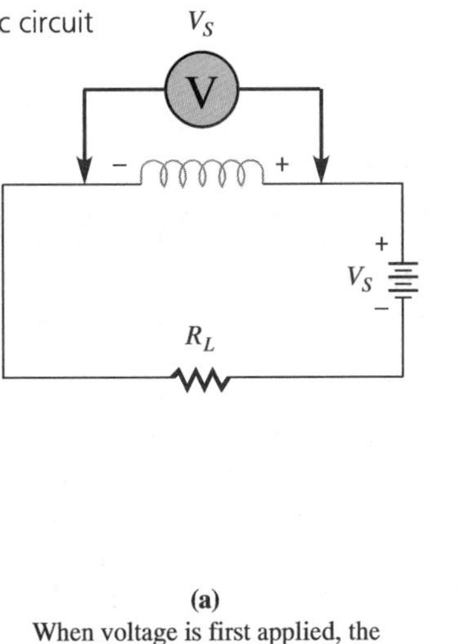

(a)
When voltage is first applied, the inductor appears as an open circuit.

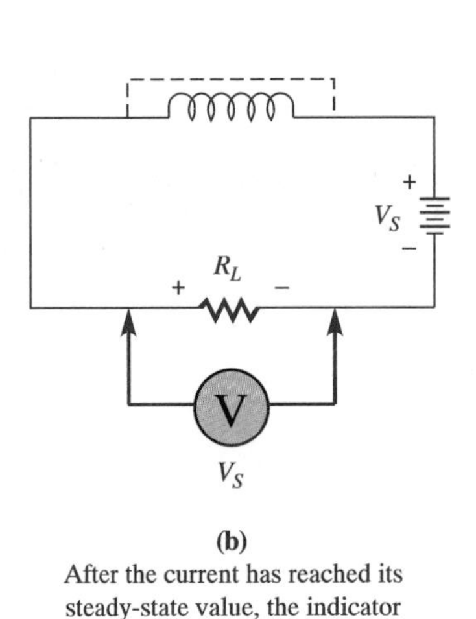

(b)
After the current has reached its steady-state value, the indicator appears as a short circuit.

You know that the time taken for the current through an inductor to reach its steady-state value is inversely proportional to the resistance and directly proportional to the inductance. In equation form, this relationship is expressed as

$$\tau = \frac{L}{R} \tag{9-7}$$

where τ (tau, from the Greek alphabet) is known as the time constant. The equation shows that the time constant in seconds is equal to the inductance in henries divided by the resistance in ohms.

EXAMPLE 9–13 For the circuit in Figure 9–18, compute the *RL* time constant.

■ **Solution:**

$$\tau = \frac{300 \text{ mH}}{20} = 15 \text{ ms}$$

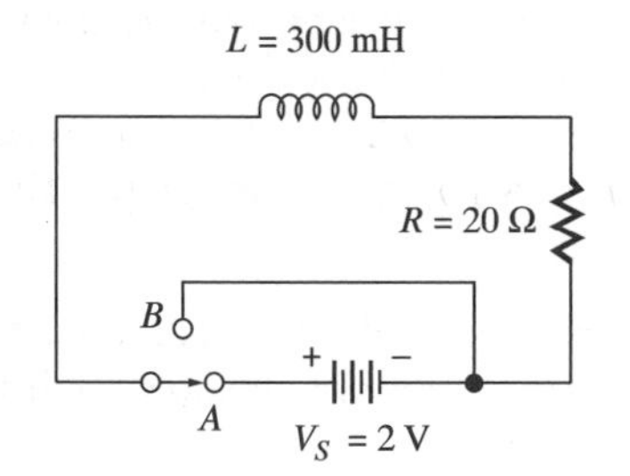

FIGURE 9–18 Circuit for Example 9–13

The ***RL* time constant** of a circuit is the time taken for the current to rise or decay by approximately 63.2%. An interesting observation here is that if the current builds up in this manner, theoretically it will *never* reach its full, steady-state value. After five time constants, however, the current reaches approximately 99.3% of its steady state value, and the inductor is considered fully energized. The following example demonstrates the energizing of an inductor with direct current.

For the circuit of Figure 9–18, the energizing current at the end of each of the first five time constants is computed and plotted on the graph in Figure 9–19.

1. The instant that the switch (Figure 9–18) is placed in position *A*, the inductor appears as an open circuit, and the circuit current is zero. The current for this circuit is represented by curve *A* in Figure 9–19. The zero point for the current is at τ-0.
2. It is now necessary to compute the final, steady-state value the current assumes after the inductor is fully energized. After it is energized, the inductor appears as a short to

FIGURE 9–19 Graph of energizing and deenergizing current in an inductor

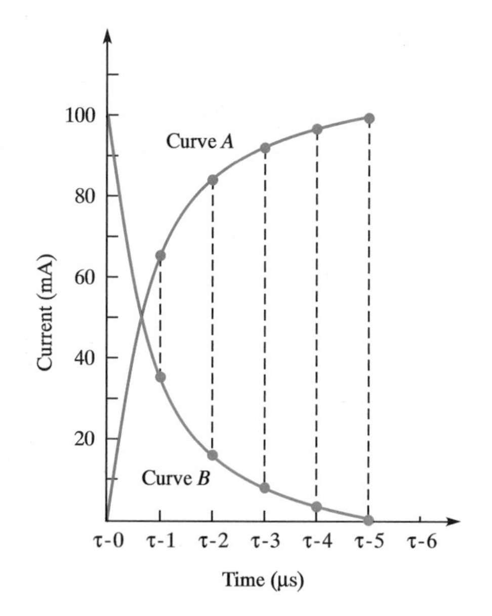

dc, and maximum current flows. At that time, it is opposed by the resistance only. Thus, the steady-state current is

$$I_F = \frac{V_S}{R}$$

$$= \frac{2\text{ V}}{20\ \Omega} = 100\text{ mA}$$

WHERE: I_F = final current

3. After the first time constant, or 15 ms, the current builds to 63.2% of 100 mA, or 63.2 mA. This value of current is plotted as point τ-1 on the graph (Figure 9–19). Current must increase another 36.8 mA for the full energizing current to be present.
4. During the second time constant, the current increases by 63.2% of the remaining 36.8 mA. Thus it increases by 23.26 mA to 86.46 mA. This value of current is at point τ-2 on the graph. A 13.54 mA increase is needed for the full energizing current to be present.
5. During the third time constant, the current increases by 63.2% of the remaining 13.54 mA. It increases by 8.56 mA to 95.02 mA, which is point τ-3 on the graph. Current must still increase another 4.98 mA to reach the full value.
6. During the fourth time constant, the current increases by 63.2% of the remaining 4.98 mA. Thus, it increases by 3.15 mA to 98.17 mA, which is point τ-4 on the graph. A 1.83 mA increase is needed for the full energizing current to be present.
7. During the fifth time constant, the current increases by 63.2% of the remaining 1.83 mA. It increases by 1.16 mA to 99.33 mA, which is point τ-5 on the graph. The inductor is considered to be fully energized at this time.

Two observations should be made about the final current value of 99.33 mA. First of all, at only 0.67 mA below the desired value, it is very close! Second, it requires another five time constants to reach 99.995 mA, which is even closer!

Question: Can a similar graph be developed for the situation where the current decays to zero?

When the switch in Figure 9–18 is moved to position *B*, the magnetic field collapses and induces a voltage across the inductor. For a period of five time constants, the inductor acts as a generator. The induced voltage now has a polarity that is the same as the source voltage. The current will decay by 63.2% at each time constant. The current at the end of each time constant is as shown in the following table:

Time Constant	Current Value
$\tau - 0$	100 mA
$\tau - 1$	36.8 mA
$\tau - 2$	13.54 mA
$\tau - 3$	4.98 mA
$\tau - 4$	1.83 mA
$\tau - 5$	0.67 mA

Notice that the current is not exactly zero but very close to it after five time constants. The curve of the current values for each time constant is shown as curve *B* in Figure 9–19.

The Exponential Equation

The curve plotted in Figure 9–19 for the energizing current is the graph of what is known as an *exponential equation*. The value of the energizing current at any instant in time after power is applied can be computed with the following equation:

(9–8) $$i_l = I_F(1 - \varepsilon^{-Rt/L})$$

This equation states that the instantaneous value of the energizing current at any instant of time *t*, is equal to the product of the final current I_F, and the quantity 1 minus ε (Greek epsilon) raised to the indicated power. The power to which ε is raised is the product of the resistance and time under consideration divided by the inductance, or Rt/L. Notice that this power will always be negative. **Epsilon** is the base of the natural or Naperian logarithm and is equal to approximately 2.7182818.

NOTE

It is not necessary to memorize the value of epsilon. It can be found by using the following calculator sequence:

Keystrokes: [1] [INV] [$\ln_x$]

Display: 1 1 2.7182818

Some calculators come equipped with a key labeled ε^x, which requires the following calculator sequence:

Keystrokes: [1] [ε^x]

Display: 1 2.7182818

EXAMPLE 9–14 An *RL* circuit is composed of a 100 Ω resistance and a 250 mH inductor across a 10 V source. Compute the value of the current 1.25 ms after the voltage is applied.

Solution:

1. First compute the final current:

$$I_F = \frac{V_S}{R}$$

$$= \frac{10\ \text{V}}{100\ \Omega} = 100\ \text{mA}$$

2. Next compute the exponent of ε:

$$\frac{-Rt}{L} = \frac{-100\ \Omega \times 1.25\ \text{ms}}{250\ \text{mH}} = -0.5$$

3. Key these values into Eq. 9–8, the exponential equation, and solve:

$$i_l = I_F(1 - \varepsilon^{-Rt/L})$$

$$= (100 \text{ mA})(1 - 2.7182818^{-0.5})$$
$$= (100 \text{ mA})(1 - 0.60653)$$
$$= (100 \text{ mA})(0.39347) = 39.347 \text{ mA}$$

(Note: Raising epsilon to the exponent $-Rt/L$ requires the use of the x^Y key on the calculator, *not* the EE key!)

EXAMPLE 9–15 For the circuit of Example 9–14, an RL circuit is composed of a 100 Ω resistor, and a 100 mh inductor across a 10 V source. Compute the current after one time constant.

■ Solution:

Since the time in question is now one time constant after the voltage is applied, the exponent of ε is recalculated to be -1. Substitute into Eq. 9–8, the exponential equation:

$$i_l = I_F(1 - \varepsilon^{-Rt/L})$$
$$= (100 \text{ mA})(1 - 2.7182818^{-1})$$
$$= (100 \text{ mA})(1 - 0.3678)$$
$$= (100 \text{ mA})(0.6322) = 63.21 \text{ mA}$$

In Example 9–15, notice the factor that the final current is multiplied by in this example, 0.6322, is 63.22% of I_F. This result gives proof of the statement that the current increases by approximately 63.2% in one time constant!

■ De-Energizing Current

The curve plotted for the de-energizing current is described by a similar exponential equation:

(9–9)
$$i_l = I_F \varepsilon^{-Rt/L}$$

This equation can be used to solve for the current at any instant as it decays from I_F to zero.

EXAMPLE 9–16 An *RL* circuit is composed of a 30 mH inductor in series with a 15 Ω resistor across a 3 V source. Compute the current 3.5 ms after the inductor begins to de-energize.

■ Solution:

1. First compute the final current for this circuit:

$$I_F = \frac{V_S}{R}$$
$$= \frac{3 \text{ V}}{15 \text{ }\Omega} = 200 \text{ mA}$$

[3] [÷] [1] [5] [=]

2. Next compute the exponent of ε:

$$\frac{-Rt}{L} = \frac{-15\ \Omega \times 3.5\ \text{ms}}{30\ \text{mH}} = -1.75$$

[1] [5] [±] [×] [3] [·] [5] [EXP] [±] [3] [÷] [3] [0] [EXP] [±] [3] [=]

3. Key these values into Eq. 9–9, the exponential equation, and solve:

$$i_l = I_F\ \varepsilon^{-Rt/L}$$
$$= (200\ \text{mA})(\varepsilon^{-Rt/L})$$
$$= (200\ \text{mA})(2.7182818^{-1.75})$$
$$= (200\ \text{mA})(0.1738) = 34.75\ \text{mA}$$

[2] [·] [7] [1] [8] [2] [8] [1] [8] [X^Y] [1] [·] [7] [5] [±] [=] [×]

[2] [0] [0] [EXP] [±] [3] [=]

SECTION 9–8 REVIEW

1. How many time constants are required in order to fully energize an inductor?
2. Is the time constant directly or inversely proportional to the inductance?
3. Is the time constant directly or inversely proportional to the resistance?
4. What is the time constant of a circuit composed of a 110 μH inductor in series with a 10 Ω resistor?
5. If 10 V dc is applied to the circuit of Problem 4, what is the value of the current after 25 μs?

9–9 APPLICATIONS OF INDUCTORS

Three properties of an inductor are made use of in its many applications. One is the magnetic field produced by a current through the inductor. The relay, a device that can be used as an electrically operated switch, is an example of this use. Another property is its opposition to a change in current. The choke, a device that smooths the rapid variations in current, is an example of this use. The third is the fact that very high voltages can be produced by the counterelectromotive force and a "reactive kick" occurs when the current through an inductor suddenly stops. *The induction coil* is an example of this use. Each of these applications are explored in the following paragraphs.

Question: How does the relay make use of the magnetic field produced by an inductor?

A **relay** is an electromechanical device that can be used to temporarily complete an electrical path through the energizing of a coil. The basic relay consists of a coil, a moveable contact known as the relay armature, and a fixed contact. The electrical layout of a relay is illustrated in Figure 9–20(a). Voltage applied to the coil produces a magnetic field around it. This field attracts the relay armature, which is made of a metal with magnetic properties. In moving downward, it makes an electrical connection with the fixed contact and completes a circuit, acting as an SPST switch.

FIGURE 9–20 Relay devices that function as switches

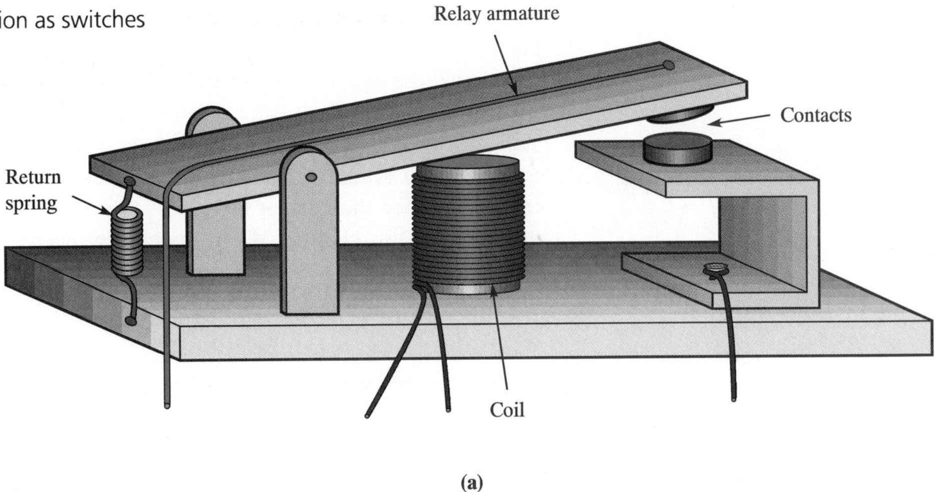

(a)

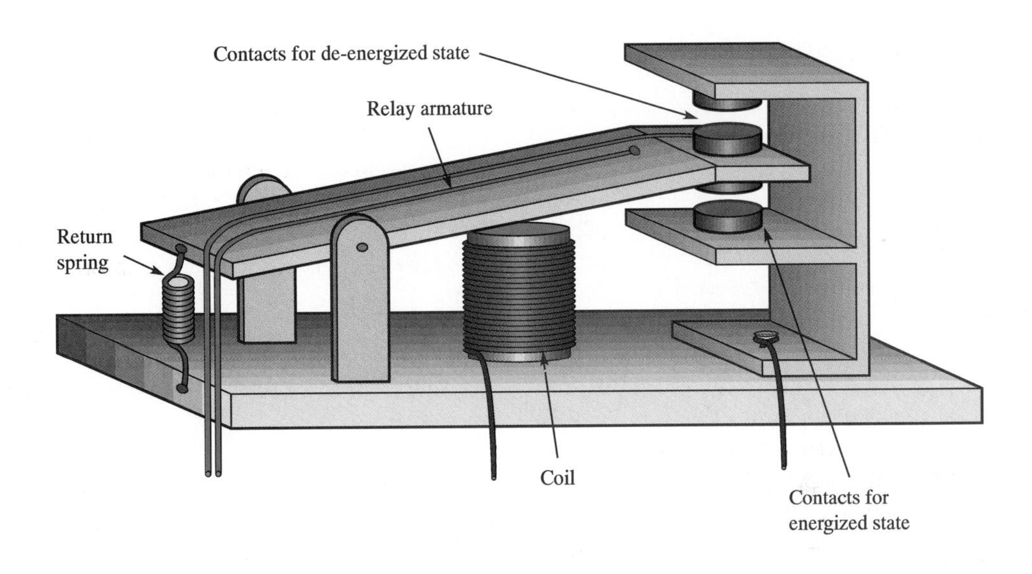

(b)

FIGURE 9–21 Types of relay devices

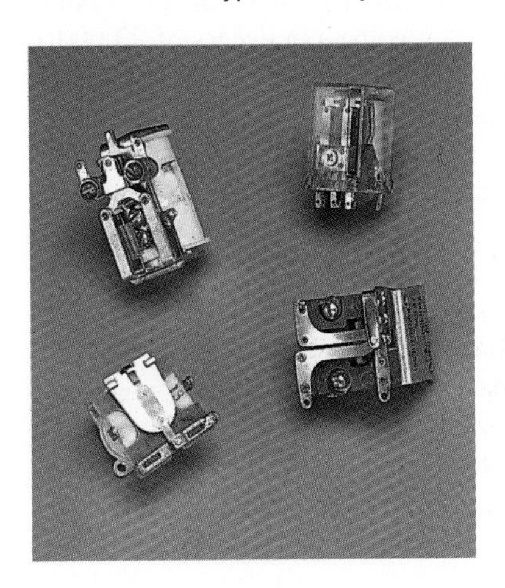

Another possibility is shown in Figure 9–20(b). In this relay, there are two sets of contacts. In the de-energized condition, the armature makes an electrical connection with the upper contact. When the relay is energized, it moves downward breaking the upper connection and making a connection with the second contact. In this way, one circuit can be energized at the same time another is de-energized, acting as a SPDT switch.

Relays make it possible to remotely switch circuits on or off through the application of a voltage. They may have multiple sets of contacts allowing them to control multiple circuits simultaneously. Several representative relays are shown in Figure 9–21.

Question: How is the inductive effect used to smooth rapid changes in current?

Recall that an inductor opposes a change in current. This property is used to smooth rapid changes in current. An inductor used for this purpose is usually known as a choke.

FIGURE 9–22 Graphs illustrating the inductive effects on current

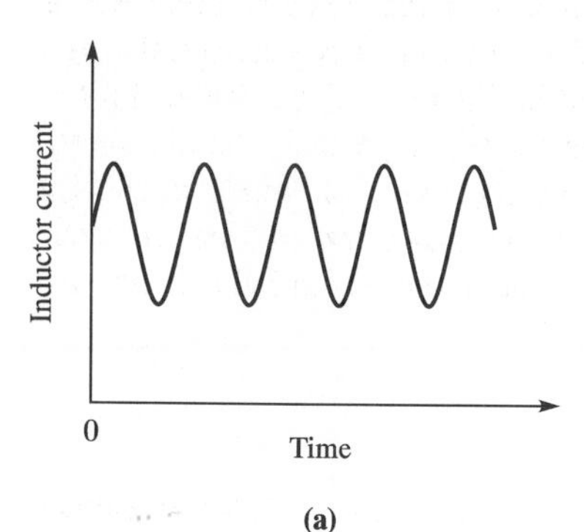

(a)
Rapidly changing direct current

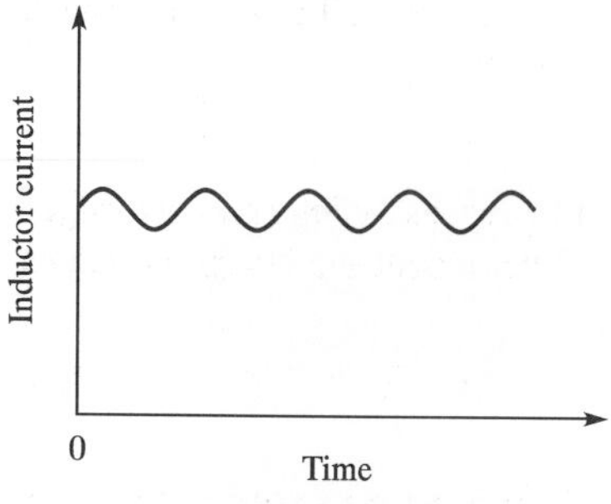

(b)
Current smoothed by an inductor used as a choke

A graph of such a current is illustrated in Figure 9–22(a). Notice that the current is dc, but over time it rises and falls with regularity, which is usually the result of the direct current having an alternating current component.

If only direct current were present, the magnetic field surrounding the inductor would stabilize when the current reaches its steady-state value. The current, being a series of increases and decreases, causes the magnetic field to first increase and then decrease in intensity. When the current increases, the change in the magnetic field induces a voltage that opposes this change. When the current decreases, the induced voltage also opposes this change. This action of the inductor upon the current of Figure 9–22(a) is illustrated in Figure 9–22(b).

Question: How is an inductor used to produce a large voltage?

Recall that the voltage induced by an inductor is equal to the product of the inductance in henries times the rate of change of current with respect to time. Thus, the inductance may be relatively small, but if the change in current is rapid enough, a large voltage can be induced. For example, if the current increased from 0 to 100 mA in 10 μs, the rate of change is

$$\text{rate of change} = \frac{\Delta i}{\Delta \text{t}}$$

$$= \frac{100 \text{ mA}}{10\ \mu\text{s}} = 10{,}000 \text{ A/s}$$

So if the inductance in this case were 2 H, the voltage induced is

$$v_l = L\frac{\Delta i}{\Delta \text{t}}$$

$$= 2\,\frac{100 \text{ mA}}{10\ \mu s}$$

$$= 2(10{,}000) = 20 \text{ kV}$$

Thus, as you can see, even though the change in current is quite small, its rate of change is large, and large voltages can be induced for a moderate value of inductance! As mentioned earlier, this property is made use of in the ignition system of an automobile. Here, although only 12 V is available from the battery, a voltage in excess of 20,000 V is produced by a component known as the ignition coil. The breaker points are closed, allowing current to flow through the coil. A cam shaft, which is rotated by the engine, opens the points each time a cylinder is to fire. The rapid collapse of the magnetic field induces the high voltage connected to the appropriate sparkplug. This large voltage is often referred to as a "reactive kick."

■ SECTION 9–9 REVIEW

1. Explain how an inductor can smooth the ripple in a direct current.
2. Explain what is meant by the rate of change of current with respect to time.
3. If a current increases from zero to 90 mA in 90 μs, compute the rate of change.
4. What voltage will be produced over this period if the inductor has a value of 5 H?

9–10 TROUBLESHOOTING

Faults that occur in an inductor are usually shorts or opens. A short may be complete or partial. That is, all the windings may short or just a few, which can be detected through a resistance check of the coil. This check is illustrated in Figure 9–23(a). If the short is complete, the ohmmeter indicates practically zero ohms. If it is partial, the ohmmeter indicates some reading lower than the winding resistance of the coil, which is usually given on schematics. An inductor that is completely or partially shorted must be replaced. As Figure 9–23(b) shows, an open coil winding causes the ohmmeter to indicate infinity on all scales. An open inductor requires replacement.

CAUTION

Be sure to begin with the low resistance range of the ohmmeter, $R \times 1$, when making the resistance check. Inductors such as those used in transformers and certain other applications may have winding resistances of only a few ohms. Use of a higher scale would make the coil appear to be shorted even though it may be perfectly all right. This check is much more valid when the winding resistance of the coil is known.

The best check of an inductor is made with an analyzer known as an *impedance bridge*. An example of this instrument is shown in Figure 9–24. The impedence bridge indicates

FIGURE 9–23 Ohmmeter checks of winding resistance

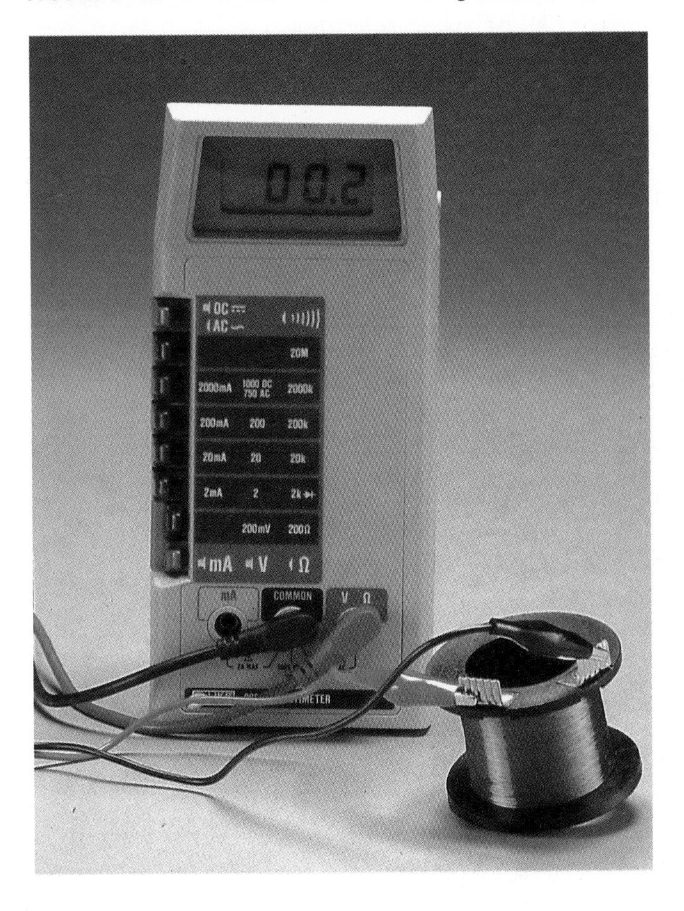

not just opens and shorts, but also the inductance value. This latter feature makes it possible to wind an inductor to a given value if a direct replacement part is not available.

FIGURE 9–24 Impedance bridge

SECTION 9–10 REVIEW

1. What types of faults usually occur in an inductor?
2. What does an ohmmeter reading of infinity indicate about an inductor?
3. Why is it important to know the approximate coil resistance when making an ohmmeter check?
4. What is an impedance bridge?

CHAPTER SUMMARY

1. Inductance is the property of a circuit that induces a voltage when the current changes.
2. Inductance is the property of a circuit that opposes a change in current.
3. Inductance is introduced into a circuit in the form of inductors.
4. Lenz' law states that the magnetic field of an induced current has a polarity that opposes the external magnetic field that produced it.
5. The polarity of the self-induced voltage in an inductor has a polarity that opposes the change in current that produces it.
6. The magnitude of the self-induced voltage is equal to the product of the inductance and the rate of change of current with respect to time.
7. When voltage is first applied to an inductor, it appears as an open circuit.
8. An inductance has no affect on a steady-state direct current.
9. An inductor does not dissipate energy, but stores it in its magnetic field.
10. The inductance value of an inductor is directly proportional to the square of the number of turns, the area of the core, and its relative permeability.
11. The inductance value of an inductor is inversely proportional to the length of the core.
12. The core of an inductor may be air, laminated iron, or ferrite.
13. The total inductance of inductors in series is equal to their sum if there is no magnetic coupling between the coils.
14. The total inductance of inductors in parallel is equal to the reciprocal of the sum of their reciprocals.
15. The current rises to its steady-state value or decays to zero in five time constants.

KEY TERMS

inductance
henry
inductor
self-induced voltage
Lenz' law
eddy currents
ferrite core
RL time constant
energizing current
de-energizing current
epsilon
relay

EQUATIONS

(9–1) $V_l = L\dfrac{\Delta i}{\Delta t}$

(9–2) $E = \dfrac{L \cdot I^2}{2}$

(9–3) $\dfrac{(N^2 A\mu)}{l}$

(9–4) $L_T = L_1 + L_2 + L_3 \cdots + L_n$

(9–5)

$$L_T = \frac{1}{(1/L_1) + (1/L_2) \cdots + (1/L_n)}$$

(9–6) $L_T = \dfrac{L}{N}$

(9–7) $\tau = \dfrac{L}{R}$

(9–8) $i_l = I_F(1 - \varepsilon^{-Rt/L})$

(9–9) $i_l = I_F\varepsilon^{-Rt/L}$

Variable Quantities

V_l = voltage induced in the inductor

L = inductance

Δ = change or difference

i = instantaneous current

t = time

E = energy stored in the inductor's magnetic field

R = resistance

N = number of inductors

τ = time constant

i_l = instantaneous inductor current

ε = approximately 2.7182818

T = total

n = any number

I_f = final value of inductor current

TEST YOUR KNOWLEDGE

1. Why is inductance called a magnetic effect?
2. How is a self-induced voltage produced?
3. What determines the polarity of a self-induced voltage?
4. What determines the magnitude of the self-induced voltage?
5. Why is the self-induced voltage referred to as a counter electromotive force?
6. What does the inductor appear as the instant that the voltage is applied?
7. Why does the inductor present no opposition to steady-state direct current?
8. What is the symbol and unit of measure for an inductor?
9. What physical factors affect the inductance of an inductor? State whether each is a direct or inverse relationship.
10. What materials may an inductor core be made of?
11. What is meant by a laminated core?
12. What is a time constant?
13. How many time constants are required for the current to build to its steady-state value?
14. List three applications for inductors in dc circuits.
15. What is an impedance bridge and what is it used for?

PROBLEM SET: BASIC LEVEL

Section 9–1

1. Will the inductor of Figure 9–25, with the graph of its current as shown, have a voltage induced? Why?

FIGURE 9–25 Inductor and current graph for Problem 1

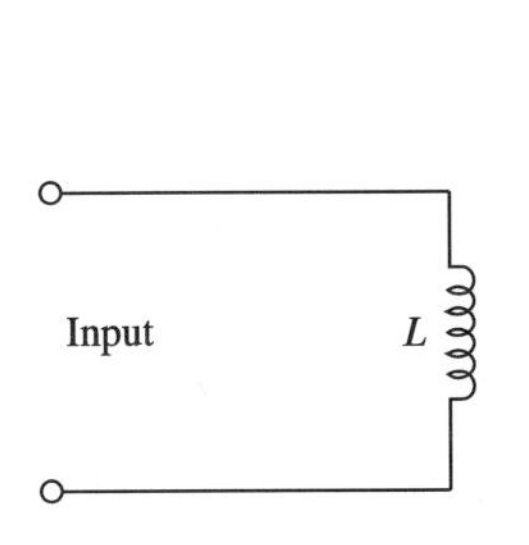

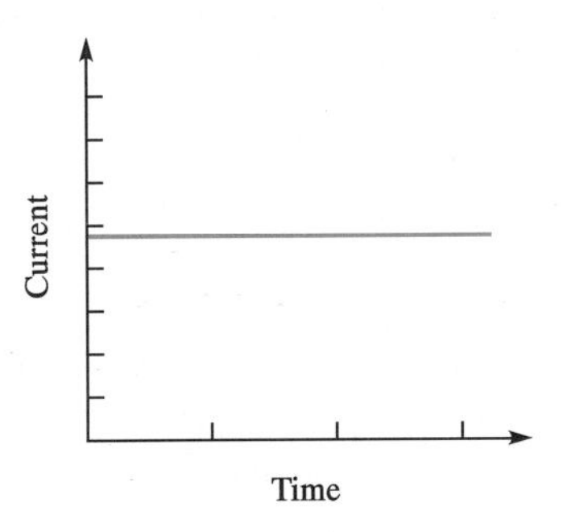

2. Will the inductor of Figure 9–26, with the graph of its current as shown, have a voltage induced? Why?

FIGURE 9–26 Inductor and current graph for Problem 2

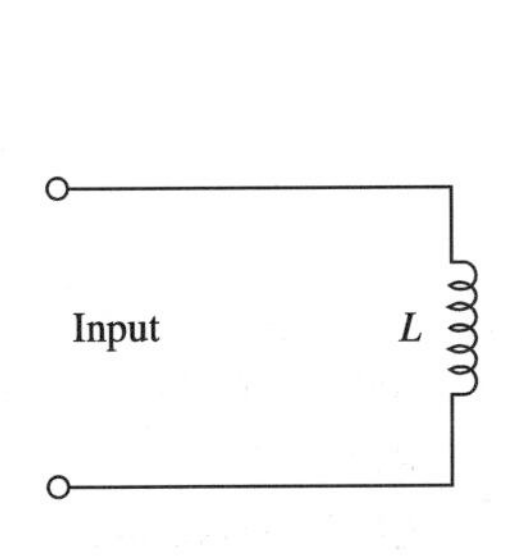

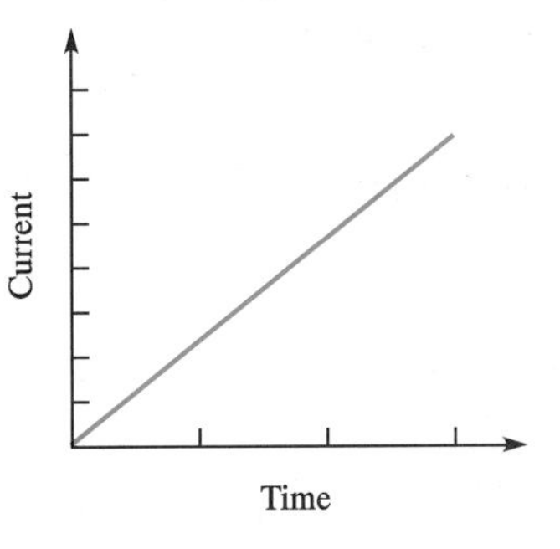

Section 9–2

3. For the circuit of Figure 9–27, what is the polarity of the induced voltage when the switch is first placed in position A?

4. For the circuit of Figure 9–27, what is the polarity of the induced voltage when the switch is placed in position B?

FIGURE 9–27 Circuit for Problems 3 and 4

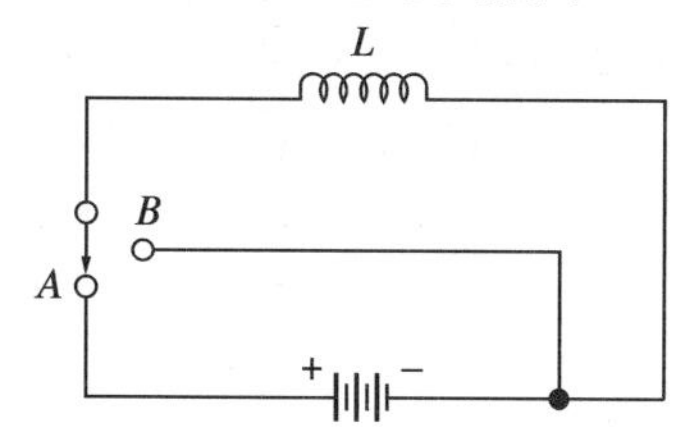

Section 9–3

5. How much voltage is induced when a change of 20 mA occurs in 100 μs across a 30 mH inductor?

6. How much voltage is induced when a change of 100 mA occurs in 20 μs across a 30 mH inductor.

7. What is the rate of change ($\Delta i/\Delta t$) if a current increases by 100 mA in 5 μs?

8. The graph in Figure 9–28 shows the change of current during the energizing and then de-energizing of the inductor. If the voltage produced during the energizing period is 15 kV, what is the value of the inductor?

FIGURE 9–28 Graph for Problem 8

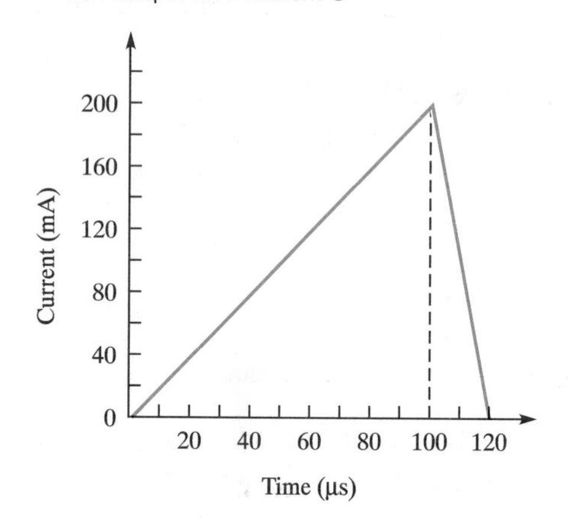

Section 9–4

9. What is the energy stored in joules if 0.5 A of current flows through a 10 H inductor?

10. An inductor stores 15 mJ of energy in its magnetic field when 200 mA of current flows. What is the value of the inductor?

Section 9–5

11. An inductor with 10 turns is wound on a core with an area of 0.03 m and a length of 0.05 m. If the absolute permeability of the core is 252 $\times$ 10^{-8} T/A · t/m, compute the value of the inductance.

12. An inductor with 1000 turns has an inductance of 200 mH. If the number of turns is reduced to 500, what is the value of the inductance?

13. An inductor with 50 turns has an inductance of 10 mH. If the number of turns is increased to 100, what is the value of the inductance?

14. Compute the inductance of the inductor shown in Figure 9–29.

FIGURE 9–29 Inductor for Problem 14

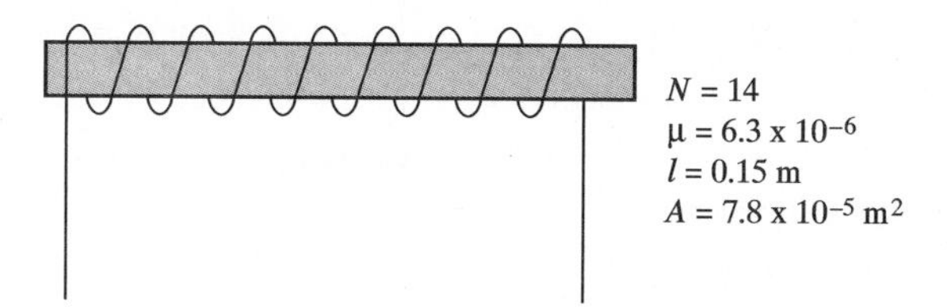

Section 9–7

15. Compute the total inductance of each of the circuits of Figure 9–30.

FIGURE 9–30 Circuits for Problem 15

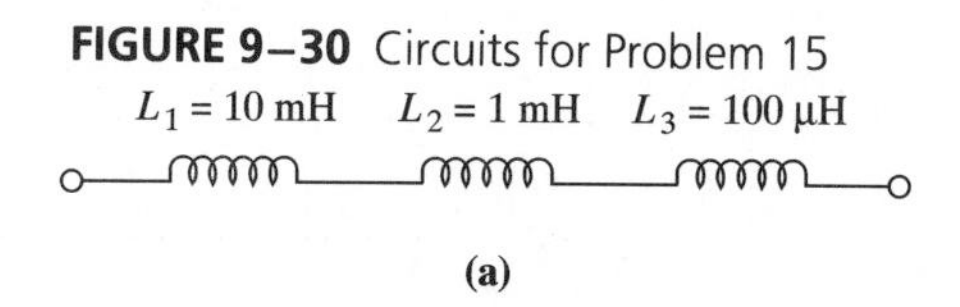

(a)

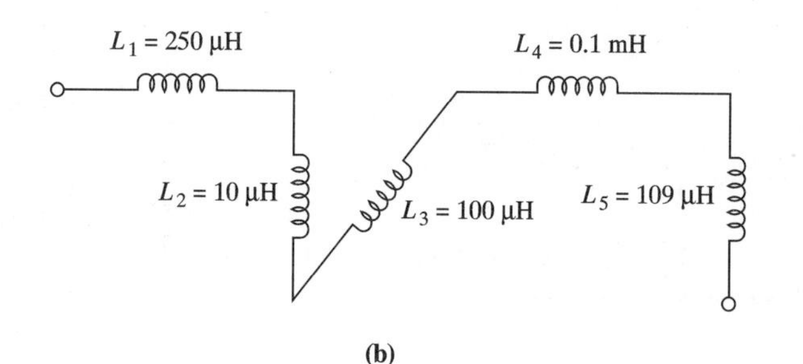

(b)

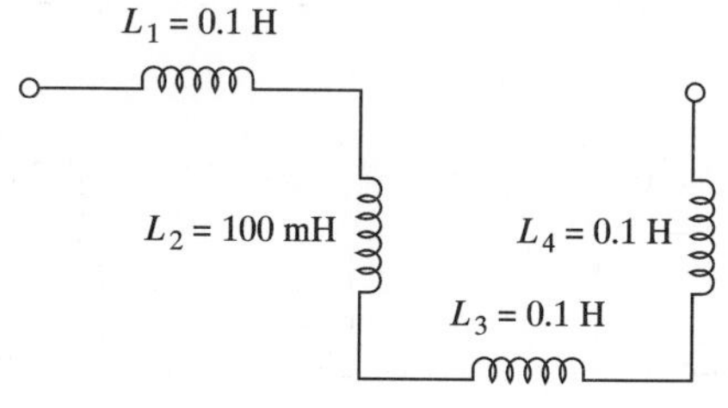

(c)

16. For the circuit of Figure 9–31, what is the value of L_4?

FIGURE 9–31 Circuit for Problem 16

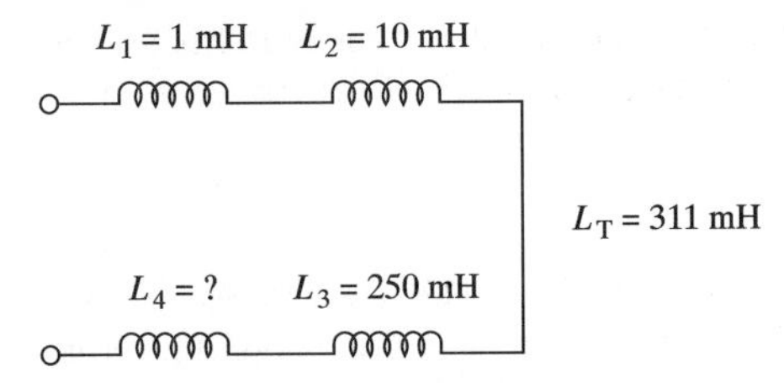

17. Six 30 mH inductors are connected in series. Compute the value of the total inductance.

Section 9–8

18. Compute the total inductance of each of the circuits of Figure 9–32.

FIGURE 9–32 Circuits for Problem 18

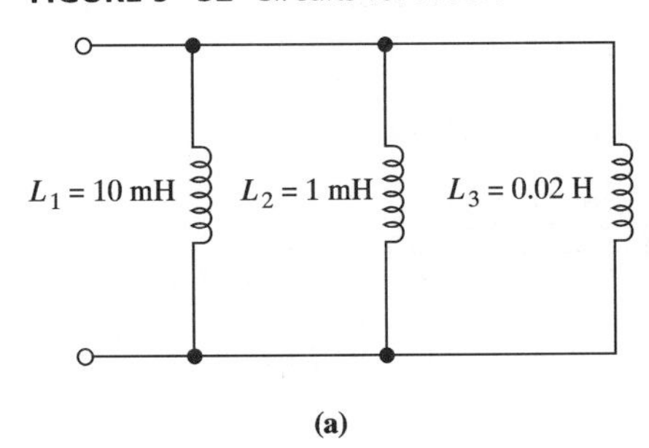

(a)

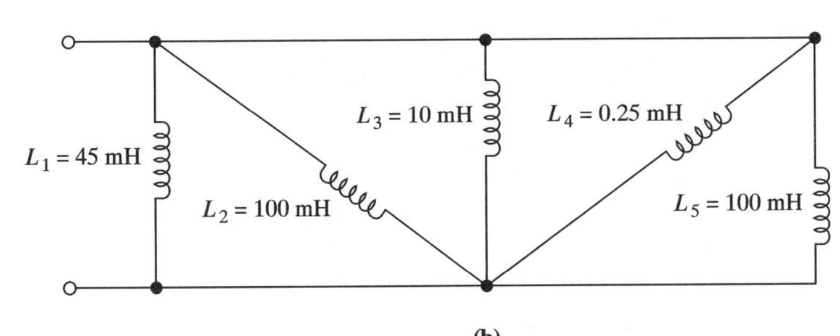

(b)

19. Using the most efficient method, compute the total inductance of the circuit of Figure 9–33.

FIGURE 9–33 Circuit for Problem 19

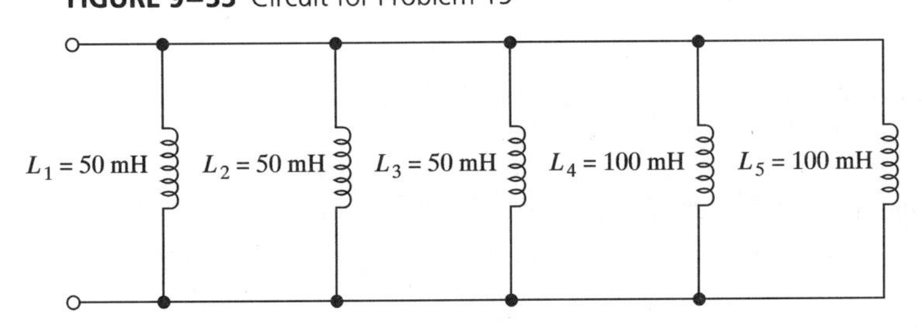

20. The total inductance of the circuit of Figure 9–34 is 10 mH. Compute the value of L_2.

FIGURE 9–34 Circuit for Problem 20

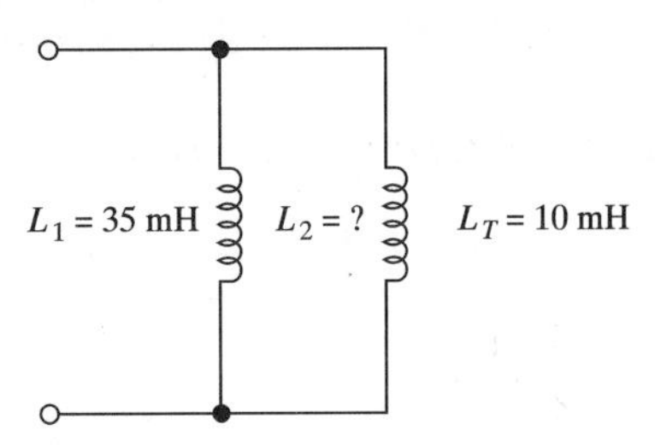

Section 9–9

21. For the circuit of Figure 9–35, compute the steady-state current that flows after the inductor is completely energized.

FIGURE 9–35 Circuit for Problem 21

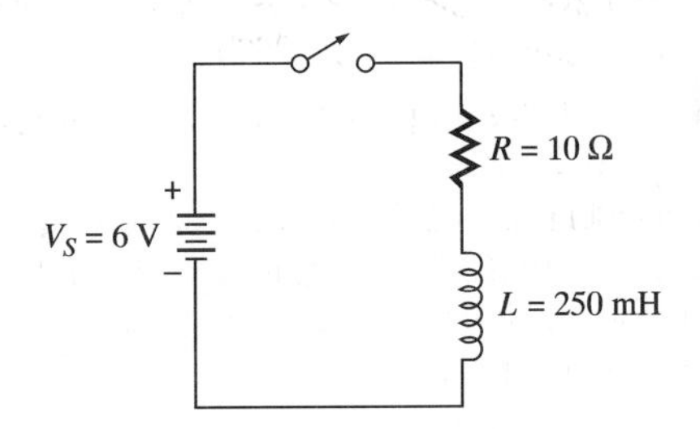

22. For the circuit of Figure 9–36, compute the time constant.

FIGURE 9–36 Circuit for Problem 22

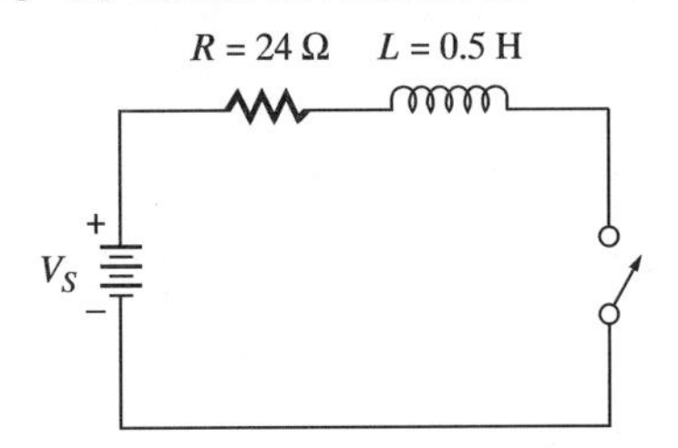

23. For the circuit of Figure 9–37, compute how long it will take the inductor to completely energize.

FIGURE 9–37 Circuit for Problem 23

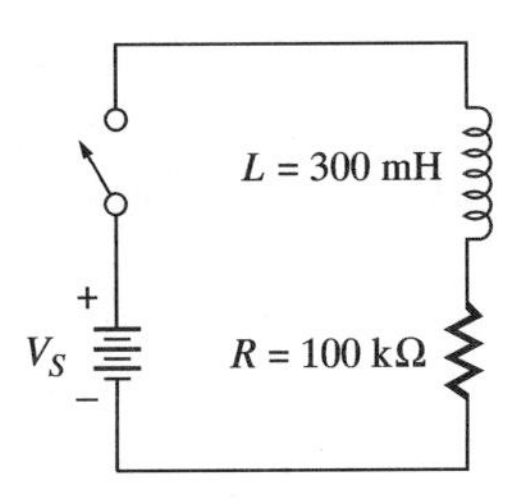

PROBLEM SET: CHALLENGING

24. An inductor has an inductance of 5 H. If the number of turns is doubled, what is the new value of inductance?

25. For the circuit of Figure 9–38, compute the current 35 ms after the switch is closed.

FIGURE 9–38 Circuit for Problem 25

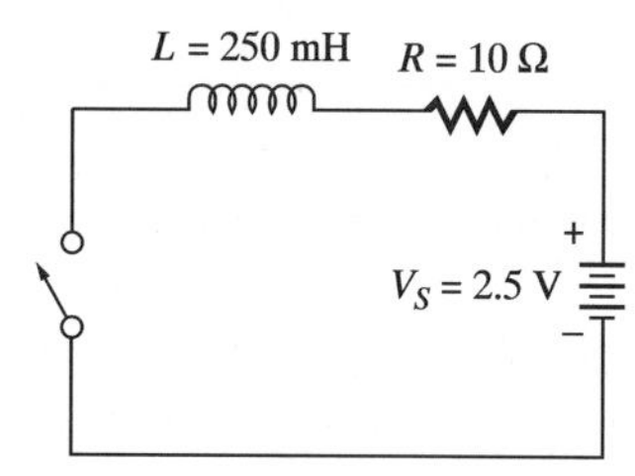

26. For the circuit of Figure 9–39, compute the total inductance.

FIGURE 9–39 Circuit for Problem 26

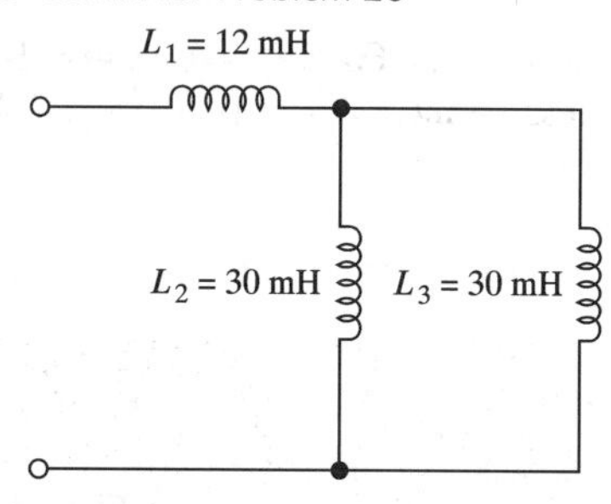

27. For the circuit of Figure 9–40, compute the total inductance.

FIGURE 9–40 Circuit for Problem 27

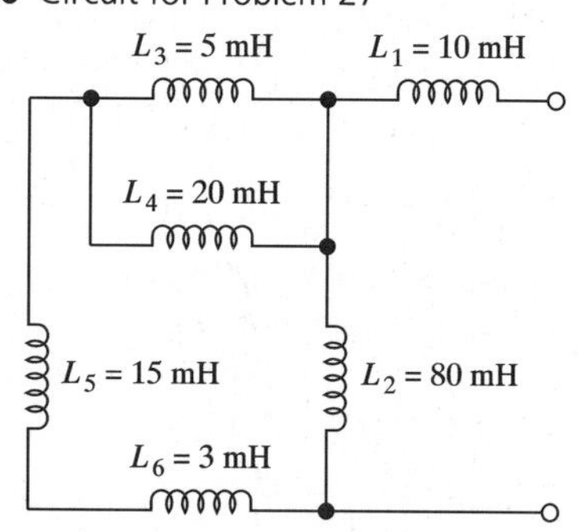

28. In Figure 9–41, if the total inductance is 5.45 mH, compute the value of L_2.

FIGURE 9–41 Circuit for Problem 28

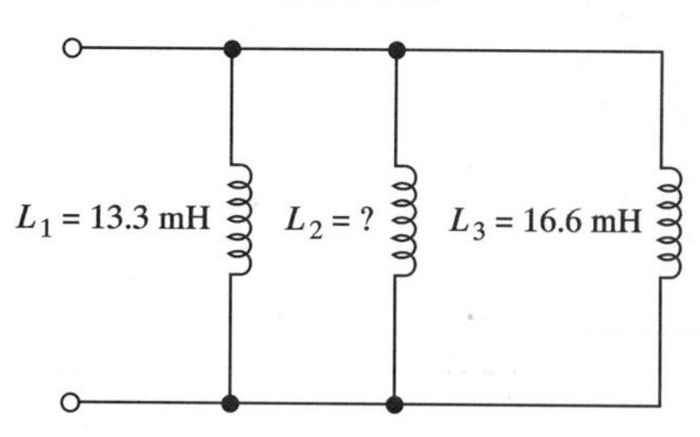

29. If the total inductance in Figure 9–42 is 498 μH, compute the value of L_1.

FIGURE 9–42 Circuit for Problem 29

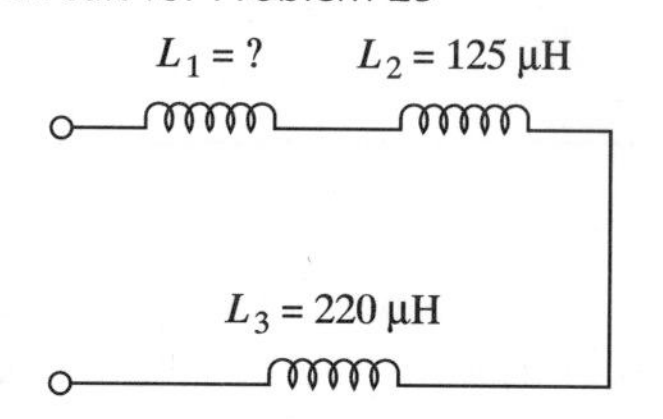

30. In the circuit of Figure 9–43, the switch is first placed in position *A* for 2 s and then placed in position *B* for 3 s. What is the final current?

FIGURE 9–43 Circuit for Problem 30

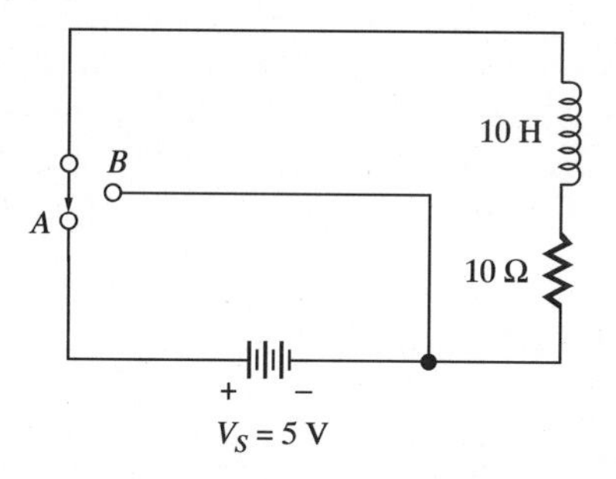

31. Figure 9–44(a) shows a graph of the current applied to the inductor of Figure 9–44(b). Compute the value of the induced voltage: (a) over the period when the current is increasing and (b) over the period when the current returns to zero.

FIGURE 9–44 Current graph and inductor for Problem 31

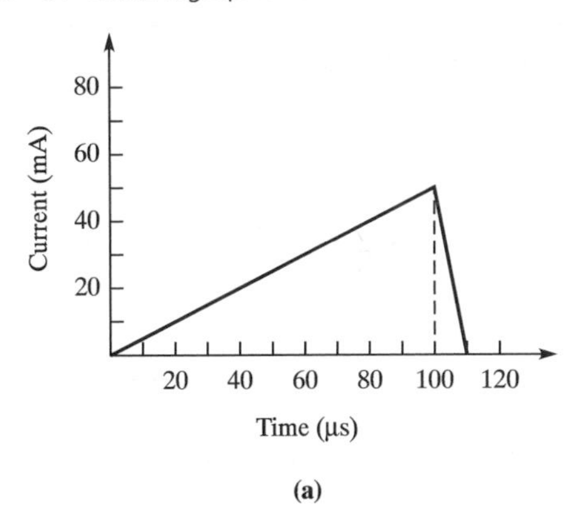

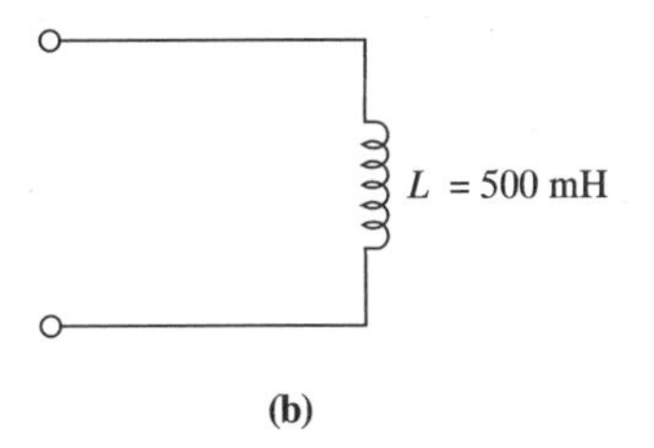

32. An inductor produces an induced voltage of 58 mV when the current varies at a rate of 29 mA/s. What is the value of the inductance?

33. An inductance of 250 μH produces an induced voltage of 1 V. What is the rate of current change in A/s?

PROBLEM SET: TROUBLESHOOTING

34. If the ohmmeter in Figure 9–45 reads as shown when placed across an inductor, what is the probable fault with this inductor?

FIGURE 9–45 Ohmmeter reading for Problem 34

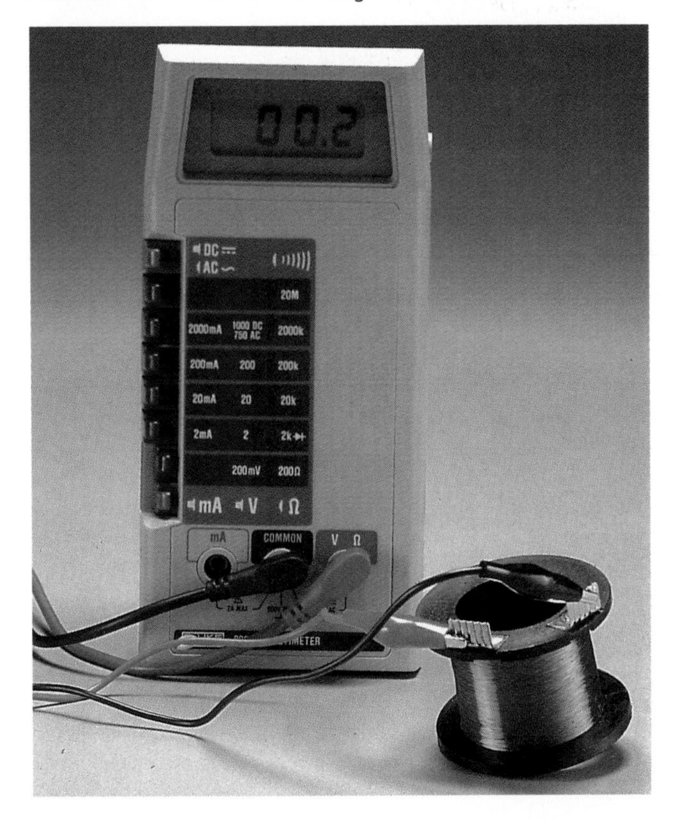

35. If the ohmmeter in Figure 9–46 reads as shown when placed across an inductor, what is the probable fault with this inductor?

FIGURE 9–46 Ohmmeter reading for Problem 35

36. What should the meter of Figure 9–47 indicate if no fault exists in the inductor?

FIGURE 9–47 Circuit for Problem 36

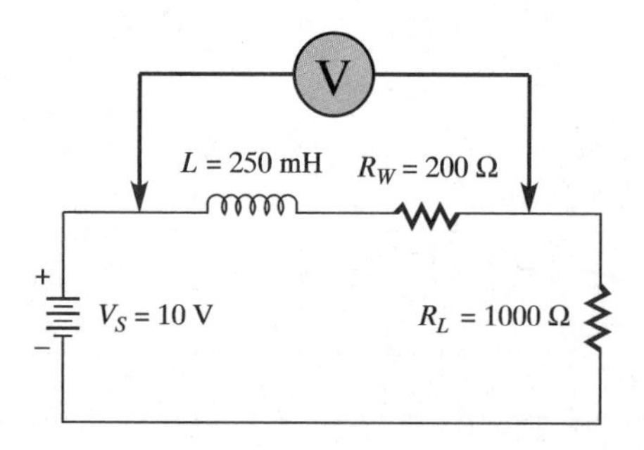

37. The total inductance of the circuit in Figure 9–48 is measured and found to be 40 mH. Is this value correct? If not, what is the probable fault?

FIGURE 9–48 Circuit for Problem 37

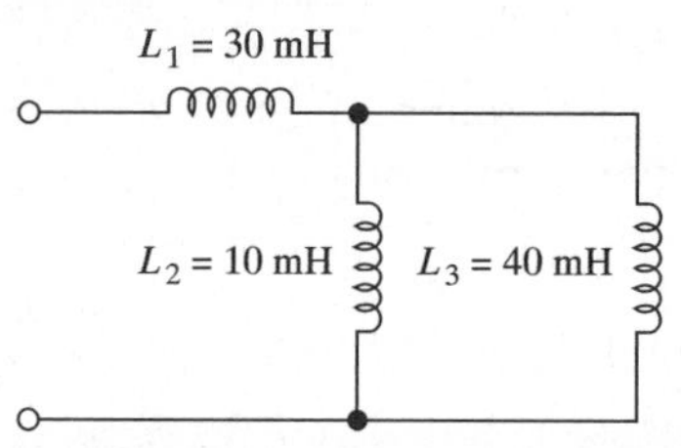

38. A 200-turn inductor with a marked value of 30 mH is measured and found to have an actual value of 15 mH. If the inductor is intact with no visible signs of damage, what is the probable fault?

39. The total inductance of the circuit of Figure 9–49 is measured and found to be 200 μH. What is the probable fault in this circuit?

FIGURE 9–49 Circuit for Problem 39

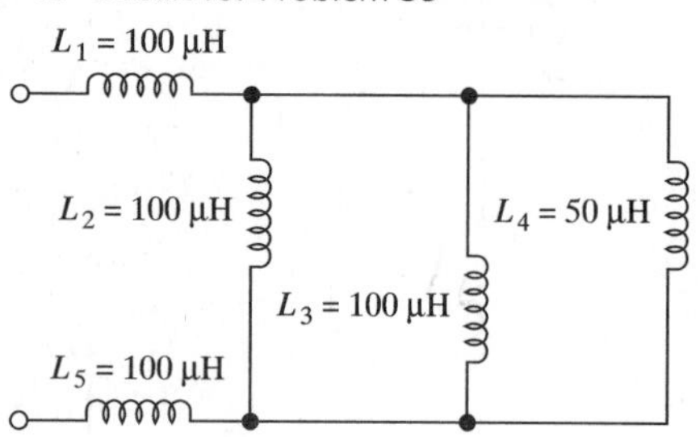

Documentation

There is the story told of a company that bought a digital computer back in the early seventies. The computer, when it arrived, was unloaded from a semitrailer using fork lifts. After it was unloaded, the driver handed the shipping and receiving clerk a small box and said: "This is the documentation." Twenty years later, the company replaced their computer with a more modern system. A semitrailer still delivered the computer. But now, the driver handed the clerk a box and said: "Here is your computer." He then proceeded to open the semitrailer and unload the documentation.

The preceding story is of course an exaggeration—but only to a degree! Electronic equipment has in fact become vastly smaller over the years while the supporting documentation has vastly increased. It is a fact that the most useful piece of equipment ever known is useless without the proper documentation. Documentation includes instructions on how to put the equipment into operation. It also explains all the options for using the equipment. It also provides support information of various types, such as mechanical drawings, schematics, and service instructions.

The material received by the user is really only the tip of the iceberg as far as documentation is concerned. Back at the manufacturer's facility, there is a tremendous amount of documentation. The documentation process begins when the equipment is first conceived and continues through the design, testing, breadboarding, and final assembly. As a technician, you can't get away from documentation

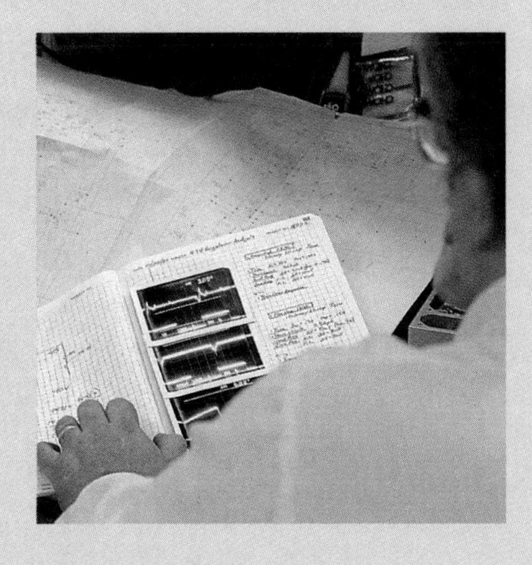

whether you work for the manufacturer or the user. Working for the manufacturer, you will develop equipment documents, and working for the customer, you will use equipment documents.

Documentation for a technician usually begins with maintaining what is known as a Technician's Notebook. If employed as a technologist or associate engineer, you may work directly under an electrical engineer in the design and testing of new systems. You will be required to maintain a notebook in which you record each test performed, the results of each test, and the date upon which the test was performed. This notebook is a bound document in which the pages are numbered consecutively and cannot be removed. All entries must be handwritten and no erasures are allowed. If a mistake is made, that entry must be lined through in a manner such that it can still be read. Each day's entries must be attested to by the signatures of two technicians who saw the test performed and the results recorded.

Why all this care in preparation of the Technician's Notebook? Any number of companies may be working on similar systems to fill some users' needs. If a new concept or technology is discovered, a patent may be issued that enhances the value of the discovery. Suppose there is a difference of opinion over who discovered the concept first? This issue will be decided through litigation in court, and the Technician's Notebook, with its dated entries, may become an exhibit in a court of law!

Entries made in the Technician's Notebook consist of more than just drawings and figures. They also include a written narrative of what transpired during testing. This information includes the items of test equipment used and their serial numbers, how the test was performed, and the results of the test. Recalling once again that the notebook could become a part of litigation, it is essential that entries be clearly written in unambiguous language that projects the desired meaning. These entries will also be used by the technical writer who completes the documentation of the system. This person does not want to be overwhelmed with data, but would like to have more than needed rather than not enough. Another purpose of the notebook is to ensure that the company does not have to perform the same work again. Anyone needing the information on such a test would have it available.

The technician must also assist in preparation of operating procedures and maintenance and service instructions. These are of utmost importance to the user. More customer frustration can result from lack of good instructions than from equipment that may fail occasionally. After all, if the documentation is good, the equipment can be quickly brought back into operation. If the documentation is poor, the equipment will be out of commission and the user may be unable to repair it. The latter will cause much more dissatisfaction than the former!

CHAPTER

10 Capacitance in DC Circuits

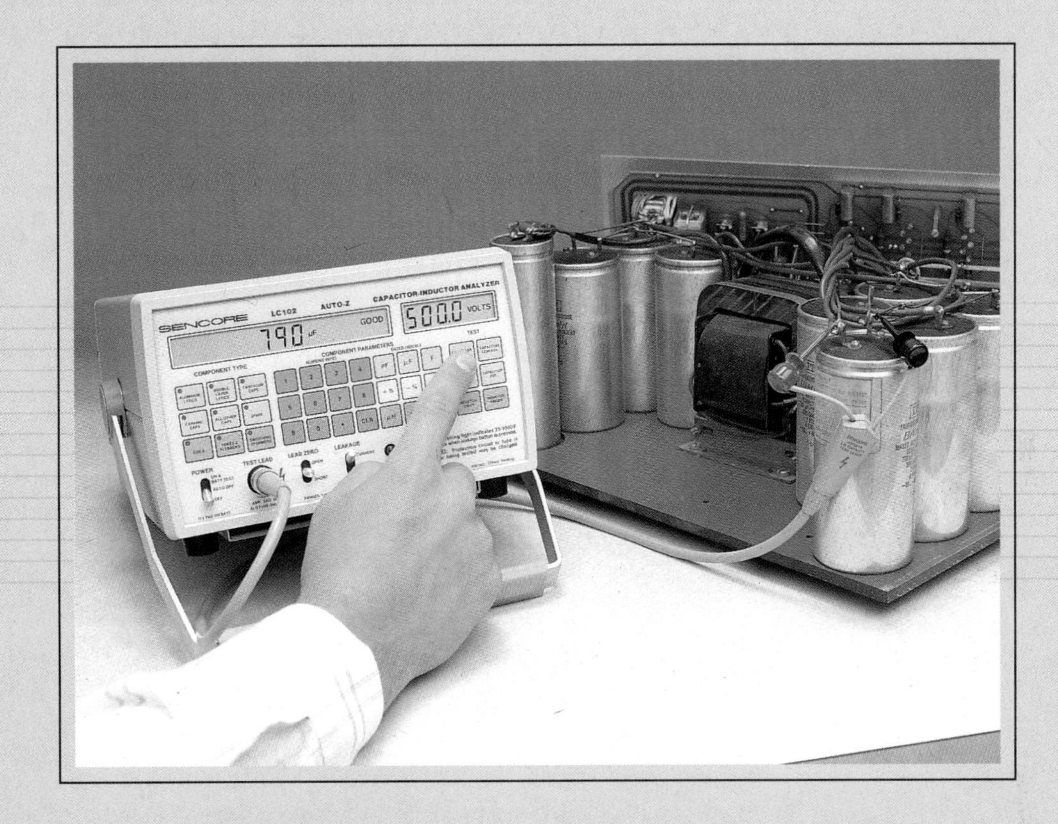

■ UPON COMPLETION OF THIS CHAPTER, YOU WILL BE ABLE TO

1. Define capacitance in two ways.
2. Sketch the construction of the basic capacitor, and describe how it is charged and discharged.
3. Write the symbol and unit of measure for capacitance.
4. Relate the amount of capacitance to the charge stored and the voltage developed across a capacitor.
5. Describe the physical characteristics of a capacitor that determine its capacitance.
6. Know the various types of capacitors, the value ranges in which they are available, and their labeling to include working voltage, temperature coefficient, and capacitance value.
7. Compute the total capacitance of capacitors in series and in parallel.
8. Define the farad in terms of the charging current and the rate of change of voltage with respect to time.
9. Compute the voltage across series capacitors using the capacitive voltage divider equation.
10. Define and compute the *RC* time constant.
11. List typical applications of capacitors.
12. Describe the process of troubleshooting capacitors.

One way of describing the content of this chapter is "Two down and one to go!" In the previous chapters, you studied the circuit properties of resistance and inductance. The circuit property of capacitance is presented in this chapter. These three circuit properties, although being vastly different in their operation, have one thing in common: They are used to control current. Thus, all electronic systems utilize resistors, inductors, and capacitors.

You will find in studying this chapter that capacitance and inductance are opposites. Recall that an inductor produces a magnetic field. In this chapter, you will learn that a capacitor produces an electrostatic field. An inductor opposes a change in current, while a capacitor opposes a change in voltage. Current must flow continuously in order to produce the magnetic field of an inductor, but the electrostatic field of a capacitor is the result of static electricity or stored charge. The inductor is a short circuit that at times appears as an open circuit; the capacitor is an open circuit that at times appears as a short circuit. Other differences between inductors and capacitors will be covered in later chapters. These two circuit devices do have one thing in common, however: Neither dissipates electrical energy, but in fact stores it.

The capacitor is an open circuit because it is basically two conductors separated by an insulator. The thought of deliberately installing an open in an electric circuit often boggles the mind of beginning students of electronics! It is done, however, and the results are at times quite intriguing. The capacitor is not an ordinary open circuit, but a very special one with very special properties. You will learn of these properties and some of the applications to which they are put in this chapter. To reiterate, capacitance is one of the three properties of an electric circuit. Understanding it is an essential part of your technical knowledge.

10–1 CAPACITANCE

Question: What is capacitance?

Charge Q, which was introduced in Chapter 2, is an excess or deficiency of electrons in a material. Recall that the result of separation of charges is an electromotive force, or voltage, produced between the charged bodies. **Capacitance** is the property of a circuit that stores electrical charge. The symbol for capacitance is C, and its unit of measure is the farad (F). Capacitance is introduced into a circuit in the form of a device known as a **capacitor.**

Capacitance exists between any two conductors separated by a **dielectric,** or insulating material between the plates of a capacitor. A dielectric is basically an insulator. Although not capable of conducting electric current, a dielectric can be electrically polarized.

FIGURE 10–1 Sources of capacitance in a circuit

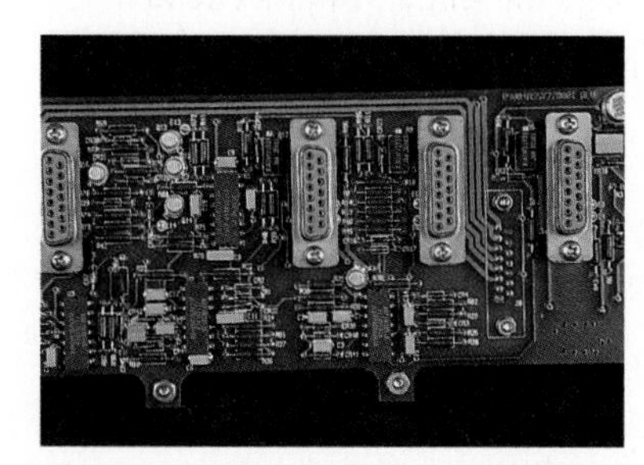

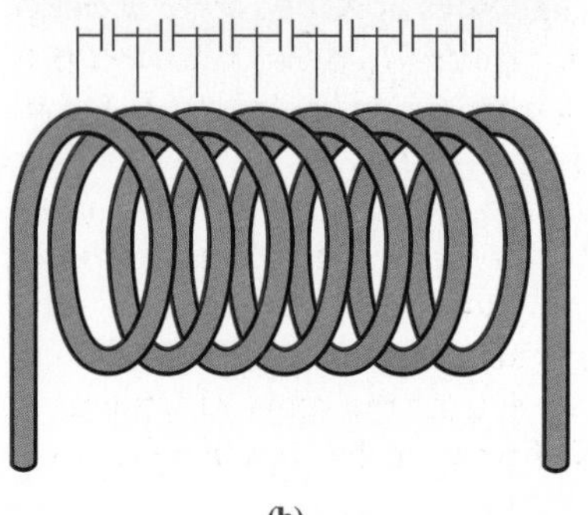

(b)

Question: What is meant by distributed capacitance?

The capacitance that exists when two conductors are separated by a dielectric can occur at many points in a circuit. For example, in the printed circuit board of Figure 10–1(a), capacitance exists between any of the traces separated by the insulation of the board. As illustrated in Figure 10–1(b), capacitance also exists between two insulated wires, such as those making up inductors. As you learned in Chapter 9, an inductor is insulated wire wound into a coil. The capacitance that occurs between each turn of an inductor is known as *distributed capacitance.* Distributed capacitance is usually unplanned for and unwanted, although some special circuits, which you will encounter later in the electronics curriculum, do make use of distributed capacitance.

FIGURE 10–2 Basic capacitor

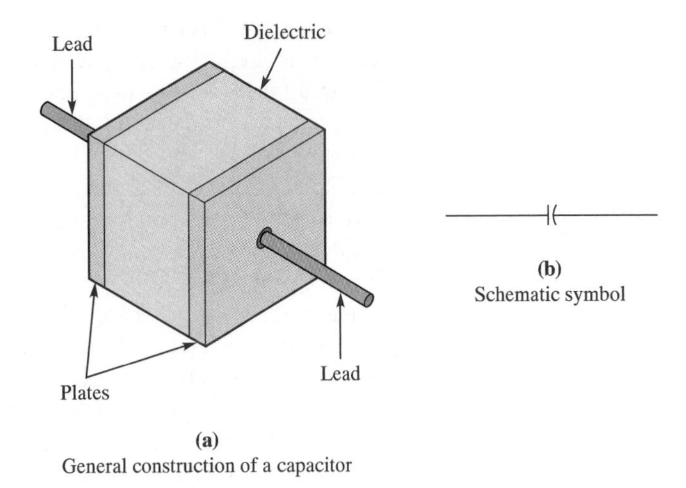

(a)
General construction of a capacitor

(b)
Schematic symbol

Question: What is meant by lumped capacitance?

The capacitance introduced into a circuit using a capacitor is known as *lumped capacitance*. The basic capacitor and its schematic symbol are shown in Figure 10–2. Two metal conductive plates with connecting leads are separated by a dielectric material. The dielectric material is an insulator such as paper, glass, mica, ceramic, or air. In actual practice, the plates are most commonly metal foil.

■ SECTION 10–1 REVIEW

1. What is capacitance?
2. What is the symbol and unit of measure for capacitance?
3. What is a dielectric?
4. What is the device called through which capacitance is introduced into a circuit?
5. What is meant by distributed capacitance and lumped capacitance?
6. What is an example of distributed capacitance?
7. Sketch the construction of a basic capacitor and its schematic symbol.

10–2 ELECTRIC CHARGES AND THE CAPACITOR

■ Charging the Capacitor

Question: How is a capacitor charged?

In charging a capacitor, electrons must be moved from one plate to the other. The movement of electrons in a circuit is known as current, so current must flow in order to charge a capacitor. This current flow ceases once the capacitor is fully charged. If current is to flow, there must be a voltage to produce it. Thus, a capacitor is charged through placing a voltage source between its plates. As shown in Figure 10–3(a), the capacitor is connected to a voltage source. With the switch open and the capacitor uncharged, the plates have equal numbers of electrons and protons. As illustrated in Figure 10–3(b), the closing of the switch places a dc voltage source across the capacitor. Electrons on the left-hand plate of the capacitor are attracted toward the positive side of the source. This loss of electrons by the left-hand plate causes electrons to move from the negative terminal of the battery to the right-hand plate. Thus, the left-hand plate of the capacitor begins to

FIGURE 10–3 Charging a capacitor

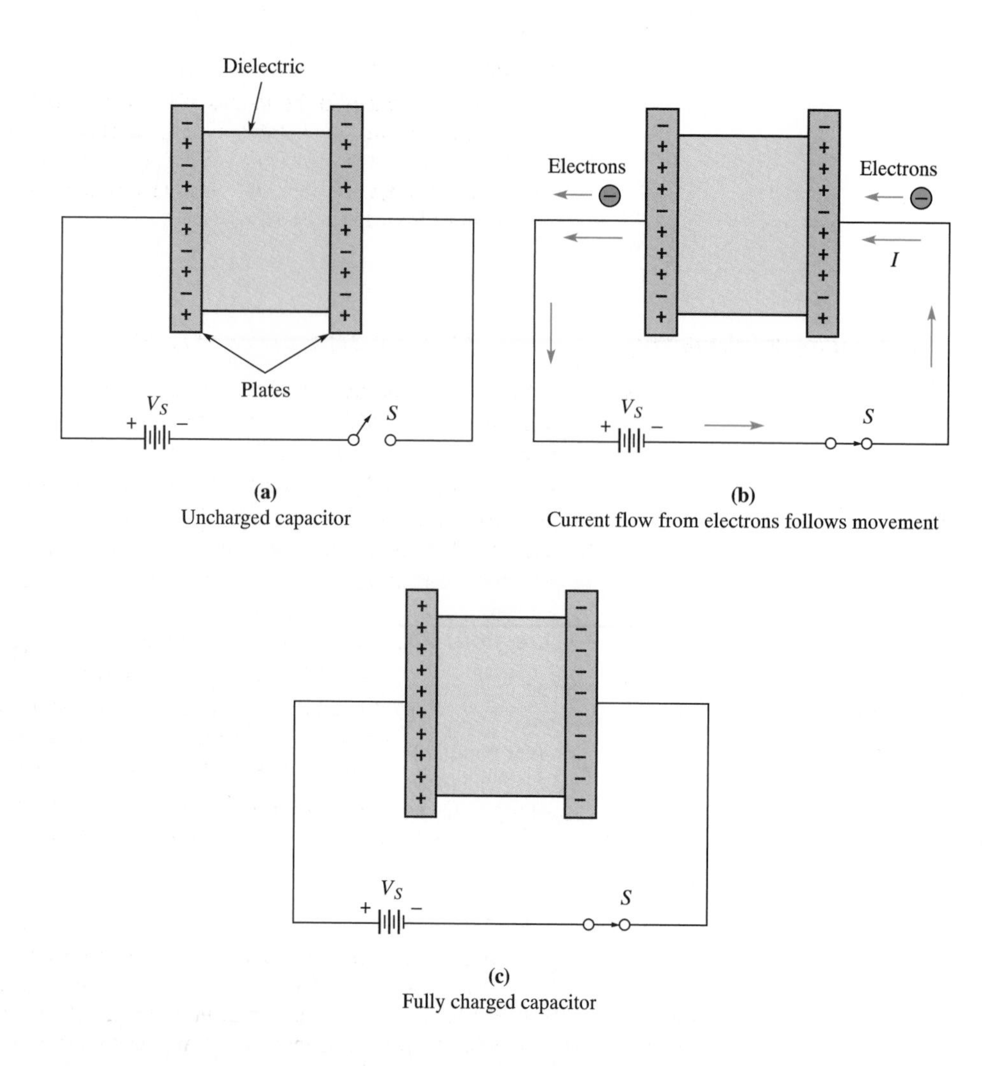

(a)
Uncharged capacitor

(b)
Current flow from electrons follows movement

(c)
Fully charged capacitor

become positively charged, while the right becomes negatively charged. With this storing of charge, a voltage begins to build across the capacitor.

Charging of the capacitor is accomplished through current flow. As electrons flow off the left-hand plate and into the right-hand plate, the voltage across the capacitor continues to build. A point is reached where the voltage across the capacitor is equal to the source voltage. As shown in Figure 10–3(c), the polarity of the voltage across the capacitor is series opposed to that of the source, so once the voltages are equal the current flow ceases. At this time, the capacitor is said to be charged. Even though current flows in the circuit during the storage of charge, it is from one plate to the other through the source, not through the capacitor itself. This current is sometimes called *displacement current* because it describes the behavior of the electrons.

Note carefully the sequence of charging the capacitor: When the switch is first closed, the voltage across the capacitor is zero because no voltage is present until charge has been stored. (At this time, the voltage is across the internal resistance of the source.) As the charging current flows into the capacitor, the voltage across it increases. A point is reached at which the capacitor is fully charged and the voltage across it is equal to the source voltage. Thus, a second and very important definition of capacitance is that capacitance is the property of a circuit that opposes a change in voltage. (Recall that inductance was defined as that property of a circuit that opposes a change in current.)

NOTE

Since differences in the actions of inductance and capacitance are being emphasized here, notice that current must flow in order to charge the capacitor and to energize the inductor. Notice, however, that once the capacitor charges, the current stops. On the other hand, when the magnetic field is formed around the inductor, the current must continue to flow in order for it to stay energized.

Discharging the Capacitor

Question: How is a capacitor discharged?

If the voltage source is removed from a charged capacitor, it remains charged as shown in Figure 10–4(a). The charge remains because there is no conducting path for the electrons to move from the right-hand to the left-hand plate. The charge remains stored as long as this condition exists. If a conducting path is provided, as shown in Figure 10–4(b), electrons flow from the left-hand to the right-hand plate in order to balance the charges. Just as in the charging process, current also flows in the discharging process. When there is no longer an excess or deficiency of electrons in the plates, the capacitor is said to be discharged, and the current flow ceases.

Care must be taken in manually discharging a capacitor. A large-value capacitor stores a large amount of electrical energy. If a short circuit is placed across such a capacitor, a large amount of current will flow. This current may be large enough to cause damage to the capacitor. The power developed could cause the leads connecting the plates to the external circuit to burn open. Thus, in the case of large capacitors it is good practice to connect a small resistor, possibly 100 Ω, across the capacitor when manually discharging it. The resistance will limit the current to a safe value.

CAUTION

Notice that when the voltage source is removed, the capacitor is capable of holding its charge, as is true of all capacitors in electronic systems. A large-value capacitor in a high voltage circuit (for example, in the high voltage section of a television receiver) can be quite dangerous. Capacitors in applications where large voltages may be present should be provided with means for automatic discharging when the system is de-energized. However, it is a good idea to manually discharge such capacitors prior to working on the circuit. Where your safety is concerned, never completely trust automatic safety devices!

FIGURE 10–4 Discharging capacitor

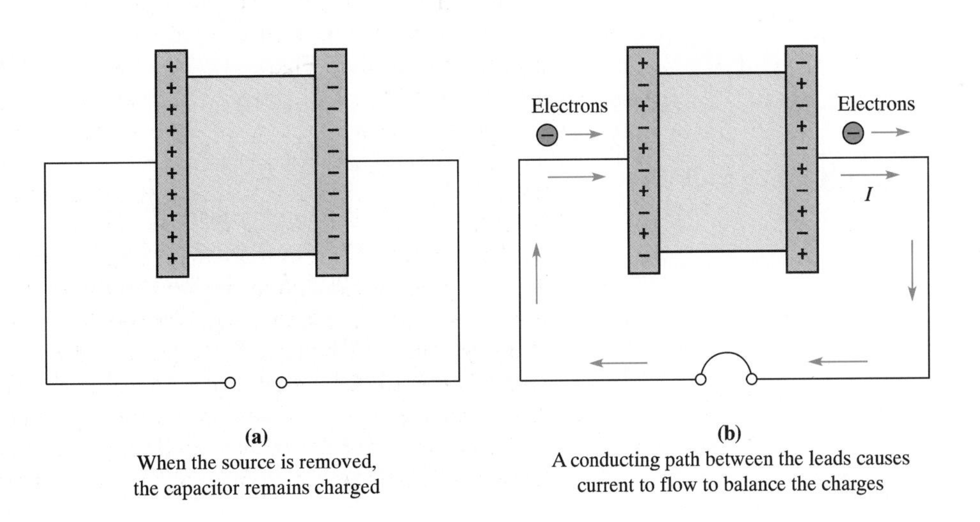

■ Formation of the Electrostatic Field

In Chapter 2, you learned that when bodies are charged, a force field known as an electrostatic field is formed between them. These lines of electrostatic force emanate from the positively charged body and terminate on the negatively charged body. They do not form loops as is the case with magnetic flux lines. As shown in Figure 10–5, this electrostatic field is formed in the dielectric material. Its strength is proportional to the amount of charge stored and the permittivity of the dielectric material. **Permittivity,** which is similar to permeability in magnetic materials, is a measure of the ease with which an electrostatic field can be established in a dielectric.

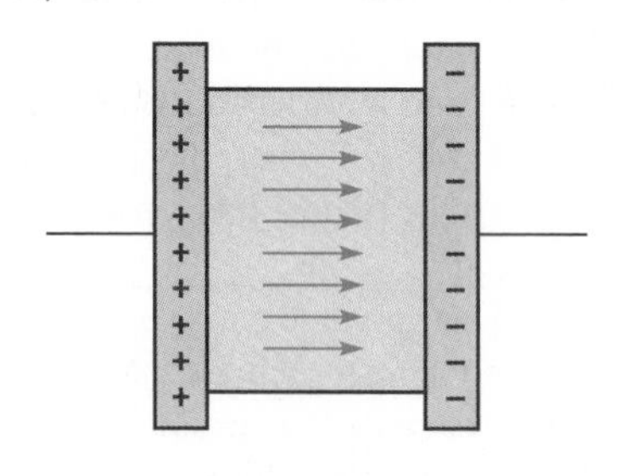

FIGURE 10–5 Electrostatic field formed in the dielectric of a charged capacitor

Question: How is the dielectric of a capacitor electrically polarized?

As shown in Figure 10–6(a), the electrons in the atoms of the dielectric material are normally in circular orbits. When charge is stored, however, the electrons enter an elliptical orbit as shown in Figure 10–6(b). They are attracted by the positively charged plate and repelled by the negatively charged plate. This distortion of electron orbits is what is known as *electrical polarization* of the dielectric.

If this attraction and repulsion become great enough, electrons are pulled completely out of orbit and become free electrons, or charge carriers. This ionization process results in electrons jumping from ion to ion, and current flow is established through the dielectric. The result is breakdown of the dielectric, and the capacitor may be destroyed because too much voltage is placed across it.

FIGURE 10–6 Electrical polarization of the dielectric

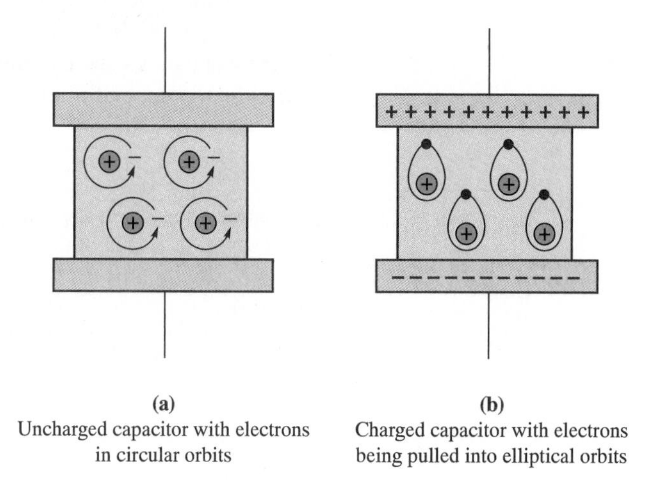

(a) Uncharged capacitor with electrons in circular orbits

(b) Charged capacitor with electrons being pulled into elliptical orbits

■ SECTION 10–2 REVIEW

1. Explain how a capacitor is charged.
2. Why does current flow cease when the capacitor has become fully charged?
3. Why does current flow in charging a capacitor even though the capacitor is an open circuit?
4. How is a capacitor discharged?
5. What is the nature of the field formed between the plates of a capacitor when it is charged?
6. What determines the strength of the electrostatic field between the plates?
7. What is meant by electrical polarization of the dielectric?
8. What can happen if the stress on the dielectric becomes too great?

10–3 CAPACITANCE AND POWER

As mentioned previously, the one thing that capacitance has in common with inductance is the ability to store, rather than dissipate, electrical energy. A "perfect" capacitor would be one in which no resistance existed and, as a result, no power was developed. The perfect capacitor does not exist in practice due to the resistance of the connecting leads and of the plates themselves. This resistance results in a small energy loss. For the most part, however, the energy is stored and returned to the source when the capacitor is discharged.

When a capacitor is charged, an electrostatic field is developed between the plates. The electrical polarization of the dielectric is the result of electrons in the atoms of the material being stressed from circular into elliptical orbits. This process requires energy, which is not dissipated, but stored in the electrostatic field. When the capacitor is discharged, the dielectric loses its electrical polarization, and the stored energy is returned to the source.

The amount of energy stored in a capacitor can be computed using the following equation:

(10–1)

$$E = Q\frac{V}{2}$$

WHERE: E = energy (J)
Q = charge (C)
V = voltage (V)

This equation states that the energy stored (in joules) is equal to the product of the charge (in coulombs) and half the applied voltage (in volts).

EXAMPLE 10–1 A capacitor stores 50 mC of charge with 6 V applied. Compute the energy stored in joules.

■ Solution:
Use Eq. 10–1:

$$E = Q\frac{V}{2}$$

$$= (50 \times 10^{-3}\text{ C})\frac{6\text{ V}}{2} = 0.15\text{ J}$$

EXAMPLE 10–2 A capacitor stores 250 mJ of electrical energy. If 10 V are applied, how much charge is stored?

■ Solution:
Transpose Eq. 10–1 to solve for Q:

$$E = Q\frac{V}{2}$$

$$2E = QV$$

$$Q = \frac{2E}{V}$$

$$= \frac{2 \times 0.25 \text{ J}}{10 \text{ V}} = 50 \text{ mC}$$

SECTION 10–3 REVIEW

1. Where and in what form is the energy stored in a capacitor?
2. Why is the "perfect" capacitor a theoretical concept only?
3. What factors determine the amount of energy stored in a capacitor?
4. A capacitor stores 600 micro C of charge with 50 mV applied. Compute the energy stored in joules.

10–4 THE UNIT OF CAPACITANCE

The unit of measure for capacitance is the farad (F), named for Michael Faraday whose work was introduced in a previous chapter. The **farad** can be described in two ways: (1) in terms of the charge stored and amount of voltage produced, or (2) in terms of the amount of current that flows when a capacitor charges.

Question: How is the farad defined in terms of the amount of charge stored and the voltage produced?

If 1 V is placed across the plates of a capacitor and it stores 1 C of charge, then its capacitance value is 1 F. Thus, capacitance in farads is found by the following equation:

(10–2)

$$C = \frac{Q}{V}$$

WHERE: C = capacitance (F)
Q = charge (C)
V = voltage (V)

Capacitors in the farad range are rare, but they are sometimes found in computer applications. In fact, the millifarad is also a quite large unit. In practice, most capacitance values encountered are expressed in micro or pico units.

EXAMPLE 10–3 If 6 V across a capacitor causes 30 μC of charge to be stored, what is the capacitance value?

■ Solution:

Use Eq. 10–2:

$$C = \frac{Q}{V}$$

$$= \frac{30 \times 10^{-6} \text{ C}}{6 \text{ V}} = 5 \ \mu\text{F}$$

EXAMPLE 10–4 A 470 μF capacitor is across a 100 V source. What is the value of the charge stored?

■ Solution:

Transpose Eq. 10–2 to solve for Q:

$$C = \frac{Q}{V}$$

$$Q = CV$$

$$= (470 \times 10^{-6}\ \text{F})(100\ \text{V}) = 47\ \text{mC}$$

EXAMPLE 10–5 A 10 μF capacitor has 1000 V applied. Compute the amount of charge that is stored.

■ Solution:

$$C = \frac{Q}{V}$$

$$Q = CV$$

$$= (10 \times 10^{-6}\ \text{F})(1000\ \text{V}) = 10\ \text{mC}$$

Question: How is the farad defined in terms of the current that flows during charging?

The current flowing into a capacitor at any instant is equal to the product of the capacitance times the rate of change of voltage with respect to time. In equation form, current flow during charge can be found by

(10–3)

$$i_C = C\frac{\Delta V}{\Delta t}$$

WHERE: i_C is the instantaneous capacitor charge current. Note that this equation is similar to that for the induced voltage produced by an inductor when the current changes. Equation 10–3 can be transposed in order to develop a definition for the farad:

(10–4)

$$C = \frac{i_C}{\Delta V/\Delta t}$$

Thus, if a change of 1 V in 1 s produces 1 A of charging or discharging current, the capacitor has a value of 1 F.

EXAMPLE 10–6 What is the charging current of a 5000 μF capacitor if the rate of change of voltage is 1000 V/s?

■ Solution:

Use Eq. 10–3:

$$i_C = C\frac{\Delta V}{\Delta t}$$

$$= (5000 \times 10^{-6}\ \text{F})(1000\ \text{V}) = 5\ \text{A}$$

EXAMPLE 10–7 The voltage across a capacitor changes from 0 to 15 V in 3.75 ms, and 60 A of charging current flows. Compute the rate of change of voltage and the value of the capacitor.

■ Solution:

1. Find the rate of change:

$$\frac{\Delta V}{\Delta t} = \frac{15 \text{ V}}{3.75 \times 10^{-3} \text{ s}} = 4000 \text{ V/s}$$

2. Use Eq. 10–4 to find the value of C:

$$C = \frac{i_C}{\Delta V/\Delta t}$$

$$= \frac{60 \text{ A}}{4000 \text{ V/s}} = 15{,}000\ \mu\text{F}$$

[6] [0] [÷] [(] [1] [5] [÷] [3] [·] [7] [5] [EXP] [±] [3] [)] [=]

SECTION 10–4 REVIEW

1. Define a farad in terms of the charge stored for a given applied voltage.
2. Define a farad in terms of the charging current produced for a given rate of change of voltage with respect to time.
3. What are the units in which capacitance is usually expressed?

10–5 CHARACTERISTICS OF CAPACITORS

■ Physical Factors Affecting Capacitance

Question: What are the physical factors affecting capacitance?

Just as was the case with inductance, physical factors determine the capacitance of a capacitor. Capacitance is determined by basically three factors: (1) the area of the plates, (2) the distance between the plates, and (3) the permittivity of the dielectric. These physical dimensions are illustrated in Figure 10–7. Each of these physical factors and their effects upon the capacitance value are discussed in the following subsections.

FIGURE 10–7 Physical factors affecting capacitance

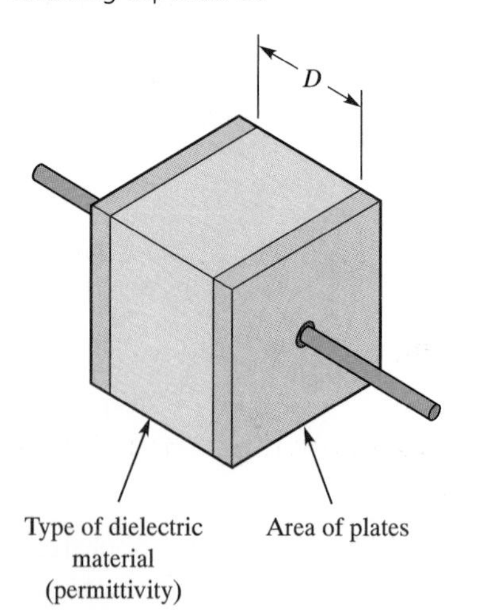

Question: How do the physical factors of plate area, distance between the plates, and dielectric permittivity affect the capacitance value?

Area of the Plates

The capacitance value is directly proportional to the area of the plates. As illustrated in Figure 10–8, larger plates give more area over which to develop an electrostatic field, similar to the way a large container holds a greater quantity of liquid than a small one.

Distance Between the Plates

Capacitance is inversely proportional to the distance between the plates. That is, the closer the plates are spaced, the stronger the electrostatic field produced. As illustrated in Figure 10–9, all other factors being equal, the capacitor in Figure 10–9(a) will have a higher value than that in (b).

Dielectric Permittivity

Just as certain magnetic materials are easier to magnetize than others, it is easier to set up electrostatic fields in some dielectric materials than in others. The **relative permittivity** ε_r, also known as the dielectric constant of a material, indicates the relative ease with which an electrostatic field can be developed in a dielectric material. The relative permit-

FIGURE 10–8 Comparison of plate areas

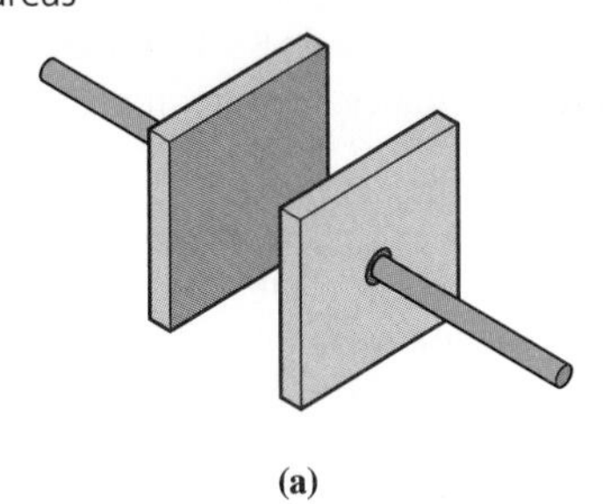

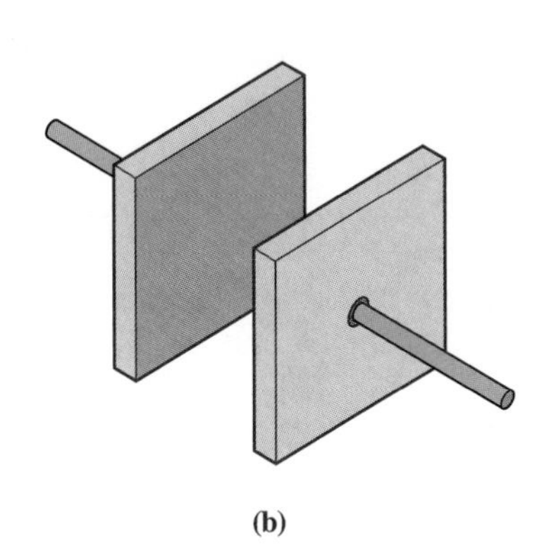

FIGURE 10–9 Comparison of distances between plates

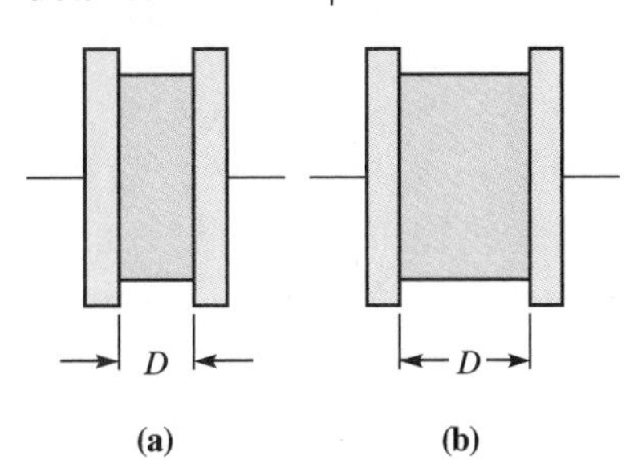

tivity of a material is referenced to that of a vacuum, which has a value of 1. Thus, a material with a relative permittivity of 5 produces 5 times the capacitance value as a vacuum. Table 10–1 contains a list of some dielectric materials and their relative permittivity values.

Computing capacitance requires the absolute permittivity ε of the dielectric. The absolute permittivity of vacuum ε_O is 8.85×10^{-12} farads per meter (F/m). The ε of any other material is found by multiplying its ε_r by ε_O. In equation form,

(10–5) $$\varepsilon = \varepsilon_r \varepsilon_O$$

EXAMPLE 10–8 Compute the absolute permittivity of mylar.

■ TABLE 10–1 Permittivity of Dielectic materials

Material	Relative Permittivity
Vacuum	1
Teflon	2
Polyethylene	2.3
Paper	3.5
Mylar	3
Glass	7
Mica	7
Ceramic	10 to 7500

■ Solution:

Table 10–1 shows the relative permittivity of mylar to be 3. Use Eq. 10–5 to determine mylar's absolute permittivity in farads per meter (F/m).

$$\varepsilon = \varepsilon_r \varepsilon_O$$
$$= 3(8.85 \times 10^{-12})$$
$$= 2.655 \times 10^{-11}\ \text{F/m}$$

Question: How can capacitance be computed using the physical factors that affect capacitance?

The physical factors affecting capacitance are related mathematically in the following equation:

(10–6) $$C = \frac{A\varepsilon}{D}$$

This equation states that the capacitance value of a capacitor in farads, is equal to the product of the area of the plates in square meters (m^2) and the absolute permittivity of the dielectric ($\varepsilon_r \times \varepsilon_O$) divided by the distance between the plates in meters. The area of the plates must be in square meters and the distance between the plates must be in meters.

EXAMPLE 10–9 A capacitor has plates with the dimensions of 0.05 m by 0.02 m. The dielectric is 0.0001 m thick and is made of polyethylene. Compute its capacitance.

■ Solution:

1. Compute the absolute permittivity of polyethylene using Eq. 10–5:

$$\varepsilon = \varepsilon_r \varepsilon_O$$

$$= 2.3(8.85 \times 10^{-12}) = 2.03 \times 10^{-11}\ \text{F/m}$$

[2] [·] [3] [×] [8] [·] [8] [5] [EXP] [±] [1] [2] [=]

2. Use Eq. 10–6 to compute the capacitance:

$$C = \frac{A\varepsilon}{D}$$

$$= \frac{(0.05\ \text{m}^2 \times 0.02)(2.03 \times 10^{-11})}{0.0001\ \text{m}} = 203\ \text{pF}$$

[(] [·] [0] [5] [×] [·] [0] [2] [×] [2] [·] [0] [3] [EXP] [±] [1] [1] [)]

[÷] [1] [EXP] [±] [4] [=]

■ Voltage Rating of Capacitors

Question: Why does a capacitor have a voltage rating?

A capacitor has a voltage rating above which dielectric breakdown may occur. The voltage at which breakdown occurs is sometimes referred to as the *working voltage dc*. As mentioned previously, voltage applied across the capacitor places a stress on the atomic structure of the dielectric. If the stress becomes great enough, electrons completely leave their orbits and break through the dielectric to the positively charged plate. When breakdown occurs in solid dielectrics, a carbon path is formed, and the capacitor is destroyed. Capacitors with air or metal film dielectrics are usually self-healing in this regard.

The safe working voltage of the capacitor is determined by the thickness of the dielectric and its dielectric strength. Table 10–2 lists the dielectric materials and their dielectric strength. They are specified in volts-per-mil (V/mil). (A mil is 1/1000 of an inch.) **Dielectric strength** is the voltage that a 1 mil thickness of a material can withstand without breaking down. Notice that glass, mica, and ceramic can have very high dielectric strength.

■ Temperature Coefficient of Capacitors

Question: What is the temperature coefficient of a capacitor and what is the significance of it?

The capacitance value of some capacitors is affected by a change in temperature. If the capacitance increases as temperature increases and decreases as temperature decreases,

■ **TABLE 10–2 Strength of Dielectic materials**

Material	Dielectric Strength (V/mil)
Vacuum	20
Teflon	1500
Polyethylene	400
Paper	1300
Mylar	400
Glass	700 to 2000
Mica	600 to 2000
Ceramic	500 to 1300

the capacitor is said to have a *positive temperature coefficient.* If, on the other hand, the capacitance decreases with an increase in temperature and increases with a decrease in temperature, it is said to have a *negative temperature coefficient.* Some capacitors remain unchanged in value over a range of temperatures and are said to have a *zero temperature coefficient.*

A capacitor marked P700 has a positive temperature coefficient of 700 parts per million (ppm) per degree Celsius change. For example, a capacitor with a value of 1 μF contains 1 million pF. Thus, the change in value is 700 pF per degree Celsius. Its capacitance value changes by this factor for each degree of temperature change above or below a reference, which is usually 25°C. A label of N500 indicates a negative temperature coefficient in which the capacitance value changes inversely by 500 ppm for every degree of change above 25°C. A label of NP0 indicates that there is no temperature coefficient and that the value should not change with a change in temperature.

EXAMPLE 10–10 A capacitor is labeled 10 μF N2000. If its temperature changes from 25°C to 40°C, what is its new value?

■ Solution:

1. Since the capacitance changes at 2000 ppm, determine the number of "millions":

$$\frac{10 \times 10^{-6}\text{ F}}{1 \times 10^{6}} = 1 \times 10^{-11}\text{ F}$$

[1] [0] [EXP] [±] [6] [÷] [1] [EXP] [6] [=]

2. Multiply this value by 2000, the parts change per million:

$$2000(1 \times 10^{-11}\text{ F}) = 0.02\ \mu\text{F}$$

[2] [EXP] [3] [×] [1] [EXP] [±] [11] [=]

3. Multiply this value by the number of degrees change:

$$(0.02\ \mu F)(15°) = 0.3\ \mu F$$

4. Since this is a negative temperature coefficient, subtract this last value from the rated value of the capacitor:

$$10\ \mu F - 0.3\ \mu F = 9.7\ \mu F$$

Leakage Resistance of Capacitor

Question: Will a charged capacitor hold its charge indefinitely after the voltage source is removed?

A charged capacitor will not hold its charge indefinitely when the source is removed. Capacitors have a rating known as **leakage resistance** that causes the charge to be lost slowly over time. The equivalent circuit of the capacitor and its leakage resistance is shown in Figure 10–10(a). Notice, in Figure 10–10(b), that the leakage resistance is effectively in parallel with the capacitor. For most capacitors, this resistance is in hundreds of megohms. For some capacitors, such as electrolytics, this resistance can be much lower—on the order of 1 to 10 MΩ. The leakage current in such capacitors is much higher. Leakage current can be measured with the impedance bridge described in Chapter 9.

FIGURE 10–10 Leakage resistance

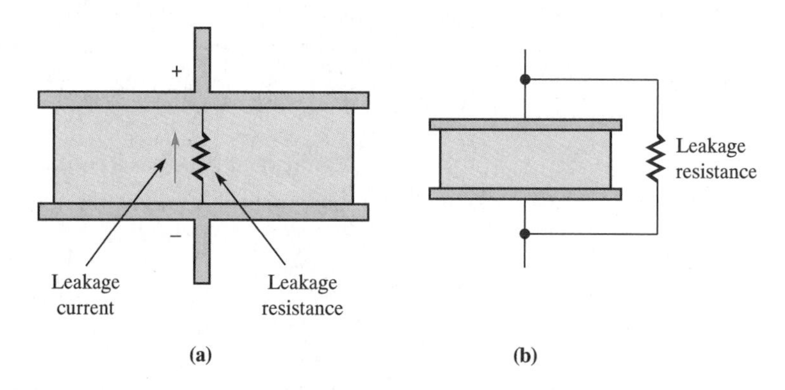

SECTION 10–5 REVIEW

1. What are the physical factors affecting capacitance?
2. How is the capacitance related, inversely or directly, to each of the physical factors?
3. What is the working voltage of a capacitor?
4. What is the significance of the labeling NP0 on a capacitor?
5. What is meant by leakage current in a capacitor?
6. A 100 μF capacitor is labeled P5000. If its temperature changes from 25°C to 45°C, what is its new value?

10–6 TYPES OF CAPACITORS

Question: How are capacitors classified?

Capacitors can be broadly classified as fixed value or variable. A **fixed-value capacitor** is one whose value cannot be easily changed. A **variable capacitor** is one whose value is easily changed usually through the rotation of a shaft. Within each classification are several subclassifications that relate mainly to the kind of dielectric used. For fixed capacitors, the mica, ceramic, paper, and electrolytic types are discussed in this section. Air, mica, ceramic, and plastic types of variable capacitors are also presented.

■ Fixed Capacitors

Mica capacitors come in two types that are related to their construction: molded and dipped mica. Typical mica capacitors are shown in Figure 10–11. They are constructed of thin sheets of metal, usually aluminum, interlaced with thin sheets of mica. The connecting leads are attached to alternate plates. The entire package is then pressed and given either a molded or dipped coating. They have relatively high breakdown voltages that are on the order of 500 kV to 20 kV. Their range of capacitance values, which is relatively low, is 10 pF through about 5000 pF. Silver mica capacitors are more temperature stable than conventional types because the metal plates are silver-plated directly on the mica sheets. They are also more expensive than capacitors made with aluminum plates.

FIGURE 10–11 Typical mica capacitors

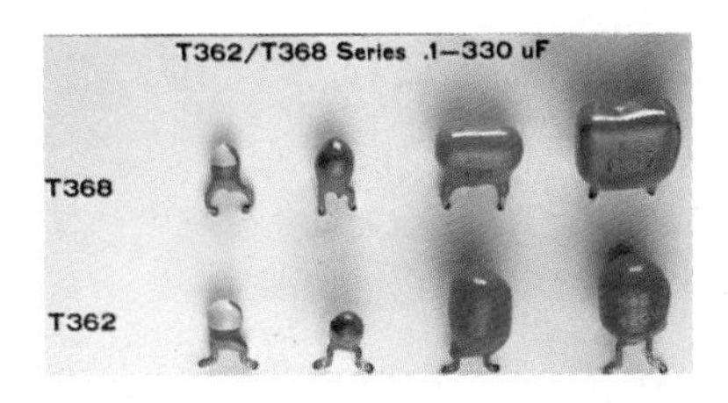

Ceramic dielectrics are made from earth materials fired under extreme heat. In the disc type, silver is fired onto both sides of the ceramic in order to form the plates. The tubular type is constructed by firing silver onto both sides of a hollow ceramic tube. Typical disc and tubular ceramic capacitors are shown in Figure 10–12. These capacitors are characterized by very large values of permittivity—on the order of 1200. As a result, larger values of capacitance—on the order of 0.01 μF—can be manufactured in small packages. Their breakdown voltage is also quite high, ranging from 500 V to 20 kV.

FIGURE 10–12 Typical ceramic capacitors

Paper capacitors use aluminum foil as the plates and paper as the dielectric. Typical paper capacitors are shown in Figure 10–13. Long sheets of foil are separated by a sheet of paper. They are rolled into tubular form and a connecting lead is attached to each piece of foil. The capacitor is then molded in plastic or encapsulated in a cardboard tube with a wax or plastic coating. Values of paper capacitors range from 0.001 μF to a few μF. The breakdown voltage range is from 200 V through 1600 V. There is a contrasting color band on the end of the capacitor near the lead attached to the outer foil. When installed, this lead should be attached to circuit common or the low-potential side of the circuit. The outer foil has a shielding effect on the rest of the circuit.

FIGURE 10–13 Typical paper capacitors

NOTE

The color-banded side of a paper capacitor does not imply that there is any polarity involved. The capacitance is the same and the capacitor operates properly no matter which way it is connected. Connecting the outer foil lead to common or the low-potential side merely takes advantage of something inherent in the capacitor construction, which is the shielding effect of the outer foil.

An *electrolytic capacitor* has a very special construction, as illustrated in Figure 10–14(a). In it, you see the electrolytic capacitor as a negative electrode of aluminum foil, an electrolyte, and a positive electrode of aluminum foil. Recall that an electrolyte is a liquid that conducts current with the charge carriers in the form of ions. In this capacitor, you notice that one essential part, the dielectric, is missing!

Question: How is the dielectric formed in an electrolytic capacitor?

The dielectric for an electrolytic capacitor is formed in the manufacturing process. A voltage is applied as shown in Figure 10–14(b). Current flows from the negative side of the source, into the negative electrode, through the electrolyte, and exits at the positive terminal. In this process, a chemical action between the electrolyte and the aluminum foil

FIGURE 10–14 Construction and formation of electrolytic capacitor

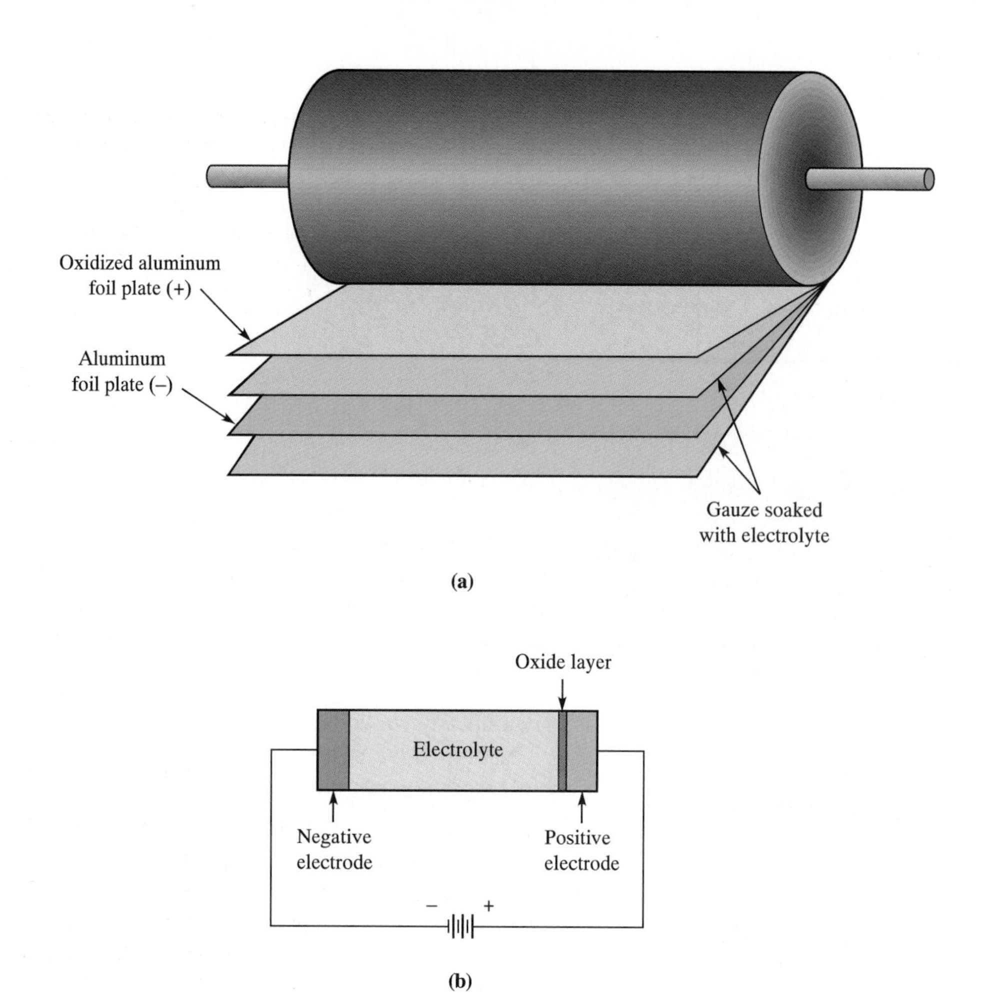

FIGURE 10–15 Electrolytic capacitors and their symbol

takes place in which a molecule-thin layer of aluminum oxide is formed on the aluminum foil. This oxide layer is an insulator and thus the last part of the capacitor, the dielectric, is formed. This extremely thin dielectric allows the manufacture of capacitors with very large values. It does, however, lower the breakdown voltage of the capacitor. They are available in values up to 5000 μF with up to 450 V breakdown voltage. Higher values (much higher) are available, but they have a lower breakdown voltage. Since chemicals are involved in this type of capacitor, they may not be usable after their given shelf-life has expired. Typical electrolytic capacitors and the symbol are shown in Figure 10–15.

Question: Is it absolutely necessary to observe polarity when connecting an electrolytic capacitor into a circuit?

The type of electrolytic capacitor described here is said to be polarized. It must be used in circuits having a dc polarizing voltage that maintains the aluminum oxide coating. The negative lead of the capacitor must be connected to the negative side of the source and the positive lead must be connected to the positive side of the source. If connected in reverse polarity, the chemical action will be reversed, the dielectric will disappear, and heavy current will flow. The heat produced by this current will produce gases that put pressure on the metal can in which the capacitor is enclosed. Needless to say, the capacitor will be destroyed and may even explode!

CAUTION

A faulty electrolytic capacitor may short and explode even though it is connected properly. It is a good idea, when working on circuits that involve large electrolytic capacitors, to wear a face shield of some type, especially if close inspection of the circuit is necessary. An exploding aluminum capacitor can throw shards of aluminum and hot electrolyte in all directions. A splash of the fluid on the face would be, at the least, painful and, if the eyes were involved, devastating! Thus, in working with large electrolytics, a little extra caution is the wise course.

FIGURE 10–16 Tantalum capacitors

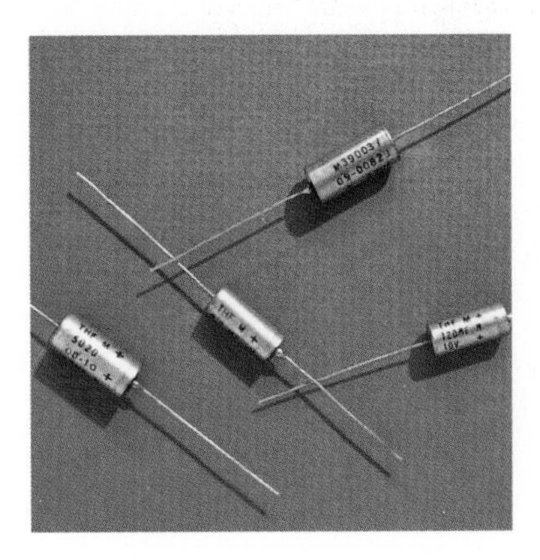

Another form of electrolytic capacitor is the type made with *tantalum*. Typical tantalum capacitors are shown in Figure 10–16. Tantalum capacitors, as compared to aluminum, have larger values of capacitance in smaller spaces, longer shelf-life, and lower leakage current. They are, however, more expensive than their aluminum counterparts. As with all types of electrolytic capacitors, special precautions should be observed when working with tantalum capacitors. When testing these devices, special attention should be paid to the polarity and value of the test instrument's output voltage. Tantalum capacitors contain strong acids in their dielectric. Should polarity be reversed or the operating voltage be exceeded, these devices can explode spraying metal particles and acid in all directions.

Variable Capacitors

FIGURE 10–17 Air dielectric capacitor

The most common type of variable capacitor is the *air dielectric*. This type and its symbol are shown in Figure 10–17. It has a moving part, known as the *rotor,* and a stationary part, known as the *stator.* To vary the capacitance, the parameter being controlled is the area of the plates. When rotated to where the plates are fully meshed, maximum plate area exists, and capacitance is maximum. When rotated so the plates are fully separated, the plate area and the capacitance are minimum. In this manner, the capacitance is continuously variable over a certain range. This type of variable capacitor is commonly used in the tuner in an AM radio receiver.

Other variable or, more accurately, adjustable capacitors are made with mica, ceramic, or plastic dielectrics. Shown in Figure 10–18, they are composed of a stationary plate, a moveable plate, a dielectric, and an adjusting screw. The adjusting screw moves the moveable plate either nearer to or farther from the stationary plate and thus controls the distance between the plates and the capacitance. These capacitors are often called *trimmer capacitors,* or just "trimmers."

Trimmer capacitors should be adjusted with a plastic tool. A metal screwdriver in close proximity to the capacitor can cause changes in the total capacitance value and thus makes the setting of an accurate capacitance very difficult.

Another type of variable capacitor is the *varactor,* or semiconductor capacitor. These capacitors are mentioned here merely for informational purposes. You will learn of their technical aspects in later studies. Their tuning is unique among capacitors. It is accomplished by varying a dc voltage that is applied across them.

SECTION 10–6 REVIEW

1. What are some common materials used as dielectrics in capacitors?
2. What is the approximate range of capacitances available in mica capacitors?
3. What does the colored band on one end of a paper capacitor signify?
4. Why must polarity be observed in connecting electrolytic capacitors into a circuit?
5. How is the capacitance varied in an air dielectric capacitor?

FIGURE 10–18 Adjustable (trimmer) capacitor

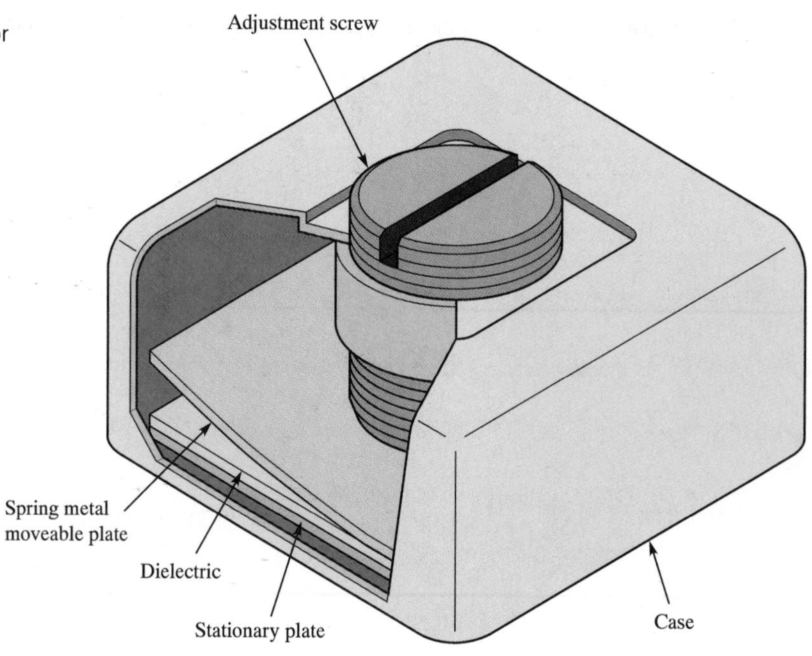

SYNOPSIS

At this point, it is a good idea to review the important facts concerning capacitors. A capacitor is an electronic device that stores charge. A capacitor is composed of two conductive plates separated by a dielectric. Thus, capacitance is present in an circuit anywhere two conductors are separated by an insulator, such as between the traces of a printed circuit board. When charge is stored in a capacitor, a voltage develops across it. Since current must flow in order to store the charge, the voltage across a capacitor builds up over time. Thus, a capacitance opposes a change in voltage. The charge polarizes the dielectric producing an electrostatic field throughout it. The electrical energy required to charge the capacitor is not dissipated, but stored in the electrostatic field. When the capacitor is discharged, the energy is returned to the source. A capacitor has ratings of maximum working voltage, temperature coefficient, and leakage current. Capacitors come in both fixed and variable varieties.

10–7 CAPACITORS IN SERIES

Question: What are the effects of placing capacitors in series?

When placed in series, the total capacitance is computed using the reciprocal equation, which is similar to the method of computing total resistance of resistors in parallel. The total capacitance of capacitors in series is smaller than that of the smallest capacitance. When capacitors are placed in series, as shown in Figure 10–19, the distance between the plates becomes effectively the total of the distances between the plates of each capacitor. The electrostatic field is not as intense, and the capacitance is lower. In equation form, total capacitance for capacitors in series is given by

FIGURE 10–19 Capacitors in series

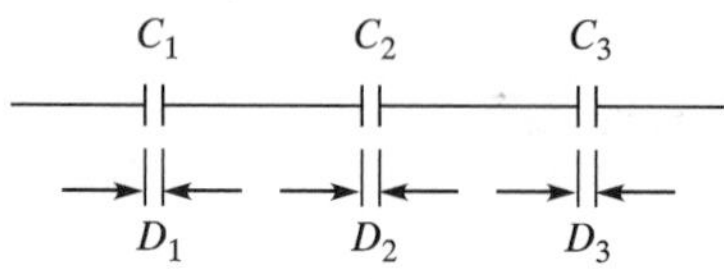

(10–7)
$$C_T = \frac{1}{(1/C_1) + (1/C_2) + (1/C_3) \cdots + (1/C_n)}$$

EXAMPLE 10–11 For the capacitors in Figure 10–20, compute the total capacitance.

FIGURE 10–20 Capacitors for Example 10–11

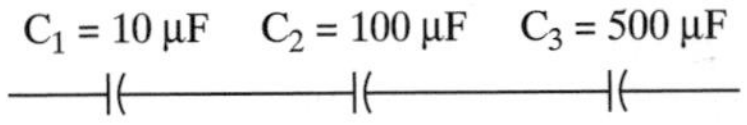

$C_1 = 10\ \mu F$ $C_2 = 100\ \mu F$ $C_3 = 500\ \mu F$

■ Solution:

Use Eq. 10–7:

$$C_T = \frac{1}{(1/C_1) + (1/C_2) + (1/C_3)}$$

$$= \frac{1}{[1/(10 \times 10^{-6})] + [1/(100 \times 10^{-6})] + [1/(500 \times 10^{-6})]} = 8.93\ \mu F$$

[1] [0] [EXP] [±] [6] [1/x] [+] [1] [0] [0] [EXP] [±] [6] [1/x] [+]

[5] [0] [0] [EXP] [±] [6] [1/x] [=] [1/x]

EXAMPLE 10–12 The total capacitance of two capacitors in series is 8 μF. If C_1 is 24 μF what is the value of C_2?

■ Solution:

Use the reciprocal form of Eq. 10–7:

$$\frac{1}{C_T} = \frac{1}{C_1} + \frac{1}{C_2}$$

Then,

$$\frac{1}{C_2} = \frac{1}{C_T} - \frac{1}{C_1}$$

and

$$C_2 = \frac{1}{(1/C_T) - (1/C_1)}$$

$$= \frac{1}{[1/(8 \times 10^{-6})] - [1/(24 \times 10^{-6})]} = 12\ \mu F$$

[8] [EXP] [±] [6] [1/x] [−] [2] [4] [EXP] [±] [6] [1/x] [=] [1/x]

For equal value capacitors in series, the similar relationship for resistors in parallel can be applied. The total capacitance of equal capacitors in series is the value of one capacitor divided by the number of capacitors in the series. In equation form, total capacitance for equal-value capacitors in series is

(10–8)
$$C_T = \frac{C}{N}$$

where N is the number of capacitors.

EXAMPLE 10–13 What is the total capacitance of four 100 μF capacitors in series?

■ Solution:

Use Eq. 10–8:

$$C_T = \frac{C}{N}$$

$$= \frac{100 \times 10^{-6}}{4} = 25\ \mu\text{F}$$

Recall that in a two-resistor parallel combination that if one is much larger than the other, the total resistance can be considered the value of the smaller. The same holds true for capacitors in series. For two capacitors in series, if one is much larger than the other, the total can be assumed to be the value of the smaller. Once again, one must be at least ten times larger than the other.

Capacitive Voltage Divider

Question: How does the total voltage divide across capacitors in series?

When a dc voltage is placed across a series capacitor string, the voltage divides in inverse proportion to the capacitance values. Thus, the smallest capacitor charges to the highest voltage, and largest capacitor charges to the smallest voltage. In this section, an equation is developed, similar to the voltage divider equation, for computing the voltage across each capacitor in a series.

FIGURE 10–21 Capacitive series circuit

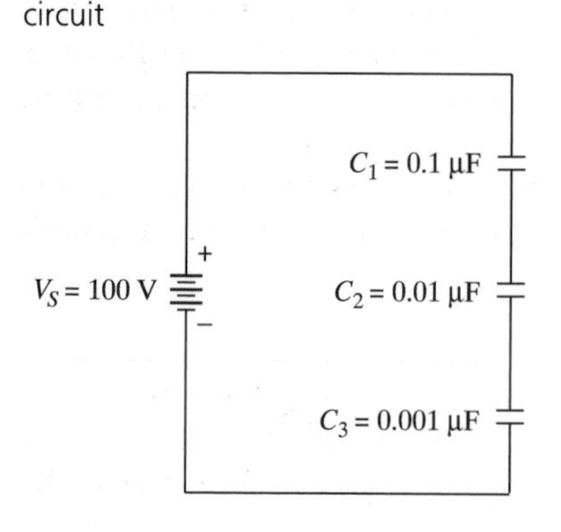

Figure 10–21 shows three capacitors in series across a dc source. Since the capacitors are in series and current is the same at all points in a series circuit, the same amount of current flows into each capacitor. Each capacitor thus stores the same amount of charge. Since there is only one source, there can only be one charge stored Q_C. Thus, the charge on any of the capacitors is Q_{Cx}, where C_x is used to designate any capacitor. The equation for the voltage developed across any of the capacitors is derived as follows:

From equation 10–2,

$$Q_T = C_T V_S$$

$$Q_{Cx} = C_S V_S \quad \text{and} \quad Q_T = Q_{Cx}$$

These lead to

$$C_T V_S = C_x V_S$$

Solving for V_x gives

(10–9)
$$V_x = \frac{C_T}{C_x} \times V_S$$

To solve for the voltages developed across each capacitor in Figure 10–22, the total capacitance is computed:

$$C_T = \frac{1}{(1/C_1) + (1/C_2) + (1/C_3)}$$

$$= \frac{1}{[1/(0.1 \times 10^{-6})] + [1/(0.01 \times 10^{-6})] + [1/(0.001 \times 10^{-6})]} = 901\ \text{pF}$$

Next, the value of each voltage is computed:

FIGURE 10–22 Capacitive voltage divider circuit

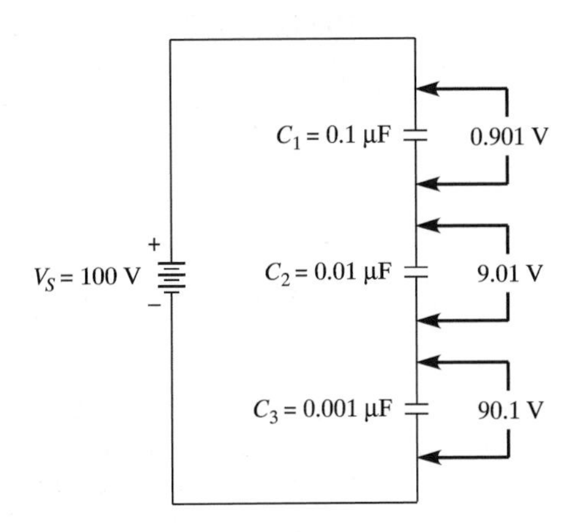

$$V_{C1} = \frac{C_T}{C_1} \times VS = \frac{901 \times 10^{-12}}{0.1 \times 10^{-6}} \times 100 = 0.901 \text{ V}$$

$$V_{C2} = \frac{C_T}{C_2} \times VS = \frac{901 \times 10^{-12}}{0.01 \times 10^{-6}} \times 100 = 9.1 \text{ V}$$

$$V_{C3} = \frac{C_T}{C_3} \times VS = \frac{901 \times 10^{-12}}{0.001 \times 10^{-6}} \times 100 = 90.1 \text{ V}$$

Note that the smallest capacitor developed the highest voltage when charged. Recall that each capacitor stores the same amount of charge. The charge Q is equal to the capacitance C times the voltage V. Thus, it will take more voltage to store the same charge in a small capacitor than in a larger one. This relationship may also be thought of in another way: If the same charge is stored in a smaller area, the resulting voltage will be greater.

Also note that Kirchhoff's voltage law applies to the capacitor as a voltage divider and is a good way of checking your work. According to KVL, the sum of the three capacitor's voltages is equal to the source voltage.

SECTION 10–7 REVIEW

1. Why does adding capacitors in series reduce the total capacitance?
2. Why does the smaller capacitor in a series string have the greatest voltage when fully charged?
3. Three capacitors with values of 50 μF, 100 μF, and 470 μF are connected in series. Compute the value of the total capacitance.
4. For the circuit of Figure 10–23, compute the value of the voltage across each capacitor.

FIGURE 10–23 Circuit for Section Review Problem 4

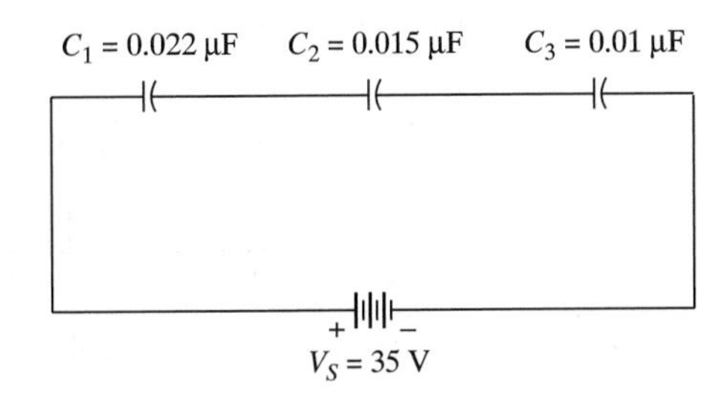

10–8 CAPACITORS IN PARALLEL

Question: What are the effects of placing capacitors in parallel?

The total capacitance of capacitors in parallel is equal to their sum, which is similar to resistors in series. As illustrated in Figure 10–24, the total capacitance increases when capacitors are placed in parallel due to the larger plate area. In equation form, total capacitance for capacitors in parallel is written as

(10–10) $$C_T = C_1 + C_2 + C_3 \cdots + C_n$$

FIGURE 10–24 Capacitors in parallel

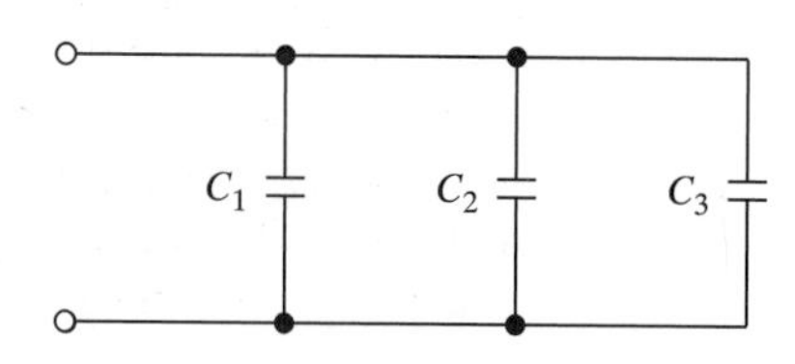

EXAMPLE 10–14 What is the total capacitance of 0.02 μF, 0.047 μF, and 0.001 μF capacitors in parallel?

■ Solution:

Use Eq. 10–10:

$$C_T = (0.02 \times 10^{-6}) + (0.047 \times 10^{-6}) + (0.001 \times 10^{-6})$$
$$= 0.068\ \mu\text{F}$$

EXAMPLE 10–15 For the circuit of Figure 10–25, C_T is equal to 0.0111 μF. Compute the value of C_2.

FIGURE 10–25 Circuit for Example 10–15

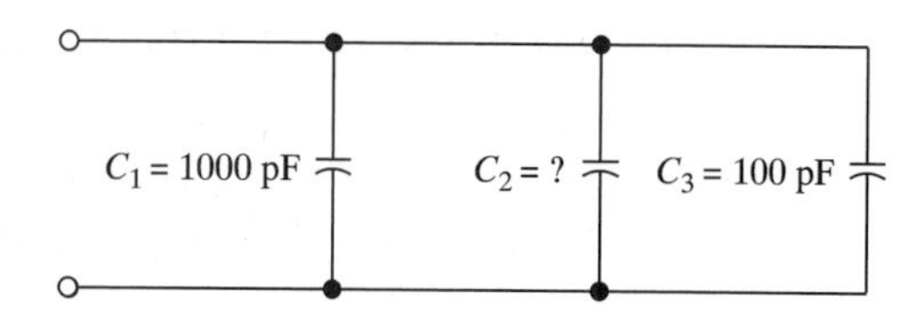

■ Solution:

Transpose Eq. 10–10 to solve for C_2:

$$C_T = C_1 + C_2 + C_3$$
$$C_2 = C_T - (C_1 + C_3)$$
$$= (0.0111 \times 10^{-6}) - [(1000 \times 10^{-12}) + (100 \times 10^{-12})] = 10{,}000\ \text{pF}$$

[.] [0] [1] [1] [1] [EXP] [±] [6] [−] [(] [1] [0] [0] [0] [EXP]
[±] [1] [2] [+] [1] [0] [0] [EXP] [±] [1] [2] [)]

SECTION 10–8 REVIEW

1. Why does total capacitance increase when capacitors are placed in parallel?
2. Will capacitors, regardless of value, have the same voltage across them when connected in parallel?
3. Three capacitors with values of 1000 pF, 0.001 μF, and 100 pF are connected in parallel. Compute the total capacitance.

10–9 THE *RC* TIME CONSTANT

When a capacitor is charged, a voltage source moves electrons from one plate, creating a deficiency, and deposits them on the other plate, creating an excess. It takes time for these electrons to move. If the source is removed and a suitable conducting path is provided, the excess electrons on one plate flow to the other in balancing the charge. Thus, a capacitor cannot change its charge instantaneously. For any values of resistance and capacitance in the circuit, a definite amount of time is required for the capacitor to charge or discharge. The length of time needed for a capacitor to charge or discharge depends upon two factors: the value of the capacitor and the value of the series resistor.

Consider the charging process for the circuit shown in Figure 10–26(a). When the switch is moved from position *B* to position *A*, the voltage across the capacitor is zero. By Kirchhoff's voltage law, the entire source voltage must be across the resistor. Thus, the capacitor is appearing as a short circuit, and maximum current flows at this time (V_S/R).

As the capacitor charges, its voltage increases. The voltage across the resistor decreases by the same amount as the voltage across the capacitor, which means that the current has decreased. The process continues until, as shown in Figure 10–26(b), all the source voltage is across the capacitor, none is across the resistor, and the current is zero. Thus, after charging, the capacitor appears as an open circuit to steady-state direct current.

After the capacitor is fully charged, the switch is moved to position *B*. As shown in Figure 10–26(c), the capacitor discharges through the resistor. It now appears as a source, and current will flow until the charge that was stored is depleted.

Question: What is meant by the RC *time constant?*

The ***RC* time constant** is the amount of time needed for the voltage across the capacitor to rise or fall by approximately 63.2%. (Notice the similarity between this and the *RL* time constant.) The *RC* time constant is given the symbol τ and is computed using the following equation:

(10–11) $$\tau = RC$$

This equation states that the *RC* time constant in seconds is equal to the product of the capacitance in farads and its series resistance in ohms.

A capacitor is considered to be charged in the interval of five time constants after the voltage source is applied, as will be demonstrated in the following example. The circuit of Figure 10–27 is used to demonstrate computation of the voltage across the capacitor from the time the switch is moved to position *A* until the capacitor is considered fully charged.

First, the *RC* time constant is computed:

$$\tau = RC$$
$$= (10 \times 10^3)(1 \times 10^{-6}) = 0.01 \text{ s}$$

FIGURE 10–26 Current flow in a capacitive circuit

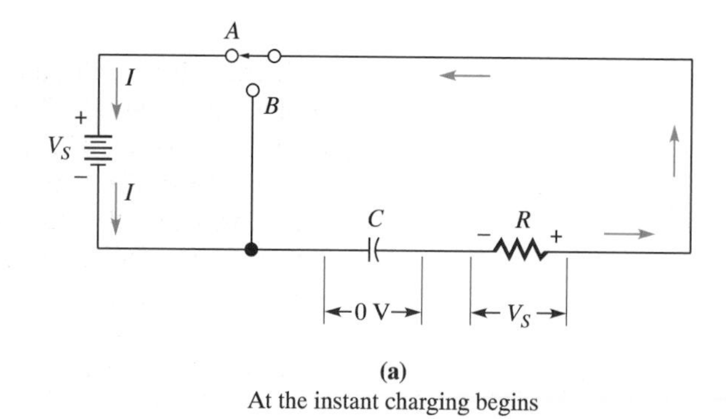

(a)
At the instant charging begins

FIGURE 10–27 Circuit for demonstration of *RC* time constant

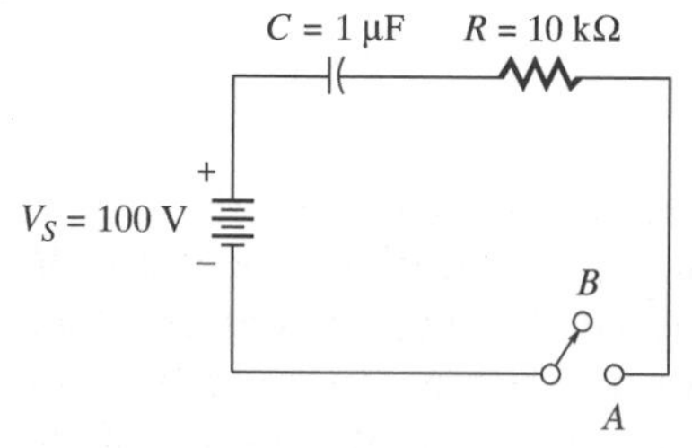

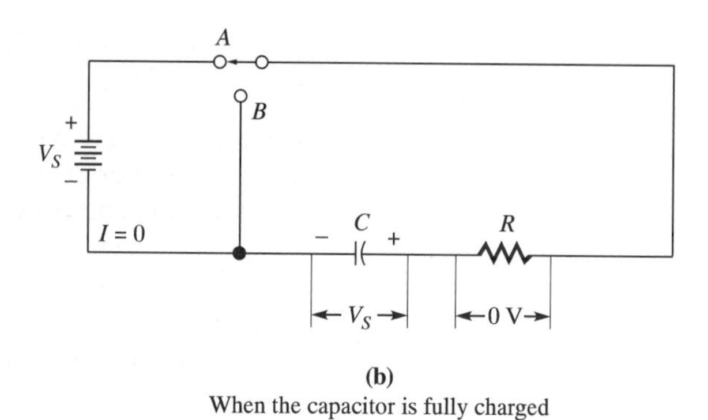

(b)
When the capacitor is fully charged

(c)
When the capacitor discharges

FIGURE 10–28 Graph of capacitor charge in five time constants

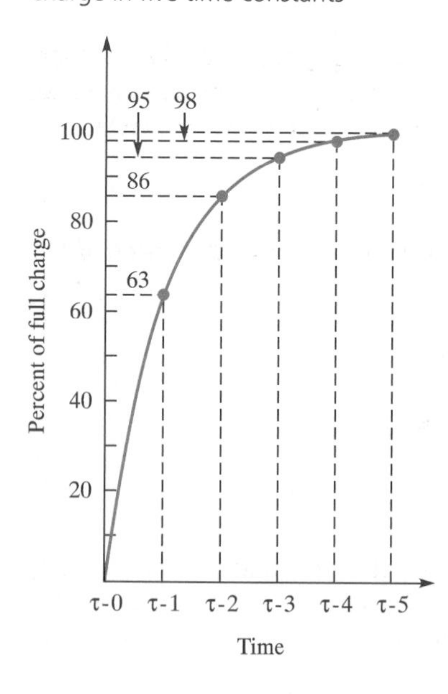

The voltages at the end of each of the first five time constants can now be computed and plotted on the graph of Figure 10–28. Recall that when a capacitor first begins to charge, it appears as a short circuit. Thus, when the switch is first moved to position *A*, the voltage across the capacitor is zero.

During the first time constant, the voltage across the capacitor increases by 63.2% of the final voltage. At this time, the capacitor has 63.2 V across it, leaving 36.8 V across the resistor. This point is plotted on the graph in Figure 10–28.

During the second time constant, the voltage across the capacitor increases by 63.2% of the remaining 36.8 V. Thus, the capacitor voltage is 63.2 + 23.3, or 86.5 V, leaving 13.5 V across the resistor. This point is plotted on the graph.

During the third time constant, the voltage across the capacitor increases by 63.2% of the 13.5 V remaining. Thus, the capacitor voltage is 86.5 + 8.5, or 95 V, with the remaining 5 V across the resistor. This point is plotted on the graph.

During the fourth time constant, the voltage across the capacitor increases by 63.2% of the remaining 5 V. Thus, the capacitor voltage is 95 + 3.2, or 98.2 V, and 1.8 V is across the resistor. This point is plotted on the graph.

During the fifth time constant, the voltage across the capacitor increases by 63.2% of the remaining 1.8 volts. Thus, the capacitor voltage is 98.2 + 1.1, or 99.3 V, and the remaining 0.7 volts is found across the resistor. This final point is plotted and the capacitor is considered fully charged.

Since five time constants must elapse for the capacitor to be considered fully charged, the time for the charging process equates to 5×0.01 s or 0.05 s. It will take the same amount of time to discharge, if the discharging takes place through the same resistance.

Question: Can a similar curve be developed for the discharge of the capacitor?

A similar curve can be developed for the discharge of the capacitor. When the switch is moved to position B, the capacitor acts like a source, and current flows until it is discharged. It discharges by approximately 63.2% in each of the five time constants. Since the capacitor is discharging through the same resistance, it takes the same 0.05 s to discharge as it did to charge. The following table contains the resistor voltages during discharge at the end of each time constant. Note that these values are the complement of the values for the charging of the capacitor.

Time Constant	Resistor Voltages
$\tau - 0$	100 V
$\tau - 1$	36.8 V
$\tau - 2$	13.5 V
$\tau - 3$	5 V
$\tau - 4$	1.8 V
$\tau - 5$	0.7 V

The Exponential Equations

There are exponential equations for charging and discharging a capacitor, just as there were for the energizing of an inductor. The equations for capacitors follow the same lines as those for inductors, but now you will be dealing with a build-up of voltage rather than current.

Question: What is the exponential equation for the charge of a capacitor?

The equation for the charge curve is

(10–12) $$v = V_F\,(1 - \varepsilon^{-t/RC})$$

WHERE: v = instantaneous value of capacitor voltage

ε = 2.7182818

V_F = final source voltage

This equation is used to compute the instantaneous value of the capacitor voltage v at any time t after the source voltage has been applied. Notice the exponent of epsilon, $-t/RC$, which is the time in question divided by the time constant.

EXAMPLE 10–16 An RC circuit consists of a 500 kΩ resistor in series with a 1 μF capacitor across a 100 V source. Compute the voltage across the capacitor 0.8 s after the switch is closed.

■ Solution:

1. Compute the time constant:

$$\tau = RC$$
$$= (500 \times 10^3)(1 \times 10^{-6}) = 0.5 \text{ s}$$

[5] [0] [0] [EXP] [3] [×] [1] [EXP] [±] [6] [=]

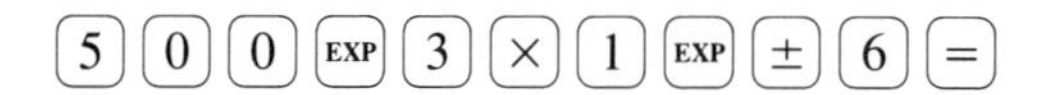

2. Compute the exponent of ε:

$$\frac{-t}{RC} = \frac{-0.8 \text{ s}}{0.5 \text{ s}} = -1.6 \text{ s}$$

[·] [8] [÷] [·] [5] [=]

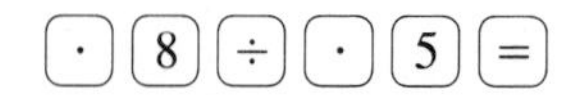

3. Key the values into Eq. 10–12, the exponential equation, and solve:

$$v = V(1 - \varepsilon^{-t/RC})$$
$$= (100 \text{ V})(1 - 2.7182818^{-1.6})$$
$$= 100(1 - 0.20189)$$
$$= 100(0.79811) = 79.811 \text{ V}$$

[1] [−] [(] [2] [·] [7] [1] [8] [2] [8] [1] [8] [x^y] [1] [·] [6] [±] [=]

[)] [×] [1] [0] [0] [=]

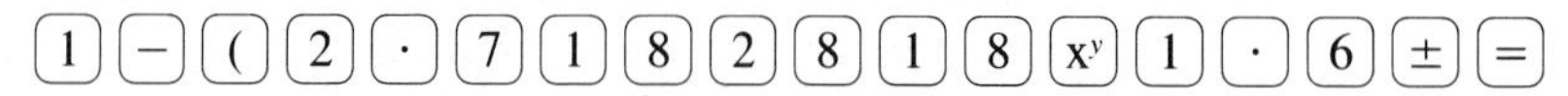

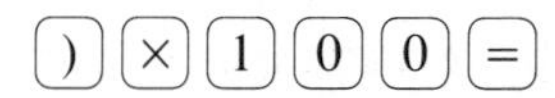

Question: What is the exponential equation for the instantaneous capacitor voltage during discharge?

There is also an exponential equation for the instantaneous capacitor voltage during discharge:

(10–13) $$v = V_I\varepsilon^{-t/RC}$$

where V_T is the initial voltage.

EXAMPLE 10–17 For the circuit of Figure 10–29, compute the capacitor voltage 0.5 s after the capacitor begins to discharge.

FIGURE 10–29 Circuit for Example 10–17

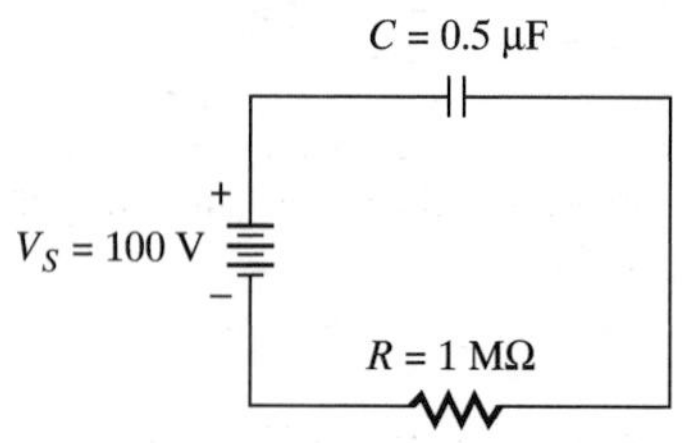

■ Solution:

Use exponential equation 10–13:

$$v = V_I(\varepsilon^{-t/RC})$$
$$= (100 \text{ V})(2.7182818^{-1})$$
$$= 100(0.36787) = 36.787 \text{ V}$$

[2] [·] [7] [1] [8] [2] [8] [1] [8] [x^y] [1] [±] [=] [×] [1] [0] [0] [=]

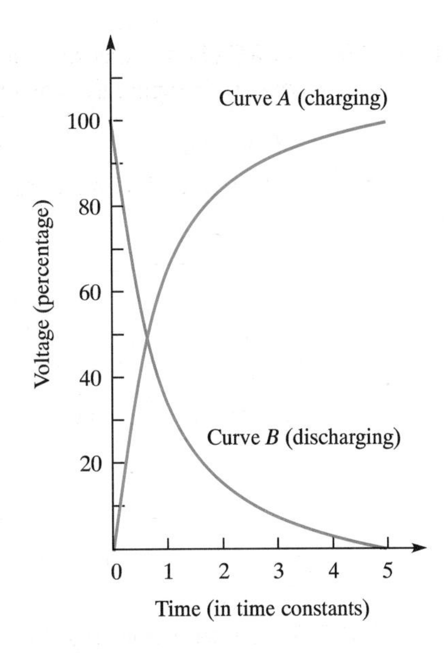

FIGURE 10–30 Graph of universal time constant

■ The Universal Time Constant Curves

The curves that were plotted in Figure 10–28 are the curves for the exponential equations. The *universal time constant curves,* shown in Figure 10–30, provide a graphical means of determining the instantaneous voltage across the capacitor at any time after either charging or discharging has begun. The percentage of the source voltage across the capacitor is plotted as a function of the time constant. For example, move upward on the graph from the first time constant until you reach the point at which it crosses the charge curve. Moving to the left from this point you see that the capacitor voltage is approximately 63% of the source voltage. Next move downward to the point at which the first time constant line crosses the discharge curve. Looking to the left, you see that the voltage across the capacitor is approximately 37% of the source voltage. These percentages are close to those already established for charge or decay during one time constant. In a similar manner, the capacitor voltage can be found as a percentage of the source voltage at any instant during charge or discharge cycles.

EXAMPLE 10–18 For the circuit of Figure 10–31, using the universal time constant curves, find the voltage across the capacitor 16 ms after charging begins.

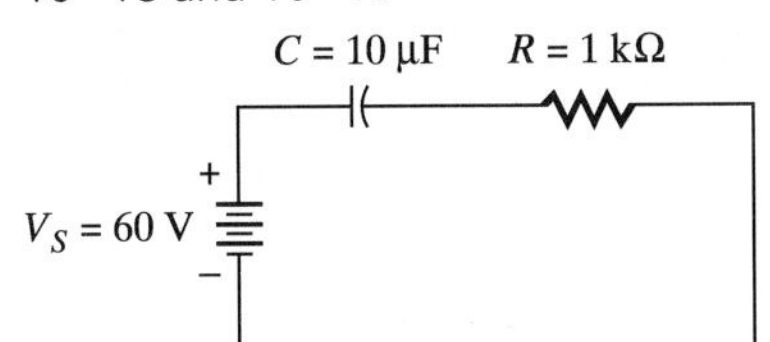

FIGURE 10–31 Circuit for Examples 10–18 and 10–19

■ Solution:

Notice that 16 ms is 1.6 time constants. Find 1.6 time constants on the graph and move upward until it crosses the charge curve. This crosses the curve at approximately the 80% point. Thus, the voltage is 80% of the final voltage at that time or 48 V.

EXAMPLE 10–19 For the same circuit as in Example 10–18, using the universal time constant curves, determine how long it will take for the voltage to reach 54 V.

■ Solution:

Computing the percentage, you find that 54 V is 90% of 60 V. Find the 90% point on the graph and move to the right until it crosses the charge curve. Moving downward, you see that this point occurs at 2.4 time constants. Since one time constant is 10 ms, 2.4 time constants would be 24 ms.

SECTION 10–9 REVIEW

1. By what percentage does a capacitor charge in one time constant?
2. How many time constants are required for a capacitor to fully charge?
3. An RC circuit consists of a 100 kΩ resistor and a 0.01 μF capacitor across a 40 V source. Compute the voltage across the capacitor 3.5 ms after the source voltage is applied.
4. If the capacitor in Problem 3 were allowed to charge for two time constants and then was discharged for three time constants, what would be the final voltage?

10–10 APPLICATIONS OF CAPACITORS

Capacitors are used very extensively in electronics but the majority of them are in alternating current circuits. Two uses in dc circuits are discussed here: those involving timing operations and those requiring a short, intense current.

Timing Operations

Timing operations take advantage of the very predictable charging time of the capacitor in an RC circuit. As you have learned, there is a definite time at which the voltage across the capacitor in an RC circuit reaches a certain percentage of the source voltage. The timing of this voltage can be used to cause something to occur in another circuit, which is often referred to as "triggering." In this type of timing circuit, once the action has occurred, the capacitor is discharged and the process begins again.

Timing circuits are known as *relaxation oscillators,* an example of which is shown in Figure 10–32. In this RC circuit, the capacitor is in parallel with a small neon lamp. A neon lamp is one containing neon gas, which ionizes when the voltage across the lamp is great enough. In ionizing, the gas becomes conductive and emits light. The action of this circuit is to make the lamp blink periodically, with the time being dependent upon the time constant of the RC circuit. Notice that when the switch is closed, the voltage begins to build across the capacitor. The voltage rises in the exponential manner, and when the voltage across the capacitor reaches the ionization potential of the lamp, the neon ionizes, emitting light. When the neon ionizes, it becomes a very low resistance path, and the capacitor rapidly discharges through it. Once below a certain voltage, the lamp is extinguished, the capacitor recharges, and the process repeats itself. The rate at which the lamp blinks is inversely proportional to the time constant. An increase in value of either the resistor or capacitor gives a longer time constant, reducing the blink rate. A decrease in the value of either the resistor or the capacitor has the opposite effect.

FIGURE 10–32 Example timing circuit

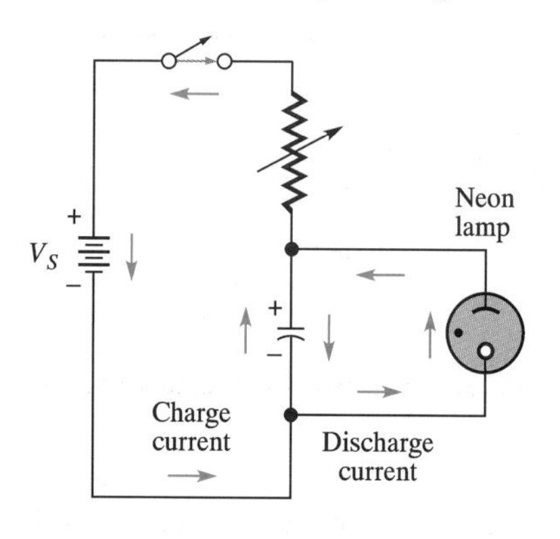

High Current Source

Question: How can a capacitor provide a short, intense current?

A capacitor, once charged, can be used as a source when it is discharged through a load. If a large-value capacitor is charged to a high voltage and then discharged through a low resistance, a very large value of current is produced. This principle is used in flash circuits in cameras. A charged capacitor is discharged through a flash bulb, providing enough current to produce a very intense light. The lamp is a very low-resistance load. The capacitor must be given time between pictures in order to recharge.

SECTION 10–10 REVIEW

1. What types of operations are appropriate for capacitors in dc circuits?
2. What characteristic of a capacitor is used in timing circuits?
3. What characteristic of a capacitor is used in circuits requiring a short, intense current?

10–11 TROUBLESHOOTING

Capacitor faults are of three types: opens, shorts, or leaks. An open capacitor is usually caused by one of the connecting wires burning open where it connects to the plate. Shorts occur when current arcs through the dielectric leaving a carbonized path that allows more current to flow. A leaky capacitor is one that is not shorted but does allow current through the dielectric. As a result, it never fully charges. Leaky capacitors become more prevalent as the equipment ages. Some electrolytics become leaky due to chemical action between the electrolyte and the electrodes while just sitting in a bin or on a shelf. Thus, such capacitors are sold with a specified shelf-life.

An ohmmeter can be used to test large-value capacitors for opens or shorts. A normal capacitor, when connected across the ohmmeter, initially shows a low resistance and then, as the capacitor charges, moves toward infinity. If the capacitor is shorted or leaky, the resistance reading stays low and does not move toward infinity. The normal resistance should be between 500 k and 1 MΩ for capacitors above 1 μF. Smaller capacitors would charge too quickly for the ohmmeter to follow and should read infinite on all scales. When checking electrolytics, the polarity of the meter *must* match the polarity of the capacitor leads.

The most comprehensive check of a capacitor is made with a capacitor analyzer or impedance bridge. These instruments not only check the value of the capacitor, but also measure the leakage current under the normal working voltage of the capacitor. A capacitor that checks normally under the low voltage of an ohmmeter may break down when a higher voltage is applied.

SECTION 10–11 REVIEW

1. What types of faults occur in capacitors?
2. What would an ohmmeter indication be for a shorted capacitor?
3. What would be an ohmmeter indication for an open capacitor?
4. Why is an ohmmeter check not really conclusive?

CHAPTER SUMMARY

1. Capacitance is the property of a circuit that stores electric charge.
2. Capacitance is the property of a circuit that opposes a change in voltage.
3. Capacitance is introduced into a circuit through devices known as capacitors.
4. A capacitor is made of conductive plates separated by a dielectric.
5. A capacitor stores charge on its plates. One plate loses electrons and becomes positive. The other gains electrons and becomes negative.
6. An electrostatic field is formed between the charged plates of a capacitor.
7. A capacitor stores electrical energy in its electrostatic field.
8. The capacitance value of a capacitor is directly proportional to the area of the plates and the permittivity of the dielectric.
9. The capacitance value of a capacitor is inversely proportional to the distance between the plates.

10. Capacitors can be either fixed or variable.
11. Some of the dielectrics used are mica, ceramic, paper, plastic, and air.
12. Polarity must be observed in connecting an electrolytic capacitor.
13. The lead nearest the colored band on a paper capacitor should be connected to the circuit common in order to take advantage of the shielding action of the outer foil.
14. The *RC* time constant is the time taken for the voltage to increase or decay by approximately 63.2%.
15. Five time constants are required for a capacitor to fully charge or fully discharge.
16. Capacitors are used in dc circuits in timing operations and where a short, intense current is needed.
17. Faults in capacitors can be opens, shorts, or leaks.
18. A capacitor analyzer or an impedance bridge are the best instruments for testing a capacitor.

KEY TERMS

capacitance
capacitor
dielectric
permittivity
farad
relative permittivity
absolute permittivity
dielectric strength
leakage current
fixed-value capacitor
variable capacitor
RC time constant

EQUATIONS

(10–1) $E = Q\dfrac{V}{2}$

(10–2) $C = \dfrac{Q}{V}$

(10–3) $i_C = C\dfrac{\Delta V}{\Delta t}$

(10–4) $C = \dfrac{i_C}{\Delta V/\Delta t}$

(10–5) $\varepsilon = \varepsilon_r \varepsilon_O$

(10–6) $C = \dfrac{A\varepsilon}{D}$

(10–7) $C_T = \dfrac{1}{(1/C_1) + (1/C_2) + (1/C_3) \cdots + (1/C_n)}$

(10–8) $C_T = \dfrac{C}{N}$

(10–9) $V_x = \dfrac{C_T}{C_x} \times V_S$

(10–10) $C_T = C_1 + C_2 + C_3 + \cdots C_n$

(10–11) $\tau = RC$

(10–12) $v = V_F(1 - \varepsilon^{-t/RC})$

(10–13) $v = V_I\varepsilon^{-t/RC}$

Variable Quantities

E = energy stored in capacitor
Q = charge stored by capacitor
V = voltage applied across capacitor
C = capacitance
i_C = capacitor charging current
$\dfrac{\Delta V}{\Delta t}$ = rate of change of voltage with respect to time
ε = absolute permittivity
ε_r = relative permittivity
ε_O = absolute permittivity of vacuum
d = distance between plates
τ = time constant in seconds
v = instantaneous capacitor voltage
V_F = final capacitor voltage
$-t$ = time in question
RC = time constant
V_I = initial voltage

TEST YOUR KNOWLEDGE

1. Define capacitance in two ways.
2. Is energy stored in a capacitor in a magnetic field or an electrostatic field?
3. Is the amount of charge stored by a capacitor directly or inversely proportional to its capacitance?
4. Does a capacitor store or dissipate electrical energy?
5. What is the unit of capacitance?
6. Is capacitance directly or inversely proportional to the area of the plates?
7. Is capacitance directly or inversely proportional to the distance between the plates?
8. List four dielectric materials.
9. What is the significance of the voltage rating of a capacitor?
10. What does a capacitor marking of N800 signify?

11. How may the polarity of an electrolytic capacitor be indicated?
12. Why may capacitors be placed in series or parallel?
13. If two capacitors are in series, which will charge to the higher value, the one with the larger or the smaller value?
14. What is meant by the *RC* time constant?
15. List two applications of capacitors in dc circuits.

PROBLEM SET: BASIC

Section 10–3

1. For the capacitor connected as shown in Figure 10–33(a), draw arrows to show the direction of current during charging of the capacitor. Mark the polarity of the voltage developed across the capacitor.
2. For the capacitor connected as in Figure 10–33(b), using arrows, show the path of the discharge current.

FIGURE 10–33 Circuits for Problems 1 and 2

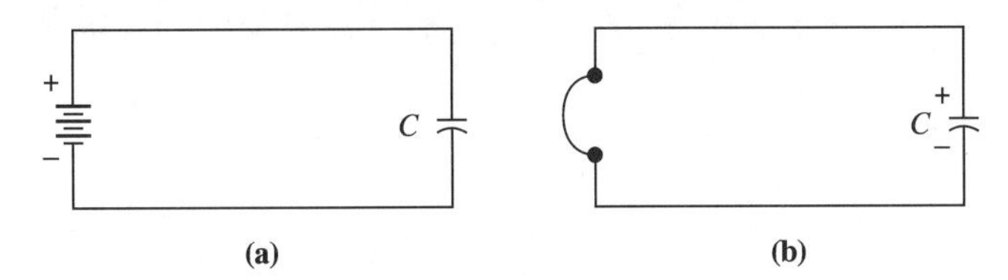

Section 10–4

3. In the diagram of the capacitor of Figure 10–34(a), draw the orbits of an electron around each nucleus for an uncharged capacitor.
4. In the diagram of the capacitor of Figure 10–34(b), draw the orbits of the electron around each nucleus for a capacitor charged as shown.

FIGURE 10–34 Circuit for Problems 3 and 4

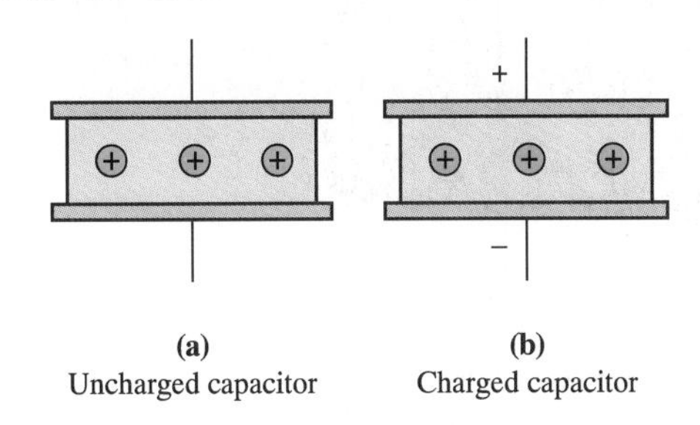

Section 10–5

5. A capacitor is charged to 12 V and stores 800 mC of charge. How many joules of energy are stored?
6. A capacitor is charged to 100 V and stores 100 mJ of electrical energy. How much charge is stored?

Section 10–6

7. A capacitor is charged to 100 V. If it contains 0.4 mC of charge, what is its capacitance value?
8. A 4 μF capacitor is charged to 15 V. How much charge is stored?
9. A 10 μF capacitor contains 200 μC of charge. What is the voltage across the capacitor?
10. If 10 V are developed across a capacitor storing 500 μC of charge, how many joules of energy are stored?

Section 10–7

11. The plates of a capacitor are 0.02 m by 0.03 m and the distance between the plates is 0.001 m. If the relative permittivity of the dielectric is 25, compute the value of the capacitor.
12. A 100 μF capacitor has a positive temperature coefficient of 1000 ppm. If its temperature increases from 25°C to 45°C, what is the value of its capacitance?

Section 10–8

13. In Figure 10–35, draw the connections, observing proper polarity, for connecting the electrolytic capacitor shown in series with the resistor.

FIGURE 10–35 Circuit for Problem 13

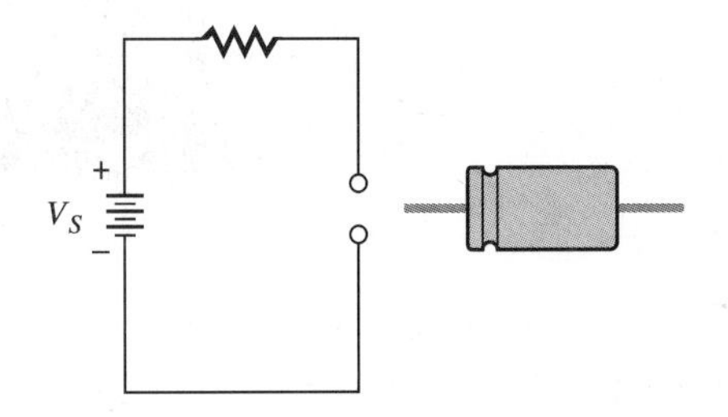

Section 10–10

14. For each circuit of Figure 10–36, compute the value of the total capacitance.

FIGURE 10–36 Circuit for Problem 14

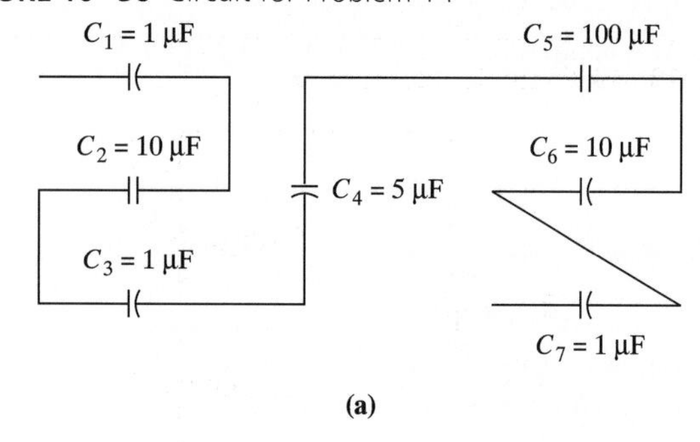

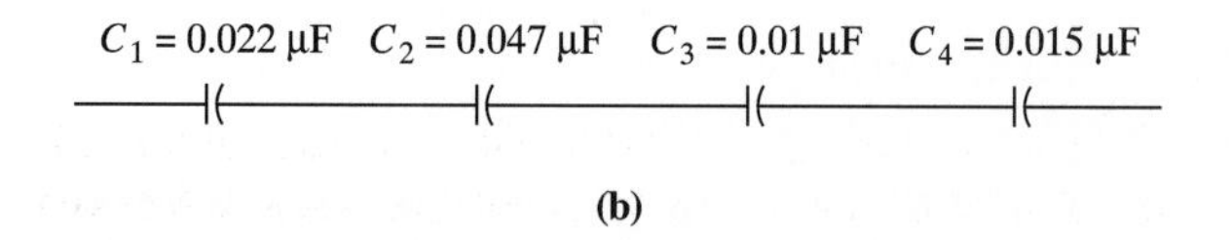

15. The total capacitance of the circuit of Figure 10–37 is 250 pF. Compute the value of C_2.

FIGURE 10–37 Circuit for Problem 15

$C_1 = 750$ pF $\quad C_2 = ?\quad C_3 = 750$ pF

16. For the circuit of Figure 10–38, compute the voltage developed across each capacitor.

FIGURE 10–38 Circuit for Problem 16

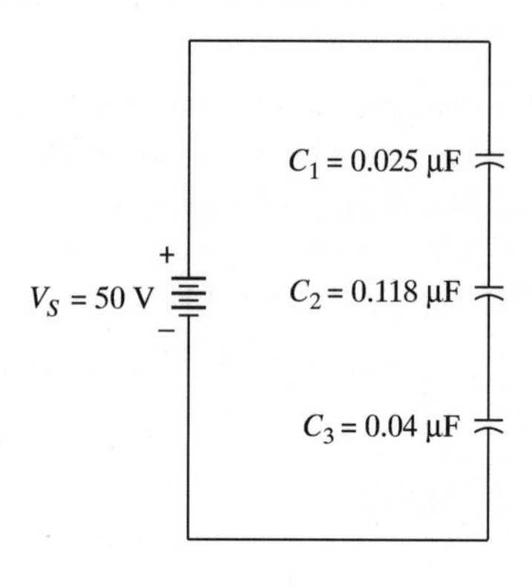

Section 10–11

17. For the circuit of Figure 10–39, compute the total capacitance.

FIGURE 10–39 Circuit for Problem 17

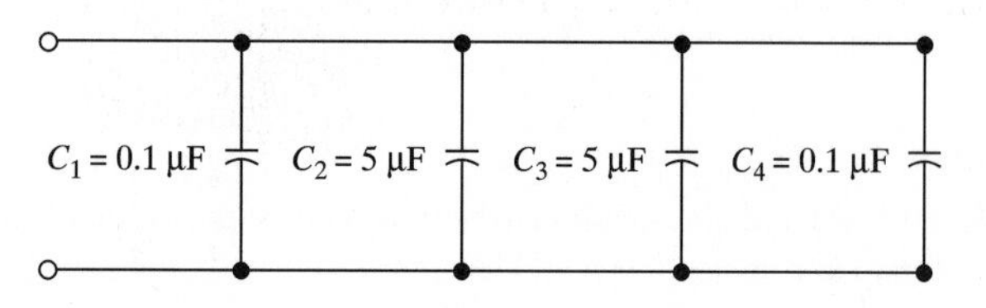

18. For the circuit of Figure 10–40, compute the total capacitance.

FIGURE 10–40 Circuit for Problem 18

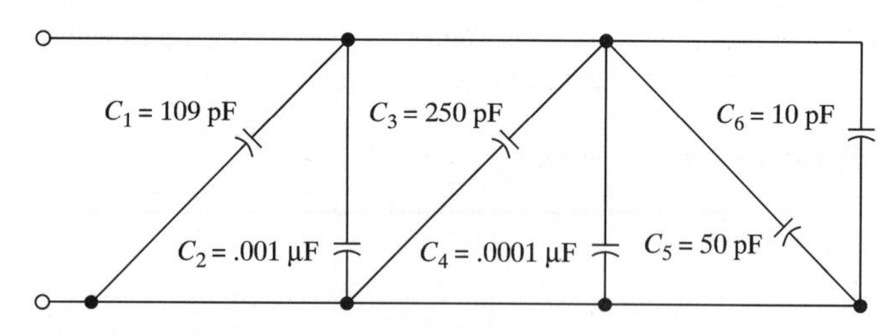

Section 10–12

19. Compute the time constant for a circuit consisting of a 10 μF capacitor and a 100 kΩ resistor.

20. How long would it take for the capacitor in Problem 19 to fully charge?

21. For the circuit of Figure 10–41, compute the voltage across the capacitor at the end of each time constant until it is fully charged?

FIGURE 10–41 Circuit for Problem 21

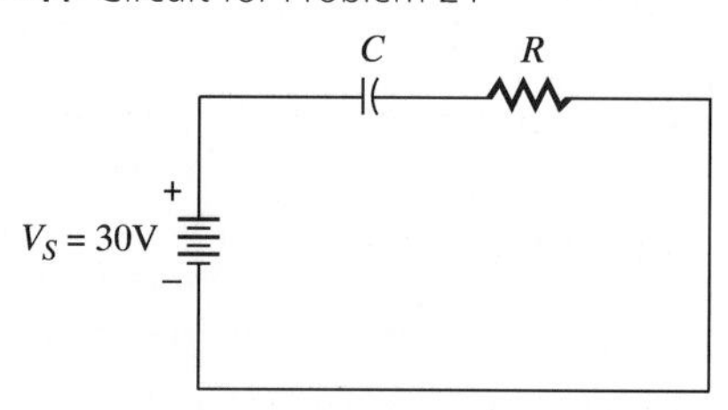

22. For the circuit of Figure 10–42, what is the capacitor voltage after 1.7 s?

FIGURE 10–42 Circuit for Problem 22

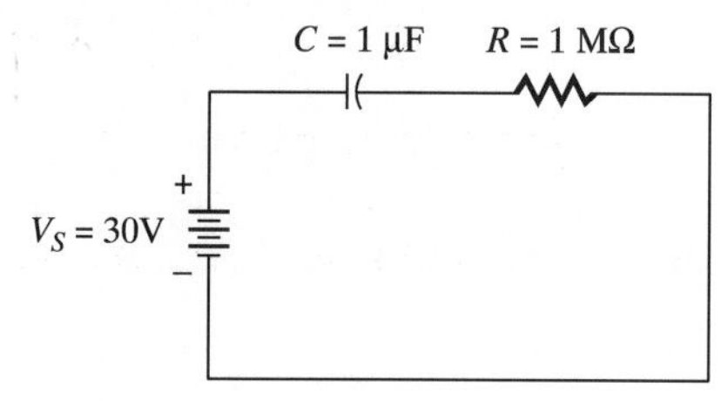

23. An *RC* circuit across a 160 V source is allowed to charge for two time constants and then discharge for four time constants. What is the final voltage?

24. A 50 μF capacitor is in series with a 1000 Ω resistor across a 90 V source. If the capacitor is fully charged, what is the capacitor voltage after it has discharged for 75 ms?

PROBLEM SET: CHALLENGING

25. What value capacitor must be placed in parallel with a 10 μF capacitor to give a total capacitance 8.33 μF?

26. For the parallel bank of Figure 10–43, compute the value of C_3 if the total capacitance is 0.08 μF.

FIGURE 10–43 Circuit for Problem 26

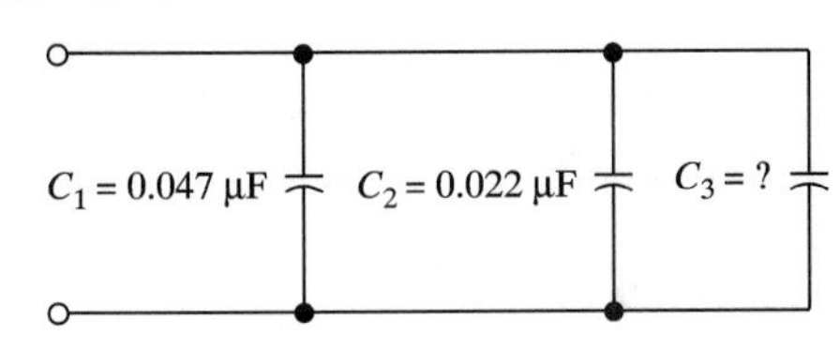

27. For the circuit of Figure 10–44, compute the total capacitance.

FIGURE 10–44 Circuit for Problem 27

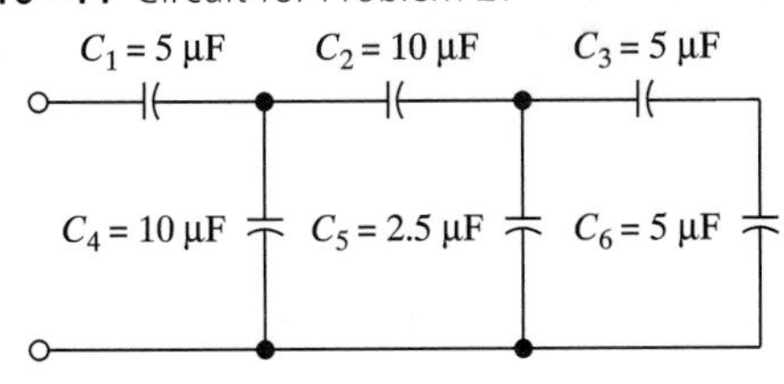

28. For the circuit of Figure 10–45, what is the voltage across C_5?

FIGURE 10–45 Circuit for Problem 28

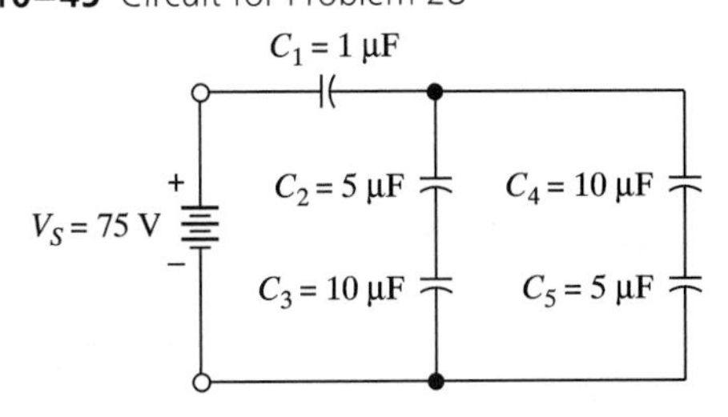

29. For the circuit of Figure 10–46, what is the voltage across C_2?

FIGURE 10–46 Circuit for Problem 29

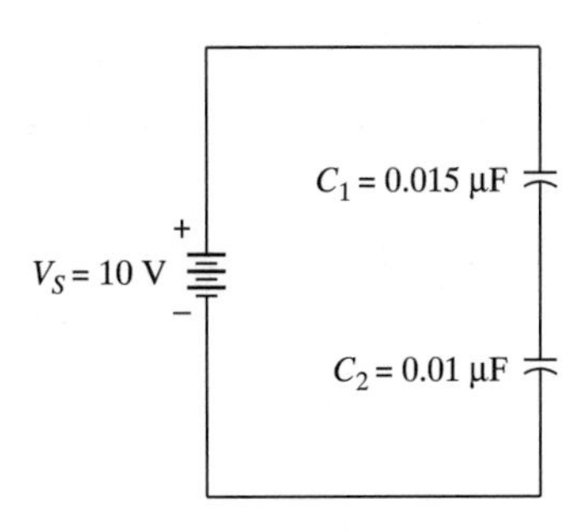

30. A capacitor is charged to 100 V and has a value of 5 μF. If the plates are moved so as to double the distance between them, what is the resulting voltage? Why?

31. For the circuit of Figure 10–47, compute the capacitance of C_2 and the voltages developed across C_1 and C_3.

FIGURE 10–47 Circuit for Problem 31

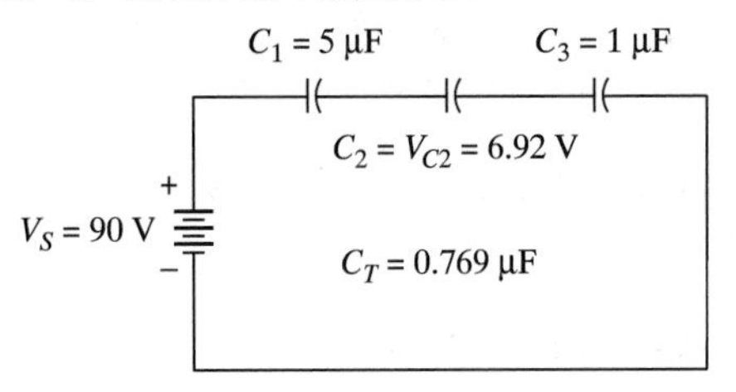

32. In the circuit of Figure 10–48, capacitor C_1 stores 0.06 μC of charge and has a value of 0.022 μF. What is the value of C_2 and the voltage developed across it?

FIGURE 10–48 Circuit for Problem 32

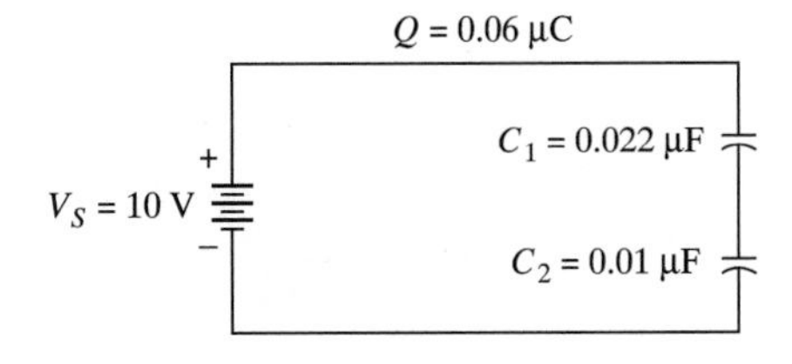

33. A mica capacitor is depicted in Figure 10–49. From the number printed upon it, what is its capacitance value?

FIGURE 10–49 Capacitor for Problem 33

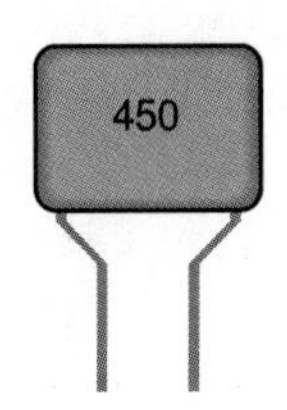

34. A ceramic disc capacitor is depicted in Figure 10–50. From the number and letters printed upon it, what is its capacitance value?

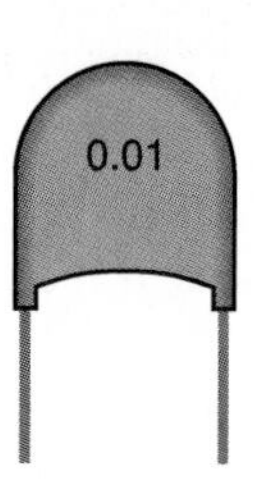

FIGURE 10–50 Capacitor for Problem 34

PROBLEM SET: TROUBLESHOOTING

35. The circuit in Figure 10–51 is measured and found to have a total capacitance of 0.014 μF. What is the probable fault?

FIGURE 10–51 Circuit for Problem 35

$C_1 = 0.047\ \mu F$ $C_2 = 0.01\ \mu F$ $C_3 = 0.02\ \mu F$

36. The circuit in Figure 10–52 is measured and found to have a total capacitance of 0.035 μF. What is the probable fault?

FIGURE 10–52 Circuit for Problem 36

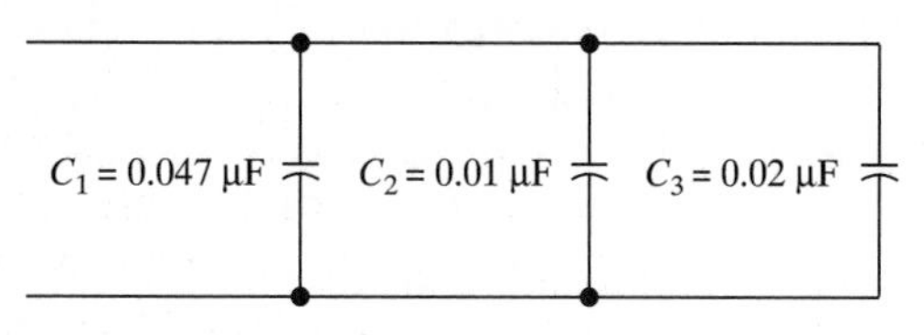

37. An ohmmeter is connected across a capacitor. After a long period of time, the resistance reading is 100Ω. Does this reading indicate a defective capacitor?

38. Figure 10–53 shows an electrolytic capacitor connected to a circuit. Is this capacitor connected correctly, or will there be some excitement when power is applied?

FIGURE 10–53 Electrolytic capacitor for Problem 38

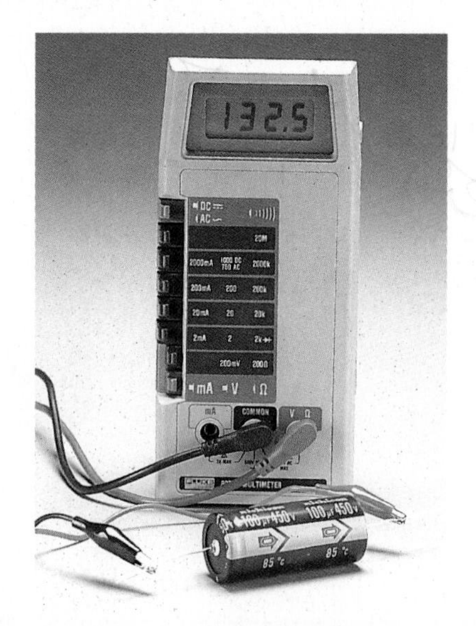

39. One trouble exists in the circuit shown in Figure 10–54. From the readings shown, what is the probable cause?

FIGURE 10–54 Circuit for Problem 39

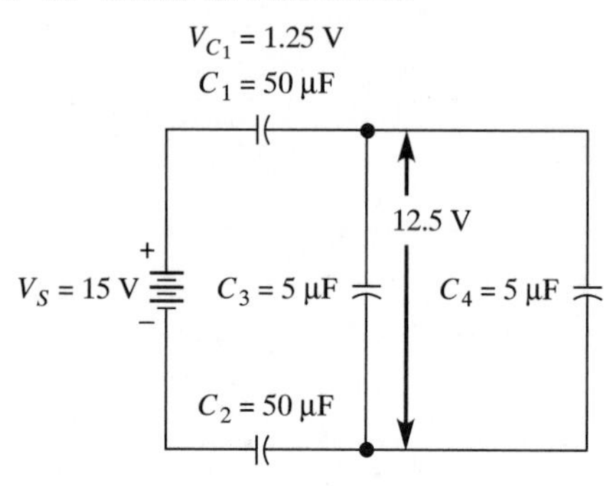

40. From the readings shown in Figure 10–55, what is the voltage across C_2?

FIGURE 10–55 Circuit for Problem 40

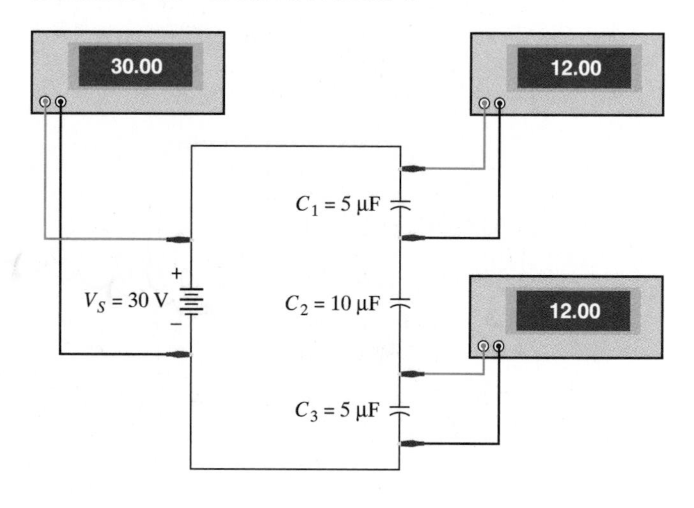

High-Energy Ultracapacitors

A reliable, cheap, long-lasting, and small electrical energy source ranks high as an important goal of science. The major emphasis of the research in this area has centered upon batteries. As you know, batteries produce electrical energy through chemical action. Once its chemical action is complete, a battery becomes useless, unless it is the type that can be recharged. In recharging, the chemical action is reversed and the battery once again becomes operational. But there is also a limit to the number of times the discharge-charge cycle can take place. Further, impurities in the battery cause a small amount of discharging to take place even when the battery is not in use, so it will have a definite shelf-life. These problems make the battery something less than the ideal energy source.

As you have learned in this chapter, the capacitor is a device capable of storing electrical energy for long periods of time. When the energy is needed, the capacitor can be made to act as a source and provide its stored energy to some load. In order to store a large amount of energy, however, the capacitor must be quite large in value. Large-value capacitors are of the electrolytic variety and are composed of metals and chemicals. There are several problems involved with electrolytic capacitors: (1) high leakage currents and (2) a definite shelf-life beyond which they are no longer usable. Thus, as energy sources, both batteries and electrolytic capacitors have the same disadvantages.

Recently, a breakthrough in electrical energy storage was announced by scientists at Eglin Airforce Base in Florida. Scientists at the Armament Laboratory were looking for a reliable energy source for a class of weapons known as hard target penetrators. These weapons must penetrate hard structures such as concrete bunkers. Once penetration has been accomplished, a voltage source must supply energy to a fuse to initiate a post-impact explosion. Thus, the energy source must be rugged enough to survive impact, small and light-weight enough to fit in the fuse, and most of all reliable. Reliable, in this case, means that the charge must be stored without loss for years and years. The scientists have developed such an electrical energy source. Known as an ultracapacitor, this device, once charged, will store electrical energy until it is needed with no need for recharging and no deterioration in its physical characteristics.

In the photograph below, the device on the right is a high-energy ultracapacitor. Notice that it is slightly smaller than a type C nicad battery. What is the capacity of the pictured ultracapacitor? Hold on to your hats, students of electronics! The capacitance of the ultracapacitor shown is 700 farads! A capacitance value of this magnitude would have been unthinkable a few years ago.

Two types are being researched: a liquid capacitor in the advanced phase of research and a dry type in the exploratory stage. Although details are sketchy at this time, the charge in the ultracapacitor is in the form of ions, which are trapped in tiny grooves within a special coating of titanium. The ions can be released upon demand to produce a flow of current, similar to the discharging of a conventional capacitor.

The ultracapacitor has several distinct advantages over other electrical energy sources. Unlike batteries, the

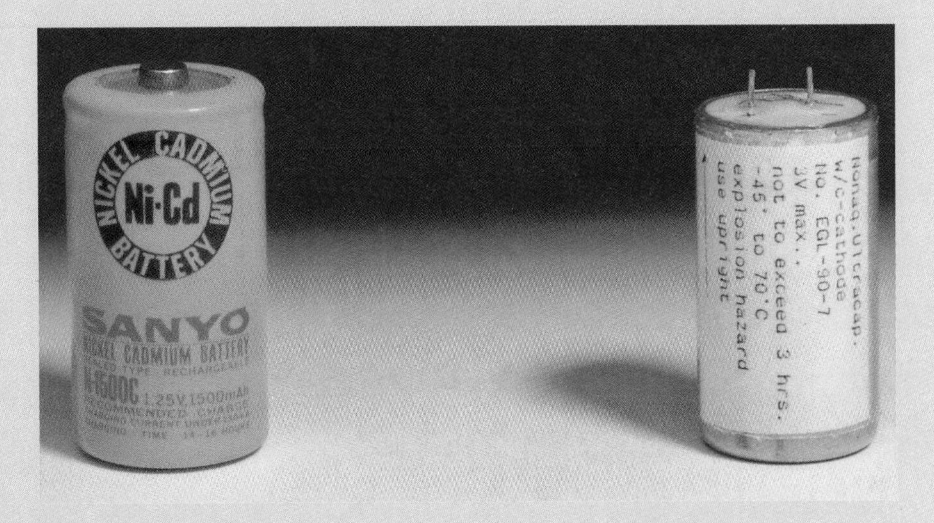

ultracapacitor can be recharged any number of times. Also unlike batteries, it has an unlimited shelf-life. Unlike the conventional capacitor, it has no leakage current. So once charged, there is nothing to stop it from storing this charge forever. Actually, no one is sure how long the charge will remain, 10 years or 100 years or what. Tests are now being conducted to determine the shelf-life and how long the ultracapacitor will hold its charge.

Possible applications of the ultracapacitor, other than its military applications, are many. Large-value capacitors of this type could be used as energy sources for the electrical systems in a car. Think of never having to buy another car battery! Ultracapacitors could also be used as the energy source for such items as electric carving knives, where leaving the capacitor in a discharged state would greatly enhance the item's safety when not in use.

CHAPTER

11 Characteristics of Alternating Current and Voltage

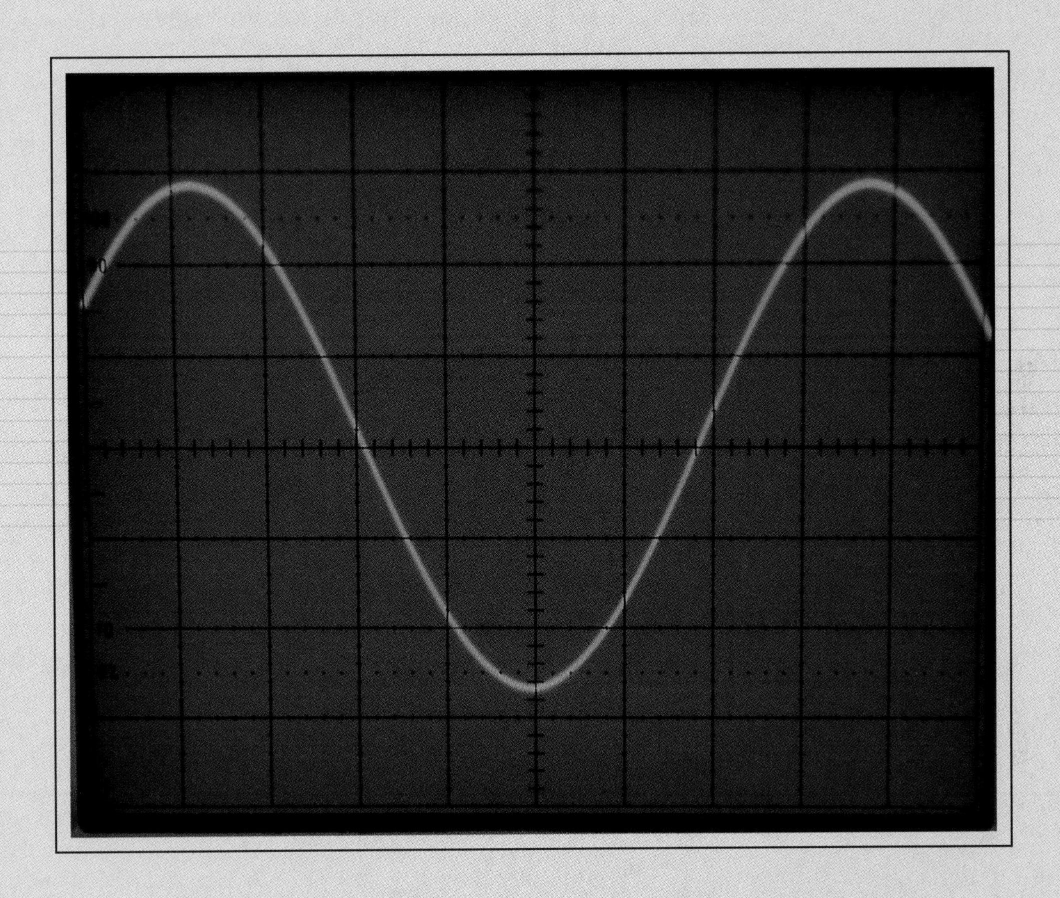

UPON COMPLETION OF THIS CHAPTER, YOU WILL BE ABLE TO

1. Explain the differences between alternating and direct current.
2. Sketch and explain the graph of the sine function.
3. Diagram and explain the generation of a sine wave.
4. Compute the instantaneous value of a sine wave of voltage.
5. Express angular measure in both degrees and radians.
6. Define cycle, alternation, period, and frequency.
7. Define and solve for the average and rms values of a sine wave.
8. Illustrate phase relationships of waveforms using phasors.
9. Define fundamental and harmonic frequencies.
10. Explain the differences between sinusoidal and nonsinusoidal waves.
11. Understand how the oscilloscope is used in the analysis of sinusoidal and nonsinusoidal waves.

Up to this point, the text has dealt with direct current, or current that flows in only one direction. This chapter introduces alternating current, or current that changes direction periodically over time. The changes in direction of alternating current are produced by an alternating voltage, which changes in polarity periodically. The way in which alternating voltages and currents are produced is introduced in this chapter.

An alternating current proceeds down a conductor in what is known as a wave-like fashion, similar to the action of an ocean wave. The most basic of these waves is known as a sine wave. The sine wave gets its name from the fact that its variations follow the values of the sine trigonometric function. Thus the level of mathematics for this chapter must be raised to include the trigonometric functions. Of all the alternating current waves, the sine wave will be given by far the most attention in the remainder of this text. All other waves are known as nonsinusoidal waves. Of these, the square wave, pulsed wave, and sawtooth wave are introduced in this chapter.

In alternating voltage and current waves, you encounter quantities of voltage and current that are continuously changing. The question comes to mind: How do you place a value on a quantity that is continuously changing? In this chapter, you learn how to express these varying quantities in meaningful ways.

The material in this chapter is a radical departure from what you have experienced up to this point. However, it is the foundation stone for the learning which is to follow in the remainder of the text.

11–1 DIFFERENCES BETWEEN ac AND dc

Question: How does alternating current differ from direct current?

In the circuit of Figure 11–1(a), a dc voltage source is connected to a load. As the graph of Figure 11–1(b) shows, the current of this circuit is a constant value over time. A variable dc voltage source is shown across the load in Figure 11–1(c). If this source voltage is varied over time, the current varies proportionately. Although the current may vary, it does not change directions. A varying dc current is illustrated in Figure 11–1(d).

In Figure 11–2(a), an ac voltage source, symbolized as shown, is applied to the load. As shown in Figure 11–2(b), the polarity of the voltage produced by an ac source reverses periodically. During period *A* of the graph, the top of the source is positive and the bottom negative. During period *B*, the top of the source is negative and the bottom is positive. Figure 11–2(c) shows the current produced by this source. During period *A*, the current through the load flows in a direction determined by the polarity of the source. During period *B*, the current flows in the opposite direction, once again under the influence of the source. Thus, an **alternating voltage** is one whose polarity changes periodically, while an **alternating current** is one that changes direction periodically.

Question: What is meant by a waveform?

The graph of the current shown in Figure 11–2(b) repeats itself over time as the polarity of the source voltage alternates. It follows a wave-like pattern. An alternating current wave may take several different forms such as the sine wave, sawtooth wave, and square waves, as illustrated in Figure 11–3. The form (shape) of the wave are referred to as waveforms.

Question: What are the differences in the uses of direct and alternating current?

Differences exist in the uses to which alternating and direct current are put. Unlike direct current, the characteristics of alternating current allow it to be transported by wire over

FIGURE 11–1 Direct voltage and current

FIGURE 11–2 Alternating voltage and current

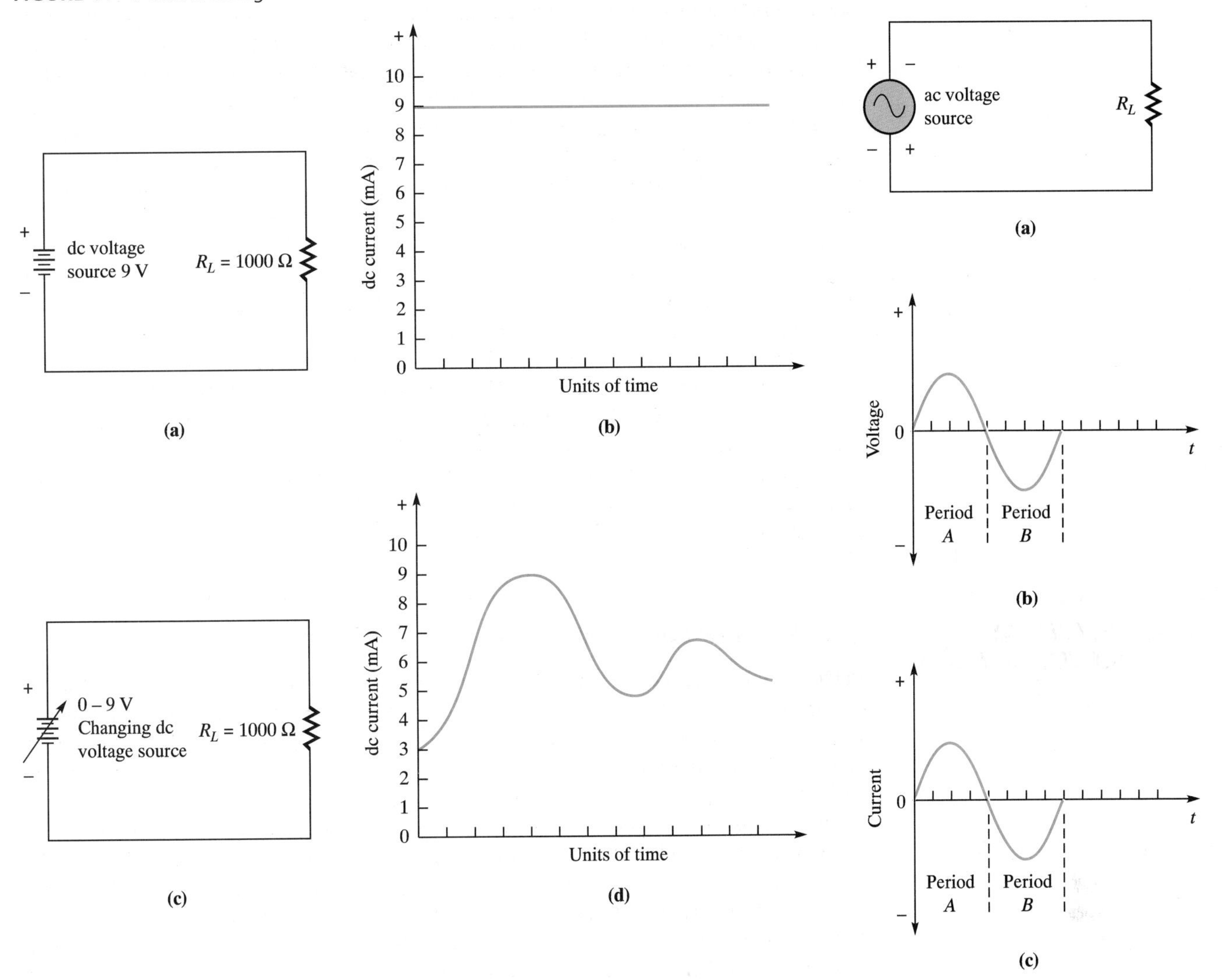

great distances with a relatively small loss in electrical energy. Since commercial production and distribution of electrical power must be done as efficiently as possible, commercial power systems utilize alternating current. Direct current is used in applications where the load is close to the source. For example, dc voltages and current are used to supply power to electronic circuits such as amplifiers and oscillators that make up electronic systems.

Another example of something that can be done with ac but not dc is in the field of electronic communications. Very rapidly alternating currents, on the order of several million alternations or more, can be made to produce electromagnetic fields capable of radiating from an antenna. Prior to radiation, intelligence in the form of voice, music, or television images are impressed upon the currents through a process known as *modulation*. In a receiving antenna, the radiated fields produce currents from which the intelligence is detected and reproduced through a process known as *demodulation*. These processes are the basis for long-distance communication systems such as radio and television.

FIGURE 11–3 Common types of waveforms

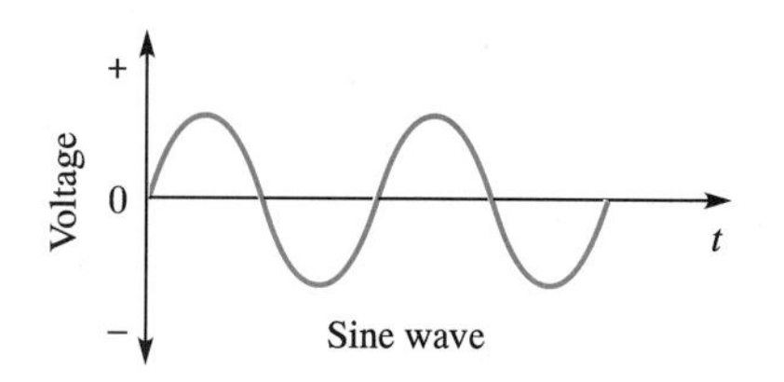

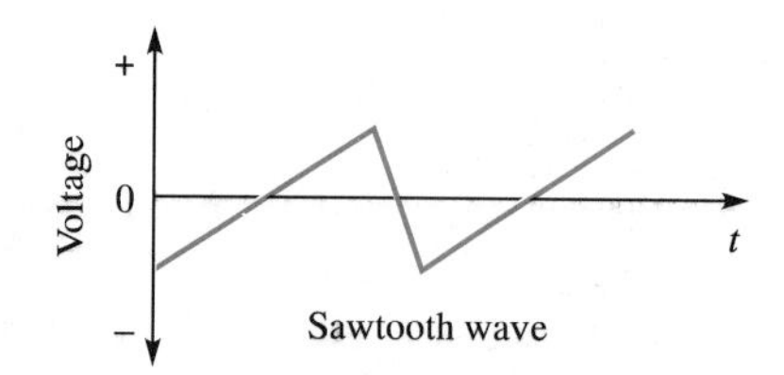

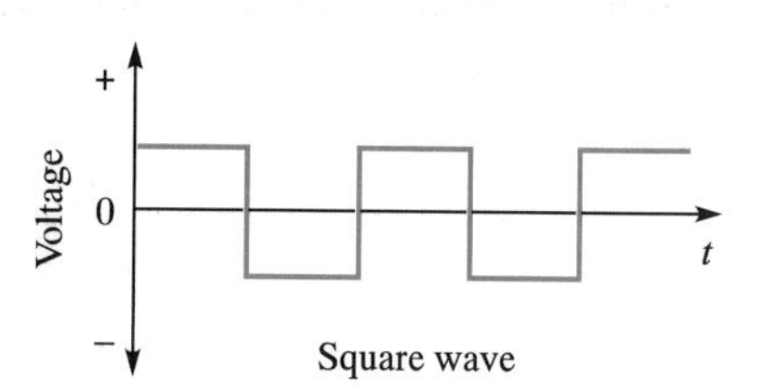

SECTION 11–1 REVIEW

1. What is alternating current?
2. What type of voltage produces alternating current in a load?
3. In Figure 11–4, is the current waveform ac or dc?
4. What is the major difference in the use of ac and dc?

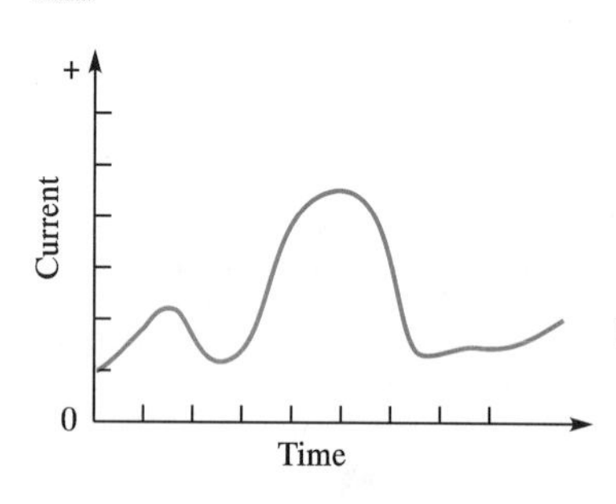

FIGURE 11–4 Waveform for Section Review Problem 3

11–2 SINE WAVES

The basic waveform for alternating current and voltage is known as the sine wave, due to its relationship to the trigonometric function known as the sine function. A brief review of this function is given next as an aid to understanding the generation of alternating voltage and current.

The Sine Function

As you may know, the trigonometric functions are numbers that relate magnitudes of the sides and angles of a right triangle. The three basic trigonometric functions are sine (sin), cosine (cos), and tangent (tan). The equations for these functions and the right triangle are shown in Figure 11–5. The sine function is covered in this section; the other two functions will be discussed in later chapters when needed.

FIGURE 11–5 Trigonometric functions

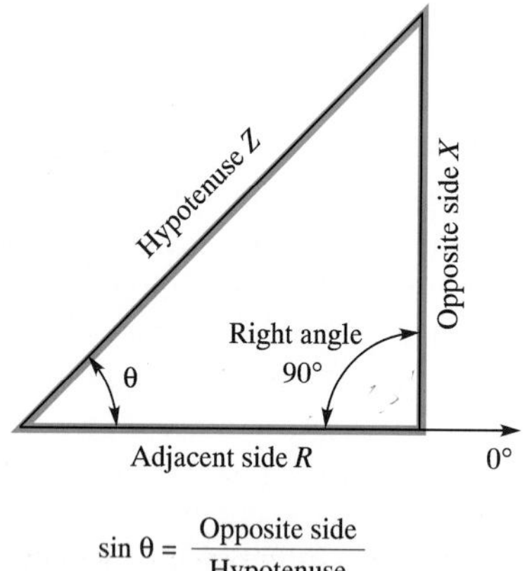

$$\sin\theta = \frac{\text{Opposite side}}{\text{Hypotenuse}}$$

$$\cos\theta = \frac{\text{Adjacent side}}{\text{Hypotenuse}}$$

$$\tan\theta = \frac{\text{Opposite side}}{\text{Adjacent side}}$$

Question: What is the sine function and what are its limits?

In Figure 11–5, one of the acute angles is labeled θ (Greek theta), and the sine function for this angle is given as the ratio of the opposite side (the side labeled X) to the hypotenuse (the side labeled Z) of a right triangle. Notice that the reference for the angle is zero degrees, which is parallel to the adjacent side R. For purposes of this discussion, side Z is considered as one unit long, although the length itself is unimportant. If side Z is rotated clockwise toward the reference, the length of side X decreases, as shown in Figure 11–6(a). When angle θ is zero degrees, side X becomes zero units long, and the sine, which is the ratio X/Z, is also zero ($\sin\theta = 0/1 = 0$). As the hypotenuse is rotated counterclockwise, the angle θ and side X both increase, as shown in Figure 11–6(b). When angle $\theta = 90°$, side X is the same length as the hypotenuse and thus the sine is ($\sin\theta = 1/1 = 1$). Continued rotation in a counterclockwise direction is shown in Figure 11–6(c). Side X becomes zero once again when angle $\theta =$ at 180° making the sine once again zero. Continuing the counterclockwise rotation in Figure 11–6(d), side X becomes longer and reaches maximum length when angle $0 = 270°$, at which it is the same length

FIGURE 11–6 Demonstration of the sine function

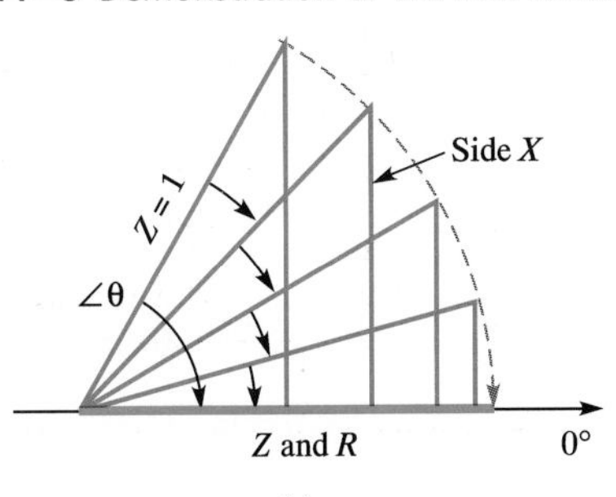

(a)
When $\angle\theta = 0°$, $X = 0$ and $\sin\theta = 0$.

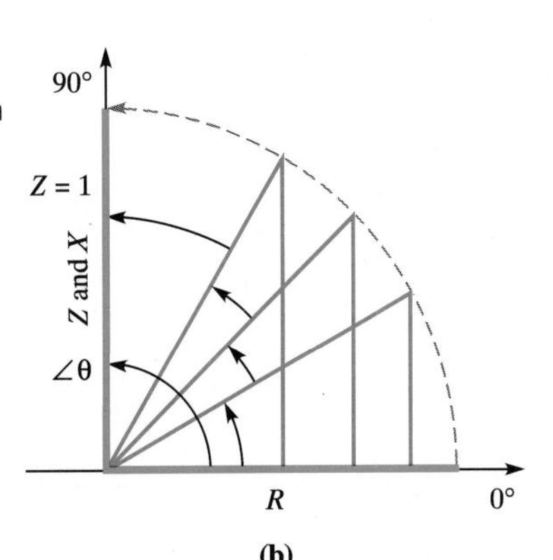

(b)
When $\angle\theta = 90°$, $X = 1$ and $\sin\theta = 1$.

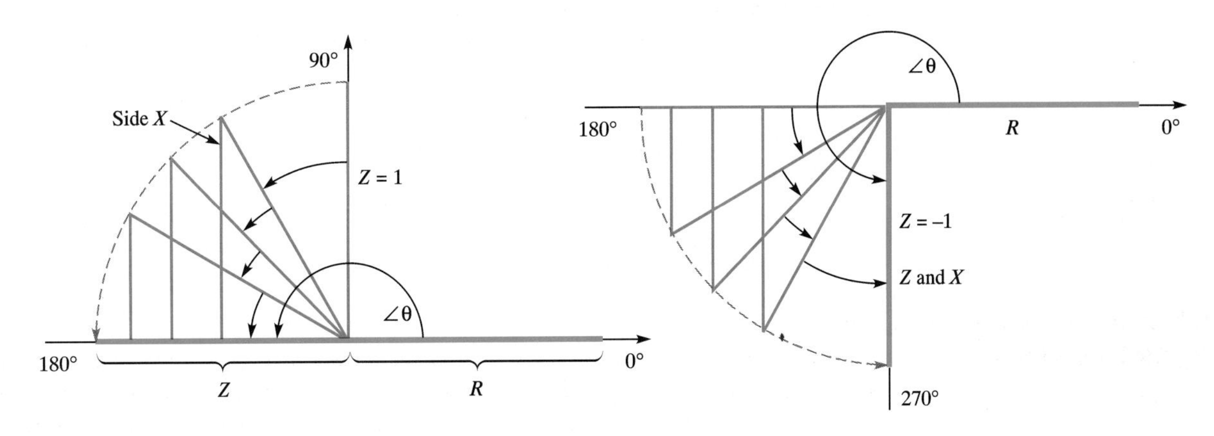

(c)
When $\angle\theta = 180°$, $X = 0$ and $\sin\theta = 0$.

(d)
When $\angle\theta = 270°$, $X = -1$ and $\sin\theta = -1$.

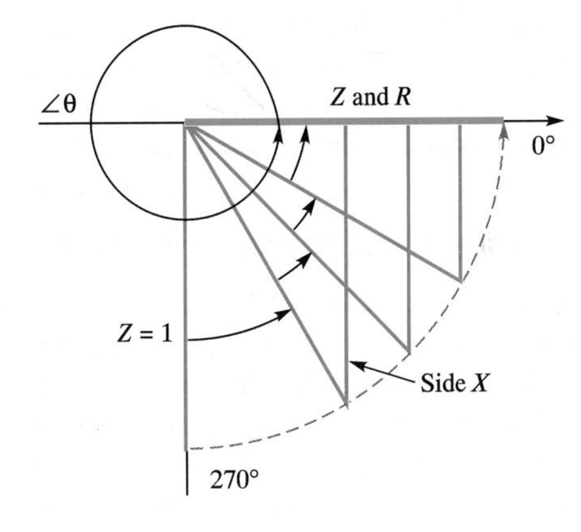

(e)
When $\angle\theta = 360°$, $X = 0$ and $\sin\theta = 0$.

as the hypotenuse and the sine is -1. ($\sin\theta = -1/1 = -1$). In Figure 11–5(e) the counterclockwise rotation continues until angle $\theta = 360°$. Angle θ becomes smaller as 360° is approached, at which point it is once again zero degrees and the sine is also zero.

The values of the sine can be plotted on a graph as shown in Figure 11–7. The sine functions for other angles have been added in order to more closely plot the function. Notice that the range of the sine is from $+1$ to -1.

FIGURE 11–7 Graph of sine function values

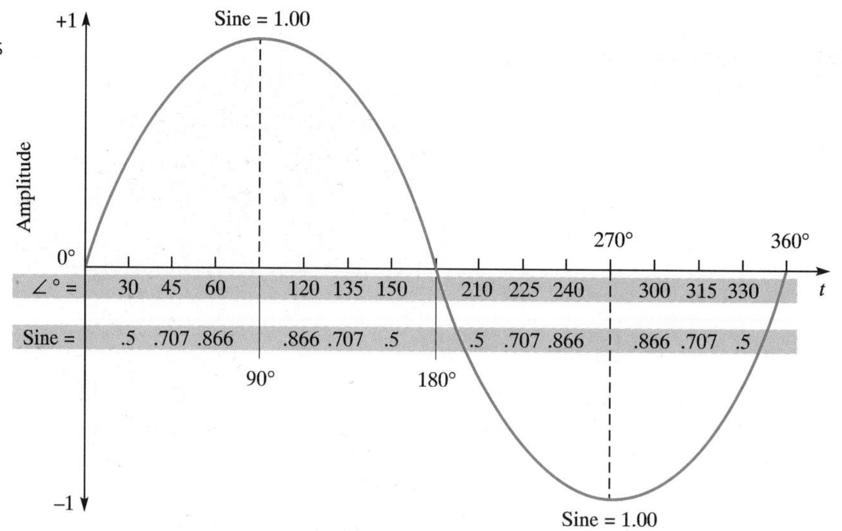

NOTE

The values of the sine, cosine, and tangent functions for any angle can be found using a scientific calculator. The value of the angle is entered into the calculator and the appropriate key (sin, cos, tan) depressed. The display then shows the value of the desired function.

■ Generation of a Sine Wave of Voltage

Recall from Chapter 8 that there are three items required for electromagnetic induction: a magnetic field, a conductor, and relative motion between the two. In generating sine wave voltages, these three items are brought together in what is known as an **alternator.** As illustrated in Figure 11–8, an **alternator** basically consists of a loop of wire (the conductor) that is rotated to create motion within a magnetic field. In understanding its operation, two facts must be kept in mind: (1) The relative magnitude of the voltage produced is *proportional* to the angle at which the conductor cuts the lines of force, and (2) its polarity depends upon the *direction* of motion.

Question: How is a sine wave of voltage produced?

The production of a sine wave of voltage is also illustrated in Figure 11–8. In Figure 11–8(a), the highlighted portion of the loop is at the reference position and the angle of rotation is at zero degrees. At the instant the loop begins to rotate, it is moving parallel to the lines of force and no voltage is induced on the loop. In Figure 11–8(b), the loop has rotated counterclockwise so that the angle of rotation is now at 90°. Notice that over this period, the angle at which the conductor cuts the lines of force also increases from zero to 90°. As the angle of cutting increases, the amount of voltage induced increases, and reaches its maximum when the angle of rotation is 90°. In Figure 11–8(c), the loop continues to rotate in a counterclockwise direction, and the angle of rotation moves from 90° to 180°. Over this period, the angle of cutting also decreases until the angle of rotation is 180°, when the loop is once again moving parallel to the lines of force. The voltage is zero at this time.

In Figure 11–8(d), as the loop rotates from 180° to 270°, a major change takes place. The motion is now in the opposite direction, and the polarity of the induced voltage

FIGURE 11–8 Generation of a sine wave with an alternator

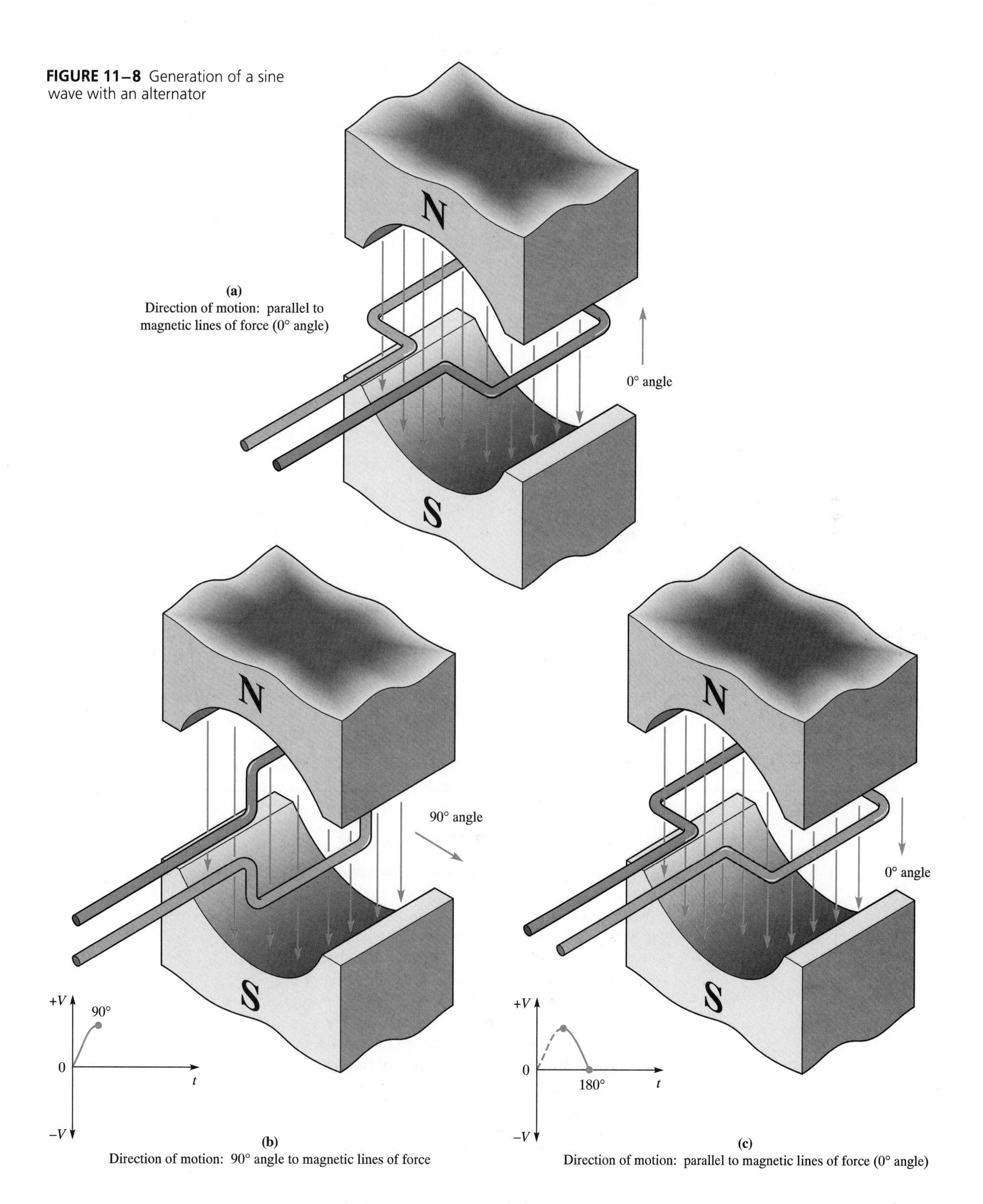

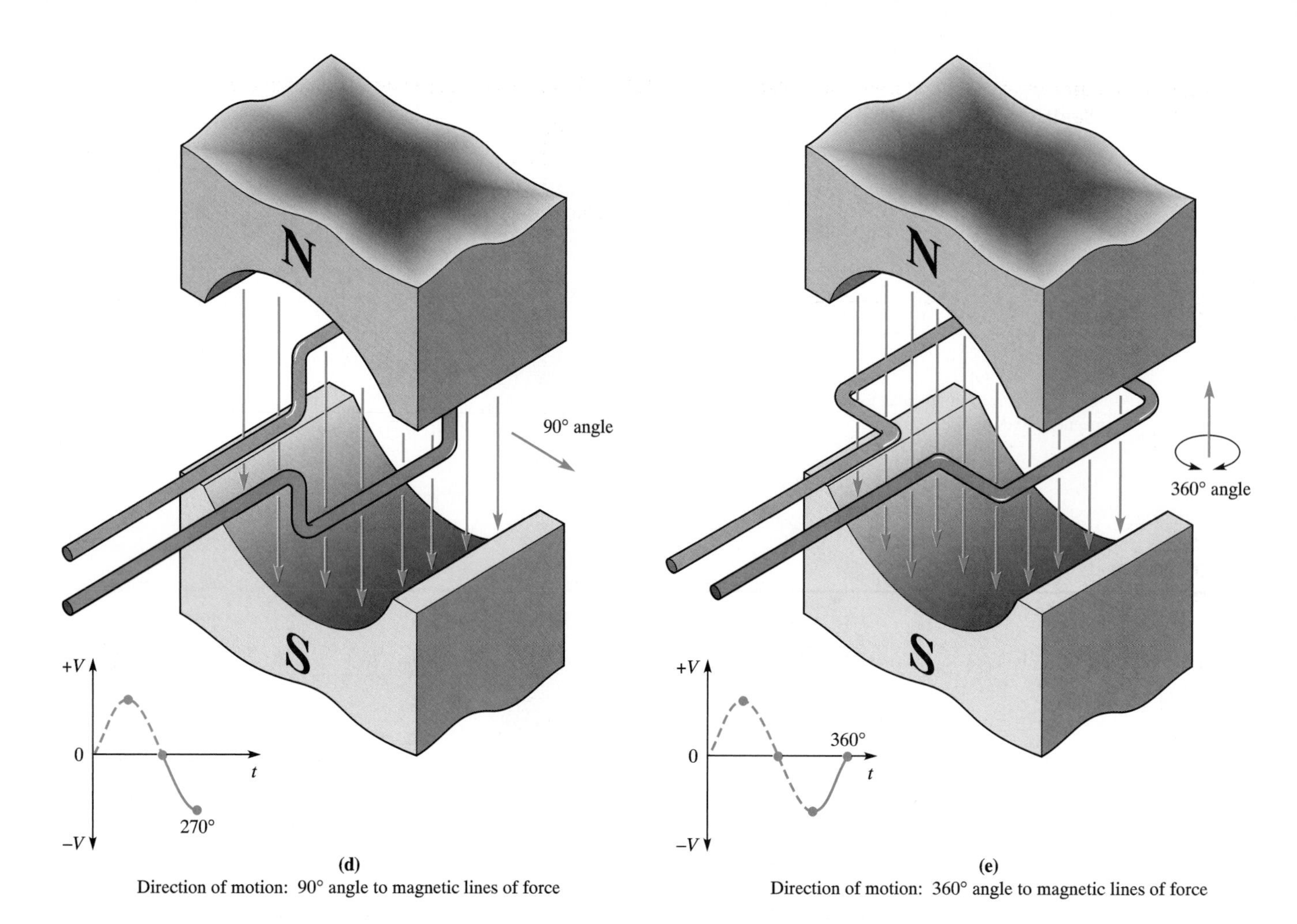

(d)
Direction of motion: 90° angle to magnetic lines of force

(e)
Direction of motion: 360° angle to magnetic lines of force

reverses. Over this period, the angle of cutting increases to 90°, and the voltage increases to its maximum value, but of the opposite polarity. In Figure 11–8(e), the loop rotates from 270° to 360°. The angle of cutting and the induced voltage decrease to zero.

Notice that for each segment of rotation of the loop in Figure 11–8, a small graph shows the voltage produced for that segment. A graph of the voltage produced for the full rotation of the loop is shown in Figure 11–9. Note the similarity of this graph to the plot of the sine function in Figure 11–7. They both reach zero and maximum at the same points. They both change polarity at the same points. In fact, the voltage produced at every point on the graph is proportional to the sine of the angle of rotation. Thus, a **sine wave** of voltage follows the shape of the sine function when plotted on a graph.

One rotation of the alternator produces what is known as one **cycle** of voltage. A cycle of a sine wave is illustrated in Figure 11–10. The portion of the cycle between 0° and 180° is known as the **positive half-cycle,** while the portion between 180° and 360° is known as the **negative half-cycle.** The line upon which the sine wave proceeds is known as the *baseline*. The baseline represents zero volts on the vertical axis and units of time on the horizontal axis. The magnitude of the voltage is represented by the vertical axis and is often referred to as the voltage **amplitude.**

Instantaneous Value of Voltage

In the previous discussion, notice that the value of the voltage changes from one instant to the next. Thus, at any instant, the sine wave has a value known as its instantaneous

FIGURE 11–12 Time periods of a sine wave

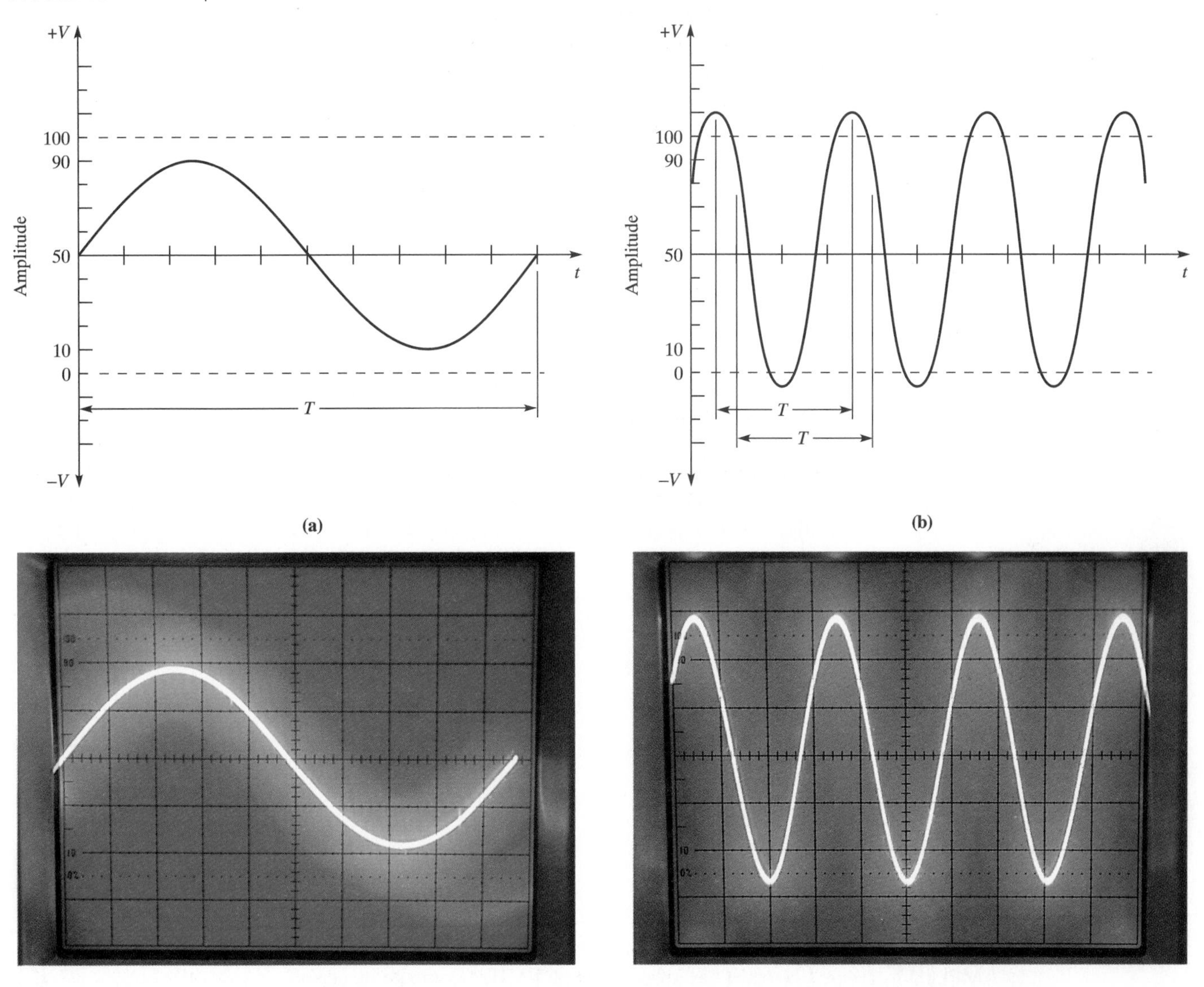

Period T is illustrated in Figure 11–12(a). Actually, period can be measured form any point on a given cycle to the corresponding point on the next, as is illustrated in Figure 11–12(b).

Frequency

Question: What is frequency and how is it determined?

The number of cycles that a waveform completes in one second is known as its **frequency,** which is symbolized by f and measured in units of cycles per second. Formerly, cycles per second was abbreviated cps. In the mid-seventies, this designation was changed to hertz (Hz) in honor of Heinrich Hertz, the German physicist who first demonstrated the production of radio waves. An example of a frequency with which almost everyone is familiar is the 60 Hz commercial power. It indicates that the power provided to homes and factories completes 60 cycles in one second.

Question: How are period and frequency related?

The period of a waveform determines its frequency. The longer the period of one cycle, the lower the frequency. Period and frequency are related mathematically in the following equations:

(11–1)
$$f = \frac{1}{T}$$

and

(11–2)
$$T = \frac{1}{f}$$

As these equations show, a reciprocal relationship exists between frequency and period. The 1 in these equations represents one second. In Eq. 11–1, dividing 1 s by the time of 1 cycle yields the number of cycles completed in 1 s. In Eq. 11–2, dividing 1 s by the frequency yields the time of 1 cycle. Thus, a *longer* period results in a *lower* frequency, while a *shorter* period results in a *higher* frequency. Also, the greater the number of cycles to be repeated in one second, the shorter the time (period) available for completing each.

EXAMPLE 11–1 A sine wave has a period of 100 μs. Compute its frequency.

■ Solution:

Use Eq. 11–1:

$$f = \frac{1}{T}$$

$$= \frac{1}{100 \times 10^{-6}} = 10 \text{ kHz}$$

EXAMPLE 11–2 A sine wave has a frequency of 2 MHz. Compute its period.

■ Solution:

Use Eq. 11–2:

$$T = \frac{1}{f}$$

$$= \frac{1}{2 \times 10^{6}} = 0.5\ \mu\text{s}$$

EXAMPLE 11–3 The waveform of Figure 11–13 covers exactly 20 ms. Compute its period and frequency.

■ Solution:

Four cycles of the waveform are included in Figure 11–13, and they are known to cover 20 mS. The time of 1 cycle, or period, can be found by dividing the time of the waveform

FIGURE 11–13 Waveform for Example 11–3

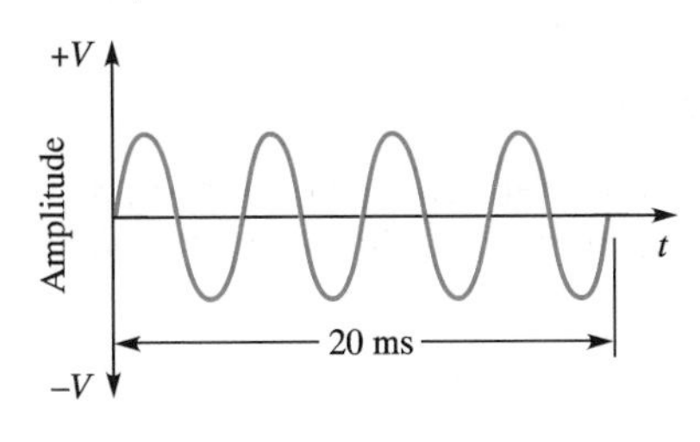

by the number of cycles included. Thus, use

1. To find the period:

$$T = \frac{20 \times 10^{-3}}{4} = 5 \text{ ms}$$

2. Now find the frequency:

$$f = \frac{1}{T}$$

$$= \frac{1}{5 \times 10^{-3}} = 200 \text{ Hz}$$

■ Rate of Change of a Sine Wave

Question: Is the rate of change of voltage constant throughout a sine wave with respect to time?

The rate of change of a sine wave is the rate at which the voltage is increasing or decreasing at any instant. The rate of change is found by the ratio of $\Delta v/\Delta t$, where v and t are instantaneous values of voltage and time, respectively. As illustrated in Figure 11–14(a), between times t_1 and t_2 the voltage changes from v_1 to v_2, so that $\Delta t = t_2 - t_1$ and $\Delta v = v_2 - v_1$. If the interval were started at an earlier point, as shown in Figure 11–14(b), the time difference (Δt) and the voltage difference (Δv) would be greater, and the rate of change ($\Delta V/\Delta t$) would thus be greater. If the interval were started at a later point, the differences and the rate of change would be smaller, as illustrated in Figure 11–14(b). If the time interval is small enough, the.rate of change approaches a single point.

Thus, the rate of change is not constant over time. At points A, C, and E in Figure 11–15, the voltage is crossing the baseline, or zero point for current or voltage, and is changing polarity, the rate of change is greatest. At points B and D, the voltage is greatest. At these two points, the voltage has stopped increasing and will soon begin to decrease, and the rate of change is zero.

FIGURE 11–14 Rate of change in voltage throughout a sine wave

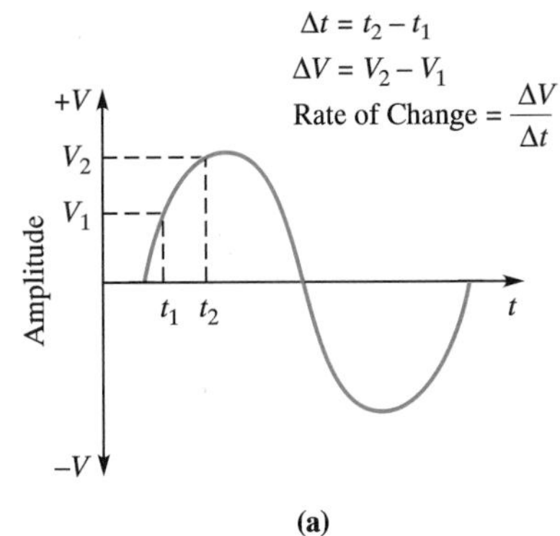

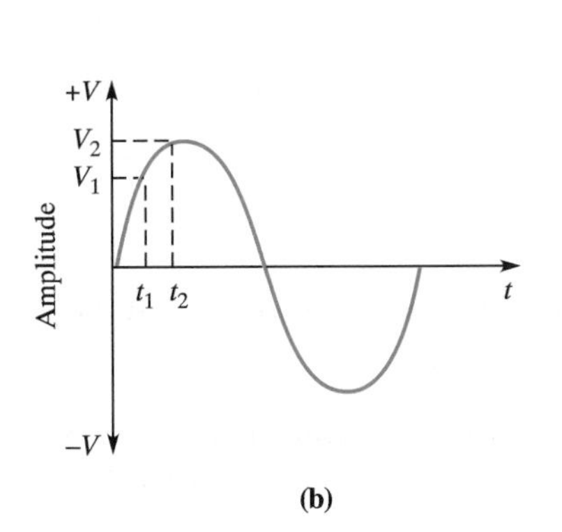

FIGURE 11–15 Variations in the rate of change: *A, C,* and *D* are points where the rate of change is maximum; *B* and *D* are points where the rate of change is zero.

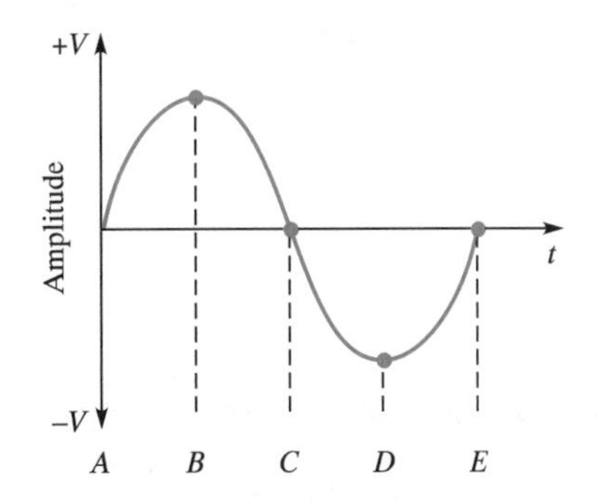

Points A, C, and E are Maximum Rate of change. B and D are points where Rate of change is zero.

SECTION 11–3 REVIEW

1. Define the period of a wave.
2. Define the frequency of a wave.
e. How are the period and frequency of a wave related?
4. What characteristics must a waveform have if it is sinusoidal?

11–4 PHASORS

Question: What is a phasor?

A **phasor** is a line, drawn from an origin, that has both magnitude and direction. Its magnitude is indicated by its length and its direction by its angular displacement from a reference, as illustrated in Figure 11–16(a). The reference, line *O-A*, is established horizontally to the right of the origin. The phasor, line *O-B*, lies directly over the reference. This phasor will rotate counterclockwise around the origin over a circular plane. The quadrants of this circular plane are labeled with the angles that they form in relation to the reference.

In Figure 11–16(b), a phasor three units in magnitude is rotated counterclockwise from the reference to form a 30° angle. In Figure 11–16(c), the phasor is five units long and is rotated counterclockwise from the reference to form a 135° angle. In Figure 11–16(d), the phasor is five units long and rotated counterclockwise to form a 225° angle. In Figure

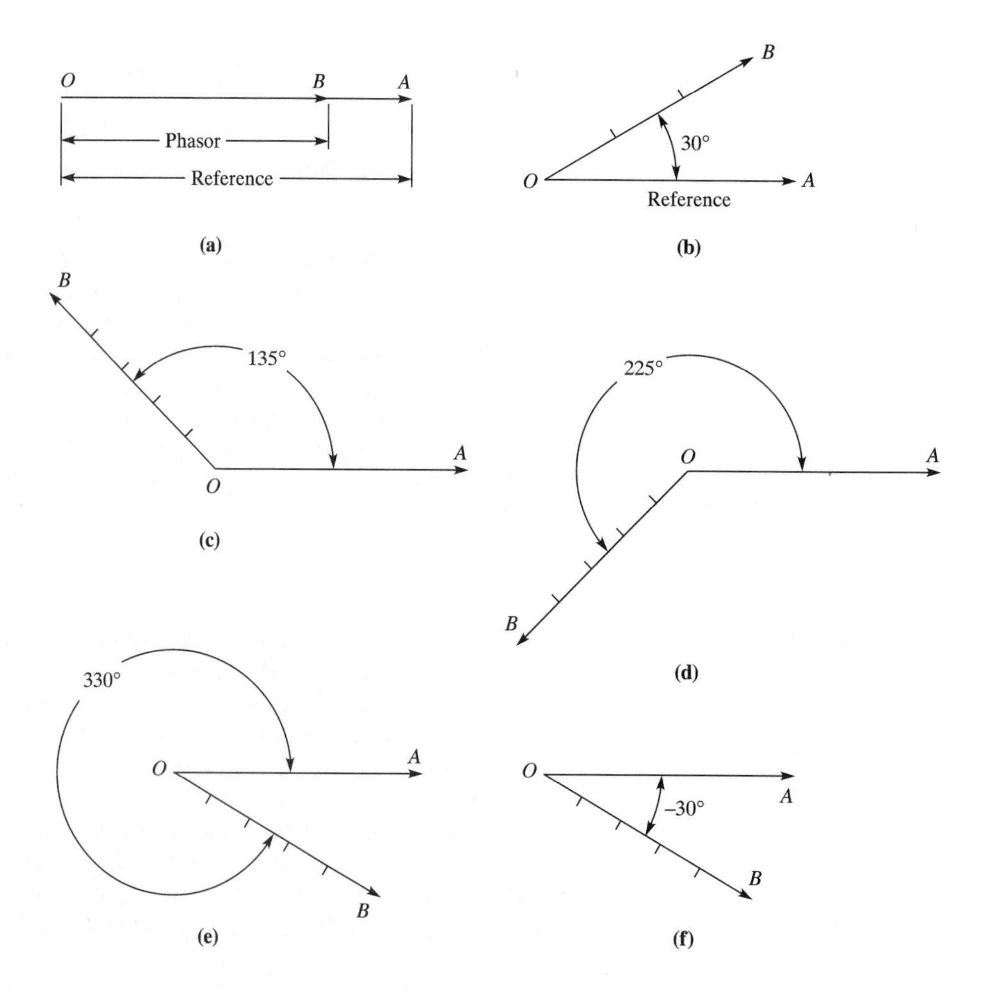

FIGURE 11–16 Direction and magnitude of phasors

11–16(e), the phasor is five units long and rotated counterclockwise to form a 330° angle. A rotation of 360° returns the phasor to the reference point once again. In Figure 11–16(f), the phasor has been rotated 30° in a clockwise direction. This position is the same as in Figure 11–16(e), but the angle is said to be −30° angle. Thus, positive angles result from counterclockwise rotation and negative angles from clockwise rotation.

Question: How are phasors used in the representation of ac quantities?

FIGURE 11–17 Phasor representation of ac quantities

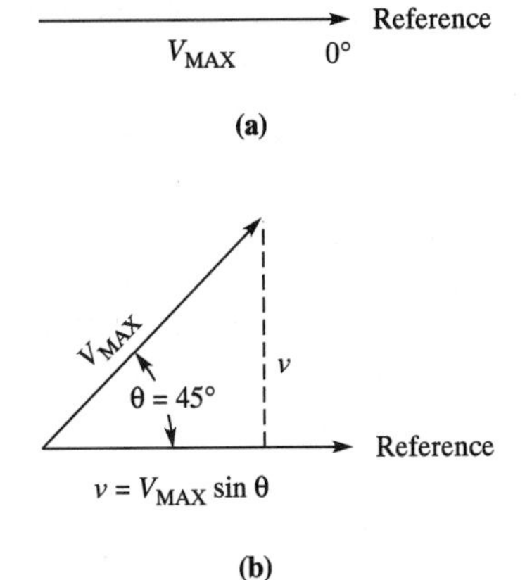

Recall that sine waves are produced by the rotation of an alternator. This rotation produces a voltage that varies continuously in amplitude over time. Phasors can be used to represent such time-varying quantities. In Figure 11–17(a), a phasor, representing the maximum voltage amplitude value of a sine wave, lies directly over the reference. Recall that in this position, no voltage is produced by the alternator. The phasor will rotate in step with the alternator. In Figure 11–17(b) the rotation is stopped after 45°. A line dropped vertically from the tip of the phasor to the baseline represents the *instantaneous value* of the sine wave. Notice that the geometric figure produced is a right triangle. The side representing the instantaneous voltage value is the side *opposite* the angle of rotation. The hypotenuse of the right triangle represents the maximum voltage. The sine of the angle of rotation is thus equal to the ratio of the instantaneous voltage v and maximum voltage V_{max} values. In equation form, the sine of the angle of rotation is found by

$$\sin \theta = \frac{v}{V_{max}}$$

Transposing gives the equations for instantaneous voltage

(11–3) $$v = V_{max} \sin \theta$$

and instantaneous current

(11–4) $$i = I_{max} \sin \theta$$

Note the lower case notation for voltage and current in Eqs. 11–3 and 11–4. This notation is used when the quantity measured has a dynamic, or changing, value. This will often be the case where ac quantities are concerned.

Eq. 11–3 states mathematically that the instantaneous voltage value of a sine wave produced by an alternator is a function of the sine of the angle of rotation. In Table 11–1, the instantaneous voltage values of a sine wave with a maximum value of 100 V are computed for several points throughout one cycle. These data are shown plotted in Figure 11–18. Note that the curve plotted does follow the general form of the sine trigonometric function.

Phase Difference

The word *phase* refers to a time displacement between two events. Two events occurring at the same time are said to be in phase, while events occurring at different times are out of phase. Figure 11–19(a) shows two sine waves of the same frequency drawn on a common baseline. Notice that waveform *B* crosses the baseline in a positive direction, point *X*, at a time earlier than waveform *A*, point *Y*. Notice further that waveform *B* crosses the baseline in a negative direction, point *X′*, at a time earlier than waveform *A*, point *Y′*. Thus, the two waveforms are out of phase. Waveform B, occurring at an earlier time, is said to *lead* waveform *A*. It may also be said that waveform *A lags* waveform *B*.

The displacement between two out-of-phase waveforms may also be expressed as an angle. In Figure 11–19(b), two sine waves with a frequency of 10 kHz are shown on a

FIGURE 11–18 Graph of instantaneous values of voltage in Table 11–1

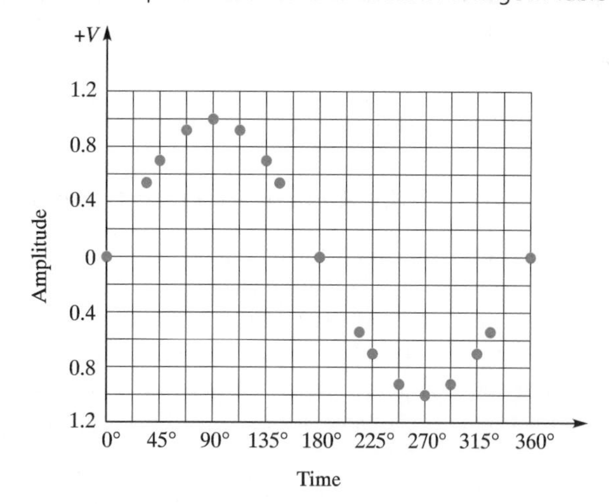

■ **TABLE 11–1 Instantaneous voltages of a sinewave produced by an alternator (V_{max} = 100 V; $v = V_{max} \sin \theta$)**

Angle of Rotation	Value of v
0°	0
30°	50 V
45°	70.7 V
60°	86.6 V
90°	100 V
120°	86.6 V
135°	70.7 V
150°	50 V
180°	0 V
210°	−50 V
225°	−70.7 V
240°	−86.6 V
270°	−100 V
300°	−86.6 V
315°	−70.7 V
330°	−50 V
360°	0

common baseline. As shown, the time difference between the waveforms is 8.31 μs. This time displacement can be converted to an angular displacement. Recall that one cycle of a sine wave contains 360°. The period of the waveforms is $1/f$ or 100 μs. The time of 1° is the ratio of the period T to 360°. In this instance, the time of 1° is 100 μs/360° or 0.278 μs. The angular displacement is then the time displacement divided by the time of 1°, or 29.9° (8.31 μs/0.278 μs). Waveform B thus leads waveform A by 29.9°.

EXAMPLE 11–4 The waveforms of Figure 11–20 are of the same frequency. What are the phase relationships between the waveforms?

■ Solution:

Waveform A is chosen as the reference and is shown to begin at 0°. Waveform B leads waveform A by 45°. Waveform C lags waveform A by 60°. Waveform B leads waveform C by 105°.

Since the sine waves of Example 11–4 are of the same frequency, they can be represented using phasor diagrams, since the phase differences, once established, remain the same throughout. For example, the sine waves of Figure 11–21(a) can be represented by the phasor diagram of Figure 11–21(b). The length of the phasors are proportional to the maximum amplitudes, V_{max}, of the waveforms. Phasor A represents sine wave A and is drawn along the reference. Phasors B and C are drawn at the appropriate angles of lead or lag from phasor A. As the phasors rotate, they maintain the same positions relative to one another.

FIGURE 11–19 Phase difference in two waveforms of the same frequency

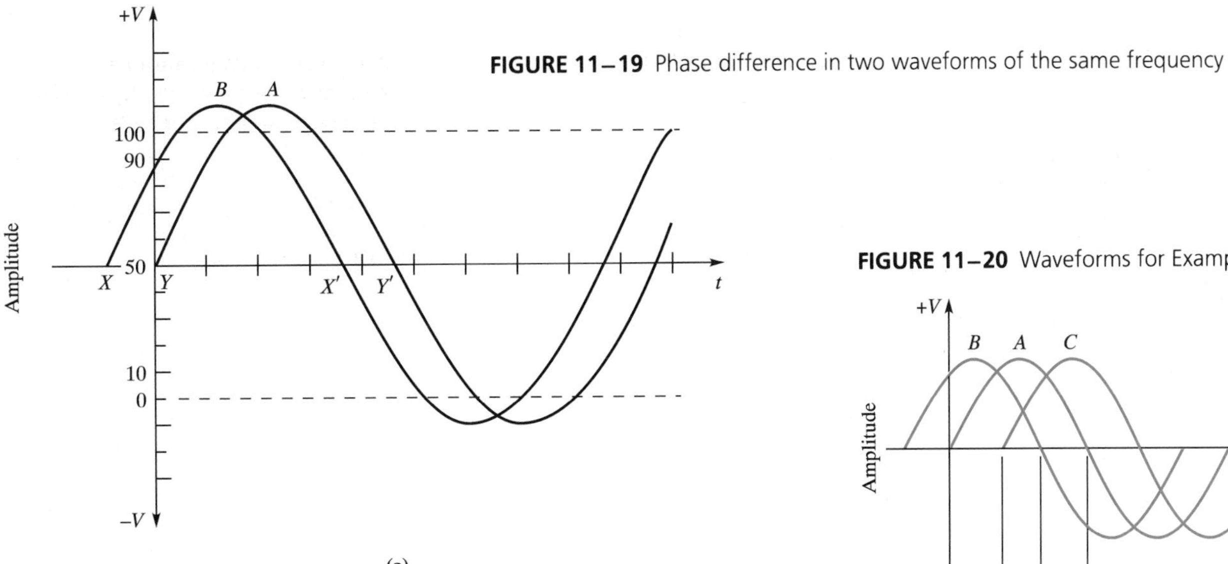

(a)

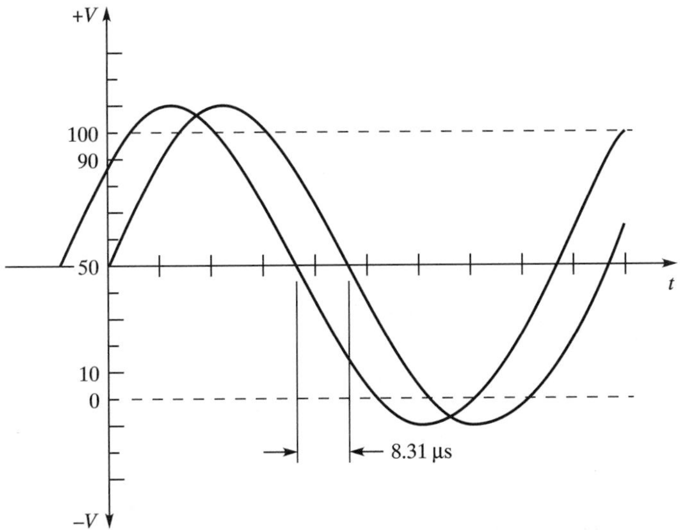

(b)

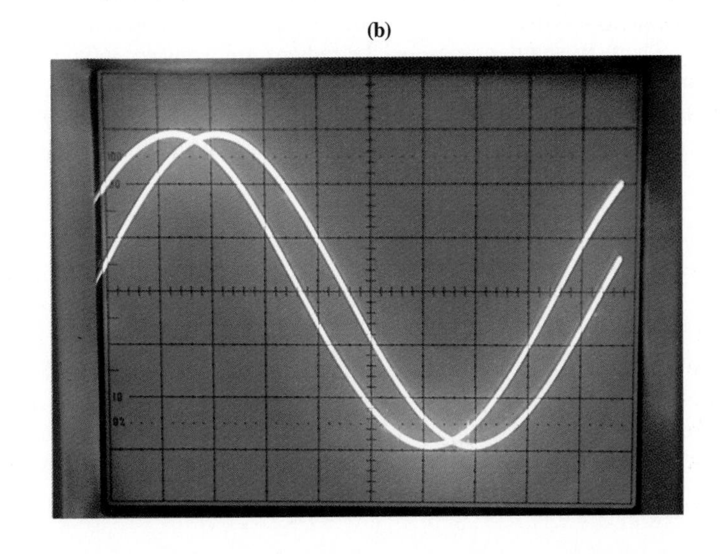

FIGURE 11–20 Waveforms for Example 11–4

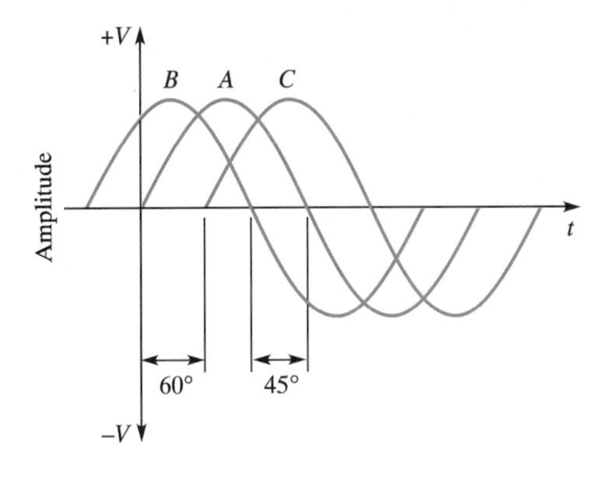

FIGURE 11–21 Phasor representation of out-of-phase waveforms

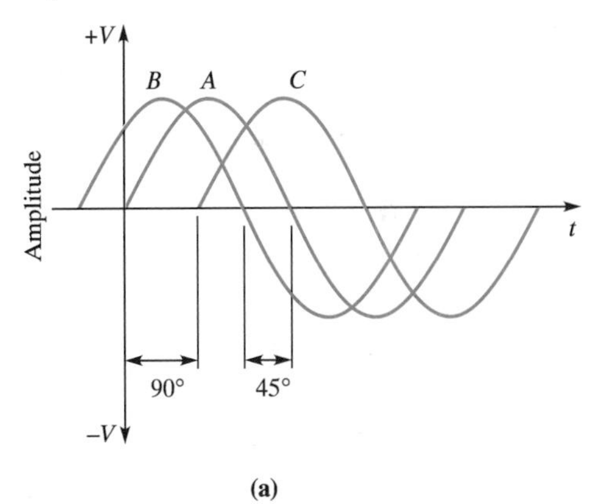

(a)

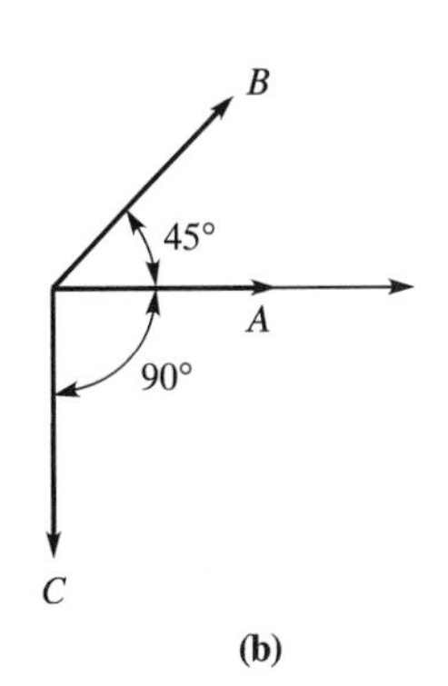

(b)

FIGURE 11–22 The radian as a measure of angular displacement

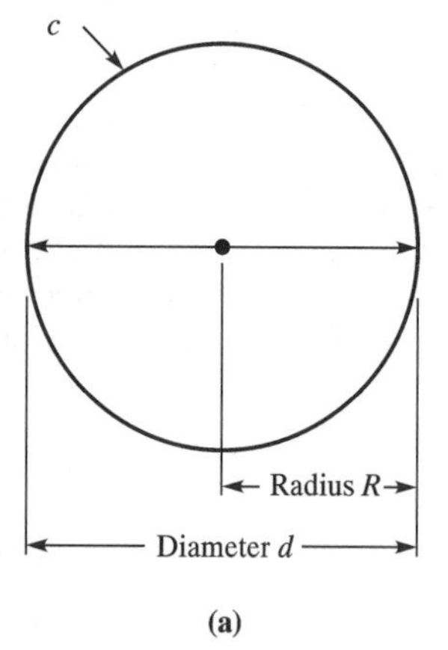

(a)

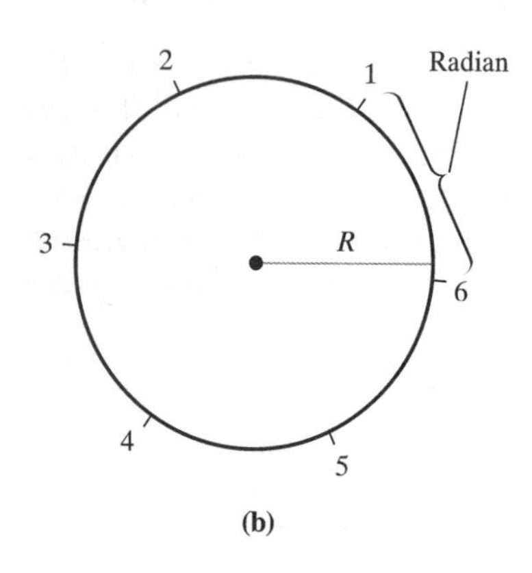

(b)

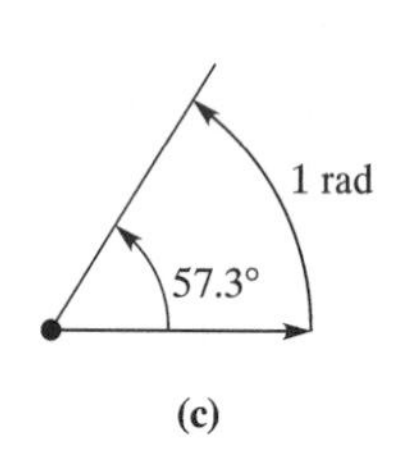

(c)

Question: Are there alternate ways to express the value of angles?

As you know, degrees are used to express the angular displacement of a phasor from the reference phasor. A complete rotation of a phasor moves it through 360°. When one thinks of rotation, however, what comes to mind is the geometric figure known as a circle. It is thus a good idea then to review the facts of circles before proceeding. As shown in Figure 11–22(a), a circle is formed by a line around a point of origin that always remains the same distance from the point. This distance from the origin is known as the *radius* of the circle. A line drawn from one side of the circle to the other, through the origin, is called the *diameter.* Thus, the diameter is twice the radius. The distance around the circle is known as the *circumference.*

Back in the classical civilization of ancient Greece, a relationship between the diameter and circumference of the circle was discovered. What they found was that if the circumference of any circle was divided by its diameter, the result was always the same. This number, equal to approximately 3.1416, was given the symbol π (pi) from the Greek alphabet. You have probably encountered this number in the past, and rest assured, you will see it ample times in the remainder of this text!

The facts about a circle are expressed by the following mathematical relationships:

$$\pi = \frac{C}{d}$$

WHERE C = circumference
d = diameter

Substituting yields

$$\pi = \frac{C}{2R}$$

WHERE R = radius

Cross multiplying gives

$$C = 2\pi R$$

The significance of this last equation is that the circumference of a circle is equal to 2π times the radius. In other words, as shown in Figure 11–22(b), the radius can be laid out exactly 2π times on the circumference of a circle. Each length of one radius along the circumference is known as a **radian,** abbreviated rad. Thus, it can be said that there are 2π radians in a circle. Then, in angular measure, 2π radians are equal to 360°. In fact, any angle can be expressed in radians. A half circle, or 180°, is equal to π radians. One radian is equal to $360/2\pi$ or approximately 57.3°. This latter angle is illustrated in Figure 11–22(c). Some other angles and their radian values are shown in the following table:

Angle	
Degrees	**Radians**
90°	$\pi/2$
45°	$\pi/4$
30°	$\pi/6$
60°	$\pi/3$
270°	$3\pi/2$

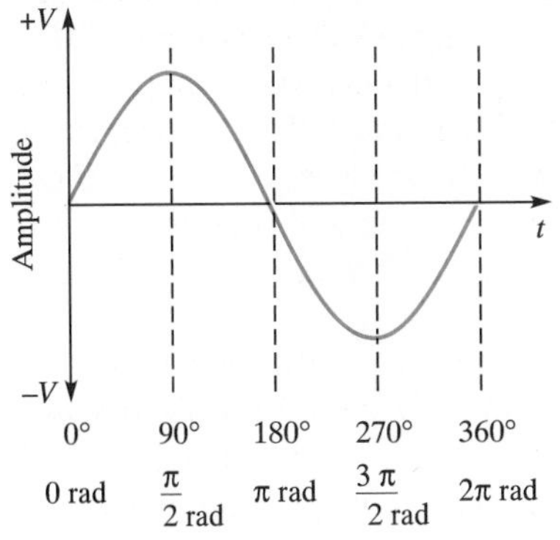

FIGURE 11–23 Sine wave produced by angles of rotation in degrees and radians

+V
Amplitude
t
−V
0° 90° 180° 270° 360°
0 rad $\frac{\pi}{2}$ rad π rad $\frac{3\pi}{2}$ rad 2π rad

The rotation of a phasor can be represented in radians just as well as degrees. In Figure 11–23, a sine wave is shown in which the angle of rotation is given in both degrees and radians.

NOTE

Most scientific calculators accept values of angle in either degrees or radians. If you wish to insert the angle in degrees, and the calculator is conditioned for radians, an incorrect answer will result. Thus, you must ensure that the calculator is conditioned for the appropriate units of angle prior to beginning a calculation. The units are usually set with a switch, and they will be displayed in one corner of the display. The display usually shows "deg" for degrees and "rad" for radians.

Angular Velocity of a Phasor

Question: What is angular velocity of a phasor and how is it computed?

As you learned earlier in this chapter, the rotation of a phasor produces a sine wave. The faster the phasor rotates, the more cycles are produced in a given period of time. The rate at which the phasor rotates is known as its **angular velocity,** which is symbolized by ω (Greek letter omega, lowercased) and measured in units of radians per second. One rotation of the phasor covers 2π radians. The number of rotations per second is the frequency of the sine wave. Thus, the angular velocity of the phasor is equal to 2π times the frequency in hertz. In equation form, angular velocity is written as

(11–5) $$\omega = 2\pi f$$

EXAMPLE 11–5 Compute the angular velocity for the waveform of Figure 11–24.

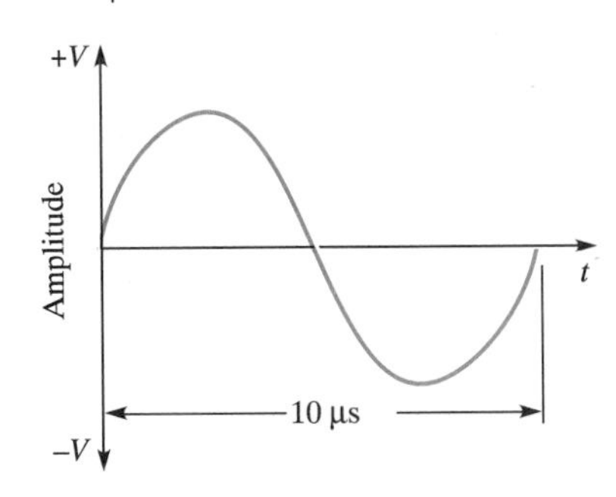

FIGURE 11–24 Waveform for Example 11–5

Solution:

1. Use Eq. 11–1 to find the frequency:

$$f = \frac{1}{T}$$

$$= \frac{1}{10 \times 10^{-6}} = 100 \text{ kHz}$$

2. Use Eq. 11–5 to find the angular velocity:

$$\omega = 2\pi f$$

$$= 2\,(3.14)(100 \times 10^{3}) = 628 \text{ krad/s}$$

SECTION 1–4 REVIEW

1. A sine wave has a maximum value of 170 V. What is its instantaneous value after 60° of rotation?
2. What two quantities does a phasor possess?
3. In Figure 11–25, does waveform A lead or lag waveform B?
4. In Figure 11–26, what is the phase difference between waveforms B and C?

FIGURE 11–25 Waveforms for Section Review Problem 3

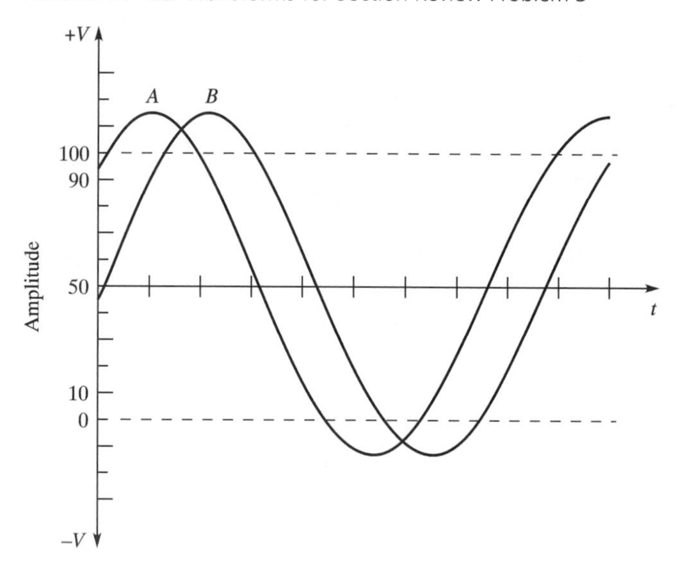

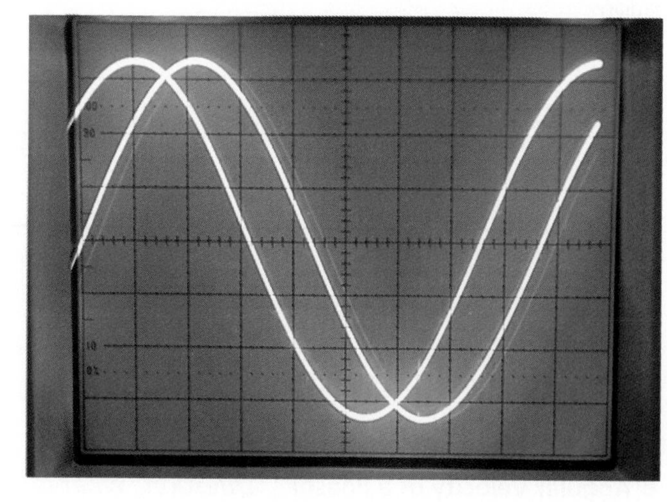

FIGURE 11–26 Waveforms for Section Review Problem 4

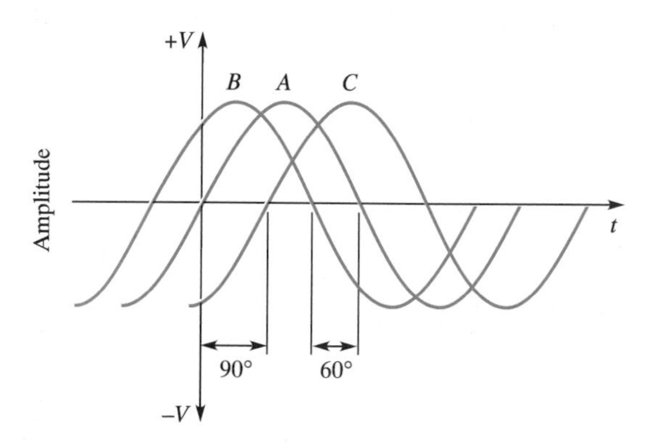

11–5 AMPLITUDE VALUES OF A SINE WAVE

Question: What problem is involved in assigning a voltage or current value to a sine wave?

Anything that is not constant poses problems when it comes to assigning it a value. So it is with the sine wave. As you have already seen, the sine wave can be assigned a value at any instant, which is its instantaneous value. There are four other ways of specifying the amplitude of both a sine wave of voltage and current. These are peak, peak-to-peak, average, and rms values. Each will be discussed in the following subsections.

■ Peak Value

The **peak value** of a sine wave, designated V_p, is the voltage measured from the baseline to its most positive or negative value. These points, which occur at the 90° and 270° points of the sine curve, are illustrated in Figure 11–27(a). (In the previous section, these points were referred to as the maximum values of the sine wave.) The two peak voltages are identical in value, differing only in time and polarity.

FIGURE 11–27 Peak values of sine wave voltage

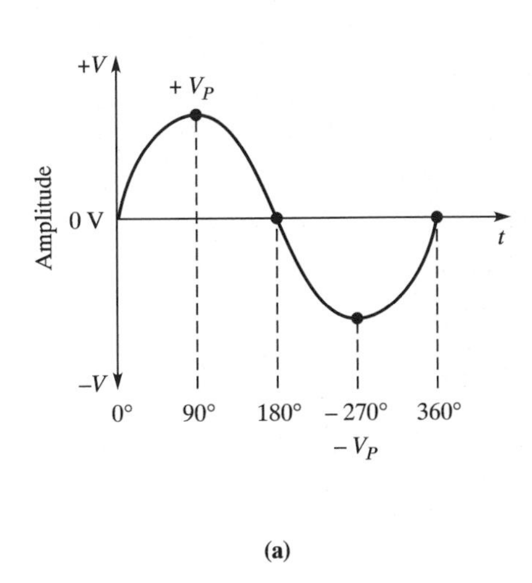

(a)

(b)

FIGURE 11–28 Oscilloscope

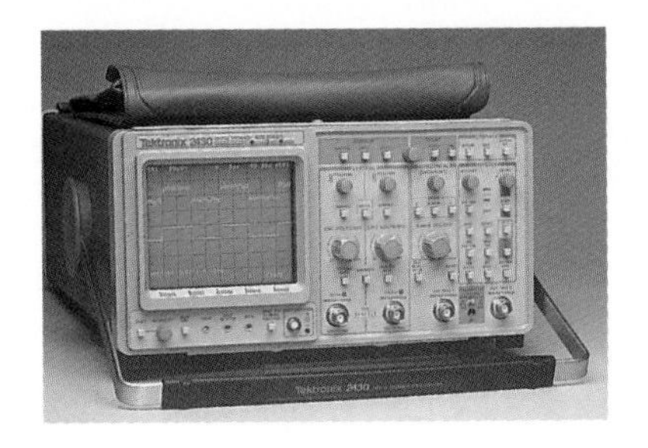

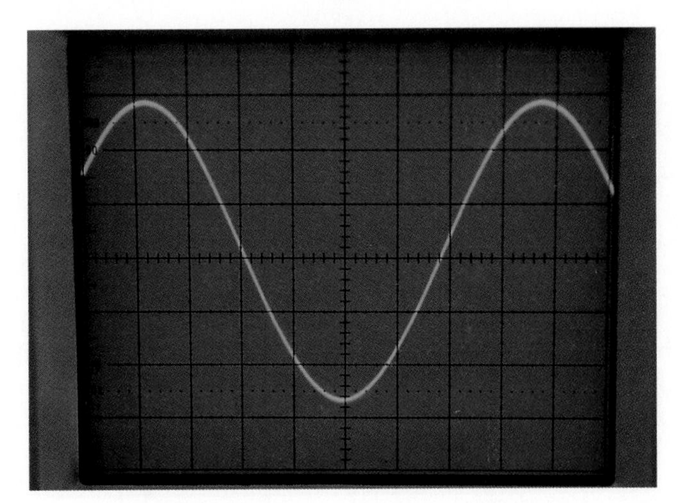

Peak-to-Peak Value

The **peak-to-peak value** of a sine wave, designated V_{p-p}, is the voltage measured between its positive and negative peaks, as illustrated in Figure 11–27(b). The peak-to-peak value is twice the peak value, or

(11–6) $$V_{p-p} = 2\ V_p$$

Although meters exist that indicate peak values, the peak and peak-to-peak values of a waveform are usually measured with an **oscilloscope,** an instrument that not only displays the value, but also the shape of the waveform. A typical oscilloscope is shown in Figure 11–28.

Average Value

As previously stated, the positive and negative alternations, or half-cycles, of a sine wave are identical in voltage amplitude and time. Thus, if all the instantaneous values assumed by a sine wave over a full-cycle are averaged, the result is zero. The **average value** of a

TABLE 11–2 Average value of sines of angles from zero to 180° in 15° increments

Angle of Rotation	Sine Function
15°	0.259
30°	0.500
45°	0.707
60°	0.866
75°	0.966
90°	1.00
105°	0.966
120°	0.866
135°	0.707
150°	0.500
165°	0.259
180°	0
Total	7.596
Average	$\frac{7.596}{12} = 0.633$

sine wave then is the average of all its instantaneous values computed over only a half-cycle. For this reason, it is sometimes referred to as the *half-cycle average.*

Recall that the peak value of the sine function is 1. Recall, further, that the instantaneous value of a sine wave is equal to its peak value times the sine of the angle of rotation. Thus, a sine wave with a peak value of 1 V, has an instantaneous value at any angle θ that is equal to the sine of that angle. The average value of this sine wave is the average of the sines of all angles from zero through 180°. Table 11–2 gives the average value of the sines for all angles from zero to 180°, taken in 15° increments. Notice that this average is 0.633. If more angles were used, the average value would be slightly higher. Using a form of higher mathematics known as integral calculus, a more accurate average value of 0.637 is obtained.

Thus, for a sine wave with any peak value, its average value is the product of its peak value and the constant 0.637. Thus, the average values of voltage V_{avg} and current I_{avg} are found by the following equations:

(11–7)
$$V_{avg} = 0.637\ V_p$$

and

(11–8)
$$I_{avg} = 0.637\ I_p$$

EXAMPLE 11–6 What is the average value of a sine wave whose peak-to-peak value is 340 V?

Solution:

1. Transpose Eq. 11–6 to find the peak value:

$$V_p = \frac{V_{p-p}}{2}$$

$$= 340 \frac{V_{p-p}}{2} = 170\ V_p$$

2. Use Eq. 11–7 to find the average value:

$$v_{\text{avg}} = V_p\ 0.637$$
$$= (170)(0.637) = 108.3\ V_{\text{avg}}$$

EXAMPLE 11–7 A sine wave has an average value of 9.555 V. Compute the peak value.

■ Solution:

Use Eq. 11–7 and transpose to solve for V_p

$$V_{\text{avg}} = 0.637\ V_p$$

$$V_p = \frac{V_{\text{avg}}}{0.637} = 15\ V_p$$

■ Root-Mean-Square Value

Question: Why aren't the ac values presented up to now suitable for metering?

Since instantaneous, peak, and peak-to-peak values occur for only a brief instant, a meter cannot respond quickly enough to follow the change (and if they could, whose eye is fast enough to follow these changes). The average value can be metered, but is of limited value since the entire sine wave is not involved. Thus, meters are not a suitable way to measure ac values. Recall that in a dc circuit, the value of the power developed can be computed from the relationship between voltage, current, and resistance. With alternating current this method poses a problem: The power can be computed, but what significance will it have if computed for only one instant of time? Thus, a way is needed to express alternating voltage and current as an equivalent steady-state dc value. Such a value exists: The root-mean-square value, known as the **rms** or *effective value,* indicates the same heating effect as obtained in steady-state direct current.

Question: How is the rms value of a sine wave computed, and what is its significance?

Like the average value, the rms value is found by averaging instantaneous values of a sine wave. Now, however, it is not the sine of each angle that is averaged, but rather its square. In squaring, all values become *positive* and the average is taken over the *entire* cycle. The square root is then taken of this average, or mean, which gives rise to the name root-mean-square value. As illustrated in Table 11–3, the number thus derived is 0.707. The root-mean-square value of a sine wave of voltage or current is the product of the peak value V_p and 0.707. In equation form, the rms value of a sine wave is found by

(11–9) $$V_{\text{rms}} = 0.707\ V_p$$

and

(11–10) $$I_{\text{rms}} = 0.707\ I_p$$

The rms value of a sine wave is actually a measure of its heating effect. It can be defined as the equivalent value of dc that will produce the same amount of heat (power)

TABLE 11–3 rms value derived from square of sine function of angles from zero to 180° in 15° increments

Angle of Rotation	(Sine Function)²
15°	0.067
30°	0.250
45°	0.500
60°	0.750
75°	0.933
90°	1.00
105°	0.933
120°	0.750
135°	0.500
150°	0.250
165°	0.067
180°	0
Total	6
Average	$\frac{6}{12} = 0.5$
rms value	$\sqrt{0.5} = 0.707$

FIGURE 11–29 Effective, or rms, value of sine wave voltage: Power produced by 170 V_p equals that produced by 120 Vdc.

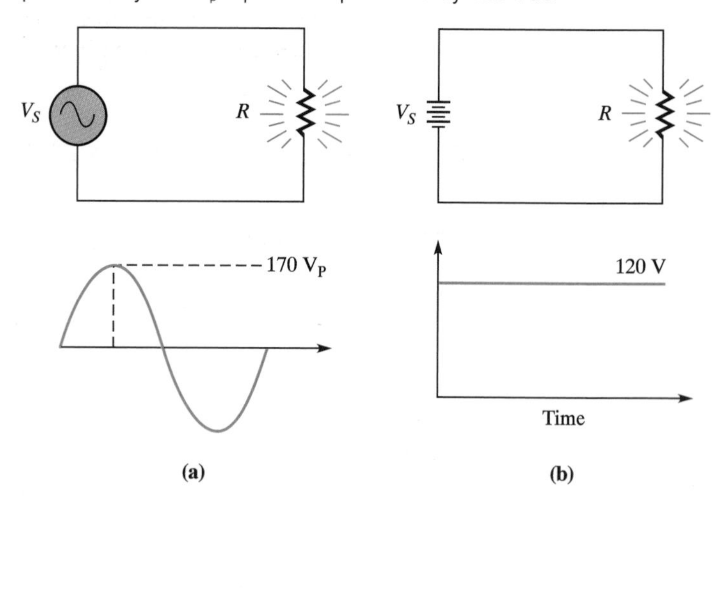

in a load as a given sine wave as illustrated in Figure 11–29. Notice in Figure 11–29(a) that a sine wave with a peak value of 170 V is applied across a resistor. A certain amount of heat is produced. Computing the rms value of this voltage yields approximately 120 V. Thus, as shown in Figure 11–29(b), 120 V of steady-state dc produces the same heating effect as a sine wave with a peak value of 170 V. For this reason, the rms value of a sine wave is often referred to as its effective value (V_{eff}).

CAUTION

As you have learned, a sine wave has certain characteristics that set it apart from all other waveforms encountered in electronics. Any waveform that is not a sine wave is said to be nonsinusoidal. These are the complex waves that will be presented in the next section. It is important to understand that the equations presented here for average and rms values apply to sine waves only! Never attempt to use them with any other waveform.

Question: Which value of the sine wave—peak, peak-to-peak, average, or rms—should be used?

At one time or another, you will use each of the values with which the amplitude of an ac signal can be expressed. Most ac meters, however, indicate the rms value. In special instances, meters may indicate peak, peak-to-peak, or average. When a voltage is expressed, such as in a text, an article, or on a schematic, *unless otherwise indicated, it is assumed to be an rms value.*

In circuit analysis, using Ohm's law equations, any value may be used. If the peak value of a current is multiplied by resistance, the peak value of the voltage source is obtained. If the peak-to-peak value of a source voltage is divided by the circuit resistance, the peak-to-peak value of the current is obtained. When solving for resistance, it is necessary to use the same values. For example, if a peak value of voltage is divided by the peak current flowing in the circuit, the total resistance is obtained.

EXAMPLE 11–8 The voltage across a 1 kΩ resistor is 12.8 V peak-to-peak. What is the value of the current?

■ **Solution:**

$$I_{p-p} = \frac{V_{p-p}}{R}$$

$$= \frac{12.8\ \mathrm{V_{p-p}}}{1000\ \Omega} = 12.8\ \mathrm{mA_{p-p}}$$

EXAMPLE 11–9 The peak value of the current through a 1.8 kΩ resistor is 17.5 mA. What is the voltage developed across the resistor?

■ **Solution:**

$$V_p = I_p R$$

$$= (17.5 \times 10^{-3}\ \mathrm{mA_p})(1.8 \times 10^3\ \Omega) = 31.5\ \mathrm{V_p}$$

EXAMPLE 11–10 The average current through a resistor is 10 mA. If the average voltage is 5V, what is the value of the resistor?

■ **Solution:**

$$R = \frac{V_{avg}}{I_{avg}}$$

$$= \frac{5\ \mathrm{V_{avg}}}{10 \times 10^{-3}\ \mathrm{mA_{avg}}} = 500\ \Omega$$

EXAMPLE 11–11 What is the rms value of a sine wave with a peak value of 170 V?

■ **Solution:**

$$V_{rms} = 0.707 V_p$$

$$= (0.707)\ (170\ \mathrm{V}) = 120\ \mathrm{V_{rms}} \qquad [120\ \mathrm{V}]$$

Remember that unless specified otherwise, all ac voltages are assumed to be rms. Therefore, the designation rms is optional.

EXAMPLE 11–12 What is the rms value of a sine wave that has an average value of 14.8 V?

■ **Solution:**

1. First, compute the peak value from the average value:

$$V_{avg} = V_p\ 0.637$$

2. Transpose:

$$V_p = \frac{V_{\text{avg}}}{0.637} = \frac{14.8\ \text{V}_{\text{avg}}}{0.637} = 23.23\ \text{V}_\text{p}$$

3. Solve:

$$V_{\text{rms}} = V_p 0.707 = (23.23\ \text{V}_p)(0.707) = 16.4\ \text{V}$$

EXAMPLE 11–13 A sine wave has an rms value of 9.9 V. What is its peak-to-peak value?

■ Solution:

1. Compute the peak value:

$$V_{\text{rms}} = V_p\ 0.707$$

2. Transpose:

$$V_p = \frac{V_{\text{rms}}}{0.707} = \frac{9.9\ \text{V}_{\text{rms}}}{0.707} = 14\ \text{V}_\text{p}$$

3. Solve:

$$V_{p-p} = 2\ V_p = (2)(14\ \text{V}_\text{p}) = 28\ \text{V}_{\text{p}-\text{p}}$$

EXAMPLE 11–14 A sine wave has a peak value of 45 V. Compute its (a) rms, (b) average, and (c) peak-to-peak values.

■ Solution:

a.

$$V_{\text{rms}} = V_p 0.707 = (45\ \text{V}_\text{p})(0.707) = 31.8\ \text{V}$$

b.

$$V_{\text{avg}} = V_p 0.637 = (45\ \text{V}_\text{p})(0.637) = 28.7\ \text{V}_{\text{avg}}$$

c.

$$V_{p-p} = 2V_p = (2)(45\ \text{V}_\text{p}) = 90\ V_{\text{p}-\text{p}}$$

TECH TIP: RMS FROM THE OSCILLOSCOPE

A common method of troubleshooting is through sine wave analysis using an oscilloscope. This presents a somewhat ambiguous situation in that sinusoidal voltages are normally stated as rms but voltages are read on the oscilloscope as peak-to-peak.

One method of converting is through the equations $V_{rms} = V_p \times 0.707$ or $V_p = V_{rms} \times 1.414$. To use this method, the peak-to-peak reading from the oscilloscope must be divided by two. A simpler way is to learn two different equations (actually these are derived from the same two original equations): $V_{rms} = V_{p-p} \times 0.3535$ (0.707/2) or $V_{p\text{-}p} = V_{rms} \times 2.828$ (1.414×2)

This Tech Tip is *not* meant to extol the merits of memorization in electronics. It is definitely not recommended in your learning the basics of electronics. You will on the job, however, learn and develop many short cut that you will use in troubleshooting (such as those presented here). In fact, it will be a necessary part of your job, where speed is often a necessity!

SECTION 11–5 REVIEW

1. A sine wave has a peak value of 30 volts. What is its average value?
2. The peak value of a sine wave is 140 mV. What is its peak-to-peak value?
3. What is the rms value of the sine wave of problem 2?
4. A sine wave has an rms value of 38.8 V. What is its average value?

11–6 NONSINUSOIDAL WAVES

Question: What is a nonsinusoidal wave?

Up to now, the only ac waveform that has been considered is the sine wave. Except for steady-state dc, the sine wave is the simplest of all waveforms. In your study of electronics, however, you will encounter many waveforms that do not have the characteristics and simple structure of the sine wave. These are known as nonsinusoidal or complex waves. A **nonsinusoidal wave** does not follow the sine curve in amplitude variations, its form is not necessarily symmetrical, and, most importantly, it is composed of more than one frequency (Recall that a sine wave contains only a single frequency). In the following subsections, several of the more common nonsinusoidal waves are presented and their structure analyzed.

The Square Wave

One of the most common nonsinusoidal waves is the square wave. Figure 11–30(a) shows the graph of a square wave, and Figure 11–30(b) shows how it appears on an oscilloscope. Recall that an oscilloscope presents a real-time graph of a waveform's voltage amplitude over a period of time, which is known as a *time domain* presentation. At point *A* of Figure 11–30(a), the voltage has risen abruptly from a negative 50 V to a positive 50 V level. The voltage remains at the positive level through the period from *A* to *B* and then abruptly drops to the negative value at point *C*. The voltage remains negative for a time identical to the period that it was positive. At point *D* the wave period is complete and the cycle begins again. The period of the square wave is the time from point *A* to point *D*. Its frequency, known as its *fundamental frequency,* is the reciprocal of the period. Like the sine wave, a square wave has a peak and a peak-to-peak value as shown.

FIGURE 11–30 Square wave

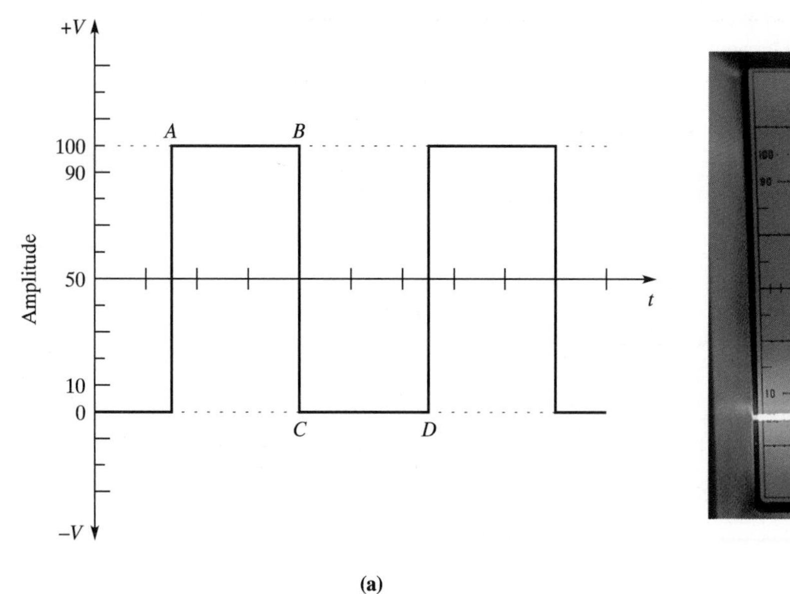

(a)
Graph of waveform

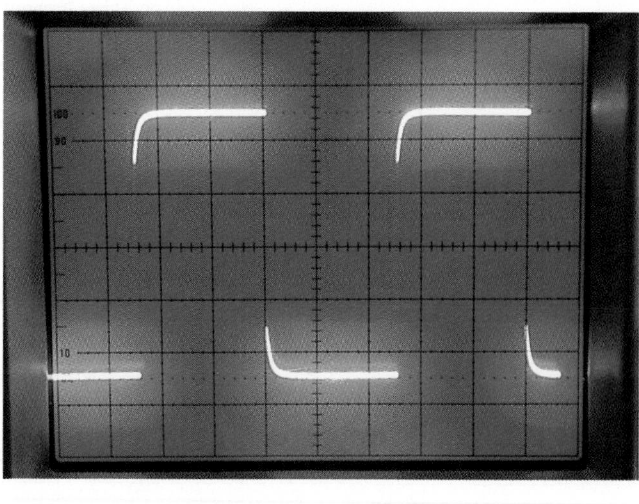

Actual waveform as viewed on the oscilloscope

Question: What are the frequency components of a square wave?

As stated previously, a sine wave contains only a single frequency. A square wave, on the other hand, contains not only its fundamental frequency, but also all of its odd harmonics. A **harmonic** is a sine wave whose frequency is an integral (whole number) multiple of the fundamental frequency. For example, for a frequency of 1 kHz, the first harmonic is 1 × 1 kHz, which is the fundamental frequency itself, the second harmonic is 2 × 1 kHz, or 2 kHz, the third harmonic is 3 kHz, the fourth harmonic is 4 kHz, the fifth harmonic is 5 kHz, and so on up through any number. The even multiples are known as *even harmonics,* and the odd multiples are known as *odd harmonics.* Thus the harmonic content of a square wave is the first, third, fifth, seventh, ninth, eleventh, and so on.

■ Harmonic Analysis of a Square Wave

The formation of a square wave can be demonstrated by adding the odd harmonics, one at a time, to a fundamental frequency. The harmonics will not have the same amplitude as the fundamental. The third harmonic's amplitude is 1/3 that of the fundamental, the fifth harmonic's amplitude 1/5 the fundamental, the seventh harmonic's amplitude 1/7 the fundamental, and so on. The square wave to be produced is shown in Figure 11–31(a). In Figure 11–31(b), a sine wave with the same frequency as the fundamental frequency of the square wave is labeled *A*. Its third harmonic is labeled *B*. The algebraic sum of the fundamental and the third harmonic is labeled *C*. Notice that the sides begin to show a sharper rise and the top is beginning to flatten. In Figure 11–31(c), the fifth harmonic is added algebraically to waveform *C* to form waveform *D,* which is the fundamental plus the third and fifth harmonics. In Figure 11–31(d), the seventh harmonic is added algebraically to waveform *D* to form waveform *E,* which is the fundamental plus the third, fifth, and seventh harmonics. Notice that the sides of waveform *D* are becoming vertical and the positive and negative peaks are flattening out. As more harmonics are added, this process will continue until a nearly perfect square wave is produced.

FIGURE 11–31 Harmonic analysis of a square wave

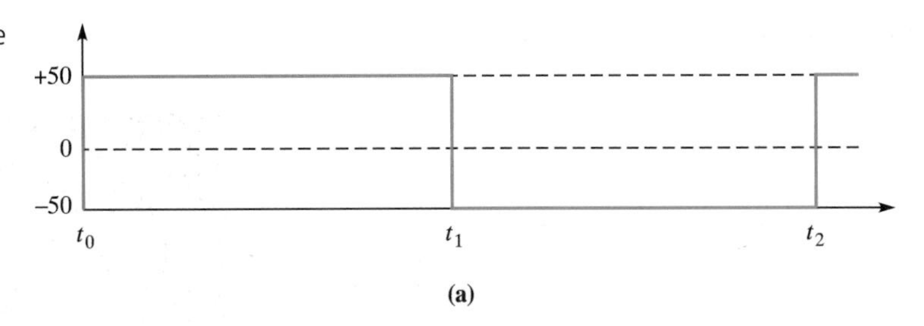

(a)

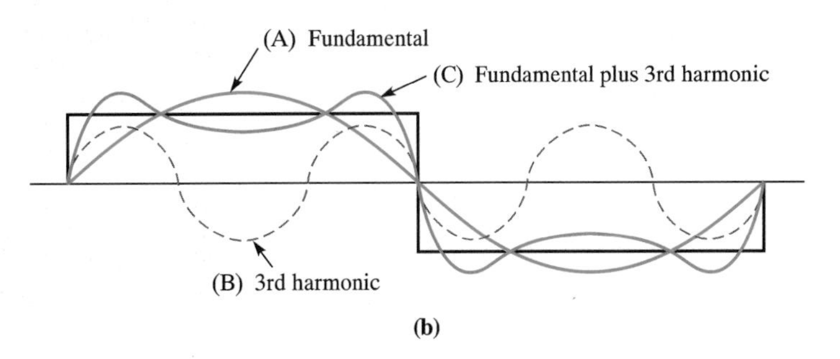

(b)

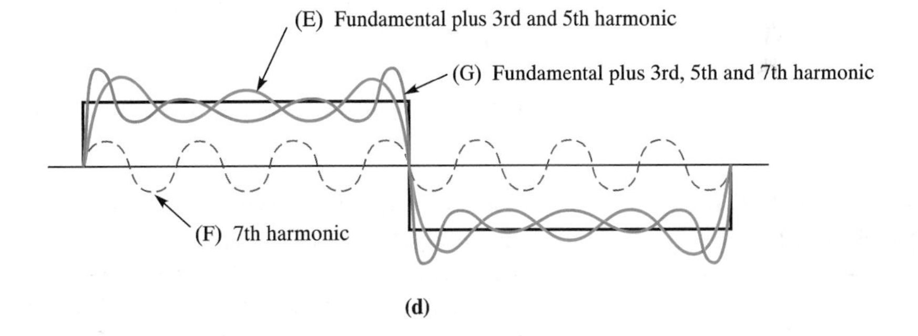

(c)

(E) Fundamental plus 3rd and 5th harmonic

(G) Fundamental plus 3rd, 5th and 7th harmonic

(F) 7th harmonic

(d)

FIGURE 11–32 Spectrum analyzer

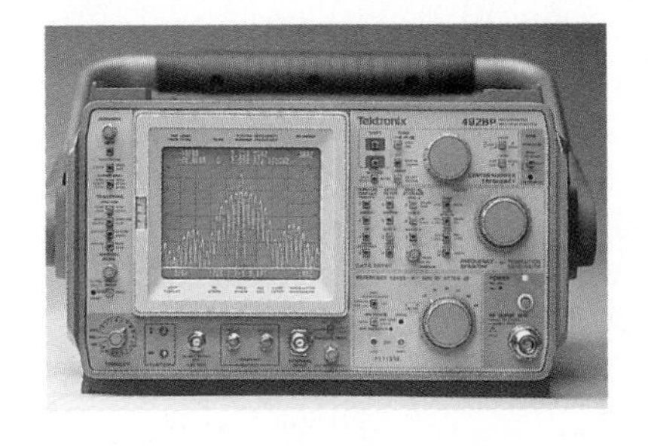

The harmonic content of a square wave, or any nonsinusoidal wave, is viewed on an instrument known as a **spectrum analyzer.** A spectrum analyzer is shown in Figure 11–37. The trace of a spectrum analyzer represents frequency; thus, its presentation is in the *frequency domain.* Figure 11–33(a) shows the graph of a sine wave in the time domain, such as would be displayed on an oscilloscope. Figure 11–33(b) shows the same sine wave applied to a spectrum analyzer. A single *spectral line* appears on the spectrum analyzer's trace at the frequency of the sine wave (Recall that a sine wave has only one frequency component). The height of this spectral line is proportional to the voltage amplitude of the sine wave. Figure 11–34(a) shows the time domain presentation of the square wave developed in the preceding paragraph. Figure 11–34(b) shows the spectrum

FIGURE 11–33 Sine wave displayed in the time domain (oscilloscope) and frequency domain (spectrum analyzer)

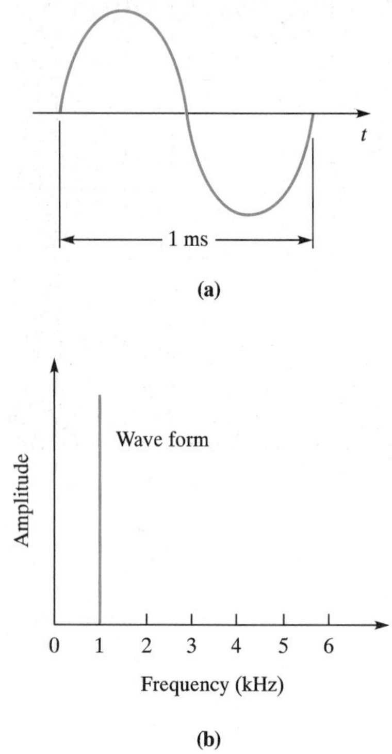

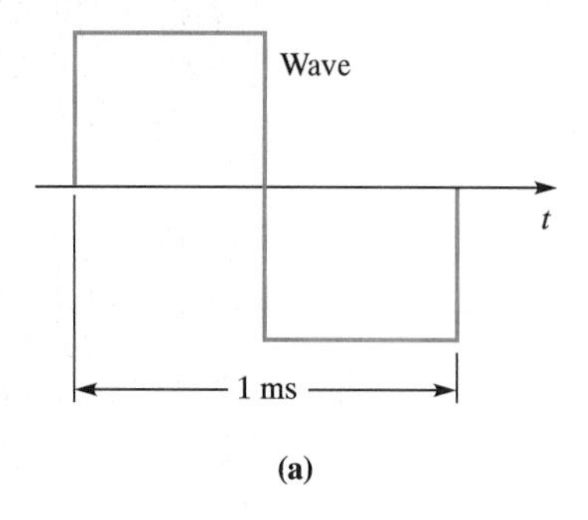

FIGURE 11–34 Square wave displayed in the time domain (oscilloscope) and frequency domain (spectrum analyzer)

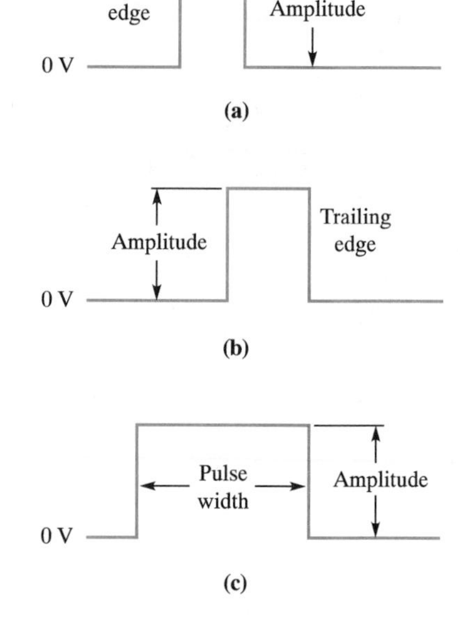

FIGURE 11–35 Ideal voltage pulses

analyzer presentation of this same square wave. Notice now that spectral lines appear at the position of the fundamental frequency and each odd harmonic. Notice also that the amplitude of the harmonics decrease as their frequency increases.

Ideal Voltage Pulses

A **pulse** is a voltage or current that momentarily makes a sharp change in amplitude, remains at this value for a period of time, and then returns to its original value. As shown in Figure 11–35, a pulse may be of a short, medium, or relatively long duration. The difference between the lower and upper voltage levels of the pulse is known as its amplitude. The waveforms in this figure are shown in the time domain, with time increasing from left to right. Thus the left rise of the pulse is known as the *leading edge,* while the right rise is known as the *trailing edge*. The *pulse width* (t_w) is the time between the leading and trailing edges. The pulse may be recurrent—that is, repeating itself over a definite period of time. The **pulse repetition time,** or PRT, is the time from the leading edge of one pulse to the leading edge of the next, as illustrated in Figure 11–36. The rate at which the pulses occur is known as the **pulse repetition frequency,** or the PRF, and is equal to the reciprocal of the PRT. (The PRF may also be referred to as the PRR, pulse repetition rate.) In equation form,

(11–11)
$$PRF = \frac{1}{PRT}$$

and

FIGURE 11–36 Pulse repetition time

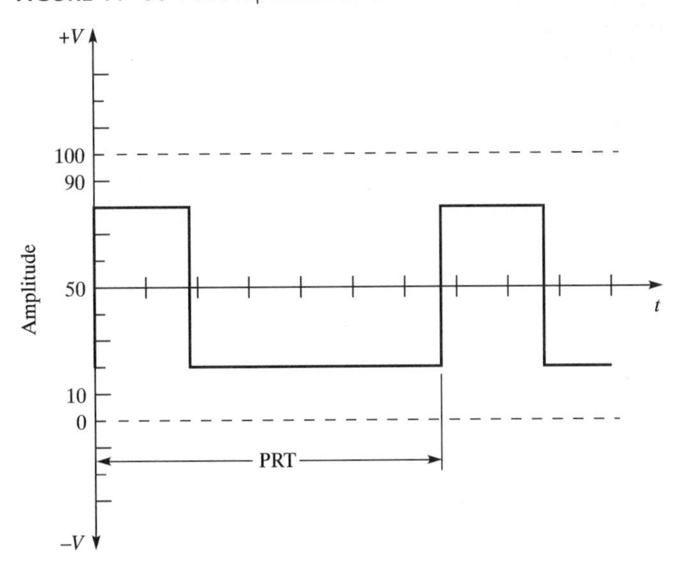

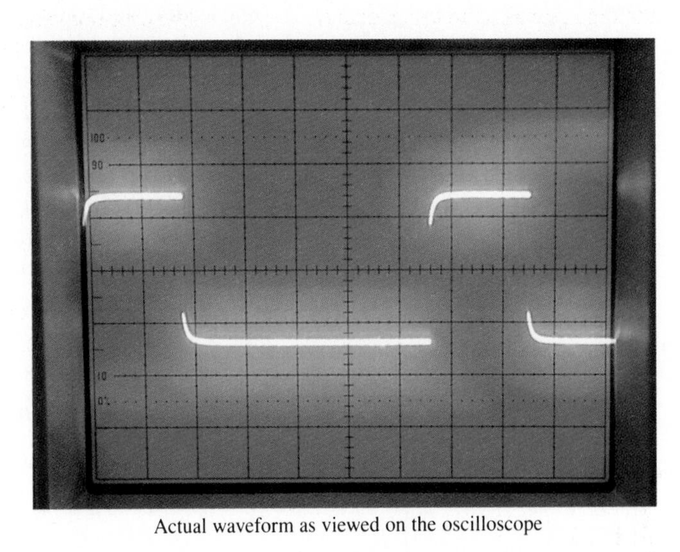

Actual waveform as viewed on the oscilloscope

(11–12)

$$\text{PRT} = \frac{1}{\text{PRF}}$$

The pulse repetition frequency is expressed in pulses-per-second (pps).

EXAMPLE 11–15 What is the pulse repetition frequency of a pulsed wave with a PRT of 1 ms?

■ Solution:

$$\text{PRF} = \frac{1}{\text{PRT}}$$

$$= \frac{1}{0.001 \text{ s}} = 1000 \text{ pps}$$

EXAMPLE 11–16 What is the PRT of a pulsed wave with a PRF of 500 pps?

■ Solution:

$$\text{PRT} = \frac{1}{\text{PRF}}$$

$$= \frac{1}{500} = 2 \text{ ms}$$

■ Nonideal Voltage Pulses

Question: What are the characteristics of a nonideal pulse?

The pulses shown in the preceding paragraph are idealized in that they are considered to increase from zero to their peak value instantaneously. A nonideal pulse is shown in

FIGURE 11–37 Nonideal voltage pulses

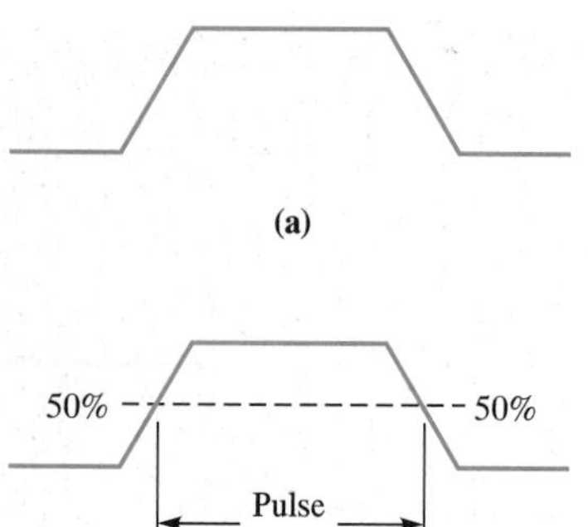

Figure 11–37(a). Notice the slope of both the leading and trailing edges. The slopes on the leading and trailing edges leads to a question: "Between what points is the pulse width measured?" Pulse width of a nonideal pulse is the time measured between the leading and trailing edges at a point where the amplitude is 50% of peak value, as is illustrated in Figure 11–37(b). The slope of the leading edge shows that there is a definite period of time involved in the waveform attaining its peak value, which is known as the **rise time** t_r. As shown in Figure 11–37(c), the rise time is the period taken for the pulse to increase from 10% to 90% of its peak value. There is also a definite period of time involved in the waveform returning from its peak value to zero, which is known as the **fall time** t_f. The fall time is the period taken for the pulse to decrease from 90% to 10% of its peak value. In practical pulses, the rise time and fall times are not necessarily equal. The fall time is sometimes referred to as the *decay time*.

Question: What is meant by duty cycle?

The **duty cycle** of a pulsed wave is the ratio of the pulse width to the pulse repetition time of the waveform. It is usually expressed as a percentage. In equation form, duty cycle is found by

(11–13)

$$\%\text{ duty cycle} = \frac{t_w}{\text{PRT}}100$$

EXAMPLE 11–17 For the pulse waveform of Figure 11–38, compute the duty cycle.

■ Solution:

$$\%\text{ duty cycle} = \frac{t_w}{\text{PRT}}100$$

$$= \frac{2 \times 10^{-3}\text{ s}}{10 \times 10^{-3}\text{ s}}100 = 20\%$$

FIGURE 11–38 Waveform for Example 11–17

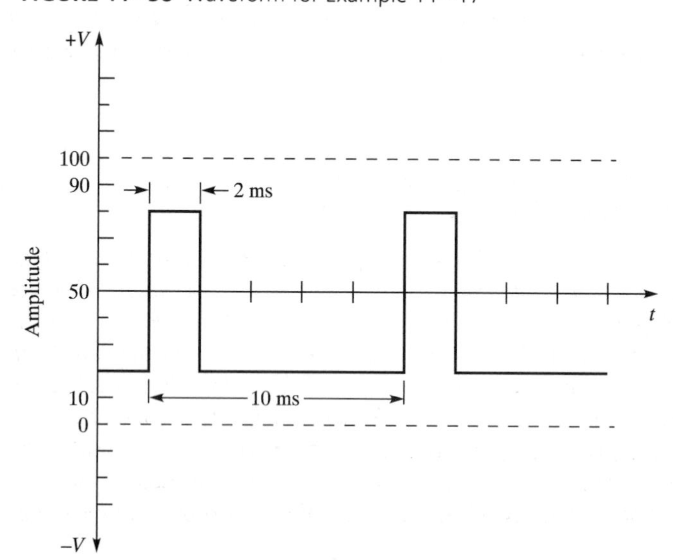

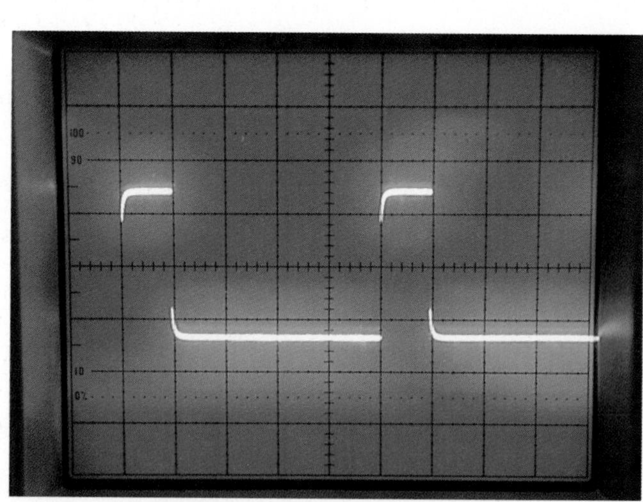

Actual waveform as viewed on the oscilloscope

FIGURE 11–39 Waveform for Example 11–18

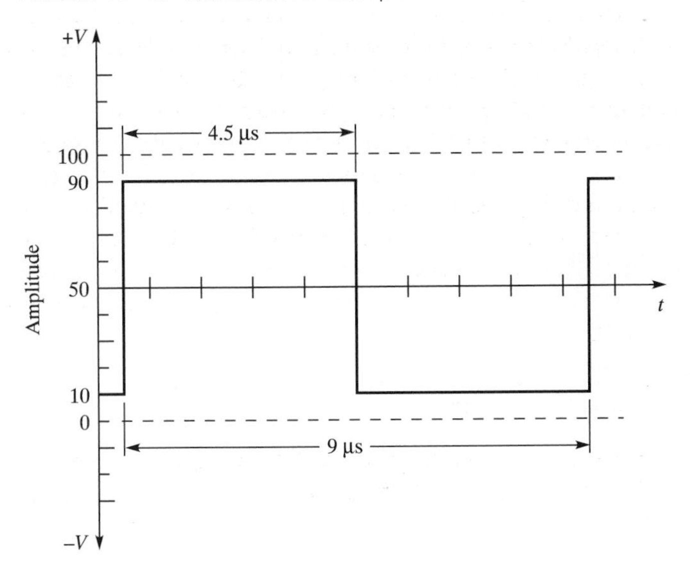

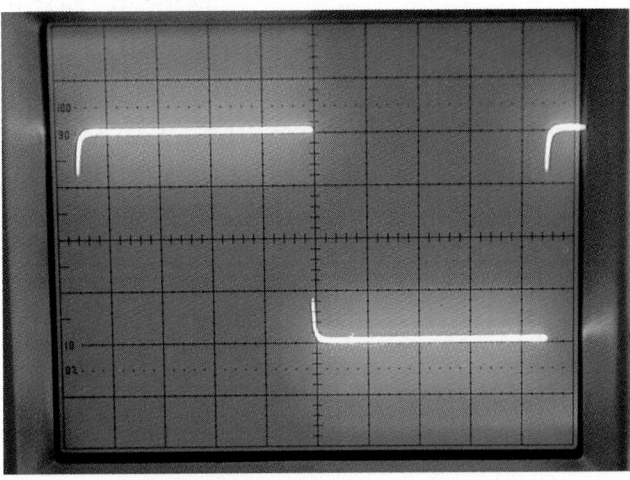

Actual waveform as viewed on the oscilloscope

EXAMPLE 11–18 For the waveform of Figure 11–39, compute the duty cycle.

■ Solution:

$$\% \text{ duty cycle} = \frac{t_w}{\text{PRT}}100$$

$$= \frac{4 \times 10^{-6}\text{ s}}{8 \times 10^{-6}\text{ s}}100 = 50\%$$

In a previous paragraph, a waveform known as a square wave was presented. It was stated that the periods of the positive and negative alternations were equal. A square wave may also be a dc waveform in which the period of peak value is one half the total period of the pulse. Thus, a **square wave** can be defined simply as a pulse waveform with a 50% duty cycle.

■ Average Value of a Pulsed Wave

Question: What is meant by the average value of a pulsed waveform, and how is it computed?

The amplitude of a pulse is defined as the difference between its lower and upper voltage values. The average voltage value is this amplitude averaged over the entire period of the wave, as illustrated in Figure 11–40. Notice that the average value is lower than the amplitude. How much lower depends upon two things: the amplitude of the waveform and the duty cycle. A higher value of duty cycle means that the pulse width is relatively long as related to the period. The amplitude is spread out over a shorter period of time so the average voltage will be greater. Thus, the average value is *directly proportional* to the duty cycle. With a higher amplitude value, there is more voltage to average; hence, the average value is greater. The average value, then, is directly proportional to the amplitude. In equation form,

FIGURE 11–40 Average value of a pulsed wave

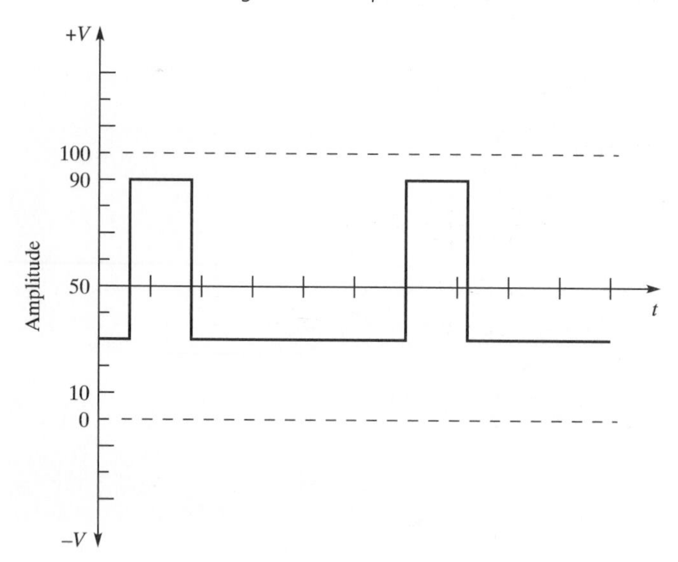

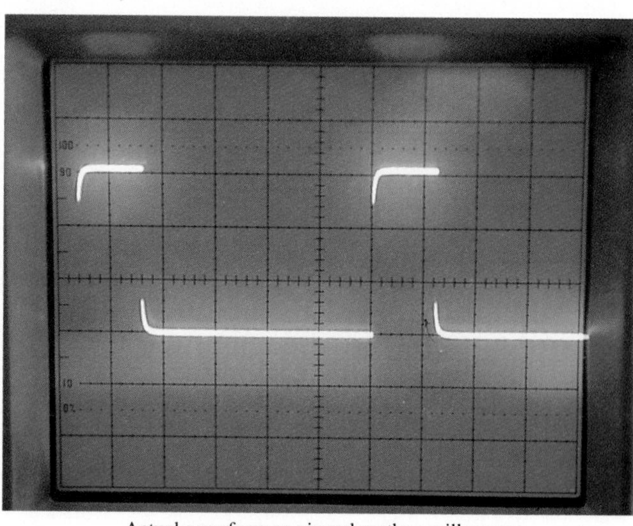

Actual waveform as viewed on the oscilloscope

(11–14) $$V_{avg} = \text{baseline} + (\text{amplitude} \times \text{duty cycle})$$

The **baseline** is considered to be the lower voltage value of the pulse. This value is zero *only* for purely dc pulses beginning at the zero volt baseline.

EXAMPLE 11–19 For the waveform of Figure 11–41, compute the average value.

■ Solution:

1. Compute the duty cycle as a decimal fraction:

$$\text{duty cycle} = \frac{t_w}{\text{PRT}}$$
$$= \frac{15 \times 10^{-3}\text{ s}}{100 \times 10^{-3}\text{ s}} = 0.15$$

2. Since baseline is zero, use Eq. 11–14:

$$V_{avg} = \text{amplitude} \times \text{duty cycle}$$
$$= 10\text{ V}_p \times 0.15 = 1.5\text{ V}_{avg}$$

Notice that it is the duty cycle as a decimal fraction, not as a percentage, that is used in this equation.

FIGURE 11–41 Waveform for Example 11–19

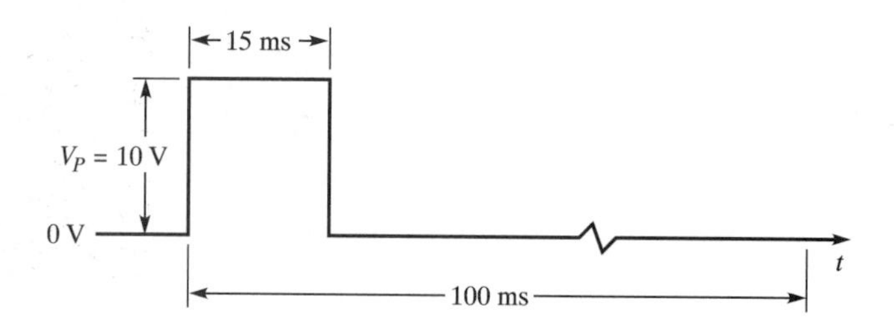

EXAMPLE 11–20 Compute the average value of the waveforms in Figure 11–42.

■ Solution:

1. Waveform (a),

$$\text{duty cycle} = \frac{t_w}{\text{PRT}} = \frac{3 \times 10^{-3}\ \text{s}}{9 \times 10^{-3}\ \text{s}} = 0.333$$

$$V_{\text{avg}} = \text{baseline} + (\text{amplitude} \times \text{duty cycle}) = -3\ \text{V} + [(1\ \text{V})(0.333)] = -2.667\ \text{V}_{\text{avg}}$$

FIGURE 11–42 Waveforms for Example 11–20

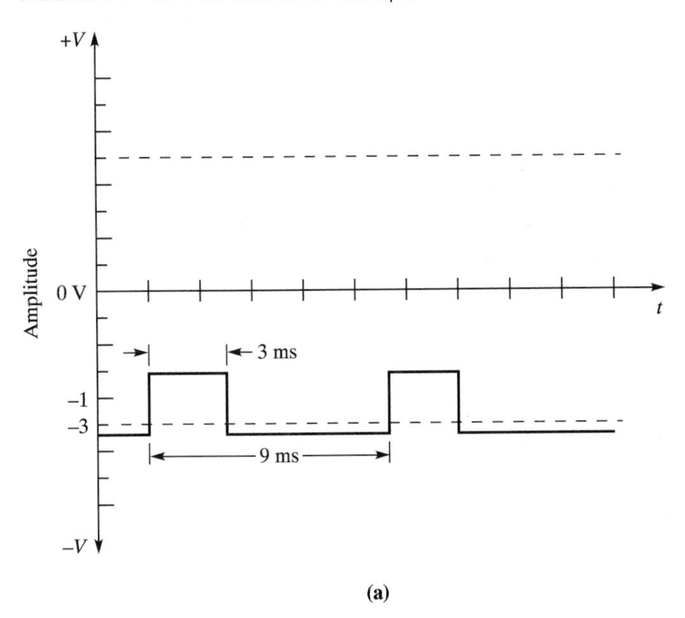

(a)

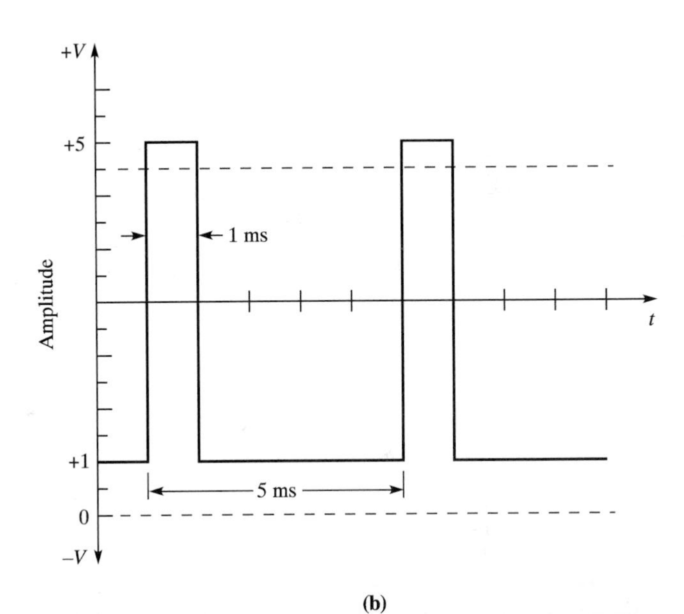

(b)

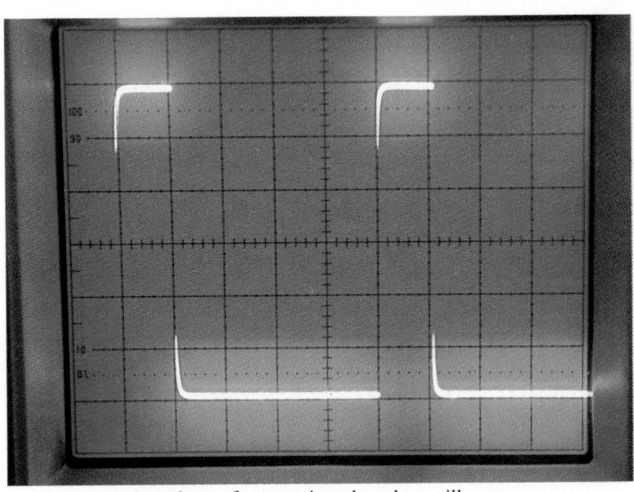

Actual waveform as viewed on the oscilloscope

2. Waveform (b),

$$\text{duty cycle} = \frac{t_w}{\text{PRT}}$$
$$= \frac{1 \times 10^{-3}\ \text{s}}{5 \times 10^{-3}\ \text{s}} = 0.2$$
$$V_{avg} = \text{baseline} + (\text{amplitude} \times \text{duty cycle})$$
$$= 1 + [(5\ \text{V})(0.2)] = 2\ \text{V}$$

Sawtooth Waves

The waveform shown in Figure 11–43 is known as a sawtooth wave. The name comes, of course, from its resemblance to the teeth of a saw blade. The voltage rises from zero in a linear manner known as a *positive ramp*. It then drops very quickly back to zero volts in what is known as a *negative ramp*. Notice that the positive ramp is relatively long compared to the negative ramp. As shown in Figure 11–43, the period T can be measured from one positive peak to another (or between any two consecutive points of the same amplitude and phase).

Sawtooth waves of current are used to sweep the electron beam across the picture tube of a television. The long positive ramp pulls the beam across the screen relatively slowly, producing the elements of the picture as it goes. The sharp negative ramp sends the beam back to the other side of the screen very quickly. (This action is known as the retrace or "flyback".) The process then repeats for another line of picture information. In a similar manner, sawtooth waves are used to sweep the beam across the display of an oscilloscope.

FIGURE 11–43 Sawtooth wave

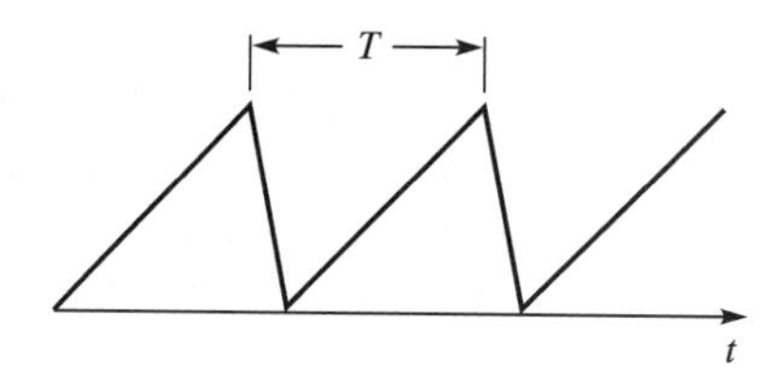

SECTION 11–6 REVIEW

1. Of what is any nonsinusoidal wave composed?
2. What is the harmonic content of a square wave?
3. Which edge of a voltage pulse is known as the leading edge?
4. For the waveform of Figure 11–44, compute the duty cycle.
5. For the waveform of Figure 11–45, compute the average value of the voltage.

FIGURE 11–44 Waveform for Section Review Problem 4

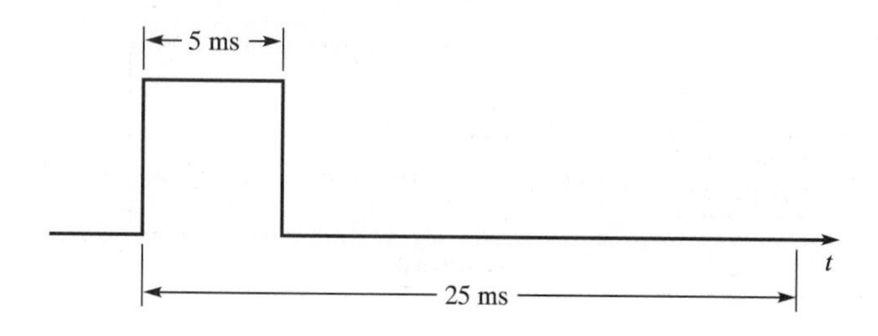

FIGURE 11–45 Waveform for Section Review Problem 5

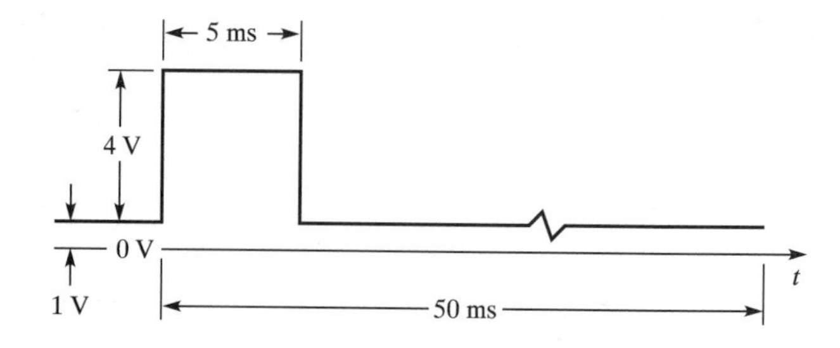

11–7 THE OSCILLOSCOPE

The oscilloscope is one of the most versatile and widely used items of electronic test equipment. With it you can determine a signal's voltage amplitude, waveshape, period, and frequency. It can also be used to determine the phase difference between signals. This application is extremely important in computer circuits where the timing of events is critical. The oscilloscope has a very high input impedance. Thus, it will cause very little, if any, loading of the circuit under test. All in all, the oscilloscope is considered by many to be indispensable in electronic testing.

At this point in the electronics curriculum, you already know a great deal about the operation of the oscilloscope. You have had practice in its use through laboratory exercises. Students sometimes experience difficulty in learning to operate the oscilloscope, which is understandable, since it truly looks like a maze of controls! In this section, the internal workings of a basic oscilloscope will be presented. The basic controls and their effect upon the display will be presented. Knowing what is happening when a control is adjusted will make you more confident in using the oscilloscope.

Question: What are the basic components of an oscilloscope?

The block diagram of a basic oscilloscope is shown in Figure 11–46. It is composed of three sections: the vertical, the horizontal, and the cathode ray tube (CRT). In the vertical section, the signal is conditioned and applied to the display. The horizontal section produces the trace by sweeping a beam of electrons across a phosphor-coated screen. A *phosphor* is a material that converts some of the energy of the electrons striking it to light. The CRT produces the electron beam, focuses it, and produces the display. Each of these sections is considered in the following subsections.

■ The Cathode Ray Tube

The cathode ray tube is the heart of the oscilloscope. It is composed of an electron gun, a control grid, a focus anode, an accelerating anode, vertical and horizontal deflection plates, and the screen. All of this is inside a glass enclosure from which a vacuum has been pulled. The mechanical layout of these items is shown in Figure 11–47.

FIGURE 11–46 Block diagram of oscilloscope

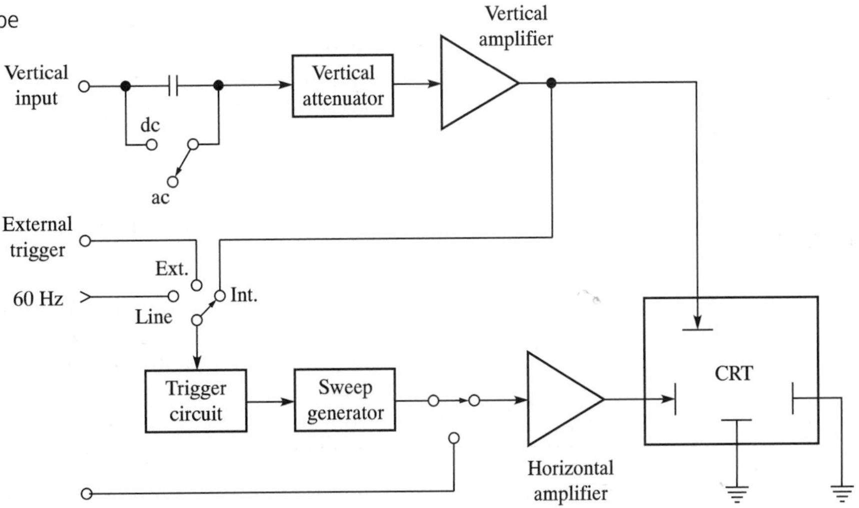

FIGURE 11–47 Mechanical layout of CRT

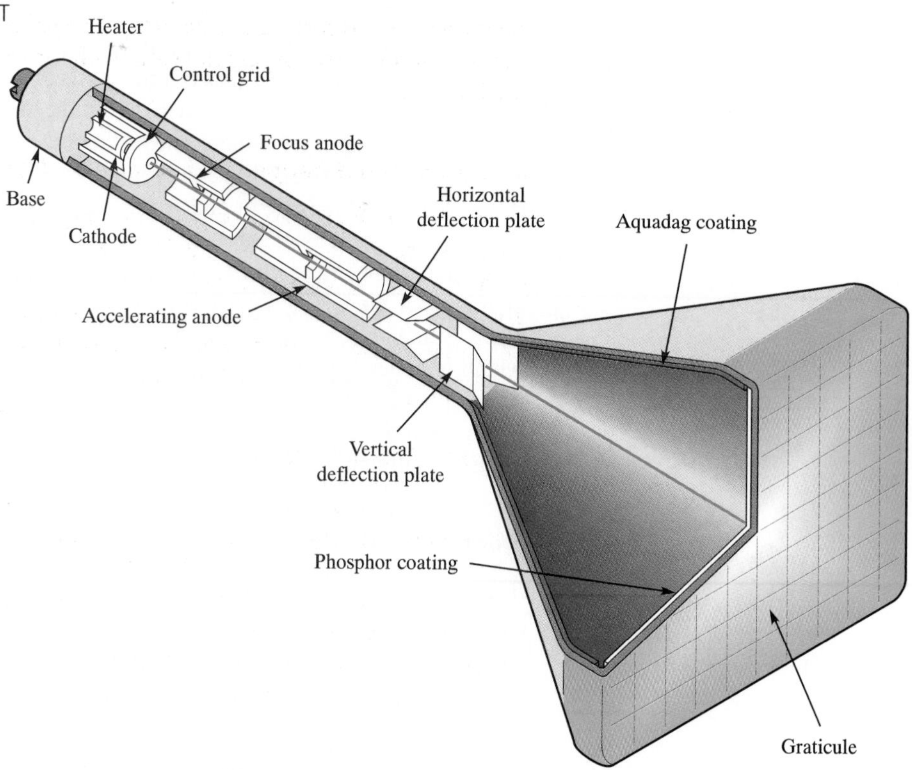

Question: How does the electron gun function?

The electron gun emits the electrons that make up the electron beam. It is composed of a heating filament surrounded by a metal cathode. Current flowing through the filament creates heat that, in turn, heats the cathode. When heated, electrons gain enough energy to leave the metal and enter the space around it. (This process is known as *thermionic emission*). The electron beam is formed from the cloud of electrons in this space.

Question: How is the electron beam developed?

The beam is formed as the electrons move through the control grid. As the name implies, the control grid controls the number of electrons in the beam. It has a negative potential with respect to the cathode, which limits the number of electrons that can pass through it. The more electrons in the beam, the brighter the display and vice versa. This potential can be adjusted using the BRIGHTNESS control on the oscilloscope. As you probably know, this adjustment can make the display very bright, or cut it off completely. (On some oscilloscopes, this adjustment may be known as the INTENSITY control). The electrons passing through the control grid must then be focused before arriving at the screen.

Question: How is the electron beam focused to a fine point on the screen?

The electrons passing through the control grid next come under the influence of the focus and accelerating anodes, and the aquadag coating. Each of these is at a relatively high positive potential with respect to the cathode; thus the electrons making up the beam are greatly accelerated. The aquadag coating is at a higher potential than the focus anode. There is a difference in charge between them, which produces electrostatic fields. These electrostatic fields focus the beam into a fine point on the screen. The fineness of the line can be controlled by varying the voltage on the focus using the FOCUS adjustment on the oscilloscope.

Question: What happens when the electron beam strikes the screen?

Upon striking the screen, the kinetic energy of the electron beam is converted to other forms of energy, one of which is light. Some of the energy of the electron beam causes electrons to be emitted by the phosphor. (This process is known as *secondary emission.*) These are not high-speed electrons and they are attracted to the high positive potential of the aquadag coating. If the emission of electrons by the phosphor did not take place, a build-up would take place on the phosphor, and the beam would eventually be repelled.

Electron Beam Deflection

The last thing to consider is the deflection of the electron beam. The horizontal trace on the oscilloscope represents time. Thus, the beam must be deflected, at a constant rate, from one side of the screen to the other. At the same time the beam is being deflected horizontally, the beam is deflected vertically by the signal to be displayed. The two deflections, operating on the beam simultaneously, display a graph of the signal on the screen. The voltages producing deflection of the beam are applied to the vertical and horizontal deflection plates within the CRT.

Question: How is the horizontal deflection and time base developed?

As shown in Figure 11–48, one of the horizontal deflection plates is grounded while the other has voltage in the form of a sawtooth waveform applied. As shown, the sawtooth begins at a negative voltage and deflects the beam to the left side of the screen. As it moves in a positive direction, the beam moves toward the center of the screen. When the sawtooth is at zero volts, there is ground on both plates, and the beam is at the center of the screen. The sawtooth now becomes positive, attracting the beam toward the right side of the screen. After reaching its maximum positive voltage, the sawtooth quickly drops

FIGURE 11–48 Electron beam deflection

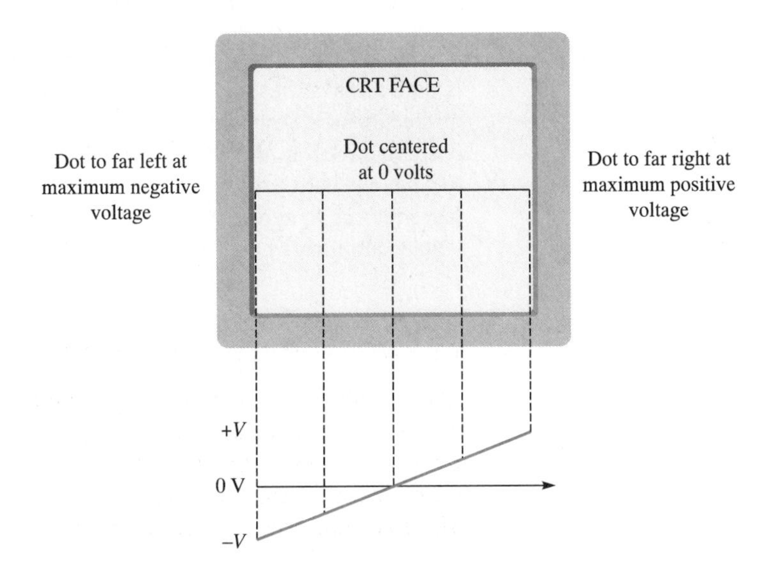

FIGURE 11–49 Signal display

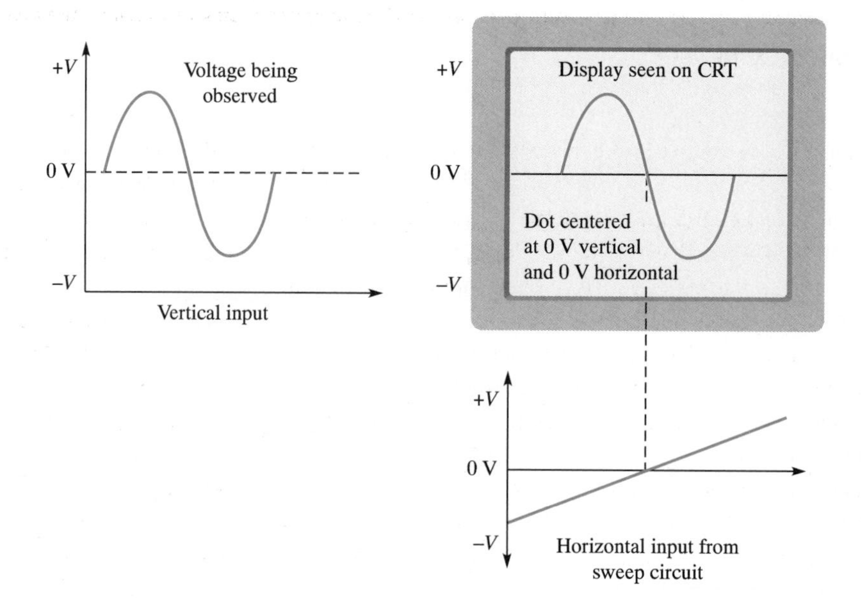

back to its maximum negative value, and the beam returns to the left of the screen. The relatively slow rise of the sawtooth is known as the *trace* and the rapid drop is known as the *retrace*.

The rate at which the sawtooth occurs establishes what is known as the *time base*. For example, if the rise of the sawtooth occurs in 100 ms, the trace represents 100 ms of time. Since the screen is divided into 10 equal divisions, each represents 10 ms. The time base is controlled by the TIME/DIVISION switch. Refer once again to the Figure 11–46. The time base is produced by the sweep generator, is amplified, and applied to the horizontal deflection plate. The trigger circuit produces a pulse that causes the start of the time base. Notice that the trigger can come from the input signal, a 60 Hz source, or an external source.

Question: How is an input signal displayed on the screen?

As shown in Figure 11–49, a sawtooth wave is applied to the horizontal deflection plate, while a sine wave is applied to the vertical. The period of the sine wave is identical to the trace time of the sawtooth. As the trace is moved to the right by the time base, the beam is pulled upward as shown by the positive half-cycle on the vertical plate. At the center of the sawtooth, the sine wave has returned to the baseline. As the sawtooth moves in a positive direction, the negative half-cycle of the sine wave on the vertical plate moves the trace in a negative direction. When the trace reaches the right edge of the screen, the sine wave returns to zero. The vertical presentation is calibrated in volts-per-division, so the voltage amplitude can be accurately determined.

■ SECTION 11–7 REVIEW

1. What components make up the CRT?
2. What is applied to the horizontal deflection plates?
3. What is applied to the vertical deflection plates?
4. What is the voltage applied to that controls the brightness of the trace?
5. What type of waveform produces the sweep on the oscilloscope?

CHAPTER SUMMARY

1. An alternating voltage is one that changes in polarity over a period of time.
2. An alternating current is one that changes direction in response to changes in the polarity of the voltage.
3. A sine wave of voltage is a time-varying periodic waveform whose amplitude follows a sine curve.
4. A sine wave of voltage has an instantaneous value that is a function of the sine of the angle of rotation.
5. A sine wave of voltage produces a sine wave of current through a load.
6. Sine waves are produced by machines known as alternators and electronic circuits known as oscillators.
7. A sine wave is made up of a positive and a negative alternation.
8. One cycle of a sine wave is 360° or 2π rad.
9. A decrease in the time period of a wave results in an increase in its frequency.
10. Two waves that do not begin simultaneously but have the same polarity are said to have a phase difference.
11. The angular velocity of a sine wave is directly proportional to its frequency.
12. Any nonsinusoidal wave is made up of a fundamental frequency plus harmonics.
13. The duty cycle of a square wave is 50%.
14. An oscilloscope presents a graph of a wave in the time domain.
15. A spectrum analyzer presents the harmonic content of a wave in the frequency domain.

KEY TERMS

alternating voltage
alternating current
alternator
sine wave
cycle
positive half-cycle
negative half-cycle
amplitude
period
frequency
phasor
radian
angular velocity
peak value
oscilloscope
average value
nonsinusoidal wave
harmonic
spectrum analyzer
pulse
pulse repetition time
pulse repetition frequency
rise time
fall time
duty cycle
square wave
baseline

EQUATIONS

(11–1) $f = \frac{1}{T}$

(11–2) $T = \frac{1}{f}$

(11–3) $v = V_{max} \sin \theta$

(11–4) $i = I_{max} \sin \theta$

(11–5) $\omega = 2\pi f$

(11–6) $V_{p-p} = 2V_p$

(11–7) $V_{avg} = 0.637\ V_p$

(11–8) $I_{avg} = 0.637\ I_p$

(11–9) $V_{rms} = 0.707\ V_p$

(11–10) $I_{rms} = 0.707\ I_p$

(11–11) $\text{PRF} = \frac{1}{\text{PRT}}$

(11–12) $\text{PRT} = \frac{1}{\text{PRF}}$

(11–13) $\%\ \text{duty cycle} = \frac{t_w}{\text{PRT}}100$

(11–14) $V_{avg} = \text{baseline} + (\text{amplitude} \times \text{duty cycle})$

Variable Quantities

f = frequency
T = time of one cycle (period)
v = instaneous voltage
i = instantaneous current
V_{max} = maximum (peak) voltage
I_{max} = maximum (peak) current
V_{avg} = half-cycle average voltage value
I_{avg} = half-cycle average current value
V_{p-p} = peak-to-peak voltage
PRF = pulse repetition frequency
PRT = pulse repetition time
t_w = pulse width

TEST YOUR KNOWLEDGE

1. Does a sine wave vary continuously in amplitude or polarity and periodically in amplitude or polarity?
2. Are frequency and period directly or inversely proportional?
3. Is the rate of change of a sine wave greatest when the voltage is passing through its maximum or zero point?
4. At what points, does the peak value V_p of a sine wave occur?
5. Which value of a sine wave is known as the effective value?
6. Is the average value of a sine wave computed over a full cycle or a half-cycle?
7. How many radians does one half-cycle rotation of a phasor represent?
8. How many radians are there in 45°?
9. How many degrees are there in $2\pi/3$ rad?
10. What is the seventh harmonic of 2 kHz?
11. What is the fourth harmonic of 3 kHz?
12. What is the harmonic content of a square wave?
13. What is meant by the pulse width of a voltage pulse?
14. Between what points on a voltage pulse is the rise time measured?
15. What is the relationship between the PRF and PRT of a pulse wave?
16. Is the duty cycle directly or inversely proportional to the pulse width?
17. Is the average value of a pulse wave directly or inversely proportional to the pulse width?
18. What is one use of a sawtooth wave?

PROBLEM SET: BASIC LEVEL

Section 11–1

1. For the graphs of Figure 11–50, state which are ac and which are for dc waveforms.

FIGURE 11–50 Waveforms for Problem 1

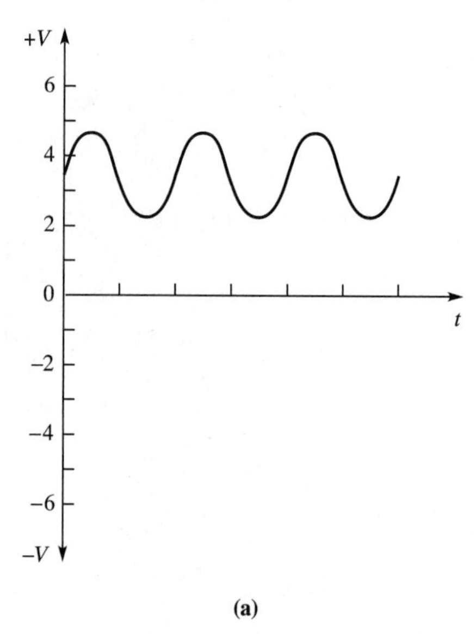

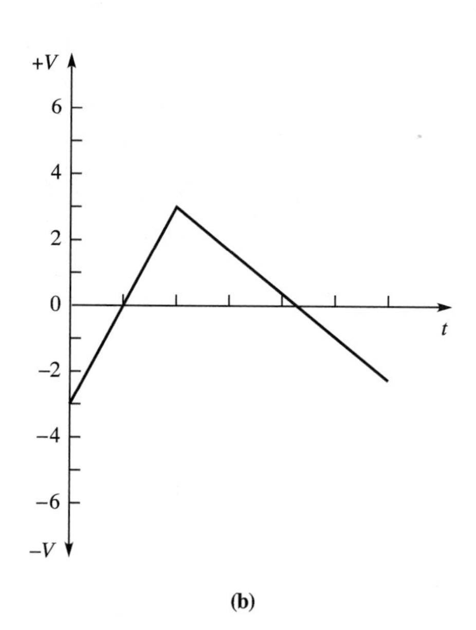

2. For the graphs of Figure 11–51, state which are ac and which are for dc waveforms.

FIGURE 11–51 Waveforms for Problem 2

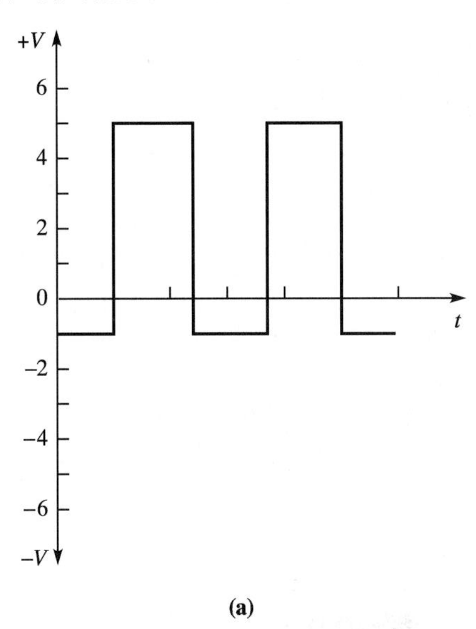

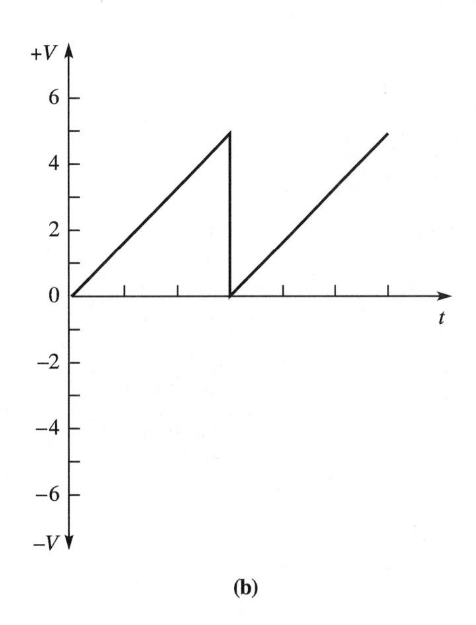

Section 11–2

3. Find the sine function of each of the following angles: (a) 45°, (b) 72°, (c) 28°, (d) −30°, and (e) −52°.

4. Plot the graph of the sine function from zero through 360°.

Section 11–3

5. For the position of the loop in Figure 11–52, will maximum or minimum voltage be induced? Why?

FIGURE 11–52 Alternator for Problem 5

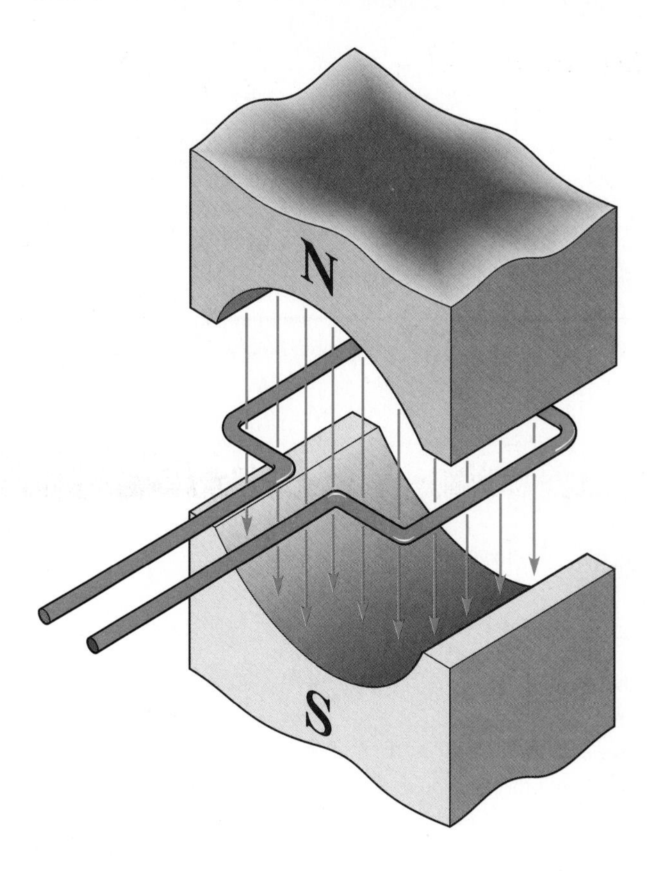

6. For the position of the loop in Figure 11–53, will maximum or minimum voltage be induced? Why?

FIGURE 11–53 Alternator for Problem 6

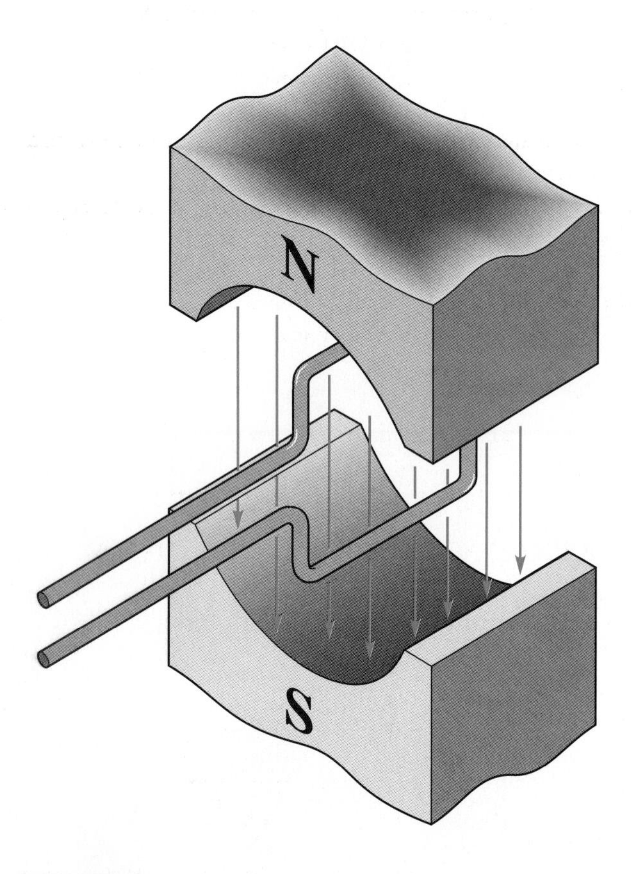

Section 11–4

7. In the circuit of Figure 11–54, how much current will flow at the indicated points on the input sine wave?

FIGURE 11–54 Circuit and sine wave for Problem 7

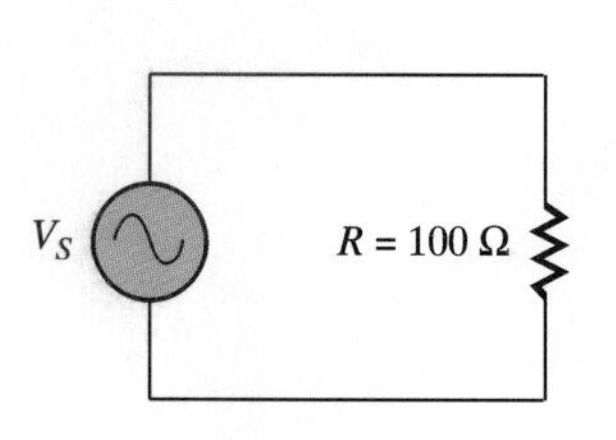

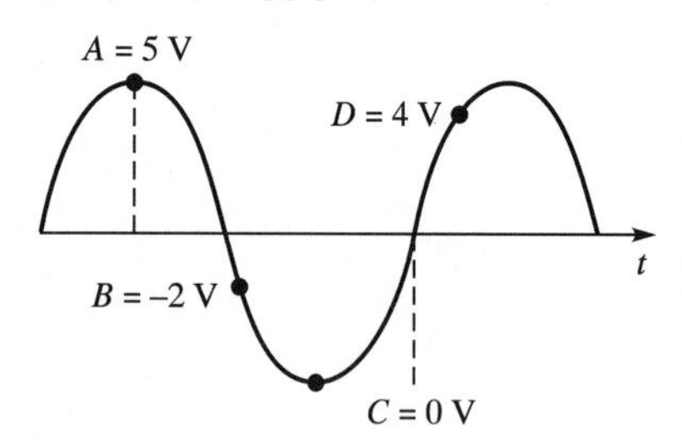

Section 11–5

8. For each of the following frequencies, compute the period: (a) 2 kHz, (b) 4 MHz, (c) 400 Hz, and (d) 60 Hz.
9. Compute the frequency of waveforms having the following periods: (a) 1 μs, (b) 10 mS, (c) 10 μs, and (d) 500 μs.
10. A sine wave contains how many frequencies?
11. A sine wave has a frequency of 500 kHz. How many cycles does it complete in 100 μs?
12. A sine wave completes 10 cycles in 50 μs. What is its frequency?
13. For the oscilloscope display shown in Figure 11–55, what is the frequency of the signal displayed?

FIGURE 11–55 Oscilloscope display for Problem 13

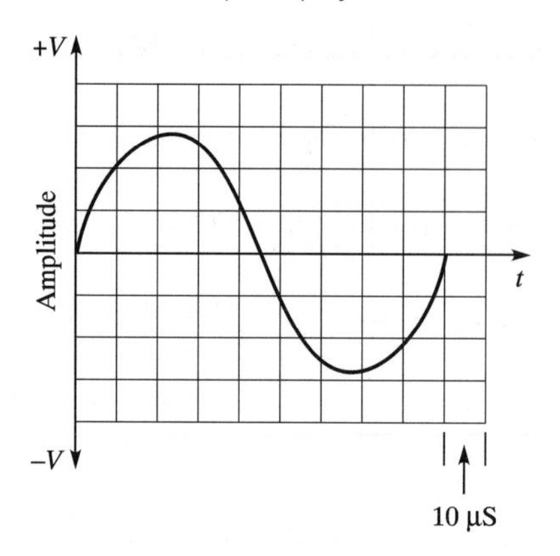

Section 11–6

14. A sine wave crosses the baseline at the reference and has an rms value of 40 volts. Determine its instantaneous value at each of the following angles: (a) 30°, (b) 45°, (c) 60°, (d) 90°, (3) 135°, (f) 145°, (g) 230°, (h) 320°, and (i) 360°.
15. Figure 11–56 shows the graph of two sine waves of voltage. Using waveform *B* as the reference, draw a phasor diagram to show their phase relationship.

FIGURE 11–56 Waveforms for Problem 15

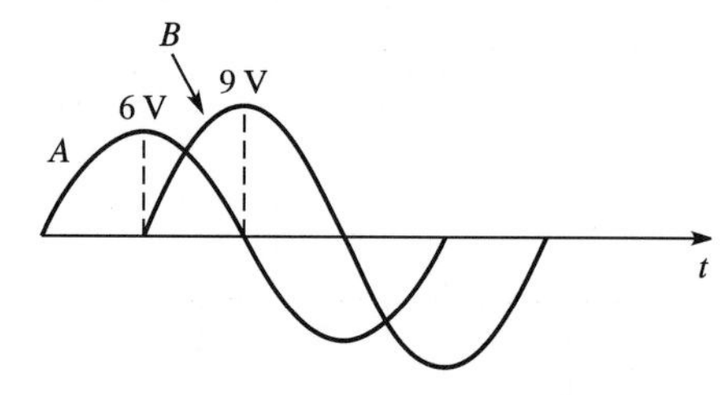

16. Figure 11–57 shows the phasor diagram of three sine waves. Draw the graph of these waveforms.

FIGURE 11–57 Phasor diagram for Problem 16

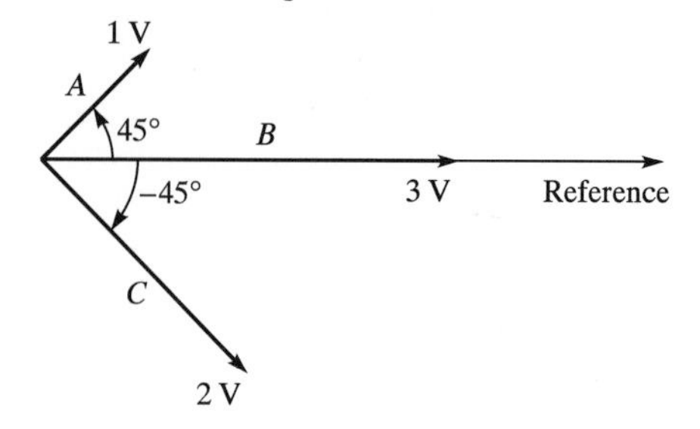

17. A sine wave has a frequency of 100 kHz. Compute its angular velocity in rad/s.

Section 11–7

18. For the following peak sine wave values, compute the rms values: (a) 12 V, (b) 170 V, (c) 35 V, and (d) 60 V.
19. Determine the peak-to-peak value of each of the voltages in Problem 18.
20. Determine the average value of each of the voltages in Problem 18.
21. A sine wave has a peak-to-peak value of 339 V. Compute its rms value.
22. A sine wave has an rms value of 70 Vs. Determine its peak value.
23. A sine wave has an rms value of 120 V. Determine its peak value.
24. A sine wave of current through a 2.4 kΩ resistor has a peak value of 8.4 mA_p. What is the peak value of the voltage across the resistor?
25. The peak value of the current through a resistance is 6.58 mA. If the voltage developed is 33.6 V_p, what is the value of the resistance?

Section 11–8

26. For each of the waveforms of Figure 11–58, compute the average value of the voltage.

FIGURE 11–58 Waveforms for Problem 26

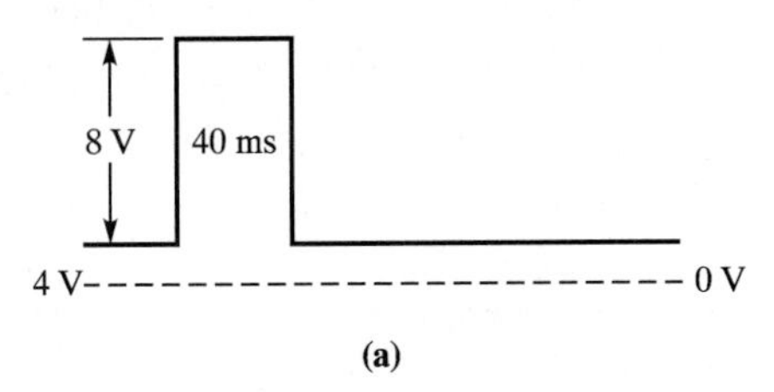

(a)

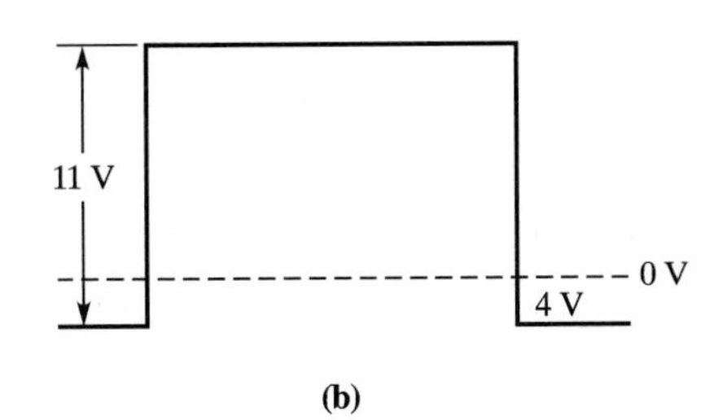

(b)

27. Determine the fundamental frequency for the square wave of Figure 11–59.

FIGURE 11–59 Square wave for Problem 27

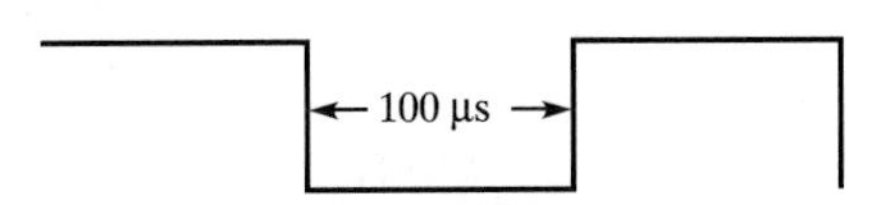

28. List the first five odd harmonics for the square wave with a fundamental frequency of 2 kHz.
29. For the pulse wave of Figure 11–60, compute the duty cycle as a percentage.

FIGURE 11–60 Pulse wave for Problem 29

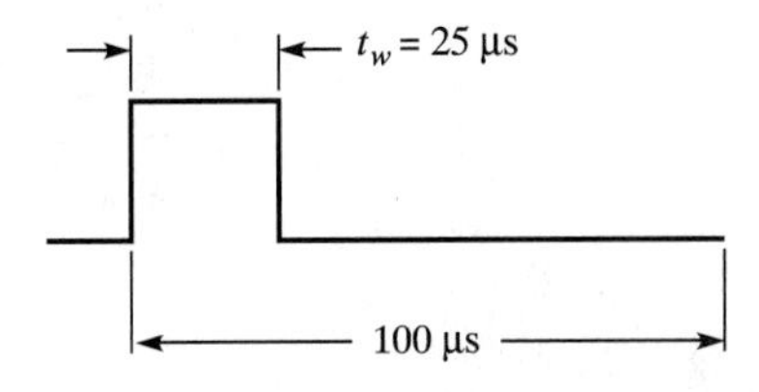

30. What is the PRF of the waveform of Figure 11–61?

FIGURE 11–61 Waveform for Problem 30

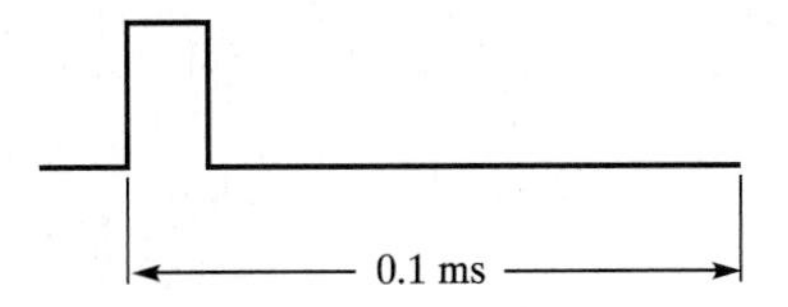

Section 11–9

31. The complete trace of an oscilloscope is 1 μs. What is the time of each major division?

PROBLEM SET: CHALLENGING

32. Determine the approximate pulse width, rise time, and fall time for the pulse of Figure 11–62.

FIGURE 11–62 Pulse wave for Problem 32

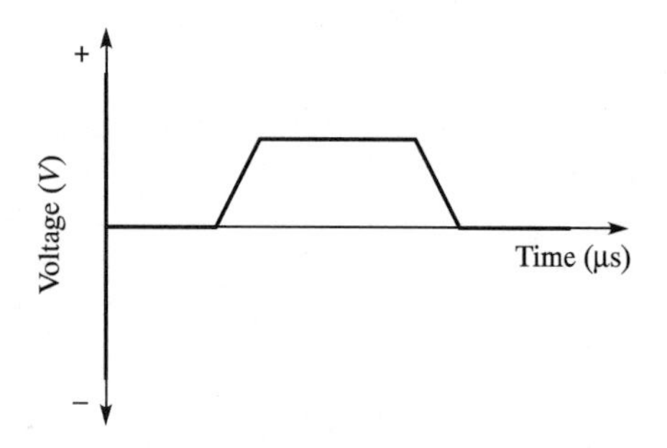

33. For the circuit of Figure 11–63, determine the current through and voltage across each resistor.

FIGURE 11–63 Circuit for Problem 33

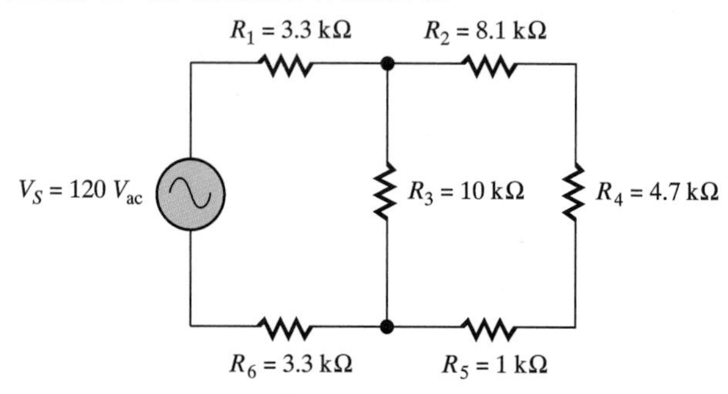

34. A pulse waveform has a duty cycle of 60%. If the pulse width is 10-mS, compute the frequency.

35. A pulsed wave has a duty cycle of 15% and a PRF of 1000 pps. Compute its pulse width.

36. Two waveforms, each with a frequency of 1 kHz, are viewed simultaneously on an oscilloscope. If the points at which they cross the baseline are as shown in Figure 11–64, determine the phase difference between them in degrees.

FIGURE 11–64 Waveforms for Problem 36

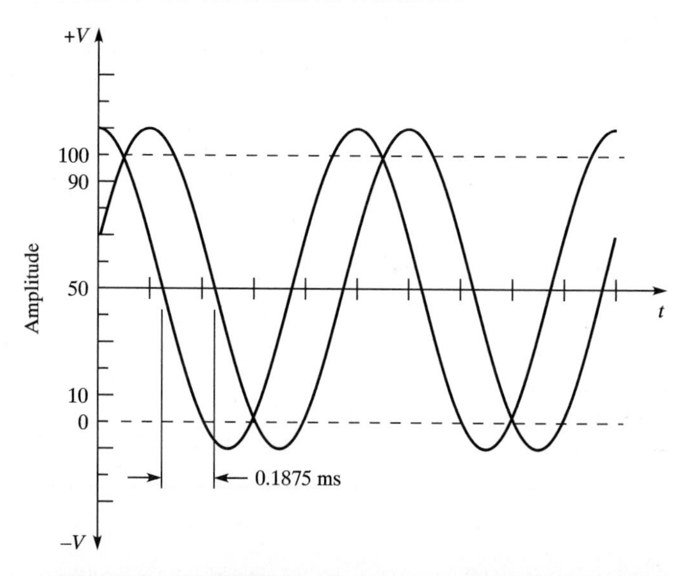

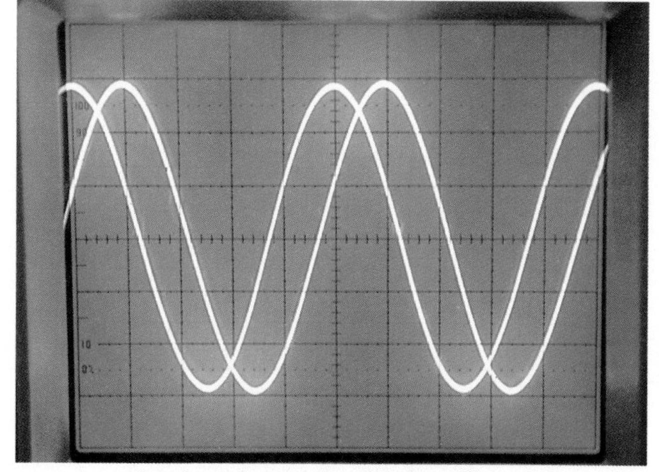

Actual waveforms as viewed on the oscilloscope

37. Sine wave A crosses the baseline in a positive direction at an angle of 45°. Waveform B crosses the baseline in a positive direction at an angle of 135°. What is the phase difference between the waveforms? Which is leading?

38. The positive peak of sine wave A occurs at 50°. The positive peak of sine wave B occurs at minus 35°. Sketch these two waveforms. What is their phase difference?

39. A sine wave has a period of 200 μs. Compute its angular velocity.

40. Determine the frequency of the sine wave from the oscilloscope display in Figure 11–65.

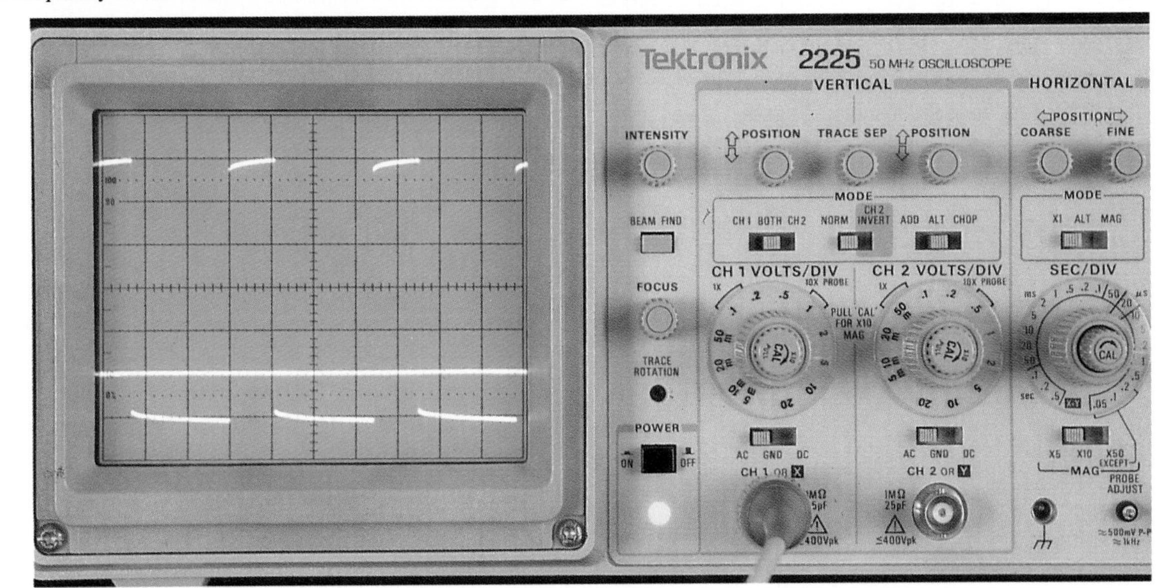

FIGURE 11–65 Oscilloscope display for Problem 40

(The X10 probe is used in the scope measurements)

41. Determine the period and frequency of the waveform of Figure 11–66. What name is given to this waveform?

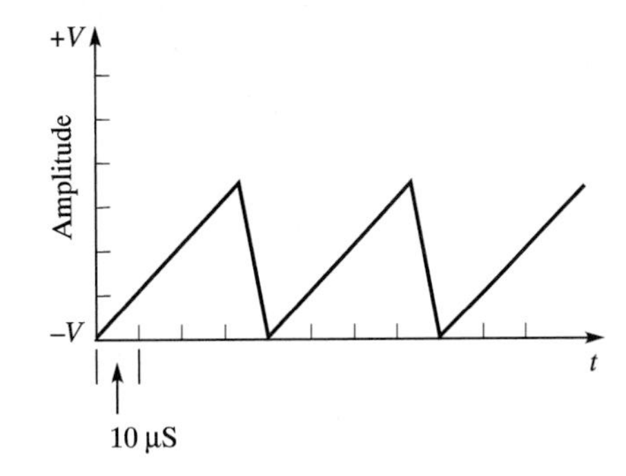

FIGURE 11–66 Waveform for Problem 41

42. What is the name given to the waveform of Figure 11–67? From the oscilloscope presentation of this figure, determine the signal's amplitude, average voltage, pulse width, and duty cycle.

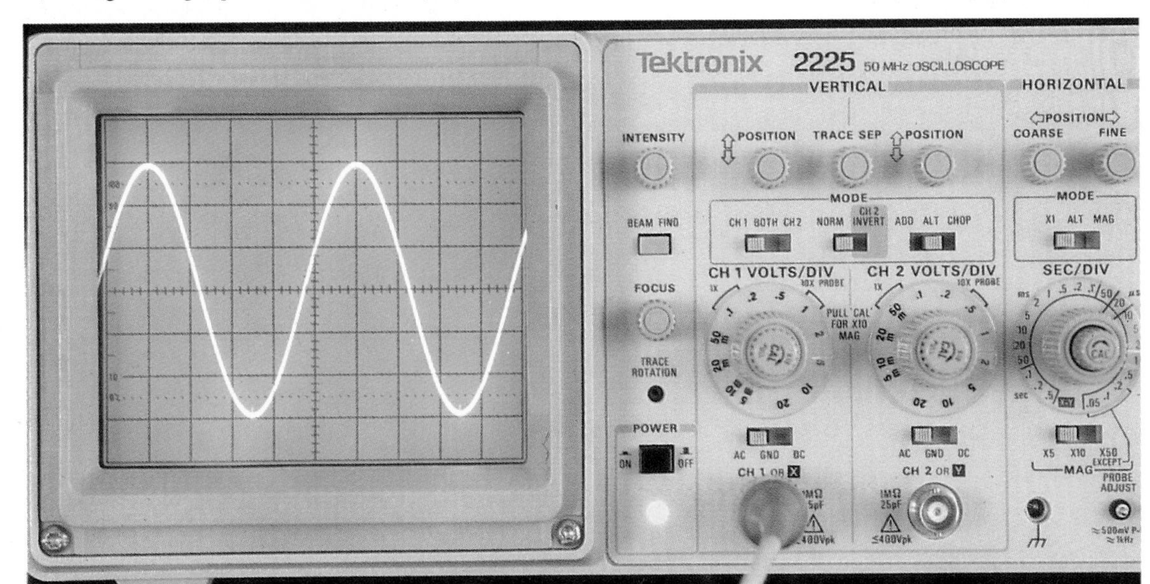

FIGURE 11–67 Oscilloscope display for Problem 42

(The X10 probe is used in the scope measurements)

PROBLEM SET: TROUBLESHOOTING

43. One fault exists in the circuit of Figure 11–68. From the readings given, determine the probable cause of the fault. (Assume V_S is correct.)

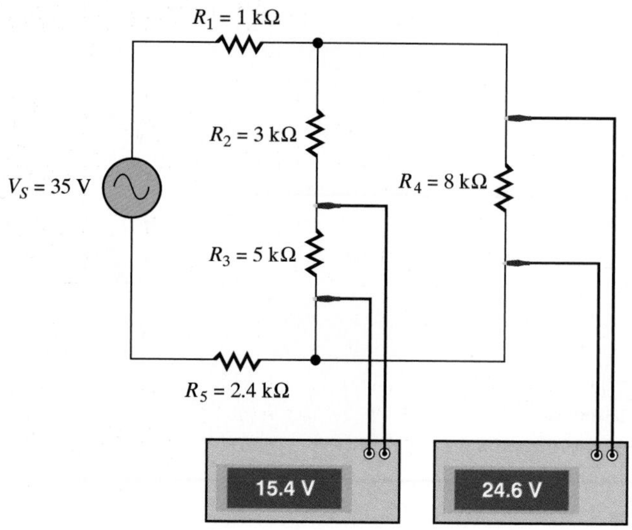

$R_1 = 1\ \text{k}\Omega$

$R_2 = 3\ \text{k}\Omega$

$V_S = 35\ \text{V}$

$R_4 = 8\ \text{k}\Omega$

$R_3 = 5\ \text{k}\Omega$

$R_5 = 2.4\ \text{k}\Omega$

15.4 V

24.6 V

FIGURE 11–68 Circuit for Problem 43

44. One fault exists in the circuit of Figure 11–69. From the readings given, what is one possible cause of the fault?

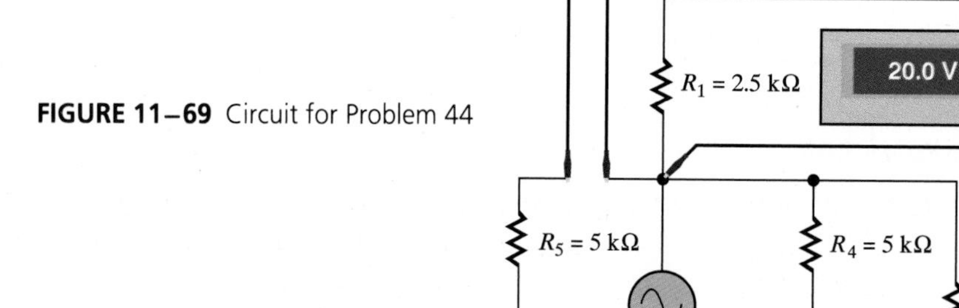

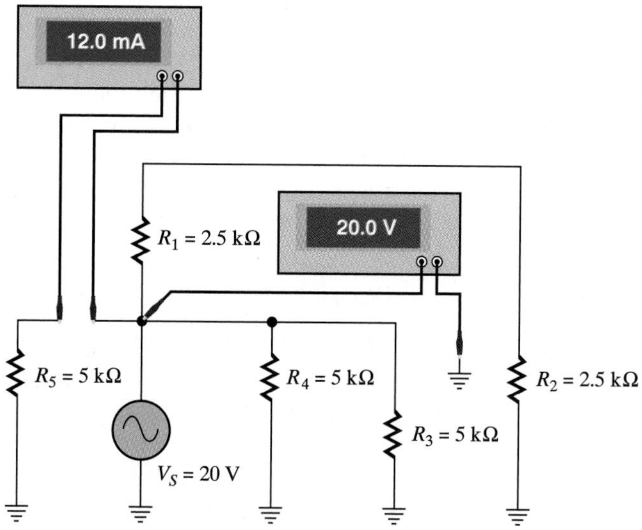

FIGURE 11–69 Circuit for Problem 44

Digital Storage Oscilloscopes

An oscilloscope is a device that provides a real-time graph of a voltage amplitude versus time, which means that you see the change in voltage as it happens. In most applications, a real-time display is just fine and is, in fact, what is really needed. There are times, however, when this type of graph is less than desirable. Suppose the voltage change in which you are interested is short in duration and only occurs after relatively long periods of time, as illustrated in Figure 1. It would be difficult to study such a waveform on a real-time display. As another example, suppose the voltage change is only a small part of a much longer, very complex, waveform, as illustrated in Figure 2. Getting the one small portion would be difficult on a real-time oscilloscope. Thus there are many times that it is advantageous to store a waveform sampled by an oscilloscope for future reference or analysis. Such storage is impossible with a standard real-time scope, but it can be accomplished using what is known as a *digital storage oscilloscope* (DSO).

With a DSO, the signal viewed can be stored in memory and later retrieved for viewing and analysis. This method of storage first digitizes the

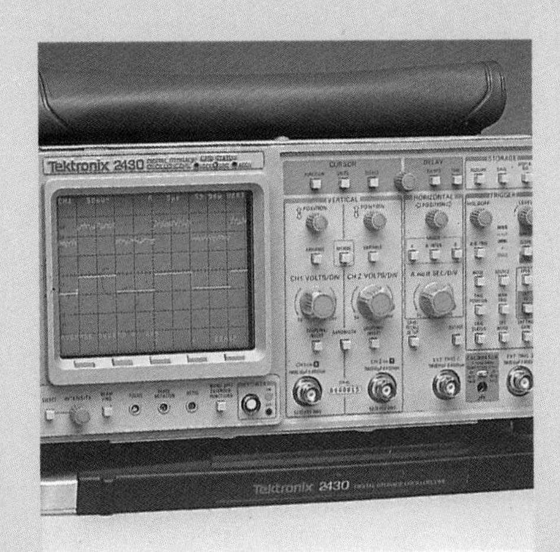

capture waveform. Digitizing means to sample an analog signal periodically, and assign each sample a number, as illustrated in Figure 3(a). In other words, the analog (continuously changing) voltage is converted to an equivalent series of digital (binary 1 or 0) numbers. Each digital number is then stored for either immediate or future display. To display the waveform,the binary numbers of the digitized waveform are used to restore the points on the graph. These points are joined by a smooth line to restore the waveform, as illustrated in Figure 3(b). Since the waveform is stored in memory it can be continuously displayed by repeatedly scanning the same digital equivalent. In addition the digitized waveform can be analyzed by computer and stored on disc indefinitely.

Figure 4 shows a block diagram of a typical digital storage oscilloscope. The input is conditioned by amplification (increasing) or attenuation (decreasing), as in any other oscilloscope. Since DSOs have basically the same input circuitry and input probes as any other scope, most will function as a conventional analog oscilloscope. Operating the scope in a manner such that it displays the waveform at the same time it is being sampled, is the real-time mode of operation.

The output of the vertical amplifier is connected to a sample and hold circuit. This circuit has the ability to measure a voltage and then remember that voltage for a brief period of time. This is where the sampling mentioned in a previous paragraph takes place. The output of the sample and hold circuit is connected to an analog-to-digital (A-to-D) converter. This circuit looks at the analog voltage that was sampled and assigns each sample an

FIGURE 1 Waveform of long duration

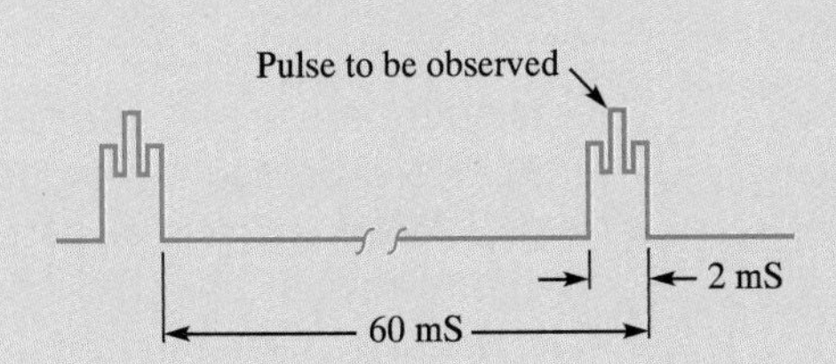

FIGURE 2 Complex waveform of long duration

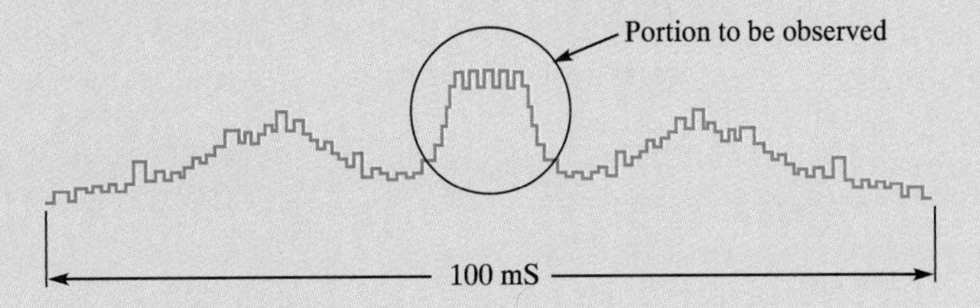

FIGURE 3 Sampling, storage, and reconstruction of a waveform by a DSO

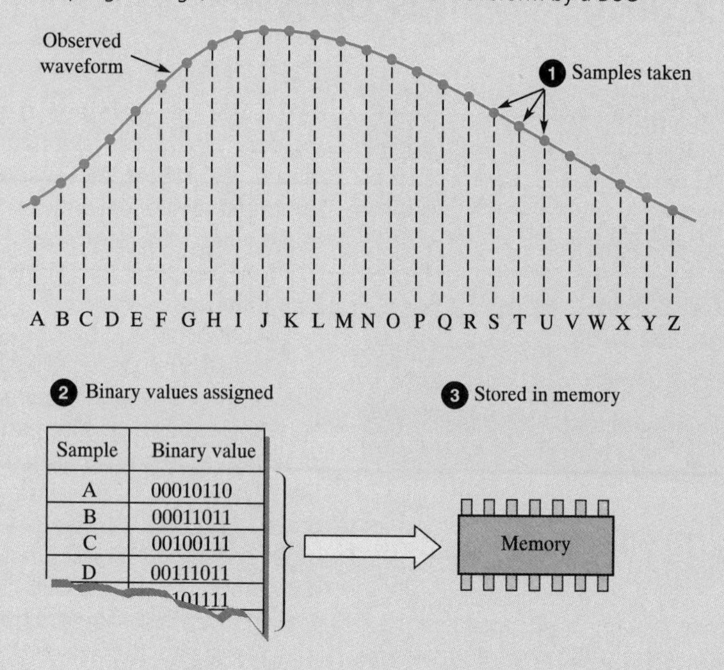

(a)

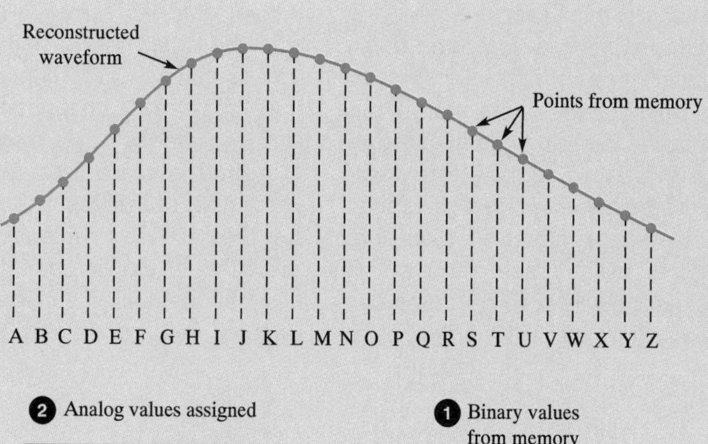

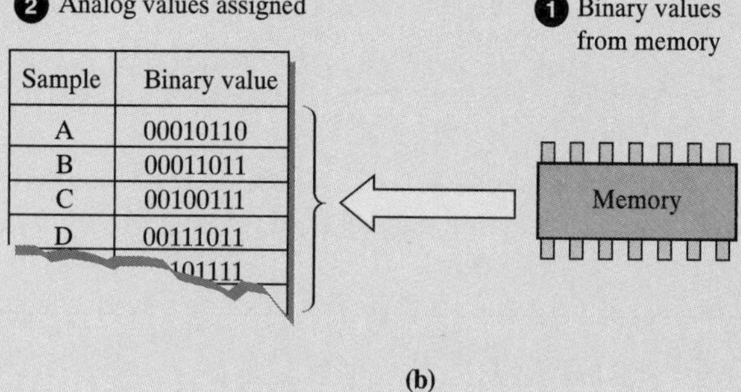

(b)

equivalent digital value. Because the input to the oscilloscope is constantly changing, the speed at which the A-to-D converter operates is extremely important.

The speed with which a DSO operates relates directly to what is known as its *resolution.* Its resolution is a measure of how small a change in the analog signal can be detected. For the average DSO, resolution for the A-to-D converter is 8 bits, which means that the voltage being sampled is divided into 256 (the value 2 raised to the eighth power) equal parts with each part being assigned a digital value. As an example, if a 10 V_{p-p} voltage is being sampled, the smallest change that can be detected and stored is 10V divided by 256, or 0.039 V. This equivalent digital value is then stored in the oscilloscope's memory for use as needed.

To display the stored waveform, the binary number that represents each portion of the digitized sample is retrieved from memory and fed in sequence into a digital-to-analog converter. The D-to-A converter does just the opposite of what the A-to-D converter does. It receives a digital number and produces an analog equivalent it. This voltage has the same value as that point in the digitized waveform. Once the D-to-A converter has reproduced the analog voltage, it is applied to the scope's vertical deflection amplifier in the conventional manner.

Another unique advantage of the DSO is its ability to look at a waveform both before and after any preselected event called the *trigger.* This trigger can be supplied as part of the waveform being analyzed or as a separate input. Operating the scope in this mode is called *pretrigger view* and can be very helpful in determining not only what happened but also why it happened.

While this introduction to the DSO is in no way complete, it will hopefully give you some idea as to how a DSO operates and how valuable it can be as a troubleshooting tool. The above photo shows a typical digital storage oscilloscope.

FIGURE 4 Block diagram of a typical digital storage oscilloscope

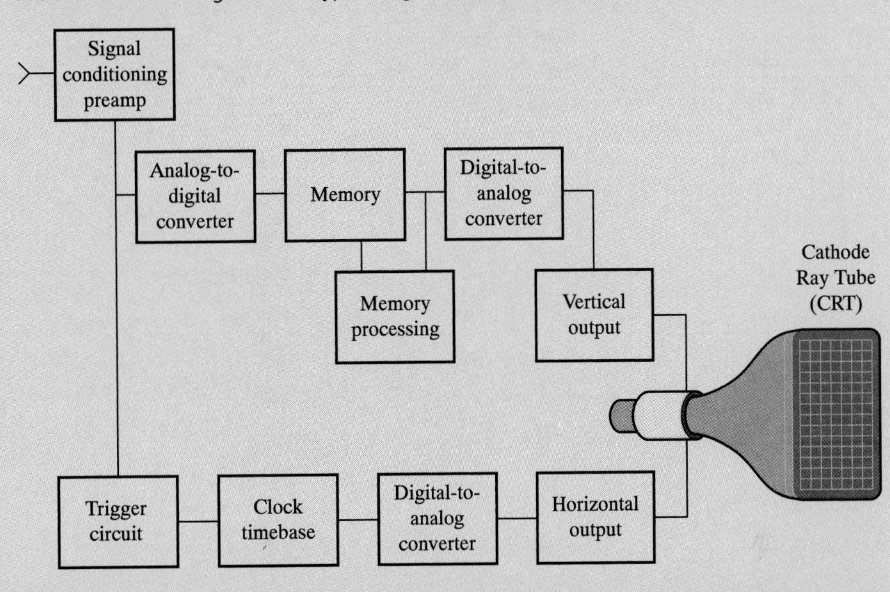

CHAPTER

12 Inductance in ac Circuits

UPON COMPLETION OF THIS CHAPTER, YOU WILL BE ABLE TO

1. Explain why inductance causes the voltage to lead the current.
2. Compute the inductive reactance and inductive susceptance of an inductor.
3. Explain the operation and perform analysis of series *RL* circuits.
4. Explain the operation and perform analysis of parallel *RL* circuits.
5. Explain how changes in frequency affect the operation of an *RL* circuit.
6. Compute and explain the significance of the quality factor of an inductor.
7. Explain what is meant by a choke.
8. List applications of *RL* circuits.
9. Perform troubleshooting on *RL* circuits.

You have learned the basics of inductors, and it is now time to put them to work in ac circuits. In this chapter, the inductor is first considered alone in its effects upon a sine wave of current. It is then combined with resistance, in what are known as *RL* circuits, and the action of these circuits is analyzed.

RL circuit analysis involves the use of Ohm's law, Kirchhoff's voltage and current laws, and the characteristics of series and parallel circuits. (Sound familiar?) In addition to the sine function, introduced in Chapter 11, the cosine and tangent functions are required in the analysis of *RL* circuits. These trigonometric functions, along with phasor algebra, are reviewed in this chapter.

As previously stated, an electronic circuit contains the effects of resistance, inductance, and capacitance only. Any electronic system, no matter how complex, can be reduced to these three effects plus current and voltage sources. This chapter presents the action of resistance and inductance in a circuit. Later chapters will cover the action of resistance and capacitance and bring all three together. Here, and in the chapters that follow, do not just memorize how to solve problems, but rather gain an understanding of how and why the circuits work they way they do. You will be glad that you did when it comes to learning the operation of electronic circuits such as amplifiers!

12–1 INDUCTANCE AND THE SINE WAVE

A Brief Review

Recall from Chapter 9 that *inductance* is the property of a circuit that induces a voltage when the current changes. The induced voltage is a result of the movement of magnetic flux lines produced by the change in current. Cutting across the windings of the inductor, the moving magnetic field induces a voltage with a polarity that opposes the change in current that produced it. Thus, inductance is the property of a circuit that opposes a change in current.

Recall further that when a dc voltage is first applied to an inductor, the induced counter voltage causes it to appear as an open circuit. The current then builds over a period of time, determined by the *LR* time constant, to a value limited only by the circuit resistance. Five of these time constants must elapse before the current reaches its full value. Once the current has reached full value, the inductor appears as a short circuit to dc. Thus, to steady-state direct current, the only opposition offered by an inductor is its winding resistance.

Inductive Effect on a Sine Wave of Current

You know from chapter 11 that a sine wave of current varies continuously over time. Thus, inductance has a continuous effect on the current. In Figure 12–1(a), a sine wave of current is shown flowing through an inductor. The varying current produces a moving magnetic field that induces a voltage across the inductor. The amount of voltage induced is a function of the inductance and the rate of change of current with respect to time ($V_L = L(\Delta i/\Delta t)$).

In Figure 12–1(b), one cycle of a sine wave of current is shown. Recall that the rate of change of a sine wave of current is greatest at the instant it crosses the zero voltage axis. At these points, the current has decreased to zero and is changing directions. As a result, the rate of change of magnetic flux and self-induced voltage are also maximum at these points. At the points where the current is maximum, its rate of change is zero. At these points the current has stopped increasing, prior to beginning to decrease. Thus, the magnetic flux is motionless, and the self-induced voltage is zero at these points.

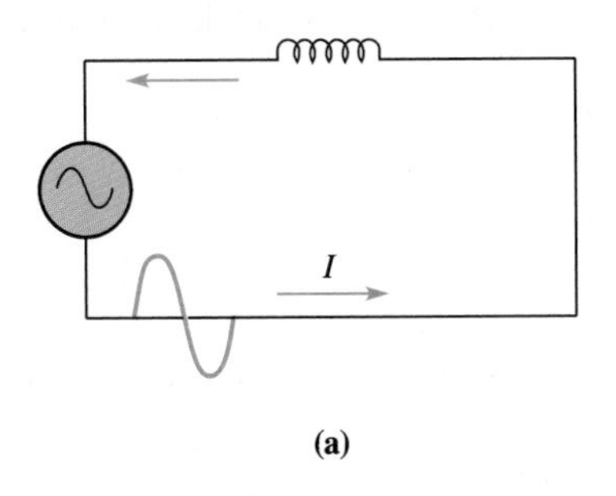

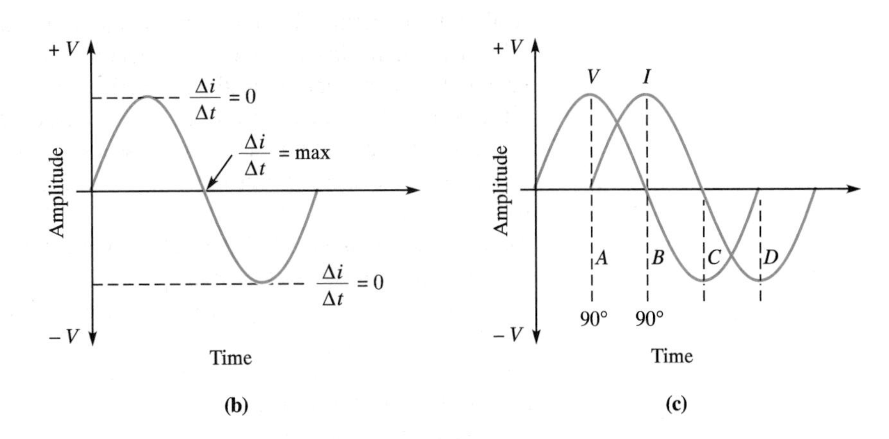

FIGURE 12–1 Effects of inductance on current in an *RL* circuit

Question: What is the effect of inductance upon the phase relationship between current and voltage?

In Figure 12–1(c), one cycle of the current and induced voltage are shown. Beginning at point *A*, the current is crossing the zero voltage axis in a positive direction. At this instant, a maximum positive voltage is induced. As the current increases, its rate of change decreases, and the magnitude of the induced voltage decreases. At point *B*, the current is maximum, its rate of change is zero, and the induced voltage is zero. As the current drops toward zero, its rate of change increases, as does the magnitude of the induced voltage. At point *C*, its rate of change is once again maximum, but now in a negative direction. The voltage produced is also maximum, but of the opposite polarity. As the current builds in the opposite direction, its rate of change decreases, and the magnitude of the induced voltage decreases. At point *D*, the current is maximum, and the induced voltage is zero. The current now moves toward zero, going in a positive direction. Its rate of change increases, as does the magnitude of the induced voltage, which reaches a maximum as the current crosses the zero axis.

Note carefully what has been demonstrated in Figure 12–1(c). The current through and the self-induced voltage across the inductor are not in phase. The inductor causes the voltage to lead the current by 90°. This concept is very important and must be thoroughly understood! The relationship of the current and voltage is shown in *phasor form* in Figure 12–2. Phasors, as you recall from Chapter 11, are lines drawn that represent two properties—its magnitude and direction—of some quantity. The current is drawn on the reference axis and the voltage is shown leading by 90°.

FIGURE 12–2 Phasor relationship of voltage and current, with V_L leading I_L by 90°

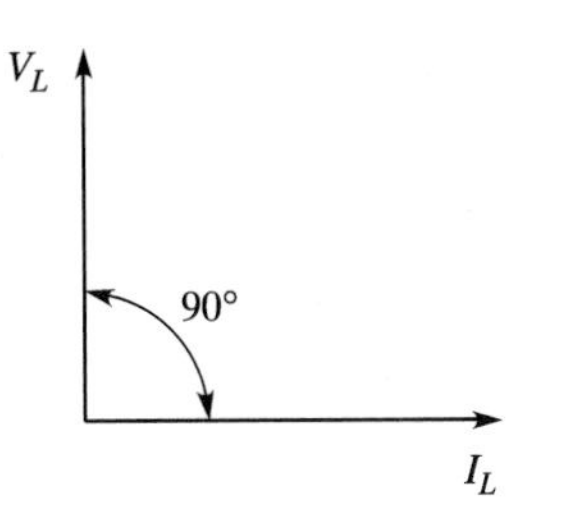

SECTION 12–1 REVIEW

1. What are the two definitions of inductance?
2. Why does inductance oppose a change in current?
3. What is the dc voltage across and the dc current through an inductor at the instant the supply voltage is applied?
4. What is the dc voltage across and the dc current through an inductor after five time constants have elapsed?
5. At what points in a sine wave of current is the rate of change the greatest, and at what points is it zero?

6. At what points on a current waveform does an inductor induce maximum voltage, and at what points does it induce zero volts?
7. State the phase shift that an inductor produces between the current and induced voltage.

12–2 INDUCTIVE REACTANCE

Question: How does an inductor affect the amplitude of a sine wave of current?

Inductance has been defined as the property of a circuit that opposes a change in current. This opposition, known as *reactance,* is symbolized by X. A continuously varying current, such as a sine wave, has continuous opposition from an inductor, known as **inductive reactance,** which is abbreviated X_L and read as "X sub L." Since inductive reactance is an opposition to current flow, its unit of measure is the ohm.

Question: What factors affect the amount of inductive reactance produced by a varying current?

An inductor with a high value of inductance is easier to magnetize than one with a lower value. Thus, the magnetic field produced is stronger, and the induced counter voltage is greater, thereby producing a greater opposition to the change in current. Inductive reactance, then, is *directly proportional* to the value of the inductance.

The induced voltage is also directly proportional to the rate of change of current with respect to time ($\Delta i/\Delta t$). Recall that a sine wave of current can be graphed from the rotation of the current phasor. If the phasor rotates 1 cycle in 1s, the frequency f is 1 Hz, and the period T is 1 s. If the speed of rotation of the phasor is increased to 60 cycles per second, the period decreases to 16.67 ms. Thus, the rate of change of current must *increase* in order to complete each cycle in a shorter time. With each increase in frequency, the period gets shorter, and the rate of change of current increases. As presented in Chapter 11, the speed of rotation of a phasor is stated in radians-per-second and is known as its *angular velocity.* It is computed as $2\pi f$. Thus, the rate of change of current can be expressed as the angular velocity of the phasor times the inductance. The equation for inductive reactance becomes

(12–1)
$$X_L = 2\pi fL$$

The symbol for angular velocity, $2\pi f$, is ω (lowercase omega from the Greek alphabet). Thus, the equation for inductive reactance is often written as

(12–2)
$$X_L = \omega L$$

EXAMPLE 12–1 Compute the reactance of a 30 mH inductor at a frequency of 10 kHz.

■ Solution:

Use Eq. 12–1:

$$X_L = 2\pi fL$$
$$= 2\,(3.14)(10\text{ kHz})(30\text{ mH}) = 1885\ \Omega$$

[2] [×] [π] [×] [1] [0] [EXP] [3] [×] [3] [0] [EXP] [±] [3] [=]

EXAMPLE 12–2 Compute the reactance of the same 30 mH inductor at a frequency of 7.5 kHz.

■ Solution:

$$X_L = 2\pi fL$$
$$= 2(3.14)(7.5\text{ kHz})(30\text{ mH}) = 1414\ \Omega$$

Notice the linear relationship in these two examples. The frequency was decreased in Example 12–2 to three-quarters its original value and the inductive reactance decreased to three-quarters its original value.

EXAMPLE 12–3 What value of inductor produces 157 Ω of reactance at a frequency of 100 kHz?

■ Solution:

Use Eq. 12–1, and transpose to solve for L:

$$X_L = 2\pi fL$$
$$L = \frac{X_L}{2\pi f}$$
$$= \frac{157\ \Omega}{2(3.14)(100\text{ kHz})} = 250\ \mu\text{H}$$

EXAMPLE 12–4 At what frequency does a 250 μH inductor produce 1.57 kΩ of reactance?

■ Solution:

Transpose Eq. 12–1 to solve for f:

$$X_L = 2\pi fL$$
$$f = \frac{X_L}{2\pi L}$$
$$= \frac{1570\ \Omega}{2(3.14)(250\ \mu\text{H})} = 1\text{ MHz}$$

■ Inductive Reactance and Current

A circuit containing only inductance is shown in Figure 12–3. Such a circuit is purely theoretical; the winding resistance R_w has been ignored in order to simplify the explanation. The amount of current that flows in this circuit is limited by the inductive reactance developed by the inductor. In the circuit of Figure 12–3, the total opposition to current flow, as computed using Equation 12–1, is 1884 Ω of inductive reactance. The current is computed using Ohm's law relationships, but X_L is substituted for resistance. In equation form, this relationship is

FIGURE 12–3 Theoretical circuit containing only inductance

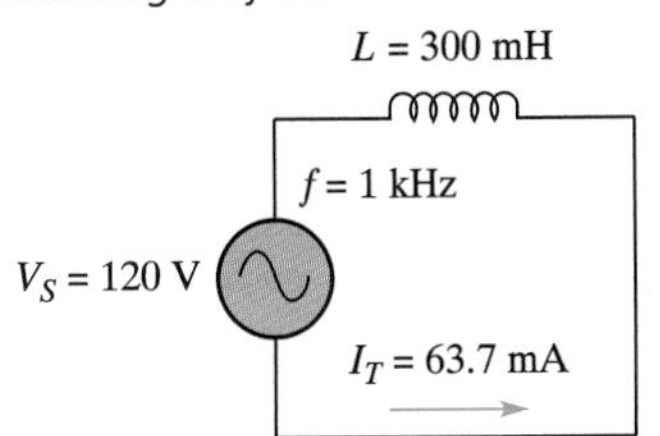

(12–3)
$$I_L = \frac{V_S}{X_L}$$

Thus, for the circuit of Figure 12–3, the current is

$$I_L = \frac{V_S}{X_L}$$

$$= \frac{120 \text{ V}}{1884 \ \Omega} = 63.7 \text{ mA}$$

The relationship of Equation 12–2 is often referred to as "Ohm's law for ac circuits." Just as is the case with the Ohm's law equation, it can be transposed in order to compute voltage or inductive reactance:

(12–4) $$V_L = IX_L \quad \text{and} \quad X_L = \frac{V_L}{I}$$ **(12–5)**

■ Inductive Reactances in Series and Parallel

Question: How is the total reactance of reactances in series computed?

The total reactance of inductive reactances in series is found in exactly the same manner as the total resistance for series resistors. The total reactance is the sum of the individual reactances:

$$X_{LT} = X_{L1} + X_{L2} + X_{L3} \cdots + X_{Ln}$$

FIGURE 12–4 Total reactance

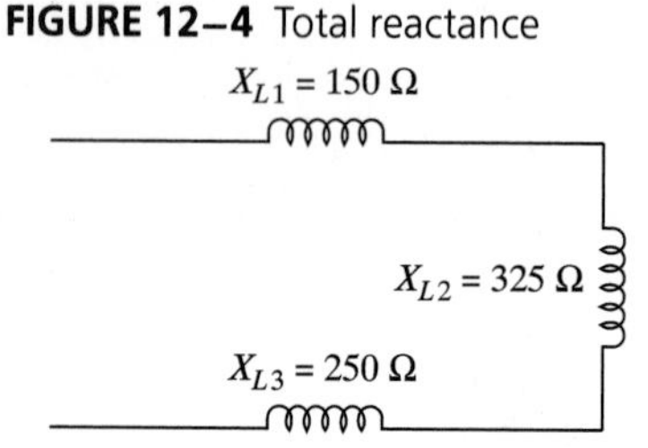

Thus, in Figure 12–4, the total reactance is

$$X_{LT} = X_{L1} + X_{L2} + X_{L3}$$

$$= 150 \ \Omega + 325 \ \Omega + 250 \ \Omega = 725 \ \Omega$$

Question: How is the total reactance of reactances in parallel computed?

The total reactance of inductive reactances in parallel is found in a manner similar to that used in finding total resistance of parallel resistors. Recall that it was conductance, the reciprocal of resistance, that was used. The reciprocal of a reactance is known as a *susceptance*. The reciprocal of inductive reactance is **inductive susceptance,** symbolized by B_L. The unit of measure for inductive susceptance is the siemen. The individual susceptances are added to find the total susceptance. The reciprocal of the total susceptance is the total reactance.

EXAMPLE 12–5 For the circuit of Figure 12–5, compute the total reactance.

■ Solution:

$$B_{LT} = B_{L1} + B_{L2} + B_{L3}$$

$$= \frac{1}{X_{L1}} + \frac{1}{X_{L2}} + \frac{1}{X_{L3}}$$

$$= \frac{1}{400 \ \Omega} + \frac{1}{1600 \ \Omega} + \frac{1}{4800 \ \Omega} = 3.33333 \text{ mS}$$

$$X_{LT} = \frac{1}{B_T}$$

$$= \frac{1}{3.33333 \text{ mS}} = 300 \ \Omega$$

FIGURE 12–5 Circuit for Example 12–5

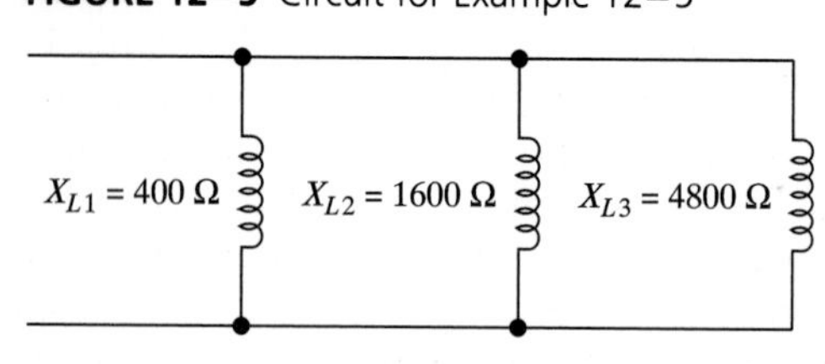

Total reactance of inductors in either series or parallel may be computed by first finding the total inductance and inserting this value in the inductive reactance equation:

(12–6) $$X_{LT} = 2\pi f L_T$$

■ SECTION 12–2 REVIEW

1. Is inductive reactance directly or inversely proportional to frequency?
2. Is inductive reactance inversely or directly proportional to inductance?
3. Compute the inductive reactance of an 1.089 mH inductor with a 500 kHz signal applied to it.
4. Compute the current in the circuit of Figure 12–6.
5. Compute the total current in the circuit of Figure 12–7.

FIGURE 12–6 Circuit for Section Review Problem 4

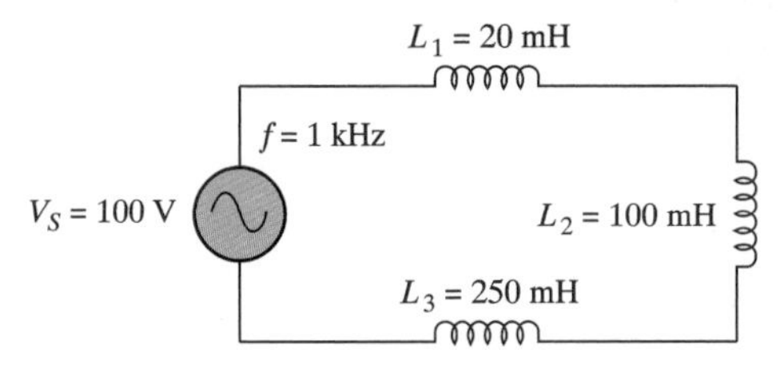

FIGURE 12–7 Circuit for Section Review Problem 5

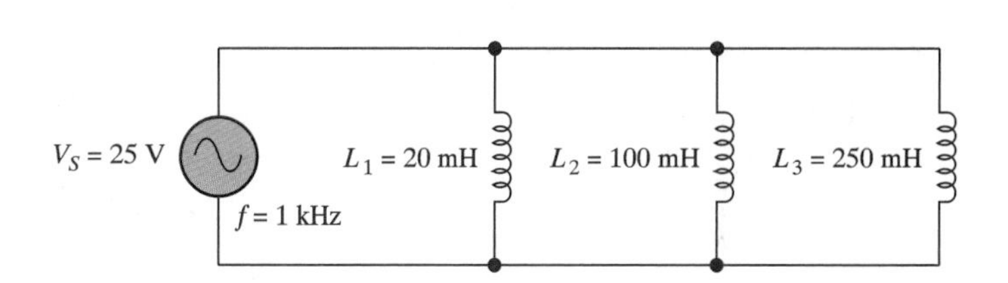

12–3 ANALYSIS OF SERIES *RL* CIRCUITS

Like any series circuit, the series *RL* circuit of Figure 12–8(a) is a voltage divider. There is a voltage drop across the resistor, V_R. This voltage and the current through it are in phase as shown in the graph and phasor diagram of Figure 12–8(b). A voltage is induced across the inductor, V_L. By the action of the inductor, this voltage leads the current by 90°, as shown in the graph and phasor diagram of Figure 12–8(c). Since the current is the same through both components, the phasor diagrams can be merged as shown in Figure 12–8(d). Thus, it can be seen that the inductor voltage V_L *leads* the resistor voltage V_R by 90°.

Question: Does Kirchhoff's voltage law apply to series RL *circuits?*

Recall that Kirchhoff's voltage law is the basic law of all series circuits. By way of review, it states that the sum of the voltages around a series loop is equal to the source voltage. Kirchhoff's voltage law can be applied to series *RL* circuits, but its application requires a special form of mathematics. The voltages that must be added, V_R and V_L, are not in phase. Thus, a special form of mathematics, known as *phasor algebra,* is necessary in series *RL* circuit analysis. Phasor algebra is reviewed briefly in the following paragraphs.

Figure 12–9(a) is a diagram of two forces acting on the same object. Notice that the two forces are acting in the same direction. They are said to be in phase. The total force on the object is the sum of the two forces. Its direction of motion is in the direction shown. Figure 12–9(b) shows the same two forces acting on the object, but at a 90° angle to one another. The object will not move in the direction of either force *A* or force *B*. It will move in a direction somewhere between the two, as shown in Figure 12–9(c). The movement will be closer to the direction of the greater force. The combination of the two forces is known as the *resultant force*.

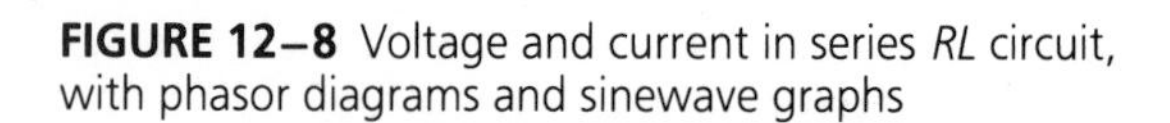

FIGURE 12–8 Voltage and current in series *RL* circuit, with phasor diagrams and sinewave graphs

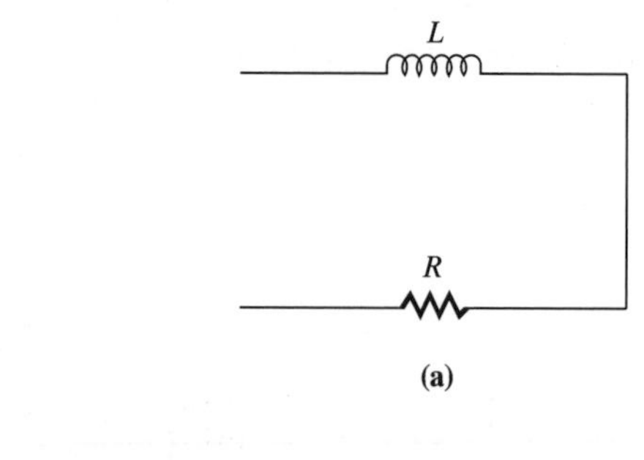

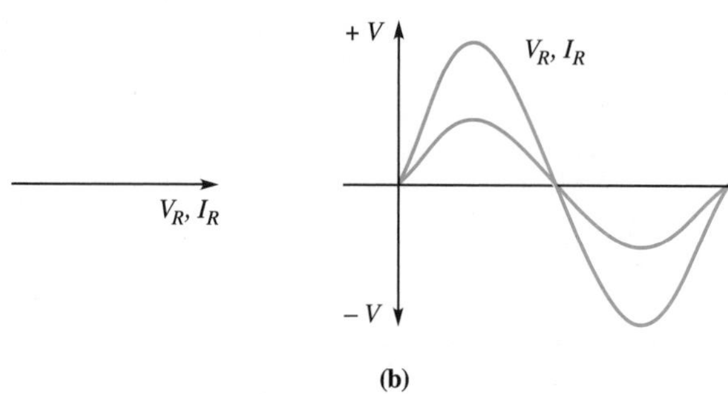

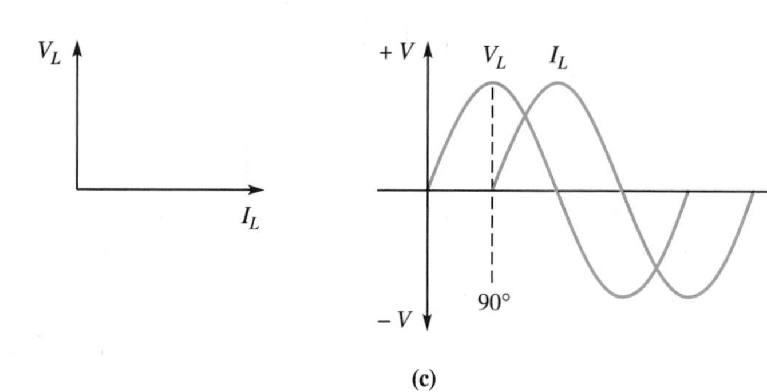

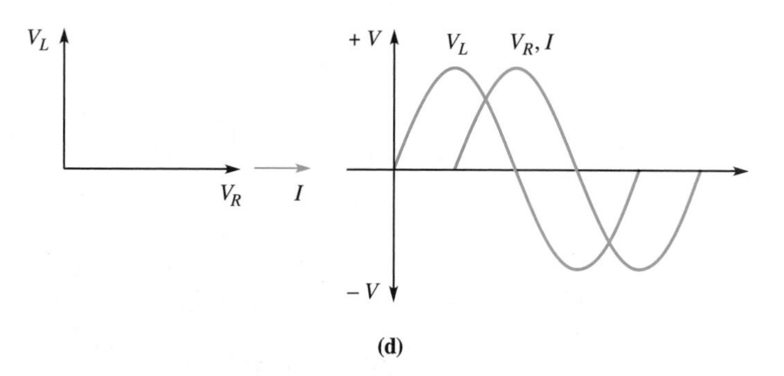

Question: How is the total force acting upon the object determined?

The two forces cannot be added directly because they are not acting in the same direction. Notice, in the phasor diagram of Figure 12–9(d), that when a vertical line is drawn from the tip of the resultant phasor to the tip of phasor B, a right triangle is formed. Recall that the square of the hypotenuse of a right triangle is equal to the sum of the squares of the other two sides. Thus,

$$(\text{resultant})^2 = A^2 + B^2$$

FIGURE 12–9 Phasor addition of two forces

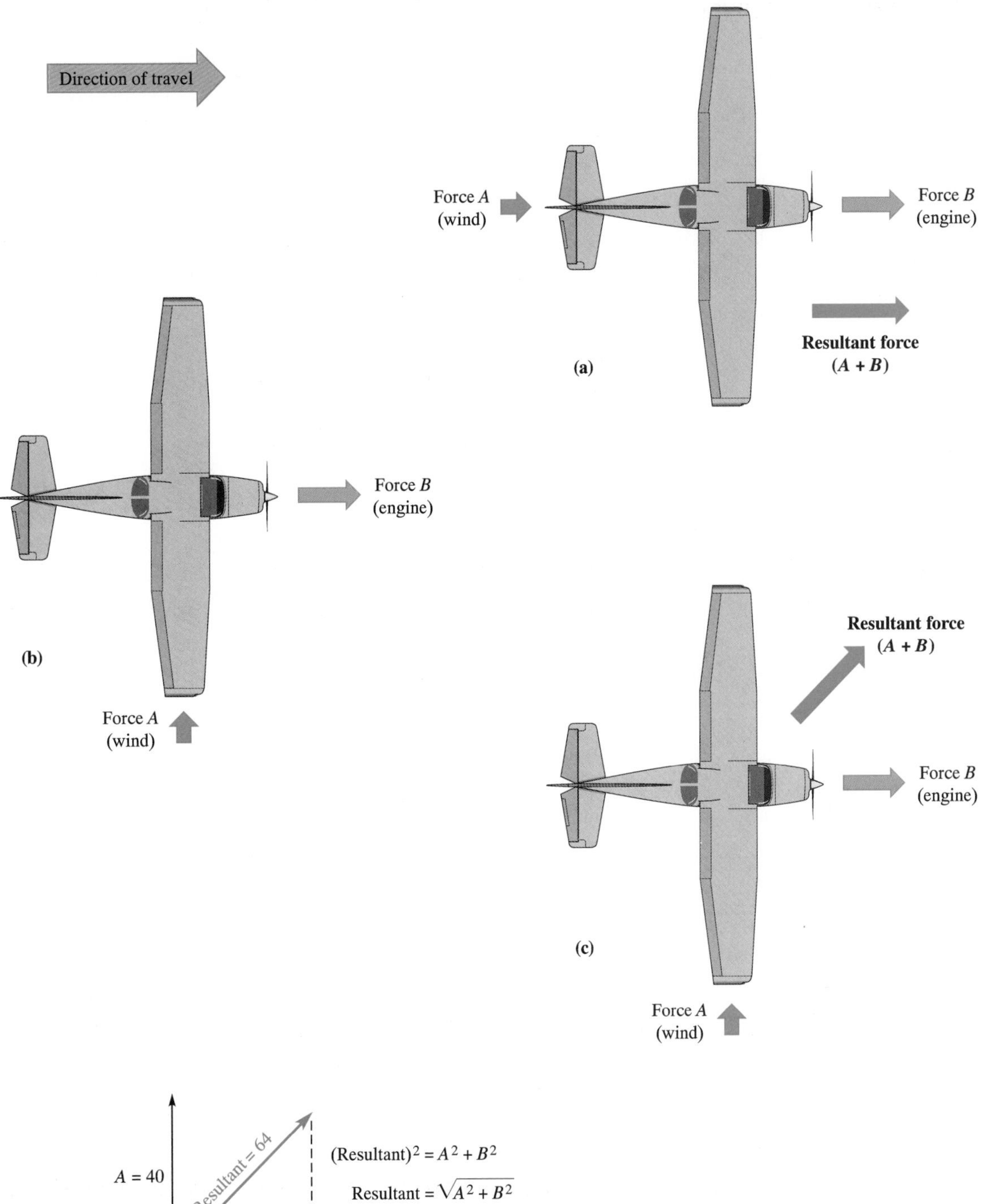

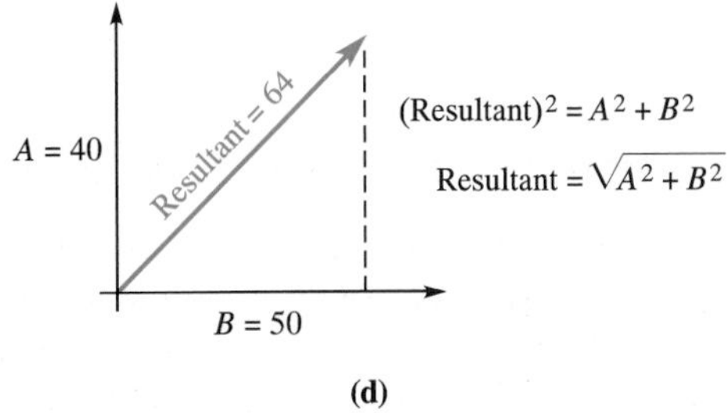

The value of the resultant can then be determined by taking the square root of both sides of the equation:

$$\text{resultant} = \sqrt{A^2 + B^2}$$

This type of addition is known as *phasor addition*. It is used to add two quantities whose phase difference is 90°.

EXAMPLE 12–6 Two forces act upon an object as shown in Figure 12–10(a). Compute the resultant force and sketch its phasor.

[4] [0] [X²] [+] [5] [0] [X²] [=] [√]

■ **Solution:**

FIGURE 12–10 Phasor diagrams for Example 12–6

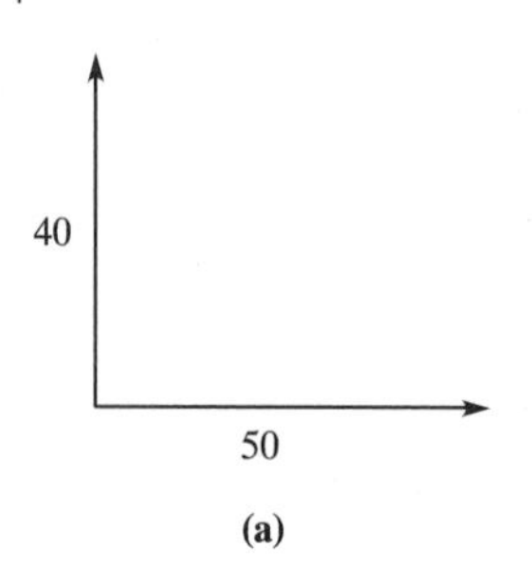

(a)

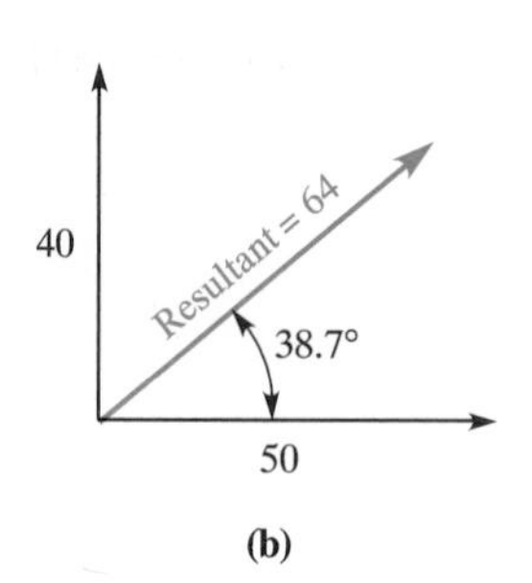

(b)

1. Compute the resultant force:

$$\text{resultant} = \sqrt{A^2 + B^2}$$

$$= \sqrt{40^2 + 50^2} = \sqrt{4100} = 64$$

Thus, the resultant has a value of 64.

2. In order to sketch the phasor, it is necessary to determine what angle θ that it makes with phasor B. The angle can be found using the trigonometric function known as the *tangent* (tan). The tangent of the angle is equal to the ratio of the opposite side to the adjacent side of the right triangle. Thus,

$$\tan \theta = \frac{A}{B}$$

$$= \frac{40}{50} = 0.8$$

The angle θ can next be determined:

$$\angle\theta = \text{arc tan } 0.8 = 38.7°$$

(Note that the function arc tan is designated $\tan^{-1}$ on many calculators.)

3. Draw the resultant phasor, as shown in Figure 12–10(b).

■ Voltages in a Series *RL* Circuit

Consider the series *RL* circuit of Figure 12–11(a). As with any series circuit, the same current flows through both components. The voltage drop across the resistor V_R is by Ohm's law the product of the current and resistance. The voltage developed across the inductor V_L is the product of the current and inductive reactance. Figure 12–11(b) is the graph and phasor diagram of V_R in relation to the current. Figure 12–11(c) is the graph and phasor diagram of V_L in relation to the current. Notice that the phasor for V_L has been placed on the reference to the point at the end of the phasor for V_R.

In Figure 12–11(d), the two phasor diagrams are merged to form what is known as a *voltage triangle*. The hypotenuse represents the source voltage, which is the phasor sum of V_R and V_L. From the voltage triangle, the following equations can be developed:

(12–7) $$V_S^2 = V_R^2 + V_L^2$$

(12–8) $$V_S = \sqrt{V_R^2 + V_L^2}$$

FIGURE 12–11 Voltage triangle for an *RL* circuit

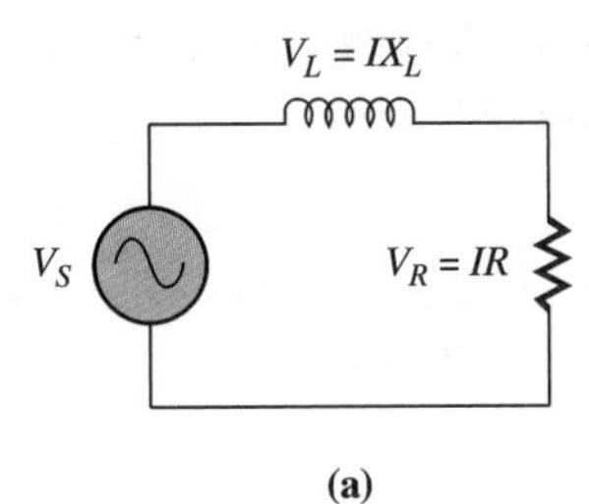

(a)

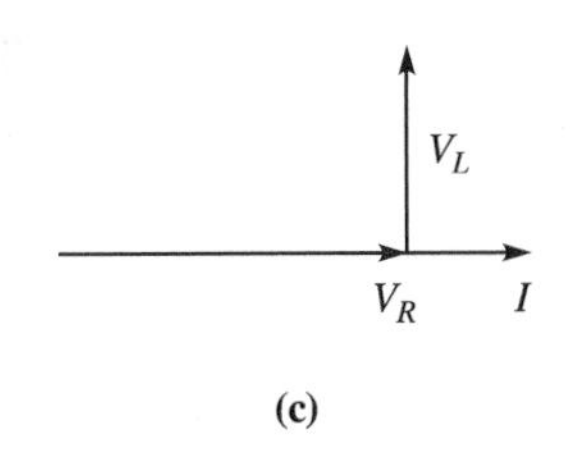

(c)

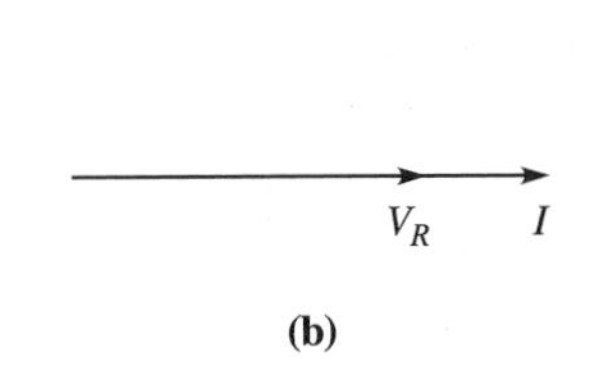

(b)

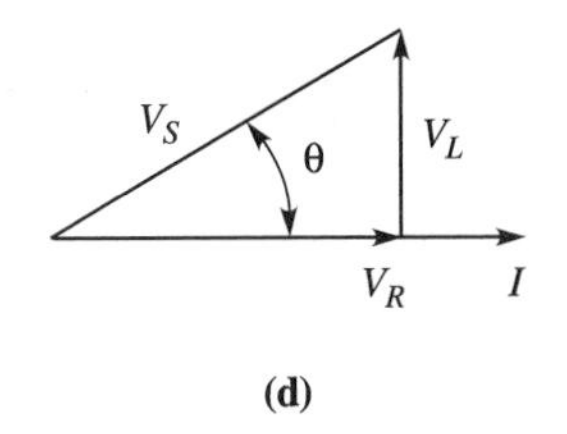

(d)

FIGURE 12–12 Circuit for Example 12–7

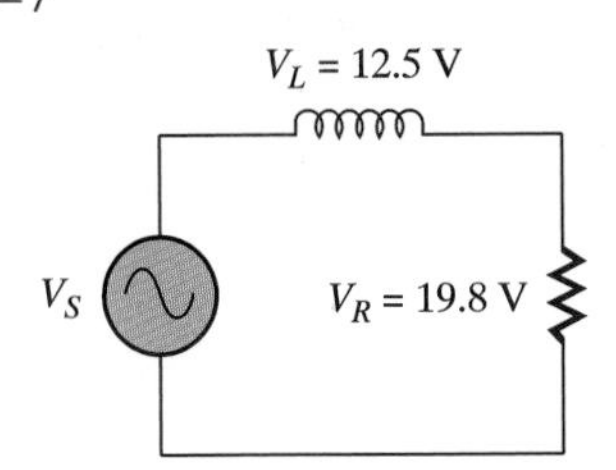

EXAMPLE 12–7 For the circuit of Figure 12–12, compute the value of V_S.

■ Solution:

Use Eq. 12–8:

$$V_S = \sqrt{V_R^2 + V_L^2}$$

$$= \sqrt{(19.8\text{ V})^2 + (12.5\text{ V})^2} = \sqrt{548\text{ V}} = 23.4\text{ V}$$

EXAMPLE 12–8 In a series *RL* circuit, a source voltage of 18 V produces a voltage drop across the resistor of 15 V. Compute the value of V_L.

■ Solution:

Use Eq. 12–7, and transpose to solve for V_L:

$$V_S^2 = V_R^2 + V_L^2$$

$$V_L^2 = V_S^2 - V_R^2$$

$$V_L = \sqrt{V_S^2 - V_L^2}$$

$$= \sqrt{(18\text{ V})^2 - (15\text{ V})^2} = \sqrt{99\text{ V}} = 9.95\text{ V}$$

EXAMPLE 12–9 For the circuit of Figure 12–13, compute the value of V_S.

■ Solution:

As the voltage phasors show, the voltages developed across the inductors are in phase and can be added directly. This results in a total inductive voltage, V_{LT}, of 30 V. The same can be done with the voltage drops across the resistors. Thus, V_R is equal to 50 V. The source voltage may now be determined:

$$V_S = \sqrt{V_R^2 + V_L^2}$$

$$= \sqrt{(50\text{ V})^2 + (30\text{ V})^2} = \sqrt{3400\text{ V}} = 58.3\text{ V}$$

FIGURE 12–13 Circuit for Example 12–9

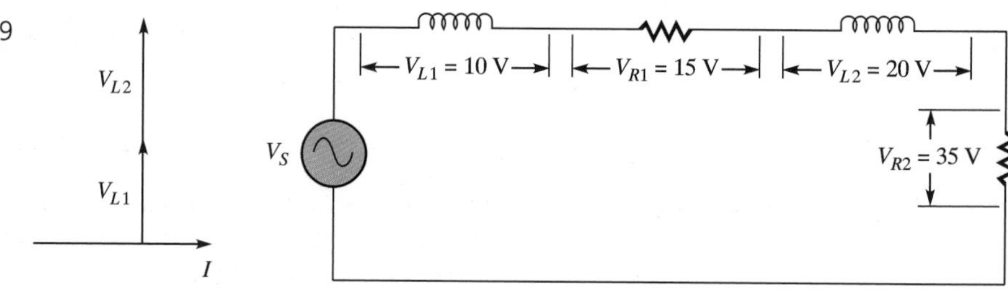

■ Impedance in a Series *RL* Circuit

Question: What is impedance?

The total opposition offered by a circuit to an alternating current is known as **impedance.** In an *RL* circuit two quantities oppose current: resistance and inductive reactance. Thus, the impedance is a combination of these two. The symbol for impedance is *Z*, and its unit of measure is the ohm. An equation for impedance can be derived using the voltage relationships in a series *RL* circuit.

EXAMPLE 12–10 For the circuit of Figure 12–14, compute the value of the impedance.

FIGURE 12–14 Circuit for Example 12–10

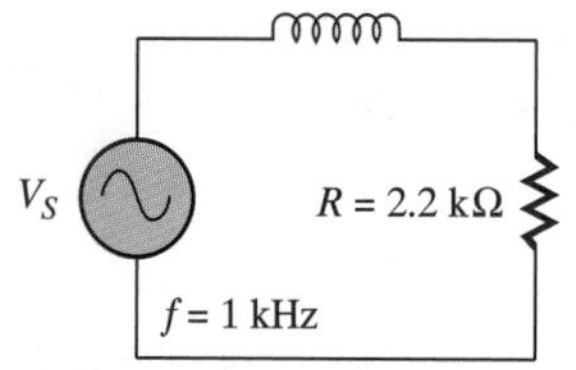

$L = 250$ mH
V_S
$R = 2.2$ kΩ
$f = 1$ kHz

■ Solution:

1. Use Eq. 12–1 to determine the value of X_L:

$$X_L = 2\pi fL$$
$$= 2(3.14)(1000 \text{ Hz})(250 \times 10^{-3} \text{ H}) = 1570\ \Omega$$

2. Use Eq. 12–9 to find the value of Z:

$$Z = \sqrt{R^2 + X_L^2}$$
$$= \sqrt{(2200)^2 + (1570)^2} = 2.703 \text{ k}\Omega$$

EXAMPLE 12–11 An *RL* circuit has an impedance of 1235 Ω. If the value of the resistor is 500 Ω, compute the reactance.

■ Solution:

Transpose Eq. 12–9 to solve for X_L:

$$Z = \sqrt{R^2 + X_L^2}$$
$$X_L = \sqrt{Z^2 - R^2}$$
$$= \sqrt{(1235)^2 - (500)^2} = 1129\ \Omega$$

By Kirchhoff's voltage law,

$$V_S^2 = V_R^2 + V_L^2$$

Substituting gives

$$(IZ)^2 = (IR)^2 + (IX_L)^2$$

Dividing by I yields

$$Z^2 = R^2 + X_L^2$$

Then,

(12–9) $$Z = \sqrt{R^2 + X_L^2}$$

FIGURE 12–15 Impedance triangle

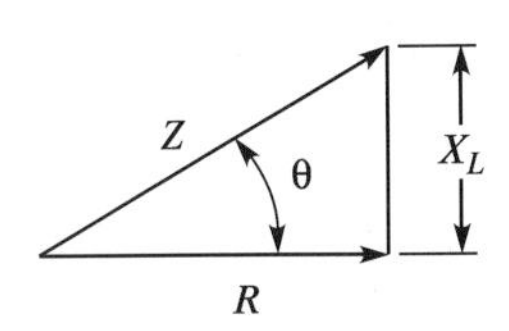

The impedance of a series *RL* circuit is the phasor sum of resistance and inductive reactance. This relationship is shown in the *impedance triangle* of Figure 12–15, in which the phasors representing resistance and inductive reactance occupy the same positions as the phasors for the voltages they produce. Ohm's law for ac circuits can be used to develop the following equations:

(12–10) $$I = \frac{V_S}{Z}$$

(12–11) $$V_S = IZ$$

(12–12) $$Z = \frac{V_S}{I}$$

■ Circuit Phase Angle

Question: What is meant by the circuit phase angle?

The **phase angle** of an ac circuit is the phase difference between the source voltage and the current. The action of the resistor in the circuit is to cause the current and voltage to be in phase. The action of the inductor is to cause the voltage to lead the current by 90°. The result is that the voltage leads the current by some angle between zero and 90°, depending upon the relative influences of the resistance and inductance.

Question: How is the phase angle determined?

FIGURE 12–16 Voltage and impedance triangles for computing the phase angle

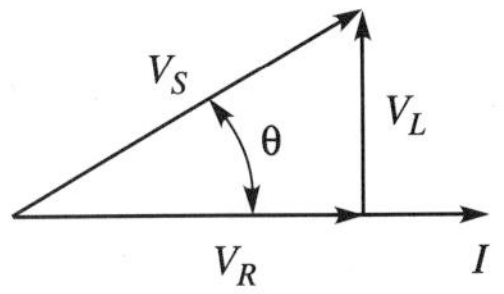

$\theta = \text{arc tan}\ \frac{V_L}{V_R}$

(a)

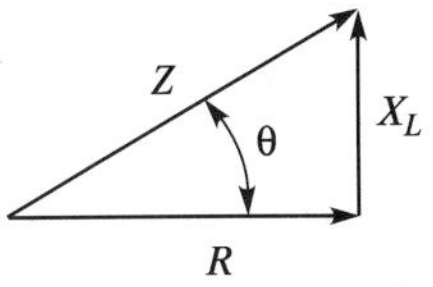

$\theta = \text{arc tan}\ \frac{X_L}{R}$

(b)

The phase angle is illustrated in Figure 12–16(a), where you will recognize a voltage triangle. The source voltage leads the current by the angle θ. This angle can be determined by applying the trigonometric functions. For example, the tangent of $\angle\theta$ is equal to V_L divided by V_R. Then, $\angle\theta$ will equal the arc tangent of this value. In equation form, this relationship is written as

(12–13) $$\theta = \text{arc tan}\ \frac{V_L}{V_R}$$

The phasors making up the impedance triangle are directly proportional to the phasors making up the voltage triangle. Thus, the impedance triangle may also be used in computing the phase angle, as illustrated in Figure 12–16(b). The equation is

(12–14) $$\theta = \text{arc tan}\ \frac{X_L}{R}$$

Question: How is the phase angle affected by frequency?

The value of the phase angle indicates the relative values of R and X_L. As Figure 12–17(a) shows, the greater the value of the phase angle, the larger X_L must be in comparison to R. Recall that X_L is directly proportional to frequency, so that as the frequency increases, the phase angle also increases. Thus, the phase angle of a series *RL* circuit is *directly proportional* to frequency. If, on the other hand, the frequency is held constant and the resistance increased, the phase angle decreases, as illustrated in 12–17(b).

FIGURE 12–17 Relationship of frequency and phase angle

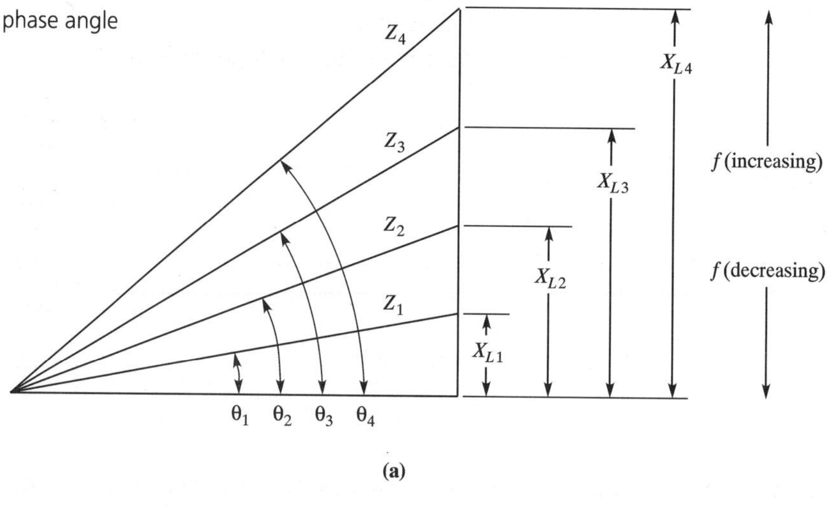

(a)

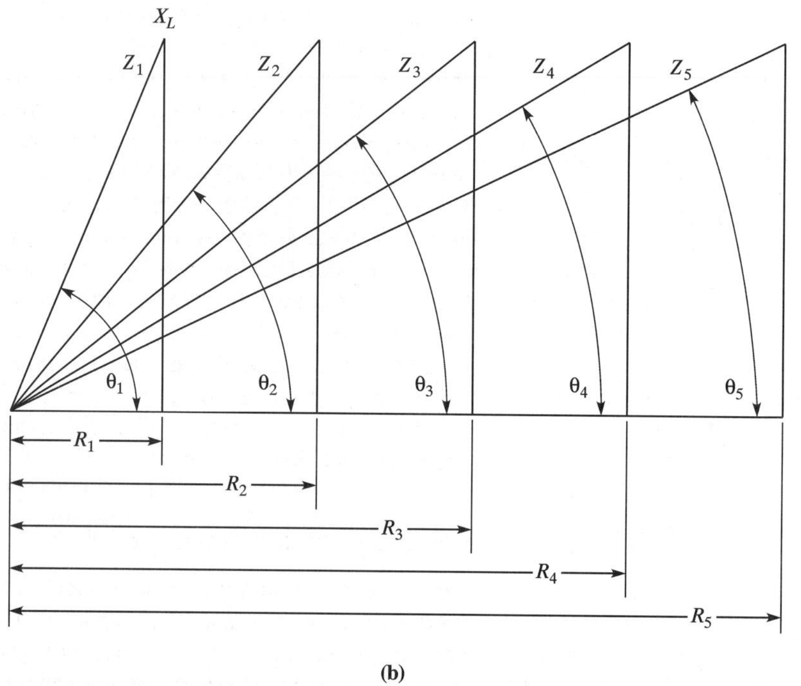

(b)

EXAMPLE 12–12 A 22 Ω resistance and a 39 Ω reactance are in series. Compute the phase angle.

■ Solution:

Use Eq. 12–14:

$$\theta = \text{arc tan } \frac{X_L}{R}$$

$$= \text{arc tan } \frac{39}{22} = \text{arc tan } 1.77 = 60.5°$$

EXAMPLE 12–13 What value resistance must be placed in series with a 400 Ω reactance in order to produce a phase angle of 14.9°?

■ Solution:

Transpose Eq. 12–14 to solve for R:

$$\tan\theta = \frac{X_L}{R}$$

$$R = \frac{X_L}{\tan\theta} = \frac{400\ \Omega}{\tan 14.9} = \frac{400\ \Omega}{0.266} = 1504\ \Omega$$

■ Power in an *RL* Circuit

Electrical energy is dissipated when current flows through a resistance. Recall that the rate at which this energy is dissipated in the form of heat is known as *power.* Thus, in the series *RL* circuit of Figure 12–18(a), the resistor develops power, the value of which is computed using the power equations presented in Chapter 3. This power developed by a resistor in the form of heat is known as **true power,** which is designated P_{true}.

Current also flows through the inductor and a voltage is developed across it. In an inductor, however, a magnetic field is formed in the space around it. The electrical energy is stored in this magnetic field. When the current ceases to flow, the magnetic field collapses, and the energy is returned to the source. When a sine wave of current flows through an inductor, the energy is alternately stored and returned on each half-cycle, as illustrated in Figure 12–18(b). Notice that as the current builds in value, energy is stored in the magnetic field. During the period when the current decreases to zero, the energy is returned to the source. During the period when the current builds to its maximum negative value, energy is stored once again. When the current falls to zero, the energy is once again returned to the source. Notice that during 1 cycle of the current, energy is stored and returned twice.

FIGURE 12–18 Power in an *RL* circuit

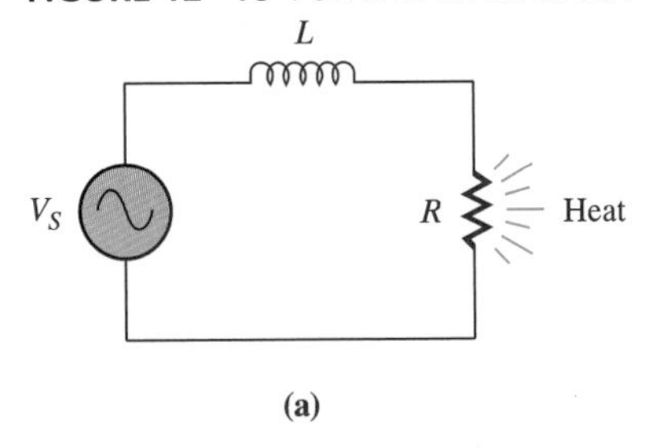

(a)

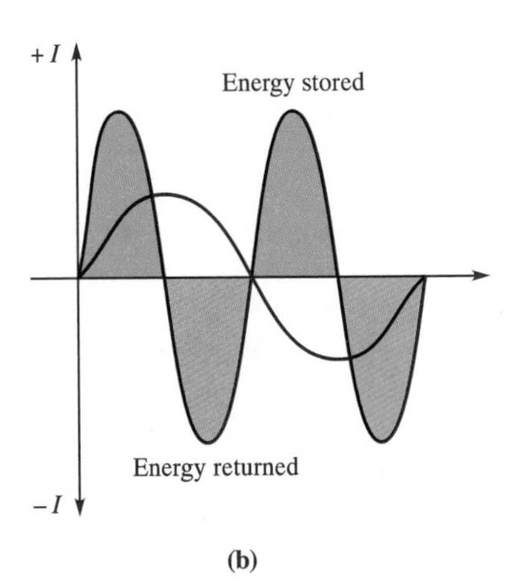

(b)

Question: What is reactive power?

No true power is developed in an ideal inductor. Since current flows and a voltage is developed across the inductor, it appears to develop power. This power that the inductor appears to develop is known as **reactive power,** which is defined as the rate at which electrical energy is alternately stored and returned to the source. Its symbol is P_{react}, and its unit of measure is the VAR (volt-ampere reactive). All the power equations may be used in computing reactive power:

(12–15) $$P_{\text{react}} = V_L I$$

(12–16) $$P_{\text{react}} = I^2 X_L$$

(12–17) $$P_{\text{react}} = \frac{V^2}{X_L}$$

■ Quality Factor of an Inductor

An ideal, or pure inductor, is one in which the only circuit effect present is inductance. As you know, such an inductor is only theoretical due to the resistance of the coil windings R_W. The presence of the winding resistance means that not all the energy delivered to the inductor is returned to the source; some will be dissipated in the winding resistance. In the preceding discussion of *RL* circuit analysis, this resistance was ignored in order to simplify the explanations. In the real world of electronics, however, it may not always be possible to ignore the effects of R_w. Whether or not the winding resistance must be considered depends upon the Q-factor, or quality factor, of the inductor.

Question: What is the quality factor of an inductor and how is it determined?

The **Q-factor** of an inductor is the ratio of its inductive reactance to the winding resistance R_W. In equation form, this relationship is

(12–18)
$$Q = \frac{X_L}{R_W}$$

Thus, the Q-factor of an inductor is a number without units that indicates how closely it approaches the ideal. As this equation demonstrates, an inductor with no resistance would have infinite quality. Higher values of winding resistance lower the value of Q. Notice also that the value of Q is dependent upon the value of X_L. Thus Q is *directly proportional* to both the frequency and the inductance value. Multiplying both the numerator and denominator of Eq. 12–18 by the square of the current yields the ratio of reactive to true power: I^2X_L/I^2R_L. As this substitution illustrates, the quality factor is also equal to the ratio of the reactive power to the true power.

Question: What value of Q is required if the winding resistance is to be ignored?

In a practical circuit, if the Q-factor is 10 or greater, it is usually permissible to ignore the winding resistance in *RL* circuit analysis. Even if the winding resistance is known, however, it is impossible to state that an inductor has a Q-factor of 10 or greater. Q also depends upon the frequency of the applied signal. An inductor with a Q-factor of 10 or greater is often referred to as a choke. This name stems from the fact that the inductor will readily pass dc current, but will effectively block, or **choke,** current at the signal frequency.

Question: What is apparent power?

The product of the source voltage and the circuit current give a value for what is known as the **apparent power,** which is the power that the circuit is "apparently" developing. Its symbol is P_{app}, and its unit of measure is the VA (volt-ampere). It is made up of two components: true power developed by the resistor and the reactive power developed by the inductor. As shown in Figure 12–19, a power triangle can be drawn using the phasors for reactive and true power. Notice that these phasors occupy the same relative position as the resistance and reactance phasors. The equation for the apparent power is

FIGURE 12–19 Power triangle

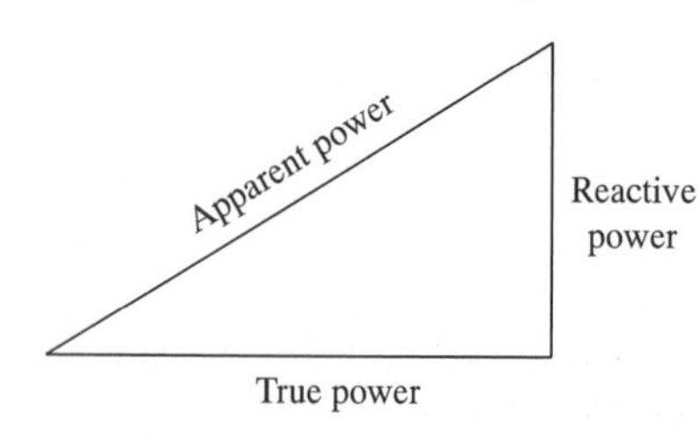

(12–19)
$$P_{app} = \sqrt{(P_{true})^2 + (P_{react})^2}$$

The apparent power may also be computed using the power equations:

(12–20)
$$P_{app} = V_S I$$

(12–21)
$$P_{app} = \frac{V_S^2}{Z}$$

(12–22)
$$P_{app} = I^2 Z$$

Power Factor

Notice in the power triangle of Figure 12–19 that the phasor representing the true power is adjacent to the phase angle. Thus, if the apparent power and phase angle of the circuit are known, an equation for true power can be derived using the trigonometric functions. (Note that if the apparent power and the true power are known the phase angle may be determined.)

The cosine of angle θ is the ratio of the adjacent side (P_{true}) and the hypotenuse (P_{app}). Thus,

(12–23) $$\cos\theta = \frac{P_{true}}{P_{app}} \quad \text{and} \quad P_{true} = P_{app}(\cos\theta)$$

True power is determined by multiplying the apparent power by the cosine of the phase angle, and for this reason, the cosine of the phase angle is known as the circuit **power factor,** which is abbreviated PF. If the circuit is purely resistive, the phase angle is zero. The cosine of zero degrees is 1, and 100% of the apparent power is true power. If the circuit is purely inductive, the phase angle is 90°. The cosine of 90° is zero, and none of the apparent power is true power.

EXAMPLE 12–14 For the circuit of Figure 12–20, compute the Q-factor.

FIGURE 12–20 Circuit for Examples 12–14 and 12–15

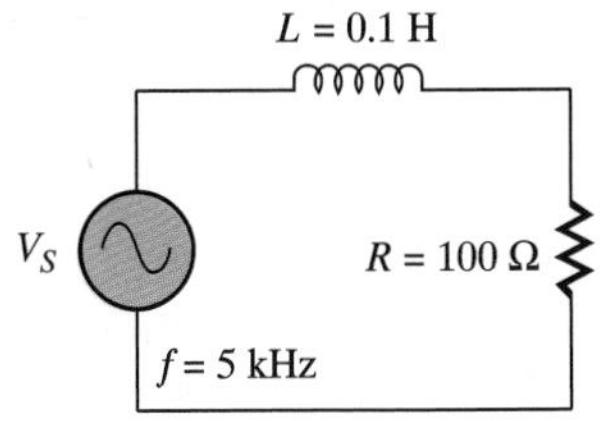

■ Solution:

$$X_L = 2\pi fL$$
$$= 2(3.14)(5000\text{ Hz})(0.1\text{ H}) = 3.140\text{ k}\Omega$$
$$Q = \frac{X_L}{R}$$
$$= \frac{3140\ \Omega}{100\ \Omega} = 31.4$$

EXAMPLE 12–15 Compute the Q-factor for the circuit of Example 12–14 if the frequency is reduced to 1 kHz.

$$X_L = 2\pi fL$$
$$= 2(3.14)(1000\text{ Hz})(0.1\text{ H}) = 628\ \Omega$$
$$Q = \frac{X_L}{R}$$
$$= \frac{628\ \Omega}{100\ \Omega} = 6.28$$

Notice that the lower frequency results in a greatly reduced value of quality factor.

EXAMPLE 12–16 For the circuit of Figure 12–21 compute: (a) the reactive power, (b) the true power, (c) the apparent power, (d) the phase angle, and (e) the power factor.

■ Solution:

First, compute the value of Z:

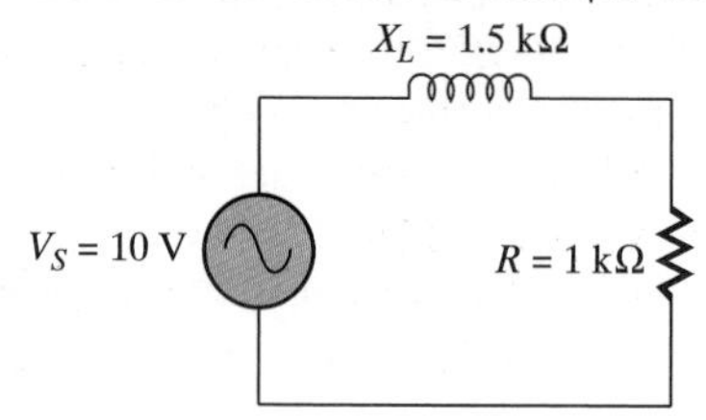

FIGURE 12–21 Circuit for Example 12–16

$$Z = \sqrt{R^2 + X_L^2}$$
$$= \sqrt{(1000\ \Omega)^2 + (1500\ \Omega)^2} = 1.803\ \text{k}\Omega$$

Next, compute the value of I:

$$I = \frac{V_S}{Z}$$
$$= \frac{10\ \text{V}}{1803\ \Omega} = 5.55\ \text{mA}$$

a. $P_{\text{react}} = I^2X_L$

$$= (5.55\ \text{mA})^2(1500\ \Omega) = 46.2\ \text{mVAR}$$

b. $P_{\text{true}} = I^2R$

$$= (5.55\ \text{mA})^2(1000\ \Omega) = 30.8\ \text{mW}$$

(Remember the note when Eq. 12–23 was derived?)

c. $P_{\text{app}} = \sqrt{(P_{\text{react}})^2 + (R_{\text{true}})^2}$

$$= \sqrt{(46.2 \times 10^{-3}\ \text{VAR})^2 + (30.8 \times 10^{-3}\ \text{W})^2} = 55.5\ \text{mVA}$$

d. $\theta = \text{arc tan}\ \dfrac{\text{P}_{\text{react}}}{\text{P}_{\text{true}}}$

$$= \text{arc tan}\ \frac{46.2 \times 10^{-3}\ \text{VAR}}{30.8 \times 10^{-3}\ \text{W}} = 56.3°$$

e. $\text{PF} = \cos\theta$

$$= \cos 56.3 = 0.555$$

■ SECTION 12–3 REVIEW

1. Which component in an *RL* circuit develops true power?
2. Which component in an *RL* circuit develops reactive power?
3. What are the components of apparent power?
4. An *RL* circuit develops 100 VA of apparent power. If the true power is 65W, what is the power factor?
5. An *RL* circuit with a phase angle of 65° develops 500 VA of apparent power. Compute the value of the true power (P_{true}) developed by this circuit.

12–4 ANALYSIS OF PARALLEL *RL* CIRCUITS

Parallel *RL* circuit analysis is similar in approach to that of any parallel circuit. As shown in Figure 12–22, the same voltage is across each branch and the current divides in inverse proportion to the resistance and reactance values. The value of the branch and total currents can be computed using Ohm's law:

$$I_R = \frac{V_S}{R} \qquad I_{\text{L}} = \frac{V^S}{X_L} \qquad I_{\text{T}} = \frac{V_S}{Z}$$

FIGURE 12–22 Currents in an *RL* circuit

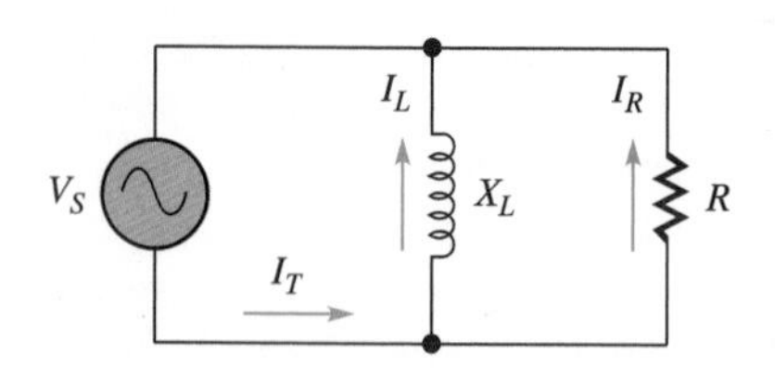

Question: Does Kirchhoff's current law apply in the analysis of parallel RL *circuits?*

Recall that Kirchhoff's current law is the basic law of all parallel circuits. As you know, it states that the sum of the branch currents in a parallel circuit is equal to the total current.

Kirchhoff's current law applies to parallel *RL* circuits, but phasor addition must be used because of the 90° phase difference between the branch currents. The current through the resistive branch is in phase with the source voltage, while the current through the inductive branch lags the source voltage by 90°. Figure 12–23 is a current triangle drawn to show these phase relationships. Notice that the source voltage, which is common in all parallel circuits, is chosen as the reference phasor. The current triangle is the starting point in the analysis of parallel *RL* circuits. The total current is the phasor sum of I_R and I_L. In equation form, this relationship is written

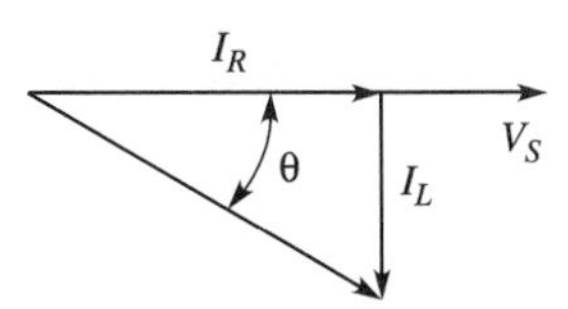

FIGURE 12–23 Current triangle for a parallel *RL* circuit

(12–24) $$I_T^2 = I_R^2 + I_L^2 \quad \text{and} \quad I_T = \sqrt{I_R^2 + I_L^2}$$

EXAMPLE 12–17 For the parallel RL circuit of Figure 12–24, compute the total current.

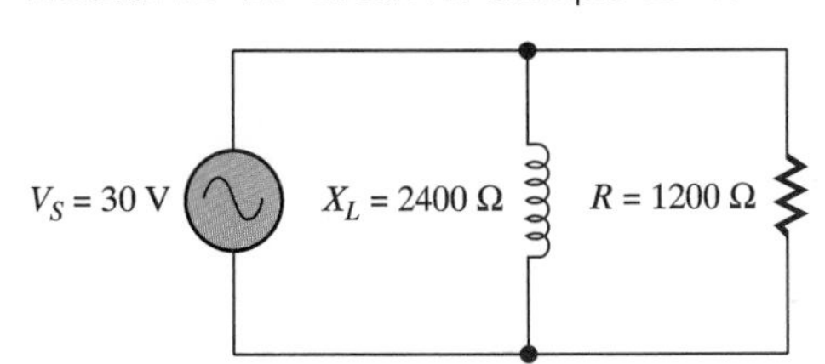

FIGURE 12–24 Circuit for Example 12–17

■ Solution:

1. Compute the branch currents:

$$I_R = \frac{V_S}{R} = \frac{30\text{ V}}{1200\ \Omega} = 25\text{ mA}$$

$$I_L = \frac{V_S}{X_L} = \frac{30\text{ V}}{2400\ \Omega} = 12.5\text{ mA}$$

2. Use Eq. 12–24 to find the total current:

$$I_T = \sqrt{I_R^2 + I_L^2} = \sqrt{(25 \times 10^{-3}\text{ A})^2 + (12.5 \times 10^{-3}\text{ A})^2} = 28\text{ mA}$$

■ Impedance in a Parallel *RL* Circuit

As was the case with the series *RL* circuit, the impedance of a parallel *RL* circuit is made up of two components: one resistive and one reactive. These components are first converted to conductance *G* and inductive susceptance B_L. The phasor sum of conductance and inductive susceptance is known as **admittance** and given the symbol *Y*. In equation form, admittance is

(12–25) $$Y^2 = G^2 + B_L^2 \quad \text{and} \quad Y = \sqrt{G^2 + B_L^2}$$

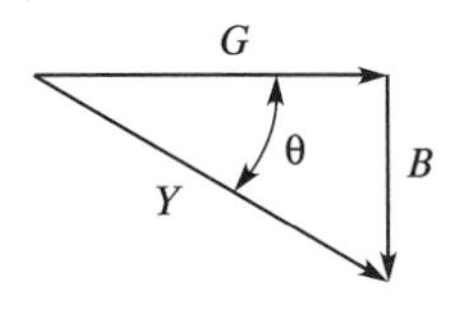

FIGURE 12–25 Admittance triangle

An admittance triangle illustrating these phase relationships is shown in Figure 12–25. Admittance is similar to total conductance in a parallel resistive circuit. Thus, the impedance of a parallel *RL* circuit is the *reciprocal* of the admittance. In equation form, this relationship is

(12–26) $$Z = \frac{1}{Y} \quad \text{and} \quad Y = \frac{1}{Z}$$

EXAMPLE 12–18 Compute the impedance for the parallel RL circuit of Figure 12–26.

FIGURE 12–26 Circuit for Example 12–18

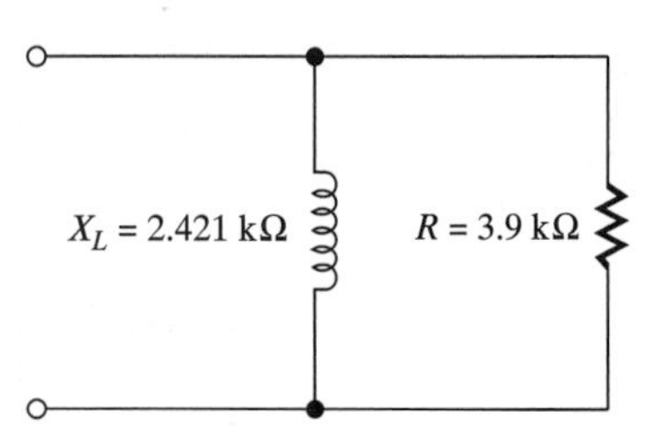

■ **Solution:**

1. Use Eq. 12–25 to compute the admittance:

$$Y = \sqrt{G^2 + B_L^2}$$
$$= \sqrt{(1/3.9\ \text{k}\Omega)^2 + (1/2.421\ \text{k}\Omega)^2} = 486\ \mu\text{S}$$

2. Use Eq. 12–26 to compute the impedance:

$$Z = \frac{1}{Y}$$
$$= \frac{1}{486 \times 10^{-6}\ \text{S}} = 2057\ \Omega$$

EXAMPLE 12–19 The impedance of a parallel RL circuit is 35.3 kΩ. If the value of the resistor is 47 kΩ, what is the reactance of the inductor?

■ **Solution:**

Simplify and transpose Eq. 12–25:

$$Y^2 = G^2 + B_L^2$$
$$B_L^2 = Y^2 - G^2$$

and

$$B = \sqrt{Y^2 - G^2}$$
$$= \sqrt{[1/(35.3 \times 10^3\ \text{S})]^2 - [1/(47 \times 10^3\ \text{S})]^2} = \frac{1}{18.7 \times 10^{-6}\ \text{S}} = 53.476\ \text{k}\Omega$$

■ Circuit Phase Angle in a Parallel *RL* circuit

The phase angle of a parallel RL circuit may be computed using the current triangle, as illustrated in Figure 12–27. Notice that the inductive current phasor is the side opposite, while the resistive current phasor is side adjacent to the phase angle. Thus,

$$\tan\theta = \frac{I_L}{I_R}$$

and

(12–27)
$$\theta = \arctan\frac{I_L}{I_R}$$

FIGURE 12–27 Current triangle for computing the phase angle (θ = arc tan I_L/I_R)

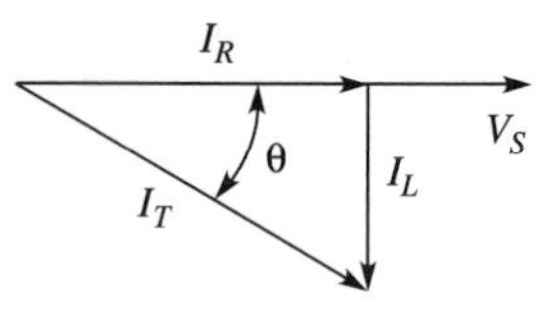

The phase angle may also be determined using the admittance triangle:

(12–28)
$$\tan\theta = \frac{B_L}{G}$$
$$\theta = \arctan\frac{B_L}{G}$$

FIGURE 12–28 Relationship of frequency and phase angle in an *RL* circuit.

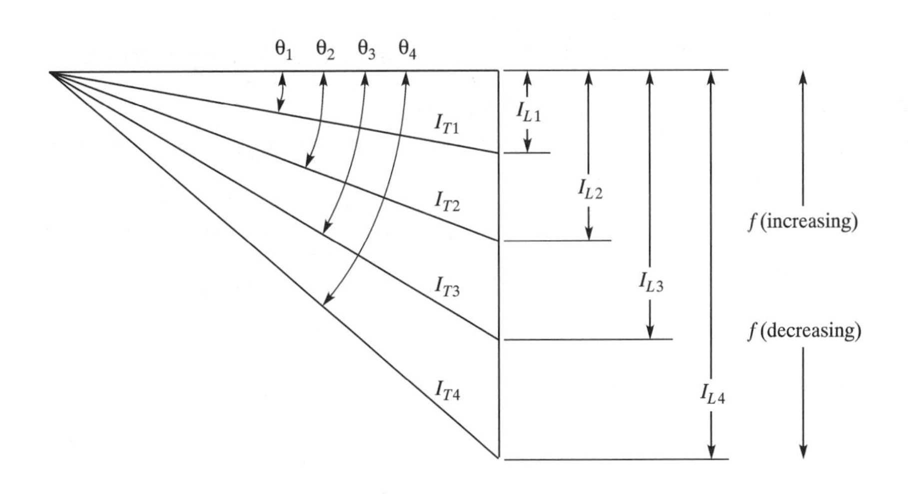

Question: How does frequency affect the phase angle of a parallel RL *circuit?*

As was the case with the series *RL* circuit, the phase angle of a parallel *RL* circuit changes as frequency changes. The reactance of the inductor is directly proportional to the frequency of the applied voltage. Thus, the current through the inductor is *inversely proportional* to frequency. As illustrated in Figure 12–28, as frequency increases, the length of phasor I_L decreases. Since the resistance is a fixed value, the phasor I_T moves toward phasor V_S and the phase angle decreases. The effect is opposite as the frequency decreases.

EXAMPLE 12–20 For the parallel *RL* circuit of Figure 12–29, calculate I_R, I_L, I_T, Z, and θ.

■ Solution:

1. Compute the value of X_L:

$$X_L = 2\pi fL$$
$$= 2(3.14)(30 \times 10^3 \text{ Hz})(30 \times 10^{-3} \text{ H}) = 5.652 \text{ k}\Omega$$

2. Compute I_R and I_L:

FIGURE 12–29 Circuit for Example 12–20

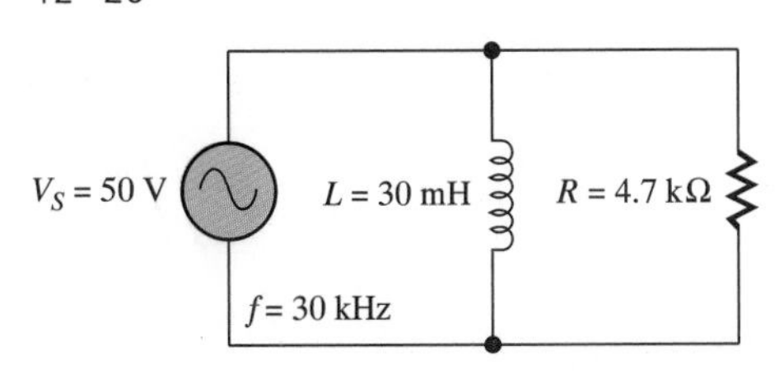

$$I_R = \frac{V_S}{R}$$
$$= \frac{50\text{V}}{4700\ \Omega} = 10.6 \text{ mA}$$
$$I_L = \frac{V_S}{X_L}$$
$$= \frac{50 \text{ V}}{5652\ \Omega} = 8.85 \text{ mA}$$

3. Compute I_T:

$$I_T = \sqrt{I_R^2 + I_L^2}$$
$$= \sqrt{(10.6 \times 10^{-3} \text{ A})^2 + (8.85 \times 10^{-3} \text{ A})^2} = 13.8 \text{ mA}$$

4. Compute Z:

$$Z = \frac{V_S}{I_T}$$

$$= \frac{50\ \text{V}}{(13.8 \times 10^{-3}\ \text{A})} = 3.623\ \text{k}\ \Omega$$

5. Compute the phase angle θ:

$$\theta = \arctan \frac{I_L}{I_R}$$

$$= \arctan \frac{8.85 \times 10^{-3}\ \text{A}}{10.6 \times 10^{-3}\ \text{A}} = 39.8°$$

■ SECTION 12–4 REVIEW

1. Which phasor is used as the reference phasor in a current triangle?
2. What is the reciprocal of impedance known as?
3. As the frequency of the applied voltage increases, does the phase angle of a parallel *RL* circuit increase or decrease?
4. Does an increase in the phase angle of a parallel *RL* circuit indicate that the frequency has increased or decreased?

12–5 EXAMPLES OF *RL* CIRCUIT ANALYSIS

In the previous Section, the procedures involved in *RL* circuit analysis were presented. In the examples that follow, these procedures will be put to work. As you have probably noticed, ac circuit analysis is a bit more complex than dc circuit analysis. One suggestion that will make your task easier is to *always* draw the voltage, current, impedance, or admittance triangles for the circuit to be analyzed. It is not necessary that they be drawn to an exact scale, but merely to show the position of the phasors making up the sides of these triangles in relation to the phase angle. The drawings will make it easier to remember the trigonometric functions and other equations used in the analysis.

EXAMPLE 12–21

For the series *RL* circuit of Figure 12–30(a), compute the following: (a) X_L, (b) Z, (c) I_T, (d) V_R, (e) V_L, (f) θ, (g) P_{true}, (h) P_{react}, and (i) P_{app}.

■ Solution:

The voltage and impedance triangles for the circuit are shown in Figure 12–30(b) and (c). They are not drawn to scale. It is their position relative to one another and the phase angle that is important.

a. Compute X_L:

$$X_L = 2\pi f L$$

$$= 2(3.14)(1 \times 10^6\ \text{Hz})(109 \times 10^{-6}\ \text{H}) = 685\ \Omega$$

b. Compute Z:

$$Z = \sqrt{R^2 + X_L^2}$$

$$= \sqrt{(810\ \Omega)^2 + (685\ \Omega)^2} = 1.061\ \text{k}\ \Omega$$

FIGURE 12–30 Circuit and phasor diagrams for Example 12–21

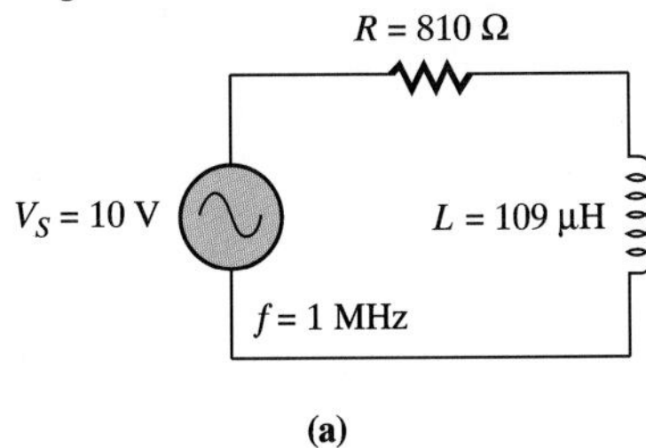

(a)

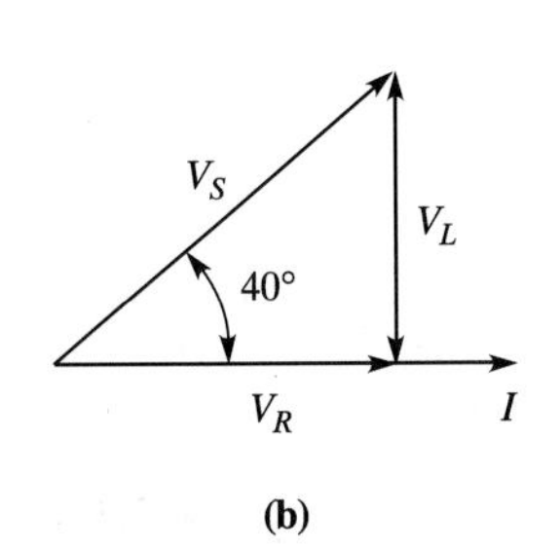

(b)

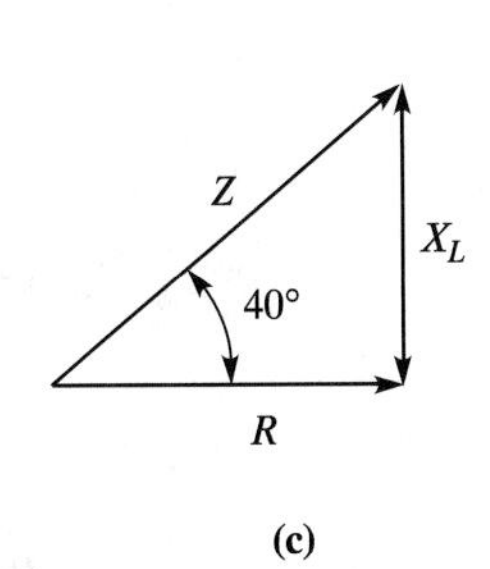

(c)

c. Compute I_T:

$$I_T = \frac{V_S}{Z} = \frac{10 \text{ V}}{1061 \ \Omega} = 9.43 \text{ mA}$$

d. Compute V_R:

$$V_R = IR$$
$$= (9.43 \times 10^{-3} \text{ A})(810 \ \Omega) = 7.64 \text{ V}$$

e. Compute V_L:

$$V_L = IX_L$$
$$= (9.43 \times 10^{-3} \text{ A})(685 \ \Omega) = 6.46 \text{ V}$$

f. Compute θ:

$$\theta = \text{arc tan } \frac{V_L}{V_R}$$
$$= \text{arc tan } \frac{6.46 \text{ V}}{7.64 \text{ V}} = \text{arc tan } 0.846 = 40.2°$$

g. Compute P_{true}:

$$P_{\text{true}} = I_R{}^2R$$
$$= (9.43 \times 10^{-3} \text{ A})^2(810 \ \Omega) = 72 \text{ mW}$$

h. Compute P_{react}:

$$P_{\text{react}} = I_L{}^2X_L$$
$$= (9.43 \times 10^{-3} \text{ A})^2(685 \ \Omega) = 60.9 \text{ mVAR}$$

i. Compute P_{app}:

$$P_{\text{app}} = \sqrt{(P_{\text{true}})^2 + (P_{\text{react}})^2}$$
$$= \sqrt{(72 \times 10^{-3} \text{ W})^2 + (60.9 \times 10^{-3} \text{ VAR})^2} = 94.3 \text{ mVA}$$

NOTE

Needless to say, this problem has involved many calculations. Each is a possible error point. There are places within the problem, however, at which you can check your work. After the values of V_R and V_L have been determined, you can use them to compute V_S.

$$V_S = \sqrt{V_R{}^2 + V_L{}^2}$$
$$= \sqrt{(7.64 \text{ V})^2 + (6.46 \text{ V})^2} = 10 \text{ V}$$

This value matches the value given for V_S. Thus, you may assume that the calculations are correct. The same may be done with the power calculations.

EXAMPLE 12–22

For the parallel *RL* circuit of Figure 12–31(a), compute the following: (a) X_L, (b) I_R, (c) I_L, (d) I_T, (e) θ, (f) PF, (g) P_{app}, and (h) P_{true}.

Solution:

The current triangle for this circuit is shown in Figure 12–31(b).

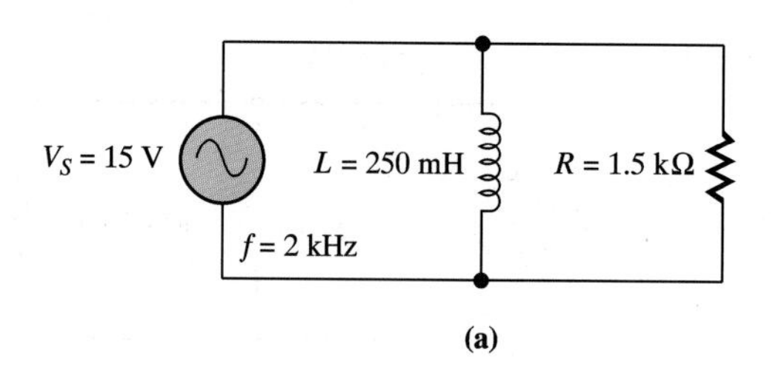

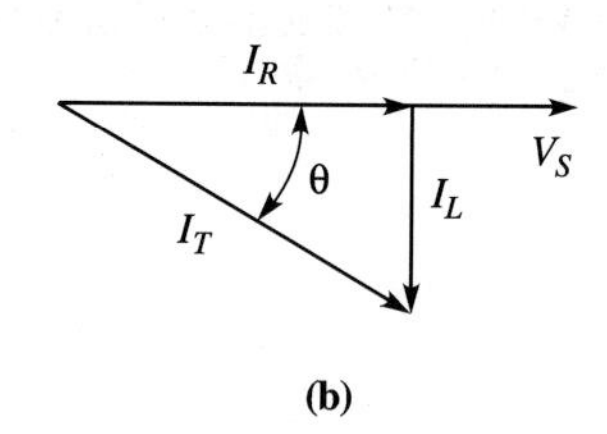

FIGURE 12–31 Circuit and phasor diagram for Example 12–22

a. Compute X_L:

$$X_L = 2\pi fL$$
$$= 2(3.14)(2 \times 10^3 \text{ Hz})(250 \times 10^{-3} \text{ H}) = 3141\ \Omega$$

b. Compute I_R:

$$I_R = \frac{V_S}{R}$$
$$= \frac{15 \text{ V}}{1500\ \Omega} = 10 \text{ mA}$$

c. Compute I_L:

$$I_L = \frac{V_S}{X_L}$$
$$= \frac{15 \text{ V}}{3140\ \Omega} = 4.78 \text{ mA}$$

d. Compute I_T:

$$I_T = \sqrt{I_R^2 + I_L^2}$$
$$= \sqrt{(10 \times 10^{-3} \text{ A})^2 + (4.78 \times 10^{-3} \text{ A})^2} = 11.1 \text{ mA}$$

e. Compute θ:

$$\theta = \arctan \frac{I_L}{I_R}$$
$$= \arctan \frac{4.78 \times 10^{-3} \text{ A}}{10 \times 10^{-3} \text{ A}} = \arctan 0.478 = 25.547°$$

f. Compute PF:

$$\text{PF} = \cos \theta$$
$$= \cos 25.5° = 0.903$$

g. Compute P_{app}:

$$P_{\text{app}} = V_S I_T$$
$$= (15 \text{ V})(11.1 \times 10^{-3} \text{ A}) = 167 \text{ mVA}$$

h. Compute P_{true}:

$$P_{\text{true}} = P_{\text{app}}\text{PF}$$
$$= (167 \times 10^{-3} \text{ VA})(0.903) = 151 \text{ mW}$$

FIGURE 12–32 Circuit and phasor diagram for Example 12–23

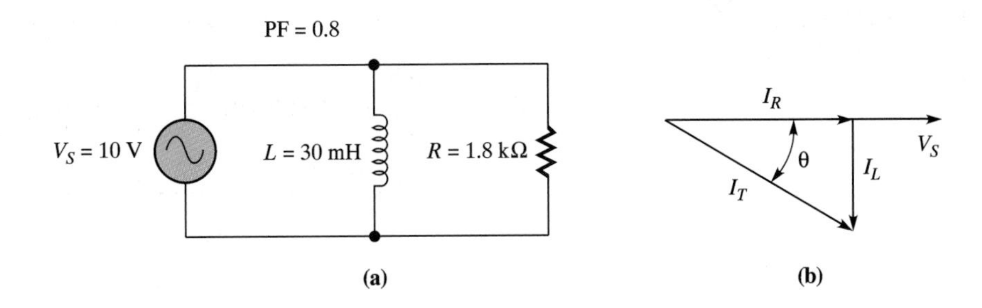

EXAMPLE 12–23 Compute the following for the circuit of Figure 12–32(a): (a) I_R, (b) I_L, (c) Z, and (d) f.

■ Solution:

The current triangle for this circuit is shown in Figure 12–32(b). First, compute the phase angle:

$$\text{PF} = \cos\theta$$

$$\theta = \text{arc}\cos \text{PF} = \text{arc}\cos 0.8 = 36.9°$$

a. Compute I_R:

$$I_R = \frac{V_S}{R} = \frac{10 \text{ V}}{1800\ \Omega} = 5.56 \text{ mA}$$

b. Compute I_L:

$$\tan 36.9° = 0.751$$

$$\tan\theta = \frac{I_L}{I_R}$$

$$I_L = I_R(\tan\theta) = (5.56 \times 10^{-3} \text{ A})(0.751) = 4.18 \text{ mA}$$

c. First compute I_T:

$$I_T = \sqrt{I_R^2 + I_L^2}$$

$$= \sqrt{(5.56 \times 10^{-3} \text{ A})^2 + (4.18 \times 10^{-3} \text{ A})^2} = 6.96 \text{ mA}$$

Then compute Z:

$$Z = \frac{V_S}{I_T}$$

$$= \frac{10 \text{ V}}{6.96 \times 10^{-3} \text{ A}} = 1.437 \text{ k}\Omega$$

d. Frequency is contained in the equation for inductive reactance ($X_L = 2\pi fL$). Thus, if the inductance and reactance are known, the equation can be transposed to compute the frequency. the inductance is given, but it is necessary to compute the value of reactance:

$$X_L = \frac{V_S}{I_L}$$

$$= \frac{10 \text{ V}}{4.18 \times 10^{-3} \text{ A}} = 2.392 \text{ k}\Omega$$

and

$$X_L = 2\pi fL$$

$$f = \frac{X_L}{2\pi L}$$

$$= \frac{2392\ \Omega}{2(3.14)(30 \times 10^{-3}\ \text{H})} = 12.689\ \text{kHz}$$

EXAMPLE 12–24 For the circuit of Figure 12–33(a), compute the following: (a) V_S, (b) V_L, (c) V_R, and (d) P_{true}.

FIGURE 12–33 Circuit and phasor diagram for Example 12–24

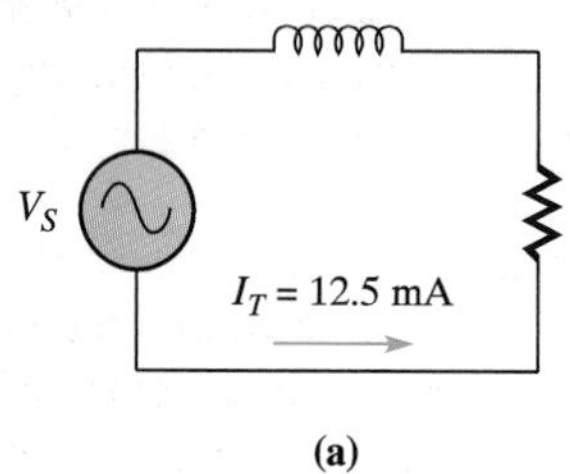

(a)

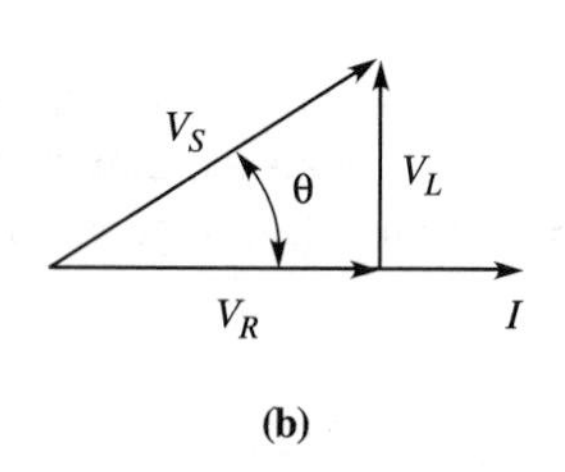

(b)

■ Solution:

The voltage triangle for this circuit is shown in Figure 12–33(b).

a. Compute V_S:

$$V_S = I_T Z$$

$$= (12.5 \times 10^{-3}\ \text{A})(2 \times 10^{3}\ \Omega) = 25\ \text{V}$$

b. With only the phase angle and source voltage known, it is necessary to use the trigonometric functions in computing V_L and V_R. Thus, compute V_L:

$$\sin\theta = \frac{V_L}{V_S}$$

$$V_L = V_S(\sin\theta)$$

$$= (25\ \text{V})(\sin 58°) = (25\ \text{V})(0.848) = 21.2\ \text{V}$$

c. Compute V_R:

$$\cos\theta = \frac{V_R}{V_S}$$

$$V_R = V_S(\cos\theta)$$

$$= (25\ \text{V})(\cos 58°) = (25\ \text{V})(0.530) = 13.25\ \text{V}$$

Note that this is a good point to check your accuracy. The phasor sum of V_L and V_R should equal 25 V:

$$V_S = \sqrt{V_R^2 + V_L^2}$$

$$= \sqrt{(13.25\ \text{V})^2 + (21.2\ \text{V})^2} = 25\ \text{V}$$

This voltage matches that computed for V_S so the calculations are correct.

d. The computation of P_{true} can proceed in several different ways. Since the current through and voltage across the resistor are known, the simplest method is to use the power equation:

$$P_{\text{true}} = V_R I_T$$

$$= (13.3\ \text{V})(12.5 \times 10^{-3}\ \text{A}) = 166\ \text{mW}$$

There is another way of computing P_{true}. See if you can find it!

12–6 APPLICATIONS OF *RL* CIRCUITS

There are two basic applications of *RL* circuits: as electronic filters or as phase-shifting devices. Both applications take advantage of the frequency-sensitive characteristic of the inductor. The circuits for accomplishing these purposes are only briefly discussed at this point; they will be covered in great detail in Chapter 18.

RL Filters

An *RL* circuit can be made to pass a certain range of frequencies and block all others. There are basically two types of *RL* filters: high-pass and low-pass filters. A **low-pass filter** is one in which all frequencies below a cutoff frequency are passed, and all those *above* blocked. A **high-pass filter** is one in which all frequencies above a cutoff frequency are passed, and all those *below* blocked.

Question: How is a series RL *circuit used as a low-pass filter?*

FIGURE 12–34 *RL* circuit connected as a low-pass filter

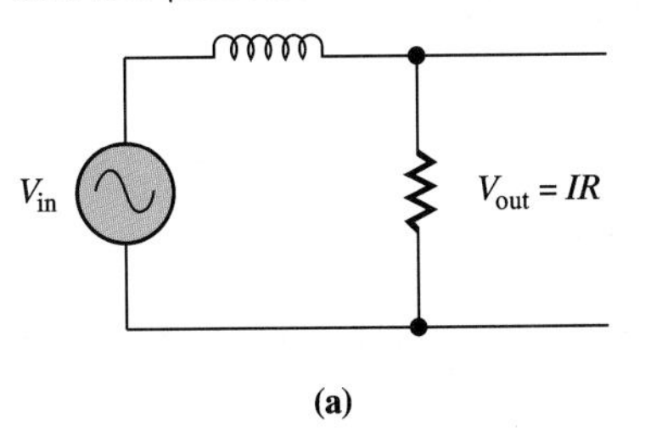

(a)

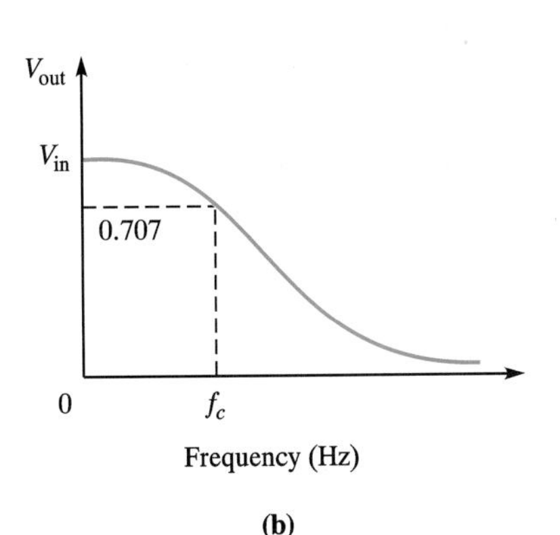

(b)

Figure 12–34(a) shows a series *RL* circuit connected as a low-pass filter. Notice that the input is across the entire circuit, while the output is taken across the resistor. Thus, the output voltage, by Ohm's law, is the resistance times the current flowing in the circuit. The resistance is a fixed value, so the output voltage is dependent on the amount of current flow. This current is controlled by the impedance, which is the phasor sum of the resistance and reactance. The reactance is directly proportional to the frequency of the applied voltage. Thus, as frequency increases, reactance and impedance both increase, thereby decreasing the current and the output voltage.

Figure 12–34(b) is a graph of the output voltage versus the frequency of the input voltage. Notice that at zero hertz, the reactance of the inductor is zero, and the entire source voltage appears across the resistor. As frequency increases, the reactance of the inductor increases causing an increase in impedance. The current decreases as does the output voltage. The voltage that formerly appeared across the resistor now appears across the inductor.

The cutoff frequency f_c of the filter is the frequency at which the voltages across the two circuit elements are equal. This point is the frequency at which the output voltage drops to 70.7% of its maximum value. Thus, all frequencies below this frequency are said to be passed while all those above it are blocked.

Question: How is a series RL *circuit used as a high-pass filter?*

Figure 12–35(a) shows a series *RL* circuit connected as a high-pass filter. Notice now that the output is taken across the inductor. Figure 12–35(b) is a graph of the output voltage verses the frequency of the input voltage. The output voltage is now the reactance of the inductor times the current flowing in the circuit (IX_L). At zero hertz the inductive reactance is zero, so the output voltage is also zero. As frequency increases, the reactance of the inductor also increases, thereby increasing the output voltage. As the graph shows, this trend continues until practically all the input voltage appears across the output.

The cutoff frequency is, once again, the frequency at which the voltages across the two circuit elements are equal. At this frequency, the output voltage has dropped to 70.7% of its maximum value. Now all frequencies above the cutoff frequency are passed, and all those below it are blocked.

RL Lead-Lag Networks

An *RL* lead-lag network is a circuit producing a phase shift between the output and input voltages. If the output voltage leads the input, the circuit is known as a **lead network.** If

FIGURE 12–35 *RL* circuit connected as a high-pass filter

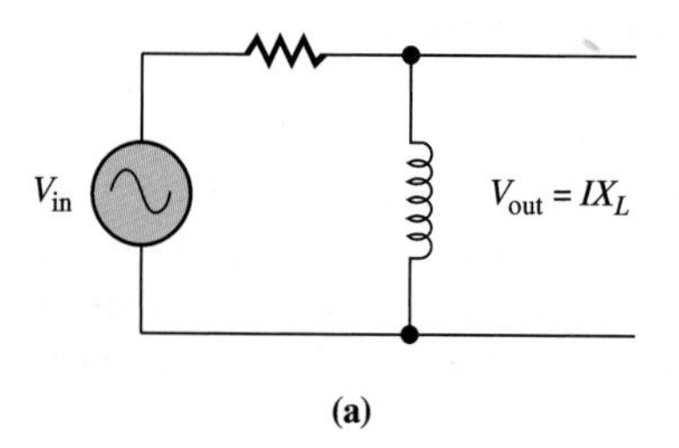

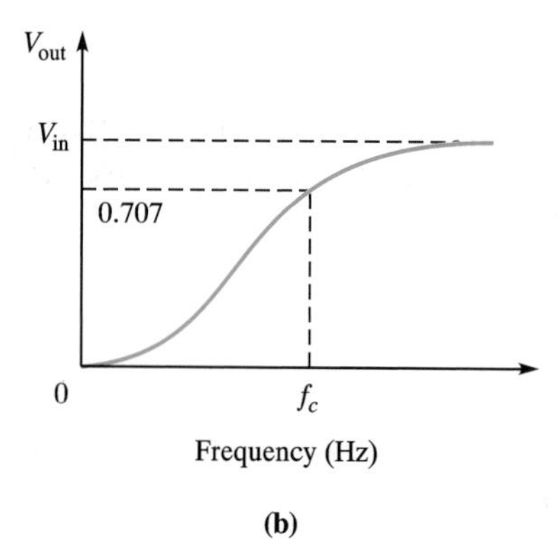

the output voltage lags the input, the circuit is known as a **lag network.** In a series *RL* lead network, the output is taken across the inductor. In a series *RL* lag network, the output is taken across the resistor.

Question: How does a series RL *lead network operate?*

The schematic for a lead network is shown in Figure 12–36(a). Notice that the output voltage is the voltage developed across the inductor. As shown in the phasor diagram of Figure 12–36(b), the output voltage V_{out} leads the input voltage V by the angle Φ. Notice that the angles θ and Φ are complementary—that is, their sum is always 90°. Thus, in equation form, the angle of lead Φ is

(12–29)
$$\Phi = 90° - \theta$$

Figure 12–36(c) is the graph illustrating the phase difference between the input and output voltages.

Question: How does a series RL *lag network operate?*

The schematic for a lag network is shown in Figure 12–37(a). Notice now that the output voltage is the voltage across the resistor. As shown in the phasor diagram of Figure

FIGURE 12–36 *RL* lead network

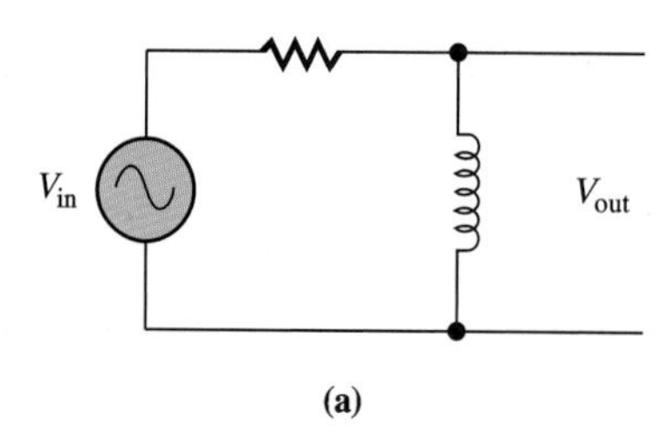

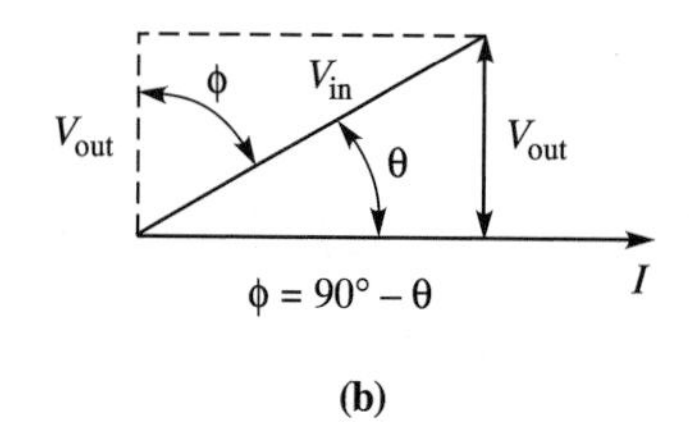

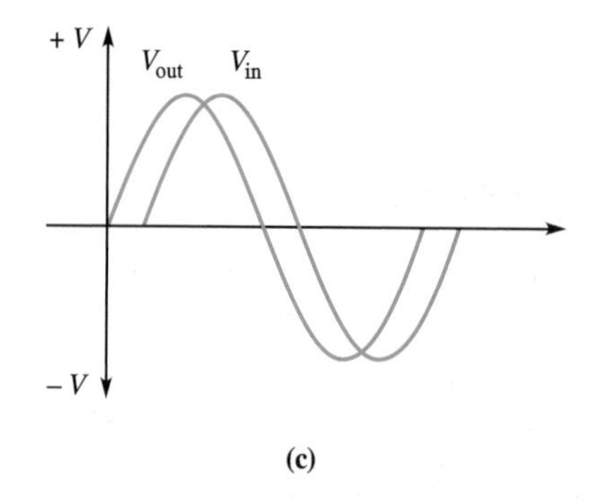

FIGURE 12–37 *RL* lag network

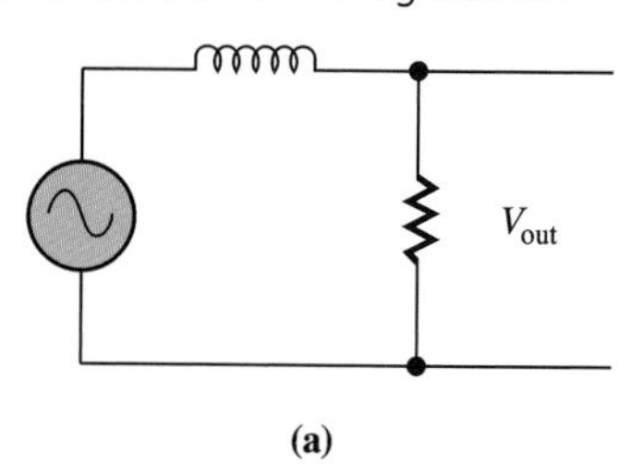

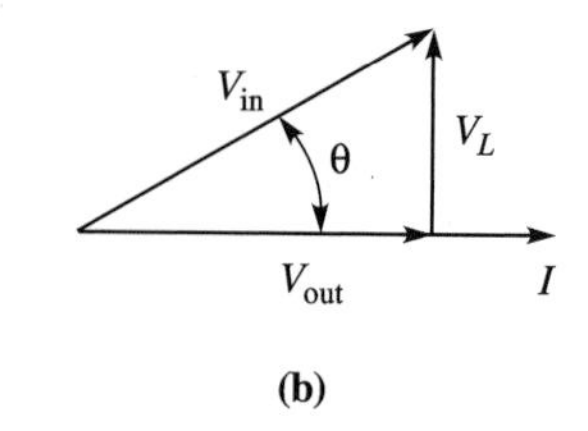

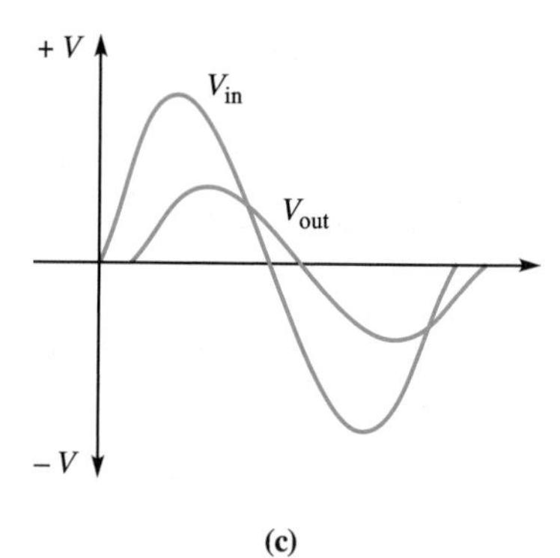

12–37(b), the output voltage V_{out} lags the input voltage V_{in} by the angle θ. Thus, the angle of lag can be found by computing the circuit phase angle. Figure 12–37(c) is a graph illustrating the phase difference between the input and output voltages.

■ Computation of the Output Voltages

The output voltages of the lead or lag network can be computed using procedures presented in the section dealing with series *RL* circuit analysis. Recall that computation of the circuit impedance, current, and the individual voltages are involved. A simpler method involves the voltage divider equation introduced in Chapter 4. For two resistors in series, this equation is

$$V_x = \frac{R_x}{R_1 + R_2} V_S$$

This equation, with one modification, applies equally well to series *RL* circuits. When applied to *RL* circuits, the phasor sum of the reactance and resistance must be used. In equation form, this relationship is

$$V_R = \frac{R}{\sqrt{R^2 + X_L^2}} V_S \tag{12–30}$$

and

$$V_L = \frac{X_L}{\sqrt{R^2 + X_L^2}} V_S \tag{12–31}$$

The use of these equations reduces the amount of work necessary in the computation of the output voltages.

EXAMPLE 12–25 For the lead network of Figure 12–38(a), compute the angle of lead and the output voltage.

FIGURE 12–38 Circuits for Examples 2–25 and 2–26

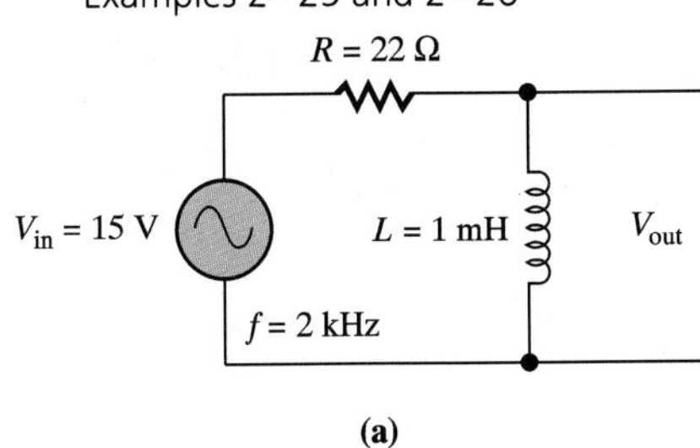

(a)

L = 1 mH

V_in = 15 V

R = 22 Ω

V_out

f = 2 kHz

(b)

■ Solution:

1. Compute X_L:

$$X_L = 2\pi f L$$
$$= 2(3.14)(2000 \text{ Hz})(1 \times 10^{-3} \text{ H}) = 12.6 \ \Omega$$

2. Compute the phase angle θ:

$$\theta = \arctan \frac{X_L}{R}$$
$$= \arctan \frac{12.6}{22} = 29.8°$$

3. Compute the angle of lead Φ:

$$\Phi = 90° - \theta$$
$$= 90° - 29.8° = 60.2°$$

4. Compute the output voltage:

$$V_L = \frac{X_L}{\sqrt{R^2 + X_L^2}} V_S$$

$$= \frac{12.6\ \Omega}{\sqrt{(22\ \Omega)^2 + (12.6\ \Omega)^2}} 15\ \text{V} = (0.497)(15\ \text{V}) = 7.44\ \text{V}$$

EXAMPLE 12–26 What change must be made to the circuit of Figure 12–38(a) in order to produce a lag network? Determine the angle of lag and the output voltage.

Solution:

In order to make the circuit of Figure 12–38(a) a lag network, the output is taken across the resistor rather than the inductor as shown in Figure 12–38(b). The lag angle is the angle between the source voltage and resistance and is the circuit phase angle θ. In Example 12–25, θ was computed to be 29.8°. Thus, computation of the output voltage is

$$V_R = \frac{R}{\sqrt{R^2 + X_L^2}}$$

$$= \frac{22\ \Omega}{\sqrt{(22\ \Omega)^2 + (12.6\ \Omega)^2}} 15\ \text{V} = (0.866)(15\ \text{V}) = 13\ \text{V}$$

SECTION 12–6 REVIEW

1. What characteristic of an inductor is taken advantage of in ac circuit applications?
2. What name is given a filter that passes all frequencies above a given cutoff frequency?
3. If the output is taken across the resistor of an *RL* circuit, will the output voltage lead or lag the input?

12–7 TROUBLESHOOTING *RL CIRCUITS*

Question: What type of faults occur in an inductor and how are they detected?

Faults associated with inductors were introduced in Chapter 9. By way of review, shorts or opens are the faults that normally occur in inductors. An open causes the inductor to be an infinite impedance and no current flows, A short may be either partial or complete. A complete short eliminates practically all inductance and resistance from the inductor. A partial short, meaning only a portion of the windings have shorted, causes the inductance and the winding resistance to be lower than normal. An open or a complete short can be detected using an ohmmeter. Any change in inductance can be detected using an impedance bridge. In the following paragraphs, troubleshooting of series and parallel *RL* circuits is considered. Keep in mind that a problem in the circuit causes one of three current conditions: (1) the current is too low, (2) the current is too high, or (3) the current is zero.

An incorrect or zero source voltage can cause any of the three fault conditions to exist. If the source voltage is low, the current is also low, and if the source voltage is high, the current is high. If the source voltage is zero, the current is zero. Since the source voltage can cause any of the current faults, it should be the *first check* made in troubleshooting. Problems caused by faulty resistors in a circuit have already been covered numerous times. Thus, only troubles introduced by the inductor are considered here.

If the current is low in an *RL* circuit, it is an indication that the impedance is high. Since impedance is directly proportional to both resistance and reactance, one must have

FIGURE 12–39 *RL* circuit for troubleshooting current that is too low

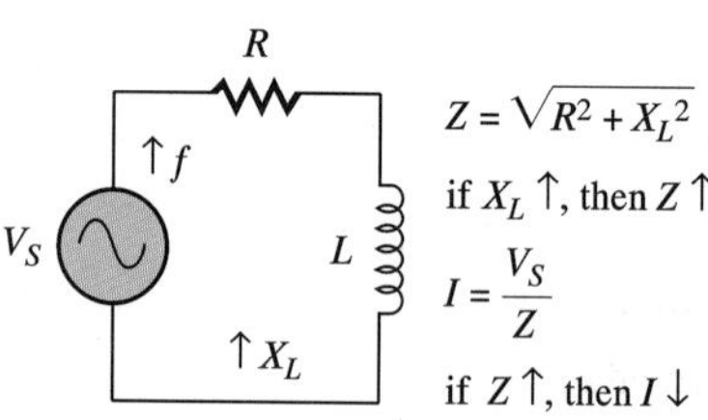

FIGURE 12–40 Typical frequency counter

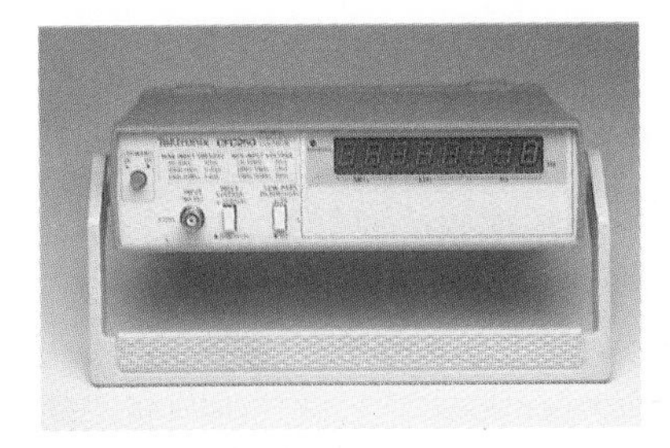

FIGURE 12–41 *RL* circuit for troubleshooting current that is too high

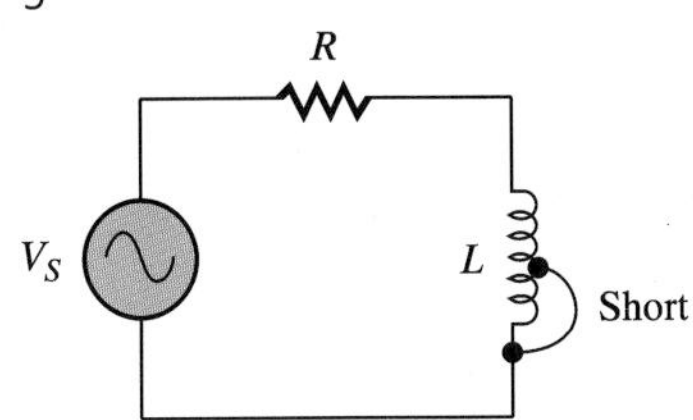

increased in value. Reactance is directly proportional to both frequency and inductance. Thus, an increase in *either* causes an increase in impedance. Although it is not inconceivable, it is unlikely that the inductor would have increased in value. Thus, too low a current in an *RL* circuit is probably the result of the frequency being too high or the resistor having increased in value, as illustrated in figure 12–39. Frequency can be measured with an oscilloscope or a *frequency counter.* A typical frequency counter is shown in Figure 12–40.

If the current is too high, then the impedance is lower than its normal value. Low impedance is an indication of either low reactance or low resistance. A low value of reactance can be caused by the frequency of the applied voltage being lower than normal, or the inductor having decreased in value. A decrease in inductance is a distinct possibility. It can be caused by a short being developed across some of the coil windings, as illustrated in Figure 12–41. An impedance bridge can be used to measure the inductance.

Bear in mind that all of the problems mentioned in the two preceding paragraphs also cause the circuit phase angle to change value. The phase angle can be determined using an oscilloscope. If the circuit is a lead-lag network, the output voltage no longer leads or lags the input by the desired amount. If being used as a low or high-pass filter, any variation in inductance causes a change in the cutoff frequency.

CHAPTER SUMMARY

1. In an inductor, the induced voltage leads the current by 90°.
2. The current through an inductor is inversely proportional to the inductive reactance.
3. Reactances in series are added in order to find the total reactance.
4. The reciprocals of individual reactances in parallel are added, and the reciprocal of this sum taken in order to the find total reactance.
5. In an *RL* circuit, the voltage induced across the inductor leads the voltage across the resistor by 90°.
6. The source voltage of a series *RL* circuit is equal to the phasor sum of V_R and V_L.
7. The impedance of a series *RL* circuit is equal to the phasor sum of R and X_L.
8. The phase angle of a series *RL* circuit is equal to the arc tangent of the ratio of V_L to V_R.
9. The total current of a parallel *RL* circuit is equal to the phasor sum of I_L and I_R.
10. The admittance of a parallel *RL* circuit is equal to the phasor sum of G and X_L.
11. The impedance of a parallel circuit is equal to the reciprocal of admittance.
12. The phase angle of a parallel *RL* circuit is the arc tangent of the ratio of I_L to I_R.
13. True power in an *RL* circuit is that developed in the resistor.
14. Reactive power in a parallel *RL* circuit is that developed in the inductor. It is imaginary.
15. Apparent power in an *RL* circuit is the total power that is "apparently" developed in an *RL* circuit.
16. Apparent power is equal to the phasor sum of true and reactive power.
17. The true power is the product of apparent power and the circuit power factor.
18. *RL* circuits are used in applications in which its frequency sensitive characteristics can be taken advantage of.

KEY TERMS

inductive reactance
inductive susceptance
impedance
phase angle
true power
reactive power

Q-factor
choke
apparent power
power factor
admittance
high-pass filter
low-pass filter
RL lead network
RL lag network

EQUATIONS

(12–1) $X_L = 2\pi fL$

(12–2) $X_L = \omega L$

(12–3) $I_L = \frac{V_S}{X_L}$

(12–4) $V_L = I_L X_L$

(12–5) $X_L = \frac{V_L}{I_L}$

(12–6) $X_{LT} = 2\pi fL_T$

(12–7) $V_S^2 = V_R^2 + V_L^2$

(12–8) $V_S = \sqrt{V_R^2 + V_L^2}$

(12–9) $Z = \sqrt{R^2 + X_L^2}$

(12–10) $I_L = \frac{V_L}{Z}$

(12–11) $V_L = IZ$

(12–12) $Z = \frac{V_S}{I_T}$

(12–13) $\theta = \arctan \frac{V_L}{V_R}$

(12–14) $\theta = \arctan \frac{X_L}{R}$

(12–15) $P_{react} = V_L I$

(12–16) $P_{react} = I^2 X_L$

(12–17) $P_{react} = \frac{V_L^2}{X_L}$

(12–18) $Q = \frac{X_L}{R_W}$

(12–19) $P_{app} = \sqrt{(P_{true})^2 + (P_{react})^2}$

(12–20) $P_{app} = V_S \times I$

(12–21) $P_{app} = \frac{V_S^2}{Z}$

(12–22) $P_{app} = I^2 Z$

(12–23) $P_{true} = P_{app}(\cos\theta)$

(12–24) $I_T = \sqrt{I_R^2 + I_L^2}$

(12–25) $Y = \sqrt{G^2 + B_L^2}$

(12–26) $Z = \frac{1}{Y}$

(12–27) $\theta = \arctan \frac{I_L}{I_R}$

(12–28) $\theta = \arctan \frac{B_L}{G}$

(12–29) $\Phi = 90 - \theta$

(12–30) $V_R = \frac{R}{\sqrt{R^2 + X_L^2}} V_S$

(12–31) $V_L = \frac{X_L}{\sqrt{R^2 + X_L^2}} V_S$

Variable Quantities

X_L = inductive reactance
X_{LT} = total inductive reactance
I_L = inductor current
V_L = inductor voltage
Z = impedance
P_{react} = reactive power
P_{app} = apparent power
P_{true} = true power
Q = quality factor
θ = the phase angle between resistance and impedance
Y = admittance
B_L = inductive suceptance
Φ = the angle between impedance and reactance

TEST YOUR KNOWLEDGE

1. At what point on a sine wave of current is maximum voltage induced in an inductor?
2. At what point on a sine wave of current is minimum voltage induced in an inductor?
3. Does the voltage induced in an inductor lead or lag the energizing current?
4. Is inductive reactance directly or inversely proportional to frequency?
5. What symbol is used for angular velocity?
6. In a series *RL* circuit, does the voltage induced in the inductor lead or lag the voltage across the resistor?
7. Of what is the impedance of a series *RL* circuit composed?
8. Is the phase angle of a series *RL* circuit inversely or directly proportional to frequency?
9. Is the phase angle of a parallel *RL* circuit inversely or directly proportional to frequency?
10. What is the reference phasor for the current triangle of a parallel *RL* circuit?
11. What is the only device that develops true power in an electric circuit?
12. What is the power factor of an ac circuit?
13. In an *RL* high-pass filter, the output is taken across which device?
14. In an *RL* lag network, the output is taken across which device?
15. What instrument, besides an oscilloscope, may be used in the measurement of frequency?

PROBLEM SET: BASIC

Sections 12–1, 12–2

1. Draw the phasor diagram showing the phase relationship between the source voltage and current for the circuit of Figure 12–42. (Consider the inductor to be perfect.)

FIGURE 12–42 Circuit for Problem 1

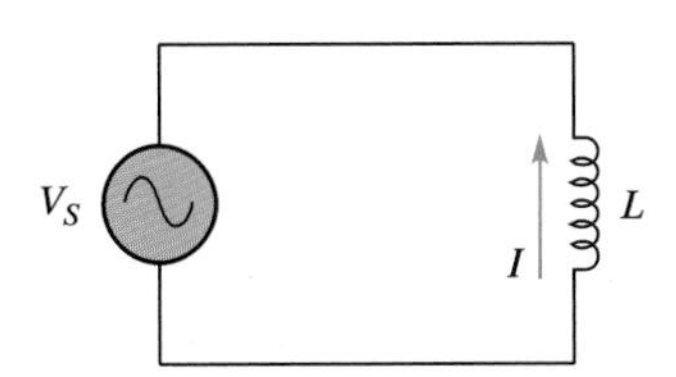

Section 12–3

2. An inductor with an inductance of 10 mH is operated by a 1000 Hz source. Compute the inductive reactance.
3. At what frequency will a 30 mH inductor produce 2 k Ω of reactance?
4. For the circuit of Figure 12–43, compute the total inductive reactance.

FIGURE 12–43 Circuit for Problem 4

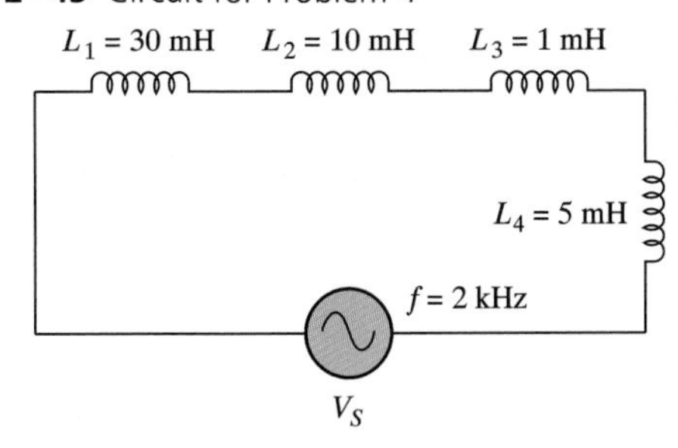

5. An inductor with a reactance of 1240 Ω is across a 300 mV source. Compute the current.
6. For the circuit of Figure 12–44, compute the total inductive reactance.

FIGURE 12–44 Circuit for Problem 6

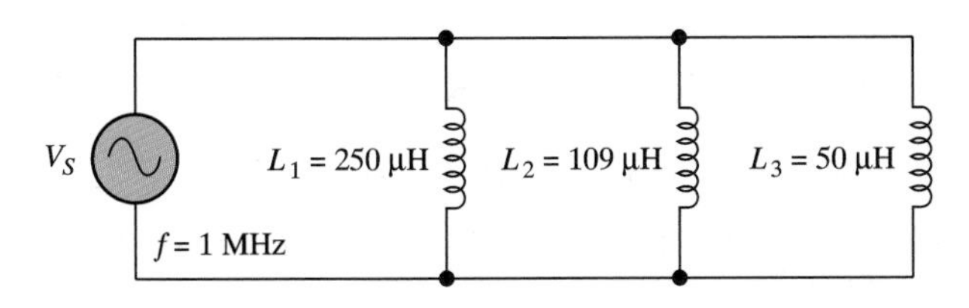

7. A current of 640 μA flows through an inductor that has a reactance of 1562 Ω. What is the value of the source voltage?
8. Compute the inductive reactance for the circuit of Figure 12–45.

FIGURE 12–45 Circuit for Problem 8

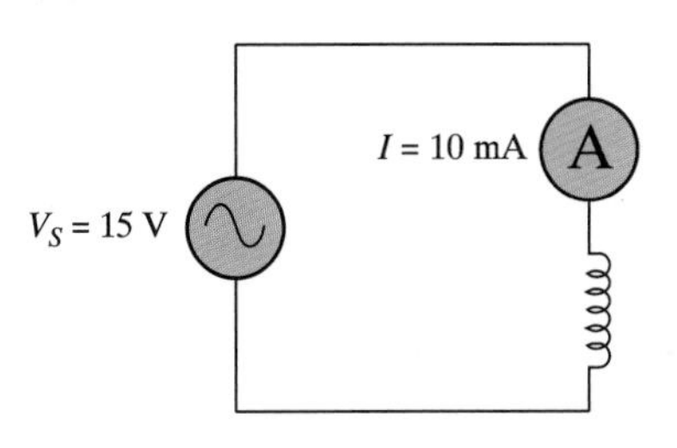

9. For the circuit of Figure 12–46, compute the following (a) the reactance of each inductor, (b) the total reactance, and (c) the total current.

FIGURE 12–46 Circuit for Problem 9

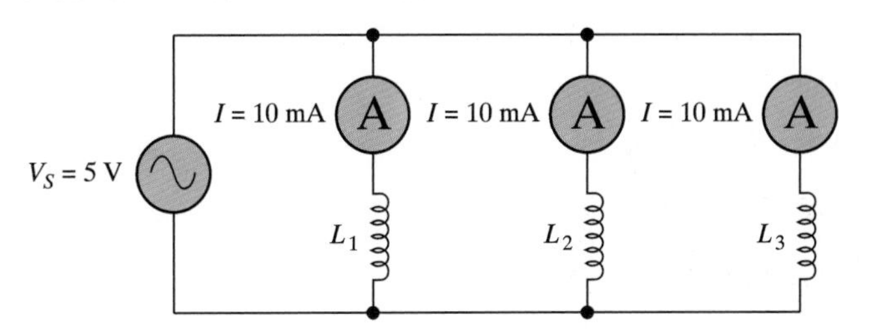

10. For the circuit of Figure 12–47, compute the value of the source voltage.

FIGURE 12–47 Circuit for Problem 10

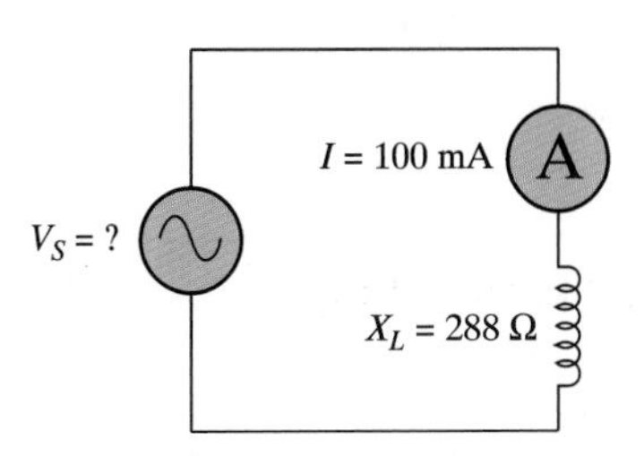

Sections 12–4, 12–5, 12–6

11. For the circuit of Figure 12–48, sketch the graphs of the voltages developed across the resistor and inductor in relation to the current. Draw the phasor diagram for each.

FIGURE 12–48 Circuit for Problem 11

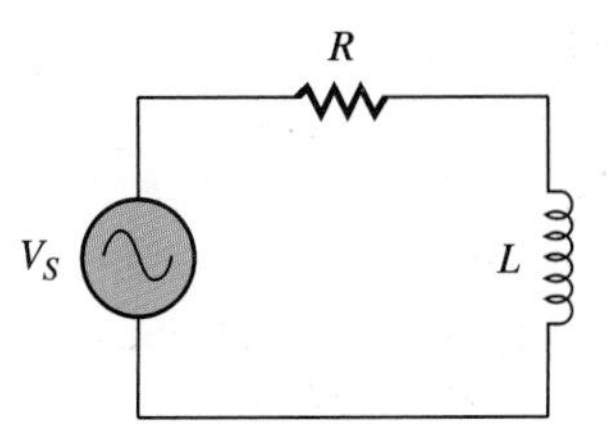

12. If a series RL circuit has 60 V across the resistor and 40 V across the inductor, compute V_S.

13. For each circuit of Figure 12–49, compute the following: (a) X_L, (b) Z, (c) I, (D) V_R, and (e) V_L.

FIGURE 12–49 Circuits for Problem 13

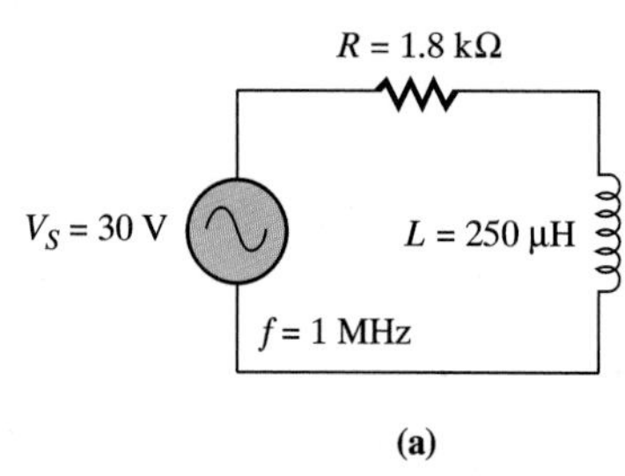

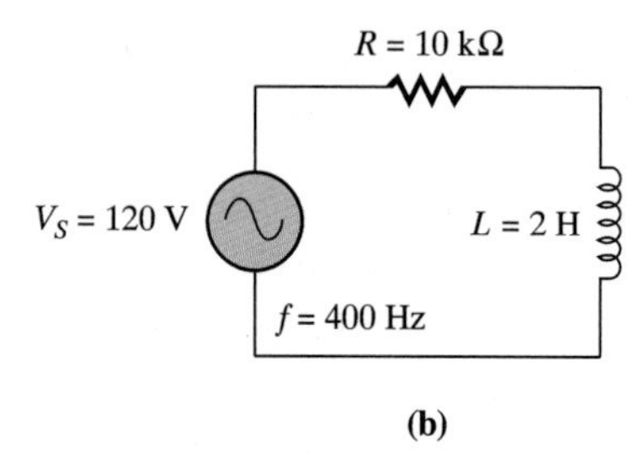

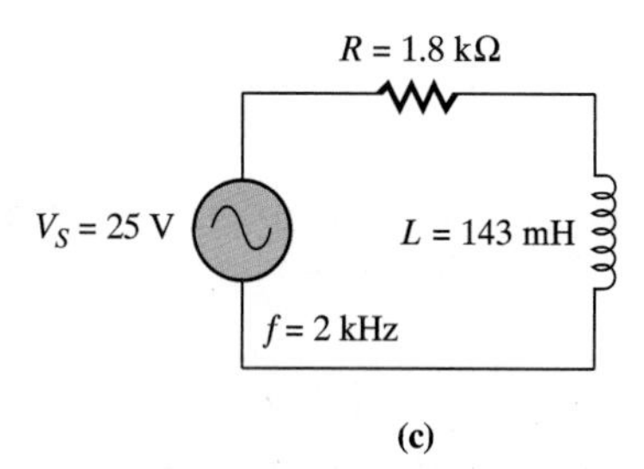

14. For each circuit of Figure 12–50, compute the following: (a) I_R, (b) I_L, and (c) I_T

FIGURE 12–50 Circuits for Problem 14

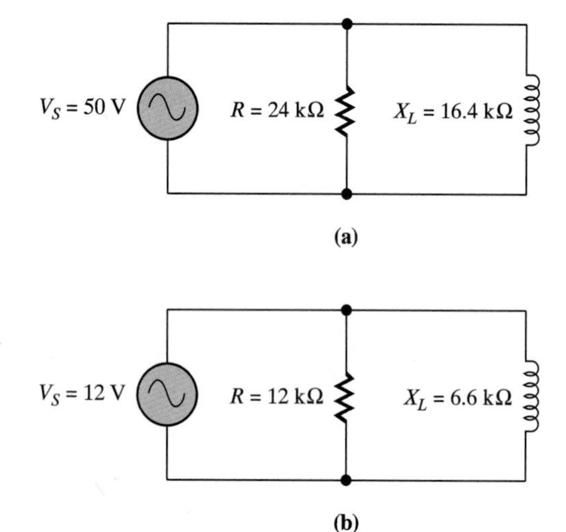

15. For the circuit of Figure 12–51, sketch the graphs of the currents through the resistor and inductor in relation to the source voltage. Draw the phasor diagrams for each.

FIGURE 12–51 Circuit for Problem 15

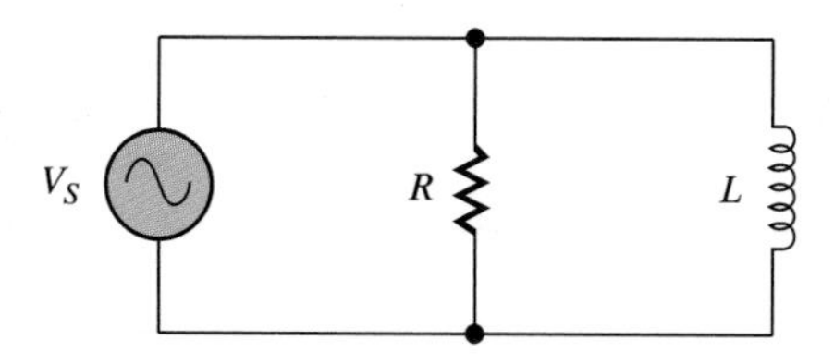

16. For the circuit of Figure 12–52, compute the conductance, inductive susceptance, and admittance.

FIGURE 12–52 Circuit for Problem 16

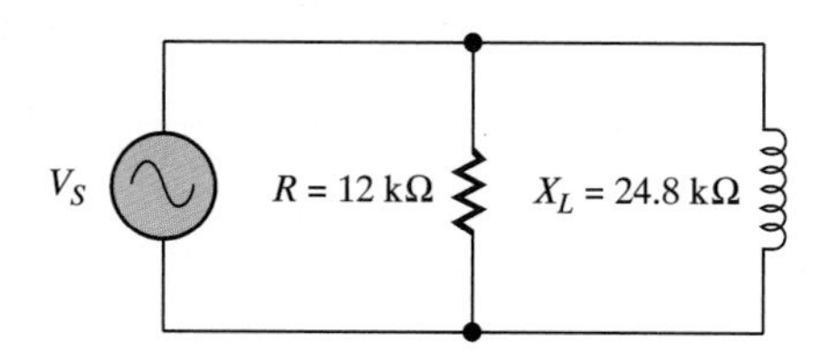

17. Compute the circuit phase angle θ for each circuit of Figure 12–53.

FIGURE 12–53 Circuits for Problem 17

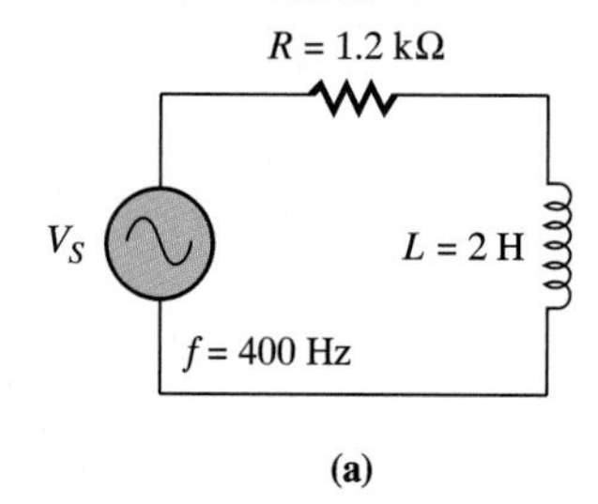

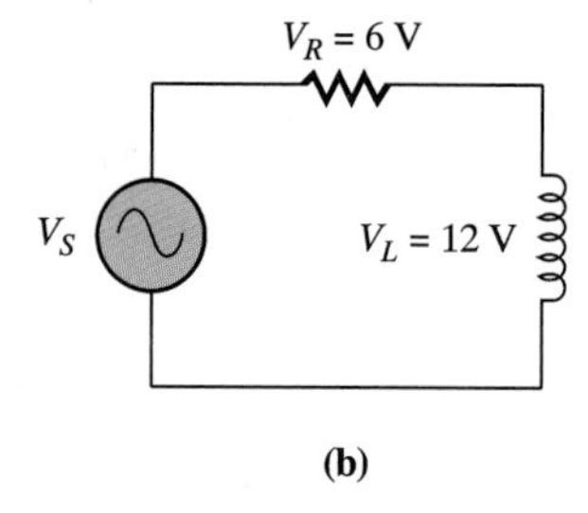

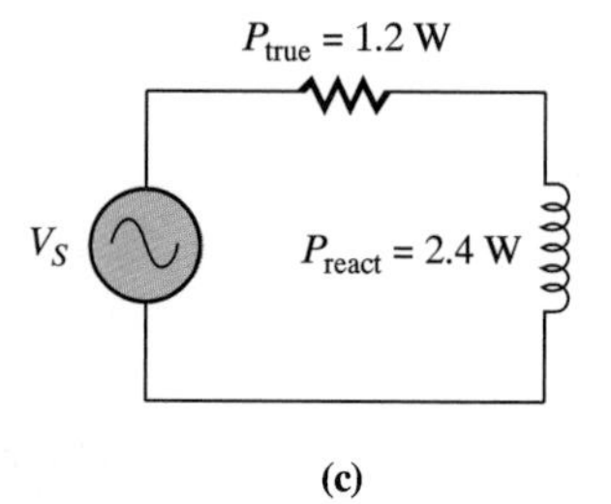

18. Compute the circuit phase angle θ for each circuit of Figure 12–54.

FIGURE 12–54 Circuits for Problem 18

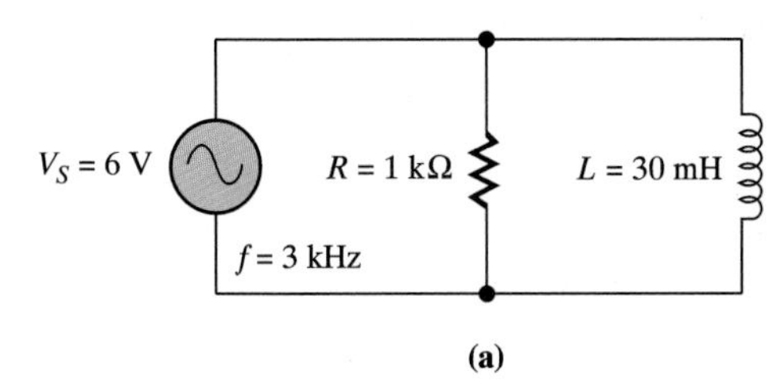

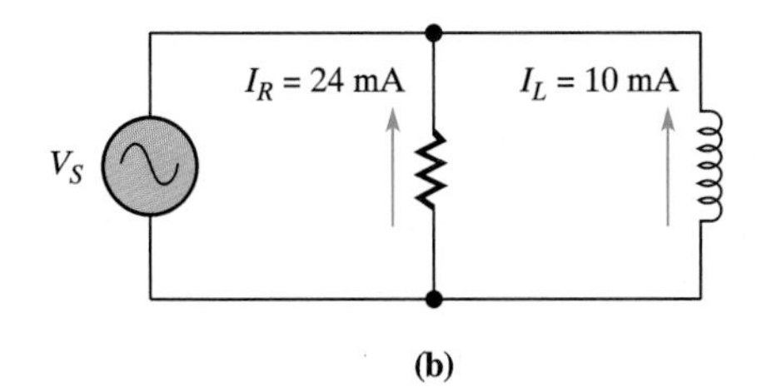

19. For the circuit of Figure 12–55, compute the true power, reactive power, and apparent power.

FIGURE 12–55 Circuit for Problem 19

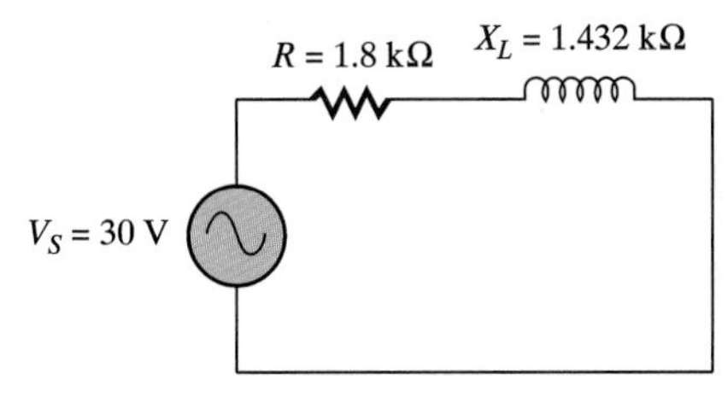

20. Determine the following quantities for the circuit of Figure 12–56: (a) X_L, (b) Z, (c) I, (d) V_L, (e) V_R, (f) θ, and (g) power factor.

FIGURE 12–56 Circuit for Problem 20

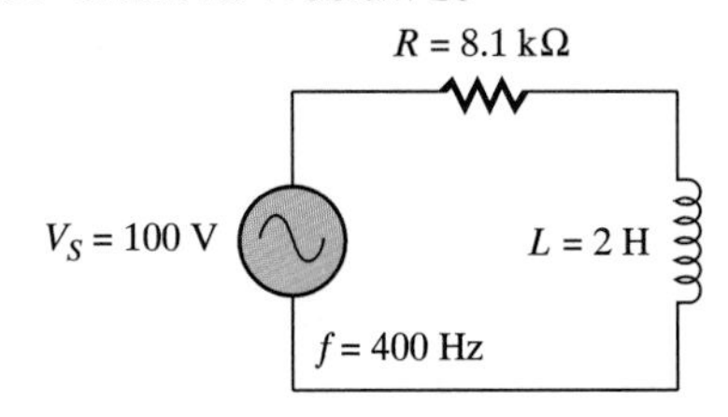

21. Determine the following quantities for the circuit of Figure 12–57: (a) X_L, (b) I_L, (c) I_R, (d) I_T, (e) θ, (f) Z, (g) P_{app}, and (h) P_{react}.

FIGURE 12–57 Circuit for Problem 21

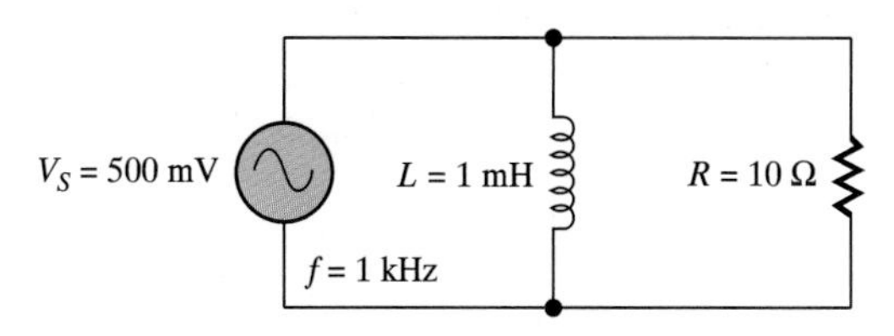

Section 12–7

22. An *RL* high-pass filter has an input voltage of 10 V. What is the output voltage at the cutoff frequency?

23. For the *RL* lag network of Figure 12–58, compute the output voltage and the angle of lag.

FIGURE 12–58 Circuit for Problem 23

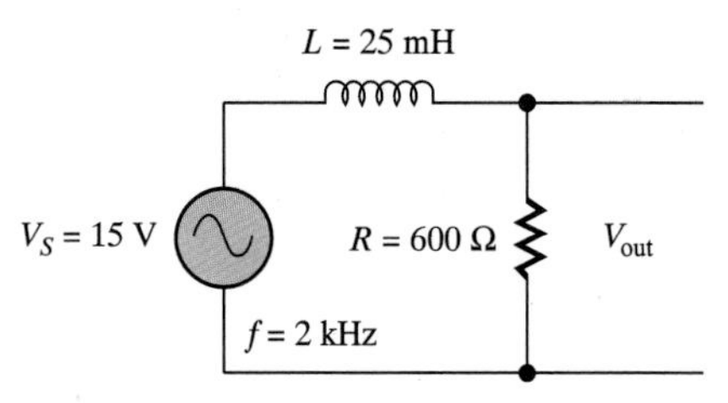

24. For the *RL* lead network of Figure 12–59, compute the output voltage and the angle of lead.

FIGURE 12–59 Circuit for Problem 24

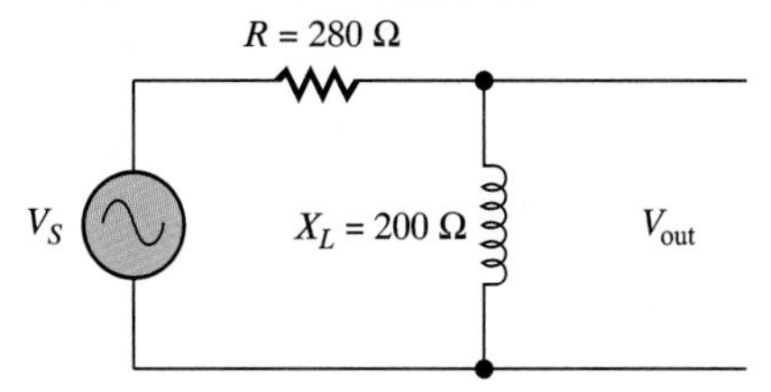

PROBLEM SET: CHALLENGING

25. For the circuit of Figure 12–60, determine the following values:

a. I_R, **b.** I_L, **c.** I_T, and **d.** PF.

FIGURE 12–60 Circuit for Problem 25

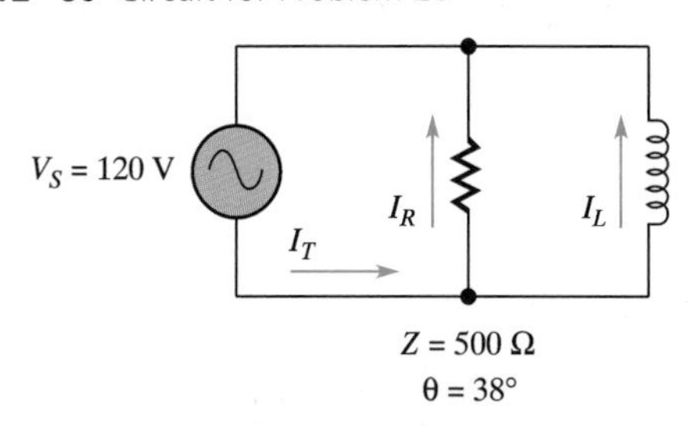

26. For the circuit of Figure 12–61, determine the following values:

a. V_R, **b.** V_L and **c.** V_S.

FIGURE 12–61 Circuit for Problem 26

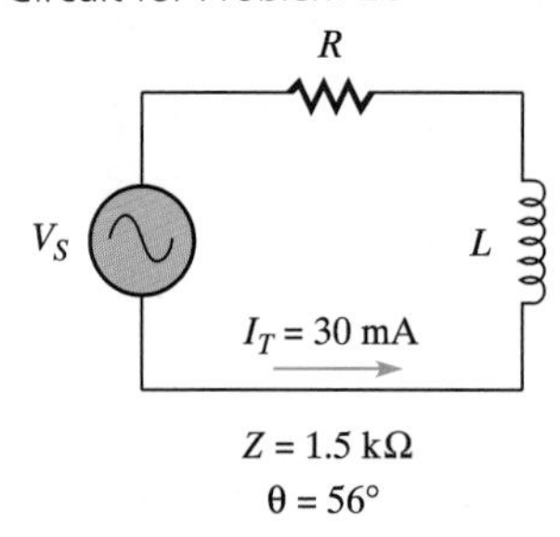

27. For the circuit of Figure 12–62, compute the following values:

a. X_L and **b.** L.

FIGURE 12–62 Circuit for Problem 27

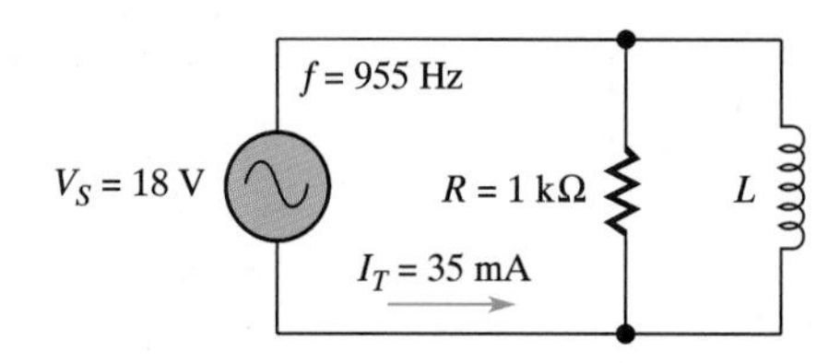

28. A series *RL* lead network is composed of a 22 Ω resistor and a 35 Ω reactance. Compute the value of the output voltage and the angle by which it leads the input voltage. Draw the schematic diagram of the circuit.

29. A series *RL* lag network is composed of a 100 Ω resistor and a 50 Ω reactance. The source voltage is 10V. Compute the value of the output voltage and the angle by which it lags the input voltage. Draw the schematic diagram of the circuit.

30. For the circuit of Figure 12–63, compute the value of V_r.

FIGURE 12–63 Circuit for Problem 30

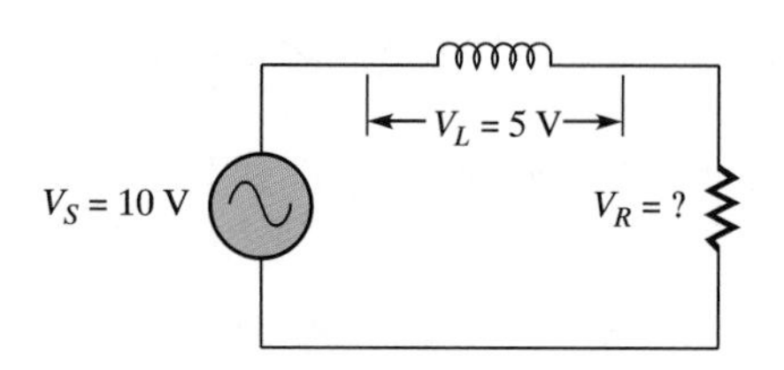

31. For the circuit of Figure 12–64, compute the value of I_L.

FIGURE 12–64 Circuit for Problem 31

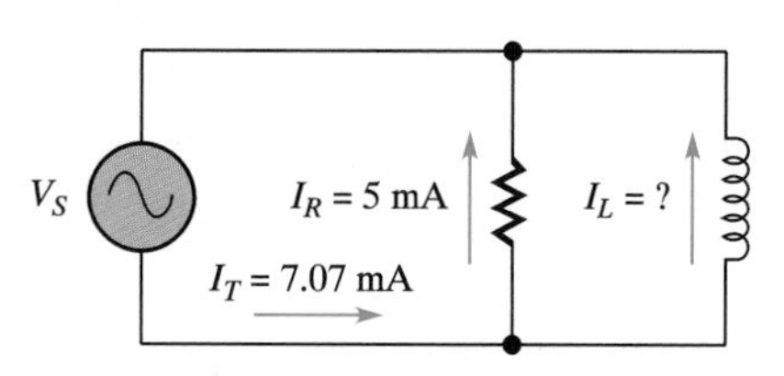

32. For the circuit of Figure 12–65, compute the following:

a. X_{LT}, **b.** Z, **c.** I,
d. voltage across each component and **e.** phase angle.

FIGURE 12–65 Circuit for Problem 32

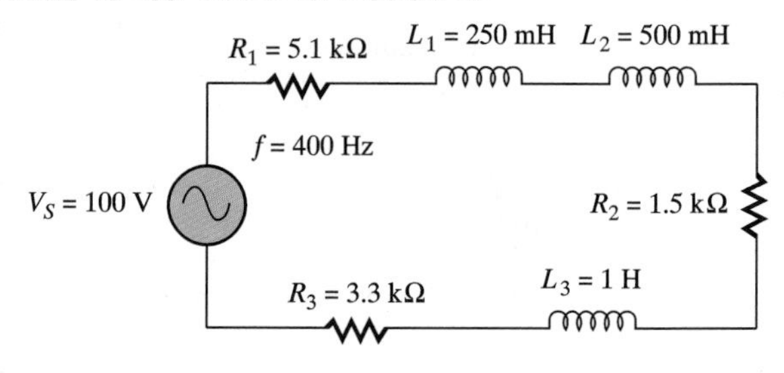

33. For the circuit of Figure 12–66, compute the following:

a. X_{LT}, **b.** Z, **c.** I_T,
d. the current through each component
and **e.** the phase angle.

FIGURE 12–66 Circuit for Problem 33

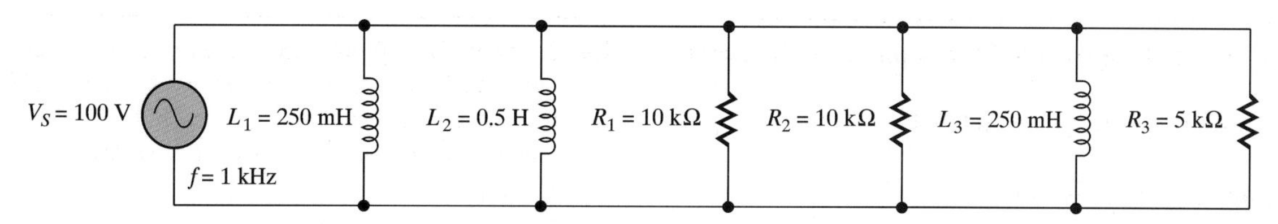

PROBLEM SET: TROUBLESHOOTING

34. In Figure 12–67, the voltage measurements are as shown. If one fault exists, what is the probable cause?

FIGURE 12–67 Circuit for Problem 34

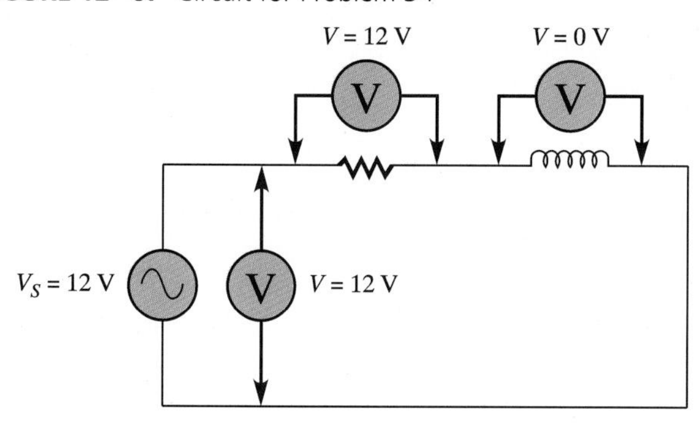

35. The phase angle of the circuit of Figure 12–68 is measured and found to be 6.84°. If the resistance is measured and found to be correct, what is the probable fault in the circuit?

FIGURE 12–68 Circuit for Problem 35

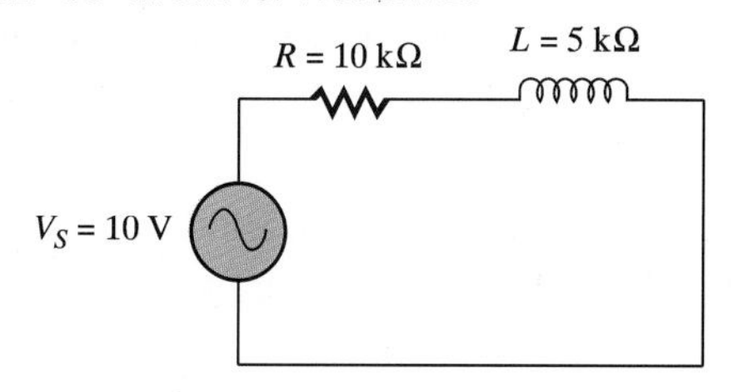

36. The phase angle of an *RL* circuit is measured and found to be zero degrees. If one fault exists, what is the probable cause?

37. In the circuit of Figure 12–69, one fault exists. From the readings shown, determine one possible cause of the fault.

FIGURE 12–69 Circuit for Problem 37

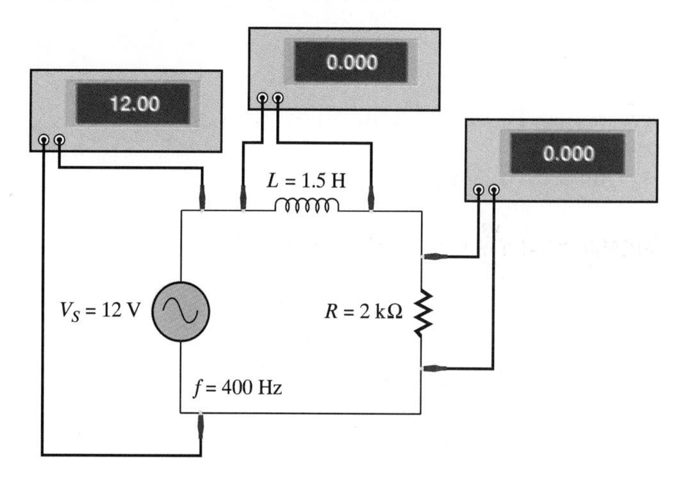

38. In the circuit of Figure 12–70, one fault exists. From the readings shown, determine one possible cause of the fault.

FIGURE 12–70 Circuit for Problem 38

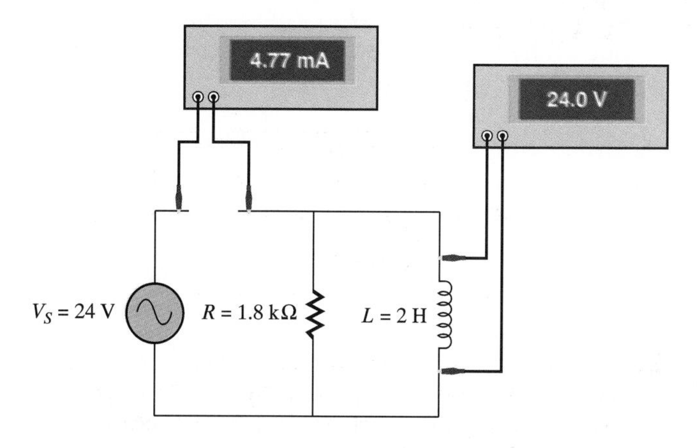

Automotive Electronics

Electronics has found its way into so many areas it is difficult to imagine anything without it. Such is the case with the automobile. Electricity has always been used in the operation of a car powered by an internal combustion engine. The lights, horn, starter, and windshield wipers all require electric current in their operation. The ignition system utilizes a transformer that builds the 6 or 12 V of the battery to the 20,000 V or so that is needed at the spark plugs. In early years, an auto mechanic needed a knowledge of electric current and simple circuits, but there were no truly electronic devices used in a car.

The first electronic system incorporated in cars was the radio. When the radio broke down, however, people did not take it to the garage for repair, but to a radio dealer. Thus, the auto mechanic still did not need a knowledge of electronics. Improved radios and sound systems followed, but once again, these items were usually installed and serviced by sound shops rather than mechanics.

One of the first truly electronic devices for which the mechanic was responsible was an improved voltage regulator, which was introduced in the late 1970s. The purpose of the voltage regulator is to ensure that the battery stays fully charged, but not overcharged. The early devices that performed this function were electromechanical and were one of the first items on a car to fail. Failure of this device had two bad outcomes: (1) If it failed to close the charging circuit, the battery would slowly discharge to a point at which there would be insufficient energy to start the car, or (2) the charging circuit could stay closed all the time, overcharging the battery and possibly destroying it.

The improved version, a solid-state regulator, is an electronic device that senses the battery voltage and goes to the ON state to connect the charging alternator or the OFF state to disconnect it. Since it needs no mechanical adjustments, it is encapsulated in plastic. It is composed of solid-state devices made from the same materials as transistors. The solid-state regulator cuts down on charging system problems, but failures do occur. At this point, thoughtful auto mechanics began to realize that they would soon have to learn the theory and operation of electronic devices and circuits.

Another application of electronics in the automobile corrects another perpetual problem: arcing at the ignition breaker points. As shown in Figure 1, when the points are closed, current flows through the coil (the transformer) producing a large magnetic field around it. When the points open, the current stops, the magnetic field collapses, developing a high voltage (20,000 V or so) across the coil in the process known as electromagnetic induction. This high voltage causes electrons to jump the spark plug gap, igniting the fuel and producing power. The continual, rapid opening and closing of the points produces arcing that can cause the contacts to scar and pit. Scarring and pitting increases the resistance of the points and, by Ohm's law, reduces the current through the coil. Reduced current results in a reduced magnetic field in the coil and lower induced voltage. A point is reached at which the voltage is not high enough to cause the electrons to jump the gap of the spark plugs, and the engine does not start.

The fix for this problem involved replacing the breaker points with a power transistor as shown in Figure 2. Closed breaker points are simulated by a transistor in the ON (conducting) condition. The opening of the breaker points is simulated by turning the transistor off for an instant. Without contacts to scar and pit, the ignition system is more reliable and requires far less maintenance. However, when this unit fails, the auto mechanic must know how to test it.

Note that drivers were often caught in a quandary by the new electronic ignitions. In the past, if a car refused to run, one of the first checks was invariably the points. If they looked pitted or scarred, a fingernail file or piece of emery cloth was often used to clean them to the point where the engine could be started. Although it may have sputtered, the car usually got the driver home or to a garage. But with the electronic ignition, it became a matter of parking the car and having it towed!

The late 1970s saw a tremendous increase in the price of gasoline and diesel fuel. Also, new federal regulations required that automobiles meet stringent emission standards. Thus, efficiency became the watchword in automobile design. Tuning the engine every 10,000 miles was not enough; it had to be done continuously. This need led to electronic computer's becoming a part of the under-the-hood equipment on most cars. This computer sensed the demands being placed upon the engine and automatically tuned it for optimum performance and minimum emissions. This computer was not only rather expensive, but difficult for dealers to test and repair. Thus, equipment was developed that could be used to test the computer itself for proper operation.

An ever-increasing amount of electronics has continued to show up in cars. Electronic controls have made the speed of windshield wipers continuously variable. Some cars are equipped with electronic timers that automatically turn off the lights when the driver forgets. The dashboards of some cars light up with an array of electronic gadgets. These include digital readouts of speed, the number of engine RPMs, fuel level, and the rate of fuel consumption. Some have a display that shows the projected distance that can be travelled at this rate of fuel consumption.

These are but a few of the many examples of electronics used in automobiles. Many more are in use, especially in high-performance racing vehicles. This proliferation of electronic devices and circuits does not, however, provide many employment opportunities for electronics technicians. Many of the devices used in the automobile are manufactured in the form of modules. A *module* is a portion of an electronic system that is easily detached and replaced. thus, when trouble occurs, it is usually localized to a module that is replaced rather than repaired. Automobile dealerships often send their auto mechanics to factory-sponsored training in which they learn to troubleshoot to these modular systems. As this trend toward more electronic control of the automobile's engine and transmission continues, however, the day when every automobile dealership will require a full time electronics technician is nearly at hand.

FIGURE 1 Ignition system utilizing breaker points

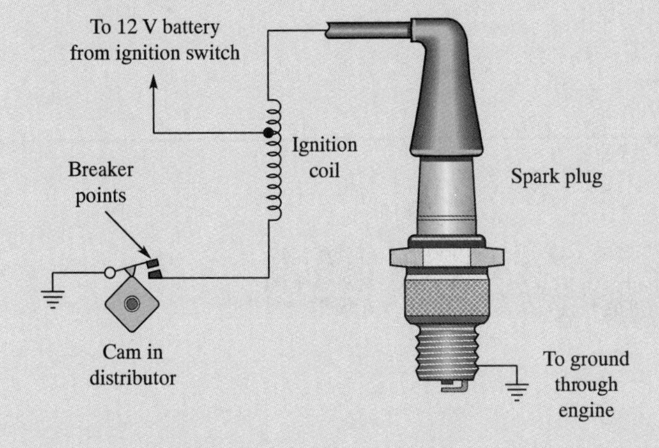

FIGURE 2 Ignition system utilizing solid-state device

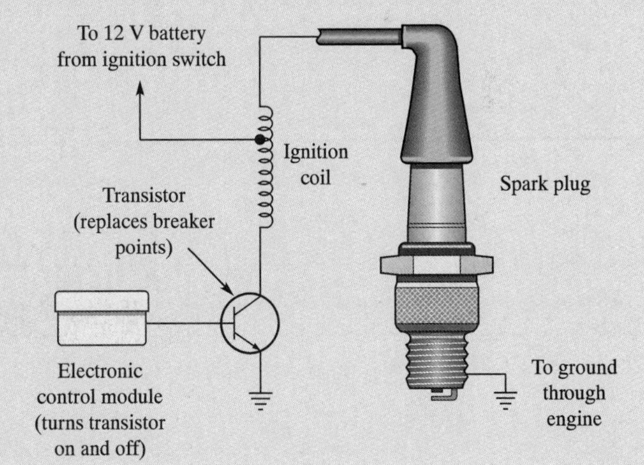

CHAPTER

13 Transformers

■ UPON COMPLETION OF THIS CHAPTER, YOU WILL BE ABLE TO

1. Explain how a transformer couples electrical energy from one circuit to another with no electrical contact between them.
2. Define mutual inductance and coupling coefficient.
3. Given the values of two inductors and the coefficient of coupling, compute their mutual inductance.
4. Define and explain the relationship between the turns ratio, transformation ratio, current ratio, and impedance ratio.
5. Explain the relationship between the power developed in the primary and secondary of a transformer.
6. List the power losses in a transformer.
7. Compute the efficiency of a transformer.
8. Draw the symbols for and describe the construction of several types of transformers.
9. List the four basic uses of transformers.
10. List and explain several practical applications of transformers.

Transformers are used in a wide variety of electronic applications. In commercial power systems, they are used to step up (increase) or step down (decrease) voltages that may be as high as 150,000 V or more. At the other end of the voltage scale, within TV and radio receivers, they are used to couple signals as low as a few microvolts. In the interest of safety, they are often used to isolate one circuit from another. There are many more applications, but these few show the true versatility of transformers.

Self-inductance, which was introduced in chapter 9, takes place within an individual inductor. The operation of a transformer, however, is based upon what is known as mutual inductance between two inductors. This principle along with the operation, construction, applications, and troubleshooting of transformers are covered in this chapter.

13–1 MUTUAL INDUCTANCE

Question: What is mutual inductance?

In Chapter 9, you learned that the circuit property known as self-inductance induces a voltage within itself when the current varies. **Mutual inductance** results when the varying magnetic flux around one inductor cuts across the windings of a nearby inductor, thus inducing voltage in the second inductor, as illustrated in Figure 13–1. Inductors L_1 and L_2 are in close proximity to one another. Assume that an increasing current is passed through L_1. As the current increases, the magnetic flux builds in magnitude. The moving flux cuts across the windings of L_2, inducing a voltage in it. The same effect occurs if the current decreases. The decreasing current causes the magnetic field to collapse, once again inducing a voltage but now of the opposite polarity. Through this principle, a **transformer** thus couples electrical energy from one circuit to another through magnetic induction, with no electrical contact between the circuits.

■ Calculation of Mutual Inductance

The amount of mutual inductance between two inductors is dependent upon their individual inductance values and the number of lines of flux coupling them. In equation form, mutual inductance is found by

(13–1) $$L_M = k\sqrt{L_1 L_2}$$

FIGURE 13–1 Mutual inductance

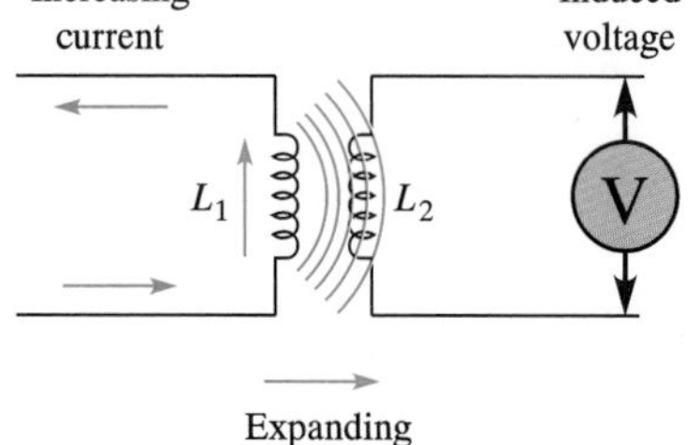

where k is the coefficient of coupling. The **coefficient of coupling** is the ratio of the number of lines of flux coupling L_1 and L_2 to the total number of lines of flux produced by L_1, or

(13–2) $$k = \frac{\text{lines of flux linking } L_1 \text{ and } L_2}{\text{lines of flux produced by } L_1}$$

The coefficient of coupling is a number between zero and one. The closer the inductors are to one another, the greater the number of lines of flux linking the two and, hence, the larger the value of k. As Eq. 13–1 shows, mutual inductance is directly proportional to the coefficient of coupling.

EXAMPLE 13–1 One thousand lines of force are produced by an inductor, and 750 of them link a second inductor. Compute the coefficient of coupling.

■ Solution:

$$k = \frac{\text{lines of flux linking}}{\text{line of flux produced}}$$

$$\frac{750}{1000} = 0.75$$

EXAMPLE 13–2 Two inductors are in close proximity one to the other. Their inductance values are 30 mH and 100 mH. If the coefficient of coupling is 0.67, compute the mutual inductance.

■ Solution:

$$L_M = k\sqrt{L_1L_2} = 0.67\sqrt{(30\text{ mH})(100\text{ mH})} = 36.69\text{ mH}$$

■ SECTION 13–1 REVIEW

1. What is the range of the values of k?
2. For the inductors of Figure 13–2, compute the mutual inductance if $k = 0.9$.
3. If the distance between two inductors is increased, will the value of k increase or decrease?
4. An inductor produces 200 lines of flux. If the coefficient of coupling is 0.8, how many flux lines link the second inductor?

FIGURE 13–2 Inductors for Section Review Problem 2

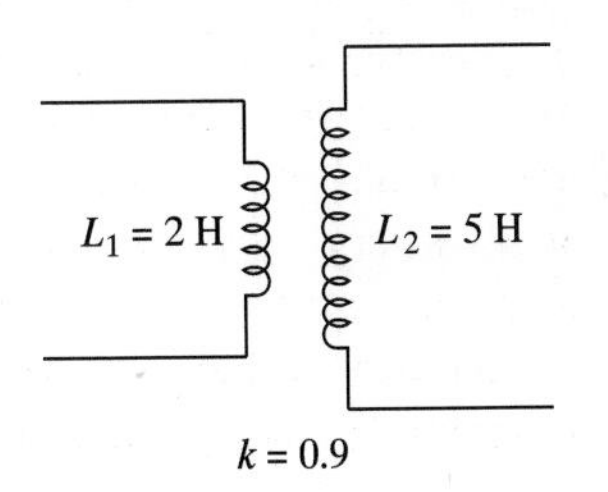

13–2 THE BASIC TRANSFORMER

The basic transformer is two inductors wound in very close proximity to one another. The schematic symbol is shown in Figure 13–3(a). As shown in Figure 13–3(b), the inductor to which the voltage source is connected is known as the *primary winding,* and the inductor to which the load is connected is known as the *secondary winding.* A changing current in the primary winding produces a moving magnetic field that cuts across the secondary.

FIGURE 13–3 Basic transformer

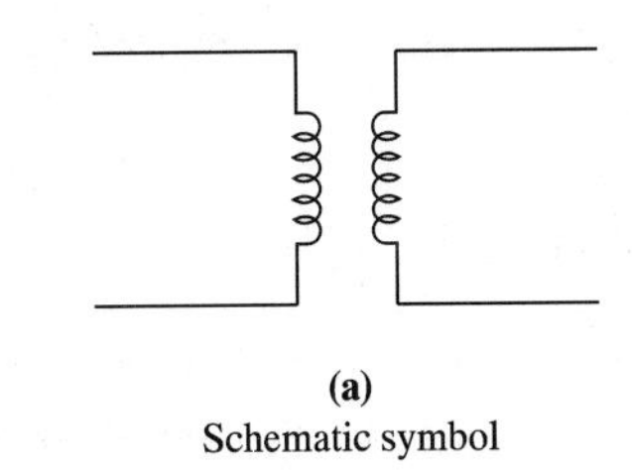

(a)
Schematic symbol

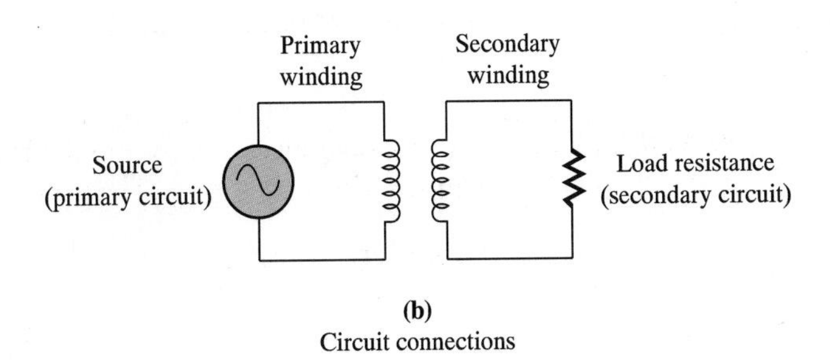

(b)
Circuit connections

The three requirements for electromagnetic induction—a conductor, a magnetic field, and relative motion between the two—are present. Thus, voltage is induced in the secondary winding. Notice that electrical energy is transferred from the primary to the secondary with no electrical contact. This *magnetic linkage* causes the secondary circuit to be isolated from the primary.

Although the winding of a transformer is covered in detail later in this chapter, it is a good idea at this point to have some idea of how it may be wound in order to place the two inductors "in close proximity." In Figure 13–4(a), the inductors are wound, side by side, on a common core. In Figure 13–4(b), they are once again wound on a common core, but now they lay one over the other. This latter method produces a larger coefficient of coupling k and a better transfer of energy from primary to secondary. They may also be wound on an iron core, as shown in Figure 13–4(c). The high permeability of the iron core provides a low-reluctance path for the lines of flux. Nearly all the flux lines produced in the primary link the secondary. Thus, the coefficient of coupling of iron core transformers is considered to be one (1).

FIGURE 13–4 Transformer windings

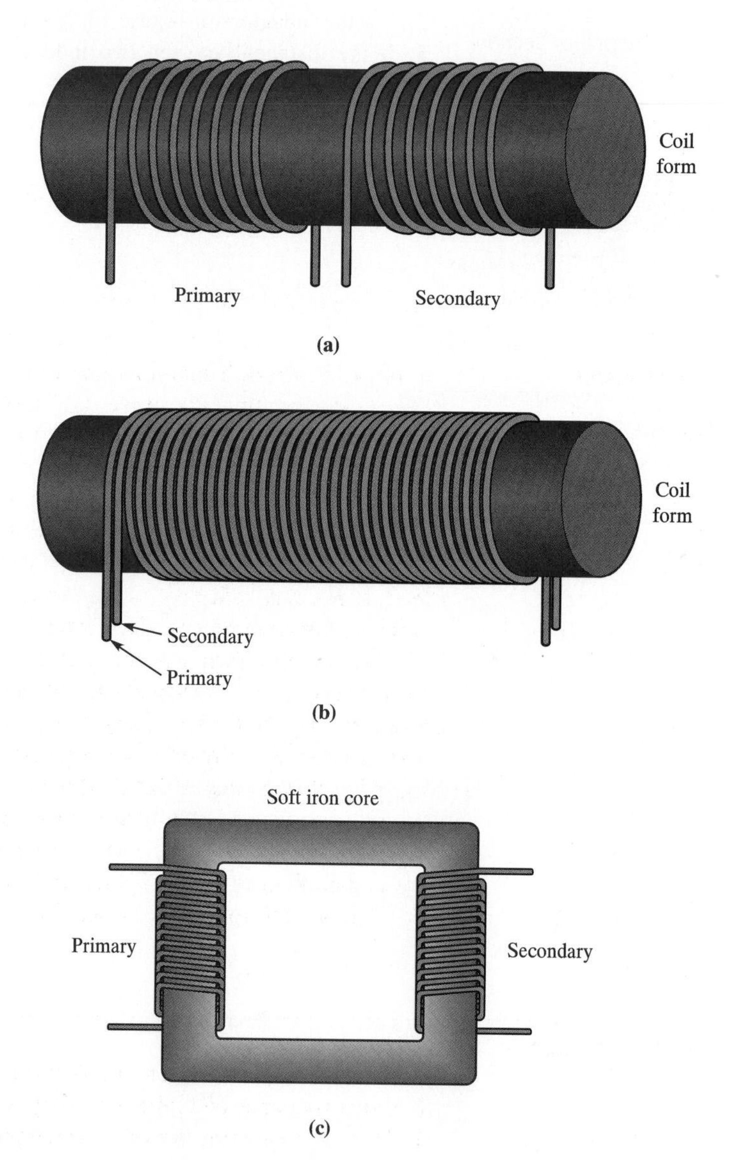

FIGURE 13–5 Transformer action

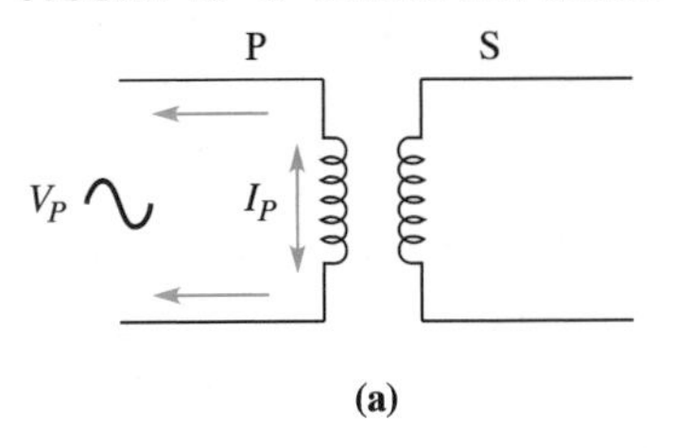

(a)

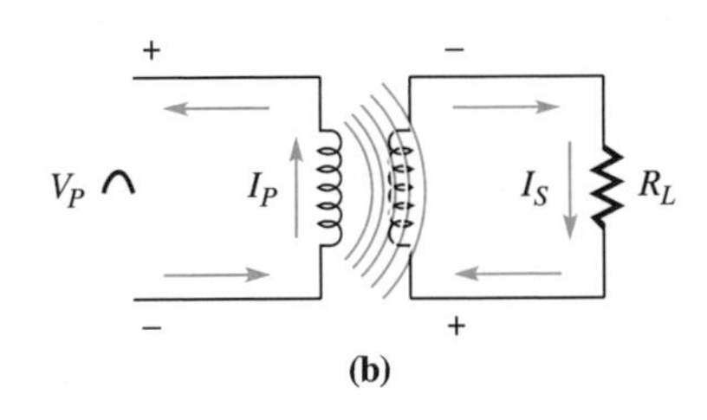

(b)

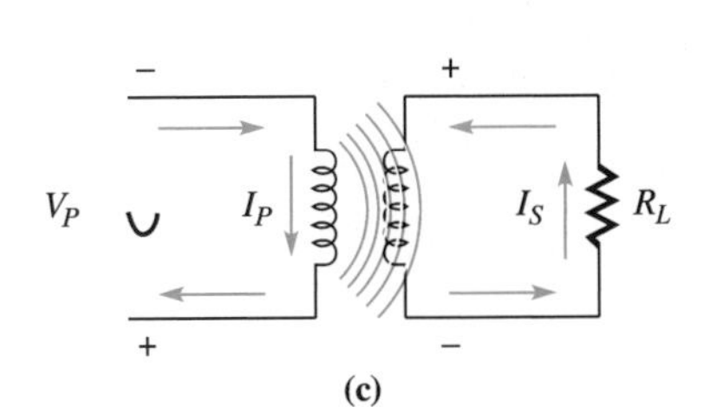

(c)

FIGURE 13–6 Transformer phase shifts

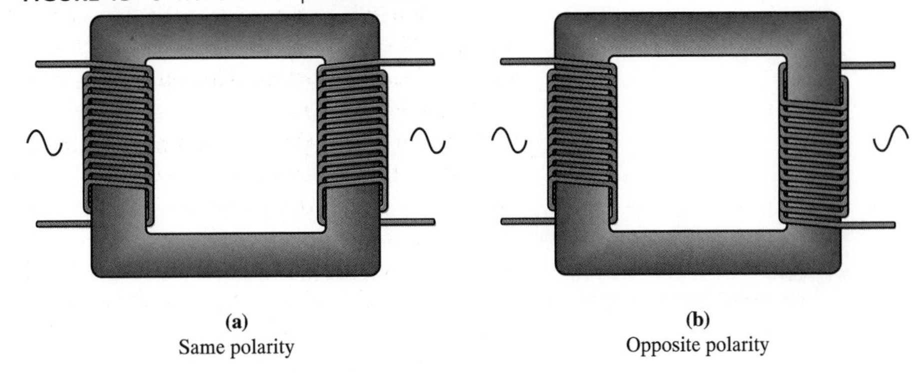

(a) Same polarity

(b) Opposite polarity

FIGURE 13–7 Dot convention, indicating points of like polarity

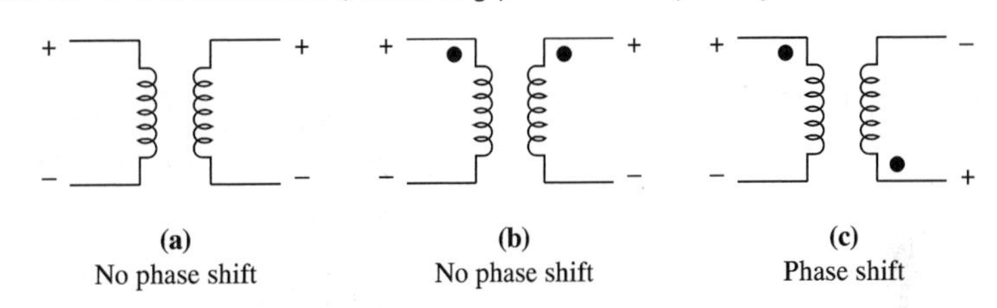

(a) No phase shift

(b) No phase shift

(c) Phase shift

Transformer Action

In Figure 13–5(a), a sine wave source is applied to the primary of a transformer, producing a corresponding sine wave of current within it. On the positive half-cycle, as the current builds to its peak value, the magnetic field expands cutting across the secondary winding. This moving magnetic field induces a voltage in the secondary and produces a current through the load in the direction shown. This action is illustrated in Figure 13–5(b). On the negative half-cycle of the primary current, a moving magnetic field is once again produced. This field, however, has the opposite magnetic polarity of that of the first and thus induces a voltage of the opposite polarity. The current through the load is in the direction shown. This action is illustrated in Figure 13–5(c). Thus, a voltage with the same frequency as the source is reproduced across the load, and a current of the same frequency flows through it. The only changes that may take place are in the amplitude and phase of the secondary voltage compared to the primary.

A transformer may or may not give a 180° phase shift in the voltage from primary to secondary, depending upon the direction in which the primary and secondary coils are wound. If wound as shown in Figure 13–6(a), no phase shift takes place. If wound as shown in Figure 13–6(b), a 180° phase shift takes place from primary to secondary. A **dot convention,** in which the dots indicate points of like polarity, as illustrated in Figure 13–7, is sometimes used to indicate whether or not a phase shift takes place.

SECTION 13–2 REVIEW

1. What is connected to the primary winding of a transformer?
2. What is connected to the secondary winding of a transformer?
3. How is energy transferred from the primary to the secondary of a transformer?

4. What are the three requirements for electromagnetic induction?
5. Will the frequencies of the primary and secondary currents of a transformer be the same?
6. Will the value of the secondary voltage be the same as the primary voltage?
7. Will the secondary voltage have the same phase as the primary voltage?
8. What do dots on the symbol for a transformer indicate?

13–3 THE TRANSFORMATION RATIO

Question: What is meant by the transformation ratio of a transformer?

As was stated previously, the secondary voltage of a transformer may be larger or smaller than the primary voltage. If the secondary voltage is larger, the transformer is said to *step up* the voltage. If it is smaller, the transformer *steps down* the voltage. The ratio of the secondary voltage to the primary voltage is known as the voltage ratio, or **transformation ratio.**

Question: What determines whether the transformer is of a step-up or step-down variety?

The voltage ratio of a transformer is directly related to what is known as the turns ratio. The **turns ratio** is equal to the ratio of the number of turns on the secondary to the number of turns on the primary. Thus, the following mathematical relationship is established:

(13–3)
$$\frac{N_S}{N_P} = \frac{V_S}{V_P}$$

According to this equation, if the number of turns on the secondary is larger than the number of turns on the primary, then the secondary voltage is larger than the primary voltage. The opposite is also true.

Question: Why do more turns on the secondary than on the primary produce a greater secondary voltage?

Recall that when current in an inductor changes, the change in magnetic flux induces a counter voltage (CEMF) equal to the applied voltage. Thus, the applied voltage is across the primary of the transformer. If the primary has 100 turns and 120 V is applied, the change in flux must produce 1.2 V across each turn. (Recall that the turns are in series so their voltages add.) By transformer action, this same flux induces voltage in the secondary winding. Thus, the same voltage, 1.2 V, is induced in each loop of the secondary. If the secondary has 500 turns, the voltage across the entire secondary is 500 × 1.2, or 600 V. Notice that the turns ratio, N_S/N_P, and the voltage ratio, V_S/V_P, are equal and in this instance, are equal to 5. The opposite effect is true for a transformer with less turns on the secondary than on the primary.

EXAMPLE 13–3 For the transformer in Figure 13–8, compute the turns ratio. Is a step-up or step-down transformer shown?

■ Solution:

$$\text{turns ratio} = \frac{N_S}{N_P}$$

FIGURE 13–8 Transformer for Example 13–3

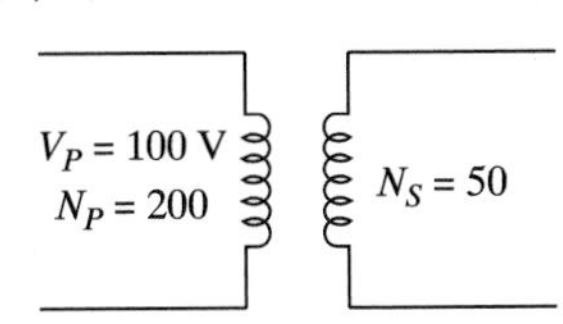

$$= \frac{900}{30} = 30$$

Since there are more turns on the secondary than on the primary, this transformer is a step-up type. The secondary voltage is 30 times greater than the primary voltage.

EXAMPLE 13–4 For the transformer in Figure 13–9, compute the value of the secondary voltage.

FIGURE 13–9 Transformer for Example 13–4

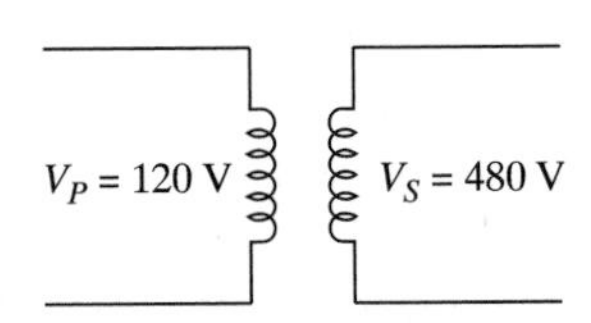

1. Use Eq. 13–3:

$$\frac{N_S}{N_P} = \frac{V_S}{V_P}$$

2. Cross multiply:

$$V_S N_P = V_P N_S$$

3. Solve:

$$V_S = \frac{V_P N_S}{N_P}$$

$$= \frac{(100 \text{ V})50}{200} = 25 \text{ V}$$

Notice in the preceding example that the transposed equation can be rewritten as:

(13–4)
$$V_S = V_P \frac{N_S}{N_P}$$

Thus, the secondary voltage is equal to the product of the primary voltage and the turns ratio.

EXAMPLE 13–5 For the transformer in Figure 13–10, compute the turns ratio.

FIGURE 13–10 Transformer for Example 13–5

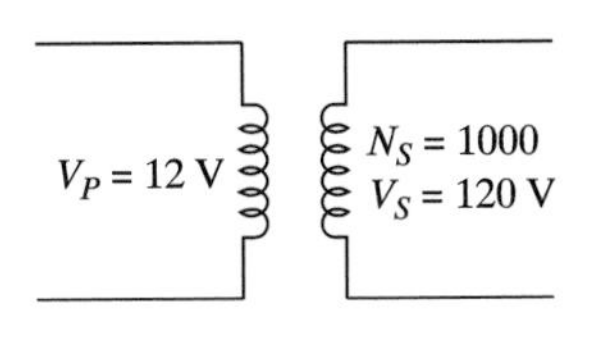

■ Solution:

The turns ratio is equal to the voltage ratio. Thus,

$$\text{turns ratio} = \frac{V_S}{V_P}$$

$$= \frac{480 \text{ V}}{120 \text{ V}} = 4$$

EXAMPLE 13–6 For the transformer in Figure 13–11, compute the number of turns on the primary.

FIGURE 13–11 Transformer for Example 13–6

$V_P = 12$ V

$N_S = 1000$

$V_S = 120$ V

■ Solution:

1. Use Eq. 13–3:

$$\frac{N_S}{N_P} = \frac{V_S}{V_P}$$

2. Cross multiply:

$$N_S V_P = N_P V_S$$

3. Solve:

$$N_p = \frac{N_S V_P}{V_S}$$

$$= \frac{1000(12\text{ V})}{120\text{ V}} = 100\text{ turns}$$

SECTION 13–3 REVIEW

1. For the transformer in Figure 13–12, compute the voltage ratio.
2. For the transformer in Figure 13–13, compute the turns ratio.
3. A transformer has a turns ratio of 4. If the primary voltage is 12 V, compute the secondary voltage.
4. A transformer has 120 V on its primary and 30 V on its secondary. If there are 40 turns on the primary, how many turns are on the secondary?

FIGURE 13–12 Transformer for Section Review Problem 1

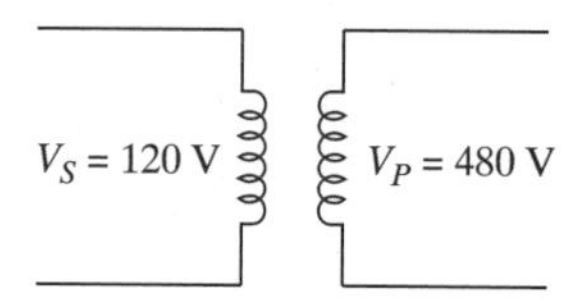

FIGURE 13–13 Transformer for Section Review Problem 2

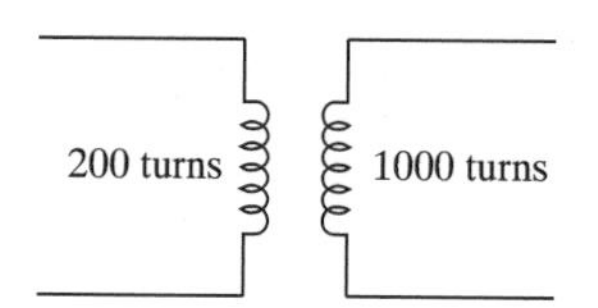

13–4 CURRENT RATIO

Before considering how current is treated from the primary to secondary of a transformer, it is necessary to understand the relationship between primary and secondary power. Recall that the electrical power developed by a circuit is equal to the product of the voltage and current. Thus, it can be seen that if the voltage is transformed, then the current must be also transformed. There is a great difference in the current transformation, however, as will be discussed in the following paragraphs.

Question: What is the relationship between the amount of power developed in the primary and secondary of a transformer?

The power developed in the primary of a transformer, P_P, is equal to the product of the primary voltage and current ($V_P I_P$). The primary power is the *input power* and the secondary power is the *output power* of the transformer. It will be shown later in this chapter that the amount of power developed in the primary is determined by the amount of power required by the load attached to the secondary. It is a fact of science, that in any system, the *total output power* can never exceed the *total input power.* Thus, the best that can be expected is that the secondary power will equal the primary power ($P_S = P_P$).

Question: How is the current ratio of a transformer related to its voltage and turns ratios?

This requirement—that the secondary power equal the primary power—causes the current ratio to be equal to the reciprocal of the voltage and turns ratios:

$$\textbf{(13–5)} \qquad \frac{N_S}{N_P} = \frac{V_S}{V_P} = \frac{I_P}{I_S}$$

The reason for this reciprocal relationship can be illustrated using the power equation:

$$P_{out} = P_{in}$$

Substituting gives

$$V_S I_S = V_P I_P$$

From this latter equation, it can be seen that if the voltage in the secondary is stepped up, then the secondary current must be stepped down by the *same* factor in order to maintain the equality. Conversely, if the voltage in the secondary is stepped down, the secondary current is stepped up by the same factor, as is illustrated in the following examples.

EXAMPLE 13–7 For the transformer in Figure 13–14, compute the current in the secondary.

■ Solution:

FIGURE 13–14 Transformer for Example 13–7

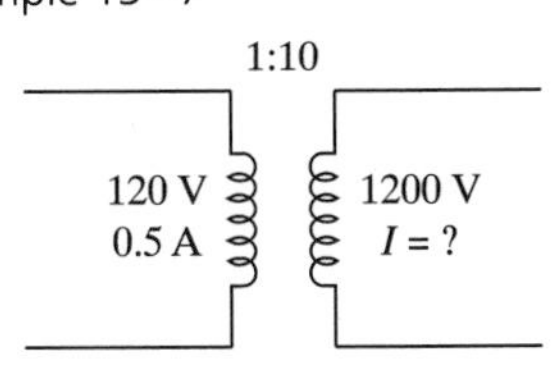

$$\frac{V_S}{V_P} = \frac{I_P}{I_S}$$

$$V_S I_S = V_P I_P$$

$$I_S = \frac{V_P I_P}{V_S}$$

$$= \frac{(120 \text{ V})(0.5 \text{ A})}{1200 \text{ V}} = 50 \text{ mA}$$

EXAMPLE 13–8 For the transformer in Figure 13–15, compute the current in the secondary.

■ Solution:

1. Compute the value of the turns ratio:

$$N = \frac{1}{5} = 0.2$$

2. Compute the value of V_S:

FIGURE 13–15 Transformer for Example 13–8

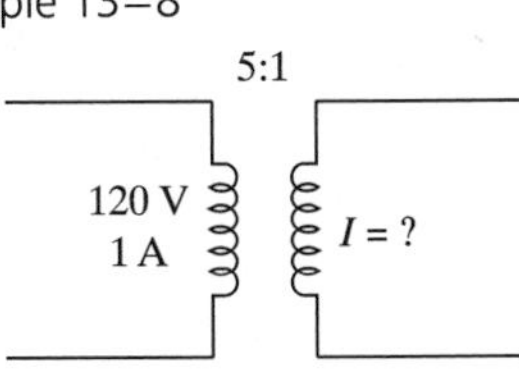

$$N = \frac{V_S}{V_P}$$

$$0.2 = \frac{V_S}{120 \text{ V}}$$

$$V_S = 0.2(120 \text{ V}) = 24 \text{ V}$$

3. Use Eq. 13–6 and solve for I_S:

$$\frac{V_S}{V_P} = \frac{I_P}{I_S}$$

$$V_S I_S = V_P I_P$$

$$I_S = \frac{V_P I_P}{V_S}$$

$$= \frac{(120\ \text{V})(1\ \text{A})}{24\ \text{V}} = 5\ \text{A}$$

SECTION 13–4 REVIEW

1. If the voltage on the secondary of a transformer is stepped up, will the secondary current be greater or less than the primary current?
2. If the current in the secondary of a transformer is greater than the current in the primary, is the transformer step up or step down?
3. Does a step-up transformer develop more power in the secondary than in the primary?

13–5 IMPEDANCE RATIO

As stated earlier, the load attached to the secondary of a transformer has a direct effect on the current in the primary. Consider first the unloaded transformer of Figure 13–16. No current flows in the secondary, so the secondary power P_S is zero. Since the secondary power is zero, no power need be developed in the primary. Under these conditions, then, it would seem that no primary current would flow. However, a small primary current does flow, opposed by the inductive reactance of the primary winding. This current produces a counter voltage, CEMF, that opposes and is approximately equal to the source voltage. Thus, the high impedance of the secondary is "seen" in the primary.

FIGURE 13–16 Unloaded transformer, with a small energizing current flowing in the primary winding

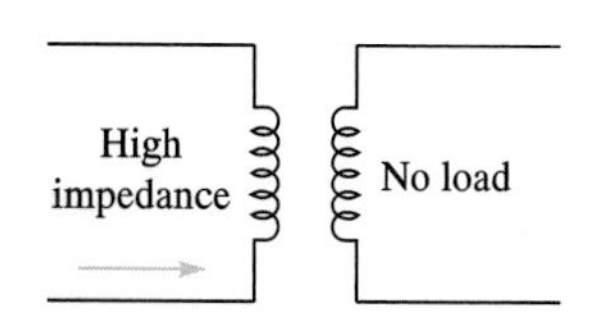

When a load is attached as shown in Figure 13–17, current flows in the secondary. The secondary now produces a magnetic field that is opposite in polarity to the field of the primary. Thus, the voltage that is self-induced in the secondary opposes the CEMF of the primary and allows the primary current to rise. This increase in primary current continues until the CEMF produced by it equals approximately the source voltage. Note carefully what happens. The primary current increases, and the higher power demand of the secondary is met. This increase in primary current indicates that the primary impedance has decreased. It is said that the impedance of the secondary has been "reflected" into the primary. So, not only are voltage and current transformed in a transformer, impedance is also transformed.

FIGURE 13–17 Loaded transformer

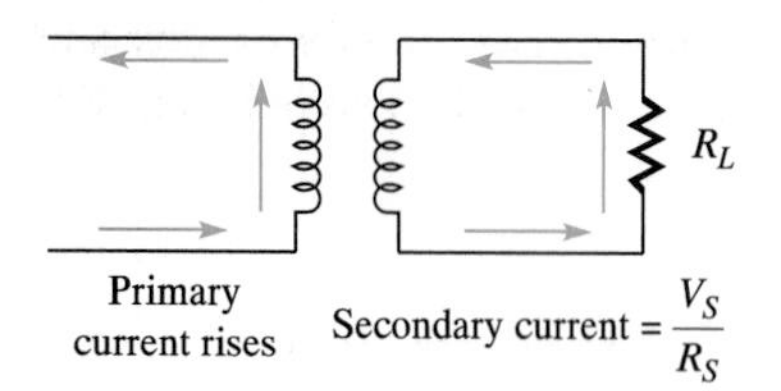

Question: How is the reflected impedance computed?

The impedance reflected into the primary of a transformer is transformed—stepped up or stepped down—by a factor proportional to the reciprocal of the turns ratio, as the following mathematical derivation demonstrates.

1. Computing secondary impedance gives

$$Z_S = \frac{V_S}{I_S}$$

2. Computing primary impedance gives

$$Z_P = \frac{V_P}{I_P}$$

3. Computing the transformation ratio gives

$$\frac{Z_S}{Z_P} = \frac{\frac{V_S}{I_S}}{\frac{V_P}{I_P}}$$

4. Solving yields

$$\frac{Z_S}{Z_P} = \frac{V_S I_S}{I_P V_P}$$

and

$$\frac{Z_S}{Z_P} = \frac{V_S I_P}{V_P I_S}$$

but

$$\frac{V_S}{V_P} = \frac{N_S}{N_P} \quad \text{and} \quad \frac{I_P}{I_S} = \frac{N_S}{N_P}$$

5. Substituting gives

$$\frac{Z_S}{Z_P} = \frac{N_S N_S}{N_P N_P}$$

and

(13–7)

$$\frac{Z_S}{Z_P} = \left(\frac{N_S}{N_P}\right)^2$$

or

(13–8)

$$\frac{Z_P}{Z_S} = \left(\frac{1}{N}\right)^2$$

Eq. 13–7 and 13–8 show that the impedance ratio is equal to the *square of the reciprocal of the turns ratio*. An equation for the reflected impedance Z_P can be derived from Eq. 13–8:

(13–9)

$$Z_P = \left(\frac{1}{N}\right)^2 Z_S$$

EXAMPLE 13–9 For the transformer in Figure 13–18, compute the impedance reflected into the primary.

FIGURE 13–18 Transformer for Example 13–9

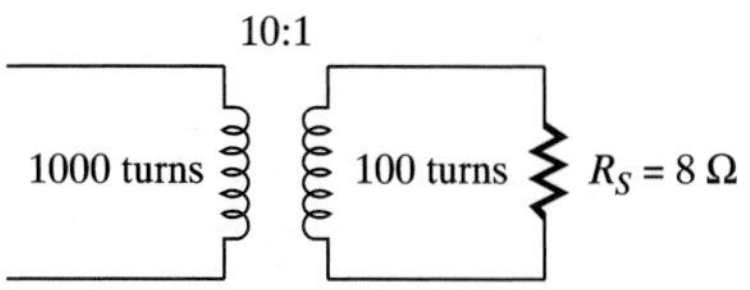

10:1

1000 turns 100 turns $R_S = 8\ \Omega$

■ **Solution:**

$$Z_P = \left(\frac{1}{N}\right)^2 Z_S$$

$$= \left(\frac{1000}{100}\right)^2 (8\ \Omega)$$

$$= 100(8\ \Omega) = 800\ \Omega$$

EXAMPLE 13–10

For the transformer in Figure 13–19, compute the turns ratio that will cause 6.25 Ω to be reflected into the primary.

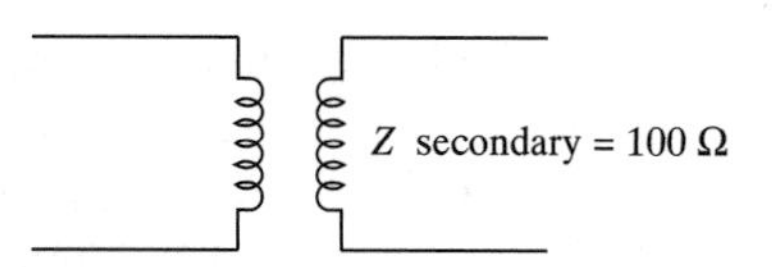

FIGURE 13–19 Circuit for Example 13–10

■ Solution:

1. Use Eq. 13–9:

$$Z_P = \left(\frac{1}{N}\right)^2 Z_S$$

2. Transpose and solve:

$$\left(\frac{1}{N}\right)^2 = \frac{Z_P}{Z_S} = \frac{6.25\ \Omega}{100\ \Omega} = 0.0625$$

$$\frac{1}{N} = \sqrt{0.0625} = 0.25$$

3. The reciprocal of the turns ratio is 0.25. The turns ratio is the reciprocal of the following equation:

$$N = \frac{1}{1/N} = \frac{1}{0.25} = 4$$

Thus, for a 100 Ω load to reflect 6.25 Ω into the primary requires a 1:4 step-up transformer.

■ SECTION 13–5 REVIEW

1. Is the impedance reflected by a step-down transformer larger or smaller than the load impedance?
2. What is the impedance reflected into the primary from a 300 Ω load if the turns ratio of the transformer is 5?
3. What turns ratio will cause 1.5 kΩ to be reflected into the primary from a 4 Ω load?

13–6 TRANSFORMER POWER LOSSES

Question: Will all output power of a transformer be useful?

The maximum total power produced at the output of a transformer is equal to the total input power, as was established for the transformer earlier in this chapter. The statement is simple and true, but it does not tell the whole story. Some of the power produced in the output is not useful power. You may have noticed, that after a period of operation, an electronic device feels warm to the touch. This heat, produced by the resistors, transistors, transformers, and so on that make up the system, represents power produced and lost in the process of operation. The useful power is the total output power minus that lost in the form of heat. The concept of efficiency was introduced in Chapter 3. As a percentage, it is equal to the ratio of the useful output power to the total input power times 100 —

$(P_{out}/P_{in})100$. It is always most desirable to have efficiency as close to 100% as possible. The factors that produce power loss and thus lower efficiency in transformer are discussed in the paragraphs that follow.

Winding Loss

Power is developed whenever current flows through a resistance. **Winding loss** of power occurs as current flows through the resistance of the windings and develops some power in the form of heat. This loss is often referred to as an I^2R (*I-squared R*) loss, or *copper loss,* and is directly proportional to the amount of current flow. Thus, the power loss increases as the current increases. The winding resistance of power transformers is usually quite small, so this type of loss does not cause a great drop in transformer efficiency.

Hysteresis Loss

Magnetic hysteresis is an effect in transformers that have an iron core. Recall that when magnetic polarity changes, the magnetic domains within the material must be realigned. Energy is expended in this process, and some heat is produced due to molecular stress and friction. **Hysteresis loss** is thus the power lost in reversing the magnetic domains within the iron core of the transformer. For this reason, iron core transformers are avoided in high frequency applications. Hysteresis loss can be minimized by constructing the core of highly permeable materials with low retentivity.

Eddy Current Loss

The iron core is also where eddy current losses take place. A solid iron core acts much like a large secondary winding with a very small resistance attached. Thus, the same magnetic induction inducing voltage in the secondary induces currents in the core. These currents are known as eddy currents. Figure 13–20(a) illustrates the lines of magnetic flux and the eddy currents that they produce. Notice that the eddy currents flow in a plane perpendicular to the direction of the flux lines. **Eddy current loss** of power results as these currents develop power in the form of heat and reduce the efficiency of the transformer.

Question: How are eddy currents reduced or eliminated?

One way of eliminating, or at least reducing, eddy currents is through use of a *laminated* iron core. As Figure 13–20(b) illustrates, the laminated core is composed of thin sheets of iron separated by an insulating·material. The insulation may be either applied as a varnish or formed as an oxide layer. In either case, the laminations are insulated from one another, and no organized eddy currents can develop. Those that do develop will be very small, and the power losses they produce will be minimal. Another way of reducing eddy currents is through the use of ferrite cores, which were introduced in chapter 9. Eddy currents are greatly reduced by a ferrite core construction consisting of many tiny specks of iron held together by a ceramic insulating material. Ferrite core transformers are used in high frequency applications, such as radio and television receivers. The symbols for iron and ferrite core transformers are shown in Figure 13–21.

SECTION 13–6 REVIEW

1. What causes power loss in the windings of a transformer?
2. What two losses occur in the iron core of a transformer?
3. How are eddy current losses reduced?

FIGURE 13–20 Eddy current

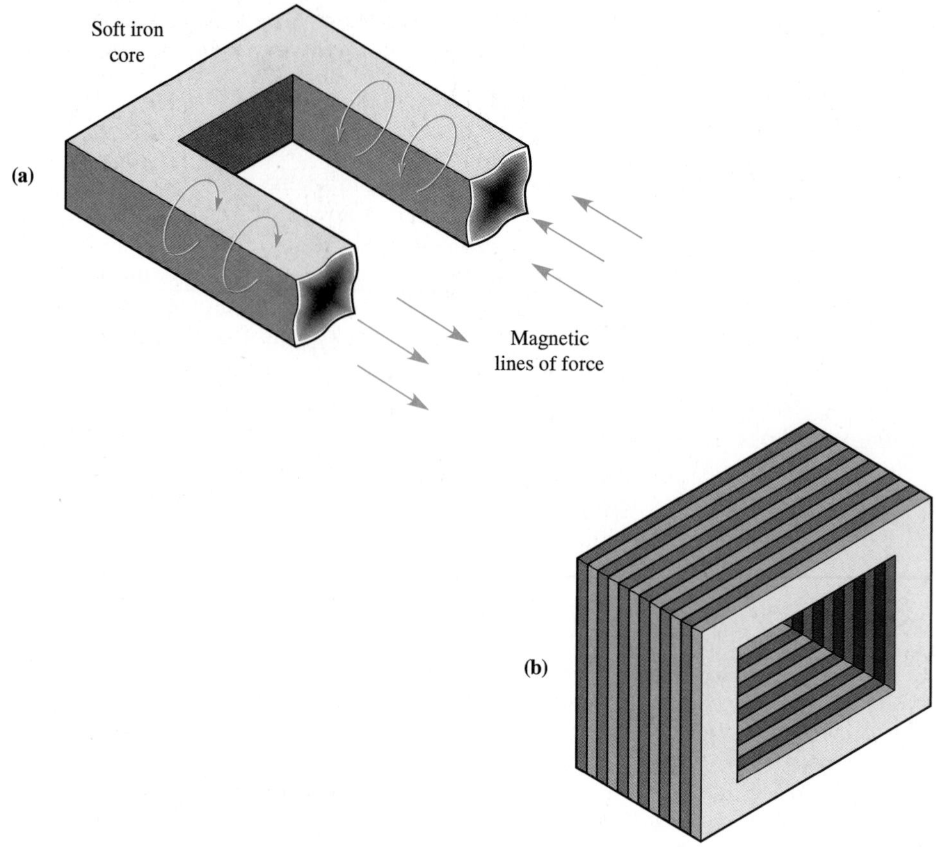

Soft iron core

Magnetic lines of force

(a)

(b)

FIGURE 13–21 Transformer core symbols

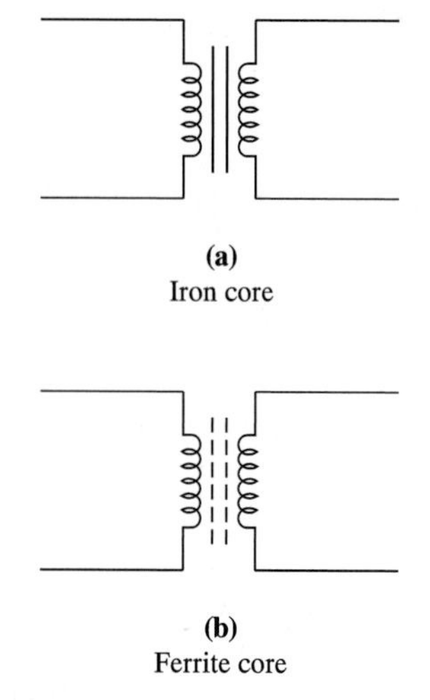

(a) Iron core

(b) Ferrite core

13–7 TRANSFORMER RATINGS

■ Transformer Power Rating

Question: What ratings are specified for a transformer?

The ratings of a typical transformer are its power rating, primary/secondary voltage ratio, and operating frequency. For example, these may be respectively 2 kVA, 400/40 V, and 60 Hz. Notice that the power rating of the transformer is the *apparent power* rating. The voltages represent the voltage transformation ratio, where the 400 V side is the primary and the 40 V side is the secondary. The current ratings can thus be determined:

$$I_S = \frac{P}{V_S}$$

$$= \frac{2\text{ kVA}}{40\text{ V}} = 50\text{ A}$$

Thus, the current rating of the secondary is 50 A. A power transformer can be reversed, however, such that the primary becomes the secondary and the secondary becomes the primary. Thus, the primary voltage could be 40 V and the secondary 400. Now the current rating is

$$I_S = \frac{P}{V_S}$$

$$\frac{2\text{ kVA}}{400\text{ V}} = 5\text{ A}$$

Under these conditions, for a step-up transformer, the current rating would be only 5 A. Care must be taken that these current ratings are not exceeded.

Question: Why is the power rating of a transformer expressed as apparent power?

In specifying the power rating, apparent power (P_{app}) is used rather than true power (P_{true}). The reason is that the load may be purely inductive, in which case no true power in watts develops in the load. For example, consider a transformer with ratings such as those in the previous paragraph (2 kVA, 400/40 V, and 60 Hz). If the load had a reactance of 40 Ω, even though no power is developed, the current would still be 10 A. This amount is double the maximum rating, and the transformer would be damaged! Thus, a true power rating is meaningless.

■ Efficiency

Recall that the percent of efficiency of a system is the ratio of the output power to the input power times 100 or

$$\% \text{ efficiency} = \frac{P_{out}}{P_{in}}100$$

As you have seen, a practical transformer has some power losses so the efficiency is less than 100%. A good power transformer, however, has an efficiency in excess of 95%.

■ SECTION 13–7 REVIEW

1. A transformer has a power rating of 1 kVA. If the secondary voltage is 50 V, what is the maximum current rating?
2. The input power to a transformer is 500 W. If the power developed in the load is 470 W, what is the percent of efficiency?
3. Why is the power rating of a transformer stated as apparent power?

13–8 TRANSFORMER APPLICATIONS

■ Power Transformer

The primary of a power transformer connects to the power line. In Figure 13–22(a), the power transformer has a single winding supplying a voltage, either larger or smaller than the source, to a load. This type of transformer always has a laminated iron core in order to provide *near-unity coupling*, or a k factor of 1. A power transformer may have what is known as a *center tap* on its secondary in order to develop equal voltages of opposite phase, as illustrated in Figure 13–22(b). The entire secondary has 480 V ac across it. Between either side and the center tap, half the secondary voltage, or 240 V, will be felt. Notice that these two voltages are 180° out of phase.

Figure 13–23 shows a power transformer with multiple secondaries. Each secondary has its own turns ratio and thus its own voltage value. In this way, a variety of load voltages can be supplied from a single transformer. Figure 13–24 shows some common examples of practical power transformers. A *shell-type transformer* is shown in Figure 13–24(a). The shell, which is made of soft iron, provides a low-reluctance path for flux, thus keeping the magnetic field inside the transformer. This type of transformer is used in equipment where circuits may be adversely affected by external magnetic fields. The *open-core transformer* is shown in Figure 13–24(b). It has no shield and is used in applications where external magnetic fields are not a problem.

FIGURE 13–22 Power transformer with secondary single winding

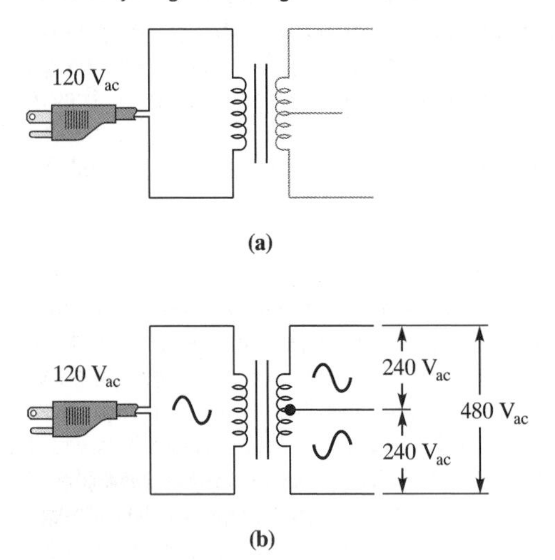

FIGURE 13–23 Transformer with multiple secondary windings

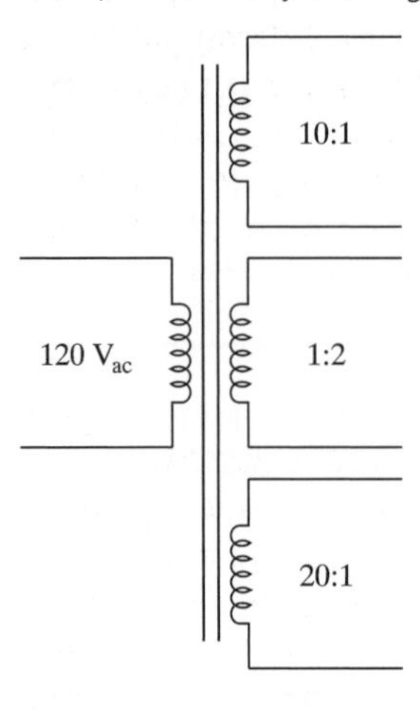

FIGURE 13–24 Practical power transformers

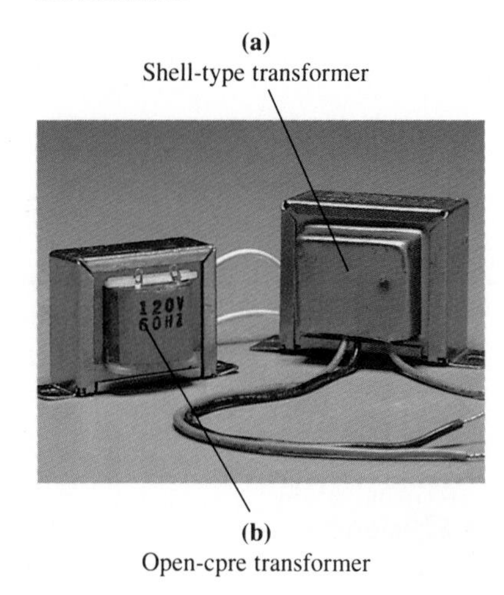

EXAMPLE 13–11 For the power transformer shown in Figure 13–25, compute the voltage of each secondary between the points shown.

■ Solution:

FIGURE 13–25 Transformer for Example 13–11

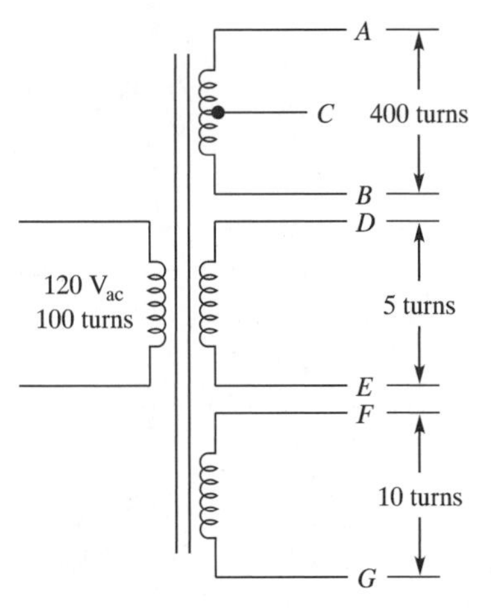

$$V_{AB} = \frac{N_S}{N_P}V_P$$

$$= \frac{400}{100}(120 \text{ V}) = 480 \text{ V}$$

$$V_{AC} = \frac{N_S}{N_P}V_P$$

$$= \frac{200}{100}(120 \text{ V}) = 240 \text{ V}$$

$$V_{CB} = \frac{N_S}{N_P}V_P$$

$$= \frac{200}{100}(120 \text{ V}) = 240 \text{ V}$$

$$V_{DE} = \frac{N_S}{N_P}V_P$$

$$= \frac{5}{100}(120 \text{ V}) = 6 \text{ V}$$

$$V_{FG} = \frac{N_S}{N_P}V_P$$

$$= \frac{10}{100}(120 \text{ V}) = 12 \text{ V}$$

■ Variable Transformers

Figure 13–26(a) shows a power transformer with connections known as taps on its primary. The secondary voltage can be varied by the placement of the top connector of the primary. If moved downward to a lower tap, the number of turns on the primary is reduced. Reducing the primary turns increases the turns ratio (N_S/N_P). Thus, the voltage across the secondary is increased. Moving the tap upward has the opposite effect. Figure 13–26(b) shows a power transformer with taps on its secondary. If the tap is moved downward, the number of turns on the secondary is reduced as is the turns ratio and the voltage across the secondary.

■ The Autotransformer

An autotransformer has only one winding that serves as both the primary and the secondary. As shown in Figure 13–27, one connection is common to both the primary and the secondary. Thus, in the *autotransformer,* there is no isolation between the primary and secondary. The taps are placed as shown in order to achieve either step-up or step-down operation. If the input voltage is between pins 1 and 3 and the output is taken between 2 and 3, the autotransformer steps down the voltage. If the input voltage is between pins 2 and 3 and the output is taken between 1 and 3, the autotransformer steps up the voltage. Autotransformers are more compact, more efficient, and cheaper than transformers with more than one coil. The wire size, however, must be capable of handling both the primary and the secondary current.

■ dc Isolation

As shown in Figure 13–28(a), if steady-state direct current flows in the primary of a transformer, no voltage is induced in the secondary. Steady-state dc does not produce the moving magnetic field required for electromagnetic induction. In Figure 13–28(b), a

FIGURE 13–26 Tapped transformers

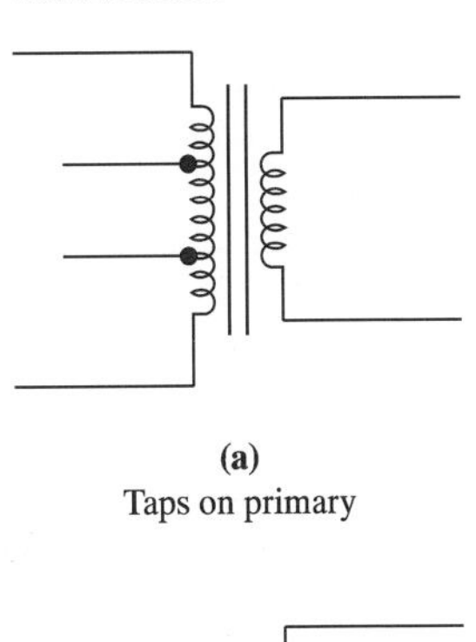

(a)
Taps on primary

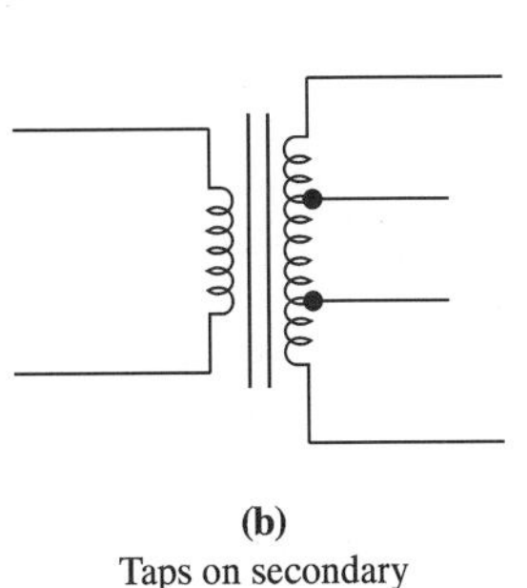

(b)
Taps on secondary

FIGURE 13–27 Autotransformer

Pin 1

Pin 2

Pin 3

FIGURE 13–28 Direct current in the transformer

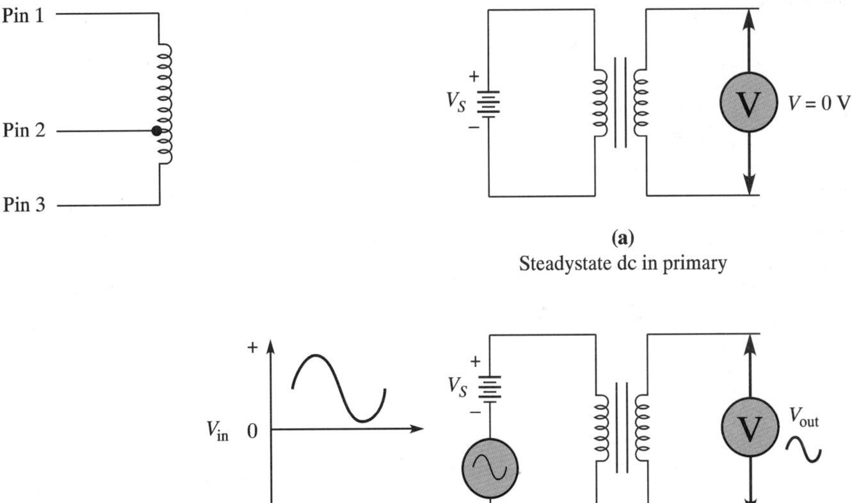

(a)
Steadystate dc in primary

(a)
Varying dc in primary

FIGURE 13–29 Power line connections

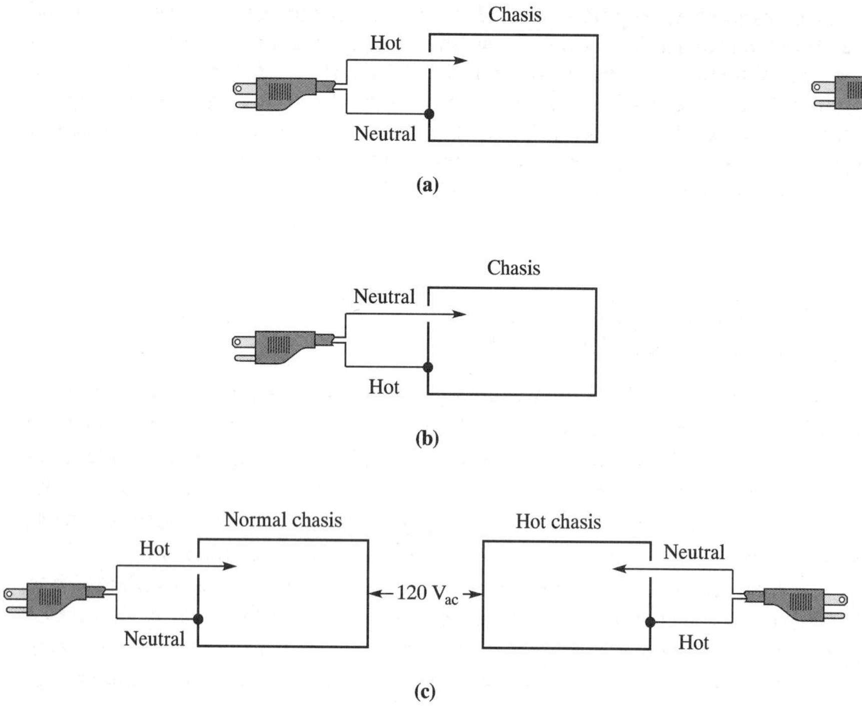

(a)

(b)

(c)

FIGURE 13–30 Power line isolation

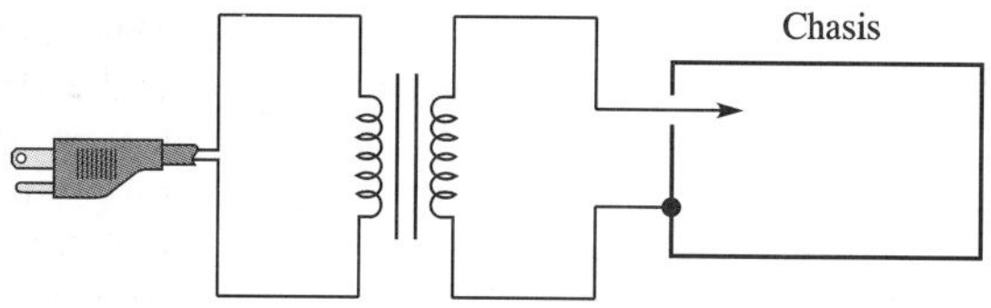

(a)
Isolation circuit

(b)
Isolation transformer

varying direct current is flowing in the primary. This direct current is said to have an ac component riding on it. It is similar to the output signal of an electronic amplifier. The varying portion of the signal induces voltage in the secondary while the dc portion does not. In this instance, the ac component is passed while the dc component is blocked.

■ Power Line Isolation

Question: Why is it good safety practice to isolate an electronic circuit from the power line?

As shown in Figure 13–29(a), in high voltage circuits, the power line ground (neutral, N) is sometimes connected to the metal chassis upon which the electronic system is built. The 120 V line (hot, H) goes directly to the circuit that converts the ac to dc. This setup is perfectly safe so long as the power plug is connected to the proper points in the receptacle. If the plug should be reversed, however, the chassis then becomes "hot" as shown in Figure 13–29(c). If another piece of equipment nearby is properly connected, then the two have a potential of 120 V between them, which is definitely not a safe or desirable situation!

If, however, the power line is brought into the equipment through a transformer, the equipment is *isolated* from the power line, as illustrated in Figure 13–30(a). Now, no matter which way the plug is connected into the receptacle, the chassis does not become "hot." Note that if the equipment has a transformer input, it is isolated. If not, it is good safety practice to use an external transformer known as an *isolation transformer*. An isolation transformer is pictured in Figure 13–30(b).

CAUTION

Many times variable transformers are used at electronic repair stations. These are useful to the service technician in that they allow an accurate input voltage to be selected for testing and calibrating equipment. Many of these variable transformers are of the auto-transformer type and as such *do not* provide dc isolation! Use caution when this type of transformer is used. Always check for the presence of voltage between the equipment being serviced and test equipment with a battery-operated voltmeter if there is any question as to whether proper isolation has been provided.

Impedance Matching

Question: Why is it necessary that the impedance of the load match the impedance of the source?

In order to transfer maximum power from a source to a load, the load and source impedances must be equal. This fact is the basis of the maximum transfer of power theorem introduced in Chapter 7. Most load impedances are small. For example, a speaker may have an impedance as low as 4 Ω or 8 Ω. The final amplifier, which is the source for the load, may have an impedance of 1000 Ω or higher. Thus, if hooked directly one to another, a *mismatch* results, and maximum power is not delivered to the load.

In a previous paragraph, it was shown that the secondary impedance is reflected into the primary. You also learned that this impedance was not reflected directly, but transformed by a factor equal to the square of the reciprocal of the turns ratio. Thus, by using the appropriate turns ratio, the load impedance can be made to appear to be equal to the source impedance, as the following example illustrates.

EXAMPLE 13–12 What turns ratio will match an amplifier with an output resistance of 1600 Ω to a 4 Ω speaker?

FIGURE 13–31 Circuit for Example 13–12

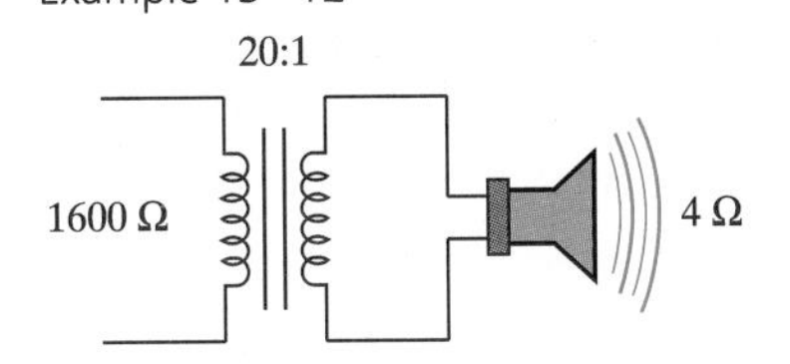

■ **Solution:**

$$\left(\frac{1}{N}\right)^2 = \frac{Z_P}{Z_S}$$

$$= \frac{1600\ \Omega}{4\ \Omega} = 400\ \Omega$$

Thus,

$$\left(\frac{1}{N}\right) = \sqrt{400} = 20$$

and

$$N = 0.05$$

The turns ratio is less than 1, which means the transformer is a step-down type. What this number means is that for every 20 turns on the primary, there is 1 turn on the secondary. This circuit is shown in Figure 13–31.

EXAMPLE 13–13 The cable that brings CATV signals into the home for television viewing has an impedance of 75 Ω. The television receiver to which it must attach has an input impedance at 300 Ω. Modern television receivers come equipped with a matching device called a "ballun" transformer whose access point is labeled "75 Ω." Since this transformer is impedance matching, what will be its turns ratio?

■ Solution:

$$\left(\frac{1}{N}\right)^2 = \frac{Z_P}{Z_S}$$

$$= \frac{75\ \Omega}{300\ \Omega} = 0.25$$

$$N^2 = \frac{1}{0.25} = 4$$

$$N = \sqrt{4} = 2$$

This transformer is a step-up type, with a turn ratio of 2.

■ SECTION 13–8 REVIEW

1. What two functions of a transformer may be made use of in a power transformer?
2. Why is it good safety practice to isolate electronic equipment from the power line?
3. Why may impedance matching be necessary between electronic circuits?
4. How are multiple voltages obtained from a single transformer?
5. Compute the turns ratio of a transformer that will match a 10 Ω load to a 60 Ω source.

13–9 TROUBLESHOOTING TRANSFORMERS

A transformer has four basic purposes: (1) to step up voltage, (2) to step down voltage, (3) to match impedances, and (4) to isolate one circuit from another. All but the last depends upon the turns ratio of the transformer. Thus, anything causing a change in the turns ratio has an effect on transformer operation. The most likely problems to occur in a transformer are shorts or opens.

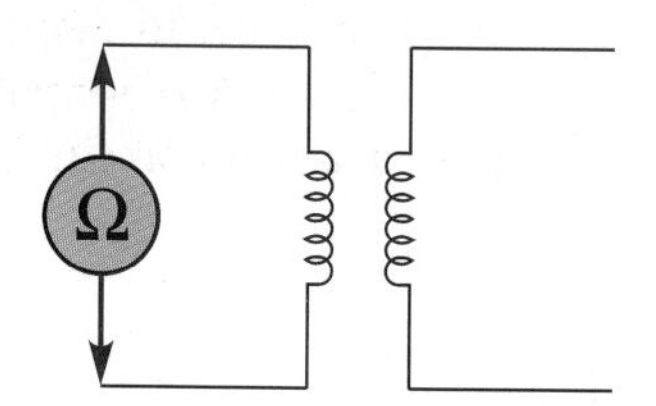

FIGURE 13–32 Detecting a shorted primary with an ohmmeter: The ohmmeter will read zero across a direct short.

Question: What instruments are used in troubleshooting of transformers?

The instruments used in troubleshooting of transformers are the ohmmeter and the impedance bridge. The service instructions often give the dc resistance of each winding of a transformer. Thus, the ohmmeter can be used to measure these resistance values, which can then be compared to the standard values. With an impedance bridge, the inductance of each winding can be measured. Once again, these values can then be compared to those contained in the service instructions.

■ Troubleshooting Shorts

Question: What are the results of a direct short across the primary or secondary of a transformer?

A direct short across the primary of a transformer removes all the inductance. Thus, with only the minute resistance of the short to oppose current, the primary current greatly increases, and the protection device, either a fuse or circuit breaker, will operate in order to protect the voltage source. A shorted primary can be detected using either an ohmmeter, as shown in Figure 13–32, or impedance bridge. The ohmmeter will read nearly zero across a direct short. Remember, though, that the primary resistance of a power transformer may be as low as one-half of an ohm. Thus, the ohmmeter must be on a scale low enough to indicate such a small value. On a higher scale, such as $R \times 100\ \Omega$, it could indicate a short across a perfectly good transformer. An impedance bridge can also be used to test the primary. If shorted, the impedance of the primary will be practically zero. A transformer with a shorted primary winding requires replacement.

A short across the secondary of a transformer, either in the winding or load, will be reflected into the primary. The primary current increases greatly, and once again the protection device will operate. If the protection device does not operate, the heat produced by the high current could destroy the primary. If disconnecting the load removes the short, then the transformer is probably not defective. If the short persists when the load is removed, the secondary winding is at fault, and the transformer must be replaced. The testing procedure, as shown in Figure 13–33, is similar to that for the primary.

FIGURE 13–33 Detecting a shorted secondary with an ohmmeter

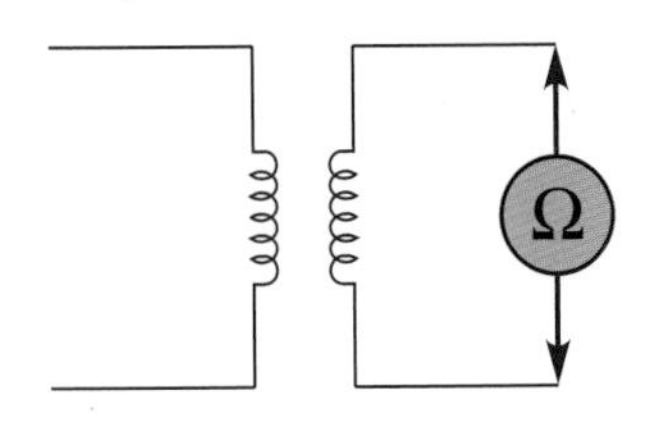

FIGURE 13–34 Partially shorted windings

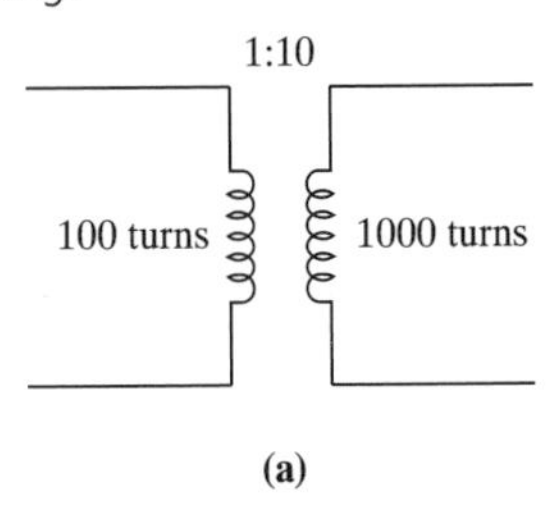

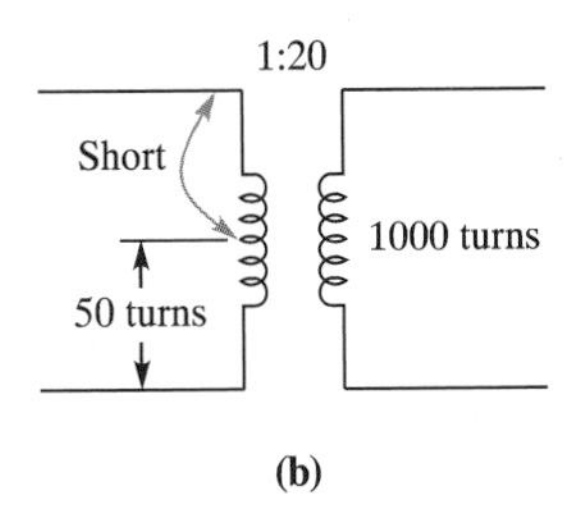

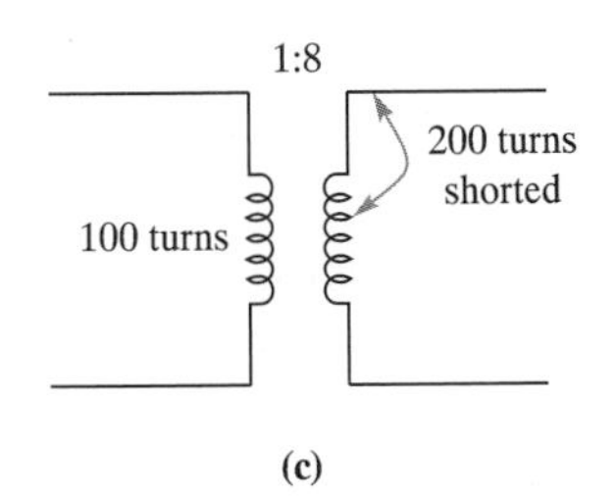

Question: What are the results of a partial short in the primary or secondary of a transformer?

A partial short means some, but not all, of the windings have shorted. A loss of windings on the primary or secondary results in a change in the turns ratio. For example, as shown in Figure 13–34, if 50 turns of the primary short, the turns ratio is no longer 10 but 20. Thus, the secondary voltage would be twice its designed value. If, on the other hand, 200 turns of the secondary were to short, the turns ratio would become 8 instead of 10. The secondary voltage would not be only 80% of its designed value. These conditions of shorted windings could be detected using an ohmmeter, but it would be necessary to know the resistance value of the winding. The same is true of using an impedance bridge; the inductance value of the windings must be known.

Question: What are the results of opens in the primary or secondary of a transformer?

If an open occurs in the primary winding, no energizing current flows and thus no magnetic field is created. No voltage is induced in the secondary under these conditions. This problem can be detected using a voltmeter, as shown in Figure 13–35. As illustrated in Figure 13–36(a), no voltage is induced in an open secondary. In a multiple secondary transformer, such as shown in Figure 13–36(b), only the open secondary has no induced voltage; the voltages on the others are normal. The same is also true of a center-tapped secondary. As shown in Figure 13–36(c), the voltage between points A and B is normal, while the voltage from points B to C is zero. Testing for an open secondary is done in the same manner as for the primary.

FIGURE 13–35 Open in primary

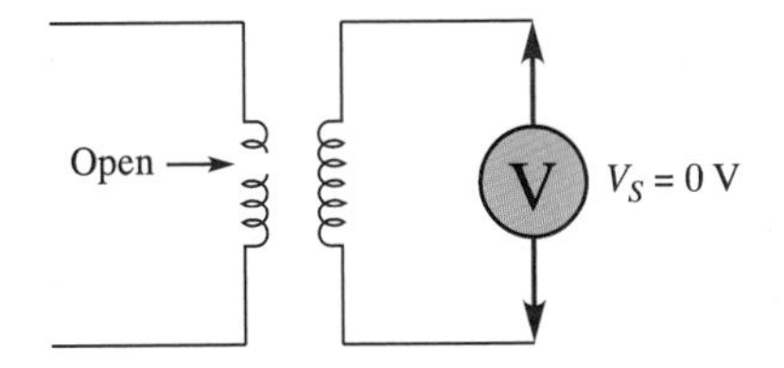

SECTION 13–8 REVIEW

1. What two types of faults are normally encountered in a transformer?
2. What are the results of a shorted primary?
3. What are the results of a shorted secondary?
4. Will an open secondary necessarily mean that there will be no output voltage? Explain.

FIGURE 13–36 Opens in secondary

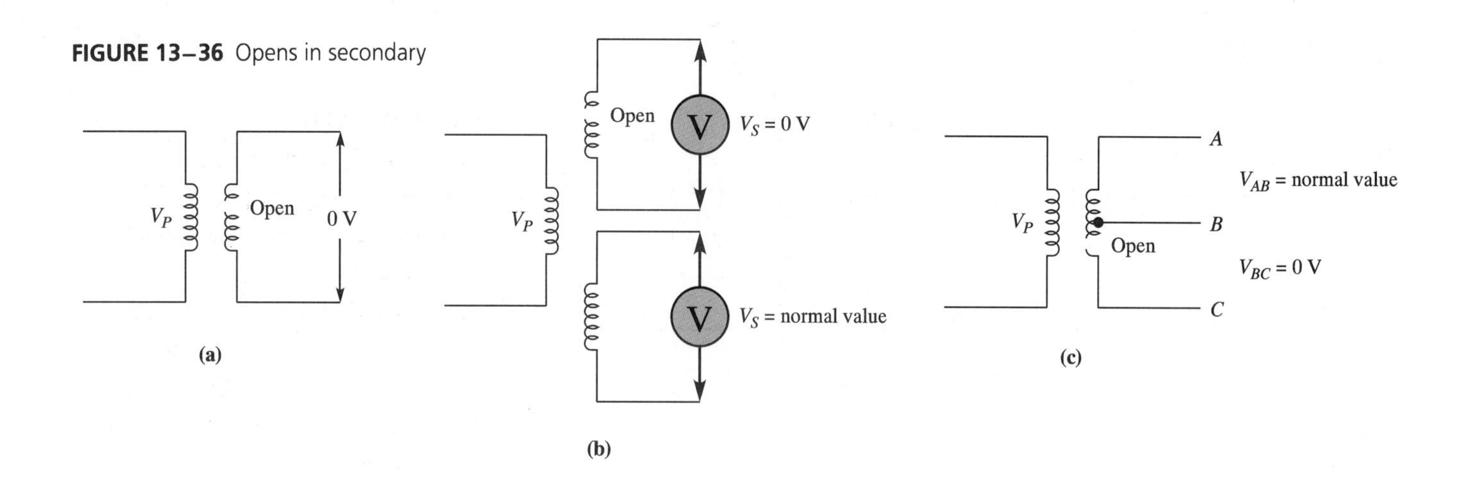

CHAPTER SUMMARY

1. A transformer transfers electrical energy from one circuit to another through magnetic linkage.
2. Mutual inductance exists between two inductors wound in close proximity one to another.
3. The coefficient of coupling k is an indication of how closely the two inductors are coupled.
4. The primary winding connects to the source voltage, and the secondary winding connects to the load.
5. The voltage induced in the secondary has the same frequency as that in the primary.
6. The turns ratio is the ratio of the number of turns on the secondary to the number of turns on the primary.
7. The voltage ratio is the ratio of the secondary voltage to the primary voltage and is equal to the turns ratio.
8. If the efficiency of a transformer is 100%, the power developed in the secondary equals the power developed in the primary.
9. The current ratio is the ratio of the current in the primary to the current in the secondary and is equal to the reciprocal of the turns ratio.
10. The impedance ratio is the ratio of the impedance of the primary to the impedance of the secondary and is equal to the square of the reciprocal the turns ratio.
11. Transformer losses are the winding loss, hysteresis loss, and eddy current loss.
12. Iron cores in power transformers are laminated in order to reduce eddy current losses.
13. Transformers are used to step up voltage, step down voltage, isolate one circuit from another, and match the impedances of two circuits.
14. A shorted secondary reflects a short into the primary, producing a high primary current that causes the current protection device to operate.

KEY TERMS

mutual inductance
transformer
coefficient of coupling
dot convention
transformation ratio
turns ratio
winding loss
hysteresis loss
eddy current loss
autotransformer

EQUATIONS

(13–1) $L_M = k\sqrt{L_1 L_2}$

(13–2) $k = \dfrac{\text{number of lines linking } L_1 \text{ and } L_2}{\text{number of lines produced in } L}$

(13–3) $\dfrac{N_S}{N_P} = \dfrac{V_S}{V_P}$

(13–4) $V_S = V_P \dfrac{N_S}{N_P}$

(13–5) $N_P = \dfrac{N_S V_P}{V_S}$

(13–6) $\dfrac{N_S}{N_P} = \dfrac{V_S}{V_P} = \dfrac{I_P}{I_S}$

(13–7) $\dfrac{Z_S}{Z_P} = \left(\dfrac{N_S}{N_P}\right)^2$

(13–8) $\dfrac{Z_P}{Z_S} = \left(\dfrac{1}{N}\right)^2$

(13–9) $Z_P = \left(\dfrac{1}{N}\right)^2 Z_S$

Variable Quantities

L_M = mutual inductance

k = coefficient of coupling

N_S = number of turns of the secondary winding

N_P = number of turns of the primary winding

Z_S = impedance of the secondary circuit

Z_P = impedance of the primary circuit

N = turns ratio

TEST YOUR KNOWLEDGE

1. What four functions can a transformer perform?
2. What is the range of values for the coefficient of coupling?
3. What is the basic transformer composed of?
4. What is the purpose of the iron core in a power transformer?
5. Is a transformer with a turns ratio of 10 a step-up type or step-down type?
6. Is a transformer with a voltage ratio of 0.2 a step-up type or step-down type?
7. Is a transformer with a current ratio of 0.5 a step-up type or step-down type?
8. A transformer has a turns ratio of 5. What is its voltage ratio?
9. A transformer has a turns ratio of 10. What is its current ratio?
10. A transformer with a turns ratio of 0.166 is used in impedance matching. Which is larger, the load or source impedance?
11. What causes power loss in the windings of a transformer?
12. How are eddy currents reduced in a transformer?
13. A transformer has 240 V across its entire secondary. What is the voltage from either side to the center tap?
14. Why is it impossible for a transformer to pass steady-state dc voltage to the secondary?
15. Give an example in which impedance matching is necessary.
16. Why will a shorted secondary reflect a short into the primary?
17. What are the results of a partial short in the primary or secondary of a transformer?

PROBLEM SET: BASIC LEVEL

Sections 13–1 and 13–2

1. The primary of a power transformer produces 500 lines of flux. If 475 lines link the secondary compute the coupling factor k.
2. The primary of a transformer produces 800 lines of flux. If the coupling factor k is 85%, how many lines link the secondary?
3. For the transformer of Figure 13–37, compute the value of the mutual inductance.

FIGURE 13–37 Transformer for Problem 3

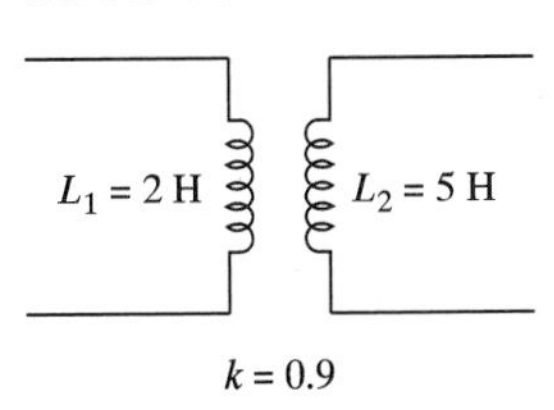

Section 13–3

4. The frequency of the current through the primary of a transformer is 400 Hz. What is the frequency of the secondary current?

5. For the transformer as shown in Figure 13–38, will points A and B have the same phase?

FIGURE 13–38 Transformer for Problem 5

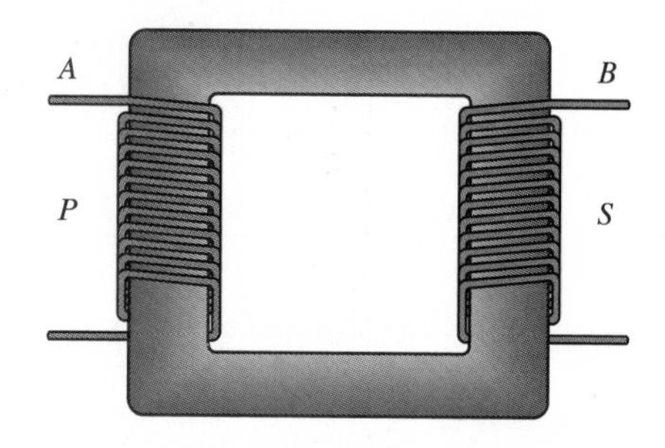

Section 13–4

6. A transformer has 40 turns on the primary and 1600 turns on the secondary. Compute the turns ratio.
7. Is the transformer in Problem 6 step up or step down?
8. A transformer has 80 turns on the secondary and 100 turns on the primary. Compute the turns ratio.
9. Is the transformer in Problem 8 step up or step down?
10. A transformer has 100 turns on its primary and 10 turns on its secondary. If 120 is applied to the primary, what is the value of the secondary voltage?
11. A transformer has a turns ratio of 4. How much voltage must be applied to the primary to produce 120 V in the secondary?
12. A transformer has 50 turns on its primary. How many turns must there be on the secondary in order to quadruple the voltage?

Section 13–5

13. A transformer has a turns ratio of 0.5. If the secondary current is 5 A, what is the value of the primary current?
14. If the turns ratio of a transformer is 0.4, what is the current ratio?
15. For the circuit of Figure 13–39, determine the following: (a) secondary voltage, (b) secondary current, (c) primary current, and (d) power developed in the load.

FIGURE 13–39 Circuit for Problem 15

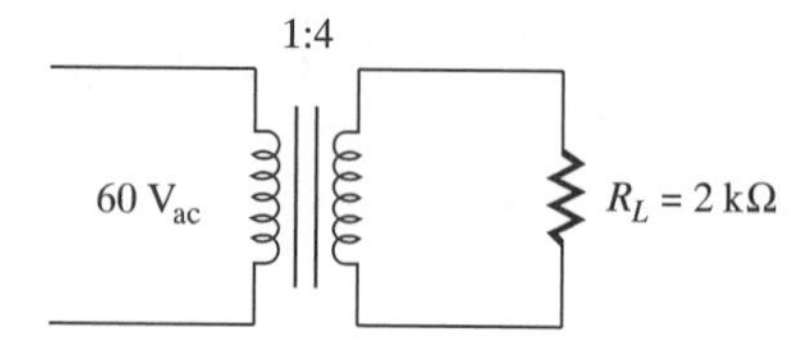

Section 13–6

16. For the circuit of Figure 13–40, what resistance is reflected into the primary?

FIGURE 13–40 Circuit for Problem 16

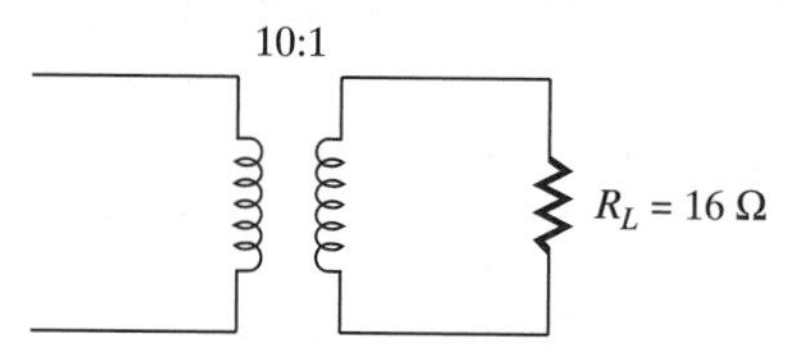

17. For the circuit of Figure 13–41, what must the turns ratio be in order to reflect 100 Ω into the primary?

FIGURE 13–41 Circuit for Problem 17

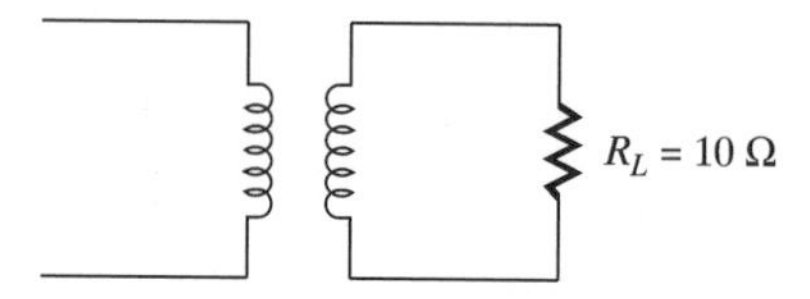

Sections 13–7 and 13–8

18. If the transformer of Figure 13–42 is 80% efficient, how much power is developed in the primary?

FIGURE 13–42 Circuit for Problem 18

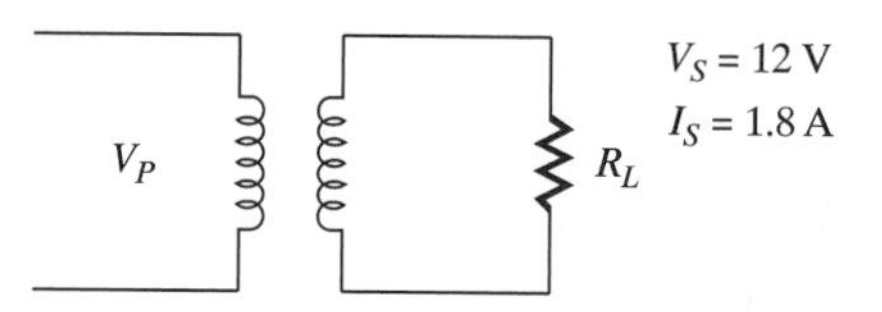

19. The input power of a transformer is 400 W. If the useful output power is 380 W, compute the efficiency.
20. The transformer circuit of Figure 13–43 has primary and secondary winding resistances as shown. For the load shown, compute the amount of power loss due to these resistances.

FIGURE 13–43 Circuit for Problem 20

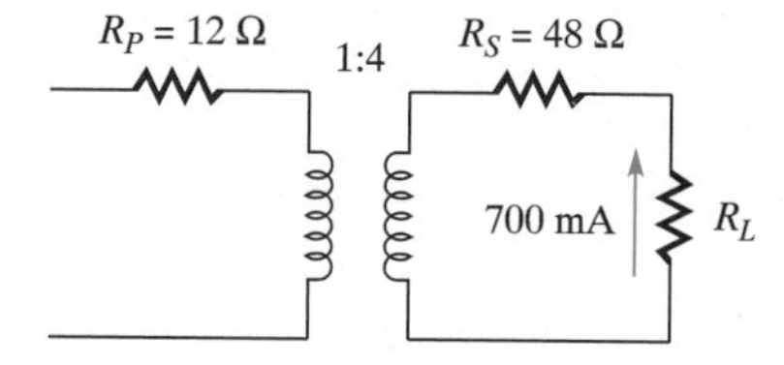

21. A transformer has the following ratings: 2 kVA, 500/50 V, and 60 Hz. If the secondary voltage is 500 V, compute the following: (a) maximum load current and (b) the smallest value of R_S.

Section 13–9

22. In Figure 13–44, determine the turns ratio between the primary and each tapped section.

FIGURE 13–44 Transformer for Problem 22

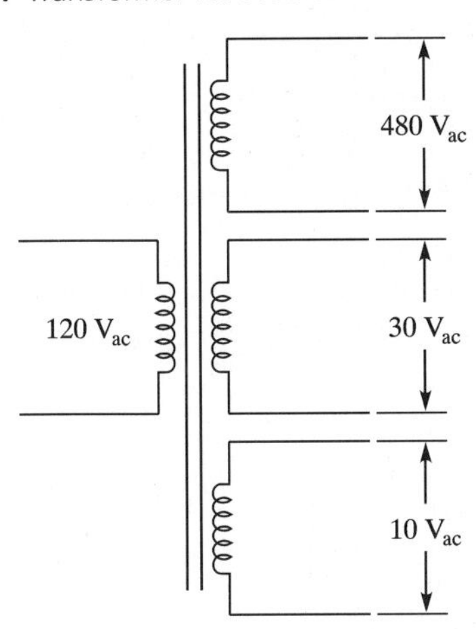

23. For the autotransformer of Figure 13–45, compute the secondary voltage.

FIGURE 13–45 Transformer for Problem 23

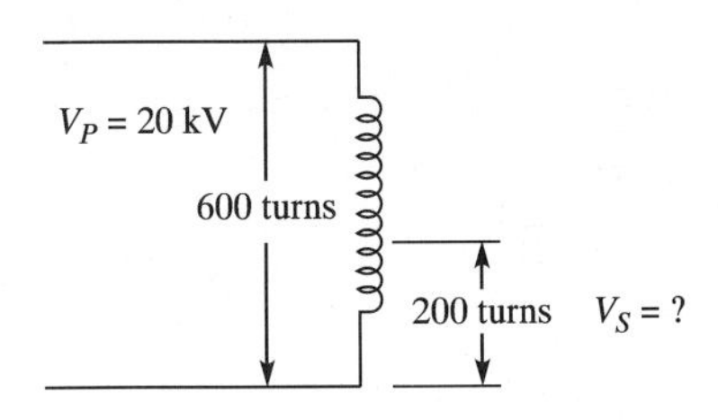

24. For the power transformer of Figure 13–46, compute the voltage on each secondary.

FIGURE 13–46 Transformer for Problem 24

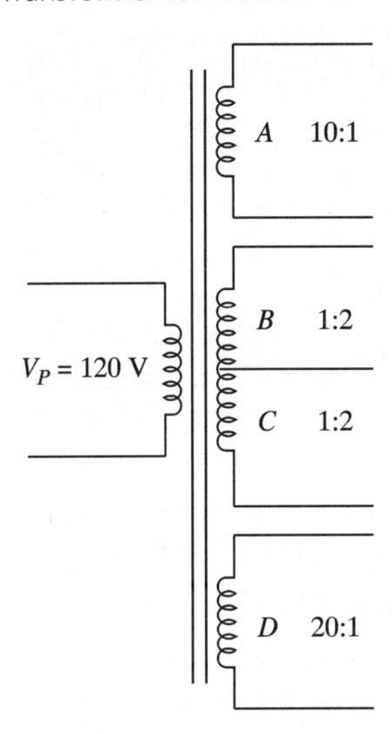

■ PROBLEM SET: CHALLENGING

25. If 2.4 kV is applied to the primary of a transformer with a 10:1 step-down turns ratio and if the primary current is 10 A, what is the value of the secondary voltage and current?

26. What is the resistance value of the load in Problem 25?

27. For the transformer in Figure 13–47, what turns ratio is required to deliver maximum power to the load?

FIGURE 13–47 Transformer for Problem 27

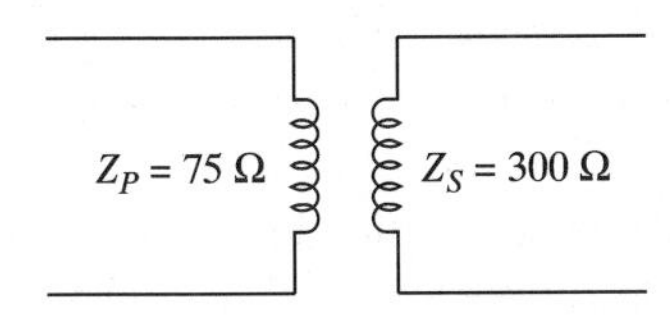

28. A transformer is rated at 3 kVA, 2400/120 V, and 60 Hz. (a) What is the turns ratio if 2400 volts is the secondary voltage? (b) What is the current rating of the secondary if 2400 volts is the primary voltage?

29. For the circuit of Figure 13–48, what must the turns ratio be in order to reflect 100 Ω into the primary?

FIGURE 13–48 Circuit for Problem 29

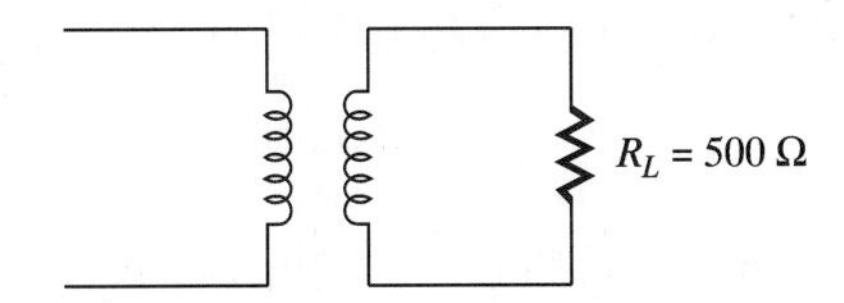

30. From the values supplied, determine if a fault exists in the circuit of Figure 13–49. What is the symptom of the fault?

FIGURE 13–49 Circuit for Problem 30

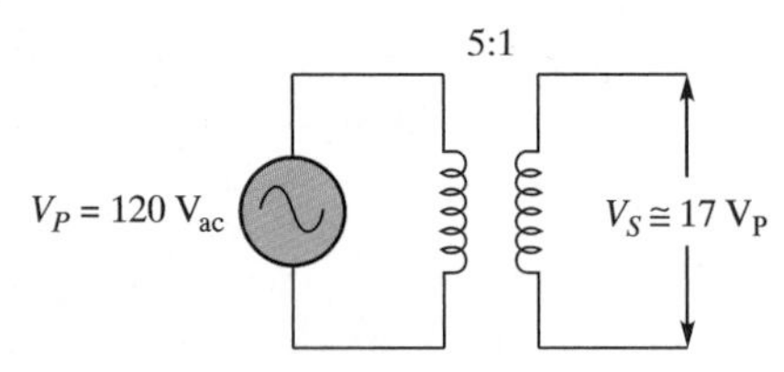

31. If one fault exists in the circuit of Figure 13–50, what is the probable cause? Why?

FIGURE 13–50 Circuit for Problem 31

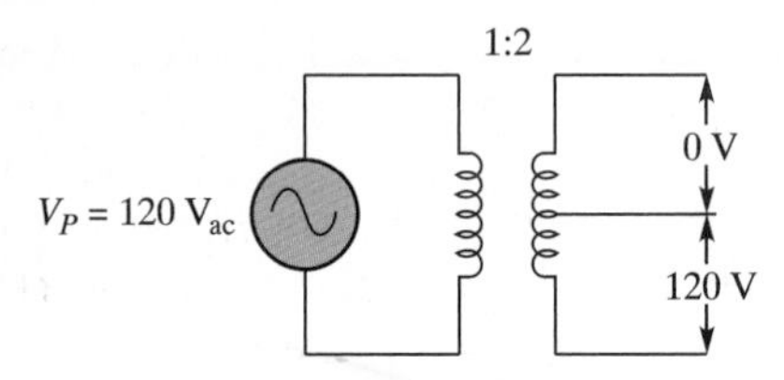

32. From the values supplied, determine if a fault exists in the circuit of Figure 13–51. What is the probable cause of the fault?

FIGURE 13–51 Circuit for Problem 32

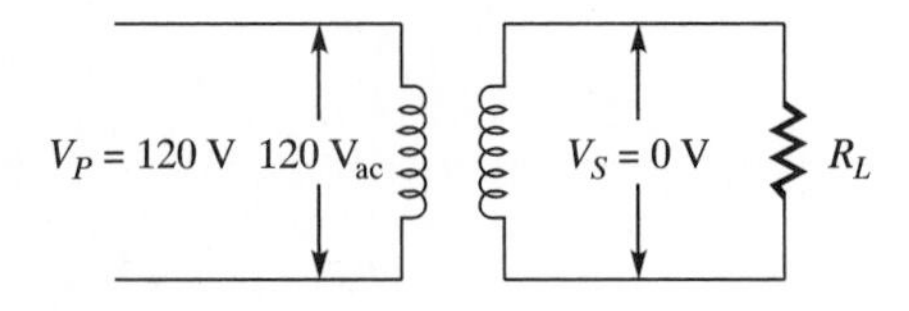

33. If one fault exists in the circuit of Figure 13–52, what is the probable cause? Why?

FIGURE 13–52 Circuit for Problem 33

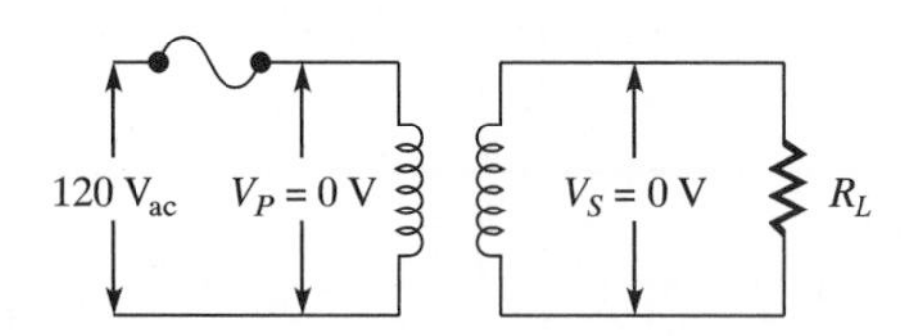

34. In the circuit of Figure 13–53, if the secondary were to short and if a fuse did not open, would the primary or the secondary burn up first? Why?

FIGURE 13–53 Circuit for Problem 34

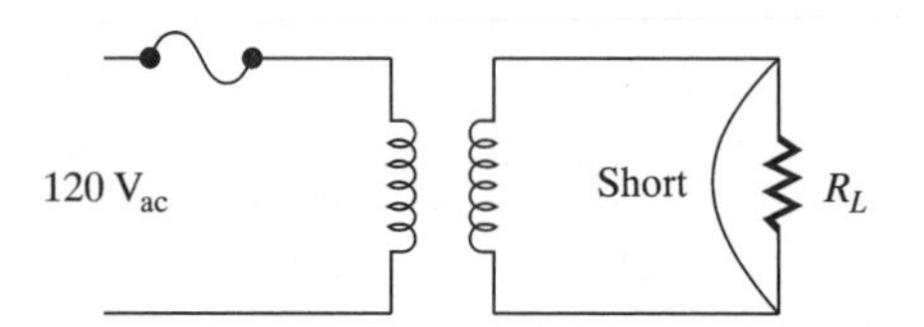

Radar

Almost everyone is familiar with radar. People know it as a device that alerts our military to the approach of hostile aircraft or missiles. They know it as a device that can see through darkness or dense rain and assist aircraft in landing. They know it as a device often used by surveyors in order to get exact measurements of land. They have seen radar screens displayed on television, showing the extent and approach of storms. They have also watched on television as radar antennas moved upward to track the movement of rockets. And they may have encountered radar in the hands of the police as it is used to enforce speed laws!

The word "radar" is a contraction of the phrase "radio detection and ranging." Thus, the basic radar system is a radio transmitter and receiver used to detect and determine the range of some distant object. (The range is the distance from the transmitter to the object.) It is, however, a very special kind of radio transmitter/receiver, deriving its information from a minute amount of electromagnetic energy that is returned from the object. This returned energy is known as a *radar echo.*

A crude form of range detection was used by early fishermen off the rocky coast of Maine. When fog obscured the coastline, they fired rifles and noted the time taken for the echo to return from the cliffs. Then, knowing that the speed of sound waves is approximately 1100 feet per second, they determined their approximate distance form the coast. For example, if the boat was 1100′ from the coast, the sound wave would take 2 s to leave and return. This two seconds, however, represents a "round trip" for the signal. Thus, the one-way time would be 1 s, indicating the boat was about 1100 feet from the coast. Sonar, a system of ranging used by submarines, is based on this principle.

The same ranging principles apply to radar. In radar, however, an electromagnetic wave, usually referred to as a radio wave, is transmitted, echoed back, and detected. Radio waves have a velocity of 328 yards per microsecond. Thus, radar gives range information much more quickly than systems based upon sound waves. Radio waves need no medium for propagation, such as air provides for sound waves. This aspect makes radar useful in the tracking of vehicles in the vacuum of space.

The basic principles of radar are shown in Figure 1. In this type, known as pulsed radar, a short burst of high frequency energy is radiated into space. The energy is reradiated by the

FIGURE 1 The principle of radar

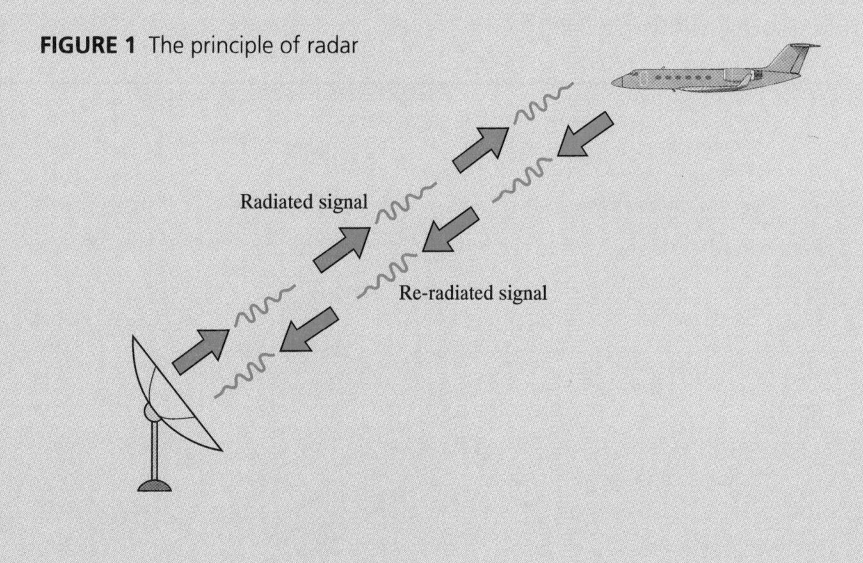

object, in this case an aircraft, back to the transmitter. This returned signal is detected and displayed upon a cathode ray tube. The returned signal is referred to as a "pip" in radar terminology. One type of display, known as an "A" scope, looks similar to the display of the oscilloscope with which you are familiar. As shown in Figure 2, at the beginning of the sweep, a portion of the transmitted signal is displayed as a reference (time zero). When the signal returns, it is displayed at some time later on the sweep. This time is automatically measured and converted to yards of range by circuits within the radar system. This type of display is most often used with precision tracking radars, an example of which is shown in Figure 3.

Another type of display is known as PPI (plan position indicator) scan, which has a rotating sweep, as illustrated in Figure 4. In this type of presentation, the beam is rotated as the energy is radiated. Objects are displayed on the CRT as they are encountered. This type of scan and presentation is most often used in long-range radar used to search for incoming aircraft. It is also used in terrain mapping. A typical search type of radar is shown in Figure 5.

What has been described up to now is known as *pulse radar.* Another type of radar is known as *Doppler radar.* It gets its name from the Doppler effect, discovered in 1842 by the Austrian physicist Christian Doppler. It relates to the change in frequency of a signal from a body that is approaching or receding from an observer. If the radiating device is approaching, the frequency of the signal increases; if it is receding, the frequency decreases. This pattern is known as a Doppler shift in frequency. You may have observed a common example of Doppler shift from a moving ambulance or fire truck with its siren sounding. The pitch (frequency) of the sound increases as the vehicle approaches and decreases as it recedes.

In Doppler radar, this Doppler shift in frequency is used to locate and follow moving objects. As shown in Figure 6, if the object is moving, the signal returned to the radar is at a different frequency than that transmitted. The faster the object is moving relative to the observer, the greater is the frequency shift. The difference in frequency can be determined, and circuits within the radar system convert this difference to the rate of motion of the object. This type of radar is used by the police in the enforcement of speed laws. Doppler radar is also used in showing the direction and speed of winds within weather systems or squall lines. A larger Doppler shift is obtained from the higher winds within the center of the storm than from those on the edge. These changes in wind speed are usually enhanced by a computer and displayed in different colors.

FIGURE 2 "A" scope presentation

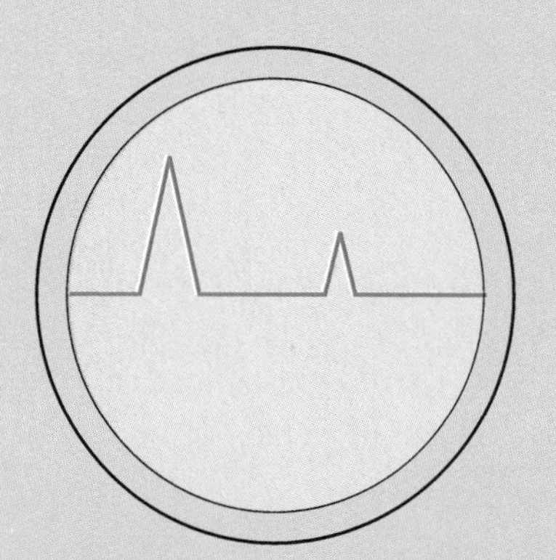

FIGURE 3 Plan-position indicator presentation

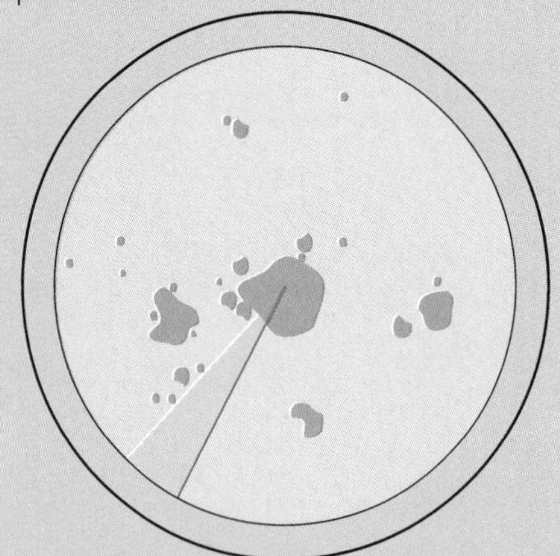

Where and in what capacity would one seek work as a radar technician? Radar, in many forms and of all types, is used extensively by the military, around commercial air terminals, in weather forecasting, by ships (large or small) at sea, and in traffic law enforcement.

The military trains its own radar technicians and maintains its own radar systems. Employment in this capacity requires a military commitment for some period of time. It is also possible to become involved in military radar through employment within the civil service. The Federal Aviation Administration (FAA) maintains the radar systems around major air terminals. The displays from these systems are used by air traffic controllers in maintaining a smooth, safe flow of air traffic. Thus, employment with the civil service is required for anyone working as a radar technician in maintenance of these systems.

Passenger and merchant ships are equipped with radar. Radar is also used in the dock areas to control entry and departures. Small and relatively cheap radars are now available for smaller fishing and pleasure boats. On commercial boats, the radar technician is usually an employee of the line. He or she may also be the radio operator. Shipboard radio operators must hold a first-class radiotelegraph operators license issued by the Federal Commissions Commission (FCC). Written tests are required for this license, which may also include the ship's radar endorsement. Successful completion of these tests usually requires a great deal of study and preparation.

No license is required to service and maintain radars used on fishing and pleasure boats. In fact, this area may be the best opportunity for someone at entry level to begin a career as a radar technician. (Entry level usually means an A.S. degree in electronics or its equivalent.) The smaller radar systems, including those used by the police, are usually marketed and maintained by small companies located close to the users. These companies usually employ an experienced technician and, depending upon the company size, one or more trainees. In addition to radar, these companies usually deal in a variety of electronic devices, such as depth finders, VHF radios, and LORAN receivers.

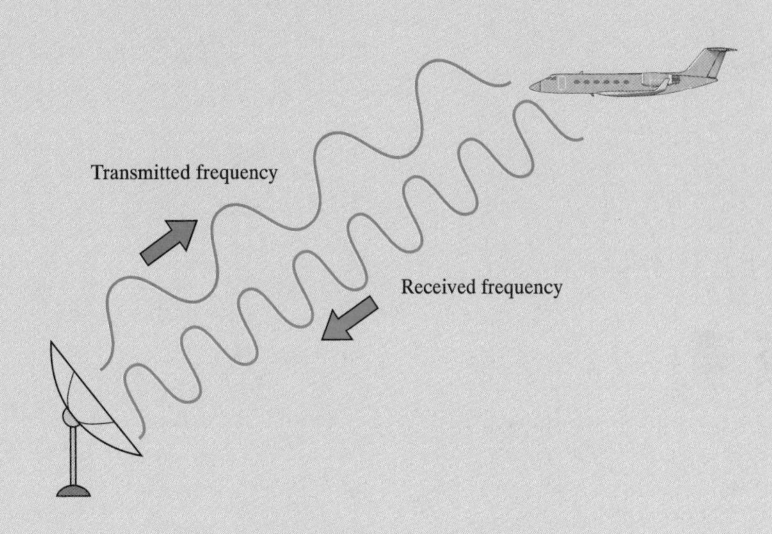

FIGURE 4 Principle of Doppler radar

CHAPTER

14 Capacitance in ac Circuits

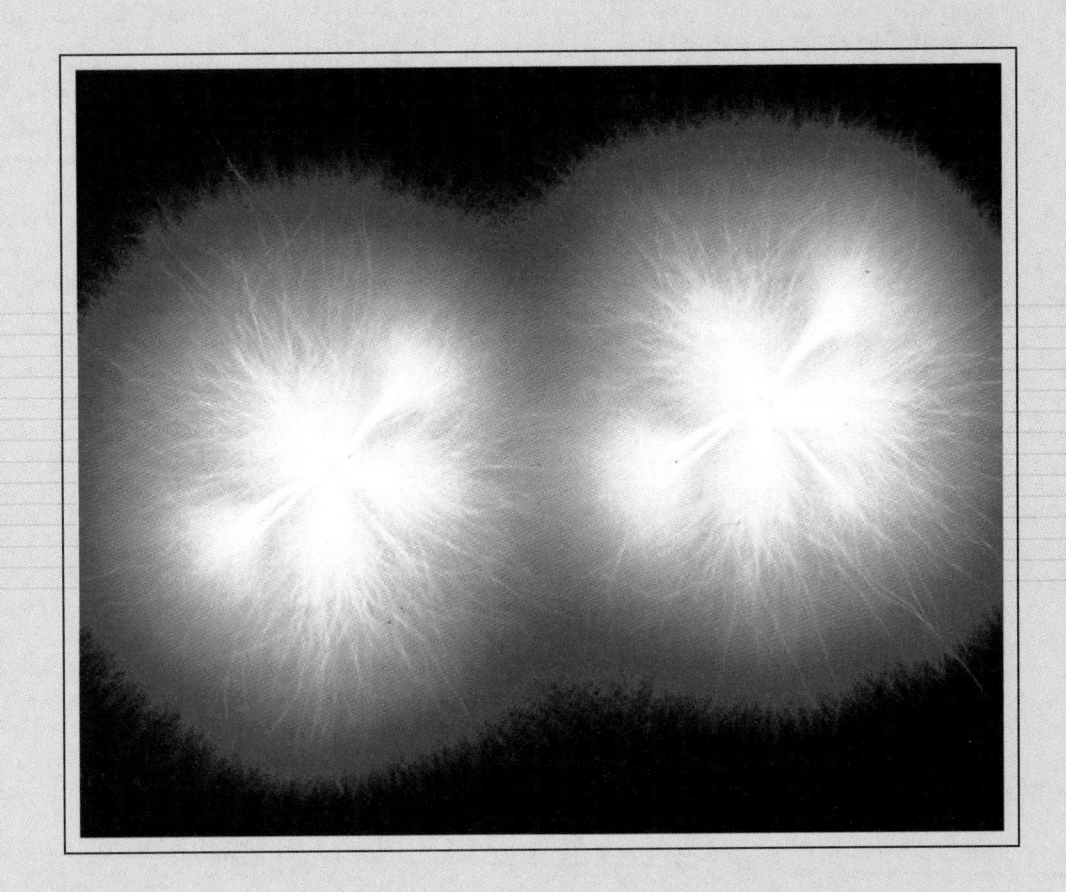

UPON COMPLETION OF THIS CHAPTER, YOU WILL BE ABLE TO

1. Explain why a capacitor causes the current to lead the voltage in a circuit.
2. Compute capacitive reactance and capacitive susceptance.
3. Explain the operation and perform analysis of series *RC* circuits.
4. Explain the operation and perform analysis of parallel *RC* circuits.
5. Explain how changes in frequency affect the response of an *RC* circuit.
6. Define and compute the power factor of an *RC* circuit.
7. List and explain applications of *RC* circuits.
8. Perform fault analysis of *RC* circuits.

Capacitance and its effects upon dc current were introduced in Chapter 10. In this chapter, the effects of capacitance upon ac current are presented. First, circuits containing only capacitance are considered. Then, capacitance is considered in combination with resistance. The action of these series and parallel *RC* circuits and their analysis is presented next. The chapter concludes with the applications and fault analysis of *RC* circuits.

Even though the effects of capacitance and inductance are opposites, the analysis of *RC* circuits proceeds in a manner very similar to that for *RL* circuits. The approach, the laws of electronics, the equations, and the mathematics used are identical. The only major differences are in the phase angles, which are of opposite polarities, and in the effect of frequency on them. If you developed a good understanding of the analysis of *RL* circuits, there will be little new learning to be done in the analysis of *RC* circuits!

As was the case with the *RL* circuit, it is essential that you develop a good understanding of the operation of the *RC* circuit. Once again, don't just memorize how to solve the problems, but develop an understanding of what is happening and why. Later, in Chapter 15, the electrical effects of resistance, inductance, and capacitance will be combined in a common circuit. A good understanding of both *RL* and *RC* circuits will greatly enhance and simplify this learning.

14–1 CAPACITANCE AND THE SINE WAVE

Recall from Chapter 10 that capacitance C is the property of a circuit that stores electric charge Q. At the instant a dc voltage is applied, the voltage across the capacitor is zero. Current, in the form of electrons leaving one plate and entering the other, must flow in order to store the charge and thus produce a potential difference between the plates. The voltage is not present immediately, but builds up over time. Thus, capacitance is the property of a circuit that opposes a change in voltage.

The charging current is heaviest when the capacitor first begins to charge and drops to zero when the capacitor is fully charged. Thus, at the instant it begins to charge, a capacitor appears to be a short circuit. The time taken for the capacitor to attain full charge is a function of the *RC* time constant. The *RC* time constant in seconds is the product of the resistance and the capacitance of the circuit. Five of these time constants must elapse in order for the capacitor to attain full charge.

Figure 14–1(a) shows a series *RC* circuit. The graph of a sine wave of voltage applied to it and the current that flows during charging of the capacitor is shown in Figure 14–1(b). At point A the voltage is moving through zero in a positive direction, and the capacitor must take on charge in order to follow. Since the voltage across the capacitor is zero, the charging current is maximum at this time. At point B, the voltage has reached

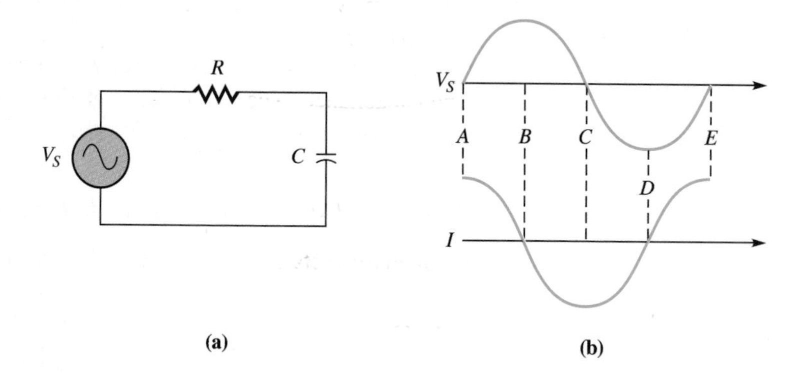

FIGURE 14–1 Capacitance in a series *RC* Circuit

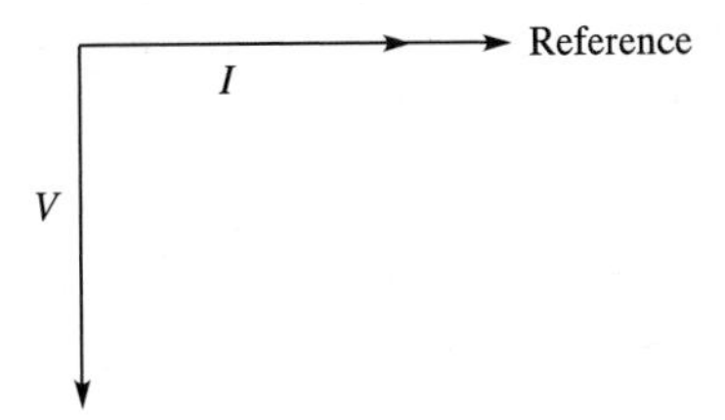

FIGURE 14–2 Phasor diagram of current and voltages in a series *RC* circuit

its maximum point and is no longer changing. The capacitor is charged, and the charging current is zero. The voltage now begins to decrease, and the capacitor must lose charge. Current again flows, but in the opposite direction. At point *C* the voltage is zero—about to change polarity—while the current is maximum in this opposite direction. The voltage increases once again, but now in a negative direction. The current increases, reaching zero when the voltage reaches its maximum negative value at point *D*. As the voltage now increases toward zero, the current builds once again, reaching its maximum value at point *E*.

Question: What is the phase relationship between the charging current and voltage in a capacitor?

Study carefully the graphs of voltage and charging current. The current and voltage are no longer in phase. The current is now leading the voltage by 90°, as shown in phasor form in Figure 14–2. This condition is the opposite of what occurs in a series *RL* circuit in which the voltage leads the current by 90°. Be sure that you understand not only what the relationship is, but also how it comes about.

SECTION 14–1 REVIEW

1. Does a capacitor opposes a change in current or voltage?
2. Does an uncharged capacitor appear to be a shorted or open circuit?
3. Does a fully charged capacitor appear to be a shorted or open circuit?
4. How many time constants are required to fully charge a capacitor?
5. When the voltage across a capacitor is maximum, is the charging current minimum or maximum?
6. When the voltage across a capacitor is zero, is the charging current minimum or maximum?
7. What is the phase relationship between the charging current and the voltage across a capacitor?

14–2 CAPACITIVE REACTANCE

Question: How does a capacitor affect the amount of sine wave current?

A capacitor reacts to a change in voltage by allowing charging current to flow. As shown previously, the amount of this current is directly proportional to the product of the capacitance and the rate of change of voltage with respect to time ($i_C = C(\Delta V/\Delta t)$). For a sine wave, which is considered in this chapter, the rate of change of voltage is *directly proportional* to the angular velocity. Recall that angular velocity is equal to $2\pi f$ and is symbolized by ω. Thus, the charging current is *directly proportional* to both frequency and capacitance. The opposition to sine wave current offered by a capacitor, is known as **capacitive reactance,** which is symbolized X_C (read "X sub C").

Question: How is the capacitive reactance computed?

Capacitive reactance is computed using the following equation:

(14–1)
$$X_C = \frac{1}{2\pi fC}$$

Since the charging current is directly proportional to the angular velocity of the sine wave and the capacitance value, the capacitive reactance is *inversely proportional* to these values, as Eq. 14–1 shows.

EXAMPLE 14–1 A capacitor has a value of 0.01 μF. If the frequency of the applied voltage is 10 kHz, compute the capacitive reactance.

■ Solution:

$$X_C = \frac{1}{2\pi fC}$$

$$= \frac{1}{2(3.14)(10 \times 10^3)(0.01 \times 10^{-6})} = 1592\ \Omega$$

[2] [×] [π] [×] [1] [0] [EXP] [3] [×] [.] [0] [1] [EXP] [±] [6] [=] [1/x]

EXAMPLE 14–2 A capacitor has a value of 1 μF. If the frequency of the applied voltage is 10 kHz, compute the value of the capacitive reactance.

■ Solution:

$$X_C = \frac{1}{2\pi fC}$$

$$= \frac{1}{2(3.14)(10 \times 10^3)(1 \times 10^{-6})} = 1.59\ \Omega$$

Notice in these examples that as the frequency and/or capacitance increased, the capacitive reactance decreased. You can confirm this fact by substituting smaller values than those used in the examples.

EXAMPLE 14–4 What value of capacitor produces 18.086 kΩ of reactance when a voltage with a frequency of 400 Hz is applied?

■ Solution:

Use Eq. 14–1 and transpose to solve for *C*:

$$X_C = \frac{1}{2\pi fC}$$

$$C = \frac{1}{2\pi fX_C}$$

$$= \frac{1}{2(3.14)(400\text{ Hz})(18086\Omega)} = 0.022\ \mu\text{F}$$

■ Capacitive Reactance and Current

As you would probably expect, Ohm's law for ac circuits applies here just the same as it does for inductive circuits. The current in an *RC* circuit can be found using the following equation:

(14–2) $$I_C = \frac{V_C}{X_C}$$

This equation can be transposed as follows:

(14–3) $$X_C = \frac{V_C}{I_C}$$

(14–4) $$V_C = I_C X_C$$

Capacitive Reactances in Series and Parallel

Question: How is the total reactance of series and parallel reactances computed?

Finding the total of capacitive reactances in series is done in a manner identical to that used for inductive reactances. If they are in series, they are merely added to find the total. Thus, total reactance in series is found by

(14–5) $$X_{ct} = X_{C1} + X_{C2} + X_{C3}$$

The total reactance of capacitive reactances in parallel is computed in a manner identical to that for inductive reactances in parallel. The individual reactances are converted to susceptances and summed to obtain the total susceptance B_T. The reciprocal of the total susceptance is the total reactance. In equation form, total reactance in parallel is found by

(14–6) $$B_T = B_1 + B_2 + B_3$$

(14–7) $$X_{CT} = \frac{1}{B_T}$$

EXAMPLE 14–5 For the circuit of Figure 14–3, compute the total capacitive reactance.

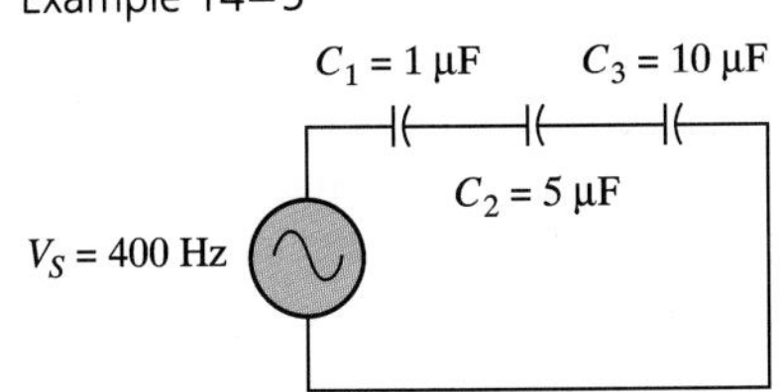

FIGURE 14–3 Circuit for Example 14–5

■ Solution:

1. Compute the individual reactances:

$$X_{C1} = \frac{1}{2\pi f C_1}$$

$$= \frac{1}{2(3.14)(400\text{ Hz})(1 \times 10^{-6})} = 398\ \Omega$$

$$X_{C2} = \frac{1}{2\pi f C_2}$$

$$= \frac{1}{2(3.14)(400\text{ Hz})(5 \times 10^{-6})} = 79.6\ \Omega$$

$$X_{C3} = \frac{1}{2\pi f C_3}$$

$$= \frac{1}{2(3.14)(400\text{ Hz})(10 \times 10^{-6})} = 39.8\ \Omega$$

2. Compute the total:

$$X_{CT} = X_{C1} + X_{C2} + X_{C3}$$

$$= 398\ \Omega + 79.6\ \Omega + 39.8\ \Omega = 517\ \Omega$$

FIGURE 14–4 Circuit for Example 14–6

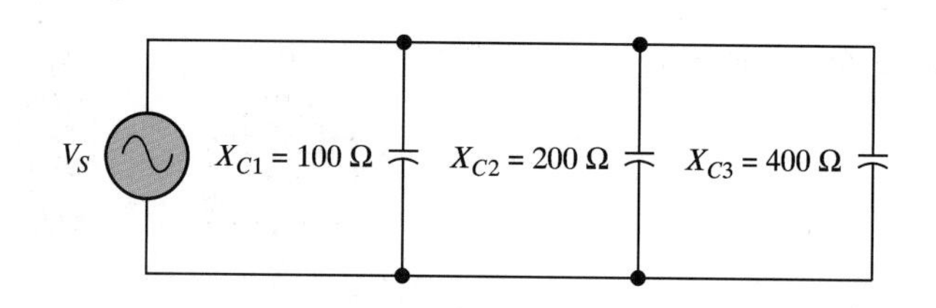

EXAMPLE 14–6 For the circuit of Figure 14–4, compute the total capacitive reactance.

■ **Solution:**

$$B_T = B_1 + B_2 + B_3$$

$$= \frac{1}{X_{C1}} + \frac{1}{X_{C2}} + \frac{1}{X_{C3}}$$

$$= \frac{1}{100\ \Omega} + \frac{1}{200\ \Omega} + \frac{1}{400\ \Omega} = 17.5\text{ mS}$$

Then,

$$X_{CT} = \frac{1}{B_T}$$

$$= \frac{1}{17.5 \times 10^{-3}} = 57.1\ \Omega$$

Notice that in a parallel circuit, the total capacitive reactance is smaller than the smallest reactance value.

■ SECTION 14–2 REVIEW

1. If the frequency of the applied voltage increases, does the capacitive reactance increase or decrease?
2. If the value of the capacitance is decreased, does the capacitive reactance increase or decrease?
3. Four reactances of 400 Ω, 800 Ω, 1200 Ω, and 1600 Ω are in series. Compute the total reactance.
4. The same four reactances of Problem 3 are placed in parallel. Compute the total reactance.

14–3 ANALYSIS OF SERIES *RC* CIRCUITS

The series *RC* circuit of Figure 14–5(a) is a voltage divider. The voltage drop across the resistor, known as V_R, is in phase with the current, as illustrated in the graph and phasor diagram of Figure 14–5(b). The voltage developed across the capacitor, known as V_C, lags the charging current by 90°, as illustrated in the graph and phasor diagram of Figure 14–5(c). Being a series circuit, the same current flows through the resistor and into the capacitor. Thus, the two phasor diagrams can be merged as shown in Figure 14–5(d). Notice that the voltage drop across the resistor leads the voltage across the capacitor by 90°. The hypotenuse of this right triangle is the phasor representing the value of the source

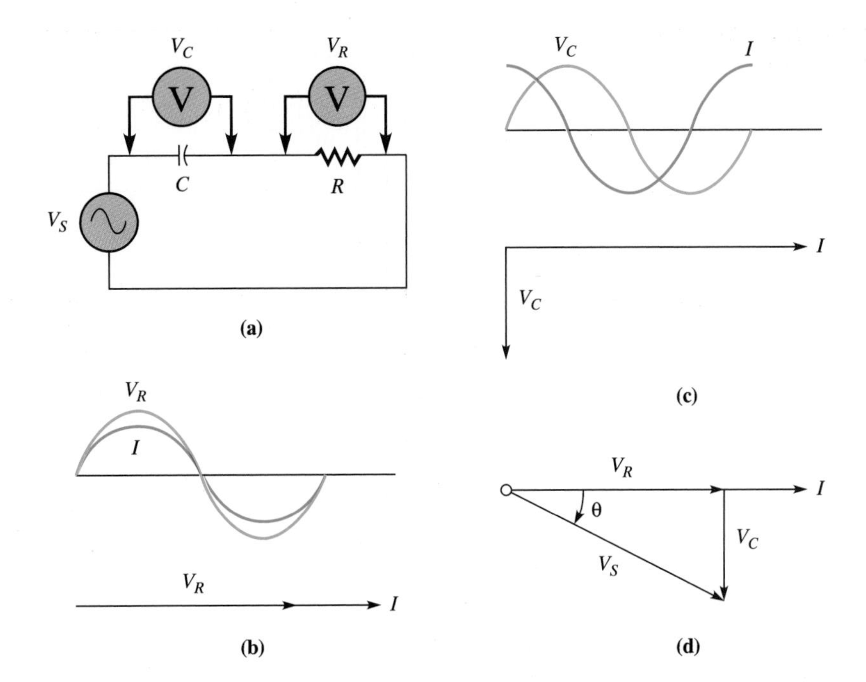

FIGURE 14–5 Series *RC* circuit and analysis diagrams

voltage. Depending upon the lengths of the phasors for V_R and V_C, the source voltage lags the current by an angle between zero and 90°. This angle is the circuit phase angle, which will be discussed in more detail later in this chapter.

Voltages in a Series *RC* Circuit

Question: Does Kirchhoff's voltage law apply to series RC *circuits?*

Kirchhoff's Voltage Law is applicable to series *RC* circuit analysis. But as was the case with the series *RL* circuit, θ (the angle between the two voltages) must be considered. Since the voltage drop across the resistor leads the voltage across the capacitor by 90°, the two cannot be added directly in order to find the source voltage, but must be added as phasors. The phasor sum of V_R and V_C is equal to the source voltage V_S. By Kirchhoff's voltage law, this relationship is given by

(14–8) $$V_S^2 = V_R^2 + V_C^2$$

(14–9) $$V_S = \sqrt{V_R^2 + V_C^2}$$

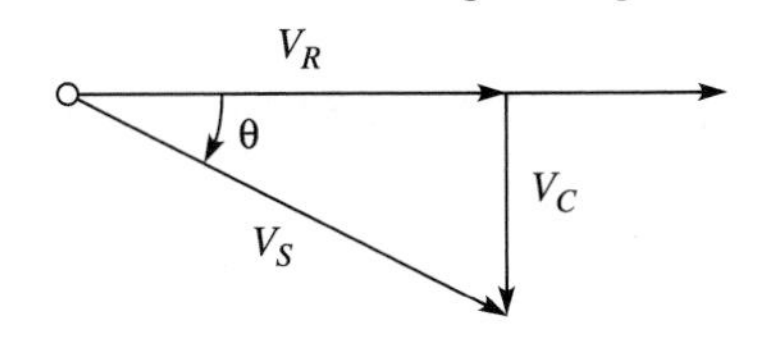

FIGURE 14–6 Voltage triangle

The phasor diagram, known as a *voltage triangle,* is shown in Figure 14–6. Angle θ is the circuit phase angle. Notice that θ opens in a negative direction, which is opposite that of the series *RL* circuit.

EXAMPLE 14–7 For the circuit of Figure 14–7(a), compute the value of V_S.

■ Solution:

$$V_S = \sqrt{V_C^2 + V_C^2}$$
$$= \sqrt{(12\text{ V})^2 + (18\text{ V})^2} = 21.633\text{ V}$$

FIGURE 14–7 Circuit and phasor diagram for Example 14–7

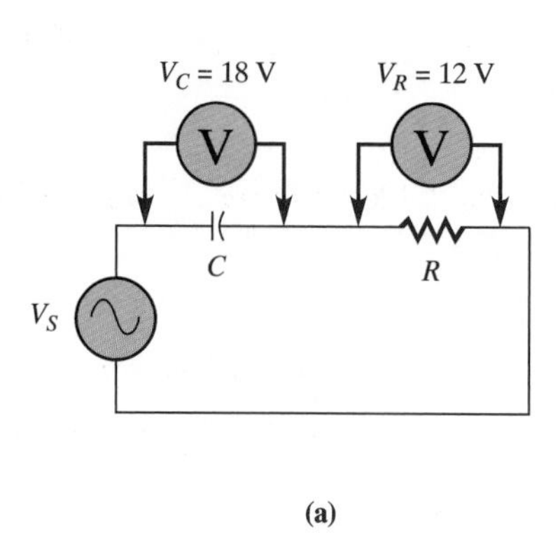

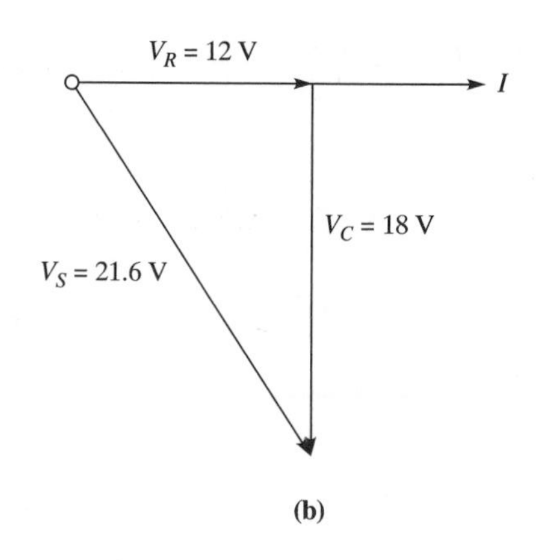

The phasor diagram for this circuit is shown in Figure 14–7(b).

EXAMPLE 14–8 The source voltage applied to a series *RC* circuit has a value of 120 V. If the capacitor voltage V_C is 60 V, compute the value of V_R.

■ Solution:

$$V_S^2 = V_R^2 + V_C^2$$
$$V_R^2 = V_S^2 - V_C^2$$
$$V_R = \sqrt{V_S^2 - V_C^2}$$
$$= \sqrt{(120\text{ V})^2 - (60\text{ V})^2} = 103.923\text{ V}$$

■ Impedance in a Series *RC* Circuit

Question: How is the impedance of a series RC circuit determined?

Recall that **impedance** is the total opposition offered by a circuit to the flow of alternating current. In an *RC* circuit, impedance is composed of resistance and reactance. Eq. 14–8 can be used to derive an equation for determining the impedance of a series *RC* circuit:

By Kirchhoff's voltage law,

$$V_S^2 = V_R^2 + V_C^2$$

Substituting gives

$$(IZ)^2 = (IR)^2 + (IX_C)^2$$

Dividing by *I* yields

(14–10) $$Z^2 = R^2 + X_C^2 \quad \text{and} \quad Z = \sqrt{R^2 + X_C^2}$$

FIGURE 14–8 Impedance triangle

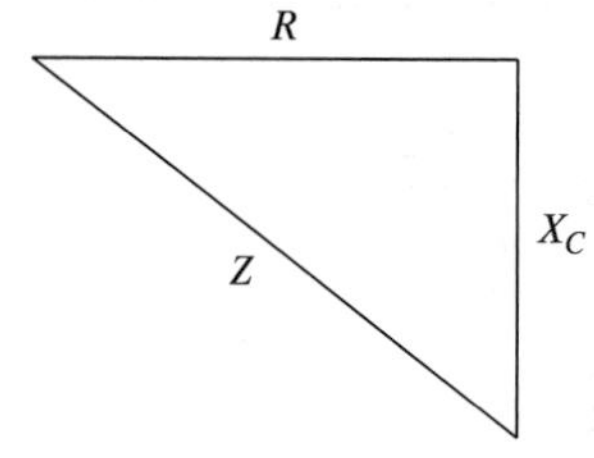

Eq. 14–10 illustrates that impedance is equal to the phasor sum of the resistance and reactance. The phasor diagram of Figure 14–8, known as an *impedance triangle,* represents the relationship between impedance, resistance, and reactance. Ohm's law for ac circuits applies here also:

(14–11) $$I = \frac{V_S}{Z}$$

(14–12) $$V_S = IZ$$

(14–13) $$Z = \frac{V_S}{I}$$

EXAMPLE 14–9 A reactance of 2.4 kΩ is in series with a resistance of 1.2 kΩ. What is the impedance of the circuit?

■ Solution:

$$Z = \sqrt{R^2 + X_C^2}$$
$$= \sqrt{(1200\ \Omega)^2 + (2400\ \Omega)^2} = 2.683\ \text{k}\Omega$$

EXAMPLE 14–10 For the circuit of Figure 14–9, compute the impedance.

FIGURE 14–9 Circuit for Example 14–10

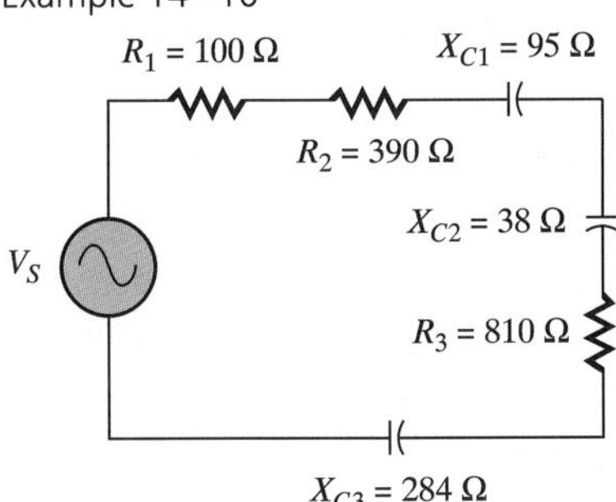

■ Solution:

1. Sum the individual resistances to determine the total resistance:

$$R_T = R_1 + R_2 + R_3$$
$$= 100\ \Omega + 390\ \Omega + 810\ \Omega = 1.3\ \text{k}\Omega$$

2. Sum the individual reactances to obtain the total capacitive reactance:

$$X_{CT} = X_{C1} + X_{C2} + X_{C3}$$
$$= 95\ \Omega + 38\ \Omega + 284\ \Omega = 417\ \Omega$$
$$Z = \sqrt{R^2 + X_C^2}$$
$$= \sqrt{(1300\ \Omega)^2 + (417\ \Omega)^2} = 1.365\ \text{k}\Omega$$

■ Circuit Phase Angle in a Series *RC* Circuit

Question: How is the phase angle of a series* RC *circuit computed?

As in the *RL* circuit, the current and voltage in an *RC* circuit are not in phase. The resistance tries to keep the current and voltage in phase, while the capacitance attempts to make the current lead the voltage by 90°. The result is a phase angle, between the source voltage and current, of some value between zero and 90°. In the voltage triangle of Figure 14–6, notice that the phasor V_C lies opposite and phasor V_R lies adjacent to the phase angle. The tangent function relates the lengths of these phasors to the value of the angle theta. In equation form, this relationship is

(14–14) $$\tan\theta = \frac{V_C}{V_R} \quad \text{and} \quad \theta = \arctan\frac{V_C}{V_R}$$

Recall that this last equation is read as "the angle whose tangent is V_C/V_R." Thus, Eq. 14–14 is the equation for the circuit phase angle. The impedance triangle of Figure 14–8 can also be used in computing the phase angle. The opposite side is now X_C and the adjacent side is R. The equation is

(14–15) $$\tan\theta = \frac{X_C}{R} \quad \text{and} \quad \theta = \arctan\frac{X_C}{R}$$

EXAMPLE 14–11 Compute the phase angle for the circuit of Figure 14–10.

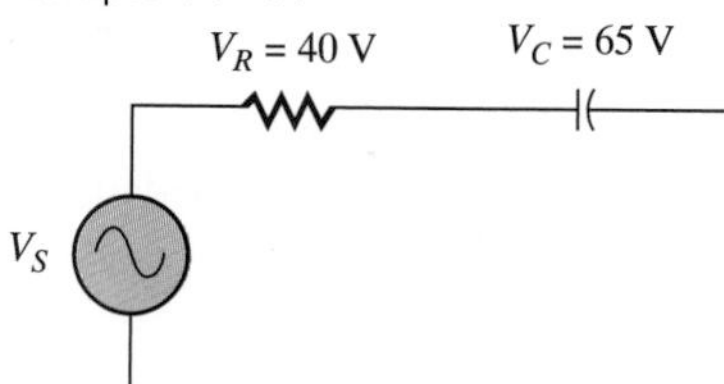

FIGURE 14–10 Circuit for Example 14–11

■ Solution:

$$\theta = \arc\tan \frac{V_C}{V}$$

$$= \arc\tan \frac{65 \text{ V}}{40 \text{ V}} = \arc\tan 1.625 = 58.392°$$

[6] [5] [÷] [4] [0] [=] [TAN⁻¹]

Question: How does the frequency of the applied voltage and the value of the capacitance affect the phase angle?

Notice in Figure 14–11 that the phase angle is directly proportional to the length of the phasor representing X_C. Capacitive reactance X_C is inversely proportional to both frequency and capacitance ($X_C = 1/2\pi fC$). As the frequency or capacitance increases, the reactance decreases, and when these quantities decrease, reactance increases. Thus, an increase in either frequency or capacitance causes both the reactance and the phase angle to decrease. A decrease in either frequency or capacitance causes both the reactance and the phase angle to increase.

If an *RC* circuit becomes more reactive, the phase angle increase, and if it becomes less reactive, the phase angle decreases. Since a series circuit is a voltage divider, anything that causes the capacitor voltage to increase will make it more reactive and cause an increase in the phase angle. The capacitor voltage is directly proportional to the reactance ($V_S = IX_C$). This fact confirms the theory of the preceding paragraph.

Another point that must be considered is that any change in current for a given reactance also causes a change in capacitor voltage and phase angle. Any change in resistance, with no other circuit changes, causes a change in current. An increase in resistance decreases the current and the voltage across the capacitor. A decrease in resistance has the opposite effect. Thus, the phase angle is *inversely proportional* to the resistance value, as illustrated in Figure 14–12.

FIGURE 14–11 Effect of changes in reactance on phase angle

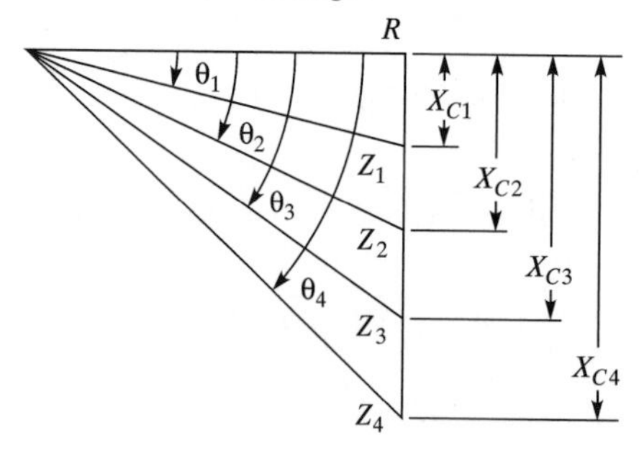

FIGURE 14–12 Effect of changes in resistance on phase angle

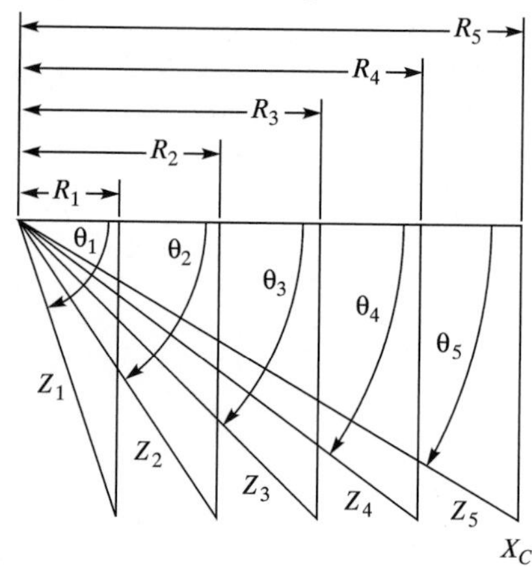

EXAMPLE 14–12 The circuit of Figure 14–13 has a phase angle of 36°. What frequency would cause the phase angle to double to 72°?

FIGURE 14–13 Circuit and phasor diagram for Example 14–12

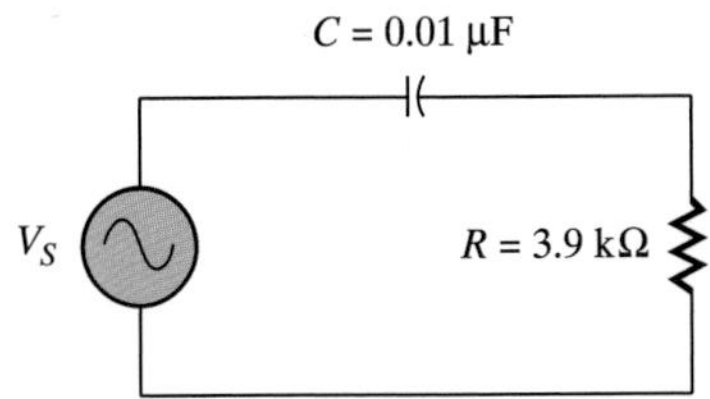

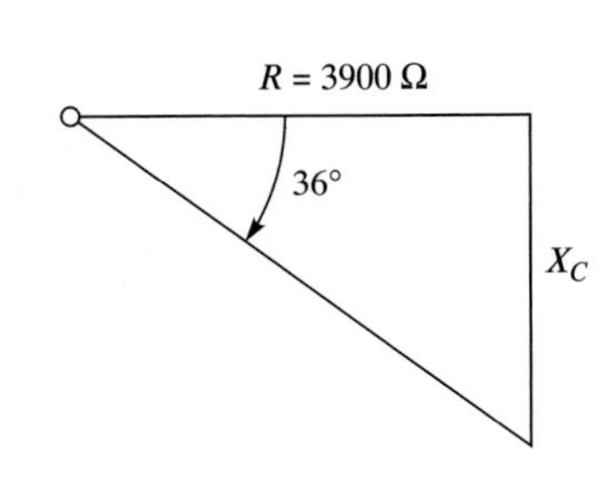

■ Solution:

1. Compute the value of X_C needed for a phase angle of 72°:

$$\tan\theta = \frac{X_C}{R}$$

$$X_C = R(\tan\theta)$$

$$X_C = (3900\ \Omega)(\tan 72) = 12.003\ \text{k}\Omega$$

2. Compute the frequency that will produce 12.003 kΩ of reactance:

$$X_C = \frac{1}{2\pi fC}$$

$$f = \frac{1}{2\pi CX_C}$$

$$= \frac{1}{2(3.14)(0.01\times10^{-6})(12003\ \Omega)} = 1.326\ \text{kHz}$$

CAUTION

In analyzing problems involving the trigonometric functions, it is sometimes easy to make a wrong assumption. In the problem of Example 14–12, the phase angle must double. You might assume this doubling of the phase angle requires doubling the length of the phasor representing X_C, but this is not the case. For example, a resistance of 3900 Ω and a reactance of 2400 Ω produce a phase angle of 31.6°. If the reactance is doubled, the phase angle does not double, but increases to 50.9°. The reason is that the trigonometric relationships are not linear, as was demonstrated previously when it was shown that the rate of change of a sine wave varied dramatically throughout 1 cycle.

EXAMPLE 14–13 Compute the phase angle for the circuit of Figure 14–14. Compute the phase angle if the capacitance is doubled.

FIGURE 14–14 Circuit for Example 14–13

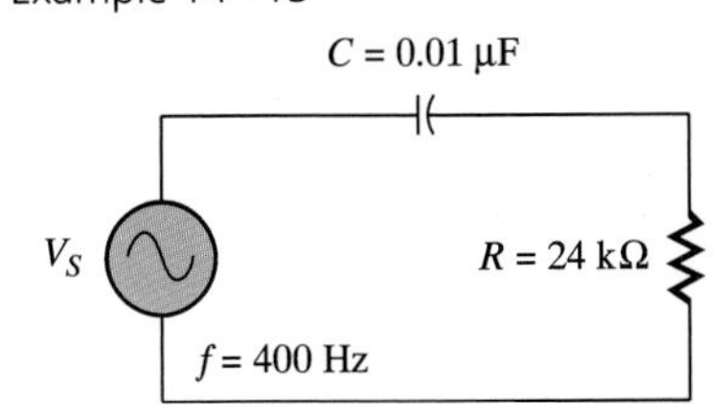

■ Solution:

1. Compute the phase angle:

$$X_C = \frac{1}{2\pi fC}$$

$$= \frac{1}{2(3.14)(400\ \text{Hz})(0.01\times10^{-6})} = 39.8\ \text{k}\Omega$$

$$\Theta = \arctan\frac{X_C}{R}$$

$$= \arctan\frac{39.8\times10^3\ \Omega}{24\times10^3\ \Omega} = \arctan 1.658 = 58.9°$$

2. Compute the phase angle for $C = 0.02\ \mu\text{F}$:

$$X_C = \frac{1}{2\pi fC}$$

$$= \frac{1}{2(3.14)(400\text{ Hz})(0.02 \times 10^{-6})} = 19.9\text{ k}\Omega$$

$$\theta = \arc\tan \frac{X_C}{R}$$

$$= \arc\tan \frac{19.9 \times 10^3\ \Omega}{24 \times 10^3\ \Omega} = \arc\tan 0.8292 = 39.664°$$

Notice that the increase in capacitance caused a decrease in phase angle.

■ Power in an *RC* Circuit

The power considerations for an *RC* circuit are similar to those for an *RL* circuit. The resistor dissipates electrical energy in the form of heat. The rate at which the energy is dissipated is the **true power** (P_{true}) developed in the circuit. True power, expressed in watts, is computed using any of the forms of the power equation.

Question: What is reactive power and how is it computed?

FIGURE 14–15 Reactive power for an applied sinusoidal voltage

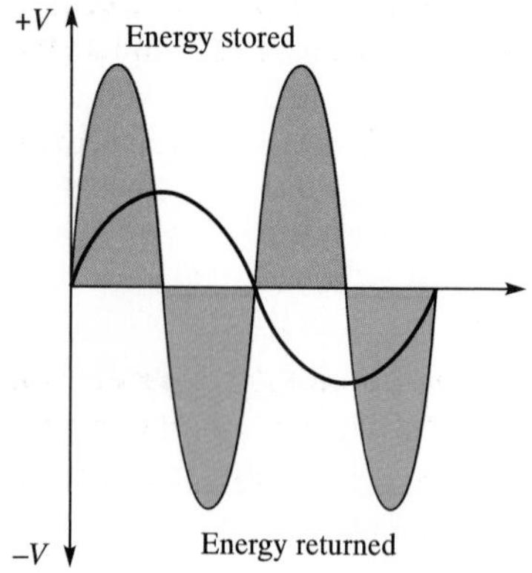

The capacitor does not dissipate electrical energy, but alternately stores and then returns it to the source. Figure 14–15 shows one cycle of a sine wave of voltage applied to a capacitor. During the first half of the positive alternation, the capacitor charges, and energy is stored in the electrostatic field. During the second half of the positive alternation, the capacitor discharges to zero, and the energy is returned to the source. During the negative alternation, the same events occur. Energy is stored during the first half of the alternation and returned during the second. The rate at which the energy is stored and returned to the source is known as **reactive power,** symbolized by P_{react} and measured by a unit called the volt-ampere reactive (VAR). The reactive power is computed using one of the following equations:

(14–16) $$P_{react} = I^2X_C$$

(14–17) $$P_{react} = V_SI$$

(14–18) $$P_{react} = \frac{V_S^2}{X_C}$$

Question: What is apparent power and how is it computed?

Apparent power is the power that the circuit is "apparently" developing. It is a combination of both the true and reactive power. The unit of measure for apparent power is the volt-ampere (VA). Figure 14–16 shows a *power triangle* for an *RC* circuit. Notice that the phasor representing the reactive power occupies the same position as reactance in the impedance triangle. The phasor representing true power occupies the same position as resistance in the impedance triangle. The phasor sum of reactive power and true power is **apparent power,** which is expressed in equation form as

FIGURE 14–16 Power triangle

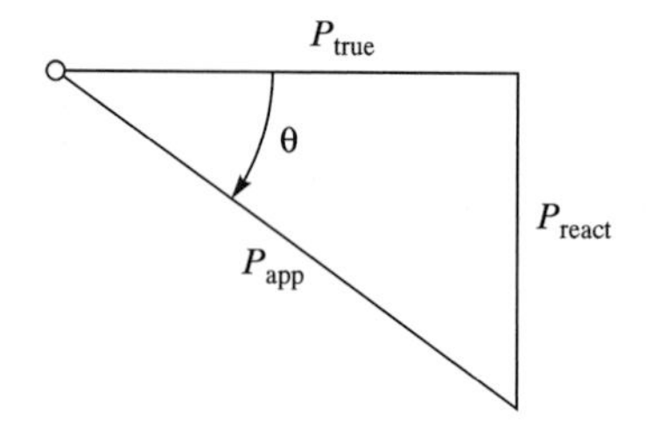

(14–19) $$P_{app} = \sqrt{(P_{true})^2 + (P_{react})^2}$$

The apparent power may also be computed using the power equations:

(14–20) $$P_{app} = I^2Z$$

(14–21) $$P_{app} = \frac{V_S^2}{Z}$$

(14–22) $$P_{app} = V_SI$$

Question: What is the power factor and how is it determined?

Notice in Figure 14–18 that the true power phasor is the side adjacent to the phase angle. If the apparent power and the phase angle are known, the true power can be determined using the cosine trigonometric function:

(14–23) $$\cos\theta = \frac{P_{true}}{P_{app}} \quad \text{and} \quad P_{true} = P_{app}(\cos\theta)$$

As the equation shows, true power is the product of the apparent power and the cosine of the phase angle. For this reason, the cosine of the phase angle is known as the **power factor** of the circuit.

SECTION 14–3 REVIEW

1. Does a capacitance cause the current to lead or lag the voltage?
2. In a series *RC* circuit, does the resistor voltage leads or lag the capacitor voltage?
3. A 450 Ω reactance is in series with a 390 Ω resistance. Compute the impedance.
4. In a series *RC* circuit, a 25 V source produces 15 V across the capacitor. Compute the value of the voltage across the resistor.
5. The phase angle of an *RC* circuit is 39°. If the apparent power is 85 mVA, compute the true power.

14–4 ANALYSIS OF PARALLEL *RC* CIRCUITS

The analysis of parallel *RC* circuits proceeds in a manner similar to that of parallel *RL* circuits. A parallel *RC* circuit is shown in Figure 14–17(a). As with any parallel circuit, the same voltage V_S is across both branches. Thus, by Ohm's law, the equations for the branch currents are

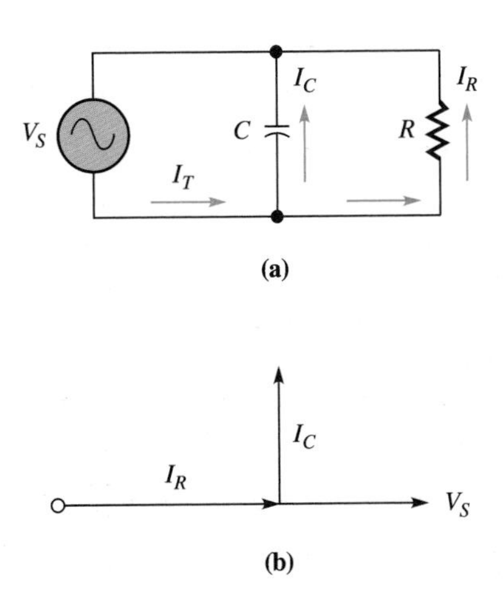

FIGURE 14–17 Parallel *RC* circuit and phasor diagram

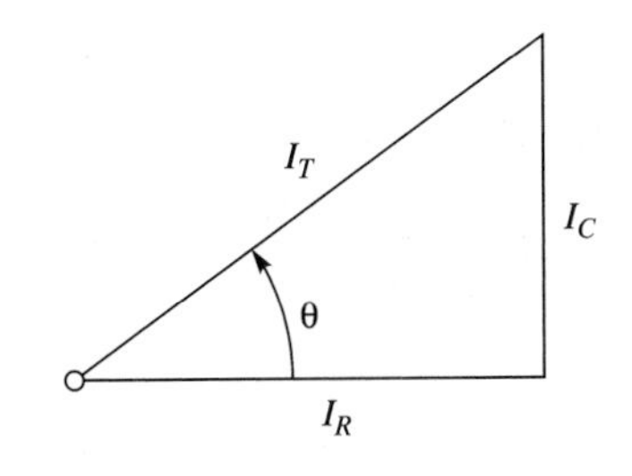

FIGURE 14–18 Current triangle

$$I_R = \frac{V_S}{R} \quad \text{and} \quad I_C = \frac{V_S}{X_C}$$

The current through the resistive branch is in phase with the source voltage. The current through the capacitive branch leads the source voltage by 90°. Thus, as shown in the phasor diagram of Figure 14–17(b), the current through the capacitive branch leads the current through the resistive branch by 90°.

Question: Is Kirchhoff's current law applicable in the analysis of parallel RC *circuits?*

Recall that Kirchhoff's current law is the basic law of all parallel circuits. As you know, it states that the sum of the branch currents in a parallel circuit is equal to the total current. Kirchhoff's current law is applicable to parallel *RC* circuits, but once again the 90° phase shift between I_R and I_C must be considered. The total current is the phasor sum of the branch currents. By Kirchhoff's current law, the total current is given by

(14–24)

$$I_T = \sqrt{I_R^2 + I_C^2}$$

Figure 14–18, known as a *current triangle,* illustrates the phase relationship of the currents.

EXAMPLE 14–14 Compute the total current for the circuit of Figure 14–19.

■ Solution:

FIGURE 14–19 Circuit for Example 14–14

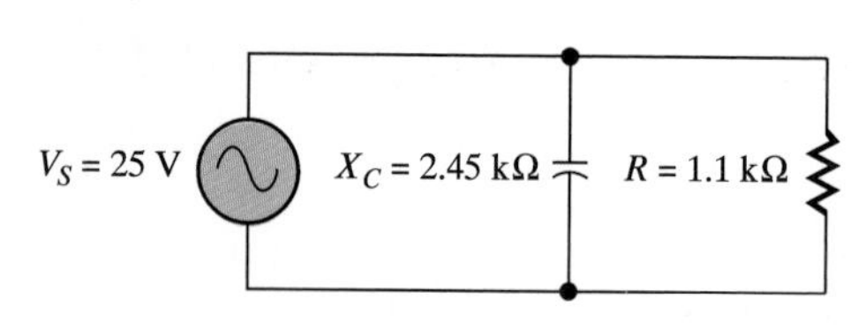

$$I_C = \frac{V_S}{X_C}$$

$$= \frac{25\text{ V}}{2450\ \Omega} = 10.2\text{ mA}$$

$$I_R = \frac{V_S}{R}$$

$$= \frac{25\text{ V}}{1100\ \Omega} = 22.7\text{ mA}$$

$$I_T = \sqrt{I_R^2 + I_C^2}$$

$$= \sqrt{(22.7 \times 10^{-3})^2 + (10.2 \times 10^{-3})^2} = 24.886\text{ mA}$$

EXAMPLE 14–15 For the circuit of Figure 14–20, compute the value of the resistance.

■ Solution:

FIGURE 14–20 Circuit for Example 14–15

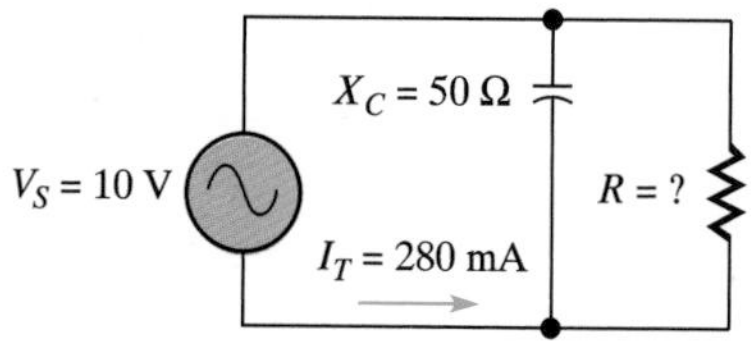

$$I_C = \frac{V_S}{X_C}$$

$$= \frac{10\text{ V}}{50\ \Omega} = 200\text{ mA}$$

$$I_T^2 = I_R^2 + I_C^2$$

Transposing,

$$I_R^2 = I_T^2 - I_C^2$$

Then,

$$I_R = \sqrt{I_T^2 - I_C^2} = \sqrt{(280 \times 10^{-3}\ \text{A})^2 - (200 \times 10^{-3}\ \text{A})^2} = 196\ \text{mA}$$

$$R = \frac{V_S}{I_R}$$

$$= \frac{10\ \text{V}}{(196 \times 10^{-3}\ \text{A})} = 51\ \Omega$$

Impedance in a Parallel *RC* Circuit

The impedance of a parallel *RC* circuit is computed in a manner similar to that for a parallel *RL* circuit. First, the **admittance** is determined by finding the phasor sum of the conductance and susceptance. The impedance is then determined by taking the reciprocal of the admittance:

(14–25) $$Y = \sqrt{G^2 + B^2}$$

and

(14–26) $$Z = \frac{1}{Y}$$

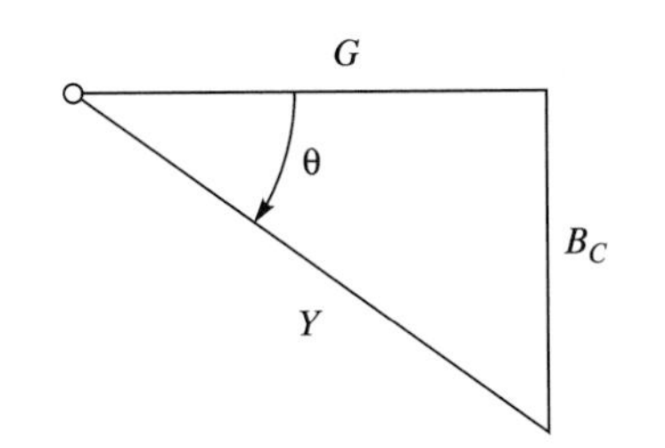

FIGURE 14–21 Admittance triangle

Since I_T, I_R, and I_C are directly proportional to admittance, conductance, and susceptance, respectively, it is possible to illustrate the phase relationships in a parallel *RC* circuit using an *admittance triangle*. An admittance triangle is shown in Figure 14–21.

EXAMPLE 14–16 A 16 kΩ resistor is in parallel with 16 kΩ capacitive reactance. Compute the impedance.

Solution:

$$Y = \sqrt{G^2 + B^2}$$

$$= \sqrt{(1/16000\ \Omega)^2 + (1/16000\ \Omega)^2} = 88.388\ \mu\text{S}$$

$$Z = \frac{1}{Y}$$

$$= \frac{1}{(88.388 \times 10^{-6})} = 11314\ \Omega$$

Circuit Phase Angle in a Parallel *RC* Circuit

Question: What is the phase angle of a parallel* RC *circuit?

The phase angle of a parallel *RC* circuit is illustrated in Figure 14–22. Notice that a current triangle is drawn on the phasor for the source voltage. The *circuit phase angle* is

FIGURE 14–22 Parallel *RC* circuit phase angle

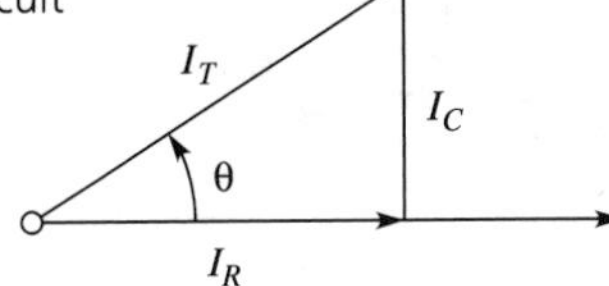

FIGURE 14–23 Affect of frequency on phase angle

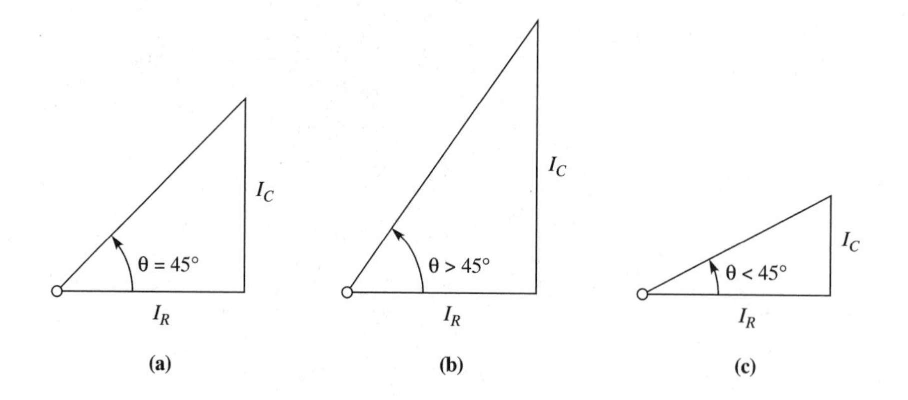

the angle between the source voltage and current phasors. Notice that this phase angle opens in a positive direction, while that of a parallel *RL* circuit opens in a negative direction.

Question: How is the circuit phase angle computed?

In the phasor diagram of Figure 14–22, notice that the phasor representing I_C is opposite the phase angle and that the phasor representing I_R is adjacent to the phase angle. Thus, the trigonometric functions are used to calculate the circuit phase angle:

(14–27)
$$\tan\theta = \frac{I_C}{I_R} \quad \text{and} \quad \theta = \arctan\frac{I_C}{I_R}$$

Question: How is the phase angle affected by changes in frequency?

FIGURE 14–24 Circuit for Example 14–17

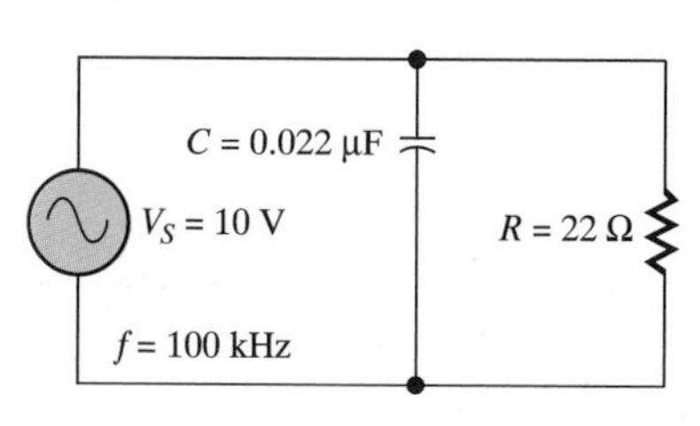

The value of the phase angle gives an indication of the relative values of I_C and I_R. In Figure 14–23(a), notice that if the current phasors are equal, the phase angle will be 45°. If the frequency of the applied voltage increases, the reactance of the capacitor decreases and causes an increase in the current through the capacitive branch. As shown in Figure 14–23(b), the phasor representing I_C is longer, and the phase angle increases. Thus, if the phase angle of a parallel *RC* circuit is greater than 45°, the reactance is smaller than the resistance. If, on the other hand, the frequency decreases, the capacitive reactance increases. The current through the capacitive branch decreases. The phasor representing I_C is shorter, and as shown in Figure 14–23(c), the phase angle decreases. If the phase angle of a parallel *RC* circuit is less than 45°, the reactance is greater than the resistance.

EXAMPLE 14–17 For the circuit of Figure 14–24, compute the phase angle.

■ Solution:

$$X_C = \frac{1}{2\pi fC}$$

$$= \frac{1}{2(3.14)(100 \times 10^3\ \text{Hz})(0.022 \times 10^{-6})} = 72.343\ \Omega$$

$$I_R = \frac{V_S}{R}$$

$$= \frac{10\text{ V}}{22\ \Omega} = 455\text{ mA}$$

$$I_C = \frac{V_S}{X_C}$$

$$= \frac{10\text{ V}}{72.3\ \Omega} = 138\text{ mA}$$

$$\theta = \text{arc tan } \frac{I_C}{I_R}$$

$$= \text{arc tan } \frac{0.138\text{ A}}{0.455\text{ A}} = \text{arc tan } 0.303 = 16.872°$$

EXAMPLE 14–18 In the circuit of Figure 14–25, what value of capacitor will produce a phase angle of approximately 51.6°?

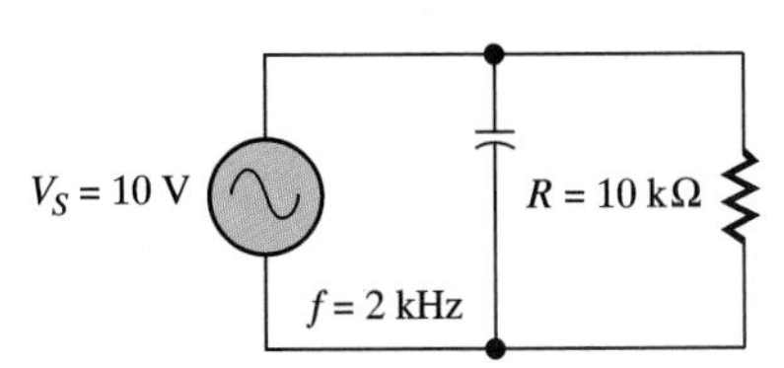

FIGURE 14–25 Circuit for Example 14–18

■ Solution:

1. Compute the current I_R:

$$I_R = \frac{V_S}{R}$$

$$= \frac{10\text{ V}}{10000\ \Omega} = 1\text{ mA}$$

2. Compute the value of I_C required to produce the desired value of phase angle:

$$\tan\theta = \frac{I_C}{I_R}$$

$$I_C = I_R \tan\theta$$

$$= (1 \times 10^{-3}\text{ A})(\tan 51.6) = (1 \times 10^{-3}\text{ A})(1.26) = 1.26\text{ mA}$$

3. Compute what reactance is required to produce 1.26 mA of current in the capacitive branch:

$$X_C = \frac{V_S}{I_C}$$

$$= \frac{10\text{ V}}{(1.26 \times 10^{-3}\text{ A})} = 7.937\text{ k}\Omega$$

4. Compute capacitance value:

$$X_C = \frac{1}{2\pi fC}$$

$$C = \frac{1}{2\pi fX_C}$$

$$= \frac{1}{2(3.14)(2000\text{ Hz})(7937\ \Omega)} = 0.01\ \mu\text{F}$$

NOTE

In presenting *RC* circuit analysis, it would have been possible just to make the following statement: "Proceed with the analysis in the same manner as for *RL* circuits but have the phase angles open in the opposite direction." This statement is true, because the methods of *RC* and *RL* circuit analysis are practically identical. The procedure is, however, so important that a second look at it in this section was well worth the time and effort! Once again, make sure that you understand these principles of *RC* and *RL* circuit analysis. This knowledge will be absolutely necessary when all three effects (resistance, inductance, and capacitance) are brought together in the practical ac circuit.

SECTION 14–4 REVIEW

1. In a parallel *RC* circuit, will the current through the capacitive branch lead or lag the current through the resistive branch?
2. Which phasor is used as the reference phasor in the current triangle?
3. In a parallel *RC* circuit, as the frequency of the applied voltage increases, does the phase angle increase or decrease?
4. In a series *RC* circuit, as the frequency decreases, does the total current increase or decrease?
5. In a parallel *RC* circuit, the capacitive current is 40 mA and the resistive current is 25 mA. What is the power factor?
6. If the apparent power is 40 mVA and the true power is 28 mW, what is the power factor?

14–5 EXAMPLES OF *RC* CIRCUIT ANALYSIS

In this section, examples of *RC* circuit analysis are presented. Two are rather straightforward, while two others require some manipulation to obtain the solution. In *RC* circuit analysis, it is always good practice to draw a representation of the phasor diagrams prior to beginning the analysis. They do not have to be drawn to exact scale; their purpose is merely to show the relationships that exist between the currents and voltages within the circuit.

EXAMPLE 14–19 For the circuit of Figure 14–26(a), compute the following: (a) X_C, (b) Z, (c) I_T, (d) V_R, (e) V_C, (f) θ, (g) P_{react}, (h) P_{true}, and (i) P_{app}.

Solution:

The impedance triangle for this circuit is shown in Figure 14–27(b).

a. Compute X_C:

$$X_C = \frac{1}{2\pi fC}$$

$$= \frac{1}{2(3.14)(1 \times 10^6 \text{ Hz})(100 \times 10^{-12} \text{ F})} = 1592\ \Omega$$

b. Compute Z:

$$Z = \sqrt{R^2 + X_C^2}$$

$$= \sqrt{(2.2 \times 10^3\ \Omega)^2 + (1.592 \times 10^3\ \Omega)^2} = 2.716\ \text{k}\Omega$$

FIGURE 14–26 Circuit and phasor diagram for Example 14–19

FIGURE 14–27 Circuit and phasor diagram for Example 14–20

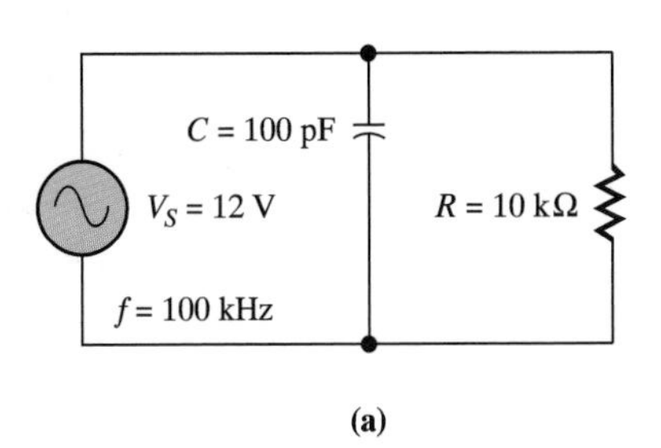

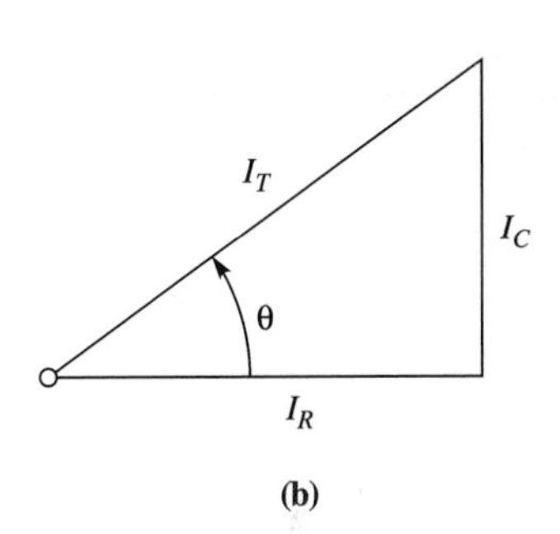

c. Compute I_T:

$$I_T = \frac{V_S}{Z}$$

$$= \frac{50 \text{ V}}{2716\ \Omega} = 18.4 \text{ mA}$$

d. Compute V_R:

$$V_R = I_R R$$

$$= (18.4 \times 10^{-3} \text{ A})(2.2 \times 10^{3}\ \Omega) = 40.5 \text{ V}$$

e. Compute V_C:

$$V_C = I_C X_C$$

$$= (18.4 \times 10^{-3} \text{ A})(1.592 \times 10^{3}\ \Omega) = 29.3 \text{ V}$$

f. Compute θ:

$$\theta = \text{arc tan } \frac{V_C}{V_R}$$

$$= \text{arc tan } \frac{29.3 \text{ V}}{40.5 \text{ V}} = \text{arc tan } 0.723 = 35.884°$$

g. Computer P_{react}:

$$P_{\text{react}} = I_C{}^2 X_C$$

$$= (18.4 \times 10^{-3} \text{ A})^2(1.592 \times 10^{3}\ \Omega) = 539 \text{ mVAR}$$

h. Computer P_{true}:

$$P_{\text{true}} = I_R^2 R$$
$$= (18.4 \times 10^{-3}\text{ A})^2(2.2 \times 10^3\ \Omega) = 745\text{ mW}$$

i. Compute P_{app}:

$$P_{\text{app}} = \sqrt{(P_{\text{true}})^2 + (P_{\text{react}})^2}$$
$$= \sqrt{(0.745\text{ W})^2 + (0.539\text{ VAR})^2} = 920\text{ mVA}$$

EXAMPLE 14–20: For the circuit of Figure 14–27(a), compute the following: (a) X_C, (b) I_R, (c) I_C, (d) I_T, (e) θ, (f) PF, and (g) P_{true}

■ Solution:

The current triangle for the circuit is shown in Figure 14–27(b).

a. Compute X_C:

$$X_C = \frac{1}{2\pi f C}$$
$$= \frac{1}{2(3.14)(100 \times 10^3\text{ Hz})(100 \times 10^{-12}\text{ F})} = 15.9\text{ k}\Omega$$

b. Compute I_R:

$$I_R = \frac{V_S}{R}$$
$$= \frac{12\text{ V}}{(10 \times 10^3\ \Omega)} = 1.2\text{ mA}$$

c. Compute I_C:

$$I_C = \frac{V_S}{X_C}$$
$$= \frac{12\text{ V}}{(15.9 \times 10^3\ \Omega)} = 0.755\text{ mA}$$

d. Compute I_T:

$$I_T = \sqrt{I_R^2 + I_C^2}$$
$$= \sqrt{(1.2 \times 10^{-3}\text{ A})^2 + (0.755 \times 10^{-3}\text{ A})^2} = 1.429\text{ mA}$$

e. Compute θ:

$$\theta = \arctan \frac{I_C}{I_R}$$
$$= \arctan \frac{(0.755 \times 10^{-3}\text{ A})}{(1.2 \times 10^{-3}\text{ A})} = \arctan 0.620 = 32.2°$$

f. Compute the power factor:

$$\text{PF} = \cos\theta$$
$$= \cos 32.2° = 0.846$$

g. Compute P_{true} (using a different approach than used in Example 4–19):

$$P_{app} = V_S I_T$$
$$= (12 \text{ v})(1.429 \times 10^{-3} \text{ A}) = 17.1 \text{ mVA}$$

Then,

$$P_{true} = P_{app}\text{PF}$$
$$= (17.1 \times 10^{-3} \text{ VA})(0.846) = 14.466 \text{ mW}$$

EXAMPLE 14–21 For the circuit of Figure 14–28(a), compute the following: (a) V_S and (b) C.

■ **Solution:**

The impedance triangle for this circuit is shown in Figure 14–28(b).

a. First compute V_R:

$$V_R = IR$$
$$= (15 \times 10^{-3} \text{ A})(1000 \ \Omega) = 15 \text{ V}$$

Then compute V_C:

$$\tan\theta = \frac{V_C}{V_R}$$
$$V_C = V_R(\tan\theta)$$
$$= (15 \text{ V})(\tan 68) = (15 \text{ V})(2.48) = 37.2 \text{ V}$$

FIGURE 14–28 Circuit and phasor diagram for Example 14–21

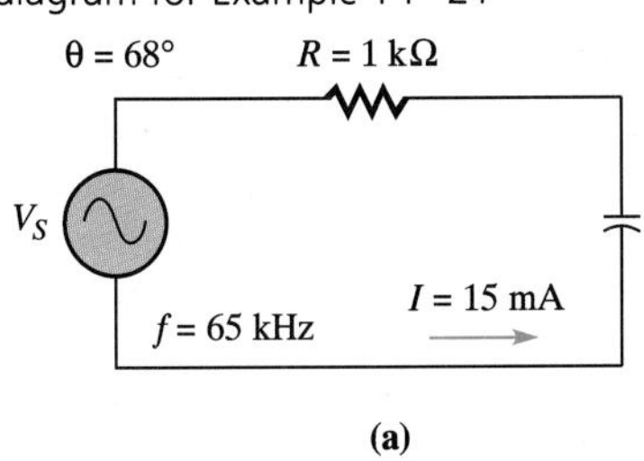

(a)

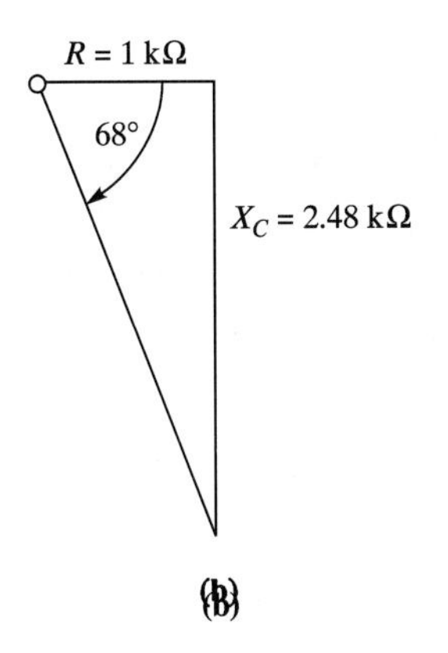

(b)

Next compute V_S:

$$V_S = \sqrt{V_R^2 + V_C^2}$$
$$= \sqrt{(15 \text{ V})^2 + (37.2 \text{ V})^2} = 40.11 \text{ V}$$

b. Compute X_C:

$$X_C = \frac{V_C}{I}$$
$$= \frac{37.2 \text{ V}}{15 \times 10^{-3} \text{ A}} = 2.48 \text{ k}\Omega$$

Compute C:

$$X_C = \frac{1}{2\pi f C}$$

and

$$C = \frac{1}{2\pi f X_C}$$
$$= \frac{1}{2(3.14)(65 \times 10^3 \text{ Hz})(2480 \ \Omega)} = 987 \text{ pF}$$

In actual practice, a standard value of 1000 pF would probably be chosen.

EXAMPLE 14–22 For the circuit of Figure 14–29(a), compute the following: (1) I_T, (b) I_R, and (c) I_C.

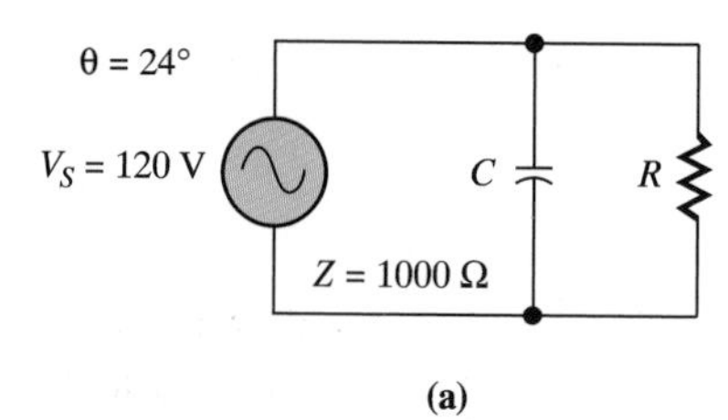

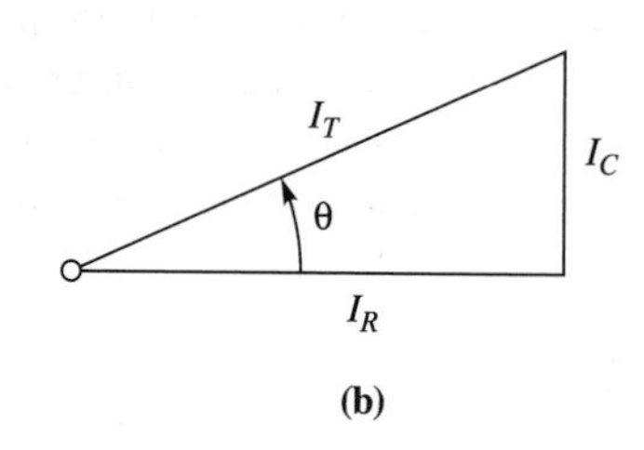

FIGURE 14–29 Circuit and phasor diagram for Example 14–22

Solution:

The current triangle for this circuit is shown in Figure 14–39(b).

a. Compute I_T:

$$I_T = \frac{V_S}{Z} = \frac{120 \text{ V}}{1000\ \Omega} = 120 \text{ mA}$$

b. Compute I_R:

$$\cos\theta = \frac{I_R}{I_T}$$

$$I_R = I_T \cos\theta$$

$$= (0.120 \text{ A})(\cos 24°) = (0.120 \text{ A})(0.914) = 110 \text{ mA}$$

c. Compute I_C:

$$\sin\theta = I_C/I_T$$

$$I_C = I_T \sin\theta$$

$$= (0.120 \text{ A})(\sin 24°) = (0.120 \text{ A})(0.407) = 48.84 \text{ mA}$$

14–6 APPLICATIONS OF *RC* CIRCUITS

The major applications of *RC* circuits follow closely the same lines as those for *RL* circuits: filters and lead or lag networks. Once again, the frequency-sensitive characteristics of the capacitor are taken advantage of. As has been mentioned several times before, the effects of inductance and capacitance are opposites, particularly in the case of the phase angle and how it responds to changes in frequency. You can note these opposites by comparing the circuits presented in the following paragraphs to similar circuits presented in Chapter 12 for *RL* circuits. Recall also that what is presented here is only an introduction; these circuits will be covered in detail in Chapter 18.

The High Pass Filter

Question: How may an* RC *circuit be used as a high-pass filter?

A high-pass filter passes all frequencies above a cutoff frequency and blocks all those below it. An *RC* high-pass filter is shown in Figure 14–30(a). Notice that the output is taken across the resistor. The value of the output voltage depends upon the amount of current flow. The amount of current is determined by the circuit impedance, which is directly proportional to the reactance. Thus, at low frequencies, where the reactance is high, both the current flow and output voltage are relatively low. As the frequency increases, the reactance decreases. The impedance decrease allows an increase in both the current and the output voltage.

The graph of this relationship between frequency and output voltage is shown in Figure 14–30(b). Notice at zero hertz, dc, the capacitor is an infinite impedance and no current flows. The output voltage at this time is zero. As the frequency increases, the output voltage rises due to the decrease in impedance and increase in current. The *cutoff fre-*

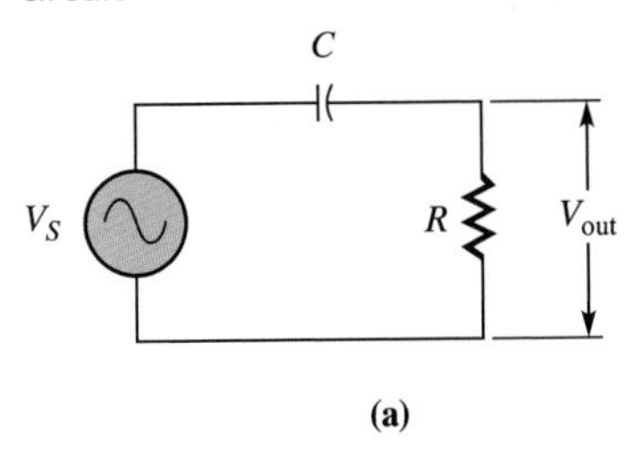

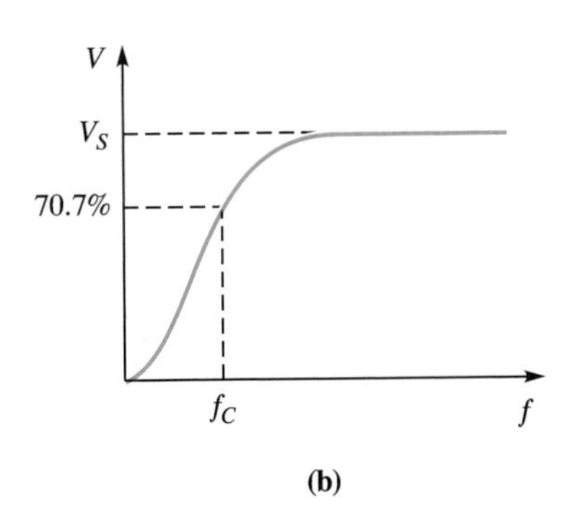

FIGURE 14–30 *RC* high-pass filter circuit

quency f_c is the frequency at which the output voltage drops to 70.7% of its maximum value ($f_C = 1/2\pi RC$). Thus, all frequencies above the cutoff are passed, and all those below it are blocked.

Question: How is the high-pass filter used in order to couple a signal from an amplifier?

The circuit shown in Figure 14–31(a) is the most common way of coupling amplifier signals. The signal from one amplifier is coupled to the input of the next. The signal to the left of the capacitor has both a dc and an ac component as shown. (This signal has been conveniently shown as a sine wave.) The dc component must be kept away form the input of amplifier 2. Figure 14–31(b) shows the situation with only the dc component present. The capacitor charges to this dc value, and no voltage is developed across the resistor. In Figure 14–31(c), the ac component is included. Notice that the dc voltage

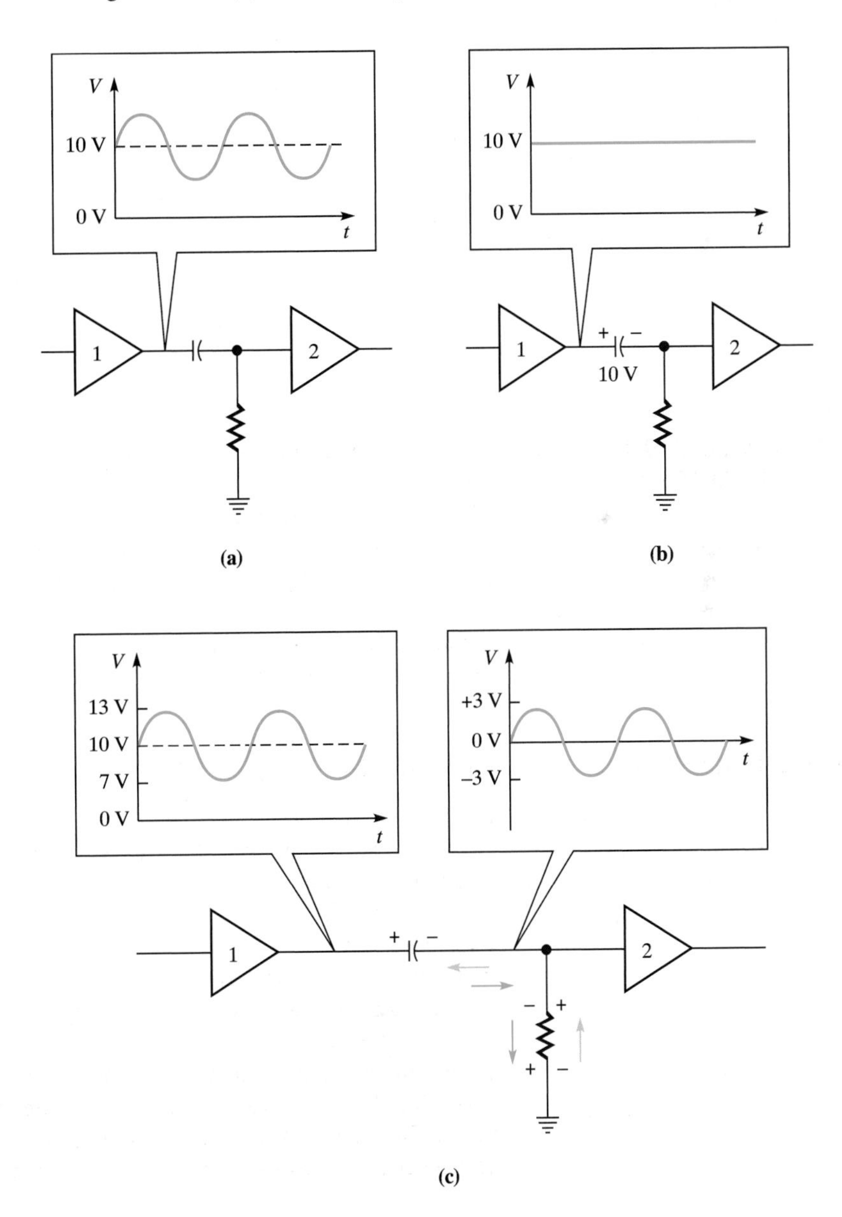

FIGURE 14–31 High-pass filter circuit used to couple amplifier signals

increases to 13 V and decreases to 7 V in a sinusoidal manner. When the voltage increases, the capacitor must take on charge, and current flows through the resistor in the direction shown. When the voltage decreases, the capacitor must lose charge, and current flows in the direction shown. Thus, the output voltage is pure ac, produced by the alternating current through the resistor.

The *RC* Low-Pass Filter

An *RC* circuit used as a low-pass filter is shown in Figure 14–32(a). Notice that the output is taken across the capacitor. The output voltage is greatest when no current flows, which occurs at zero hertz when the capacitor is an infinite impedance. As the frequency increases, the reactance of the capacitor decreases, and current flows. Thus, some of the input voltage is now dropped across the resistor, leaving less voltage across the output capacitor. As the frequency continues to increase, the current increases, more voltage is dropped across the resistor, and the output voltage decreases.

A graph of this relationship between output voltage and frequency is shown in Figure 14–32(b). Notice that at zero hertz, the reactance of the capacitor is infinite, and the entire input voltage appears across the output. As the frequency increases, the decrease in reactance allows current to increase. The drop across the resistor now reduces the output voltage. Once again, the cutoff frequency f_c occurs at the frequency at which the output voltage has dropped to 70.7% of its maximum value. All frequencies below this cutoff frequency are passed, while all those above are blocked.

FIGURE 14–32 *RC* low-pass filter circuit

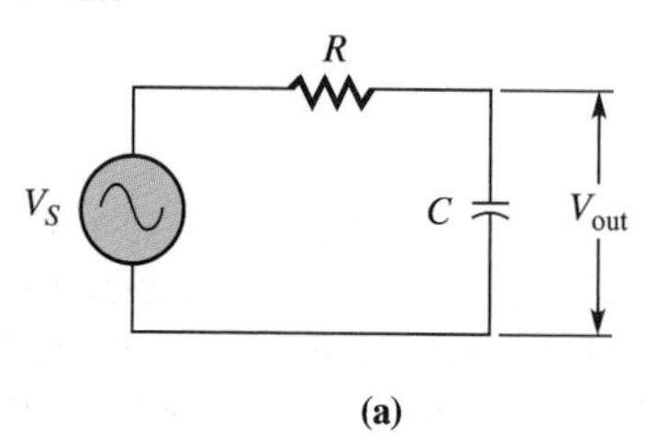

(a)

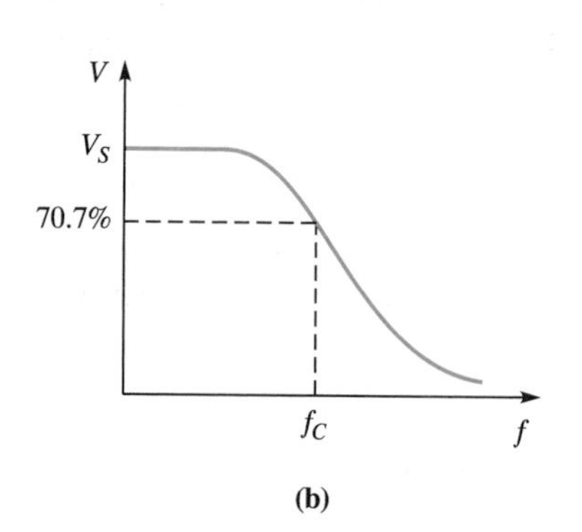

(b)

Question: How is an RC *low-pass filter used to smooth the output voltage of a rectifier?*

A **rectifier** is a circuit that converts a sine wave of voltage to dc, as illustrated in Figure 14–33. The input to the rectifier may be the sine wave from the commercial power line. The output, as shown, is the peak value of the input, with the negative alternation inverted. This output is a dc voltage, but the tremendous amount of ripple makes it worthless for many applications. An *RC* low-pass filter can be used to remove the ripple frequency from the voltage. As shown in Figure 14–33, the capacitor charges very quickly on the first positive-going half-cycle. This rapid charge time is due to a very short time constant in this direction. When the voltage from the rectifier drops to zero, the output voltage cannot follow due to the charge on the capacitor. The capacitor attempts to discharge through the load. If this time constant is long enough, the output voltage will not change too greatly. Thus, as shown, the output voltage is smoothed, making it useful as the dc supply voltage in circuits such as amplifiers.

The *RC* Lead Network

Question: How does a lead network cause the output voltage to lead the input?

The output voltage of a **lead network** leads the input by some angle. An *RC* lead network is shown in Figure 14–34(a). Notice that the input is across the series combination of resistance and capacitor, while the output is taken across the resistor. The angle of lead is illustrated in the phasor diagram of Figure 14–34(b). As you know, the voltage across the resistor is in phase with the current. Since this current leads the source voltage by the phase angle, the resistor voltage does so as well. Notice that the angle by which the output voltage leads the source voltage is the circuit phase angle. Thus, the angle of lead can be determined using the analysis procedures for series *RC* circuits.

FIGURE 14–33 *RC* low-pass filter used as a rectifier

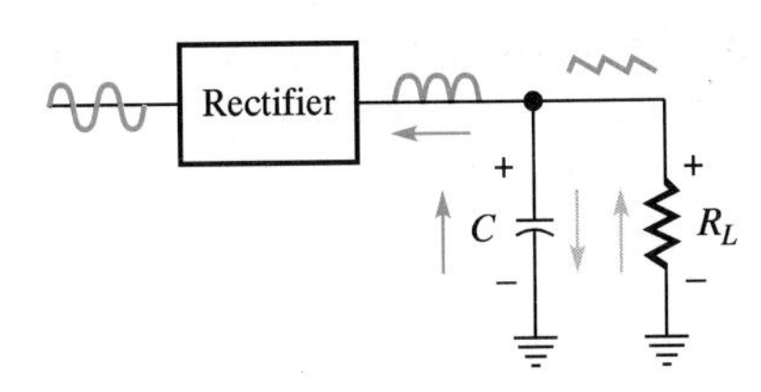

FIGURE 14–34 *RC* lead network

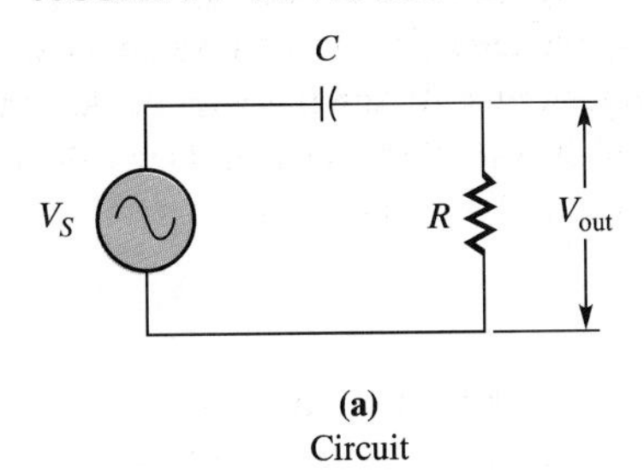

(a) Circuit

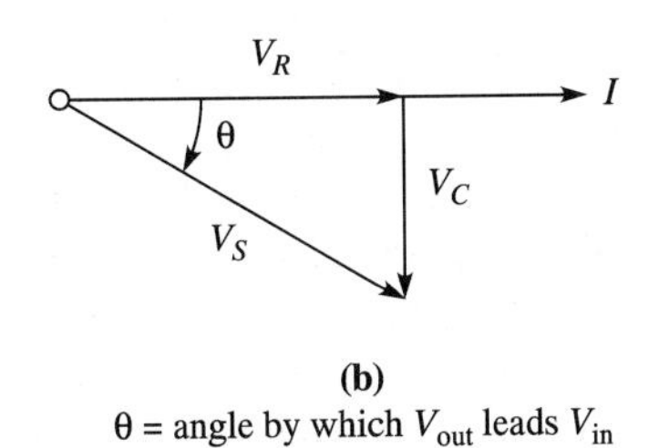

(b) θ = angle by which V_{out} leads V_{in}

FIGURE 14–35 *RC* lag network

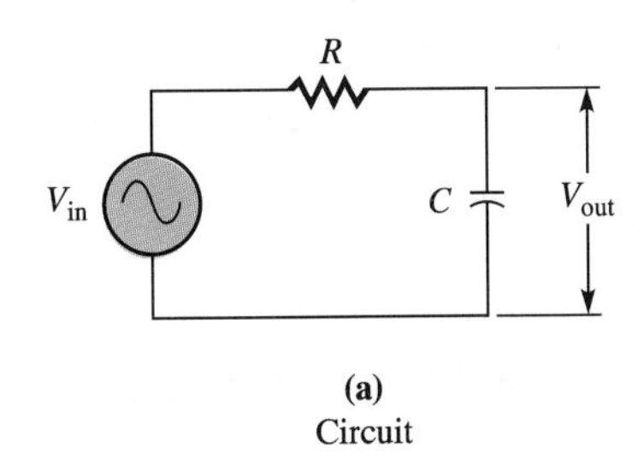

(a) Circuit

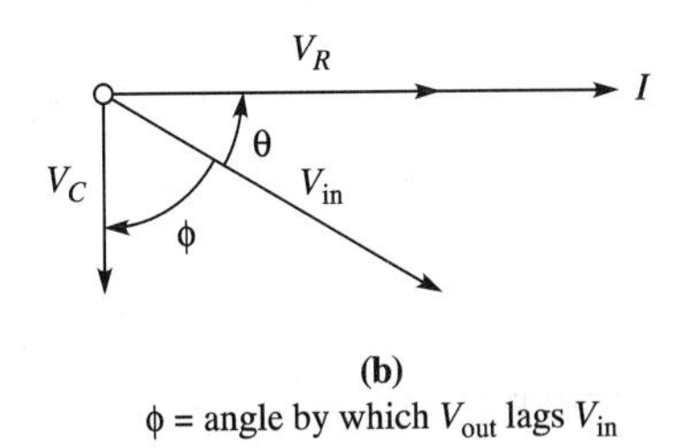

(b) ϕ = angle by which V_{out} lags V_{in}

The *RC* Lag Network

Question: How does an* RC *lag network cause the output voltage to lag the input?

An *RC* **lag network** is one in which the output voltage lags the input by some angle. An *RC* lag network is shown in Figure 14–35(a). The input is across the series combination of resistance and capacitance, while the output is taken across the capacitor. The angle of lag is illustrated in the phasor diagram of Figure 14–35(b). Through capacitor action, its voltage lags the current by some angle. The output voltage V_{out} lags the input voltage by the angle Φ (phi). To find Φ, it is first necessary to determine the value of θ. The angles θ and Φ are complementary, so

(14–28)
$$\Phi = 90° - \theta$$

EXAMPLE 14–23 An *RC* lead network is composed of a 39 kΩ resistance in series with a 14.4 kΩ reactance. Compute the angle of lead.

Solution:

The angle of lead is equal to the value of the circuit phase angle θ.

$$\theta = \arc \tan \frac{X_C}{R}$$

$$= \arc \tan \frac{14400\ \Omega}{39000\ \Omega} = \arc \tan 0.369 = 20.265°$$

EXAMPLE 14–24 If, in Example 14–23, the output is taken across the capacitor, what is the angle of lag?

Solution:

$$\Phi = 90° - \theta$$

$$= 90° - 20.3° = 69.7°$$

Computing the Output Voltages

The output voltage of a lead or lag network may be computed using the principles of series RC analysis presented previously. A much simpler way, however, is in the use of the voltage divider equation. These equations were introduced in Chapter 1.

For the lead network,

(14–29)

$$V_{out} = V_S \frac{R}{\sqrt{(R^2 + X_C^2)}}$$

For the lag network,

(14–30)

$$V_{out} = V_S \frac{X_C}{\sqrt{(R^2 + X_C^2)}}$$

EXAMPLE 14–25 For the lead network of Figure 14–36, compute the output voltage.

FIGURE 14–36 Circuit for Example 14–25

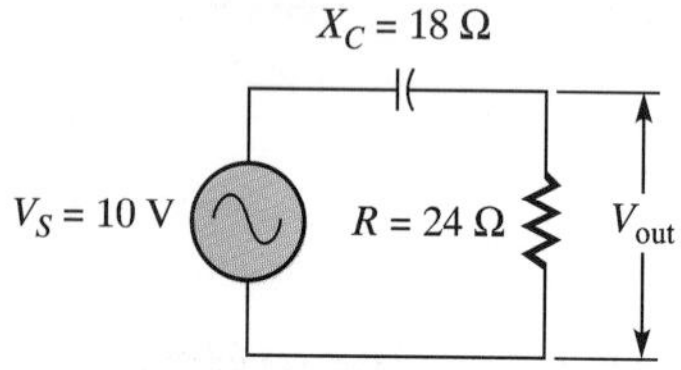

Solution:

$$V_{out} = V_S \frac{R}{\sqrt{(R^2 + X_C^2)}}$$

$$= (10\text{ V})\frac{24\ \Omega}{\sqrt{(24\ \Omega)^2 + (18\ \Omega)^2}} = (10\text{ V})(.8\ \Omega) = 8\text{ V}$$

14–7 TROUBLESHOOTING *RC* CIRCUITS

Capacitors can fail in several ways. They can develop two of the problems most common in electronics: opens and shorts. An open occurs when the lead that connects the plates to the external circuit burns open. A short is caused by the dielectric breaking down and allowing current flow between the plates. The most common fault, however, is the "leaky" capacitor. It is not completely shorted, but it does allow some current flow between the plates. Once again, circuit faults can cause the current to be too high, too low, or zero.

Leaky Capacitor

Recall in Section 14–6 where it was stated that an RC high-pass filter is one of the most common ways of coupling signals from one amplifier to another. The circuit is shown again in Figure 14–37(a). Recall that the purpose of the capacitor is to block the dc component and pass the ac component along to the next amplifier. It does this by charging to the average dc level and then taking on charge as the voltage increases and discharging as the voltage decreases. This charging and discharging produces ac current through the load, and the ac portion of the signal is passed to the next amplifier. Figure 14–37(b) shows how the signal should appear on the display of an oscilloscope.

Question: What is leakage current and what is the effect of it?

If, however, the capacitor allows current to leak through the dielectric, a dc voltage component will be present across the resistor. Notice the difference in the oscilloscope presentation of Figure 14–37(c). The signal is now *above* the baseline, which indicates that it

FIGURE 14–37 High-pass filter circuit with leaky capacitor

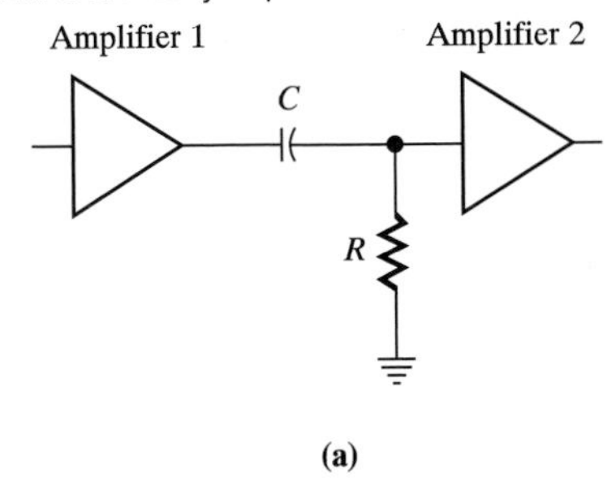

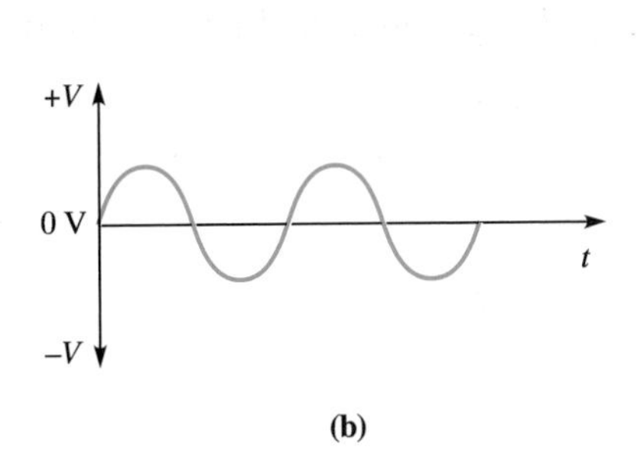

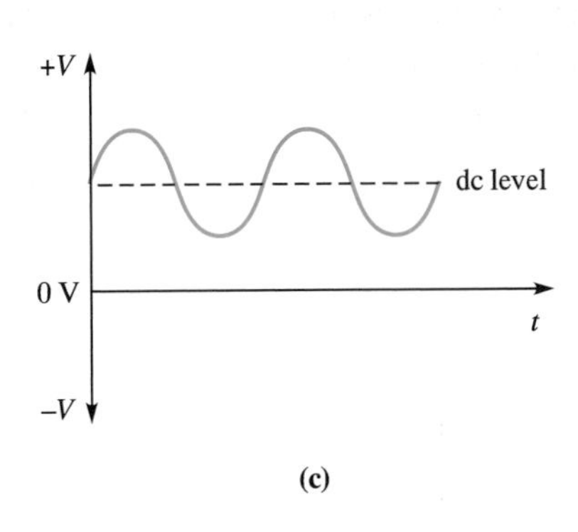

FIGURE 14–38 Oscilloscope display when capacitor is open

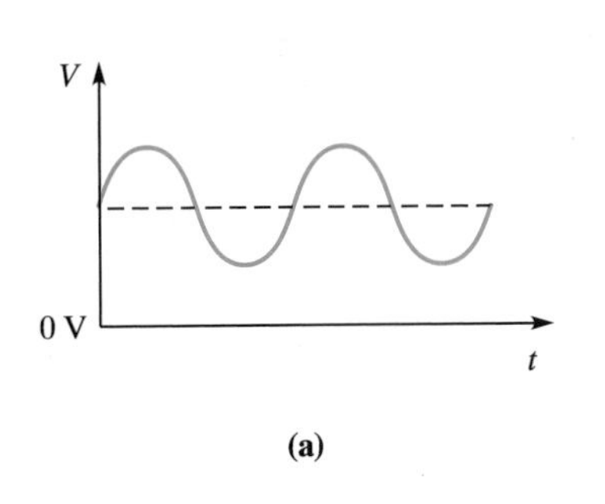

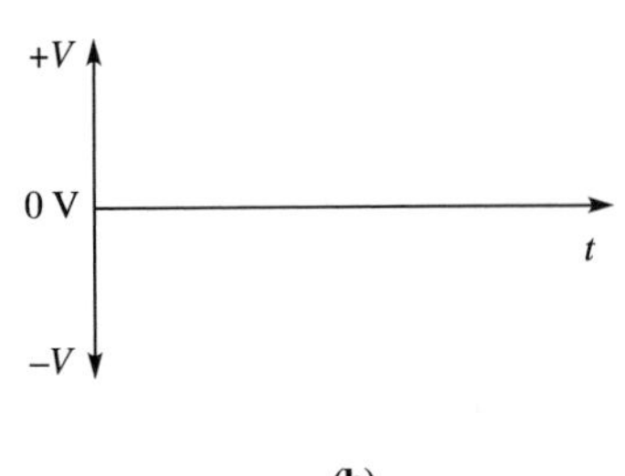

(b)
Open capacitor — no signal into amplifier 2

contains a dc component. If a capacitor is suspected of current leakage, it is removed from the circuit and tested with an impedance bridge. With this instrument, the full working voltage is placed across the capacitor, and the amount of leakage current is indicated. A high leakage current is an indication of a leaky capacitor.

Open Capacitor

When a capacitor is open, it loses all ability to store charge. Thus, if the capacitor does not charge, no current flows. Looking once again at the circuit of Figure 14–37(a), if the capacitor opens, no charging or discharging current flows. Thus, no current flows through the resistor, and no signal is transferred to amplifier 2. Whatever system this circuit is a part of will be open at this point. This is illustrated by the oscilloscope displays of Figure 14–38. Once again, the best method for checking for an open capacitor is using an impedance bridge. If the capacitor is of a large enough value, on the order of 1 μF, opens can be tested with an ohmmeter. Across a normal capacitor, the ohmmeter should first indicate a short and then move toward infinity as the capacitor charges. Across an open capacitor, an ohmmeter will indicate infinity.

Shorted Capacitor

If a capacitor shorts, current flows through it rather than into and out of it. Thus, in Figure 14–39, the ac and dc components will be present on both sides of the capacitor. This condition causes problems in what are known as the *bias circuits* in the amplifier. Recall that an *RC* circuit can be used as a filter in the output of a dc power supply. If the capacitor should short in this application, a heavy current would flow and the current protection device would open in order to protect the source. Across a shorted capacitor, an ohmmeter will indicate zero ohms.

FIGURE 14–39 dc component on both sides of capacitor when capacitor is shorted

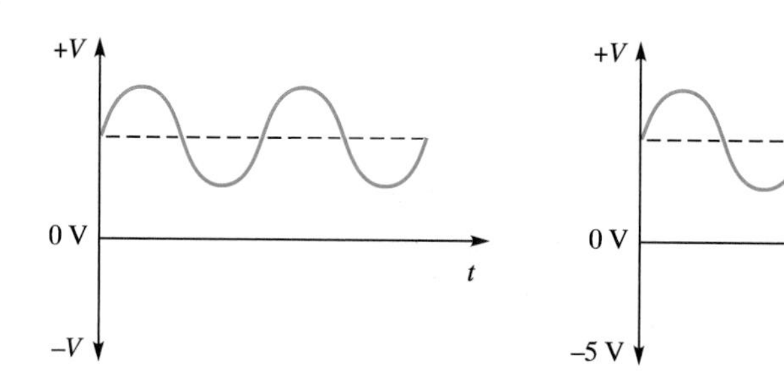

CAUTION

The ohmmeter usually supplies only a small portion of the voltage for which a capacitor is rated. A capacitor that is found to be "good" with an ohmmeter check can break down in the circuit when higher voltages are applied. Thus, an ohmmeter only indicates that a capacitor is "bad," not that it is necessarily "good."

SECTION 14–7 REVIEW

1. What is the best instrument for detecting leakage current in a capacitor?
2. What is the circuit indication of leakage current?
3. What is the circuit indication of a shorted capacitor?
4. What is the circuit indication of an open capacitor?

CHAPTER SUMMARY

1. A pure capacitance causes the current to lead the voltage by 90°.
2. In a series *RC* circuit, the voltage drop across the resistor leads the voltage across the capacitor by 90°.
3. The source voltage in a series *RC* circuit is the phasor sum of the voltages across the resistor and capacitor.
4. The impedance of a series *RC* circuit is the phasor sum of the resistance and capacitive reactance.
5. The total current in an parallel *RC* circuit is the phasor sum of the currents through the resistive and capacitive branches.
6. The admittance of an *RC* circuit is the phasor sum of the conductance and capacitive susceptance.
7. The impedance of a parallel *RC* circuit is the reciprocal of the admittance.
8. In a series *RC* circuit, the phase angle is inversely proportional to frequency.
9. In a series *RC* circuit, the phase angle is inversely proportional to capacitance.
10. In a parallel *RC* circuit, the phase angle is directly proportional to frequency.
11. In a parallel *RC* circuit, the phase angle is directly proportional to capacitance.
12. A capacitor stores electrical energy and returns it to the source on each half-cycle.
13. *RC* circuits find use as electronic filters and as lead-lag networks.
14. Common faults occurring in a capacitor are opens, shorts, and leakage current.
15. The impedance bridge is the best instrument for testing capacitors.

KEY TERMS

capacitive reactance
impedance
true power
reactive power
apparent power
power factor
admittance
lead network
lag network

EQUATIONS

(14–1) $X_C = \frac{1}{2\pi fC}$

(14–2) $I_C = \frac{V_C}{X_C}$

(14–3) $X_C = \frac{V_C}{I_C}$

(14–4) $V_C = I_C X_C$

(14–5) $X_{CT} = X_{C1} + X_{C2} + X_{C3}$

(14–6) $B_T = B_1 + B_2 + B_3$

(14–7) $X_{CT} = 1/B_T$

(14–8) $V_S^2 = V_R^2 + V_C^2$

(14–9) $V_S = \sqrt{V_R^2 + V_C^2}$

(14–10) $Z = \sqrt{R^2 + X_C^2}$

(14–11) $I = \frac{V_S}{Z}$

(14–12) $V_S = IZ$

(14–13) $Z = \frac{V_S}{I}$

(14–14) $\theta = \arctan \frac{V_C}{V_R}$

(14–15) $\theta = \arctan \frac{X_C}{R}$

(14–16) $P_{react} = IX_C$

(14–17) $P_{react} = V_S I$

(14–18) $P_{react} = \frac{V_S^2}{X_C}$

(14–19) $P_{app} = \sqrt{(P_{true})^2 + (P_{react})^2}$

(14–20) $P_{app} = I^2 Z$

(14–21) $P_{app} = \frac{V_S^2}{Z}$

(14–22) $P_{app} = V_S I$

(14–23) $P_{true} = P_{app}(\cos \theta)$

(14–24) $I_T = \sqrt{I_R^2 + I_C^2}$

(14–25) $Y = \sqrt{G^2 + B^2}$

(14–26) $Z = \frac{1}{Y}$

(14–27) $\theta = \arctan \frac{I_C}{I_R}$

(14–28) $\Phi = 90° - \theta$

(14–29) $V_{out} = V_S \frac{R}{\sqrt{R^2 + X_C^2}}$

(14–30) $V_{out} = V_S \frac{X_C}{\sqrt{R^2 + X_C^2}}$

Variable Quantities

X_C = capacitive reactance
I_C = capacitor current
X_{CT} = total capacitive reactance
X_{C1} = reactance of capacitor C_1
X_{C2} = reactance of capacitor C_2
Z = impedance
B_C = capacitive susceptance
Y = admittance

TEST YOUR KNOWLEDGE

1. At the instant a capacitor begins to charge, is the charging current maximum or zero?

2. Is capacitive reactance inversely or directly proportional to frequency?

3. Is capacitive reactance inversely or directly proportional to capacitance?

4. Does the charging current of a capacitor lead or lag the voltage by 90°?

5. In a series *RC* circuit, does the capacitor voltage lead or lag the voltage across the resistor by 90°?

6. Is impedance directly or inversely proportional to capacitive reactance?

7. The circuit phase angle is the angle between what two quantities?

8. Apparent power is the phasor sum of what two quantities?

9. The circuit power factor is the ratio of what two quantities?

10. In a parallel *RC* circuit, does the current through the resistive branch lead or lag the current through the capacitive branch?

11. In a series *RC* circuit, as frequency increases, does the phase angle increase or decrease?

12. In a parallel *RC* circuit, as frequency decreases, does the phase angle increase or decrease?

13. In a series *RC* circuit, if the phase angle is greater than 45°, does V_R or V_C have the larger value?

14. In a parallel *RC* circuit, if the phase angle is less than 45 degrees, does I_R or I_C have the larger value?

15. I a series *RC* lag network is the output is taken across the resistor or the capacitor?

PROBLEM SET: BASIC LEVEL

Section 14–1 and 14–2

1. Draw the phasor diagram showing the phase relationship between the source voltage and current for the circuit of Figure 14–40.

FIGURE 14–40 Circuit for Problem 1

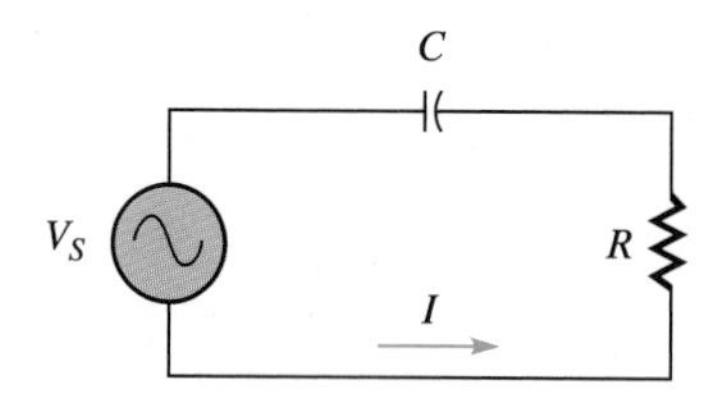

Section 14–3

2. For each circuit of Figure 14–41, compute the capacitive reactance.

FIGURE 14–41 Circuits for Problem 2

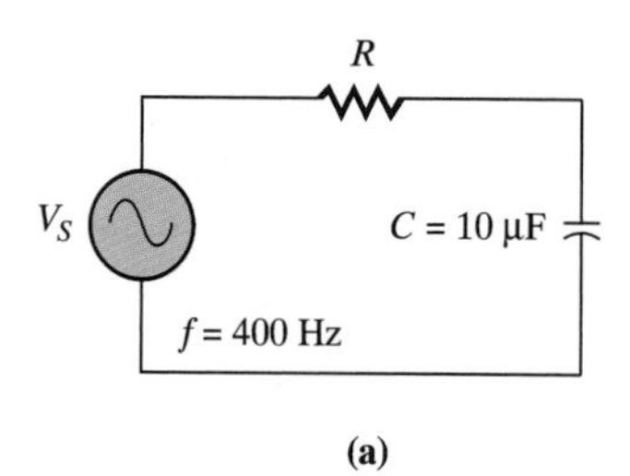

(a)

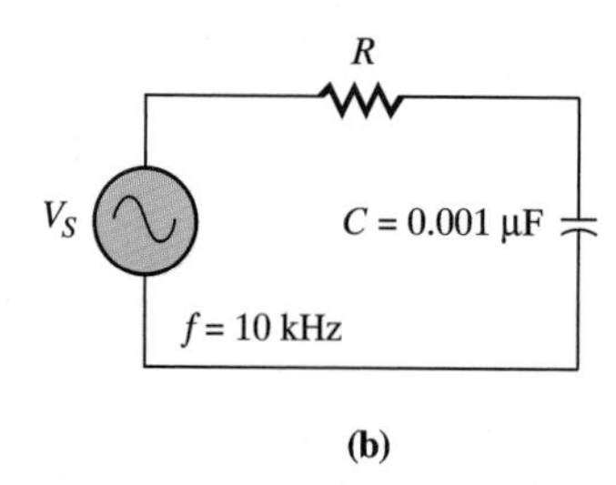

(b)

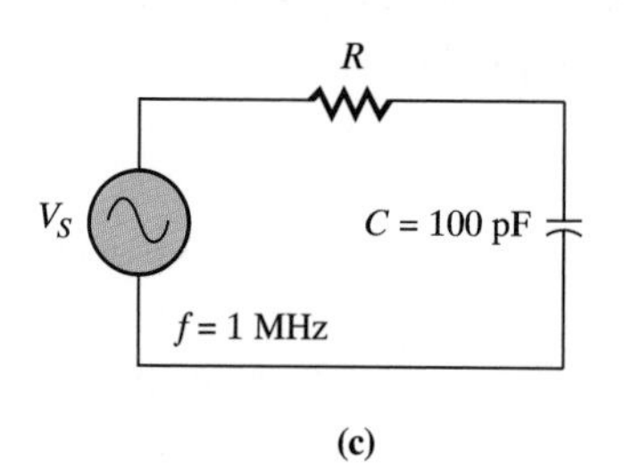

(c)

3. Compute the total capacitive reactance for the circuit of Figure 14–42.

FIGURE 14–42 Circuit for Problem 3

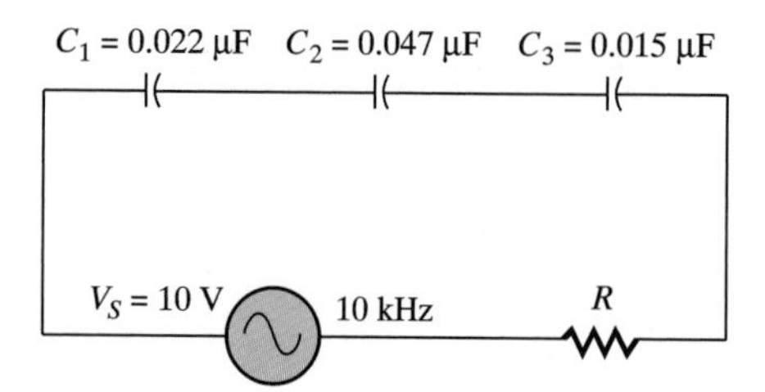

4. Compute the total capacitive reactance for the circuit of Figure 14–43.

FIGURE 14–43 Circuit for Problem 4

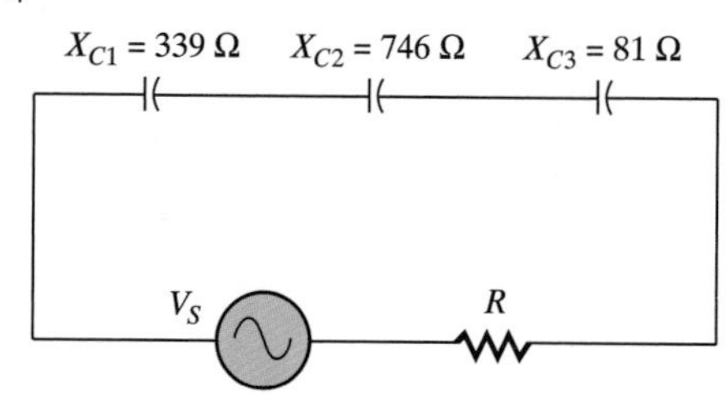

5. For the circuit of Figure 14–44, compute (a) the reactance of each capacitor, (b) the susceptance of each capacitor, (c) the total susceptance of the circuit, and (d) the total reactance.

FIGURE 14–44 Circuit for Problem 5

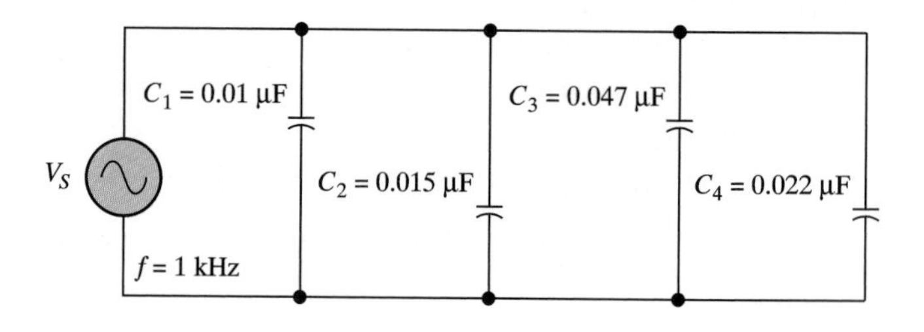

Sections 14–4, 14–5, 14–6

6. For each circuit of Figure 14–45, compute the impedance.

FIGURE 14–45 Circuits for Problem 6

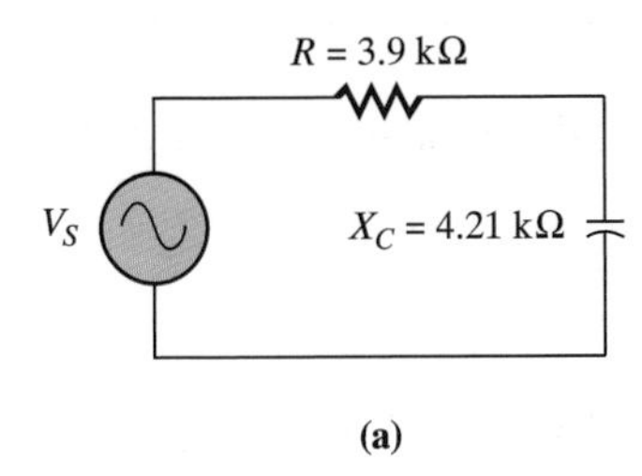

(a)

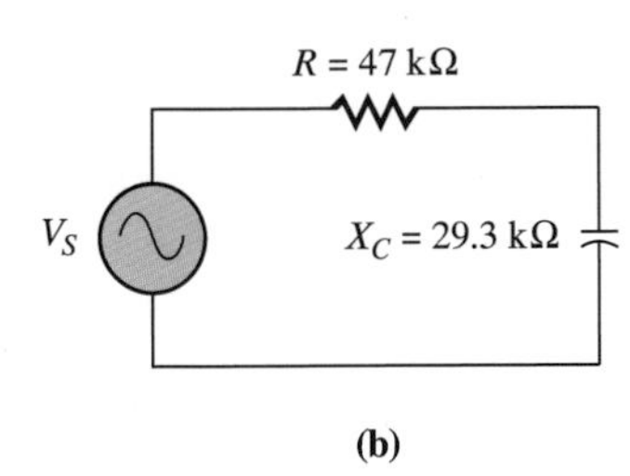

(b)

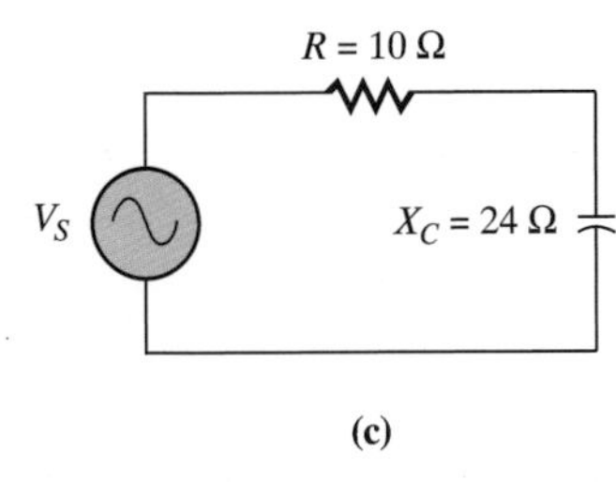

(c)

7. A series *RC* circuit has an impedance of 450 Ω. If the value of the resistor is 100 Ω, compute the value of the capacitive reactance.

8. A parallel *RC* circuit has an impedance of 186 Ω. If the value of the capacitive reactance is 500 Ω, compute the value of the resistance.

9. For each circuit of Figure 14–46, compute the impedance.

FIGURE 14–46 Circuits for Problem 9

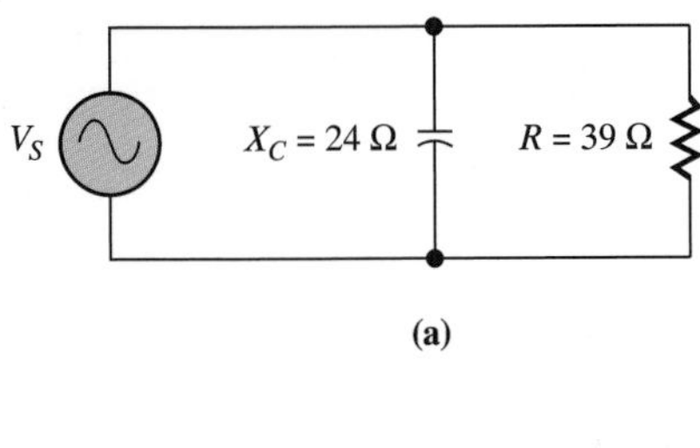

(a)

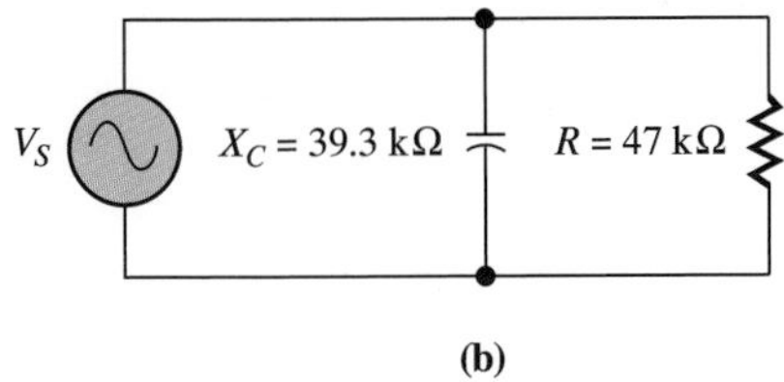

(b)

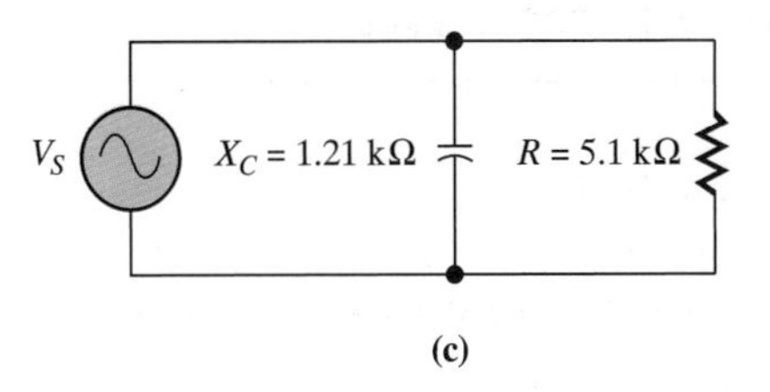

(c)

10. For each circuit of Figure 14–47, compute the value of the source voltage and phase angle.

FIGURE 14–47 Circuits for Problem 10

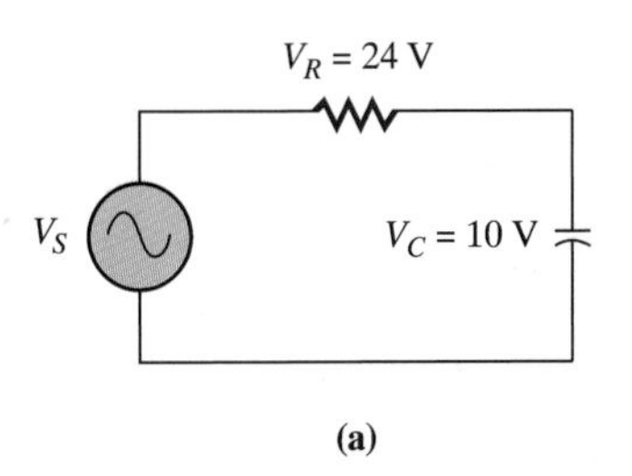

(a)

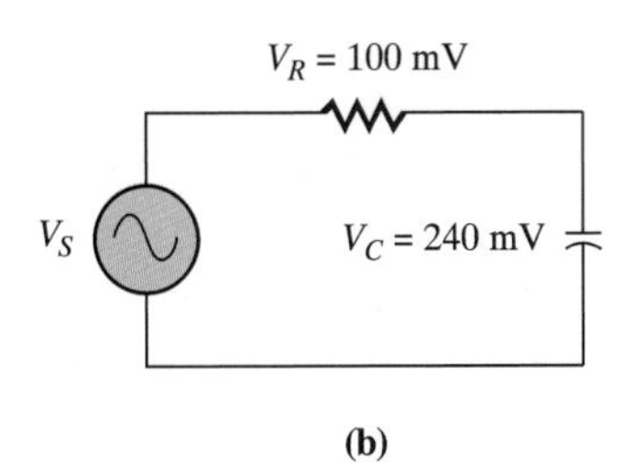

(b)

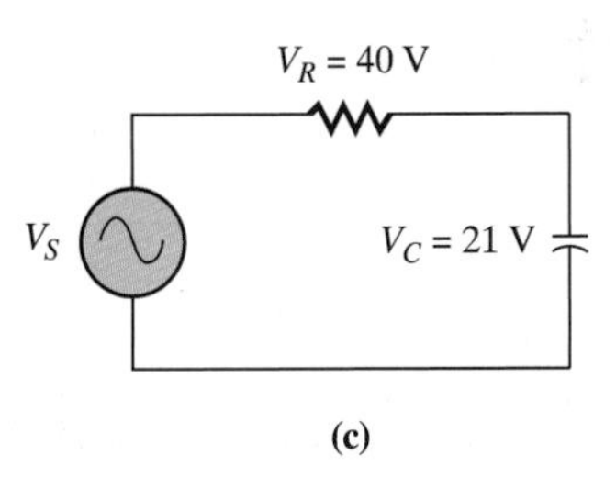

(c)

11. A series *RC* circuit has a source voltage of 15 V. If 7.5 V is produced across the resistor, compute the value of the voltage across the capacitor.

12. For each circuit of Figure 14–48, compute the value of the total current and phase angle.

FIGURE 14–48 Circuits for Problem 12

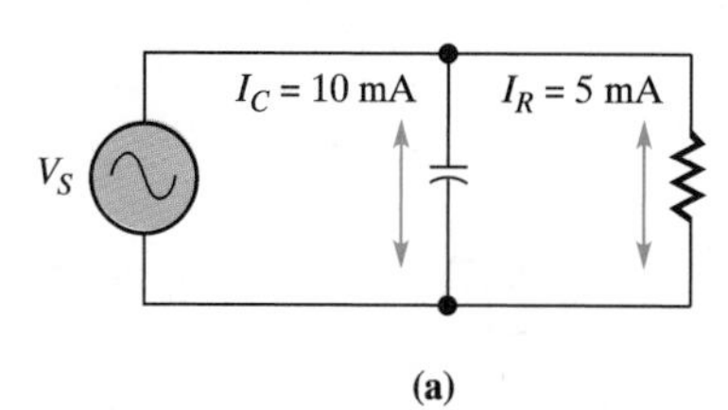

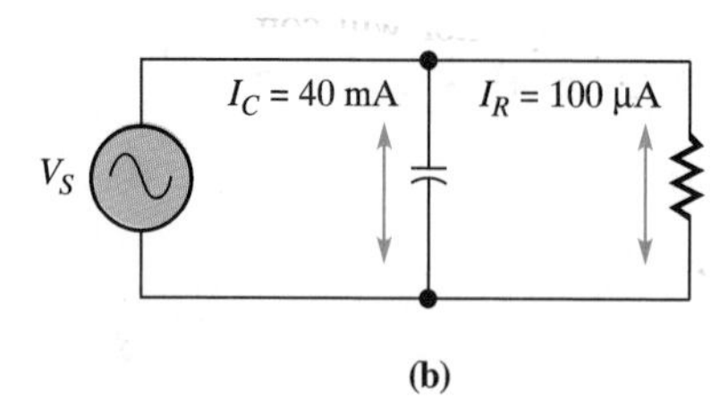

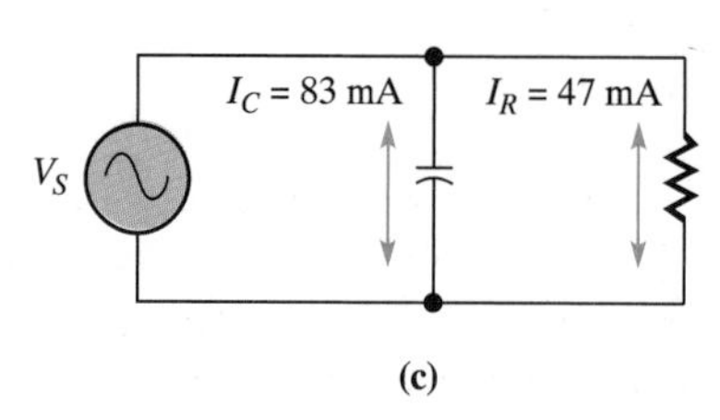

13. For the circuit of Figure 14–49, compute P_{true}, P_{react}, and P_{app}.

FIGURE 14–49 Circuit for Problem 13

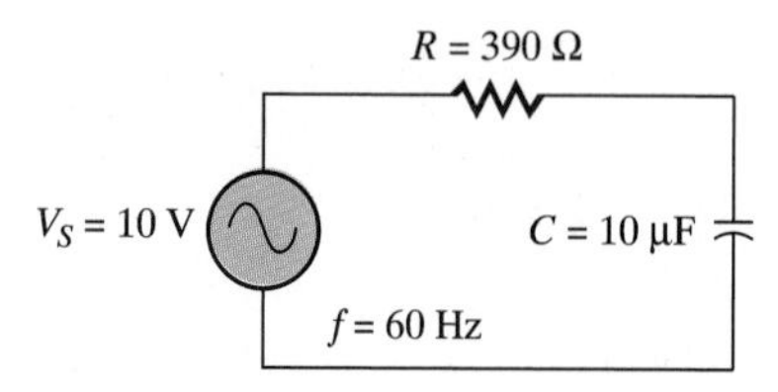

14. A series *RC* circuit has a phase angle of 28°. If the apparent power is 300 mVA, compute the true power.

15. For the series *RC* circuit of Figure 14–50, compute the following: (a) X_C, (b) *Z*, (c) *I*, (d) V_R, (e) V_C, (f) θ, (g) P_{true}, (h) P_{react}, and (i) P_{app}.

FIGURE 14–50 Circuit for Problems 15 and 16

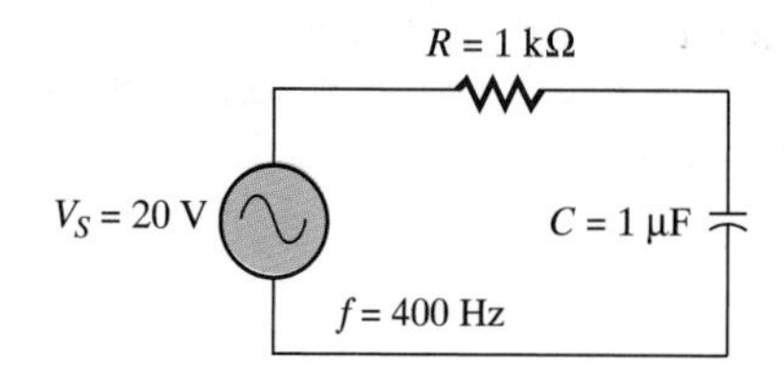

16. Draw the phasor diagrams for the circuit of Figure 14–50.

17. For the parallel *RC* circuit of Figure 14–51, compute the following: (a) X_C, (b) I_R, (c) I_C, (d) I_T, (e) θ, and (f) power factor.

18. Draw the phasor diagrams for the circuit of Figure 14–51.

FIGURE 14–51 Circuit for Problems 17 and 18

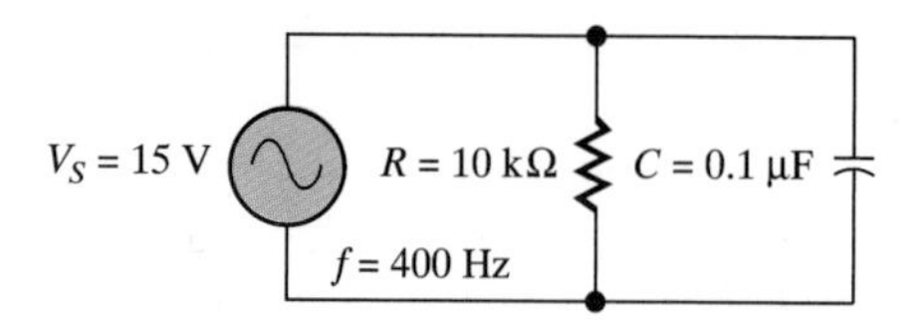

Section 14–7

19. For the series *RC* lead network of Figure 14–52, compute the angle of lead and the output voltage.

FIGURE 14–52 Circuit for Problem 19

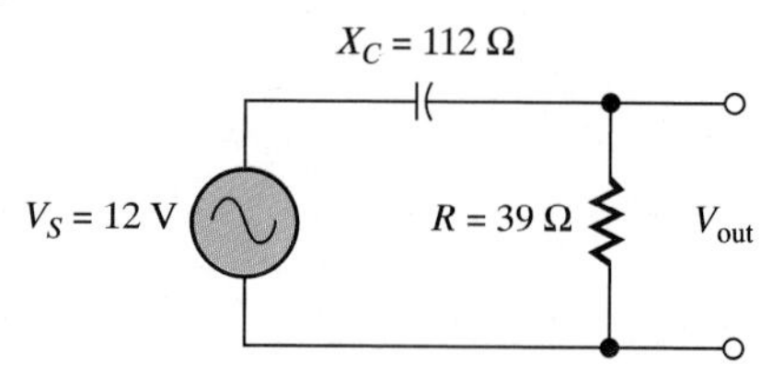

20. For the series *RC* lag network of Figure 14–53, compute the angle of lag and the output voltage.

FIGURE 14–53 Circuit for Problem 20

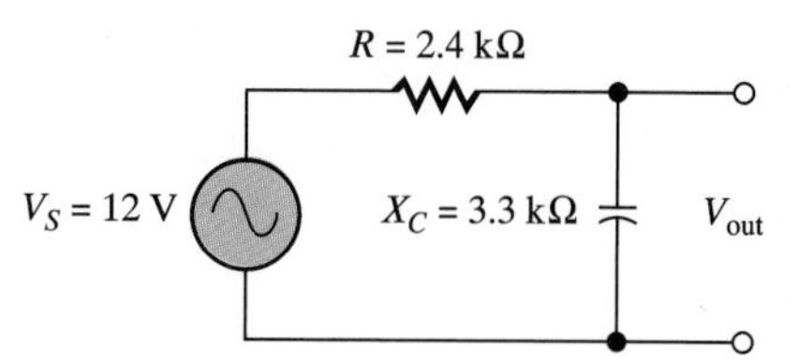

PROBLEM SET: CHALLENGING

21. A 0.001 μF capacitor produces 31.8 kΩ of reactance. Compute the frequency of the voltage applied.

22. For the circuit of Figure 14–54, compute I_R, I_C, and I_T.

FIGURE 14–54 Circuit for Problem 22

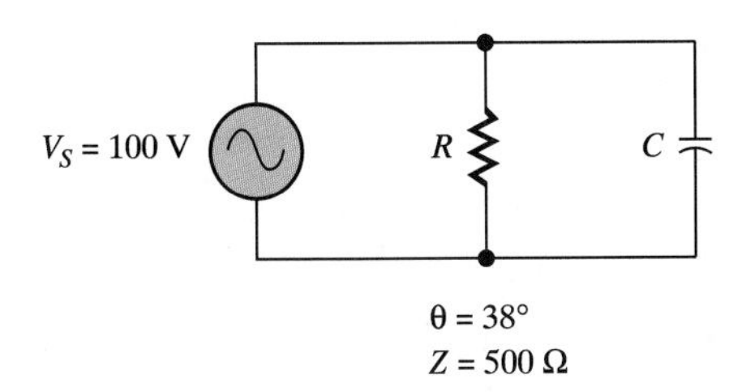

23. In the circuit of Figure 14–55, what value of capacitance will cause the phase angle to be 45°?

FIGURE 14–55 Circuit for Problem 23

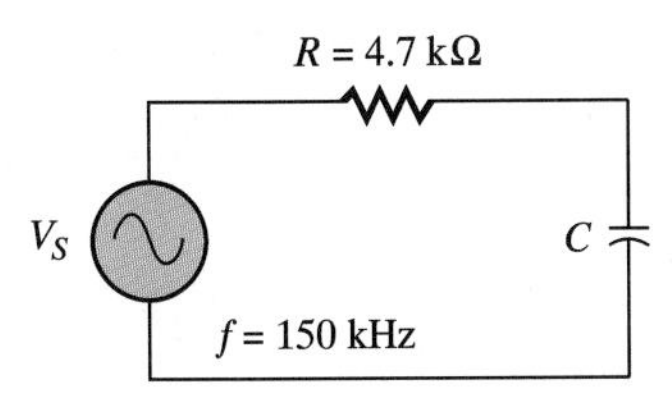

24. A series RC circuit has a source voltage of 10 V and a phase angle of 30°. Compute the values of V_R and V_C.

25. In the circuit of Figure 14–56, compute the value of V_R.

FIGURE 14–56 Circuit for Problem 25

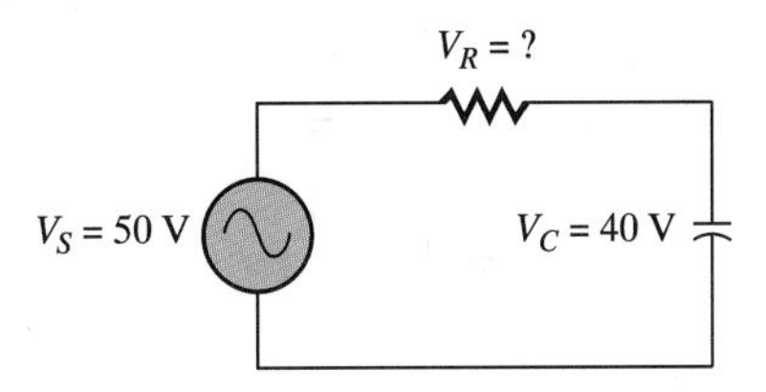

26. What value of capacitor will combine with the resistor in the circuit of Figure 14–57 to produce a lead angle of 58°?

FIGURE 14–57 Circuit for Problem 26

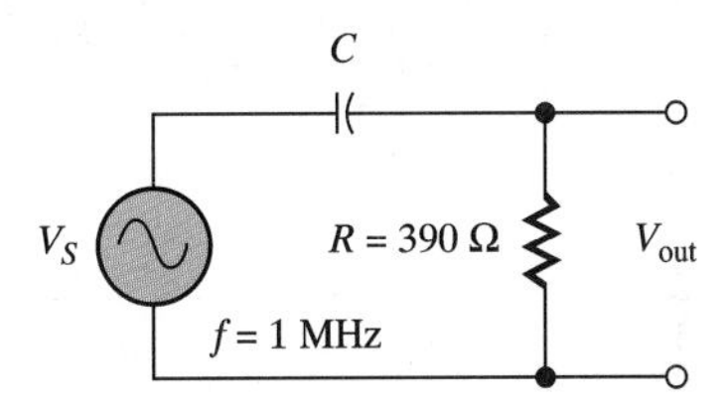

27. A series RC circuit has a phase angle of 36° and a source voltage of 10 V. Compute the values of V_R and V_C.

28. What value capacitor must be added in series with 0.01 μF in order to reduce the total capacitance to 0.006 μF?

PROBLEM SET: TROUBLESHOOTING

29. One fault exists in the circuit of Figure 14–58. The source voltage is checked and found to be correct. You suspect that the frequency is incorrect. If this in fact is the fault, what value would a counter indicate for the frequency?

FIGURE 14–58 Circuit for Problem 29

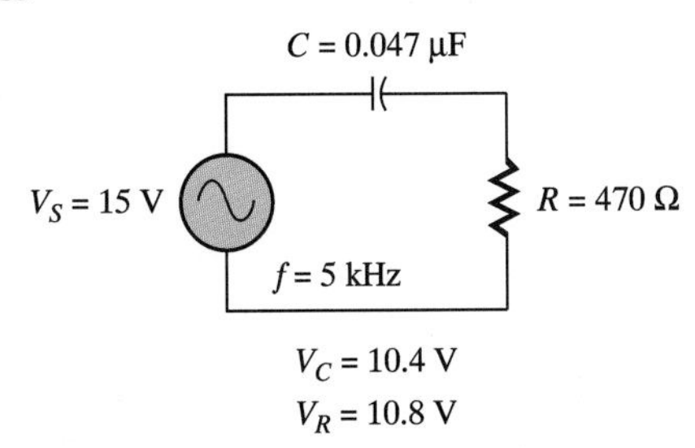

30. The correct phase angle of the circuit in Figure 14–59 is 31.1°. It is found to be 50.4°. The source voltage, frequency, and resistor value are found to be correct. The capacitor is removed and tested for value with an impedance bridge. What value would you expect to find for the capacitor?

FIGURE 14–59 Circuit for Problem 30

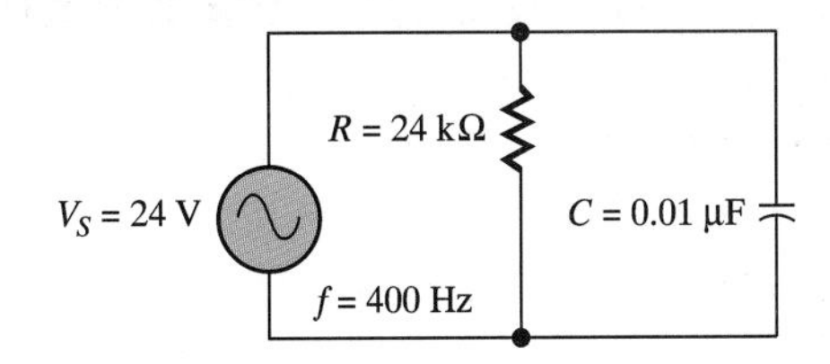

31. The current in the circuit of Figure 14–60 is measured and found to be 10 mA. If one fault exists in this circuit, what is its probable cause?

FIGURE 14–60 Circuit for Problem 31

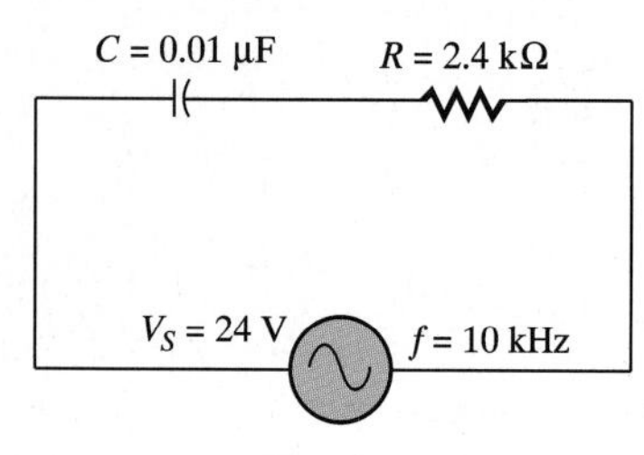

32. The current in the circuit of Figure 14–61 is measured and found to be 38.4 mA. If one fault exists in this circuit, what is the probable cause?

FIGURE 14–61 Circuit for Problem 32

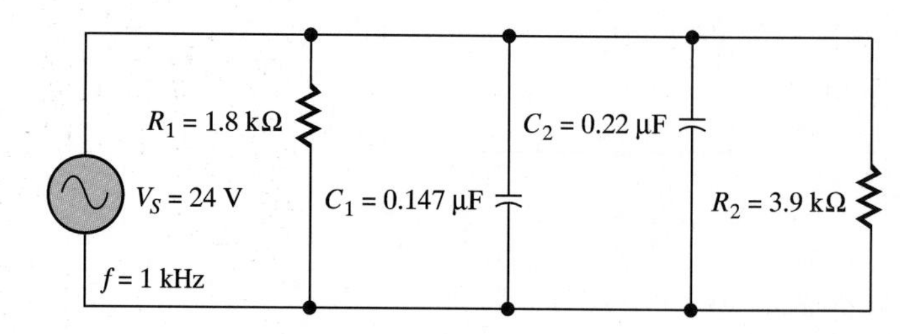

Automated Test Equipment (ATE) Systems

In quality assurance programs in the manufacturing process, inspections at various points in the process are made to ensure that standards are met and that the items or systems will perform as designed. Even in the recent past, testing was a was slow and costly process.

One test that is often made involves checking the frequency response of an amplifier. The test consists of applying a constant amplitude signal to the input of an amplifier and measuring its output over a band of frequencies in order to ensure faithful reproduction of the input signals. The data is collected and then plotted on a graph as part of the documentation. Often as many as five pieces of test equipment must be set up, calibrated, and made to work together in gathering data.

As illustrated in Figure 1, a power supply provides the dc voltages required in the operation of the circuit under test. It must be manually set to the proper voltage and then monitored periodically to ensure that it does not change. A signal generator is used to provide the input signal to the amplifier, and an oscilloscope is used to measure the value of the output signal. The amplitude of the input signal must be held constant over the band of frequencies for which output data is taken. Thus, it must be checked each time the frequency is changed. (The output of some signal generators varies as frequency is changed.)

The oscilloscope presentation must then be read and the data recorded. These data are then plotted and the

parameters of frequency are determined. The curve may look as shown in Figure 1. The upper-half and lower-half power points are then determined. You will recall that these occur at the points where the voltage drops to 70.7% of maximum.

This manual process has several disadvantages. The pieces of equipment used in this test are not always compatible and are, in fact, sometimes made by different manufacturers. This sometimes makes it difficult for the test technician to make the equipment work together successfully. The technician must constantly check the test setup parameters to make sure that the frequency or applied signal voltage has not changed. Thus, the data collection process is slow and tedious. The chance of human error is always a possibility. When data collection is complete, the data must be graphed manually and the rest of the documentation completed. Thus, it is no surprise that the electronics industry was waiting for a better testing system!

Automated Test Equipment (ATE) systems are available today. They are composed of compatible test devices that automatically run the test setup and even automatically graph and document the results. The ATE is a very sophisticated system composed of a digital storage oscilloscope, a logic analyzer, an arbitrary function generator, a programmable power supply, a digital multimeter, a frequency counter, a series of programmable switches, and a control computer. It is highly automated in the measurement of signal parameters such as amplitude and period. The logic analyzer is used in testing such equipment as digital computers. You will probably encounter this device later in the electronics curriculum. The arbitrary function generator can be used to generate any waveform, sinusoidal or nonsinusoidal, as selected by the technician. The programmable power supply can be set to a desired voltage that will be maintained within tight tolerances. Voltage, current, and

FIGURE 1 Amplifier frequency response curve

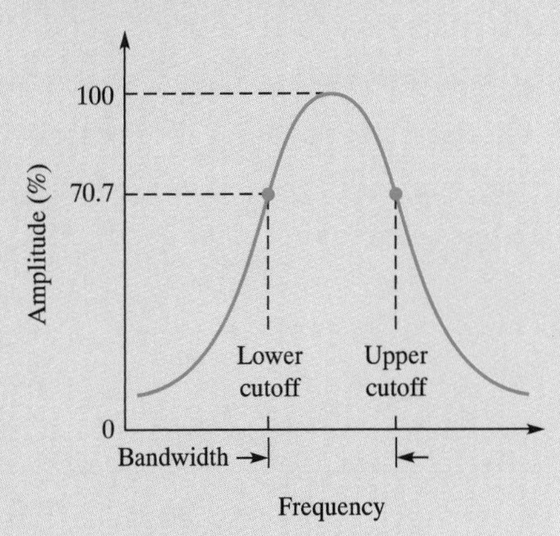

resistance measurements can be made by the digital multimeter, and accurate frequency measurements are made by the frequency counter.

The items making up this system are manufactured so as to be compatible in every way. For example, the input and output impedances are matched so that no loading of the signals takes place. (Recall that the problem of loading was introduced in Chapter 7.) They are also configured so as to be capable of communicating with one another through the system controller. The system controller is based upon a digital computer that can be programmed to run a wide variety of tests. Once programmed and placed in operation, the test is entirely automatic. The equipment is thus operated by the computer that also stores and analyzes the data taken. This particular system can also support a four-color printer, which will print out the graph of the data to include documentation such as pulse width, frequency bandwidth, and voltage amplitude. The test data can also be saved in a disc file for permanent record or for future use.

More and more manufacturers of electronic systems are acquiring ATEs in spite of their cost, which is considerable at this time. They have found that ATEs quickly pay for themselves in increased productivity and in customer satisfaction. Tests are made much more quickly than is the case with manual systems. Another benefit of ATE systems is the fact that tests are performed in an identical manner each time. Thus, uniformity is achieved in the performance of the systems. Also the chance of human error is greatly reduced.

This is not to imply that the technician in such situations is now superfluous. Quite the contrary, the technician is even more valuable when he or she is trained in the operation of ATEs. Among the duties of the technician are devising and entering tests into the system. The technician must then verify the tests and periodically checked to ensure that they are valid. Further, if a tested item does not meet specifications, the technician must repair it. So once again, as has been the case with all advances in technology, the technician has not been replaced, but made more productive and valuable!

CHAPTER

15 Nonresonant ac Circuits

UPON COMPLETION OF THIS CHAPTER, YOU WILL BE ABLE TO

1. Draw the phasor diagrams and explain the operation of series *RLC* circuits.
2. Explain the effects of frequency variations upon series *RLC* circuits.
3. Perform analysis of series *RLC* circuits.
4. Draw the phasor diagrams and explain the operation of parallel *RLC* circuits.
5. Explain the effects of frequency variations upon parallel *RLC* circuits.
6. Perform analysis of parallel *RLC* circuits.
7. Perform analysis of series-parallel *RLC* circuits.
8. List applications of *RLC* circuits.
9. Explain troubleshooting procedures for RLC circuits.

This chapter introduces composite circuits, which are composed of all three electrical effects: resistance, inductance, and capacitance. As you have learned, each, in its own unique way, presents opposition to current flow. Resistance is the simplest of the three. Its opposition is the same, regardless of whether the current is alternating or direct. Inductance and capacitance, on the other hand, are frequency sensitive in their opposition to current flow. Another point, which has been made several times in previous chapters, is that inductance and capacitance are basically opposites in their effects.

As you will learn in this chapter, in any electric circuit, resistance, inductance, and capacitance seem to vie for control. A change in frequency that causes an increase in the effect of inductance causes a corresponding decrease in the effect of capacitance. The opposite is also true. You will learn that a series *RLC* circuit is inductive at higher frequencies and capacitive at lower frequencies, while the opposite is true of the parallel *RLC* circuit. This type of circuit response finds important applications in electronic systems such as radar and FM radio. The circuit resistance is a fixed value. Thus, its influence is determined by the relative values of the reactances.

RLC circuits are analyzed using the principles of circuit analysis that were presented in earlier chapters for *RL* and *RC* circuits. These procedures are combined and applied to the practical circuit containing all three effects. Thus, though there is little in this chapter that is new, it provides you with an opportunity to take another look at ac circuit analysis and reinforce your knowledge of it. Make maximum use of this opportunity!

15–1 ANALYSIS OF SERIES RLC CIRCUITS

Figure 15–1(a) shows a series *RLC* circuit. As with any series circuit, it is a voltage divider, and the current is the same at all points. The voltage drop across the resistor V_R is in phase with the current. The graph for V_R and its phasor diagram are shown in Figure 15–1(b). The voltage induced in the inductor leads the current by 90°. The graph of V_L and its phasor diagram are shown in Figure 15–1(c). The voltage developed across the

FIGURE 15–1 Series *RLC* circuit

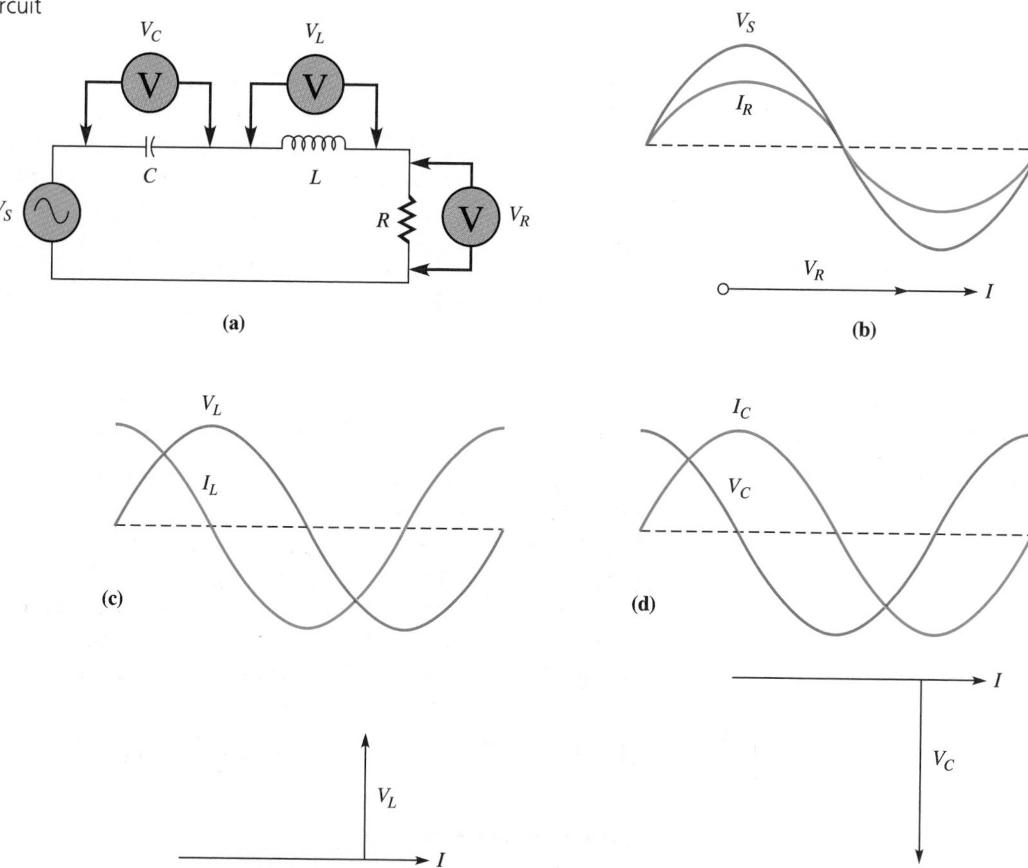

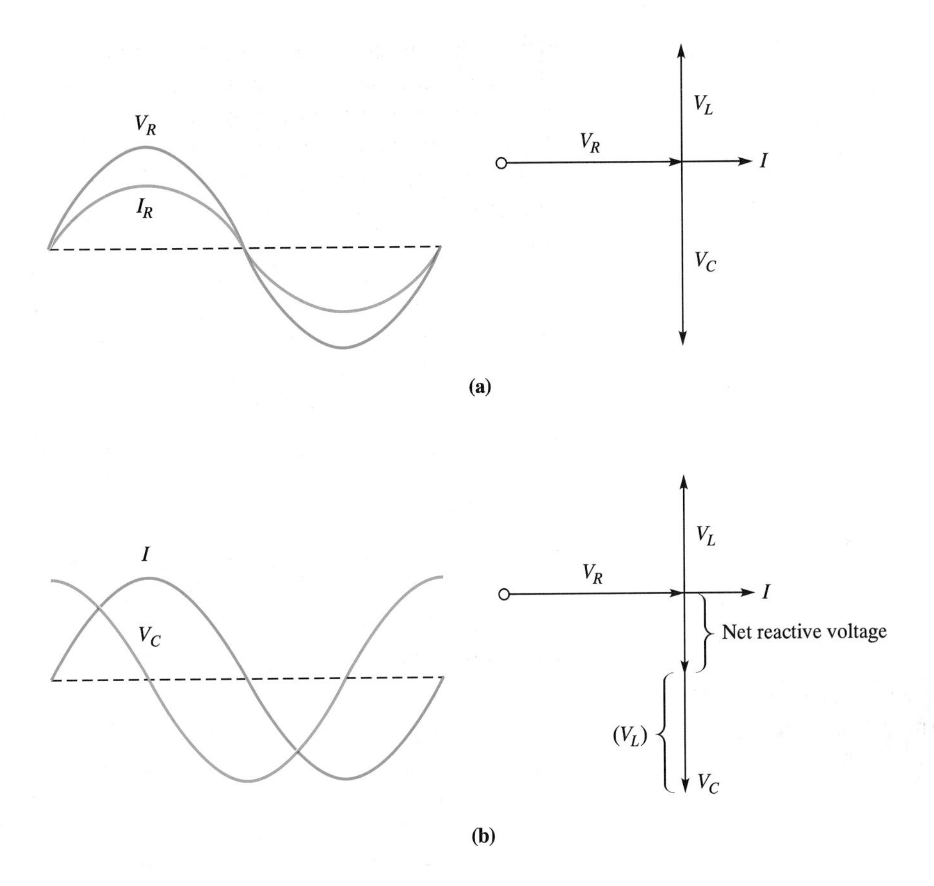

FIGURE 15–2 Phasor diagrams for the circuit of Figure 15–1, showing net reactive voltage

capacitor V_C lags the current by 90°. The graph of V_C and its phasor diagram are shown in Figure 15–1(d).

Since the current phasor is common, the phasor diagrams are shown merged in Figure 15–2(a). From this merged diagram, notice that the phasors for V_L and V_C are 180° *out of phase*. Once again, this bears out the contention that the effects of inductance and capacitance are opposites. Thus, the **net reactive voltage** is the difference between V_L and V_C. The net reactive voltage is illustrated in the phasor diagram of Figure 15–2(b). The only effect of the smaller reactive voltage, in this case V_L, is to reduce the effective value of the larger.

Voltage in a Series *RLC* circuit

Question: Does Kirchhoff's voltage law applies to a series RLC *circuit?*

Recall that the basic law of series circuits is Kirchhoff's voltage law. By way of review, it states that the sum of the voltages around a series circuit is equal to the source voltage. This law does apply to series *RLC* circuits, but since the voltages to be added are not in phase, phasor addition must be used. Figure 15–3(a) shows a series *RLC* circuit and the voltages that are developed. Figure 15–3(b) shows the phasor diagram for these voltages. Notice that the phasor sum of the resistive and net reactive voltages is the source, or total, voltage. Together, the three voltages form a *voltage triangle*. Thus, the equation for the source voltage of a series RLC circuit is

(15–1)
$$V_S = \sqrt{V_R^2 + (V_L - V_C)^2}$$

FIGURE 15–3 Voltage development in a series *RLC* circuit

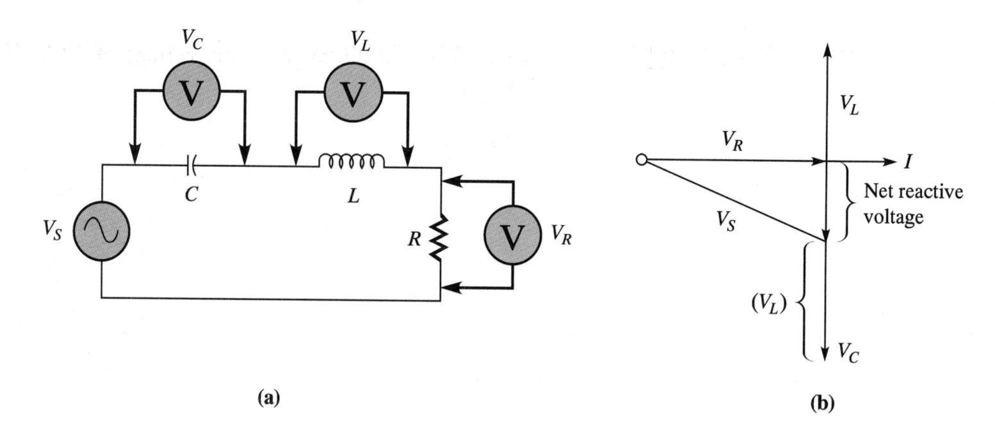

FIGURE 15–4 Circuit and phasor diagram for Example 15–1

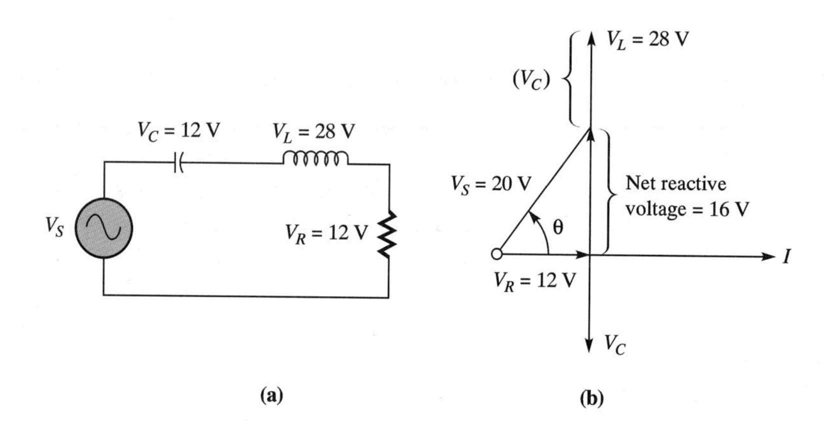

Notice that in the term for net reactive voltage, it does not matter whether V_L or V_C is subtracted from the other quantity. Since the difference is squared, the result will be positive.

EXAMPLE 15–1 In a series *RLC* circuit, the voltages are $V_R = 12$ V, $V_L = 28$ V, and $V_C = 12$ V. Compute the value of V_S, and draw the phasor diagram.

■ Solution:

$$V_S = \sqrt{V_R^2 + (V_C - V_L)^2}$$
$$= \sqrt{(12\text{ V})^2 + (12\text{ V} - 28\text{ V})^2}$$
$$= \sqrt{(12\text{ V})^2 + (-16\text{ V})^2} = \sqrt{144\text{ V} + 256\text{ V}} = 20\text{ V}$$

The circuit and phasor diagram are shown in Figure 15–4.

EXAMPLE 15–2 A series *RLC* circuit has a source voltage of 60 V. If the net reactive voltage is 40 V, compute V_R.

■ Solution:

$$V_S^2 = V_R^2 + (V_L - V_C)^2$$
$$V_R = \sqrt{V_S^2 - (V_L - V_C)^2} = \sqrt{60^2 - 40^2} = 44.721\text{ V}$$

EXAMPLE 15–3 A series *RLC* circuit has a source voltage of 12 V. If V_R is 8 V and V_L is 11 V, find the value of V_C.

■ Solution:

$$V_S^2 = V_R^2 + (V_L - V_C)^2$$
$$(V_L - V_C)^2 = V_S^2 - V_R^2$$
$$(V_L - V_C) = \sqrt{V_S^2 - V_R^2}$$
$$= \sqrt{(12 \text{ V})^2 - (8 \text{ V})^2} = 8.94 \text{ V}$$

Then,

$$V_C = 11 \text{ V} - 8.94 \text{ V} = 2.06 \text{ V}$$

■ Impedence in a Series *RLC* Circuit

Question: How is the impedance of a series RLC *circuit computed?*

The impedance of a series *RLC* circuit is equal to the phasor sum of the resistance and net reactance. Note in Figure 15–5 that the phasors representing X_L and X_C lie in the same relative positions as V_L and V_C, respectively. Thus, the **net reactance** is the difference of X_L and X_C. The impedance is

(15–2) $$Z = \sqrt{R^2 + (X_L - X_C)^2}$$

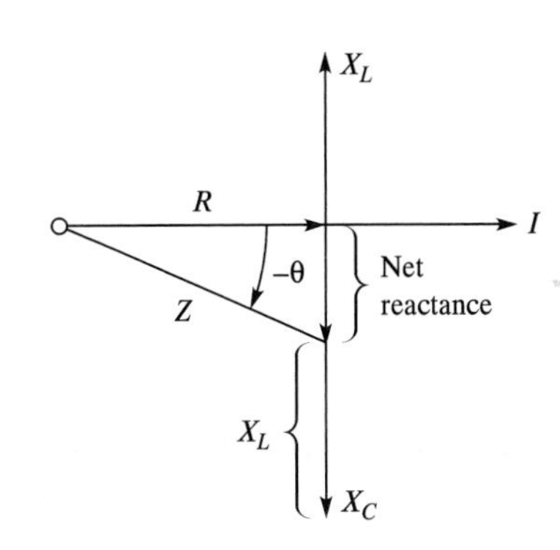

FIGURE 15–5 Phasor diagram for a Series *RLC* circuit, showing net impedance

EXAMPLE 15–4 Compute the impedance for the circuit of Figure 15–6.

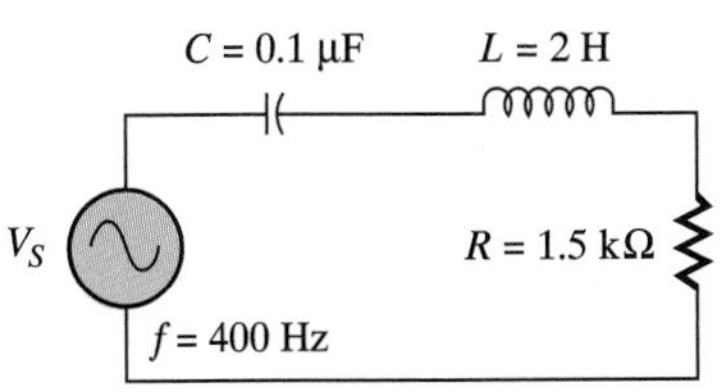

FIGURE 15–6 Circuit for Example 15–4

■ Solution:

1. Compute X_L:

$$X_L = 2\pi fL$$
$$= 2(3.14)(400 \text{ Hz})(2 \text{ H}) = 5.024 \text{ k}\Omega$$

2. Compute X_C:

$$X_C = \frac{1}{2(3.14)(400 \text{ Hz})(0.1 \times 10^{-6} \text{ F})}$$
$$= \frac{1}{2.513 \times 10^{-4}} = 3.981 \text{ k}\Omega$$

3. Compute Z:

$$Z = \sqrt{R^2 + (X_L - X_C)^2}$$
$$= \sqrt{(1500\ \Omega)^2 + (5024\ \Omega - 3981\ \Omega)^2} = \sqrt{(1500\ \Omega)^2 + (1043\ \Omega)^2}$$
$$= 1.827\ \text{k}\Omega$$

Phase Angle of a Series *RLC* Circuit

Question: How is the phase angle of a series* RLC *circuit computed?

The phase angle of a series *RLC* circuit is the angle between the phasors representing the source voltage and the circuit current. In Figure 15–7(a), V_C is greater than V_L. Thus, the phase angle opens in a negative direction. Since the current leads the voltage, the circuit is said to be *capacitive*. In Figure 15–7(b), V_L is greater than V_C, and the angle opens in a positive direction. Since the voltage now leads the current, the circuit is said to be *inductive*. Notice that the net reactive voltage is the opposite side and the resistor voltage the adjacent side of the voltage triangle. Thus, the equation for computing the circuit phase angle is

(15–3)
$$\tan\theta = \frac{V_L - V_C}{V_R}$$
$$\theta = \arctan\frac{V_L - V_C}{V_R}$$

EXAMPLE 15–5 Compute the phase angle from the phasor diagram of Figure 15–8.

■ Solution:

$$\theta = \arctan\frac{V_L - V_C}{V_R}$$
$$= \arctan\frac{18\ \text{V} - 36\ \text{V}}{25\ \text{V}} = \arctan\frac{-18\ \text{V}}{25\ \text{V}} = \arctan -0.72 = -35.753^\circ$$

FIGURE 15–7 Phase angles in a series *RLC* circuit

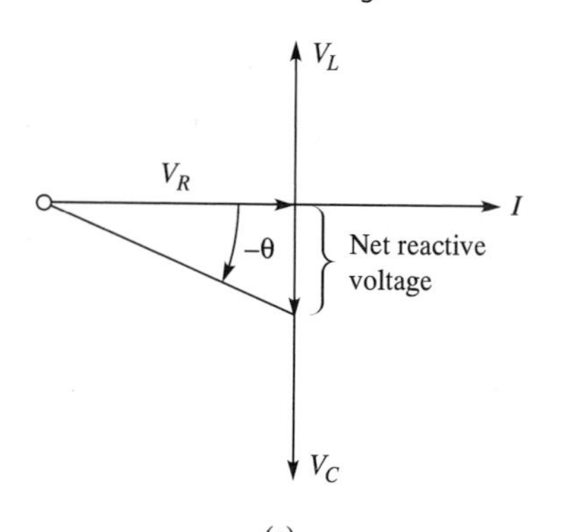

(a)
When the phase angle is negative, the circuit is capacitive.

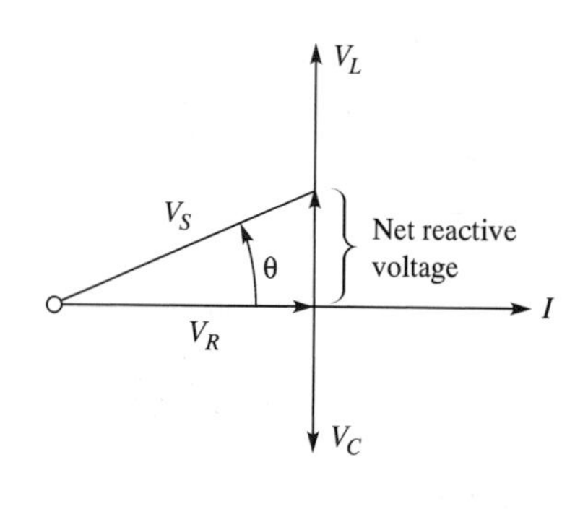

(b)
When the phase angle is positive, the circuit is inductive.

FIGURE 15–8 Phasor diagram for Example 15–5

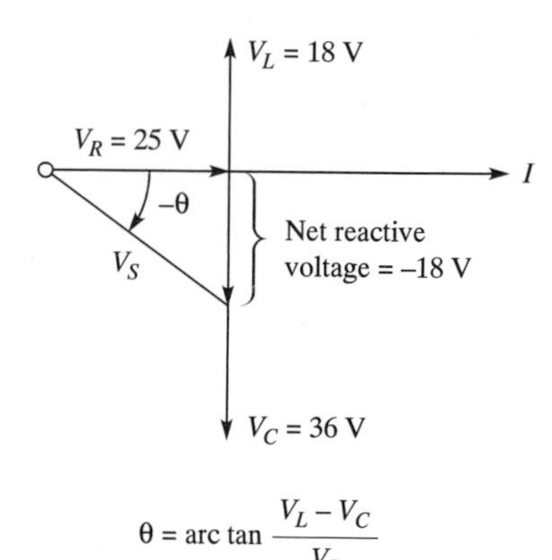

FIGURE 15–9 Impedance triangle for a series *RLC* circuit

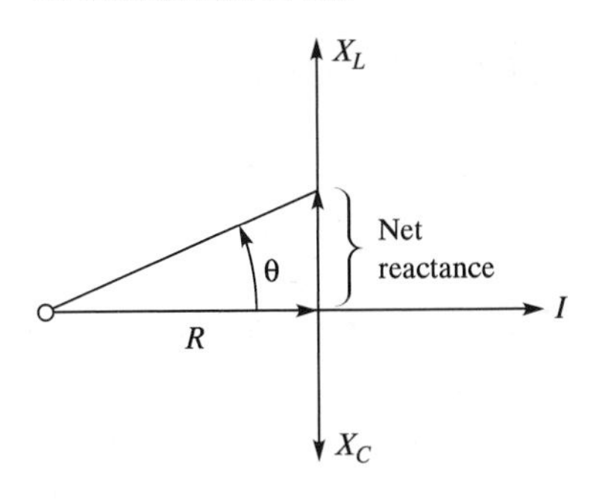

The *impedance triangle*, shown in Figure 15–9, can also be used in determining the phase angle. The net reactance is the opposite side and the resistance is the adjacent side of the impedance triangle. The equation for the phase angle is

(15–4)

$$\tan \theta = \frac{X_L - X_C}{R}$$

$$\theta = \arctan \frac{X_L - X_C}{R}$$

Notice that the series *RLC* circuit is either inductive or capacitive. The effects of inductance and capacitance cannot exist simultaneously in a circuit; one or the other will be predominant. In the case of the series *RLC* circuit, the one with the greatest reactance completely cancels the affect of the other.

NOTE

It is possible for the values of inductive and capacitive reactance to be equal. If so, they cancel one another, leaving only the resistance to oppose current. This circuit condition, known as *resonance,* produces some very interesting results that will be the subject of the next chapter.

Power in an *RLC* Circuit

Power considerations for an *RLC* circuit are identical to those for *RL* and *RC* circuits. As you know, the resistor is the only device that develops *true power* in the circuit. The reactive components store the electrical energy and return it to the source. The product of the voltage across the reactive element and the circuit current is known as *reactive power* and is the vertical leg of the power triangle. The product of the source voltage and the circuit current is known as the apparent power and forms the hypotenuse of the power triangle. *Apparent power* is also the phasor sum of the true and reactive power.

EXAMPLE 15–6 For the circuit of Figure 15–10, compute the true power, reactive power, and apparent power.

FIGURE 15–10 Circuit for Example 15–6

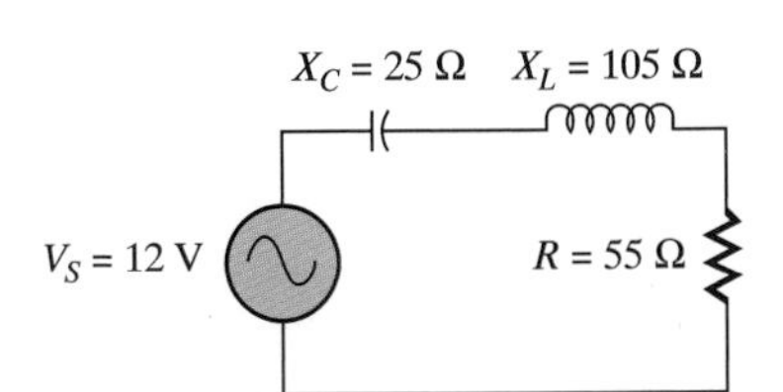

Solution:

1. Compute Z:

$$Z = \sqrt{R^2 + (X_L - X_C)^2}$$

$$= \sqrt{(55\ \Omega)^2 + (105\ \Omega - 25\ \Omega)^2} = \sqrt{(55\ \Omega)^2 + (80\ \Omega)^2} = 97.08\ \Omega$$

2. Compute I:

$$I = \frac{V_S}{Z}$$

$$= \frac{12\text{ V}}{97.08\ \Omega} = 124\text{ mA}$$

3. Compute true power:

$$P_{\text{true}} = I^2R$$

$$= (0.124\text{ A})^2(55\ \Omega) = 0.846\text{ W}$$

4. Compute reactive power:

$$P_{react} = I^2X$$
$$= (0.124)^2(80\ \Omega) = 1.23\ \text{VAR}$$

5. Compute apparent power:

$$P_{app} = \sqrt{(P_{true})^2 + (P_{react})^2}$$
$$= \sqrt{(.846\ \text{W})^2 + (1.23\ \text{VAR})^2} = 1.49\ \text{VA}$$

Question: What is meant by a leading or lagging power factor?

Depending upon whether the circuit is inductive or capacitive, one of two *power triangles* can be used. The angle opens in a positive direction for an inductive circuit and in a negative direction for a capacitive circuit, as illustrated in Figure 15–11. Recall that the *power factor* of the circuit is the cosine of the phase angle. If the series *RLC* circuit is inductive, the power factor is said to be *lagging*. If the circuit is capacitive, the power factor is said to be *leading*.

FIGURE 15–11 Power triangles for a series *RLC* circuit

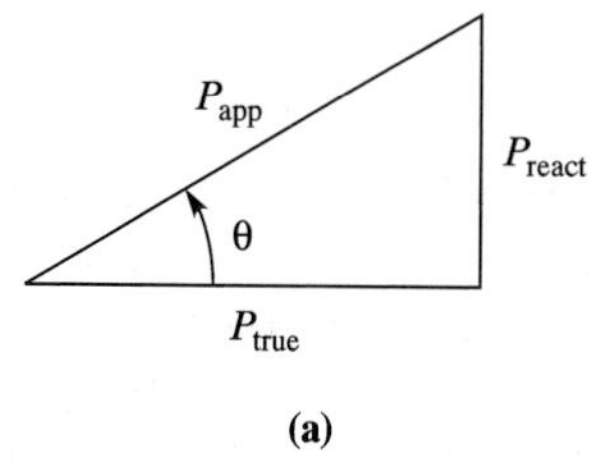

(a)
In an inductive circuit, θ is positive, and the power factor is leading.

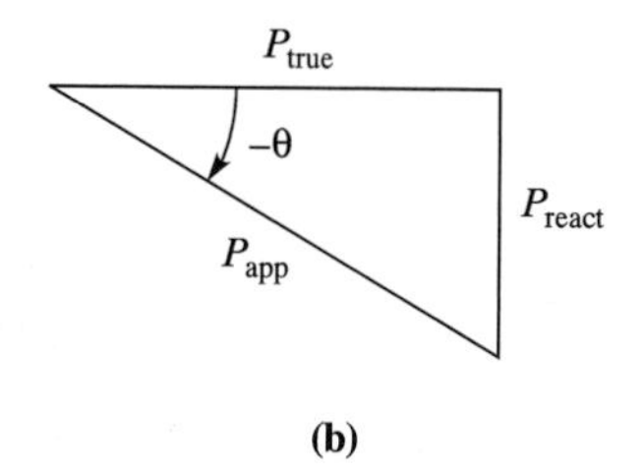

(b)
In a capacitive circuit, θ is negative, and the power factor is lagging.

SECTION 15–1 REVIEW

1. If V_C is 12 V and V_L is 18 V, is a series *RLC* circuit inductive or capacitive?
2. If X_L is 42 Ω and X_C is 27 Ω, is a series *RLC* circuit inductive or capacitive?
3. What is the net reactive voltage in Problem 1?
4. What is the net reactance in Problem 2?
5. What is meant by a leading or lagging power factor?

15–2 SERIES *RLC* CIRCUIT RESPONSE

Frequency Changes

Question: What effect do changes in frequency have on the response of series RLC *circuits?*

As with any series circuit, the series *RLC* circuit is a voltage divider. Thus, it is the *relative* sizes of the voltages developed across each component that determine its response. As you learned in the preceding paragraphs, if the voltage developed across the inductor V_L is greater than that across the capacitor V_C, then the phase angle opens in a positive direction. The source voltage leads the current and the circuit is *inductive*. If the voltage across the capacitor is greater than that across the inductor, the phase angle opens in a negative direction. The source voltage now lags the current, and the circuit is *capacitive*.

Recall that current is the same at all points in a series circuit. Thus, the relative values of V_L and V_C are directly proportional to the inductive and capacitive reactances (V_L =

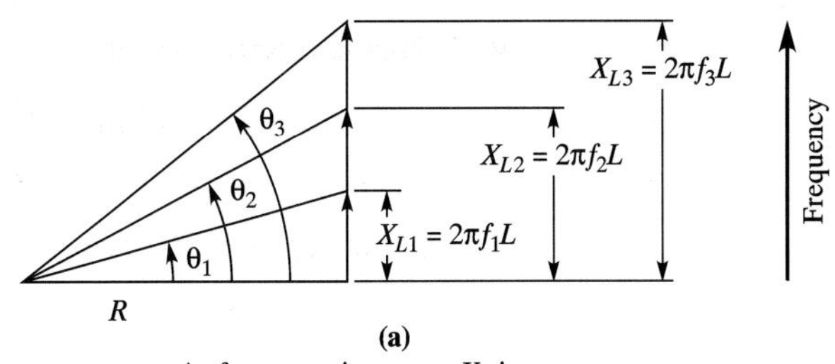

(a)
As frequency increases, X_L increases, and θ increases in a positive direction.

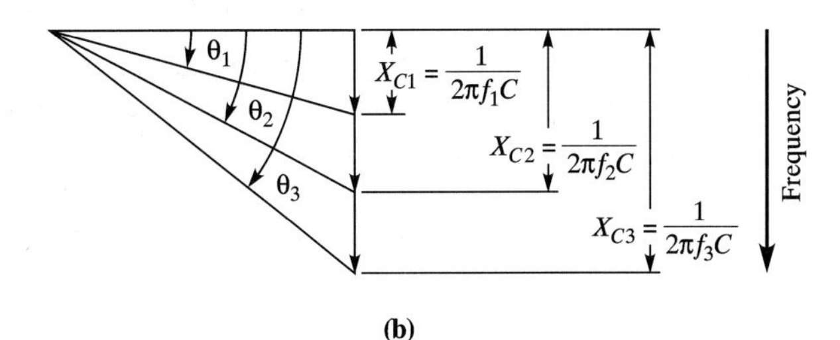

(b)
As frequency decreases, X_C increases, and θ increases in a negative direction.

FIGURE 15–12 Effects of frequency changes in a series *RLC* circuit

IX_L and $V_C = IX_C$). Inductive reactance is *directly proportional* to frequency ($X_L = 2\pi fL$) and capacitive reactance is *inversely proportional* to frequency ($X_C = 1/(2\pi fC)$. These mathematical relationships show that an increase in frequency increases X_L, causing the phase angle to increase in a positive direction. By the same token, a decrease in frequency increases X_C, causing the phase angle to increase in a negative direction. Thus, an increase in frequency makes the series *RLC* circuit more inductive, while a decrease in frequency makes it more capacitive, as illustrated in Figure 15–12.

EXAMPLE 15–7 For the circuit of Figure 15–13, compute the values of X_L, X_C, and θ, for the following frequencies: (a) 8 kHz, (b) 16 kHz, and (c) 32 kHz.

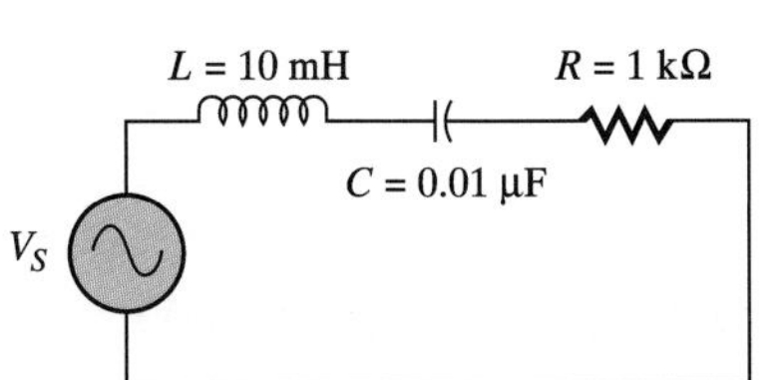

FIGURE 15–13 Circuit for Example 15–7

■ **Solution:**

a. For 8 kHz,

$$X_L = 2\pi fL = 502\ \Omega$$

$$X_C = \frac{1}{2\pi fC} = 1989\ \Omega$$

$$\text{net reactance} = X_L - X_C = 1448\ \Omega \quad \text{(capacitive)}$$

$$\theta = \text{arc tan}\,\frac{\text{net reactance}}{R}$$

$$= \text{arc tan}\,\frac{1448\ \Omega}{1000\ \Omega} = -55.37°$$

b. For 16 kHz,

$$X_L = 2\pi fL = 1005\ \Omega$$

$$X_C = \frac{1}{2\pi fC} = 995\ \Omega$$

FIGURE 15–14 Phasor diagrams for Example 15–7

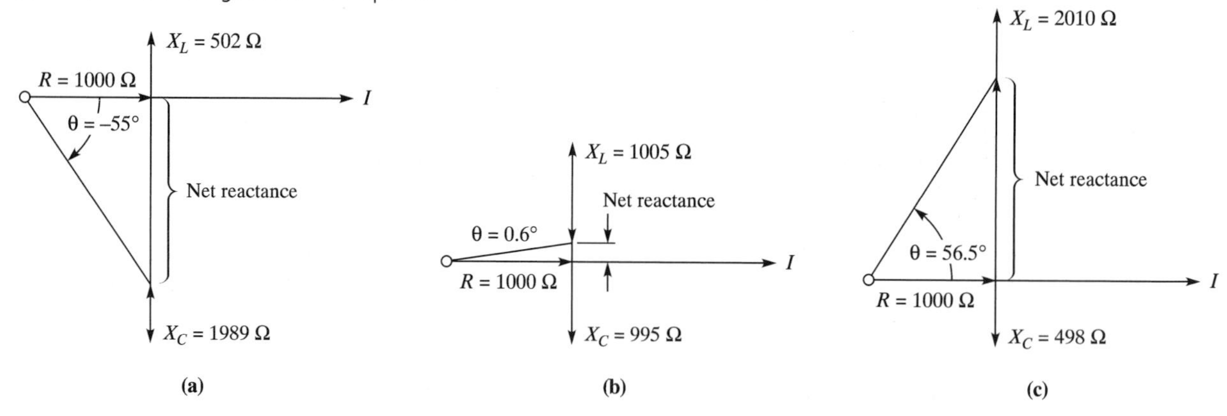

$$\text{net reactance} = X_L - X_C = 10\ \Omega \quad \text{(inductive)}$$

$$\theta = \text{arc tan}\ \frac{\text{net reactance}}{R}$$

$$= \text{arc tan}\ \frac{10\ \Omega}{1000\ \Omega} = 0.573°$$

c. For 32 kHz,

$$X_L = 2\pi f L = 2010\ \Omega$$

$$X_C = \frac{1}{2\pi f C} = 498\ \Omega$$

$$\text{net reactance} = X_L - X_C = 1512\ \Omega \quad \text{(inductive)}$$

$$\theta = \text{arc tan}\ \frac{\text{net reactance}}{R}$$

$$= \text{arc tan}\ \frac{1512\ \Omega}{1000\ \Omega} = 56.52°$$

The phasor diagrams for the circuit at these three frequencies are shown in Figure 15–14. These calculations bear out the point made that at higher frequencies the circuit is inductive, while at lower frequencies it is capacitive. Notice the approximate point (16 kHz) at which the circuit moves from inductive to capacitive and vice versa.

■ Changes in Inductance or Capacitance

Question: What effect do changes in inductance or capacitance have on series* RLC *circuit response?

In electronic systems, you will often find variable inductors or capacitors. They provide adjustable values of inductance and capacitance in order to make the circuit respond to a specific frequency. For example, in Figure 15–15(a), if the value of the inductance is increased, the value of X_L increases. The latter causes V_L to increase, and the circuit becomes more inductive. Figure 15–15(b) shows that as the value of the capacitance is

FIGURE 15–15 Effects of changes in inductance and capacitance

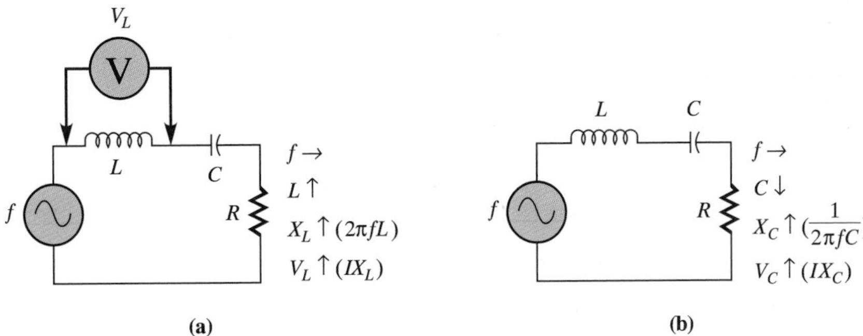

(a)
An increase in inductance causes the circuit to become more inductive.

(b)
A decrease in capacitance causes the circuit to become more capacitive.

decreased, both X_C and V_C increase, making the circuit more capacitive. Thus, the effects of changes in inductance and capacitance are identical to those of changes in frequency.

EXAMPLE 15–8 In Figure 15–16, compute the phase angle for the given values of inductance and capacitance. Next compute the phase angle for double and half the given values.

FIGURE 15–16 Circuit for Example 15–8

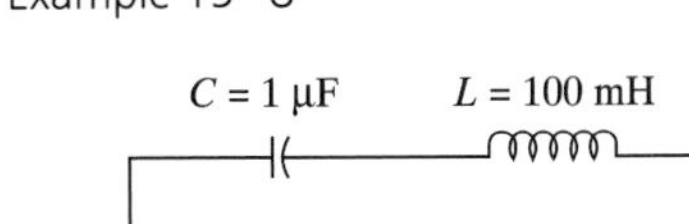

■ Solution:

1. Use the original values:

$$X_L = 2\pi fL = 251\ \Omega$$

$$X_C = \frac{1}{2\pi fC} = 398\ \Omega$$

$$\theta = \arctan\frac{(X_L - X_C)}{R} = -16.4°$$

2. Double X_L:

$$X_L = 502\ \Omega$$

$$X_C = 398\ \Omega$$

$$\theta = 11.7°$$

The phase angle increases in a *positive* direction.

3. Halve X_L:

$$X_L = 126\ \Omega$$

$$X_C = 398\ \Omega$$

$$\theta = -28.5°$$

The phase angle increases in a *negative* direction.

4. Double X_C:

$$X_L = 251\ \Omega$$

$$X_C = 796\ \Omega$$

$$\theta = -47.5°$$

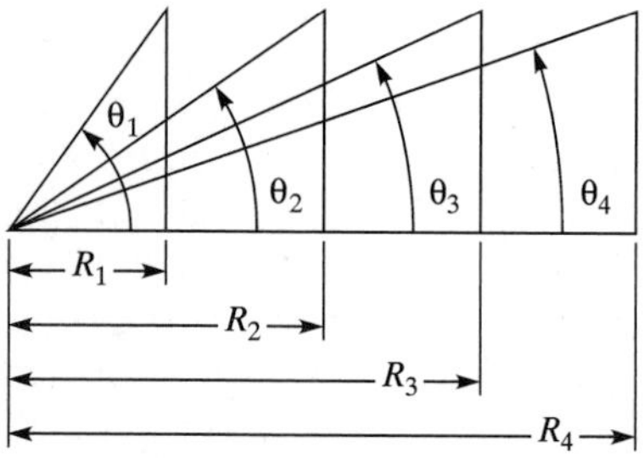

FIGURE 15–17 Effects of changes in resistance

The phase angle increases in a *positive* direction.

5. Halve X_C:

$$X_L = 251\ \Omega$$

$$X_C = 199\ \Omega$$

$$\theta = 5.93°$$

The phase angle increases in a *negative* direction.

Question: Will a change in resistance have any effect on series* RLC *circuit response?

Even though resistance is not frequency sensitive, changes in its value have a definite effect upon *RLC* circuit response. As previously shown, the phase angle of the circuit is a function of the ratio of the net reactance to the resistance: $\theta = \text{arc tan}\ (V_L - V_C)/\text{R}$. The arc tangent function assumes values from zero to extremely large numbers over the range of zero to 90°. (It is undefined at 90°. Thus, the phase angle, for given values of reactance, is *inversely proportional* to the resistance, as illustrated in Figure 15–17.

EXAMPLE 15–9 A certain *RLC* circuit has a net reactance of 1600 Ω capacitive. Compute the phase angle for the following values of resistance: (a) 500 Ω, (b) 1600 Ω, and (c) 2400 Ω.

Solution:

a. For 500 ohms,

$$\theta = \text{arc tan}\ \frac{\text{net reactance}}{R} = -72.6°$$

b. For 1600 ohms,

$$\theta = \text{arc tan}\ \frac{\text{net reactance}}{R} = -45°$$

c. For 2400 ohms,

$$\theta = \text{arc tan}\ \frac{\text{net reactance}}{R} = -33.7°$$

Notice that as resistance increased, the phase angle decreased in this example.

SECTION 15–2 REVIEW

1. As the frequency of a series *RLC* circuit decreases, does it become more inductive or capacitive?
2. A series *RLC* circuit must be made more inductive. Should the inductor be replaced with one of a larger or smaller value?
3. The resistance in a series *RLC* circuit is decreased in value. Will the phase angle increase or decrease?
4. Why is it that the voltages across the components determine whether a series *RLC* circuit is inductive or capacitive?

15–3 ANALYSIS OF PARALLEL *RLC* CIRCUITS

Question: What are the voltage and current phase relationships in a parallel RLC *circuit?*

A parallel *RLC* circuit is shown in Figure 15–18(a). Since it is a current divider, the branch currents are considered in circuit analysis. The same voltage V_S is across each branch, so the phasor representing it is used as the reference. You will find that the mathematics used is identical to that used in series *RLC* analysis.

The graph and phasor diagram relating the source voltage and resistor current are shown in Figure 15–18(b). Notice that the current I_R and the source voltage V_S are in phase. Figure 15–18(c) shows the graph and phasor diagram relating the source voltage and the current through the inductive branch. Here, the source voltage leads the current

FIGURE 15–18 Parallel *RLC* circuit

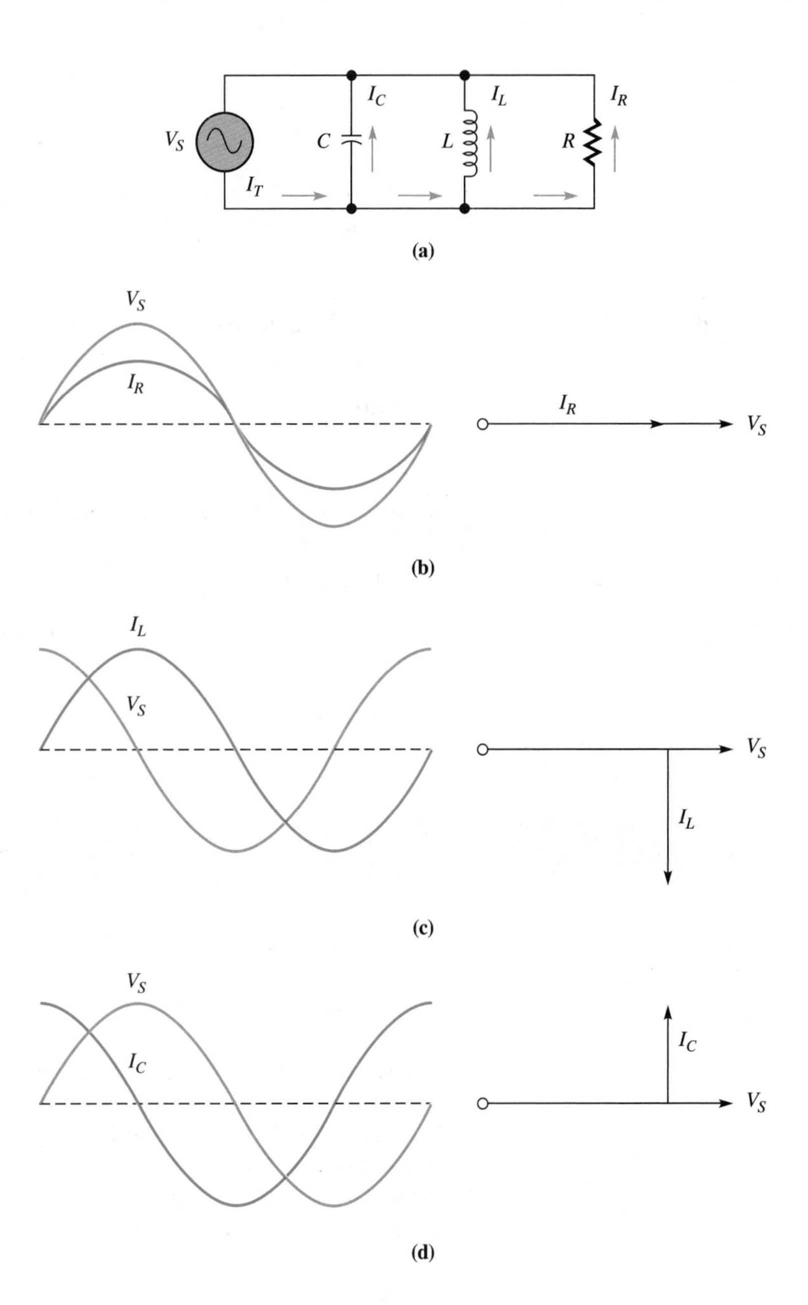

FIGURE 15–19 Merged phasor diagrams for the circuit of Figure 15–18

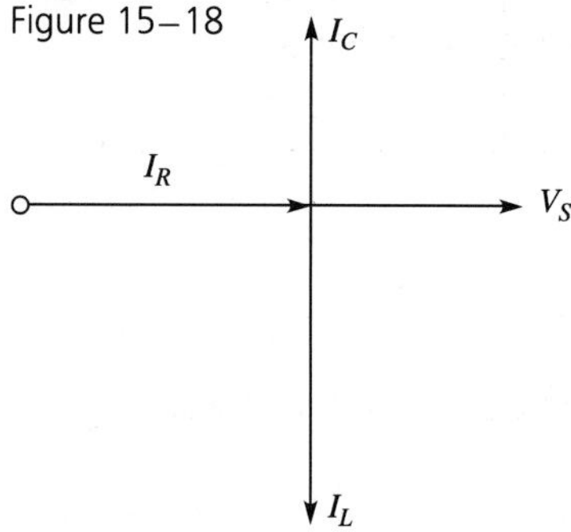

I_L by 90°. Figure 15–18(d) shows the graph and phasor diagram relating the source voltage and current through the capacitive branch. As shown, the current I_C leads the source voltage by 90°.

The three phasor diagrams are merged in Figure 15–19. Notice its similarity to the phasor diagram for the series *RLC* circuit. The difference here lies in the use of current rather than voltage phasors. From this diagram, you can note several things. The phasors for I_L and I_C are 180° out of phase. Thus, the phasor sum of the reactive currents is their difference ($I_{net} = I_L - I_C$) and is known as the **net reactive current.** The circuit is inductive or capacitive according to which current, that through the inductor or that through the capacitor, is larger. Also notice that the sense of the phase angle is reversed for parallel *RLC* circuits. A phase angle opening in a positive direction indicates a capacitive circuit, while one opening in a negative direction indicates an inductive circuit.

Current in a Parallel *RLC* Circuit

Question: Does Kirchhoff's current law apply to parallel* RLC *circuits?

Recall that Kirchhoff's current law is the basic law for all parallel circuits. As you know, it states that the sum of the branch currents in a parallel circuit is equal to the total current. Kirchhoff's current law applies to parallel *RLC* circuits, but once again, phasor addition of the currents must be used. The total current in a parallel *RLC* circuit is the phasor sum of the current through the resistive branch and the net reactive current. By Kirchhoff's current law, the equation is

(15–5)
$$I_T = \sqrt{I_R^2 + (I_L - I_C)^2}$$

The *current triangle* for a parallel *RLC* circuit is shown in Figure 15–20. Notice that the triangle is formed using the resistance and net reactive current phasors. The hypotenuse of this right triangle represents the total current. The total current is often referred to as the "line current."

EXAMPLE 15–10 For the circuit of Figure 15–21, compute the value of the total current.

Solution:

1. Compute X_L,

$$X_L = 2\pi fL = 377\ \Omega$$

2. Compute X_C:

$$X_C = \frac{1}{2\pi fC} = 796\ \Omega$$

3. Compute I_L:

$$I_L = \frac{V_S}{X_L} = 31.83\text{ mA}$$

4. Compute I_C:

$$I_C = \frac{V_S}{X_C} = 15.07\text{ mA}$$

5. Compute I_R:

$$I_R = \frac{V_S}{R} = 30.76\text{ mA}$$

FIGURE 15–20 Current triangle for a parallel *RLC* Circuit. Inductive current is greater than capacitive; thus, θ is negative, and the circuit is inductive.

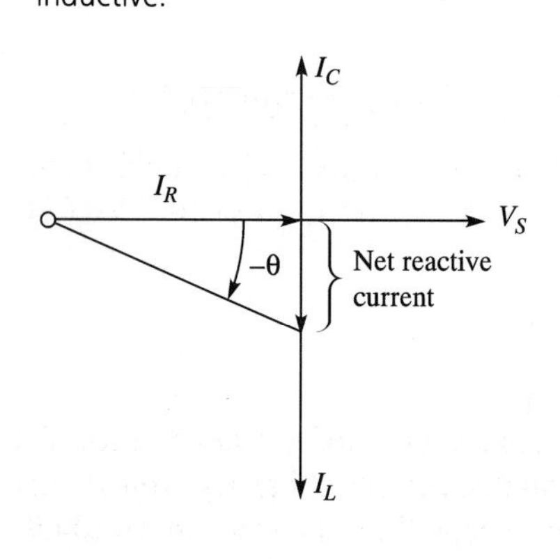

FIGURE 15–21 Circuit for Example 15–10

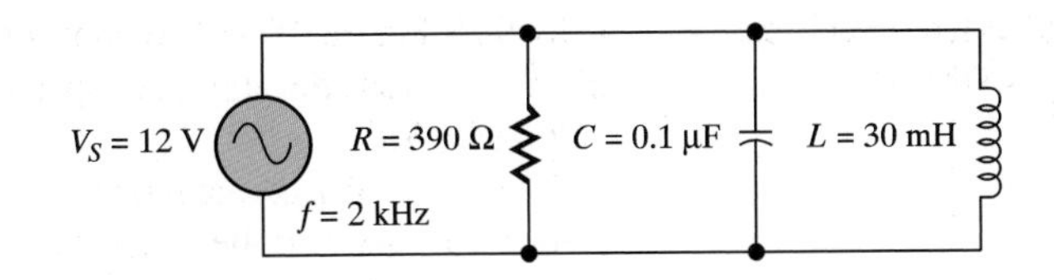

6. Compute I_T:

$$\begin{aligned} I_T &= \sqrt{I_R^2 + (I_L - I_C)^2} \\ &= \sqrt{(30.76 \text{ mA})^2 + (31.83 \text{ mA} - 1507 \text{ mA})^2} = 35.03 \text{ mA} \end{aligned}$$

EXAMPLE 15–11 If the total current of a parallel *RLC* circuit is 120 mA and the net reactive current is 45 mA, compute the resistive current I_R.

■ **Solution:**

1. Use Eq. 15–5:

$$I_T = \sqrt{I_R^2 + (I_L - I_C)^2}$$

2. Transpose:

$$\begin{aligned} I_R &= \sqrt{I_T^2 - (I_L - I_C)^2} \\ &= \sqrt{(120 \text{ mA})^2 - (45 \text{ mA})^2} = 111 \text{ mA} \end{aligned}$$

EXAMPLE 15–12 In a parallel *RLC* circuit the total current is 90 mA, and the resistive current is 48 mA. If the inductor current is 35 mA, what is the value of the capacitor current?

■ **Solution:**

$$\begin{aligned} I_T &= \sqrt{I_R^2 + (I_L - I_C)^2} \\ (I_C - I_L) &= \sqrt{I_T^2 - I_R^2} \\ &= \sqrt{(90 \text{ mA})^2 - (48 \text{ mA})^2} = 76.13 \text{ mA} \end{aligned}$$

Then,

$$I_C - 35 \text{ mA} = 76.13 \text{ mA}$$

$$I_C = 76.13 \text{ mA} + 35 \text{ mA} = 111 \text{ mA}$$

■ Impedance in a Parallel *RLC* Circuit

Question: How is the impedance of a parallel* RLC *circuit computed?

As is the case with any parallel circuit, the impedance of a parallel *RLC* circuit is not simply the phasor sum of resistance and reactance. The *impedance* of a parallel *RLC* circuit is found as the reciprocal of its admittance:

(15–6)
$$Z = \frac{1}{Y}$$

The admittance Y is equal to the phasor sum of the conductance and net susceptance. An *admittance triangle* is shown in Figure 15–22. Recall that current is directly proportional to conductance and susceptance. Thus, their phasors occupy the same place in the admit-

FIGURE 15–22 Admittance triangle for a parallel *RLC* circuit

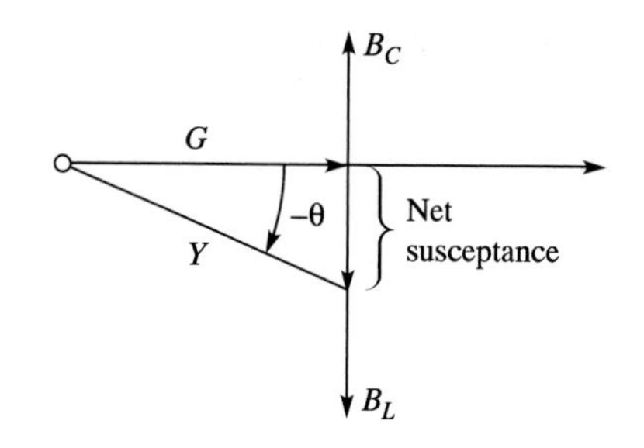

tance triangle as their currents did in the current triangle. The **net susceptance** is the difference between the inductive and capacitive susceptance. Thus, in equation form, admittance is

(15–7) $$Y = \sqrt{G^2 + (B_L - B_C)^2}$$

EXAMPLE 15–13 In a parallel *RLC* circuit the resistance and reactances are $R = 1.2\ \text{k}\Omega$, $X_L = 1655\ \Omega$, and $X_C = 1210\ \Omega$. Compute the impedance Z of the circuit.

■ Solution:

1. Compute B_L:

$$B_L = \frac{1}{X_L} = 604\ \mu\text{S}$$

2. Compute B_C:

$$B_C = \frac{1}{X_C} = 826\ \mu\text{S}$$

3. Compute G:

$$G = \frac{1}{R} = 833\ \mu\text{S}$$

4. Compute Y:

$$\begin{aligned} Y &= \sqrt{G^2 + (B_L - B_C)^2} \\ &= \sqrt{(833\ \mu\text{S})^2 + (826\ \mu\text{S} - 604\ \mu\text{S})^2} = 862\ \mu\text{S} \end{aligned}$$

5. Compute Z:

$$Z = \frac{1}{Y} = 1159.99\ \Omega$$

EXAMPLE 15–14 A parallel *RLC* circuit has an impedance of 1240 Ω. If the net susceptance is 1860 Ω capacitive, calculate the value of the resistance.

■ Solution:

$$Y = \sqrt{G^2 + B^2}$$

and

$$\begin{aligned} G &= \sqrt{Y^2 - B^2} \\ &= \sqrt{(1/1240\ \Omega)^2 - (1/1860\ \Omega)^2} = \sqrt{(806\ \mu\text{S})^2 - (538\ \mu\text{S})^2} = 600\ \mu\text{S} \end{aligned}$$

Then,

$$R = \frac{1}{G}$$

$$= \frac{1}{600\ \mu S} = 1667\ \Omega$$

■ Phase Angle of the Parallel *RLC* Circuit

Question: How is the phase angle of a parallel* RLC *circuit computed?

As in any electric circuit, the circuit phase angle is the angle between the source voltage and total current. As illustrated in the current triangle of Figure 15–23, the side opposite the phase angle is the phasor representing the net reactive current. The side adjacent to the phase angle is the phasor representing the resistive current. Thus, the phase angle is equal to the arc tangent of the ratio of the net reactive current to the resistive current:

(15–8)
$$\theta = \text{arc tan} \frac{I_C - I_L}{I_R}$$

The admittance triangle may also be used in computing the phase angle. As shown in Figure 15–24, the net susceptance is the opposite side and the conductance is the adjacent side. Thus, the phase angle may also be computed as

(15–9)
$$\theta = \text{arc tan} \frac{B_L - B_C}{G}$$

EXAMPLE 15–15 For the circuit of Figure 15–25, compute the phase angle.

■ Solution:

$$\theta = \text{arc tan} \frac{I_C - I_L}{I_R}$$

$$= \text{arc tan} \frac{28\text{ mA} - 12\text{ mA}}{24\text{ mA}} = \text{arc tan } 0.667 = 33.69°$$

FIGURE 15–23 Computation of the phase angle with the current triangle

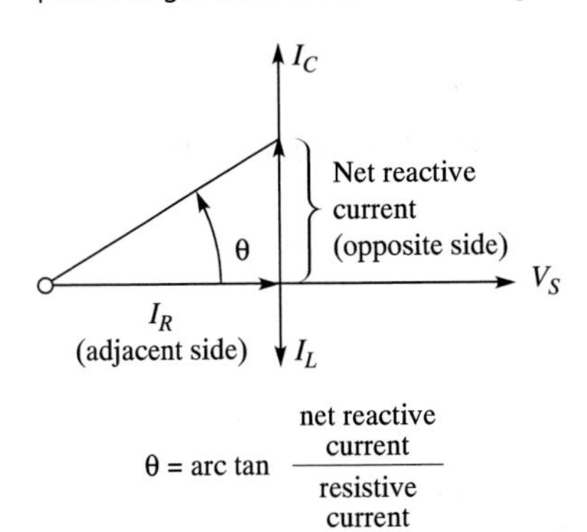

FIGURE 15–24 Computation of phase angle with the admittance triangle

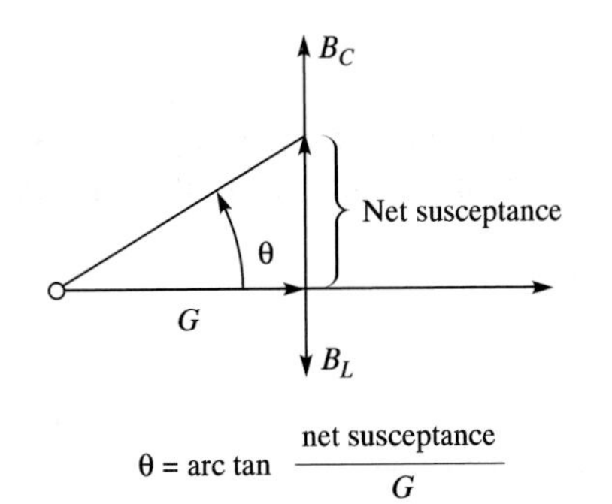

FIGURE 15–25 Circuit for Example 15–15

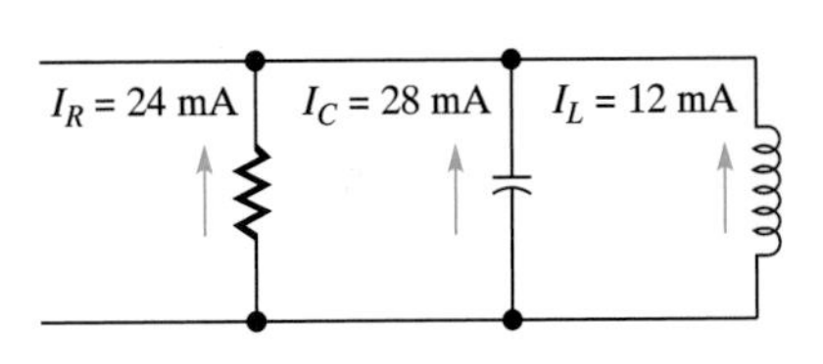

EXAMPLE 15–16 In a parallel *RLC* circuit, the inductive current is 10 mA and the capacitive current is 35 mA. If the phase angle of the circuit is 52°, compute the value of I_R.

■ Solution:

$$\tan\theta = \frac{I_C - I_L}{I_R} = \tan 52° = 1.279$$

$$1.279 = \frac{35\text{ mA} - 10\text{ mA}}{I_R} = \frac{25\text{ mA}}{I_R}$$

$$(1.279)I_R = 25\text{ mA}$$

$$I_R = \frac{25\text{ mA}}{1.279} = 19.546\text{ mA}$$

■ SECTION 15–3 REVIEW

1. If I_C is 4.48 mA and I_L is 9.33 mA is a parallel *RLC* circuit inductive or capacitive?
2. If B_L is 600 μS and B_C is 960 μS, is a parallel *RLC* circuit inductive or capacitive?
3. Compute the total current I_T in Problem 1.
4. Compute the impedance Z in Problem 2.
5. A parallel *RLC* circuit consists of a 39 Ω resistance, a 24 Ω inductive reactance, and a 6 Ω capacitive reactance. If the source voltage is 6 V, compute the circuit phase angle θ.

15–4 PARALLEL *RLC* CIRCUIT RESPONSE

Question: What effect does frequency changes have on the response of parallel* RLC *circuits?

■ Effects of Changes in Frequency

Since the parallel *RLC* circuit is a current divider, the branch currents must be considered when determining circuit response. If the current through the inductive branch I_L is greater than the current through the capacitive branch I_C, the phase angle opens in a negative direction. The source voltage leads the current, and the circuit is inductive. The opposite is also true. If the capacitive current I_C is greater than inductive current I_L, the phase angle opens in a positive direction. The current leads the source voltage, and the circuit is capacitive.

The voltage across each branch is the same, so the branch currents are inversely proportional to the reactance ($I_L = V_S/X_L$ and $I_C = V_S/X_C$). Any increase in frequency causes an increase in X_L and a decrease in X_C. Thus, the current I_C increases and I_L decreases. The phase angle increases in a positive direction and the circuit becomes more capacitive. A decrease in frequency causes a decrease in X_L and an increase in X_C. Now the current I_L increases while I_C decreases. The phase angle increases in a negative direction and the circuit becomes more inductive. Thus, in a parallel *RLC* circuit, an increase in frequency makes the circuit more capacitive, while a decrease makes it more inductive, as illustrated in Figure 15–26.

EXAMPLE 15–17 For the circuit of Figure 15–27, compute the value of the phase angle for the following frequencies: (a) 1 kHz, (b) 2 kHz, and (c) 4 kHz.

FIGURE 15–26 Effects of frequency changes in a parallel *RLC* circuit

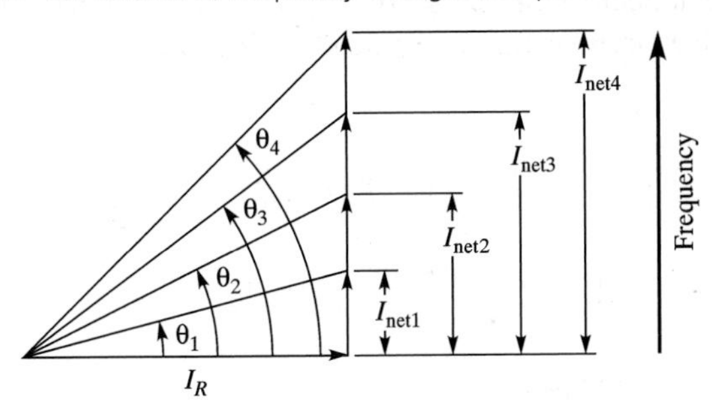

(a)
As frequency increases, X_C decreases, and the net capacitive current increases, increasing θ in a positive direction.

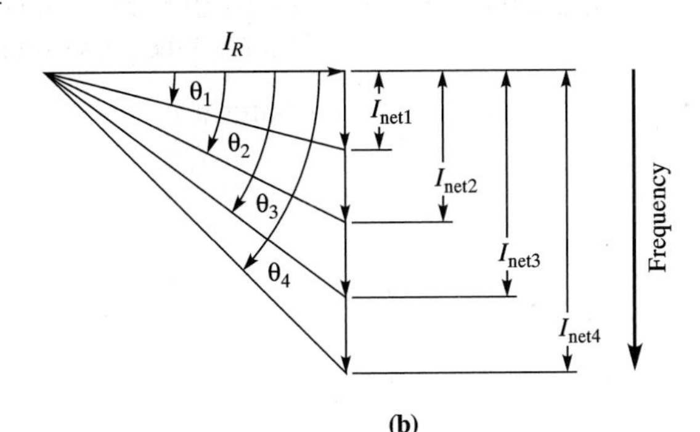

(b)
As frequency decreases, X_L decreases, and the net inductive current increases, increasing θ in a negative direction.

FIGURE 15–27 Circuit for Example 15–17

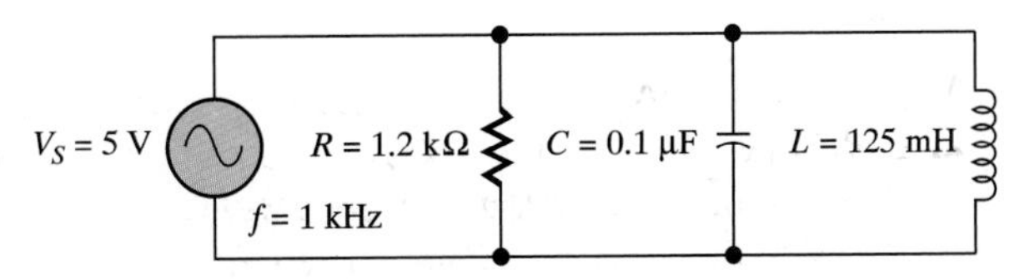

■ Solution:

a. For 1 kHz,

$$X_L = 2\pi fL = 785\ \Omega$$

$$X_C = \frac{1}{2\pi fC} = 1592\ \Omega$$

$$I_L = \frac{V_S}{X_L} = 6.37\text{ mA}$$

$$I_C = \frac{V_S}{X_C} = 3.14\text{ mA}$$

$$I_R = \frac{V_S}{R} = 4.17\text{ mA}$$

$$\theta = \arctan\frac{I_C - I_L}{I_R}$$

$$\theta = \arctan\frac{3.14\text{ mA} - 6.37\text{ mA}}{4.17\text{ mA}} = \arctan -0.775 = -37.76°$$

b. For 2 kHz,

$$X_L = 2\pi fL = 1570\ \Omega$$

$$X_C = \frac{1}{2\pi fC} = 796\ \Omega$$

$$I_L = \frac{V_S}{X_L} = 3.184\text{ mA}$$

$$I_C = \frac{V_S}{X_C} = 6.281 \text{ mA}$$

$$\theta = \text{arc tan} \frac{I_C - I_L}{I_R} = 36.6°$$

c. For 4 kHz

$$X_L = 2\pi fL = 3140 \ \Omega$$

$$X_C = \frac{1}{2\pi fC} = 398 \ \Omega$$

$$I_L = \frac{V_S}{X_L} = 1.592 \text{ mA}$$

$$I_C = \frac{V_S}{X_C} = 12.562 \text{ mA}$$

$$\theta = \text{arc tan} \frac{I_C - I_L}{I_R} = 69.186°$$

Notice in this example that the phase angle increased in a positive direction as the frequency increased. Practice now by substituting decreasing values of frequency in this example, and confirm that the phase angle decreases.

■ Changes in Inductance and Capacitance

Question: How do changes in inductance or capacitance affect the response of a parallel RLC?

As you know, inductive reactance is directly proportional to inductance, and capacitive reactance is inversely proportional to capacitance. Thus, these two quantities, inductance and capacitance, affect parallel *RLC* circuit response in the same sense as frequency. An increase in inductance value increases X_L and reduces I_L. Thus, the circuit becomes less inductive and more capacitive. An increase in capacitance has the opposite effect, as illustrated in Figure 15–28. Decreases in inductance and capacitance have the opposite effects.

Question: Does a change in resistance affect the phase angle of a parallel RLC *circuit?*

Once again, even though resistance is not frequency sensitive, changes in resistance for given values of net inductive current cause changes in circuit response. A decrease in

FIGURE 15–28 Effects of inductance and capacitance

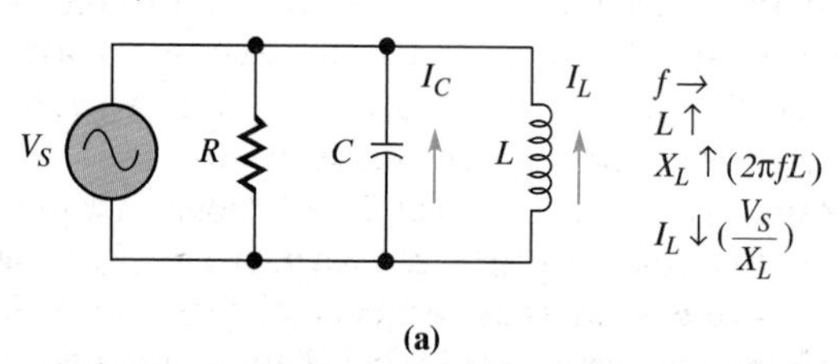

(a)
An increase in inductance causes X_L to increase and I_L to decrease; thus, the circuit becomes less inductive.

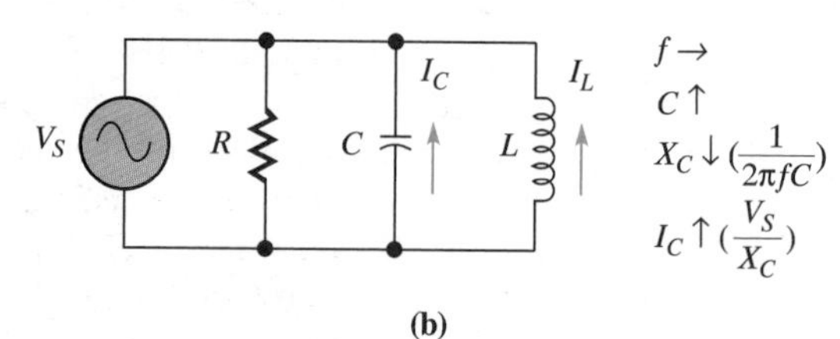

(b)
An increase in capacitance causes X_C to decrease and I_C to increase; thus, the circuit becomes more capacitive.

resistance increases the resistive current. This moves the phase angle toward zero whether the circuit is inductive or capacitive. An increase in resistance lowers the resistive current and causes the phase angle to increase.

SECTION 15–4 REVIEW

1. Does a decrease in frequency cause a parallel *RLC* circuit to become more inductive or more capacitive?
2. In a parallel *RLC* circuit, if I_C is greater than I_L, is the circuit capacitive or inductive?
3. As frequency increases, does the phase angle of parallel *RLC* circuit increase in a positive or negative direction?
4. Does an increase in the value of capacitance cause a parallel *RLC* circuit to become more inductive or capacitive?
5. Does an increase in resistance cause the phase angle of a parallel *RLC* circuit to increase or decrease?

15–5 APPLICATIONS OF NONRESONANT *RLC* CIRCUITS

Question: What are the applications of* RLC *circuits based upon?

The applications of nonresonant *RLC* circuits are based upon the fact that the effects of inductance and capacitance cancel one another. The one with the strongest effect—on voltage in a series circuit or current in a parallel circuit—combines with resistance to determine the circuit's response. As you have learned, the relative strengths of the inductive and capacitive effects are dependent upon two things: (1) the frequency of the applied voltage and (2) the values of inductance and capacitance. The circuit response can be varied through changes in any of these. This variation in circuit response will be utilized in the applications discussed in the following paragraphs.

Correction of Power Factor

As you know, the power factor of a circuit indicates two things: (1) the percentage of apparent power that is true power and (2) whether the circuit is inductive or capacitive. The ideal condition is one in which the power factor is 1, which indicates that the circuit phase angle is zero and that the circuit is completely resistive. Thus, all power developed is true power. The appliances in a home are predominantly resistive (electric stove, lights, resistance heater, and so forth). Some appliances, such as the refrigerator and furnace, contain electric motors that are inductive. Thus, the power distribution circuit for the home can be reduced to an *RL* circuit. The resistance, however, is great compared to the inductance, and the power factor is close to 0.9. The power factor could be corrected to zero, but under current circumstances, it would not be economical to do so.

A factory, on the other hand, may have many more electric motors and fewer resistive devices. Thus, its power factor may be much lower—on the order of 0.6. As illustrated in Figure 15–29(a), the motors are shown as parallel inductances, while the resistive devices are shown as resistors. Recall that the phase angle is determined by the ratio of the inductive current to the capacitive current. The net inductive current is the difference between the inductive and capacitive currents. (Even though inductance predominates in most power circuits, there will be some capacitance present.) The phase angle is illustrated in Figure 15–29(b). Notice that the phase angle is relatively large, resulting in a small power factor. Since it is the apparent power for which the owner pays, this situation is unacceptable from an economic standpoint. Also, if the power factor were 1, the net

FIGURE 15–29 Power distribution circuit

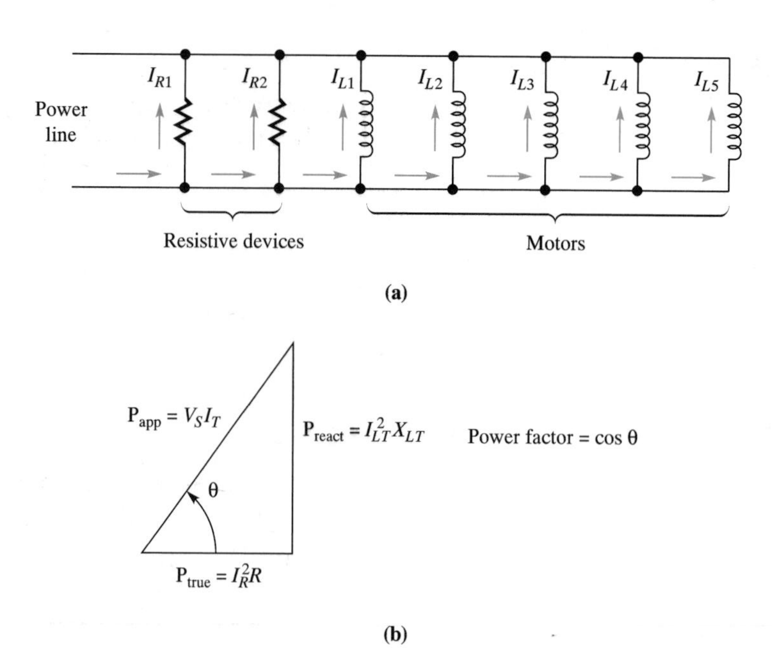

current would be that through the resistive branches only. Thus, the current supplied by the generator would be smaller and the power lines would run cooler. So a small power factor value is unacceptable to the power company also!

FIGURE 15–30 Power factor correction: A capacitor added to the circuit produces the same current as inductance

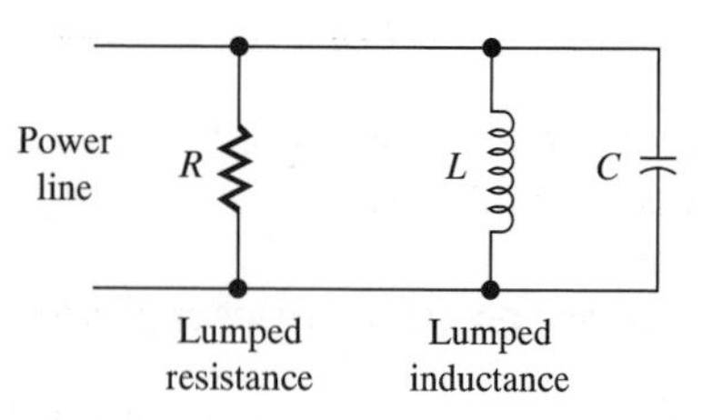

Question: What is meant by power factor correction, and how is it accomplished?

If the power factor is equal to 1, all the apparent power is true power ($P_{true} = P_{app} \times$ power factor). Thus, correcting the power factor means making circuit changes that produce a power factor of as close to 1 as possible, as shown in Figure 15–30, where the resistances and inductances have been combined. A physical capacitance is added in parallel with the other circuit elements. If this reactance produces the same current as the inductive components, the net reactive current will be zero. The phase angle will be zero, and the power factor will be 1. It is not necessary for the power factor to be exactly 1. Usually, 0.85 or above is considered adequate.

■ Elimination of Unwanted High Frequencies

The power cord of a radio or television often acts as an antenna, with signals from high-frequency sources being induced in it. These voltages can get into the dc power supply, mix with the desired signals in the amplifiers, and cause a very objectionable deterioration in system performance. These signals can be eliminated, or at least greatly reduced, using what are known as *wavetraps.*

FIGURE 15–31 Wavetrap circuit

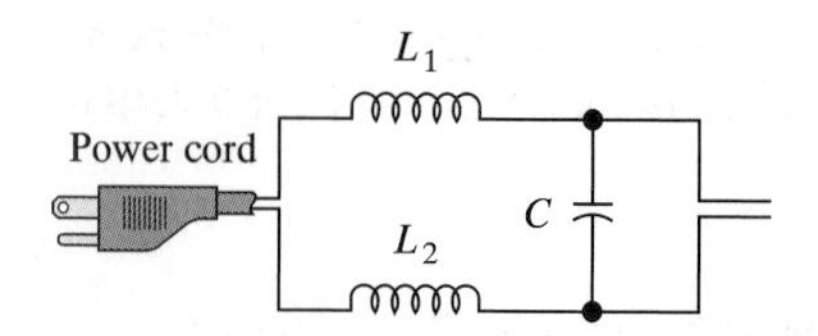

A wavetrap is shown in Figure 15–31. Notice that an inductance is placed in series with the power line and that a capacitance shunts it. The series inductances produce high reactance at higher frequencies, greatly reducing their currents. The capacitance, on the other hand, is a low reactance at these frequencies. The capacitance thus shunts the higher frequencies around the power supply. The result of these two actions is a great reduction in high-frequency pickup by the power lines and cords.

■ Detection of Frequency Variations

An *RLC* circuit can be used to detect changes in frequency above or below some standard. The standard is usually the frequency at which the effects of inductance and capacitance

FIGURE 15–32 Circuit for detecting frequency changes

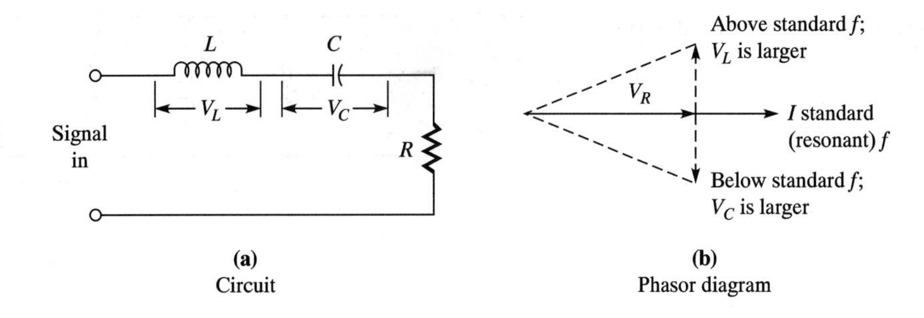

FIGURE 15–33 Series *RLC* circuit with an open resistor

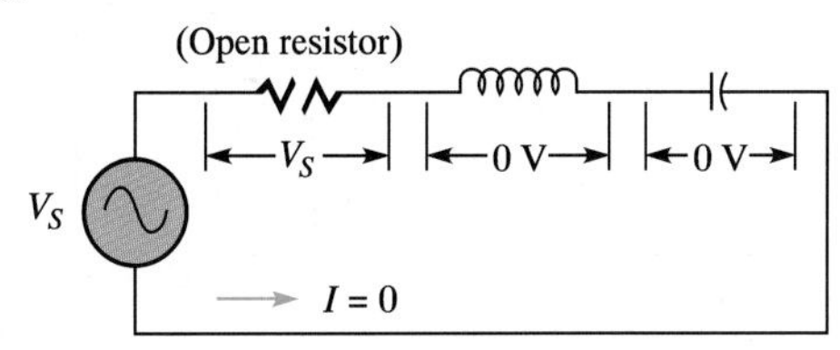

just cancel one another, leaving the circuit totally resistive. As illustrated in Figure 15–32, above the standard frequency, the circuit is inductive, and below it, the circuit is capacitive. The change in phase angle from negative to positive, or vice versa, indicates when the frequency moves above or below the standard. This type of circuit is used in conjunction with others in FM receivers. They convert frequency variations into audio output signals for reproduction by a speaker.

SECTION 15–5 REVIEW

1. What characteristics of an *RLC* circuit are made use of in their applications?
2. What characteristic is made use of in the correction of power factor?
3. What is considered a reasonable value of power factor?
4. In suppressing high frequencies from the power line, is the inductor or capacitor placed in series?
5. When the effects of inductance and capacitance just cancel one another, is the circuit power factor zero, 1, or 0.5?

15–6 TROUBLESHOOTING

Troubleshooting an *RLC* circuits proceeds in a manner similar to that presented for *RL* and *RC* circuits. Recall that the problems that can occur in inductors and capacitors are shorts and opens. In an inductor, the short may be partial or complete. In a capacitor, the dielectric may break down completely or just "leak" current through it. The effects of opens or shorts depends upon the type of circuit—series or parallel—and the relative values of the reactances.

Question: What are the effects of opens and shorts in series RLC *circuits?*

The effect of an open in a series *RLC* circuit is quite simple: The current ceases to flow, and no voltage is developed across the components. The source voltage is found across the open component, as illustrated in Figure 15–33. A short, on the other hand, may

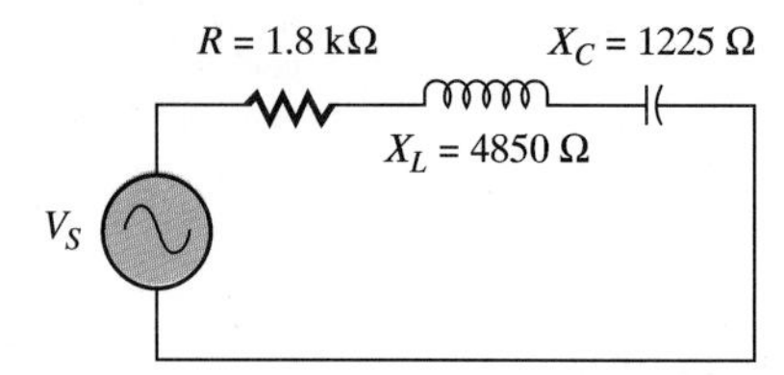

FIGURE 15–34 Effects of shorts in series *RLC* circuits

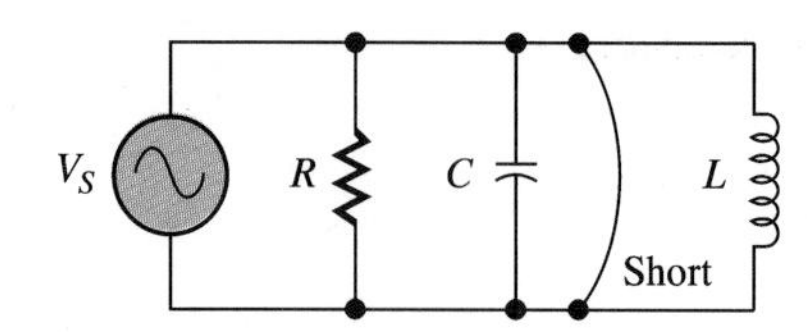

FIGURE 15–35 Parallel *RLC* circuit with a shorted branch

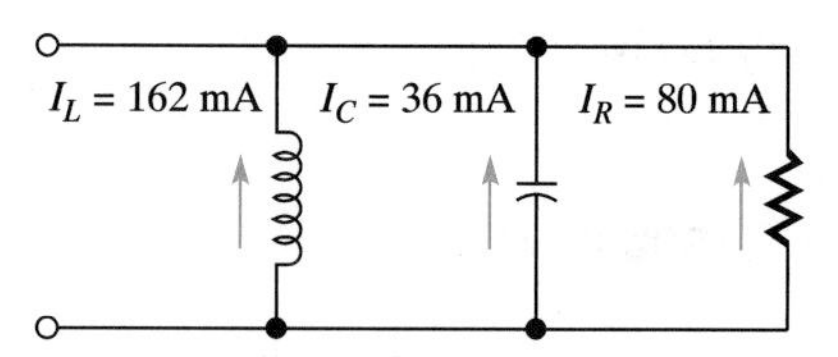

FIGURE 15–36 Effects of opens in parallel *RLC* circuits

have a rather subtle effect. It may seem logical to think that any shorted component will remove it from the circuit and reduce the impedance, but this is not necessarily the case. In Figure 15–34, the normal impedance for the circuit is 4047 Ω. If the capacitor shorts, the net reactance becomes the value of X_L. Thus, the impedance *increases* to 5173 Ω. If the inductor should short, making the net reactance the value of X_C, the impedance *decreases* to 2177 Ω.

Question: What are the effects of opens and shorts in a parallel RLC *circuit?*

As shown in Figure 15–35, a short across any branch of a parallel *RLC* circuit shorts all branches. The short carries the entire current, and no current flows through any branch. Once again, it would seem that an open circuit would remove one branch and reduce the total current, but this is not always the case. An open has an effect that is dependent upon the relative value of the reactances. In Figure 15–36, the normal total current is 149 mA. If the capacitor opens, the current will rise to 181 mA. On the other hand, if the inductor opens, the current will decrease to 88 mA.

CHAPTER SUMMARY

1. An *RLC* circuit contains all three circuit effects: resistance, inductance, and capacitance.
2. The source voltage in a series *RLC* circuit is equal to the phasor sum of the resistor voltage and net reactive voltage.
3. The impedance of a series *RLC* circuit is equal to the phasor sum of the resistance and net reactance.
4. The phase angle of a series *RLC* circuit is equal to the arc tangent of the ratio of net reactance to the resistance.
5. True power in an *RLC* circuit is developed by the resistive components.
6. Reactive power in an *RLC* circuit is produced by the reactive components.
7. Apparent power in an *RLC* circuit is the phasor sum of true and reactive power.
8. The power factor of an *RLC* circuit is the cosine of the phase angle.
9. The phase angle of a series *RLC* circuit is directly proportional to the inductance and inversely proportional to the capacitance.
10. An increase in frequency makes a series *RLC* circuit more inductive, while a decrease in frequency makes it more capacitive.
11. The total current in a parallel *RLC* circuit is the phasor sum of the resistive and net reactive currents.
12. The net susceptance is the difference of B_L and B_C.
13. The admittance of a parallel *RLC* circuit is equal to the phasor sum of the net susceptance and the conductance.
14. The phase angle of a parallel *RLC* circuit is equal to the arc tangent of the ratio of the net reactive current and the resistive current.
15. A parallel *RLC* circuit is made more inductive by a decrease in frequency and is made more capacitive by an increase in frequency.
16. A parallel *RLC* circuit is made more capacitive by an increase in capacitance and more inductive by a decrease in inductance.

KEY TERMS

net reactive voltage

net reactance

net susceptance

EQUATIONS

(15–1) $V_S = \sqrt{V_R^2 + (V_L - V_C)^2}$

(15–2) $Z = \sqrt{R^2 + (X_L - X_C)^2}$

(15–3) $\theta = \arctan \dfrac{V_L - V_C}{V_R}$

(15–4) $\theta = \arctan \dfrac{X_L - X_C}{R}$

(15–5) $I_T = \sqrt{I_R^2 + (I_L - I_C)^2}$

(15–6) $Z = \dfrac{1}{Y}$

(15–7) $Y = \sqrt{G^2 + (B_L - B_C)^2}$

(15–8) $\theta = \arctan \dfrac{I_C - I_L}{I_R}$

(15–9) $\theta = \arctan \dfrac{B_L - B_C}{G}$

Variable Quantities

B_L = inductive suceptance

B_C = capacitive suceptance

Z = impedance

X_L = inductive reactance

X_C = capacitive reactance

y = admittance

f = frequency

L = inductance

C = capacitance

TEST YOUR KNOWLEDGE

1. In a series *RLC* circuit, if the phase angle opens in a negative direction, is the circuit inductive or capacitive?
2. In a parallel *RLC* circuit, if the phase angle opens in a negative direction, is the circuit inductive or capacitive?
3. If the inductive reactance of a series *RLC* circuit is greater than the capacitive reactance, is the circuit inductive or capacitive?
4. The power factor of a series *RLC* circuit is lagging. Is the circuit inductive or capacitive?
5. Will adding inductance or capacitance to the circuit correct the power factor in Problem 4?
6. Does an increase in frequency cause a series *RLC* circuit to become more inductive or more capacitive?
7. Will a decrease in frequency cause a parallel *RLC* circuit to become more capacitive or more inductive?
8. Will a decrease in the capacitance cause a series *RLC* circuit to become more inductive or more capacitive?
9. Does an increase in inductance cause a parallel *RLC* circuit to become more inductive or more capacitive?
10. What is considered a reasonable value for power factor?
11. Is an *RLC* circuit with a power factor of 1 inductive, capacitive, or resistive?
12. Does an increase in resistance cause the phase angle of a parallel *RLC* circuit to increase or decrease?

PROBLEM SET: BASIC LEVEL

Section 15–1

1. For each circuit of Figure 15–37, compute the impedance. State whether each is inductive or capacitive.

FIGURE 15–37 Circuits for Problem 1

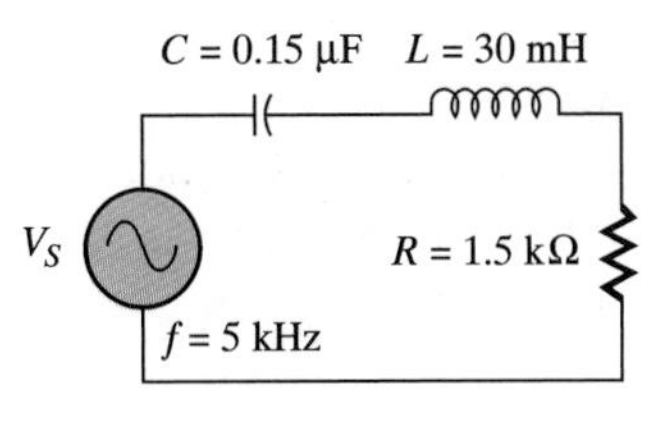

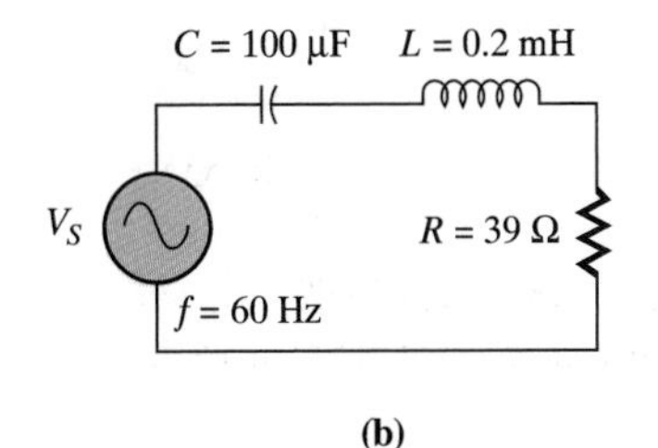

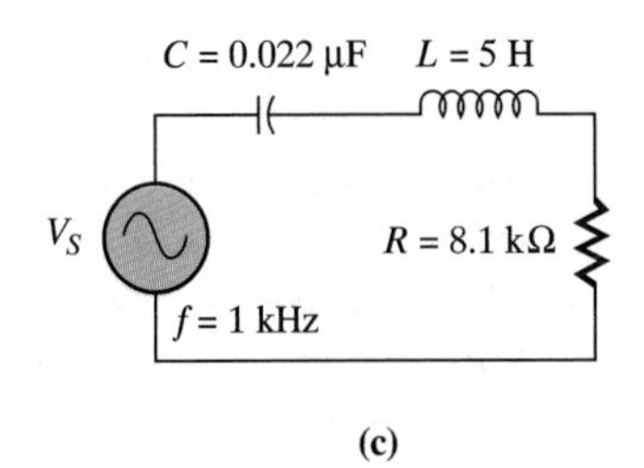

2. For each circuit of Figure 15–38, compute of the source voltage. State whether each is inductive or capacitive.

FIGURE 15–38 Circuits for Problem 2

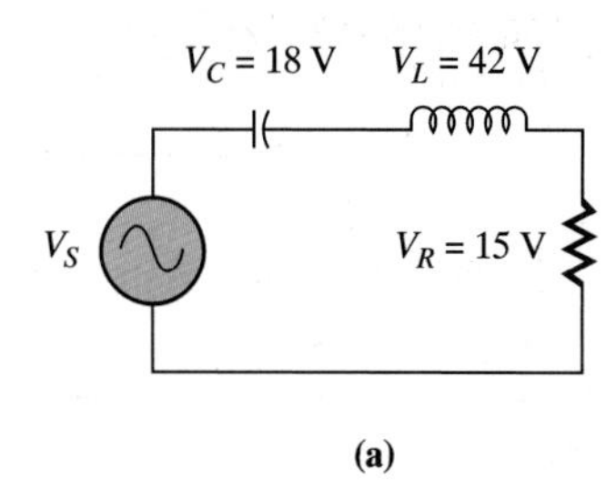

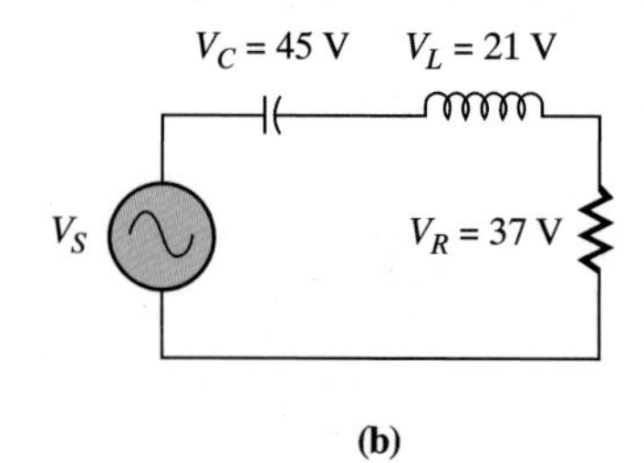

3. For each circuit of Figure 15–39, compute the apparent power P_{app}. State whether each is inductive or capacitive.

FIGURE 15–39 Circuits for Problem 3

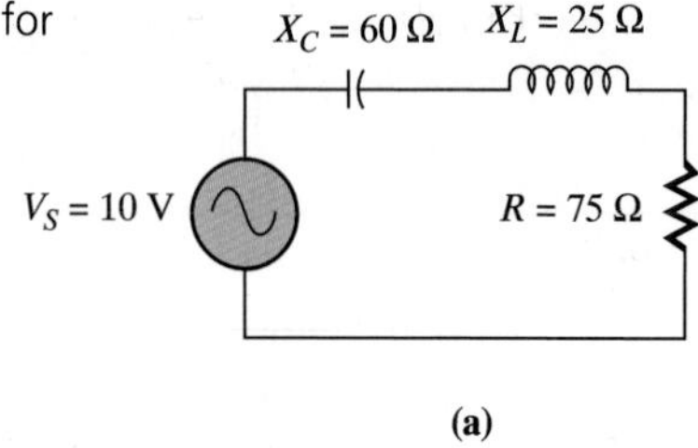

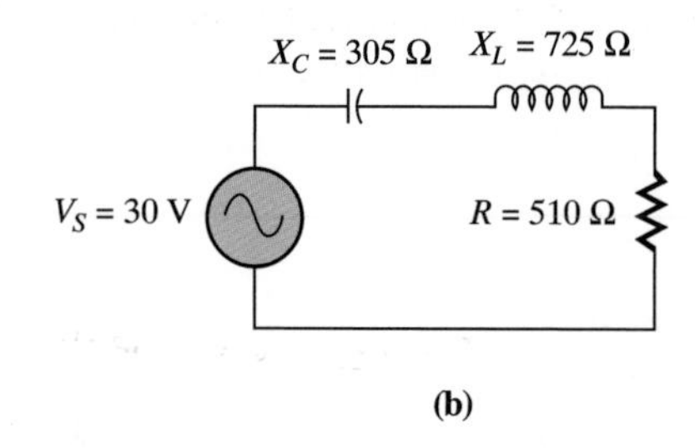

4. For the circuit of Figure 15–40, compute: (a) X_L, (b) X_C, (c) Z, (d) I_T, (e) V_L, (f) V_C, and (g) θ.

FIGURE 15–40 Circuit for Problem 4

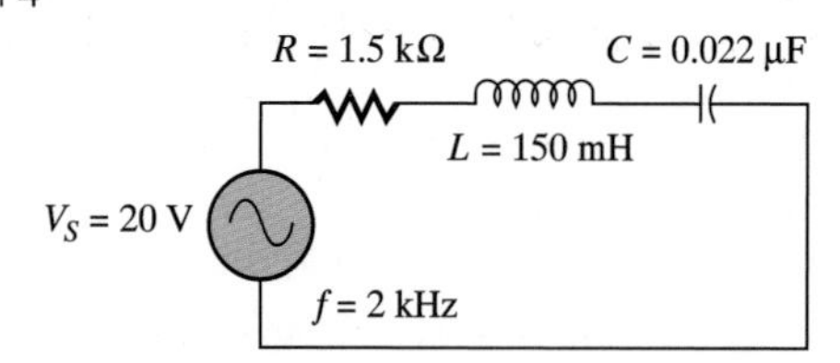

5. For the circuit of Figure 15–41, compute the power factor. Is the power factor leading or lagging?

FIGURE 15–41 Circuit for Problem 5

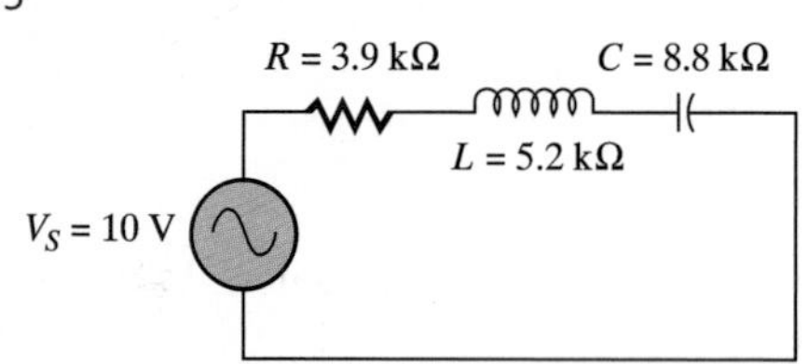

Section 15–2

6. Compute the phase angle for the circuit of Figure 15–42. Recompute the phase angle for the following conditions: (a) the value of the frequency is doubled and (b) the value of the frequency is halved.

FIGURE 15–42 Circuit for Problem 6

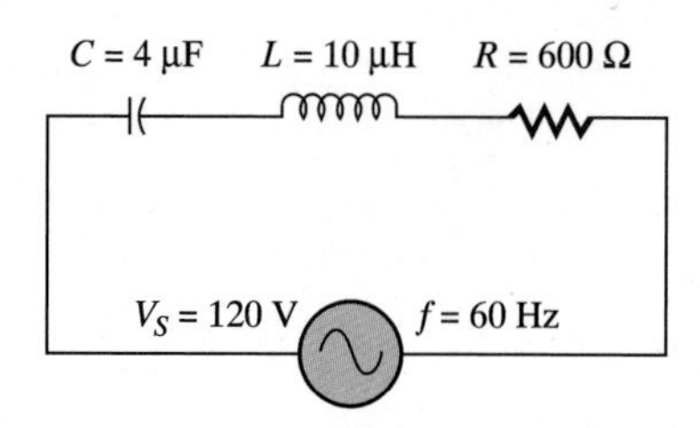

7. Compute the phase angle for the circuit of Figure 15–43. Recompute the phase angle for the following conditions: (a) the value of the capacitance is doubled and (b) the capacitance value is halved.

FIGURE 15–43 Circuit for Problem 7

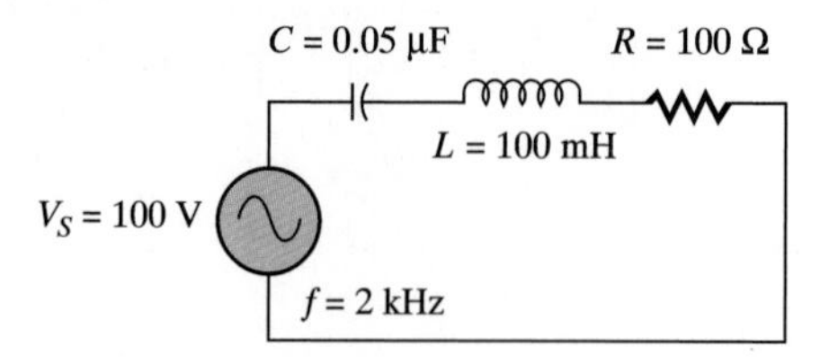

8. Compute the phase angle for the circuit of Figure 15–44. Recompute the phase angle for the following conditions: (a) the value of the inductance is doubled and (b) the value of the inductance is halved.

FIGURE 15–44 Circuit for Problem 8

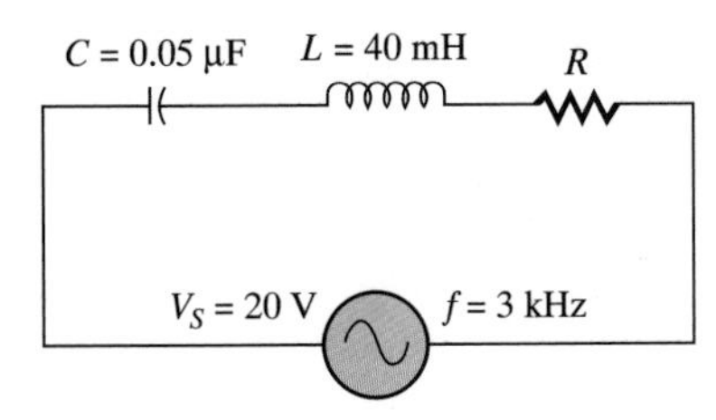

Section 15–3

9. For each circuit of Figure 15–45, compute the impedance. State whether each is inductive or capacitive.

FIGURE 15–45 Circuits for Problem 9

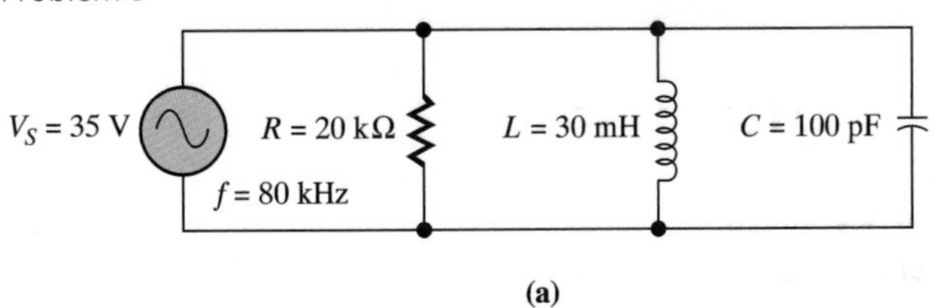

(a)

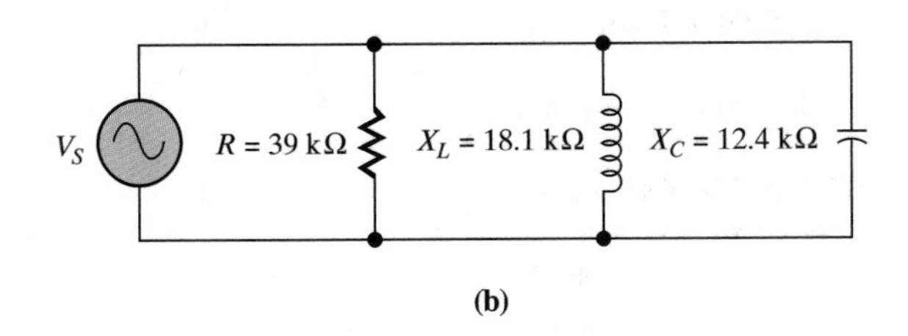

(b)

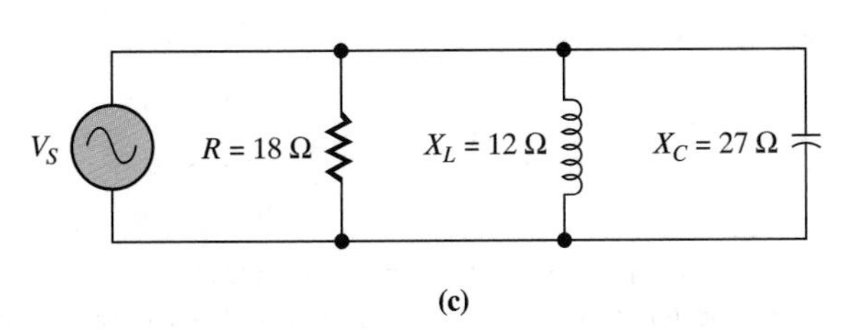

(c)

10. For each circuit of Figure 15–46, compute the total current I_T. State whether each is inductive or capacitive.

FIGURE 15–46 Circuits for Problem 10

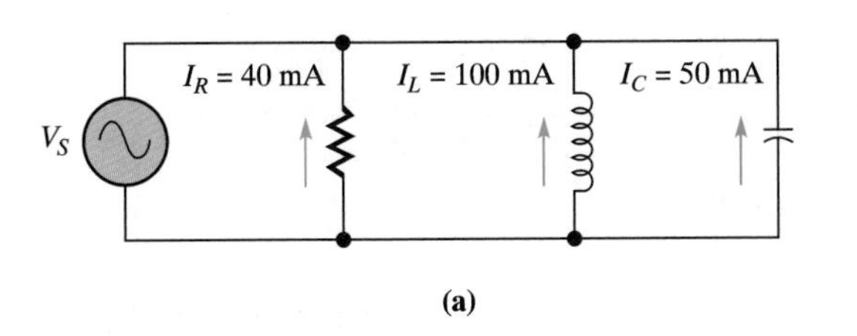

(a)

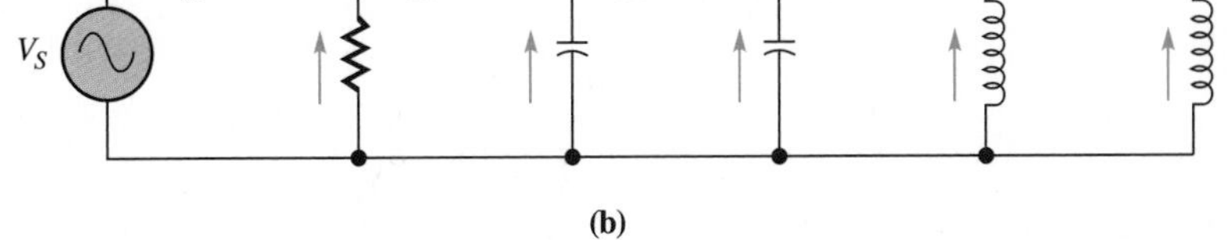

(b)

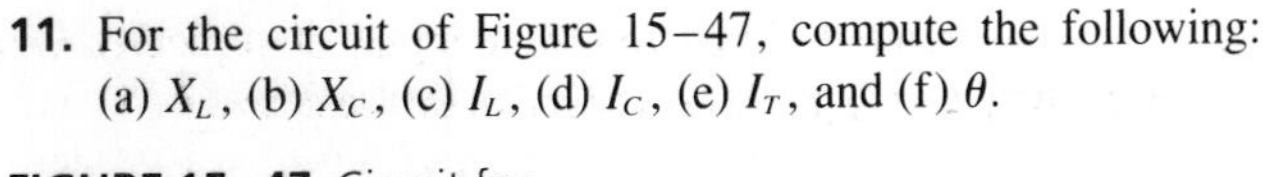

11. For the circuit of Figure 15–47, compute the following: (a) X_L, (b) X_C, (c) I_L, (d) I_C, (e) I_T, and (f) θ.

FIGURE 15–47 Circuit for Problem 11

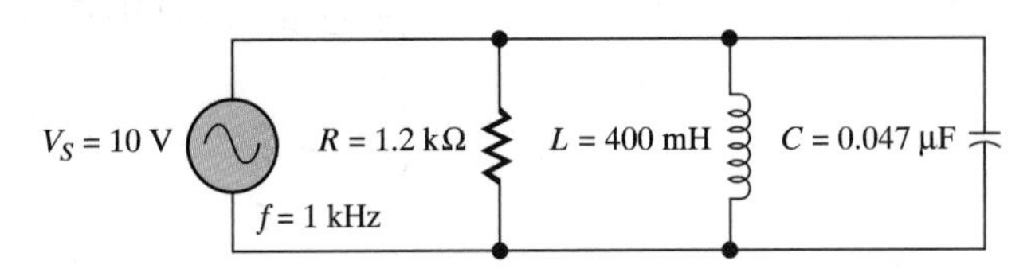

Section 15–4

12. Compute the phase angle for the circuit of Figure 15–48. Recompute the phase angle under the following conditions: (a) the value of the frequency is doubled and (b) the value of the frequency is halved.

FIGURE 15–48 Circuit for Problem 12

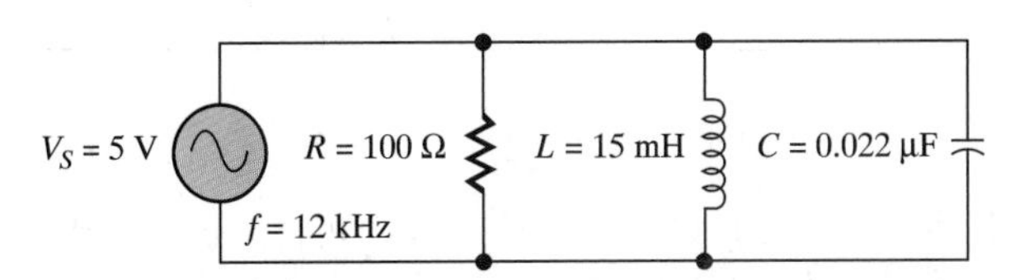

13. Compute the phase angle for the circuit of Figure 15–49. Recompute the phase angle under the following conditions: (a) the capacitance value is doubled and (b) the capacitance value is halved.

FIGURE 15–49 Circuit for Problem 13

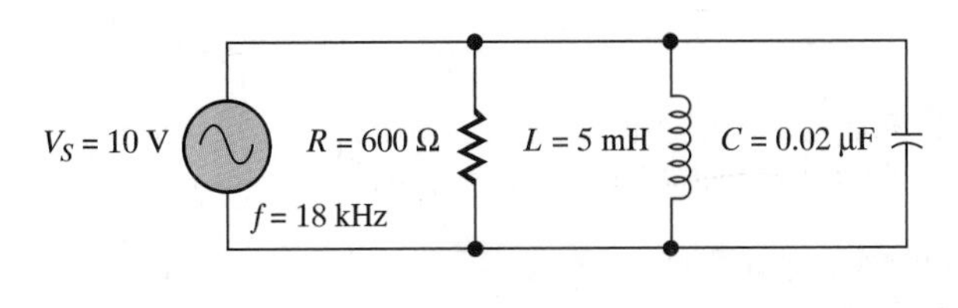

14. Compute the phase angle for the circuit of Figure 15–50. Recompute the phase angle under the following conditions: (a) the inductance value is doubled and (b) the inductance value is halved.

FIGURE 15–50 Circuit for Problem 14

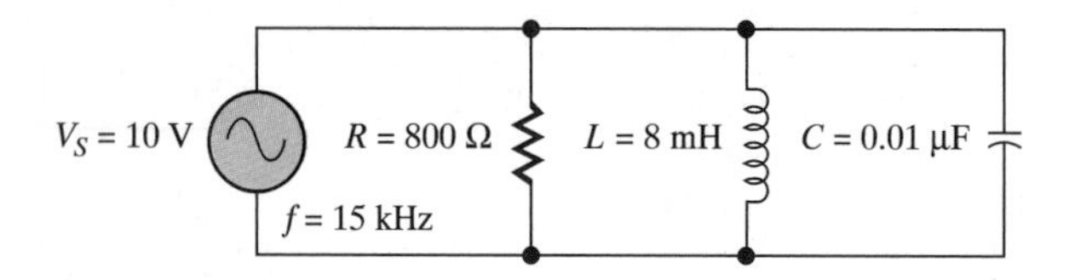

Section 15–5

15. In the circuit of Figure 15–51, compute the value of capacitor, which when added in parallel, will cause the power factor to be 1.

FIGURE 15–51 Circuit for Problem 15

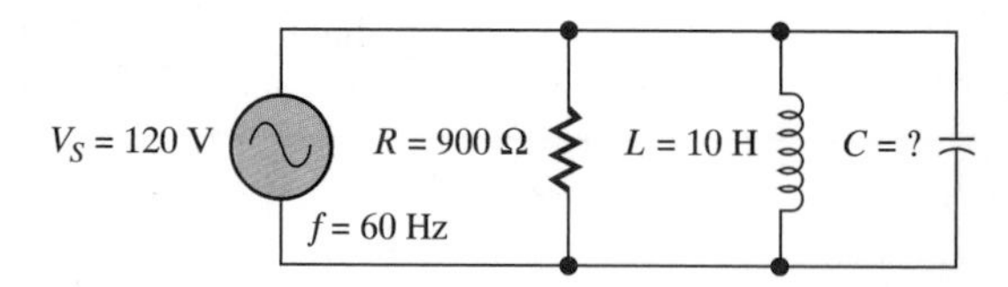

PROBLEM SET: CHALLENGING

16. A series *RLC* circuit with an impedance of 1.55 kΩ has a current flow of 60 mA. If the circuit phase angle is a $-29°$ compute (a) V_S and (b) the net reactive voltage.

17. A parallel *RLC* circuit has a source voltage of 120 V and an impedance of 500 Ω. If the phase angle is 58°, compute (a) I_T and (b) the net reactive current.

18. An *RLC* circuit has a 0.75 power factor. If the circuit draws 100 A from a 120 V line compute (a) the true power and (b) the reactive power.

19. For the series *RLC* circuit of Figure 15–52, compute the following: (a) V_R, (b) V_L, (c) V_C, (d) V_S, (e) Z, (f) C, (g) θ, and (h) L.

FIGURE 15–52 Circuit for Problem 19

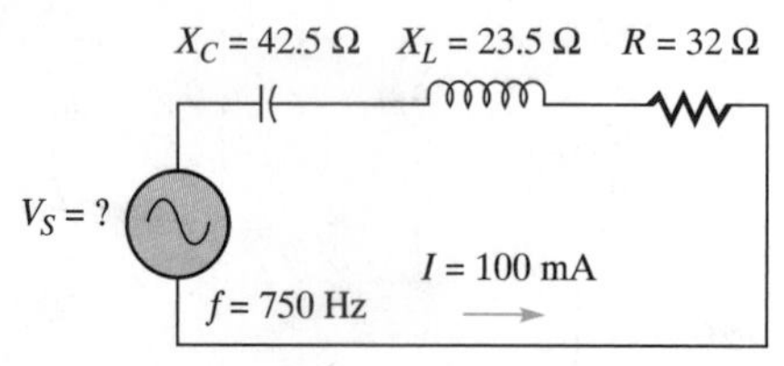

20. For the series *RLC* circuit of Figure 15–53, compute the following: (a) f, (b) X_C, (c) Z, (d) V_R, (e) V_L, (f) V_C, (g) R, and (h) θ.

FIGURE 15–53 Circuit for Problem 20

$C = 0.1\ \mu$F $L = 5$ mH R
$X_L = 314\ \Omega$
$V_S = 20$ V
$I = 123$ mA

21. For the parallel *RLC* circuit of Figure 15–54, compute the following: (a) f, (b) X_L, (c) R, (d) Z, (e) I_L, (f) I_C, (g) I_T, and (h) θ.

FIGURE 15–54 Circuit for Problem 21

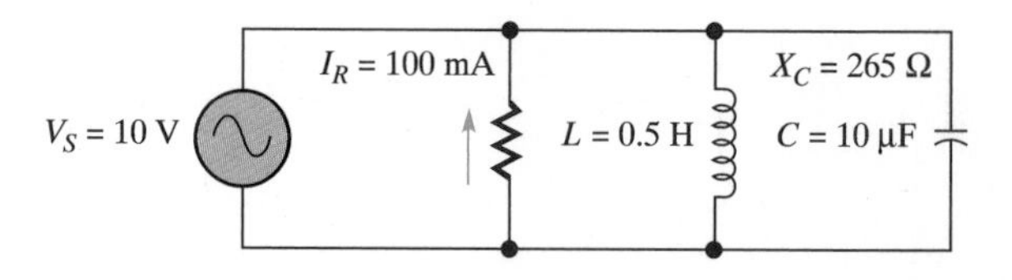

22. For the parallel *RLC* circuit of Figure 15–55, compute the following: (a) V_S, (b) X_L, (c) X_C, (d) Z, (e) I_R, (f) I_L, (g) I_C, and (h) θ.

FIGURE 15–55 Circuit for Problem 22

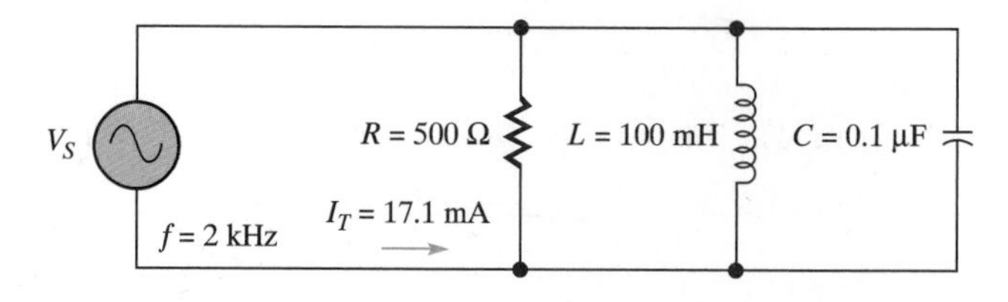

23. For the parallel *RLC* circuit of Figure 15–56, compute the following: (a) R, (b) X_L, (c) X_C, (d) I_T, (e) Z, (f) C, (g) L, and (h) θ.

FIGURE 15–56 Circuit for Problem 23

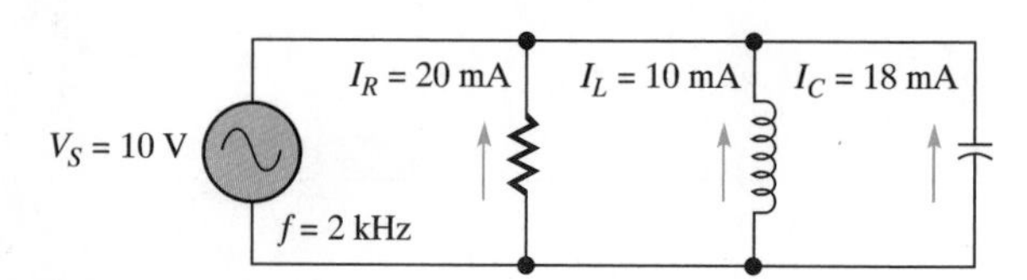

24. For the circuit of Figure 15–57, what value resistance must be added in parallel with L and C to make the phase angle 30°?

FIGURE 15–57 Circuit for Problem 24

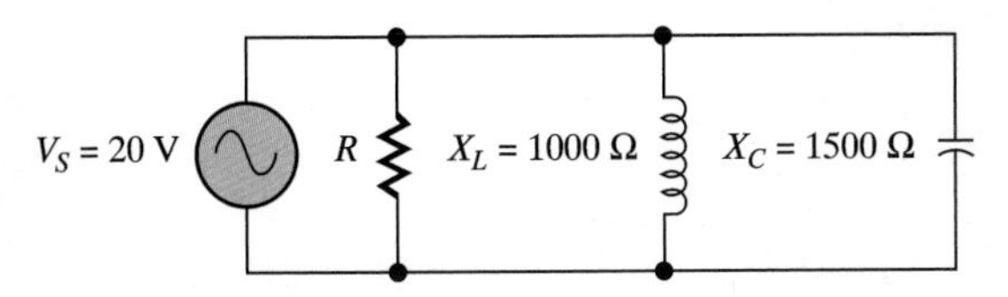

PROBLEM SET: TROUBLESHOOTING

25. The current in the circuit of Figure 15–58 is measured and found to be high. List two faults that might cause the current to be high.

26. In the circuit of Figure 15–58, what would be the indication of a slight decrease in frequency?

FIGURE 15–58 Circuit for Problems 25 and 26

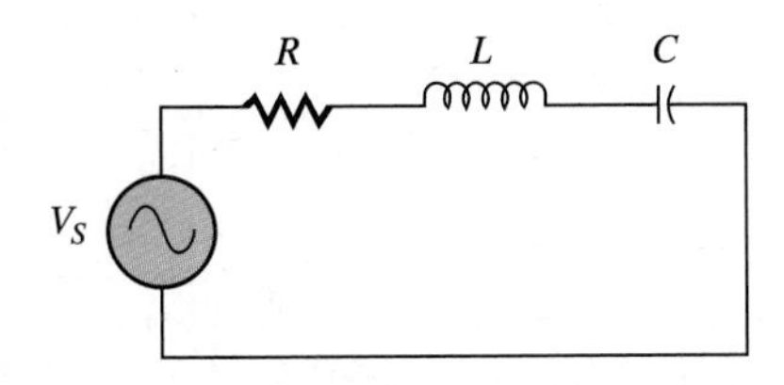

27. The current in the circuit of Figure 15–59 is measured and found to be 39 mA. A single fault exists in this circuit. The source voltage is measured and found to be 19.7 V. The frequency is measured and found to be 399 Hz. What is probable cause of this fault?

FIGURE 15–59 Circuit for Problem 27

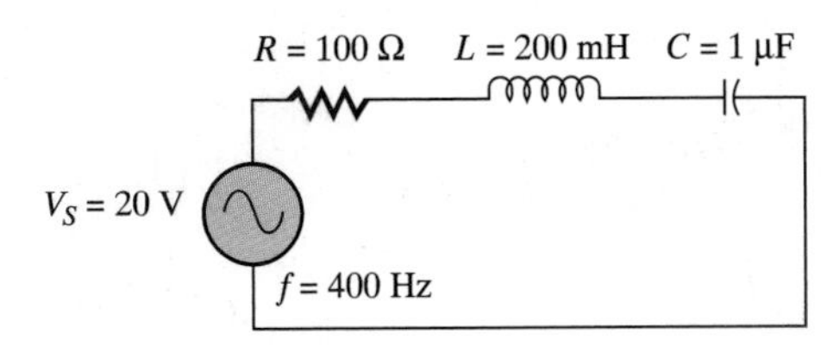

28. In the circuit of Figure 15–60, the voltage measured from point A to point B is 15 V. List three possible causes for such a reading.

29. The voltage across the resistor in the circuit of Figure 15–60 is measured and found to be 7.66 V. The voltage across the capacitor measures 12.98 V. The source voltage and frequency are both correct. If one fault exists, what is the probable cause?

30. If half of the turns on the inductor in Figure 15–60 were to short, what would be the effect on the circuit current and the voltages across C and R?

FIGURE 15–60 Circuit for Problems 28, 29, and 30

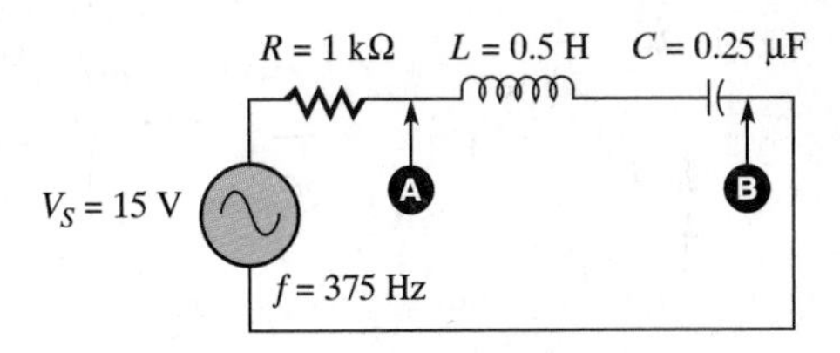

Radiation

With the advent of the use of atomic energy in the mid-1940s, the term *radiation* became a part of almost everyone's vocabulary. This term means different things to different persons. To the average non-technical person, the term *radiation* brings to mind terrible images of the radiation sickness suffered by people exposed to the detonation of the atomic bomb. It also brings to mind stories of those exposed to radiation produced through accidents in the handling of radioactive materials. They also think of the horror stories of the results of atomic war between super powers and the philosophy "if the blast doesn't get you, the radiation will." Thus, some would ask: "Who needs radiation anyway?" As you will see in the next paragraph, the answer is "Everyone!"

To physicists and engineers, radiation is merely one method of transferring energy from one point to another. This type of energy, known as *radiant energy,* is in the form of electromagnetic waves. There are many forms of these waves including heat, light, radio waves, X-rays, and gamma rays. The waves making up these various types of energy differ basically in their frequency. X-rays and gamma rays are emitted naturally and spontaneously by radioactive elements such as radium and uranium. Another example of a radiator is the source of all earthly energy, the sun. The sun is a giant thermonuclear reaction that continuously produces huge amounts of heat and light. This energy, which powers all life on earth, is passed to us through the vast vacuum of space in the form of electromagnetic waves. Thus, it is through radiation that life can exist! (That really answers the question of who needs it.)

In electronics technology, however, the major interest is in radio waves. It is probably from the term radiation that the word "radio" evolved. Radio waves are electromagnetic waves through which radio and television signals are transmitted. They are also the transmitted and received signals in radar. Radio waves are produced and transmitted over a wide range of frequencies. At the transmitting antenna, electrical currents are converted to electromagnetic waves, and at the receiving antenna, the waves are converted back into currents.

As discussed in Chapters 9 and 10 of this text, there are two types of fields, magnetic and electrostatic. Magnetic fields are produced by current flow, while electrostatic fields are the result of stored charge. An electromagnetic field, then, is one containing both magnetic and electrostatic components. Under the proper conditions, these fields will radiate into space as electromagnetic waves. In order to accomplish radiation, the frequency of the currents produced must be relatively high, beginning somewhere near 30 kHz. There must also be a radiating element known as an antenna. The antenna must be a conductor and be of a given length in relation to the wavelength of the currents exciting it.

As shown in Figure 1, the radio transmitter produces high frequency signals with intelligence impressed upon them. These signals are conducted to the antenna in which they produce alternating currents. Figure 2 shows the electrostatic field produced around the antenna. Notice in Figure 2(a) that the field is positive at the top and negative at the bottom. On the next half-cycle, as shown in Figure 2(b), the electrostatic field changes polarity. Figure 3 shows the magnetic field produced by the currents flowing in the antenna. Notice that the magnetic field is greatest in the center where the current is greatest, and zero at the ends where the current is zero. The magnetic field also changes polarity on alternate half-cycles.

As shown in Figure 4, when the current changes directions the electrostatic and magnetic fields that had been produced must fully collapse back onto the antenna. If the frequency of the current is high enough, the new fields being produced will be on their way outward before the previous fields have time to fully collapse, as illustrated in Figure 5(a). As shown in Figure 5(b), the new field will repel the previous field and force it off into space. If you were sitting out in space and could see this field, it would appear as shown in Figure 6. Notice that the lines of electrostatic force are vertical, while the lines of magnetic force are horizontal. The two fields move through space together at the speed of light. The energy contained in this field is radiant energy.

At the radio receiver, the electromagnetic field cuts across the receiving antenna, as illustrated in Figure 6. Notice that the antenna is so placed as to have the magnetic portion of the field cut across it at as close to an angle of 90° as possible. Three items are required for electromagnetic induction: a magnetic field, a conductor, and relative motion between the two. Thus, the moving magnetic field induces voltages in the receiving antenna of the same frequency as those that produced radiation at the transmitting antenna.

When you consider all the sources of radiation, it is easy to see that we are awash in an ocean of radiation. All radio transmitters, (commercial broadcast stations, CB radios, amateur radio stations, taxi dispatch radios, pagers, police and ambulance radios, and so on) contribute to low-level radiation. So do television stations, radars, and the sources of celestial radiation. Is this radiation dangerous to life? No one knows for sure, but the short-term effects seem to be negligible. The long-term effects are unknown, and this is what worries some health experts. Studies are now under way to determine the long-term effects of low-level radiation.

FIGURE 1 Intelligence impressed on carrier

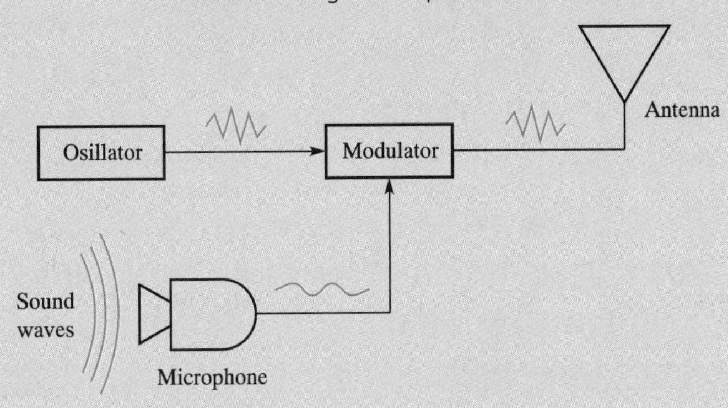

FIGURE 2 Electrostatic field format around antenna

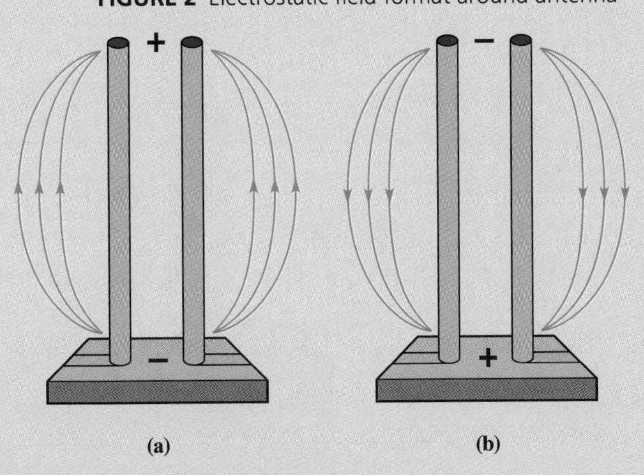

FIGURE 3 Magnetic field formed around antenna

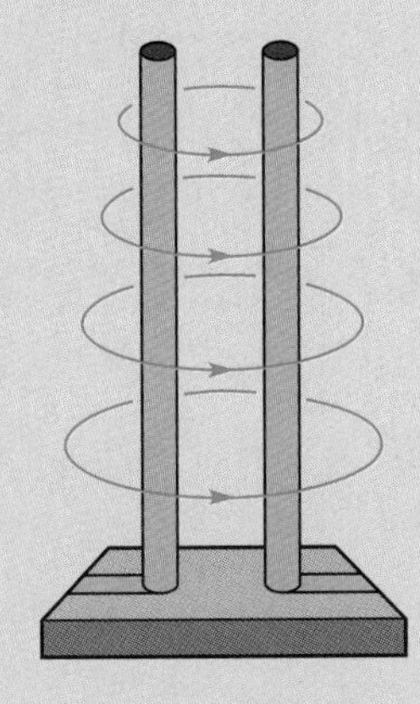

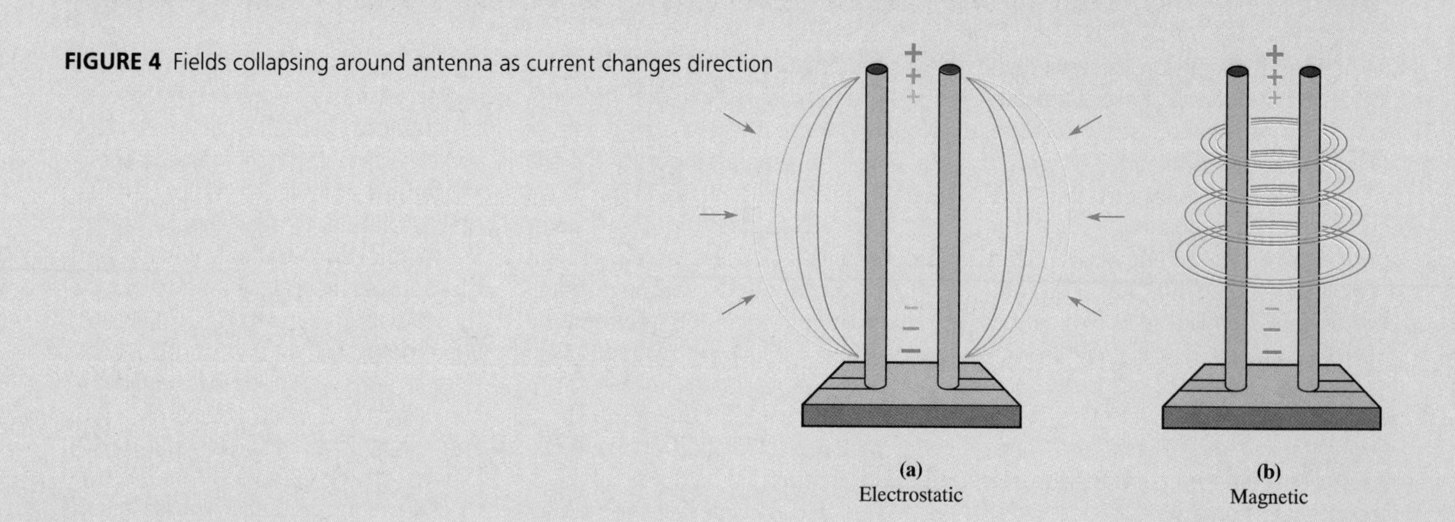

FIGURE 4 Fields collapsing around antenna as current changes direction

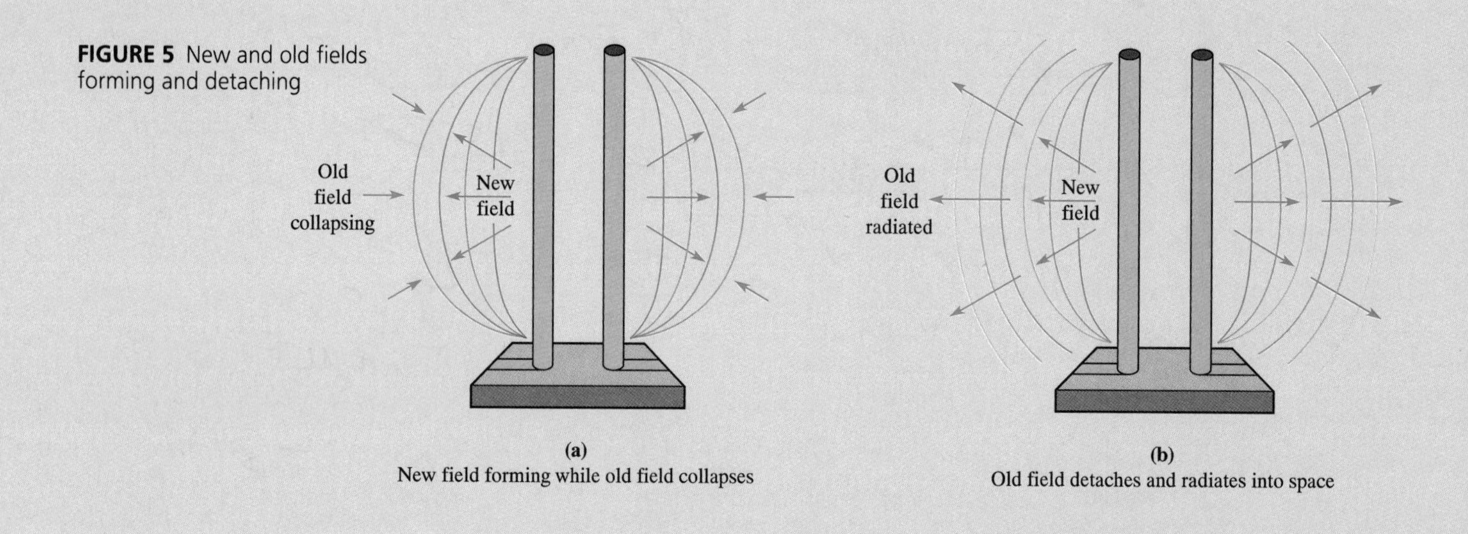

FIGURE 5 New and old fields forming and detaching

FIGURE 6

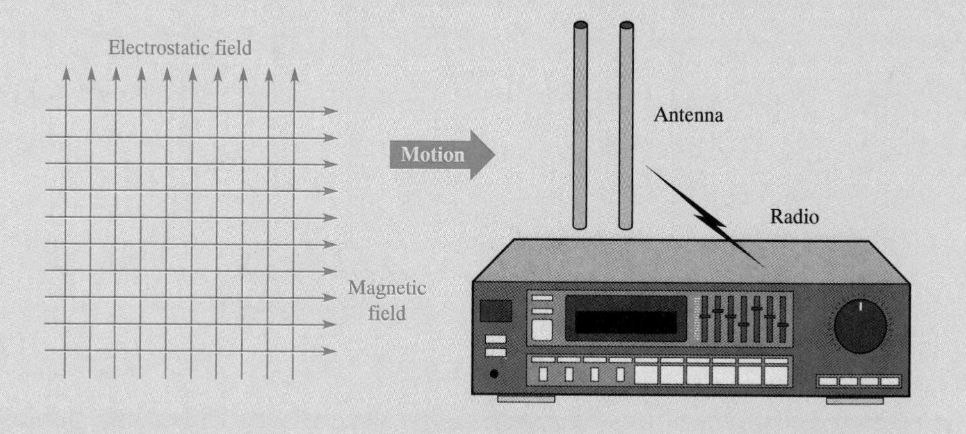

CHAPTER

16 Resonant ac Circuits

UPON COMPLETION OF THIS CHAPTER, YOU WILL BE ABLE TO

1. Explain how resonance occurs in a circuit.
2. Compute the resonant frequency of an *LC* circuit.
3. List the characteristics of series resonance.
4. Explain the rise in reactive voltage at series resonance.
5. List the characteristics of parallel resonance.
6. Explain the rise in impedance at parallel resonance.
7. Explain the effects of circuit quality factor Q.
8. Perform analysis of series resonant circuits.
9. Perform analysis of parallel resonant circuits.
10. List and explain applications of resonant circuits.
11. Perform troubleshooting of resonant circuits.

The preceding chapter dealt with the characteristics and analysis of circuits composed of resistance, inductance, and capacitance. As you learned, the effects of inductance and capacitance are opposites. Thus, one cancels the effects of the other. A circuit is either inductive or capacitive, depending upon the frequency of the applied signal. It was noted that at some frequency the effects of inductance and capacitance are equal and completely cancel one another, leaving only resistance to oppose the current. This circuit condition, known as resonance, is the subject of this chapter.

The resonant circuit is one of the most important in the electronics industry. Resonant circuits make possible the telecommunication industry as it exists today. Using resonant circuits, you can tune a radio to one of the thousands of radio signals impinging upon the antenna. The same can be said for television. This selection is done through a filtering action that passes a relatively narrow band of frequencies, which is quite an accomplishment when you consider the number of radio transmitters in use today (commercial broadcast, citizens' band, amateur radio, police, taxis, emergency medical services, and so on). Within the radio receiver, resonant *LC* circuits are used to couple the selected signal from one amplifier to the next. Resonant circuits may also be used to block a narrow band of frequencies within a system.

This introduction to the characteristics and uses of resonant circuits certainly demonstrates their importance. Knowledge of them is absolutely essential to any electronics technician. But beyond this, they are at the least very interesting and at times quite fascinating. This chapter should thus be challenging, interesting, and essential to your success as an electronics technician!

16–1 HOW RESONANCE OCCURS

Question: What is resonance?

Before beginning a study of electrical resonance, it might be a good idea to consider something similar, with which you may already be familiar—mechanical resonance. Recall the TV commercial that shows a glass being shattered by sound waves reproduced from an audio tape. At some frequency, known as the resonant frequency, the sound waves striking the glass reinforce one another. The result is violent vibrations that cause the glass to shatter. Another example of mechanical resonance is in the wood making up a piano. The vibrating strings set up vibrations in the wood, which once again reinforce one another, amplifying the sound and giving it more "timber." Once again, there is a specific relationship between the desired frequency and the mechanical qualities of the wood. Other examples are resonators on sound systems and car exhausts that amplify the sounds and make them more pleasing to the ear. What should be noted in all these cases is that certain conditions must exist in order for mechanical resonance to take place.

Question: Under what conditions does resonance occur in RLC *circuits?*

The same conditions just described for mechanical resonance also exist for electrical resonance. Within an *LC* circuit, some very interesting things happen at the resonant frequency. In one case, voltages many times greater than the applied voltage may be measured across the reactive components. But when the voltage across the reactive combination of *L* and *C* is measured, zero volts is found. Under another circumstance, the line current may drop all the way to zero, even while relatively large currents are circulating in the *LC* combination. Thus, while it may appear that the resonant circuit defies Kirchhoff's laws, in fact, it does not, as you will learn.

Certain conditions must exist if electrical resonance is to occur. First, a circuit must

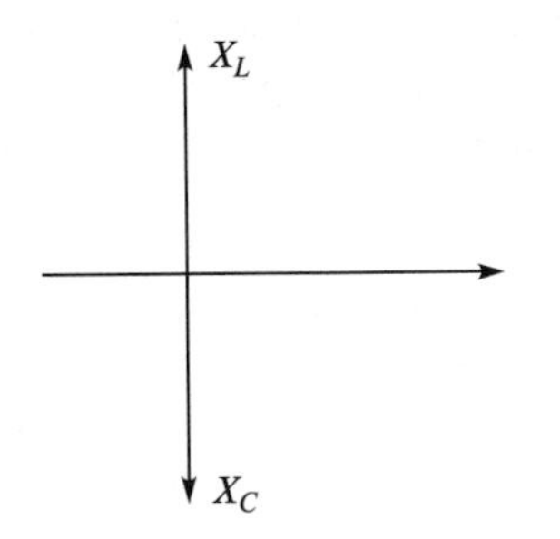

FIGURE 16–1 Zero net reactance in a series *RLC* circuit

include the properties of inductance and capacitance in order to be resonant. Second, the frequency of the applied voltage must be such that the inductive and capacitive reactances are equal. How does this equality cause the reinforcement spoken of in mechanical resonance? Recall that in a series *RLC* circuit the net reactance is the difference of X_L and X_C. Thus, as illustrated in Figure 16–1, if these quantities are equal, the net reactance is zero, and only the circuit resistance opposes current. If the resistance is relatively small, the current responds by becoming quite high at resonance, as illustrated in Figure 16–2(a). The opposite effect takes place in a parallel *RLC* circuit. Here, inductive and capacitive susceptance are used in computing impedance. These also cancel, but the result of zero susceptance is infinite (or at least extremely high) impedance. As illustrated in Figure 16–2(b), the source current now drops to a very low value.

Question: How is the value of the resonant frequency computed?

At resonance, the inductive and capacitive reactances are equal. An equation for the resonant frequency f_r can thus be derived in the following manner:

1. Begin with the equality of inductive and capacitive reactances:

$$X_L = X_C$$

2. Substitute:

$$2\pi fL = \frac{1}{2\pi fC}$$

3. Cross multiply:

$$4\pi^2 f_R^2 LC = 1$$

4. Solve for f_r:

$$f_r^2 = \frac{1}{4\pi^2 LC}$$

(16–1)

$$f_r = \frac{1}{2\pi\sqrt{LC}}$$

From Eq. 16–1, it can be seen that the resonant frequency is inversely proportional to the square root of the product of *L* and *C*. It also shows that for any given values of inductance and capacitance, there is *only one frequency* at which the circuit is resonant. Thus, the resonant frequency is a function of the *LC* product. Many combinations of inductance and capacitance will satisfy the equation for a given resonant frequency, as is demonstrated in the following example.

FIGURE 16–2 Circuit conditions for resonance in series and parallel RLC circuits

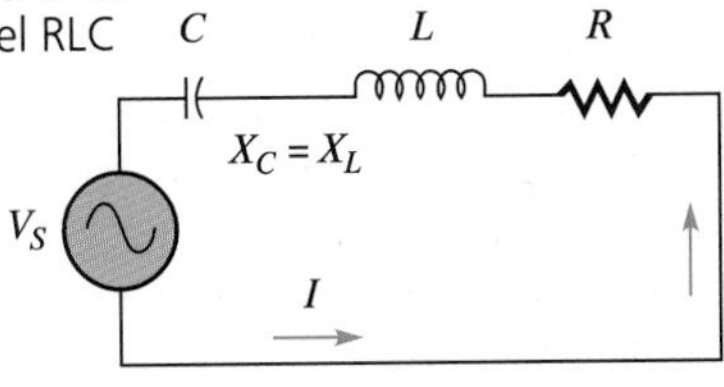

(a)
Net reactance is zero, impedance is value of resistance, and current is maximum.

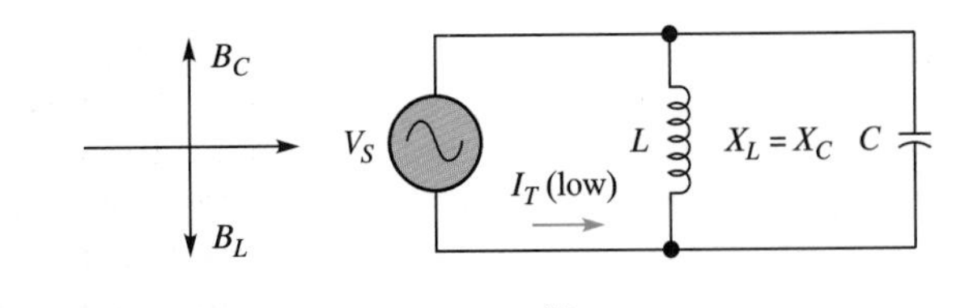

(b)
When $X_L = X_C$, net susceptance equals zero, resulting in very high impedance and a low current value.

EXAMPLE 16–1 Compute the resonant frequency for the following combinations of inductance and capacitance: (a) 250 μH and 0.01 μF and (b) 100 mH and 25 pF.

■ **Solution:**

a.
$$f_r = \frac{1}{2\pi\sqrt{LC}}$$
$$= \frac{1}{2(3.14)\sqrt{(250 \times 10^{-6}\text{ H})(0.01 \times 10^{-6}\text{ F})}} = 100.7\text{ kHz}$$

b.
$$f_r = \frac{1}{2\pi\sqrt{LC}}$$
$$= \frac{1}{2(3.14)\sqrt{100 \times 10^{-3}\text{ H})(25 \times 10^{-12}\text{ F})}} = 100.7\text{ kHz}$$

In Example 16–1, notice that both combinations of L and C give the same resonant frequency. If you multiply the inductance by the capacitance in each case, you will find that the products are *identical*. Thus, any combination of L and C with the same product, known appropriately enough as the LC product, produces the same resonant frequency in an LC circuit.

EXAMPLE 16–2 What value of capacitance produces resonance with a 100 μH inductor at 1 MHz?

■ **Solution:**

1. Square both sides of the equation to remove the radical:

$$f_r^2 = \frac{1}{4\pi^2 LC}$$

2. Solve for C:

$$C = \frac{1}{4\pi^2 f_r^2 L}$$
$$= \frac{1}{4(9.86)(1 \times 10^6)^2(100 \times 10^{-6})} = 254\ \mu\text{F}$$

EXAMPLE 16–3 What value of inductance produces resonance with a 100 pF capacitor at 1.592 MHz?

$$L = \frac{1}{4\pi^2 f_r^2 C}$$
$$= \frac{1}{4(9.86)(1.592 \times 10^6)^2(1000 \times 10^{-12})} = 100\ \mu\text{H}$$

■ **SECTION 16–1 REVIEW**

1. What is the meaning of resonance?
2. What condition must exist in an RLC circuit for resonance to occur?
3. What is the difference in the resonant effect in series and parallel RLC circuits?

4. Is the resonant frequency directly or inversely proportional to inductance?
5. Is the resonant frequency directly or inversely proportional to capacitance?
6. How many resonant frequencies are there for any given values of inductance and capacitance?
7. How many combinations of inductance and capacitance will produce a given resonant frequency?

16–2 SERIES RESONANCE

■ Voltages in a Series *RLC* Circuit at Resonance

Analysis of the voltages in a series resonant circuit proceeds in exactly the same manner as for the series *RLC* circuits of the last chapter. A series *RLC* circuit at resonance is shown in Figure 16–3(a). Since X_L equals X_C, the voltages V_L and V_C are also equal ($V_L = IX_L$ and $V_C = IX_C$). The result is a net reactive voltage of zero volts. Remember that the source voltage is the phasor sum of the resistor and net reactive voltages. Thus, in equation form,

$$V_S = \sqrt{V_R^2 + (V_L - V_C)^2}$$

Since V_L equals V_C,

(16–2) $$V_S = \sqrt{V_R^2 + 0} \quad \text{and} \quad V_S = V_R$$

As this equation shows, the entire source voltage is across the resistor at series resonance. The voltage measured across the inductor and capacitor combination is zero. Thus, by Kirchhoff's voltage law, the algebraic sum of the voltages developed is equal to the source voltage. The phasor diagram for the voltages at resonance is shown in Figure 16–3(b).

■ Impedance in a Series *RLC* Circuit at Resonance

Question: What is the impedance of a series* RLC *circuit at resonance?

Once again, the impedance of a series *RLC* circuit at resonance is computed in the same manner as for any series *RLC* circuit. At resonance, the inductive and capacitive reactances are equal and 180° out of phase. Thus, at resonance, the impedance of a series *RLC* circuit is equal to the resistance value, as can be illustrated using the equation for impedance:

$$Z = \sqrt{R^2 + (X_L - X_C)^2}$$

Since $X_L = X_C$,

(16–3) $$Z = \sqrt{R^2 + 0} \quad \text{and} \quad Z = R$$

FIGURE 16–3 Series *RLC* circuit at resonance

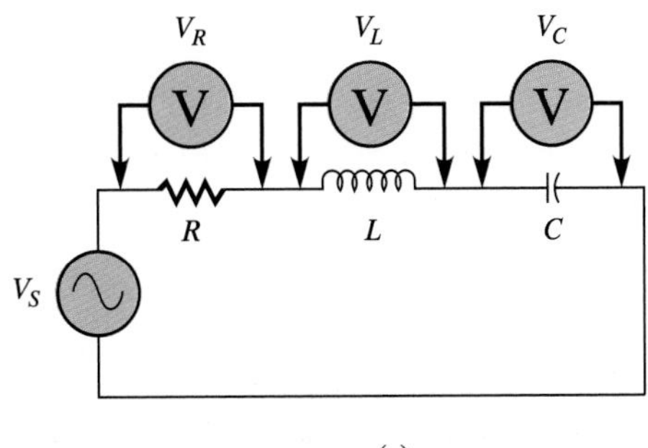

(a)
$X_L = X_C$, $V_L = V_C$, and $V_R = V_S$.

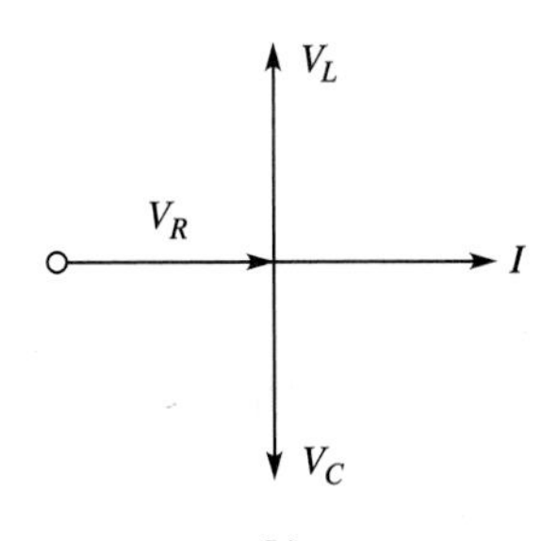

(b)
$V_L = V_C$, and net reactive voltage is zero.

The phasor diagram for the circuit impedance is shown in Figure 16–4.

EXAMPLE 16–4 For the circuit of Figure 16–5, compute the following: (a) the resonant frequency, (b) X_L, (c) X_C, (d) Z, (e) I, (f) V_L, (g) V_C, (h) θ, and (i) power factor (PF).

■ Solution:

a.

$$f_r = \frac{1}{2\pi\sqrt{LC}}$$

$$= \frac{1}{2(3.14)\sqrt{(250 \times 10^{-3}\text{ H})(25 \times 10^{-12}\text{ F})}} = 63.694\text{ kHz}$$

b.

$$X_L = 2\pi fL$$

$$= 2(3.14)(63.694\text{ KHz})(250\text{ mH}) = 100\text{ k}\Omega$$

c.

$$X_C = \frac{1}{2\pi fC} = \frac{1}{2(3.14)(63.694\text{ KHz})(25\text{ pF})} = 100\text{ k}\Omega$$

d.

$$Z = \sqrt{R^2 + (X_L - X_C)^2}$$

$$= \sqrt{(500\ \Omega^2) + (100\text{ k}\Omega - 100\text{ k}\Omega)^2} = \sqrt{250\text{ k}\Omega + 0} = 500\ \Omega$$

e.

$$I = \frac{V_S}{Z}$$

$$= \frac{500\text{ mV}}{500\ \Omega} = 1\text{ mA}$$

f.

$$V_L = IX_L$$

$$= (1\text{ mA})(100\text{ k}\Omega) = 100\text{ V}$$

g.

$$V_C = IX_C$$

$$= (1\text{ mA})(100\text{ k}\Omega) = 100\text{ V}$$

h.

$$\theta = \arctan\frac{X_L - X_C}{R}$$

$$= \arctan\frac{0}{500} = \arctan 0 = 0°$$

i.

$$\text{PF} = \cos\theta$$

$$= \cos 0° = 1$$

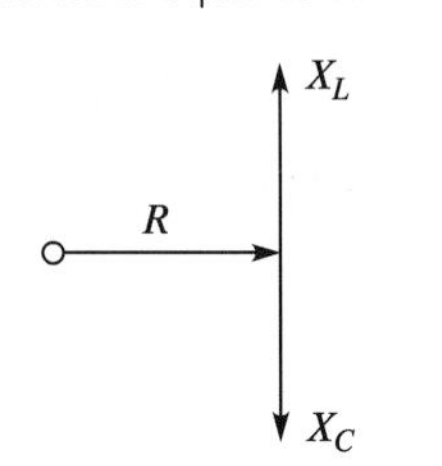

FIGURE 16–4 Phasor diagram for circuit impedance. At resonance, X_L equals X_C, which cancel, and impedance is equal to R.

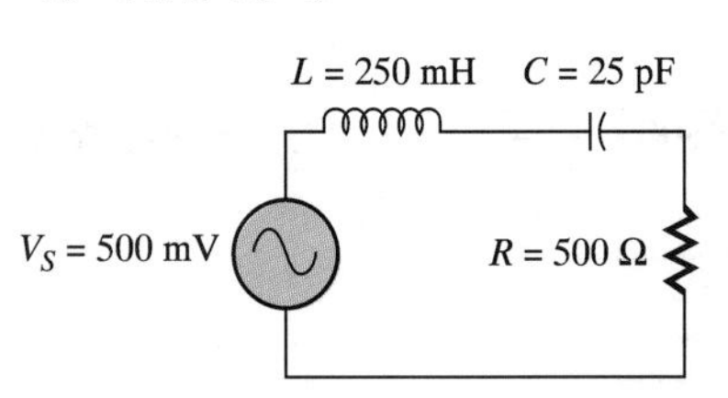

FIGURE 16–5 Circuit for Examples 16–4 and 16–5

From this example, several facts concerning series resonance should be noted:

1. The reactive voltages, V_L and V_C, and the impedances, X_L and X_C, are in fact equal.
2. Since Z equals R, it is as low as it will ever be. Thus, the current is at its largest value.
3. Since the impedance is equal to the resistance, the phase angle is zero and the power factor is one.

It is important that you not only know these facts, but that you understand why they are so. If in doubt, take time now to review the circuits and phasor diagrams of the preceding paragraphs.

In the presentation of these facts, it is assumed that the inductor and capacitor are both ideal; that is to say, neither contains resistance. Remember, however, that in a practical circuit, the leads and plates making up the capacitor do have some resistance, even though it is usually not a large amount. The wire making up the windings of the inductor also has some resistance, the amount of which is determined by the number of turns and size of the wire used. Thus, the inductor is actually a series *RL* circuit with a phase angle of less than 90°. If this resistance is small in proportion to the reactance of the inductor, it can be ignored and the circuit treated as if it were ideal. In the examples that follow, the circuits will be considered ideal. The nonideal resonant circuit will be considered in a later section.

EXAMPLE 16–5 In the circuit of Figure 16–5, if the frequency is increased by one third (84,925 Hz), recompute the values of (a) Z, (b) V_L, (c) V_C, and (d) V_R. Is the circuit now inductive or capacitive?

■ **Solution:**

a. Compute the values for X_L and X_C at the new frequency:

$$X_L = 2\pi fL$$

$$= 2(3.14)(84{,}925)(250 \times 10^{-3}) = 133.3\ \text{k}\Omega$$

$$X_C = \frac{1}{2\pi fC}$$

$$= \frac{1}{2(3.14)(84{,}925)(25\ \text{pF})} = 75\ \text{k}\Omega$$

b. Compute the value for Z:

$$Z = \sqrt{R^2 + (X_L - X_C)^2}$$

$$= \sqrt{(500\ \Omega)^2 + (133.3\ \text{k}\Omega - 75\ \text{k}\Omega)^2} = \sqrt{(500\ \Omega)^2 + (58.3\ \text{k}\Omega)^2} = 58.3\ \text{k}\Omega$$

c. Compute I:

$$I = \frac{V_S}{Z}$$

$$= \frac{0.5\ \text{V}}{58{,}300\ \Omega} = 8.58\ \mu\text{A}$$

d. Compute V_L, V_C, and V_R:

$$V_L = IX_L$$

$$= (8.58\ \mu\text{A})(133.3\ \text{k}\Omega) = 1.14\ \text{V}$$

$$V_C = IX_C$$

$$= (8.58\ \mu\text{A})(75\ \text{k}\Omega) = 0.64\ \text{V}$$

$$V_R = IR$$

$$= (8.58\ \mu\text{A})(500\ \Omega) = 4.29\ \text{mV}$$

The circuit is *inductive* because the voltage drop across the inductor is larger than that across the capacitor. Notice the tremendous drop in the response of the circuit, as evidenced by the much lower voltages across L and C, when it is not at the resonant frequency!

The Resonant Rise in Voltage

Question: Why are the voltages across the inductor and capacitor so high at resonance?

The voltages developed across the inductor and capacitor are equal to the product of their common current and the reactances. Notice that in the circuit of Figure 16–5, these voltages are 200 times greater than the source voltage! This **resonant rise in voltage** across the capacitor and the inductor is due to the rise in current that occurs at resonance (IX_L and IX_C). As shown in Example 16–5, above resonance, X_L is greater than X_C, and the net reactance is no longer zero. The impedance increases, and the current decreases. Below resonance, the same events occur with X_C being larger. Thus, above and below resonance, the series *RLC* circuit is analyzed as in Chapter 15.

SECTION 16–2 REVIEW

1. At series resonance, what is the algebraic sum of V_L and V_C?
2. At series resonance, what voltage is found across the resistor?
3. What is the circuit impedance at series resonance?
4. What is the phase angle at series resonance?
5. At series resonance, are V_L and V_C relatively large or small?

16–3 SERIES RESONANT CIRCUIT RESPONSE

Question: What is the response of the series* RLC *circuit over a band of frequencies from below to above resonance?

The resonant response, as noted by the rise in current, is greatest at the resonant frequency. This response drops both above and below resonance. Thus, the circuit responds over a given band of frequencies. The circuit illustrated in Figure 16–6(a) is resonant at a frequency of 1000 Hz. The output is taken across the resistor. At the resonant frequency, the output voltage equals the source voltage. Output voltages for frequencies both above and below the resonant frequency are computed and shown in Table 16–1. The graph of the data of Table 16–1 is shown in Figure 16–6(b), which plots the output voltage produced when the frequency of the input signal is swept across a band of frequencies containing the resonant frequency. Notice that the output voltage drops both above and below resonance.

TABLE 16–1 Output voltages for frequencies above and below resonance in a series *RLC* circuit

FREQUENCY	X_L	X_C	NET X	Z	I	$V_{out} = IR$
400 Hz	6.48 Ω	39.8 Ω	33.3 Ω	34.80 Ω	0.345 A	3.45 V
600 Hz	9.57 Ω	26.5 Ω	16.9 Ω	19.70 Ω	0.61 A	6.10 V
800 Hz	12.8 Ω	19.9 Ω	7.1 Ω	12.26 Ω	0.978 A	9.78 V
1000 Hz	15.9 Ω	15.9 Ω	0 Ω	10.00 Ω	1.2 A	12.00 V
1200 Hz	19 Ω	13.3 Ω	5.7 Ω	11.50 Ω	1.04 A	10.40 V
1400 Hz	22.3 Ω	11.4 Ω	10.9 Ω	14.80 Ω	0.811 A	8.11 V
1600 Hz	25.5 Ω	9.95 Ω	15.5 Ω	18.40 Ω	0.651 A	6.51 V

FIGURE 16–6 Series resonant circuit response

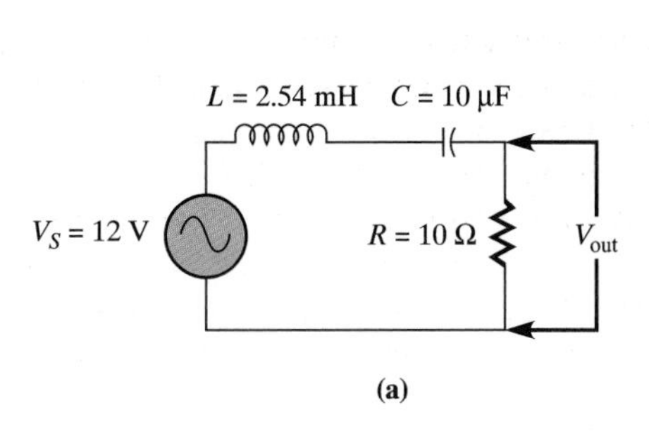

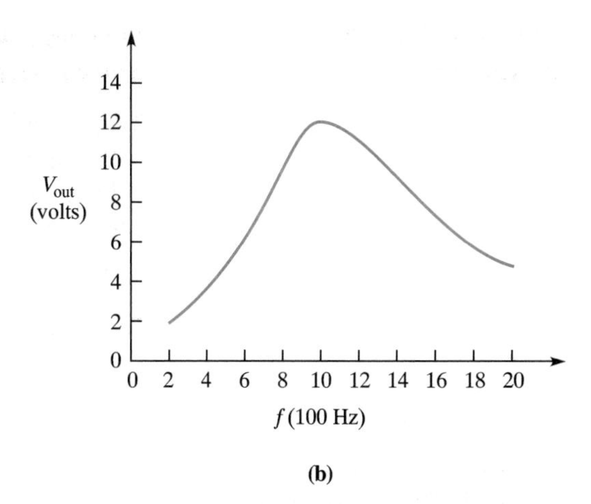

FIGURE 16–7 *RLC* circuit bandwidth

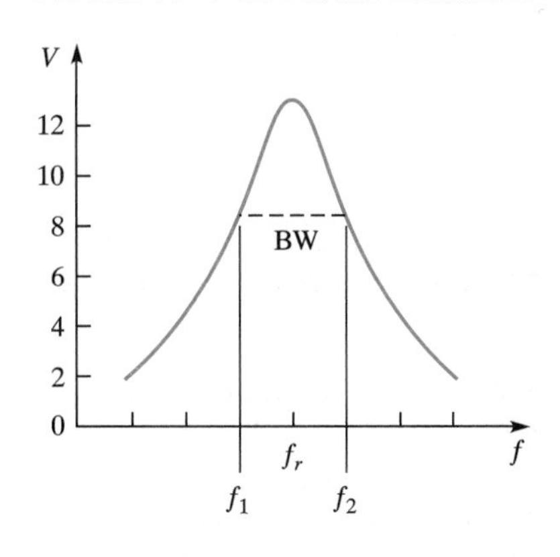

FIGURE 16–8 Response curve for Example 16–6

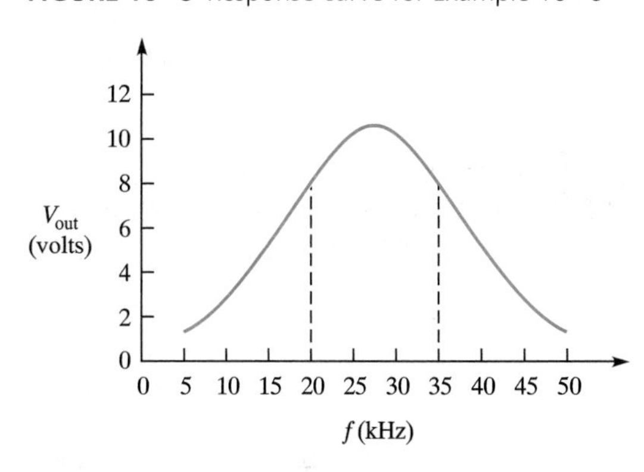

Bandwidth of a Series Resonant Circuit

There is a band of frequencies over which the circuit's response is considered to be useful, which is known as the **bandwidth** of the circuit. The bandwidth of an *RLC* circuit is illustrated in Figure 16–7, which shows not one but two cutoff frequencies. They occur at the frequencies at which the output voltage drops to 70.7% of its maximum value. One of these cutoff frequencies, f_1, occurs below the resonant frequency, and the other, f_2, occurs above. The frequency range between the lower and upper cutoff points is considered to be the bandwidth of the circuit. In equation form, the bandwidth (BW or Δf) is equal to

(16–4)

$$\text{BW} = f_2 - f_1$$

EXAMPLE 16–6 From the response curve of Figure 16–8, determine the frequency range over which the circuit's output is useful.

Solution:

The bandwidth is the frequencies between the points where the output voltage has dropped to 70.7% of its maximum value. The lower cutoff f_1 is at 20 kHz and the upper cutoff f_2 is at 35 kHz. Thus,

$$\text{BW} = f_2 - f_1 = 35\text{ kHz} - 20\text{ kHz} = 15\text{ kHz}$$

EXAMPLE 16–7 A circuit is resonant at 455 kHz. If the bandwidth is 10 kHz, what are the values of f_1 and f_2?

■ Solution:

Ideally, the bandwidth is centered upon the resonant frequency. Thus, 50% of the bandwidth is above the resonant frequency and 50% below.

$$f_2 = f_r + \frac{\text{BW}}{2} = 460 \text{ kHz}$$

$$f_1 = f_r - \frac{\text{BW}}{2} = 450 \text{ kHz}$$

■ Question: What determines the bandwidth of an *RLC* circuit?

As previously stated, the degree to which an *RLC* circuit responds at resonance depends upon the amount of current that flows. Since the reactances cancel at resonance, the amount of this current is determined by the value of the source voltage and the total circuit resistance. (The total resistance is the sum of the winding resistance and any series resistors.) The smaller this resistance, the greater the current flow at resonance and the greater the circuit response.

Recall that the figure of merit of an inductor, symbolized as Q, is the ratio of inductive reactance to winding resistance ($Q = X_L/R_W$). A series *RLC* circuit has a figure of merit, known as circuit Q, which is the ratio of X_L to the total circuit resistance.

(16–5)
$$Q_{\text{ckt}} = \frac{X_L}{R_T}$$

The bandwidth of an *RLC* circuit is the ratio of the resonant frequency to the circuit Q:

(16–6)
$$\text{BW} = \frac{f_R}{Q_{\text{ckt}}}$$

As previously stated, the ideal condition has the bandwidth centered on the resonant frequency. Thus, the lower cutoff frequency f_1 and the upper cutoff frequency f_2 can be computed as follows:

(16–7)
$$f_1 = f_r - \frac{\text{BW}}{2}$$

(16–8)
$$f_2 = f_r + \frac{\text{BW}}{2}$$

NOTE

It is important to understand that Eq. 16–7 and Eq. 16–8 are *approximations* only. They assume that the bandwidth curve is symmetrical, which is the case in narrow bandwidth circuits. Since narrow bandwidths result from a high value of circuit Q, these equations may be considered adequate for *LC* circuits with a Q of 10 or greater.

The resonant rise in voltages across the inductor and capacitor may result in quite high value. The magnitude of these values is determined by the circuit Q. In addition to being equal to the ratio of X_L to R_T, circuit Q is also equal to the ratio of V_L to V_S. Thus, circuit Q may be derived using Equation 16–5:

1. Begin with 16–5:

$$Q_{ckt} = \frac{X_L}{R_T}$$

2. Multiply by I/I:

$$Q_{ckt} = \frac{IX_L}{IR_T}$$

3. Since IX_L is V_L and IR_T is V_{RT}, and at resonance $V_{RT} = V_S$, find Q_{ckt} by

(16–7)
$$Q_{ckt} = \frac{V_L}{V_S}$$

Eq. 16–7 can be transposed to yield an equation for V_L at resonance:

(16–8)
$$V_L = QV_S$$

Thus, the resonant rise in voltage is *directly proportional* to the circuit Q. Eq. 16–5 can also be transposed to yield an equation for X_L at resonance:

$$Q_{ckt} = \frac{X_L}{R_T} \quad \text{and} \quad X_L = QR_T$$

The rise in the value of X_L at resonance is also directly proportional to the value of circuit Q. For this reason the circuit Q is often referred to as the *amplification factor* of a resonant circuit.

EXAMPLE 16–8 For the series *RLC* circuit of Figure 16–9, determine the bandwidth and the upper and lower cutoff frequencies.

FIGURE 16–9 Circuit for Example 16–8

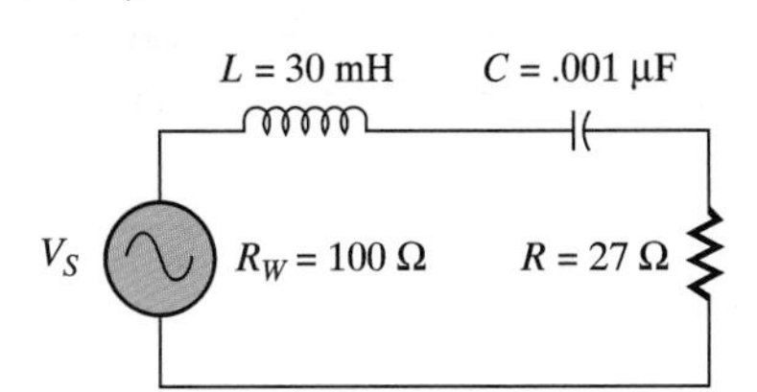

■ Solution:

1. Compute the resonant frequency:

$$f_r = \frac{1}{2\pi\sqrt{LC}}$$

$$= \frac{1}{2(3.14)\sqrt{(30\text{ mH})(0.001\ \mu\text{F})}} = 29.072\text{ kHz}$$

2. Compute X_L at the resonant frequency:

$$X_L = 2\pi fL$$

$$= 2(3.14)(29.072\text{ kHz})(30\text{ mH}) = 5.477\text{ k}\Omega$$

3. Compute total resistance:

$$R_T = R_L + R_W$$

$$= 27\ \Omega + 100\ \Omega = 127\ \Omega$$

4. Compute circuit Q:

$$Q_{ckt} = \frac{X_L}{R_T}$$

$$= \frac{5477\ \Omega}{127\ \Omega} = 43\ \Omega$$

5. Compute BW:

$$\text{BW} = \frac{f_r}{Q_{ckt}}$$

$$= \frac{29072 \text{ Hz}}{43} = 676 \text{ Hz}$$

6. Compute f_2 and f_1:

$$f_2 = f_r + \frac{\text{BW}}{2}$$

$$= 29072 \text{ Hz} + \frac{676 \text{ Hz}}{2} = 29.410 \text{ kHz}$$

$$f_1 = f_r - \frac{\text{BW}}{2}$$

$$= 29072 \text{ Hz} - \frac{676 \text{ Hz}}{2} = 28.734 \text{ kHz}$$

The bandwidth of a series *RLC* circuit at resonance is inversely proportional to the circuit Q. Thus, in a high-Q circuit, the response is large, and the bandwidth is relatively narrow. The opposite is true of a low-Q circuit, as is illustrated in the following example.

EXAMPLE 16–9 For the circuit of Figure 16–10(a), compute the Q and bandwidth for R_L values of (a) 50 Ω, (b) 100 Ω, and (c) 200 Ω. Ignore any inductor winding resistance.

■ Solution:

1. Compute the resonant frequency:

$$f_r = \frac{1}{2\pi\sqrt{LC}}$$

$$= \frac{1}{2\pi\sqrt{(30 \text{ mH})(0.01 \ \mu\text{F})}} = 9.193 \text{ kHz}$$

2. Compute X_L at the resonant frequency:

$$X_L = 2\pi f L$$

$$= 2(3.14)(9193 \text{ Hz})(30 \text{ mH}) = 1.732 \text{ k}\Omega$$

3. Using the equation $Q_{ckt} = X_L/R_T$, compute Q for each value of R_L:

a. 50 ohms, $Q_{ckt} = \frac{1732 \ \Omega}{50 \ \Omega} = 35$

b. 100 ohms, $Q_{ckt} = \frac{1732 \ \Omega}{100 \ \Omega} = 17$

c. 200 ohms, $Q_{ckt} = \frac{1732 \ \Omega}{200 \ \Omega} = 9$

4. Compute the value of circuit current at resonance ($I = V_S/R_T$) for each value of R_L:

FIGURE 16–10 Circuit and response graph for Example 16–9

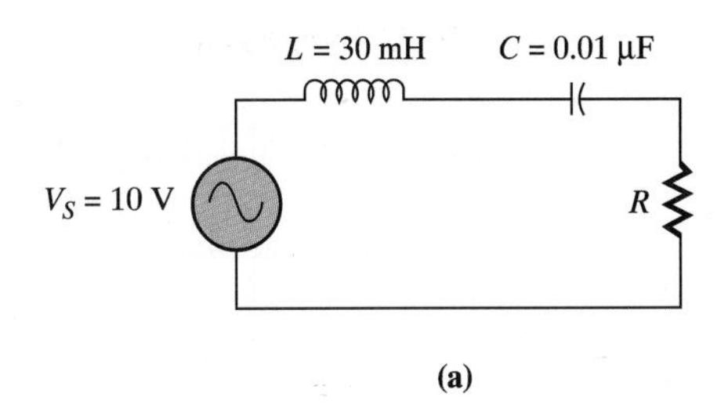

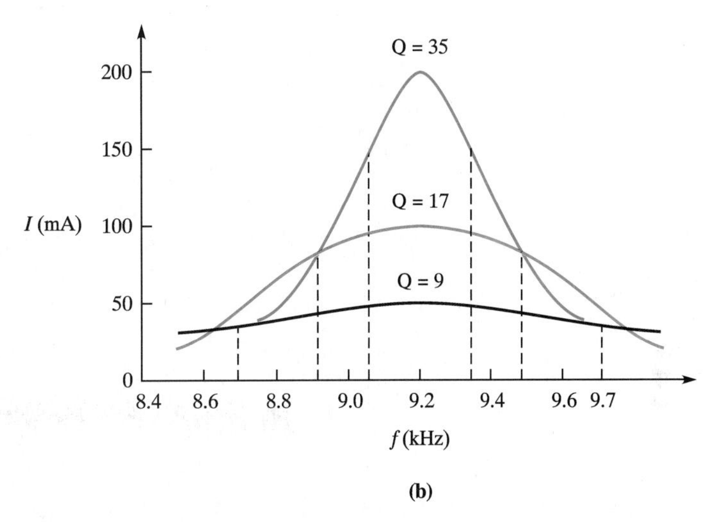

a. 50 ohms, $I = \dfrac{10\text{ V}}{50\ \Omega} = 200\text{ mA}$

b. 100 ohms, $I = \dfrac{10\text{ V}}{100\ \Omega} = 100\text{ mA}$

c. 200 ohms, $I = \dfrac{10\text{ V}}{200\ \Omega} = 50\text{ mA}$

5. Compute the bandwidth for each value of Q using the equation $\text{BW} = f_r/\text{Q}_{\text{ckt}}$:

For $\text{Q}_{\text{ckt}} = 35$, $\text{BW} = \dfrac{9193\text{ Hz}}{35} = 263\text{ Hz}$

For $\text{Q}_{\text{ckt}} = 17$, $\text{BW} = \dfrac{9193\text{ Hz}}{17} = 541\text{ Hz}$

For $\text{Q}_{\text{ckt}} = 9$, $\text{BW} = \dfrac{9193\text{ Hz}}{9} = 1021\text{ Hz}$

Figure 16–10(b) is the graph of the series *RLC* response for each value of resistance. Notice that the response does in fact drop as resistance is increased.

SECTION 16–3 REVIEW

1. Why is the response of a series *RLC* circuit greatest at resonance?
2. What is meant by the bandwidth of a series resonant circuit?

3. Is the bandwidth of a series *RLC* circuit directly or inversely proportional to the circuit Q?
4. If the total resistance of a series *RLC* circuit is increased, does the bandwidth increase or decrease?
5. Why is the circuit Q often referred to as the amplification factor of the resonant circuit?

16–4 PARALLEL RESONANCE

A parallel *LC* circuit at resonance is shown in Figure 16–11(a). Once again, X_C and X_L are equal, and the values of I_L and I_C are also equal. Figure 16–11(b) shows the phasor diagram for the currents. Notice that the source voltage is the reference phasor. The phase difference between the source voltage and currents is 90°, with the capacitive current leading and the inductive current lagging. Thus, the inductive and capacitive currents are equal and 180° out of phase. The result is that they cancel one another, and the total, or line current, is zero. With zero line current, the circuit impedance is infinite.

This situation may seem, on the surface, to be a little strange. On the one hand, current flows: I_L and I_C. On the other hand, the canceling effect of I_L and I_C causes no current to flow from the source into the line. If no current flows from the source, then where do the inductive and capacitive currents come from? The answer lies with the fact that both the inductor and capacitor store electrical energy and that they interact in a very special way when in parallel.

■ The *LC* Tank Circuit

Question: What is an LC tank circuit?

A **tank circuit** is a parallel resonant *LC* circuit. It takes its name from its ability to store electrical energy. As you know, electrical energy is stored in the magnetic field of the inductor and the electrostatic field of the capacitor. Most of this energy, except for a small amount dissipated in the winding resistance of the inductor, is returned to the source when the circuit is de-energized. As previously stated, the line current is zero at parallel resonance due to the canceling effect of the reactive currents. The currents within the tank circuit, however, may be quite high. This current flow is a result of oscillations within the tank circuit.

FIGURE 16–11 Parallel *LC* circuit at resonance

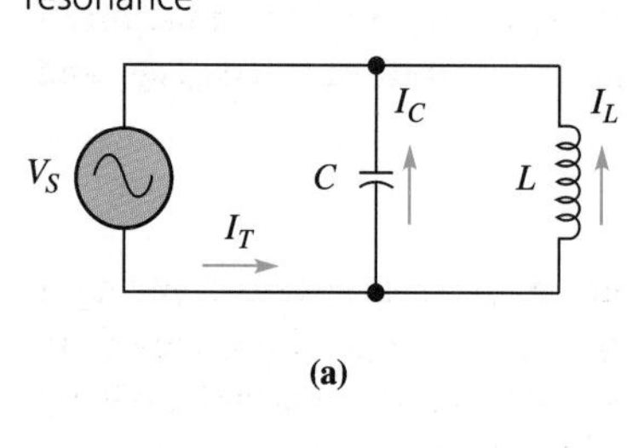

(a)

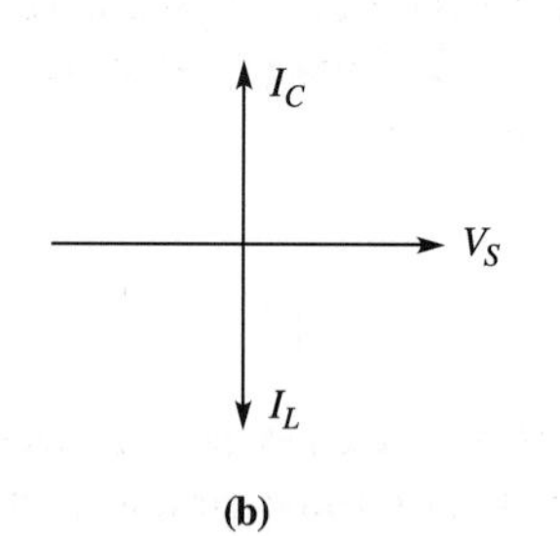

(b)

Question: What are tank circuit oscillations, and how are they produced and sustained?

Oscillations within a tank circuit are the currents that flow as the electrical energy is passed between the inductor and capacitor. In Figure 16–12(a), switch S_1 is closed. Instantaneously, the capacitor appears as a short circuit and begins to charge, while the inductor opposes the change in current and appears as an open. Thus, initially the current flows to charge the capacitor, storing the energy required in its electrostatic field. Current then ceases in the capacitive branch and builds up in the inductor, as its opposition to the change in current decreases. The energy required to energize the inductor is stored in its magnetic field. This static condition exists as long as the voltage source is applied through S_1.

In Figure 16–12(b), the switch is opened. With no voltage source applied, the current drops to zero. With no current flow there is nothing to sustain the magnetic field around the inductor. The magnetic field collapses, inducing a voltage in the inductor with a polarity that opposes the decrease in current. The inductor now acts as a source, supplying charging current to the capacitor in the direction shown. Thus, the energy that was stored in the magnetic field has been transferred to the electrostatic field of the capacitor.

FIGURE 16–12 *LC* tank circuit

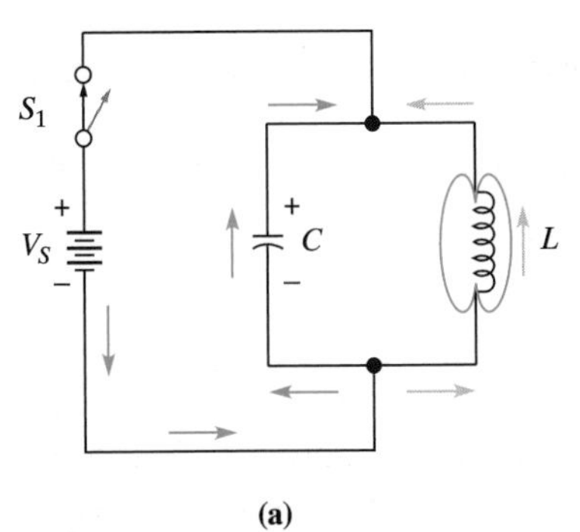

(a)
Closing switch S_1 allows current flow to charge the capacitor and energize the inductor.

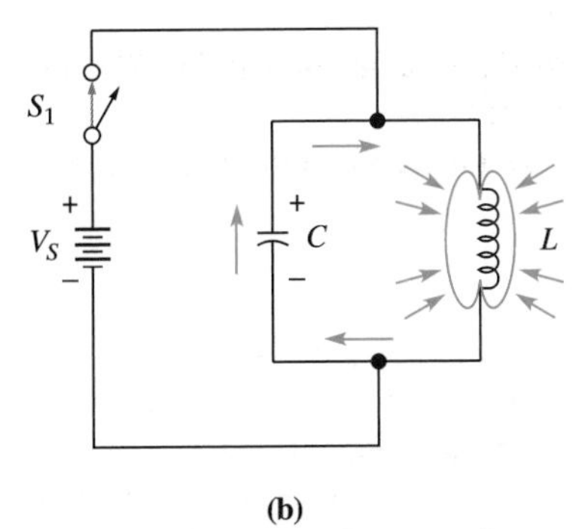

(b)
Opening switch S_1 causes the current flow to stop and the magnetic field to collapse. Voltage induced in the inductor is stored in the capacitor.

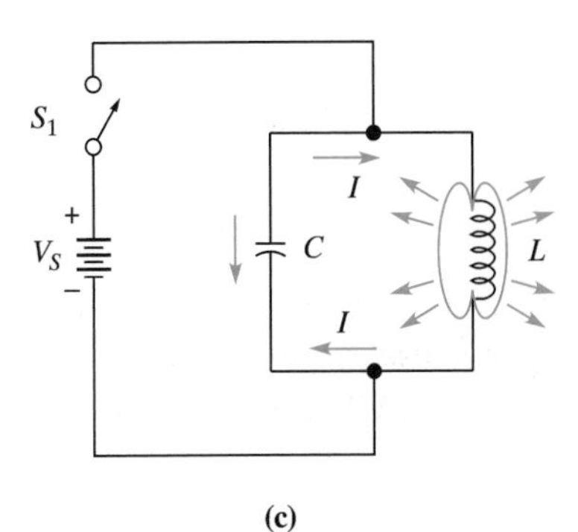

(c)
Magnetic field is re-established as the capacitor discharges through the inductor. Energy is transferred from the electrostatic to the magnetic field.

After the magnetic field has collapsed, there is no induced voltage to maintain the charge upon the capacitor. The capacitor now acts as a voltage source across the inductor. As shown in Figure 16–12(c), the capacitor discharges through the inductor re-establishing the magnetic field. The total energy is transferred to the inductor. When the capacitor has fully discharged, there is once again no current to support the magnetic field. The magnetic field collapses, making the inductor once again a voltage source to recharge the capacitor. Thus, in this manner, the stored energy is passed back and forth between the capacitor and inductor.

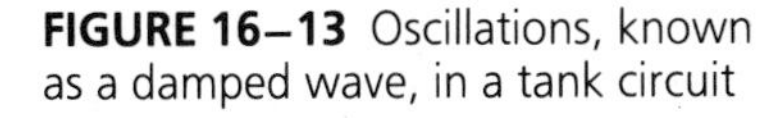

FIGURE 16–13 Oscillations, known as a damped wave, in a tank circuit

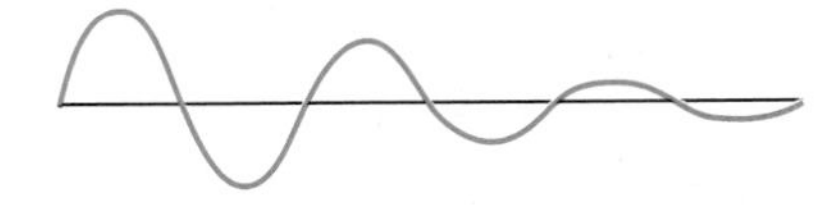

Note that the current flowing within the tank is not two currents, I_L and I_C, but actually only one current. In the previous paragraphs, when capacitive current was referred to, it was the current supplied by the inductor to charge the capacitor. Also the inductive current was the current supplied by the capacitor to energize the inductor.

The waveform produced in the preceding sequence is a sine wave. Its frequency is the resonant frequency of the tank circuit. This frequency is computed using the same equation as was presented previously $\{f_r = 1/(2\pi\sqrt{LC})\}$. If there were no resistance in the circuit, no energy would be dissipated and the oscillations would continue. All *LC* circuits, however, contain some resistance. Thus, the amplitude of the oscillations decreases over time and eventually stop altogether, as illustrated in Figure 16–13. When the oscillations in a tank circuit cease due to circuit losses, the process is known as **ringing,** and the waveform is often referred to as a **damped wave.**

■ The Tank Circuit at Resonance

As previously stated, at parallel resonance, the line current is zero. Thus, applying a signal at the resonant frequency to a tank circuit is the equivalent of opening switch S_1 in Figure 16–12. As Figure 16–14 shows, the capacitor is charged during one half-cycle by the voltage produced by the collapsing magnetic field of the inductor. On the next half-cycle, the inductor is energized with current supplied by the discharging capacitor. This process continues as the tank circuit oscillates at its resonant frequency.

Question: In a practical circuit, will the line current actually be zero at resonance?

As noted in the discussion of series resonance, the ideal is never reached in resonant circuits. The tank circuit does not oscillate continuously with no line current flowing.

FIGURE 16–14 Tank circuit at resonance

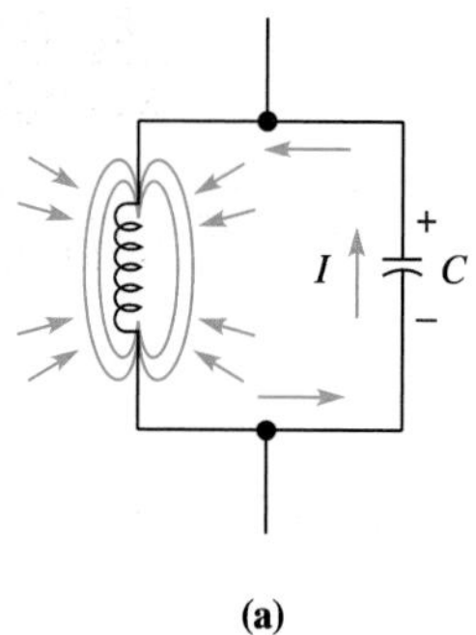

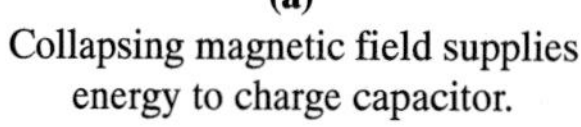
(a)
Collapsing magnetic field supplies energy to charge capacitor.

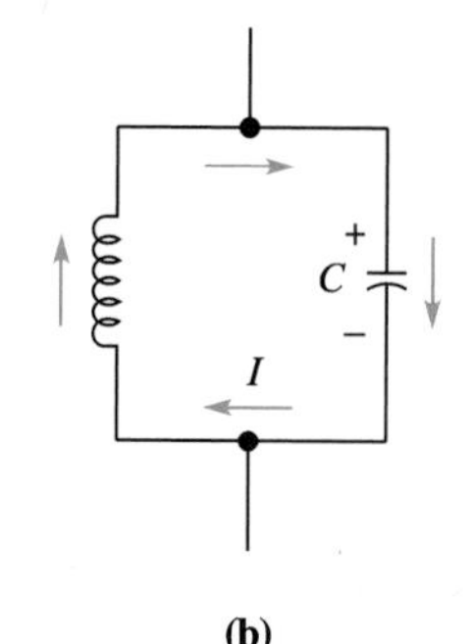

(b)
Discharging capacitor supplies current to energize inductor.

Where the circuit contains no parallel resistor, the only appreciable resistance present is in the windings of the inductor. Some line current is required in order to overcome the losses due to this winding resistance. The Q of a circuit such as this is equal to X_1/R_T. Thus, a high-Q circuit would require a relatively small value of line current in order to sustain oscillations in the tank circuit. A circuit with a Q-factor of 10 or greater is usually considered to be a high-Q circuit. If the Q-factor produced by the inductor and its winding resistance is 10, the phase angle will be approximately 85°. Thus, cancellation of the inductive and capacitive current will be nearly complete and the circuit will approach the ideal.

Impedance in an *LC* Circuit at Resonance

As you have seen by the dramatic drop in current in a parallel LC circuit at resonance, there is a rapid increase in the impedance known as a **resonant rise in impedance.** In fact, if the inductor and capacitor were ideal, the impedance would rise to infinity. Such a rise is not the case, however, due largely to the winding resistance of the inductor. Thus, the impedance will be quite large, but not infinite. Using complex algebra, a relationship between the impedance at resonance and the circuit elements of resistance, inductance, and capacitance can be derived:

(16–9)

$$Z_T = \frac{L}{CR}$$

Another expression for impedance at resonance, involving inductance only, can be obtained from Eq. 16–9. Recall that $X_L = 2\pi fL$ and $X_C = 1/(2\pi fC)$. Thus,

$$L = \frac{X_L}{2\pi f_r} \quad \text{and} \quad C = \frac{1}{2\pi f_r X_C}$$

But since at resonance $X_L = X_C$, it is possible to write

$$L = \frac{X_L}{2\pi f_r} \quad \text{and} \quad C = \frac{1}{2\pi f_r X_L}$$

These expressions can be substituted in Eq. 16–9:

$$Z = \frac{L}{CR_w}$$

$$= \frac{X_L/(2\pi f_r)}{1/(2\pi f_r X_L)(\mathrm{R}_w)}$$

(16–10)

$$= \frac{X_L/(2\pi f_r)}{R_w/(2\pi f_r X_L)}$$

$$= \frac{(X_L)}{\{(2\pi f_r)(2\pi f_r X_L/R_w)\}} = \frac{(X_L) \times (X_L)}{R_w}$$

Thus,

$$Z = \frac{X_{L\hat{L}}^2}{R_w}$$

CAUTION

Eq. 16–9 and Eq. 16–10 are applicable only when the source is at the resonant frequency of the LC combination. Above or below resonance, the questions and methods introduced in Chapter 15 must be used.

EXAMPLE 16–10 For the LC circuit of Figure 16–15, compute the following: (a) f_r, (b) Q, (c) Z, (d) I_L, (e) I_C, and (f) I_T.

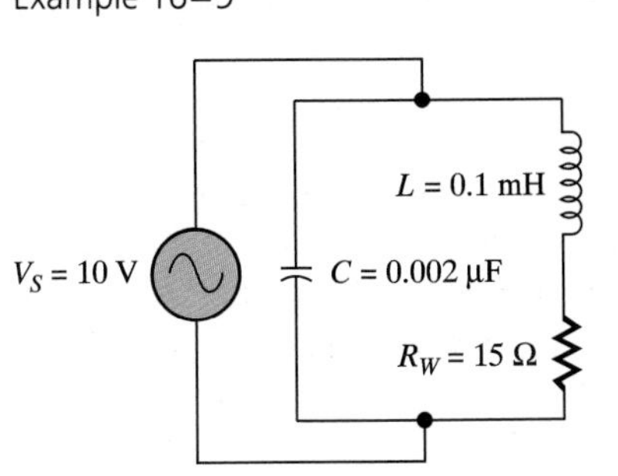

FIGURE 16–15 Circuit for Example 16–9

■ Solution:

a.

$$f_r = \frac{1}{2\pi\sqrt{LC}}$$

$$= \frac{1}{2(3.14)\sqrt{(0.1 \text{ mH})(0.002\ \mu\text{F})}} = 356.061 \text{ kHz}$$

b.

$$X_L = 2\pi fL$$

$$2(3.14)(356.061 \text{ kHz})(0.1 \text{ mH}) = 224\ \Omega$$

$$\text{Q} = \frac{X_L}{R_w}$$

$$= \frac{224\ \Omega}{15\ \Omega} = 14.9$$

c.

$$Z = \frac{L}{CR}$$

$$= \frac{0.1 \text{ mH}}{(0.002\ \mu\text{F})(15\ \Omega)}$$

$$= \frac{0.1 \text{ mH}}{3 \times 10^{-8}} = 3.333 \text{ k}\Omega$$

d.

$$I_T = \frac{V_S}{Z}$$

$$= \frac{10 \text{ V}}{3.333 \text{ k}\Omega} = 3.0 \text{ mA}$$

e.

$$I_L = \frac{V_S}{X_L}$$

$$= \frac{10 \text{ V}}{224\ \Omega} = 44.6 \text{ mA}$$

f.

$$I_C = \frac{V_S}{X_C}$$

$$= \frac{10\ \text{V}}{224\ \Omega} = 44.6\ \text{mA}$$

Example 16–10 shows several things about parallel resonance:

1. At the resonant frequency, the impedance of the tank circuit rises.
2. As a result of this rise in impedance, the line current is minimum.
3. In spite of the fact that the line current is low, the current circulating within the tank circuit is relatively high.

Notice that the line current in Example 16–10 is small compared to a rather large circulating current in the tank. In fact, the tank current is approximately equal to the line current times the Q, or 14.9×3.0 mA = 45.1 mA. There is only a slight difference between this value and that computed. Further, notice that the impedance is approximately equal to the product of X_L and Q, or $14.9 \times 224 = 3338\ \Omega$. For this reason, the Q-factor of the tank circuit is known as the *magnification factor.*

■ Phase Angle and Power Factor

As was stated in the previous paragraphs, the reactive currents within a parallel *LC* circuit cancel one another. This leaves only the effect of resistance within the circuit. Thus, the circuit phase angle is zero, producing a power factor of 1. The cancellation between the currents is not exact, but can be considered as such if the Q-factor is 10 or greater.

■ SECTION 16–4 REVIEW

1. At the resonant frequency, is the line current of a parallel *LC* circuit relatively large or small?
2. At the resonant frequency, are the currents circulating within a tank circuit relatively large or small?
3. At the resonant frequency, is the impedance of a parallel *LC* circuit large or small?
4. What causes the oscillations within a tank circuit?
5. What is the phase angle of a parallel *LC* circuit at resonance?
6. What is the power factor at the resonant frequency?

16–5 PARALLEL RESONANT CIRCUIT RESPONSE

Question: What is the response of a parallel resonant circuit over a band of frequencies from below to above resonance?

As was the case with the series *RLC* circuit, the response of a parallel *LC* circuit is greatest at the resonant frequency. The response drops both above and below resonance. The circuit of Figure 16–16(a) is used as an example for computing the circuit conditions of I_T and Z for frequencies above and below resonance. These computed values are contained in Table 16–2. The response curve for this circuit is shown in Figure 16–16(b). As the response curve shows, this circuit responds to a band of frequencies centered around resonance.

FIGURE 16–16 Parallel resonant circuit response

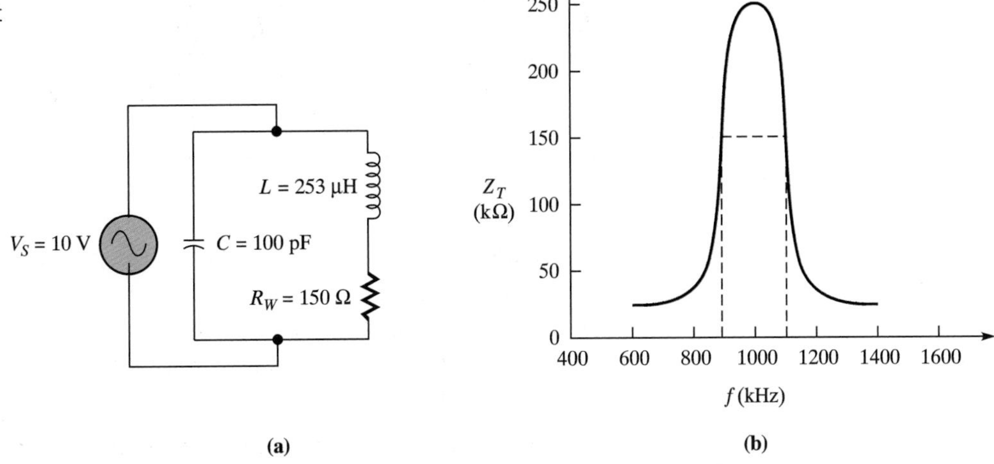

■ **TABLE 16–2 Circuit conditions at frequencies above and below resonance in a parallel LC circuit**

FREQUENCY	X_L	X_C	Z_T
600 kHz	953 Ω	2653 Ω	1487 Ω
800 kHz	1271 Ω	1990 Ω	3517 Ω
1000 kHz	1588 Ω	1588 Ω	253 kΩ
1200 kHz	1906 Ω	1327 Ω	4368 Ω
1400 kHz	2224 Ω	1137 Ω	2326 Ω

FIGURE 16–17 Bandwidth in a parallel LC circuit at resonance

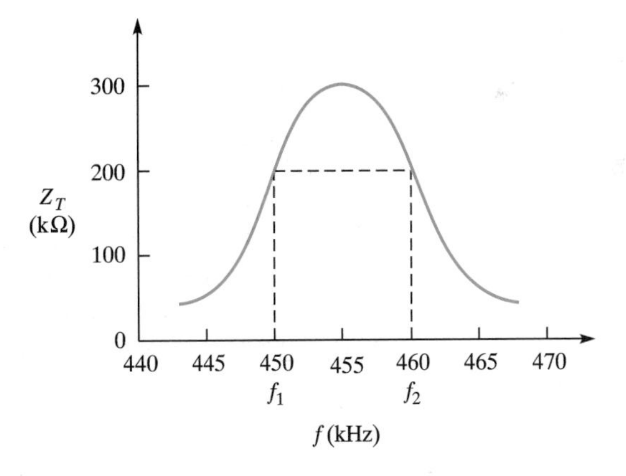

■ Bandwidth of a Parallel Resonant Circuit

The considerations for bandwidth of a parallel resonant circuit are similar to those for series resonance. The circuit response is useable over a band of frequencies from a lower to an upper cutoff frequency. Recall that in an LC circuit the impedance peaks at resonance and drops both above and below it. The cutoff frequencies are those at which the impedance drops to 70.7% of its peak value, as illustrated in the response curve of Figure 16–17. The bandwidth is found by subtracting f_1 from f_2. Thus,

$$\text{BW} = f_1 - f_2$$
$$= 460\text{ kHz} - 450\text{ kHz} = 10\text{ kHz}$$

Question: How is the bandwidth of a parallel* LC *resonant circuit computed?

The bandwidth of a parallel LC circuit is inversely proportional to the circuit Q-factor. Thus, Eq. 16–3 is used in the computation of bandwidth ($BW = f_r/Q$). The ideal situation, once again, is for the bandwidth to be centered on the resonant frequency. If bandwidth is known, the upper and lower cutoff frequencies, f_2 and f_1, can be computed using Eq. 16–4 and Eq. 16–5: $f_2 = f_r + (BW/2)$ and $f_1 = f_r - (BW/2)$.

EXAMPLE 16–11 For the circuit of Figure 16–18, compute the bandwidth and the upper and lower cutoff frequencies.

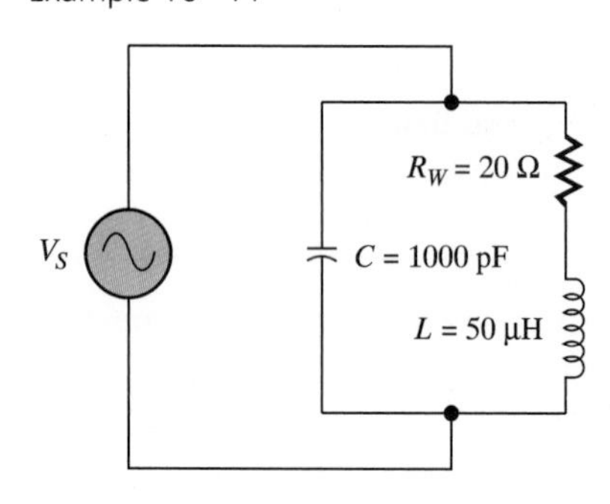

FIGURE 16–18 Circuit for Example 16–11

■ Solution:

1. Compute f_r:

$$f_r = \frac{1}{2\pi\sqrt{LC}}$$

$$= \frac{1}{2(3.14)\sqrt{(50\ \mu H)(1000\ pF)}} = 712\ kHz$$

2. Compute X_L at the resonant frequency:

$$X_L = 2\pi fL$$

$$= 2(3.14)(712\ kHz)(50\ \mu H) = 224\ \Omega$$

3. Compute the Q-factor:

$$Q = \frac{X_L}{R}$$

$$= \frac{224\ \Omega}{20\ \Omega} = 11$$

4. Compute the bandwidth:

$$BW = \frac{f_r}{Q}$$

$$= \frac{712\ KHz}{11} = 64.7\ kHz$$

5. Compute the upper and lower cutoff frequencies:

$$f_2 = f_r + \frac{BW}{2}$$

$$= 712\ KHz + \frac{64.7\ KHz}{2} = 744.35\ kHz$$

$$f_1 = f_r - \frac{BW}{2}$$

$$= 712\ KHz - \frac{64.7\ KHz}{2} = 679.65\ kHz$$

Once again, the bandwidth and the sharpness of the response are dependent upon the Q-factor of the circuit.

A high Q produces a sharp response and a narrow bandwidth. The Q-factor is inversely proportional to the circuit resistance. Thus, the more resistance there is in the circuit, the lower the response and the wider the bandwidth.

EXAMPLE 16–12 For the circuit of Figure 16–19(a), compute the bandwidth for the following values of R_W: (a) 100 Ω, (b) 50 Ω, and (c) 10 Ω. Sketch the response curve for each on the same graph.

■ **Solution:**

1. Compute f_r:

$$f_r = \frac{1}{2\pi\sqrt{LC}}$$

$$= \frac{1}{2(3.14)\sqrt{(109\ \mu\text{H})(232\ \text{pF})}}$$

$$= 1\ \text{MHz}$$

2. Compute X_L at the resonant frequency:

$$X_L = 2\pi f L$$

$$= 2(3.14)(1\ \text{MHz})(109\ \mu\text{H}) = 685\ \Omega$$

3. Compute Q ($Q = X_L/R_W$) for each value of R_W:

a. 100 Ω, $\quad Q = \dfrac{685\ \Omega}{100\ \Omega} = 6.85$

b. 50 Ω, $\quad Q = \dfrac{685\ \Omega}{50\ \Omega} = 13.7$

c. 10 Ω, $\quad Q = \dfrac{685\ \Omega}{10\ \Omega} = 68.5$

4. Compute the bandwidth ($BW = f_r/Q$) for each value of Q:

FIGURE 16–19 Circuit and response graph for Example 16–12

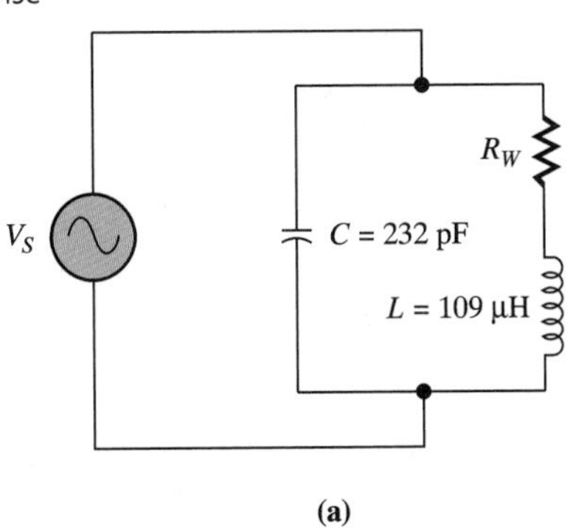

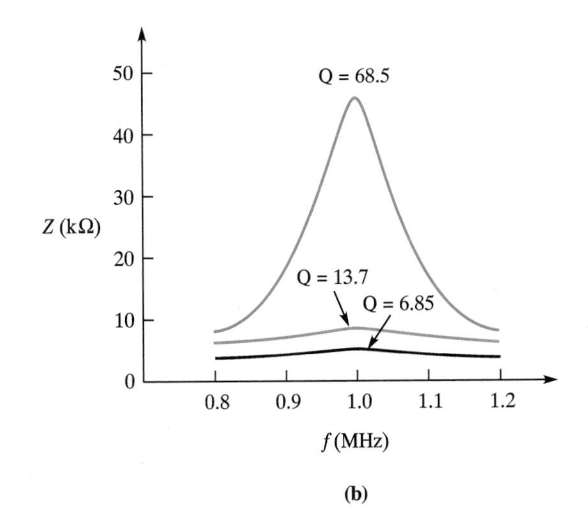

$$\text{For Q} = 6.85, \qquad \text{BW} = \frac{1 \text{ MHz}}{6.85} = 146 \text{ kHz}$$

$$\text{For Q} = 13.7, \qquad \text{BW} = \frac{1 \text{ MHz}}{13.7} = 73 \text{ kHz}$$

$$\text{For Q} = 68.5, \qquad \text{BW} = \frac{1 \text{ MHz}}{68.5} = 14.6 \text{ kHz}$$

The graph of the response curves are shown in Figure 16–19(b).

Damping of Parallel Resonant Response

Up to this point, the circuits considered have contained inductance, capacitance, and winding resistance only. Although they are sometimes used in this manner, they are most often shunted by a resistive load. The shunting action of this load will have a definite effect upon the resonant response. In fact, a shunt resistor is sometimes added for the very purpose of altering the response at resonance. In this section, the effects of shunt resistance are considered.

Question: What are the effects of a shunt resistance in the response of a parallel LC *circuit?*

FIGURE 16–20 Shunt resistance in a parallel *LC* circuit

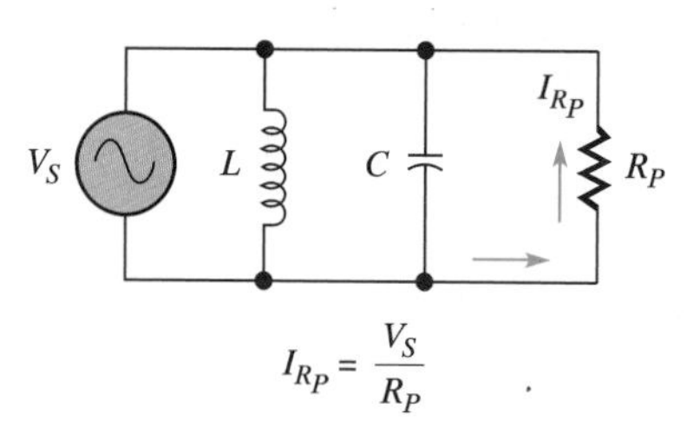

In the circuit of Figure 16–20, a resistor is placed across a parallel *LC* tank circuit. The current through the parallel resistor is, by Ohm's law, V_S/R_P. This value is not affected by frequency. At resonance, even though the reactive currents still cancel one another, the current through the resistor remains constant. Thus, the line current does not decrease below the resistive current value, and the line current does not drop as much or as sharply as before. The effect is to limit the resonant response and make it occur over a wider band of frequencies.

It can be seen that with the shunt resistor in place, the bandwidth is wider. Thus, the circuit must be lowered by the shunt resistor. An equation for the Q-factor of the circuit of Figure 16–20 can be derived in the following manner. Recall that a high-Q circuit is mostly reactive and very little true power is developed. In fact, if no true power were developed, the circuit Q would be infinite. The Q-factor can be related to the following ratios:

$$Q = \frac{\text{energy stored}}{\text{energy dissipated}}$$

$$Q = \frac{\text{reactive power}}{\text{true power}}$$

WHERE: $\text{reactive power} = \dfrac{V_S^2}{X_L}$

$\text{true power} = \dfrac{V_S^2}{R_P}$

Substituting gives

(16–11)

$$Q = \frac{V_S^2/X_L}{V_S^2/R_P} = \frac{V_S^2 R_P}{X_L V_S^2}$$

$$= \frac{V_S^2 R_P}{X_L V_S^2}$$

Thus,

$$Q = \frac{R_P}{X_L}$$

The significance of this equation is that any decrease in the value of R_P reduces the circuit Q. This effect, you will recall, is opposite that of the winding resistance of the inductor. A smaller winding resistance yields a larger Q. This effect of the value of R_P on the bandwidth is often made use of in what is known as *broadbanding*. With the shunt resistor in place, the circuit responds over a wider, or broader, band of frequencies, and the effect is to *dampen* the resonant response, so that R_P is sometimes referred to as a **damping resistor.**

EXAMPLE 16–13 For the circuit of Figure 16–21, compute the resonant frequency and the bandwidth.

■ Solution:

1. Compute f_r:

$$f_r = \frac{1}{2\pi\sqrt{LC}}$$

$$= \frac{1}{2(3.14)\sqrt{(200\ \mu\text{H})(50\ \text{pF})}} = 1.592\ \text{MHz}$$

2. Compute X_L at resonance:

$$X_L = 2\pi f L$$

$$= 2(3.14)(1.59\ \text{MHz})(200\ \mu\text{H}) = 2\ \text{k}\Omega$$

3. Compute Q:

$$Q = \frac{R_P}{X_L} = 19.5$$

4. Compute bandwidth:

$$\text{BW} = \frac{f_r}{Q} = 81.6\ \text{kHz}$$

FIGURE 16–21 Circuit for Example 16–13

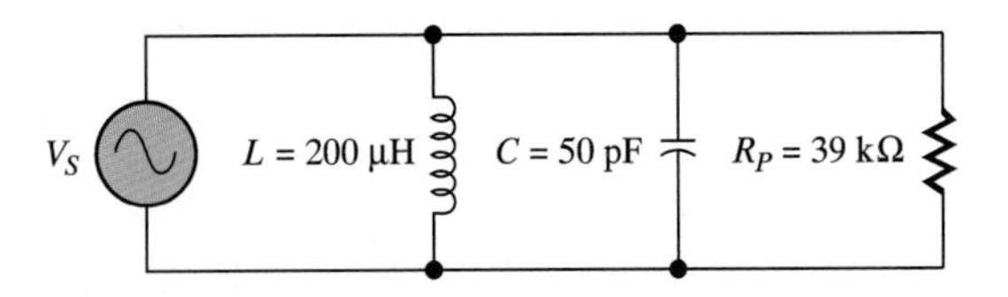

■ Nonideal Parallel *LC* Resonant Circuits

An ideal *LC* circuit would be one that contained no resistance. This situation is, of course, impossible; all inductors have some winding resistance. As you know, the degree to which this resistance affects the operation of the circuit depends upon the Q-factor. As has been established, if the Q-factor is 10 or greater, the circuit can be analyzed as though it were ideal. Through modern technology, achieving a Q-factor of 10 or greater is not too difficult. Thus, in most instances, assuming the ideal is adequate.

Question: What is the effect upon resonance of nonideal conditions?

There are three general conditions that exist at resonance. These are (1) the frequency at which X_L equals X_C, (2) the frequency at which the total impedance is maximum and line current is minimum, and (3) the frequency at which the power factor is 1. This last condition makes the circuit totally resistive and the phase angle zero. In the high-Q circuits considered previously, these three conditions occur at the same frequency. In low-Q circuits, however, there is appreciable resistance in the inductive branch, making it an impedance rather than a reactance. Thus, the three conditions occur at three separate frequencies!

FIGURE 16–22 Resonance under nonideal conditions

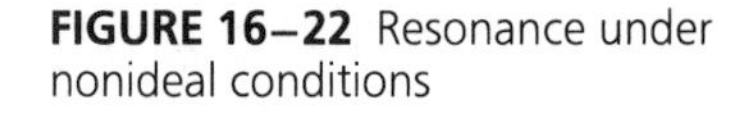

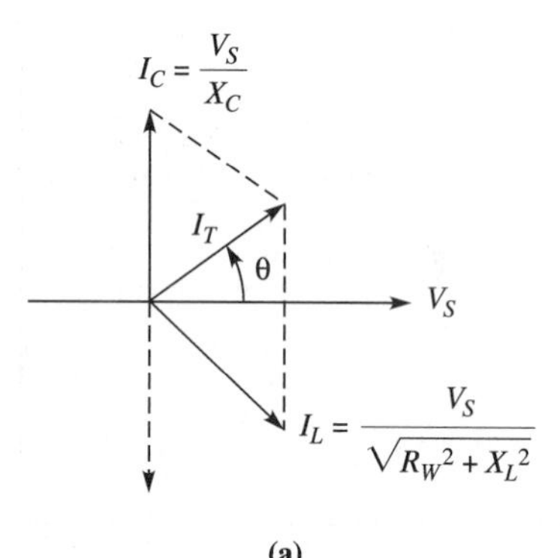

(a)
$X_L = X_C$, but the winding resistance in L is large, which causes I_L to be less than I_C.

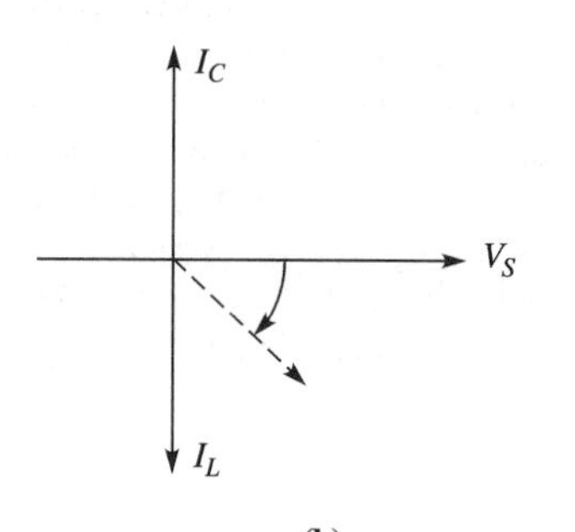

(b)
Lowering the frequency reduces Z, increases I_L, and decreases I_C.

Under condition 1, even with X_L and X_C equal, the impedance of the inductive branch is greater than X_C ($\sqrt{R_W^2 + X_L^2}$). Thus, I_L is less than I_C and lags the source voltage by some value much less than 90°, as illustrated in Figure 16–22(a). The line current leads the source voltage, and the circuit is capacitive. Even though X_L and X_C are equal, the line current is not minimum, and the phase angle is not zero. To obtain the minimum line current of condition 2, it is necessary to reduce the frequency of the applied signal. The reduction in frequency reduces X_L and the impedance of the inductor. The reduction in impedance increases the current through the inductive branch I_L. The value of X_C increases, decreasing the capacitive current I_C. As this continues, more cancellation takes place between I_L and I_C. The line current and the phase angle both decrease. At some frequency, the line current reaches its minimum, and any further reduction in frequency causes it to increase once again, as illustrated in Figure 16–22(b).

A further reduction in current causes the inductive and capacitive components of the total current to be equal. Thus, these two cancel, and the circuit is purely resistive, which, together with the resulting phase angle zero, satisfies condition 3.

Question: Which is the true resonant frequency of a low-Q parallel LC circuit?

Resonance is usually thought to occur at the frequency at which the reactive effects cancel one another, leaving the circuit totally resistive (phase angle = 0 and power factor = 1). As you have seen, these characteristics exist in condition 3. This resonant frequency, which is less than that for a corresponding high-Q circuit, is usually considered the resonant frequency of a low-Q *LC* circuit. It is often referred to as anti-resonance in order to distinguish it from the series resonance or high-Q parallel resonance.

■ SECTION 16–5 REVIEW

1. What value of Q qualifies a circuit as a high-Q circuit?
2. Will a parallel *LC* circuit with a relatively large Q have a wide or narrow bandwidth?
3. Will any increase in the value of the resistance in either the inductive or capacitive branch raise or lower the circuit Q?
4. Will a resistor shunting a parallel *LC* circuit widen or narrow the bandwidth?
5. Why is the resonant frequency of a low-Q circuit less than that computed for a high Q circuit?

■ **TABLE 16–3 Comparison of series and parallel resonant circuits**

	SERIES RESONANCE	PARALLEL RESONANCE
EQUATION	$f_r = \dfrac{1}{2\pi\sqrt{LC}}$	$f_r = \dfrac{1}{2\pi\sqrt{LC}}$ (high Q)
CIRCUIT CURRENT	Maximum: $\dfrac{V_S}{R_T}$	Minimum: $\dfrac{V_S}{Z\ (\text{tank})}$
IMPEDANCE	Minimum: $Z = R_T$	Maximum: $\dfrac{L}{CR}$ or $\dfrac{X_L^2}{R}$
BANDWIDTH	$\text{BW} = \dfrac{f_r}{\text{Q}}$	$\text{BW} = \dfrac{f_r}{\text{Q}}$
QUALITY FACTOR	$\text{Q} = \dfrac{X_L}{R_W}$ or $\text{Q} = \dfrac{X_L}{R_T}$	$\text{Q} = \dfrac{X_L}{R_W}$ or $\text{Q} = \dfrac{R_P}{X_L}$
RESONANT RISE	Voltage: $\text{Q}V_S$	Impedance: $\text{Q}X_L$
PHASE ANGLE	$\angle\theta = 0°$	$\angle\theta = 0°$
POWER FACTOR	PF = 1	PF = 1

SYNOPSIS

Before considering the applications and troubleshooting of resonant circuits, it would be a good idea to review the concepts of each type. For the most part, as was the case with any series or parallel circuit, series and parallel resonant circuits are opposites. In a series resonant circuit, the impedance is minimum, and the current is maximum. In a parallel resonant circuit, the impedance is maximum, and the line current is minimum. In a series resonant circuit, large voltages are developed across both the inductor and capacitor. In a parallel resonant circuit, a large circulating current is developed within the tank circuit. In both, the reactive elements cancel one another, leaving the circuit purely resistive. Thus, in a resonant circuit, the phase angle is zero and the power factor is one. The complete comparison of series and parallel resonance is contained in Table 16–3.

16–6 APPLICATIONS OF RESONANT CIRCUITS

■ Resonant Filters

Resonant filters operate in a manner similar to the *RL* and *RC* filters with which you are already familiar. The major difference is that instead of one cutoff frequency, a resonant filter has two. Thus, a resonant filter is capable of passing or blocking a specific, narrow band of frequencies. Resonant filters that pass a band of frequencies are known as **band-pass filters,** while those blocking a band of frequencies are known as **band-reject filters,** or band-stop filters. They find many uses in sound systems, radio, and television. Both types will be discussed in the following subsections. Remember, only an introduction is presented here; they will be discussed in detail in Chapter 17.

■ Band-Pass Filters

Question: How is a series resonant circuit used as a band-pass filter?

As shown in Figure 16–23(a), a series resonant circuit must be placed in series with the load in order to perform as a band-pass filter. As you know, at series resonance, the

FIGURE 16–23 Series resonant circuit used as a band-pass filter

FIGURE 16–24 Parallel resonant circuit used as a band-pass filter

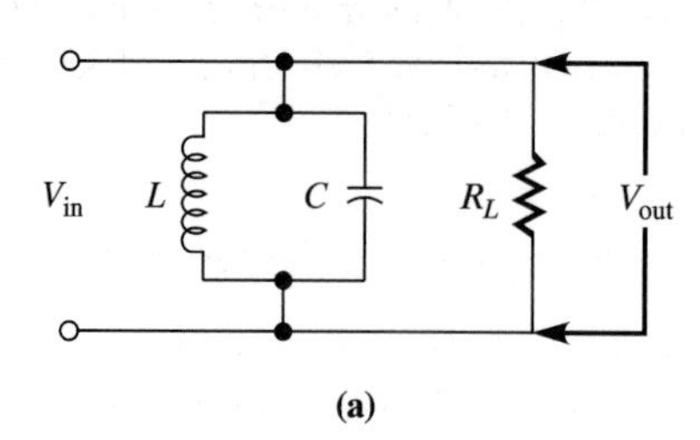

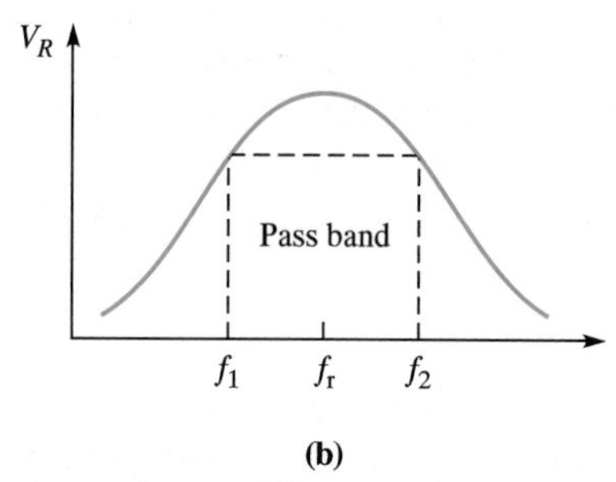

FIGURE 16–25 Series resonant circuit used as a band-reject filter

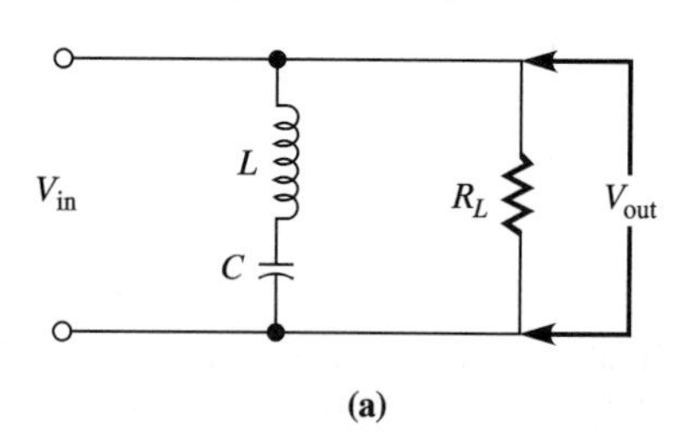

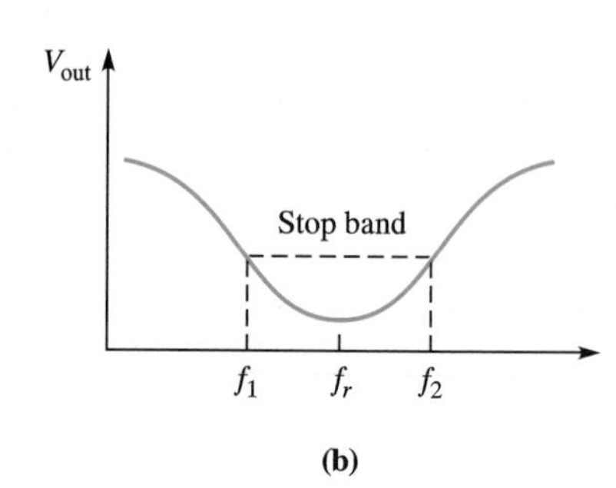

current is maximum. Thus, at the resonant frequency, maximum current flows through the load, and maximum voltage is developed across it. The value of both the load current and voltage declines at frequencies above and below resonance. A point is reached at which the voltage produced is no longer useable. The frequencies at which this occurs are known as the cutoff frequencies. By definition, the cutoff frequencies are those at which the output voltage has decreased to 70.7% of its maximum value. The response curve for this filter is shown in Figure 16–23(b). The bandwidth of the filter is all the frequencies between the upper and lower cutoff. Thus, this band of frequencies is passed, and all others are discriminated against.

Question: How is a parallel resonant circuit used as a band-pass filter?

As shown in Figure 16–24(a), a parallel resonant circuit must be placed in parallel with the load in order to act as a band-pass filter. Recall that at the resonant frequency, impedance is maximum, and current is minimum through the parallel resonant circuit. The resonant circuit has maximum voltage developed across it at this time (IZ). This maximum voltage is also felt across the load. Both above and below resonance, the impedance of the tank circuit decreases. The output voltage, which is directly proportional to the impedance of the tank circuit, drops accordingly. The frequencies at which the output voltage declines to 70.7% of its maximum value are the cutoff frequencies. The response curve for this circuit is the same as that shown in Figure 16–24(b). Notice its similarity to that for the series resonant band-pass filter. The bandwidth is once again the frequencies between the upper and lower cutoff.

■ Band-Reject Filters

Question: How is a series resonant circuit used as a band-reject filter?

As shown in Figure 16–25(a), a series resonant circuit must be placed in parallel with the load in order to perform as a band-reject filter. Whatever voltage is developed across the resonant circuit is also across the load. At the resonant frequency, the low impedance of the resonant circuit develops only a small voltage. At frequencies both above and below resonance, the impedance of the LC circuit increases. The voltage developed across it, and thus the load, increases. The cutoff frequencies are those at which the output voltage has reached 70.7% of its maximum value. Thus, all frequencies between the cutoff frequencies are rejected, and all those outside this band are passed. The response curve is shown in Figure 16–25(b).

Question: How is a parallel resonant circuit used as a band-reject filter?

FIGURE 16–26 Parallel resonant circuit used as a band-reject filter

As shown in Figure 16–26, a parallel resonant circuit must be placed in series with the load in order to perform as a band-reject filter. At the resonant frequency, the high impedance of the tank circuit allows only a small current flow through the load. Thus, the output voltage is minimum at this time. At frequencies above or below resonance, the decrease in impedance of the tank circuit allows more current to flow. The cutoff frequencies are those at which the output voltage has reached 70.7% of its maximum value. Once again, the frequencies between the upper and lower cutoff are rejected, and those outside this band are passed. The response curve is the same as that shown in Figure 16–25(b).

NOTE

In electronics, some things often seem so obvious that a technician often makes a snap judgement that he or she lives to regret! A case in point is the circuit of Figure 16–27(a). This simplified tuning circuit could be used to select a particular radio station for listening. The LC circuit looks for all the world like a parallel resonant circuit. This is not the case, however. Notice that L_2 is the secondary of a transformer that couples the signal from the antenna into the amplifier. The voltage is induced in the windings of L_2 and is situated as shown in Figure 16–27(b). As illustrated, the voltage induced in L_2 acts as a source in series with the inductor and capacitor. Thus, L_2 and C form a series resonant circuit.

Frequency-Determining Devices

An *oscillator* is an electronic device that produces a repetitive waveform with a constant amplitude and frequency. The waveform may be sinusoidal or nonsinusoidal. All oscillators have some component, or components, that determine the frequency of oscillation, which, in the case of sinusoidal oscillators, is often a tank circuit. In this type of oscillator, the tank circuit is shocked into oscillation in a manner similar to that described in Section 16–4. Recall that these oscillations rapidly dampen out due to resistance losses in the tank. In an oscillator, this is overcome as shown in Figure 16–28. A small amount of the tank circuit energy is fed back into the input of an amplifier. The output of the

FIGURE 16–27 Simplified tuning circuit

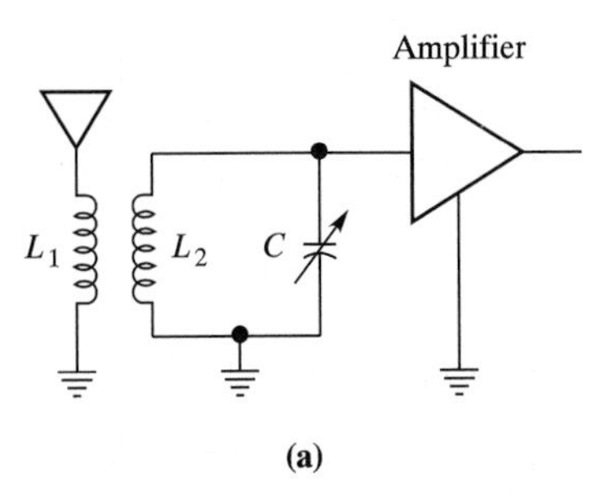

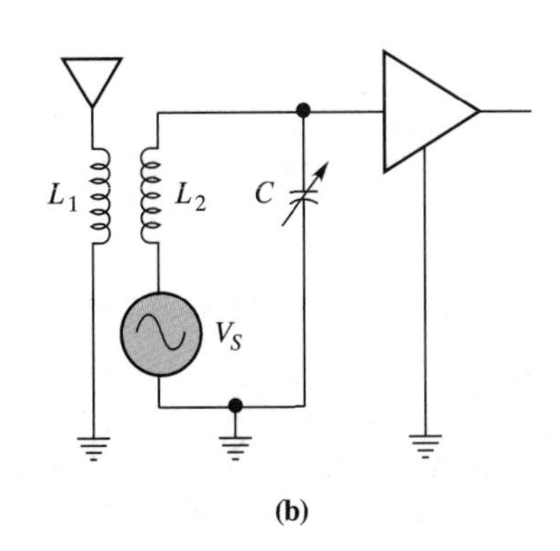

FIGURE 16–28 Tank circuit used as a frequency determining device

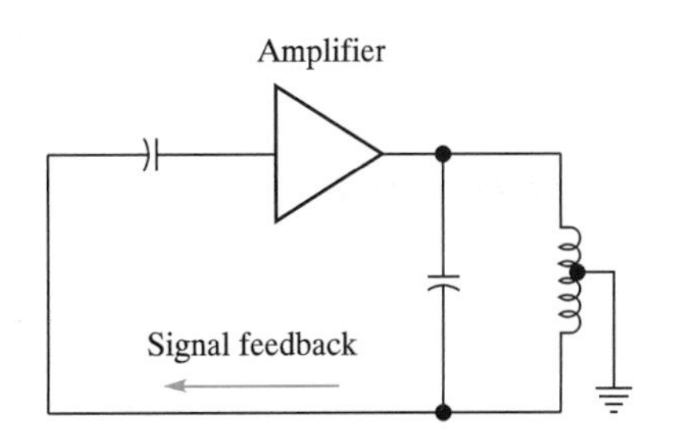

amplifier is then fed to the tank circuit. If adjusted to the proper phase and amplitude, this feedback signal just overcomes the tank circuit losses, and it has continuous oscillations. The frequency of these oscillations are the resonant frequency of the tank circuit.

SECTION 16–6 REVIEW

1. In a band-pass filter, is a parallel resonant circuit placed in series or parallel with the load?
2. In a band-reject filter, is a series resonant circuit placed in series or parallel with the load.
3. If a series resonant circuit is in series with a load, is the filter formed a band-pass or band-reject filter?
4. How many cutoff frequencies does a resonant filter have?

16–7 TROUBLESHOOTING

The purpose of a resonant filter is to pass or reject a narrow band of frequencies. Its two most important characteristics are its resonant frequency and Q-factor. A variation in either of these quantities could cause the system in which it is incorporated to fail. In the following subsections, the factors effecting these characteristics are considered.

As you know, the resonant frequency of an *LC* circuit is determined by the values of inductance and capacitance. Any change in the resonant frequency, then, would be the result of a change in one or both of these values. The resonant frequency is inversely proportional to changes in both *L* and *C*, decreasing as they increase and increasing as they decrease.

Changes in the Resonant Frequency

Question: What factors can cause changes in the inductance value, and how do they affect the resonant frequency?

As you have learned, an inductor's value is directly proportional to the square of the number of turns of wire. Although the number of turns could not increase, it can decrease due to shorts that may develop between the windings. Shorts reduce the number of turns, and the inductance decreases in value accordingly. Another factor that can cause reduced inductance is an increase in the spacing of the windings. This spacing could be caused by physical damage or from environmental factors. As illustrated in Figure 16–29, a reduced inductance value causes the resonant frequency to increase. Thus, the bandwidth is now centered at a higher frequency than that desired, and the system may fail to operate properly.

FIGURE 16–29 Effect of change in inductance: Reduced inductance value causes the resonant frequency to rise.

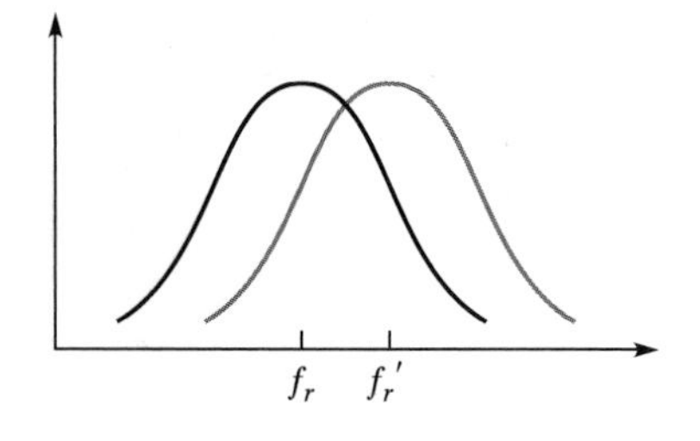

Question: What factors can cause changes in the capacitance value, and how do they affect the resonant frequency?

In Chapter 10, you studied the affect of temperature upon the value of capacitors. You learned that the value of the capacitor may increase or decrease in proportion to the amount of temperature change. This temperature coefficient is either positive or negative and is stated in parts-per-million change per degree Celsius. Although the change is usually not great, it must be considered in an environment where large temperature variations are expected. Notice that the capacitance can both increase or decrease, depending upon the direction of the temperature change. As illustrated in Figure 16–30, an increase in capacitance causes a decrease in resonant frequency, while a decrease in capacitance causes an increase. Once again, the bandwidth would be centered at the wrong frequency, and the system may fail to operate properly.

Changes in Bandwidth

Changes in the bandwidth of an *LC* circuit can also be critical in the operation of an electronic system. For example, *LC* band-pass filters are used in radios to pass a narrow band of frequencies from a specific radio station. The bandwidth must be just wide enough to pass all the intelligence in the stations signal and also narrow enough to keep signals from adjoining stations from interfering. If the bandwidth is too narrow, intelligence will be lost, and the audio (especially music) will be of poor quality. If the bandwidth is too wide, the signals from adjacent stations in the frequency spectrum will be reproduced, and the result is what is known as "babble," as illustrated in Figure 16–31.

FIGURE 16–30 Effects of a change in capacitance: An increase in capacitance causes the resonant frequency to decrease, while a decrease in capacitance causes the resonant frequency to increase.

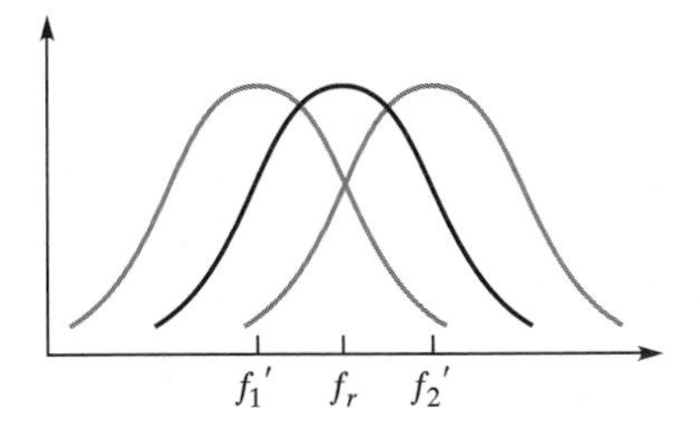

FIGURE 16–31 Result of bandwidth that is too wide

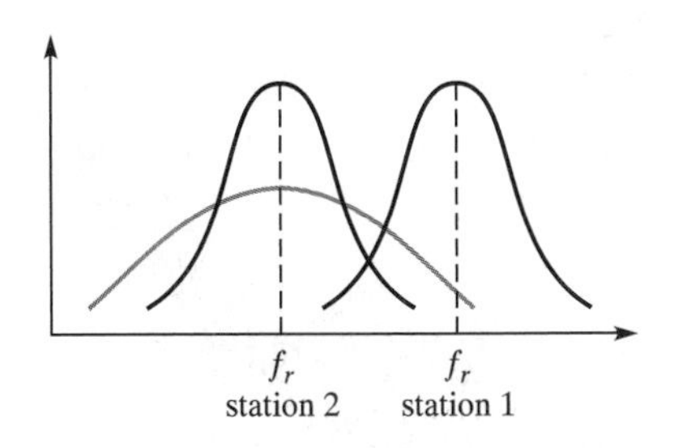

Bandwidth values are just as critical in other applications. Recall that bandwidth is the ratio of the resonant frequency to the quality factor of the circuit (BW $= f_r/\mathrm{Q}$). Since the resonant frequency is fixed by the values of L and C, it is changes in the circuit Q that causes changes in the bandwidth.

Question: What factors cause variations in the Q-factor of an LC circuit?

As you have learned, the Q of an unloaded *LC* circuit is inversely proportional to the winding resistance of the inductor (Q $= X_L/R_W$). Several things could cause a change in R_W. Shorts between the windings of the inductor not only reduce inductance but also the winding resistance, which results in an increase in Q and a reduction of bandwidth. The system would have already failed, however, due to the accompanying change in resonant frequency. A change in ambient temperature may also affect the winding resistance. As you know, the resistance of a conductor is directly proportional to temperature. Thus, an increase in temperature causes R_W to rise, lowering the Q. This lowered value of Q causes the bandwidth to widen, creating the possibility of interference from adjoining channels. Control of temperature variations have to be considered in the design state of the system. This control must be tight enough to keep the operation of the *LC* circuit within the design parameters.

A more likely cause of variations in the Q-factor is shown in Figure 16–32(a). Notice that the load is in parallel with the *LC* circuit. Thus, it acts in a manner identical to the damping resistor discussed in a previous paragraph. Many loads do not have a constant value of impedance; their values varying according to circuit conditions. Recall that the Q of this type circuit is the ratio of the parallel resistance to the inductive reactance (Q =

FIGURE 16–32 Cause and solution to variations in Q-factor

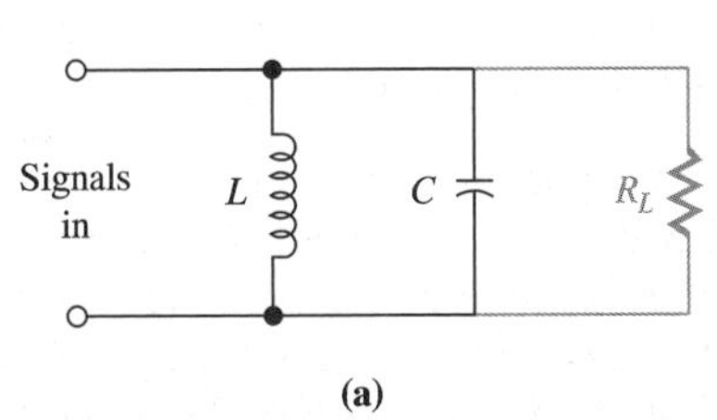

(a)

An added load appears as a resistance in parallel with the tank; Q decreases and BW increases.

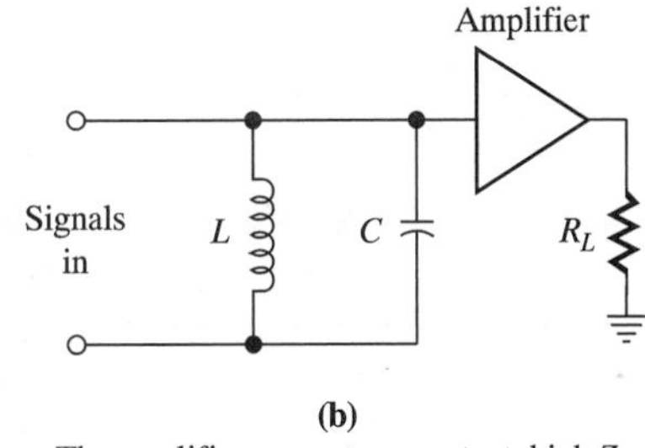

(b)

The amplifier presents a constant, high Z load to the tank circuit.

R_P/X_L). Thus a decrease in load resistance also causes a decrease in Q and a widening of the bandwidth. The opposite effect occurs for an increase in load resistance. A method of overcoming this problem is shown in Figure 16–32(b). Here you see a *buffer amplifier* between the *LC* circuit and the load. The function of the amplifier, which you will study in a later portion of the electronics curriculum, is to present a constant load to the *LC* circuit. Thus, the Q and bandwidth are stabilized.

ac Resistance of an Inductor

Before concluding the discussion of troubleshooting of resonant circuits, it is necessary to consider the effects of frequency upon the resistance of the inductor. As you know, the winding resistance, R_W is a dc resistance value that can be measured with an ohmmeter. The **ac resistance,** on the other hand, is frequency sensitive and is made up of three things: the core losses of magnetic hysteresis and eddy currents and skin effect.

NOTE

Recall that both eddy currents and hysteresis were covered in previous chapters. Recall that they occur in the core of an inductor. Thus, an air core inductor will not be affected by them. If a core is necessary, as is often the case where tuning is required, then it is made of some highly permeable material. Ferrite and powered iron cores are examples of such materials.

FIGURE 16–33 Techniques for eliminating skin effect

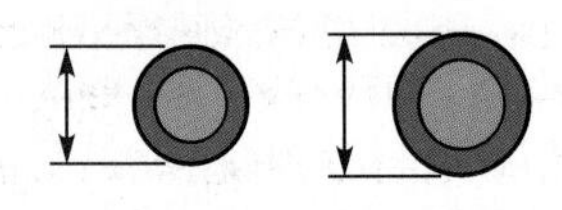

(a)
Larger diameter wire is used to increase conducting area.

(b)
Area of wire's core is increased, increasing inductance.

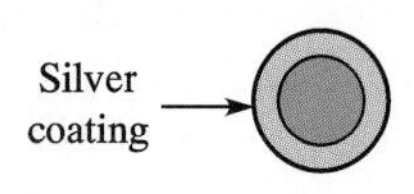

(c)
Silver coated wire makes surface more conductive.

Question: What is skin effect?

The resistance of a conductor is not the same for a low-frequency current as for a high-frequency current. When high-frequency current flows, the lines of magnetic flux are concentrated in its center. Thus, the current encounters more inductance in the center of the conductor than at the edge, where some of the magnetic flux is in the air. Thus, at high frequencies, the current tends to flow on the surface, or *skin,* of a conductor. This tendency of high-frequency signals to flow on the surface of the conductor is known as **skin effect.** The reduction in the cross-sectional area of the conductor increases its effective resistance, reducing the circuit Q and widening the bandwidth.

Question: How is skin effect minimized?

There are three techniques used to reduce skin effect. First the diameter of the wire used is made large in order to reduce resistance. In some cases, since the current flows on the surface, copper tubing is used rather than wire. Second, the diameter of the inductor is made larger. The area of the inductor is thus increased, and fewer turns of wire are needed. Third, the surface of the inductor may be silver coated to give it greater conductivity. These three techniques are illustrated in Figure 16–33.

SECTION 16–7 REVIEW

1. What two basic problems can be associated with a resonant circuit?
2. What environmental factor may cause changes in the resonant frequency?
3. What problem with the inductor may cause the resonant frequency to increase?
4. What environmental factor may cause the bandwidth to increase?
5. What is skin effect?
6. What are three techniqued to minimize skin effect?

CHAPTER SUMMARY

1. In order to produce electronic resonance, a circuit must contain inductance and capacitance.
2. At the resonant frequency, X_L equals X_C.
3. At series resonance, the reactances cancel, and the circuit is resistive.
4. At series resonance the current is maximum.
5. At parallel resonance, the capacitive and inductive susceptances cancel.
6. At parallel resonance, the line current is minimum.
7. The resonant frequency of an *LC* circuit is a function of the *LC* product.
8. At resonance, the circuit phase angle is zero and the power factor is one.
9. At series resonance, there is a resonant rise in the voltages across the inductor and capacitor.
10. The bandwidth of a series resonant circuit is the band of frequencies between the points where the output voltage drops to 70.7% of its maximum value.
11. At parallel resonance, there is a resonant rise in the impedance of the tank circuit.
12. The bandwidth of a parallel resonant circuit is the frequencies between the points at which the impedance drops to 70.7% of its maximum value.
13. Bandwidth is inversely proportional to the circuit Q.
14. The effects of parallel resonance can be damped by placing a resistor in parallel with the tank circuit.
15. In low Q circuits ($Q < 10$), the three conditions of resonance will not occur at the same frequency.
16. A series resonant circuit in parallel with the load forms a band-reject filter.
17. A series resonant circuit in series with the load forms a band-pass filter.
18. A parallel resonant circuit in parallel with the load forms a band-pass filter.
19. A parallel resonant circuit in series with the load forms a band-reject filter.
20. Anything causing changes in the value of inductance or capacitance will cause variations in the resonant frequency.
21. A varying value of load resistance causes changes in circuit Q, which causes the bandwidth to vary.
22. The ac resistance of a coil is the result of eddy currents, hysteresis loss, and skin effect.

KEY TERMS

resonance
resonant frequency
LC product
resonant rise in voltage
resonant rise in impedance
bandwidth
tank circuit
oscillations
ringing
damped wave
damping resistor
band-pass filter
band-reject filter
ac resistance
skin effect

EQUATIONS

(16–1) $f_r = \dfrac{1}{2\pi\sqrt{LC}}$

(16–2) $V_s = V_R$ (series resonance)

(16–3) $Z = R$ (series resonance)

(16–4) $\text{BW} = f_2 - f_1$

(16–5) $Q_{\text{ckt}} = \dfrac{X_L}{R_T}$

(16–6) $\text{BW} = \dfrac{f_r}{Q}$

(16–7) $f_1 = f_r - \dfrac{\text{BW}}{2}$

(16–8) $f_2 = f_r + \dfrac{\text{BW}}{2}$

(16–9) $Z = L/CR$ (parallel resonance)

(16–10) $Z = \dfrac{X_L^2}{R}$ (parallel resonance)

(16–11) $Q = \dfrac{R_P}{X_L}$ (parallel damping)

Variable Quantities

BW = (Δf) bandwidth
Q = quality factor
Q_{ckt} = quality factor of the circuit
f_1 = lower cutoff frequency
f_2 = upper cutoff frequency
R_P = parallel resistor

TEST YOUR KNOWLEDGE

1. What conditions must exist in a circuit for resonance to occur?
2. Is the resonant frequency directly or inversely proportional to the capacitance?
3. Is the resonant frequency directly or inversely proportional to the inductance?
4. At series resonance, is the impedance maximum or minimum?
5. At series resonance, is the voltages across both the inductor and capacitor relatively high or low?
6. At parallel resonance, is the impedance maximum or minimum?
7. At parallel resonance, is the line current minimum or maximum?
8. Does a series resonant circuit in parallel with the load forms a band-pass or band-reject filter?
9. Does a parallel resonant circuit in series with the load forms a band-pass or band-reject filter?
10. What is the band of frequencies passed or rejected by a filter known as?
11. What is skin effect?
12. What is meant by a low Q circuit?
13. If the resistor placed in parallel with a tank circuit is increased in value, will the Q increase or decrease?
14. If the resistance in series with a series *LC* circuit is increased in value, will the Q increase or decrease?
15. Will an increase in Q cause an increase or decrease in bandwidth?

PROBLEM SET: BASIC LEVEL

Sections 16–1 and 16–2

1. Compute the resonant frequencies for the following combination of inductors and capacitors: (a) $L = 100$ mH and $C = 0.1\ \mu$F, (b) $L = 50$ mH and $C = 0.2\ \mu$F, (c) $L = 109\ \mu$H and $C = 100$ pF, and (d) $L = 200$ mH and $C = 10\ \mu$F.
2. Compute the resonant frequency of the circuit of Figure 16–34. Recompute the resonant frequency if the inductance is (a) doubled and (b) halved.

FIGURE 16–34 Circuit for Problem 2

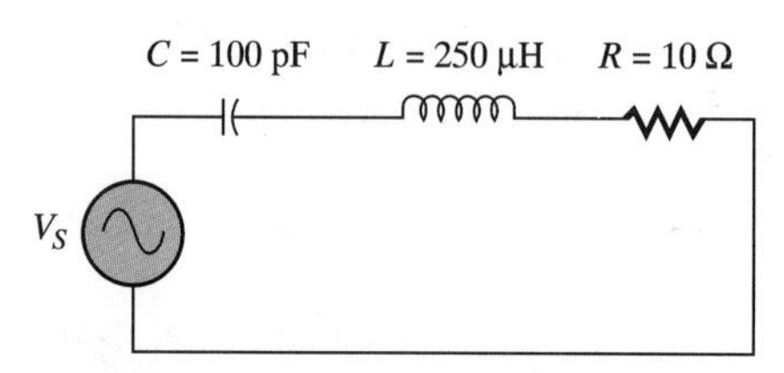

3. Compute the resonant frequency of the circuit of Figure 16–35. Recompute the resonant frequency if the capacitance is (a) doubled and (b) halved.

FIGURE 16–35 Circuit for Problem 3

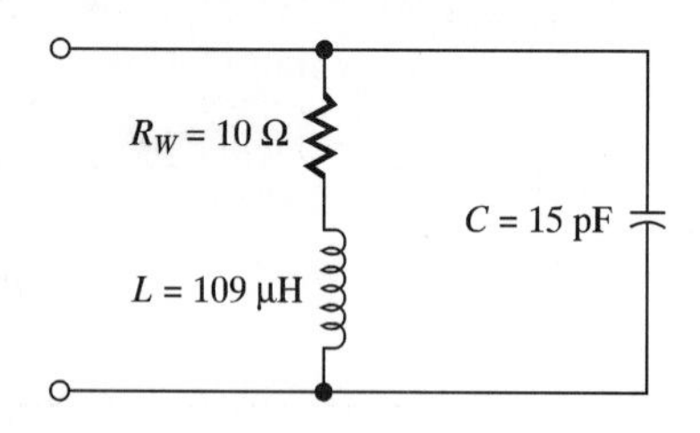

Sections 16–3 and 16–4

4. For the circuit of Figure 16–36, compute the following: (a) f_r, (b) V_L, (c) V_C, and (d) V_R and I (at resonance).

FIGURE 16–36 Circuit for Problems 4–7

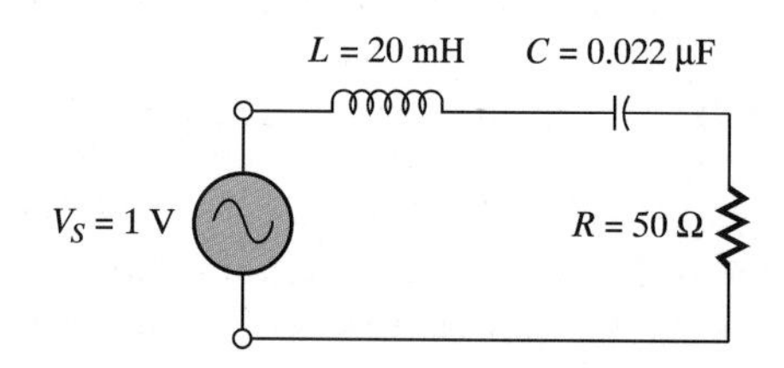

5. For the circuit of Figure 16–36, compute the bandwidth and the upper and lower cutoff frequencies.
6. For the circuit of Figure 16–36, compute the impedance at the upper cutoff frequency. Is the circuit inductive or capacitive at this frequency?
7. For the circuit of Figure 16–36, compute the impedance at the lower cutoff frequency. Is the circuit inductive or capacitive at this frequency?
8. A series *RLC* circuit is resonant at 15 MHz. It has a bandwidth of 100 kHz. If the resistance is halved, what will be the new upper and lower cutoff frequencies?
9. A series *RLC* circuit has an output voltage of 80 V at the resonant frequency. What is its output voltage at the upper and lower cutoff frequencies?
10. A series *RLC* circuit has a Q-factor of 35. If the applied voltage is 100 mV, what is the voltage across the capacitor at resonance?

Sections 16–5 and 16–6

11. For the circuit of Figure 16–37, compute the following: (a) f_r, (b) I_L, (c) I_C, (d) Z, and (e) BW (currents and impedance at the resonant frequency).

12. For the circuit of Figure 16–37, what are the upper and lower cutoff frequencies?

13. A 39 kΩ damping resistor is added to the circuit of Figure 16–37. Compute the new circuit Q.

14. Recompute the bandwidth of the circuit of Figure 16–37 with the damping resistor installed.

FIGURE 16–37 Circuit for Problems 11–14

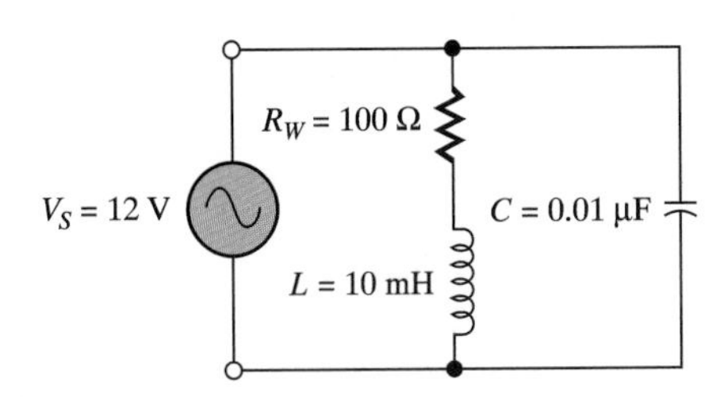

15. A parallel *RLC* circuit is resonant at 10 MHz. It has a bandwidth of 50 kHz. If the value of the damping resistor is halved, what will be the new upper and lower cutoff frequencies?

16. A parallel *LC* circuit has an impedance of 25 kΩ at its resonant frequency. What is its impedance at the upper and lower cutoff frequencies?

17. A parallel *LC* circuit has a Q-factor of 15. If X_L at resonance is 500 Ω, what is the impedance of the circuit?

Section 16–8

18. Draw the schematic of a series resonant band-pass filter connected to a 10 Ω resistive load. If the value of the inductance is 10 mH and the value of the capacitance is 0.01 μF, what band of frequencies will be passed?

19. Draw the schematic of a parallel resonant band-reject filter connected across a 30 kΩ resistive load. If the value of the capacitance is 0.022 μF and the value of the inductance is 1 mH, what band of frequencies will be rejected?

PROBLEM SET: CHALLENGING

20. Calculate the value of capacitance required to be resonant at 2400 Hz with an inductor of 200 mH.

21. Calculate the value of inductor required to be resonant at 1 MHz with a 100 pF capacitor.

22. A parallel *LC* circuit is to be resonant at 500 kHz. The inductor used has a value of 0.25 mH and a winding resistance of 20 ohms. (a) Compute the value of the capacitor required. (b) Compute the impedance of the circuit at resonance. (c) If V_S is 1 V, compute I_L, I_C, and I_R at resonance.

23. At the resonance, the reactance of the inductor of a series *RLC* circuit is 1550 Ω. The circuit Q is 15 Ω. (a) What is the value of the circuit resistance? (b) If the source voltage is 5 V, what are the voltages across the resistor, inductor, and capacitor at resonance?

24. A variable capacitor is used to tune a 109 μH inductor to cover the tuning range of the standard AM radio broadcast band (540 kHz through 1600 kHz.). What must the minimum and maximum capacitance be?

25. A series *RLC* circuit is resonant at 100 MHz. If the inductance is decreased 50 times and the capacitance increased 8 times, what is the new resonant frequency? (Hint: Recall that the resonant frequency varies inversely in proportion to the square root of the *LC* product.)

26. The circuit of Figure 16–38 has a resonant frequency of 1 MHz. If the bandwidth is to be 10 kHz, to what value must the variable resistor be set?

FIGURE 16–38 Circuit for Problem 26

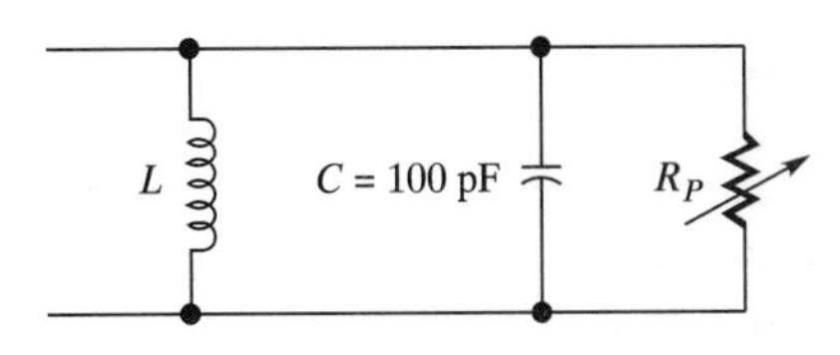

PROBLEM SET: TROUBLESHOOTING

27. The frequency response of the circuit of Figure 16–39 is checked, and the bandwidth is found to be one-third the design value. The resonant frequency is correct. What is the probable fault in this circuit? The normal setting of R_L is 60 Ω.

$L = 100\ \mu$H
$C = 100$ pF
$R = 60\ \Omega$

FIGURE 16–39 Circuit for Problem 27

28. The circuit of Figure 16–40 is found to resonate at 225.192 kHz. If the capacitance is correct, what is the probable fault in this circuit?

FIGURE 16–40 Circuit for Problem 28

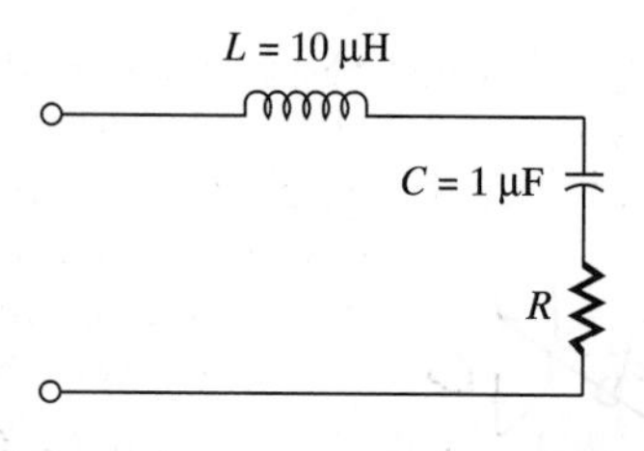

29. The bandwidth of the circuit of Figure 16–41 is measured and found to be one quarter its designed value. If the resonant frequency is correct, what is the probable fault?

FIGURE 16–41 Circuit for Problem 29

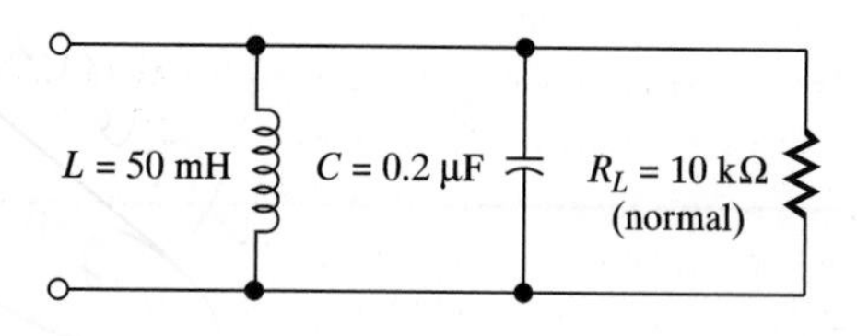

A Day in the Life of a Field Service Technician

Equipment manufacturers often establish service companies in local areas where their products are marketed and used. Staffed by engineers and technicians, these companies service and repair equipment in the local area, usually at the users' location. This policy relieves the user of the responsibility of shipping an item back to the plant for repair, and thus avoids costly downtime. Field service technicians are those that make short trips in these local areas in answer to trouble calls. Another function of such a company is to watch for recurring troubles that must be reported to the manufacturer's home office. If a trend in breakdowns is established, modifications can be designed and installed to prevent them. In the case of computer systems, the service company personnel often provide initial training in system operation to customers.

At the beginning of each day, trouble calls are logged and apportioned to the field service technicians who utilize company transportation and equipment to travel to customers' locations within the service company's region. If possible, they perform the service or repair on site. If the problem is major in nature, the equipment sometimes must be removed to the shop for repair. The customer is then billed for the service call and any parts that are replaced.

Tommy has worked for a service company of computer manufacturer for the past three years, after receiving his A.S. degree in electronics from an area community college. Tommy began his employment by attending an intensive five-week course covering the computers for which he would be responsible. He has become quite proficient in his work and has advanced steadily in the technician grades as well as in the pay scale.

On this particular day, as he parks his car, he is thinking of an event that transpired a few days before. He was on a service call when his company car inexplicably quit running. Anything that delays answering a service call is a worry because it has a way of perturbing the customer, not to mention lengthening the work day! Feeling sorry that he would be late for his service calls, Tommy informed the office and then called a local car dealer for road service. While at the garage waiting for the repairs to be completed, Tommy was approached by the owner who asked about the company name and logo on the car. When informed that he worked for a computer service company, the owner became quite interested and wished to know more about the computers and how they were used in a business. Before leaving the garage, about two hours later, Tommy phoned the office again and arranged for a sales representative to contact the garage owner. The result, as he was informed later, was that one of their computers would probably be installed at this business.

As Tommy enters his office, he sees the neat stack of service calls on his desk. He quickly shuffles through them and finds that today he will be working the local area of the company region. He recognizes some of the local businesses that he has visited previously and decides that none appear to be emergencies. As he develops a plan for placing the service calls in the most efficient order, he notices the MESSAGE light blinking on his phone. Accessing phone-mail, he listens to a series of messages from the previous day. His supervisor needed to see him. Performance evaluation time, Tommy thinks. Next, Shelly, an engineer, wanted to discuss a report he had made regarding a problem with a printer.

Tommy calls his supervisor, who does, in fact, want to set a time for Tommy's performance appraisal. This annual review will involve a comparison of the goals he set for himself the previous year with what he had actually achieved, the determination of a raise in pay, as well as the establishment of new goals for the coming year. The supervisor tells Tommy to have his goals for the next year ready for an interview on the following Monday.

Next, Tommy goes to the office of the engineer who wanted to see him. As he suspected, Shelly is not pleased with his report on the problem in a printer at a local business. The trouble call indicated that the printer was not responding to the computer's commands. To confirm that the trouble was not in the computer, he attached another printer that had worked. When he tested the original printer, however, it, too, worked fine! Thus, he had reconnected it without having made any repairs and submitted a report to that effect.

Shelly has the report on her desk and expresses her concern that the problem could occur again since the cause had not been found. She assures him, however, that she agreed with the procedures he used to test the equipment and that she didn't have any further suggestions on how he might have determined the cause of the problem.

After completing the conference with Shelly, Tommy gathers up his equipment and leaves on the first trouble call for the day. It is at a real estate office in a shopping center where a customer's personal computer will not "boot up." His tests prove that the disc drive is out of alignment. Removing the computer to an area provided in the rear of the offices, he procedes to further test and align the drive. The trouble is cleared, and the computer operates normally.

His next stop is at city hall. Here, a small computer is not responding to any command, covering the screen with "garbage" only. His first test is of the power supply. With a voltmeter, the dc voltage measures lower than normal, but not zero. Tommy knows that one cause of low dc voltage is failure of the filter capacitor on the output the rectifier. This was confirmed when he viewed the dc voltage on an oscilloscope. Other tests of the power supply indicated a particular filter capacitor to be open. Tommy replaces the capacitor, and the computer operates normally.

On his last service call, late in the afternoon at a machine shop north of town, Tommy finds another printer that refuses to speak with the computer. He fervently hopes that this printer would not be like the contrary printer he encountered several days ago, and he is relieved to find a mechanical problem. He quickly repairs the printer by tightening a set-screw on the motor shaft.

Having completed all his service calls, Tommy returns to the office. He thinks that it has been a pretty good day! He had plenty to keep him busy, but nothing that caused him problems. The conference with Shelley had been unproductive, but at least she had agreed with his methods. He has an appointment for his performance appraisal and expects that his supervisor will be satisfied with his progress. Tonight he plans to work on the goals he wants to set for the coming year and perhaps look at the community college catalog to decide on some new courses to take. As in the past year, the company will pay for the tuition and books for courses he successfully completes.

CHAPTER

17 Filters

■ UPON COMPLETION OF THIS CHAPTER, YOU WILL BE ABLE TO

1. Define, draw the schematic, and explain the characteristics of *RC* and *RL* low-pass filters.
2. Define, draw the schematic, and explain the characteristics of *RC* and *RL* high-pass filters.
3. Compute the cutoff frequency of *RC* and *RL* low- and high-pass filters.
4. Define, draw the schematic, and explain the characteristics of band-pass and band-reject filters.
5. Compute the bandwidth and upper and lower cutoff frequencies of band-pass and band-reject filters.
6. List the applications of low-pass, high-pass, band-pass, and band-reject filters.
7. Define the decibel.
8. Compute the power, voltage, and current gain of filters in decibels.
9. Troubleshoot low-pass, high-pass, band-pass, and band-reject filters.
10. Compute the value of a bypass capacitor.

As you have learned, electronics involves the systematic control of electric current, utilizing devices that are poor conductors of electricity. Among these devices are resistors, capacitors, and inductors. Previous chapters have shown the effects of each, both individually and in combination. Each device has unique characteristics that are taken advantage of in specific current control applications.

A resistor may be fixed or variable. Its effects, however, are the same regardless of whether the current is direct or alternating. Resistance opposes the flow of current and in so doing produces two effects: a voltage drop and power in the form of heat. The voltage drop and power developed are directly proportional to both the current and resistance values.

Capacitors and inductors also oppose the flow of electric current. This opposition, known as reactance, is frequency sensitive. It produces a voltage drop across each device that is directly proportional to the current and reactance. Recall that unlike resistance, these two devices store electrical energy rather than dissipate it, and thus no power is developed. Capacitive reactance is infinite to direct current and decreases as frequency increases. Inductive reactance, on the other hand, is zero to direct current and increases as frequency increases. This frequency sensitivity is taken advantage of in circuits known as electronic filters.

As you already know, there are basically four types of electronic filters: (1) those that pass all frequencies below a specified frequency and block all those above, (2) those that pass all frequencies above a specified frequency and block all those below, (3) those that pass a band of frequencies between specified upper and lower frequencies and block all others, and (4) those that block a band of frequencies between specified upper and lower frequencies and pass all others. Previous chapters have introduced each type of filter; each will be considered in detail in this chapter.

Along with the theory of operation, the circuit applications of each type filter are discussed in this chapter. These applications should convince you of the importance of filter circuits in electronics technology. Mastery of the material in this chapter requires a thorough knowledge of inductance and capacitance. It might be a good idea to review both prior to proceeding.

17–1 LOW-PASS FILTER

Question: What is a low-pass filter?

A **low-pass filter** is a circuit that passes all frequencies below a specified frequency and blocks all those above. As you may recall from a previous chapter, this specified frequency, known as the cutoff frequency, is the frequency at which the output voltage decreases to 70.7% of the input.

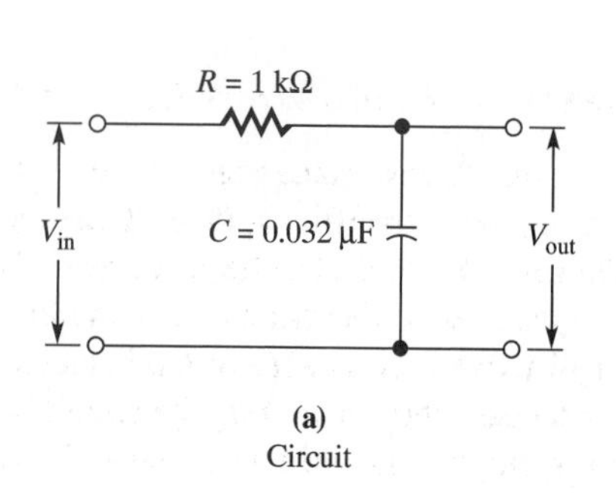

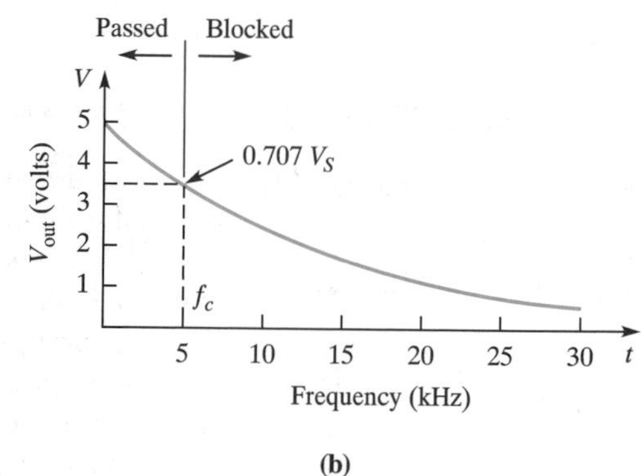

FIGURE 17–1 *RC* low-pass filter

■ **TABLE 17–1 Output voltages for various frequencies of an *RC* low-pass filter**

Frequency	X_c	Z	I	$V_{out} = IX_c$
0	∞	∞	0	5 V
1 kHz	4976 Ω	5074 Ω	0.985 mA	4.92 V
2 kHz	2488 Ω	2679 Ω	1.87 mA	4.64 V
3 kHz	1659 Ω	1934 Ω	2.59 mA	4.29 V
4 kHz	1244 Ω	1592 Ω	3.14 mA	3.9 V
5 kHz	995 Ω	1407 Ω	3.55 mA	3.53 V
6 kHz	829 Ω	1295 Ω	3.86 mA	3.2 V
7 kHz	710 Ω	1222 Ω	4.1 mA	2.9 V
10 kHz	497 Ω	1112 Ω	4.5 mA	2.24 V
20 kHz	248 Ω	1025 Ω	4.88 mA	1.21 V
30 kHz	166 Ω	1008 Ω	4.96 mA	0.823 V

■ *RC* Low-Pass Filter

Question: What is an RC *low-pass filter, and how does it operate?*

An *RC* low-pass filter is shown in Figure 17–1(a). Notice that it is a series *RC* circuit with the output taken across the capacitor. As with any series circuit, it is a voltage divider. The output voltage is the product of capacitive reactance and circuit current: $V = IX_C$. Reactance varies inversely with frequency, so the output voltage also varies inversely with frequency. At a frequency of 0 Hz (dc), the reactance is nearly infinite, and virtually all the input voltage appears across the output. As the frequency is increased, the reactance and output voltage decreases, while the voltage across the resistor increases. As this process continues, a point is reached at which the output voltage and the voltage across the resistor are equal. The frequency at which they are equal is known as the cutoff frequency (f_c).

In Table 17–1, the output voltage is computed for a series of frequencies. These points are shown plotted in the graph of Figure 17–1(b). This graph is the typical characteristic curve of a low-pass filter. All frequencies below the cutoff are passed, while all above are considered to be blocked (which may also be referred to as stopped, rejected, or attenuated). Notice that the cutoff point is not abrupt, but gradual. Thus, some output voltage is present for frequencies above the cutoff, but it is greatly *attenuated* (decreased). Notice further that at the cutoff frequency, the output voltage has decreased to 70.7% of the input.

■ *RL* Low-Pass Filter

FIGURE 17–2 *RL* low-pass filter

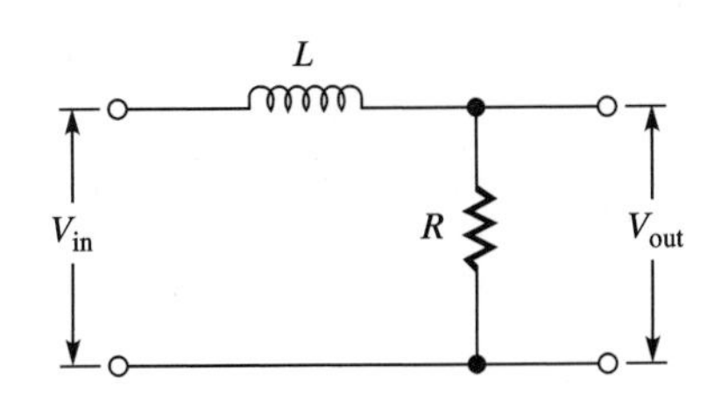

Question: What is an RL *low-pass filter, and how does it operate?*

An inductor may also be used as the reactive element in a low-pass filter. As Figure 17–2 shows, the positions of the resistive and reactive elements are reversed, and the output is taken across the resistor. At 0 Hz, the reactance of the inductor is zero. The resistor is thus the sole impedance in the circuit, and the entire input voltage appears across it, ignoring the small winding resistance of the inductor. As frequency increases, the inductive reactance increases, thus increasing the circuit impedance. The result is lower circuit current and lower output voltage. Once again, the cutoff frequency is that at which the output voltage and the voltage across the inductor are equal. The pass characteristic is

FIGURE 17–3 Types of low-pass filters

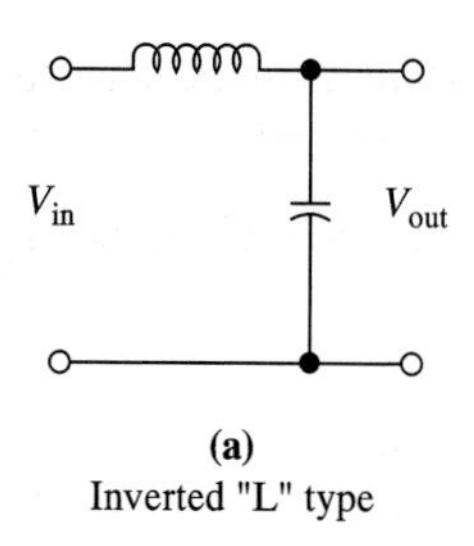

(a)
Inverted "L" type

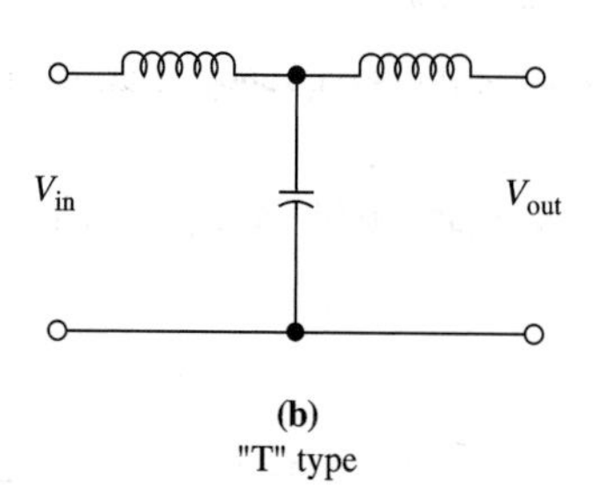

(b)
"T" type

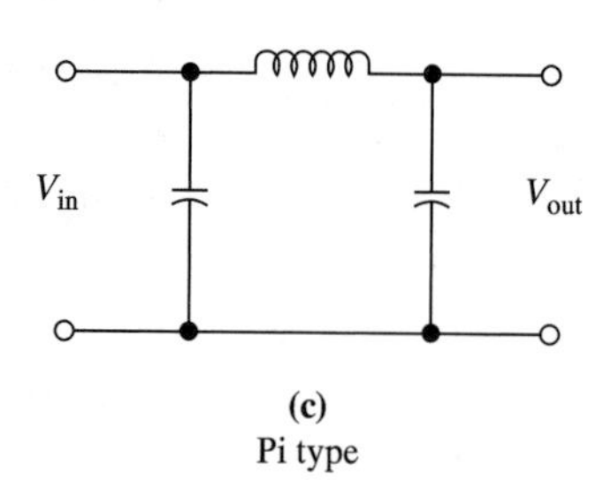

(c)
Pi type

identical to that shown in 17–1(b). As in the case of the *RC* low-pass filter, at the cutoff frequency, the output voltage has decreased to 70.7% of the input. (It is left for the student to compute the output voltage for a series of frequencies, and verify the response of this filter.)

Other Low-Pass Filters

Figure 17–3 shows low-pass filters made up of inductors and capacitors in combination. Notice in each that the capacitors are parallel elements, while the inductors are series elements. Frequencies below the cutoff encounter a high reactance in the capacitive branch and are shunted through the relatively low impedance of the inductor and load. Frequencies above the cutoff bypass the load through the relatively low reactance of the capacitive branch.

The filters of Figure 17–3 each operate on the same basic principle. For example, in the inverted "L" filter, the reactance of the capacitor is much less than that of the inductor (at least ten times less). Thus, higher frequencies are greatly attenuated in the inductor and produce only small output voltages across the capacitor. The "T" type is an inverted "L" filter followed by another series inductor. Here the higher frequencies are further attenuated. The pi type filter is the opposite of the "T" type. In this type, some filtering is accomplished by the first capacitor, prior to the inverted "L" filter.

SECTION 17–1 REVIEW

1. In an *RC* low-pass filter, is the output taken across the resistor or capacitor?
2. In an *RL* low-pass filter, is the output taken across the resistor or inductor?
3. Give the definition of a low-pass filter.
4. The cutoff frequency occurs at the point where the output voltage drops to what percentage of its maximum value?

17–2 THE CUTOFF FREQUENCY

As you know, the **cutoff frequency** is the frequency at which the output has decreased to 70.7% of its maximum value. Recall that at the cutoff frequency, the voltages across the resistive and reactive elements are equal. The phasor relationship between these voltages and the source voltage is shown in Figure 17–4, which gives a phasor diagram for an *RC* filter. (A phasor diagram for an *RL* filter would be the same with angles reversed.) Since the voltages are equal, the phase angle is 45°. Figure 17–4 also reviews the mathematical solutions for the voltages. Notice that the voltages developed across the resistive and reactive elements are equal and that they are the product of the input voltage and the factor 0.707.

FIGURE 17–4 Phasor diagram and voltage equations for an *RC* filter

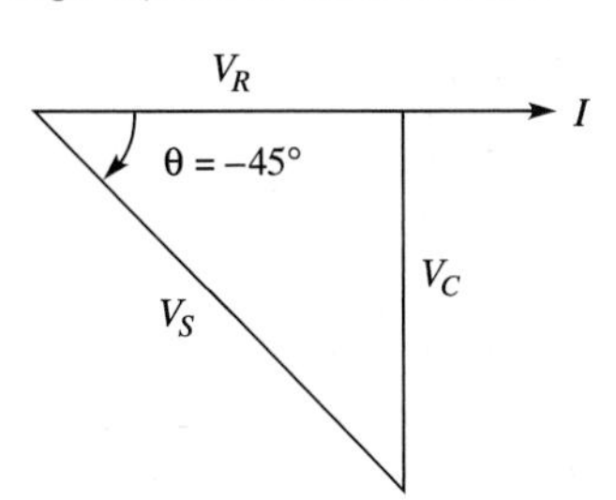

$$\sin\theta = \frac{V_C}{V_S} \qquad \cos\theta = \frac{V_R}{V_S}$$

$$V_C = \sin\theta\, V_S = \sin 45°\, V_S = 0.707\, V_S$$

$$V_R = \cos\theta\, V_S = \cos 45°\, V_S = 0.707\, V_S$$

Question: What is the equation for the cutoff frequency, and how is it derived?

An equation for the cutoff frequency of an *RC* low-pass filter can be derived in the following manner:

1. Begin with the equation for equal voltages across the resistor and the capacitor:

$$V_C = V_R$$

2. Substitute:

$$IX_C = IR \quad \text{and} \quad X_C = R$$

3. Substitute again

$$\frac{1}{2\pi fC} = R$$

4. Equate:

$$2\pi fRC = 1$$

5. Solve:

(17–1)

$$f_c = \frac{1}{2\pi RC}$$

The same rationale may be used in deriving the equation for the cutoff frequency of an *RL* low-pass filter. While the derivation will be left to the student, the equation for the cutoff frequency of an *RL* low-pass filter is as follows:

(17–2)

$$f_c = \frac{R}{2\pi L}$$

EXAMPLE 17–1 An *RC* low-pass filter is composed of a 0.01 μF capacitor and an 8.1 kΩ resistor. Compute the cutoff frequency.

■ Solution:

$$f_c = \frac{1}{2\pi RC} = \frac{1}{2(3.14)(8.1\text{ k}\Omega)(0.01\ \mu\text{F})} = 1.966\text{ kHz}$$

EXAMPLE 17–2 What value of resistor will combine with a 0.022 μF capacitor to produce a cutoff frequency of 72 Hz?

■ Solution:

$$R = \frac{1}{2\pi Cf} = \frac{1}{2(3.14)(0.022\ \mu\text{F})(72\text{ Hz})} = 100.5\text{ k}\Omega$$

EXAMPLE 17–3 What value capacitor will combine with a 10 kΩ resistor to produce a cutoff frequency of 100 Hz?

■ Solution:

$$C = \frac{1}{2\pi Rf}$$

$$= \frac{1}{2(3.14)(10\ \text{k}\Omega)(100\ \text{Hz})} = 0.159\ \mu\text{F}$$

■ SECTION 17–2 REVIEW

1. What is the relationship between the inductive and resistive voltages at the cutoff frequency?
2. In an RC low-pass filter, if the value of the resistor is increased, will the cutoff frequency increase or decrease?
3. Why are the resistive and capacitive voltages equal to the source voltage times 0.707 at the cutoff frequency?

17–3 APPLICATIONS OF LOW-PASS FILTERS

Question: How is a low-pass filter used to smooth the output voltage of a rectifier?

■ Power Supply Filter

A common application of a low-pass filter is in smoothing the output of a dc power supply. Power supplies convert the 120 V ac line voltage into the lower-value dc voltages required by most consumer electronics products. In Figure 17–5(a), the block labeled *rectifier* contains the circuit that converts ac to dc. The devices in the rectifier are known as diodes and they perform the rectification. You will study them later in the electronics curriculum. Notice in the graph of its output that the dc voltage increases from zero to a peak value and then drops back to zero in a periodic manner. Few electronic systems will operate successfully with this type of power!

FIGURE 17–5 Low-pass filter application: Smoothing dc output

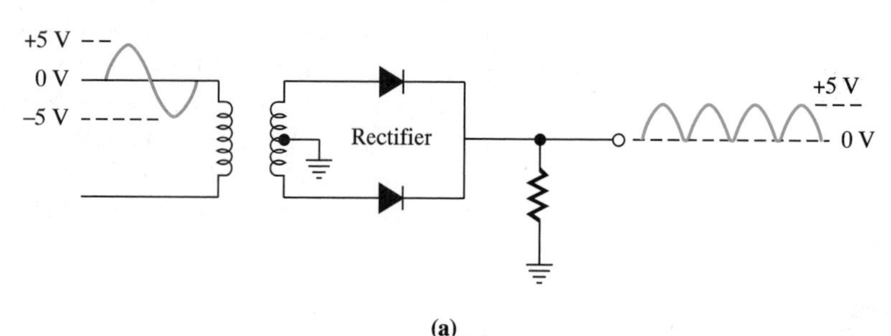

(a)
Full-wave rectifier with pulsating dc output

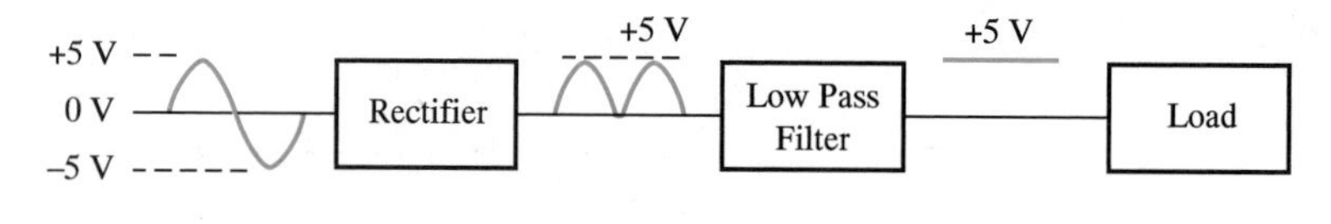

(b)
Low-pass filter between rectifier and load

This *pulsating dc output* is applied to a low-pass filter installed between the rectifier and load. The filtering action of the capacitor reduces the ripple voltage to a level that is within the operating limits of the load. During the period when the output of the rectifier increases from zero to peak, the capacitor charges quickly to this peak value since the time constant through the rectifier circuit is very brief. When the dc drops back to zero, the diodes in the rectifier prevent the capacitor from discharging back through the short time constant path. The capacitor is thus forced to discharge through the long, time-constant path of the load. As a result, it may discharge by only a few tenths of volts. The next dc pulse recharges it to peak very quickly, and the cycle is repeated. Figure 17–5(b) demonstrates this action. Note that the average dc level is now very close to the peak value.

■ Capacitive Bypass

Question: What is a bypass capacitor, and how is its value determined?

If a pulsating direct current flows through a resistor, the voltage across it varies. In some circuits, this voltage variation is objectionable. It can be overcome if a capacitor of appropriate value, known as *bypass capacitor,* is placed in parallel with the resistor as shown in Figure 17–6. In this circuit, the load is driven by dc and ac sources in series, the waveform of which is also shown in Figure 17–6. (You will encounter this type of waveform in amplifiers.) The capacitor charges to the average dc value and then becomes an infinite impedance to the dc. The ac component finds a low-impedance path around the resistor through the capacitive branch. The ac component is said to *bypass* the resistor, and hence, the voltage across the resistor remains constant.

Another way of explaining the bypassing action is to recall the definition of capacitance: the property of a circuit that opposes a change in voltage. The voltage across a capacitor cannot change, regardless of any variation in the source, until the capacitor either loses or gains charge. This loss or gain cannot happen instantaneously, since current must flow when a capacitor charges or discharges. If a large-value capacitor is used, the resulting time constant is relatively long. This long time constant prevents the capacitor from charging or discharging any appreciable amount, resulting in a nearly constant voltage across its plates and, in turn, across the parallel resistor.

The two explanations of bypassing have one thing in common: The value of the capacitor must be relatively large. This large value is necessary in order to present a low reactance to the ac component and to have a long time constant so that it cannot follow the amplitude variations. One other consideration that must be taken into account is the frequency range of the ac component to be bypassed. For most practical applications, successful bypassing can be achieved if the capacitive reactance is *one-tenth or less* of the resistance that it must bypass. The value of the capacitor can then be computed using the lowest-frequency component expected in the signal.

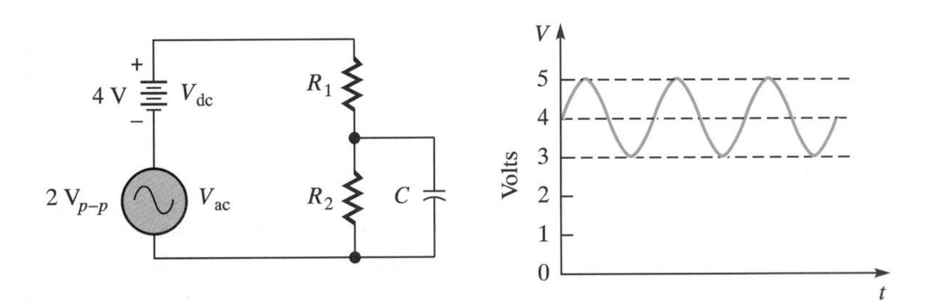

FIGURE 17–6 Low-pass filter application: Bypass capacitor circuit, in which voltage across R_1 is pulsating dc and voltage across R_2 is essentially pure dc

EXAMPLE 17–4 A signal with a frequency range of 500Hz through 3400 Hz must be bypassed around a 330 Ω resistor. Compute the value of the capacitor required.

■ Solution:

1. Transpose the equation for capacitive reactance in order to solve for capacitance:

$$C = \frac{1}{2\pi f X_C}$$

2. Substitute the lower frequency limit, 500 Hz, for f and one-tenth the resistance value for X_C:

$$C = \frac{1}{(2\pi)(500 \text{ Hz})(33 \ \Omega)} = 9.64 \ \mu\text{F}$$

3. Select a standard value of 10 μF electrolytic capacitor.

■ Power Line Filter

The ac power line acts much like an antenna, picking up radio frequency (RF) interference from electric motors, florescent lights, RF equipment, and other sources. If this interference gets into the power supply of a radio or television receiver, the result may be objectionable. Figure 17–7 shows the arrangement for a low-pass power line filter. The series inductors offer a low reactance to the 60 Hz power frequency but a high reactance to RF. The parallel capacitor offers a high reactance to the power line, but its low reactance to radio frequencies shunts them around the power supply.

■ SECTION 17–3 REVIEW

1. In low-pass filters, are capacitors in series or in parallel with the load?
2. What characteristic of a capacitor is taken advantage of in filtering the output of a rectifier?
3. The reactance of a bypass capacitor must be what value relative to the resistor that it bypasses?

17–4 HIGH-PASS FILTER

A **high-pass filter** is a circuit that passes all frequencies above the cutoff frequency and blocks all those below. As was the case with the low-pass filter, the cutoff frequency is that at which the output voltage decreases to 70.7% of its maximum value. The rationale for computing the cutoff frequency is exactly the same as for low-pass filters. Thus, the same equations used in computing the cutoff frequency of low-pass filters apply to high-pass filters also.

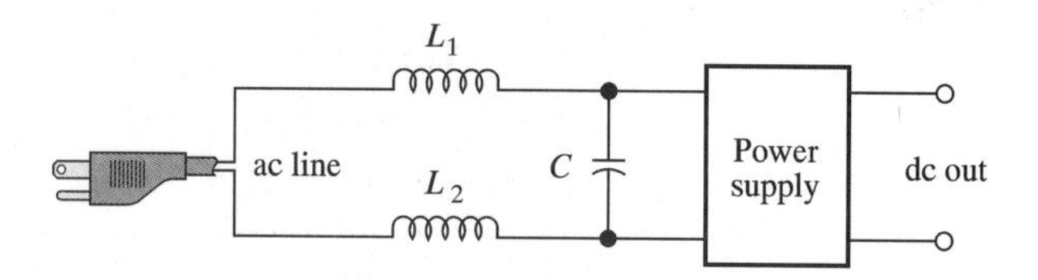

FIGURE 17–7 Low-pass filter application: Power line filter

■ *RC* High-Pass Filter

Question: What is an RC *high-pass filter, and how does it operate?*

Figure 17–8(a) shows an *RC* high-pass filter. Note that it is a series *RC* circuit with the output taken across the resistor. At 0 Hz, the reactance of the capacitor is infinite, and the circuit current is zero. Thus, the output voltage, taken across the resistor is zero: $V = IR$. As the frequency is increased, the reactance decreases. Impedance decreases, resulting in an increase in circuit current and output voltage. When the output voltage becomes 70.7% of its maximum value, the cutoff frequency is reached, and all frequencies above this point are passed. Table 17–2 contains the output voltage computed for a series of frequencies. The graph of this data is shown in Figure 17–8(b), which is the pass characteristic for a high-pass filter. The cutoff frequency can be computed using the same equation as for the *RC* low-pass filter.

■ *RL* High-Pass Filter

Question: What is an RL *high-pass filter, and how does it operate?*

An *RL* high-pass filter is shown in Figure 17–9. Notice that it is a series *RL* circuit with the output taken across the inductor. At 0 Hz, the inductor is practically a short circuit, and no voltage is developed across it or the parallel load: $V_L = IX_L$. As the frequency is

FIGURE 17–8 *RC* high-pass filter

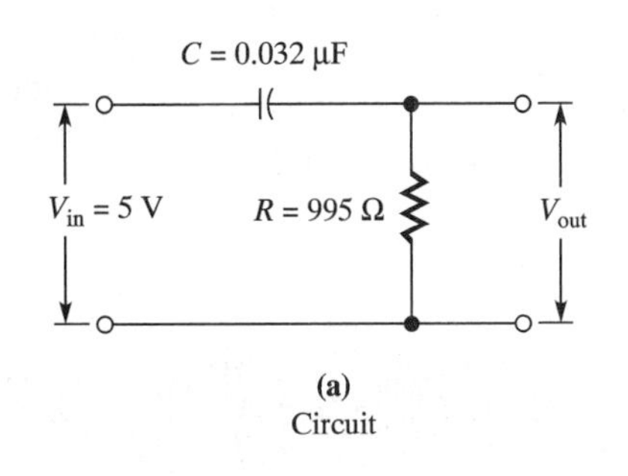

(a) Circuit

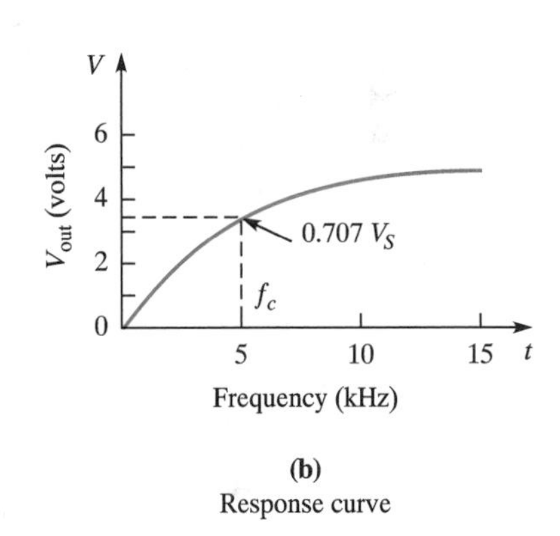

(b) Response curve

FIGURE 17–9 *RL* high-pass filter circuit

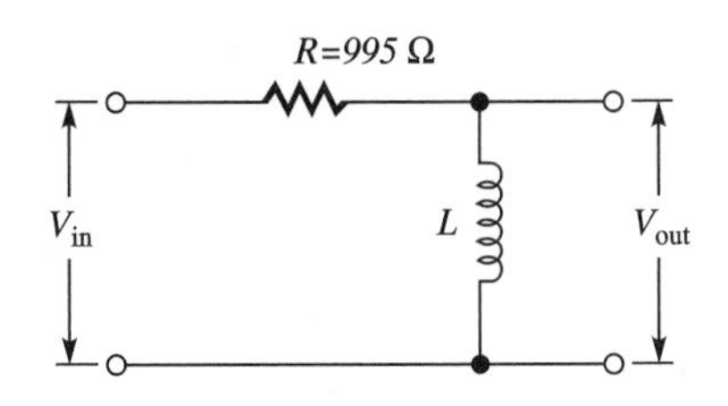

■ **TABLE 17–2 Output voltages for various frequencies of an *RC* high-pass filter**

Frequency	X_c	Z	I	$V_{out} = IR$
0	∞	∞	0	0
1 kHz	4976 Ω	5074 Ω	0.985 mA	0.98 V
2 kHz	2488 Ω	2679 Ω	1.87 mA	1.86 V
3 kHz	1659 Ω	1934 Ω	2.59 mA	2.58 V
4 kHz	1244 Ω	1592 Ω	3.14 mA	3.12 V
5 kHz	995 Ω	1407 Ω	3.55 mA	3.53 V
6 kHz	829 Ω	1295 Ω	3.86 mA	3.84 V
7 kHz	710 Ω	1222 Ω	4.1 mA	4.1 V
10 kHz	497 Ω	1112 Ω	4.5 mA	4.48 V
20 kHz	248 Ω	1025 Ω	4.88 mA	4.86 V
30 kHz	165 Ω	1008 Ω	4.96 mA	4.94 V

Note: $R = 995\ \Omega$.

increased, the reactance of the inductor increases, as does the voltage drop across the inductor and parallel load. The cutoff frequency is reached when the load voltage is 70.7% of its maximum value. All frequencies above the cutoff are passed, while all those below are blocked. The pass characteristic is the same as that shown in Figure 17–8(b). The cutoff frequency can be computed using the same equation as for the *RL* low-pass filter.

EXAMPLE 17–5 An *RL* high-pass filter is composed of a 100 Ω resistor and a 100 mH inductor. Compute the cutoff frequency.

■ Solution:

$$f_c = \frac{R}{2\pi L}$$

$$= \frac{100\ \Omega}{2(3.14)(100\ \text{mH})} = 159\ \text{Hz}$$

■ Other High-Pass Filters

Figure 17–10 shows combinations of inductors and capacitors used as high-pass filters. Notice that now the capacitors are series elements and inductors are parallel elements. The reader should verify the operation of these filters using the same rationale as was applied in the previous section on *RC* and *RL* low-pass filters.

■ SECTION 17–4 REVIEW

1. In an *RL* high-pass filter, is the inductor in series or in parallel with the load?
2. In an *RC* high-pass filter, is the resistor in series or in parallel with the load?
3. A high-pass filter consists of a 100 Ω resistor and a 1 μF capacitor. Compute the cutoff frequency.
4. What value capacitor combines with a 100 Ω resistor to form a high-pass filter with a cutoff frequency of 10 kHz?

17–5 APPLICATIONS OF HIGH-PASS FILTERS

Question: How is a high-pass filter used to pass the ac component of a signal and block the dc?

■ Coupling Circuit

A common application of a high-pass filter is a circuit that couples a signal into an amplifier. (Note that it is not the intention to present the operation of transistors or amplifiers in this section, but merely to show how an ac signal can be coupled into an amplifier while the dc component is blocked!) Figure 17–11(a) is the schematic of an amplifier being fed by a source containing both an ac and a dc component. Resistors R_1 and R_2 form a voltage divider that produces the dc bias voltage, or operating voltage, for the amplifier. The voltage from point *A* to ground is critical to the operation of the transistor and must be maintained or the amplifier's performance will be degraded.

Figure 17–11(b) shows the ac equivalent circuit of the amplifier input. Notice that C_1 and R_{IN} form a high-pass filter. If capacitor C_1 is made large enough, the ac portion of the signal encounters a low reactance and be developed across R_{in}, which for practical purposes is the input of the amplifier. The dc component, on the other hand, encounters an infinite impedance in C_1 and is eliminated. Thus, the bias voltage will be maintained.

FIGURE 17–10 Types of high-pass filters

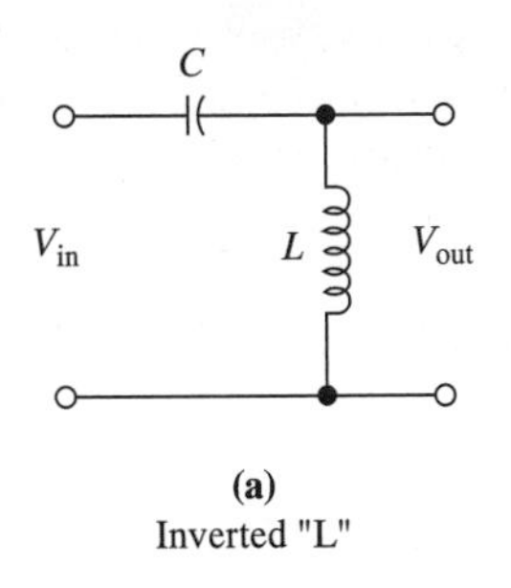

(a)
Inverted "L"

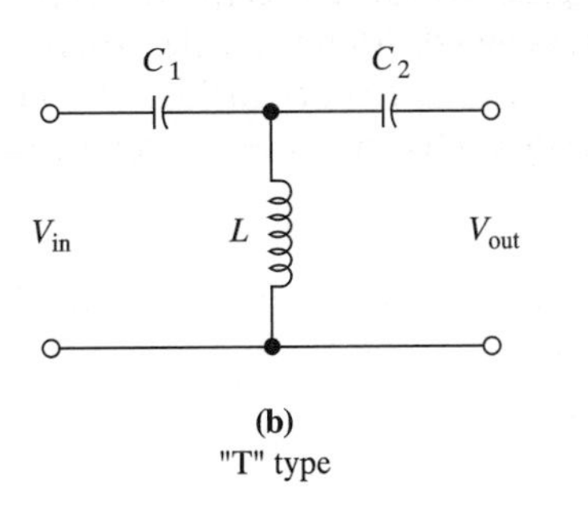

(b)
"T" type

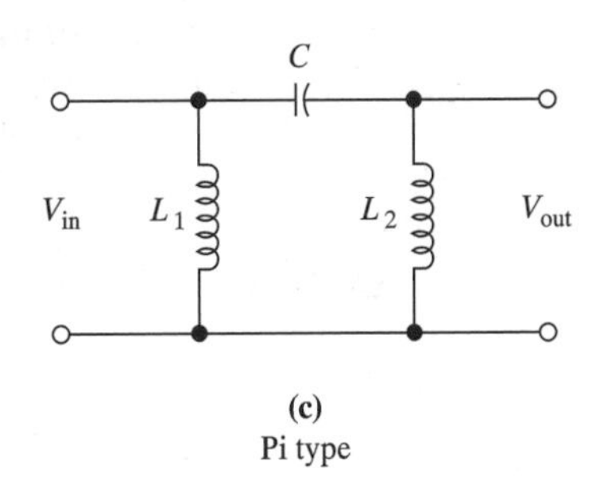

(c)
Pi type

FIGURE 17–11 High-pass filter application: Coupling circuit

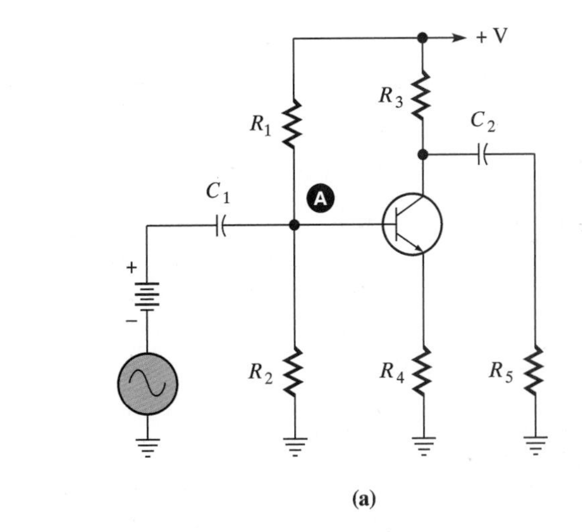

(a)

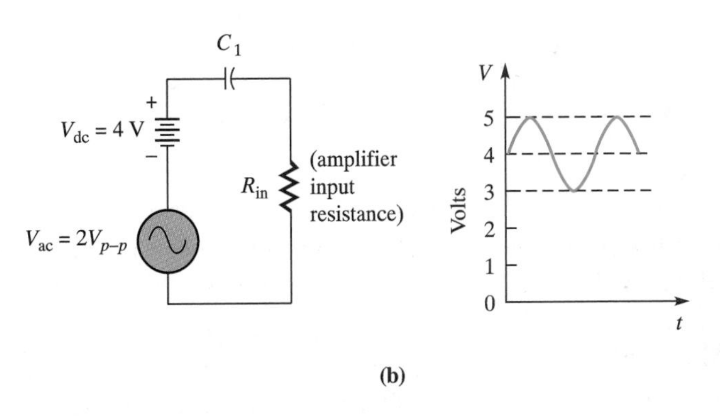

(b)

■ Signal Injection

Amplification in an electronic system is not normally accomplished in a single amplifier, but with what is known as a *chain* of amplifiers, as shown in Figure 17–12. A fault occurring in any one of the amplifiers, each of which is known as a stage of amplification, would cause the output to be zero. The fault, if in fact a single fault exists, must be localized to a specific stage. As the figure shows, test signals are injected at various points, and the output or lack of output is noted. In this way, the stage with proper input, but no output, is determined.

As Figure 17–12 shows, a capacitor must be inserted between the signal generator and the circuit under test. It forms a high-pass filter with the input resistance of the amplifier. This is necessary in order to block any dc component that may be part of the test signal.

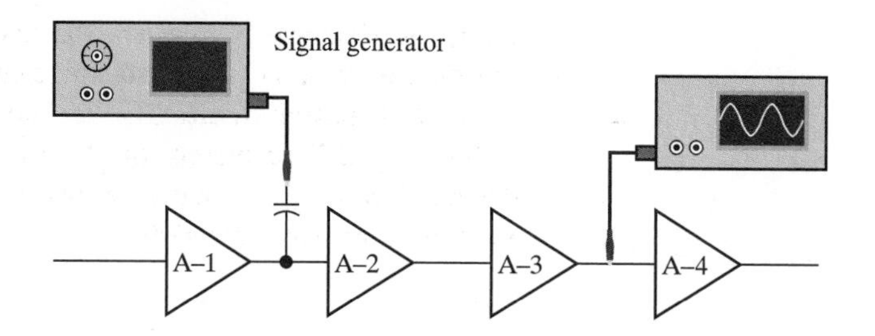

FIGURE 17–12 High-pass filter application: Signal injection to test amplifier chain. A capacitor is inserted between the signal generator and amplifier A-2 to block any dc component, thus preventing it from becoming part of the test signal viewed at A-4.

A dc level reaching the input of the amplifier would upset the bias, or operating voltage, and degrade amplifier performance. A 0.1 μF, or larger, capacitor could be used for audio frequencies, while a 0.001 μF would suffice for radio frequencies.

SECTION 17–5 REVIEW

1. How is a high-pass filter used as a coupling circuit between amplifier stages?
2. If the frequency coupled is relatively low, will the capacitor be a relatively low or high value?
3. When injecting an RF test signal into an amplifier, what value capacitor should be used?

17–6 HALF-POWER POINT

Question: What is meant by the half-power point?

At the cutoff frequency, with the load voltage having decreased to 70.7% of its maximum value, the load power will have decreased by 50%, as demonstrated in the following example.

EXAMPLE 17–6a

For the circuit of Figure 17–13, compute the power developed in the load at maximum load voltage and at the cutoff frequency.

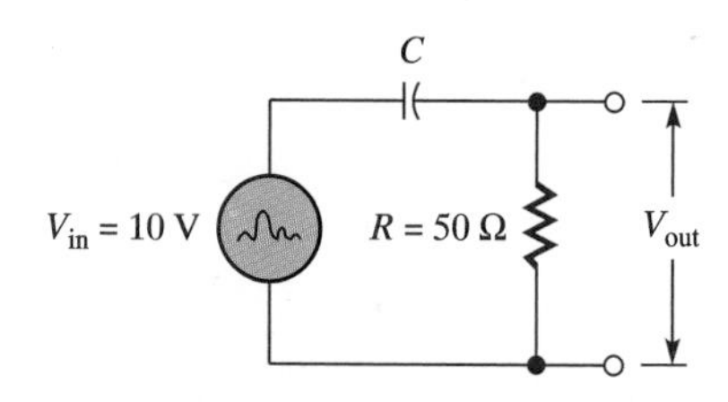

FIGURE 17–13 Circuit for Example 17–6a

■ Solution:

As you will notice, the circuit of Figure 17–13 is a high-pass filter. At frequencies well above the cutoff frequency, the load voltage is virtually the entire source voltage. So, the maximum power developed in the load is

$$P = \frac{V^2}{R}$$

$$= \frac{(10\text{ V})^2}{50\ \Omega} = 2\text{ W}$$

At the cutoff frequency, the load voltage is 70.7% of the source voltage, or 7.07 V. So, at the cutoff frequency, the power developed in the load is

$$P = \frac{V^2}{R}$$

$$= \frac{(7.07\text{ V})^2}{50\ \Omega} = 1\text{ W}$$

Notice in Example 17–6a that the load power has in fact decreased by 50%. For this reason, the point on the characteristic curve at which the cutoff frequency occurs is known as the **half-power point.** The half-power point may also be known as "3 dB down," "minus 3 dB," or merely the "3 dB" point. The term "dB" is the abbreviation for decibel. The decibel is a unit of measure that can be used to indicate the gain or loss of a filter or other electronic systems. It will be discussed in the next section.

SECTION 17–6 REVIEW

1. What are the other names given the half-power points?
2. When the voltage has decreased to 70.7% of maximum, by what percentage has the power decreased?

17–7 THE DECIBEL

Gain

Question: What is gain, and how is it computed?

The **gain** of an electronic system, which may be expressed in terms of power, voltage, or current, is equal to the output quantity divided by the input quantity. Gain is given the symbol A (amplification) with a subscript indicating which type: voltage v, power p, or current i. For example, if an amplifier has an input power of 300 mW and an output power of 1.8 W the power gain would be

$$A_P = \frac{P_{out}}{P_{in}}$$

$$= \frac{1.8 \text{ W}}{300 \text{ mW}} = 6$$

The figure for gain is a *dimensionless* number, which means the output power will be 6 times larger than the input power. The same can be done using the input/output voltage or input/output current. The equations for each are as follows:

(17–3) $$A_p = \frac{P_{out}}{P_{in}}$$

(17–4) $$A_v = \frac{V_{out}}{V_{in}}$$

(17–5) $$A_i = \frac{I_{out}}{I_{in}}$$

EXAMPLE 17–6b The input voltage of a system is 135 mV. The output is measured and found to be 6 V. What is the voltage gain of the system?

Solution:

$$A_v = \frac{V_{out}}{V_{in}}$$

$$= \frac{6 \text{ V}}{135 \text{ mV}} = 44.4$$

EXAMPLE 17–7 An amplifier has a power input of 5 mW. If the power gain is 12, what is the output power?

■ Solution:

$$A_p = \frac{P_{out}}{P_{in}}$$

Thus,

$$P_{out} = P_{in}A_p$$
$$= (5 \text{ mV})(12) = 60 \text{ mW}$$

EXAMPLE 17–8 An amplifier has a voltage gain of 15. If an output voltage of 8 V is required, what input voltage is required?

■ Solution:

$$A_v = \frac{V_{out}}{V_{in}} \quad \text{and} \quad V_{out} = V_{in}A$$

$$V_{in} = \frac{V_{out}}{A}$$
$$= \frac{8 \text{ V}}{15} = 533 \text{ mV}$$

■ **The Bel**

Question: What is a bel, and how is it used to express gain?

Another way of expressing gain is through use of the common logarithm of the computed ratio. If the common, or base-10, logarithm of the ratio is taken, the gain is now expressed in **bels,** named for Alexander Graham Bell, inventor of the telephone and early pioneer in radio. For example, a gain figure of 6 is expressed in bels as follows:

$$A_P = \log_{10} 6$$
$$= 0.77815 \text{ bel}$$

Note that this number is quite small, indicating that the bel is a rather large unit. For this reason the **decibel,** or one-tenth of a bel, is used in order to get a more manageable number. Decibel is abbreviated dB.

$$A_P = 10(\log_{10} 6)$$
$$= 10(0.77815) = 7.7815 \text{ dB}$$

EXAMPLE 17–9 The input power of a filter is 68 mW and the output power is 34 mW. Compute the power gain in dB.

■ Solution:

$$A_p = 10\log_{10} \frac{P_{out}}{P_{in}}$$

$$= 10\log_{10} \frac{34 \times 10^{-3}}{68 \times 10^{-3}}$$

$$= 10\log_{10} 0.5 = 10\,(-0.3) = -3 \text{ dB}$$

Notice that the ratio of power out to power in is 0.5 and the gain is -3 dB; hence, the use of "minus 3 dB" or "3 dB down" to designate the half-power point. If the power out of a system were to double, the gain would then be $+3$ dB. Verify this by substituting appropriate values in the equation.

NOTE

In Example 17–9, even though there is a "loss" in power, the factor arrived at is still called "gain." Use of this term avoids a requirement for another set of symbols to indicate loss. Thus, a gain figure of less than one, or a negative value of dB, indicates a loss in power.

Question: Can the gain ratios for voltage and current also be expressed in dB?

Most of the measurements made in electronics are not power but voltage. For this reason, it is convenient to have an equation for gain in decibels containing output and input voltages. With one condition and one minor modification, the equation for power in dB can be made to accommodate voltage. The condition is that the input and output voltages must be taken across identical resistances. The modification is that the multiplication factor must be 20 rather than 10. Beginning with the basic equation for power gain, an equation for voltage gain in dB is derived as follows:

$$A_P = 10\log_{10} \frac{P_{out}}{P_{in}}$$

Substituting gives

$$A_p = 10\log_{10} \frac{V_{out}^2/R_{out}}{V_{in}^2/R_{in}}$$

and

$$A_p = 10\log_{10} \frac{V_{out}^2 R_{in}}{R_{out} V_{in}^2}$$

Since $R_{out} = R_{in}$,

$$A_p = 10\log_{10} \frac{V_{out}^2}{V_{in}^2}$$

Mathematically, the log of a quantity raised to a power is equal to that power times the log of the quantity. Thus,

$$A_v = 20\log_{10} \frac{V_{out}}{V_{in}}$$

EXAMPLE 17–10 The input voltage of a system is 300 mV and the output voltage is 600 mV. Compute the voltage gain in dB.

■ Solution:

$$A = 20\log_{10} \frac{V_{out}}{V_{in}}$$

$$= 20\log_{10} \frac{600 \text{ mV}}{300 \text{ mV}}$$

$$= 20\log_{10} 2$$

$$= 20(0.301) = 6 \text{ dB}$$

In this example, the voltage has doubled. Notice that the voltage gain is 6 dB. If the voltages were halved, the gain would be -6 dB.

EXAMPLE 17–11 The input voltage of a system is 10 V. If the voltage gain is a -3 dB, what is the value of the output voltage?

■ Solution:

$$A_v = 20\log_{10} \frac{V_{out}}{V_{in}}$$

$$-3 = 20\log_{10} \frac{V_{out}}{10}$$

$$-0.15 = \log_{10} \frac{V_{out}}{10}$$

Then,

$$\text{antilog } -0.15 = \frac{V_{out}}{10}$$

$$0.707 = \frac{V_{out}}{10}$$

$$V_{out} = (0.707 \text{ V})10 = 7.07 \text{ V}$$

In Example 17–11, notice that the output voltage has decreased from 10 V to 7.07 V. Thus, the output voltage had decreased to 70.7% of the input for a gain of -3dB. This result verifies the definition given previously for the cutoff frequency and the half-power point. Notice further that if any two of the quantities are known, the third can be computed!

The condition and modification of the dB equation for power can also make it compatible with a current ratio:

$$A_i = 20\log \frac{I_{out}}{I_{in}}$$

Thus, the equations for gain in decibels are as follows:

(17–6)
$$A_p = 10\log \frac{P_{out}}{P_{in}}$$

(17–7)
$$A_v = 20\log \frac{V_{out}}{V_{in}}$$

(17–8) $$A_i = 20\log \frac{I_{out}}{I_{in}}$$

EXAMPLE 17–12 The output current of an amplifier must be 60 mA. If the current gain is 5 db, what input current is required?

■ Solution:

$$A_i = 20\log \frac{I_{out}}{I_{in}}$$

$$5 = 20\log \frac{60 \text{ mA}}{I}$$

$$0.25 = \log \frac{60 \text{ mA}}{I}$$

Then,

$$\text{antilog } 0.25 = 1.778$$

and

$$1.778 = \frac{60 \text{ mA}}{I}$$

$$1.778I = 60 \text{ mA}$$

$$I = \frac{60 \text{ mA}}{1.778} = 33.75 \text{ mA}$$

■ SECTION 17–7 REVIEW

1. Is a decibel a linear or logarithmic ratio?
2. What is the definition of gain?
3. What is the stipulation placed upon the input and output resistances in the gain equations for voltage and current?
4. A system has an input power of 150 mW and an output power of 2.5 W. Compute the power gain in dB.
5. A system has an input voltage of 100 mV. If its voltage gain is 18 dB, what is the output voltage?
6. A system has an output power of 5 W. If the power gain is 4 dB, what is the input power?

17–8 BAND-PASS FILTERS

A **band-pass filter** is defined as one that passes all frequencies between an upper and a lower cutoff frequency and rejects all others. As shown in Chapter 16, the response of an *RLC* circuit is greatest at the resonant frequency. As demonstrated, this response drops both above and below resonance—not precipitously, but gradually in a manner determined by the circuit Q. Thus, a resonant circuit can be made to pass a given band of frequencies. Both series and parallel resonant circuits may be used in band-pass filter applications. The important characteristics are their center, or resonant, frequency and their bandwidth.

Series Resonant Band-Pass Filter

Question: What is the theory of operation of a series resonant band-pass filter?

Figure 17–14(a) shows a series *LC* circuit used as a band-pass filter. Notice that the *LC* circuit is in series with the load. The impedance of the *LC* circuit determines the current through the load and hence the output voltage. The impedance of a series circuit at resonance in minimum, making the current maximum. Thus, at the resonant frequency, the output voltage is maximum. Above or below resonance, the impedance increases, causing a decrease in circuit current and output voltage. The half-power points and cutoff frequencies are reached when the output voltage has decreased to 70.7% of its maximum value.

Table 17–3 contains data of the output voltage over a range of frequencies centered at resonance. Figure 17–14(b) shows the graph of these data, which is the response curve for a band-pass filter. Note the upper and lower cutoff frequencies, f_2 and f_1, which occur at the half-power points. As previously stated, the filter does not cutoff abruptly but gradually. Although the cutoff is not complete at the half-power points, those frequencies above and below are greatly attenuated. Any frequencies greater than 3 dB below the peak at resonance are not considered to be passed. The band of frequencies passed are those between the upper and lower cutoff frequencies. The bandwidth is the difference of the upper and lower cutoff frequencies.

EXAMPLE 17–12

For the response curve shown in Figure 17–14(b), determine the pass band and the bandwidth.

FIGURE 17–14 Series resonant band-pass filter

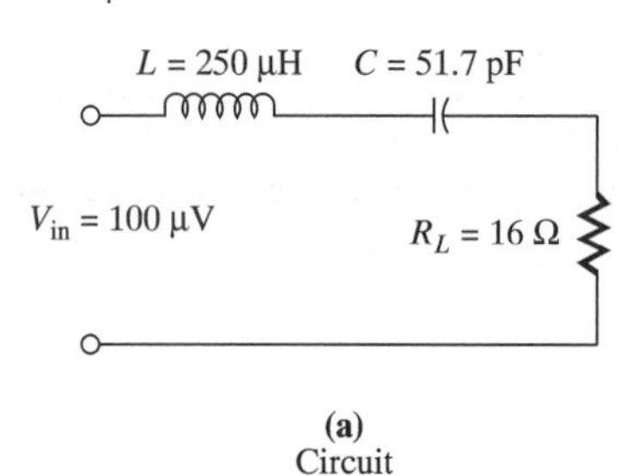

(a)
Circuit

(b)
Response curve

Solution:

The upper and lower cutoff frequencies are 1405 kHz and 1395 kHz, respectively. Therefore, the pass band is all frequencies between 1395 kHz and 1405 kHz. The bandwidth, which is abbreviated BW (sometimes Δf), is the difference of the upper and lower cutoff frequencies. Thus,

TABLE 17–3 Output voltages for various frequencies of a series resonant band-pass filter

Frequency	X_L	X_c	Z	I	$V_{out} = IR$
1392 kHz	2185 Ω	2212 Ω	31.4 Ω	3.18 μA	50.8 μV
1394 kHz	2189 Ω	2209 Ω	25.6 Ω	3.9 μA	62.6 μV
1396 kHz	2192 Ω	2206 Ω	21.3 Ω	4.69 μA	75 μV
1398 kHz	2195 Ω	2203 Ω	17.9 Ω	5.59 μA	89.4 μV
1400 kHz	2198 Ω	2198 Ω	16 Ω	6.25 μA	100 μV
1402 kHz	2201 Ω	2196 Ω	16.8 Ω	5.95 μA	95.2 μV
1404 kHz	2204 Ω	2193 Ω	19.4 Ω	5.15 μA	82.4 μV
1406 kHz	2207 Ω	2190 Ω	23.3 Ω	4.29 μA	68.6 μV
1408 kHz	2211 Ω	2187 Ω	28.8 Ω	3.47 μA	55.5 μV

$$BW = f_2 - f_1$$
$$= 1405 \text{ kHz} - 1395 \text{ kHz} = 10 \text{ kHz}$$

Parallel Resonant Band-Pass Filter

A parallel *LC* circuit, when connected as shown in Figure 17–15, operates as a band-pass filter. Notice that now the output is taken across the filter. At resonance, the parallel *LC* circuit has maximum impedance, resulting in maximum voltage being developed across the output. Above or below resonance, the impedance and consequently the output voltage decreases. The half-power points and cutoff frequencies are reached when the output voltage has decreased to 70.7% of its maximum value. The response curve is the same as that in Figure 17–14(b).

Bandwidth and Selectivity

A band-pass filter is used to select a band of frequencies in the frequency spectrum and reject all others. If a narrow band of frequencies is passed, the filter is said to be *highly selective*. The selectivity of the filter decreases as the bandwidth widens. Later, when applications are considered, it will be shown that the issue of selectivity is very important in radio and television receivers.

As previously stated, the bandwidth of a filter is dependent upon the Q-factor of the circuit. The Q-factor is in turn dependent upon the resistance of the circuit. The Q-factor is the ratio of the inductive reactance, at resonance, to the total circuit resistance. A high-Q circuit has a narrow bandwidth, while a low-Q circuit has a wider bandwidth.

Figure 17–16 shows the response curve of a band-pass filter with several different bandwidths. The narrow bandwidth has a sharper response curve than the wider one. It also has a greater selectivity. As the bandwidth widens, notice that the response curve is not as steep, and there is less selectivity. The circuit application determines what bandwidth and what degree of selectivity is required.

EXAMPLE 17–13 For the series *RLC* circuit shown in Figure 17–17, compute the following: (a) the resonant frequency, (b) the Q-factor, (c) the bandwidth, and (d) the upper and lower cutoff frequencies.

FIGURE 17–15 Parallel resonant band-pass filter

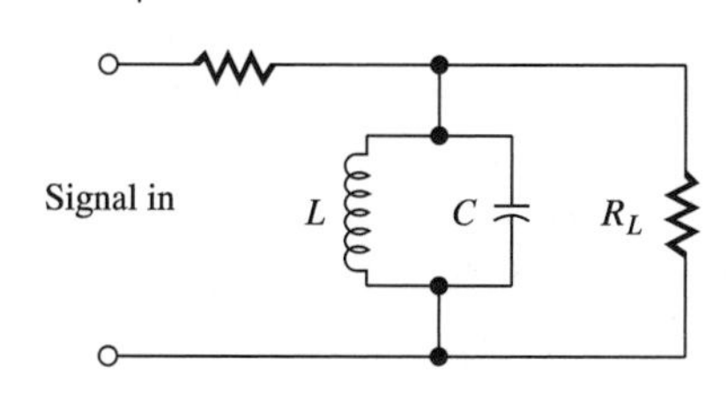

FIGURE 17–16 Bandwidths in a band-pass filter

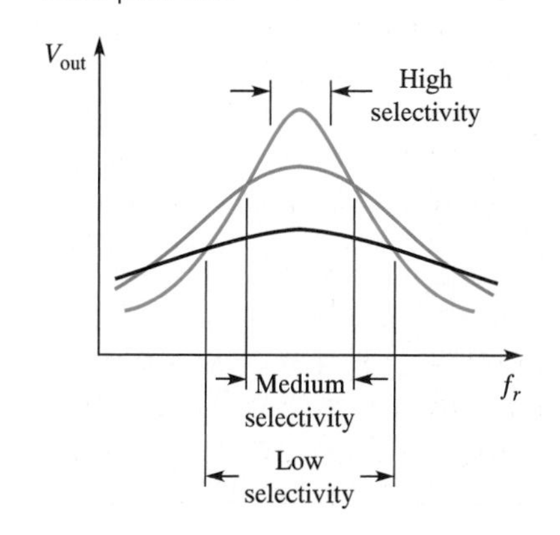

FIGURE 17–17 Circuit for Example 17–13

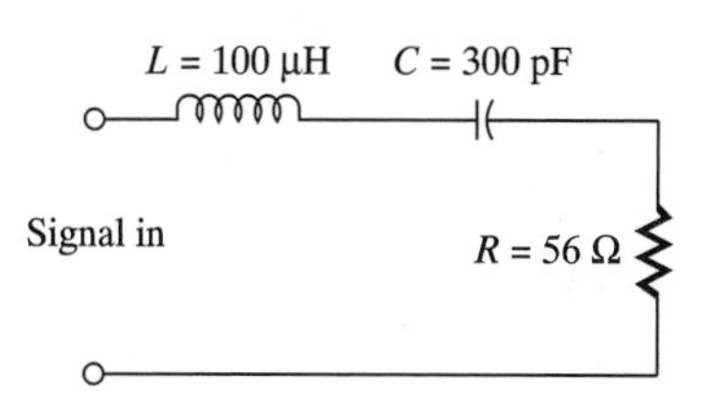

■ Solution:

a.
$$f_r = \frac{1}{2\pi\sqrt{LC}}$$
$$= \frac{1}{2(3.14)\sqrt{(100\ \mu\text{H})(300\ \text{pF})}}$$
$$= 919.3\ \text{kHz}$$

b.
$$X_L = 2\pi fL$$
$$= 2(3.14)(919.3\ \text{kHz})(100\ \mu\text{H}) = 577\ \Omega$$
$$Q = \frac{X_L}{R}$$
$$= \frac{577\ \Omega}{56\ \Omega} = 10.3$$

c.
$$\text{BW} = \frac{f_r}{Q}$$
$$= \frac{919.3\ \text{kHz}}{10.3} = 89.3\ \text{kHz}$$

d.
$$f_2 = f_r + \frac{\text{BW}}{2}$$
$$= 919.3\ \text{kHz} + \frac{89.3\ \text{kHz}}{2} = 964\ \text{kHz}$$
$$f_1 = f_r - \frac{\text{BW}}{2}$$
$$= 919.3\ \text{kHz} - \frac{89.3\ \text{kHz}}{2} = 874.7\ \text{kHz}$$

■ SECTION 17–8 REVIEW

1. Is a parallel resonant band-pass filter placed in series or in parallel with the load?
2. Is a series resonant band-pass filter placed in series or in parallel with the load.
3. What determines the selectivity of a band-pass filter?
4. The upper and lower cutoff frequencies occur when the output voltage is how many dB down from maximum?
5. How is the bandwidth of a band-pass filter computed?

17–9 APPLICATIONS OF BAND-PASS FILTERS

■ Radio Signal Selection

One application of a band-pass filter is in the tuning of a radio receiver. For example, in AM (amplitude modulation) commercial radio broadcasting, each station is assigned a 10 kHz channel, centered upon a carrier frequency over which its intelligence is broadcast. These station assignments begin at 540 kHz and continue in 10 kHz steps through 1600 kHz. One such channel, of the many impinging upon the antenna, must be selected and all others blocked.

Figure 17–18 shows a filter used in this application. Notice that the capacitance is variable for tuning a particular station. When it is adjusted to 133.5 pF, the tank circuit is resonant at approximately 1320 kHz. The radio station broadcasting on this frequency will be selected. In an application such as this, the bandwidth is critical. It must be wide enough to pass all the intelligence of the station's broadcast but narrow enough to prevent signals from adjoining channels being passed. If too narrow, the signal reproduced, es-

FIGURE 17–18 Band-pass filter application: Selecting AM radio stations

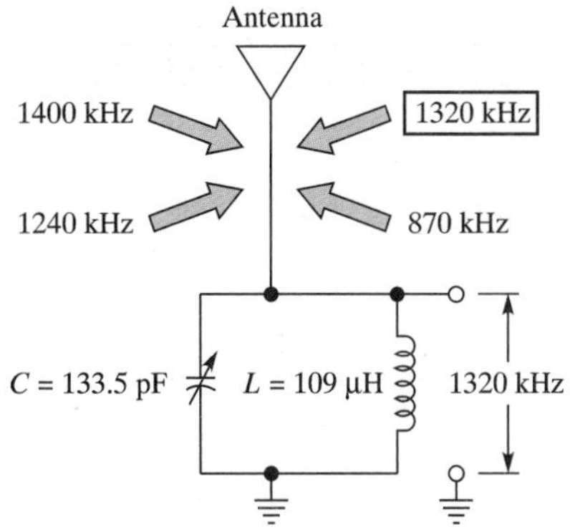

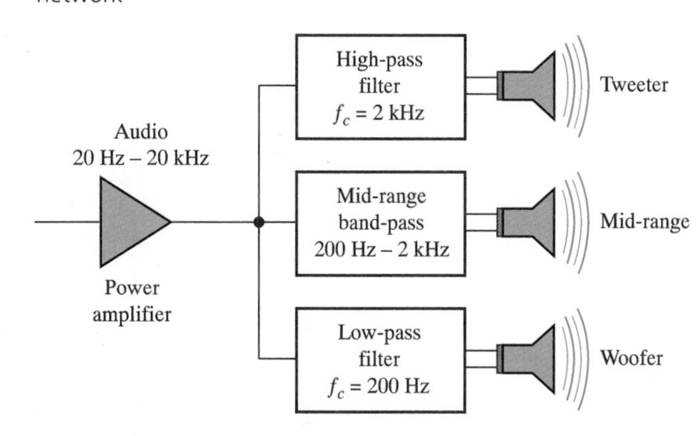

FIGURE 17–19 High-pass filter application: Speaker crossover network

pecially music, will have poor fidelity. If too wide, signals passed from adjoining channels will cause chaotic listening.

■ Speaker Crossover Network

A single loud speaker cannot reproduce the entire range of audio frequencies with equal efficiency. In a quality audio system, there are often three separate loudspeakers: a small one, known as a *tweeter,* for reproducing the high frequencies; a large one, known as a *woofer,* for reproducing the low frequencies; and a midsize one that reproduces the *mid-range frequencies.* A system of filters, known as a *crossover network,* routes the appropriate signal frequencies to each loudspeaker.

A block diagram of this arrangement is shown in Figure 17–19. The signal from the amplifier contains frequencies over a band from 20 Hz to 20 kHz. Notice that signal frequencies above 2 kHz are routed to the tweeter through a high-pass filter, while those below 200 Hz are routed to the woofer through a low-pass filter. The midrange frequencies are routed to the midrange speaker through a band-pass filter with appropriate bandwidth.

■ SECTION 17–9 REVIEW

1. List two applications of band-pass filters.
2. Is bandwidth important in these applications? Explain.

17–10 BAND-REJECT FILTER

A **band-reject filter** blocks a band of frequencies between an upper and lower cutoff frequency and passes all others. If this band of frequencies is very narrow, the filter may be referred to as a *notch filter.* As in the case of band-pass filters, they are implemented with LC resonant circuits.

Figure 17–20(a) shows a series LC circuit connected as a band-reject filter. The output voltage is that developed by the filter. At the resonant frequency, the impedance of the filter is low, thus producing a low output voltage. Above or below resonance, the impedance and output voltage increases. When the output voltage is equal to 70.7% of its maximum value, the half-power and cutoff frequencies are achieved. Table 17–4 contains data of the output voltage computed over a band of frequencies centered at resonance.

FIGURE 17–20 Series resonant band-reject filter

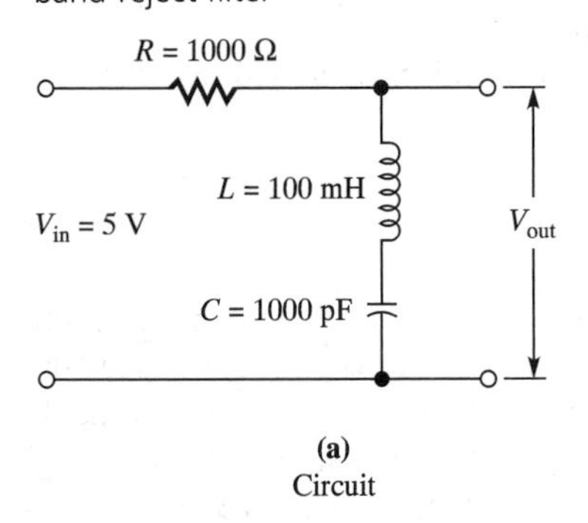

(a)
Circuit

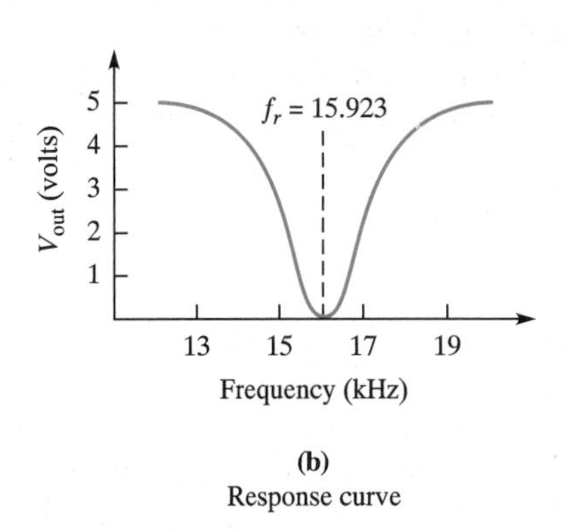

(b)
Response curve

FIGURE 17–21 Parallel resonant band-reject filter

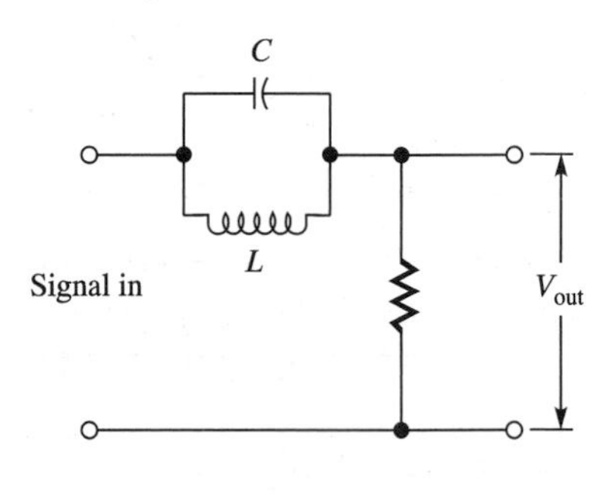

■ **TABLE 17–4 Output voltages for various frequencies of a series resonant band-reject filter**

Frequency	X_L	X_c	Z	I	$V_{out} = IZ$
14 kHz	8792 Ω	11373 Ω	2768 Ω	1.81 mA	4.66 V
14.5 kHz	9106 Ω	10981 Ω	2125 Ω	2.35 mA	4.41 V
15 kHz	9420 Ω	10615 Ω	1558 Ω	3.21 mA	3.83 V
15.5 kHz	9734 Ω	10273 Ω	1136 Ω	4.4 mA	2.37 V
15.923 kHz	10000 Ω	10000 Ω	1000 Ω	5 mA	0 V
16.5 kHz	10367 Ω	9650 Ω	1227 Ω	4.07 mA	2.89 V
17 kHz	10676 Ω	9366 Ω	1648 Ω	3.03 mA	3.97 V
17.5 kHz	10990 Ω	9099 Ω	2139 Ω	2.34 mA	4.44 V
18 kHz	11304 Ω	8846 Ω	2650 Ω	1.89 mA	4.65 V

Figure 17–20(b) is the graph of this data, which is the response curve of a band-reject filter. The bandwidth and cutoff frequencies are computed in a like manner to those of the band-pass filter. Notice now, however, that the frequencies between the half-power points are rejected rather than passed.

■ Parallel Resonant Band-Reject Filter

Figure 17–21 shows a parallel resonant band-reject filter. Notice that now the filter is in series with the load. The output voltage is determined by the amount of current permitted to flow by the filter. At the resonant frequency, the impedance of the filter is high, and the current and output voltage is low. Above and below resonance, the impedance decreases, resulting in a rise in current and output voltage. Once again, the half-power points and cutoff frequencies are reached when the output voltage is 70.7% of its maximum value. The response curve is the same as in Figure 17–20(b).

■ SECTION 17–10 REVIEW

1. In order to function as a band-reject filter, is a parallel resonant circuit placed in series or in parallel with the load?

2. In order to function as a band-reject filter, is a series resonant circuit placed in series or in parallel with the load?
3. How is the bandwidth of a band-reject filter computed?
4. At the half-power points, the output voltage of a band-reject filter will be what percentage of its maximum value?

17–11 APPLICATIONS OF BAND REJECT FILTERS

60 Hz Notch Filter

The commercial 60 Hz power line is the source of energy for most electronics applications. It is rectified or converted to dc for some applications. For others, it provides the power for the moving parts such as motors. In either case, it must be kept out of the dc voltage supply. If allowed to enter the supply, the dc voltage varies at a 60 Hz rate, and an objectionable hum results. A band-reject filter with a 60 Hz center frequency and a narrow bandwidth will eliminate this 60 Hz hum. This circuit is often referred to as a 60 Hz notch filter.

Broadband VHF Amplifier FM Trap

In areas where television antennas are still a necessity, it is often necessary to enhance the received signal for good viewing. To accomplish this, a wide-band RF amplifier is installed on the antenna at the point where the signal is received. This arrangement sometimes poses a problem because the FM broadcast band is located in the frequency spectrum between television Channels 6 and 7. Thus, sometimes the FM signal interferes with the television signals, which can be avoided if an FM trap is installed in the amplifier. An *FM trap* is basically a band-reject filter that blocks all frequencies of the standard FM broadcast band. These are the frequencies between 88 MHz and 108 MHz.

NOTE

Electronic devices can be categorized as being either active or passive. *Active devices* are capable of amplifying electric signals such as those produced by a tape head or antenna. In this category are bi-polar and field-effect transistors. Passive devices, such as inductors, capacitors, and resistors, are not capable of amplification. Electronic amplifiers are circuit elements made up of both active and passive devices.

The filters presented in this chapter are known as *passive filters*. They're made of passive elements, and their outputs can never be greater than their inputs. Later studies in electronics will introduce filter networks made up of both active and passive devices. The active devices will be integrated circuits known as operational amplifiers. These active filters can be constructed to provide a flatter response over the pass band and to cut off more sharply at the cutoff frequencies.

17–12 TROUBLESHOOTING

Troubleshooting electronic systems must begin with a knowledge of failures that may occur in their individual components. As mentioned in previous chapters, resistors, inductors, and capacitors are subject to failure from opens, shorts, and changes in value. A short means that the impedance of the device is zero, circuit current is high, and no voltage is developed across the component. An open means infinite impedance, no current flow, and source voltage developed across the component. A change in value results in a change in the component's influence in the circuit. Whether the latter is significant or not depends on the amount of variation and the circuit application.

To understand how these component failures affect filtering action, recall the position

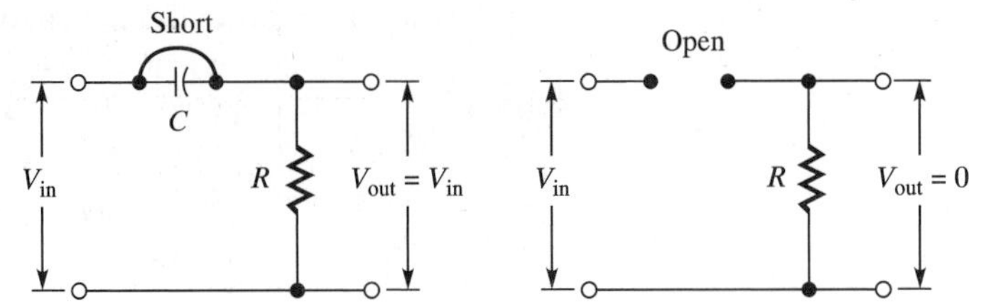

FIGURE 17–22 Effect of open and short on series element

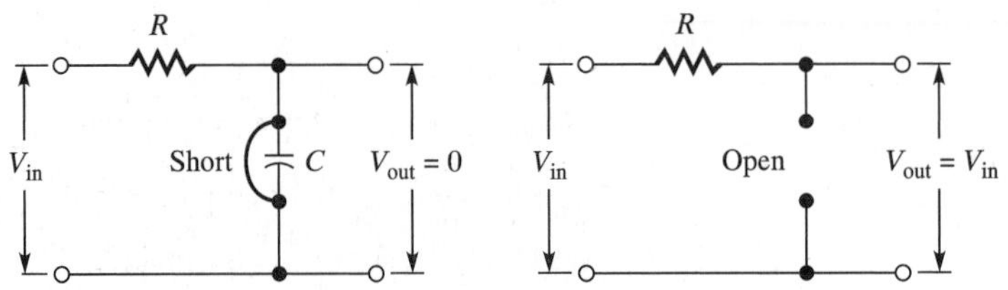

FIGURE 17–23 Effect of open and short on parallel element

of each in the various filters. As shown in Figure 17–22, if the component is a series element, a short will cause all the input voltage to appear across the output, while an open will result in no current and, thus, no output voltage. If, on the other hand, the component is a parallel element, a short will cause the output voltage to be zero, and an open will cause the output to be its maximum value, as demonstrated in Figure 17–23. Filter element measurements should be made with a high-impedance instrument such as an oscilloscope. Manufacturers specifications should be checked to ensure that the bandwidth of the oscilloscope used is adequate for the filter frequencies.

One of the most common problems in an *RC* filter is deterioration of the capacitor's ability to hold charge stored on its plates, which is especially true in electrolytic capacitors that are commonly used as filters in electronic power supplies. Over a period of time, the dielectric material loses some of its insulating properties and becomes conductive. The stored charge begins to "leak" through the dielectric from one plate to another. Thus, the capacitor's efficiency both as a low- or high-pass filter is reduced. If used as a high-pass filter, this conductance will allow some dc current to pass and some dc voltage to appear across the load. If used as a low-pass filter, frequencies above the cutoff will appear across the load. Testing a capacitor with an ohmmeter probably will not indicate a leakage problem. The capacitor should be tested for both value and leakage using an appropriate capacitor analyzer.

A change in either inductance or capacitance value causes a change in the center, or resonant, frequency of a band-pass or band-reject filter. If the filter were part of a radio receiver, the frequencies selected would not represent the channel of a particular station. The result would be a loss in the sensitivity and selectivity of the receiver. The filters mentioned are equipped with variable inductors that allow a small adjustment in order to retune the filter to the appropriate frequency. Test instruments used in this adjustment inject a signal which sweeps across the band of frequencies which the filter must pass. At the output, the instrument provides a graph of the filter response and markers to indicate the pass band.

■ Troubleshooting Practical Circuits

Power Supply Filter

Refer to Figure 17–24(a). In most modern low-voltage power supplies, the low-pass filter is an *RC* combination circuit. As mentioned in the previous section, a filter element may short, open, or change value over a period of use. These problems are more likely to occur in the capacitor than in the resistor.

FIGURE 17–24 Troubleshooting power supply low-pass filter circuit

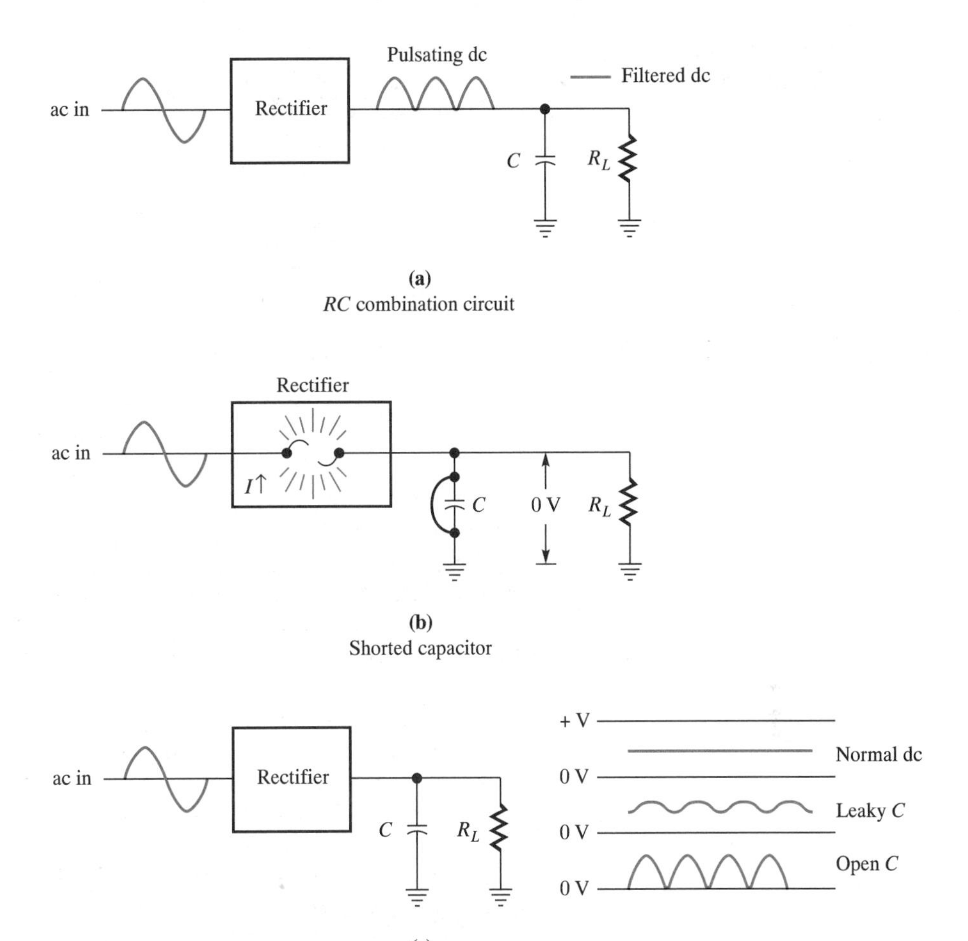

Recall that in a low-pass filter, the capacitor is in parallel with the load. A shorted capacitor would cause a drastic reduction in resistance and an equally drastic increase in current. The fuse protecting the power supply would probably fail. Thus, in a system where the input power fuse fails, a possible cause is a shorted filter capacitor, as illustrated in Figure 17–24(b). An ohmeter test may indicate a shorted capacitor, but a more definitive test could be made with a capacitor analyzer.

An open capacitor will cause the power supply to deliver pulsating rather than pure dc voltage to the load. In a radio, the result would be a severe hum in the reproduced sound. In a television, both the sound and picture would be affected. A personal or business computer may be put completely out of service by this problem. A "leaky" capacitor, one in which the charge begins to leak through the dielectric from one plate to the other, would give a similar indication. Figure 17–24(c) illustrates both leaky and open capacitors. Recall from a previous chapter that a large-value capacitor can be tested for charging using an ohmmeter. Leakage, however, is best checked with a capacitor analyzer.

Capacitive Bypass Circuit

Refer to Figure 17–25(a). If either the resistor or capacitor short, the results are identical: No voltage develops across the combination. This condition is easily detected with a simple voltmeter measurement. The results are different if either component opens. If the resistor opens, there is no path for dc current. As a result, all the dc component appears

FIGURE 17–25 Opens and shorts in a capacitor bypass circuit

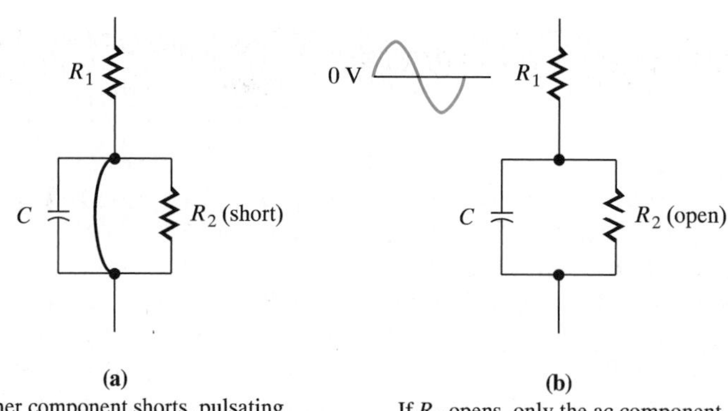

(a)
If either component shorts, pulsating dc appears across R_1, and zero voltage appears across R_2 and C

(b)
If R_2 opens, only the ac component appears across R_1

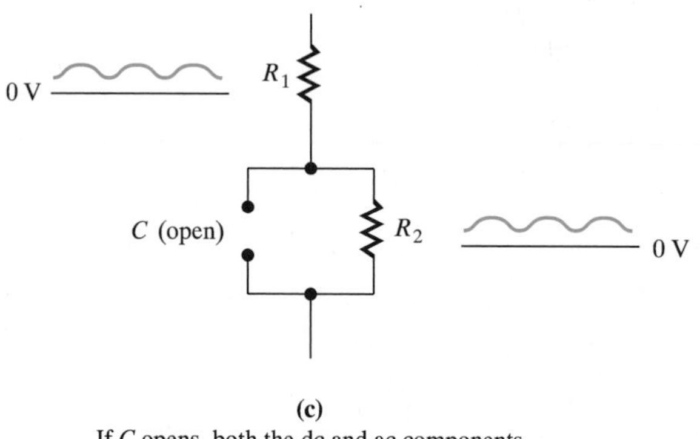

(c)
If C opens, both the dc and ac components appear across each series device

FIGURE 17–26 Troubleshooting amplifier input circuit

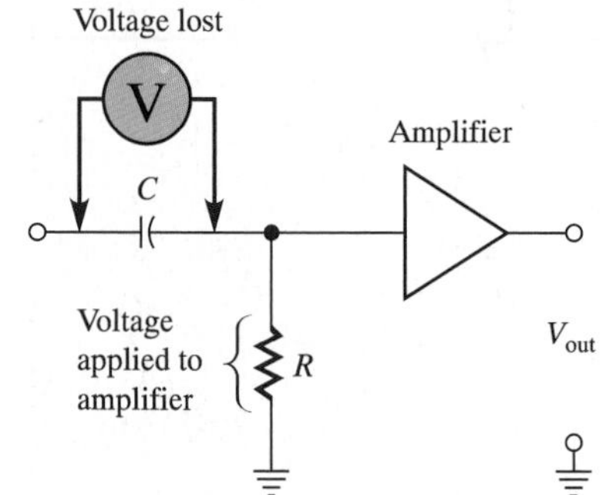

across the capacitor, and none appears across the resistor or any series component that may be in the circuit. Once again, a simple voltmeter measurement will detect this problem. An open capacitor causes both the dc and ac components to be developed across the resistor. These problems are illustrated in Figures 17–25(b) and (c), respectively. Either problem will adversely affect the operation of the circuit of which it is a part.

A change in value of the capacitor may or may not affect circuit operation. An increase in capacitor value results in a lower reactance and an actual improvement in bypass operation. A decrease in value has the opposite effect. The reactance increases, and some of the ac component begins to be developed across the combination. This condition is once again easily detected by viewing the voltage across the combination with an oscilloscope.

Coupling Circuit

Refer to Figure 17–26. The signal voltage enters the amplifier across the series circuit composed of the capacitor and resistor. The actual voltage that is amplified is that developed across the resistor. Any change in value of either the capacitor or resistor affects amplifier performance. Any increase in the value of the resistance or decrease in the reactance of the capacitor increases the amount of signal voltage entering the amplifier. A decrease in the value of the resistor or increase in the reactance of the capacitor has the opposite effect. A decrease in the value of the capacitor causes an increase in its reactance and thus provides less voltage into the amplifier for amplification. The result is less output from the amplifier.

TROUBLESHOOTING PRACTICE

The circuit of Figure 17–27 is a crossover network such as that described in Section 17–9. In the audio system of which it is a part, the frequencies above 2 kHz and below 200 Hz are reproduced normally. The midrange frequencies, however, are reproduced at extremely reduced volume. The problem has been localized to the crossover network. Your task is to determine the fault and what component is involved.

On the job, you would make the necessary measurements of component values, capacitor leakage, or inductor winding resistance one at a time. For this practice, after deciding on a measurement, you can determine its value from the table given in Figure 17–27. Try to determine the fault in as few steps as possible. Rate yourself as follows:

Supertech: 2 or less steps
Technician: 3 steps
Apprentice: 4 steps
Need review: 5 or more steps

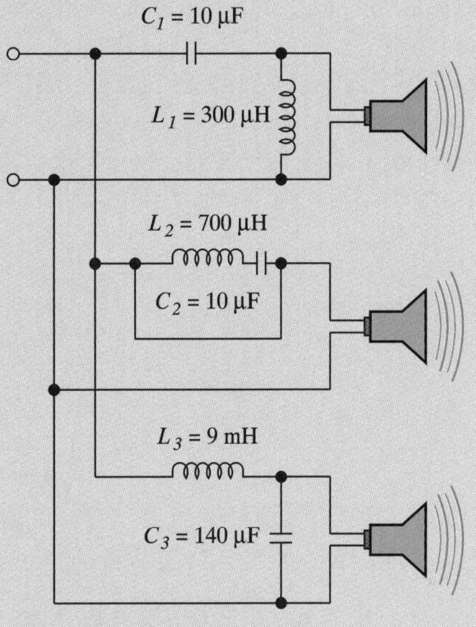

Component	Measured value	Leakage/Ω test
L_1	300 μH	0.5 Ω
L_2	700 μH	125.0 Ω
L_3	9 mH	0.3 Ω
C_1	10 μF	Leakage normal
C_2	30 μF	Leakage normal
C_3	140 μF	Leakage normal

Cutoff frequencies
Low pass: 20 Hz – 200 Hz
Mid-range: 200 Hz – 2 kHz
High pass: 2 kHz – 20 kHz

FIGURE 17–27

A shorted or "leaky" capacitor has a different effect. Now the dc component of the signal enters the amplifier, and as mentioned previously, the operating point of the amplifier is changed, resulting in a distorted output signal.

SECTION 17–13 REVIEW

1. What would be the symptom of a shorted filter capacitor in a power supply?
2. What is meant by a "leaky" capacitor?
3. What is the symptom of an open bypass capacitor?
4. What is the appropriate instrument for checking for leakage in a capacitor?
5. What would be the effect of a change of inductance or capacitance in a band-pass filter?

CHAPTER SUMMARY

1. Electronic systems operate through the systematic control of electric current.
2. Resistors, inductors, and capacitors are devices used in the control of electric current.
3. The effects of inductors and capacitors are frequency sensitive while the effects of resistance are not.
4. A low-pass filter passes all frequencies below the cutoff frequency and blocks all those above.
5. A high-pass filter passes all frequencies above the cutoff frequency and blocks all those below.
6. In a high-pass filter, inductors are parallel elements while capacitors are series elements.
7. A band-pass filter passes a band of frequencies between an upper and a lower cutoff frequency and blocks all others.
8. A band-pass filter may consist of a series *LC* circuit in series with the load or a parallel *LC* circuit in parallel with the load.
9. A band-reject filter blocks a band of frequencies between an upper and a lower cutoff frequency and passes all others.
10. A band-reject filter may consist of a parallel *LC* circuit in series with the load or a series *LC* circuit in parallel with the load.
11. The cutoff frequency of a filter is the frequency at which the output voltage has decreased to 70.7% of maximum.
12. The point at which the output voltage of the filter has decreased to 70.7% of maximum is known as the half-power or 3-dB-down point.
13. The bel is the logarithm of a power, voltage, or current ratio.
14. The decibel is one-tenth of a bel and is the unit normally used to express a power, voltage, or current ratio.

KEY TERMS

low-pass filter
cutoff frequency
high-pass filter
half-power point
band-pass filter
band-reject filter
decibel
bell
gain

EQUATIONS

(17–1) $f_c = \frac{1}{2\pi RC}$

(17–2) $f_c = \frac{R}{2\pi L}$

(17–3) $A_P = \frac{P_{\text{out}}}{P_{\text{in}}}$

(17–4) $A_v = \frac{V_{\text{out}}}{V_{\text{in}}}$

(17–5) $A_i = \frac{I_{\text{out}}}{I_{\text{in}}}$

(17–6) $A_p = 10\log \frac{P_{\text{out}}}{P_{\text{in}}}$

(17–7) $A_v = 20\log \frac{V_{\text{out}}}{V_{\text{in}}}$

(17–8) $A_i = 20\log \frac{I_{\text{out}}}{I_{\text{in}}}$

Variable Quantities

f_c = cutoff frequency
A_p = power gain
A_v = voltage gain
A_i = current gain

TEST YOUR KNOWLEDGE

1. List the four types of electronic filters.
2. In an *RC* low-pass filter, what component is the output taken across? Why?
3. In an *RC* high-pass filter, what component is the output taken across? Why?
4. What determines the cutoff frequency of an *RC* or *RL* filter?
5. What is the equation for the cutoff frequency of an *RL* filter?
6. What is the phase angle of an *RC* or *RL* filter at the cutoff frequency?
7. Why does the cutoff frequency of an *RC* filter occur at the frequency where the reactance equals the resistance.
8. Describe the need for a capacitive bypass.
9. Explain why the point at which the output voltage of a filter is equal to 70.7% of the input is known as the half-power point.

10. Define the bel.
11. Define the decibel.
12. What characteristics of a tuned circuit are made use of in band-pass and band-reject filters?
13. How does the Q-factor of the filter circuit affect the bandwidth?
14. What is meant by the selectivity of the filter?
15. Explain one application of each type of filter presented in this chapter.

PROBLEMS SET: BASIC LEVEL

Sections 17–1, 17–2, and 17–3

1. For the *RC* low-pass filter of Figure 17–28, compute the cutoff frequency.

FIGURE 17–28 Circuit for Problem 1

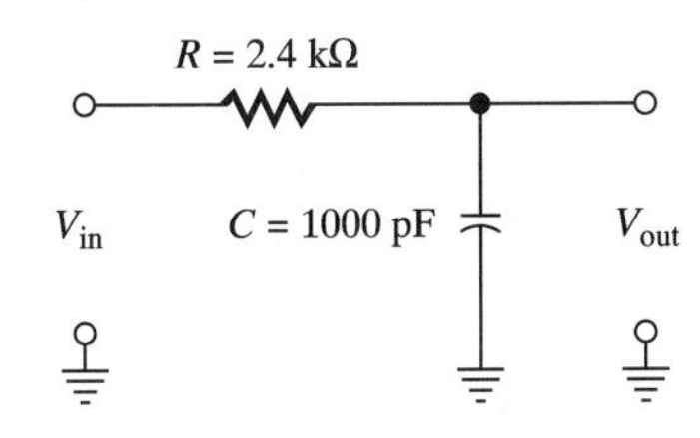

2. For an *RC* low-pass filter consisting of a 100 pF capacitor and 16 kΩ resistor, (a) draw the circuit, and (b) compute the cutoff frequency.
3. Compute the cutoff frequency for the *RL* low-pass filter of Figure 17–29.

FIGURE 17–29 Circuit for Problem 3

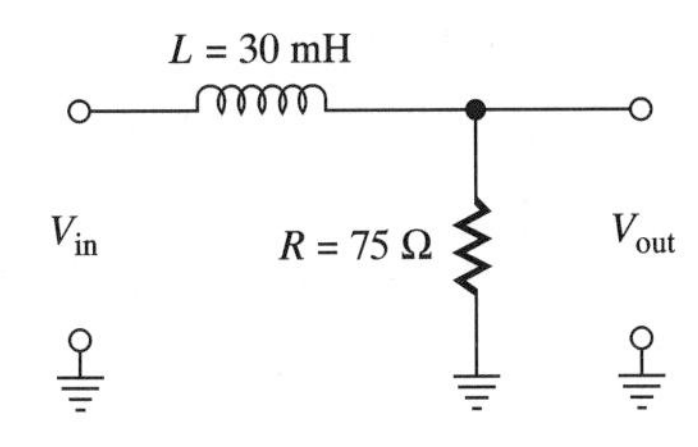

4. In the circuit of Figure 17–30, compute the value of bypass capacitor required for a band of frequencies between 20 Hz and 20 kHz.

FIGURE 17–30 Circuit for Problem 4

+ Source voltage

R_1 R_3 C_1 R_2 $R_4 = 500\ \Omega$ C_2

What value of C_2 will bypass R_4 for frequencies ranging from 20 Hz through 20 KHz?

Sections 17–4 and 17–5

5. For an *RC* high-pass filter consisting of a 10 k Ω resistor and a 0.01 μF capacitor, (a) draw the circuit, and (b) compute the cutoff frequency.
6. Compute the cutoff frequency for the *RL* high-pass filter of Figure 17–31.

FIGURE 17–31 Circuit for Problem 6

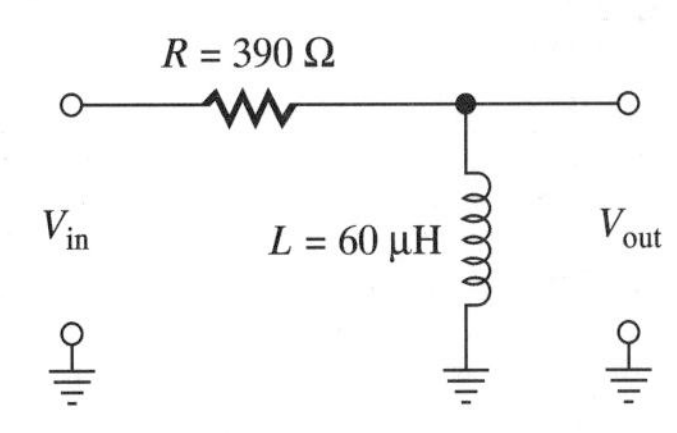

Section 17–6

7. For the circuit of Figure 17–32, compute the power developed by the load at the half-power point.

FIGURE 17–32 Circuit for Problem 7

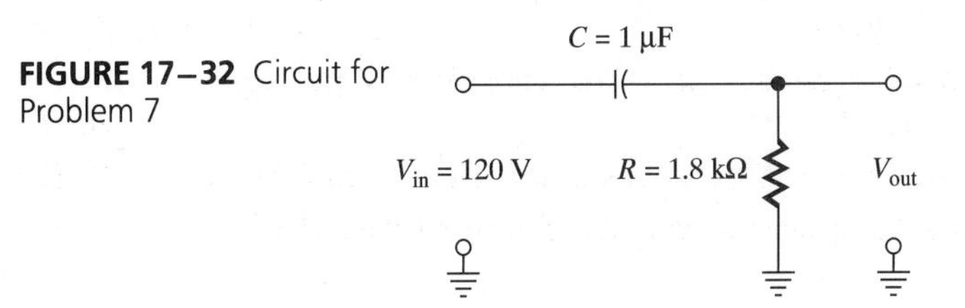

8. For the circuit of Figure 17–33, what is the value of the output voltage at the half power point?

FIGURE 17–33 Circuit for Problem 8

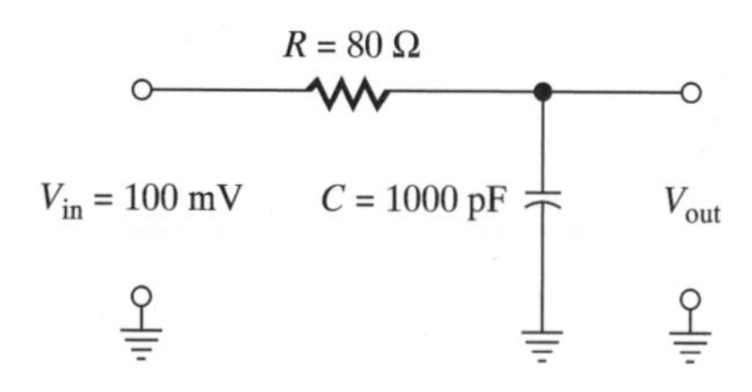

Section 17–7

9. An amplifier has an input power of 100 mW. If the output power is 1 W, compute the power gain in dB.
10. For the amplifier of Figure 17–34 with inputs and outputs as shown, compute the voltage gain in dB.

FIGURE 17–34 Circuit for Problem 10

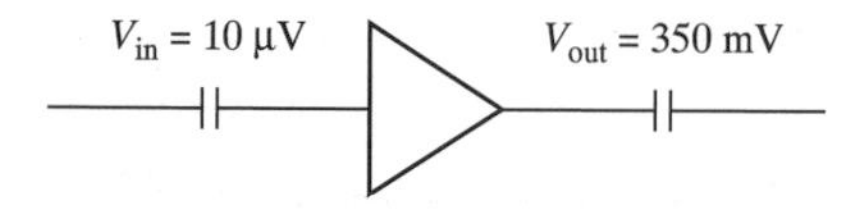

11. An amplifier has an input voltage of 300 mV. If its voltage gain is 15 dB, compute the output voltage.
12. It is required that the output power of an amplifier be 1.3 W. If its power gain is 20 dB, what is the required input power?
13. The amplifier of Figure 17–35 has input and output currents as shown. Compute the current gain in dB.

FIGURE 17–35 Circuit for Problem 13

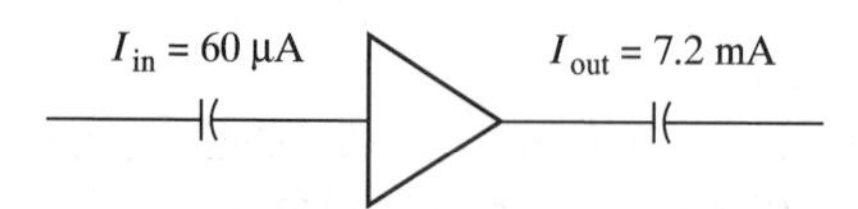

14. What is the voltage gain ratio of an amplifier that has a voltage gain of -6 dB?

Section 17–8 and 17–9

15. A series LC band-pass filter has the following values: $L =$ 250 uH, $C =$ .01 uF, and $R = 15\ \Omega$. Draw the circuit, and compute the following: (a) f_r, (b) BW, (c) f_c (upper), and f_c (lower).
16. For the series LC band-pass filter of Figure 17–36, compute the center and the upper and lower cutoff frequencies.

FIGURE 17–36 Circuit for Problem 16

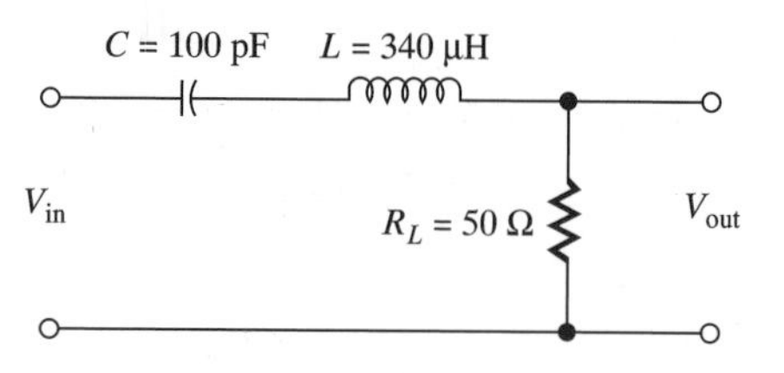

17. A parallel LC band-pass filter has the following values: $L = 125\ \mu$H and $C = 1000$ pF. The winding resistance of the inductor is 30 Ω. Draw the circuit and compute the following: (a) f_r, (b) BW, (c) f_c (upper), and (d) f_c (lower).
18. In the series LC band-reject filter of Figure 17–37, compute the output voltages at the upper and lower cutoff frequencies.

FIGURE 17–37 Circuit for Problem 18

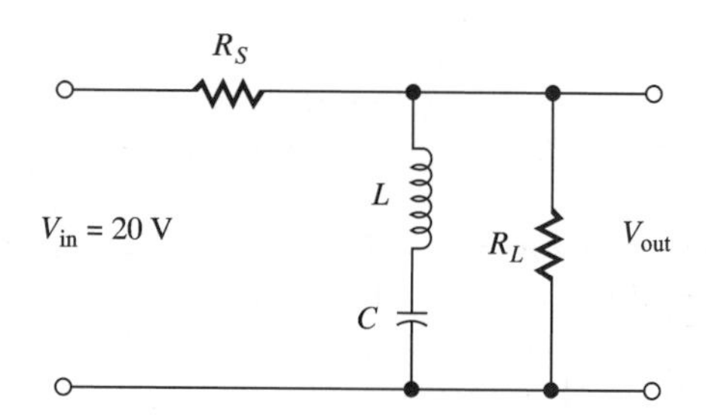

Sections 17–10 and 17–11

19. For the parallel LC band-reject filter of Figure 17–38, consider R_W to be insignificant and compute: (a) Q-factor, (b) BW, (c) the upper and lower cutoff frequencies.

FIGURE 17–38 Circuit for Problem 19

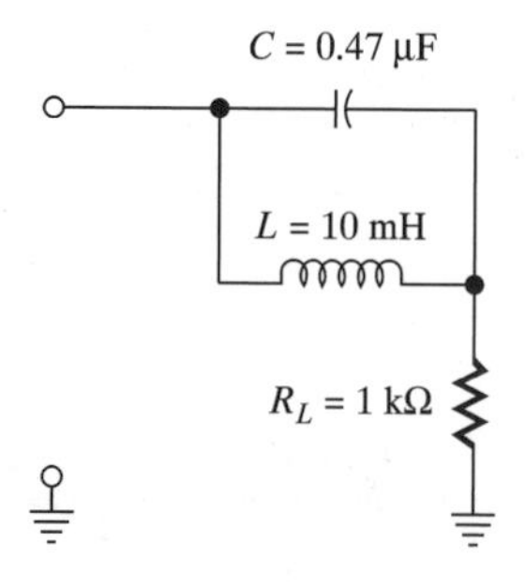

20. For the series LC band-reject filter of Figure 17–39, compute: (a) Q-factor, (b) BW, (c) the upper and lower cutoff frequencies.

FIGURE 17–39 Circuit for Problem 20

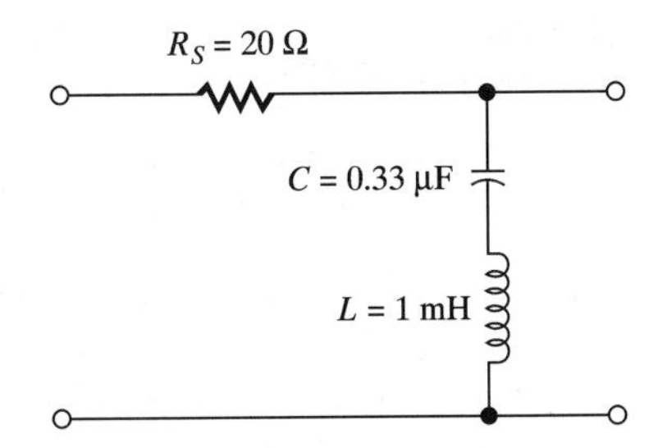

FIGURE 17–40 Response curve for Problem 21

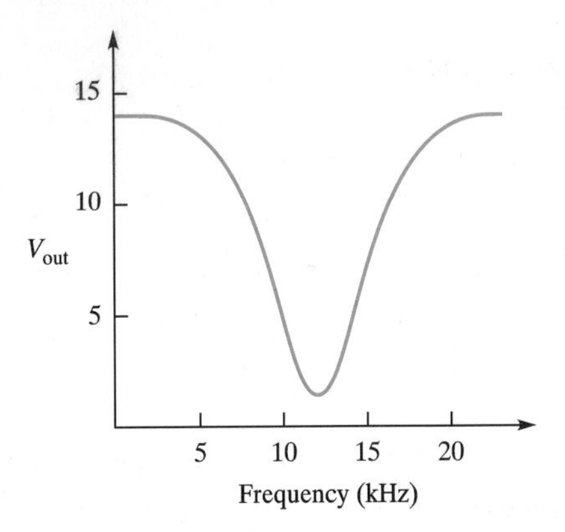

21. For the band-reject filter response curve of Figure 17–40, determine the bandwidth and the voltages at the upper and lower cutoff frequencies.

PROBLEM SET: CHALLENGING

22. An *RC* low-pass filter consists of a 2400 Ω resistor and a 0.01 μF capacitor. Compute its output voltage in 1 kHz steps from 0 Hz through 9 kHz. Plot these points on a graph to demonstrate the response curve of this filter. Mark the 3-dB-down point.

23. For the high-pass filter of Figure 17–41, compute the output voltage at 80 kHz and 120 kHz.

FIGURE 17–41 Circuit for Problem 23

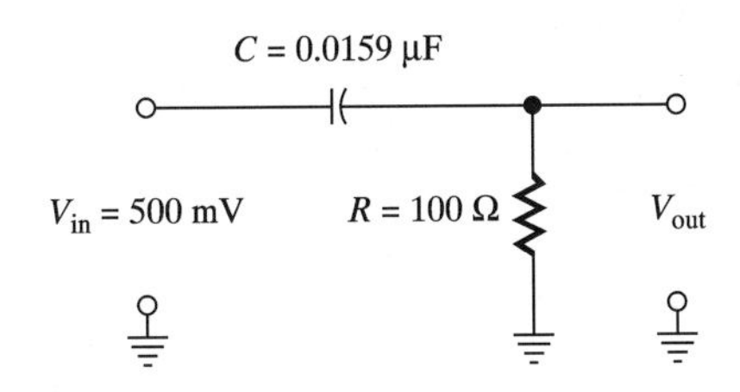

24. For the low-pass filter of Figure 17–42, compute the cutoff frequency. Compute the output voltage at 6 kHz and 10 kHz.

FIGURE 17–42 Circuit for Problem 24

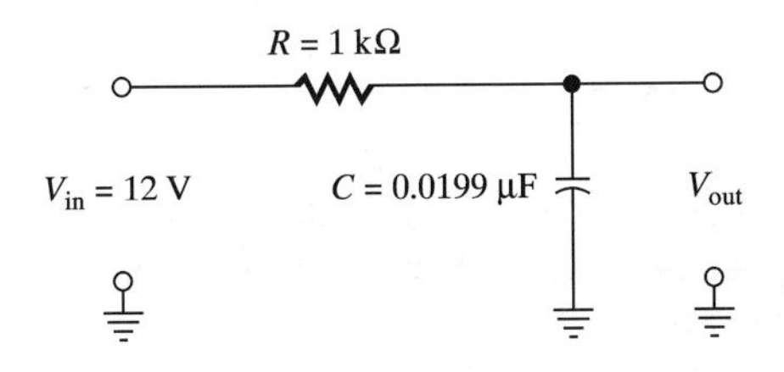

25. Figure 17–43 capacitor C_2 bypasses R_4. Compute the lowest frequency at which the by-pass is effective.

FIGURE 17–43 Circuit for Problem 25

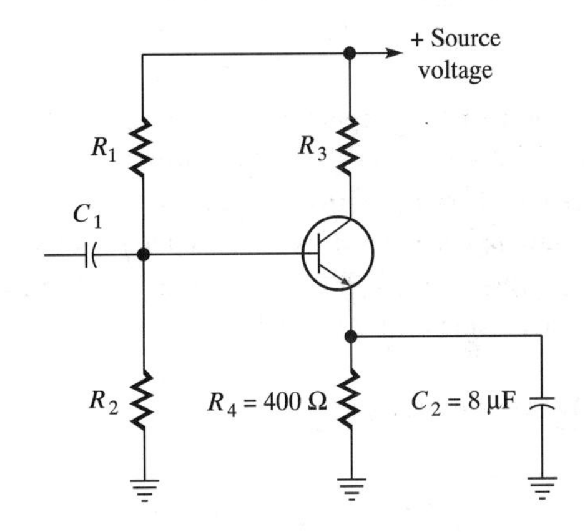

26. As the frequency of the signal applied to a filter increases from 400 Hz to 10 kHz, its output voltage increases from 180 mV to 1.4 V. Compute the gain in dB.

27. As the frequency of the signal applied to a filter increases from 400 Hz to 10 kHz, the power developed in the load drops from 560 mW to 36 mW. Compute the gain in dB.

28. What is the ratio of output to input voltage for a filter with a gain of −12 dB?

29. A filter has an output of 1.2 V and a gain of −8.5 dB. Compute its input voltage.

30. If the output of the high-pass filter of Figure 17–44 is as shown, what are two possible causes of this fault?

FIGURE 17–44 Circuit and graphs of input/output voltages for Problem 30

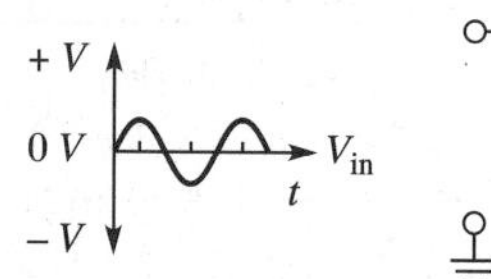

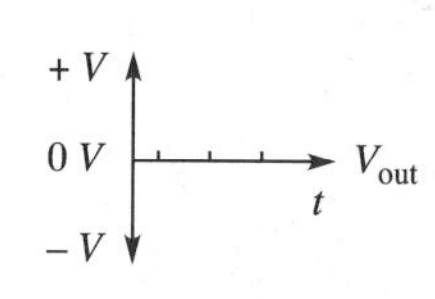

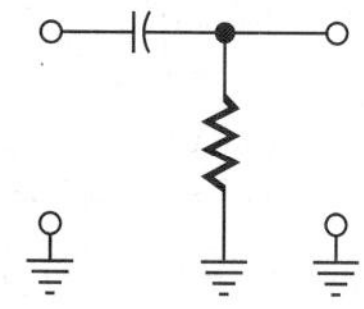

31. If the output of the low-pass filter of Figure 17–45 is as shown, and if one fault exists, what is the probable cause?

FIGURE 17–45 Circuit and graphs of input/output voltages for Problem 31

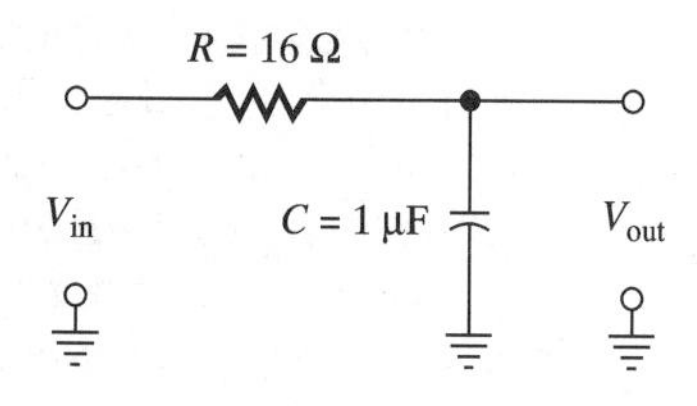

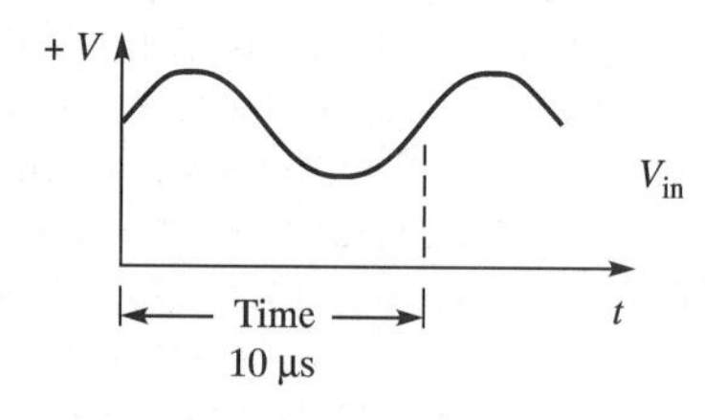

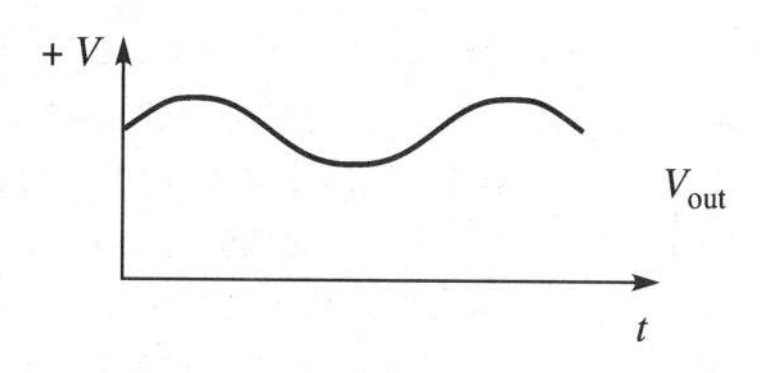

32. There is no output from the pi filter of Figure 17–46. List all the possible causes of this fault.

FIGURE 17–46 Circuit for Problem 32

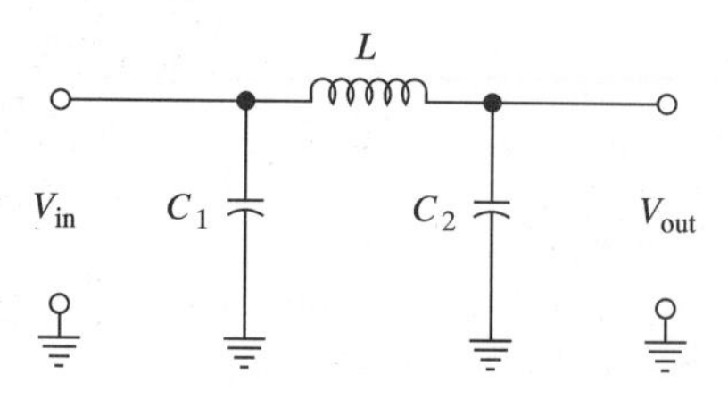

A Day in the Life of an Associate Engineer

An associate engineer is a technician who is employed to work closely with engineers in the design and development of new electronic systems. As design concepts are developed, the associate engineer builds the circuit, performs tests on it, collects test data, and writes a report on the outcome. In addition, he or she may suggest design changes to overcome problems. The associate engineer is usually involved in the project from its inception through the early manufacturing stages.

Bill is an associate engineer with a company that specializes in electronic power supplies and power control equipment. He has been employed in this capacity for almost a year. It is not his intention to remain a technician; he wants to become an electrical engineer. Thus, while at the community college earning an associate of science degree in electronic engineering technology, he also earned an associate of arts degree in pre-engineering. He is saving money in order to be able to attend the engineering college at the university. He feels that the experience he is gaining now will make him a much more productive engineer when he does become one.

In his present position, Bill works closely with most of the members of the engineering staff. On Mondays, he attends the staff meeting at which all current developmental projects are discussed and their progress tracked. At the end of each week, he must have prepared a report on the up-to-date status of each project's tasks for which he is responsible. He currently has three tasks in progress. One is in

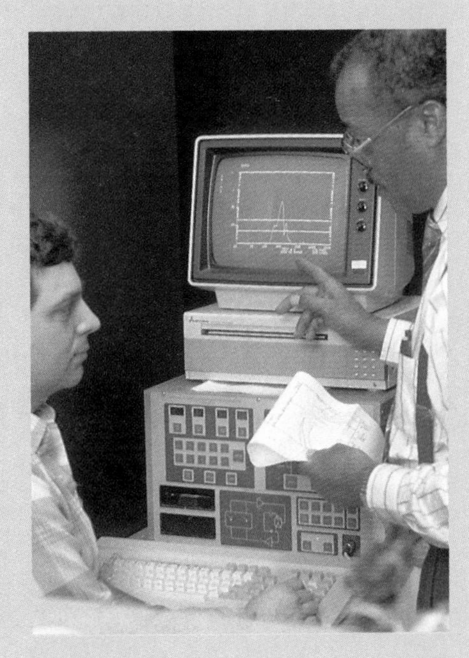

suspension, awaiting equipment to be furnished by a subcontractor. Another is nearing completion as the problems of manufacturing are being worked out. The one he will be working on today is a new design of a "crowbar" circuit, which will be a part of a very sophisticated power supply being produced under a contract with the government of Taiwan.

A crowbar is a circuit that projects the load on the power supply from an overvoltage condition. When the output voltage increases beyond a certain set value, the crowbar circuit will trip and remove it. (It gets its name from the act of placing a steel crowbar across the output, thus shorting out the load). It is one of the circuits of this power supply that Bill has helped to perfect. The engineers have been quite pleased with his performance, relying more and more upon him for answers to problems. He has begun to feel that this is "his" power supply.

The problem he is experiencing with the crowbar circuit is false triggering, which means that, for seemingly no reason, the circuit trips and removes the voltage—in spite of the fact that a recorder on the output voltage indicates no change in value! False triggering is a problem that is sometimes associated with one component of the circuit. Today Bill is considering using a piece of test equipment to see if a "transient" is involved. Transients are extremely short duration voltage pulses that are often caused when electronic equipment is turned on or off. The rapid change in current produces the transient. The transient could be coming from some external point in the shop, totally unrelated to the power supply under test. Regardless of the source, it is Bill's job to track it down and look for a solution.

The morning passes quickly for Bill as he is engrossed in the problem. The engineer in charge of the project stops by to check progress. After being brought up to date, he offers a few suggestions and leaves. Later the quality assurance supervisor stops by to discuss the project now in production. What could be an overheat condition has been found in the item, and Bill is to look into it as soon as possible—today! The company has a sophisticated, computer-controlled environmental test facility for learning of such problems during the design and test stages. Having personally supervised these tests, Bill doubts that a problem exists, but agrees to look into it.

After a quick lunch, Bill wraps up the work he started in the morning and decides what to do about the project in production. He feels the best way to alleviate the concerns of the production people is to check the environmental test data to see what it shows. He quickly scans the data he had collected during the testing to refresh his mind before going to the production area. The item had been tested under full load at both very high and very low temperatures. There were no failures recorded and no indication of overheating on the circuit boards. With this information on the record, overheating problems are not likely to occur, he believes, and he writes a short memo relating the information and his opinion to the engineer in charge of the project.

Bill puts in the rest of the day working on and thinking about the false-triggering problem. No transients have shown up yet on the recorder, and no false triggering has occurred.

In his job, he has found that there are no short cuts or easy solutions to most problems in electronics. Thus, it often takes a great deal of perseverance in order to find solutions to problems. He remembers how in school he thought everything should always work exactly the way Mr. Ohm and Mr. Kirchhoff said they would. It does, of course, but sometimes becomes quite subtle when devices are pushed to the limits demanded in the modern electronics industry! He wonders if these intelligent pioneers of the electronics world could have forseen the lengths to which their simple laws would be pushed in the control of current.

He spends the last few minutes of the day completing his log book, recording the circuit schematic, each test performed, and the results. He carefully fills in the entries for the day and signs the log, knowing that failure to stringently follow the recording procedures would bring management's displeasure down on him as quickly as anything!

As he leaves the plant, Bill wishes he had made some progress on the task. Maybe I have, he thinks. No transients showed up on the recorder and no false triggers occurred. He decides to pursue this line further the next day. Bill longs for the day when he will be the engineer, but in the meanwhile, he enjoys the challenges of his present job.

CHAPTER

18 Pulse Response of *RC* and *RL* Circuits

■ UPON COMPLETION OF THIS CHAPTER, YOU WILL BE ABLE TO

1. Describe the response of an *RC* circuit to pulse type signals.
2. Draw and describe the current and voltage waveforms in an *RC* circuit when a pulse is applied.
3. Explain how the length of the time constant in relation to the pulse width affects the action of an *RC* circuit.
4. Draw the schematic of an *RC* integrator, and describe its operation.
5. Draw the schematic of an *RC* differentiator, and describe its operation.
6. Describe the response of an *RL* circuit to pulse type signals.
7. Draw the schematic of an *RL* integrator, and describe its operation.
8. Draw the schematic of an *RL* differentiator, and describe its operation.
9. List and explain applications of integrators and differentiators.

In Chapters 9 and 10, you learned of the effects of inductance and capacitance on direct currents and voltages. In Chapters 12 and 14, you learned of the response of *RL* and *RC* circuits to the simplest of all ac waveforms, the sine wave. In this chapter, you will learn of the response of *RC* and *RL* circuits to pulses and pulse waves. Recall that pulses are complex waves that are very rich in harmonics. Thus, you would expect capacitance and inductance, which are frequency sensitive, to have quite an effect upon pulse signals.

As you will learn, the shape of a pulse can be greatly altered by the effects of inductance or capacitance. Sometimes the shape of pulses are altered to fulfill a specific purpose. Two such circuits, the integrator and differentiator, are presented in this chapter. These circuits take their names from functions in the branch of mathematics known as calculus. This shaping is accomplished by changing the harmonic content of the pulse. An integrator reduces the number of higher-order harmonics, while a differentiator reduces the number of lower-order harmonics. Pulse shaping has many uses in electronic circuits and systems.

Sometimes pulse shaping takes place that is not desirable. It is not necessary to have inductors and capacitors physically present in order to experience their effects. For example, distributed capacitance and inductance are present in most transmission lines. Thus, if a pulse is transmitted down such a line, its amplitude and shape will be changed to some degree by the reactances present. A typical circuit board contains many components, all of which contribute small values of distributed capacitance and inductance. As with the transmission line, the amplitude and shape of a pulse may be changed by passing a signal through the board. This consideration is very important in digital systems where numbers are represented by the presence or absence of a pulse.

The practical uses of the circuits in this chapter are many. An integrator can be used as a pulse counter. A differentiator can produce narrow pulses used to trigger actions in a circuit. They are also used in motor speed control circuits. Thus, the information of this chapter must become a part of your technical knowledge.

18–1 REVIEW OF TIME CONSTANTS AND PULSE PRINCIPLES

Time Constants

A thorough knowledge of time constants is essential in the study of pulse response of *RC* and *RL* circuits. Thus, in this section, both the *RC* and *RL* time constant are briefly reviewed. If, after this review, you feel uncertain of your knowledge of time constants, it would be a good idea to review the material in detail. Refer to Chapter 9 for information on the *RL* time constant and to Chapter 10 for the *RC* time constant.

The *RC* Time Constant

Recall that a capacitor cannot change its charge condition instantaneously. Charging current must flow in order to move electrons from one plate to the other. A series *RC* circuit with a dc voltage source is shown in Figure 18–1(a). At the instant the switch is closed, the charging current is maximum, limited only by the resistance. Thus, all the source voltage is across the resistor. As charge is stored in the capacitor, the voltage across it increases and the charging current decreases. The decrease in charging current results in a decrease in voltage across the resistor. The increase in capacitor voltage follows an exponential curve as shown in Figure 18–1(b). Once the capacitor is fully charged, the current ceases to flow. In Figure 18–1(c) the dc voltage source is removed and replaced with a short. Current must now flow in order to neutralize the charge on the capacitor. As the capacitor discharges, its voltage decreases, following an exponential curve as shown in Figure 18–1(d).

The time taken for the capacitor to fully charge or discharge is a function of the *RC* time constant of the circuit. The *RC* time constant is equal to the product of the resistance

FIGURE 18–1 Charging and discharging of a capacitor

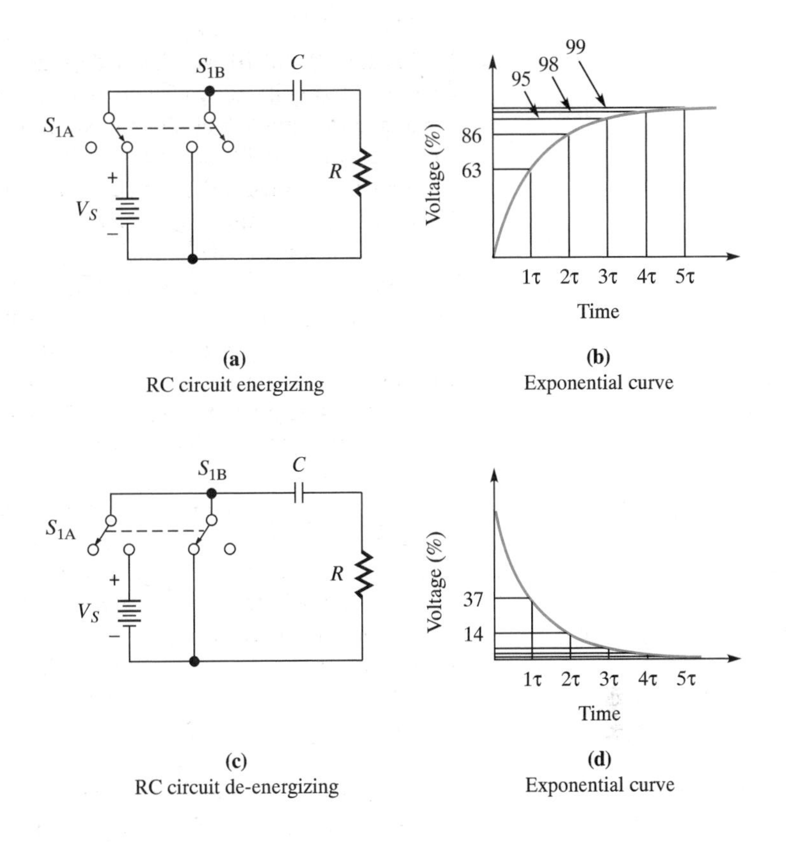

and capacitance ($R \times C$). In one time constant, a capacitor charges to approximately 63% of the applied voltage. During discharge, in the same one time constant, the voltage across the capacitor decreases to approximately 37% of its value. In order for the capacitor to fully charge or discharge, approximately five time constant periods must elapse.

■ The *RL* Time Constant

When the switch of Figure 18–2(a) is placed in the position shown, the energizing current for the inductor is not immediately present. The switch closure produces an expanding magnetic field that induces a counter voltage (CEMF), whose polarity opposes the build-up in current. As the magnetic field expands, the induced counter voltage decreases and the current increases. When the magnetic field ceases to expand, there is no counter voltage induced, and the current has a value limited by the resistance only. The current builds up following an exponential curve as illustrated in Figure 18–2(b). When the position of the switch is reversed, as in Figure 18–2(c), the source is removed and replaced with a short. There is no current to support the magnetic field. The magnetic field collapses, once again inducing a counter voltage that keeps the current flowing for a period of time. The current decreases, following an exponential curve as shown in Figure 18–2(d).

The time taken for the current to build to maximum or decay to zero is a function of the *RL* time constant. The *RL* time constant is the ratio of the inductance to the resistance. In one time constant, the current builds to approximately 63% of its maximum value or decreases to 37% of its maximum value. Once again, approximately five time constant periods must elapse in order for the current to build to its maximum value or decay to zero.

FIGURE 18–2 Energizing and de-energizing an inductor

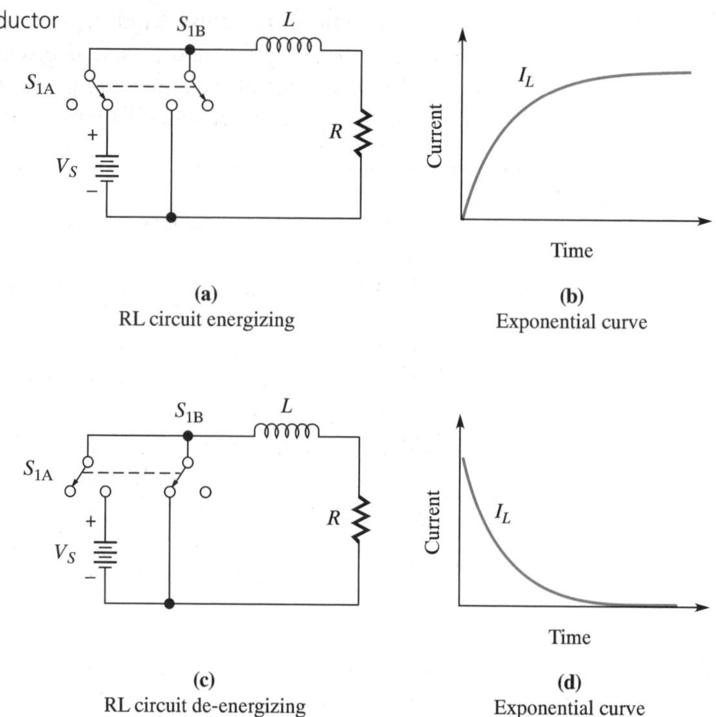

FIGURE 18–3 Voltage pulse wave

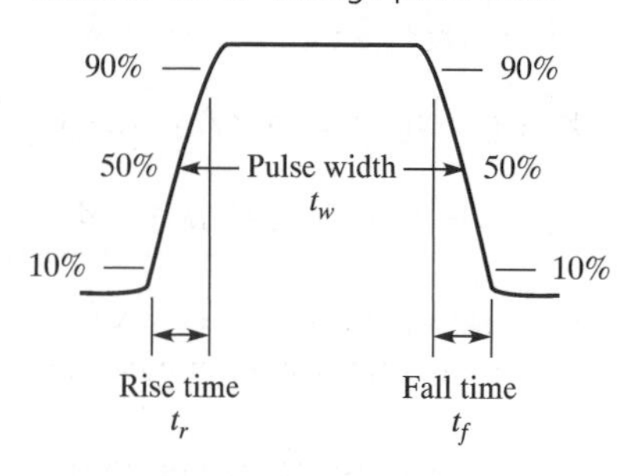

FIGURE 18–4 Idealized pulse

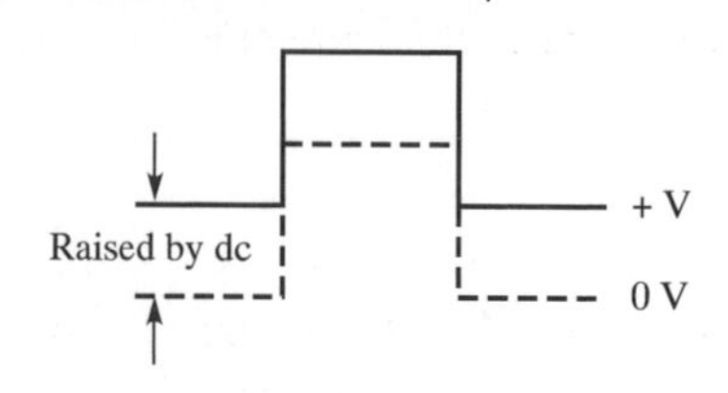

NOTE

Notice the careful use of the word "approximate" in the preceding paragraphs. It must be understood that the 63% and 37% values mentioned are only approximate. A more accurate value would be obtained using the exponential equations introduced in Chapters 9 and 10. Further notice that if anything increases or decreases by some percentage of a maximum value, it will theoretically never reach the maximum value. However, for practical purposes, it is assumed that the capacitor will be fully charged or discharged in five time constants, and the inductor will be fully energized or deenergized in five time constants.

■ Pulse Principles

A typical voltage pulse is illustrated in Figure 18–3. The pulse parameters, which were introduced in Chapter 11, are pulse width, rise time, and fall time. The leading and trailing edges are as shown. The *pulse width* is the time of one pulse. It is measured from the leading edge of the pulse to its trailing edge, at a point where the voltage is 50% of maximum value. The *rise time* is the time taken by the pulse to rise from 10% to 90% of the maximum value. The *fall time* is the time taken to fall from 90% to 10% of the maximum value.

Recall that any nonsinusoidal wave can be constructed from a fundamental sine wave and its harmonics. If a dc voltage is added to raise the reference level, then the pulse wave of Figure 18–4 is produced. This wave consists of the fundamental frequency plus its odd harmonics. The sharpness of the sides of the pulse is a measure of how many of the odd harmonics are contained in it. The shorter the rise or fall time, the higher is the harmonic content.

The ratio of the pulse width (t_w) to the pulse recurrence time (PRT) is known as the *duty cycle* of a pulse. Duty cycle is usually expressed as a percentage. Thus, duty cycle is the percentage of the total pulse period that the pulse is at a high state. For example, in a square wave, the on and off times of the pulse are equal. Thus, its duty cycle is 50%.

SECTION 6–1 REVIEW

1. How many time constant periods must elapse in order for a capacitor to fully charge or discharge?
2. An *RC* circuit is composed of a 1 kΩ resistor and a 100 μF capacitor. Compute the time constant.
3. At the instant a capacitor begins to charge, is the voltage across it zero or maximum?
4. An *RL* circuit is composed of a 100 kΩ resistor and a 30 mH inductor. Compute the time constant.
5. How many time constant periods must elapse for the inductor energizing current to reach maximum once the source voltage is applied?
6. What is the harmonic content of a square wave?
7. Is the harmonic content of a pulse higher or lower as the rise time shortens?
8. What is the duty cycle of a square wave?

18–2 PULSE RESPONSE OF AN *RC* CIRCUIT

Figure 18–5(a) shows a pulse wave applied to an *RC* circuit.* There will be a period when the pulse is at its peak value, and a period between pulses when the voltage is zero. As illustrated in Figure 18–5(b), the presence of the pulse voltage effectively places a dc source across the circuit. Current flows during this period charging the capacitor, and

*Unless otherwise specified, in the circuit analyses in the remainder of this chapter, the input pulses will be considered to be square waves.

FIGURE 18–5 Pulse response of an *RC* circuit

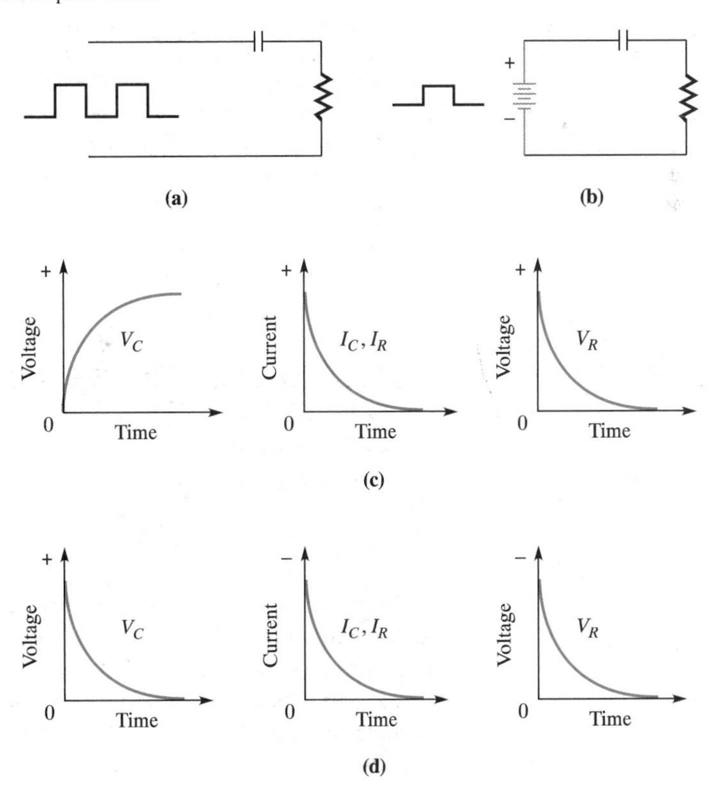

the voltage across the capacitor rises in an exponential manner. The voltage across the resistor is initially maximum and decreases in an exponential manner as the capacitor charges. The waveforms of the current and voltages for the pulse period are shown in Figure 18–5(c).

When the pulse drops to zero, it is equivalent to replacing the voltage source with a short. At this time, the capacitor becomes the voltage source for the circuit. It discharges through the resistor and the internal resistance of the original voltage source. (In order to simplify the explanations throughout this chapter, the internal resistance will be considered to be zero.) Once again, the voltage across the resistor is initially maximum and drops to zero as the capacitor discharges. Notice, however, that the current is in the opposite direction, and the voltage produced is of the opposite polarity. The waveforms of current and voltage for the period between pulses are shown in Figure 18–5(d).

Question: What determines the shape of the current and voltage waveforms when a pulse is applied to an RC circuit?

The width of the input pulse is the time allowed for the capacitor to charge. The time between pulses is the time allowed for the capacitor to discharge. Thus, the shape of the waveforms is determined by the relative length of the circuit time constant in relation to the width of the input pulse. For purposes of discussion, three relative lengths of pulse period are considered in this section: (1) medium, (2) short, and (3) long. A *medium pulse* period is one which is equal to five time constants. A *short pulse* period is one which is less than five time constants. A *long pulse* period is one which is greater than five time constants.

Pulse Width Equal to Five Time Constants

Question: What is the shape of the waveforms of current and voltage developed in the circuit of Figure 18–6?

If the pulse width is equal to five time constants, the capacitor has just sufficient time to fully charge during the pulse and fully discharge in the time between pulses (assuming a 50% duty cycle), as illustrated in Figure 18–6(a). The capacitor voltage waveform is shown in Figure 18–6(b). Notice the exponential rise on the leading edge, and the exponential decay on the trailing edge of the pulse. Both the rise and decay of the capacitor voltage require one pulse period.

As illustrated in Figures 18–6(c) and (d), the waveshapes of the circuit current and resistor voltage are similar to one another. At the instant the capacitor begins to charge, the current is maximum, and the voltage across the resistor equals the source voltage. As the capacitor charges, the current decreases exponentially, producing a similar drop in

FIGURE 18–6 Development of voltage and current in a medium pulse period

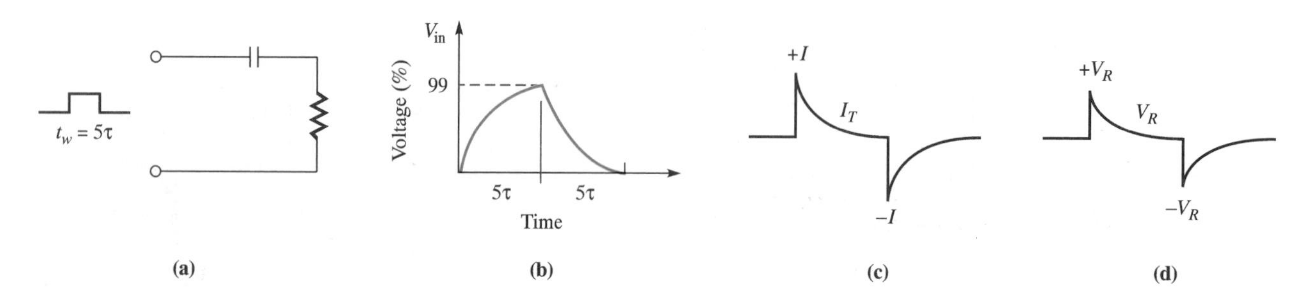

voltage across the resistor. Both reach zero near the end of the input pulse. At the end of the pulse period, the capacitor begins to discharge. The discharge current produces another voltage pulse across the resistor. Now, however, it is of opposite polarity. Both the discharge current and the voltage across the resistor decay in an exponential manner, reaching zero near the end of the between-pulse period.

Pulse Width Greater Than Five Time Constants

Question: What is the effect upon the circuit waveforms if the pulse width is greater than five time constants?

In Figure 18–7(a), the pulse is adjusted to have a width of 20 μs. The ratio of the pulse width to the time constant is now 10. Thus, the capacitor now has a period of ten time constants in which to charge or discharge. As shown in Figure 18–7(b), the capacitor fully charges in half the pulse width (five time constants) and remain in this condition through the rest of the pulse. It also discharges in half the pulse width. Thus, with more time allowed for charging, the capacitor voltage begins to resemble the input pulse. If the pulse width were made even longer or the time constant made shorter, the capacitor would charge in a shorter percentage of the input pulse, as illustrated in Figure 18–7(c). Notice that in each instance the width of the capacitor voltage pulse increases, approaching that of the input pulse.

Since the capacitor charges in half the pulse period, the current becomes zero in the same period. Thus, the resistor voltage is as shown in Figure 18–8(a). It forms a sharp positive-going spike on the leading edge of the pulse that decreases to zero in half the pulse period. At the trailing edge of the pulse, the resistor voltage forms a sharp negative-going spike that decreases to zero as the capacitor discharges over half the period between pulses. Thus, the wider input pulse width has effectively decreased the width of the pulse developed across the resistor. If the pulse width is increased, the capacitor charges and discharges in even a smaller percentage of the input pulse period. If the time constant is shortened, either by reducing the value of the capacitor or resistor, the capacitor needs less time to charge or discharge. Thus, for a given pulse width, the spikes produced at the leading and trailing edges are narrower, as illustrated in Figure 18–8(b).

FIGURE 18–7 Development of voltage and current in a long pulse period

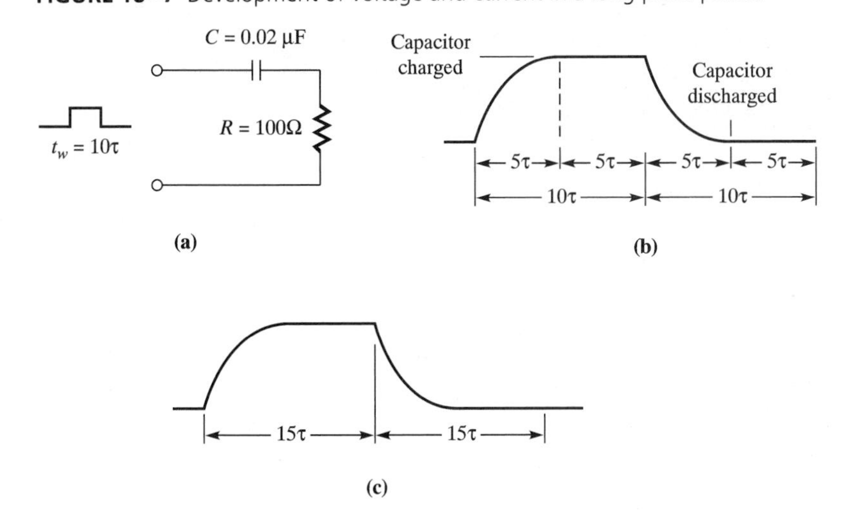

FIGURE 18–8 Examples of resistor voltage waveforms for $t_w > 5\tau$

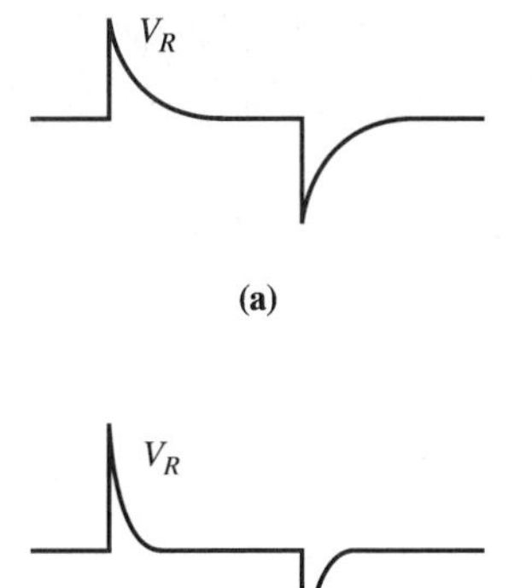

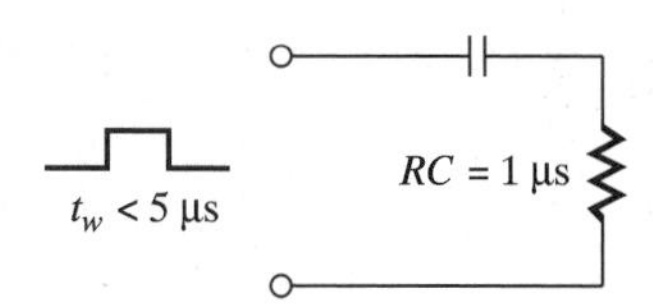

FIGURE 18–9 *RC* circuit driven by pulse with $t_w < 5\tau$

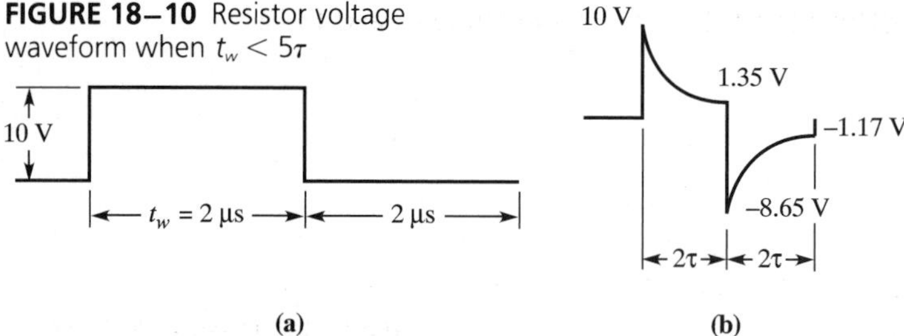

FIGURE 18–10 Resistor voltage waveform when $t_w < 5\tau$

■ Pulse Width Less Than Five Time Constants

If the width of the input pulse is made less than five time constants, the capacitor now does not have sufficient time to fully charge or discharge. The *RC* circuit of Figure 18–9 has a time constant of 1 μs. Thus, if the pulse width were adjusted to any value less than 5 μs, the capacitor would not fully charge or discharge. For example, as shown in Figure 18–10(a), the width of the input signal is 2 μs and its amplitude is 10 V. The capacitor has two time constants in which to charge or discharge. In order to plot the output signal, it is necessary first to compute the voltage across the capacitor after two time constants. The exponential equation for charge of the capacitor gives

$$\begin{aligned} V &= V_F\,(1 - e^{-t/RC}) \\ &= (10\text{ V})(1 - 2.7182818^{-2}) \\ &= (10\text{ V})(0.865) = 8.65\text{ V} \end{aligned}$$

Thus, during the period when the pulse is high, the output voltage rises, in an exponential manner, to 8.65 V.

During the period when the pulse is low, the capacitor discharges for two time constants. This discharge begins from the 8.65 V to which it has charged. The exponential equation for discharge of the capacitor gives

$$\begin{aligned} V &= V_i(e^{-t/RC}) \\ &= (8.65)(2.718^{-2}) \\ &= (8.65)(0.135) = 1.17\text{ V} \end{aligned}$$

Probably the easiest way of determining the voltage waveform across the resistor is through the application of Kirchhoff's voltage law. The resistor and capacitor form a series circuit. Thus, the sum of the capacitor and resistor voltages equals the source voltage. At the leading edge of the pulse, the entire source voltage is across the resistor. At the end of two time constants, with the capacitor charged to 8.65 V, 1.35 V is across the resistor. On the trailing edge of the input pulse, the capacitor voltage is placed across the circuit, and −8.65 V appear across the resistor. The capacitor discharges in the period between pulses and the resistor voltage rises to −1.17 V. The output waveform is shown in Figure 18–10(b).

Note what has happened to the capacitor and resistor voltage waveforms relative to the input pulse. The capacitor voltage waveform now bears less resemblance to the input pulse than with the longer pulse width. Both the leading and trailing edges cover less of the charge and discharge curves. The resistor waveform, though still converted to an ac signal, begins to resemble the input pulse.

EXAMPLE 18–1 For the circuit of Figure 18–11 with input as shown, compute the output voltage at the end of the high and low periods of the pulse.

FIGURE 18–11 Circuit for Examples 18–1 and 18–2

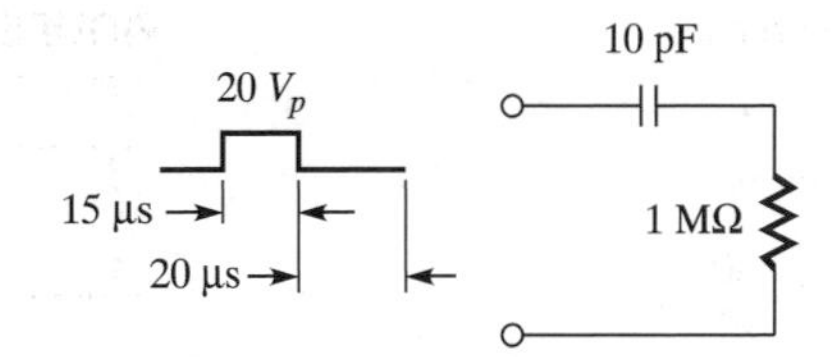

■ Solution:

First compute the time constant of the circuit:

$$\tau = RC$$
$$= (1\ \text{M}\Omega)(10\ \text{pF}) = 10\ \mu\text{s}$$

The period of the pulse is 15 μs. Using the exponential equation for the charge curve, compute the voltage at the end of the pulse period:

$$V = V_F(1 - e^{-t/RC})$$
$$= (20\ \text{V})(1 - 2.7182818^{-1.5})$$
$$= (20\ \text{V})(0.7768) = 15.5\ \text{V}$$

Notice that the period between pulses is longer than the pulse period. During the period between pulses, the capacitor discharges from the 15.5 V to which it has charged:

$$V = V_i(e^{-t/RC})$$
$$= (15.5\ \text{V})(2.7182818^{-2})$$
$$= (15.5\ \text{V})(0.135) = 2.1\ \text{V}$$

EXAMPLE 18–2 For the same circuit and input as in Example 18–1, compute the voltage across the resistor.

■ Solution:

At the leading edge of the pulse, the entire source voltage is across the resistor. During the period of the pulse, the resistor voltage decreases to 4.5 V (20 V − 15.5 V). At the trailing edge of the pulse, the resistor voltage drops to −15.5 V (the capacitor becomes the source) and drops to −2.1 volts during the period between pulses.

Question: What is the effect of making the ratio of pulse width to time constant smaller?

In Example 18–2, if the pulse width and pulse off-time were each reduced to 10 μs, the capacitor would have even less time to charge or discharge. The voltage across the capacitor at the end of the first pulse period would be only 12.642 V. During the period between pulses, the capacitor would discharge from this lower voltage to 4.65 V. (You should use the exponential equations to verify these voltages.) Thus, the capacitor charges less and also discharges less on each PRT of the input. The voltage waveform across the resistor would more closely resemble the input pulse.

■ SECTION 18–2 REVIEW

1. What determines the shape of the voltage and current waveforms when a pulse is applied to an *RC* circuit?

2. In an *RC* circuit, under what conditions will the output voltage be of a lower value than the input pulse?
3. Under what condition will the output voltage be equal to the value of the input pulse?

18–3 THE *RC* INTEGRATOR

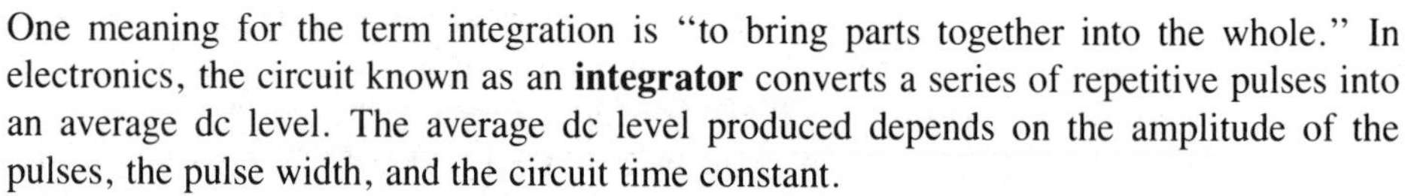

One meaning for the term integration is "to bring parts together into the whole." In electronics, the circuit known as an **integrator** converts a series of repetitive pulses into an average dc level. The average dc level produced depends on the amplitude of the pulses, the pulse width, and the circuit time constant.

If the pulse and time between pulse periods of the waveform are equal to or greater than five time constants, the capacitor will fully charge and discharge during each, as illustrated in Figure 18–12. The input to the integrator, illustrated in Figure 18–12(a), is a dc square wave with a pulse width of 10 ms. The time constant of the integrator is 1 ms. Thus, the time allowed by the waveform for charging and discharging is ten time constants. The capacitor charges or discharges in one-half this period. The output waveform is shown in Figure 18–12(b). The average dc level is nearly that of the input pulse.

FIGURE 18–12 *RC* integrator circuit with $t_w > 5\tau$

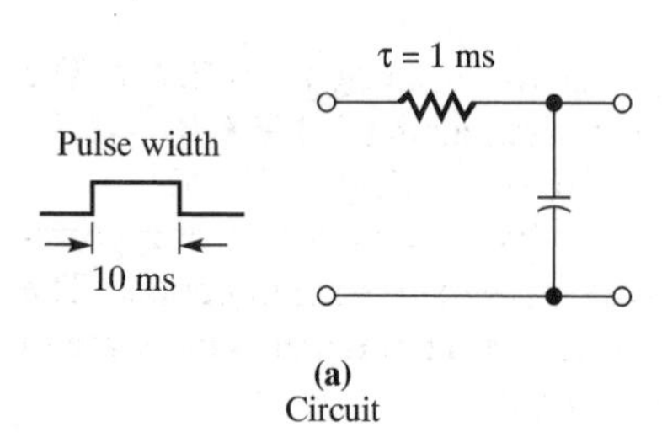

(a)
Circuit

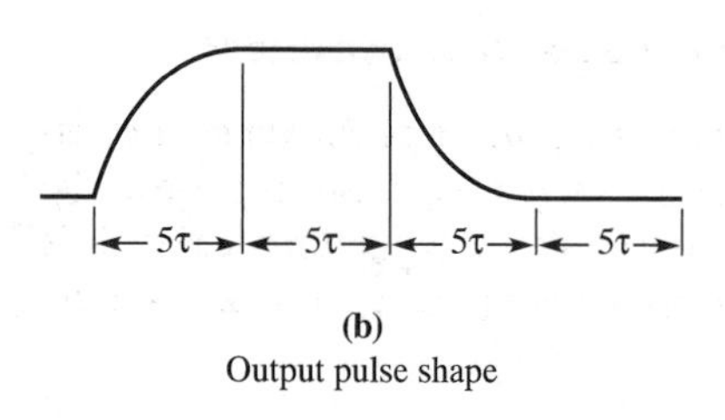

(b)
Output pulse shape

Question: What is the action of an integrator if the capacitor does not have sufficient time to charge and discharge?

If the pulse width and time between pulses is less than five time constants, the capacitor will not have sufficient time to charge. Near the end of the last section, the integrator response to a single pulse with a width less than five time constants was considered. This section will consider the action of the integrator with not one, but a series of such pulses.

The input to the *RC* integrator of Figure 18–13 is a pulsed wave with an amplitude of 10 V. The width of the pulses and the time between pulses have been made equal to one time constant. Thus, the approximations of 63% on the charge cycle and 37% on the discharge cycle can be used rather than the exponential equations. Figure 18–14(a) is a graph of the input waveform, and Figure 18–14(b) is the integrated output waveform. In the analysis that follows, it is assumed that the capacitor is initially uncharged. The capacitor takes on charge during the pulse periods, and it loses charge during the periods between pulses.

Pulse 1: During this period the capacitor charges to 63% of the total voltage. Thus at the end of the pulse, the output voltage is 6.3 V (10 × 0.63) as shown in Figure 18–14(b).

During the period between pulses 1 and 2, the capacitor discharges to 37% of the 6.3 V present at the beginning of this interval. The voltage at the end of this interval is 2.33 V (6.3 × 0.37).

Pulse 2: At the beginning of this pulse, the voltage across the capacitor is 2.33 V. The capacitor begins charging from this voltage to 63% of the remaining 7.67 V (10 − 2.33). The capacitor voltage at the end of pulse 2 is (7.67 V × 0.63) + 2.33 V = 7.16 V.

FIGURE 18–13 *RC* integrator circuit with $t_w < 5\tau$

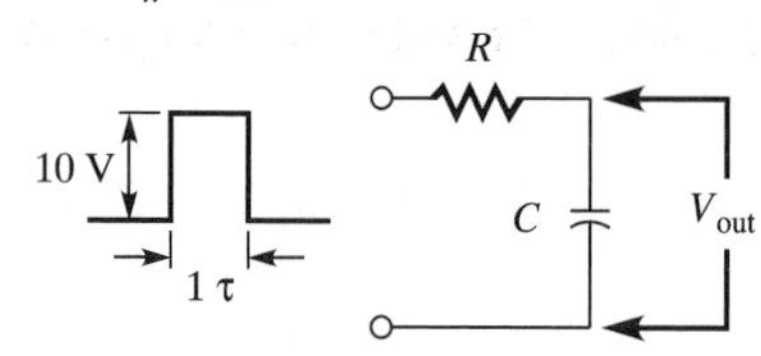

FIGURE 18–14 Input and output waveforms for *RC* integrator circuit of Figure 18–13

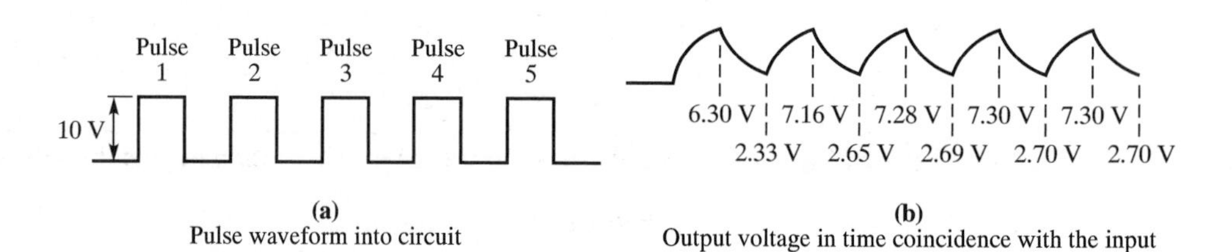

(a)
Pulse waveform into circuit

(b)
Output voltage in time coincidence with the input

During the period between pulses 2 and 3, the capacitor discharges to 37% of the 7.16 V present at the beginning of the interval. The voltage at the end of this interval is 2.65 V (7.16 × 0.37).

Pulse 3: At the beginning of this pulse, the voltage across the capacitor is 2.65 V. The capacitor begins charging from this voltage to 63% of the remaining 7.35 V (10 − 2.65). The capacitor voltage at the end of pulse 3 is (7.35 V × 0.63) + 2.65 V = 7.28 V.

During the period between pulses 3 and 4, the capacitor discharges to 37% of the 7.28 V present at the beginning of the interval. The voltage at the end of this interval is 2.69 V (7.28 × 0.37).

Pulse 4: At the beginning of this pulse, the voltage across the capacitor is 2.69 V. The capacitor begins charging from this voltage to 63% of the remaining 7.31 V (10 − 2.69). The capacitor voltage at the end of pulse 4 is (7.31 V × 0.63) + 2.69 V = 7.3 V.

During the period between pulses 4 and 5, the capacitor discharges to 37% of the 7.3 V present at the beginning of the interval. The voltage at the end of this interval is 2.7 V (7.3 × 0.37).

Pulse 5: At the beginning of this pulse, the voltage across the capacitor is 2.7 V. The capacitor begins charging from this voltage to 63% of the remaining 7.3 V (10 − 2.7). The capacitor voltage at the end of pulse 5 is (7.3 V × 0.63) + 2.7 V = 7.3 V.

Notice that after five pulses, the voltage became steady, varying between approximately 2.7 and 7.3 V, which is known as the **steady-state response** of the integrator. The output voltage will remain between these limits, so long as the input remains the same in amplitude and frequency and the time constant is not changed.

FIGURE 18–15 Output voltage of integrator circuit with extremely long time constant

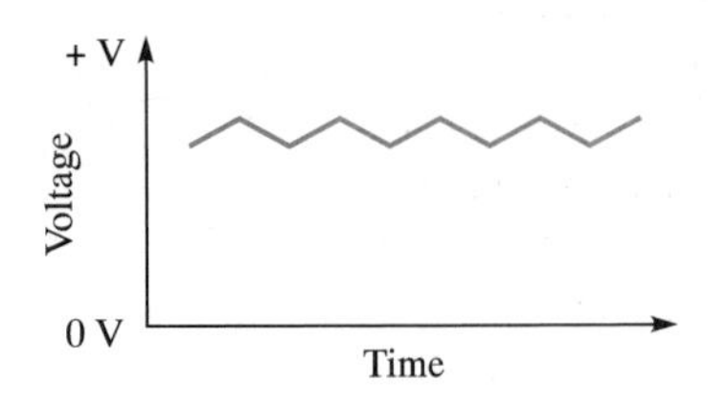

Question: What would be the effect of increasing the time constant?

If the time constant of the integrator is increased, such as through a larger value of resistance, the capacitor charges less during the pulse and discharges less between pulses. The result is less fluctuation in the voltage around its average value. If the time constant is made extremely long, the output of the integrator approaches a pure dc voltage, as illustrated in Figure 18–15.

The Integrator As a Low-Pass Filter

As you have learned, a pulse wave is made up of a fundamental frequency plus all the odd harmonics. In fact, the greater the number of harmonics the "sharper" the sides of the pulse. Actually, the sharpness of the sides of the pulse is related to its rise time (t_r). Recall that the rise time of a pulse is the time taken to increase from 10% to 90% of its maximum value. The shorter this rise time, the higher the harmonic content of the pulse, and the sharper are its sides. The equation that relates the rise time and highest frequency component (f_h) of a pulse is as follows:

(18–1)
$$f_h = \frac{0.35}{t_r}$$

and

(18–2)
$$t_r = \frac{0.35}{f_h}$$

EXAMPLE 18–2 A pulse has a rise time of 5 μs. Compute the highest frequency component.

■ Solution:

$$f_h = \frac{0.35}{t_r}$$

$$= \frac{0.35}{5\ \mu s} = 70\ \text{kHz}$$

EXAMPLE 18–3 A square wave has a fundamental frequency of 2 kHz. If the highest harmonic present is the fifteenth, what is its rise time?

■ Solution:

The fifteenth harmonic of 2 kHz is 30 kHz. Thus,

$$t_r = \frac{0.35}{f_h}$$

$$= \frac{0.35}{30\ \text{kHz}} = 11.7\ \mu s$$

Question: What effect does an integrator have on the harmonic content of a pulse waveform?

Figure 18–16(a) shows the input pulse, and Figure 18–16(b) shows the output of an *RC* integrator. Notice, in Figure 18–16(a), that the voltage between the leading and trailing edges of the input pulse is dc. Thus, this portion of the pulse contains the low-frequency components. The sharpness of the leading and trailing edges of the pulse is directly proportional to the rise time, which is directly proportional to the highest frequency component in the pulse. Thus, the high frequency components are contained in the edges of the pulse.

The shape of the output pulse shows the rounding of the edges typical of an integrator. The rise time is now longer, and the highest frequency contained in the pulse is now lower. The top of the pulse, except for leading edge rounding, is practically intact. Thus, the integrator has discriminated against the high-frequency components and passed the lower. This action is explained when you recall that an *RC* circuit, with the output taken across the capacitor, is a low-pass filter.

FIGURE 18–16 Input and output pulses of *RC* integrator circuit

■ SECTION 18–3 REVIEW

1. In an integrator, across what element is the output taken?
2. What will be the shape of the output waveform of an integrator if the time constant is very long in comparison to the pulse width?
3. What is meant by the steady state response of an integrator?
4. In what part of a pulse are the high-frequency components contained?
5. In what part of a pulse are the low-frequency components contained?
6. An *RC* integrator is what type of filter?

18–4 THE *RC* DIFFERENTIATOR

Question: What is the basic* RC *differentiator circuit, and how does it operate?

The term differentiation means "to separate or make distinct." In electronics, a **differentiator** is used to convert pulse waves into a series of short positive and negative pulses. The basic *RC* differentiator is a series *RC* circuit with the output taken across the resistor. As demonstrated in previous paragraphs, the time constant must be very short in relation to the pulse width in order to produce short, sharp pulses. The ratio of pulse width to time constant must be at least ten in order for differentiation to take place.

The *RC* circuit of Figure 18–17(a) has a time constant of 100 μs. The input is a dc square wave with a pulse width of 1 ms. The circuit time constant is one-tenth the pulse width, so the input waveform will be differentiated. When the leading edge of the pulse occurs, the capacitor begins to charge, and all the voltage is across the resistor. The current decreases exponentially as the capacitor charges, causing the voltage across the resistor to decrease in the same manner. The capacitor is fully charged in 0.5 ms or one-half the pulse period. The input wave is shown in Figure 18–17(b), and the output waveform is shown in Figure 18–17(c). Notice that the output waveform is an ac signal with a peak-to-peak value of twice the peak value of the pulse.

The Differentiator As a High-Pass Filter

Recall from the previous section that the low-frequency components are contained in the flat top of the pulse, while the high-frequency components are contained in the sides. Notice that the sides of the input pulse are present in the output, while the dc component of the top is not. Thus, the high frequencies have been passed, while the low frequencies have been blocked. As shown in Chapter 14, a series *RC* circuit with the output taken across the resistor is a high-pass filter. Thus, the action of the differentiator upon a pulse wave can be explained in terms of the action of a high-pass filter.

FIGURE 18–17 *RC* differentiator circuit

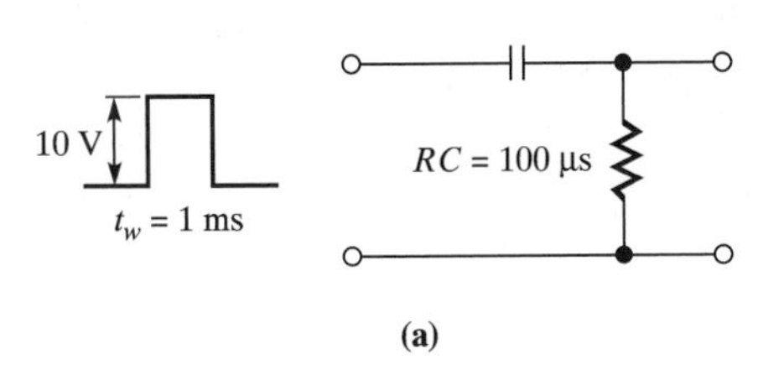

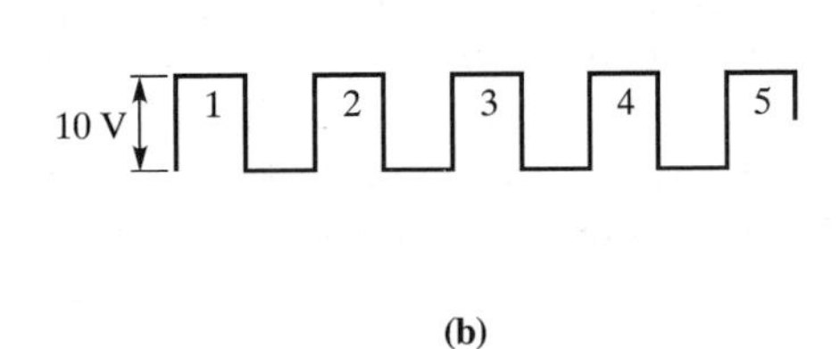

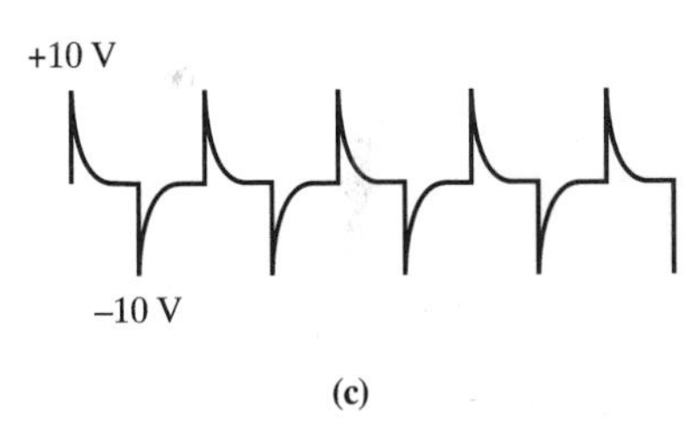

SECTION 18–4 REVIEW

1. Across which component is the output taken in an *RC* differentiator?
2. What must be the relationship between the width of the input pulse and circuit time constant in an *RC* differentiator?
3. What type of filter is an *RC* differentiator?
4. The input of a differentiator is a dc square wave. Is the output a dc or ac signal?

18–5 PULSE RESPONSE OF AN *RL* CIRCUIT

When a voltage pulse is applied to an *RL* circuit, a counter-voltage is induced in the inductor. The counter-voltage has a polarity that opposes the build up of circuit current. As the current increases, the counter-voltage decreases in an exponential manner, reaching zero after five *RL* time constants, as illustrated in Figure 18–18(a). When the pulse drops to zero, it is equivalent to replacing the voltage source with a short (assuming the internal resistance of the source is zero). The magnetic field around the inductor collapses, inducing a counter-voltage of the opposite polarity. The counter-voltage drops to zero in five *RL* time constants, as illustrated in Figure 18–18(b). This voltage induced in the inductor is identical to that produced across the resistor in the *RC* circuit.

Question: What effect does the counter-voltage have upon the circuit current and the voltage developed across the resistor?

The counter-voltage induced in the inductor opposes the change in current. Thus, since it requires five *RL* time constants for the counter-voltage to decrease to zero, it also requires five time constants for the current to build to its maximum value, as illustrated in Figure 18–19(a). Since it is this same current that produces the voltage drop across the resistor, the resistor voltage will also build exponentially over five time constants, as illustrated in Figure 18–19(b). The maximum current is by Ohm's law equal to the pulse voltage divided by the resistance. When the pulse drops to zero, the induced counter-voltage keeps the current flowing in the same direction. Thus, as illustrated in Figure 18–20(a), it also requires five time constants for the current to drop to zero. Once again, the resistor voltage follows the current and requires five time constants to decrease to zero, as illustrated in Figure 18–20(b).

Notice that the voltage produced across the inductor is similar to the voltage produced across the resistor in the *RC* circuit. Further notice that the voltage produced across the resistor is similar to the voltage produced across the capacitor in the *RC* circuit. These differences bear out the contention made previously that inductive and capacitive circuits are, for the most part, opposites!

FIGURE 18–18 Counter voltage and current waveforms produced by an *RL* circuit.

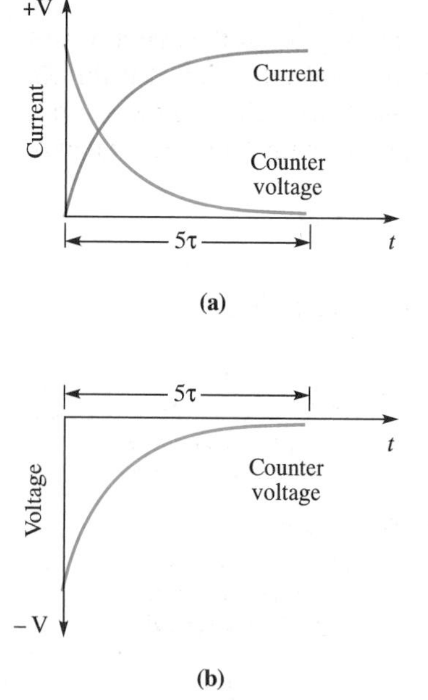

FIGURE 18–19 Effects of counter voltage

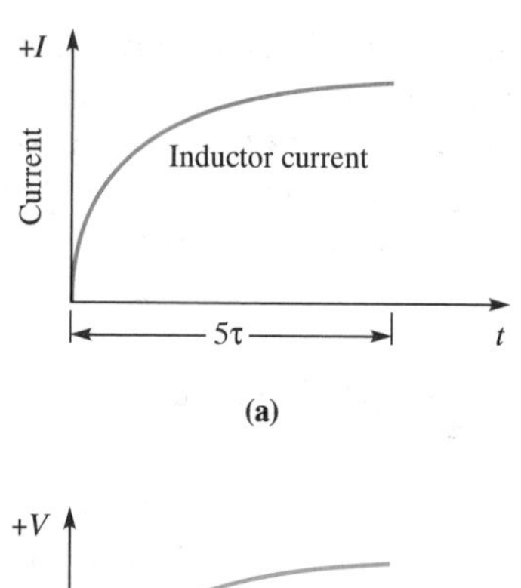

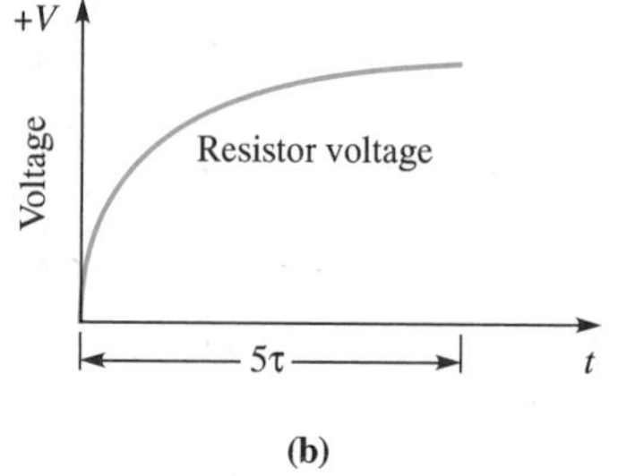

FIGURE 18–20 Current and voltage waveforms produced by collapse of magnetic field

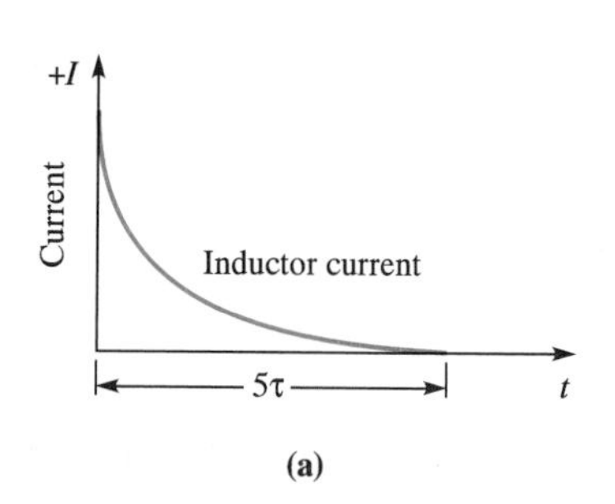

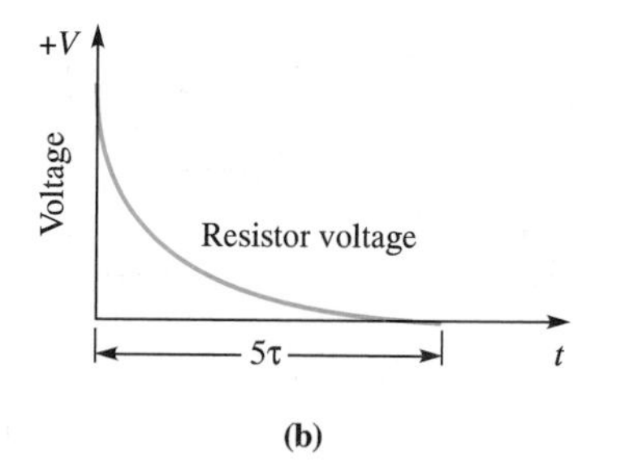

Question: What determines the shape of the waveforms in an RL *circuit?*

Identical factors affect the shape of the waveforms in an *RL* circuit as did those in the *RC* circuit. These factors are the width of the input voltage pulse and the length of the time constant. A medium pulse width—one that is equal to five time constants—allows sufficient time for the energizing current to just reach its maximum value. Thus, at the leading and trailing edges of the pulse, the voltage induced in the inductor just equals the source voltage. The resistor voltage rises to the value of the source at the trailing edge of the pulse and just drops to zero at the end of the period between pulses, as illustrated in Figure 18–21.

FIGURE 18–21 Inductor voltage waveform when $t_w = 5\tau$

If the pulse width is greater than five time constants, the time allowed for attaining full energizing current is greater than required. The inductor voltage decreases to zero in less than one pulse width, and the voltage pulses developed across the inductor are a smaller percentage of the input pulse. The current reaches its full value in less time, thus developing a wider voltage pulse across the resistor. The voltage developed across the resistor more closely resembles the input pulse. These trends continue as the pulse width is made wider. If the time constant is shortened, the inductor needs less time to energize, and the pulses developed across it are narrower, as illustrated in Figure 18–22.

FIGURE 18–22 Voltage waveforms produced when $t_w > 5\tau$

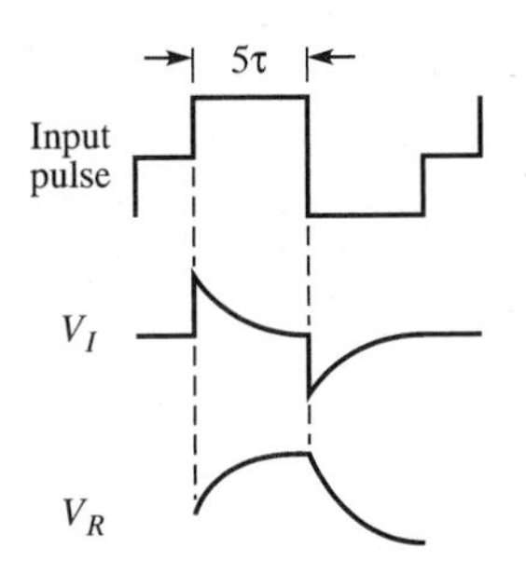

If the pulse width is less than five time constants, then insufficient time is allowed for the energizing current to reach its full value. The decrease in the counter-voltage across the inductor is not complete by the end of the pulse nor in the period between pulses. Thus, the current and resistor voltage do not reach their full values during the pulse nor drop to zero in the period between pulses, as illustrated in Figure 18–23.

The *RL* Integrator

The *RL* and *RC* integrators have identical functions. Both are used to convert a series of input pulses to an average dc level. The circuit for an *RL* integrator is shown in Figure 18–24(a). Notice that it is a series *RL* circuit with the output taken across the resistor. (In the *RC* integrator, the output was taken across the reactive component.) The time constant must be at least ten times longer than the period of the input pulse. Thus, the output is a voltage that builds up, from pulse to pulse, to some average dc value. The longer the time constant is made, the closer this average value is to dc, as illustrated in Figure 18–24(b).

Notice once again that the portion of the input pulse that appears at the output is that containing the low-frequency components, which is explained by the fact that the *RL* integrator is a low-pass filter. The component developing the output voltage is the resistor. Its voltage is greatest when the current flow is highest. Thus, the output voltage is greatest when the reactive effect is smallest.

FIGURE 18–23 Voltage waveforms produced when $t_w < 5\tau$

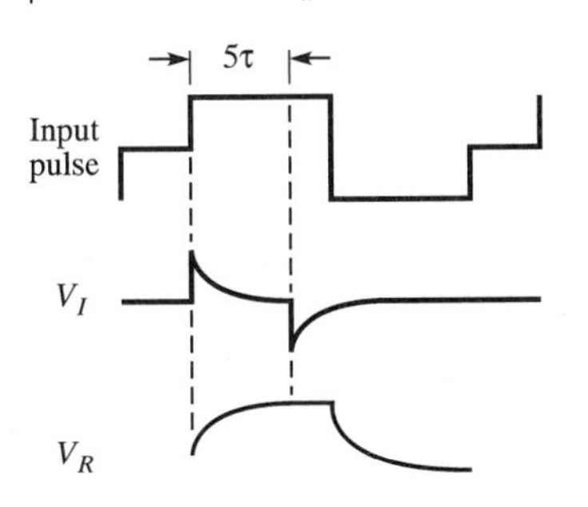

FIGURE 18–24 *RL* integrator and output voltage waveforms

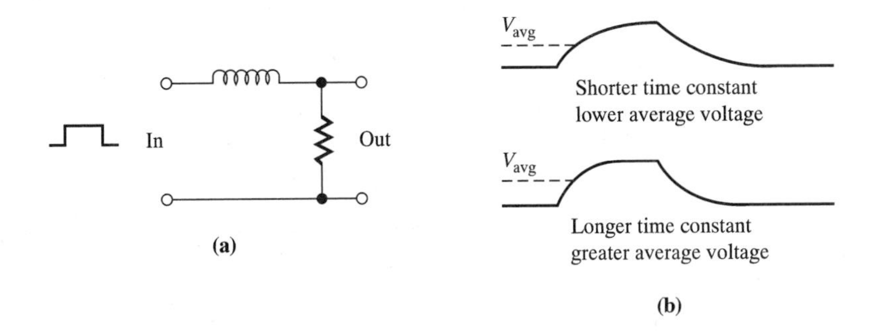

FIGURE 18–25 *RL* differentiator circuit and pulse waves

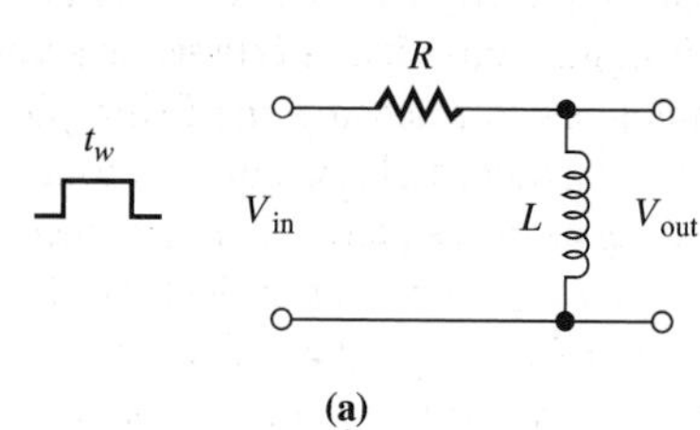

(a)
Pulse width t_w of input pulse must be at least 10 times the time constant.

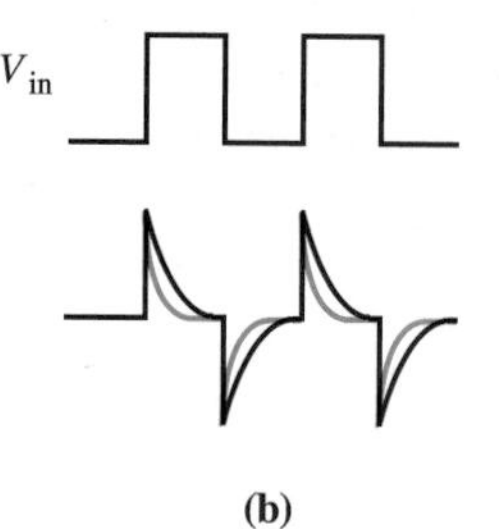

(b)
A shorter time constant or longer period results in narrower pulses.

The *RL* Differentiator

Once again, *RL* and *RC* differentiators have the same basic function. Both are used to convert an input pulse to series of short positive and negative voltage pulses. The circuit for an *RL* differentiator is shown in Figure 18–25(a). Notice that it is a series *RL* circuit with the output taken across the inductor. The width of the input pulse must be greater than ten time constants. Thus, the energizing current reaches its maximum value in one-half the pulse period and decreases to zero in one-half the period between pulses. If the pulse width is made wider or the time constant is made shorter, the pulse developed is narrower, as illustrated in Figure 18–25(b).

The output of a differentiator is the portion of the pulse (the sides) that contains the high-frequency components, which is explained by the fact that the RL differentiator is a high-pass filter. The component developing the output voltage is the inductor. Thus, its effect will be greatest at higher rather than lower frequencies.

SECTION 18–5 REVIEW

1. How many time constants are required for the energizing current to reach its maximum value?
2. In order to integrate a series of input pulses, must the time constant be long or short in relation to the width of the input pulses.
3. In an *RL* differentiator, is the output taken across the resistor or the inductor?
4. Is an *RL* integrator a low- or high-pass filter?

18–6 APPLICATIONS

The applications of *RC* and *RL* circuits, in which the signals involved are pulses, are many and varied. They are found in power supplies, motor speed controls, computer circuits, and a host of other applications. An *RC* or *RL* circuit may be used to remove a dc level from a series of repetitive pulses and pass them to another circuit without changing their shape. They may also, in the case of the differentiator and integrator, greatly change the shape of the pulses before passing them to the next circuit. Several representative applications are presented in this section.

NOTE

You will find both *RL* and *RC* circuits used in pulse applications. It must be stated, however, that *RC* circuits are by far the most common, due primarily to the relatively lower cost of capacitors as compared to inductors.

■ Triggering

One major application of the differentiator is to produce short, sharp pulses for use as triggers in digital circuits. A **trigger** is a voltage that causes an event to happen at a given time (analogous to releasing the firing pin in a rifle with a trigger). In a digital system such as a computer, a clock signal is used to trigger, or initiate, actions in a given order. The *clock signal* is a square wave with a constant amplitude and frequency. Certain things happen within the computer at the time the pulse makes a high-to-low transition. Thus, a trigger must be developed at the trailing edge of each clock pulse. The triggering effect can be accomplished using a circuit similar to that in Figure 18–26(a).

The clock signal, as shown in Figure 18–26(b), is applied to an *RC* differentiator. The time constant of the circuit is made very short in relation to the width of the clock pulses. Thus, as you have learned, a very short positive pulse is developed at the leading edge of the pulse, and a very short negative pulse is developed at the trailing edge, as illustrated in Figure 18–26(c). Since the trigger must occur at the trailing edge of the clock, the positive pulse must be discarded, and the negative pulse must be *inverted,* or changed in polarity. The rest of the circuit accomplishes this change. The component following the differentiator is known as a *diode*. When connected as shown, the diode allows the negative pulse to pass, but blocks the positive pulse, as illustrated in Figure 18–26(d). The next device is known as an *inverter.* As its name implies, it inverts the polarity of the input pulse. Figure 18–26(e) shows the final output, which is a very short pulse occurring at the trailing edge of the clock signal.

NOTE

Do not concern yourself with the operation of diodes or inverters at this time. Just accept what has been said about them for the time being. They will be covered in great detail later in the electronics curriculum. The purpose at this time is merely to show how a differentiator may fit into a circuit in order to accomplish a particular function.

■ Pulse Counting

As previously stated, pulses are often used as triggers in electronic circuits. When a trigger occurs, it often indicates that some event has taken place. It is often necessary to count these events and, after a certain number of them have occurred, cause some related event to happen. An integrator can be used for this purpose. Such a circuit is shown in Figure 18–27. If the time constant is extremely long compared to the width of the input

FIGURE 18–26 Trigger generator circuit utilizing a differentiator

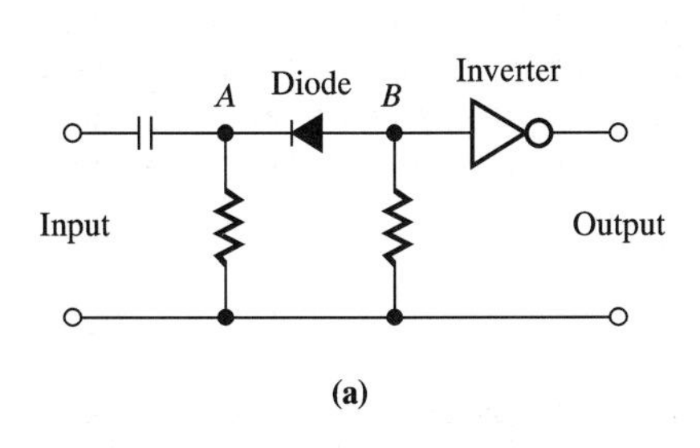

FIGURE 18–27 Integrator application: Pulse counting

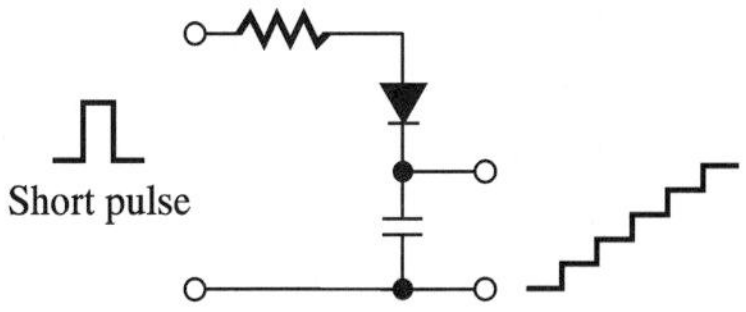

pulse, the capacitor charges only a small amount during each pulse period. The diode, as it is installed, does not allow the capacitor to discharge through the source during the period between pulses. Thus, the capacitor voltage builds up from pulse to pulse. The voltage, which builds up across the capacitor, can be computed for any number of pulses. For example, some related event may be required to occur after 262 pulses have arrived. The voltage across the capacitor can be computed for this number of pulses and used to drive a subsequent circuit that will trigger the related event. Once the related event has occurred, some means is necessary in order to discharge the capacitor.

Pulse Width Motor Speed Control

An inductor is a current-averaging device when pulses are applied to it. The armature of a dc motor is an inductor. Thus, as shown in Figure 18–28, if the motor is driven by a dc pulsed supply, the inductance of the motor integrates, or averages, the current produced by the pulses. The amount of averaging is proportional to the pulse width of the current. If the width of the current pulse is increased, the average current is higher, and the motor speed increases. If the width of the pulse is decreased, the average current is lower, and the motor speed decreases. Thus, the armature of the motor acts as an *RL* integrator and controls the speed of the motor.

Coupling Circuit

An *RC* circuit is often used to couple the ac component of a signal to a succeeding circuit stage, while it blocks the dc component. The signal shown in Figure 18–29(a) represents the output of a transistor amplifier. Notice that the voltage varies between + 8 and + 12 V. The average dc voltage is 10V, and the peak-to-peak value of the ac component is 4 V. If this signal is applied to the circuit of Figure 18–29(b), the capacitor charges to the average dc value of 10 V. When the voltage increases to 12 V, the capacitor must take on charge. To do this, current flows upward through the resistor, developing a positive going voltage across it. When the voltage drops back to 8 V, the capacitor now must lose charge. Current flows downward through the resistor and back through the source, developing a negative going voltage across it. Thus, the voltage developed across the resistor is the 4 V peak-to-peak ac component of the signal. For this circuit to be effective in coupling the signal, the time constant must be long in relation to the period of the signal.

FIGURE 18–28 Motor speed control utilizing *RL* integrator

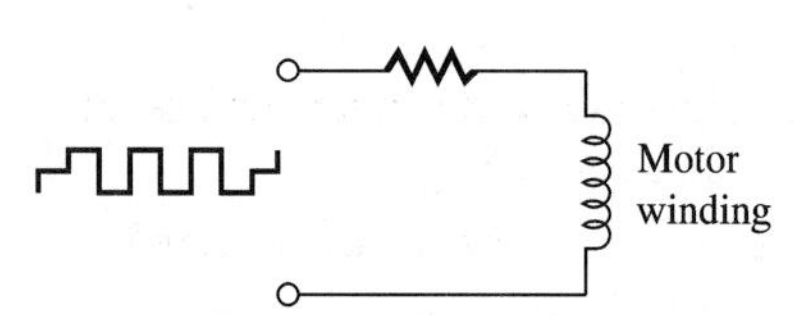

FIGURE 18–29 Differentiator application: Coupling circuit

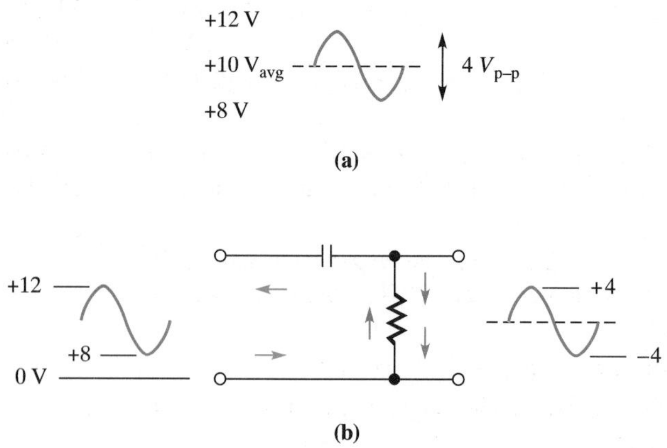

CHAPTER SUMMARY

1. During the time a voltage pulse is high, it is equivalent to placing a dc source across a circuit.
2. When a pulse drops to zero volts, it is equivalent to removing the source and replacing it with a short.
3. In an *RC* integrator, the output is taken across the capacitor.
4. In an *RC* differentiator, the output is taken across the resistor.
5. In an *RL* integrator, the output is taken across the resistor.
6. In an *RL* differentiator, the output is taken across the inductor.
7. The ratio of the width of the input pulse to the circuit time constant determines the shape of the output waveforms.
8. An *RC* or *RL* integrator is a low-pass filter.
9. An *RC* or *RL* differentiator is a high-pass filter.
10. The high-frequency components of a pulse are contained in its sides.
11. The low-frequency components of a pulse are contained in its top.
12. The shorter the rise and fall time, the greater the number of harmonics contained in a pulse.
13. The rounding of the leading and trailing edges of a pulse illustrates that an integrator is a low-pass filter.
14. The loss of the top of a pulse illustrates that a differentiator is a high-pass filter.

KEY TERMS

integrator
differentiator
steady-state response
trigger

EQUATIONS

(18–1) $$f_h = \frac{0.35}{t_r}$$

(18–2) $$t_r = \frac{0.35}{f_h}$$

Variable Quantities
f_h = high-frequency component
t_r = rise time of the pulse

TEST YOUR KNOWLEDGE

1. What is the equation for the time constant of an *RC* circuit?
2. What is the equation for the time constant of an *RL* circuit?
3. Between what two points of a pulse is the pulse width measured?
4. How many time constants must elapse in order to fully charge a capacitor?
5. If the time constant is short in relation to the width of the input pulse, does the output of an integrator resemble the input pulse?
6. If the time constant is short in relation to the width of the input pulse, does the output of a differentiator resemble the input pulse?
7. Is an integrator a low-pass or high-pass filter?
8. In an *RL* differentiator, is the output taken across the inductor or resistor?
9. In an *RC* integrator, is the output taken across the capacitor or resistor?
10. In a differentiator, what is the minimum value for the ratio of the width of the input pulse to the time constant?
11. What portion of a pulse contains the high-frequency components?
12. Which harmonics, the higher or lower order, will be discriminated against by a differentiator?
13. What are three applications of *RC* or *RL* pulse circuits?
14. Is the output of a differentiator a dc or ac waveform?

PROBLEMS: BASIC LEVEL

Sections 18–1 and 18–2

1. For the circuit of Figure 18–30, compute the value of the time constant. How much time is required for the capacitor to completely charge?

FIGURE 18–30 Circuit for Problem 1

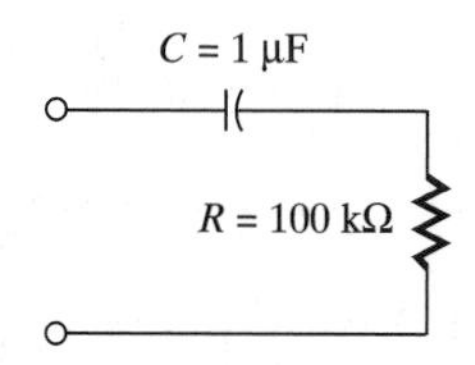

2. For the circuit of Figure 18–31, compute the value of the time constant. How much time is required for the inductor to be fully energized?

FIGURE 18–31 Circuit for Problem 2

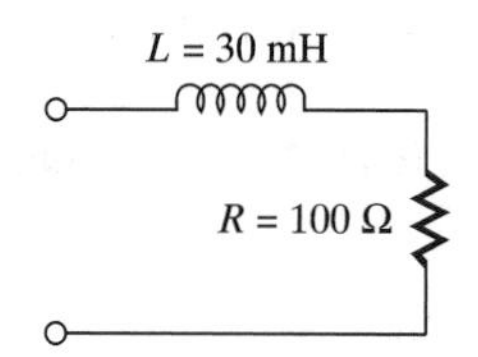

3. For the pulse of Figure 18–32, determine the approximate values of rise time, fall time, and pulse width.

FIGURE 18–32 Pulse wave for Problem 3

Section 18–3

4. For the circuit of Figure 18–33, compute the output voltage. Sketch the output waveform in time coincidence with the input. As shown, the input is a single pulse.

FIGURE 18–33 Circuit for Problem 4

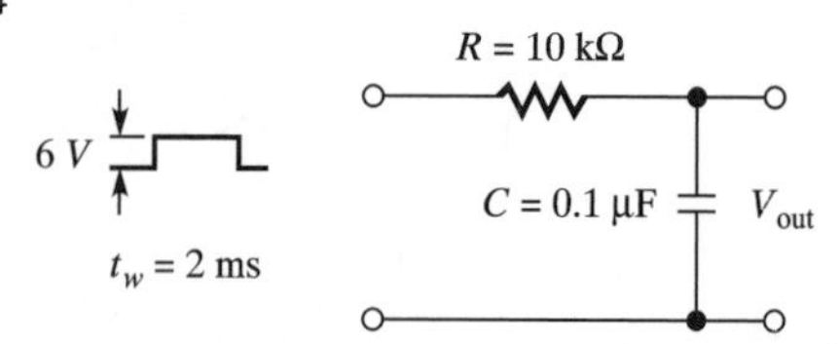

5. For the circuit of Figure 18–34, compute the value of the output voltage. Sketch the output waveform in time coincidence with the input. As shown, the input is a single pulse.

FIGURE 18–34 Circuit for Problem 5

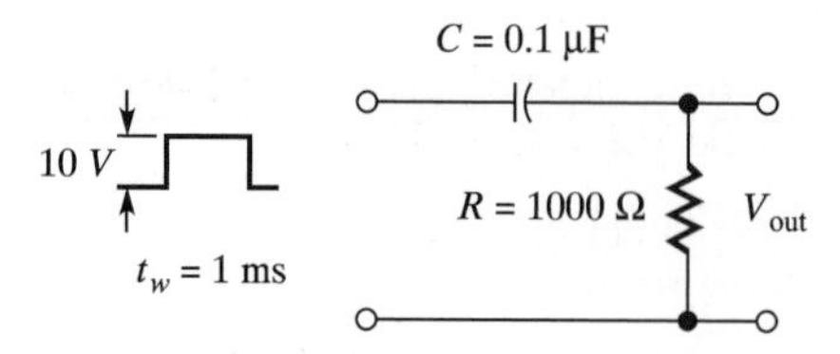

6. For the circuit of Figure 18–35, compute the output voltage. Sketch the output waveform in time coincidence with the input. As shown, the input is a single pulse.

FIGURE 18–35 Circuit for Problem 6

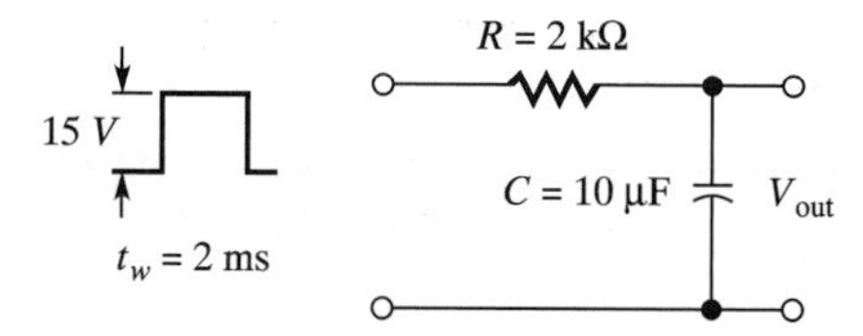

7. For the circuit of Figure 18–36, compute the output voltage. Sketch the output waveform in time coincidence with the input. As shown, the input has a single pulse.

FIGURE 18–36 Circuit for Problem 7

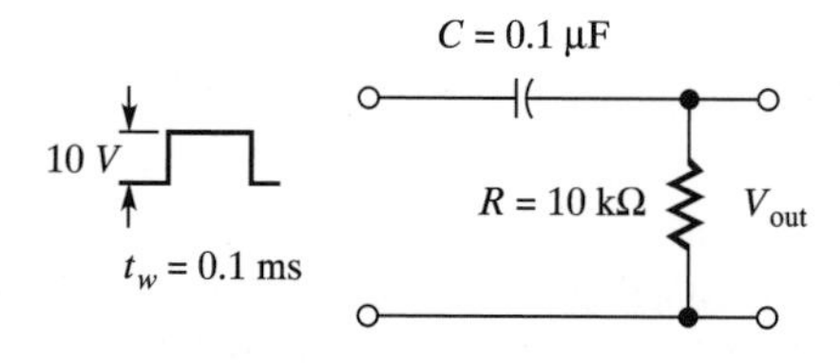

Section 18–4

8. For the circuit of Figure 18–37, compute the output voltage at the end of the second pulse period.

FIGURE 18–37 Circuit for Problem 8

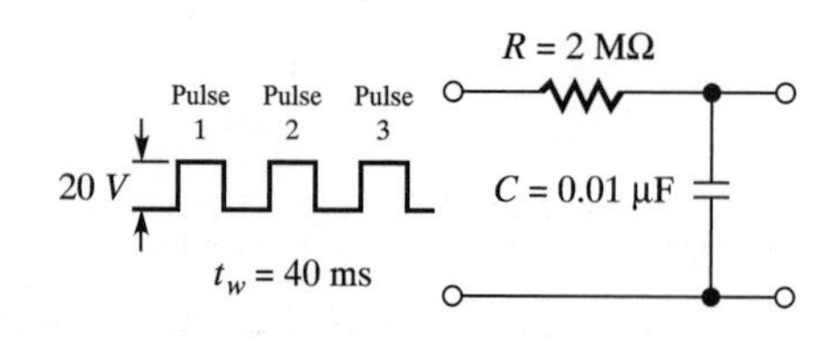

9. In Problem 8, what would be the resistor voltage at the end of the same two pulse periods?

10. Compute the highest-frequency component of a pulse whose width is 10 μs.

11. The highest-frequency component of a pulse is 500 kHz. Compute the width of the pulse.

Section 18–5

12. For the circuit of Figure 18–38, sketch the output waveform in time coincidence with the input.

FIGURE 18–38 Circuit for Problems 12 and 13

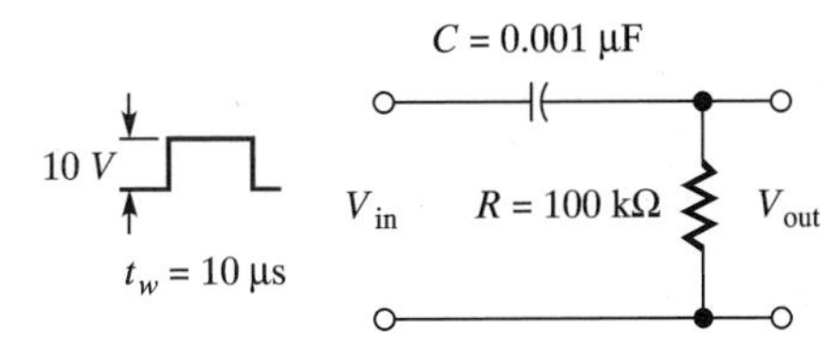

13. In Problem 12, reduce the time constant by 50%, and redraw the output waveform.

14. For the circuit of Figure 18–39, draw the waveforms across both the resistor and capacitor in time coincidence with the input.

FIGURE 18–39 Circuit for Problem 14

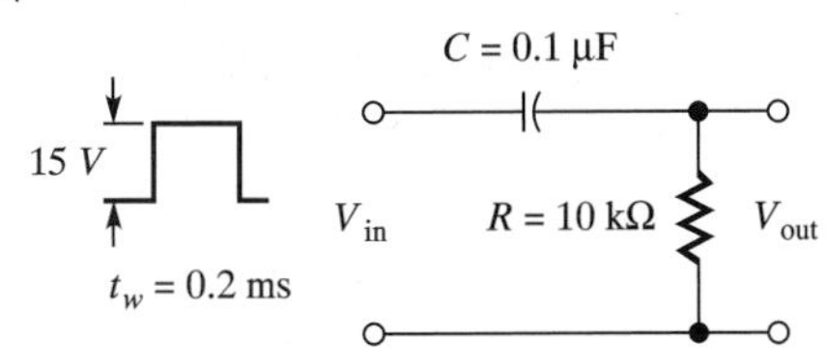

Section 18–6

15. For the circuit of Figure 18–40, what is the voltage across the resistor at the trailing edge of the first pulse?

FIGURE 18–40 Circuit for Problems 15 and 16

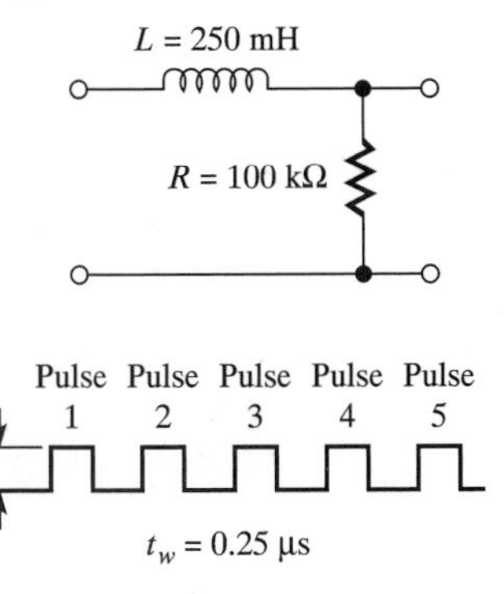

16. In Problem 15, what is the voltage across the capacitor at the trailing edge of the first pulse?

17. For the circuit of Figure 18–41, compute the steady-state response.

FIGURE 18–41 Circuit for Problems 17 and 18

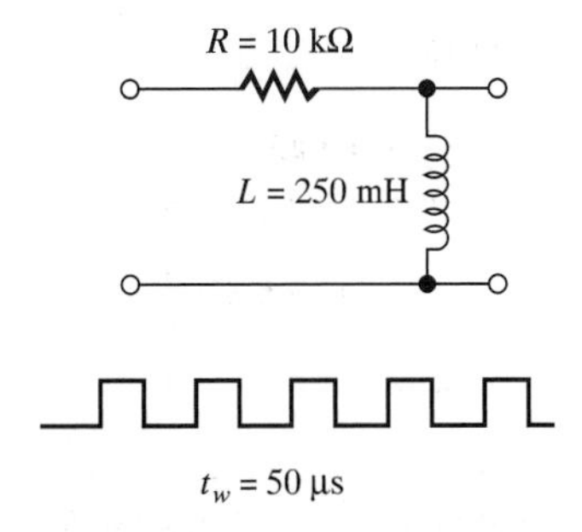

18. How many pulses does it require in Problem 17 to reach the steady-state response?

19. It is necessary to convert the circuit of Problem 17 to a differentiator. What component change(s) could be made to accomplish this?

20. For the circuit of Figure 18–42, compute the steady-state response.

FIGURE 18–42 Circuit for Problem 20

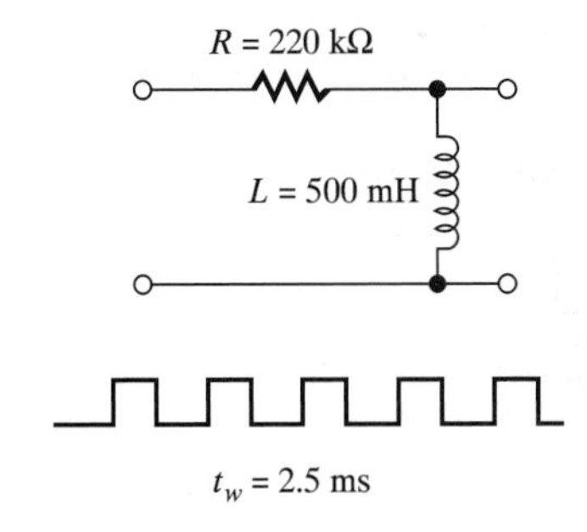

PROBLEM SET: CHALLENGING

21. For the circuit of Figure 18–43, compute the time constant.

FIGURE 18–43 Circuit for Problems 21 and 22

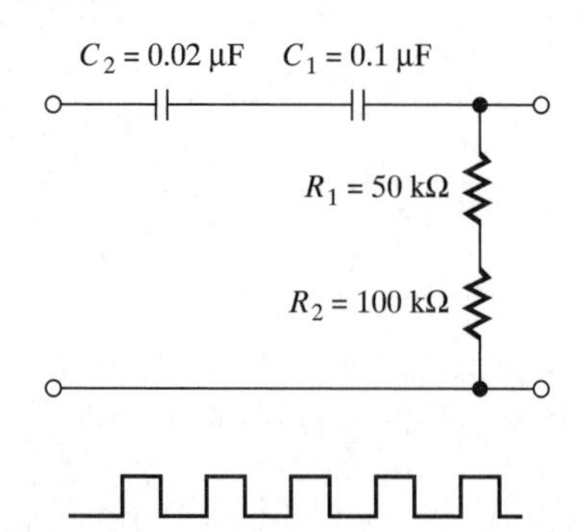

22. For the circuit of Figure 18–43, compute the voltage across the resistor at the end of the first five pulse periods.

23. For the circuit of Figure 18–44, compute the time constant.

24. For the circuit of Figure 18–44, compute the steady-state response.

FIGURE 18–44 Circuit for Problems 23 and 24

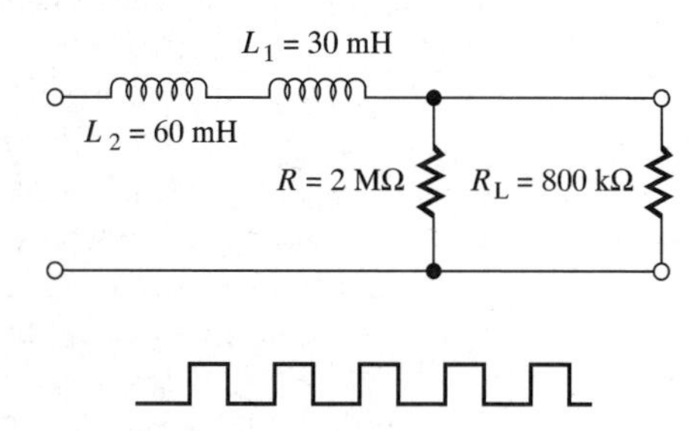

25. For the pulse of Figure 18–45, compute the highest-frequency component present.

FIGURE 18–45 Circuit for Problem 25

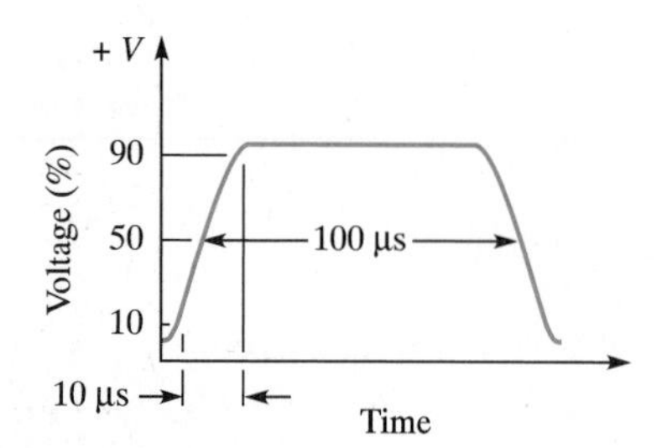

26. The highest-frequency component in a pulse is 80 kHz. Compute the rise time.

Medical Electronics

In the past twenty years or so, electronic systems and devices have become pervasive in medical procedures. Operating rooms, intensive care units (ICU), coronary care units, and recovery rooms are replete with electronic devices and systems. Today, some surgical procedures, which would otherwise be impossible, are performed with such equipment. Some internal problems can be detected using sophisticated electronic systems rather than surgery. In other uses, patients recovering from surgery or a heart attack are continuously monitored using electronic systems; pacemakers are implanted in the heart to regulate the pulse; and portable systems that continuously monitor vital signs are worn by heart patients.

Electronic devices are also used outside the hospital/clinical setting. Many health centers today have testing facilities in which tests are made with the results fed to a computer that analyzes the data and prepares a printout of the results. This list could go on, but these examples show the pervasiveness of electronics in medicine.

What was the impetus for this union of medicine and electronics? First of all, the body functions are controlled by precise electrical signals. Transmitted from the brain through the nervous system, these minute signals tell a muscle when to contract, an eyelid when to blink, or the heart when to pump. Second, the tissues of the body are made up of atoms containing electrons in motion. Thus, magnetic fields are produced by the organs. These magnetic fields can be detected and used to indicate the condition of a body organ, especially the brain. Thus, the electrical nature of body tissues and control of them make electrical detection and stimulation possible.

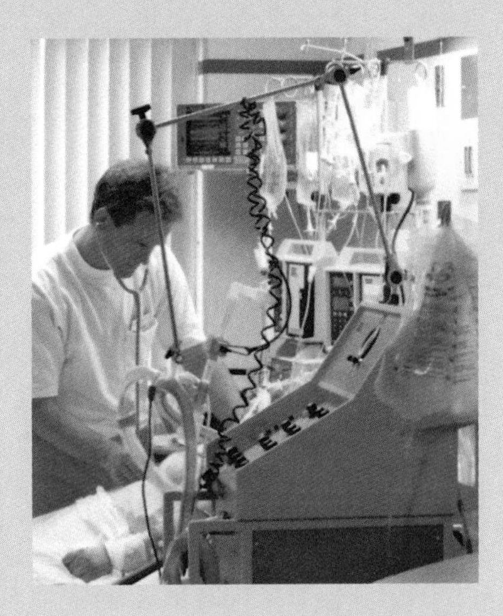

Another impetus was the advent of the space program, where the monitoring of vital signs of people under extreme physical and mental stress became necessary. At such great distances, an electronic solution was not only desirable but necessary. The vital signs of the astronauts had to be detected, converted to scaled electrical signals, impressed on an electromagnetic wave carrier, carried to earth by a radio link, detected, and once again displayed as heart-beats-per-minute, degrees of temperature, or breaths per minute—all with a high degree of accuracy. The direct spin-off of space technology to medical technology has been tremendous!

One such spin-off is the patient monitoring system used in hospitals. In the past, it was necessary for a nurse or doctor to make regular visits to each patient in order to monitor vital signs. However, during periods when health personnel were not present, a patient could have a seizure and no one would know about it until it was too late. A monitor system, such as illustrated in the above photograph, has greatly increased the amount of surveillance a patient receives. A nurse can continuously monitor the vital signs of all patients on the ward. With closed-circuit television, the same nurse can monitor all that happens in the ward. This type of system affords the patient better care and allows the hospital or clinic to operate with a reduced staff.

Emergency medical services make great use of electronics in their business of saving lives of accident and heart attack victims. When dispatched to an accident, they first must make sure that the patient is stable enough to endure the fast ride to the hospital. They carry portable monitoring equipment through which the victims vital signs can be telemetered to a doctor at the hospital. (*Telemetering* means measuring at a distance.) The doctor can then prescribe treatment or medication to make the victim stable prior to being moved.

What does all this mean to you and your career as an electronics technician? First of all, it opens doors to employment in the calibration and repair of the electronic equipment used. Every hospital, clinic, or emergency medical service becomes a possible employer. Repair and calibration of medical electronics equipment is very serious business. Used as it is in the health care of an individual, it must be in perfect working order. After all, if a doctor sees an irregularity on the electrocardiogram of a patient, he/she wants to know for sure that it not just a problem with an amplifier in the equipment! Further, diagnoses are often made based upon the results of tests made with electronic systems. Thus, the wrong diagnosis could result in the wrong treatment and have an adverse effect on the health of the patient. The health facility then would be in danger of a malpractice suit! Thus, regular calibration of these systems by a qualified technician is in the interest of both the patient and the hospital.

There are other compelling reasons for health facilities employing electronics technicians to service, repair, and calibrate their electronic equipment. If a piece of equipment fails, the hospital must have backup equipment available that it can bring up on a moments notice. Without a technician to perform immediate repairs, the hospital must keep a large and costly inventory of what is usually idle equipment. With a full time technician on the staff, an average hospital can usually reduce its inventory by as much as 50%. A surgeon often wants a technician in the operating room in the event of trouble in the equipment. Thus, the technician would have to scrub, don operating room gear, and stand by during the operation.

Another function of a medical electronics technician is to ensure the safety of patients attached, through sensors, to several systems at the same time. A ground fault in such a system could be devastating to the patient. It would be terrible, would it not, to survive open heart surgery only to be electrocuted by a faulty electronic monitoring system. Thus, safety checks of the electronic systems are made daily in the ICU and CCU.

To get into this field as a technician, you need, first of all, the good grounding in the basics of electronics that a course such as the one you are taking provides. You may need specialized training on specific items of equipment, especially as new products enter the market. The manufacturer, who has just as much interest as the hospital in the proper use of the equipment, will provide seminars for technician training. The employer usually finances these seminars and provides time for technicians to attend. You will, of course, have to become a member of the professional organizations that keep their members abreast of new developments in the medical electronics field. As the responsible person on the staff, the technician must ensure that he/she stays current in all phases of the medical electronics field, especially anything pertaining to safety.

Whether or not you are cut out for the medical electronics field depends to a great degree upon your emotional and physical make up. Remember that there may be occasions when you must be in the operating room. Can you take the tension and stand the sight of blood? Can you work around an ICU or CCU where there is usually a great deal of human suffering? Can you work in harmony with the doctors and nurses on the staff? Can you handle the stress of knowing that the equipment for which you are responsible may mean life or death to a patient? Only you can answer these questions.

CHAPTER

19 Complex Quantities in AC Calculations

■ **UPON COMPLETION OF THIS CHAPTER, YOU WILL BE ABLE TO**

1. Write any number using real or imaginary number notation.
2. Explain what is meant by a complex number.
3. Write complex numbers in rectangular form.
4. Write complex numbers in polar form.
5. Perform addition and subtraction of complex numbers in rectangular form.
6. Perform multiplication and division of complex numbers in polar form.
7. Write the values of voltage and impedance of ac circuits using complex numbers.
8. Perform analysis of *RL* and *RC* circuits using complex algebra.

Basic algebra is not sufficient for ac circuit analysis. The reason is that the currents, voltages, and impedances are no longer in phase but separated by a phase angle. The source voltage is composed of two components: the resistive voltage and the reactive voltage. The impedance is composed of a resistive and a reactive component. These components are separated by an angle of 90°.

In this chapter, you will learn of a notation, known as a complex number, that contains both the magnitude and the angle. Complex numbers come in two forms: polar form and rectangular form. (The rectangular form is sometimes referred to as *Cartesian form.*) The polar form consists of a magnitude and an angle. Thus, the magnitude and angle are expressed directly. The rectangular form consists of two components, one of which is rotated from the other by 90°, which should sound familiar to you. This 90° rotation is performed by an operator which is given the designation j.

The impedance, voltage, or current in a circuit can be expressed in either polar or rectangular form. The real benefit in expressing them in this manner is that arithmetic can be done directly with these numbers since the magnitude and angle are both present. The operations of mathematics can be done in either polar or rectangular form. You will find, however, that addition and subtraction are done most conveniently in rectangular form, while multiplication and division are more easily done in polar form.

In this chapter, the background and mathematics of complex numbers is presented. Their use is demonstrated in the analysis of representative ac circuits. Even if you never have the occasion to solve real-world problems using complex numbers, you will need a knowledge of them. Many technical journal articles and text books use them in their presentations. Thus, a knowledge of their theory and application will allow you to read these articles with understanding!

19–1 REAL AND IMAGINARY NUMBERS

The **real number system**, the system of numeration with which most people are familiar, is the set of numbers composed of the positive and negative integers, fractions, mixed numbers, rational, and irrational numbers. The number of this set represent all the points on a number line, a small portion of which is shown in Figure 19–1. As shown, this line is horizontal and moves both to the right and left of an origin.

Notice on the number line of Figure 19–2(a) the point labeled $+5$. This length represents a phasor 5 units in magnitude and at an angle of zero degrees. If this phasor were rotated 180°, the magnitude would now be -5 at an angle of 180°. What mathematical operation will cause this 180° rotation to take place? If the original value of $+5$ were multiplied by a -1, the result would be -5; that is, $5 \times (-1) = -5$. Thus, minus one

FIGURE 19–1 Real number line

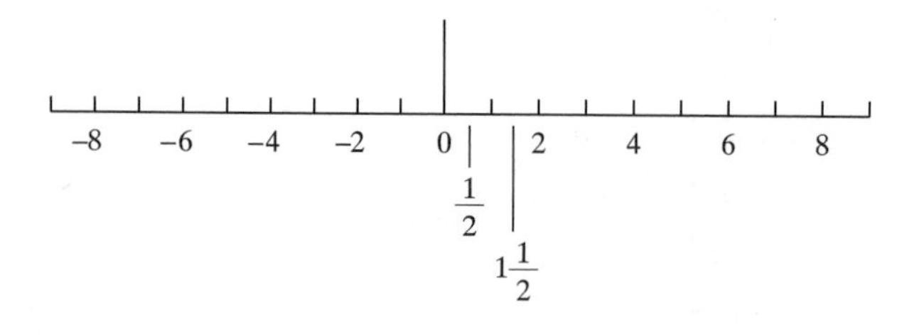

FIGURE 19–2 Rotating a phasor through 180°

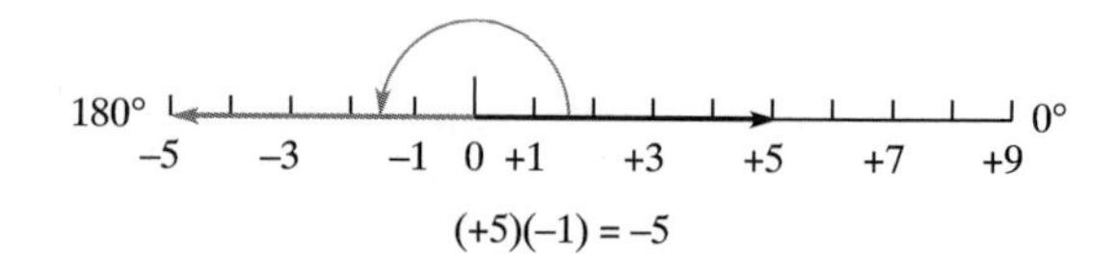

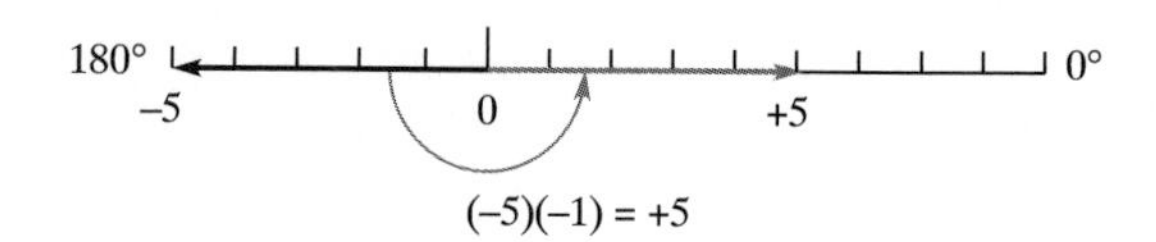

(−1) is an operator that rotates a phasor 180°, as illustrated in Figure 19–2(b). The rotation of 180° occurs each time the phasor is multiplied by a −1. (You can prove this by multiplying once again by a minus one; the phasor returns to zero degrees).

Question: Is there an operator that will rotate a phasor by 90°?

First, assume that there is such an operator and give it the designation j. (The reason j was selected will be made apparent later.) Refer to the number line of Figure 19–3(a). If the phasor with a magnitude of +5 is multiplied by j, it will rotate as shown. Notice that the phasor magnitude is now specified as $j5$. If the phasor were again multiplied by j, it would rotate another 90°, or a total of 180°. Now the phasor magnitude is specified as $j^2 5$, as illustrated in Figure 19–3(b). Thus, 180° rotation of the phasor is accomplished through multiplication by j^2. A comparison of Figure 19–2 and 19–3(b) shows that they are identical. That is to say, the operators −1 and j^2 yield the same results, which leads to the following relationships:

$$j^2 = -1 \qquad \text{and} \qquad j = \sqrt{-1}$$

FIGURE 19–3 Phasor diagram of the *j*-operator

j5

90°

0 +5

(a)
$+5 \times j = j5$

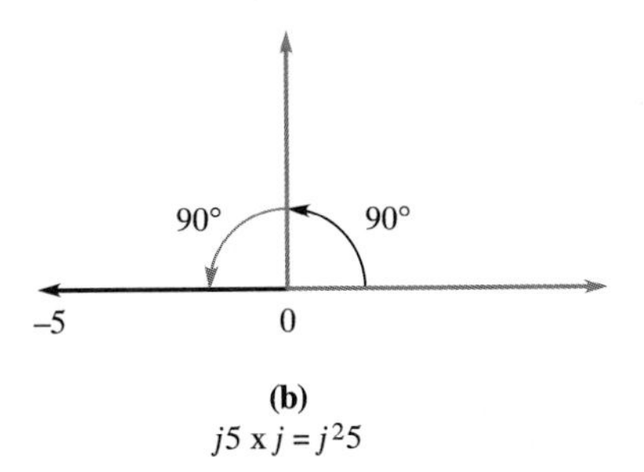

(b)
$j5 \times j = j^2 5$

These relationships show that if a phasor is multiplied by the square root of a minus one, it will be rotated 90°. The answer to the question, What is the value of $\sqrt{-1}$, is that there is no such value. It is known as an **imaginary number**.

As shown in Figure 19–4, if all the values on the real number line are multiplied by the operator j, the entire number line will be rotated 90°. Thus, the ***j*-operator** rotates a number 90° each time it is applied. Since these values are multiplied by an imaginary number, the entire set is imaginary. This *imaginary number line* contains the same values as the real number line, but their notation is preceded by j. The term *imaginary* is not meant to imply that the numbers do not exist. It is merely a convenient way of differentiating between the two number systems.

NOTE

In mathematics, a number is preceeded by the letter i in order to indicate that it is imaginary. When used in electronics, however, the letter i means an instantaneous value of current. Thus, to avoid confusion, the next letter in the alphabet, j, is used to denote the imaginary number in electronics.

FIGURE 19–4 Imaginary number line

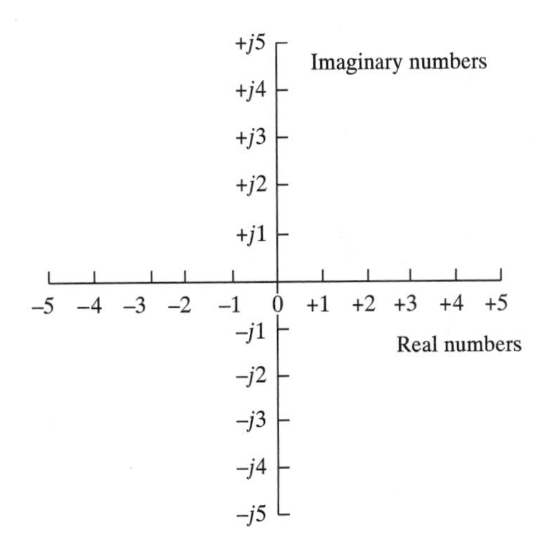

FIGURE 19–5 Complex plane: All points in all directions around the origin

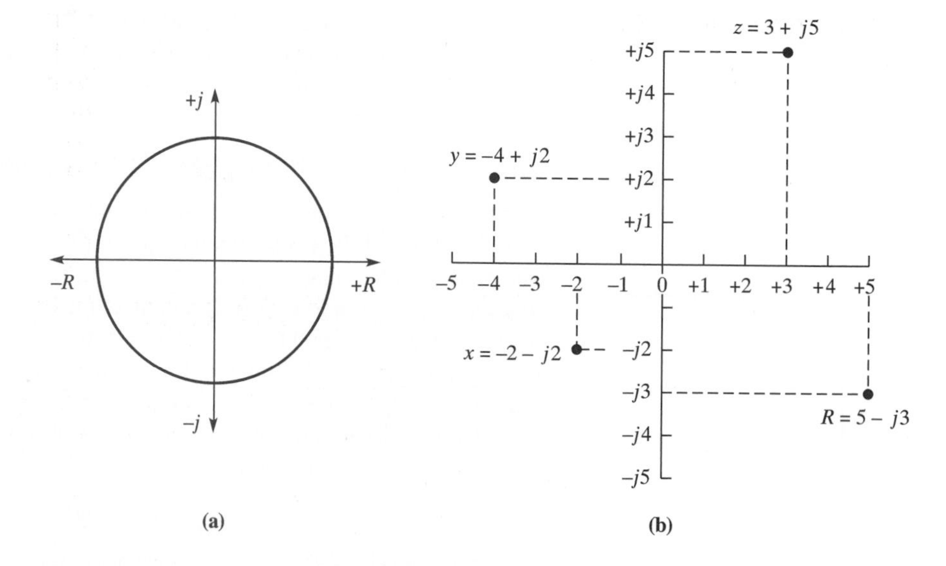

SECTION 19–1 REVIEW

1. What operator rotates a phasor through an angle of 180°?
2. What operator rotates a phasor through an angle of 90°?
3. What is the numerical value of the *j*-operator?
4. What is an imaginary number?

19–2 COMPLEX NUMBERS

Question: What is a complex number?

A **complex number** is one composed of two components: one real and one imaginary. As illustrated in Figure 19–5(a), the area in all directions around the origin is known as the *complex plane*. Any point in this plane can be assigned a number consisting of two components: one real and one imaginary. Consider for example point *Z*. As illustrated in Figure 19–5(b), its value is found by drawing lines that intersect it, with one line parallel to the real axis and one parallel to the imaginary axis. The real portion is the distance measured along the real axis, while the imaginary portion is that measured along the imaginary axis. Thus, the real part of this number is 3 and the imaginary part is 5. This number can be written as $3 + j5$. As another example, consider point *R*. Notice that it is 5 units to the right of the origin and 3 units below it. The real part of the number is 5 and the imaginary part is -3. This number can be written as $5 - j3$.

EXAMPLE 19–1 Write the complex number associated with each point shown in Figure 19–6.

■ Solution:

1. Point *A* lies 6 units to the right of the origin and 4 units above it. Thus,

$$A = 6 + j4$$

2. Point *B* lies 5 units to the left of the origin and 4 units above it. Thus,

$$B = -5 + j4$$

3. Point *C* lies 8 units to the left of the origin and 6 units below it. Thus,

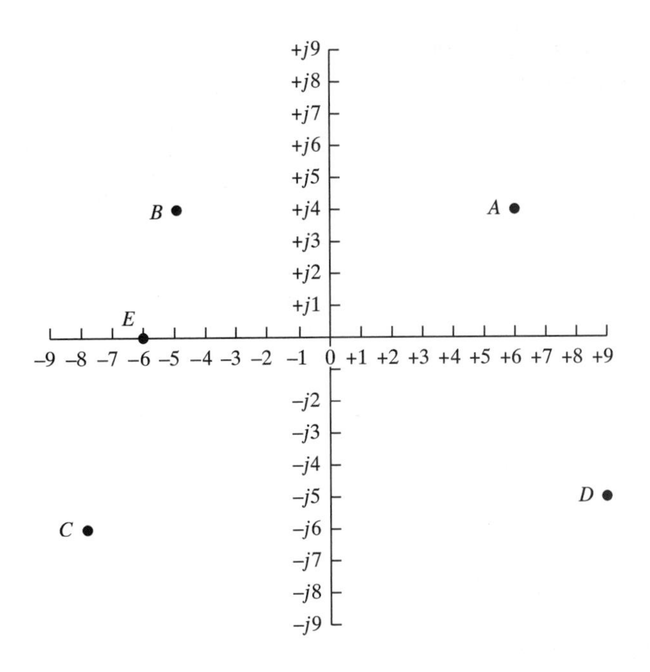

FIGURE 19–6 Number lines for Example 19–1

$$C = -8 - j6$$

4. Point D lies 9 units to the right of the origin and 5 units below it. Thus,

$$D = 9 - j5$$

5. Point E lies 6 units to the left of the origin. Thus the real portion is -6 and the imaginary portion is 0:

$$E = -6 + j0$$

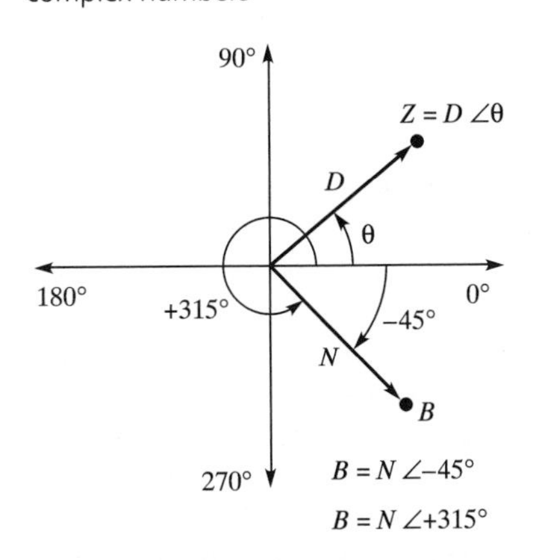

FIGURE 19–7 Polar form of complex numbers

This method of expressing complex numbers—a real part plus an imaginary part—is known as the **rectangular form,** in which the complex number is composed of the sum of two phasors with an indicated phase difference of 90°. This name stems from the rectangle formed by the lines drawn in Figure 19–5(b). This form is sometimes referred to as *Cartesian form* named for the French mathematician, physicist, and philosopher, Rene Descartes. The number is actually composed of two phasors, with the j-operator indicating a phase difference of 90°.

Complex Numbers in Polar Form

Question: How are complex numbers expressed in polar form?

There is another method of expressing complex numbers known as **polar form,** in which the complex number is expressed as a magnitude and an angle, as illustrated in Figure 19–7. Point Z can be found by facing in the direction of the reference, turning through angle θ, and moving the distance required to reach point Z. Thus, the value of Z can be expressed as

$$Z = D\angle\theta$$

The angles formed are measured from the reference, which is usually a horizontal line drawn to the right of the origin. These angles may be measured either in a counterclockwise or clockwise direction from the reference line. Thus, point B could be expressed in two ways:

$$B = N\angle -45^\circ$$

or

$$B = N\angle +315^\circ$$

EXAMPLE 19–2 Express the numbers associated with the points in the complex plane of Figure 19–8 in polar form.

FIGURE 19–8 Complex plane for Example 19–2

■ Solution:

1. Point A is 7 units long and forms an angle of 45° with the reference. Thus,

$$A = 7\angle 45^\circ$$

2. Point B is 3 units long and forms an angle of -30° with the reference. Thus,

$$B = 3\angle -30^\circ$$

3. Point C is 3 units long and forms an angle of 135° with the reference. Thus, point C may be expressed as

$$C = 3\angle 135^\circ$$

SECTION 19–2 REVIEW

1. What is a complex number?
2. What is meant by rectangular form of complex notation?
3. What is meant by polar form of complex notation?
4. For the graph of Figure 19–9, state the value of each point in rectangular form.
5. For the graph of Figure 19–10, state the value of each point in polar form.

19–3 CONVERSION BETWEEN POLAR AND RECTANGULAR FORMS

It is often desirable to convert from one form of complex notation to the other. As you will learn in the next section, it is easier to add or subtract in rectangular form than it is in polar form. And by the same token, it is easier to multiply or divide in polar form than it is in rectangular form. In this section, you will learn methods of converting from one complex form to the other.

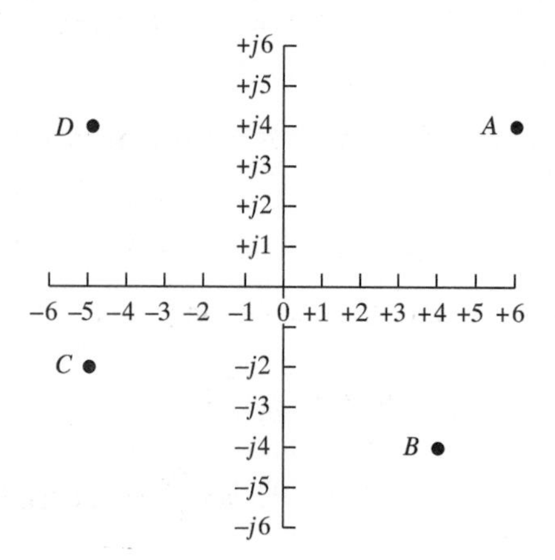

FIGURE 19–9 Number lines for Section Review Problem 4

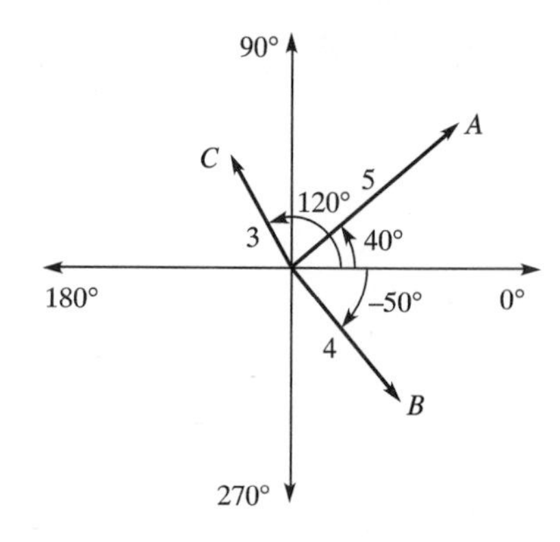

FIGURE 19–10 Complex plane for Section Review Problem 5

■ Converting from Rectangular to Polar Form

Recall that in rectangular form a complex number is composed of two phasors separated by an angle of 90°. A complex number Y is represented in phasor form in Figure 19–11(a). In rectangular form, Y is equal to $R + jX$. Notice that the phasors R and X form two legs of a right triangle. In order to convert this to polar form, it is necessary to compute the value of the hypotenuse of this triangle and the angle θ adjacent to R, which can be accomplished using the Pythagorean theorem and the trigonometric functions.

FIGURE 19–11 Converting rectangular form of a complex number to the polar form

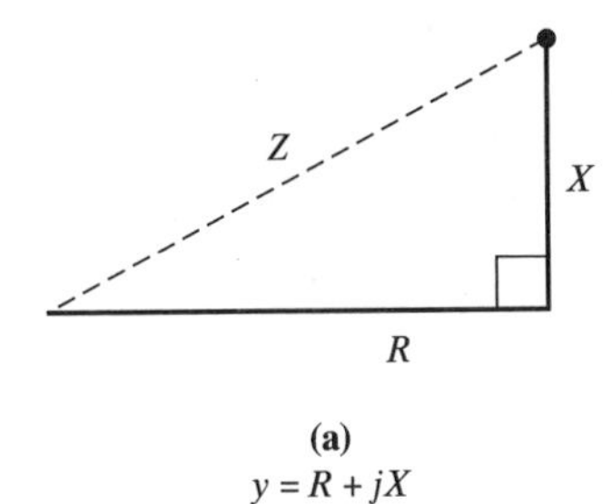

(a)
$y = R + jX$

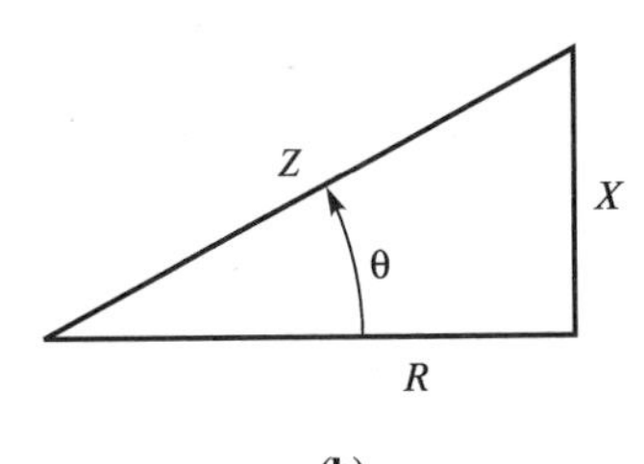

(b)
$Z = \sqrt{R^2 + X^2}$

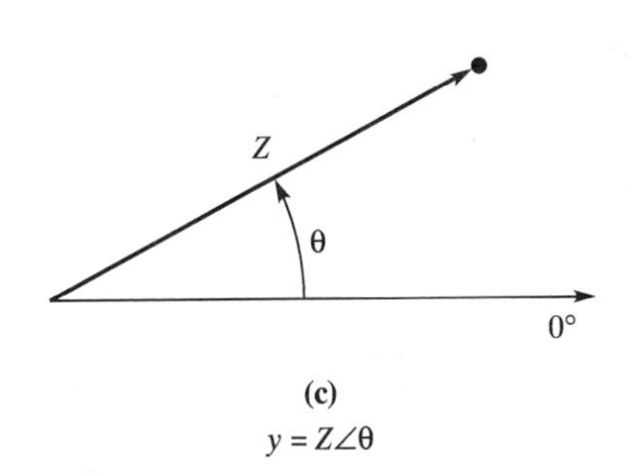

(c)
$y = Z\angle\theta$

Question: How is the conversion from rectangular to polar form accomplished?

Recall that the Pythagorean theorem states that the sum of the squares of the sides of a right triangle is equal to the square of the hypotenuse. Thus, in Figure 19–11(b), Z^2 is equal to the sum of R^2 plus X^2:

(19–1) $$Z^2 = R^2 + X^2$$

Extracting the square root of both sides of this equation yields

(19–2) $$Z = \sqrt{R^2 + X^2}$$

The trigonometric functions are then used in order to compute the value of the angle θ. The function most often used in order to find the angle θ is the tangent function. The tangent function is equal to the side opposite the angle X divided by the side adjacent to the angle R:

(19–3) $$\tan\theta = \frac{X}{R}$$

This equation can be transposed to yield an equation for the angle θ:

(19–4) $$\theta = \arctan\frac{X}{R}$$

As shown in Figure 19–11(c), the elements making up a complex number in polar form have been determined. Thus, in polar form, Y is equal to Z at an angle of θ degrees:

$$Y = Z\angle\theta$$

EXAMPLE 19–3 A complex number Z is equal to $40 + j30$. Convert Z to the polar form.

■ Solution:

1. The magnitude of Z is computed using the Pythagorean theorem:

$$Z = \sqrt{40^2 + 30^2}$$
$$= \sqrt{2500} = 50$$

2. The angle is computed using the tangent function:

$$\tan\theta = \frac{30}{40}$$

$$= \arctan\frac{30}{40} = \arctan 0.75 = 36.9°$$

3. Thus, Z in polar form is

$$Z = 50\angle 36.9^\circ$$

The calculator sequence for Example 19–3 is

KEY	DISPLAY
40	40
X^2	1600
+	1600
30	30
X^2	900
=	2500
$\sqrt{\ }$	50

In Example 19–3 both the real and the imaginary part of the complex number were positive. The following examples show how the conversion is done if one or both of the components of the complex number is negative.

EXAMPLE 19–4 A complex number X is equal to $-50 + j20$. Convert X to polar form.

Solution:

1. First draw a simple phasor diagram of $-50 + j20$, shown in Figure 19–12. Notice that the angle of the phasor is between 90° and 180°.
2. Solve for the magnitude of the phasor:

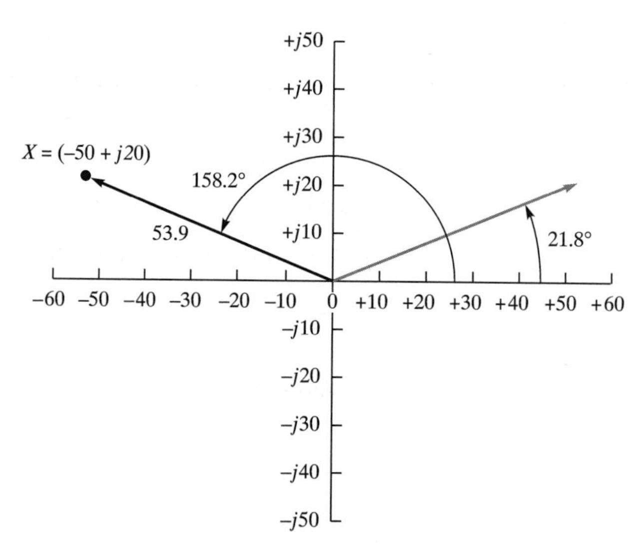

FIGURE 19–12 Complex plane for Example 19–4

$$X = \sqrt{50^2 + 20^2}$$
$$= \sqrt{2900} = 53.9$$

3. Solve for the angle:

$$\theta = \arctan \frac{20}{50}$$
$$= \arctan 0.4 = 21.8°$$

This angle would place the phasor at a position indicated by the dashed line, which is not the correct position. The true angle is 180° − 21.8°–158.2°. This angle places the phasor in the correct position:

$$X = 53.9\angle 158.2°$$

Without the phasor diagram, this result may not have been apparent. Thus, it is a good idea to draw a phasor diagram when making conversions.

EXAMPLE 19–5 A complex number Y is equal to $-12 - j7$.
Convert Y to polar form.

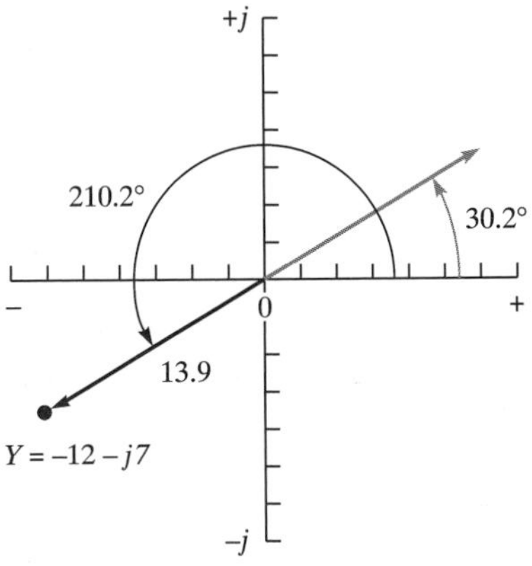

FIGURE 19–13 Complex plane for Example 19–5

1. Once again, draw a simple phasor diagram as shown in Figure 19–13. Notice that Y will be at some angle between 180°–360°.
2. Solve for the magnitude of Y;

$$Y = \sqrt{12^2 + 7^2}$$
$$= \sqrt{193} = 13.9$$

3. Solve for the angle;

$$\theta = \arctan \frac{7}{12}$$
$$= \arctan 0.583 = 30.2°$$

Note that this angle places the phasor as shown by the dashed line in Figure 19–13.

4. Find the correct angle by adding 180° (180° + 30.2° = 210.2°):

$$Y = 13.9\angle 210.2°$$

EXAMPLE 19–6 A complex number Z is equal to $15 - j10$.
Convert Z to polar form.

■ Solution:
The phasor diagram for Z is shown in Figure 19–14.

1. Solve for the magnitude of Z:

$$Z = \sqrt{15^2 + 10^2}$$
$$= \sqrt{325} = 18$$

2. Solve for the angle:

$$\theta = \arctan \frac{10}{15}$$
$$= \arctan 0.667 = 33.7°$$

3. Correct the angle by subtracting it from 360° (360° − 33.7° = 326.3°) to find Z:

$$Z = 18\angle 326.3°$$

Recall that angles in a counterclockwise direction are positive, while those measured in a clockwise direction are negative. Thus, the angle in Example 19–6 could be expressed as a −33.7°. Thus,

$$Z = 18\angle -33.7°$$

Conversion from Polar to Rectangular Form

Converting from polar to rectangular form involves the computation of the two sides of a right triangle when the hypotenuse and one angle are known, as illustrated in Figure 19–15(a). As you know, side R is the real portion of the number, and side X is the imaginary portion. These values must be determined in order to express the complex number in rectangular form, which is accomplished using the trigonometric functions.

Question: How are complex numbers converted from polar to rectangular form?

The sine function of an angle is equal to the ratio of the side opposite the angle to the hypotenuse. Thus, in Figure 19–15(b), the sine of angle θ is the ratio of X to Z:

(19–5)
$$\sin\theta = \frac{X}{Z}$$

This equation can be transposed to yield an equation for X:

$$X = Z\sin\theta$$

FIGURE 19–14 Complex plane for Example 19–6

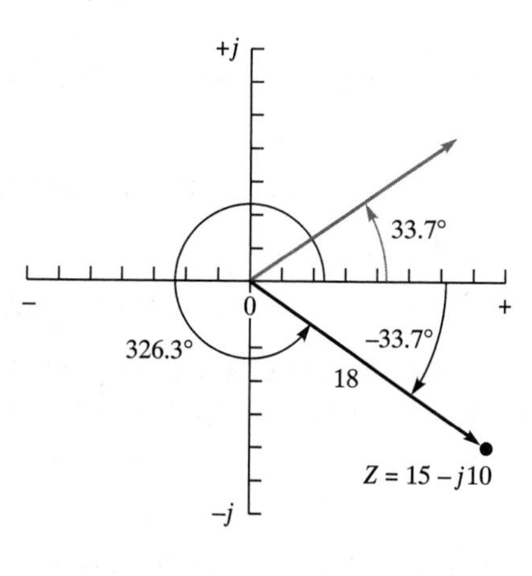

FIGURE 19–15 Converting polar form of a complex number to rectangular form

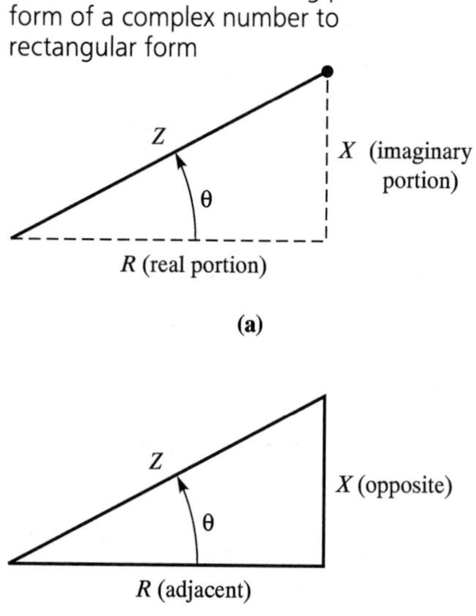

The cosine function of an angle is equal to the ratio of the side adjacent the angle to the hypotenuse. In Figure 19–15(b), the cosine of angle θ is the ratio of R to Z:

(19–6)
$$\cos \theta = \frac{R}{Z}$$

This equation can be transposed to yield an equation for R:

$$R = Z \cos \theta$$

EXAMPLE 19–7 A complex number (A) in polar form has a value of $100\angle 36°$. What is its value in rectangular form? (See Figure 19–16.)

FIGURE 19–16 Diagram for Example 19–7

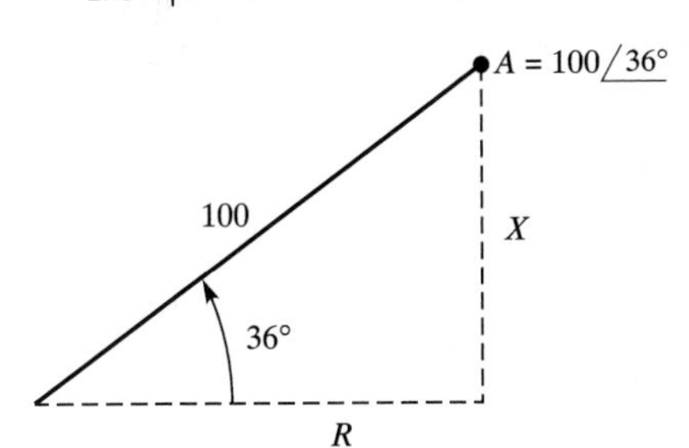

■ Solution:

$$X = Z \sin \theta$$
$$= 100(\sin 36°) = 100(0.588) = 58.8\ \Omega$$
$$R = Z \cos \theta$$
$$= 100(\cos 36°) = 100(0.809) = 80.9\ \Omega$$

Thus, in rectangular form,

$$A = 80.9\ \Omega + j58.8\ \Omega$$

EXAMPLE 19–8 A complex number (B) has a value in polar form of $550\angle -50°$. What is its value in rectangular form? (See Figure 19–17.)

■ Solution:

As is the case with negative angles, phasor Z is rotated in a clockwise direction. Notice though that this also represents a positive angle of 310° (360° − 50° = 310°). The same values of X and R are obtained using either angle value:

FIGURE 19–17 Diagram for Example 19–8

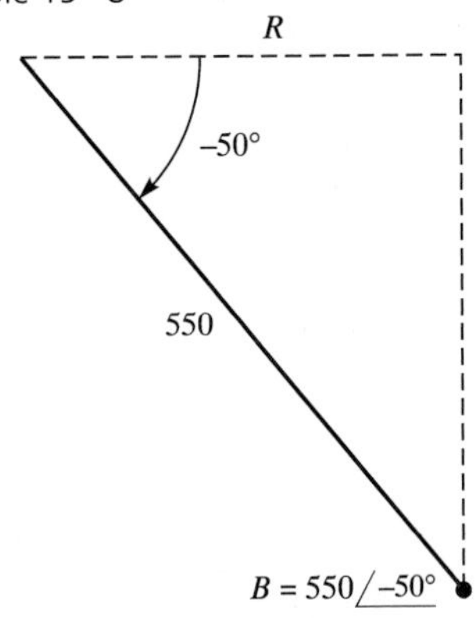

$$X = Z \sin \theta$$
$$= 550(\sin -50°) = 550(-0.766) = -421.3$$
$$R = Z \cos \theta$$
$$= 550(\cos -50°) = 550(0.643) = 353.7$$

Thus, in rectangular form,

$$B = 353.7 - j421.3$$

You should substitute the value of 310° for angle θ and confirm that the values obtained for X and R are identical to those given for the $-50°$ angle.

■ Conversion Using the Calculator

Conversions between the forms of complex numbers can also be made using the scientific calculator. In order to do so, the calculator must be equipped with the keys $R \rightarrow P$, $P \rightarrow R$, and $X \leftrightarrow Y$. (To see if these procedures apply, check the instruction book for your calculator.) To convert from rectangular to polar form, proceed as follows:

1. Enter the real portion of the complex number.
2. Depress the $R \rightarrow P$ key.

3. Enter the imaginary portion of the complex number.
4. Depress the = key. The magnitude of the phasor is now displayed.
5. Depress the $X \leftrightarrow Y$ key. The value of the angle is now displayed.
6. Repeated pressing of the $X \leftrightarrow Y$ will alternately display the magnitude and the angle.

To convert from polar to rectangular form, proceed as follows:

1. Enter the magnitude of the phasor.
2. Depress the $P \rightarrow R$ key.
3. Enter the value of the angle.
4. Depress the = key. The real portion of the complex number is now displayed.
5. Depress the $X \leftrightarrow Y$ key. The imaginary portion of the complex number is now displayed.
6. Repeated pressing of the $X \leftrightarrow Y$ will alternately display the real and imaginary portions of the complex number.

19–4 ARITHMETIC OPERATIONS WITH COMPLEX NUMBERS

In this section, the arithmetic operations of addition, subtraction, multiplication, and division of complex numbers is reviewed. As previously stated, all arithmetic operations can be performed on complex numbers in either polar or rectangular form. Arithmetic operations can be performed on complex numbers in either polar or rectangular form. It is much more convenient, however, to perform addition and subtraction in rectangular form and to perform multiplication and division in polar form. This practice is observed in the explanation and applications that follow.

Addition and Subtraction of Complex Numbers

Question: How is addition and subtraction of complex numbers performed?

Addition of two complex numbers is accomplished by adding their real parts and their imaginary parts. Algebraic addition is performed, and the rules of signs must be observed. Subtraction is performed by changing the sign of the subtrahend and proceeding as in addition.

EXAMPLE 19–9 Add the following complex numbers in rectangular form: (a) $35 + j28$ and $15 + j7$, (b) $-53 + j12$ and $43 - j28$, and (c) $-10 + j16$ and $-25 + j14$

Solution:

a.
$$\begin{array}{r} 35 + j28 \\ 15 + j7 \\ \hline 50 + j35 \end{array}$$

b.
$$\begin{array}{r} -53 + j12 \\ 43 - j28 \\ \hline -10 - j16 \end{array}$$

c.
$$\begin{array}{r} -10 + j16 \\ -25 + j14 \\ \hline -35 + j30 \end{array}$$

EXAMPLE 19–10 Subtract the following complex numbers in rectangular form: (a) $48 + j5$ minus $35 + j20$, (b) $128 - j85$ minus $-24 - j12$, and (c) $-96 + j51$ minus $53 + j33$.

■ Solution:

a.

$$\begin{array}{r} 48 + j5 \\ -35 - j20 \\ \hline 13 - j15 \end{array}$$ (signs changed)

b.

$$\begin{array}{r} 128 - j85 \\ 24 + j12 \\ \hline 152 - j73 \end{array}$$ (signs changed)

c.

$$\begin{array}{r} -96 + j51 \\ -53 - j33 \\ \hline -149 + j18 \end{array}$$ (signs changed)

■ Multiplication and Division of Complex Numbers

Question: How is multiplication and division of complex numbers performed?

Multiplication of complex numbers is performed by multiplying the magnitude of the phasors and algebraically adding the angles. Division of complex numbers is performed by dividing the phasors and algebraically subtracting the angle of the divisor from the angle of the dividend.

EXAMPLE 19–11 Multiply the following complex numbers in polar form: (a) $15\angle 60°$ times $5\angle 10°$, (b) $7\angle -38°$ times $8\angle 12°$, and (c) $12\angle -17°$ times $8\angle -25°$.

■ Solution:

a. $15 \times 5 = 75$
$60° + 10° = 70°$
$75\angle 70°$

b. $7 \times 8 = 56$
$-38° + 12° = -26°$
$56\angle -26°$

c. $12 \times 8 = 96$
$-17° + (-25°) = -42°$
$96\angle -42°$

EXAMPLE 19–12 Divide the following complex numbers in polar form: (a) $72\angle 46°$ divided by $12\angle 30°$, (b) $125\angle -27°$ divided by $5\angle -13°$, and (c) $39\angle 27°$ divided by $3\angle -12°$.

■ Solution:

a. $72 \div 12 = 6$
$46° - 30° = 16°$
$6\angle 16°$

b. $125 \div 5 = 25$
$-27° - (-13°) = -14°$
$25\angle -14°$

c. $39 \div 3 = 13$
$27° - (-12°) = 39°$
$13\angle 39°$

Squaring and Extracting Square Roots of Complex Numbers

Once again, it is easier to square or extract the square root of a number in polar form than in rectangular. In order to square a complex number in polar form, square the magnitude and double the size of the angle. In order to extract the square root of a complex number in polar form, take the square root of the magnitude and take half the angle. That is,

$$(15\angle 25°)^2 = 225\angle 50°$$
$$(12\angle 18°)^2 = 144\angle 36°$$
$$\sqrt{25\angle 30°} = 5\angle 15°$$
$$\sqrt{49\angle 12°} = 7\angle 6°$$

SECTION 19–4 REVIEW

1. What are the rules for addition and subtraction of complex numbers in rectangular form?
2. What are the rules for multiplication and division of complex numbers in polar form?
3. Add $45 + j37$ and $12 + j19$.
4. Subtract $12 - j9$ from $29 + j0$.
5. Multiply $17\angle -39°$ and $5\angle 7°$.
6. Divide $42\angle 17°$ by $7\angle -7°$.
7. Find the square root of $900\angle 30°$.
8. Square $20\angle 28°$.

19–5 REPRESENTING ac CIRCUIT QUANTITIES WITH COMPLEX NUMBERS

Complex numbers are well suited for expressing the quantities of voltage, current, and impedance in an ac circuit. As you have learned, a complex number consists of two parts: one real and one imaginary. The imaginary part is rotated 90° from the real. This 90° phase difference is exactly what is needed to express currents and voltages that are affected by reactances.

Question: How are voltages and impedances of series RL *and* RC *circuits expressed as complex numbers?*

Recall that in a series *RL* circuit the resistor voltage is in phase with the current, while the voltage across the inductor leads the current by 90°. Figure 19–18 shows the phasor diagram for this relationship, but notice that the phasors are drawn upon the real and imaginary number axes. The phasor for the resistor voltage is drawn along the real number axis. The voltage developed across the inductor is drawn 90° ahead of the current phasor along the positive imaginary axis. Notice that the number representing the source voltage is in the complex plane at the points V_R and jV_L:

FIGURE 19–18 Voltage phase relationships in a series RL circuit

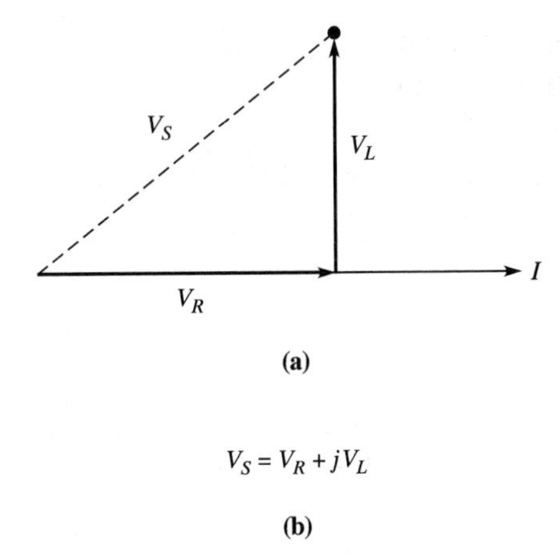

FIGURE 19–19 Phasor diagram for a series RC circuit

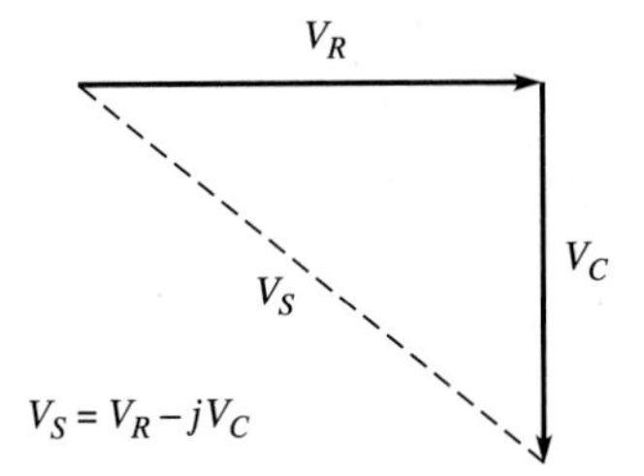

(19–7)
$$V_S = V_R + jV_L$$

The complex number for the voltages in an *RC* circuit are written in a similar manner. Figure 19–19 shows the phasor diagram for a series *RC* circuit. The resistor voltage is drawn once again along the real number axis, in phase with the current. The capacitor voltage is drawn along the imaginary axis, but now in a negative direction. Thus, the equation for source voltage in this circuit is

(19–8)
$$V_S = V_R - jV_C$$

The $-j$ in this equation indicates that the imaginary term is rotated 90° in a negative direction from the real term.

EXAMPLE 19–13 For the circuits of Figure 19–20, express the source voltage in rectangular form.

■ Solution:

For circuit (a),

$$V_S = 35V_R + j12V_L$$

For circuit (b),

$$V_S = 10V_R - j20V_C$$

Impedances are expressed as complex numbers in a manner similar to voltages. The resistance is the real part of the number and the reactance is the imaginary part. The equations for impedance in series *RL* and *RC* circuits are

(19–9) Series *RL*, $Z = R + jX_L$

(19–10) Series *RC*, $Z = R - jX_C$

EXAMPLE 19–4 For the circuits of Figure 19–21, express the impedances in rectangular form.

■ Solution:

For circuit (a),

$$Z = 100\ \Omega + j36\ \Omega$$

FIGURE 19–20 Circuits and phasor diagrams for Example 19–13

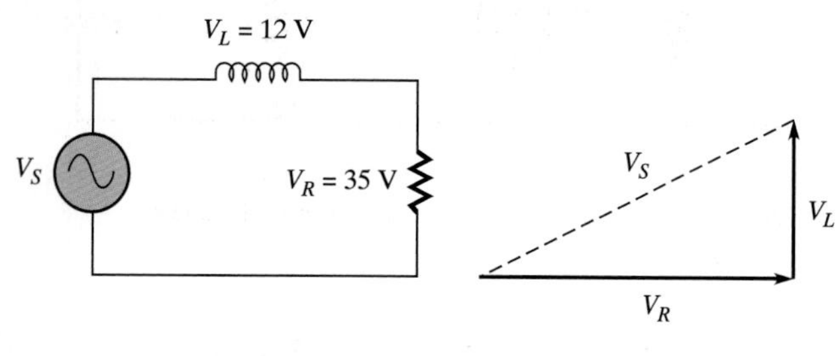

(a)
$V_S = V_R + jV_L$ and $V_S = 35\text{ V} + j12\text{ V}$

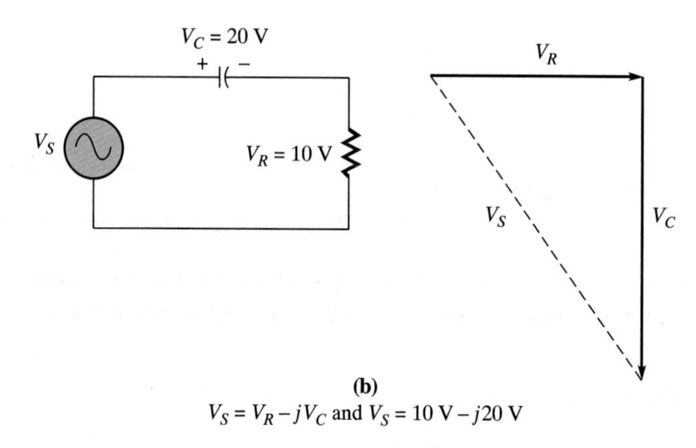

(b)
$V_S = V_R - jV_C$ and $V_S = 10\text{ V} - j20\text{ V}$

FIGURE 19–21 Circuits for Example 19–14

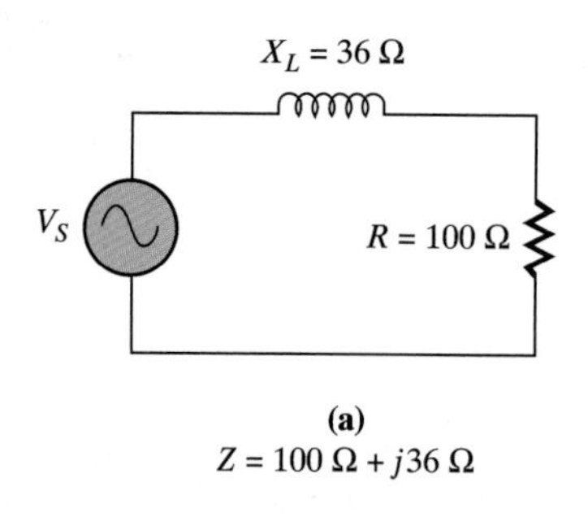

(a)
$Z = 100\ \Omega + j36\ \Omega$

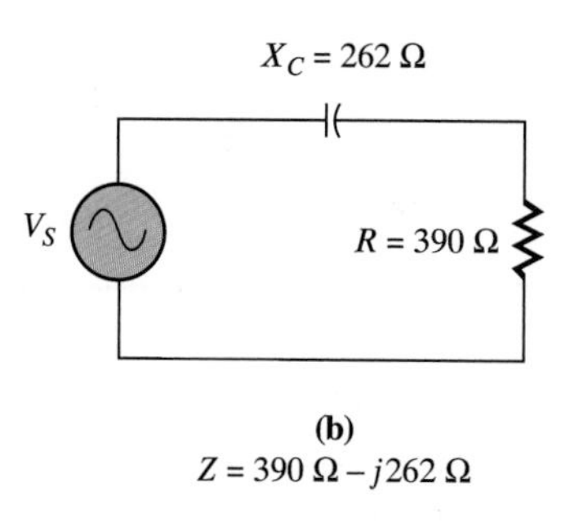

(b)
$Z = 390\ \Omega - j262\ \Omega$

For circuit (b),

$$Z = 390\ \Omega - j262\ \Omega$$

Question: How are the currents and impedances of parallel RL and RC circuits expressed as complex numbers?

Recall that the phasor diagram for a parallel *RL* circuit uses the source voltage phasor as the reference. The current through the resistor is in phase with the reference and drawn along the real axis. Thus, the resistor current will be the real part of the complex number. The current through the inductor lags the source voltage by 90° and is drawn along the imaginary axis, as illustrated in Figure 19–22. The equation is

FIGURE 19–22 Current relationships in a parallel RL circuit

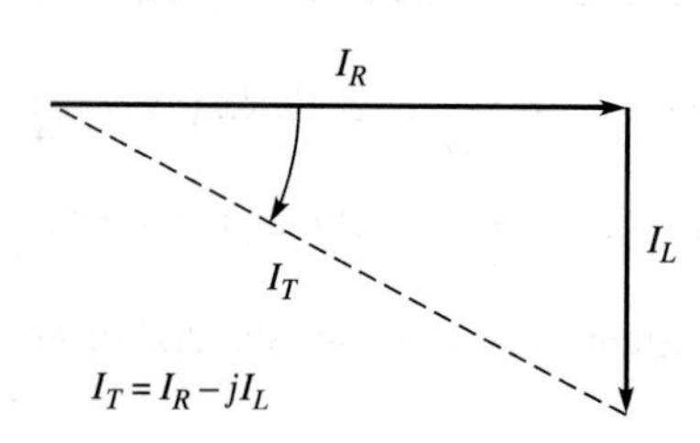

(19–11) $$I_T = I_R - jI_L$$

The resistor current is also the real part of the complex number for a parallel *RC* circuit. The only difference is in the capacitor current, which will lead the source voltage by 90°. Thus, the equation is

(19–12) $$I_T = I_R + jI_C$$

FIGURE 19–23 Circuits for Example 19–15

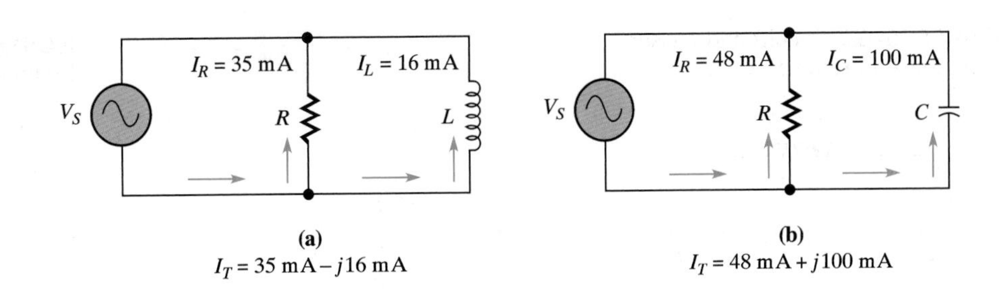

EXAMPLE 19–15 For the circuits of Figure 19–23, express the total current in rectangular form.

■ Solution:

For circuit (a),

$$I_T = 35\text{ mA} - j16\text{ mA}$$

For circuit (b),

$$I_T = 48\text{ mA} + j100\text{ mA}$$

Recall that the impedance of a parallel circuit is found by first solving for the admittance and then taking its reciprocal. Thus, admittance is the value that is expressed in complex form:

(19–13) Parallel RL, $Y = G - jB_L$

(19–14) Parallel RC, $Y = G + jB_C$

■ Expressing ac Quantities in Polar Form

Any of the quantities in rectangular form can be converted to polar using procedures introduced earlier in the chapter. The series RL and RC circuits are used as an example. The first step is to solve for the magnitude, which in this case is V_S.

$$\text{Series } RL, \quad V_S = \sqrt{V_R^2 + V_L^2}$$

$$\text{Series } RC, \quad V_S = \sqrt{V_R^2 + V_C^2}$$

The angle which the source voltage makes with the reference is then computed using the tangent function. The result is an equation in polar form:

$$V_R + jV_L = V_S\angle\theta \quad \text{and} \quad V_R - jV_C = V_S\angle-\theta$$

The other quantities in rectangular form are converted to polar form in a like manner.

■ SECTION 19–5 REVIEW

1. A resistance of 24 Ω is in series with an inductive reactance of 15 Ω. Express the impedance in both rectangular and polar form.
2. Express the source voltage of the circuit of Figure 19–24 in both rectangular and polar form.
3. For the circuit of Figure 19–25, express the total current in both rectangular and polar form.

FIGURE 19–24 Circuit for Section Review Problem 2

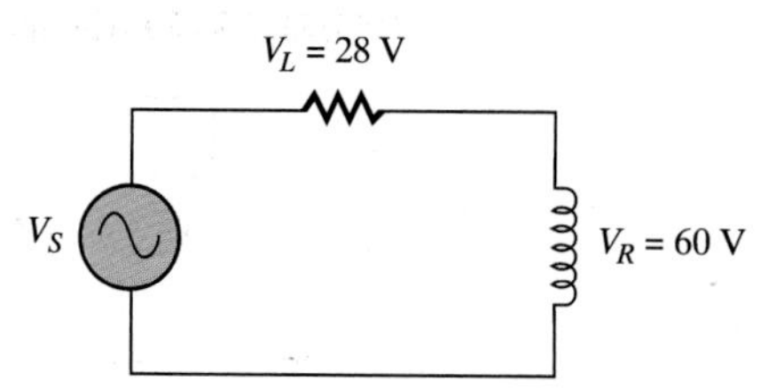

FIGURE 19–25 Circuit for Section Review Problem 3

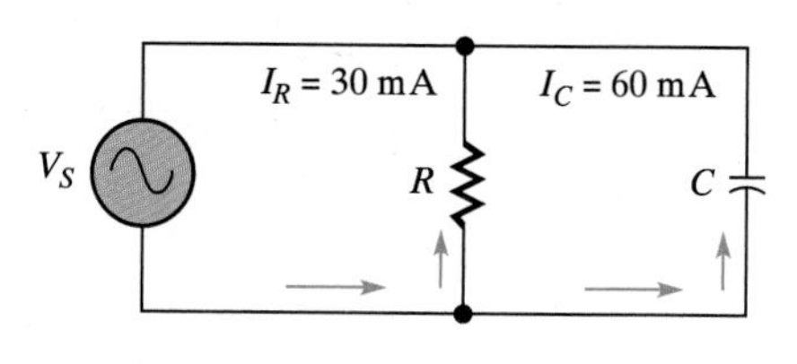

19–6 ac CIRCUIT ANALYSIS USING COMPLEX NOTATION

In this section, representative ac circuits are analyzed using complex notation. You already know the procedures for ac circuit analysis; the only differences will be in the arithmetic notation. It might be a good idea to make a quick review of the material in Chapters 12, 14, and 15.

EXAMPLE 19–16 For the circuit of Figure 19–26, compute: (a) Z_T, (b) I, (c) V_{Z1}, and (d) V_{Z2}.

■ **Solution:**

The load is made up of two complex impedances in series with values as shown. They must be added to find Z_T, so first convert them to rectangular form:

$$Z_1 = (180\ \Omega)(\cos 30°) + j(180\ \Omega)(\sin 30°)$$
$$= 155.9\ \Omega + j90\ \Omega$$
$$Z_2 = (395\ \Omega)(\cos -40°) - j(395)(\sin -40°)$$
$$= 302.6\ \Omega - j253.9\ \Omega$$

Then,

$$Z_T = Z_1 + Z_2$$
$$= 458.5\ \Omega - j163.9\ \Omega$$

Z_T is now converted to polar form so that I may be computed:

$$\theta = \arctan \frac{-163.9\ \Omega}{458.5\ \Omega}$$
$$= \arctan -0.357 = -19.6°$$
$$Z_T = \sqrt{458.5\ \Omega^2 + 163.9\ \Omega^2}$$
$$= 487.3\ \Omega$$
$$Z_T = 487.3\ \Omega\angle -19.6°$$

Calculate I:

$$I = V_S/Z_T$$
$$= \frac{(8\ \text{V})\angle 0°}{487.3\angle -19.6°}$$
$$= (16.4\ \text{mA})\angle 19.6°$$

FIGURE 19–26 Circuit for Example 19–16

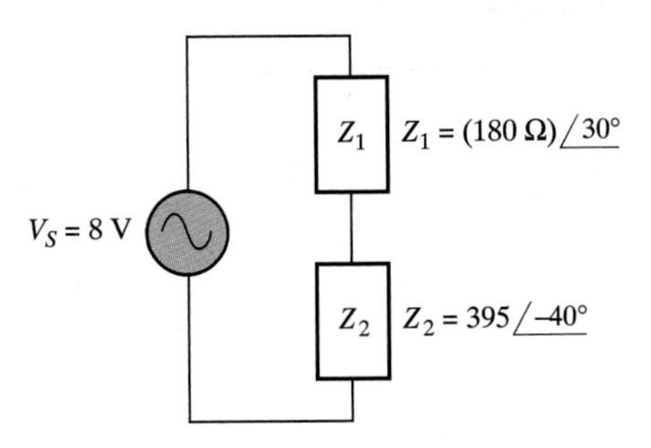

FIGURE 19–27 Circuit for Example 19–17

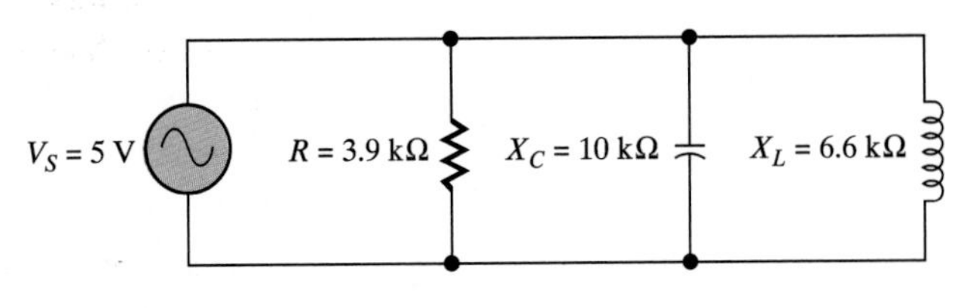

Calculate V_{Z1} and V_{Z2}

$$\begin{aligned}V_{Z1} &= IZ_1\\ &= [(16.4\text{ mA})\angle -19.6°][(180\ \Omega)\angle 30°]\\ &= (2.95\text{ V})\angle 49.6°\\ V_{Z2} &= IZ_2\\ &= [(16.4\text{ mA})\angle 19.6°][(395\ \Omega)\angle -40°\\ &= (6.48\text{ V})\angle -20.4°\end{aligned}$$

EXAMPLE 19–17 For the circuit of Figure 19–27, compute Z and I_T.

■ Solution:

1. Compute G:

$$G = \frac{1}{R} = 256\ \mu\text{S}$$

2. Compute B_C:

$$B_C = \frac{1}{X_C} = \frac{1}{(1/2\pi fC)} = 100\ \mu\text{S}$$

3. Compute B_L:

$$B_L = \frac{1}{X_L} = \frac{1}{2\pi fL} = 152\ \mu\text{S}$$

4. Compute the value of admittance Y:

$$\begin{aligned}Y &= G + (jB_C - jB_L)\\ &= 256\ \mu\text{S} + (j100\ \mu\text{S} - j152\ \mu\text{S})\\ &= 256\ \mu\text{S} - j52\ \mu\text{S}\end{aligned}$$

5. Convert Y to polar form:

$$\begin{aligned}Y &= 256\ \mu\text{S} - j52\ \mu\text{S}\\ &= (261.2\ \mu\text{S})\angle -11.5°\end{aligned}$$

6. Compute value of Z_T:

$$Z = \frac{1}{Y}$$

$$= \frac{1}{(261.2\ \mu\text{S})\angle -11.5^\circ}$$

$$= (3828\ \Omega)\angle 11.5^\circ$$

7. Compute value of I_T:

$$I_T = \frac{V_S}{Z}$$

$$= \frac{(5\ \text{V})\angle 0^\circ}{(3828\ \Omega)\angle 11.5^\circ}$$

$$= (1.31\ \text{mA})\angle -11.5^\circ$$

EXAMPLE 19–18 For the circuit of Figure 19–28, compute V_{Z_1} and V_{Z_2}. Use the voltage divider equation.

■ **Solution:**

The load, once again. is two complex impedances with values as shown.

1. Convert Z_1 and Z_2 to rectangular form:

$$Z_1 = (485\ \Omega)\angle 36^\circ = 392.4\ \Omega + j285.1\ \Omega$$

$$Z_2 = (786\ \Omega)\angle 14^\circ = 762.7\ \Omega + j190.2\ \Omega$$

2. Compute Z_T:

$$Z_T = Z_1 + Z_2$$

$$= (392.4\ \Omega + j285.1\ \Omega) + (762.7\ \Omega + j190.2\ \Omega)$$

$$= 1155.1\ \Omega + j475.3\ \Omega$$

3. Convert Z_T to polar form:

$$Z_T = 1155.1\ \Omega + j475.3\ \Omega = (1249.1\ \Omega)\angle 22.4^\circ$$

4. Compute V_{Z1} and V_{Z2}:

$$V_{Z1} = \frac{Z_1}{Z_T} V_S$$

$$= \frac{(485\ \Omega)\angle 36^\circ}{(1249.1\ \Omega)\angle 22.4^\circ}\ 12\ \text{V}$$

$$= (4.66\ \text{V})\angle 13.6^\circ$$

$$V_{Z2} = \frac{Z_2}{Z_T} V_S$$

$$= \frac{(786\ \Omega)\angle 14^\circ}{(1249.1\ \Omega)\angle 22.4^\circ}\ 12\ \text{V}$$

$$= (7.55\ \text{V})\angle -8.4^\circ$$

FIGURE 19–28 Circuit for Example 19–18

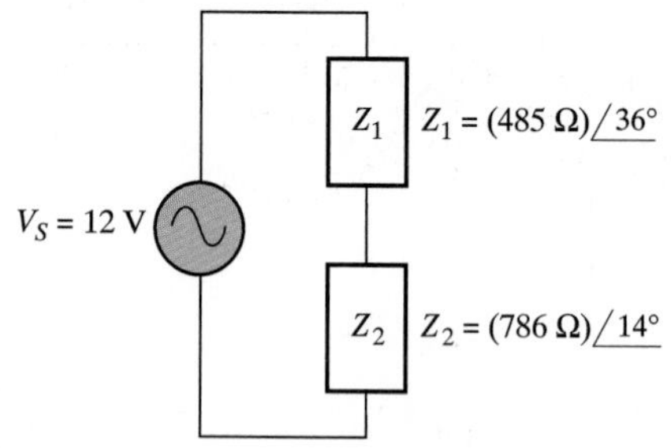

Z_1 $Z_1 = (485\ \Omega)\angle 36^\circ$

$V_S = 12$ V

Z_2 $Z_2 = (786\ \Omega)\angle 14^\circ$

EXAMPLE 19–19 For the circuit of Figure 19–29(a), compute
(a) Z, (b) I, (c) V_{R1}, (d) V_{R2}, (e) V_L, and (f) V_C.

■ Solution:

1. Compute Z:

$$
\begin{aligned}
Z &= R_1 + R_2 + jX_L - jX_C \\
&= 10\ \Omega + 5\ \Omega + j38\ \Omega - j18\ \Omega \\
&= 15\ \Omega + j20\ \Omega
\end{aligned}
$$

2. Convert value of Z to polar form:

$$
\begin{aligned}
Z &= 15\ \Omega + j20\ \Omega \\
&= (25\ \Omega)\angle 53.1^\circ
\end{aligned}
$$

3. Compute value of I:

$$
\begin{aligned}
I &= \frac{V_S}{Z} \\
&= \frac{(15\ \text{V})\angle 0^\circ}{25\ \Omega\angle 53.1^\circ} \\
&= (0.6\ \text{A})\angle -53.1^\circ
\end{aligned}
$$

This angle of -53.1° is the circuit phase angle. It indicates that the current lags the voltage by this angle. Thus the circuit is inductive.

4. Compute the value of the voltage drops:

$$
\begin{aligned}
V_{R_1} &= IR_1 \\
&= [(0.6\ \text{A})\angle -53.1^\circ][(10\ \Omega)\angle 0^\circ] = (6\ \text{V})\angle -53^\circ \\
V_{R_2} &= IR_2 \\
&= [(0.6\ \text{A})\angle -53.1^\circ][(5\ \Omega)\angle 0^\circ] = (3\ \text{V})\angle -53.1^\circ
\end{aligned}
$$

FIGURE 19–29 Circuit for Example 19–19

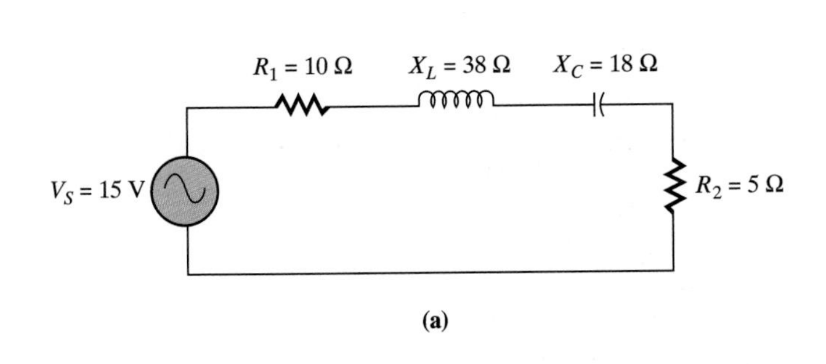

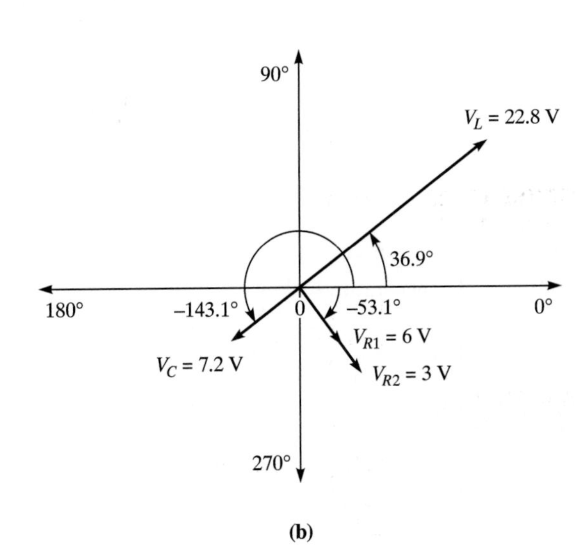

$$V_L = IX_L$$
$$= [(0.6 \text{ A})\angle -53.1°][(38 \text{ }\Omega)\angle 90°] = 22.8 \text{ V}\angle 36.9°$$
$$V_C = IX_C$$
$$= [(0.6 \text{ A})\angle -53.1°][(18 \text{ }\Omega)\angle -90°] = (10.8 \text{ V})\angle -143.1°$$

The phasor diagram of the voltages is shown in Figure 19–29(b).

One of the major advantages of using complex numbers in ac analysis is demonstrated in Example 19–16. Using the approach used in Chapter 16 would have resulted in the same values for the voltages and circuit phase angle, but it would not have shown the phase relationships of Figure 19–29. Thus, complex numbers yield more details of the phase relationships of the voltages and current.

SECTION 19–5 REVIEW

1. For the circuit of Figure 19–30, compute the following: V_{R_1}, V_{R_2}, V_L, V_C, and V_S.
2. What is the phase angle of V_L with respect to I?
3. What is the phase angle of V_C with respect to V_S?
4. Is this circuit inductive or capacitive as indicated by the phase angle?

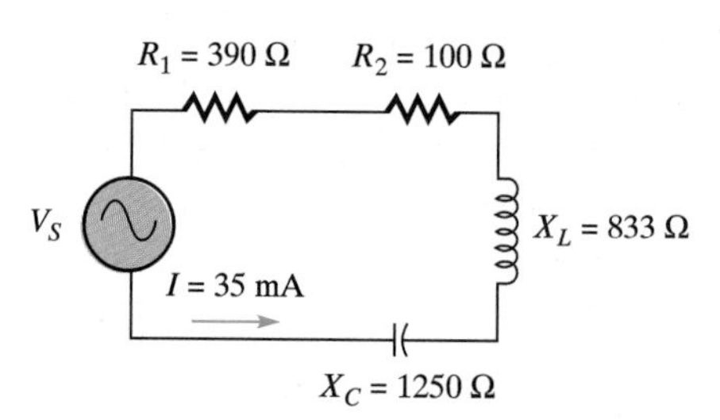

FIGURE 19–30 Circuit for Section Review Problems 1–4

CHAPTER SUMMARY

1. The imaginary number axis is rotated 90° from the real number axis.
2. The operator that rotates the imaginary axis is known as the *j*-operator.
3. The *j*-operator is equal to $\sqrt{-1}$.
4. The entire plane surrounding an origin is known as the complex plane.
5. Complex numbers may be expressed in either polar or rectangular form.
6. Complex numbers are more easily added or subtracted in rectangular form.
7. Complex numbers are more easily multiplied or divided in polar form.
8. Squaring or taking the square root of a complex number is more easily accomplished in polar form.
9. Complex numbers can be easily and quickly converted from one form to the other.
10. Quantities in ac circuits can be expressed as complex numbers.
11. An imaginary value of $-jX$ is rotated 90° in a negative direction, while a value of jX is rotated 90° in a positive direction.
12. Using complex numbers in the analysis of ac circuits yields more information of the phase angles between circuit quantities than other means.

KEY TERMS

imaginary number
j-operator
complex number
rectangular form of a complex number
polar form of a complex number

EQUATIONS

(19–1) $Z^2 = R^2 + X^2$

(19–2) $Z = \sqrt{R^2 + X^2}$

(19–3) $\tan \theta = \frac{X}{R}$

(19–4) $\theta = \arctan \frac{X}{R}$

(19–5) $\sin \theta = \frac{X}{Z}$

(19–6) $\cos \theta = \frac{R}{Z}$

(19–7) $V_S = V_R + jV_L$

(19–8) $V_S = V_R + jV_C$

(19–9) $Z = R + jX_L$

(19–10) $Z = R - jX_C$

(19–11) $I_T = I - jI_L$

(19–12) $I_T = I + jI_C$

(19–13) $Y = G - jB_L$

(19–14) $Y = G + jB_C$

TEST YOUR KNOWLEDGE

1. What is meant by a real number?
2. What is meant by an imaginary number?
3. What is meant by a complex number?
4. In what two forms may complex numbers be expressed?
5. Are addition and subtraction more easily accomplished in rectangular form or polar form?
6. Are multiplication and division more easily accomplished in polar form or rectangular form?
7. What are the rules for addition and subtraction of complex numbers in rectangular form?
8. What are the rules for multiplication and division of complex numbers in polar form?
9. How is a complex number in polar form squared?
10. How is the square root of a complex number in polar form taken?

PROBLEM SET, BASIC LEVEL

Section 19–1

1. In the following complex numbers, indicate which is the real and which is the imaginary part: (a) $100 + j60$, (b) $230 - j15$, and (c) $400 + j1000$.
2. Draw a graph of the number lines for the real and imaginary numbers.

Section 19–2

3. Draw a graph of the complex plane and locate the position of the following complex numbers: (a) $12 + j15$, (b) $7 - j5$, (c) $-18 - j12$, and (d) $-10 + j12$.
4. Draw a graph of the complex plane and locate the position of the following complex numbers: (a) $10\angle 45^\circ$, (b) $15\angle -52^\circ$, (c) $-18\angle 25^\circ$, and (d) $-14\angle -30^\circ$.

Section 19–3

5. Convert the following complex number from rectangular to polar form: (a) $40 + j33$, (b) $-60 + j85$, (c) $35 - j12$, and (d) $-95 - j60$.
6. Convert the following complex numbers from rectangular to polar form: (a) $3600 + j6500$, (b) $10300 - j2500$, and (c) $2000 + j2000$.
7. Convert the following complex numbers from polar to rectangular form: (a) $65\angle 28^\circ$, (b) $188\angle -50^\circ$, (c) $300\angle 37^\circ$, and (d) $8\angle -30^\circ$.
8. Convert the following complex numbers from polar to rectangular form: (a) $4500\angle 138^\circ$, (b) $5000\angle 235^\circ$, and (c) $12000\angle 336^\circ$.

Section 19–4

9. Perform the following additions of complex numbers: (a) $(5 + j12) + (-18 - j7)$, (b) $(85 - j20) + (35 - j28)$, and (c) $(-36 + j18) + (12 - j8)$.

10. Perform the following subtractions of complex numbers: (a) $(148 - j68) - (39 + j12)$, (b) $(125 + j155) - (-18 - j20)$, and (c) $(-40 + j39) - (40 - j1)$.

11. Perform the following multiplications of complex numbers: (a) $(40\angle 52°)(5\angle 10°)$, (b) $(-30\angle -35°)(-3\angle 17°)$, and (c) $(-15\angle 17°)(5\angle -29°)$.

12. Perform the following divisions of complex numbers: (a) $(39\angle 68°)/(3\angle 18°)$, (b) $(125\angle -36°)/(10\angle 12°)$, and (c) $(45\angle -12°)/(9\angle -39°)$.

13. Square the following complex numbers: (a) $9\angle 17°$, (b) $12\angle -23°$, and (c) $15\angle 36°$.

14. Find the square root of the following complex numbers: (a) $36\angle -12°$, (b) $144\angle 38°$, and (c) $900\angle 40°$.

15. Perform the following multiplications of complex numbers: (a) $(12 + j5)(6 + j8)$ and (b) $(7 + j9)(4 - j6)$.

16. Perform the following additions of complex numbers: (a) $16\angle 36° + 14\angle 12°$ and (b) $30\angle 60° + 40\angle 30°$.

Section 19–5

17. For each circuit of Figure 19–31, state the impedance in complex form.

FIGURE 19–31 Circuits for Problem 17

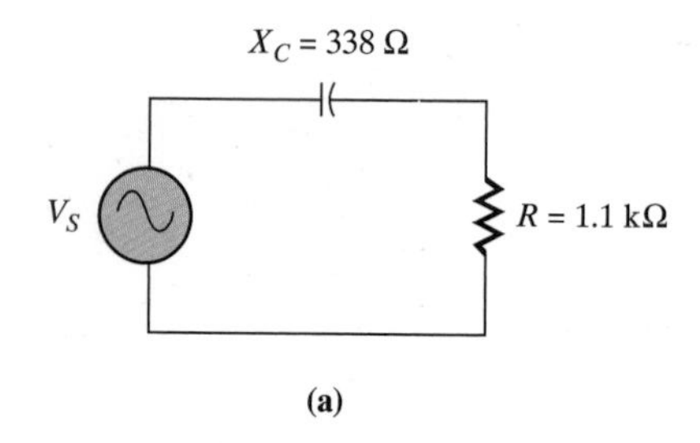

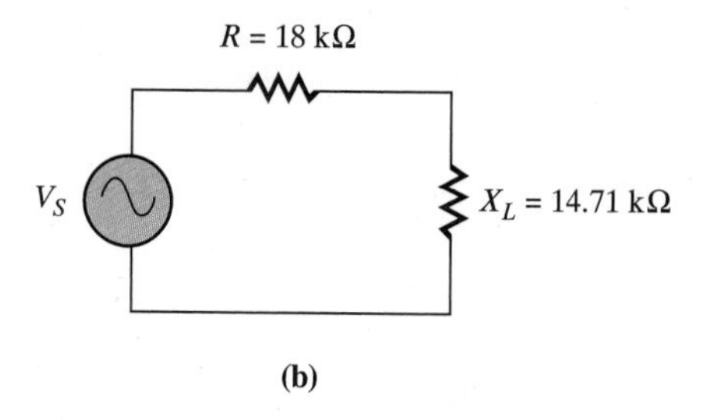

18. For each circuit of Figure 19–32, state the source voltage in complex form.

FIGURE 19–32 Circuits for Problem 18

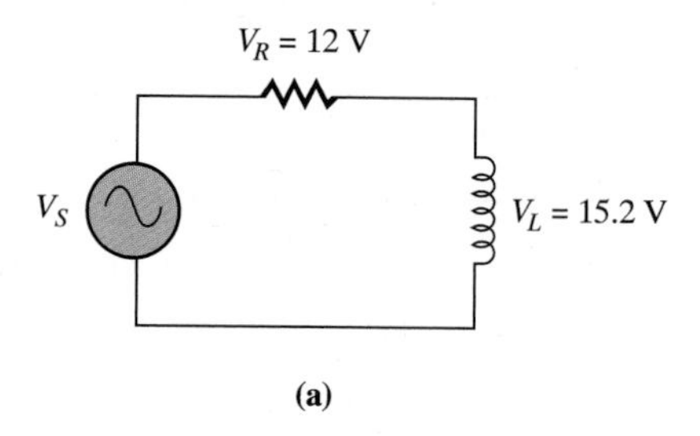

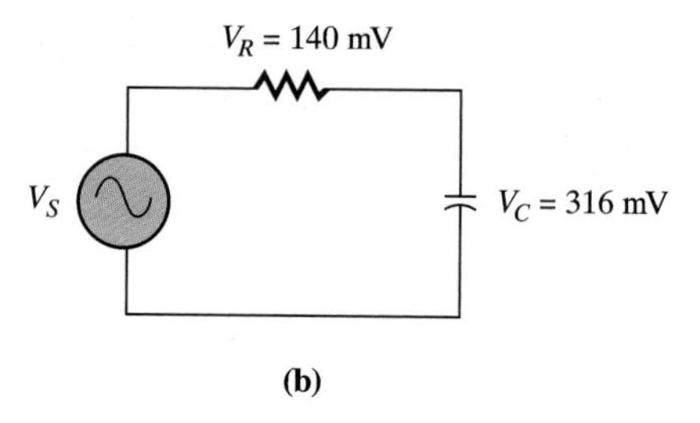

19. For each circuit of Figure 19–33, state the total current in complex form.

FIGURE 19–33 Circuits for Problem 19

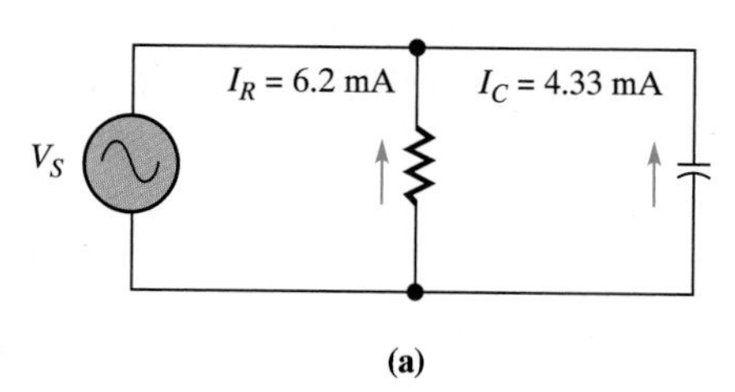

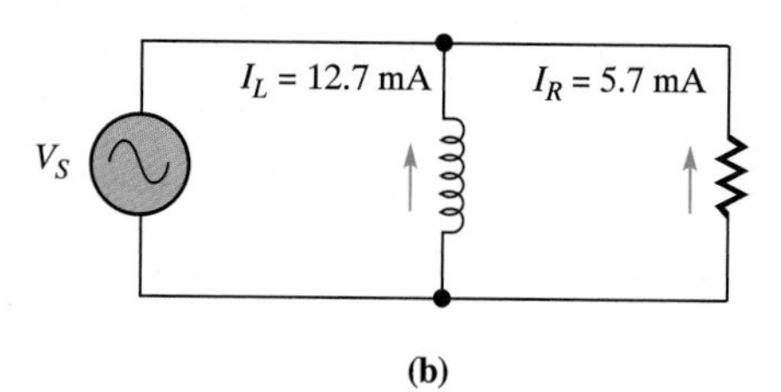

Section 19–6

20. For each circuit of Figure 19–34, compute I_R and I_X using complex numbers, and express I_T in complex form.

FIGURE 19–34 Circuits for Problem 20

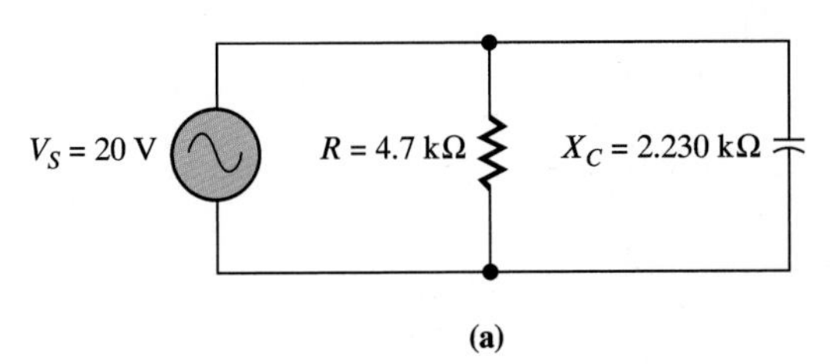

(a)

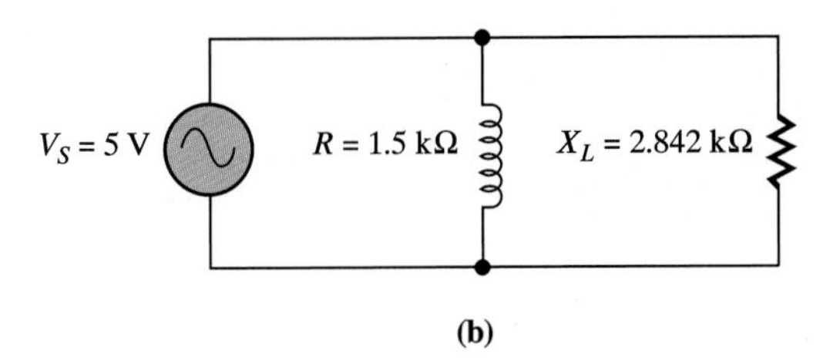

(b)

21. For the circuit of Figure 19–35, compute the following using complex notation: (a) Z, (b) I, (c) V_R, and (d) V_L.

FIGURE 19–35 Circuits for Problem 21

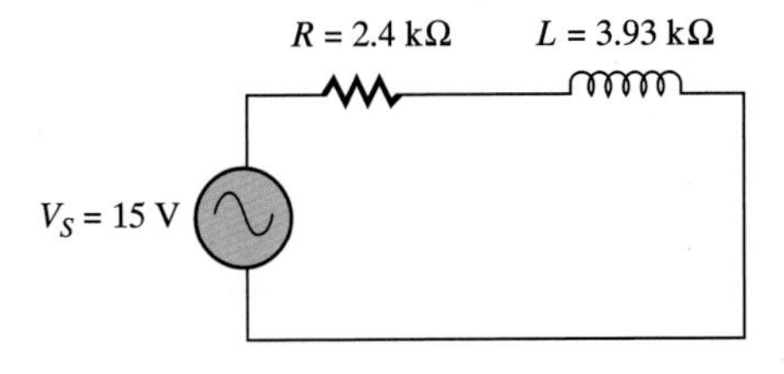

22. For the circuit of Figure 19–36, compute the following using complex notation: (a) I_R, (b) I_L, (c) I_C, and (d) Z.

FIGURE 19–36 Circuits for Problem 22

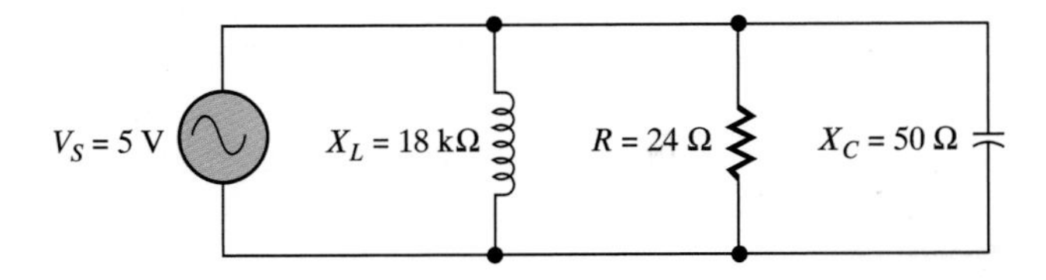

23. For the circuit of Figure 19–37, compute the impedance.

FIGURE 19–37 Circuit for Problem 23

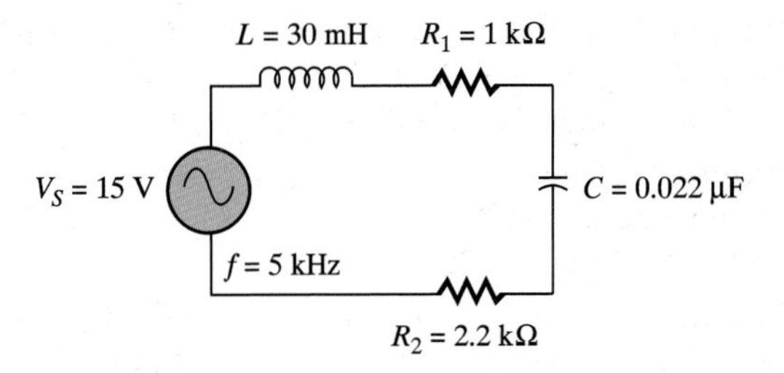

24. For the circuit of Figure 19–38, compute the following: (a) X_{L_1}, (b) X_{L_2}, (c) X_{L_T}, (d) Z, (e) I, (f) V_R, (g) V_{L_1}, (h) V_{L_2}, and (i) V_C.

FIGURE 19–38 Circuit for Problem 24

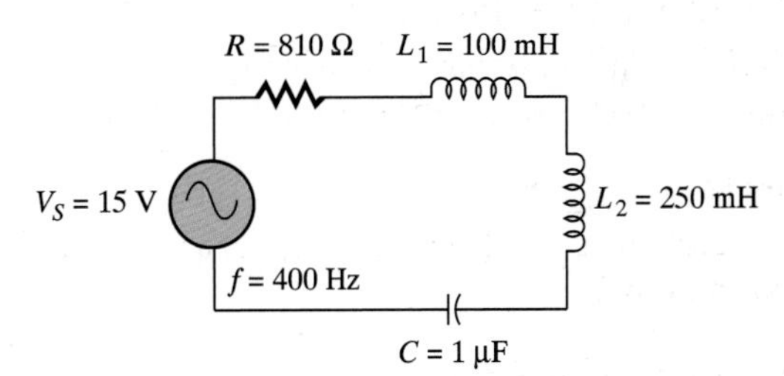

Job Search

Before you know it, you will have completed the electronics curriculum, earned your degree, and be ready to take your place as an electronics technician. As you have learned, there are a large variety of careers to choose from in electronics. There are also a great number of companies, located in all parts of the world, with which to pursue them. With this wide variety of careers, the large number of firms, and worldwide locations from which to choose, it would seem that your only problem will be sorting out the best offer. It is, of course, not quite that simple. Although some companies do send recruiters to community colleges and vocational schools, the vast majority do not. Thus, you will probably have to find the job, not wait until it finds you!

The first step is to make up your mind on a work location. This decision is strongly influenced by concerns for family, friends, personal preferences, and tradition. These concerns are not to be taken lightly in job search! The choice of work location is also influenced by the availability of jobs in a particular area. In some areas, the electronics industry is strong, while in others it is weak. Thus, a good first step is to collect data on electronics firms in as many areas as possible.

Most large cities or metropolitan areas will have electronics manufacturing or service companies, ranging in size from large manufacturing plants to small units with only a few employees. However, the fact that a certain company is located in a city does not mean that it will have employment opportunities available there. Information on companies located in a given area can be obtained from the local chamber of commerce. These chambers, which may represent a city, metropolitan area, or county, furnish information that includes the name of the company, its product or service, the number of people it employs, and the name of the contact person. These listings include all firms in the area, not only those dealing in electronics. Another source of information on electronics firms in an area is the Yellow Pages of the telephone book. Although not containing nearly the amount of information in the chambers' listings, they do give you an idea of the number of firms in the city or area.

You can also gain employer information through the help wanted advertisements in newspapers. Although you may not at the moment be in a position to respond to a particular ad, it will give you the name of the firm, its address, and a contact person. The best edition of the newspaper for this purpose is the Sunday edition. Employers prefer to run their ads in the Sunday paper because it has the widest circulation and coverage. You can also learn of electronics firms on a national basis through the U.S. Department of Labor. State Department of Labor offices can be a source for local and state employers. Another source of employer information is the job fairs sponsored by local civic groups. Employers from all parts of the country attend these fairs to pass out literature, spread the word of their employment needs, interview, and sometimes even hire employees. You should attend these if for no other reason than to pick up literature.

Once you have obtained the names and addresses of a number of firms, write a letter requesting brochures, pamphlets, or other advertising materials. In this letter, it is a good idea to mention the source (newspaper ad, Yellow Pages, Chamber of Commerce) from which you learned of the company. The materials you receive will contain, at a minimum, information on the product or service performed, work locations, and often the name of the person to contact for more information. With this material and information received from other sources, you can start a file of prospective employers that will be quite useful when your job search begins in earnest.

Prior to making application for employment, it is necessary to develop two documents: a resume and a cover letter to accompany the resume. The cover letter is actually a type of letter of introduction. It should include a brief statement of your purpose in writing, your source in learning about the company, a request for an employment application, and mention that your resume is enclosed. Experience has shown that if twenty such packages (letter of introduction and resume) are sent, five answers will be received, three of which are serious possibilities!

An up-to-date and correctly prepared resume is a very important part of your job search. Your resume is a summary of who you are, your schooling, and the work experiences that have been a part of your life. It should cover the necessary information but not be so extensive as to cause the reader to lose interest. A rule of thumb here is that it should be no longer than two pages in length. This amount of space may seem impossibly small when you consider all the schools, classes, jobs, and qualifications that you would like to put on it. The small space of two pages, however, makes it necessary to decide what is important and what is not. For example, a special project performed while in technical school should be included, while having been an eagle scout should not. Guidance in resume preparation can be found in books at your local library.

The application you submit is your first serious contact with the employment manager of the company. (Today these persons are often referred to as human resources managers.) Their opinions of you will begin to be formed by not only what is in this document, but also in the manner in which it is prepared. First of all, it must be complete. Do not assume that any question asked is trivial; it would not be asked if it were not important or didn't have a purpose. Each must have an honest, serious answer, including the question of salary you require. Many applicants write the word "open" in this entry, which is a poor practice! You cannot go to work for less than your salary needs, and the company must know if they can afford you. If you ask less than the going rate for a position, the vast majority of employers will offer you the higher wage.

Another feature of the application from which an opinion will be formed of you is how neatly it is completed. An application need not necessarily be typed. A neatly done application, written in ink, is actually a "plus" for the applicant. If you are not gifted in penmanship, however, the application should be typed. A poorly written, messy, application may give the impression of someone whose work habits may be similar. This applies, of course, to all communications with the prospective employer. As the old cliche goes: "You never have a second chance to make a first impression!"

Most employers view persistence as a virtue in an employee. Thus after the application is submitted it is a good idea to follow up with a phone call to ensure that it was received and in the hands of the right persons. In fact, some employers wait for this call before responding!

Resistivity of Materials: The Wire Gage Table

Even though materials such as copper are good conductors of electricity, they do offer some resistance to the flow of electric current. How much resistance a specific conducting material offers depends upon two factors: the length of the conductor and its cross-sectional area. These factors are illustrated in Figure A–1. The longer a conductor is, the more work it takes to cause electrons to drift through it and thus the higher its resistance is. The larger the cross-sectional area, the more free electrons there are available for conduction and the lower its resistance is. Thus, the resistance of a conductor is directly proportional to its length and inversely proportional to its cross-sectional area. In the American system of gaging wire, cross-sectional area is expressed in circular mils. A mil is 1/1000th of an inch, or 1 milli-inch. As shown in Figure A–2, the cross-sectional area is found by squaring the diameter in mils.

The system for sizing wire is known as the American Wire Gage (AWG) system. It is shown in Table A–1. This particular table is for solid copper wire at a temperature of 20°C. The wire is sized according to its cross-sectional area in descending order. Notice, however, that the AWG numbers are in ascending order. As the gage number increases, the cross-sectional area decreases and the resistance for a given length increases. Thus, AWG 20 wire is much thinner and has greater resistance than AWG 10. As illustrated in the following example, Table A–1 can be used to determine the resistance of a given length of copper wire.

EXAMPLE A–1 What is the resistance of 400′ of AWG 26 solid copper wire?

■ Solution:

From the Table A–1, AWG 26 copper wire has 40.81 Ω of resistance per 1000 feet. Thus the resistance for 400′ is

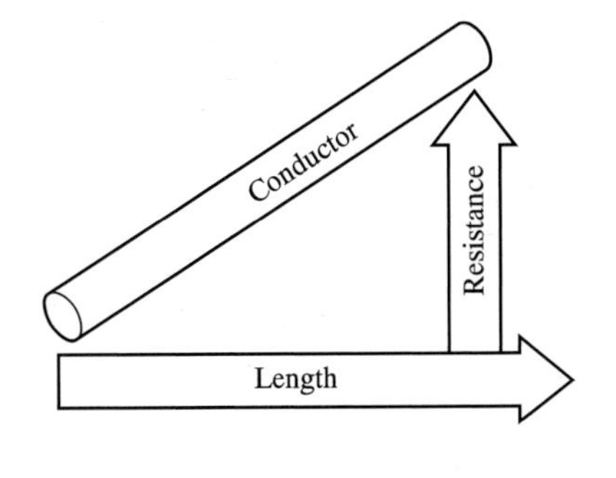

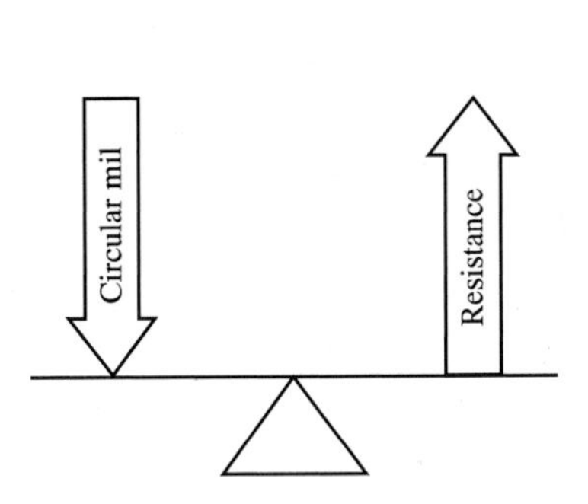

FIGURE A–1 Factors influencing resistance of a wire

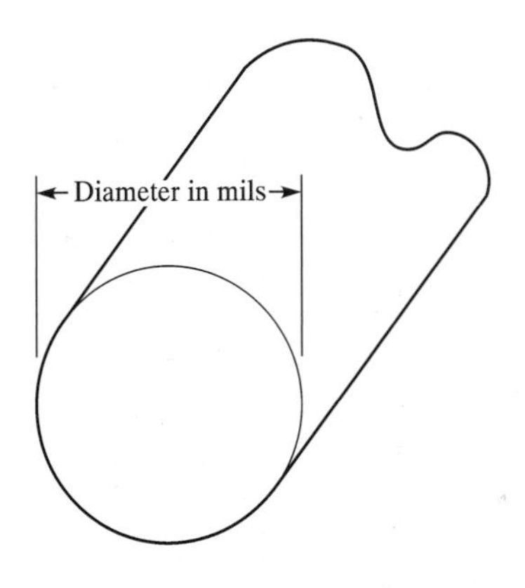

FIGURE A–2 Cross-sectional area of a in circular mils

■ **TABLE A–1 Copper Wire Table**

AWG GAGE	STANDARD METRIC SIZE (mm)	DIAMETER IN MILS	CROSS-SECTIONAL AREA CIRCULAR MILS	CROSS-SECTIONAL AREA SQUARE INCHES	OHMS/ 1000 Ft AT 20°C (68°F)	POUNDS/ 1000 Ft	FEET/POUND
0000	11.8	460.0	211,600	0.1662	0.04901	640.5	1.561
000	11.0	409.6	167,800	0.1318	0.06180	507.9	1.968
00	9.0	364.8	133,100	0.1045	0.07793	402.8	2.482
0	8.0	324.9	105,500	0.08289	0.09827	319.5	3.130
1	7.1	289.3	83,690	0.06573	0.1239	253.3	3.947
2	6.3	257.6	66,370	0.05213	0.1563	200.9	4.977
3	5.6	229.4	52,640	0.04134	0.1970	159.3	6.276
4	5.0	204.3	41,740	0.03278	0.2485	126.4	7.914
5	4.5	181.9	33,100	0.02600	0.3133	100.2	9.980
6	4.0	162.0	26,250	0.02062	0.3951	79.46	12.58
7	3.55	144.3	20,820	0.01635	0.4982	63.02	15.87
8	3.15	128.5	16,510	0.01297	0.6282	49.98	20.01
9	2.80	114.4	13,090	0.01028	0.7921	39.63	25.23
10	2.50	101.9	10,380	0.008155	0.9989	31.43	31.82
11	2.24	90.74	8,234	0.006467	1.260	24.92	40.12
12	2.00	80.81	6,530	0.005129	1.588	19.77	50.59
13	1.80	71.96	5,178	0.004067	2.003	15.68	63.80
14	1.60	64.08	4,107	0.003225	2.525	12.43	80.44
15	1.40	57.07	3,257	0.002558	3.184	9.858	101.4
16	1.25	50.82	2,583	0.002028	4.016	7.818	127.9
17	1.12	45.26	2,048	0.001609	5.064	6.200	161.3
18	1.00	40.30	1,624	0.001276	6.385	4.917	203.4
19	0.90	35.89	1,288	0.001012	8.051	3.899	256.5
20	0.80	31.96	1,022	0.0008023	10.15	3.092	323.4
21	0.71	28.46	810.1	0.0006363	12.80	2.452	407.8
22	0.63	25.35	642.4	0.0005046	16.14	1.945	514.2
23	0.56	22.57	509.5	0.0004002	20.36	1.542	648.4
24	0.50	20.10	404.0	0.0003173	25.67	1.223	817.7
25	0.45	17.90	320.4	0.0002517	32.37	0.9699	1,031.0
26	0.40	15.94	254.1	0.0001996	40.81	0.7692	1,300
27	0.355	14.20	201.5	0.0001583	51.47	0.6100	1,639
28	0.315	12.64	159.8	0.0001255	64.90	0.4837	2,067
29	0.280	11.26	126.7	0.00009953	81.83	0.3836	2,607
30	0.250	10.03	100.5	0.00007894	103.2	0.3042	3,287
31	0.224	8.928	79.70	0.00006260	130.1	0.2413	4,145
32	0.200	7.950	63.21	0.00004964	164.1	0.1913	5,227
33	0.180	7.080	50.13	0.00003937	206.9	0.1517	6,591
34	0.160	6.305	39.75	0.00003122	260.9	0.1203	8,310
35	0.140	5.615	31.52	0.00002476	329.0	0.09542	10,480
36	0.125	5.000	25.00	0.00001964	414.8	0.07568	13,210
37	0.112	4.453	19.83	0.00001557	523.1	0.06001	16,660
38	0.100	3.965	15.72	0.00001235	659.6	0.04759	21,010
39	0.090	3.531	12.47	0.000009793	831.8	0.03774	26,500
40	0.080	3.145	9.888	0.000007766	1049.0	0.02993	33,410

$$R = \frac{400}{1000} 40.81\ \Omega$$
$$= 0.4(40.81\ \Omega) = 16.32\ \Omega$$

The resistance of a conductor is an important consideration in electricity and electronics. In electronics, connecting leads and traces are kept short and usually have a very low resistance that can be ignored. This stray resistance may however be a consideration in inductors. Depending upon the application, the winding resistance of an inductor may have to be considered. Wire resistance is of far greater concern in high power circuits such as in the commercial power distribution system. Recall that power is developed in the form of heat and that the amount of heat in watts is equal to the product of the square of the current and the resistance. Thus, a wire produces heat in direct proportion to its resistance. Since all metals fuse, or melt, at some temperature, there is a certain current that a given size conductor can safely carry. This value varies, depending upon the type of material used.

Specific Resistance

The resistance of other conductive materials can be determined using what is known as their specific resistance. *Specific resistance,* which is also referred to as resistivity, is the resistance of a one foot section of the material with a cross-sectional area of 1 circular mil. It is given the symbol ρ, which is the letter rho from the Greek alphabet. Table A–2 contains the values of specific resistance for some of the common conductive materials. The resistance of a length of a conductor is directly proportional to its specific resistance. The three factors in the resistance of a conductor—its length, cross-sectional area, and specific resistance—are related in the following equation:

$$R = \rho \frac{l}{A}$$

in which l is the length of the conductor and A is its cross-sectional area in circular mils.

TABLE A–2 Specific resistance for conductive materials

MATERIAL	SPECIFIC RESISTANCE (OHMS/ CIRCULAR MIL/ FOOT)
Aluminum	17
Cadmium	46
Copper	10.4
Gold	14.5
Iron	58.2
Lead	130
Silver	9.8
Tin	69

Thus, if you know the specific resistance of a material, you can find the value of its resistance for any given length without the use of a table.

EXAMPLE A–2 What is the resistance of 2500 feet of AWG 14 copper wire?

■ **Solution:**

$$R = \rho \frac{l}{A}$$

$$= (10.4)\left(\frac{2500}{4107}\right) = 6.33\ \Omega$$

EXAMPLE A–3 What is the resistance of 2500 feet of AWG 14 aluminum wire?

■ **Solution:**

$$R = \rho \frac{l}{A}$$

$$= (17)\frac{2500}{4107} = 10.35\ \Omega$$

As these two examples show, aluminum wire of the same physical size as copper has a higher resistance than the copper wire. Thus, if aluminum wire is used to replace copper, a larger size would be required.

Preferred Values of Carbon Composition Resistors

Carbon composition resistors are manufactured with wattage ratings of 1/8 W, 1/4 W, 1/2 W, 1 W, and 2 W. They are available in tolerances of 5%, 10%, and 20%. To simplify the manufacturing process, not all values of resistance are available in each tolerance. Notice in Table B–1 that more values are available in 5% than in 10%. Also notice that more values are available in 10% than in 20%. Multiples of these values are also available. For example, 33 Ω, 330 Ω, 3,300 Ω, and 33,000 Ω values are available in 20%, 10%, and 5% tolerances. The values 39 Ω, 390 Ω, 3,900 Ω, and 39,000 Ω are only available in 5% or 10%. The values 36 Ω, 360 Ω, 3,600 Ω, and 36,000 Ω are available only in 5% tolerance. The tolerance value chosen is dictated by the voltage and current tolerances required in the circuit.

TABLE B–1 Preferred values of resistors.

0.1% 0.25% 0.5%	1%	2% 5%	10%	0.1% 0.25% 0.5%	1%	2% 5%	10%	0.1% 0.25% 0.5%	1%	2% 5%	10%	0.1% 0.25% 0.5%	1%	2% 5%	10%	0.1% 0.25% 0.5%	1%	2% 5%	10%	0.1% 0.25% 0.5%	1%	2% 5%	10%
10.0	10.0	10	10	14.7	14.7	—	—	21.5	21.5	—	—	31.6	31.6	—	—	46.4	46.4	—	—	68.1	68.1	68	68
10.1	—	—	—	14.9	—	—	—	21.8	—	—	—	32.0	—	—	—	47.0	—	47	47	69.0	—	—	—
10.2	10.2	—	—	15.0	15.0	15	15	22.1	22.1	22	22	32.4	32.4	—	—	47.5	47.5	—	—	69.8	69.8	—	—
10.4	—	—	—	15.2	—	—	—	22.3	—	—	—	32.8	—	—	—	48.1	—	—	—	70.6	—	—	—
10.5	10.5	—	—	15.4	15.4	—	—	22.6	22.6	—	—	33.2	33.2	33	33	48.7	48.7	—	—	71.5	71.5	—	—
10.6	—	—	—	15.6	—	—	—	22.9	—	—	—	33.6	—	—	—	49.3	—	—	—	72.3	—	—	—
10.7	10.7	—	—	15.8	15.8	—	—	23.2	23.2	—	—	34.0	34.0	—	—	49.9	49.9	—	—	73.2	73.2	—	—
10.9	—	—	—	16.0	—	16	—	23.4	—	—	—	34.4	—	—	—	50.5	—	—	—	74.1	—	—	—
11.0	11.0	11	—	16.2	16.2	—	—	23.7	23.7	—	—	34.8	34.8	—	—	51.1	51.1	51	—	75.0	75.0	75	—
11.1	—	—	—	16.4	—	—	—	24.0	—	24	—	35.2	—	—	—	51.7	—	—	—	75.9	—	—	—
11.3	11.3	—	—	16.5	16.5	—	—	24.3	24.3	—	—	35.7	35.7	—	—	52.3	52.3	—	—	76.8	76.8	—	—
11.4	—	—	—	16.7	—	—	—	24.6	—	—	—	36.1	—	36	—	53.0	—	—	—	77.7	—	—	—
11.5	11.5	—	—	16.9	16.9	—	—	24.9	24.9	—	—	36.5	36.5	—	—	53.6	53.6	—	—	78.7	78.7	—	—
11.7	—	—	—	17.2	—	—	—	25.2	—	—	—	37.0	—	—	—	54.2	—	—	—	79.6	—	—	—
11.8	11.8	—	—	17.4	17.4	—	—	25.5	25.5	—	—	37.4	37.4	—	—	54.9	54.9	—	—	80.6	80.6	—	—
12.0	—	12	12	17.6	—	—	—	25.8	—	—	—	37.9	—	—	—	56.2	—	—	—	81.6	—	—	—
12.1	12.1	—	—	17.8	17.8	—	—	26.1	26.1	—	—	38.3	38.3	—	—	56.6	56.6	56	56	82.5	82.5	82	82
12.3	—	—	—	18.0	—	18	18	26.4	—	—	—	38.8	—	—	—	56.9	—	—	—	83.5	—	—	—
12.4	12.4	—	—	18.2	18.2	—	—	26.7	26.7	—	—	39.2	39.2	39	39	57.6	57.6	—	—	84.5	84.5	—	—
12.6	—	—	—	18.4	—	—	—	27.1	—	27	27	39.7	—	—	—	58.3	—	—	—	85.6	—	—	—
12.7	12.7	—	—	18.7	18.7	—	—	27.4	27.4	—	—	40.2	40.2	—	—	59.0	59.0	—	—	86.6	86.6	—	—
12.9	—	—	—	18.9	—	—	—	27.7	—	—	—	40.7	—	—	—	59.7	—	—	—	87.6	—	—	—
13.0	13.0	13	—	19.1	19.1	—	—	28.0	28.0	—	—	41.2	41.2	—	—	60.4	60.4	—	—	88.7	88.7	—	—
13.2	—	—	—	19.3	—	—	—	28.4	—	—	—	41.7	—	—	—	61.2	—	—	—	89.8	—	—	—
13.3	13.3	—	—	19.6	19.6	—	—	28.7	28.7	—	—	42.2	42.2	—	—	61.9	61.9	62	—	90.9	90.9	91	—
13.5	—	—	—	19.8	—	—	—	29.1	—	—	—	42.7	—	—	—	62.6	—	—	—	92.0	—	—	—
13.7	13.7	—	—	20.0	20.0	20	—	29.4	29.4	—	—	43.2	43.2	43	—	63.4	63.4	—	—	93.1	93.1	—	—
13.8	—	—	—	20.3	—	—	—	29.8	—	—	—	43.7	—	—	—	64.2	—	—	—	94.2	—	—	—
14.0	14.0	—	—	20.5	20.5	—	—	30.1	30.1	30	—	44.2	44.2	—	—	64.9	64.9	—	—	95.3	95.3	—	—
14.2	—	—	—	20.8	—	—	—	30.5	—	—	—	44.8	—	—	—	65.7	—	—	—	96.5	—	—	—
14.3	14.3	—	—	21.0	21.0	—	—	30.9	30.9	—	—	45.3	45.3	—	—	66.5	66.5	—	—	97.6	97.6	—	—
14.5	—	—	—	21.3	—	—	—	31.2	—	—	—	45.9	—	—	—	67.3	—	—	—	98.8	—	—	—

Precision Resistors

FIGURE C–1 Color coded resistor

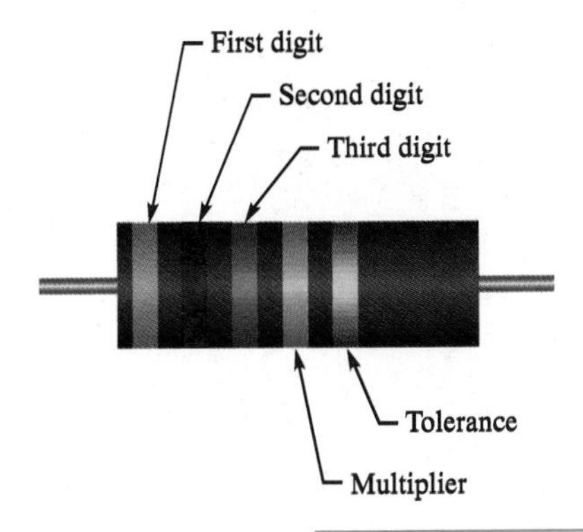

Resistors with tolerances of better than 5% are available. They are found in tolerances of 0.1%, 0.25%, 0.5%, 1%, and 2%. There is a five band color code system used to indicate the value of those with 1% or 2% tolerance. It is very similar to the four band system, but instead of only two significant digits, three are used. As shown in Figure C–1, the first color band is the first digit, the second band is the second digit, the third band is the third digit, the fourth band is the multiplier, and the fifth band is the tolerance. The color code for the three digits and the multiplier are the same as those for the four band system. The tolerance bands are

1% = brown

2% = red

The values are decoded as shown in the following examples.

EXAMPLE C–1 What is the value of a resistor color coded in Figure C–1 brown-red-orange-brown-red?

■ Solution:

First digit: 1
Second digit: 2
Third digit: 3
Multiplier: 10

The value is 1.230 kΩ ± 2%.

EXAMPLE C–2 What is the value of a resistor color coded in Figure C–1 brown-violet-yellow-orange-brown?

■ Solution:

First digit: 1
Second digit: 7
Third digit: 4
Multiplier: 1000

The value is 174 kΩ ± 1%.

Answers to Section Review Questions

Chapter 1

■ Section 1–1

1. Science dealing with the control of electon flow **2.** Computer, microwave oven, television, video camera, radio

■ Section1–2

1. Study of the theory of electricity and electronics **2.** Service store technician, field service technician, engineering assistant, technical writer, sales representative **3.** Communication skills, interpersonal relationships skills, safety

■ Section 1–4

1. (a) 1.1×10^3; **(b)** 8.5×10^{-4}; **(c)** 2.25×10^6 **2. (a)** 3,000,000; **(b)** 0.000,000,002,08; **(c)** 2,250,000,000,000 **3.** 9.33235×10^6 **4.** 4.7944×10^{-6} **5.** 4.53×10^8 **6.** 7×10^3

■ Section 1–6

1. 150 kV **2.** 10 MHz **3.** 50 μA **4.** 15.5 mS

Chapter 2

■ Section 2–2

1. Smallest particle of an element that retains its characteristics **2.** Electron: negative; proton: positive **3.** Number of protons in the nucleus **4.** 8 **5.** Chemical combination of two or more elements forming a different substance

■ Section 2–3

1. Electricity that is not in motion **2.** Unlike charges attract one another; like charges repell **3.** The material has gained an excess of electrons **4.** Friction due to movement of the styrofoam causes them to gain or lose electrons

■ Section 2–4

1. Excess or deficiency of electrons in a material **2.** Q, the coulomb **3.** An electrostatic field is developed between the charged bodies. **4.** 2.5×10^{19} electrons

■ Section 2–5

1. Energy is the ability to do work; work is a force acting through a distance. **2.** Electrical force produced by the separation of charges; symbol, V; unit of measure, volt (V) **3.** Joule

■ Section 2–6

1. (a) Fossil fuel is burned; **(b)** producing heat; **(c)** turning water to steam; **(d)** steam drives a turbine engine; **(e)** turbine rotates generator producing electricity. **2.** Heat, light, chemical activity **3.** Two dissimilar metals immersed in an electrolyte

■ Section 2–7

1. The lower the number of electrons in the valence shell, the higher the conductivity. **2.** Electron that has left the valence shell and that moves freely throughout the material **3.** Its valence electrons are tightly bound to the nucleus, resulting in few free electrons. **4.** Semiconductors are neither good conductors nor insulators. **5.** Shared electron bond between valence electrons of certain elements

■ Section 2–8

1. No **2.** Flow of electrons in a conductor under the influence of a voltage **3.** Symbol, I; unit, ampere (A) **4.** The flow of 1 C of charge is 1 s **5.** The control of electric current

■ Section 2–9

1. Opposition to current flow; symbol, R; unit, ohm (Ω) **2.** 2.2 MΩ **3.** 493.5 kΩ **4.** NEED INFO FOR FIGURE 2–25 **5.** Rheostat: variable resistor; potentiometer: variable voltage divider **6.** Measure of a materials ability to pass electric current; symbol, G; unit, siemen (S).

■ Section 2–10

1. Voltage source, fuse, switch, connecting wires, load **2.** A pictorial shows components as they actually are; a schematic shows components as symbols. **3.** To protect the voltage source and components in the event of an overload **4.** A DPDT switch has two poles for inputs and can be placed in two ON positions **5.** To indicate the voltage at which arcover may occur in the event of a blown fuse

■ Section 2–11

1. Analog and digital **2.** Because the current path must be opened in order to install the meter **3.** Series **4.** Remove the voltage source.

■ Section 2–12

1. Ensure the voltage source value is correct. **2. (a)** No applied voltage; **(b)** an open circuit in the current path

3. (a) Low source voltage; **(b)** a high resistance in the current path such as a poor connection or faulty switch **4.** Switch ON: $R = 0\ \Omega$ (approximately); Switch OFF: R = infinity **5.** Heating so intense that the material glows white, producing light

Chapter 3

Section 3–1

1. Increase **2.** Decrease **3.** Current is directly proportional to voltage and inversely proportional to resistance.

Section 3–2

1. $R = 12.5\ \Omega$ **2.** $I = 140$ mA **3.** $R = 16.7\ \Omega$ **4.** $I = 20$ mA **5.** $V = 82.5$ V

Section 3–3

1. $P = 25$ W **2.** Work **3.** 306 mW **4.** 500 mA **5.** Energy: ability to do work; power: rate at which work is done.

Section 3–4

1. 95% **2.** 725 mW **3.** Ratio of usable output power to total input power

Section 3–5

1. 250 mW **2.** 1/4 W (Use 1/2 W.)

Section 3–6

1. (a) Determine that a fault exists; **(b)** check source voltage; **(c)** determine the cause of zero, high, or reduced current; **(d)** Make necessary repairs; (e) test system for proper operation. **2. (a)** A charred or darkened surface; **(b)** a pungent odor **3. (a)** Current is greater than design value; **(b)** a short circuit

Chapter 4

Section 4–1

1. They are connected end-to-end. **2.** Multiple loads **3.** No

Section 4–2

1. They are in series if the same current flows through them. **2.** 100 mA **3.** 1 A

Section 4–3

1. The sum of the voltage drops around a series circuit is equal to the source voltage. **2.** 9 V **3.** 20 V

Section 4–4

1. The total voltage is their sum. **2.** The total voltage is their difference. **3.** $V = 40$ V; see Figure D–1. **4.** $V = 10$ V; see Figure D–2.

FIGURE D–1

V_1 V_2
15 V 25 V

FIGURE D–2

V_1 V_2
15 V 25 V

Section 4–5

1. 4.39 kΩ **2.** 150 Ω **3.** Decreased; total resistance is greater.

Section 4–7

1. 810 Ω **2.** $V_{R1} = 288$ mV; $V_{R2} = 2.39$ V; $V_{R3} = 7.36$ V; $V_{R4} = 4.97$ V **3.** $P = 6.25$ W; use standard 10 W value. **4.** 1 V **5.** $R_2 = 350\ \Omega$; $R_3 = 400\ \Omega$; $R_4 = 100\ \Omega$

Section 4–8

1. 6 (assuming the voltage source is not one of the voltages) **2.** Since resistors are in series **3.** Assuming $I = 10$ mA, $R_1 = 450\ \Omega$; $R_2 = 800\ \Omega$; $R_3 = 2$ kΩ

Section 4–9

1. The wiper divides the resistance into two resistances whose ratio can be varied. **2.** 375 kΩ and 125 kΩ **3.** 75 V and 25 V

Section 4–10

1. See Figure D–3. **2.** See Figure D–4. **3.** Not all circuit references are the same electrical point. **4.** See Figure D–5.

FIGURE D–3

FIGURE D–4

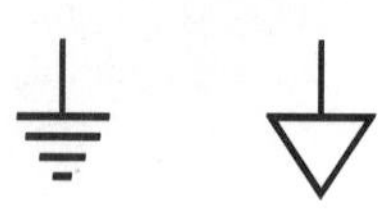

FIGURE D–5

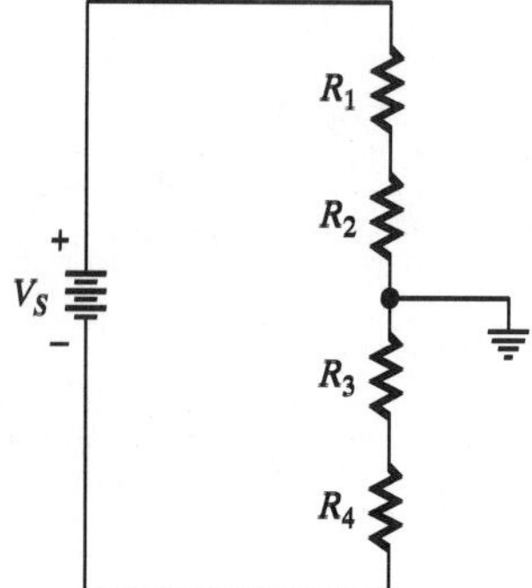

Section 4–12

1. Meter resistance, R_M, and full-scale current, I_{FS} **2.** Ohms-per-volt **3.** 249.7 kΩ **4.** 499.4 kΩ

Section 4–13

1. Verify the value of the source voltage. **2.** The current may be too great, to small, or zero. **3.** Across a short, the voltage is zero. **4.** Measure voltage across a known resistor, and find the current using Ohm's law. **5.** Replace with the same ohmic value and an equal or greater wattage rating.

Chapter 5

Section 5–1

1. Components mounted side-by-side such that they are connected to the same electrical points **2.** Connected in a manner such that there is no resistance between them **3.** A group of resistor in parallel **4.** They are in parallel if they share the same voltage source. **5.** The components are connected to common electrical points and thus develop the same voltage.

Section 5–2

1. (a) The sum of the branch currents is equal to the total current; **(b)** the algebraic sum of the currents into and out of a point is zero. **2.** Current through a particular branch of a parallel circuit **3.** 100 mA **4.** I_{R1} = 7.5 mA; I_{R2} = 5 mA; I_{R3} = 3 mA **5.** 6 mA

Section 5–3

1. Branch currents are independent of one another.
2. Relationship in which an increase or decrease in one quantity causes the opposite effect in another **3.** Current is directly proportional to conductance. **4.** 192 μS

Section 5–4

1. Adding parallel branches increass total current, reducing total resistance. **2.** 638 μS **3.** 1567 Ω **4.** 783.3 Ω **5.** 10 Ω **6.** 10 kΩ

Section 5–5

1. The development of power is the same in both circuits.
2. Use the power equations and values of I_T, R_T, and V_S.

Section 5–6

1. I_{R1} = 65.2 mA; I_{R2} = 25.1 mA; I_{R3} = 9.77 mA **2.** I_{R1} = 97.7 mA; I_{R2} = 37.6 mA; I_{R3} = 14.7 mA **3.** 20 A **4.** 1.5 A

Section 5–7

1. If one lamp fails, the other will still illuminate. **2.** To provide individual control of devices **3.** Each additional load increases total current, and the circuit may overload.

Section 5–8

1. Shunt resistors **2.** Parallel

Section 5–9

1. The source voltage is too high, or a r value. **2.** The source voltage is too lo opened. **3.** Disconnect one lead in orde paths. **4.** The source voltage is across al

Chapter 6

Section 6–1

1. A circuit in which both voltage and current division takes place **2.** Series devices carry the same current; parallel devices have the same voltage across them. **3.** R_1 is in parallel with R_2; R_4, R_5, and R_6 are in parallel, and this combination is in series with R_7. This latter combination is in parallel with R_3. These two combinations of resistors is in series with R_8.

Section 6–2

1. The series and parallel combinations are more clearly shown. **2.** At the left of the circuit **3.** Place all components vertically. **4.** See Figure D–6.

FIGURE D–6

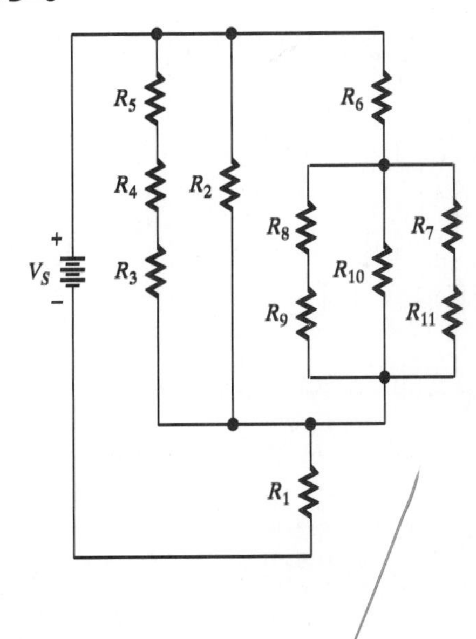

Section 6–3

1. (a) Redraw the circuit if necessary; **(b)** combine all series components; **(c)** combine all parallel components; **(d)** continue this procedure until only a single equivalent resistance is left. **2.** Resistance that could replace a combination of resistors without changing the total current **3.** 55.5 kΩ
4. *A*-to-*B*: 29.328 kΩ; *B*-to-*C*: 26.832 kΩ; *A*-to-*D*: 19.392 kΩ

6–4

mpute the current through and the voltage across each onent **2. (a)** Redraw circuit if necessary; **(b)** determ e the total equivalent resistance; **(c)** work backward, com ting current through and voltage across each component; **(d)** compute power developed by each resistor; **(e)** compute total power. **3.** R_T = 10 kΩ, I_T = 10 mA; $V_{R1} = V_{R2}$ = 50 V, $I_{R1} = I_{R2}$ = 5 mA; $V_{R3} = V_{R4} = V_{R5}$ = 15 V, I_{R3} = 5 mA, $I_{R4} = I_{R5}$ = 2.5 mA; V_{R6} = 10 V, V_{R7} = 25 V, $I_{R6} = I_{R7}$ = 10 mA

■ Section 6–5

1. A loaded voltage divider has loads across the various outputs while the unloaded does not. **2.** So that their currents may be taken into account when computing the series resistors of the divider **3. (a)** To determine the value of unknown resistances; **(b)** in the measurement of quantities such as heat, light, and pressure **4.** When the ratio of the resistances making up the legs of the bridge are equal **5.** 1.51 V **6.** 3.3 kΩ

■ Section 6–7

1. The ohmmeter scale is nonlinear. **2.** To limit the meter current to a value that produces full-scale deflection **3.** To adjust full-scale deflection to compensate for aging of the battery **4.** To provide the current for the meter movement **5.** 6 kΩ

■ Section 6–8

1. Open or shorted components **2.** A short usually causes an increase in current, while an open results in a decrease in current. **3.** Shorts: components may heat to the point where their connectors fuse, or conductive debris may collect on circuit boards; opens: components may burn open or solder connections may break. **4. (a)** No source voltage, R_3 or R_4 open, any broken conductor in R_3–R_4 line; **(b)** source voltage greater than normal, either R_5 or R_6 open; **(c)** no fault exists.

Chapter 7

■ Section 7–1

1. Source whose voltage does not vary with changes in load resistance **2.** Distributed resistance within a battery or power supply; internal resistance **3.** It contains resistance from the components of which it is made. **4.** When the value of the source resistance is no more than 1/10 the load resistance

■ Section 7–2

1. Source whose current does not vary with changes in load resistance **2.** Its internal resistance is not infinite. **3.** When the parallel internal resistance of the source is at least 10 times greater than the load resistance

■ Section 7–3

1. $I_S = V_S/R_S$; R_S is the same in both. **2.** $V_S = I_S R_S$; R_S is the same in both. **3.** The circuits are equivalent if they produce the same current and voltage for identical loads. **4.** I_S = 1.25 A; R_S = 20 Ω. **5.** Both produce V_L = 20.8 V and I_L 208 mA.

■ Section 7–4

1. The current in any branch of a multiple source circuit is the algebraic sum of the currents that would be produced by each source acting separately. **2. (a)** Short all voltage sources but one: **(b)** compute the current that the remaining voltage source produces in each branch, and mark the direction of these currents; **(c)** repeat step b for the remaining sources: **(d)** find the branch currents by algebraically adding the computed currents. **3.** V_{R1} = 5.94 V, I_{R1} = 212 mA; V_{R2} = 108 V, I_{R2} = 1.08 A; V_{R3} = 33.9 V, I_{R3} = 869 mA

■ Section 7–5

1. That any network may be reduced to a single voltage source, V_{Th}, and a single series resistor, R_{Th} **2.** It simplifies circuit analysis where multiple loads are involved. **3. (a)** Remove the load resistor; **(b)** compute the open circuit voltage, V_{Th}; **(c)** short the source and determine total resistance looking into circuit where the load was removed (R_{Th}). **4.** The load has infinite resistance, and maximum voltage is produced. **5.** V_L = 8.33 V; I_L = 5.55 mA

■ Section 7–6

1. That any network may be reduced to a current source, I_N, and a parallel resistance, R_N **2.** It has the same advantage as Thevenin's theorem. **3. (a)** Short the load; **(b)** compute the current through the short (I_N); **(c)** remove the short and compute total resistance looking into the circuit from the open (R_N). **4.** The load is zero and maximum current is produced. **5.** V_L = 4.89 V; I_L = 4.08 mA

■ Section 7–7

1. Maximum power is developed in a load when $R_L = R_S$. **2.** At the output of an electrical system **3.** $R_L = R_S$ **4.** 281 mA, 50%

■ Section 7–8

1. Some is dropped across the source resistance. **2.** When voltage amplification is needed **3.** When R_L is at least 10 times larger than R_S

■ Section 7–9

1. Circuit analysis **2. (a)** Remove the load; **(b)** measure the open circuit voltage, **(c)** place a rheostat across the output and adjust until the voltage is half the open circuit voltage; **(d)** remove the rheostat and measure its resistance. This measured value is R_S. **3.** This is the resistance that will be in series with any load.

Chapter 8

■ Section 8–1

1. In relation to their interaction with the earth's magnetic field **2. (a)** Form complete loops; **(b)** never cross one another. **3.** Unlike poles attract, like pole repell.

■ Section 8–2

1. Maxwell and Weber **2.** Flux is the number of lines of force; flux density is the number of lines per unit area. **3.** At the poles where the flux lines are most concentrated **4.** The entire group of flux lines around a magnet

■ Section 8–3

1. Magnetic dipoles within magnetic materials such as iron **2.** Permeability, reluctance, retentivity **3.** High reluctance **4.** It is similar to resistance in an electric circuit. **5.** Inducing magnetism in a material by placing it in a strong magnetic field

■ Section 8–4

1. Amount of current, number of turns, length of the core, permeability of the core **2.** If the fingers of the left hand are placed around the coil in the direction of current flow, the thumb points toward the north pole. **3.** $\mu_r = \mu/\mu_0$; $\mu = B/H$ **4.** 750 A·t **5.** 15,000 A·t/m

■ Section 8–5

1. The effect lags the cause. **2.** All available magnetic domains are aligned. **3.** Core permeability. **4.** Coercive force **5.** Energy lost in reversing the magnetic domains

■ Section 8–7

1. Some of the magnetic domains are misaligned. **2.** It results in a decrease in flux density. **3.** Some of the turns may short. **4.** It results in a decrease in flux density. **5.** Low source voltage or a poor electrical connection

Chapter 9

■ Section 9–1

1. Property of a circuit that induces a voltage when the current changes **2.** Inductance present due to conductors in the circuit **3.** Varying current producing a moving magnetic field that cuts across the windings of an inductor **4.** Coil of wire **5.** See Figure D–7; henry.

FIGURE D–7 

■ Section 9–2

1. The magnetic field of an induced current has a polarity that opposes the external magnetic field producing it. **2.** Moving magnetic field cutting across the coil windings **3.** Direction of motion of the magnetic field relative to the inductor **4.** The self-induced voltage opposes the establishment of energizing current in the inductor. **5.** It opposes these changes. **6.** Only when it changes in value

■ Section 9–3

1. Inductance and the rate of change of current **2.** Induced voltage will decrease. **3.** Induced voltage will decrease. **4.** Induced voltage will increase. **5.** $V_i = L(\Delta i/\Delta t)$

■ Section 9–4

1. $R_W = 0\ \Omega$. **2.** In the magnetic field **3.** Only in the winding resistance **4.** 938 μJ

■ Section 9–5

1. Number of turns, area of core, μ, length of core **2.** μ_r: relative ease with which a material can be magnetized; μ: product of μ_r and μ_0 (1.26×10^{-6}) **3.** Very little energy is required to reverse the magnetic field. **4.** Powered iron and ceramic **5.** They have higher hysteresis losses. **6.** 640 μH

■ Section 9–6

1. To increase total inductance **2.** 95.5 mH **3.** 20 mH

■ Section 9–7

1. To reduce total inductance **2.** 3.94 μH **3.** 50 mH

■ Section 9–8

1. 5 **2.** Directly **3.** Inversely **4.** 11 μs **5.** 897 mA

■ Section 9–9

1. As the dc current varies, the self induced voltage will oppose the change **2.** Rate at which the current varies, $\Delta i/\Delta t$ **3.** 1000 A/s **4.** 5 kV

■ Section 9–10

1. Shorts and opens **2.** That it is open **3.** A reading of the winding resistance R_W indicates that no windings are shorted. **4.** Device used in analyzing inductors

Chapter 10

■ Section 10–1

1. Property of a circuit that stores electric charge **2.** See Figure D–8; farad (F) **3.** Insulating material between the plates of a capacitor **4.** Capacitor **5.** Distributed capacitance is that which exists between insulated conductors; lumped capacitance is capacitance introduced by a capacitor. **6.** Capacitance between the windings of a coil **7.** See Figure D–9.

FIGURE D–8 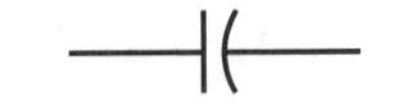

FIGURE D–9

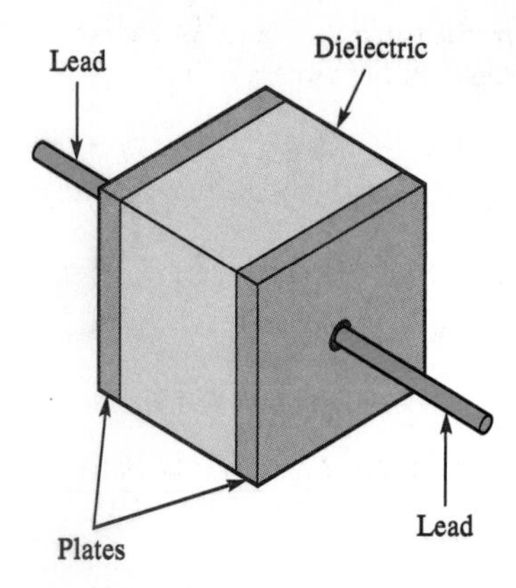

Section 10–2

1. The voltage source moves charge from one plate to the other. **2.** V_C is equal and opposite in polarity to V_S. **3.** Current flows into the capacitor, not through it. **4.** The source is removed and a conducting path placed between the plates. **5.** It is an electrostatic field formed due to polarization of the dielectric. **6.** Amount of charge stored and dielectric permitivity **7.** The stored charge forces the orbiting electrons of the atoms of the dielectric into elliptical orbits. **8.** Dielectric breakdown can occur.

Section 10–3

1. In the dielectric in the form of an electrostatic field **2.** Its plates and leads contain some resistance. **3.** Amount of charge stored and dielectric permittivity **4.** 0.75 μJ

Section 10–4

1. If 1 V stores 1 C of charge, $C = 1$ F. **2.** If a change of 1 V in 1 s produces 1 A of 1_c, $C = 1$ F. **3.** pF or μF

Section 10–5

1. Area of plates, absolute permittivity, distance between the plates **2.** Directly related to plate area and absolute permittivity; inversely related to distance between plates **3.** Voltage that may not be exceeded without risking dielectric breakdown **4.** Its value does not change with temperature changes. **5.** Amount of current that leaks through the dielectric **6.** 110 μF

Section 10–6

1. Mica, ceramic, teflon, paper, air, plastic **2.** 10 pF – 5000 pF **3.** Outer foil that should be connected to neutral **4.** Reversed polarity will destroy the capacitor. **5.** By varying the area of the plates

Section 10–7

1. The effective distance between the plates is increased. **2.** It takes more voltage to store same charge on a smaller value capacitor ($Q = CV$). **3.** 31.1 μF **4.** $V_{C1} = 7.5$ V; $V_{C2} = 11$ V; $V_{C3} = 16.5$ V

Section 10–8

1. The effective plate area is increased. **2.** Yes **3.** 2100 pF

Section 10–9

1. 63.2% **2.** 5 **3.** 38.8 V **4.** 1.72 V

Section 10–10

1. Triggering and timing **2.** Its predictable charging time **3.** High current that is produced when discharged through a low resistance

Section 10–11

1. Opens, shorts, leakage **2.** 0 Ω **3.** Infinity **4.** The capacitor under test does not have its rated voltage applied.

Chapter 11

Section 11–1

1. Current that periodically changes directions **2.** Alternating voltage **3.** dc **4.** ac: used when the load is a far distance from the source; dc: used when the load is near the source **5.** Impressing intelligence upon a current

Section 11–2

1. +1 through −1. **2.** Zero and 180° **3.** 90 and 270° **4.** The basic ac waveform follows the sine curve.

Section 11–3

1. Angle at which the conductor cuts the magnetic field changes **2.** 90 and 270° **3.** Zero and 180° **4.** The value of a sine wave at any instant is a function of the sine of the angle of rotation.

Section 11–4

1. When a sine wave of voltage is applied to a load **2.** Through the application of Ohm's law **3.** When the source voltage changes polarity **4.** No

Section 11–5

1. Time of one cycle **2.** Number of cycles in one second **3.** They are inversely proportional. **4. (a)** It follows a sine curve; **(b)** it contains only one frequency; **(c)** the two alternations must be identical, varying only in polarity.

Section 11–6

1. 147.2 V **2.** Magnitude and direction **3.** Waveform *A* leads waveform *B*. **4.** 150°

Section 11–7

1. 19.1 V_{avg} **2.** 280 mV_{pp} **3.** 99 mV **4.** 34.96 V_{avg}

Section 11–8

1. Fundamental frequency and its harmonics **2.** Fundamental and its odd harmonics **3.** The left edge **4.** 0.2 or 20% **5.** 1.4 V

Section 11–9

1. Electron gun, control grid, focus anode, accelerating anode, vertical deflection plates, horizontal deflection plates, screen **2.** Waveform that produces the time base **3.** Input signal **4.** Control grid **5.** Sawtooth wave

Chapter 12

Section 12–1

1. (a) Property of a circuit that induces a voltage when the current varies; **(b)** property of a circuit that opposes a change in current. **2.** The induced voltage is opposite in polarity to that producing the change. **3.** $V(\text{induced}) = V_S; I_L = 0$ **4.** $V(\text{induced}) = 0; I_L = V_S/R$

Section 12–2

1. At zero and 180° **2.** At 90 and 270° **3.** When it is changing most rapidly **4.** When its rate of change is zero **5.** The voltage leads the current by 90°

Section 12–3

1. Directly **2.** Directly **3.** 3.419 k Ω **4.** 43.04 mA **5.** 254.8 mA

Section 12–4

1. Resistance **2.** Inductor **3.** True and reactive power **4.** 49.5° **5.** 211.3 W

Section 12–5

1. Source voltage **2.** Admittance **3.** Decreases **4.** Decreased

Section 12–7

1. Its effects are frequency sensitive. **2.** High-pass filter **3.** Lag

Chapter 13

Section 13–1

1. Zero through 1 **2.** 2.85 H **3.** Decreases **4.** 160

Section 13–2

1. Source voltage **2.** Load **3.** Through the moving magnetic field produced by the primary **4.** A conductor, a magnetic field, and relative motion between the two

Section 13–3

1. They will be the same. **2.** May not be **3.** May not have **4.** Points of the same phase

Section 13–4

1. 1,4 **2.** 5 **3.** 48 V **4.** 10 turns

Section 13–5

1. Less than **2.** Step-down **3.** False

Section 13–6

1. Larger than **2.** 12 Ω **3.** 0.0516

Section 13–7

1. Resistance of the wire **2.** Magnetic hysteresis and eddy current losses **3.** Laminated core

Section 13–8

1. 20 A **2.** 94% **3.** Because the load may not be purely resistive

Section 13–9

1. Step up or step down the voltage. **2.** To avoid the possibility of having a high potential between to electronic systems **3.** In order to match the load to the source for maximum transfer of power **4.** By using multiple secondaries **5.** 0.408

Section 13–10

1. Opens and shorts **2.** The circuit protection device (fuse) will open. **3.** The primary may be damaged by the high power demand of the secondary. **4.** Not if the secondary is center tapped

Chapter 14

Section 14–1

1. Voltage **2.** Short **3.** Open **4.** Five

Section 14–2

1. Minimum **2.** Maximum **3.** The charging current leads the voltage by 90°

Section 14–3

1. Decreases **2.** Increases **3.** 4 k Ω **4.** 192 Ω

Section 14–4

1. Lead **2.** Leads **3.** 595 Ω **4.** 20 V **5.** 66.1 mW

Section 14–5

1. Lead **2.** Source voltage **3.** Increases **4.** Increases **5.** 0.530 **6.** 0.7

■ Section 14–6

1. (a) $X_c = 1592\ \Omega$; **(b)** $Z = 2716\ \Omega$; **(c)** $I = 9.2$ mA; **(d)** $V_R = 20.2$ V; **(e)** $V_C = 14.6$ V **2. (a)** $X_c = 199\ \Omega$; **(b)** $Z = 177\ \Omega$; **(c)** $I_R = 25.6$ mA; **(d)** $I_c = 50.3$ mA; **(e)** $I_T = 56.4$ mA; **(f)** PF = 0.454; **3. (a)** $X_c = 995\ \Omega$; **(b)** $Z = 364\ \Omega$; **(c)** $I_R = 25.6$ mA; **(d)** $I_C = 10.1$ mA; **(e)** $I_T = 27.5$ mA; **(f)** PF = 0.93 mA

■ Section 14–7

1. Resistor **2.** Circuit that passes all frequencies below a cutoff frequencies and rejects all those above **3.** Capacitor

■ Section 14–8

1. Capacitor analyzer **2.** Direct current flows through the load. **3.** The same voltage is found on both sides of the capacitor. **4.** No signal is found past the capacitor.

Chapter 15

■ Section 15–1

1. Capacitive **2.** Inductive **3.** 6 V **4.** 150 Ω
5. If the phase angle is positive, the phase angle is leading; if negative, the phase angle is lagging.

■ Section 15–2

1. Capacitive **2.** Larger **3.** Increase **4.** V_L and V_C are 180° out of phase; thus the larger will determine whether the circuit is inductive or capacitive.

■ Section 15–3

1. Capacitive **2.** Inductive **3.** 15 mA

■ Section 15–4

1. Inductive **2.** Inductive **3.** Positive
4. Capacitive **5.** Increase

■ Section 15–5

1. They are frequency sensitive. **2.** The fact that inductive and capacitive currents are 180° out of phase. **3.** 0.85
4. Inductor **5.** One

Chapter 16

■ Section 16–1

1. Resonance is a condition in which the response of a device or circuit increases greatly. **2.** $X_L = X_C$ **3.** Frequency at which there is a large rise in current in a series *LC* circuit and a large rise in impedance in a parallel *LC* circuit

■ Section 16–2

1. Inversely **2.** Inversely **3.** One **4.** Many

■ Section 16–3

1. Zero volts **2.** Ideally, the source voltage **3.** $Z = R$
4. Zero degrees **5.** Large

■ Section 16–4

1. The impedance is minimum. **2.** Band of frequencies over which it is considered to respond **3.** Inversely
4. Increases **5.** Q is the factor by which V_L increases at resonance.

■ Section 16–5

1. Small **2.** Large **3.** Large **4.** Energy of the tank being passed between the inductor and capacitor **5.** Zero degrees **6.** One

■ Section 16–6

1. 10 or greater **2.** Narrow **3.** Lower **4.** Wider **5.** The frequency must be lowered to increase I_L in order to make it equal to I_C.

■ Section 16–8

1. Parallel **2.** Parallel **3.** Band-pass **4.** Two

■ Section 16–9

1. Variations in f_r or Q. **2.** Ambient temperature
3. Shorts between some of the windings **4.** Changes in ambient temperature **5.** Tendency of high frequency currents to flow on the surface of a conductor rather than through its cross section **6. (a)** Use as large diameter conductor as possible; **(b)** make area of inductor is as large as possible; **(c)** silver coat wire for better conductivity

Chapter 17

■ Section 17–1

1. Capacitor **2.** Resistor **3.** Circuit that passes all frequencies below a cutoff frequency and rejects all those above **4.** 70.7%

■ Section 17–2

1. They are equal. **2.** Decrease **3.** Indicates a phase angle of 45°

■ Section 17–3

1. Parallel **2.** Opposes a change in voltage **3.** $X_c = R/10$.

■ Section 17–4

1. Parallel **2.** Parallel **3.** 1592 Hz **4.** 0.159 uF

■ Section 17–5

1. It passes the ac component (signal) and blocks the dc component. **2.** Large **3.** 0.001 uF

■ Section 17–6

1. −3dB, 3dB, and 3dB down. **2.** 50%

■ Section 17–7

1. Logarithmic **2.** Ratio of output to input voltage, current, or power **3.** The input and output voltages must be taken across identical resistances. **4.** 12.2 dB **5.** 794 mV **6.** 1.99 W

■ Section 17–8

1. Parallel **2.** Series **3.** Quality factor **4.** 3dB **5.** Difference between the upper and lower cutoff frequencies

■ Section 17–9

1. (a) Radio tuning; **(b)** speaker crossover network **2.** Yes; the complete band of frequencies must be passed for fidelity, but adjoining channels or frequencies must be rejected.

■ Section 17–10

1. Series **2.** Parallel **3.** Difference between the upper and lower cutoff frequencies **4.** 70.7%

■ Section 17–12

1. High current causing the fuse to fail **2.** Some of the charge "leaks" through the dielectric. **3.** Both the dc and ac components of the signal will be developed across the resistor **4.** Capacitor analyzer **5.** Change in the center frequency and possibly a change in bandwidth

Chapter 18

■ Section 18–1

1. Five **2.** 0.1 s **3.** Zero **4.** 0.3 μs **5.** Five

■ Section 18–2

1. Fundamental sine wave and all of its odd harmonics **2.** Higher **3.** Ratio of the pulse width, t_w, to the pulse repetition time, PRT

■ Section 18–3

1. The width of the input pulse and the time constant of the *RC* circuit. **2.** The pulse width is greater than the time constant. **3.** The pulse width is much less than the time constant.

■ Section 18–4

1. Capacitor **2.** The output voltage will resemble the input. **3.** With a pulse train input, after a period of time the output of the integrator will vary between an upper and a lower value voltage. **4.** Rising and falling edges of the pulse **5.** Top of the pulse **6.** Low-pass filter

■ Section 18–5

1. Resistor **2.** The time constant must be short in relation to the pulse width. **3.** High-pass filter **4.** ac

■ Section 18–6

1. Five **2.** The time constant must be short in relation to the pulse width. **3.** Inductor **4.** High-pass filter

Chapter 19

■ Section 19–1

1. -1 **2.** j **3.** -1 **4.** Product of a real number and j.

■ Section 19–2

1. Number made up of a real and an imaginary component **2.** Ordered pair of numbers—one real and one imaginary **3.** A magnitude and an angle **4. (a)** $A = 6 + j4$; **(b)** $B = 4 - j4$; **(c)** $C = -5 - j2$; **(d)** $D = -5 + j4$ **5. (a)** $A = 5, 40$; **(b)** $B = 4, -50$; **(c)** $C = 3, 120$

■ Section 19–3

1. It often simplifies mathematical operations with complex numbers. **2. (a)** 161.6, 21.8; **(b)** 88, −31.5; **(c)** 393.7, 229.6 **3. (a)** $145.4 + j399$; **(b)** $530.3 - j530.3$; **(c)** $44.9 - j21.9$

■ Section 19–4

1. Find the algebraic sum or difference of the real and imaginary numbers. **2.** Multiplication: multiply the magnitudes and algebraically add the angles; (divide: divide the magnitudes and algebraically subtract the angle of the divisor from the angle of the dividend.) **3.** $57 + j56$ **4.** $17 + j9$ **5.** 85, −32 **6.** 6, 24 **7.** 30, 15

■ Section 19–5

1. (a) $24\ \Omega + j15\ \Omega$; **(b)** 28 Ω, 32 **2. (a)** $60\ \text{V} + j28\ \text{V}$; **(b)** 66.2 V, 25 **3. (a)** $30\ \text{mA} + j60\ \text{mA}$; **(b)** 67 mA, 63.4

■ Section 19–6

1. $V_{R1} = 13.65$ V, 0; $V_{R2} = 3.5$ V, 0; $V_L = 29.2$ V, 90; $V_C = 43.75$ V, −90; $V_S = 22.49$ V, −40.3 **2.** V_L leads I by 90° **3.** V_C lags V_S by 49.7°. **4.** Capacitive

Batteries

■ The Basic Wet Cell

A basic wet cell consists of two dissimilar metals, known as *electrodes,* immersed in an electrolyte. Contacts on the electrodes provide points for connecting an external load. An *electrolyte* is a liquid that conducts electric current in the form of ions. Ions are formed when the electrolyte dissolves in water. Chemical action between the metals and the electrolyte causes one electrode to gain electrons while the other loses electrons. Thus, charge is stored and voltage produced.

■ Cell Action

In the simple cell of Figure E–1, the metals are copper and zinc, and the electrolyte is a solution of sulfuric acid, H_2SO_4, and water. The acid dissolves in the water and in the process forms positive hydrogen ($+H$) and negative sulfate ($-SO_4$) ions. The zinc is the more active of the two metals, and it slowly dissolves in the electrolyte in the form of Z^{++} ions. The zinc ions have a charge of $+2$, which means they have left behind two electrons in the zinc electrode. Thus, the zinc electrode begins to take on a negative charge. Each zinc ion displaces two hydrogen ions in the electrolyte forming zinc sulfate. The displaced hydrogen ions migrate to the copper electrode, where they take on two electrons and form a molecule of hydrogen gas, H_2. The copper electrode, in losing two electrons, takes on a positive charge. Thus, in storing charges of opposite polarity, a potential difference is developed between the electrodes. If, as shown in Figure E–2, a load is placed between the electrodes, electrons leave the zinc electrode and flow to the copper. The electrode losing electrons to the external circuit (zinc) is known as the anode, while the electrode gaining electrons (copper) is the cathode.

FIGURE E–1 Basic wet cell

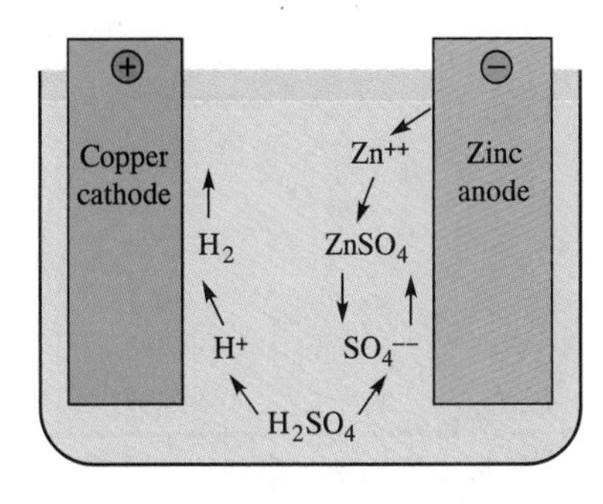

The cell continues to store charge and produce voltage as long as the chemical reaction continues. Several things happen to reduce this chemical activity. The hydrogen gas produced at the cathode acts as an insulator. This process is known as *polarization.* The insulating effect of the hydrogen gas raises the internal resistance of the source and lowers the electrode potential. Polarization is overcome by the inclusion of an agent that combines with the hydrogen and removes it from the electrode. This agent is often manganese dioxide, an element rich in oxygen, which combines with the hydrogen to form water. The zinc electrode is gradually eaten away as the zinc enters the electrolyte. Its surface area is reduced, and the terminal voltage drops.

These actions—polarization, erosion of the zinc electrode, and the resulting decrease in voltage—create what is known as *discharging.* In some cells, when discharging has progressed to the point at which the cell is no longer useable, it is discarded. These cells are known as *primary cells.* There are cells that can be recharged by reversing the chemical process, thus reconstituting the dissolved electrode. Recharging is accomplished by passing a current through the cell in the opposite direction. These cells are known as *secondary cells.*

FIGURE E–2 Basic wet cell with load attached

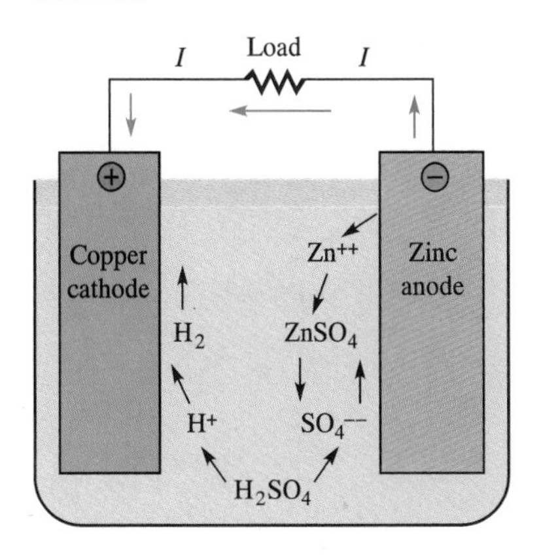

Another way of classifying cells is according to the consistency of their electrolyte. Some are truly wet cells in that the electrolyte is a liquid. These cells usually must be used in an upright position in order to avoid loss of electrolyte. The storage battery used in automobiles is an example. Some cells, classified as *dry cells* are not really "dry," but have an electrolyte that is moist in the form of a paste. An example of this type cell is the D cell used in flashlights, which are not rechargable. There are certain types of newer dry cells that are rechargeable.

■ Battery Ratings

A group of cells connected series-aiding are known as a battery. An example is the automobile battery which is six 2 V cells connected in series to provide 12 V. The capacity of a battery for supplying its rated voltage is expressed in *ampere hours,* which is abbreviated Ah. The ampere hour rating of a battery states the number of hours that it will supply its rated voltage for a given current drain. For example, a battery rated at 100 Ah could supply 1 ampere for 100 hours, 2 amperes for 50 hours, 10 amperes for 10 hours, and so on.

There are many different types of both wet and dry cells. They are made of a variety of metals and electrolytes, each having their own characteristics, advantages, and disadvantages. A representative list of these are shown in Table E–1.

■ TABLE E–1 Common cells

CELL	TYPE	NOMINAL VOLTAGE	ANODE/ CATHODE	ELECTRO-LYTE
Carbon-zinc	Primary	1.5 V	Zinc/carbon	Ammonium chloride
Manganese dioxide	Primary/ secondary	1.5 V	Zinc/ manganese dioxide	Potassium hydroxide
Mercuric oxide	Primary	1.35 V	Zinc/ mercuric oxide	Potassium hydroxide
Silver oxide	Primary	1.5 V	Zinc/silver oxide	Potassium hydroxide
Lead-acid	Secondary	2.1 V	Spongy lead/ lead peroxide	Sulfuric acid
Nickel-cadmium	Secondary	1.25 V	Nickel/ cadmium	Potassium hydroxide
Silver-zinc	Secondary	1.5 V	Zinc/silver oxide	Potassium hydroxide
Silver-cadmium	Secondary	1.1 V	Silver/ cadmium	Potassium hydroxide

Answers to Odd-Numbered Problems

Chapter 1

1. (a) 6.5×10^4 **(b)** 3.2×10^2 **(c)** 8.57×10^3 **(d)** 3.7×10^1

3. (a) 0.0000025 **(b)** 300,000 **(c)** 0.000000001 **(d)** 25,000,000

5. (a) 7.55×10^5 **(b)** 4.916×10^{-3} **(c)** 178.5×10^3

7. (a) 18×10^{-12} **(b)** 36.8235×10^{-3} **(c)** 0.099×10^3

9. (a) 3.1 GHz **(b)** 1.8 MΩ **(c)** 1.5 kV

11. (a) 9×10^9 Hz **(b)** 1.5×10^3 V **(c)** 4.8×10^9 Hz

13. $18\text{–}03 = 18 \times 10^3 = 18{,}000$

15. $35\text{–}03 = 35 \times 10^{-3} = 0.035$ **17.** EE or EXP

19. (a) 1,400 kHz **(b)** 1,400,000 Hz **(c)** 0.0014 GHz

Chapter 2

1. Silicon **3.** $Q = -3$ C **5.** $Q = +1$ C

7. 15 J **9.** $Q = 5$ C **11.** $t = 666.7\ \mu$s

13. (a) 950 Ω – 1050 Ω **(b)** 2.97 kΩ – 3.63 kΩ **(c)** 53.2 kΩ – 58.8 kΩ **(d)** 6.12 kΩ – 7.48 kΩ

15. (a) yellow-violet-black-gold **(b)** brown-blue-brown-silver **(c)** yellow-violet-brown **(d)** red-red-black-gold **(e)** brown-orange-brown-gold **(f)** orange-white-brown-gold

17. (a) 190 Ω – 210 Ω **(b)** 90 kΩ – 110 kΩ **(c)** 42.3 Ω – 51.7 Ω **(d)** 64.6 kΩ – 71.4 kΩ **(e)** 0.198 Ω – 0.242 Ω **(f)** 950 kΩ – 1.05 MΩ

19. 33 kΩ

21. (a) no source voltage **(b)** switch may be faulty **(c)** lamp may be burned out **(d)** open in the wiring

Chapter 3

1. The current will increase.

3. (a) 50 mA **(b)** 666.7 μA **(c)** 50 mA **(d)** 400 μA

5. 429 mA **7.** 2.78 mA

9. (a) 50 V **(b)** 150 V **(c)** 25 V **(d)** 90 mV

11. 75 V

13. (a) 80 Ω **(b)** 1 MΩ **(c)** 473.68 kΩ **(d)** 679.89 kΩ

15. 24 Ω

17. (a) 2.88 W **(b)** 25 μW **(c)** 0.32 W **(d)** 2.25 W

19. 25% **21.** 332.5 W

23. (a) 1.5 kΩ **(b)** 1.5 MΩ **(c)** 555 μA **(d)** 10,400,000,000 Hz **(e)** 10 kW **(f)** 0.001 μF **(g)** 30 kV **(h)** 0.6 μA **(i)** 6 mA

25. 60 V **27.** 2.16 MJ **29.** 15 V

31. Current incorrect **33.** Current incorrect

35. (a) no source voltage **(b)** lamp burned out **(c)** fuse blown

Chapter 4

1. See Figure F–1. **3.** See Figure F–2.

FIGURE F–1

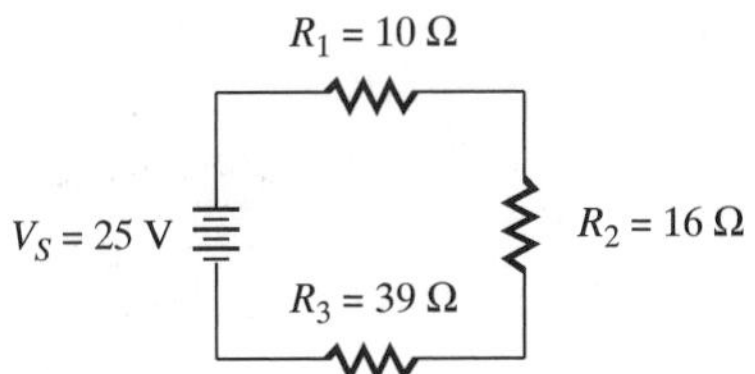

FIGURE F–2

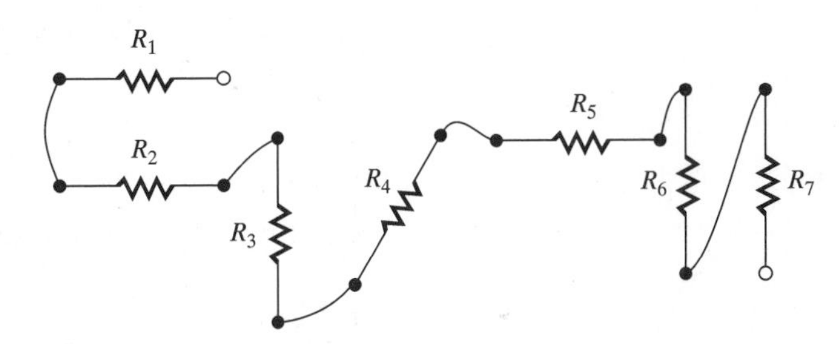

5. 150 mA **7.** 20 V **9.** 40.5 V; $V_S = 60$ V

11. 3 V; see figure F–3 for polarity.

FIGURE F–3

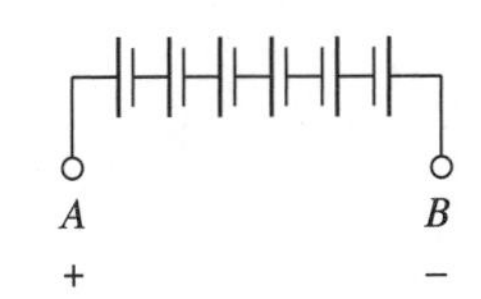

13. 22.1 kΩ **15.** $R_1 = 1/2$ W; $R^2 = 1$ W; $R_3 = 2$ W **17.** $P_T = 600$ mW

19. $V_A = 30$ V; $V_B = 26.9$ V; $V_C = 22.3$ V; $V_D = 14.8$ V; $V_E = 13.3$ V; $V_F = 12.1$ V

21. $R_4 = 25.5$ kΩ (1%); $R_3 = 30.5$ kΩ (0.5%); $R_2 = 30.5$ kΩ (0.5%); $R_1 = 33.2$ kΩ (1%); see Figure F–4.

FIGURE F–4

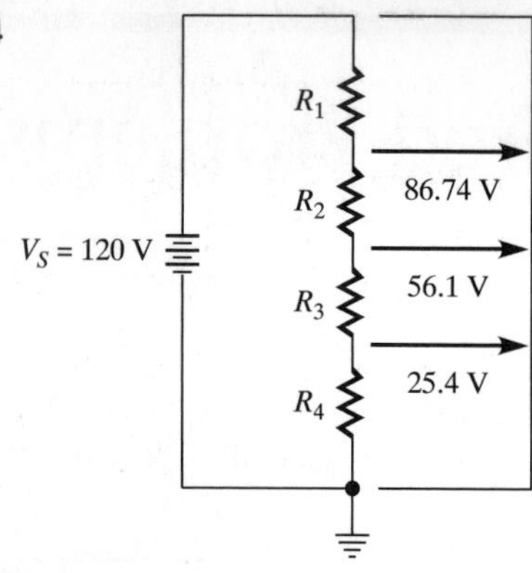

23. $R_1 = 25.3\%$; $R_2 = 21\%$; $R_3 = 53.8\%$

25. $V_{BC} = 81.65$ V **27.** $R_{CB} = 1.6$ kΩ; $R_{CA} = 3.4$ kΩ

29. $P_A = 247$ mW; $P_B = 150$ mW; $P_C = 75$ mW; $P_D = 29$ mW

31. 2.5 V, scale = 2 kΩ; 10 V, scale = 9.5 kΩ; 50 V, scale = 49.5 kΩ; 500 V, scale = 499.5 kΩ

33. $V_S = 119.98$ V; $R_T = 258.021$ kΩ; $R_1 = 10$ kΩ; $R_2 = 47.01$ kΩ; $R_3 = 82$ kΩ; $R_4 = 51.01$ kΩ; $R_5 = 68$ kΩ; $P_1 = 2.16$ mW; $P_2 = 10.2$ mW; $P_3 = 17.7$ mW; $P_4 = 11$ mW; $P_5 = 14.7$ mW; $P_T = 55.8$ mW

35. $R_1 = 8.204$ kΩ **37.** R_3 will burn up.

39. (a) Each resistor is 5 Ω. **(b)** 1.2V

41. R_3 shorted **43.** R_2 open **45.** R_2 shorted

Chapter 5

1. See Figure F–5. **3.** 10 V

FIGURE F–5

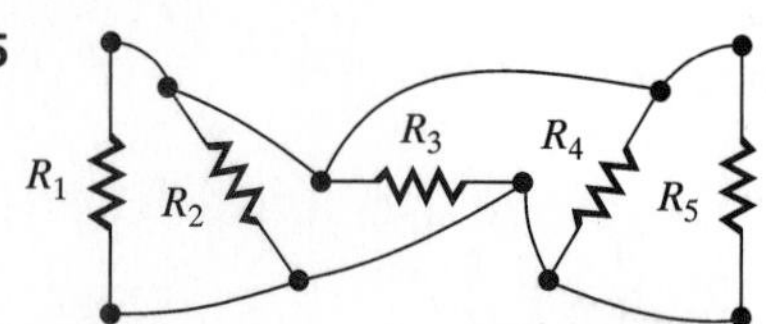

5. $I_1 = 12$ mA **7.** $I_T = 71.8$ mA **9.** $I_1 = 15$ mA

11. All conductances will be halved. **(a)** 300 μS **(b)** 23.8 μS **(c)** 55.5 μS

13. (a) 7.2 kΩ **(b)** 158.873 kΩ **(c)** 6 kΩ

15. $R_T = 1$ kΩ (approximately) **17.** 4 kΩ

19. $P_T = 1.47$ W **21.** 51.4 mV

23. $I_1 = 10$ mA; $I_2 = 4.55$ mA; $I_3 = 29.55$ mA

25. 2.4 kΩ **27.** 22.3 mA

29. $I_1 = 9.44$ mA; $I_2 = 5.41$ mA; $I_3 = 20.2$ mA

31. 1 mA, range = 1 kΩ; 10 mA, range = 52.6 Ω; 100 mA, range = 5.03 Ω

33. Maximum: 3 mA minimum: 2 mA

35. (a) $R_1 = 100$ Ω; $R_2 = 1$ kΩ; $R_3 = 500$ Ω
(b) $I_1 = 0.385$ A; $I_2 = 38.5$ mA; $I_3 = 77$ mA
(c) $P_1 = 14.8$ W; $P_2 = 1.48$ W; $P_3 = 2.96$ W
(d) $P_T = 19.2$ W

37. 50% **39.** $R_1 = 0.18$ Ω; $R_2 = 0.129$ Ω

41. (a) 240 Ω **(b)** 0.5 A **(c)** 2.5 A **(d)** 300 W

43. $V_S = 50$ V; $R_2 = 50$ kΩ; $R_3 = 5$ kΩ; $R_4 = 14.993$ kΩ; $I_2 = 1$ mA; $I_3 = 10$ mA; $I_4 = 3.33$ mA

45. R_2 open **47.** R_1 open **49.** Short across resistor bank

Chapter 6

1. (a) $(R_1 + R_2)\|R_3 + R_4 + R_5\|R_6 + R_7$
(b) $[(R_3 + R_5)\|(R_4 + R_6) + R_2]\|R_1$
(c) $(R_1 + R_2)\|R_3 + R_4 + (R_5 + R_6)\|R_7\|(R_8 + R_9)$

3. See Figure F–6.

FIGURE F–6

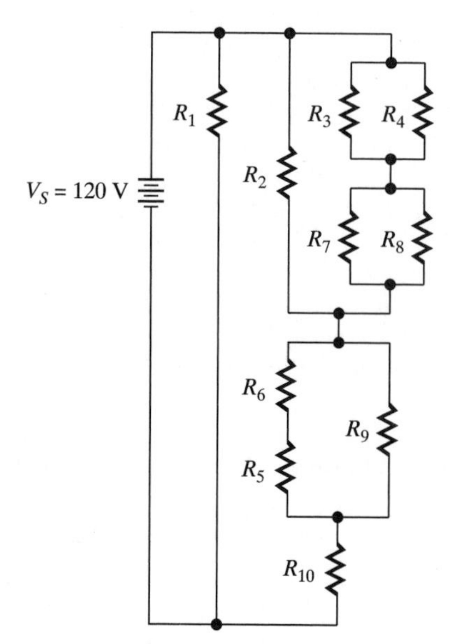

5. (a) 750 Ω **(b)** 12.625 kΩ

7. $R_{AB} = 6.667$ Ω; $R_{AC} = 7.5$ kΩ

9. $V_1 = 23.4$ V; $I_1 = 3.9$ mA
$V_2 = 11.34$ V; $I_2 = 1.89$ mA
$V_3 = 13.31$ V; $I_3 = 1.89$ mA
$V_4 = 1.51$ V; $I_4 = 1.005$ mA
$V_5 = 1.51$ V; $I_5 = 1.005$ mA
$V_6 = 3.22$ V; $I_6 = 1.61$ mA
$V_7 = 3.22$ V; $I_7 = 0.403$ mA
$V_8 = 20$ V; $I_8 = 1$ micro A
$V_9 = 20$ V; $I_9 = 2$ mA

11. $V_A = 80$ V; $V_B = 32$ V; $V_C = 6.4$ V

13. $V_A = 3.81$ V; $V_B = 9.4$ V; $V_C = 15$ V

15. $R_1 = 1.8\ \text{k}\Omega$; $R_2 = 800\ \Omega$

17. 0.238 V **19.** 5 Ω

21. $V_2 = 46.15$ V; $V_S = 100$ V

23. 100 Ω **25.** 4.7 kΩ

27. $V_1 = 2.86$ V; $I_1 = 715\ \mu\text{A}$
$V_2 = 2.86$ V; $I_2 = 179\ \mu\text{A}$
$V_3 = 7.14$ V; $I_3 = 179\ \mu\text{A}$
$V_4 = 7.14$ V; $I_4 = 714\ \mu\text{A}$
$V_5 = 10$ V; $I_5 = 5$ mA
$V_6 = 10$ V; $I_6 = 1.25$ mA
$V_7 = 10$ V; $I_7 = 417\ \mu\text{A}$

29. $V_A = 65.92$ V; $V_B = 33.02$ V; $V_C = 3.75$ V;
$V_D = 30.48$ V

31. $V_A = 54.16$ V; $V_B = 4.86$ V; $V_C = 0.552$ V;
$V_D = 4.86$ V

33. $V_1 = 25$ V; $I_1 = 25$ mA
$V_2 = 62.5$ V; $I_2 = 12.5$ mA
$V_3 = 37.5$ V; $I_3 = 12.5$ mA
$V_4 = V_5 = 62.5$ V; $I_4 = I_5 = 6.25$ mA
$V_6 = 31.875$ V; $I_6 = 6.25$ mA
$V_7 = 62.5$ V; $I_7 = 6.25$ mA
$V_8 = 30.625$ V; $I_8 = 6.25$ mA
$V_9 = V_{10} = 15$ V; $I_9 = I_{10} = 6.25$ mA
$V_{11} = 10$ V; $I_{11} = 10$ mA
$V_{12} = 10$ V; $I_{12} = 2.5$ mA
$V_{13} = 62.5$ V; $I_{13} = 25$ mA

35. R_3 open

37. A_1 incorrect; A_2 incorrect; R_6 open

39. Branch containing R_2 and R_4 is open.

Chapter 7

1. No; the load is too small in relation to the source resistance.

3. No; the load resistance is too large in relation to the source resistance.

5. (a) $I_S = 3$ A; $R_S = 12\ \Omega$ **(b)** $I_S = 120$ mA; $R_S = 100\ \Omega$

7. 610 mA; 19.5 V **9.** $I_6 = 3$ mA; $V_6 = 12$ V

11. $V_{Th} = 10.36$ V; $R_{Th} = 1.482\ \text{k}\Omega$; $V_L = 8.74$ V;
$I_L = 1.09$ mA

13. See Figure F–7. **15.** See Figure F–8.

FIGURE F–7

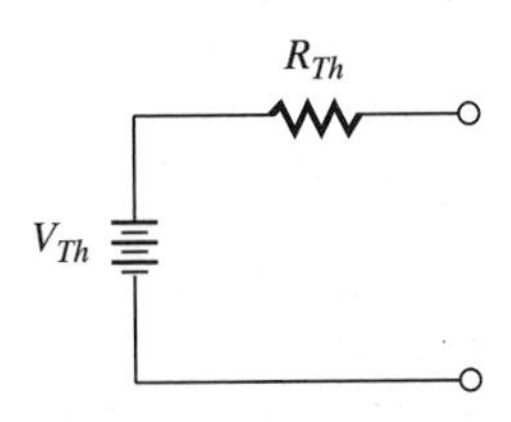

FIGURE F–8

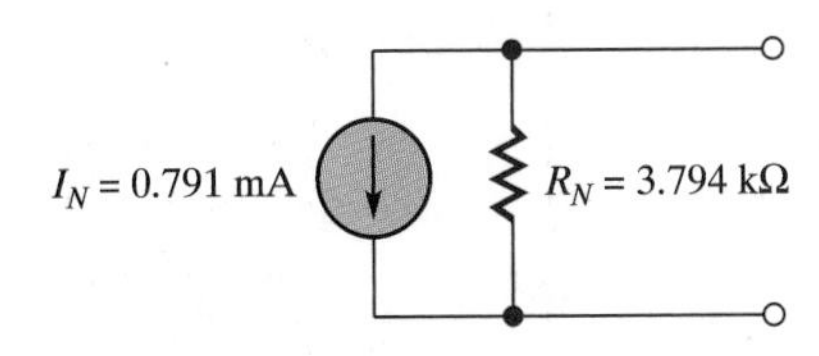

17. $I_N = 3.86$ mA; $R_N = 7.2\ \text{k}\Omega$; $I_L = 2.26$ mA;
$V_L = 11.53$ V

19. See Figure F–9. **21.** 2.25 W; efficiency = 50%

FIGURE F–9

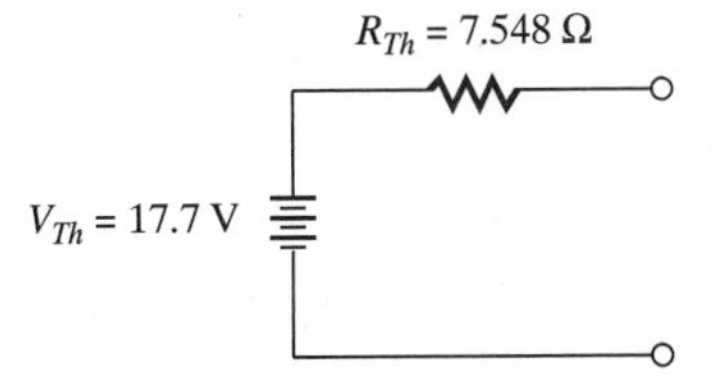

23. 9.9 V **25.** 400 Ω **27.** 8 Ω

29. 5.137 kΩ **31.** 2.79 V **33.** $I_5 = 2.01$ mA

35. See Figure F–10. **37.** R_3 shorted **39.** R_2 open

FIGURE F–10

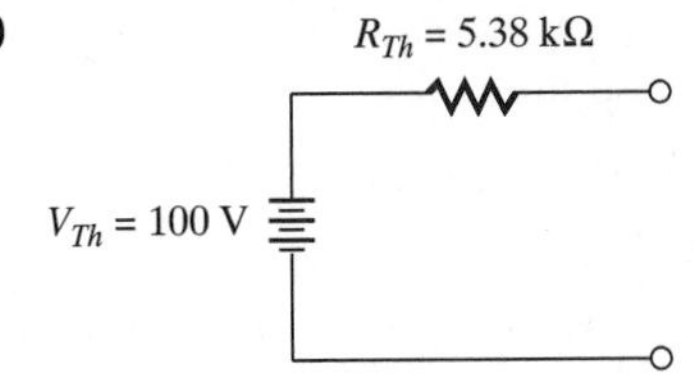

Chapter 8

1. See Figure F–11.

FIGURE F–11

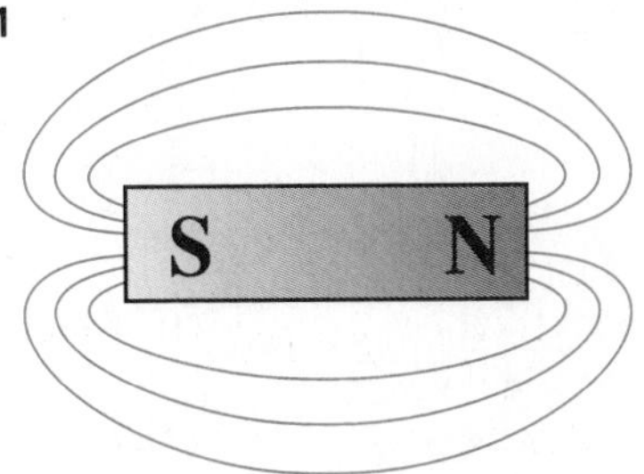

3. 25,000 Mx **5.** 350 G **7.** 0.5 T

9. 900 At·M **11.** 30 kV **13.** 80 V

15. 105 At **17.** 0.02 T

19. $I = 4.8$ A; mmf = 57.6 At; $H = 757.9$ At·m;
$B = 9.54 \times 10^{-4}$ T/At·m

21. 0.2 T

23. (a) It may have been dropped or struck a hard blow.
(b) It may have been heated excessively.

Chapter 9

1. No; the current does not change over time.

3. Negative (−) at the left of the inductor and positive (+) at the right

5. 6 V **7.** 20,000 A/s **9.** 1.25 J

11. 151.2 μH **13.** 40 mH

15. (a) 11.1 mH **(b)** 569 μH **(c)** 0.4 H

17. 180 mH **19.** 12.5 mH **21.** 600 mA **23.** 15 μs

25. 188 mA **27.** 27.3 mH **29.** 153 μH

31. Current increasing: 250 V; current decreasing: 2.5 kV

33. Inductor shorted **35.** 1.67 V

37. A portion of the turns on the inductor have shorted.

Chapter 10

1. See Figure F–12. **3.** See Figure F–13.

FIGURE F–12

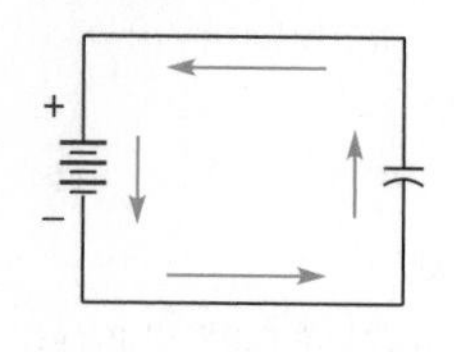

FIGURE F–13

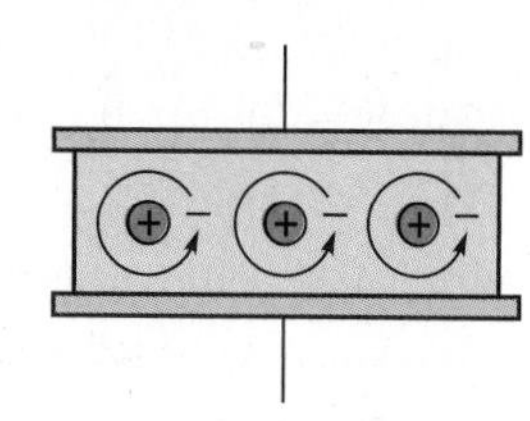

5. 4.8 J **7.** 4 μF **9.** 20 V

11. 132.8 pF **13.** See Figure F–14.

FIGURE F–14

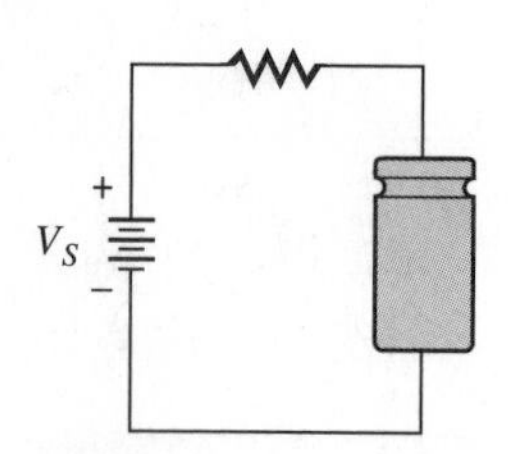

15. 750 pF **17.** 10.2 μF **19.** 1 s

21. $T_0 = 0$ V; $T_1 = 18.96$ V; $T_2 = 25.94$ V; $T_3 = 28.51$ V; $T_4 = 29.45$ V; $T_5 = 29.78$ V (considered fully charged after five time constants)

23. 2.53 V **25.** 50 μF **27.** 3.64 μF **29.** $V_2 = 6$ V

31. $C_2 = 10\ \mu$F; $V_1 = 13.84$ V; $V_3 = 69.21$ V

33. 450 pF **35.** C_2 shorted **37.** Yes

39. Either C_3 or C_4 is open.

Chapter 11

1. (a) dc **(b)** ac

3. (a) 0.707 **(b)** 0.951 **(c)** 0.469 **(d)** −0.5 **(e)** −0.788

5. Minimum; the loop is moving parallel to the magnetic lines of force.

7. (a) Point *A:* 50 mA **(b)** point *B:* 20 mA
(c) point *C:* 0 mA **(d)** point *D:* 40 mA

9. (a) 1 MHz **(b)** 100 Hz **(c)** 100 kHz **(d)** 25 kHz

11. 50 cycles **13.** 1 MHz

15. See Figure F–15. **17.** 628 krad/s

FIGURE F–15

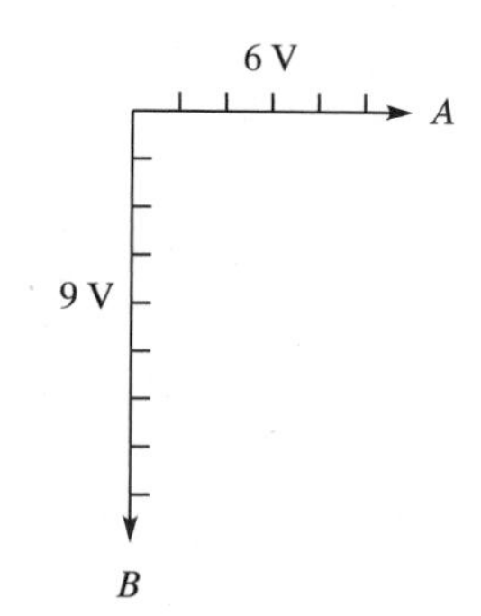

19. (a) 24 V_{PP} **(b)** 340 V_{PP} **(c)** 70 V_{PP} **(d)** 120 V_{PP}

21. 119.8 V_{RMS} **23.** 169.7 V_P **25.** 5.106 kHz

27. 5 kHz **29.** 25% **31.** 0.1 μs

31. 0.1 μs

33. $V_{R1} = 31.9$ V; $I_{R1} = 9.68$ mA
$V_{R2} = 32.9$ V; $I_{R2} = 4.07$ mA
$V_{R3} = 56.1$ V; $I_{R3} = 5.61$ mA
$V_{R4} = 19.1$ V; $I_{R4} = 4.07$ mA
$V_{R5} = 4.07$ V; $I_{R5} = 4.07$ mA
$V_{R6} = 31.9$ V; $I_{R6} = 9.68$ mA

35. 0.15 ms **37.** 90°; waveform *A* leads waveform *B*.

39. 1400 rad/s

41. $T = 40\ \mu$s; $f = 25$ kHz; sawtooth

43. R_4 is open.

Chapter 12

1. See Figure F–16. **3.** 10.616 kHz

FIGURE F–16

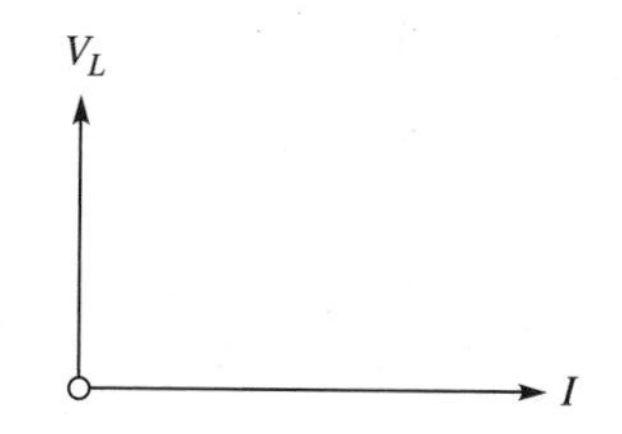

5. 242 μA **7.** 1 V

9. (a) $X_{L1} = 250\ \Omega$; $X_{L2} = 1$ kΩ; $X_{L3} = 500\ \Omega$
(b) 142.86 Ω **(c)** 35 mA

11. See Figure F–17.

FIGURE F–17

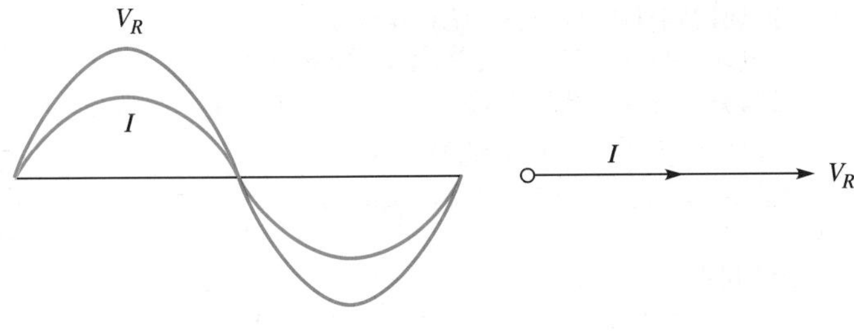

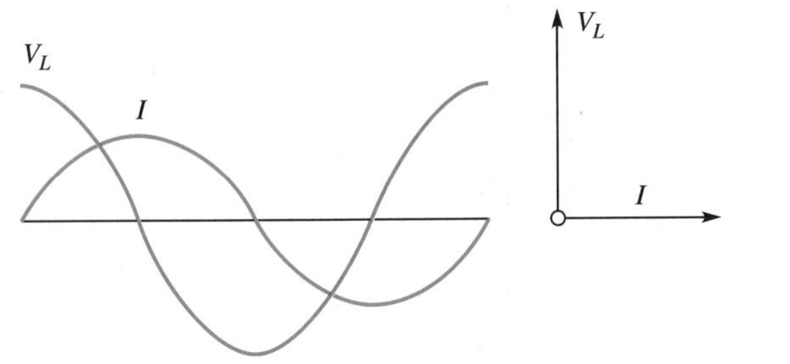

13. Circuit a: **(a)** 1.57 kΩ **(b)** 2.388 kΩ **(c)** 12.6 mA
(d) V_R = 22.68 V **(e)** V_L = 19.78V
Circuit b: **(a)** 5.024 kΩ **(b)** 11.191 kΩ **(c)** 10.7 mA
(d) V_R = 107 V **(e)** V_L = 53.76 V
Circuit c: **(a)** 1.796 kΩ **(b)** 2.543 kΩ **(c)** 9.83 mA
(d) V_R = 17.69 V **(e)** V_L = 17.65 V

15. See Figure F–18.

FIGURE F–18

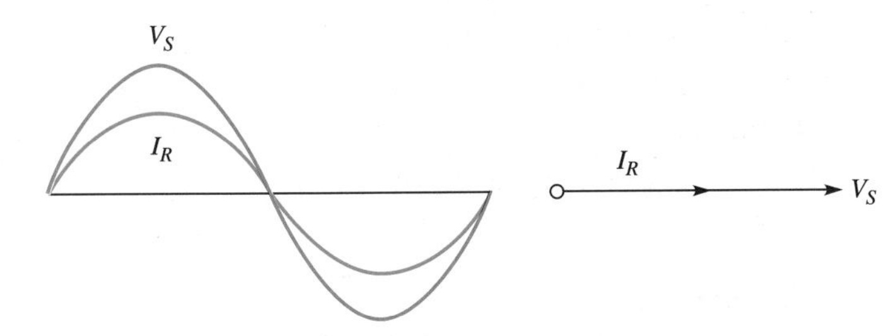

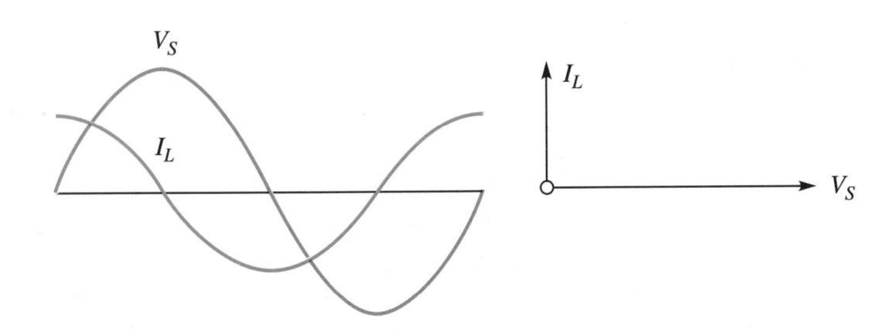

17. (a) 76.6° **(b)** 63.4° **(c)** 63.4°

19. (a) 306 mW **(b)** 243 mVAR **(c)** 391 mVA

21. (a) X_L = 6.28Ω **(b)** I_R = 50 mA **(c)** I_L = 79.6 mA
(d) I_T = 94 mA **(e)** Θ = 57.9° **(f)** Z = 5.32 Ω
(g) P_{app} = 47 mVA **(h)** P_{react} = 39.8 mVAR

23. (a) 13.29 V **(b)** 27.6°

25. (a) I_R = 189 mA **(b)** I_L = 148 mA **(c)** I_T = 240 mA
(d) PF = 0.788

27. X_L = 600 Ω **(b)** L = 100 mH

29. Lag angle = 26.6°; V_{out} = 8.94 V **31.** I_L = 5 mA

33. X_{LT} = 628 Ω; Z = 2578 Ω; I_T = 38.8 mA; I_{R1} = 10 mA; I_{R2} = 10 mA; I_{R3} = 20 mA; I_{L1} = 63.7 mA; I_{L2} = 31.8 mA; I_{L3} = 63.7 mA; Θ = 75.9°

35. (a) Frequency could be lower than correct value.
(b) Inductance has decreased in value.

37. Open in the circuit (inductor, resistor, or wiring)

Chapter 13

1. 0.95 **3.** 2.85 H **5.** No

7. Step-up **9.** Step-down **11.** 30 V **13.** 2.5 A

15. V_S = 240 V; I_S = 120 mA; I_P = 480 mA; P_L = 28.8 W

17. 0.316 **19.** 95%

21. (a) 4 A **(b)** 125 Ω

23. 6.667 kV

25. (a) V_S = 240 V **(b)** I_S = 100A

27. 2 **29.** 2.236

31. Open in secondary above the center-tap

33. Fuse has opened; source voltage is present but not present after fuse.

Chapter 14

1. See Figure F–19. **3.** 2.124 kΩ

FIGURE F–19

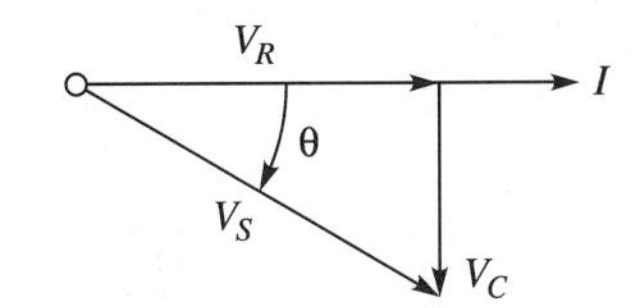

5. (a) X_{C1} = 15.924 kΩ; X_{C2} = 10.616 kΩ; X_{C3} = 3.388 kΩ; X_{C4} = 7.238 kΩ **(b)** B_{C1} = 62.8 μS; B_{C2} = 94.2 μS; B_{C3} = 295 μS; B_{C4} = 138 μS **(c)** B_{CT} = 590 μS
(d) X_{CT} = 1.695 kΩ

7. X_C = 438.7 Ω

9. (a) Z = 20.4 Ω **(b)** 30.194 kΩ **(c)** 1.177 kΩ

11. V_C = 13 V

13. P_{true} = 175 mW; P_{react} = 119 mVAR; P_{app} = 212 mVA

15. (a) X_C = 398 Ω **(b)** Z = 1.076 kΩ **(c)** I = 18.6 mA
(d) V_R = 18.6 V **(e)** V_C = 7.4 V **(f)** Θ = 21.7°
(g) P_{true} = 346 mW **(h)** P_{react} = 138 mVAR
(i) P_{app} = 372 mVA

17. (a) $X_C = 3.981$ kΩ **(b)** $I_R = 1.5$ mA **(c)** $I_C = 3.77$ mA **(d)** $I_T = 4.06$ mA **(e)** $\Theta = 68.3°$ **(f)** PF = 0.37

19. (a) 70.8° **(b)** 3.94 V

21. 5 kHz **23.** 225.8 pF **25.** $V_R = 30$ V

27. $V_R = 8.09$ V; $V_C = 5.88$ V

29. $f = 7.492$ kHz **31.** Capacitor shorted

Chapter 15

1. (a) $Z = 1.688$ kΩ (inductive) **(b)** $Z = 47.1$ Ω (capacitive) **(c)** $Z = 9.811$ kΩ (capacitive)

3. (a) $P_{app} = 1.21$ VA (capacitive) **(b)** $P_{app} = 1.36$ VA (inductive)

5. 0.735 (lagging)

7. $\Theta = -73.4°$ **(a)** $\Theta = 77.7°$ (inductive) **(b)** $\Theta = -87°$ (capacitive)

9. (a) $Z = 19.021$ kΩ (inductive) **(b)** $Z = 27.777$ kΩ (capacitive) **(c)** $Z = 14.2$ Ω (inductive)

11. $X_L = 2.512$ kΩ; $X_C = 3.388$ kΩ; $I_L = 3.98$ mA; $I_C = 2.95$ mA; $I_T = 8.39$ mA; $\Theta = 7.04°$

13. $\Theta = 16.4°$ (capacitive) **(a)** $\Theta = -58.7°$ (capacitive) **(b)** $\Theta = 21°$ (inductive)

15. 0.704 μF **17.** $I_T = 240$ mA; $I\ \text{Net}_{react} = 203.5$ mA

19. $V_R = 3.2$ V; $V_L = 2.35$ V; $V_C = 4.25$ V; $V_S = 3.72$ V; $Z = 37.2$ Ω; $C = 5$ μF; $\Theta = -30.7°$; $L = 4.99$ mH

21. $f = 60$ Hz; $X_L = 188$ Ω; $R = 100$ Ω; $Z = 99$ Ω; $I_L = 53.2$ mA; $I_C = 37.7$ mA; $I_T = 101$ mA; $\Theta = 8.3°$

23. $R = 500$ Ω; $X_L = 1$ kΩ; $X_C = 556$ Ω; $I_T = 21.5$ mA; $Z = 465$ Ω; $C = 0.143$ μF; $L = 79.6$ mH; $\Theta = 21.8°$

25. (a) Source voltage higher than normal **(b)** resistance value lower than normal

27. Capacitor shorted **29.** Inductor shorted

Chapter 16

1. (a) 1.592 kHz **(b)** 1.592 kHz **(c)** 1.5252 MHz **(d)** 112.6 Hz

3. 3.938 MHz **(a)** 2.7846 MHz **(b)** 5.5692 MHz

5. BW = 400 Hz; F_c (upper) = 7.791 kHz; f_c (lower) = 7.391 kHz

7. 70.7 Ω (capacitive) **9.** 56.56 V (assuming circuit is high-Q)

11. $f_r = 15.924$ kHz; $I_L = 12$ mA; $I_C = 12$ mA; $Z = 10$ kΩ; BW = 1592.4 Hz

13. $Q = 39$

15. f_c (upper) = 10.05 MHz; f_c (lower) = 9.95 MHz

17. 7.5 kΩ

19. f_c (upper) = 34.023 kHz; f_c (lower) = 33.840 kHz

21. 253.5 μH

23. (a) 103.3 Ω **(b)** $V_R = 5$ V; $V_L = 75$ V; $V_C = 75$ V

25. 250 MHz

27. The load value has changed to 20Ω.

29. The load has increased to a value of 40 kΩ.

Chapter 17

1. $f_c = 66.348$ kHz **3.** $f_c = 398$ Hz

5. $f_c = 1592$ Hz; see figure F–20. **7.** P_L (max) = 4 W

FIGURE F–20

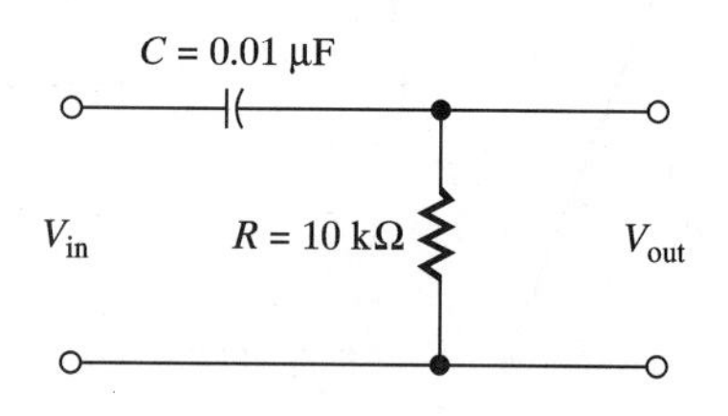

9. $A_p = 10$ dB **11.** $V_{out} = 1.686$ V

13. $A_I = 41.58$ dB

15. $f_r = 100.7$ kHz; BW = 9.59 kHz; f_c (upper) = 105.495 kHz; f_c (lower) = 95.905 kHz

17. $f_r = 450.386$ kHz; BW = 38.168 kHz; f_c (upper) = 469.47 kHz; f_c (lower) = 431.302 kHz

19. $f_r = 2.323$ kHz; $Q = 12.2$; BW = 190 Hz; f_c (upper) = 2.418 kHz; f_c (lower) = 2.228 kHz

21. BW = 10 kHz; voltages at cutoff frequencies = 10 V

23. At 80 kHz, V (out) = 0.3125 V; at 120 kHz, V (out) = 0.3846 V

25. 498 Hz **27.** −11.9 dB **29.** $V_{in} = 3.19$ V

31. Capacitor leaking

Chapter 18

1. (a) T = 0.1 s **(b)** fully charged after 0.5 s

3. (a) $t_r = 1.5$ s **(b)** $t_f = 0.9$ s **(c)** $t_w = 5.7$ s

5. See Figure F–21. **7.** See Figure F–22.

FIGURE F–21

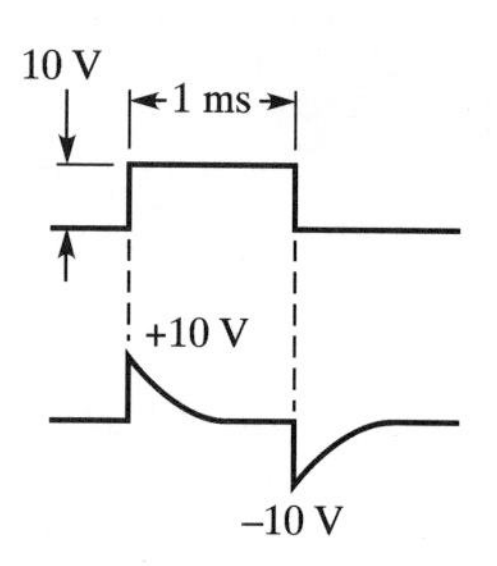

FIGURE F–22

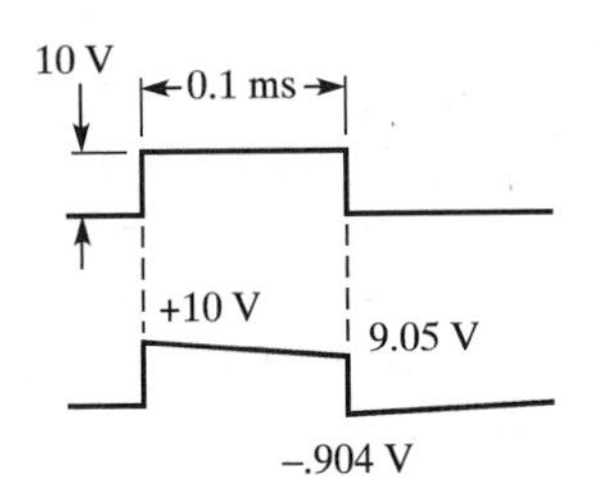

9. 5.68 V **11.** 2 μs **13.** See Figure F–23.

FIGURE F–23

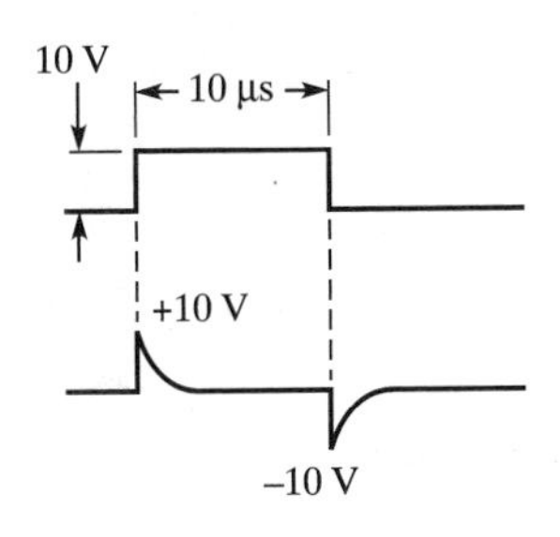

15. 4.65 V **17.** 17.6 V − 2.4 V

19. See Figure F–24. **21.** 2.5 ms **23.** 0.158 μs

FIGURE F–24

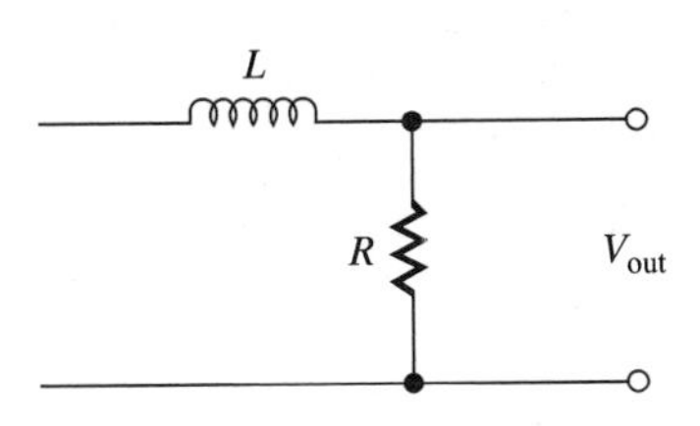

25. 35 kHz

Chapter 19

1. (a) Real 100; imaginary $j60$ **(b)** Real 230; imaginary $j15$ **(c)** Real 400; imaginary $j1000$

3. Figure F–25.

FIGURE F–25

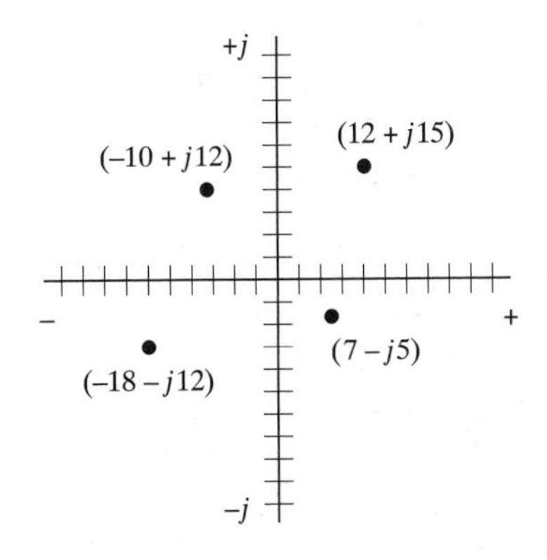

5. (a) $51.9\angle 39.5^\circ$ **(b)** $104\angle 125.2^\circ$ **(c)** $37\angle 341.1^\circ$ or $37\angle -18.9^\circ$ **(d)** $112.4\angle 212.3^\circ$

7. (a) $57.4 + j30.5$ **(b)** $120.8 - j144$ **(c)** $239.6 + j180.5$ **(d)** $6.93 - j4$

9. (a) $-13 + j5$ **(b)** $120 - j48$ **(c)** $-24 + j10$

11. (a) $200\angle 62^\circ$ **(b)** $90\angle -18^\circ$ **(c)** $-75\angle -12^\circ$

13. (a) $81\angle 34^\circ$ **(b)** $144\angle -46^\circ$ **(c)** $225\angle 72^\circ$

15. (a) $13\angle 22.6^\circ \times 10\angle 53.1^\circ = 130\angle 75.7^\circ$ **(b)** $11.4\angle 52.1^\circ \times 7.2\angle -56.3^\circ = 82.1\angle -4.2^\circ$

17. (a) $Z = 1.1\text{ k}\Omega - j338\ \Omega$ **(b)** $Z = 18\text{ k}\Omega\ j14.71\text{ k}\Omega$

19. (a) $I_T = 6.2\text{ mA} + j4.33\text{ mA}$ **(b)** $I_T = 5.7\text{ mA} - j12.7\text{ mA}$

21. $Z = 4605\ \Omega\angle 58.6^\circ$; $I = 3.26\text{ mA}\angle -58.6^\circ$; $V_R = 7.82\text{ V}\angle 0^\circ$; $V_L = 12.8\text{ V}\angle 90^\circ$

23. $Z = 3.2\text{ k}\Omega - j506\ \Omega$

Glossary of Terms

Absolute permeability (æ) Actual value of permeability measured in teslas per ampere-turn per meter or gauss per oersted.

Ac resistance Resistance of a coil made up of core losses and skin effect.

Admittance The phasor sum of conductance and susceptance.

ALNICO An alloy made of aluminum, nickel, cobalt, and iron used in the manufacture of high quality permanent magnets.

Alternation One half of a complete cycle.

Angular velocity The rate, in radians per second, at which a voltage phasor rotates.

Ammeter An instrument used to measure the flow of electrical current.

Ampere The unit of current flow; equal to a flow of one coulomb per second.

Amplitude The value of an electrical current or voltage.

Analog Continuously variable, able to respond to an infinite number of variables.

Apparent power The power "apparently" developed in an ac circuit. The phasor sum of the reactive and true power.

Arc tangent A trigonometric function meaning "the angle whose tangent is."

Atom The smallest particle of an element which will retain that element's properties and characteristics.

Attenuation The process whereby a signal is reduced in amplitude.

Audio A range of frequencies which can be detected by the human ear. Typically from 20 Hz. to 20 Khz.

Autotransformer A type of transformer with only one winding providing no isolation between primary and secondary.

Average Value of a sine wave derived from the average of all its instantaneous values over one half cycle.

AWG (American Wire Gauge) A grouping of standard wire sizes, according to diameter.

Bank A group of loads in a parallel circuit.

Band-pass filter A filter which passes all frequencies between an upper and lower cutoff frequency and blocks or greatly attenuates all others.

Band-reject filter A filter which rejects or greatly attenuates all frequencies between an upper and a lower cutoff frequency and passes all others.

Bandwidth The band of frequencies between the points at which the output voltage of a series resonant circuit drops to 70.7% of maximum. In a parallel resonant circuit, it is the band of frequencies between the points at which the impedance drops to 70.7% of its maximum value.

Battery A source of electrical energy which uses chemical action to separate charge.

Branch Any one of the loads in a parallel circuit.

Branch current The current that flows through any particular load in a parallel circuit.

Capacitance The property of a circuit that opposes a change in voltage.

Capacitive reactance The opposition offered to sine wave current by a capacitor.

Capacitive susceptance The reciprocal of capacitive reactance.

Capacitor is a device for introducing capacitance into a circuit.

Center tap A connection at the electrical center of a transformer's secondary winding.

Charge is an excess or deficiency of electrons in a material.

Chassis The metal framework which supports the various circuits and assemblies of an electronic system.

Choke A common name used for inductors with a "Q" of greater than 10. Normally applied to power supply and RF applications.

Circuit breaker A reusable protective device that opens when excessive current flows in a circuit.

Circuit common is the reference point for voltage measurements in a circuit.

Circular mil (CM) A measure of the cross-sectional area of a conductor. One circular mil equals $(.001 \text{ inch.})^2$

Closed circuit A circuit in which there is a complete path for current from the negative side to the positive side of the source.

Coefficient of coupling The ratio of the number of flux lines linking mutual coils to the total number of lines produced.

Coercive force The amount of reverse magnetizing force required to reduce the magnetic fields to zero.

Coil A name commonly used to describe an inductor.

Complex number A number which contains both a real part and an imaginary part.

Common A term used to describe a central or common low level point in an electrical circuit. Often used to represent circuit ground.

Conductance A circuit's ability to allow the flow of electric current. The reciprocal of resistance. Measured in Siemens (S).

Conductor A material with many free electrons. They are metals with less than 4 valence electrons.

Copper loss The I^2R losses in the coil windings of a transformer. Also called winding loss.

Core The material within the center of a coil or the framework around which a transformer's windings are wound.

Coulomb The basic unit of charge; equal to the storage of 6.25×10^{18} electrons or protons.

Counter electromotive force (CEMF) The voltage induced across an inductor by a change in current.

Current The flow of electrons under the influence of a voltage.

Current source One in which the source current does not change appreciably as the load resistance varies.

Cycle The completion of a repetitive event.

Damped wave Oscillations of a tank circuit which subsequently cease due to circuit losses.

Damping resistor A resistor placed in parallel with an LC circuit in order to reduce the resonant effect.

Decibel (db) A unit expressing a logarithmic ratio such as gain.

De-energizing current is the current which flows when the voltage source is removed and the magnetic field collapses around the inductor.

Degree A unit used to express angular measurements. One 1/360 of a complete circle.

Dielectric The insulating material between the plates of a capacitor.

Differentiation To separate or make distinct.

Differentiator A circuit that converts a dc pulse train to a series of short pulses, one occurring at the leading and one at the trailing edge of each pulse.

Diode A semiconductor device which allows current flow in only one direction.

Dissipate To lose energy, usually by conversion to heat.

DMM Digital Multimeter. A measuring instrument which may be configured to measure current, voltage, or resistance.

Duty cycle The ratio of the pulse width to the pulse repetition time.

DVM Digital voltmeter.

Eddy current A circulating current produced in the iron core of an inductor when a magnetic field cuts across it.

Effective value The equivalent value of steady-state dc that will give the same heating effect as a given AC.

Efficiency The ratio of output to input power.

Electromagnet A device which gains its magnetism from the flow of an electric current in a coil wound upon a magnetic core.

Electron A particle of negative charge and is found in orbit around the nucleus of the atom.

Energy The capacity for doing work.

Energizing current is the current which flows in an inductor to produce and maintain the magnetic field.

Epsilon The base of the natural of Naperian logarithm; equal to approximately 2.7182818.

Equivalent resistance One which will produce the same voltage or current when used to replace a series or parallel resistance combination.

Exponent The number of times a number is multiplied by itself.

Falling edge The negative-going transition of a pulse waveform also called trailing edge.

Fall time The time required for a pulse to fall from 90% to 10% of its maximum amplitude.

Farad The unit of capacitance.

Ferrite core A material made of ferrous oxide and ceramic. It is characterized by high permeability and low losses due to hysteresis and eddy currents.

Field An invisible force that exists between charged bodies or between magnetic poles.

Field intensity The flux developed at a given point on the magnet. Its unit of measure is the ampere-turn per meter (At/m).

Filter An electrical circuit designed to allow certain frequencies to pass and block or reject all others.

Flux density The number of lines of flux per unit area.

Flux lines The lines of magnetic force surrounding a magnet.

Free electron An electron which has broken free of an atom and can readily move within the atomic structure of a material.

Frequency The number of cycles that a waveform completes in one second.

Function generator A test instrument able to produce several types of electrical waveforms, i.e., sine, triangle, square, sawtooth.

Fuse A disposable protective device that opens when excessive current flows in a circuit.

Gain (A) The amount whereby an electrical signal is increased or decreased.

Gauss A unit of measure for flux density which equals one maxwell per centimeter.

Giga A metric prefix used to denote 10^9.

Ground A common reference point in an electrical circuit normally indicating chassis or earth ground.

Harmonics Odd or even multiples of a fundamental frequency.

Henry is the unit of inductance.

Hertz The unit of measure of frequency. One Hertz equals one cycle per second.

High-pass filter A circuit which passes all frequencies above a cutoff frequency and blocks or greatly attenuates all those below.

Hypotenuse The side of a right trangle opposite to the 90° angle. The longest side of a right triangle.

Hysteresis loss The power lost in reversing the magnetic domains within the core of the transformer.

Imaginary number is one rotated 90 degrees from the real number axis.

Impedance The total opposition offered to the flow of ac current by a circuit.

Impedance bridge A piece of precision measuring equipment that, among other things, will indicate the inductance value of an inductor.

Inductance The property of a circuit that opposes a change in current.

Inductive reactance The opposition offered by an inductor to sine wave current.

Inductive susceptance The reciprocal of inductive reactance.

Inductor A coil used for introducing inductance into a circuit.

Infinite Without limits or bounds.

Insulator A material with few free electrons. They are elements with more than 5 valence electrons.

Integration To bring together or make whole.

Integrator A circuit whose output is the dc average of a series of input pulses.

Inverse Directly opposite, inverted.

Ion An atom which loses or gains an electron.

IR drop The same as a voltage drop.

Joule The basic unit of energy.

j operator Imaginary number used to rotate a number 90 degrees each time it is applied.

Kilo The metric prefix used to denote 10^3.

Kilowatt One thousand Watts.

Kilowatt hour The product of a unit of power (kilowatts) multiplied by a unit of time (hours) to obtain a total of one thousand.

Kirchhoff's Current Law States in general that the total current in a parallel circuit is equal to the sum of the branch currents.

Kirchhoff's Voltage Law States that the sum of the voltage drops around a closed loop is equal to the source voltage.

Lag network A circuit in which the phase of the output voltage lags the input voltage.

LC product Determines the resonant frequency of an LC circuit.

Lead network A circuit in which the phase of the output voltage leads the input voltage.

Leading edge The positive going transition of a pulse waveform also called the rising edge.

Linear A relationship between two variable quantities whose graph is a straight line.

Linear circuit One in which a change in voltage will produce a proportional change in current.

Load The amount of current being supplied by a voltage source to a load device.

Load device A resistive or reactive device driven by an energy source.

Loaded voltage divider One in which loads are placed across the voltage sources developed by the divider.

Low-pass filter A circuit which passes all frequencies below a cutoff frequency and blocks or greatly attenuates all those above.

Magnetic Flux is the entire group of flux lines surrounding a magnet.

Magnetomotive force (mmf) The force producing a magnetic field and is similar to voltage in an electric circuit. Its unit of measure is the ampere-turn (At).

Magnetic hysteresis The lagging of the magnetic flux behind the magnetizing force.

Magnetic induction The process of inducing magnetism in a material by placing it in a magnetic field.

Magnitude The value of a given quantity such as vosltage or current.

Maxwell A unit of measure of magnetic flux lines which equals one line of flux.

Mega A metric prefix used to denote 10^6.

Micro A metric prefix used to denote 10^{-6}.

Milli A metric prefix used to denote 10^{-3}.

Monopole A single north or south pole. They are thought to not exist.

Multimeter A measuring instrument for voltage, current, and resistance.

Mutual inductance A measure of the inductive coupling between two coils, and is related to the flux which links them.

Nano A metric prefix used to denote 10^{-9}.

Neutron A particle with no charge and is found in the nucleus of the atom.

Net reactance The difference of X_L and S_C in a series RLC circuit.

Net reactive voltage is the difference of V_L and V_C in a series RLC circuit.

Net susceptance The difference of B_L and B_C in a series RLC circuit.

Nonsinusoidal All waveforms that are not sine waves.

Norton equivalent circuit One that has been reduced to a single current source and a single parallel resistor.

Ohm The unit of resistance; equal to the amount of resistance present when 1 ampere of current flows under a pressure of 1 volt.

Ohmmeter An instrument used to measure resistance.

Open branch One whose resistance is infinite.

Open circuit A circuit in which there is a break in the conducting path which results in no current flow.

Oscilloscope An instrument which displays visually a graph of voltage plotted over time.

Parallel circuit One with more than one path for current and the same voltage across all loads.

Parallel A relationship in which two or more paths exist for the flow of current from a common source.

Peak The maximum value that a sine wave reaches in a cycle.

Peak-to-peak The value of a sine wave measured between its negative and positive peak; equal to twice the value of V_p.

Period The time required for a waveform to complete one cycle.

Permeability A measure of the ease with which a material can be magnetized.

Permittivity A measure of the ease with which an electrostatic field may be established in a dielectric.

Phase angle The phase difference between the source voltage and the current.

Phasor A line whose direction indicates a phase angle and whose length represents the amplitude of a waveform.

Pico A metric prefix used to denote 10^{-12}.

Polar form A complex number composed of a single magnitude and an angle.

Potentiometer A variable voltage divider or two-component variable series circuit.

Power The rate at which work is accomplished.

Power factor The ratio of true power to apparent power. It is equal to the cosine of the phase angle.

PRF (Pulse Repetition Frequency) The number of pulses occurring in one second. Also known as PRR (Pulse Repetition Rate).

Primary The input winding of a transformer.

Proportional A comparative relationship in amount, size, etc. between quantities having a constant or equal ratio.

Proton A particle of positive charge found in the nucleus of the atom.

PRT (pulse repetition time) The time between two successive leading or trailing edges of a pulse wave.

Pulse wave A voltage that makes an abrupt change in amplitude, remains at this value for a period of time, and then returns to its original value.

Pulse width The time between a pulse waveform's leading and trailing edge at the 50% amplitude points.

Q factor The ratio of an inductor's reactive power to its true power.

Radian A measurement of angle equivalent to approximately 57.3 degrees.

Ramp A waveform characterized by a linear increase or decrease in amplitude.

Reactance The opposition offered by an inductor or capacitor to the flow of sinusoidal current and measured in ohms.

Reactive power The power alternately stored and returned to the source by an inductor or capacitor.

Rectangular form A complex number composed of the sum of two phasors with an indicated phase difference of 90 degrees.

Relay An electromechanical device which is used as an electromagnetically operated switch.

Reluctance A measure of a material's opposition to being magnetized.

Residual magnetism The magnetism which remains in a material when the magnetizing force is removed.

Resistance The opposition to current flow.

Resistance bridge Another name for a Wheatstone bridge.

Resistor An electrical component which offers the same opposition to both AC and DC.

Resonance The condition of a circuit in which large rise in either current or impedance takes place.

Resonant frequency The frequency at which the inductive and capacitive reactances are equal.

Resonant rise in impedance The rise in impedance that takes place across the tank circuit at parallel resonance due to the canceling effect of the circuit susceptances.

Resonant rise in voltage The rise in voltage that takes place across the capacitor and inductor at series resonance due to the rise in current.

Retentivity The measure of how much magnetism remains in a material when the magnetizing force is removed.

Rise time The time required to rise from 10% of its maximum amplitude to 90% of its maximum amplitude.

Rising edge The positive going transition of a pulse waveform also called the leading edge.

Rms The equivalent value of steady-state dc that will give the same heating effect as a given AC.

Sawtooth A waveform whose leading edge is formed by a slowly rising ramp and trailing edge by a rapidly falling ramp.

Secondary the output winding of a transformer.

Semiconductor A material that is neither a good conductor nor insulator. They are characterized by atoms with 4 valence electrons.

Series circuit One in which there is one path for current, and the same current flows through each component.

Series aiding Voltage sources connected in a manner (− to +, − to +, etc) such that they add.

Series opposing Voltage sources connected in a manner (− to −, + to +, etc) such that they subtract.

Series-parallel circuit A circuit containing both series and parallel combinations.

Short circuit A low resistance path around a component through which a larger-than-normal current can take a "short cut" back to the source.

Shorted branch One in which the resistance is zero.

Siemen The unit of measure of conductance.

Sine wave A waveform whose value is a function of the sine of the angle of rotation.

Skin effect Tendency of high-frequency currents to flow only on the skin of a wire rather than through the entire cross-section.

String A group of resistors connected in series.

Source A device which provides electrical energy to a circuit.

Square wave A symmetrical pulse waveform having a 50% duty cycle.

Steady-state response of an integrator is the average dc voltage produced for a series of input pulses.

Switch A mechanical device that provides a convenient way of opening the circuit.

Tangent A trigonometric function indicating the ratio of the opposite side of a right triangle to the adjacent side.

Tank circuit A parallel LC circuit which gets its name from its ability to store electrical energy.

Tesla A unit of measure for flux which equals one weber per meter.

Thermistor A special type of resistor whose value changes with a change in temperature.

Thevanin equivalent circuit One that has been reduced to a simple series circuit consisting of a single voltage source and a single series resistor.

Time constant The time required for the current in an RC or RL circuit to rise or decay by 63.2%.

Transformer A device which couples electrical energy from input to output by magnetic induction.

Triangle wave A waveform formed by ramps of the same slope on the leading and trailing edges.

Troubleshooting A methodical system of locating, correcting, and testing electrical circuits.

True power The power developed by a resistor in which electrical energy is converted to heat or motion.

Turns ratio A ratio of the number of turns in the primary winding of a transformer as compared to its secondary.

Valence The number of electrons contained in the outer most shell of an atom.

Volt The unit of measure of voltage and is equal to the potential difference when one joule of energy is used to move one coulomb of charge from one point to another.

Voltage The electrical pressure resulting from stored charge.

Voltage drop The difference in potential developed across a resistor when current flows through it.

Voltage rise Has the opposite polarity of that of a voltage drop.

Voltage source One in which the load voltage does not change appreciably as the load resistance varies.

Watt 1 joule of energy converted in one second.

Watthour The product of a unit of power (watts) multiplied by a unit of time (hours).

Waveform The shape which results when a voltage or current is graphed over a period of time.

Wheatstone bridge A device used for making resistance measurements.

Weber A unit of measure of magnetic flux lines which equals 100 million maxwells.

Winding loss The I^2R losses in the coil windings of a transformer. Also called copper loss.

Wiper The moveable contact of a potentiometer or rheostat.

Work The process of converting energy from one form to another form.

Index

Voltage = Current x Resistance

Current = Voltage / Resistance

Resistance = Voltage / Current

Power = Voltage x Current

Voltage = Power / Current

Current = Power / Voltage

$Power = Current^2 \times Resistance$

24 V

.4 Amps

60 Ω

.028571A

.04

.3290473

180

.02

.025

.0333

.044

1/1

1/1

.095

10.52

3.039